Lecture Notes in Computer Science 2607

Edited by G. Goos, J. Hartmanis, and J. van Leeuwen

T0189429

Springer
Berlin
Heidelberg
New York
Hong Kong
London
Milan
Paris
Tokyo

Helmut Alt   Michel Habib (Eds.)

STACS 2003

20th Annual Symposium
on Theoretical Aspects of Computer Science
Berlin, Germany, February 27 – March 1, 2003
Proceedings

 Springer

Series Editors

Gerhard Goos, Karlsruhe University, Germany
Juris Hartmanis, Cornell University, NY, USA
Jan van Leeuwen, Utrecht University, The Netherlands

Volume Editors

Helmut Alt
Freie Universität Berlin, Institut für Informatik
Takusstr. 9, 14195 Berlin, Germany
E-mail: alt@inf.fu-berlin.de

Michel Habib
LIRMM, 161, Rue Ada
34392 Montpellier Cedex 5, France
E-mail: habib@lirmm.fr

Cataloging-in-Publication Data applied for

A catalog record for this book is available from the Library of Congress

Bibliographic information published by Die Deutsche Bibliothek
Die Deutsche Bibliothek lists this publication in the Deutsche Nationalbibliographie;
detailed bibliographic data is available in the Internet at <http://dnb.ddb.de>.

CR Subject Classification (1998): F, E.1, I.3.5, G.2

ISSN 0302-9743
ISBN 3-540-00623-0 Springer-Verlag Berlin Heidelberg New York

Springer-Verlag Berlin Heidelberg New York
a member of BertelsmannSpringer Science+Business Media GmbH

http://www.springer.de

© Springer-Verlag Berlin Heidelberg 2003
Printed in Germany

Typesetting: Camera-ready by author, data conversion by PTP-Berlin, Stefan Sossna e.K.
Printed on acid-free paper   SPIN: 10872807   06/3142   5 4 3 2 1 0

Preface

The Symposium on Theoretical Aspects of Computer Science (STACS) is alternately held in France and Germany. The latest conference, February 27 to March 1, 2003 at the Institute of Computer Science, Freie Universität Berlin is the twentieth in this series. The previous meetings took place in Paris (1984), Saarbrücken (1985), Orsay (1986), Passau (1987), Bordeaux (1988), Paderborn (1989), Rouen (1990), Hamburg (1991), Cachan (1992), Würzburg (1993), Caen (1994), München (1995), Grenoble (1996), Lübeck (1997), Paris (1998), Trier (1999), Lille (2000), Dresden (2001), and Antibes/Juan-les-Pins (2002).

Unlike some other important theory conferences STACS covers the whole range of theoretical computer science including algorithms and data structures, automata and formal languages, complexity theory, semantics, logic in computer science, and current challenges like biological computing, quantum computing, and mobile and net computing.

The interest in STACS has been increasing continuously during recent years and has turned it into one of the most significant conferences in theoretical computer science. The STACS 2003 call for papers led to a record number of 253 submissions from all over the world.

It is needless to say that the job of the program committee was not easy. In a very intense meeting in Berlin on November 1 and 2, 2002, its members spent a total of 20 hours selecting the papers to be presented at the conference. The decision was based on more than 1100 reviews done by external referees in the weeks before the meeting. We would like to thank our fellow committee members and all external referees (see the list below) for the valuable work they put into the reviewing process of this conference.

After the submission deadline, the steering committee decided to extend the length of the conference to three full days. Nevertheless, with a maximum of 58 papers only, just 22 percent of the submissions could be accepted. Although it is regrettable that many good papers had to be rejected, the strict selection guaranteed the really very high scientific quality of the conference.

We would like to thank the three invited speakers Ernst W. Mayr (TU Munich), Alain Viari (INRIA, Grenoble), and Victor Vianu (UC San Diego) for presenting their results to the audience of STACS 2003.

Special thanks for the local organization go to Christian Knauer, who spent a lot of time and effort and did most of the organizational work. We also thank the other members of the work group Theoretical Computer Science at the Free University of Berlin involved in the organization of the conference, in particular Tamara Knoll and Astrid Sturm.

We acknowledge the substantial financial support STACS 2003 received from the Deutsche Forschungsgemeinschaft (DFG).

Berlin, January 2003   Helmut Alt
                                                                      Michel Habib

Program Committee

H. Alt (Berlin), Chair
A. Bouajjani (Paris)
B. Durand (Marseille)
P. Fraigniaud (Orsay)
M. Goldwurm (Milan)
M. Grohe (Edinburgh)
R. Grossi (Pisa)
M. Habib (Montpellier) Co-chair
R. Impagliazzo (San Diego)
M. Krause (Mannheim)
P.B. Miltersen (Aarhus)
D. Niwinski (Warsaw)
G. Senizergues (Bordeaux)
H.U. Simon (Bochum)
R. Wanka (Paderborn)

Organizing Committee

H. Alt
B. Broser
F. Hoffmann
C. Knauer, Chair
T. Knoll
K. Kriegel
T. Lenz
A. Sturm

External referees[1]

Pankaj Agarwal
Gagan Aggarwal
Stefano Aguzzoli
Susanne Albers
Laurent Alfandari
Eric Allender
Jean-Paul Allouche
Stephen Alstrup
Ernst Althaus
Rajeev Alur
Carme Alvarez
Amihood Amir
Bolette
  Ammitzbøll Madsen
James Anderson
Arne Andersson
Leslie Ann Goldberg
Lars Arge
Frederik Armknecht
Stefan Arnborg
André Arnold
Kristoffer Arnsfelt Hansen
Sanjeev Arora
Eugene Asarin
David Aspinall

Vincenzo Auletta
Giorgio Ausiello
Baruch Awerbuch
Maxim Babenko
Christine Bachoc
Christel Baier
Marie-Pierre Béal
Evripidis Bampis
Amotz Bar-Noy
David Barrington
Hannah Bast
Paul Beame
Luca Becchetti
Arnold Beckmann
Lev Beklemishev
Radim Belohlavek
Marcin Benke
Petra Berenbrink
Anne Bergeron
Luca Bernardinello
Anna Bernasconi
Jean Berstel
Alberto Bertoni
Claudio Bettini
Gianfranco Bilardi

Dario Bini
Johannes Blömer
Luc Boasson
Jean-Daniel Boissonnat
Mikolaj Bojanczyk
Paolo Boldi
Andreas Bomke
Paola Bonizzoni
Olaf Bonorden
Alberto Borghese
Vincent Bouchitté
Andreas Brandstädt
Peter Brass
Peter Bürgisser
Gerth Brodal
Britta Broser
Jehoshua Bruck
Danilo Bruschi
Véronique Bruyère
Nader Bshouty
Johannes Buchmann
Adam Buchsbaum
Harry Buhrmann
Peter Buneman
Marco Cadoli

---

[1] This list has been automatically compiled from the conference's database. We apologize for any omissions or inaccuracies.

Peter Hoyer
Juraj Hromkovic
Marianne Huchard
Thore Husfeldt
John Iacono
Andreas Jacoby
Florent Jacquemard
David Janin
Klaus Jansen
Mark Jerrum
Jens Jägersküpper
Bengt Jonsson
Stasys Jukna
Marcin Jurdzinski
Tomasz Jurdzinski
Yan Jurski
Valentine Kabanets
Jarkko Kari
Marek Karpinski
Michael Kaufmann
Johannes Köbler
Mark Keil
Claire Kenyon
Delia Kesner
Sanjeev Khanna
Eike Kiltz
Ralf Klasing
Hartmut Klauck
Jon Kleinberg
Teodor Knapik
Christian Knauer
Hirotada Kobayashi
Pascal Koiran
Petr Kolman
Jean-Claude Konig
Guy Kortsarz
Elias Koutsoupias
Lukasz Kowalik
Dariusz Kowalski
Miroslaw Kowaluk
Vladik Kreinovich
Klaus Kriegel
Michael Krivelevich
Jens Krokowski
Piotr Krysta
Lukasz Krzeszczakowski
Ralf Küsters
Manfred Kunde
Maciej Kurowski
Eyal Kushilevitz

Miroslaw Kutylowski
Salvatore La Torre
Yassine Lakhnech
Francois Lamarche
Klaus-Jörn Lange
Sophie Laplante
Denis Lapoire
Slawomir Lasota
Matthieu Latapy
Clemens Lautemann
Thomas Lücking
Thierry Lecroq
Hanno Lefmann
Stefano Leonardi
Bertrand Lesaec
Slawomir Leszczynski
Leonid Levin
Art Liestman
Anh Linh Nguyen
Giuseppe Liotta
Violetta Lonati
Luc Longpre
Krzysztof Lorys
Hsueh-I Lu
Fabrizio Luccio
Stefan Lucks
Denis Lugiez
Tamas Lukovszki
Arnaud Maes
Jean Mairesse
Mila Majster-Cederbaum
Konstantin Makarychev
Yuri Makarychev
Jesper Makholm
Dario Malchiodi
Adam Malinowski
Guillaume Malod
Roberto Mantaci
Heiko Mantel
Alberto
  Marchetti-Spaccamela
Jerzy Marcinkowski
Luciano Margara
Vincenzo Marra
Paolo Massazza
Giancarlo Mauri
Ernst Mayr
Richard Mayr
Jacques Mazoyer
Catherine Meadows

Kurt Mehlhorn
Carlo Mereghetti
Wolfgang Merkle
Jochen Messner
Yves Metivier
Friedhelm
  Meyer auf der Heide
Christian Michaux
Victor Mitrana
Anton Mityagin
Dirk Müller
Mark Moir
Paul Molitor
Chris Moore
Maria Morgana
Rémi Morin
Michel Morvan
Achour Moustefaoui
Marcin Mucha
Andrei Muchnik
Ian Munro
Kaninda Musumbu
Mats Naslund
Dana Nau
Stefan Näher
Santoro Nicola
Rolf Niedermeier
Mogens Nielsen
Till Nierhoff
Brigitte Oesterdiekhoff
Anna Oestlin
Patrice Ossona De Mendez
Andre Osterloh
Deryk Osthus
Sven Ostring
Vangelis Pachos
Rasmus Pagh
Linda Pagli
Beatrice Palano
Alessandro Panconesi
Michael Paterson
Christophe Paul
Arnaud Pecher
Andrzej Pelc
David Peleg
Wojciech Penczek
Elisa Pergola
Giuseppe Persiano
Antonio Piccolboni
Andrea Pietracaprina

Pascal Weil   Gerhard Woeginger   Jiawei Zhang
Carola Wenk   David Wood   An Zhu
Matthias Westermann   Thomas Worsch   Wieslaw Zielonka
Thomas Wilke   Sergio Yovine   Rosalba Zizza
Carsten Witt   Jean-Baptiste Yunes   Alexandre Zvonkin
Phillip Wölfel   Erik Zenner

Sponsoring Institutions

Deutsche Forschungsgemeinschaft (DFG)
Freie Universität Berlin

Table of Contents

Logic as a Query Language: From Frege to XML

Victor Vianu*

Univ. of California at San Diego, CSE 0114, La Jolla, CA 92093-0114, USA

Abstract. This paper surveys some of the more striking aspects of the connection between logic and databases, from the use of first-order logic as a query language in classical relational databases all the way to the recent use of logic and automata theory to model and reason about XML query languages. It is argued that logic has proven to be a consistently effective formal companion to the database area.

1 Introduction

The database area is concerned with storing, querying and updating large amounts of data. Logic and databases have been intimately connected since the birth of database systems in the early 1970's. Their relationship is an unqualified success story: indeed, first-order logic (FO), which can be traced all the way to Frege [15], lies at the core of modern database systems. Indeed, the standard query languages such as *Structured Query Language* (SQL) and *Query-By-Example* (QBE) are syntactic variants of FO. More powerful query languages are based on extensions of FO with recursion, and are reminiscent of the well-known fixpoint queries studied in finite-model theory. Lately, logic and automata theory have proven to be powerful tools in modeling and reasoning about the Extended Markup Language (XML), the current *lingua franca* of the Web. The impact of logic on databases is one of the most striking examples of the effectiveness of logic in computer science.

In this paper we survey some aspects of the connection between logic and databases. In Section 2 we discuss the use of FO as a query language in relational databases. We then look briefly at query languages beyond FO and point out the connection between the theory of query languages and finite model theory. In Section 3 we discuss how logic and automata theory provide the formal foundation for XML query languages and their static analysis.

A good introduction to the database area may be found in [40], while [42] provides a more in-depth presentation. The first text on database theory is [24], followed more recently by [2]. The latter text also described database query languages beyond FO, including Datalog and fixpoint logics. An excellent survey of early relational database theory is provided in [23].

---

* Work supported in part by the National Science Foundation under grant number IIS-9802288.

H. Alt and M. Habib (Eds.): STACS 2003, LNCS 2607, pp. 1–12, 2003.

2   FO as a Query Language

This section[1] discusses the following question: Why has FO turned out to be so successful as a query language? We will focus on three main reasons:

- FO has syntactic variants that are easy to use. These are used as basic building blocks in practical languages like SQL and QBE.
- FO can be efficiently implemented using *relational algebra*, which provides a set of simple operations on relations expressing all FO queries. Relational algebra as used in the context of databases was introduced by Ted Codd in [12]. It is related to Tarski's Cylindric Algebras [18]. The algebra turns out to yield a crucial advantage when large amounts of data are concerned. Indeed, the realization by Codd that the algebra can be used to efficiently implement FO queries gave the initial impetus to the birth of relational database systems[2].
- FO queries have the potential for "perfect scaling" to large databases: if massive parallelism is available, FO queries can *in principle* be evaluated in *constant time*, independent of the database size.

A relational database can be viewed as a finite relational structure. Its signature is much like a relational FO signature, with the minor difference that relations and their coordinates have names. The name of a coordinate is called an *attribute*, and the set of attributes of a relation R is denoted $att(R)$. For example, a "beer drinker's" database might consist of the following relations:

frequents	drinker	bar		serves	bar	beer
	Joe	King's			King's	Bass
	Joe	Molly's			King's	Bud
	Sue	Molly's			Molly's	Bass
	...	...			...	...

The main use of a database is to query its data: Which passengers are connecting from flight 173 to flight 645? List all students who take calculus and orchestra. Find the drinkers who frequent only bars serving Bass. It turns out that each query expressible in FO can be broken down into a sequence of very simple subqueries. Each subquery produces an intermediate result, which may be used by subsequent subqueries. A subquery is of the form:

$$R := \exists \boldsymbol{x} \ (L_1 \wedge \ldots \wedge L_k),$$

where L_i is a literal $P(\boldsymbol{y})$ or $\neg P(\boldsymbol{y})$, P is in the input signature or is on the left-hand side of a previous subquery in the sequence, and R is not in the input signature and does not occur previously in the sequence. The meaning of such

---

[1] This section is based on portions of [17].

[2] Codd received the Turing Award for his work leading to the development of relational systems.

a subquery is to assign to R the result of the FO query $\exists x\ (L_1 \wedge \ldots \wedge L_k)$ on the structure resulting from the evaluation of the previous subqueries in the sequence. The subqueries provide appealing building blocks for FO queries. This is illustrated by the language QBE, in which a query is formulated as just described. For example, consider the following query on the "beer drinker's" database:

Find the drinkers who frequent some bar serving Bass.

This can be expressed by a single query of the above form:

(†) $answer := \exists b\ (frequents(d, b) \wedge serves(d, Bass))$.

In QBE, the query is formulated in a visually appealing way as follows:

$$\frac{answer \mid drinker}{\mid d} \leftarrow \frac{frequents \mid drinker\ bar}{\mid d\qquad b} \quad \frac{serves \mid bar\ beer}{\mid b\quad Bass}$$

Similar building blocks are used in SQL, the standard query language for relational database systems.

Let us consider again the query (†). The naive implementation would have us check, for each drinker d and bar b, whether $frequents(d, b) \wedge serves(d, Bass)$ holds. The number of checks is then the product of the number of drinkers and the number of bars in the database, which can be roughly n^2 in the size of the database. This turns out to be infeasible for very large databases. A better approach, and the one used in practice, makes use of relational algebra. Before discussing how this works, we informally review the algebra's operators. There are two set operators, $\cup$ (union) and $-$ (difference). The *selection* operator, denoted $\sigma_{cond}(R)$ extracts from R the tuples satisfying a condition *cond* involving (in)equalities of attribute values and constants. For example, $\sigma_{beer=Bass}(serves)$ produces the tuples in *serves* for which the beer is *Bass*. The *projection* operator, denoted $\pi_X(R)$, projects the tuples of relation R on a subset X of its attributes. The *join* operator, denoted by $R \bowtie Q$, consists of all tuples t over $att(R) \cup att(Q)$ such that $\pi_{att(R)}(t) \in R$ and $\pi_{att(Q)}(t) \in Q$. A last unary operator allows to *rename* an attribute of a relation without changing its contents.

Expressions constructed using relational algebra operators are called relational algebra queries. The query (†) is expressed using relational algebra as follows:

(‡) $\pi_{drinker}(\sigma_{beer=Bass}(frequents \bowtie serves))$.

A result of crucial importance is that *FO and relational algebra express precisely the same queries*.

The key to the efficient implementation of relational algebra queries is twofold. First, individual algebra operations can be efficiently implemented using data structures called *indexes*, providing fast access to data. A simple example of such a structure is a binary search tree, which allows locating the tuples with a given attribute value in time $log(n)$, where n is the number of tuples. Second, algebra queries can be simplified using a set of *rewrite rules* applied by the query processor. The query (‡) above can be rewritten in the equivalent but more efficient form:

$$\pi_{drinker}[frequents \bowtie \pi_{bar}(\sigma_{beer=Bass}(serves))].$$

The use of indexes and rewriting rules allows to evaluate the above query at cost roughly $n \ log(n)$ in the size of the database, which is much better than n^2. Indeed, for very large databases this can make the difference between infeasibility and feasibility.

The FO queries turn out to be extremely well-behaved with respect to *scaling*: given sufficient resources, response time can *in principle* be kept constant as the database becomes larger. They key to this remarkable property is parallel processing. Admittedly, *a lot* of processors are needed to achieve this ideal behavior : polynomial in the size of the database. This is unlikely to be feasible in practice any time soon. The key point, however, is that FO query evaluation admits *linear* scaling; the speed-up is proportional to the number of parallel processors used.

Once again, relational algebra plays a crucial role in the parallel implementation of FO. Indeed, the algebra operations are *set oriented*, and thus highlight the intrinsic parallelism in FO queries. For example, consider the projection $\pi_X(R)$. The key observation is that one can project the tuples in R independently of each other. Given one processor for each tuple in R, the projection can be computed in constant time, independent of the number of tuples. As a second example, consider the join $R \bowtie Q$. This can be computed by joining all *pairs* of tuples from R and Q, independently of each other. Thus, if one processor is available for each pair, the join can be computed in constant time, independent on the number of tuples in R and Q.

Since each algebra operation can be evaluated in constant parallel time, each algebra query can also be evaluated in constant time. The constant depends only on the query, and is independent of the size of the database. Of course, more and more processors are needed as the database grows.

In practice, the massive parallelism required to achieve perfect scaling is not available. Nevertheless, there are algorithms that can take optimal advantage of a given set of processors. It is also worth noting that the processors implementing the algebra need not be powerful, as they are only required to perform very specific, simple operations on tuples. In fact, it is sufficient to have processors that can implement the basic Boolean circuit operations. This fact is formalized by a result due to Immerman [21] stating that FO is included in AC_0, the class of problems solvable by circuits of constant depth and polynomial size, with unbounded fan-in.

Beyond FO. The inability of FO to express certain useful queries, such as connectivity of finite graphs, was noted early on by database theoreticians ([14], and independently [4]). Such facts have led to the introduction of a variety of database languages extending FO with recursion. The proposed paradigms are quite diverse and include SQL extensions, fixpoint logics, imperative programming, logic programming, and production systems (e.g., see [2]).

The wide diversity of query language paradigms beyond FO may be disconcerting at first. It may appear that there is no hope for unity. However, despite the wide variations, many of the languages are equivalent. In fact, two central classes of queries emerge, known as the *fixpoint* queries and the *while* queries. The *fixpoint* queries are equivalent to the well-known fixpoint logics FO+IFP and FO+LFP, and the *while* queries are equivalent to FO+PFP. Recall that FO+LFP extends FO with an operator LFP that iterates a positive formula up to a fixpoint; FO+IFP extends FO with an operator IFP that iterates an arbitrary FO formula up to a fixpoint. Convergence is ensured by forcing the iteration to be cumulative (whence the name of inflationary fixpoint operator). And, FO+PFP extends FO with an operator PFP that iterates an arbitrary FO formula, and may or may not reach a fixpoint (whence the name of partial fixpoint operator). The connection of the database languages to the logics FO+IFP (FO+LFP) and FO+PFP underscores the intimate connection between practical query languages and logics studied in finite-model theory. The relationship between finite-model theory and databases is discussed in more detail in [44].

3 XML

Much of the recent research in databases has been motivated by the Web scenario. As the standard for data exchange on the Web, the Extended Markup Language (XML) plays a central role in current databases (see [1]). As it turns out, logic, particularly automata theory, plays a crucial role in modeling and reasoning about XML query languages. As an aside, we note that automata theory has long interacted with database theory (see [43] for a survey).

An XML document consists of nested elements, with ordered sub-elements. Each element has a name (also called tag or label). The full XML has many bells and whistles, but its simplest abstraction is as a labeled ordered tree (with labels on nodes), possibly with data values associated to the leaves. For example, an XML document holding ads for used cars and new cars is shown in Figure 1 (left), together with its abstraction as a labeled tree (right, data values omitted).

DTDs and XML Schemas. The basic typing mechanism for XML documents is called *Document Type Definition (DTD)*. Let Σ be a finite alphabet of labels. A DTD consists of a set of rules of the form $e \to r$ where $e \in \Sigma$ and r is a regular expression over Σ. There is one such rule for each e, and the DTD also specifies the label of the root. An XML document satisfies a DTD if, for each rule $e \to r$, and each node labeled e, the sequence of labels of its children spells a word in r. For example, a DTD might consist of the rules (with ϵ-rules omitted):
root : *section*;
section $\to$ *intro*, *section**, *conc*
An example of a labeled tree satisfying the above DTD is:

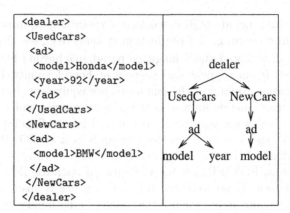

```
<dealer>
 <UsedCars>
  <ad>
   <model>Honda</model>
   <year>92</year>
  </ad>
 </UsedCars>
 <NewCars>
  <ad>
   <model>BMW</model>
  </ad>
 </NewCars>
</dealer>
```

Fig. 1. Dealer XML document (data values omitted from tree representation)

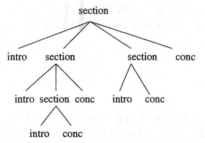

Thus, each DTD d defines a set of labeled ordered trees, denoted by $sat(d)$. It turns out that basic DTDs have many limitations as schema languages. Some are addressed in extensions that have been proposed, such as XML Schema, that are still in a state of flux. One important limitation of DTDs is the inability to separate the *type* of an element from its *name*. For example, consider the dealer document in Figure 1. It may be natural for used car ads to have different structure than new car ads. There is no mechanism to do this using DTDs, since rules depend only on the name of the element, and not on its context. To overcome this limitation, XML Schema provides a mechanism to decouple element names from their types and thus allow context-dependent definitions of their structure. One way to formalize the decoupling of names from types is by the notion of *specialized DTD*, studied in [37] and equivalent to formalisms proposed in [6,11]. Specialization is present in the XQuery typing system and also (in a restricted form) in XML-Schema. The idea is to use whenever necessary "specializations" of element names with their own type definition. For example, this allows us to use in the DTD two variants of ads with different structure: used car ads, and new car ads.

Many basic questions arise in connection to schemas for XML. How hard is it to check validity of an XML document with respect to a schema? When can a set of XML documents be characterized by a schema? Is there always a most precise schema describing a given set of XML documents? Can the union, difference, intersection of sets of valid documents specified by schemas be in turn described by another schema? If yes, how can that schema be computed? It turns out that there is a unified approach to such questions based on a powerful connection between DTDs and tree automata, discussed next. Indeed, specialized DTDs define precisely the *regular tree languages*.

DTDs and tree automata. We informally review the notion of regular tree language and tree automaton. Tree automata are devices whose function is to accept or reject their input, which in the classical framework is a complete binary tree with nodes labeled with symbols from some finite alphabet Σ. There are several equivalent variations of tree automata. A non-deterministic top-down tree automaton over Σ has a finite set Q of states, including a distinguished initial state q_0 and an accepting state q_f. In a computation, the automaton labels the nodes of the tree with states, according to a set of rules, called *transitions*. An internal node transition is of the form $(a, q) \rightarrow (q', q'')$, for $a \in \Sigma$. It says that, if an internal node has symbol a and is labeled by state q, then its left and right children may be labeled by q' and q'', respectively. A leaf transition is of the form $(a, q) \rightarrow q_f$ for $a \in \Sigma$. It allows changing the label of a leaf with symbol a from q to the accepting state q_f. Each computation starts by labeling the root with the start state q_0, and proceeds by labeling the nodes of the trees non-deterministically according to the transitions. The input tree is accepted if *some* computation results in labeling all leaves by q_f. A set of complete binary trees is *regular* iff it is accepted by some top-down tree automaton. Regular languages of finite binary trees are surveyed in [16].

There is a *prima facie* mismatch between DTDs and tree automata: DTDs describe unranked trees, whereas classical automata describe binary trees. There are two ways around this. First, unranked trees can be encoded in a standard way as binary trees. Alternatively, the machinery and results developed for regular tree languages can be extended to the unranked case, as described in [8] (an extension for unranked infinite trees is described in [3]). Either way, one can prove a surprising and satisfying connection between specialized DTDs and tree automata: *they are precisely equivalent* [8,37].

The equivalence of specialized DTDs and tree automata is a powerful tool for understanding XML schema languages. Properties of regular tree languages transfer to specialized DTDs, including closure under union, difference, complement, decidability of emptiness (in PTIME) and inclusion (in EXPTIME), etc. Moreover, automata techniques can yield algorithmic insight into processing DTDs. For example, the naive algorithm for checking validity of an XML document with respect to a specialized DTD is exponential in the size of the document (due to guessing specializations for labels). However, the existence of a bottom-up deterministic automaton equivalent to the specialized DTD shows that validity can be checked in linear time by a single bottom-up pass on the document.

XML query languages and tree transducers. There has been a flurry of proposals for XML query languages, including XML-QL [13], XSLT (W3C Web site), XQL [38], XDuce [19,20], and Quilt [10]. Recently, the language XQuery has been adopted by the W3C committee as the standard query language for XML.

XML query languages take trees as input and produce trees as output. Despite their diversity, it turns out that their tree manipulation capabilities are subsumed by a single model of tree transducer, called *k-pebble transducer* [26]. This provides a uniform framework for measuring the expressiveness of XML languages, and it is instrumental in developing static analysis techniques. In particular, as we shall see, the transducers can be used for typechecking XML queries. The k-pebble transducer uses up to k pebbles to mark certain nodes in the tree. Transitions are determined by the current node symbol, the current state, and by the existence/absence of the various pebbles on the node. The pebbles are ordered and numbered $1, 2, \ldots, k$. The machine can place pebbles on the root, move them around, and remove them. In order to limit the power of the transducer the use of pebbles is restricted by a stack discipline: pebbles are placed on the tree in order and removed in reverse order, and only the highest-numbered pebble present on the tree can be moved. The transducer works as follows. The computation starts by placing pebble 1 on the root. At each point, pebbles $1, 2, \ldots, i$ are on the tree, for some $i \in \{1, \ldots, k\}$; pebble i is called the *current pebble*, and the node on which it sits is the *current node*. The current pebble serves as the head of the machine. The machine decides which *transition* to make, based on the following information: the current state, the symbol under the current pebble, and the presence/absence of the other $i - 1$ pebbles on the current node. There are two kinds of transitions: *move* and *output* transitions. *Move* transitions are of four kinds: they can place a new pebble, pick the current pebble, or move the current pebble in one of the four directions *down-left, down-right, up-left, up-right* (one edge only). If a move in the specified direction is not possible, the transition does not apply. After each move transition the machine enters a new state, as specified by the transition. An *output* transition emits some labeled node and does not move the input head. There are two kinds of output transitions. In a *binary* output the machine spawns two computation branches computing the left and right child respectively. Both branches inherit the positions of all pebbles on the input, and do not communicate; each moves the k pebbles independently of the other. In a *nullary* output the node being output is a leaf and that branch of computation halts.

Looking at the global picture, the machine starts with a single computation branch and no output nodes. After a while it has constructed some top fragment of the output tree, and several computation branches continue to compute the remaining output subtrees. The entire computation terminates when all computation branches terminate. It turns out that all transformations over unranked trees over a given finite alphabet expressed in existing XML query languages (XQuery, XML-QL, Lorel, StruQL, UnQL, and XSLT) can be expressed as k-pebble transducers. This does *not* extend to queries with joins on data values, since these require an infinite alphabet. Details, as well as examples, can be found in [26].

Typechecking using tree automata and transducers. Typechecking by means of tree automata and k-pebble transducers is explored in [26]. As discussed above, this applies to the tree manipulation core of most XML languages. Typechecking can be done by means of *inverse type inference*. Suppose d is an input specialized DTD (or, equivalently, a tree automaton), and d' an output specialized DTD. Consider a k-pebble transducer T. It can be shown that $T^{-1}(sat(d'))$ is always a regular tree language, for which a tree automaton can be effectively constructed from T and d'. Then typechecking amounts to checking that $sat(d) \subseteq T^{-1}(sat(d'))$, which is decidable.

There are several limitations to the above approach. First, the complexity of typechecking in its full generality is very high: a tower of exponentials of height equal to the number of pebbles, so non-elementary. Thus, general typechecking appears to be prohibitively expensive. However, the approach can be used in restricted cases of practical interest for which typechecking can be reduced to emptiness of automata with *very few pebbles*. Even one or two pebbles can be quite powerful. For example, typechecking selection XML-QL queries without joins (i.e., queries that extract the list of bindings of a variable occurring in a tree pattern) can be reduced to emptiness of a 1-pebble automaton with exponentially many states. Another limitation has to do with data values. In general, the presence of data values leads to undecidability of typechecking. For example, if k-pebble transducers are extended with equality tests on the data values sitting under the pebbles, even emptiness is undecidable. However, the approach can be extended to restricted classes of queries with data value joins [5]. An overview of typechecking for XML is provided in [41].

Other models. Another transducer model for XML queries, called *query automaton*, is described in [35]. This work was the first to use MSO to study query languages for XML. Query automata, however, differ significantly from k-pebble transducers: they take an XML input tree and return a set of nodes in the tree. By contrast a k-pebble transducer returns a new output tree. Several abstractions of XML languages are studied in [25], and connections to extended tree-walking transducers with look-ahead are established. Various static analysis problems are considered, such as termination, emptiness, and usefulness of rules. It is also shown that ranges of the transducers are closed under intersection with *generalized DTDs* (defined by tree regular grammars). Tree-walking automata and their relationship to logic and regular tree languages are further studied in [34].

Another computation model for trees, based on *attribute grammars*, is considered in [32]. These capture queries that return sets or tuples of nodes from the input trees. Two main variants are considered. The first expresses all unary queries definable by MSO formulas. The second captures precisely the queries definable by first-order inductions of linear depth. Equivalently, these are the queries computable on a parallel random access machine with polynomially many processors. These precise characterizations in terms of logic and complexity suggest that attribute grammars provide a natural and robust querying mechanism for labeled trees.

To remedy the low expressiveness of pattern languages based on regular path expressions, a guarded fragment of MSO that is equivalent to MSO but that can be evaluated much more efficiently is studied in [33,39]. For example, it is shown that this fragment of MSO can express FO extended with regular path expressions. In [7] a formal model for XSLT is defined incorporating features like modes, variables, and parameter passing. Although this model is not computational complete, it can simulate k-pebble transducers, even extended with equality tests on data values. Consequently, and contrary to conventional wisdom, XSLT can simulate all of XML-QL!

Feedback into automata theory. The match between XML and automata theory is very promising, but is not without its problems. The classical formalism sometimes needs to be adapted or extended to fit the needs of XML. For example, tree automata are defined for ranked trees, but XML documents are unranked trees. This required extending the theory of regular tree languages to unranked trees [8], and has given rise to a fertile line of research into formalisms for unranked trees. This includes extensions of tree transducers [25], push-down tree automata [27], attribute grammars [28], and caterpillar expressions [9]. Another mismatch arises from the fact that XML documents have data values, corresponding to trees over *infinite* alphabets. Regular tree languages over infinite alphabets have not been studied, although some investigations consider the string case [22,36]. Tree-walking transducers accessing data values of XML documents are considered in [31]. Informative surveys on logic and automata-theoretic approaches to XML are provided in [30,29].

4   Conclusion

We first discussed why logic has proven to be a spectacularly effective tool in classical relational databases. FO provides the basis for the standard query languages, because of its ease of use and efficient implementation via relational algebra. FO can achieve linear scaling, given parallel processing resources. Thus, its full potential as a query language remains yet to be realized.

Beyond FO, there is a wide variety of database languages extending standard query languages with recursion. Despite the seemingly hopeless diversity, logic unifies the landscape. Indeed, most languages converge around two central classes of queries: the *fixpoint* queries, defined by the logics FO+LFP and FO+IFP, and the *while* queries, defined by FO+PFP. This underscores an intimate connection between the theory of query languages and finite-model theory, that has led to much fruitful cross-fertilization.

Recently, a lot of research has focused on XML, the new standard for data exchange on the Web. Again, logic, particularly automata theory, has provided crucial tools for modeling and reasoning about XML query languages.

Altogether, the impact of logic on databases is one of the most striking examples of the effectiveness of logic in computer science.

References

1. S. Abiteboul, P. Buneman, and D. Suciu. *Data on the Web*. Morgan Kauffman, 1999.
2. S. Abiteboul, R. Hull, and V. Vianu. *Foundations of Databases*. Addison-Wesley, 1995.
3. S. Abiteboul and P. C. Kanellakis. Object identity as a query language primitive. *JACM*, 45(5):798–842, 1998. Extended abstract in SIGMOD'89.
4. A. V. Aho and J. D. Ullman. Universality of data retrieval languages. In *Proc. ACM Symp. on Principles of Programming Languages*, pages 110–117, 1979.
5. N. Alon, T. Milo, F. Neven, D. Suciu, and V. Vianu. XML with data values: typechecking revisited. In *Proc. ACM PODS*, 2001.
6. C. Beeri and T. Milo. Schemas for integration and translation of structured and semi-structured data. In *Int'l. Conf. on Database Theory*, pages 296–313, 1999.
7. G. Bex, S. Maneth, and F. Neven. A formal model for an expressive fragment of XSLT. In *Proc. DOOD*, pages 1137–1151, 2000.
8. A. Bruggemann-Klein, M. Murata, and D. Wood. Regular tree and regular hedge languages over unranked alphabets, 2001. Technical Report HKUST-TCSC-2001-0, Hong-Kong University of Science and Technology.
9. A. Brüggemann-Klein and D. Wood. Caterpillars: a context specification technique. *Markup Languages*, 2(1):81–106, 2000.
10. D. Chamberlin, J. Robie, and D. Florescu. Quilt: An XML query language for heterogeneous data sources. In *WebDB (Informal Proceedings)*, pages 53–62, 2000.
11. S. Cluet, C. Delobel, J. Simeon, and K. Smaga. Your mediators need data conversion! In *Proc. ACM SIGMOD Conf.*, pages 177–188, 1998.
12. E. F. Codd. A relational model for large shared databank. *Communications of the ACM*, 13(6):377–387, June 1970.
13. A. Deutsch, M. Fernandez, D.Florescu, A. Levy, and D. Suciu. A query language for XML. In *WWW8*, pages 11–16, 1999.
14. R. Fagin. Monadic generalized spectra. *Z. Math. Logik*, 21:89–96, 1975.
15. G. Frege. *Begriffsschrift, eine der arithmetischen nachgebildete Formelsprache des reinen Denkens*. L. Nebert, Halle a. S., 1879.
16. F. Gécseg and M. Steinby. Tree languages. In G. Rozenberg and A. Salomaa, editors, *Handbook of Formal Languages*, volume 3, chapter 1, pages 1–68. Springer, 1997.
17. J. Halpern, R. Harper, N. Immerman, P. Kolaitis, M. Vardi, and V. Vianu. On the unusual effectiveness of logic in computer science. *Bulletin of Symbolic Logic*, 7(2):213–236, 2001.
18. L. Henkin, J. D. Monk, and A. Tarski. *Cylindric Algebras: Part I*. North Holland, 1971.
19. H. Hosoya and B. Pierce. XDuce: A typed XML processing language (Preliminary Report). In *WedDB (Informal Proceedings)*, pages 111–116, 2000.

20. H. Hosoya, J. Vouillon, and B. Pierce. Regular expression types for XML. In *Int. Conf. on Functional Programming*, pages 11–22, 2000.
21. N. Immerman. Languages that capture complexity classes. *SIAM Journal of Computing*, 16:760–778, 1987.
22. M. Kaminski and N. Francez. Finite-memory automata. *Theoretical Computer Science*, 134(2):329–363, 1994.
23. P. Kannelakis. Elements of relational database theory. In J. V. Leeuwen, editor, *Handbook of Theoretical Computer Science*, pages 1074–1156. Elsevier, 1991.
24. D. Maier. *The Theory of Relational Databases*. Computer Science Press, Rockville, Maryland, 1983.
25. S. Maneth and F. Neven. Structured document transformations based on XSL. In *Proc. DBPL*, pages 79–96. LNCS, Springer, 1999.
26. T. Milo, D. Suciu, and V. Vianu. Typechecking for XML transformers. In *Proc. ACM PODS*, pages 11–22, 2000. Full paper to appear in special issue of JCSS.
27. A. Neumann and H. Seidl. Locating matches of tree patterns in forests. In *Proc. Foundations of Software Technology and Theoretical Computer Science*, pages 134–145. LNCS, Springer, 1998.
28. F. Neven. Extensions of attribute grammars for structured document queries. In *Proc. DBPL*, pages 97–114. LNCS, Springer, 2000.
29. F. Neven. Automata, logic, and XML. In *Proc. Computer Science Logic*, pages 2–26. Springer LNCS, 2002.
30. F. Neven. Automata theory for XML researchers. *SIGMOD Record*, 31(3):39–46, 2002.
31. F. Neven. On the power of walking for querying tree-structured data. In *Proc. ACM PODS*, pages 77–84, 2002.
32. F. Neven and J. V. den Bussche. Expressiveness of structured document query languages based on attribute grammars. *JACM*, 49(1), 2002. Extended abstract in PODS 1998.
33. F. Neven and T. Schwentick. Expressive and efficient pattern languages for tree-structured data. In *Proc. ACM PODS*, pages 145–156, 2000.
34. F. Neven and T. Schwentick. On the power of tree-walking automata. In *Proc. ICALP*, pages 547–560, 2000.
35. F. Neven and T. Schwentick. Query automata on finite trees. *Theoretical Computer Science*, 275(1-2):633–674, 2002.
36. F. Neven, T. Schwentick, and V. Vianu. Towards regular languages over infinite alphabets. In *Proc. MFCS*, pages 560–572, 2001.
37. Y. Papakonstantinou and V. Vianu. DTD inference for views of XML data. In *Proc. ACM PODS*, pages 35–46, 2000.
38. J. Robbie, J. Lapp, and D. Schach. XML query language (XQL). In *The Query Languages Workshop (QL'98)*, 1998.
39. T. Schwentick. On diving in trees. In *Proc. MFCS*, pages 660–669, 2000.
40. A. Silberschatz, H. Korth, and S. Sudarshan. *Database System Concepts*. Mc Graw Hill, 1997.
41. D. Suciu. The XML typechecking problem. *SIGMOD Record*, 31(1):89–96, 2002.
42. J. D. Ullman. *Principles of Database and Knowledge-Base Systems*. Computer Science Press, Rockville, MD 20850, 1989.
43. M. Vardi. Automata theory for database theoreticians. In *Proc. ACM PODS*, pages 83–92, 1989.
44. V. Vianu. Databases and finite-model theory. *DIMACS Series in Discrete Mathematics and Theoretical Computer Science*, 31, 1997.

How Does Computer Science Change Molecular Biology?

Alain Viari

INRIA - Rhône-Alpes - 655 Av. de l'Europe 38334 Saint Ismier Cedex - France

One of the most frequent opinion about the advent of 'computational biology' states that the flood of data emerging from the genomic and post-genomic projects created a growing demand of computational resources for their management and analysis thus giving rise to an interdisciplinary field between computer sciences and molecular biology. This point of view is not completely satisfactory. First, it precludes the historical background of a field that appears, in the 70's, at a time where the sequence data were not flooding at all. Second, it emphasizes the quantitative aspect of the data produced by (new) experimental technologies whereas the real difficulties probably rest in the diversity and complexity of the biological data. For instance, all the human chromosomes totalize about 3 Gb in size, which is not such a large quantity of information (by comparison to, let's say, meteorological records). By contrast, the internal structure of these chromosomes such as the location of gene or regulatory sites or their large-scale organization remains a huge puzzle. This situation is still more complex in the so-called 'post-genomic' area. The biological concepts involved in the analysis of regulatory or metabolic networks, for instance, ask for much more sophisticated representations than sequence of symbols. Third, and most importantly, this statement gives to computer science and biology two distinct roles : the latter providing (nice) problems to the former that, in turn, is supposed to solve them. Unfortunately, the situation is not so clear-cut. First, it hardly ever appends that biology can provide a problem under a form directly suitable for computer analysis. The most subtle and difficult work in computational biology is actually to (re)formulate a biological problem under such a form. This is precisely where the role of computer science does not reduce to its technical aspects and where some of its key concepts could migrate to biology. From this point of view, for instance, a successful gene prediction algorithm is at least as important because of its predictive power than because its underlying model reflects some biological reality. One of the more exciting challenge of today's biology is therefore to build such operational models to help understanding complex biological systems, from chromosomal organization up to metabolic or physiological processes.

In this talk I would like to develop the idea that computer science may really change the way molecular and cellular biologists do work by bringing them practical and theoretical tools to formalize some biological questions, to formulate new hypothesis and to devise test experiments. The talk will be divided in two parts. In the first part, I will give an overview of some biological concepts needed to understand the main issues in computational biology. In the second part, I will illustrate the previous idea with several selected examples from the computational biology literature.

H. Alt and M. Habib (Eds.): STACS 2003, LNCS 2607, p. 13, 2003.

Improved Compact Visibility Representation of Planar Graph via Schnyder's Realizer

Ching-Chi Lin, Hsueh-I. Lu*, and I.-Fan Sun

Institute of Information Science, Academia Sinica, Taiwan

Abstract. Let G be an n-node planar graph. In a visibility representation of G, each node of G is represented by a horizontal segment such that the segments representing any two adjacent nodes of G are vertically visible to each other. In this paper, we give the best known compact visibility representation of G. Given a canonical ordering of the triangulated G, our algorithm draws the graph incrementally in a greedy manner. We show that one of three canonical orderings obtained from Schnyder's realizer for the triangulated G yields a visibility representation of G no wider than $\lfloor \frac{22n-42}{15} \rfloor$. Our easy-to-implement $O(n)$-time algorithm bypasses the complicated subroutines for four-connected components and four-block trees required by the best previously known algorithm of Kant. Our result provides a negative answer to Kant's open question about whether $\lfloor \frac{3n-6}{2} \rfloor$ is a worst-case lower bound on the required width. Moreover, if G has no degree-5 node, then our output for G is no wider than $\lfloor \frac{4n-7}{3} \rfloor$. Also, if G is four-connected, then our output for G is no wider than $n-1$, matching the best known result of Kant and He. As a by-product, we obtain a much simpler proof for a corollary of Wagner's Theorem on realizers, due to Bonichon, Saëc, and Mosbah.

1 Introduction

In a *visibility representation* of a planar graph G, the nodes of G are represented by non-overlapping horizontal segments, called *node segments*, such that the node segments representing any two adjacent nodes of G are vertically visible to each other. (See Figure 1.) Computing compact visibility representations of graphs is not only fundamental in algorithmic graph theory [29,8] but also practically important in VLSI layout design [25].

Without loss of generality, one may assume that the input G is an n-node plane triangulation. Under the convention of placing the endpoints of node segments on the grid points, one can easily see that any visibility representation of G can be made no higher than $n-1$. Otten and van Wijk [23] gave the first known algorithm for visibility representations of planar graphs, but no width bound was provided for the output. Rosenstiehl and Tarjan [24], Tamassia and

---

* Corresponding author. Address: 128 Academia Road, Section 2, Taipei 115, Taiwan. Email: `hil@iis.sinica.edu.tw`. URL: `www.iis.sinica.edu.tw/~hil/`. This author's research is supported in part by NSC grant NSC-91-2213-E-001-028.

H. Alt and M. Habib (Eds.): STACS 2003, LNCS 2607, pp. 14–25, 2003.

Tollis [28], and Nummenmaa [22] independently proposed $O(n)$-time algorithms whose outputs are no wider than $2n-5$. Kant [14,16] improved the required width to at most $\lfloor \frac{3n-6}{2} \rfloor$ by decomposing G into its four-connected components and then combining the representations of the four-connected components into one drawing. Kant also left open the question of whether the upper bound $\lfloor \frac{3n-6}{2} \rfloor$ on the width is also a worst-case lower bound. In this paper, we provide a negative answer by presenting an algorithm whose output visibility representation is no wider than $\lfloor \frac{22n-42}{15} \rfloor$.

Our algorithm, just like that of Nummenmaa [22], is based upon the concept of canonical ordering for plane triangulations. Specifically, our algorithm draws G incrementally in a naive greedy manner according to any given canonical ordering of G. Although an arbitrary canonical ordering of G may yield a visibility representation with large width, we show that the required width can be bounded by $\lfloor \frac{22n-42}{15} \rfloor$, if we try three different canonical orderings obtained from Schnyder's realizer [27,26] for G. (Rosenstiehl and Tarjan [24] conjectured that selecting a good numbering of nodes to minimize the area of the corresponding visibility representation is NP-hard.) Our algorithm can easily be implemented to run in $O(n)$ time, bypassing the complicated subroutines of finding four-connected components and four-block trees [13] required by the best previously known algorithm of Kant [14,16]. Moreover, for the case that G has no degree-five node, the output of our algorithm is no wider than $\lfloor \frac{4n-7}{3} \rfloor$. Also, for the case that G is four-connected, the output of our algorithm is no wider than $n-1$, matching the best known result due to Kant and He [17,18].

Schnyder's realizer [27,26] for plane triangulation was invented for compact straight-line drawing of plane graph. Researchers [7,15,18,6,11,12,5,10] also obtained similar and other graph-drawing results using the concept of canonical ordering for tri-connected plane graph. Nakano [21] attempted to explain the hidden relation between these two concepts. Recently, Chiang, Lin, and Lu [4] presented a new algorithmic tool *orderly spanning tree* that extends the concept of st-ordering [9] (respectively, canonical ordering and realizer) for plane graphs unrequired to be biconnected (respectively, triconnected and triangulated). Orderly spanning tree has been successfully applied to obtain improved results in compact graph drawing [4,19,3], succinct graph encoding with query support [4], and design of compact routing tables [20]. Very recently, Bonichon, Gavoille, and Hanusse [1] obtained the best known upper bounds on the numbers of distinct labeled and unlabeled planar graphs based on *well orderly spanning tree*, a special case of orderly spanning tree. As a matter of fact, we first successfully obtained the results of this paper using orderly spanning tree, and then then found out that Schnyder's realizer suffices.

Suppose R is a realizer of G. Our analysis requires an equality (see Lemma 2) relating the number of internal nodes in all three trees of R and the number of faces of G intersecting with all three trees of R. The equality was proved very recently by Bonichon, Saëc, and Mosbah [2] as a corollary of the so-called Wagner's Theorem [30] on Schnyder's realizers. Their proof requires a careful

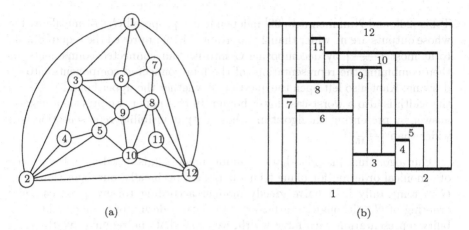

Fig. 1. A plane triangulation and one of its visibility representations.

case analysis for 32 different configurations. As a by-product, we give a simpler proof for the equality without relying on Wagner's Theorem on realizers.

The remainder of the paper is organized as follows. Section 2 gives the preliminaries. Section 3 describes and analyzes our algorithm. Section 4 discusses the tightness of our analysis.

2   Preliminaries

Let G be the input n-node *plane triangulation*, a planar graph equipped with a fixed planar embedding such that the boundary of each face is a triangle. $R = (T_1, T_2, T_3)$ is a *realizer* of G if

- the internal edges of G are partitioned into three edge-disjoint trees T_1, T_2, and T_3, each rooted at a distinct external node of G; and
- the neighbors of each internal node v of G form six blocks U_1, D_3, U_2, D_1, U_3, and D_2 in counterclockwise order around v, where U_j (respectively, D_j) consists of the parent (respectively, children) of v in T_j for each $j \in \{1, 2, 3\}$.

Schnyder [27,26] showed that a realizer of G can be obtained in linear time. For each $i \in \{1, 2, 3\}$, let ℓ_i be the node labeling of G obtained from the counterclockwise preordering of the spanning tree of G consisting of T_i plus the two external edges of G that are incident to the root of T_i. Let $\ell_i(v)$ be the label of v with respect to ℓ_i. For example, Figure 2 shows a realizer of the plane triangulation shown in Figure 1(a) with labeling ℓ_1.

Lemma 1 (see, e.g., [4,22]). *The following properties hold with respect to labeling ℓ_i for each $i \in \{1, 2, 3\}$. Let u_1 and u_2 be the nodes with labels 1 and 2 with respect to ℓ_i.*

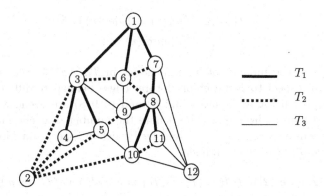

Fig. 2. A realizer for the plane triangulation shown in Figure 1(a), where $(3, 9, 9)$ and $(6, 9, 8)$ are the only two cyclic faces with respect to this realizer.

1. The subgraph G_k of G induced by the nodes with labels $1, 2, \ldots, k$ is 2-connected. The boundary of G_k's external face is a cycle C_k containing u_1 and u_2.
2. The node with label k is in the external face of G_{k-1} and its neighbors in G_{k-1} form an at least 2-element subinterval of the path $C_{k-1} - \{(u_1, u_2)\}$.
3. The neighbors of v form the following four blocks in counterclockwise order around v: (1) the parent of v in T_i, (2) the node set consists of the neighbors u in $G - T_i$ with $\ell_i(u) < \ell_i(v)$, (3) the children of v in T_i, and (4) the node set consists of the neighbors u in $G - T_i$ with $\ell_i(u) > \ell_i(v)$.

A labeling ℓ of G that labels the external nodes by 1, 2, and n and satisfies Lemmas 1(1) and 1(2) is a *canonical ordering* of G (e.g., see [22]). The *score* of an internal node v with respect to a canonical ordering ℓ of G is

$$\text{score}_\ell(v) = \min\{|N_\ell(v)^+|, |N_\ell(v)^-|\},$$

where $N_\ell(v)^-$ (respectively, $N_\ell(v)^+$) consists of the the neighbors u of v with $\ell(u) < \ell(v)$ (respectively, $\ell(u) > \ell(v)$). For example, if ℓ_1 is the labeling obtained from the tree T_1 consisting of the thick edges shown in Figure 2, then we have $\text{score}_{\ell_1}(v_8) = 2$, $\text{score}_{\ell_1}(v_9) = 1$, $\text{score}_{\ell_1}(v_{10}) = 2$, and $\text{score}_{\ell_1}(v_{11}) = 1$. The *score* of an internal node v of G with respect to realizer R of G is defined to be

$$\text{score}_R(v) = \text{score}_{\ell_1}(v) + \text{score}_{\ell_2}(v) + \text{score}_{\ell_3}(v).$$

For each $i \in \{1, 2, 3\}$, let $\text{score}(T_i)$ be the sum of $\text{score}_{\ell_i}(v)$ for all internal nodes v of G.

Let $\text{inter}(v)$ to be the number of trees in $R = (T_1, T_2, T_3)$ in which v is an internal (i.e., non-leaf) node. That is, $\text{inter}(v) = \sum_{i=1,2,3}[v \text{ is not a leaf of } T_i]$, where $[c]$ is 1 (respectively, 0) if condition c is true (respectively, false). For convenience to prove our main result, we define three subsets A, B and C of the internal nodes of G as follows.

$$A = \{v \mid \text{inter}(v) = 0\};$$

$$B = \{v \mid \text{inter}(v) = 2, \deg(v) = 5\};$$
$$C = \{v \notin B \mid \text{inter}(v) \geq 1\}.$$

Let ξ_i be the number of internal nodes in T_i. An internal face f of G is *cyclic* with respect to R if each of its three edges belongs to different trees of R. For example, in Figure 2, faces $(3,5,9)$ and $(6,9,8)$ are cyclic with respect to the underlining realizer. Let $c(R)$ denote the number of cyclic faces in R. The following lemma was recently proved by Bonichon, Saëc, and Mosbah [2]. Our alternative proof is much simpler.

Lemma 2 (see [2]). *If $R = (T_1, T_2, T_3)$ is a realizer of G, then $\xi_1 + \xi_2 + \xi_3 = n + c(R) - 1$.*

Proof. An internal face of G is *acyclic* with respect to R if it is not cyclic with respect to R. Let $\text{leaf}(T_i)$ consist of the leaves in T_i. Let $p_i(v)$ denote the parent of v in T_i. We claim that $v \in \text{leaf}(T_i)$ if and only if there exists an acyclic face $F_i(v)$ of G containing v, $p_{i_1}(v)$, and $p_{i_2}(v)$, where $\{i_1, i_2\} = \{1, 2, 3\} - \{i\}$. According to Euler's formula, the number of internal faces of G is $2n - 5$. Therefore, $\sum_{i \in I} |\text{leaf}(T_i)| = 2n - c(R) - 5$. It follows that $\xi_1 + \xi_2 + \xi_3 = 3(n - 2) - (2n - c(R) - 5) = n + c(R) - 1$.

Now it remains to prove that $v \in \text{leaf}(T_i)$ if and only if there exists a unique acyclic face $F_i(v)$ of G containing v, $p_{i_1}(v)$, and $p_{i_2}(v)$. Since (T_1, T_2, T_3) is a realizer of G, for each node $v \in \text{leaf}(T_i)$, there is a unique internal triangle face $F_i(v)$ of G containing v, $p_{i_1}(v)$, and $p_{i_2}(v)$. Obviously, edges $(v, p_{i_1}(v))$ and $(v, p_{i_2}(v))$ belong to T_{i_1} and T_{i_2}, respectively. By definition of realizer, $(p_{i_1}(v), p_{i_2}(v))$ should belong to T_{i_1} or T_{i_2}. Therefore, $F_i(v)$ is acyclic. Let $f = (u_1, u_2, u_3)$ be an acyclic face of G with $u_3 = p_i(u_1) = p_i(u_2)$. Since (T_1, T_2, T_3) is a realizer of G, one of the following two conditions holds (i) $u_1 = p_{i_1}(u_2)$ and $u_2 \in \text{leaf}(T_{i_2})$; and (ii) $u_2 = p_{i_2}(u_1)$ and $u_1 \in \text{leaf}(T_{i_1})$. Thus, each acyclic face f contains exactly one node v such that v is a leaf of T_i and the other two nodes in f are are $p_{i_1}(v)$ and $p_{i_2}(v)$, where $\{i_1, i_2\} = \{1, 2, 3\} - \{i\}$. □

Lemma 3. *If $v_1, v_2, \ldots, v_n$ are the nodes of G, where v_1, v_2, v_3 are external in G, then the following statements hold.*

1. $\xi_1 + \xi_2 + \xi_3 - 3 = \sum_{i=4}^{n} \text{inter}(v_i) \geq 2|B| + |C|$.
2. $\sum_{i=1}^{3} \text{score}(T_i) = \sum_{j=4}^{n} \text{score}_R(v_j)$.
3. $\text{score}_R(v) \geq 2 + 2 \cdot \text{inter}(v) + [v \in B]$.

Proof. It is not difficult to verify statements 1 and 2 by definitions of $\text{inter}(v)$ and $\text{score}(T_i)$. Statement 3 can be verified as follows. By definition of realizer and Lemma 1(3), we have $\text{score}_{\ell_1}(v) = \min\{|U_1| + |D_3| + |U_2|, |D_1| + |U_3| + |D_2|\} = \min\{|D_3| + 2, |D_1| + |D_2| + 1\}$. Similarly, it is not difficult to see $\text{score}_{\ell_2}(v) = \min\{|D_1| + 2, |D_2| + |D_3| + 1\}$ and $\text{score}_{\ell_3}(v) = \min\{|D_2| + 2, |D_1| + |D_3| + 1\}$. Clearly, statement 3 follows. □

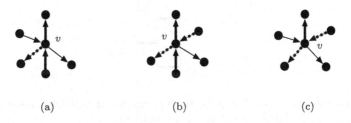

Fig. 3. Three different kinds of nodes in B. The orientation of an edge is from a child to its parent.

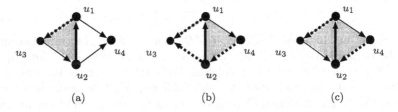

Fig. 4. At least one of (u_1, u_2, u_3) and (u_1, u_2, u_4) is cyclic.

Lemma 4. *If $R = (T_1, T_2, T_3)$ is a realizer of G, then*

$$\sum_{i=1}^{3} \text{score}(T_i) \geq \frac{23n - 78}{5}.$$

Proof. By Lemma 3, we have

$$\sum_{i=1}^{3} \text{score}(T_i) = \sum_{j=4}^{n} \text{score}_R(v_j)$$

$$\geq \sum_{j=4}^{n} \{3 + 2 \cdot \text{inter}(v_j)\} - |B|$$

$$= 3(n - 3) + 2(n + c(R) - 4) - |B|$$

$$= 5n + 2c(R) - 17 - |B|. \tag{1}$$

It remains to find an appropriate upper bound on $|B|$. Let $G[B]$ be the subgraph of G induced by B. Suppose that $G[B]$ has k connected components. Let B_i consists of the nodes of B in the i-th connected component of $G[B]$. Clearly, $B_1 \cup B_2 \cup \cdots \cup B_k = B$. We have the following two observations: (1) any two distinct nodes of A are not adjacent in G; and (2) each node in A is adjacent to at most one B_i. From above two observations, we know that the cardinality of the connected components of $G[A \cup B]$ is at least k. One can easily verify that the number of internal faces of $G - (A \cup B)$ is at least the cardinality of connected components of $G[A \cup B]$. According to Euler's formula, the number of internal faces of $G - (A \cup B)$ is at most $2(n - |A| - |B|) - 5 = 2(|C| + 3) - 5 = 2|C| + 1$.

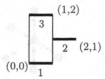

Fig. 5. A visibility representation for G_3. The number of uncovered points on the node segment of v_1, v_2, and v_3 are $0, 1$ and 2, respectively.

We have

$$k \leq 2|C| + 1. \tag{2}$$

Focus on the induced subgraph $G[B]$. Let u_1 and u_2 be two adjacent nodes of B. Without loss of generality, let u_1 be the parent of u_2 in T_1. Let (u_1, u_2, u_3) and (u_1, u_4, u_2) be the two triangular faces of G which is incident with the edge (u_1, u_2). Note that there are three different kinds of nodes in B as shown in Figure 3. By definition of realizer, one of the three conditions as shown in Figure 4 holds for u_1 and u_2. That is, at least one of (u_1, u_2, u_3) and (u_1, u_4, u_2) is cyclic. Consider a spanning forest F of $G[B]$ with $|B| - k$ edges. Each cyclic face contains at most two tree edges on F and each edge on F is incident to at least one cyclic face. Thus, we have

$$|B| - k \leq 2c(R). \tag{3}$$

It follows from Equations (1), (2), and (3) and Lemma 3(1) that $\text{score}(T_1) + \text{score}(T_2) + \text{score}(T_3) \geq \frac{23n + 6c(R) - 78}{5} \geq \frac{23n - 78}{5}$. □

3   Our Drawing Algorithm

Our algorithm computes a visibility representation for any n-node plane triangulation in a greedy manner. Given a canonical ordering ℓ of G, let v_i to be the node with $\ell(v) = i$. Clearly, v_1, v_2, and v_n are the external nodes of G. Let G_k be the subgraph of G induced by $v_1, v_2, \ldots, v_k$. Let $y(v)$ denote the y-coordinate of the node segment $\Gamma(v)$ of node v. A point (x', y') on the plane is *covered* by a node segment $\Gamma(v)$ drawn between $(x_1, y(v))$ and $(x_2, y(v))$ if $y(v) > y'$ and $x_1 \leq x' \leq x_2$. A point on the plane is *uncovered* if it is not covered by any node segment. See Figure 5 for an example.

Algorithm Visibility(G)

1. Let $\Gamma(G_3)$ be as shown in Figure 5.
2. For $k = 4$ to n, do the following steps.
    (a) For each neighbor v of v_k in G_{k-1} and $\Gamma(v)$ does not have any uncovered point in G_{k-1}, widen our current drawing to make one more uncovered point on $\Gamma(v)$.

(b) Let $y(v_k) = \max\{y(v) \mid v \text{ is a neighbor of } v_k \text{ in } G_{k-1}\} + 1$. We draw $\Gamma(v_k)$ from $(x_1, y(v_k))$ to $(x_2, y(v_k))$, where x_1 is the x-coordinate of the rightmost uncovered point on the node segment of the leftmost neighbor in G_{k-1} and x_2 is the x-coordinate of the leftmost uncovered point on the node segment of the rightmost neighbor in G_{k-1}.

When drawing the node segment $\Gamma(v_k)$, we visit each node segment $\Gamma(v_i)$ with $i < k$. Since it is not difficult to get a linear-time implementation by modifying Nummenmaa's linear-time algorithm [22], we leave out the linear-time implementation of our algorithm due to the page limit. For the rest of the section, we prove the correctness of our algorithm and show that the output drawing is no wider than $\lfloor \frac{22n-42}{15} \rfloor$.

Lemma 5. *The output of* Visibility(G) *is a visibility representation of* G.

Proof. Suppose $\Gamma(G_k)$ is the output drawing of Visibility for G_k. We prove it by induction on k. Clearly, $\Gamma(G_3)$ is a visibility representation of G_3. Suppose $\Gamma(G_{k-1})$ is a visibility representation for G_{k-1}. By Lemma 1(1), we know that v_k and its neighbors in G_{k-1} are all on the external face of G_{k-1}. Step 2(a) ensures that if v is a neighbor of v_k in G_{k-1}, then $\Gamma(v)$ has at least one uncovered point. After drawing $\Gamma(v_k)$ in Step 2(b), one can see that the node segments of v_k and its neighbors in G_{k-1} are horizontally visible to each other in $\Gamma(G_k)$. Thus, $\Gamma(G_k)$ is a visibility representation of G_k. $\square$

Theorem 1. *Let G be an n-node plane triangulation with $n > 3$. Then, there exists an output drawing of* Visibility(G) *which is a visibility representation for G of width at most* $\lfloor \frac{22n-42}{15} \rfloor$.

Proof. In Step 2(a) of Visibility(G), each neighbor v of v_k in G_{k-1} is examined to ensure having at least one uncovered point on $\Gamma(v)$ as $4 \leq k \leq n$. It will clearly be examined $3n - 9$ times. We claim that the number of times that Step 2(a) does not widen the drawing is at least $1 + \sum_{k=3}^{n-1} \text{score}_\ell(v_k)$.

Observe a drawing property in Step 2(b): v_i is a neighbor of v_k and $i < k$, the number of uncovered points on the node segment of v_i decreases by exactly one if v_k is not the last neighbor of v_i. Clearly, the number of uncovered points on the $\Gamma(v)$ is at least $|N_\ell(v)^-|$ after drawing $\Gamma(v)$ and it will be examined $|N_\ell(v)^+|$ times throughout the execution. Consider the following two cases: (i) if $|N_\ell(v)^-| \geq |N_\ell(v)^+|$, then widening $\Gamma(v)$ is skipped all the $|N_\ell(v)^+|$ times after drawing $\Gamma(v)$; (ii) if $|N_\ell(v)^-| < |N_\ell(v)^+|$, then widening $\Gamma(v)$ is skipped in the first $|N_\ell(v)^-|$ times after drawing $\Gamma(v)$. Since $\Gamma(v_2)$ has one uncovered point in $\Gamma(G_3)$, the widening is skipped in the first time of examining. Thus, the widening In Step 2(a) of Visibility(G) is skipped at least $1 + \sum_{k=3}^{n-1} \text{score}_\ell(v_k)$ times throughout the execution. We complete the claim.

We know that the number of times to widen our drawing is at most $(3n - 9) - (1 + \sum_{k=3}^{n-1} \text{score}_\ell(v_k))$ and the width of $\Gamma(G_3)$ is 2. The width of the output drawing is $3n - 8 - \sum_{k=3}^{n-1} \text{score}_\ell(v_k)$. By Lemma 4 and the fact that a realizer is obtainable in linear time, one ordering ℓ obtained from the given realizer with

$\text{score}(T_\ell) = \sum_{k=3}^{n-1} \text{score}_\ell(v_k) \geq \frac{23n-78}{15}$ is obtainable in linear time. Thus, the width of the output drawing is at most $(3n-8) - \lceil \frac{23n-78}{15} \rceil \leq \lfloor \frac{22n-42}{15} \rfloor$. □

Kant and He [17,18] proved that a visibility representation of 4-connected graph G is obtainable on a grid of width at most $n-1$. We give an alternative proof which is easy to check by Visibility(G).

Corollary 1. *If G is an n-node 4-connected plane triangulation, then there exists an $O(n)$-time obtainable visibility representation for G no wider than $n-1$.*

Proof. Kant and He [17,18] proved that a 4-canonical labeling ℓ of 4-connected triangulation graph G is derivable in linear time such that ℓ is a canonical labeling and v_k has at least two neighbors in G_{k-1} and has at least two neighbors in $G - G_{k-1}$ for each $3 \leq k \leq n-2$. We use this labeling ℓ for Visibility(G). Clearly, $\text{score}_\ell(v_k) = \min\{|N_\ell(v_k)^+|, |N_\ell(v_k)^-|\} \geq 2$ for each $3 \leq k \leq n-2$ and $\text{score}_\ell(v_{n-1}) = 1$. Therefore, $\sum_{k=3}^{n-1} \text{score}_\ell(v_k) \geq 2n-7$. The width of the output drawing is $3n - 8 - \sum_{k=3}^{n-1} \text{score}_\ell(v_k) \leq 3n - 8 - (2n-7) = n-1$. □

Corollary 2. *If G is an n-node graph without degree-5 nodes, then there exists an $O(n)$-time obtainable visibility representation for G that is no wider than $\lfloor \frac{4n-7}{3} \rfloor$.*

Proof. If G is a graph without degree-5 nodes, then $|B| = 0$. By Equation (1) in Lemma 4, we have $\text{score}(T) \geq \frac{5n-17}{3}$ for the tree T of a realizer. Thus, the width of the output drawing is at most $(3n-8) - \lceil \frac{5n-17}{3} \rceil \leq \lfloor \frac{4n-7}{3} \rfloor$. □

Corollary 3. *If G is an n-node graph without degree-3 nodes, then there exists an $O(n)$-time obtainable visibility representation for G that is no wider than $\lfloor \frac{4n-9}{3} \rfloor$.*

Proof. If G is a graph without degree-3 nodes, then $\text{score}_R(v) \geq 5$ for each internal node in G. It follows that we have $\text{score}(T) \geq \frac{5n-15}{3}$. Thus, the width of the output drawing is at most $(3n-8) - \lceil \frac{5n-15}{3} \rceil \leq \lfloor \frac{4n-9}{3} \rfloor$. □

Suppose a visibility representation of G is given and f is an internal face of G. Let G_f be the graph obtained by adding a new degree-3 node in f. As pointed out by a reviewer of this paper, an improved width would have been achieved via Corollary 3, if a visibility representation of G_f one unit wider could always be obtained from that of G. Unfortunately, this nice property does not always hold, as illustrated by Figure 6.

4   Tightness of Our Analysis

We have the following theorem regarding the tightness of our analysis on the required width.

Theorem 2. *There is an n-node plane triangulation G such that any visibility representations of G obtained from our algorithm with respect to any canonical ordering has width at least $\lfloor \frac{4n-9}{3} \rfloor$.*

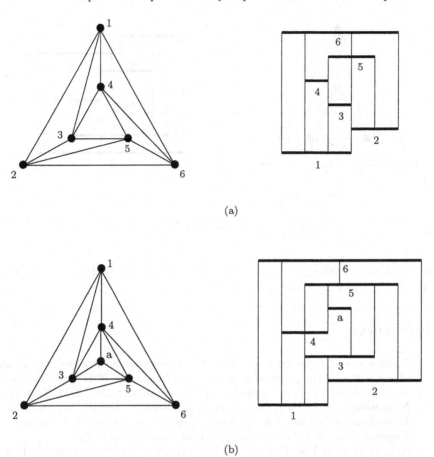

(a)

(b)

Fig. 6. An example of adding a degree-3 node increases the width by two.

Proof. Suppose $G(k)$ is the graph of k nested triangles as shown in Figure 7(a). Let ℓ be a canonical ordering of $G(k)$ and the nodes on the external face have labels 1, 2, and $3k$. By Lemma 1(1), it is not difficult to observe that the graph removing the nodes on the external face is $G(k-1)$ and the resulting ordering is still a canonical ordering which labels nodes from 3 to $3k-1$. It follows that we can prove the theorem by induction on k. Clearly, the visibility representation drawings for $G(2)$ have width at least 5. By the induction hypothesis, $\Gamma(G(k))$ has width at least $4k+1$. To make $\Gamma(1), \Gamma(2)$, and $\Gamma(3)$ visible to each other in $\Gamma(G(k+1))$, we have to increase the width by at least two. To make each of $\Gamma(1), \Gamma(2)$, and $\Gamma(3)$ visible to the corresponding node segments of its neighbors in $G(k)$, we also have to increase the width by at least two. Hence, each triangle needs to increase the width by at least 4. The visibility representation drawings for $G(k+1)$ have width at least $\frac{4n-9}{3}$. If n is not a multiple of 3, it is not difficult to construct G such that drawing each node segment of the corresponding node

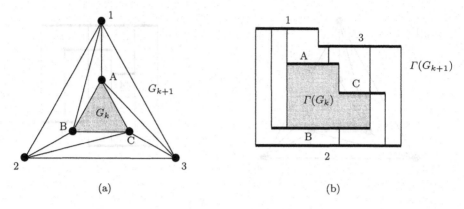

(a)   (b)

Fig. 7. A worse-case lower bound example.

which does not belong to nested triangles increases the width by at least one. Hence, the width is at least $\left\lfloor \frac{4n-9}{3} \right\rfloor$. □

References

1. N. Bonichon, C. Gavoille, and N. Hanusse. An information-theoretic upper bound of planar graphs using triangulation. In *Proceedings of the 20th Annual Symposium on Theoretical Aspects of Computer Science*, Berlin, Germany, 2003. To appear.
2. N. Bonichon, B. L. Saëc, and M. Mosbah. Wagner's theorem on realizers. In *Proceedings of the 29th International Colloquium on Automata, Languages, and Programming*, LNCS 2380, pages 1043–1053, Málaga, Spain, 2002.
3. H.-L. Chen, C.-C. Liao, H.-I. Lu, and H.-C. Yen. Some applications of orderly spanning tree in graph drawing. In *Proceedings of the 10th International Symposium on Graph Drawing*, LNCS 2528, pages 332–343, Irvine, California, 2002.
4. Y.-T. Chiang, C.-C. Lin, and H.-I. Lu. Orderly spanning trees with applications to graph encoding and graph drawing. In *Proceedings of the 12th Annual ACM-SIAM Symposium on Discrete Algorithms*, pages 506–515, Washington, D. C., USA, 7–9 Jan. 2001.
5. M. Chrobak and G. Kant. Convex grid drawings of 3-connected planar graphs. *International Journal of Computational Geometry & Applications*, 7(3):211–223, 1997.
6. H. de Frayseix, P. Ossona de Mendez, and P. Rosenstiehl. On triangle contact graphs. *Combinatorics, Probability and Computing*, 3:233–246, 1994.
7. H. de Frayseix, J. Pach, and R. Pollack. How to draw a planar graph on a grid. *Combinatorica*, 10:41–51, 1990.
8. G. Di Battista, R. Tamassia, and I. G. Tollis. Constrained visibility representations of graphs. *Information Processing Letters*, 41:1–7, 1992.
9. S. Even and R. E. Tarjan. Computing an *st*-numbering. *Theoretical Computer Science*, 2:436–441, 1976.
10. U. Fößmeier, G. Kant, and M. Kaufmann. 2-visibility drawings of planar graphs. In S. North, editor, *Proceedings of the 4th International Symposium on Graph Drawing*, LNCS 1190, pages 155–168, California, USA, 1996.

11. X. He. On floor-plan of plane graphs. *SIAM Journal on Computing*, 28(6):2150–2167, 1999.
12. X. He. A simple linear time algorithm for proper box rectangular drawings of plane graphs. *Journal of Algorithms*, 40(1):82–101, 2001.
13. A. Kanevsky, R. Tamassia, G. Di Battista, and J. Chen. On-line maintenance of the four-connected components of a graph. In *Proceedings of the 32nd Annual Symposium on Foundations of Computer Science*, pages 793–801, San Juan, Puerto Rico, 1991. IEEE.
14. G. Kant. A more compact visibility representation. In *Proceedings of the 19th Workshop on Graph-Theoretic Concepts in Computer Science*, LNCS 790, pages 411–424, Utrecht, Netherlands, 1994.
15. G. Kant. Drawing planar graphs using the canonical ordering. *Algorithmica*, 16(1):4–32, 1996.
16. G. Kant. A more compact visibility representation. *International Journal of Computational Geometry & Applications*, 7(3):197–210, 1997.
17. G. Kant and X. He. Two algorithms for finding rectangular duals of planar graphs. In *Proceedings of the 19th Workshop on Graph-Theoretic Concepts in Computer Science*, LNCS 790, pages 396–410, Utrecht, Netherlands, 1994.
18. G. Kant and X. He. Regular edge labeling of 4-connected plane graphs and its applications in graph drawing problems. *Theoretical Computer Science*, 172(1-2):175–193, 1997.
19. C.-C. Liao, H.-I. Lu, and H.-C. Yen. Floor-planning via orderly spanning trees. In *Proceedings of the 9th International Symposium on Graph Drawing*, LNCS 2265, pages 367–377, Vienna, Austria, 2001.
20. H.-I. Lu. Improved compact routing tables for planar networks via orderly spanning trees. In O. H. Ibarra and L. Zhang, editors, *Proceedings of the 8th International Conference on Computing and Combinatorics*, LNCS 2387, pages 57–66, Singapore, August 15–17 2002.
21. C.-i. Nakano. Planar drawings of plane graphs. *IEICE Transactions on Information and Systems*, E83-D(3):384–391, Mar. 2000.
22. J. Nummenmaa. Constructing compact rectilinear planar layouts using canonical representation of planar graphs. *Theoretical Computer Science*, 99(2):213–230, 1992.
23. R. Otten and J. van Wijk. Graph representations in interactive layout design. In *Proceedings of the IEEE International Symposium on Circuits and Systems*, pages 914–918, 1978.
24. P. Rosenstiehl and R. E. Tarjan. Rectilinear planar layouts and bipolar orientations of planar graphs. *Discrete & Computational Geometry*, 1(4):343–353, 1986.
25. M. Schlag, F. Luccio, P. Maestrini, D. T. Lee, and C. K. Wong. A visibility problem in VLSI layout compaction. In F. P. Preparata, editor, *Advances in Computing Research*, volume 2, pages 259–282. JAI Press Inc. Greenwich, CT, 1985.
26. W. Schnyder. Planar graphs and poset dimension. *Order*, 5:323–343, 1989.
27. W. Schnyder. Embedding planar graphs on the grid. In *Proceedings of the First Annual ACM-SIAM Symposium on Discrete Algorithms*, pages 138–148, 1990.
28. R. Tamassia and I. G. Tollis. A unified approach to visibility representations of planar graphs. *Discrete & Computational Geometry*, 1(4):321–341, 1986.
29. R. Tamassia and I. G. Tollis. Planar grid embedding in linear time. *IEEE Transactions in Circuits and Systems*, 36:1230–1234, 1989.
30. K. Wagner. Bemerkungen zum Vierfarbenproblem. *Jahresber Deutsche Math. -Verein*, 46:26–32, 1936.

Rectangle Visibility Graphs: Characterization, Construction, and Compaction

Ileana Streinu[1]* and Sue Whitesides[2]**

[1] Computer Science Department, Smith College, Northampton, MA 01063, USA
streinu@cs.smith.edu
[2] School of Comp. Sci., McGill Univ., Montreal, Quebec H3A 2A7, CANADA
sue@cs.mcgill.ca

Abstract. Non-overlapping axis-aligned rectangles in the plane define visibility graphs in which vertices are associated with rectangles and edges with visibility in either the horizontal or vertical direction. The recognition problem for such graphs is known to be NP-complete. This paper introduces the *topological rectangle visibility graph*. We give a polynomial time algorithm for recognizing such a graph and for constructing, when possible, a realizing set of rectangles on the unit grid. The bounding box of these rectangles has optimum length in each dimension. The algorithm provides a compaction tool: given a set of rectangles, one computes its associated graph, and runs the algorithm to get a compact set of rectangles with the same visibility properties.

1 Introduction

The problem of constructing visibility representations of graphs has received a lot of attention in the graph drawing literature, motivated, for example, by research in VLSI and visualization of relationships between database entities.

One particularly successful approach represents graphs with horizontal bars for the vertices and vertical, unobstructed visibility lines between bars for the edges. The class of graphs allowing such a representation (*bar visibility graphs*) has been studied by several authors ([26], [23], [27], [28], [20]), and its properties are well understood: it consists of planar st-graphs (i.e., those planar graphs having an embedding for which all cut vertices lie on the same face). Several authors have studied the properties of generalizations of such graphs to higher dimensions ([1], [4], [5], [6], [7], [8], [10], [11], [15], [16], [22], [24] and [25]), but so far, no simple characterizations have been found.

For *rectangle visibility graphs*, vertices correspond to non-overlapping isothetic rectangles, and edges correspond to unobstructed, axis-aligned lines of sight between rectangles (see section 2).

Any set of non-overlapping isothetic rectangles contains much more information of a topological, combinatorial nature than simply the visibility information

---

* Partially supported by NSF RUI grant CCR-0105507.
** Partially supported by NSERC and FCAR.

H. Alt and M. Habib (Eds.): STACS 2003, LNCS 2607, pp. 26–37, 2003.

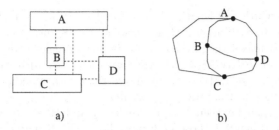

Fig. 1. a) A set $\mathcal{R}$ of rectangles. b) Its rectangle visibility graph.

in the horizontal and vertical directions. For example, walking the boundary of a rectangle in, say, the counterclockwise sense, gives rise to a cyclic ordering on those rectangles that are visible from the given one. In this paper, we make use of this kind of additional information to define the notion of a *topological rectangle visibility graph* (TRVG), defined in section 2. In brief, it is a graph arising from the horizontal and vertical visibilities among a set of non-overlapping isothetic rectangles in the plane, where the graph records ordering and direction information for these visibilities. Thus, every such set of rectangles in the plane gives rise to a TRVG.

Unlike bar visibility graphs, no combinatorial characterization of rectangle visibility graphs is known, and furthermore, it has been shown by Shermer ([24], [25]) that the problem of recognizing them is NP-complete.

The vertical visibilities among horizontal segments are parallel, so bar visibility graphs are planar. However, the axis-aligned visibilities of axis-aligned rectangles cross. Our characterization of visibility graphs of axis-aligned rectangles weaves together the properties of four planar graphs, namely, the planar vertical and horizontal visibility graphs arising from the rectangles, and the duals of these graphs.

Results. We give a complete combinatorial characterization of TRVG's. For any graph structure that satisfies the necessary and sufficient conditions to be a TRVG, we give an algorithm that constructs a set of rectangles on the unit grid realizing the graph. The bounding box of this grid realization is optimum.

Our construction provides a compaction method for axis-aligned rectangles drawn on a grid. A new set of such rectangles is produced that has the same visibility properties as the original set, such that both the width and the height of the bounding box of the new set is optimum for visibility preserving compactions of the original set. Compaction algorithms typically are heuristics that alternate between processing one direction, then the other. Our approach considers both directions simultaneously and produces an optimal result.

2   Preliminaries

We consider sets $\mathcal{R}$ of non-overlapping, isothetic (i.e., axis-aligned) rectangles in the plane and the horizontal and vertical visibilities among them. The rectangles have positive width and length. Two rectangles are said to be horizontally (vertically) visible to each other if there exists an unobstructed horizontal (vertical) line of sight between them. Such a set $\mathcal{R}$ of rectangles gives rise to a *rectangle visibility graph* (RVG), which associates a vertex with each rectangle and an edge between two vertices if and only if their corresponding rectangles "see" each other either horizontally or vertically.

A *line of sight* between two rectangles A and B is a closed line segment, possibly of zero length, whose endpoints lie in the boundaries of A and B. A line of sight must be unobstructed, i.e., its relative interior cannot intersect the closure of any of the rectangles. A line of sight between A and B is *thick* if it can be extended to an open *visibility rectangle* between A and B, that is, a rectangle with one side of positive length contained in the boundary of A and the opposite side contained in the boundary of B, where this open visibility rectangle does not intersect the interior of any rectangle in $\mathcal{R}$. While the sides of a visibility rectangle between rectangles A and B must have a pair of opposite, positive length sides contained in the boundaries of A and B, the other pair of sides is allowed to have zero length; thus if rectangles A and B share a boundary segment of positive length, then they "see" each other by a thick line of sight of zero length.

In this paper, the rectangles in $\mathcal{R}$ are open and may share boundary points. Lines of sight are *thick*. This is a standard model of visibility.

Topological Rectangle Visibility Graphs (TRVG's). We adopt the convention that vertical lines of sight (visibilities) are directed upward and that horizontal lines of sight are directed from left to right.

Note that the graph in Figure 1(b) fails to record many visibility properties of $\mathcal{R}$, such as the directions and multiplicities of visibilities. For example C has upwardly directed visibility lines to A on both sides of B. The cyclic order in which incoming and outgoing visibility lines are encountered in traversing the boundary of a rectangle is not recorded.

The graph of Figure 1(b) also does not record how and whether a given rectangle can see infinitely far. This can be facilitated by placing a *frame* around $\mathcal{R}$ consisting of long, wide rectangles N and S at the top and bottom of the picture, and long, tall rectangles W and E at the left and right of the picture (W and N should not see each other, and likewise for W and S, and so on).

The following definition captures the visibility properties of $\mathcal{R}$.

Definition 1. *Let $\mathcal{R}$ be a set of axis-aligned rectangles. The **topological rectangle visibility graph** of $\mathcal{R}$ consists of a pair (D_V, D_H) of graphs. Here, D_V records, with multiplicities, the upwardly directed, cyclically ordered vertical visibilities of $\mathcal{R} \cup \{S, N\}$; D_H records, with multiplicities, the left-to-right directed, cyclically ordered horizontal visibilities of $\mathcal{R} \cup \{W, E\}$.*

Figure 2(a) shows the horizontal and vertical visibilities among a set $\mathcal{R}$ of rectangles and its frame. Parts (b) and (c) show the neighborhood of vertex A in D_V and in D_H, respectively. Note that in (b), the visibilities are recorded with multiplicity, and in the correct cyclic order. Starting on the top edge of rectangle A and traversing the boundary clockwise, one encounters an outgoing

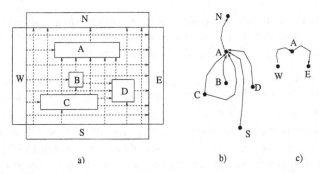

Fig. 2. a) Visibilities in a framed set $\mathcal{R}$. b) The neighborhood of A in D_V. c) The neighborhood of A in D_H.

vertical visibility to N, an outgoing horizontal visibility to E, incoming vertical visibilities from D, S, C, B, and C again, and an incoming horizontal visibility from W. Of course the graphs in (b) and (c) can be drawn together to produce a single graph (hence the term topological rectangle visibility "graph"); when this is done, the outgoing vertical visibilities are followed in cyclic order by the outgoing horizontal visibilities, then by the incoming vertical visibilities and the incoming horizontal visibilities. Note that the graph of vertical visibilities is planar, as is the graph of horizontal visibilities. However, when the two graphs are combined, the resulting graph may have many edge crossings.

We shorten the expression "topological rectangle visibility graph" to TRVG and write $\text{TRGV}(\mathcal{R}) = (D_V, D_H)$. It is easy to see that D_V has a single source S and a single sink N, and that its underlying undirected graph is 2-connected. Graph D_H has similar properties, with source W and sink E.

Basic Definitions. First we recall some standard definitions (see [9], chapter 4). A directed acyclic graph with a single source s and a single sink t is called an *st-graph*. Given an *st*-graph with n vertices and m edges, topological sorting yields a labelling $n(u)$ of the vertices with labels in $\{1, \ldots, n\}$ so that $n(s) = 1$; $n(t) \leq n$; and if there is a directed path from u to v, then $n(u) < n(v)$. This labelling is called an *st-numbering* and can be computed in $O(n + m)$. For example, we can choose $n(u)$ to be the length of a longest directed path from s to u.

An *embedded* planar graph is a planar graph together with cyclic orderings (called *rotations*) of edges at each vertex such that these orderings arise from a drawing of the graph in the plane with vertices drawn as points and edges drawn as noncrossing curves; we assume that the infinite face is specified.

A set of rotations can be constructed for a planar graph in $O(n)$ time using linear time planarity testing algorithms ([3], [12], [14], [18]). Given the rotations, the edge cycles bounding the faces can be computed in $O(n)$ time.

A *planar st-graph* is an *st*-graph that is both planar and embedded, with s and t on the external face. Thus the graphs D_H and D_V of a $\text{TRVG}(\mathcal{R})$ are planar *st*-graphs.

Definition 2. *A pseudo TRVG is a pair (D_V, D_H) of planar st-graphs, where the vertex set of D_V is $V \cup \{S, N\}$, and the vertex set of D_H is $V \cup \{W, E\}$. Both*

D_V and D_H are directed, acyclic planar st-graphs whose underlying undirected graphs are 2-connected. The source and sink of D_V are S and N, respectively, and the source and sink of D_H are W and E, respectively.

Definition 3. Let (D_V, D_H) be a pseudo TRVG. The TRVG recognition problem is to determine whether there exists a set $\mathcal{R}$ of axis-aligned rectangles such that $TRVG(\mathcal{R}) = (D_V, D_H)$.

The problem shown by Shermer ([24] and [25]) to be NP-complete is that of recognizing rectangle visibility graphs of the type shown in Fig. 1, where the given graph does not specify directions, multiplicities, cyclic orderings of edges at vertices, or axis-aligned lines of sight to infinity. By contrast, the graphs we study have much more visibility information.

Basic Properties. Here we recall some basic properties (see, for example, [9]). Let G be a planar st-graph whose underlying graph is 2-connected, e.g., the D_H or D_V of a TRVG($\mathcal{R}$). The following properties hold (see [26]).

1. All the entering edges of any vertex v appear consecutively in the rotation for v, as do all the leaving edges.

2. Each vertex other than s and t has both incoming and outgoing edges.

3. At any vertex having both incoming and outgoing edges, there are two distinguished faces that split the incoming edges from the outgoing edges. Traversing the rotation of edges at a vertex v in the cw sense, we call the face that splits the incoming edges from the outgoing edges the *left face* of v, denoted $L(v)$; the face that splits the outgoing edges from the incoming edges is the *right face* of v, denoted $R(v)$. See Fig.3(a).

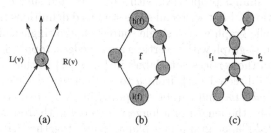

(a)   (b)   (c)

Fig. 3. a) Faces $L(v)$ and $R(v)$ of v. b) Vertices $h(f)$ and $l(f)$ of a face f. c) The dual edge (f_1^*, f_2^*) from face f_1 to face f_2.

4. The boundary of each face f has exactly one local source or "low" vertex $l(f)$ and exactly one local sink or "high" vertex $h(f)$, and the boundary consists of exactly two directed paths from $l(f)$ to $h(f)$, called the left and right chains of f. Note that for any st-numbering of the graph, among the vertices on the boundary of f, $l(f)$ receives the smallest label and $h(f)$ receives the largest label (hence the terms "low" and "high"). See Fig.3(b).

For a planar st-graph whose underlying undirected graph is 2-connected, it is convenient to draw an imaginary edge through the external face between the source and the sink, thus splitting the external face into a "left" part and a "right" part. These parts have separate representatives, s^* and t^*, respectively, in the dual graph. The dual of a planar st-graph with a 2-connected underlying undirected graph has vertices corresponding to the faces of the original graph. An internal face f in the primal graph becomes a vertex f^* in the dual. Note that the right face $R(u)$ of a vertex u becomes a vertex $(R(u))^*$ in the dual. We denote $(R(u))^*$ by $R^*(u)$. Similarly, we denote $(L(u))^*$ by $L^*(u)$. Traversing a directed edge $e = (u, v)$ determines a face $L(e)$ on the left side of e and a face $R(e)$ on the right side of e. In the dual graph, which contains vertices $L^*(e)$ and $R^*(e)$ corresponding to primal faces $L(e)$ and $R(e)$, this generates a dual edge, which we direct $(L^*(e), R^*(e))$ when the primal graph is D_V; for D_H, we direct the dual edge of e as $(R^*(e), L^*(e))$. (see Fig. 3(c)).

Starting at the source in the primal graph, one can follow the left-most outgoing edge at each encountered vertex to determine a left-most directed st-path. This path bounds a portion of the external face, and that portion is associated with a source vertex s^* in the dual. Similarly, the other portion of the external face is bounded by a right-most st-path and is associated in the dual with a sink vertex t^*. The directed edges in the left-most st-path are crossed from left to right by the edges in the dual leaving s^*. Similarly, the right-most chain is crossed from left to right by the edges directed into t^*.

We note one additional property from [9].

5. The dual graph, defined as above, of a planar st-graph whose underlying undirected graph is 2-connected is a planar s^*t^*-graph whose underlying undirected graph is 2-connected.

Lemma 1 (Tamassia and Tollis [26]). *Let D be a planar st-graph whose underlying undirected graph is 2-connected, and let D^* be its dual. Let u and v be vertices of D. Then *exactly* one of the following four conditions holds.*

1. *D has a path from u to v.*
2. *D has a path from v to u.*
3. *D^* has a path from $R^*(u)$ to $L^*(v)$.*
4. *D^* has a path from $R^*(v)$ to $L^*(u)$.*

3   Main Result

We will be working with two primal graphs D_H and D_V and their respective duals, D_H^* and D_V^*. It is crucial to keep in mind that D_H^* is in general *not* the same as D_V. We choose notation to emphasize this point. For example, the face to the left of edge e in D_H is denoted $L_H(e)$ and corresponds to a vertex of D_H^* denoted $L_H^*(e)$ (see Fig. 4a),b),c)).

Theorem 1. *(Main Theorem) Let (D_V, D_H) be a pseudo TRVG. Then there is a set $\mathcal{R}$ of rectangles such that $TRVG(\mathcal{R}) = (D_V, D_H)$ if and only if both the two following combinatorial conditions hold.*

1. If D_V has a directed path from vertex u to vertex v then D_H^* has a directed path from vertex $L_H^*(u)$ to vertex $R_H^*(v)$.

2. If D_H has a directed path from vertex u to vertex v, then D_V^* has a directed path from vertex $R_V^*(u)$ to vertex $L_V^*(v)$.

The *necessity* of these conditions is not difficult to prove, by considering some imagined set $\mathcal{R}$ of rectangles realizing (D_V, D_H), so we focus on sketching the proof of sufficiency by constructing a realizing set of rectangles when the conditions 1 and 2 both hold.

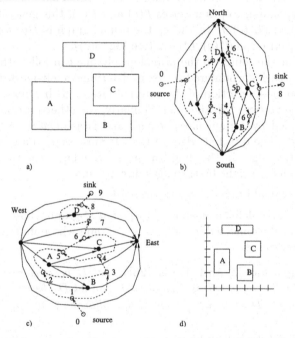

Fig. 4. a) A set of rectangles (frame not shown). b) Its graphs D_V and D_V^*. c) Its graphs D_H and D_H^*. d) The new set of rectangles that will be constructed.

Construction of the Rectangles. Given a pair (D_V, D_H) satisfying the conditions of Theorem 1, we describe now the construction of the set of rectangles R (Fig. 4). Later we prove correctness, i.e, that the interiors of the rectangles do not intersect and that $\mathrm{TRVG}(\mathcal{R}) = (D_V, D_H)$.

1. Compute D_H^* from D_H.
2. Perform a topological sorting on D_H^*. Denote it by n_H^*.
3. Compute D_V^* from D_V.
4. Perform a topological sorting on D_V^*. Denote it by n_V^*.
5. For each vertex u in the shared vertex set V, define its horizontal interval $hor(u)$ to be $(n_V^*(L_V^*(u)), n_V^*(R_V^*(u)))$; define its vertical interval $vert(u)$ to be $(n_H^*(R_H^*(u)), n_H^*(L_H^*(u)))$.
6. Define the box $B(u)$ associated with vertex u to be the open box $B(u) = hor(u) \times vert(u)$.

Proof of Theorem 1. To prove that the conditions 1 and 2 are sufficient, we prove that the rectangles produced by the construction do not overlap, that they realize all the visibilities as described in the pair (D_V, D_H) of the given pseudo TRVG, and that no additional visibilities occur. We have broken the proof of sufficiency into a series of lemmas which we now give.

Lemma 2. *Let u and v be vertices in the set V of shared vertices of D_V and D_H. If $hor(u) \cap hor(v) \neq \emptyset$ then there exists a directed path in D_V from u to v or from v to u.*

proof sketch: To get a contradiction, assume that $hor(u) \cap hor(v)$ is not empty, but that there is no directed path in D_V from u to v or from v to u. Then by Lemma 1, there exists a path in D_V^* from $R_V^*(u)$ to $L_V^*(v)$ or from $R_V^*(v)$ to $L_V^*(u)$. Since the topological sorting n_V^* on the dual D_V^* has to be compatible with this path, it must be true that $n_V^*(R_V^*(u)) < n_V^*(L_V^*(v))$ or that $n_V^*(R_V^*(v)) < n_V^*(L_V^*(u))$. But this would mean, respectively, that $hor(u)$ is strictly left of $hor(v)$, or that $hor(u)$ is strictly right of $hor(v)$, contradicting the assumption that $hor(u) \cap hor(v)$ is not empty.   △

Lemma 3. *Let u and v be vertices in the set V of shared vertices. If $vert(u) \cap vert(v) \neq \emptyset$ then there exists a directed path in D_H from u to v or from v to u.*

proof sketch: Similar to that of Lemma 2.   △

Lemma 4. *The constructed rectangles do not overlap.*

proof sketch: To get a contradiction, assume that $B(u) \cap B(v) \neq \emptyset$ for some u, v in the shared vertex set V of D_V and D_H. Then $hor(u) \cap hor(v) \neq \emptyset$ and $vert(u) \cap vert(v) \neq \emptyset$. Then by Lemma 2, there exists a directed path in D_V from u to v or from v to u, and by Lemma 3, there exists a directed path in D_H from u to v or from v to u. Suppose, for instance, that there is a directed path in D_H from u to v. Then by the assumption that condition 1 of Theorem 1 holds, there exists a path in D_V^* from $R_V^*(u)$ to $L_V^*(v)$. Then by Lemma 1, it can't be true that there also exists a path in D_V from u to v or from v to u. Contradiction.   △

Lemma 5. *No new visibilities are added.*

proof sketch: Assume that D_V, say, does not contain a directed path from u to v. By Lemma 1, exactly one of the following must happen:

1. D_V has a directed path from v to u. But then by condition 1 of Theorem 1 (interchanging the roles of u and v), graph D_H^* has a directed path from $L_H^*(v)$ to $R_H^*(u)$, which implies that the y-coordinate of the top side of $B(v)$, namely $n_H^*(L_H^*(v))$, is less than the y-coordinate of the bottom side of $B(u)$, namely $n_H^*(R_H^*(u))$. By step 5 of the construction in Subsection 3, this implies that $vert(v)$, regarded as a vertical segment, is strictly lower than $vert(u)$, regarded as a vertical segment, so there can be no horizontal visibility between rectangles $B(u)$ and $B(v)$.

2. D_V^* has a directed path from $R_V^*(u)$ to $L_V^*(v)$. Then the x-coordinate of the right side of $B(u)$, namely $n_V^*(R_V^*(u))$, is less than the x-coordinate of the

left side of $B(v)$, namely $n_V^*(L_V^*(v))$; hence $hor(u)$, regarded as a horizontal segment, is strictly left of $hor(v)$, regarded as a horizontal segment. Therefore these intervals do not overlap, and there can be no vertical visibility between rectangles $B(u)$ and $B(v)$.

3. D_V^* has a directed path from $R_V^*(v)$ to $L_V^*(u)$. This case is treated similarly: the right side of $B(v)$ is to the left of the left side of $B(u)$. △

Lemma 6. *If D_V contains a directed edge (u_i, u), then in the set $\mathcal{R}$ of rectangles produced by the construction, there is a directed vertical visibility from $B(u_i)$ to $B(u)$.*

proof sketch: Let $u_1, u_2, \cdots, u_k$ be all the vertices of D_V with outgoing edges to vertex u. For $i = 1, \cdots, k-1$, let f_i denote the face of D_V with high vertex $h(f_i)$ $= u$. See Fig.5(a).

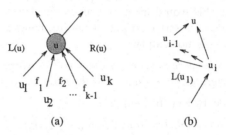

(a)   (b)

Fig. 5. (a) Faces of D_V with high vertex u. (b) The left face $L(u_i)$ of u_i.

Then D_V^* has a directed path $L_V^*(u) \to f_1^* \to f_2^* \to \cdots \to f_{k-1}^* \to R_V^*(u)$, which implies the inequalities: $n_V^*(L_V^*(u)) < n_V^*(f_1^*) < \cdots < n_V^*(f_{k-1}^*) < n_V^*(R_V^*(u))$. It follows that the horizontal interval $hor(u)$ for vertex u is partitioned into a sequence of intervals $H_1, H_2, \cdots, H_k$ where:

$$H_1 = (n_V^*(L_V^*(u)), n_V^*(f_1^*))$$
$$H_i = (n_V^*(f_{i-1}^*), n_V^*(f_i^*))$$
$$H_k = (n_V^*(f_{k-1}^*), n_V^*(R_V^*(u)))$$

We claim that for each u_i, $1 \le i \le k$, $(hor(u_i) \cap hor(u)) - (\cup_{j \ne i} hor(u_j)) \supseteq H_i$, where the interval H_i was just defined above.

To see this, consider the outgoing edges at vertex u_i. It is possible that u_i has outgoing edges ccw of the edge (u_i, u). If so, D_V has a directed path from u_i to u_{i-1}; $R_V(u_{i-1}) = f_{i-1}$; and finally, there is a directed path from $L_V^*(u_i)$ to f_i^* in D_V^*. See Fig.5(b). (The case of an outgoing edge leaving u_i cw of edge (u_i, u) is similar.) But then the st-number $n_V^*(L^*(u_i))$ is less then the st-number $n_V^*(f_i^*)$ of the dual vertex associated with face f_i. Also $n_V^*(R_V^*(u_{i-1})) = n_V^*(f_{i-1})$. Thus the left end of $hor(u_i)$, regarded as a horizontal segment, lies at or to the left of the left end of H_i, regarded as a horizontal segment, depending on whether u_i has outgoing edges ccw of (u_i, u); also the right end of interval $hor(u_i)$ lies at or right of the right end of H_i, depending on whether u_i has outgoing edges cw of (u_i, u). Thus H_i is contained in $hor(u_i)$. Clearly H_i belongs to $hor(u)$ and has no overlap with $\cup_{j \ne i} hor(u_j)$.

We are about to claim that H_i can serve as the horizontal interval for a rectangular region (not one of the rectangles in $\mathcal{R}$ produced by the construction) containing lines of sight from $B(u_i)$ to B_u. We must also produce a vertical interval for such a region. To do this, note that in the primal graph D_V we have a directed path (just an edge) from u_i to u so condition 1 of Theorem 1 implies that there is a path in D_H^* from $L_H^*(u_i)$ to $R_H^*(u)$. Therefore $n_H^*(L_H^*(u_i)) < n_H^*(R_H^*(u))$, $\forall u_i$.

We claim now that there exists an empty rectangle of visibility between $B(u_i)$ and $B(u)$, namely $H_i \times (n_H^*(L_H^*(u_i)), n_H^*(R_H^*(u)))$. By Lemma 5, the rectangles in the constructed set $\mathcal{R}$ have no axis-aligned visibilities not given in (D_V, D_H). Thus the rectangular region (not one of the rectangles in $\mathcal{R}$) whose horizontal interval equals H_i and whose top and bottom edges lie in $B(u)$ and $B(u_i)$, respectively, is empty of other rectangles as well as empty of the rectangles $B(u_i)$. Therefore, this region contains vertical lines of sight from $B(u_i)$ to $B(u)$, establishing the claim of the lemma.   △

Lemma 7. *If D_H contains a directed edge (u_i, u), then in the set $\mathcal{R}$ of rectangles produced by the construction, there is a directed horizontal visibility from $B(u_i)$ to $B(u)$.*

Complexity and Optimality. The proof of Theorem 1 yields a simple quadratic time algorithm for recognizing a topological rectangle visibility graph TRVG specified by a pair (D_V, D_H) and for drawing a configuration of rectangles compatible with (D_V, D_H). The duals of the planar st-graphs D_V and D_H can be constructed in linear time. Verifying the path conditions can be done in quadratic time: for each directed visibility edge in D_V and D_H (and there are linearly many), perform a depth-first search in the dual of the other primal graph to test reachability. A topological sorting performed on each dual takes linear time and gives the coordinates of the rectangles for the drawing.

Claim: The set $\mathcal{R}$ of rectangles produced by the construction has the minimum possible extent in each dimension.

proof sketch: Consider the horizontal dimension first. Partition the bounding box of any realizing set $\mathcal{R}'$ into the rectangles themselves and rectangular visibility regions for the vertical visibilities.

Since the rectangles have vertices with integer coordinates, the rectangular visibility regions for vertical visibilities each have width at least one. Consider a vertical sweep line moving from left to right across the bounding box of $\mathcal{R}'$. Define an *event* for this line to be an x-coordinate at which the sweep line encounters a vertical side of one or more rectangular visibility regions for vertical visibilities. Thus the horizontal extent of the bounding box of $\mathcal{R}'$ must be at least as great as the number of events encountered by the line. If we choose n_V^* so that for each u, $n_V^*(u)$ is equal to the length of the longest directed path in D_V^* from s_V^* to t_V^*, then the horizontal extent of the bounding box of our constructed set $\mathcal{R}$ will be equal to the length of the longest directed path from s_V^* to t_V^*, minus 2 (to account for the two extremal edges at s_V^* and t_V^*). This achieves the lower bound given by the events for the sweep line. The argument for the optimality of the vertical extent of the bounding box is similar. $\triangle$

4   Conclusion

The addition of extra combinatorial information has allowed us to define a family of rectangle visibility graphs that can be recognized efficiently. It remains an open problem to see how much information can be dropped from the given graph while retaining the feasibility of the recognition algorithm.

We believe that our combinatorial characterization can be used to find classes of graphs that are not rectangle visibility graphs, as is done for instance in [24], [4] and [8]. Of course other visibility models can be explored, as well as extensions to higher dimensional boxes.

References

1. H. Alt, M. Godau, and S. Whitesides. Universal 3-dimensional visibility representations for graphs. In F. J. Brandenburg, editor, *Graph Drawing (Proc. GD '95)*, vol. 1027 of *Lecture Notes Comput. Sci.*, pp. 8–19. Springer-Verlag, 1996.
2. G. Brightwell and P. Winkler. Counting linear extensions is #P-complete. *Proc. STOC 23*, pp. 175–181, 1991.
3. K.S. Booth and G.S. Lueker. Testing for the consecutive ones property, interval graphs, and graph planarity using PQ-tree algorithms. J. Comput.System Sci. 13, pp. 335–379, 1976.
4. P. Bose, A. Dean, J. Hutchinson, and T. Shermer. On rectangle visibility graphs I: k-trees and caterpillar forests. Tech. report, DIMACS and Simon Fraser U., May 1996.
5. P. Bose, H. Everett, S. P. Fekete, A. Lubiw, H. Meijer, K. Romanik, T. Shermer, and S. Whitesides. On a visibility representation for graphs in three dimensions. In D. Avis and P. Bose, eds., *Snapshots in Computational and Discrete Geometry, Vol. III*, pp. 2–25. McGill U., July 1994. McGill tech. report SOCS-94.50.
6. P. Bose, H. Everett, Michael E. Houle, S. Fekete, A. Lubiw, H. Meijer, K. Romanik, G. Rote, T. Shermer, S. Whitesides, and Christian Zelle. A visibility representation for graphs in three dimensions. *J. Graph Algorithms and Applications*, vol. 2, no. 3, pp. 1–16, 1998.
7. P. Bose, A. Josefczyk, J. Miller, and J. O'Rourke. K_{42} is a box visibility graph. In *Snapshots of Computational and Discrete Geometry*, vol. 3, pp. 88–91. School Comput. Sci., McGill U., Montreal, July 1994. Tech. report SOCS-94.50.
8. A. M. Dean and J. P. Hutchinson. Rectangle-visibility representations of bipartite graphs. In R. Tamassia and I. G. Tollis, eds., *Graph Drawing (Proc. GD '94)*, vol. 894 of *Lecture Notes Comput. Sci.*, pp. 159–166. Springer-Verlag, 1995.
9. G. Di Battista, P. Eades, R. Tamassia, and I. Tollis. Graph Drawing: Algorithms for the visualization of graphs. Prentice Hall, Upper Saddle River NJ, 1999.
10. Jeong-In Doh and Kyung-Yong Chwa. Visibility problems for orthogonal objects in two or three dimensions. *Visual Comput.*, vol. 4, no. 2, pp. 84–97, July 1988.

11. S. P. Fekete, M. E. Houle, and S. Whitesides. New results on a visibility representation of graphs in 3-d. In F. Brandenburg, ed., *Graph Drawing '95*, vol. 1027 of *Lecture Notes Comput. Sci.*, pp. 234–241. Springer-Verlag, 1996.

12. H. de Fraysseix and P. Rosenstiehl. L'algorithme gauche-droite pour le plongement des graphes dans le plan.

13. Xin He and Ming-Yang Kao. Regular edge labelings and drawings of planar graphs. In *Proc. Graph Drawing conf.*, pp. 96–103, 1994.

14. J. Hopcroft and R. Tarjan. Efficient planarity testing. *J. Assoc. Comput. Machin.* 21, pp. 549–568, 1974.

15. J. P. Hutchinson, T. Shermer, and A. Vince. On representations of some thickness-two graphs. In F. Brandenburg, ed., *Graph Drawing '95*, vol. 1027 of *Lecture Notes Comput. Sci.*, pp. 324–332. Springer-Verlag, 1995.

16. A. Josefczyk, J. Miller, and J. O'Rourke. Arkin's conjecture for rectangle and box visibility graphs. Tech. Report 036, Dept. Comput. Sci., Smith College, July 1994.

17. D. G. Kirkpatrick and S. K. Wismath. Weighted visibility graphs of bars and related flow problems. In *Proc. 1st Workshop Algorithms Data Struct.*, vol. 382 of *Lecture Notes Comput. Sci.*, pp. 325–334. Springer-Verlag, 1989.

18. A. Lempel, S. Even and I. Cederbaum. An algorithm for planarity testing of graphs. In *(P. Rosenstiehl, ed., Theory of Graphs* (Int. Symposium, Rome, 1966), pp. 215–232, Gordon and Breach, New York, 1967.

19. E. Lodi and L. Pagli. A VLSI solution to the vertical segment visibility problem. *IEEE Trans. Comput.*, vol. C-35, no. 10, pp. 923–928, 1986.

20. F. Luccio, S. Mazzone, and C. Wong. A note on visibility graphs. *Discrete Math.*, vol. 64, pp. 209–219, 1987.

21. M. H. Overmars and D. Wood. On rectangular visibility. *J. Algorithms*, vol. 9, pp. 372–390, 1988.

22. K. Romanik. Directed rectangle-visibility graphs have unbounded dimension. In *Proc. 7th Canad. Conf. Comput. Geom.*, pp. 163–167, 1995.

23. Pierre Rosenstiehl and Robert Tarjan. Rectiliniar planar layouts and bipolar orientations of planar graphs. *Discrete Comput Geom.*, vol. 1, pp. 343–353, 1986.

24. T. Shermer. On rectangle visibility graphs III: External visibility and complexity. Tech. report, DIMACS and Simon Fraser U., April 1996.

25. T. C. Shermer. On rectangle visibility graphs, III: External visibility and complexity. In *Proc. 8th Canad. Conf. Comput. Geom.*, pp. 234–239, 1996.

26. R. Tamassia and I. G. Tollis. A unified approach to visibility representations of planar graphs. *Discrete Comput. Geom.*, vol. 1, no. 4, pp. 321–341, 1986.

27. S. K. Wismath. Characterizing bar line-of-sight graphs. In *Proc. 1st Annu. ACM Sympos. Comput. Geom.*, pp. 147–152, 1985.

28. S. K. Wismath. *Bar-Representable Visibility Graphs and Related Flow Problems.* Ph.D. thesis, Dept. Comput. Sci., U. British Columbia, 1989.

Approximating Geometric Bottleneck Shortest Paths[*]

Prosenjit Bose[1], Anil Maheshwari[1], Giri Narasimhan[2],
Michiel Smid[1], and Norbert Zeh[3]

[1] School of Computer Science, Carleton University, Ottawa, Canada K1S 5B6.
{jit,maheshwa,michiel}@scs.carleton.ca
[2] School of Computer Science, Florida International University, Miami, FL 33199.
giri@cs.fiu.edu
[3] Department of Computer Science, Duke University, Durham, NC 27708-0129.
nzeh@cs.duke.edu

Abstract. In a geometric bottleneck shortest path problem, we are given a set S of n points in the plane, and want to answer queries of the following type: Given two points p and q of S and a real number L, compute (or approximate) a shortest path in the subgraph of the complete graph on S consisting of all edges whose length is less than or equal to L. We present efficient algorithms for answering several query problems of this type. Our solutions are based on minimum spanning trees, spanners, the Delaunay triangulation, and planar separators.

1 Introduction

We consider *bottleneck shortest path* problems in geometric graphs. Given a set S of n points in the plane, we consider queries of the following type: Given any two points p and q of S and any real number L, compute or approximate a shortest path in the subgraph of the complete graph on S consisting of all edges whose length is less than or equal to L.

To define these problems more precisely, given $L \in \mathbb{R}$, let $K^{(\leq L)}$ be the graph with vertex set S, in which any two distinct vertices p and q are connected by an edge if and only if their Euclidean distance $|pq|$ is less than or equal to L. Furthermore, we denote by $\delta^{(\leq L)}(p,q)$ the Euclidean length of a shortest path between p and q in the graph $K^{(\leq L)}$.

In a *bottleneck connectedness query*, we are given two points p and q of S and $L \in \mathbb{R}$, and have to decide if there exists a path between p and q in $K^{(\leq L)}$. In a *bottleneck shortest path length query*, we are given two points p and q of S and $L \in \mathbb{R}$, and have to compute $\delta^{(\leq L)}(p,q)$. In a *bottleneck shortest path query*, we are given two points p and q of S and $L \in \mathbb{R}$, and have to compute a path in $K^{(\leq L)}$ whose length is equal to $\delta^{(\leq L)}(p,q)$.

These problems are motivated by the following scenario. Imagine the points of S to be airports. Then we would like to answer queries in which we are given

---

[*] Bose, Maheshwari, and Smid were supported by NSERC.

H. Alt and M. Habib (Eds.): STACS 2003, LNCS 2607, pp. 38–49, 2003.
© Springer-Verlag Berlin Heidelberg 2003

two airports p and q and an airplane that can fly L miles without refueling, and have to compute shortest path information for this airplane to fly from p to q.

Throughout the rest of this paper, we denote by $L_1 < L_2 < \ldots < L_{\binom{n}{2}}$ the sorted sequence of distances determined by any two distinct points of S. (We assume for simplicity that all these distances are distinct.) For any i with $1 \leq i \leq \binom{n}{2}$, we write $K^{(i)}$ instead of $K^{(\leq L_i)}$, and $\delta^{(i)}(p,q)$ instead of $\delta^{(\leq L_i)}(p,q)$.

In Section 2, we show that, after an $O(n \log n)$–time preprocessing, bottleneck connectedness queries can be answered in $O(1)$ time. The data structure is a binary tree that reflects the way in which Kruskal's algorithm computes the minimum spanning tree of S.

In Section 3, we consider bottleneck shortest path length queries. We present a simple data structure of size $O(n^2 \log n)$ that supports ϵ-approximate bottleneck shortest path length queries in $O(\log n)$ time. A simple extension of this data structure allows ϵ-approximate bottleneck shortest path queries to be answered in $O(\log n + \ell)$ time, where ℓ is the number of edges on the reported path. This data structure uses $O(n^3 \log n)$ space.

In Section 4, we give a general approach for solving the approximate bottleneck shortest path query problem. The idea is to approximate the sequence $K^{(i)}, 1 \leq i \leq \binom{n}{2}$, of graphs by a collection of $O(n)$ sparse graphs. We show that the bottleneck versions of the Yao-graph [7] and the Delaunay triangulation are examples of such collections of sparse graphs. The latter claim is obtained by extending the proof of Keil and Gutwin [5] that the Delaunay triangulation has stretch factor less than or equal to $2\pi/(3 \cos(\pi/6))$. We prove that for any two points p and q of a given point set S, there exists a path in the Delaunay triangulation of S whose length is at most $2\pi/(3 \cos(\pi/6)) \approx 2.42$ times the Euclidean distance $|pq|$ between p and q, and all of whose edges have length at most $|pq|$. (In [5], there is no guarantee on the lengths of the individual edges on the path.)

Finally, in Section 5, we give a data structure of size $O(n^{5/2})$ that can be used to answer bottleneck shortest path queries in planar graphs in $O(\sqrt{n} + \ell)$ time, where ℓ is the number of edges on the reported path. This data structure uses a result of Djidjev [3] to obtain a recursive separator decomposition of the planar graph. By applying this result to those of Section 4, we obtain an efficient solution to the approximate bottleneck shortest path query problem.

To our knowledge, the type of bottleneck shortest path problems considered in this paper have not been considered before. There is related work by Narasimhan and Smid [6], who consider the following problem: Given a real number L, approximate the *stretch factor* of the graph $K^{(\leq L)}$, which is defined as the maximum value of $\delta^{(\leq L)}(p,q)/|pq|$ over all distinct points p and q of S. They present a data structure of size $O(\log n)$, that can be built in roughly $O(n^{4/3})$ time, and that can be used to answer approximate stretch factor queries (with an approximation factor of about 36) in $O(\log \log n)$ time.

Our results are based on *t-spanners*, which are sparse graphs having stretch factor less than or equal to t; see the survey by Eppstein [4].

2 Bottleneck Connectedness Queries

Let $MST(S)$ be the Euclidean minimum spanning tree of the point set S. We define a binary tree $T(S)$ as follows. If $|S| = 1$, then $T(S)$ consists of one node storing the only point of S. Assume that $|S| \geq 2$, and let e be the longest edge in $MST(S)$. Removing e partitions $MST(S)$ into two trees. Let S_1 and S_2 be the vertex sets of these trees. Then $T(S)$ consists of a root that stores the edge e and pointers to its two children, which are roots of recursively defined trees $T(S_1)$ and $T(S_2)$. Observe that the leaves of $T(S)$ are in one-to-one correspondence with the points of S, and the internal nodes are in one-to-one correspondence with the edges of $MST(S)$. Computing $T(S)$ according to the above definition corresponds to tracing back the execution of Kruskal's minimum spanning tree algorithm. The following lemma shows how the tree $T(S)$ can be used to answer bottleneck connectedness queries.

Lemma 1. *Let p and q be two distinct points of S, and let L be a real number. Let e be the edge stored at the lowest common ancestor (LCA) of the leaves of $T(S)$ storing p and q. Then p and q are connected by a path in the graph $K^{(\leq L)}$ if and only if the length of e is less than or equal to L.*

Proof. Assume that the length of e is less than or equal to L. Let u be the node of $T(S)$ that stores e. We may assume w.l.o.g. that p is stored in the left subtree of u (and, hence, that q is stored in the right subtree of u). Let S_p and S_q be the sets of points that are stored at the leaves of the left and right subtrees of u, respectively. Let x and y be the endpoints of e, where $x \in S_p$ and $y \in S_q$. Consider what happens when Kruskal's algorithm computes the minimum spanning tree of S. Immediately before e is added, all points in S_p are connected by edges in the partially constructed minimum spanning tree, and the length of each of these edges is less than the length of e. The same holds for the set S_q. Hence, immediately after edge e has been added to the partially constructed minimum spanning tree, p and q are connected by a path, all of whose edges have a length that is at most the length of e. This path is contained in the graph $K^{(\leq L)}$. To prove the converse, assume that the length of e is larger than L. Let S_1 and S_2 be the partition of S obtained by deleting e from $MST(S)$. Since the unique path in $MST(S)$ between p and q contains e, we have (i) $p \in S_1$ and $q \in S_2$, or (ii) $p \in S_2$ and $q \in S_1$. By a well known property of minimum spanning trees, the length of e is equal to the minimum distance between any point in S_1 and any point of S_2. If there is a path in $K^{(\leq L)}$ between p and q, then this path must contain an edge between some point of S_1 and some point of S_2. Since the length of any such edge is larger than L, it follows that such a path cannot exist.   □

Lemma 1 implies that a bottleneck connectedness query can be answered by answering an LCA-query in the tree $T(S)$. This tree can be computed in $O(n \log n)$ time. Given $T(S)$, we preprocess it in $O(n)$ time, so that LCA-queries can be answered in $O(1)$ time. (See Bender and Farach-Colton [1].)

Theorem 1. *We can preprocess a set of n points in the plane in $O(n \log n)$ time into a data structure of size $O(n)$, such that bottleneck connectedness queries can be answered in $O(1)$ time.*

3   Bottleneck Shortest Path Length Queries

Recall that $K^{(i)}$ denotes the graph $K^{(\leq L_i)}$. We define $K^{(0)}$ to be the graph $(S, \emptyset)$. Also recall that we write $\delta^{(i)}(p, q)$ instead of $\delta^{(\leq L_i)}(p, q)$.

Let $0 < \epsilon \leq 3$ be a fixed real constant. In this section, we show how to preprocess the points of S into a data structure of size $O(n^2 \log n)$, such that ϵ-approximate bottleneck shortest path length queries can be answered in $O(\log n)$ time. To be more precise, we show that a query of the following type can be answered in $O(\log \log n)$ time: Given two points p and q of S and an index $0 \leq i \leq \binom{n}{2}$, compute an ϵ-approximation to $\delta^{(i)}(p, q)$, i.e., a real number Δ, such that $\delta^{(i)}(p, q) \leq \Delta \leq (1 + \epsilon) \cdot \delta^{(i)}(p, q)$. Using an additional amount of $O(n^2)$ space, we will extend this solution to solve general ϵ-approximate bottleneck shortest path length queries (in which an arbitrary real number L is part of the query, rather than the distance L_i) in $O(\log n)$ time. Our solution is based on an approach by Narasimhan and Smid [6].

We fix two distinct points p and q of S, and observe that

$$|pq| = \delta^{(\binom{n}{2})}(p, q) \leq \ldots \leq \delta^{(1)}(p, q) \leq \delta^{(0)}(p, q) = \infty.$$

Let $k := \min\{i \geq 0 : \delta^{(i)}(p, q) < \infty\}$. Since p and q are not connected by a path in the graph $K^{(k-1)}$, we have $|pq| > L_{k-1}$ and, hence, $|pq| \geq L_k$. On the other hand, since p and q are connected by a path in $K^{(k)}$, and since any such path contains at most $n - 1$ edges, we have $\delta^{(k)}(p, q) \leq (n-1)L_k \leq (n-1)|pq|$. Based on this observation, we partition the set $\{k, k+1, \ldots, \binom{n}{2}\}$ into $O(\log n)$ subsets, in the following way. For any $0 \leq j \leq \log_{1+\epsilon/3}(n - 1)$, let

$$I_{pq}^j := \left\{ i : k \leq i \leq \binom{n}{2} \text{ and } (1 + \epsilon/3)^j |pq| \leq \delta^{(i)}(p, q) < (1 + \epsilon/3)^{j+1}|pq| \right\}.$$

We store for each integer j for which $I_{pq}^j \neq \emptyset$, (i) a value ℓ_j, which is the smallest element of the set I_{pq}^j, and (ii) a value $\Delta^{(j)}(p, q)$ which is equal to $(1 + \epsilon/3) \cdot \delta^{(\ell_j)}(p, q)$. Let us see how we can use this information to answer an ϵ-approximate bottleneck shortest path length query for p and q. Let i be an integer with $0 \leq i \leq \binom{n}{2}$. We start by showing how the value of $\delta^{(i)}(p, q)$ can be approximated. First compute the integer j for which $\ell_j \leq i < \ell_{j-1}$. Then return the value $\Delta := \Delta^{(j)}(p, q)$.

To prove the correctness of this query algorithm, first observe that $i \in I_{pq}^j$. This implies that $\delta^{(i)}(p, q) < (1 + \epsilon/3)^{j+1}|pq|$. Similarly, since $\ell_j \in I_{pq}^j$, we have $\delta^{(\ell_j)}(p, q) \geq (1 + \epsilon/3)^j |pq|$. It follows that $\delta^{(i)}(p, q) \leq \Delta$. In a completely symmetric way, we obtain $\Delta \leq (1 + \epsilon/3)^2 \cdot \delta^{(i)}(p, q)$, which is at most $(1 + \epsilon) \cdot \delta^{(i)}(p, q)$. This proves that Δ is an ϵ-approximation to the length of a shortest path between p and q in the graph $K^{(i)}$. By storing the values ℓ_j in sorted order in an array, Δ can be computed in $O(\log \log n)$ time.

We store all this information for each pair of points. Additionally, we store the sequence $L_1 < L_2 < \ldots < L_{\binom{n}{2}}$ of distances. Given two query points p and

q of S and an arbitrary query value $L \in \mathbb{R}$, we first use binary search to find the index i for which $L_i \leq L < L_{i+1}$. Since $\delta^{(\leq L)}(p,q) = \delta^{(i)}(p,q)$, we then answer the query as described above.

The amount of space used by this solution is $O(n^2 \log n)$, and the query time is $O(\log n)$. Let us consider the preprocessing time. It clearly suffices to solve the all-pairs-shortest-path problem for each graph $K^{(i)}$, $0 \leq i \leq \binom{n}{2}$. Using the Floyd-Warshall algorithm, one such problem can be solved in $O(n^3)$ time. Hence, the overall preprocessing time is $O(n^5)$.

If we store with each value $\Delta^{(j)}(p,q)$ a path in $K^{(\ell_j)}$ having length $\delta^{(\ell_j)}(p,q)$, then we can use this additional information to answer approximate bottleneck shortest path queries: Let L and i be as above. Then the path P stored with $\Delta^{(j)}(p,q)$ has length $\delta := \delta^{(\ell_j)}(p,q)$ satisfying $\delta^{(i)}(p,q) \leq \delta \leq (1 + \epsilon/3)\delta^{(i)}(p,q)$. (Observe that P is a path in $K^{(i)}$.)

Theorem 2. *For any real constant $\epsilon > 0$, we can preprocess a set of n points in the plane in $O(n^5)$ time into*

1. *a data structure of size $O(n^2 \log n)$, such that ϵ-approximate bottleneck shortest path length queries can be answered in $O(\log n)$ time,*
2. *a data structure of size $O(n^3 \log n)$, such that ϵ-approximate bottleneck shortest path queries can be answered in $O(\log n + \ell)$ time, where ℓ is the number of edges on the reported path.*

4   The Bottleneck Shortest Path Problem

In this section, we introduce a general approach for the approximate bottleneck shortest path problem. The idea is to approximate the sequence $K^{(i)}$, $1 \leq i \leq \binom{n}{2}$, by a "small" collection of sparse graphs, i.e., with "few" edges. This notion is formalized in the definition below. For any graph G and any two vertices p and q, we denote the length of a shortest path in G between p and q by $\delta^{(G)}(p,q)$.

Definition 1. *Let S be a set of n points in the plane, let $t \geq 1$ be a real number, let J be a subset of $\{1, 2, \ldots, \binom{n}{2}\}$ and, for each $j \in J$, let $G^{(j)}$ be a graph with vertex set S all of whose edges have length at most L_j. We say that the collection $\mathcal{G} = \{G^{(j)} : j \in J\}$ is a collective bottleneck t-spanner of S, if the following holds: for any i with $1 \leq i \leq \binom{n}{2}$, there is an index $j \in J$, such that $j \leq i$ and*

$$\delta^{(i)}(p,q) \leq \delta^{(G^{(j)})}(p,q) \leq t \cdot \delta^{(i)}(p,q)$$

holds for all pairs of points p and q in S.

In order to approximate a bottleneck shortest path between p and q in the possibly dense graph $K^{(i)}$, we compute a shortest path P between p and q in the graph $G^{(j)}$. Since $j \leq i$, P is a path in $K^{(i)}$; it is a t-approximate shortest path between p and q in $K^{(i)}$.

The goal is to define $\mathcal{G}$ in such a way that shortest path queries on them can be answered efficiently. Further goals are to minimize (i) the value of t, (ii) the size of the index set J, and (iii) the number of edges in the graphs in $\mathcal{G}$.

Lemma 2. *The size of the index set J is greater than or equal to $n - 1$.*

Proof. Let (p, q) be any edge of the Euclidean minimum spanning tree $MST(S)$ of S, and let i be the index such that $|pq| = L_i$. Observe that $\delta^{(i)}(p, q) = |pq| < \infty$. We claim that $i \in J$. To prove this, assume that $i \notin J$. By Definition 1, there is an index $j \in J$ such that $j < i$ and $\delta^{(G^j)}(p, q) \leq t \cdot \delta^{(i)}(p, q) < \infty$. In particular, we have $\delta^{(j)}(p, q) < \infty$. By well known properties of minimum spanning trees, however, $\delta^{(j)}(p, q) = \infty$, contradicting our assumption that $i \notin J$. Hence, each edge of $MST(S)$ contributes an index to J. $\square$

In the rest of this section, we give some examples of collective bottleneck spanners.

The Yao-graph [7]: Let S be a set of n points in the plane, and let θ be an angle such that $2\pi/\theta$ is an integer. We partition the plane into a collection $\mathcal{C}$ of $2\pi/\theta$ cones of angle θ, all having their apex at the origin. For any point $p \in S$ and any cone $C \in \mathcal{C}$, let C_p be the cone obtained by translating C by the vector p. (Hence, C_p has p as its apex.)

The Yao-graph $Y(S, \theta)$ has S as its vertex set. Let p be any point of S, let C be any cone of $\mathcal{C}$ such that $C_p \cap (S \setminus \{p\}) \neq \emptyset$, and let q_p be the point of $C_p \cap (S \setminus \{p\})$ whose Euclidean distance to p is minimum. The edge set of $Y(S, \theta)$ consists of all edges (p, q_p) obtained in this way. Chang et al. [2] have shown how to construct the graph $Y(S, \theta)$ in $O(n \log n)$ time.

Given two points p and q, we construct a path between p and q in $Y(S, \theta)$ in the following way. If $p = q$, then there is nothing to do. Assume that $p \neq q$. Let C be the cone in $\mathcal{C}$ such that $q \in C_p$. The graph $Y(S, \theta)$ contains an edge (p, r), where $r \in C_p$ and $|pr| \leq |pq|$. We follow this edge, and recursively construct a path between r and q. It can be shown that for any constant $\epsilon > 0$, there is an angle θ such that the path constructed in this way has length at most $(1 + \epsilon)|pq|$ and the length of each edge on this path is less than or equal to $|pq|$; see [5].

If we denote the number of edges of $Y(S, \theta)$ by m, then $m = O(n)$. Let $j_1 < j_2 < \ldots < j_m$ be the indices such that $L_{j_1} < L_{j_2} < \ldots < L_{j_m}$ are the edge lengths of $Y(S, \theta)$. For any k with $1 \leq k \leq m$, denote by $Y^{(j_k)}$ the graph consisting of all edges of $Y(S, \theta)$ whose lengths are at most L_{j_k}.

Let $J := \{j_k : 1 \leq k \leq m\}$. We claim that $\mathcal{G} := \{Y^{(j_k)} : 1 \leq k \leq m\}$ is a collective bottleneck $(1 + \epsilon)$-spanner of S. To prove this claim, consider any two points p and q of S and any integer $1 \leq i \leq \binom{n}{2}$. We may assume that p and q are connected by a path in $K^{(i)}$. Let k be the integer such that $L_{j_k} \leq L_i < L_{j_{k+1}}$, and let δ be the length of a shortest path between p and q in the graph $Y^{(j_k)}$. It is clear that $j_k \leq i$. It remains to show that $\delta^{(i)}(p, q) \leq \delta \leq (1 + \epsilon) \cdot \delta^{(i)}(p, q)$. The first inequality follows from the fact that $j_k \leq i$. This implies that $L_{j_k} \leq L_i$ and, therefore, $Y^{(j_k)}$ is a subgraph of $K^{(i)}$. To prove the second inequality, consider a shortest path P between p and q in $K^{(i)}$. Hence, the length of P is equal to $\delta^{(i)}(p, q)$. Consider an arbitrary edge (x, y) on P. Observe that $|xy| \leq L_i$. In the graph $Y(S, \theta)$, there is a path P'_{xy} between x and y whose length is at most $(1 + \epsilon)|xy|$ and all of whose edges are of length at most $|xy|$. Hence, P'_{xy} is a path in the graph $Y^{(j_k)}$. By concatenating the paths P'_{xy}, over all edges (x, y) of P, we obtain a path between p and q in $Y^{(j_k)}$ having length at most $(1 + \epsilon)$ times the length of P. This proves the second inequality.

Theorem 3. *Let S be a set of n points in the plane, and let $\epsilon > 0$ be a constant. There exists a collective bottleneck $(1+\epsilon)$-spanner of S, consisting of $O(n)$ graphs.*

Observe that the bottleneck graphs in the collection $\mathcal{G}$ are not planar (except for small values of k). We next show how to obtain a collective bottleneck spanner where the collection consists of $O(n)$ planar graphs.

The Delaunay triangulation: Let S be a set of n points in the plane, and consider the Delaunay triangulation $DT(S)$ of S. In this section, we show that for any two points p and q of S, there exists a path $P_{DT(S)}(p, q)$ between p and q in $DT(S)$ such that (i) no edge on this path has length more than $|pq|$ and (ii) the length of $P_{DT(S)}(p, q)$ is less than or equal to $2\pi/(3\cos(\pi/6)) \cdot |pq|$. We will prove this claim by modifying Keil and Gutwin's proof of the fact that $DT(S)$ is a $(2\pi/(3\cos(\pi/6)))$-spanner of S; see [5]. (The proof in [5] may produce a path between p and q that contains an edge whose length is larger than $|pq|$.) We start with the following key lemma, which is similar to Lemma 1 in [5].

Lemma 3. *Let p and q be two points of S on a horizontal line L such that p is to the left of q. Let C be a circle having p and q on its boundary, let r and O be the radius and center of C, respectively, and let θ be the upper angle $\angle pOq$. Assume that no point of S lies inside C below L. Then there exists a path $P_{DT(S)}(p, q)$ between p and q in $DT(S)$ such that (i) no edge on this path has length more than $|pq|$, and (ii) the length of $P_{DT(S)}(p, q)$ is less than or equal to $r\theta$.*

Proof. We may assume w.l.o.g. that p and q are on the x-axis. If the circle C contains no points of S in its interior, then the lemma holds since by the empty circle property of Delaunay triangulations, (p, q) is an edge of $DT(S)$.

In the rest of the proof, we assume that C contains at least one point of S in its interior (implying that (p, q) is not an edge of $DT(S)$). Observe that, by the assumptions in the lemma, all points of $C \cap S$ must be above the x-axis.

Our goal is to prove the lemma by induction on the angle θ. In order to do this, we will *normalize* the circle C so that, over all possible pairs p and q of points in S, there are only a finite number of normalized circles.

We move the center O of C downwards along the bisector of p and q. During this movement, we change C so that the points p and q stay on its boundary. (Hence, the radius r and the upper angle θ of C change. It is easy to see that $r\theta$ decreases.) We stop this process at the moment when the part of C below the x-axis hits at a point, say z, of S. Observe that z exists, because otherwise (p, q) is an edge of the convex hull of S, and thus an edge of $DT(S)$, which would contradict our earlier assumption.

We refer to this circle as a normalized circle. Hence, from now on, C is a normalized circle with center O and radius r, containing p, q, and z on its boundary, where z is below the x-axis. There are no points of S in the interior of C below the x-axis, and there is at least one point of S in the interior of C.

Each pair of points defines at most two normalized circles (depending on whether the circle is empty above or below the line through the two points). Therefore, there are at most $O(n^2)$ possible normalized circles. We proceed by induction on the rank of the angle θ of these circles.

For the base case, consider the circle C with the smallest angle θ. For any point t of $C \cap S$, let D_t be the circle through p, q, and t. Choose the t for which no point of S lies in the interior of D_t above the x-axis. (Observe that such a t must exist.) We will write D instead of D_t.

Let C_1 be the circle through p and t whose center lies on the segment pO. We denote the center of C_1 by O_1. By the definition of D, no point of S lies in C_1 below the line through pt. However, our construction implies that $\angle pO_1t < \angle pOq$, which contradicts the assumption that θ is the smallest angle. We conclude that the circle C must be empty and, hence, (p,q) is an edge of $DT(S)$. Thus, $|pq| \leq r\theta$ as required. This proves the base case of the induction.

For the inductive step, we consider a given normalized circle C and assume that the lemma holds for all normalized circles with smaller interior upper angle. Define the point t and circle D as above. Observe that the upper angle $\theta = \angle pOq$ lies in the open interval $(0, 2\pi)$.

We will treat the cases $0 < \theta \leq \pi$ and $\pi < \theta < 2\pi$ separately. Before proceeding, we outline a few constructions which will be useful in the sequel.

Let C_1 be the circle through p and t whose center lies on the segment pO. We denote the center and radius of C_1 by O_1 and r_1, respectively. Similarly, let C_2 be the circle through t and q whose center lies on the segment qO. We denote the center and radius of C_2 by O_2 and r_2, respectively. Finally, circle D implies that no point of S lies in C_1 below the line through p and t, and no point of S lies in C_2 below the line through q and t.

Consider the two intersection points between C_1 and the x-axis. One of these intersection points is p; we denote the other one by a_1. Similarly, let a_2 be the intersection point between C_2 and the x-axis that is not equal to q.

Let C_3 be the circle through a_1 and a_2 whose center is the intersection between the line through O_1 and a_1 and the line through O_2 and a_2. We denote the center and radius of C_3 by O_3 and r_3, respectively.

We observe that the following four triangles are all similar isoceles triangles with two equal base angles, denoted by ϕ: $\triangle(p, O, q)$, $\triangle(p, O_1, a_1)$, $\triangle(a_2, O_2, q)$, and $\triangle(a_2, O_3, a_1)$.

By construction, θ is larger than both angles $\theta_1 = \angle pO_1t$ and $\theta_2 = \angle tO_2q$. Therefore, by the induction hypothesis, there exists a path $P_{DT(S)}(p, t)$ between p and t in $DT(S)$, all of whose edges have length at most $|pt|$, and $|P_{DT(S)}(p, t)| \leq r_1\theta_1$. Similarly, there exists a path $P_{DT(S)}(t, q)$ between t and q in $DT(S)$, all of whose edges have length at most $|tq|$, and $|P_{DT(S)}(t, q)| \leq r_2\theta_2$.

Case 1: $0 < \theta \leq \pi$. Let $P_{DT(S)}(p, q)$ be the concatenation of the paths $P_{DT(S)}(p, t)$ and $P_{DT(S)}(t, q)$. We have

$$
\begin{aligned}
|P_{DT(S)}(p, q)| &= |P_{DT(S)}(p, t)| + |P_{DT(S)}(t, q)| \\
&\leq r_1\theta_1 + r_2\theta_2 \\
&= r_1\theta + r_2\theta - [r_1(\theta - \theta_1) + r_2(\theta - \theta_2)].
\end{aligned}
$$

Since $0 < \theta \leq \pi$, the point a_2 is to the left of a_1. Therefore, the part of C_3 that is above the x-axis is contained in the part of $C_1 \cap C_2$ that is above the x-axis. This, in turn, implies that $r_1(\theta - \theta_1) + r_2(\theta - \theta_2) \geq r_3\theta$. Hence, we have $|P_{DT(S)}(p, q)| \leq (r_1 + r_2 - r_3)\theta$. Finally, a straightforward analysis yields

$$\begin{aligned}
|P_{DT(S)}(p, q)| &\leq (r_1 + r_2 - r_3)\theta \\
&= (|pa_1|/2\cos\phi + |a_2q|/2\cos\phi - |a_2a_1|/2\cos\phi) \\
&= |pq|/2\cos\phi = r\theta.
\end{aligned}$$

Since $0 < \theta \leq \pi$, we have $|pt| \leq |pq|$ and $|tq| \leq |pq|$. Therefore, the length of each edge on $P_{DT(S)}(p, q)|$ is less than or equal to $|pq|$.

Case 2.a: $\pi < \theta < 2\pi$ and the point t is inside the circle having p and q as diameter. Let $P_{DT(S)}(p, q)$ be the concatenation of the paths $P_{DT(S)}(p, t)$ and $P_{DT(S)}(t, q)$. We have $|P_{DT(S)}(p, q)| \leq |P_{DT(S)}(p, t)| + |P_{DT(S)}(t, q)| \leq r_1\theta_1 + r_2\theta_2$. If the point a_2 is to the right of a_1, then θ_1 and θ_2 are both less than or equal to θ. Therefore, we have $|P_{DT(S)}(p, q)| \leq (r_1 + r_2)\theta$. A straightforward analysis gives

$$\begin{aligned}
|P_{DT(S)}(p, q)| &\leq (r_1 + r_2)\theta \\
&= (|pa_1|/2\cos\phi + |a_2q|/2\cos\phi) \\
&\leq |pq|/2\cos\phi = r\theta.
\end{aligned}$$

If a_2 is to the left of a_1, then $|P_{DT(S)}(p, q)| \leq r\theta$ by the same argument as in Case 1.

Since t is contained in the circle with p and q as diameter, both $|pt|$ and $|tq|$ are less than $|pq|$. As a result, the length of each edge on $P_{DT(S)}(p, q)$ is less than or equal to $|pq|$.

Case 2.b: $\pi < \theta < 2\pi$ and the point t lies outside the circle R with p and q as diameter. Let C_4 be the circle through p and z whose center is on the x-axis. We denote the center and radius of C_4 by O_4 and r_4, respectively. Let θ_4 be the lower angle $\angle pO_4z$. Similarly, let C_5 be the circle through q and z whose center is on the x-axis. We denote the center and radius of C_5 by O_5 and r_5, respectively. Let θ_5 be the lower angle $\angle zO_5q$.

The fact that no point of S is contained in the part of R that is above the x-axis, implies that there are no points of S in the part of C_4 that is above the line through p and z. Similarly, there are no points of S in the part of C_5 that is above the line through q and z. Furthermore, since both angles θ_4 and θ_5 are less than π and, hence, less than θ, we can apply the induction hypothesis. That is, there exists a path $P_{DT(S)}(p, z)$ between p and z in $DT(S)$, all of whose edges have length at most $|pz|$, and $|P_{DT(S)}(p, z)| \leq r_4\theta_4$. Similarly, there exists a path $P_{DT(S)}(z, q)$ between z and q in $DT(S)$, all of whose edges have length at most $|zq|$, and $|P_{DT(S)}(z, q)| \leq r_5\theta_5$.

Let $P_{DT(S)}(p, q)$ be the concatenation of $P_{DT(S)}(p, z)$ and $P_{DT(S)}(z, q)$. Then

$$\begin{aligned}
|P_{DT(S)}(p, q)| &\leq |P_{DT(S)}(p, z)| + |P_{DT(S)}(z, q)| \\
&\leq r_4\theta_4 + r_5\theta_5 \leq (r_4 + r_5)\theta \leq r\theta,
\end{aligned}$$

where the last inequality follows from the fact that both r_4 and r_5 are less than or equal to $|pq|/2$. Finally, since both $|pz|$ and $|zq|$ are less than $|pq|$, the length of each edge on $P_{DT(S)}(p,q)|$ is less than or equal to $|pq|$.   □

Theorem 4. *Let S be a set of n points in the plane, let $DT(S)$ be the Delaunay triangulation of S, and let p and q be two points of S. There is a path between p and q in $DT(S)$ whose length is less than or equal to $2\pi/(3\cos(\pi/6)) \cdot |pq|$, and all of whose edges have length at most $|pq|$.*

Proof. The proof is similar to the proof of Theorem 1 in [5].   □

Let m be the number of edges of $DT(S)$, and let $j_1 < j_2 < \ldots < j_m$ be the indices such that $L_{j_1} < L_{j_2} < \ldots < L_{j_m}$ are the edge lengths of $DT(S)$. Observe that $m \leq 3n - 6$. For any k with $1 \leq k \leq m$, let $DT^{(j_k)}$ be the graph with vertex set S consisting of all edges of $DT(S)$ whose lengths are at most L_{j_k}.

Corollary 1. *The collection $\mathcal{G} := \{DT^{(j_k)} : 1 \leq k \leq m\}$ of planar graphs constitutes a collective bottleneck t-spanner of S for $t = 2\pi/(3\cos(\pi/6))$.*

5 Bottleneck Shortest Path Queries in Planar Graphs

In this section, we address the following problem: Given a set S of n points in the plane and a planar graph G with vertex set S, build a data structure that can answer bottleneck queries of the following type: Given two points p and q of S and a real number L, decide whether there is a path between p and q in G all of whose edges are of length at most L, and report the shortest such path if such a path exists. We will present a structure of size $O(n^{5/2})$ that answers existence queries in $O(\log n)$ time and allows the shortest path whose edges have length at most L to be reported in $O(\sqrt{n} + \ell)$ time, where ℓ is the number of edges on the reported path. We start in Section 5.1 with reviewing Djidjev's data structure [3] for answering general shortest path queries in the planar graph G. In Section 5.2, we show how to answer bottleneck shortest path queries in G.

5.1 Shortest Path Queries in G

Djidjev's shortest path data structure [3] is based on a recursive separator partition of the planar graph G, which we represent by a *separator tree* $\mathcal{T}(G)$. This tree is defined as follows. If G has only one vertex p, then $\mathcal{T}(G)$ contains only one node, say α. We store with α the set $S(\alpha) := \{p\}$, and associate with α the graph $G(\alpha) := G$. Otherwise, we compute a set C of $O(\sqrt{n})$ vertices in G such that each of the connected components $G_1, \ldots, G_k$ of $G \setminus C$ has size at most $2n/3$. We store the set $S(\alpha) := C$ with the root α of $\mathcal{T}(G)$, and associate the graph $G(\alpha) := G$ with α. Then we recursively compute separator trees $\mathcal{T}(G_1), \ldots, \mathcal{T}(G_k)$ for the graphs $G_1, \ldots, G_k$, and make the roots of those trees the children of α.

It is easy to see that each point of S is stored exactly once in $\mathcal{T}(G)$, so that we can associate a unique node $\alpha(p)$ of $\mathcal{T}(G)$ with every point p of S.

For any two points p and q of S, every path in G between p and q must contain at least one vertex stored at a common ancestor of $\alpha(p)$ and $\alpha(q)$. Thus, we associate a set $A(p)$ with every point p of S. This set $A(p)$ consists of all pairs $(x, \delta^{(G(\alpha(x)))}(p, x))$, where x ranges over all points of S that are stored at some ancestor of $\alpha(p)$ in the separator tree $\mathcal{T}(G)$, and $\delta^{(G(\alpha(x)))}(p, x)$ is the length of a shortest path between p and x in $G(\alpha(x))$. Each such set $A(p)$ can be shown to have size $O(\sqrt{n})$, so that the total size of all sets $A(p)$, $p \in S$, is $O(n^{3/2})$. Since every path between two points p and q must contain at least one vertex stored at an ancestor of $\alpha(p)$ and $\alpha(q)$, the value of $\delta^{(G)}(p, q)$ is equal to the minimum value of $\delta^{(G(\alpha(x)))}(p, x) + \delta^{(G(\alpha(x)))}(q, x)$, where x ranges over all vertices for which $(x, \delta^{(G(\alpha(x)))}(p, x)) \in A(p)$ and $(x, \delta^{(G(\alpha(x)))}(q, x)) \in A(q)$.

Thus, we can answer a shortest path-length query in G between two vertices p and q in $O(\sqrt{n})$ time by scanning the lists $A(p)$ and $A(q)$. In [3], it is shown how to augment the data structure with shortest path trees rooted at the separator vertices, so that the actual shortest path between p and q can be reported in $O(\ell)$ additional time, where ℓ is the number of edges on the reported path.

Lemma 4 ([3]). *Let S be a set of n points in the plane and let G be a planar graph with vertex set S. We can preprocess G in $O(n^{3/2})$ time into a data structure of size $O(n^{3/2})$ such that the shortest path in G between any two query points can be computed in $O(\sqrt{n} + \ell)$ time, where ℓ is the number of edges on the reported path.*

5.2   Bottleneck Shortest Path Queries in G

Consider again the planar graph G with vertex set S. Let $e_1, e_2, \ldots, e_m$ be the m edges of G, sorted by their lengths. For any i with $1 \le i \le m$, let $|e_i|$ denote the Euclidean length of edge e_i, and let $G^{(i)}$ be the graph consisting of all edges of G having length at most $|e_i|$.

In order to answer bottleneck shortest path queries in G, for each of the graphs $G^{(i)}$, we build the shortest path data structure of Lemma 4. We also compute a labeling of the vertices of each $G^{(i)}$ so that two vertices have the same label if and only if they are in the same connected component of $G^{(i)}$.

Observation 1 *Let p and q be two points of S, let L be a real number, and let i be the integer such that $|e_i| \le L < |e_{i+1}|$. There is a path between p and q in G all of whose edges have length at most L if and only if p and q are in the same connected component of $G^{(i)}$. Furthermore, the shortest path between p and q in $G^{(i)}$ is the same as the shortest path between p and q in G all of whose edges have length at most L.*

Thus, we build a binary search tree T over the sorted edge set of G. In $O(\log n)$ time, we can find the index i such that $|e_i| \le L < |e_{i+1}|$. Given that every node of T stores a pointer to the corresponding graph $G^{(i)}$ and the shortest path data structure for $G^{(i)}$, it now takes constant time to retrieve the two labels of the vertices p and q in $G^{(i)}$, and compare them to decide whether p and q are in the same connected component of $G^{(i)}$. If they are, we query the shortest path data structure to report the shortest path, which takes $O(\sqrt{n} + \ell)$ time, where ℓ is the number of edges in the reported path, using the data structure of Lemma 4. Since G is planar, we build $O(n)$ shortest path data structures of size $O(n^{3/2})$ each, one per graph $G^{(i)}$. Each of these data structures can be constructed in $O(n^{3/2})$ time. Hence, the total preprocessing time and amount of space used by our data structure is $O(n^{5/2})$.

Theorem 5. *Let S be a set of n points in the plane and let G be a planar graph with vertex set S. We can preprocess G in $O(n^{5/2})$ time into a data structure of size $O(n^{5/2})$ such that the following type of bottleneck queries can be answered: Given any two points p and q of S and any real number L, decide whether there is a path between p and q in G all of whose edges have length at most L. If such a path exists, report the shortest such path. The decision part of the query takes $O(\log n)$ time, whereas reporting the shortest path takes $O(\sqrt{n} + \ell)$ time, where ℓ is the number of edges on the reported path.*

If we combine Theorems 4 and 5, then we obtain the following result.

Theorem 6. *Let S be a set of n points in the plane. We can preprocess S in $O(n^{5/2})$ time into a data structure of size $O(n^{5/2})$ such that t-approximate bottleneck shortest path queries, for $t = 2\pi/(3\cos(\pi/6))$, can be answered in $O(\sqrt{n} + \ell)$ time, where ℓ is the number of edges on the reported path.*

References

1. M. A. Bender and M. Farach-Colton. The LCA problem revisited. *Proc. 4th LATIN*, LNCS, 1776, pp. 88–94, 2000.
2. M. S. Chang, N.-F. Huang, and C.-Y. Tang. An optimal algorithm for constructing oriented Voronoi diagrams and geographic neighborhood graphs. *Information Processing Letters*, 35:255–260, 1990.
3. H. N. Djidjev. Efficient algorithms for shortest path queries in planar digraphs. *Proc. 22nd WG*, LNCS, 1197, pp. 151–165, 1996.
4. D. Eppstein. Spanning trees and spanners. In *Handbook of Computational Geometry*, pages 425–461. Elsevier, 2000.
5. J. M. Keil and C. A. Gutwin. Classes of graphs which approximate the complete Euclidean graph. *Discrete & Computational Geometry*, 7:13–28, 1992.
6. G. Narasimhan and M. Smid. Approximation algorithms for the bottleneck stretch factor problem. *Nordic Journal of Computing*, 9:13–31, 2002.
7. A. C. Yao. On constructing minimum spanning trees in k-dimensional spaces and related problems. *SIAM Journal on Computing*, 11:721–736, 1982.

Optimization in Arrangements

Stefan Langerman[1]* and William Steiger[2]

[1] School of Computer Science, McGill University
sl@cgm.cs.mcgill.ca
[2] Computer Science, Rutgers University
steiger@cs.rutgers.edu

Abstract. Many problems can be formulated as the optimization of functions in R^2 which are implicitly defined by an arrangement of lines, halfplanes, or points, for example linear programming in the plane. We present an efficient general approach to find the optimum exactly, for a wide range of functions that possess certain useful properties. To illustrate the value of this approach, we give a variety of applications in which we speed up or simplify the best known algorithms. These include algorithms for finding robust geometric medians (such as the Tukey Median), robust regression lines, and ham-sandwich cuts.

1 Introduction

Given a set L of n non-vertical lines $\ell_1, \ldots, \ell_n$, where line ℓ_i has equation $y = \ell_i(x) = a_i x + b_i$, many problems can be formulated as the optimization of some function $f_L : R^2 \to R$ implicitly defined by L. Similarly, given a set P of n points $p_1, \ldots, p_n$, many problems can be formulated as the optimization of some function $f_P : R^2 \to R$ implicitly defined by P. Without loss of generality, we will restrict ourselves to minimization problems.

We outline a general approach to design simple and efficient algorithms to optimize functions that possess certain good properties. First, these functions should be somehow connected to the combinatorial structure of the arrangement of lines or of points on which they are defined. For a function f_L on an arrangement of lines, we assume it is known how to find an optimum point inside a cell of the arrangement. For a function f_P on a set of points, the combinatorial structure is more complex. Consider the $\binom{n}{2}$ lines joining every pair of points. These lines decompose R^2 into $O(n^4)$ cells. We assume that it is known how to find the optimum point inside each of these cells. Note that in each cell, the ordering of the slopes of the lines from q to the points in P is the same for every q in the cell. The time to find the optimum inside a cell will be denoted $T_C(n)$. We also assume that within the same time bounds, we can find an optimum of f_P restricted to a line segment, or to a small convex polygon Q that does not intersect any line of the arrangement. We call the function that returns an optimum inside a cell the *cell tool*. Moreover, we assume we have access to one of the following tools:

---

* Chargé de recherches du FNRS à l'Université Libre de Bruxelles

H. Alt and M. Habib (Eds.): STACS 2003, LNCS 2607, pp. 50–61, 2003.
© Springer-Verlag Berlin Heidelberg 2003

Sidedness: Given a line ℓ, the sidedness tool decides in time $T_S(n)$ which of the two closed halfplanes bounded by ℓ contains an optimum point for f.

Restricted sidedness: This applies to a function f_L on an arrangement of lines. Given a line ℓ and a convex polygon Q (given as a list of its vertices) known to contain an optimum point, restricted sidedness performs sidedness on ℓ in time $T_R(m)$, where m is the number of lines in L that intersect Q.

Witness: Given a point $p \in R^2$, the witness tool returns in time $T_W(n)$ a halfplane h containing p such that $f(q) \geq f(p)$ for all $q \in h$. If the function f is *level convex*, i.e. the set $\{q \in R^2 | f(q) \leq t\}$ is convex for all t, then a witness is guaranteed to exist for every point p.

The main results of this paper are stated in terms of these primitive operations:

Theorem 1. *Suppose we have a restricted sidedness tool for a function f_L on a set L of n lines, as well as a cell tool for f_L. Then we can find an optimum of f_L in time $O(n + T_R(n) + T_C(n))$. If we only have a sidedness tool the complexity to optimize f_L is $O((n + T_S(n)) \log n + T_C(n))$*

Theorem 2. *Suppose we have a function f_P on a set P of n points, and a sidedness tool for f_P. We can obtain an optimum in time $O(n \log^3 n + T_S(n) \log n + T_C(n))$.*

Theorem 3. *Suppose we have a level convex function f_P on a set P of n points, and a witness tool for f_P. We can obtain an optimum in time $O((n \log n + T_W(n)) \log^2 n + T_C(n) \log n)$.*

We prove these statements via algorithms shown to have the asserted complexity. For arrangements of points, the algorithms we describe are probabilistic, but they can be derandomized within the same time bounds using parametric search. Note that the randomized algorithms are extremely simple and should be quite easy to implement.

These theorems were motivated by several specific geometric optimization problems. We present some of these applications in Section 2 as an important part of the contribution of the paper, and to illustrate the potential value and range of applicability of our methods. Some of the applications significantly improve previously known algorithms for those problems. We also present applications which, though they only achieve the time bounds of known methods, do so using much simpler algorithms. The proofs of the theorems appear in Section 3 and Section 4, and contain some results of independent interest.

Throughout the paper we apply the duality transform where a point $q = (a, b)$ maps to the line $D_q = \{(x, y) : y = ax + b\}$ and the non-vertical line $\ell = \{(x, y) : y = mx + b\}$ maps to the point $D_\ell = (-m, b)$ (and a vertical line $x = t$ would map to the point (t) at infinity, the point incident with all lines of slope t; the point (∞) at vertical infinity is incident with all vertical lines). It is familiar that if q is (i) above, (ii) incident with, or (iii) below line ℓ then also D_q is (i) above, (ii) incident with, (iii) below D_ℓ, and also that the vertical distance of q from ℓ is preserved under D.

2 Some Applications

We show how to use the three main theorems to solve a variety of geometric optimization problems. In addition to presenting these algorithms themselves, we want to illustrate the applicability and utility of our methods. For each application we give a brief background and state the best current solution. Then we describe our new algorithm via the relevant theorem, give its complexity, and try to indicate how the particular sidedness, witness, and cell tools can be constructed. The full details will appear in the final paper.

The first four pertain to functions defined by arrangements of points and rely on Theorems 2 and 3. The resulting algorithms improve and simplify previous algorithms for these tasks. They also constitute the first known efficient algorithms for these problems that are simple enough to be implemented.

1. **Tukey median for points:** A given set P of n points in R^2 is used to define depth. The *Tukey depth* [27], or *location depth* of a point $q \in R^2$ is the smallest number of points of P in any halfplane containing q. That is,

$$\tau_P(q) = \min_{\text{halfspace } h \ni q} |h \cap P| \qquad (1)$$

A Tukey median is a point of maximum depth.

A well known consequence of Helly's Theorem (e.g., [14]) is that there is a point $x \in \mathbb{R}^2$ (not necessarily in P) of depth at least $\lceil n/3 \rceil$. Such a point is called a *centerpoint*. The *center* is the set of all centerpoints.

Cole, Sharir and Yap [10] described an $O(n(\log n)^5)$ algorithm to construct a centerpoint, and subsequent ideas of Cole [9] could be used to lower the complexity to $O(n(\log n)^3)$. Recently, Jadhav and Mukhopadhyay [17] described a linear time algorithm to find a centerpoint.

Matoušek [20] attacked the harder problem of computing a Tukey median and presented an $O(n(\log n)^5)$ algorithm for that task. The algorithm uses two levels of parametric search and ε-nets, and would be quite difficult to implement. Despite the fact that the Tukey median is of genuine interest in statistical applications, there is no really usable implementation. The fastest algorithms actually implemented for this task have complexity $\Theta(n^2)$ [22]. We can prove

Lemma 1. *Given a set P of n points in R^2, a Tukey median can be found in $O(n \log^3 n)$ time.*

Proof. Given $q \in R^2$, we sort the points in P in radial order from q. Thus the depth of q and a witness halfplane is obtained in $T_W(n) = O(n \log n)$. Every cell in the arrangement of the $\binom{n}{2}$ lines joining pairs of points in P has all its points of the *same* depth, so the cell tool can return any point in $T_C(n) = O(1)$. Finally since τ_P is level convex, Theorem 3 may be applied. $\square$

Besides being much faster than other known methods, this algorithm could be implemented easily.

2. **A Tukey median for convex polygons:** Applications in computer vision motivated a variant of Tukey depth that is defined with respect to the area within a given polygon Q. The depth (in Q) of $p \in R^2$ is defined by

$$\tau_Q(p) = \min_{\text{halfspace } h \ni p} Area(h \cap Q) \tag{2}$$

A median is a point in Q of maximal depth.

Diaz and O'Rourke [11,12,13] developed an $O(n^6 \log^2 n)$ algorithm for finding the Tukey median, n being the number of vertices of Q. If Q is convex Brass and Heinrich-Litan were able to compute a median in time $O((n^2 \log^3 n\alpha(n))$ [6]. We can show that

Lemma 2. *Given a convex polygon $Q \subset R^2$ with n vertices, a median may be found in time $O(n \log^3 n)$.*

Proof. For a convex polygon with its points given in clockwise order, a witness can be found in $O(n)$ time, and it can be shown that the optimum of τ_Q inside a cell can be found in $O(n)$ time. The statement now follows from Theorem 3.   □

Recently, Braß, Heinrich-Litan and Morin gave a randomized linear time algorithm for this task [7].

3. **Oja median:** Given a set P of n points, the Oja depth of a point $q \in R^2$ is the sum of the areas of the triangles formed by q and every pair of points in P. An Oja median is a point if minimum depth.

The first algorithm for finding the Oja median was presented by Niinimaa, Oja and Nyblom [24] and ran in time $O(n^6)$. This was then improved to $O(n^5 \log n)$ by Rousseeuw and Ruts, and then to $O(n^3(\log n)^2)$ and $O(n^2)$ space by Aloupis, Soss and Toussaint [2].

The Oja median has to lie on the intersection between two lines of $\binom{P}{2}$ [24], the Oja depth function can be shown to be level convex [2], and a witness can be computed in $T_W(n) = O(n \log n)$ time, and so, using Theorem 3, it is shown in [1] that optimum can be found in $O(n \log^3 n)$ time.

4. **Ham-sandwich cut in arrangements:** A given set $L = \{\ell_1, \ldots, \ell_n\}$ of non-parallel, non-vertical lines in the plane defines $\binom{n}{2}$ vertices $V = \{\ell_i \cap \ell_j, i < j\}$. Consider subsets A and B of those points each specified by some convex polygonal region, or by the union of a constant number of such regions. A ham-sandwich cut for A and B is a line that simultaneously splits the vertices in both A and B exactly in half. We can show that using a duality transform, this problem corresponds to minimizing a function for an arrangement of points for which sidedness can be decided in $T_S(n) = O(n \log^2 n)$ time. Applying Theorem 2 we get

Lemma 3. *A ham-sandwich cut for subsets A and B of the vertices of an arrangement of n lines can be found in $O(n \log^3 n)$ time.*

This compares to $O(n^2)$ complexity of the optimal planar ham sandwich cut algorithm when applied to $O(n^2)$ points in R^2 [19]. Details are omitted here.

The next three applications pertain to functions defined on line arrangements. Although they do not improve the time bounds of known algorithms, they are simple, and they are of interest in illustrating the range of applicability of the technique.

5. **Huber M-estimators:** Huber (see [16]) proposed an interesting and useful method of robust regression: Given n points (x_i, y_i) find a line $y = mx + b$ to minimize
$$\sum_{i=1}^{n} \rho_k(y_i - (mx_i + b)),$$
where $\rho_k(t) = t^2/2$ if $|t| \leq k$ and $k|t| - k^2/2$ otherwise; k is a given non-negative constant. In the dual we seek $p = (p_x, p_y) \in R^2$ to minimize
$$f_L(p) = \sum_{i=1}^{n} \rho_k(p_y - (m_i p_x + b_i)),$$
given lines $\ell_i = \{(x, y) : y = m_i x + b_i\}$. Restricted sidedness can be performed in $O(m)$ time and the optimum in a cell of the arrangement of the n given lines can be found in $O(n)$ time. Using Theorem 1 we obtain:

Lemma 4. *The Huber M-estimator for n given points in the plane can be computed in time $O(n)$.*

The first finite algorithm for the Huber M-estimator ([8]) was not even known to be polynomial. If we take $k = 0$ we also get a linear time algorithm for L_1 regression. A linear time algorithm for computing the L_1 regression line was first discovered by Zemel [29] by reducing the problem to an instance of the Multiple Choice L.P. Problem (MCLPP), which Megiddo's algorithm for linear programming [21] can solve in linear time.

6. **Fermat-Torricelli for lines:** The Fermat-Torricelli points for a set $L = \{\ell_1, \ldots, \ell_n\}$ of lines are defined to be the points that minimize:
$$f_L(p) = \sum_{i=1}^{n} d_\perp(p, \ell_i)$$
where $d_\perp(p, \ell)$ denotes the perpendicular distance between point p and line ℓ.

Roy Barbara [5] showed that a Fermat-Torricelli point for a set of n lines can always be found at a vertex of the arrangement of L. He then proposed to evaluate f_L at every vertex of the arrangement, obtaining an $O(n^3)$ algorithm. Using Theorem 1, Aloupis et. al. [1] show:

Lemma 5. *Given a set L of n lines, a Fermat-Torricelli point for L can be found in $O(n)$ time.*

In their proof they show that restricted sidedness can be performed in $O(m)$ time. The statement now follows easily by applying Theorem 1.

7. Hyperplane depth: The hyperplane depth $\delta_L(p)$ of a point p with respect to a set L of lines is the minimum number of lines that intersects any ray from p. The Hyperplane median is a point p that maximizes $\delta_L(p)$. Note that the depth of all the points inside a cell is the same.

Hyperplane depth was introduced by Rousseeuw and Hubert [25], motivated by problems in robust regression. Using the duality transform, the lines in L are mapped to points and the *regression depth* of a line in dual space is the minimum number of points the line must meet in a rotation to vertical. This corresponds exactly to the hyperplane depth of the corresponding point in the original problem.

Rousseeuw and Hubert point out that $\delta_L(p)$ can be computed in time $O(n \log n)$ for any p, and since there are $O(n^2)$ vertices in the arrangement, the hyperplane median can be found in $O(n^3 \log n)$. They also mention an $O(n^3)$ algorithm [26]. In Amenta et. al. [3] it was observed that the arrangement of L can be constructed in $O(n^2)$ and then, using breadth-first-search on the graph of adjacent cells, the depth of every cell is obtained in the same $O(n^2)$ time. Van Kreveld et.al. [28] described an $O(n \log^2 n)$ algorithm for finding the hyperplane median, using an $O(n \log n)$ algorithm to perform sidedness on a line.

In [18], we showed that restricted sidedness could be performed in $O(m \log m)$ time. Using Theorem 1, we obtained:

Lemma 6. *Given a set L of n lines, a cell of maximum depth can be found in time $O(n \log n)$.*

3   Arrangements of Lines

3.1   Equipartitions

To do efficient pruning in the line arrangement, we needed a partitioning method. Given lines $L = \{\ell_1, \ldots, \ell_n\}$ in general position in R^2 and two other non-parallel lines ℓ_A and ℓ_B, not in L, nor parallel to any line in L, we write $C = \ell_A \cap \ell_B$. Every line in L crosses both ℓ_A and ℓ_B. Partition the lines in L into 4 sets, one for each quadrant ($I = +, +$, $II = +, -$, $III = -, -$, $IV = -, +$), depending on whether the line crosses ℓ_A above or below C, and whether it crosses ℓ_B above or below C. A line that belongs to one of these sets (quadrants) avoids the opposite quadrant. We say that ℓ_A and ℓ_B form an *α-partition* of L if each of the four groups contains at least α lines. We prove the following result, of independent interest.

Lemma 7. *Let $L = \{\ell_1, \ldots, \ell_n\}$ be a set of n lines in general position in R^2. There exists an $\lfloor n/4 \rfloor$-partition of L and it can be found in time $O(n)$.*

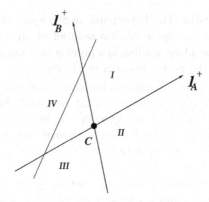

Proof. Let μ_j denote the jth smallest slope of the lines in L and let $\mu = (\mu_{\lfloor n/2 \rfloor} + \mu_{1+\lfloor n/2 \rfloor})/2$. The median level of the lines in L with slope less than μ (call this set L_1) and the median level of the lines in L with slope greater than μ (call this set L_2) cross at a unique point $C \in R^2$. Take ℓ_A to be the line of slope μ incident with C and ℓ_B to be the vertical incident with C. They form an $\lfloor n/4 \rfloor$-partition of L since half the lines in L_1 meet ℓ_A to the left of C and ℓ_B below C, and half meet ℓ_B above C and meet ℓ_A to the right of C. Similarly the lines of L_2 split into the two remaining quadrants.

In fact the construction could choose ℓ_A to be a line of arbitrary slope. The set L_1 would be the next $n/2$ lines in the (clockwise) radial ordering, and L_2 the remaining $n/2$ lines. The plane is then rotated so that vertical direction separates the last line in L_1 (in clockwise ordering) from the first line of L_2. Finally we point out that C is the dual of the ham-sandwich cut for the (separated) sets of points to which L_1 and L_2 are dual and so can be found in $O(n)$ time [19]. $\square$

There are analogous existence statements about equipartitions of *points* in R^d by d hyperplanes when $d \leq 3$. When $d > 4$ equipartitions of points do not always exist; the case $d = 4$ is open (see e.g., Avis [4]). It would be interesting to know if there exist three hyperplanes that equipartition n given hyperplanes in R^3.

Even though there is an algorithm that finds a $n/4$-partition in linear time, it is not trivial to implement. However there is an extremely easy way to randomly generate a $n/8$-partition with positive constant probability. Pick a line ℓ uniformly at random in L. Then pick randomly in L one line with slope smaller than ℓ and one line with slope larger than ℓ and let C be the intersection of those two lines. Then pick ℓ_A to be the line with the same slope as ℓ through C, and ℓ_B to be the vertical line through C. We can show that with constant probability $p > 0$, ℓ_A and ℓ_B form a $n/8$-partition. This then gives a Las Vegas algorithm in the usual way.

3.2 Line Pruning

With the notion of equipartitions, the algorithm for Theorem 1 is quite easy to state.

Algorithm LineOpt(L)
$\hat{L} \leftarrow L$
$Q \leftarrow R^2$
while $|\hat{L}| > 10$
    Find an $|\hat{L}|/c$-partition ℓ_A, ℓ_B, for some constant $c > 1$
    Perform a *sidedness test* on ℓ_A, ℓ_B restricted to Q
    Let h_A and h_B be the two halfplanes returned
        by the sidedness test.
    $Q \leftarrow Q \cap h_A \cap h_B$
    $\hat{L} \leftarrow \{\ell \in \hat{L} | \ell \cap h_A \cap h_B \neq \phi\}$
endwhile
perform a sidedness test on all the lines in $\hat{L}$.
return the optimum in the cell that contains the intersection of all the
    halfplanes returned and Q.

For correctness, the invariant that Q contains an optimum point is maintained throughout the algorithm, and at the end, all the cells remaining in Q are searched for the optimum point. Since at every step $\hat{L}$ is reduced by a constant factor $\geq 1/4$, the total number of steps is $O(\log n)$ and the running time is $O((n + T_S(n)) \log n + T_C(n))$ if sidedness is used. If restricted sidedness is used, the running time inside the loop forms a geometric progression, and the total running time is $O(n + T_R(n) + T_C(n))$. If the randomized algorithm is used to find the partition, then the running times of the algorithm are the same in the expected sense. This completes the proof of Theorem 1. $\qquad\square$

4   Arrangements of Points

4.1   Best of Candidates

Functions for arrangements of points seem to be more difficult to optimize. We first show an algorithm for a slightly simpler problem: suppose we are given a set A^* of candidate points, the algorithm returns a point p such that $f_P(p) \leq f_P(q)$ for every $q \in A^*$. The algorithm maintains a set A of points "to beat" and uses the witness tool. We will again need a pruning tool, but this time for points. Recall (from the introduction) that the Tukey depth $\tau_P(q)$ of a point q with respect to a set P of points is the minimum number of points of P that are contained in any halfplane that contains q. It is known that a point of depth $|P|/3$ always exists and can be found in linear time [17]. There is again a very simple way to obtain a point of depth $|P|/8$ with positive constant probability: pick a point at random in P and draw a vertical line through it. Then pick at random a point to the left, and a point to the right of that line. Join the two points by a line and call p the intersection of that line and the vertical line. It can be shown that with positive constant probability, $\tau_P(p) \geq |P|/8$.

Algorithm Bestof(A^*)
    $A \leftarrow A^*$; $p^* \leftarrow$ **some arbitrary point of A.**
    $Q \leftarrow R^2$

    while $A \neq \phi$ **repeat**
        find a point p **such that** $\tau_A(p) \geq c|A|$ **for some constant** $c \leq 1/3$.
        compute $f_P(p)$. **Let** h **be a witness for** p.
        if $f_P(p) \leq f_P(p^*)$ **then** $p^* \leftarrow p$ **endif.**
        $Q \leftarrow Q - h$
        $A \leftarrow A - (A \cap h)$
    endwhile

The invariant of the algorithm is that $f_P(p^*) \leq f_P(q)$ for all $q \in A^* - A$. The invariant is true at the beginning since $A^* - A = \phi$, and the invariant is preserved after each step of the while loop: $f_P(p^*) \leq f_P(p) \leq f_P(q)$ for all $q \in (A \cap h)$. As for the running time of the algorithm, we know that $|A \cap h| \geq c|A|$ because $\tau_A(p) \geq c|A|$. Thus the size of A reduces by a constant factor at every step of the loop, and so the number of loop steps is bounded by $O(\log |A^*|)$. Each step of the loop takes time $O(T_D(A) + T_W(n))$ where $T_D(A^*)$ is the time needed to compute a point p of depth $\tau(p) \geq |A^*|/c$ for some constant $c \leq 1/3$. The overall running time is $O((T_D(A^*) + T_W(n)) \log |A^*|)$ or $O(T_D(A^*) + T_W(n) \log |A^*|)$ if $T_D(A^*) = \Omega(|A^*|^\varepsilon)$ for some $\varepsilon > 0$. This is also $O(|A^*| + T_W(n) \log |A^*|)$ since we may take $T_D(A) = O(|A|)$ (e.g. use the algorithm of Jadhav and Mukhopadhyay [17] for finding a centerpoint, or the randomized method shown above).

For example, if we set A^* to the set of all $O(n^4)$ optima for each cell of the arrangement of lines in $\binom{P}{2}$, the algorithm will find the optimum for f_P, but the running time would be $O(n^4)$ if we have to enumerate all of A^* to use the centerpoint algorithm. One could think of designing a faster algorithm to find a point of Tukey depth $\geq c|A|$ for the candidate points inside Q without explicitly maintaining A, but this seems difficult even if the candidates are the vertices of the arrangement of lines in $\binom{P}{2}$, because even deciding whether Q contains any vertex of the arrangement is 3SUM-hard (for a definition of 3SUM-hardness, see [15]). Details of this fact appear in the full version.

4.2   Witness to Sidedness

In this section, we show that the witness tool can be used to perform a sidedness test on any given line.

Lemma 8. *If a witness tool is available for a function f_P on an arrangement of n points P, then a sidedness test for a line ℓ can be computed in time $T_S(n) = O((n \log n + T_W(n)) \log n + T_C(n))$.*

Proof. Consider the set of lines $\binom{P}{2}$ and let A^* be the set of the $\binom{n}{2}$ intersections between those lines and ℓ. Those intersections divide ℓ into $O(n^2)$ regions. If we run the algorithm of the previous section, at the end of the algorithm, $s = Q \cap \ell$ is a segment that does not intersect any of the $\binom{n}{2}$ lines. We can then use the cell tool to find the optimum point p^* on s in time $T_C(n)$. Neither the witness infinitesimally to the left of p^* nor the one infinitesimally to its right contain p^*, so they exclude one side of the line.

Since $|A^*| = \Theta(n^2)$, using the usual centerpoint construction algorithm for the set A in the main loop of the algorithm would take $O(n^2)$ time. We will reduce this to $O(n \log n)$ by exploiting the structure of the problem.

First notice that all the points in A^* lie on the straight line ℓ. This implies that there is a point on ℓ with Tukey depth $|A|/2$ with respect to any $A \subseteq A^*$. At any moment of the high depth algorithm, $A = A^* \cap R$ for some convex polygon R since every step eliminates a halfplane. In our case, this means that A corresponds to the points of A^* that lie inside a segment on ℓ. We will construct a point of Tukey depth $|A|/2$ with respect to A using the duality transform. First, rotate the plane so that ℓ is vertical, then apply the duality transform. P becomes a set of lines, and ℓ becomes a point T_ℓ at infinity. Rotate the dual plane so that that point becomes the point at vertical infinity. Any point a in A^* is at the intersection between ℓ and a line ℓ_{ij} that connects the two points p_i and p_j of P. Thus $T_{\ell_{ij}}$ is the vertex of the dual arrangement at the intersection of the lines T_{p_i} and T_{p_j}, and a is the vertical line passing through this vertex and the point T_v at vertical infinity. From this, we conclude that the line segment in v corresponding to A at some point of the algorithm is a vertical slab in the transformed space, and a point of Tukey depth $|A|/2$ with respect to A can be found using vertex selection inside that slab. This can thus be done in $O(n \log n)$ time. More easily we can generate a point at random inside that slab by counting the number of inversions between the ordering of the lines along the left boundary of the slab, and the ordering of the lines along the right boundary of the slab, in $O(n \log n)$ time. $\quad\square$

4.3 Line Pruning

Now that we have a sidedness test, we could use the algorithm from Section 3 to prune the lines, except that we have to be able to generate a $|\hat{L}|/c$-partition quickly even though $\hat{L}$ might be $\Theta(n^2)$. For this, we maintain $\hat{L}$ implicitly as the set of lines in $\binom{P}{2}$ that intersect Q, and we maintain Q.

As we saw before, generating a good partition randomly only requires to be able to generate random lines from $\hat{L}$. Using the duality transform, one can notice that this is equivalent to generating a random vertex in an arrangement amongst the ones inside a certain region. But this can be done in $O(n \log n)$ time using a vertex counting algorithm of Mount and Netanyahu [23]. Their algorithm is itself quite simple and easy to implement. This completes the proof of Theorems 2 and 3. $\quad\square$

References

1. G. Aloupis, S. Langerman, M. Soss, and G. Toussaint. Algorithms for bivariate medians and a fermat-torricelli problem for lines. In *Proc. 13th Canad. Conf. Comput. Geom.*, 2001.

2. G. Aloupis, M. Soss, and G. Toussaint. On the computation of the bivariate median and the fermat-torricelli problem for lines. Technical Report SOCS-01.2, School of Computer Science, McGill University, Feb. 2001.

3. N. Amenta, M. Bern, D. Eppstein, and S.-H. Teng. Regression depth and center points. *Discrete Comput. Geom.*, 23(3):305–323, 2000.
4. D. Avis. On the partitionability of point sets in space. In *Proc. 1st Annu. ACM Sympos. Comput. Geom.*, pages 116–120, 1985.
5. R. Barbara. The fermat-torricelli points of n lines. *Mathematical Gazette*, 84:24–29, 2000.
6. P. Brass and L. Heinrich-Litan. Computing the center of area of a convex polygon. Technical Report B 02-10, Freie Universität Berlin, Fachbereich Mathematik und Informatik, March 2002.
7. P. Brass, L. Heinrich-Litan, and P. Morin. Computing the center of area of a convex polygon. Technical report, 2002.
8. D. I. Clark and M. R. Osborne. Finite algorithms for Huber's M-estimator. *SIAM J. Sci. Statist. Comput.*, 7(1):72–85, 1986.
9. R. Cole. Slowing down sorting networks to obtain faster sorting algorithms. *J. ACM*, 34(1):200–208, 1987.
10. R. Cole, M. Sharir, and C. K. Yap. On k-hulls and related problems. *SIAM J. Comput.*, 16:61–77, 1987.
11. M. Díaz and J. O'Rourke. Computing the center of area of a polygon. In *Proc. 1st Workshop Algorithms Data Struct.*, volume 382 of *Lecture Notes Comput. Sci.*, pages 171–182. Springer-Verlag, 1989.
12. M. Díaz and J. O'Rourke. Chord center for convex polygons. In B. Melter, A. Rosenfeld, and P. Bhattacharyai, editors, *Computational Vision*, pages 29–44. American Mathematical Society, 1991.
13. M. Díaz and J. O'Rourke. Algorithms for computing the center of area of a convex polygon. *Visual Comput.*, 10:432–442, 1994.
14. H. Edelsbrunner. *Algorithms in Combinatorial Geometry*, volume 10 of *EATCS Monographs on Theoretical Computer Science*. Springer-Verlag, Heidelberg, West Germany, 1987.
15. A. Gajentaan and M. H. Overmars. On a class of $O(n^2)$ problems in computational geometry. *Comput. Geom. Theory Appl.*, 5:165–185, 1995.
16. P. Huber. *Robust Statistics*. John Wiley, NY, 1981.
17. S. Jadhav and A. Mukhopadhyay. Computing a centerpoint of a finite planar set of points in linear time. *Discrete Comput. Geom.*, 12:291–312, 1994.
18. S. Langerman and W. Steiger. An optimal algorithm for hyperplane depth in the plane. In *Proc. 11th ACM-SIAM Sympos. Discrete Algorithms*, 2000.
19. C.-Y. Lo, J. Matoušek, and W. L. Steiger. Algorithms for ham-sandwich cuts. *Discrete Comput. Geom.*, 11:433–452, 1994.
20. J. Matoušek. Computing the center of planar point sets. In J. E. Goodman, R. Pollack, and W. Steiger, editors, *Computational Geometry: Papers from the DIMACS Special Year*, pages 221–230. American Mathematical Society, Providence, 1991.
21. N. Megiddo. Linear-time algorithms for linear programming in R^3 and related problems. *SIAM J. Comput.*, 12:759–776, 1983.
22. K. Miller, S. Ramaswami, P. Rousseeuw, T. Sellarès, D. Souvaine, I. Streinu, and A. Struyf. Fast implementation of depth contours using topological sweep. In *Proceedings of the twelfth annual ACM-SIAM symposium on Discrete algorithms*, pages 690–699. ACM Press, 2001.
23. D. M. Mount and N. S. Netanyahu. Efficient randomized algorithms for robust estimation of circular arcs and aligned ellipses. Technical report, Dec. 1997.
24. A. Nniinimaa, H. Oja, and J. Nyblom. The oja bivariate median. *Applied Statistics*, 41:611–617, 1992.

25. P. J. Rousseeuw and M. Hubert. Depth in an arrangement of hyperplanes. *Discrete Comput. Geom.*, 22(2):167–176, 1999.
26. P. J. Rousseeuw and M. Hubert. Regression depth. *J. Amer. Statist. Assoc.*, 94(446):388–402, 1999.
27. J. W. Tukey. Mathematics and the picturing of data. In *Proceedings of the International Congress of Mathematicians (Vancouver, B. C., 1974), Vol. 2*, pages 523–531. Canad. Math. Congress, Montreal, Que., 1975.
28. M. van Kreveld, J. Mitchell, P. Rousseeuw, M. Sharir, J. Snoeyink, and B. Speckmann. Efficient algorithms for maximum regression depth. In *Proc. 15th ACM Symp. Comp. Geom.*, pages 31–40, 1999.
29. E. Zemel. An $O(n)$ algorithm for the linear multiple choice knapsack problem and related problems. *Inform. Process. Lett.*, 18(3):123–128, 1984.

Complete Classifications for the Communication Complexity of Regular Languages

Pascal Tesson and Denis Thérien*

School of Computer Science, McGill University
{ptesso,denis}@cs.mcgill.ca

Abstract. We show that every regular language L has either constant, logarithmic or linear two-party communication complexity (in a worst-case partition sense). We prove a similar trichotomy for simultaneous communication complexity and a "quadrichotomy" for probabilistic communication complexity.

1   Introduction

Communication complexity has grown to be one of the most ubiquitous tools of theoretical computer science [5] and the work pioneered by Babai, Frankl and Simon in [1] (see also [4]) has provided a rich structure of communication complexity classes to understand the relationships between various models of communication complexity.

The focus of this paper is the communication complexity of regular languages which, in some sense, are the simplest languages in the usual Time/Space complexity framework. Perhaps surprisingly, some regular languages have very high communication complexity even in powerful extensions of the deterministic model: There exist regular languages that are complete (in the sense of [1]) for every level of PH^{cc}, PSPACE^{cc} (using a result of [2]) and for $\mathrm{MOD}_q\mathrm{P}^{cc}$ (see [4]). Moreover, a majority of the most well-studied examples in the field such as Equality, Disjointness or Inner Product modulo m, are equivalent, from a communication complexity point of view, to regular languages (in these examples, equivalent to the regular languages $\{00 \cup 11\}^*$, $\{01 \cup 10 \cup 00\}^*$ and $\{\{\{01 \cup 10 \cup 00\}^*11\{01 \cup 10 \cup 00\}^*\}^m\}^*$ respectively).

The algebraic theory of finite monoids and semigroups provides a powerful framework and tools to analyze and classify regular languages [6] and has also been used to uncover connections between algebraic properties of finite monoids and classes of Boolean circuits using the program over monoid formalism of Barrington and Thérien [2,3]. Szegedy found a surprising connection between monoid programs and communication complexity by showing that a language has constant two-party deterministic communication complexity in the worst-case partition of the input if and only if it can be recognized by a program over some finite commutative monoid [9].

---

* Research supported in part by NSERC and FCAR grants.

H. Alt and M. Habib (Eds.): STACS 2003, LNCS 2607, pp. 62–73, 2003.

It was established in [8] that the class of regular languages having $O(f)$ communication complexity forms a variety and so the question of the communication complexity of regular languages has an algebraic answer. Amazingly, this point of view allows us to determine the complexity of every regular language in the deterministic, probabilistic and simultaneous models of communication. This complete classification unexpectedly features complexity gaps in all three cases: a regular language has either $O(1)$, $\Theta(\log n)$ or $\Theta(n)$ communication complexity in the deterministic two-party model, $O(1)$, $\Theta(\log n)$ or $\Theta(n)$ complexity in the simultaneous model and $O(1)$, $\Theta(\log \log n)$, $\Theta(\log n)$ or $\Theta(n)$ in the bounded-error probabilistic model. The classes of languages thus defined have fairly simple algebraic and combinatorial descriptions and are thus decidable. Moreover, some of these classes coincide: a regular language has $O(\log n)$ probabilistic complexity only if it has $O(\log n)$ deterministic complexity and further has $\Theta(\log \log n)$ probabilistic complexity if and only if it has $\Theta(\log n)$ simultaneous complexity. In proving these theorems we use rectangular reductions to and from only four different problems: Disjointness, Inner Product modulo m, Greater Than and Index which both highlights and, in retrospect, explains their importance as fundamental examples in communication complexity. We should note that, in contrast with the regular case, it is easy to artificially define, for any function $1 \leq f \leq n$, some non-regular language of communication complexity $\Theta(f)$ (in any of these models).

The main technical content of our paper is algebraic in nature and is presented in Section 2 along with some background on finite monoids. Next, Section 3 introduces the definitions and results in communication complexity needed to prove the bounds of our main theorems in Section 4. Due to space limitations, the proofs of some technical results are omitted although a full version of the paper can be obtained from the first author's web page.

2   Algebraic Background and Results

We refer the reader to e.g. [6] for further background on algebraic automata theory and semigroup theory.

A monoid M is a set together with a binary associative operation and a distinguished identity element 1_M for this operation. Throughout this paper, M will denote a finite monoid and A some finite alphabet.

The following equivalence relations on M, called Green's relations, describe whether two elements generate the same ideals: $x \mathcal{J} y$ if $MxM = MyM$, $x \mathcal{L} y$ if $Mx = My$, $x \mathcal{R} y$ if $xM = yM$ and $x \mathcal{H} y$ if both $x \mathcal{R} y$ and $x \mathcal{L} y$. It is known that $\mathcal{L}$ and $\mathcal{R}$ commute and that for finite monoids $\mathcal{L} \circ \mathcal{R} = \mathcal{J}$. We will denote by $\mathcal{J}_x$ the $\mathcal{J}$-class of x and similarly for the other relations.

If we instead ask whether element's ideals are contained in one another, we can define natural pre-orders $\leq_{\mathcal{J}}, \leq_{\mathcal{R}}, \leq_{\mathcal{L}}$ on M with e.g. $x \leq_{\mathcal{J}} y$ if and only if $MxM \subseteq MyM$. We will say that "x is (strictly) $\mathcal{J}$-above y" if $x \geq_{\mathcal{J}} y$ (resp. $x >_{\mathcal{J}} y$), and so on. Note that $x \leq_{\mathcal{J}} y$ if and only if there exists u, v such that $x = uyv$ and similarly $x \leq_{\mathcal{R}} y$ if and only if there is u with $x = yu$.

An element $m \in M$ is *idempotent* if $m^2 = m$ and since M is finite there must exist a smallest integer ω such that m^ω is idempotent for all $m \in M$. It can be shown that a subset T of M is a maximal subgroup in M if and only if T is an $\mathcal{H}$-class containing an idempotent. We call a $\mathcal{J}$-class regular if it contains at least one idempotent.

For a group G, the exponent of G is the smallest q such that $g^q = 1_G$ for all $g \in G$. More generally, we will call the *exponent of M* the least common multiple of the exponents of all subgroups of M.

A monoid N *divides* M if it is the morphic image of a submonoid of M and we denote this as $N \prec M$. A class $\mathbf{V}$ of monoids is said to form a *(pseudo-)variety* if it is closed under direct product and division. If $\mathbf{H}$ is a variety of finite groups, we denote by $\overline{\mathbf{H}}$ the variety of monoids whose subgroups all belong to $\mathbf{H}$. We will in particular mention the variety $\overline{\mathbf{Ab}}$ of monoids in which all subgroups are Abelian.

We say that $L \subseteq A^*$ can be *recognized* by M if there exists a morphism $\phi : A^* \to M$ and a subset $F \subseteq M$ such that $L = \phi^{-1}(F)$. A trivial variant of Kleene's Theorem states that a language L can be recognized by some finite monoid if and only if it is regular. Every such L admits some minimal monoid recognizer which we call L's *syntactic monoid* and denote $M(L)$. In particular, $M(L)$ recognizes L and for any N also recognizing L, we have $M(L) \prec N$. Moreover, for any class of regular languages $\mathbf{L}$ closed under Boolean operations, left and right quotients and inverse morphisms from A^* to B^*, there exists a variety of monoids $\mathbf{V}$ such that $L \in \mathbf{L}$ if and only if $M(L) \in \mathbf{V}$.

For a finite M, we say that $u, v \in A^*$ are *M-equivalent* if for all morphisms $\phi : A^* \to M$ we have $\phi(u) = \phi(v)$. For example, ab and ba are M-equivalent for any commutative M.

2.1 Combinatorial Descriptions of Subvarieties of DO

We denote $\mathbf{DS}$ the variety of monoids where every regular $\mathcal{J}$-class J is closed under multiplication. This is provably equivalent to the requirement that every $\mathcal{H}$-class within a regular J contain an idempotent. We denote by $\mathbf{DO}$ the subvariety of $\mathbf{DS}$ where every monoid satisfies $(xy)^\omega (yx)^\omega (xy)^\omega = (xy)^\omega$ for all x, y. Alternatively, one can characterize $\mathbf{DO}$ as the class of monoids such that the product of any two $\mathcal{J}$-related idempotents e, f is itself idempotent and $\mathcal{J}$-related to e and f. The following two lemmas, whose proofs we omit here, describe some of the useful properties of monoids in $\mathbf{DS}$ and $\mathbf{DO}$.

Lemma 1. *Let $M \in \mathbf{DS}$ then*

- *If e, f are $\mathcal{J}$-related idempotents then $(ef)^\omega e = e$ (but $efe = e$ if and only if ef is idempotent).*
- *for any $a, b \in M$ such that $a \leq_{\mathcal{J}} b$ and a is regular (i.e. $\mathcal{J}_a$, the $\mathcal{J}$-class of a contains an idempotent) then $a \mathcal{J} ab \mathcal{J} ba$.*
- *for any $x, y \in M$ such that $xy \mathcal{R} x$ we have in fact $R_x y \subseteq R_x$.*

Lemma 2. *Suppose* $M \in \mathbf{DO}$ *and let* $w_1, w_2 \in A^*$ *be G-equivalent for any subgroup G of M. For any morphism* $\phi : A^* \to M$ *and any* $x \in M$ *such that* $x\phi(w_1) \,\mathcal{H}\, x\phi(w_2) \,\mathcal{R}\, x$ *we have in fact* $x\phi(w_1) = x\phi(w_2)$.

For a word $u \in A^*$, we denote $\alpha(u)$ the set of letters occurring in u. For $a \in \alpha(u)$, the *a-left* (resp. *a-right*) *decomposition* of u is the unique factorization $u = u_0 a u_1$ with $a \notin \alpha(u_0)$ (resp. $a \notin \alpha(u_1)$). For a finite group G, we define $\sim_{n,k}^{G}$ on A^* where $n = |A|$ by induction on $n + k$. First, we have $x \sim_{n,0}^{G} y$ if and only if x, y are G-equivalent. Next, we let $x \sim_{n,k}^{G} y$ when and only when:

1. $x \sim_{n,k-1}^{G} y$;
2. $\alpha(x) = \alpha(y)$;
3. For any $a \in \alpha(x) = \alpha(y)$, if $x = x_0 a x_1$ and $y = y_0 a y_1$ are the a-left decompositions of x and y then $x_0 \sim_{n-1,k}^{G} y_0$ and $x_1 \sim_{n,k-1}^{G} y_1$;
4. For any $a \in \alpha(x) = \alpha(y)$, if $x = x_0 a x_1$ and $y = y_0 a y_1$ are the a-right decompositions of x and y then $x_0 \sim_{n,k-1}^{G} y_0$ and $x_1 \sim_{n-1,k}^{G} y_1$.

This equivalence relation is well-defined since $|\alpha(x_0)| < |\alpha(x)|$ in (3) and $|\alpha(x_1)| < |\alpha(x)|$ in (4). It is easy to check that $\sim_{n,k}^{G}$ is in fact a congruence of finite index.

For $a_1, \ldots, a_s \in A$ and $K_0, \ldots, K_s \subseteq A^*$, we say that the concatenation $K = K_0 a_1 K_1 \ldots a_s K_s$ is *unambiguous* if for all $w \in K$ there exists a unique factorization $w = w_0 a_1 \ldots a_s w_s$ with $w_i \in K_i$. The following theorem due to Schützenberger gives a combinatorial description of languages recognized by monoids in $\mathbf{DO} \cap \overline{\mathbf{H}}$:

Theorem 1. *Let* $L \subseteq A^*$, *then* $M(L) \in \mathbf{DO} \cap \overline{\mathbf{H}}$ *if and only if L is the disjoint union of languages of unambiguous concatenations* $K = K_0 a_1 K_1 \ldots a_s K_s$ *where each* K_i *can be recognized by the direct product of a group of* $\mathbf{H}$ *and an idempotent and commutative monoid.*

In the full version of the paper, we prove a slight refinement of this theorem by relating these languages with the congruences $\sim_G$; specifically:

Theorem 2. *Let* $M = A^*/\gamma$, *with* $|A| = n$. *Then* $M \in \mathbf{DO} \cap \overline{\mathbf{H}}$ *iff* $\sim_{n,k}^{G} \subseteq \gamma$ *for some* $k \in \mathbb{N}$ *and* $G \in \mathbf{H}$.

Thus, a language is recognized by a monoid in $\mathbf{DO} \cap \overline{\mathbf{H}}$ if and only if it is the disjoint union of $\sim_{n,k}^{G}$-classes for some $G \in \mathbf{H}$ and some $n, k \in \mathbb{N}$. It is not hard to see that each $\sim_{n,k}^{G}$ class can be described by an unambiguous concatenation as in the statement of Schützenberger's Theorem.

We will also be interested in the following small subclass of $\mathbf{DO}$.

Definition 1. *We call* $\mathbf{W}$ *the variety of monoids M satisfying:*

1. *If p is the exponent of M then* $exw^p yf = exyf$ *for all* $w, e, f, x, y \in M$ *such that* e, f *are idempotents lying* $\mathcal{J}$-*below w. (i.e. M satisfies the identity* $(swt)^\omega x w^p y (uwv)^\omega = (swt)^\omega xy (uwv)^\omega$*)*
2. $ewxf = exwf$ *for all* $w, e, f, x, y \in M$ *such that* e, f *are idempotents lying* $\mathcal{J}$-*below w. (i.e. M satisfies* $(swt)^\omega wx(uwv)^\omega = (swt)^\omega xw(uwv)^\omega$*)*

In particular, condition 1 implies $\mathbf{W} \subseteq \mathbf{DO}$ since for all $x, y \in M$:

$$(xy)^\omega (yx)^\omega (xy)^\omega = (xy)^\omega (yx)^{p\omega} (xy)^\omega = (xy)^\omega (xy)^\omega = (xy)^\omega.$$

On the other hand condition 2 implies that any two elements w, x of the maximal subgroup with identity e commute since $xw = exwe = ewxe = wx$ and so $\mathbf{W} \subseteq \overline{\mathbf{Ab}}$. Also from 2 we get $exwyf = ewxyf$ for e, f idempotent $\mathcal{J}$-below w.

For a word $u \in A^*$ and $a \in A$ we denote by $|u|_a$ the number of occurrences of a in u and let $RED_t(u)$ be the unique word of A^* obtained by keeping in u only the first and last t occurrences of each letter a with $|u|_a \geq 2t$ and all occurences of letters a with $|u|_a < 2t$. For example, $RED_2(abcbabbababba) = abcbaabba$. We will show that languages recognized by monoids in $\mathbf{W}$ have a useful combinatorial characterization: we set $u \approx_{t,p} v$ if and only if:

1. $RED_t(u) = RED_t(v)$;
2. For all $a \in A$ we have $|u|_a \equiv |v|_a \pmod{p}$.

Alternatively, we could define $RED_{t,p}(u)$ as the word obtained from u by the following process: For every $a \in \alpha(u)$ with $|u|_a \geq 2t$ mark the first and last t occurrences of a then move all *other* occurrences of a, if any, next to the tth one and then reduce that block of a's modulo p. If $|u|_a < 2t$, all occurrences of a are left untouched. Note that we clearly have $RED_{t,1}(u) = RED_t(u)$ and $u \approx_{t,p} v$ if and only if $RED_{t,p}(u) = RED_{t,p}(v)$.

Theorem 3. *Let $M = A^*/\gamma$, then $M \in \mathbf{W}$ if and only if $\approx_{t,p} \subseteq \gamma$ for some t, p.*

Proof. For one direction, we need to show that $M = A^* / \approx_{t,p}$ satisfies both conditions in the definition of $\mathbf{W}$. We start with the second: Consider the words $q = (uwv)^{tp} wx(ywz)^{tp}$ and $r = (uwv)^{tp} xw(ywz)^{tp}$. For any $a \in A$, $|q|_a \equiv |x|_a + |w|_a \equiv |r|_a \pmod{p}$ and

$$RED_t((uwv)^{tp} wx(ywz)^{tp}) = RED_t((uwv)^{tp} x(ywz)^{tp})$$
$$= RED_t((uwv)^{tp} xw(ywz)^{tp})$$

since for any letter a occurring in w, the first t occurrences of a lie in $(uwv)^{tp}$ and its last t occurrences lie in $(ywz)^{tp}$. Thus, $q \approx_{t,p} r$ so M satisfies condition 3. Similarly, $(uwv)^{tp} xw^p y(swz)^{tp} \approx_{t,p} (uwv)^{tp} xy(swz)^{tp}$ and M thus satisfies the first condition.

Conversely, suppose M is in $\mathbf{W}$. We need to show that there exist t, p such that for any morphism $\phi : A^* \to M$ we have $\phi(q) = \phi(r)$ for any $q \approx_{t,p} r$ and it is in fact sufficient to establish $\phi(q) = \phi(RED_{t,p}(q))$ since $RED_{t,p}(q) = RED_{t,p}(r)$. In particular, we choose p as the exponent of M and t as $|M| + 1$.

Recall that to obtain $RED_{t,p}(q)$, one successively considers all $a \in A$ with $|q|_a \geq 2t$, "groups" together the "middle" a's and reduces their number modulo p. We will show that the image under ϕ is preserved by this process. Consider a word $u = u_0 a u_1 a \ldots a u_t$ with at least t occurences of a. Since $t = |M| + 1$, there must exist $1 \leq i < j \leq t$ such that $s_i = \phi(u_0 a u_1 a \ldots u_i) = \phi(u_0 a u_1 a \ldots u_j) = s_j$. This means that $s_i = s_i \phi(au_{i+1} a \ldots au_j) = s_i \phi(au_{i+1} a \ldots au_j)^\omega$. Therefore there exist $g, h \in M$ such that $\phi(u)$ can be written as geh where $e = \phi(au_{i+1}a \dots au_j)^\omega$ is an idempotent lying $\mathcal{J}$-below a.

Suppose now that q contains at least $2t + 1$ occurrences of a. We can thus factor q as $q = u_0 a \dots u_{t-1} a x a y a v_{t-1} a \dots a v_0$ where the u_i's and v_i's do not contain a. From the remarks of the preceding paragraph, we can now use condition 2 in Definition 1 to obtain $\phi(q) = \phi(u_0 a \dots u_{t-1} a \, a \, xy \, a v_{t-1} a \dots a v_0)$. Repeating the same process for all occurrences of a in x or y we can get $\phi(q) = \phi(u_0 a \dots u_{t-1} a \, a^{kp+d} z \, a v_{t-1} a \dots a v_0)$ where $a \notin \alpha(z)$ and from condition 1: $\phi(q) = \phi(u_0 a \dots u_{t-1} a \, a^d z \, a v_{t-1} a \dots a v_0)$ where $0 \leq d < p$ is such that $|q|_a - 2t \equiv d \pmod{p}$. If the same manipulation is made for every $a \in A$, we obtain $\phi(q) = \phi(RED_{t,p}(q))$ as we needed. $\square$

3 Communication Complexity

We only provide here a short introduction to three communication complexity models but refer the reader to the excellent book of Kushilevitz and Nisan [5] for further details and formal definitions.

In the deterministic model, two players, Alice and Bob, wish to compute a function $f : S^{n_A} \times S^{n_B} \to T$ where S and T are finite sets. Alice is given $x \in S^{n_A}$ and Bob $y \in S^{n_B}$ and they collaborate in order to obtain $f(x, y)$ by exchanging bits using a common blackboard according to some predetermined *communication protocol* $\mathcal{P}$. This protocol determines whose turn it is to write, furthermore what a player writes is a function of that player's input and the information exchanged thus far. When the protocol ends, its output $\mathcal{P}(x, y) \in T$ is a function of the blackboard's content. We say that $\mathcal{P}$ computes f is $\mathcal{P}(x, y) = f(x, y)$ for all x, y and define the *cost* of $\mathcal{P}$ as the maximum number of bits exchanged for any input. The *deterministic communication complexity* of f, denoted $D(f)$ is the cost of the cheapest protocol computing f. We will be interested in the complexity of functions $f : S^* \times S^* \to T$ and will thus consider $D(f)$ as a function from $\mathbb{N} \times \mathbb{N}$ to $\mathbb{N}$ and study its asymptotic behavior.

In a *simultaneous protocol* $\mathcal{P}$, we disallow any interaction between Alice and Bob: Each of them *simultaneously* sends a message to a trusted referee which has access to none of the input and the referee produces the output $\mathcal{P}(x, y) \in T$. We denote $D^{\parallel}(f)$ the simultaneous communication complexity of f, i.e. the cost of the cheapest simultaneous protocol computing f.

In a *probabilistic communication protocol* $\mathcal{P}$, Alice and Bob have access to private random bits which determine their behavior. The protocol is said to compute f if for all x, y, the probability over the choices of these random bits that $\mathcal{P}(x, y) = f(x, y)$ is at least 3/4. We denote $R(f)$ the randomized communication complexity of f.

All three of these models have been extensively studied. We will use classical communication complexity lower bounds for the following functions [5]:

- For $x, y \in \{0, 1\}^n$, we define Disjointness as: $DISJ(x, y) = 1$ if and only if $\bigvee_{1 \leq i \leq n} x_i y_i = 0$;

- For $x, y \in \{0,1\}^n$, and any $q \in \mathbb{N}$ we define Inner Product (mod q) as: $IP_q(x, y) = 1$ if and only if $\sum\limits_{1 \le i \le n} x_i y_i \equiv 0 \pmod{q}$;

- For two n-bit numbers $x, y \in [2^n]$ we define Greater Than as: $GT(x, y) = 1$ if and only if $x \ge y$.

- For $x \in \{0,1\}^n$ and a $\log n$-bit number $p \in [n]$ we define $INDEX(x, p) = x_p$;

The known communication complexity bounds for these problems can be summed up in the following table:

Theorem 4.

	D	R	$D^{\parallel}$
$DISJ$	$\Theta(n)$	$\Theta(n)$	$\Theta(n)$
IP_q	$\Theta(n)$	$\Theta(n)$	$\Theta(n)$
GT	$\Theta(n)$	$\Theta(\log n)$	$\Theta(n)$
$INDEX$	$\Theta(\log n)$	$\Theta(\log n)$	$\Theta(n)$

It should be noted that non-trivial work is necessary to prove the GT randomized upper bound and the $DISJ$ and IP randomized lower bounds.

In general, we want to study the communication complexity of functions which do not explicitly have two inputs. In the case of regular languages and monoids we will use a form of *worst-case partition* definition. Formally, the deterministic (resp. randomized, simultaneous) *communication complexity of a finite monoid M* is the communication complexity of evaluating in M the product $m_1 \cdot m_2 \cdot \ldots \cdot m_{2n}$ where the odd-indexed $m_i \in M$ are known to Alice and the even-indexed m_i are known to Bob.

Similarly, the *communication complexity of a regular language $L \subseteq A^*$* is the communication complexity of the following problem: Alice and Bob respectively receive $a_1, a_3, \ldots a_{2n-1}$ and $a_2, a_4, \ldots, a_{2n}$ where each a_i is either an element of A or the empty word ϵ and they want to determine whether $a_1 a_2 \ldots a_{2n}$ belongs to L.

The following basic facts are proved in [8].

Lemma 3. *Let $L \subseteq A^*$ be regular with $M(L) = M$. We have $D(M) = \Theta(D(L))$ and similarly for $D^{\parallel}$ and R.*

In particular the deterministic (resp. simultaneous, randomized) complexity of a monoid M is, up to a constant, the maximal communication complexity of any regular language that it can recognize.

Lemma 4. *For any increasing $f : \mathbb{N} \to \mathbb{N}$ the class of monoids such that $D(M)$ (resp. $D^{\parallel}(M)$, $R(M)$) is $O(f)$ forms a variety.*

In order to compare the communication complexity of two languages K, L in different models, Babai, Frankl and Simon [1] defined *rectangular reductions* from K to L which are, intuitively, reductions which can be computed privately by Alice and Bob without any communication cost. We give here a form of this definition which specifically suits our needs:

Definition 2. *Let $f : \{0,1\}^{n_A} \times \{0,1\}^{n_B} \to \{0,1\}$ and M some finite monoid. A rectangular reduction of length t from f to M is a sequence of $2t$ functions $a_1, b_2, a_3, \ldots, a_{2t-1}, b_{2t}$, with $a_i : \{0,1\}^{n_A} \to M$ and $b_i : \{0,1\}^{n_B} \to M$ and such* *that for every* $x \in \{0,1\}^{n_A}$ *and* $y \in \{0,1\}^{n_B}$ *we have* $f(x,y) = 1$ *if and only if* $eval_M(a_1(x)b_2(y)\ldots b_{2t}(y)) \in T$ *for some target subset* T *of* M.

Such a reduction transforms an input (x,y) of the function f into a sequence of $2t$ monoid elements $m_1, \ldots, m_{2t}$ where the odd-indexed m_i are obtained as a function of x only and the even-indexed m_i are a function of y.

It should be clear that if f has communication complexity $\Omega(g(n_A, n_B))$ and has for any n_A, n_B a reduction of length $t(n_A, n_B)$ to M then M has complexity $\Omega(g(t^{-1}(n)))$.

We will write $f \leq_r^t M$ to indicate that f has a rectangular reduction of length t to M and will drop the t superscript whenever $t = O(n)$.

4   Communication Complexity Bounds for Regular Languages and Monoids

Our goal is to prove tight bounds on the communication complexity of any regular language or, equivalently using Lemma 3, of any finite monoid in the deterministic, randomized and simultaneous models. We start with a number of useful upper bounds, the first of which is an easy observation due to [8].

Lemma 5. *If M is commutative then $D^{\|}(M) = O(1)$.*

Lemma 6. *Let $L \subseteq A^*$ be such that $M(L) \in \mathbf{DO} \cap \overline{\mathbf{Ab}}$. Then $D(L) = O(\log n)$.*

Proof. By Lemma 2, L is a union of $\sim_{|A|,k}^{G}$-classes for some Abelian group G. We claim that any such class has logarithmic communication complexity and argue by induction on $t = |A| + k$. For $t = 1$ there is nothing to prove. For $t > 1$, let $u \in A^*$ be some predetermined representative of the class. Given the input $x = x_1 x_2 \ldots x_{2n} \in A^*$, Alice and Bob can check whether $\alpha(x) = \alpha(u)$ by exchanging $|A| + 1$ bits. Next, they need to verify that u and x are G-equivalent. It is well known that for any Abelian group G of exponent p, we get u and x G-equivalent if and only if $|u|_a \equiv |x|_a \pmod{p}$ for all $a \in A$. The latter condition can easily be verified with communication cost about $|A|\lceil \log p \rceil$, a constant. Let $u = vaw$ be the a-left decomposition of u and i, j denote the locations of the leftmost occurrence of a in x that is seen respectively by Alice and Bob. These indices can be exchanged at logarithmic communication cost so that if e.g. i is smaller than j Alice and Bob can conclude that $x = x_1 \ldots x_{i-1} a x_{i+1} \ldots x_n$ is the a-left decomposition of x and, by induction, can verify that $x_1 \ldots x_{i-1} \sim_{|A|-1,k}^{G} v$ and $x_{i+1} \ldots x_n \sim_{|A|,k-1}^{G} w$ using only $O(\log n)$ communication. Left-right symmetry completes the proof.

□

Lemma 7. *Let $L \subseteq A^*$ be such that $M(L) \in \mathbf{W}$. Then $D^{\|}(L) = O(\log n)$ and $R(L) = O(\log \log n)$.*

Proof. As in the previous proof, we obtain these upper bounds for the $\approx_{t,p}$ classes. Let u be some representative of the target class and x the common input of Alice and Bob. Checking whether $|u|_a \equiv |x|_a \pmod{p}$ is easily done at constant cost so we only need to show that verifying $RED_t(x) = RED_t(u)$ can be done efficiently. For the simultaneous case, the players send to the referee the locations of the first and last t occurrences that they see of each letter $a \in A$. Given this information, the referee can reconstruct $RED_t(x)$ and compare it to $RED_t(u)$.

For the probabilistic case we use a subprotocol of cost $O(\log\log n)$ to determine for any $k \leq t$ which of Alice or Bob holds the kth (or symmetrically the kth to last) occurrence of some letter a in x, provided of course that $|x|_a \geq k$. We argue by induction on k: For $k = 1$, let i, j be the positions of the first occurrence of a seen by Alice and Bob respectively. Of course Alice holds the first occurrence of a if and only if $i < j$ and by Theorem 4 this can be tested by a randomized protocol at cost $O(\log\log n)$ since i, j are only $\log n$-bits long. For $k > 1$ we can assume from induction that Alice and Bob have marked, in their respective inputs, the occurrences of a which are among the first $k - 1$ of a in x. The kth occurrence must be either the first unmarked a that Alice sees or the first unmarked a that Bob sees, whichever comes first in x. Once again, Alice and Bob are left with comparing two $\log n$-bit numbers and apply the $O(\log\log n)$ cost protocol.

For $i, j \leq t$, the ith occurrence of a in x comes before the jth occurrence of b in x if and only if the ith occurrence of a in $RED_t(x)$ comes before the jth occurrence of b in $RED_t(x)$. This means that Alice and Bob can check $RED_t(x) = RED_t(u)$ by verifying that for all $i, j \leq t$ and all $a, b \in A$ the ith occurrence of a precedes the jth occurrence of b in $RED_t(u)$ if and only if the ith occurrence of a precedes the jth occurrence of b in x. Since they can determine which of them holds these occurrences, they can check precedence either privately (when one player holds both occurrences) or by using once more the $O(\log\log n)$ randomized protocol to compare two $\log n$ bit numbers.

It should be noted that in any event, the GT protocol is used only a constant number of times (depending on t and $|A|$) so we need not worry about the dwindling of the overall probability of correctness in the protocol. □

Lemma 8. *1. If M is non-commutative then $GT \leq_r^{2^n} M$ (Notice that the reduction has exponential length);*
*2. If M is not in **DS** then $DISJ \leq_r M$.*
*3. If M lies in **DS** but is not in **DO** then $IP_q \leq_r M$ for some integer q.*
4. If G is a non-commutative group then $IP_q \leq_r G$ for some integer q.
*5. If M is in **DO** but not in **W** then $INDEX \leq_r M$.*

Proof. Since 2 and 4 are already proved explicitly in [8], we only sketch their proofs in the full version of this paper.

1- Let $a, b \in M$ be such that $ab \neq ba$. We obtain an exponential length reduction from $GT(x, y)$ by building $m_1 m_2 \ldots m_{2^n+1}$ where $m_i = a$ for $i = m_{2x}$; $m_i = b$ for $i = m_{2y-1}$ and $m_i = 1_M$ otherwise. The product of the m_i is then ba if and only if $x \geq y$ and is ab otherwise.

3- Since M is in **DS** but not **DO**, there must exist two $\mathcal{J}$-related idempotents e, f such that ef is not idempotent. Recall from Lemma 1 that we thus have $efe \neq e = (ef)^\omega e$. Let q be minimal such that $(ef)^q e = e$: The reduction produces elements $m_1 \ldots m_{2n}$ where $m_{2i-1} = e$ if $x_i = 1$ and $m_{2i-1} = e(ef)^\omega$ otherwise and $m_{2i} = fe$ if $y_i = 1$ and $m_{2i} = (ef)^\omega e$ otherwise. In particular the product $m_{2i-1}m_{2i}$ is efe if and only if $x_i = y_i = 1$ and is e otherwise and so the product $m_1 \ldots m_{2n}$ equals $(ef)^{\sum_{1 \leq i \leq n} x_i y_i \equiv 0 \ (\mathrm{mod}\ q)} e$ which equals e if and only if $IP_q(x, y) = 1$.

5- Suppose first that there exist e, f, u, v, w with e, f idempotent and $\mathcal{J}$-below w such that $euw^p vf \neq euvf$. Since we assume membership in **DO**, we can use Lemma 2 to show that $ew^p e = e$ and $fw^p f = f$. We obtain a rectangular reduction from $INDEX(x, s)$ by creating $m = m_1 m_2 \ldots m_{2n+1}$ as follows:

$$
m_i = \begin{cases}
e & \text{for } i = 1, 3, \ldots, 2s - 3 \text{ (the first } s - 1 \text{ odd-indexed } m_i\text{'s)}; \\
(eu) & \text{for } i = 2s - 1; \\
(vf) & \text{for } i = 2s + 1; \\
f & \text{for } i = 2s + 3, \ldots, 2n + 1 \text{ (all other odd indexed } m_i\text{'s)}; \\
1_M & \text{for } i = 2j \text{ and } x_j = 0; \\
w^p & \text{for } i = 2j \text{ and } x_j = 1.
\end{cases}
$$

The values of the odd indexed and even indexed m_i depend respectively on s and x as required and one can see that since $ew^p e$ is e and $fw^p f$ is f, the product of the m_i's is equal to $euvf$ when x_s is 0 and $euw^p vf$ when x_s is 1 which shows the correctness of the reduction.

Suppose now that there exist $e, f, u, w \in M$ with e, f idempotent $\mathcal{J}$-below w but $ewuf \neq euwf$. On the other hand, we can assume that M satisfies the first condition and consequently that $ewuf = ew^{p+1}uf$ and $euf = ew^p uf = euw^p f$. We obtain a rectangular reduction from $INDEX(x, s)$ (assuming w.l.o.g. that $x_n = 1$) as follows:

$$
m_i = \begin{cases}
e & \text{for } i = 1 \\
f & \text{for } i = 2n + 1 \\
u & \text{for } i = 2s + 1; \\
1_M & \text{for all other odd-indexed } i; \\
w^p & \text{for even } i = 2j \text{ such that } x_j = x_{j-1}; \\
w & \text{for even } i = 2j \text{ such that } x_j = 1 \text{ and } x_{j-1} = 0; \\
w^{p-1} & \text{for even } i = 2j \text{ such that } x_j = 0 \text{ and } x_{j-1} = 1.
\end{cases}
$$

Again, the values of the odd indexed and even indexed m_i's depend respectively on s and x. The value of the even indexed m_i's are such that the product $m_2 m_4 \ldots m_{2i}$ is w^{kp+1}, for some k, if and only if x_i is 1 and w^{kp} if x_i is 0. Similarly, using the fact that $x_n = 1$, the product $m_{2i+1} \ldots m_{2n}$ is $w^{k'p}$ if and only if $x_i = 1$ and $w^{k'p+1}$ otherwise. Using the values assigned by the reduction to the odd-indexed m_i's we have

$$
m_1 m_2 \ldots m_{2n+1} = em_2 m_4 \ldots m_{2s} u m_{2s+1} \ldots m_{2n} f
$$

and by our previous remark this is $ew^{kp+1}uw^{k'p}f = ewuf$ if $x_s = 1$ and $ew^{kp}uw^{k'p+1}f = ewuf$ if $x_s = 0$ so our reduction is correct. □

In particular, if M does not lie in $\mathbf{DO} \cap \overline{\mathbf{Ab}}$ then it either admits a linear length reduction from $DISJ$ (if it is outside $\mathbf{DS}$) or from IP_q for some q (if it is either in $\mathbf{DS}$ but not in $\mathbf{DO}$ or if it is outside $\overline{\mathbf{Ab}}$). Combining our last lemma with the upper bounds above we can obtain the three following theorems.

Theorem 5. *Let $L \subseteq A^*$ be a regular language with $M = M(L)$. Then*

$$D(L) = \begin{cases} O(1) & \text{if and only if } M \text{ is commutative;} \\ \Theta(\log n) & \text{if and only if } M \text{ is in } \mathbf{DO} \cap \overline{\mathbf{Ab}} \text{ but not commutative;} \\ \Theta(n) & \text{otherwise.} \end{cases}$$

Proof. We know $D(L) = O(1)$ if M is commutative and $D(L) = \Omega(\log n)$ otherwise since in that case $GT \leq_r^{2^n} M$. Lemma 6 gives the upper bound when $M \in \mathbf{DO} \cap \overline{\mathbf{Ab}}$. Finally, when M is not in $\mathbf{DO} \cap \overline{\mathbf{Ab}}$, then it admits a linear length reduction from $DISJ$ or IP_q which yields the last $\Omega(n)$ lower bound. □

Theorem 6. *Let $L \subseteq A^*$ be a regular language with $M = M(L)$. Then*

$$R(L) = \begin{cases} O(1) & \text{if and only if } M \text{ is commutative;} \\ \Theta(\log \log n) & \text{if and only if } M \text{ is in } \mathbf{W} \text{ but not commutative;} \\ \Theta(\log n) & \text{if and only if } M \text{ is in } \mathbf{DO} \cap \overline{\mathbf{Ab}} \text{ but not in } \mathbf{W}; \\ \Theta(n) & \text{otherwise.} \end{cases}$$

Proof. When M is in $\mathbf{W}$ but not commutative we put together Lemma 7 and part 1 of Lemma 8 to get the tight $\log \log n$ bound. Similarly, if M in $\mathbf{DO} \cap \overline{\mathbf{Ab}}$ but not in $\mathbf{W}$ then it admits a reduction from $INDEX$ which proves the $\Omega(\log n)$ lower bound matching Lemma 6 and when M is not in $\mathbf{DO} \cap \overline{\mathbf{Ab}}$ we again use the linear lower bounds on the probabilistic complexity of $DISJ$ and IP_q. □

Theorem 7. *Let $L \subseteq A^*$ be a regular language with $M = M(L)$. Then*

$$D^{\|}(L) = \begin{cases} O(1) & \text{if and only if } M \text{ is commutative;} \\ \Theta(\log n) & \text{if and only if } M \text{ is in } \mathbf{W} \text{ but not commutative;} \\ \Theta(n) & \text{otherwise.} \end{cases}$$

Proof. When M is in $\mathbf{W}$ but not commutative we combine the upper bound of Lemma 7 with the lower bound obtained from part 1 of Lemma 8. When M is not in $\mathbf{W}$ then it admits a linear length reduction from $INDEX$, $DISJ$ or IP_q which all have $\Omega(n)$ simultaneous complexity. □

5   Concluding Remarks

We have proved that the communication complexity of a regular language in the deterministic, simultaneous and probabilistic two-party models can be decided by simple and decidable algebraic properties of its syntactic monoid and that, in each of the three cases, this classification only defines a small number of classes. Because regular languages are so well-behaved in this context, they are a natural starting point for further investigations in communication complexity. For instance, it is natural to ask whether such gaps in the complexity of regular languages also occur in other communication models. Recently, we have considered the MOD_p-counting communication complexity model (see e.g. [4] for definitions), where a similar constant to logarithmic to linear trichotomy seems to hold, and the non-deterministic case. The latter requires some extra algebraic machinery since we need to deal with ordered monoids and positive varieties to reflect the fact that a language might have a non-deterministic complexity exponentially larger than that of its complement. There is also ongoing work concerning the multiparty (number in the forehead) model where one might still hope to prove a generalization of the theorem of Szegedy mentioned in our introduction. Some partial results along those lines appear in [8] and [7].

References

1. L. Babai, P. Frankl, and J. Simon. Complexity classes in communication complexity theory. In *Proc. 27th IEEE FOCS*, pages 337–347, 1986.
2. D. A. Barrington. Bounded-width polynomial-size branching programs recognize exactly those languages in NC^1. *J. Comput. Syst. Sci.*, 38(1):150–164, Feb. 1989.
3. D. A. M. Barrington and D. Thérien. Finite monoids and the fine structure of NC^1. *Journal of the ACM*, 35(4):941–952, Oct. 1988.
4. C. Damm, , M. K. C. Meinel, and S. Waack. On relations between counting communication complexity classes. To appear in Journal of Computer and Systems Sciences. Currently available e.g. from www.num.math.uni-goettingen.de/damm/ and preliminary version in proceedings of STACS'92.
5. E. Kushilevitz and N. Nisan. *Communication Complexity*. Cambridge University Press, 1997.
6. J.-E. Pin. *Varieties of formal languages*. North Oxford Academic Publishers Ltd, London, 1986.
7. P. Pudlák. An application of Hindman's theorem to a problem on communication complexity. Draft, 2002.
8. J.-F. Raymond, P. Tesson, and D. Thérien. An algebraic approach to communication complexity. *Lecture Notes in Computer Science (ICALP'98)*, 1443:29–40, 1998.
9. M. Szegedy. Functions with bounded symmetric communication complexity, programs over commutative monoids, and ACC. *J. Comput. Syst. Sci.*, 47(3):405–423, 1993.

The Commutation with Codes and Ternary Sets of Words*

Juhani Karhumäki[1], Michel Latteux[2], and Ion Petre[3]**

[1] Department of Mathematics, University of Turku and
Turku Centre for Computer Science (TUCS)
Turku 20014, Finland
karhumak@cs.utu.fi
[2] LIFL, URA CNRS 369, Université des Sciences et Technologie de Lille
F-59655 Villeneuve d'Ascq, France
michel.latteux@lifl.fr
[3] Department of Mathematics, University of Turku and
Turku Centre for Computer Science (TUCS)
Turku 20520, Finland
ipetre@cs.utu.fi

Abstract. We prove several results on the commutation of languages. First, we prove that the largest set commuting with a given code X, i.e., its *centralizer* $\mathcal{C}(X)$, is always $\rho(X)^*$, where $\rho(X)$ is the *primitive root* of X. Using this result, we characterize the commutation with codes similarly as for words, polynomials, and formal power series: a language commutes with X if and only if it is a union of powers of $\rho(X)$. This solves a conjecture of Ratoandromanana, 1989, and also gives an affirmative answer to a special case of an intriguing problem raised by Conway in 1971. Second, we prove that for any nonperiodic ternary set of words $F \subseteq \Sigma^+$, $\mathcal{C}(F) = F^*$, and moreover, a language commutes with F if and only if it is a union of powers of F, results previously known only for ternary codes. A boundary point is thus established, as these results do not hold for all languages with at least four words.

Topics: *Regular Languages, Combinatorics on Words.*

1 Introduction

The centralizer $\mathcal{C}(L)$ of a language L is the largest set of words commuting with L, i.e., the maximal solution of the language equation $XL = LX$. As it can be readily seen, the notion of centralizer of L is well defined for any language L; as a matter of fact, $\mathcal{C}(L)$ is the union of all languages commuting with L. The best known problem with respect to the notion of centralizer is the intriguing question raised by Conway [9] more than thirty years ago.

---

* Work supported by Academy of Finland under grant 44087
** Current address: Department of Computer Science, Åbo Akademi University, Turku 20520, Finland, ipetre@abo.fi

H. Alt and M. Habib (Eds.): STACS 2003, LNCS 2607, pp. 74–84, 2003.

Conway's Problem: Is it true that for any rational language, its centralizer is rational?

Surprisingly enough, very little is known on Conway's problem. In fact, a much weaker question than Conway's is unanswered up to date: *it is not known whether or not the centralizer of any finite language is even recursively enumerable !*

A closely related problem is that of characterizing the commutation of languages. It is an elementary result of Combinatorics on Words that two words commute if and only if they have the same primitive root, see, e.g., [4] and [17]. A similar result holds also for the commutation of two polynomials and of two formal power series in noncommuting variables, with coefficients in a commutative field: two polynomials/formal power series commute if and only if they are combinations of a third one; these results are due to Bergman and Cohn, respectively, see [2], [7], and [8]. Characterizing the commutation of two languages appears to be a very difficult problem in general and certainly a similar result as above does not hold. E.g., if $X = \{a, a^3, b, ab, ba, bb, aba\}$ and $Y = X \cup \{a^2\}$, then $XY = YX$, but X and Y cannot be written as unions of powers of a third set. Nevertheless, it has been conjectured by Ratoandromanana, [21], that the commutation with a code can be characterized as in free monoids:

Conjecture 1 ([21]). For any code X and any language Y commuting with X, there is a language $R \subseteq \Sigma^+$ such that $X = R^n$ and $Y = \cup_{i \in I} R^i$, for some $n \in \mathbb{N}$, $I \subseteq \mathbb{N}$.

The first major result on the conjecture was achieved by Ratoandromanana [21], in the case of prefix codes, using ingenious (and involved) techniques on codes and prefix sets. Conjecture 1 remained open however in its general form.

We say that a language X satisfies the *BTC-property*, i.e., the Bergman-type of characterization, if the commutation with X can be characterized as in the statement of Conjecture 1, that is, similarly as in Bergman's theorem, see [13]. Thus, Conjecture 1 proposes that all codes satisfy the BTC-property. This property has been established for all singletons, as well as for all two-word (or *binary*) languages, and it has been proved that it does not hold for four-word languages, see [6]. It is a conjecture of [13] and [14] that the BTC-property holds also for all ternary sets of words.

Conjecture 2 ([13],[14]). For any ternary language $F \subseteq \Sigma^+$, the following hold:

(i) If $F \subseteq u^+$, for some primitive word $u \in \Sigma^+$, then a language X commutes with F if and only if $X = \cup_{i \in I} u^i$, for some $I \subseteq \mathbb{N}$.

(ii) If $F \nsubseteq u^+$, for all $u \in \Sigma^+$, then a language X commutes with F if and only if $X = \cup_{i \in I} F^i$, for some $I \subseteq \mathbb{N}$.

A language L is called *periodic* if $L \subseteq u^*$, for some $u \in \Sigma^*$. The first part of Conjecture 2 has been proved already in [19]: all periodic sets satisfy the BTC-property. The second part however remained open.

These three problems recently received some well deserved attention and a handful of different approaches have been investigated: the combinatorial approach, the equational method, the branching point approach, and the multiplicity approach, see [12], [15], and [20] for some surveys. Thus, it has been proved in [19] and [6] that Conway's problem has an affirmative answer for all periodic and binary languages, respectively, and in [13] and [14] an affirmative answer has been given for ternary sets, proving also that Conjectures 1 and 2 hold for ternary codes, see also [16] for a different approach. Using different techniques, it has been proved in [11] that Conjecture 1 holds for all ω-codes. Moreover, it has been established that any code has a unique *primitive root*, a notion of [1] and [23], see [20] for details, and that two codes commute if and only if they have the same primitive root. These results of [11] also led to an affirmative answer for Conway's problem for rational ω-codes.

In this paper, we solve both Conjectures 1 and 2, characterizing the commutation with codes and with ternary sets of words. We also give an affirmative answer for Conway's problem in the case of rational codes: for any rational code X, both $\rho(X)$ and $\mathcal{C}(X)$ are rational and moreover, $\mathcal{C}(X) = \rho(X)^*$, where $\rho(X)$ denotes the primitive root of X; this is the most general result known up to date on Conway's problem. Our results also lead to a much simpler proof than that of [14] for Conway's problem for ternary languages; as a matter of fact, we give here a sharper result, proving that for any ternary language $F \subseteq \Sigma^+$, either $\mathcal{C}(F) = F^*$, or $\mathcal{C}(F) = u^*$, for some primitive word $u \in \Sigma^+$ and thus, $\mathcal{C}(F)$ is rational.

2   Preliminaries

We recall here several notions and results needed throughout the paper. For basic notions and results of Combinatorics on Words we refer to [4], [17], and [18] and for those of Theory of Codes, we refer to [3]. For details on the notion of centralizer and the commutation of languages we refer to [14], [15], and [20].

In the sequel, Σ denotes a finite alphabet, Σ^* the set of all finite words over Σ and Σ^ω the set of all (right) infinite words over Σ. We denote by 1 the empty word and by $|u|$ the length of $u \in \Sigma^*$. For a word $u \in \Sigma^*$, u^ω denotes the infinite word $uuu\ldots$, while for $L \subseteq \Sigma^*$, $L^\omega = \{u_1 u_2 u_3 \ldots \mid u_n \in L, n \geq 1\} \subseteq \Sigma^\omega$.

We say that a word u is a *prefix* of a word v, denoted as $u \leq v$, if $v = uw$, for some $w \in \Sigma^*$. We denote $u < v$ if $u \leq v$ and $u \neq v$. We say that u and v are *prefix comparable* if either $u \leq v$, or $v \leq u$. For a word $u \in \Sigma^+$, $\mathrm{pref}_1(u)$ denotes the first letter of u. For $L \subseteq \Sigma^+$, we denote $\mathrm{pref}_1(L) = \{\mathrm{pref}_1(u) \mid u \in L\}$. The word u is a *root* of v if $v = u^n$, for some $n \in \mathbb{N}$; v is *primitive* if it has no root other than itself.

A language L of cardinal two (three) is called *binary (ternary*, resp.). L is called *periodic* if $L \subseteq u^*$, for some $u \in \Sigma^*$.

For a word u and a language L, we say that $v_1 \ldots v_n$ is an *L-factorization* of u if $u = v_1 \ldots v_n$ and $v_i \in L$, for all $1 \leq i \leq n$. For an infinite word α, we say that $v_1 v_2 \ldots v_n \ldots$ is an L-factorization of α if $\alpha = v_1 v_2 \ldots v_n \ldots$ and $v_i \in L$, for all $i \geq 1$. A *relation* over L is an equality $u_1 \ldots u_m = v_1 \ldots v_n$, with $u_i, v_j \in L$, for all $1 \leq i \leq m$, $1 \leq j \leq n$; the relation is *trivial* if $m = n$ and $u_i = v_i$, for all $1 \leq i \leq m$.

We say that L is a *code* if any word of Σ^* has at most one L-factorization. Equivalently, L is a code if and only if all relations over L are trivial.

Let Σ be a finite alphabet, and Ξ a finite set of unknowns in a one-to-one correspondence with a set of nonempty words $X \subseteq \Sigma^*$, say $\xi_i \leftrightarrow x_i$, for some fixed enumeration of X. A (constant-free) *equation* over Σ with Ξ as the set of unknowns is a pair $(u, v) \in \Xi^\omega \times \Xi^\omega$, usually written as $u = v$. The subset X *satisfies* the equation $u = v$ if the morphism $h : \Xi^\omega \to \Sigma^\omega$, $h(\xi_i) = x_i$, for all $i \geq 0$, verifies $h(u) = h(v)$. These notions extend in a natural way to *systems of equations*.

We define the *dependence graph* of a system of equations S, as the nondirected graph G, whose vertices are the elements of Ξ, and whose edges are the pairs $(\xi_i, \xi_j) \in \Xi \times \Xi$, with ξ_i and ξ_j appearing as the first letters of the left and right handsides of some equation of S, respectively. The following basic result on combinatorics of words ([4]), known as *Graph Lemma*, is very useful and efficient in our later considerations. Note that in Graph Lemma it is crucial that *all words are nonempty*.

Lemma 1 ([4], Graph Lemma). *Let S be a system and let $X \subset \Sigma^+$ be a subset satisfying it. If the dependence graph of S has p connected components, then there exists a subset F of cardinality p such that $X \subseteq F^*$.*

For languages L, R, we say that R is a *root* of L if $L = R^n$, for some $n \in \mathbb{N}$. We say that L is *primitive* if for any R such that $L = R^n$, $n \in \mathbb{N}$, we have $L = R$ and $n = 1$. The following is a result of [11], extending central properties of words to codes.

Theorem 1 ([11]). *Any code has a unique primitive root. Moreover, two codes commute if and only if they have the same primitive root.*

3 Ternary Sets of Words

As it is well-known, two words commute if and only if they have the same primitive root, or equivalently, if and only if they are powers of another word. Based on this, it is not difficult to prove, see [19], that a set of words X commutes with a word $u \in \Sigma^+$ if and only if $X \subseteq \rho(u)^*$, where $\rho(u)$ denotes the primitive root of u. Consequently, for any word $u \in \Sigma^+$, $\mathcal{C}(\{u\}) = \rho(u)^*$.

If instead of a singleton $\{u\}$, we consider a language $L \subseteq u^+$, i.e., a periodic set, then the situation is not much different than that of a singleton: a language X commutes with L if and only if $X \subseteq \rho(u)^*$ and moreover, $\mathcal{C}(L) = \rho(u)^*$.

The above results extend to binary sets of words as well. As it is well-known, a binary set F is either a periodic set, or a code. If F is a code, then it is proved in [6], see also [16] for a simpler proof, that $\mathcal{C}(F) = F^*$ and any set X commuting with F is of the form $X = \cup_{i \in I} F^i$, for some $I \subseteq \mathbb{N}$. We recall these results in the following theorem.

Theorem 2 ([6]). *Let F be a language over the alphabet Σ.*

(i) *If $1 \in F$, then $\mathcal{C}(F) = \Sigma^*$.*

(ii) *If $F \subseteq u^+$, for some primitive word $u \in \Sigma^+$, then $\mathcal{C}(F) = u^*$. Thus, a language X commutes with F if and only if $X = \cup_{i \in I} u^i$, for some $I \subseteq \mathbb{N}$.*

(iii) *If F is a binary code, then $\mathcal{C}(F) = F^*$. Moreover, a language X commutes with F if and only if $X = \cup_{i \in I} F^i$, for some $I \subseteq \mathbb{N}$.*

Consequently, for any binary language $F \subseteq \Sigma^+$, either $\mathcal{C}(F) = F^*$, or $\mathcal{C}(F) = u^*$, for some primitive word u. On the other hand, the above results do not hold for sets with at least four words. E.g., for $F = \{a, ab, ba, bb\}$, the set $X = F \cup F^2 \cup \{bab, bbb\}$ commutes with F, but it is not a union of powers of F, see [6]. Thus, the case of ternary sets is a boundary point for the commutation of languages. We prove in Section 3.2 that the commutation with ternary sets can be characterized as for words, polynomials, and power series, thus solving Conjecture 2. To this aim, a central result is achieved in Section 3.1, regarding the centralizer of a ternary language $F \subseteq \Sigma^+$. It has been proved in [13] and [14], using some involved techniques of equations on languages that the centralizer of any ternary set is rational. However, the general form of the centralizer is known only in the case of codes: for a ternary code F, $\mathcal{C}(F) = F^*$. It has been conjectured in [13], [14] that $\mathcal{C}(F) = F^*$, for all nonperiodic ternary sets $F \subseteq \Sigma^+$. We solve in Section 3.1 this problem, using only elementary techniques of Combinatorics on Words. In particular, this gives a new, elementary solution for Conway's problem in the ternary case, much simpler than the original solution of [14].

3.1   Conway's Problem for Ternary Sets of Words

The next result ([16]) shows that we can always reduce Conway's problem to the so-called branching sets of words. We say that a language L is *branching* if there are two words $u, v \in L$ such that u and v start with different letters. This simplification turns out to be essential in our results. The intuitive idea behind this result is that having a language L and a letter $a \in \Sigma$, Conway's problem has the same answer for languages aL and La. Thus, if all words in a language start with the same letter, we can "shift" the letter in the end, without essentially changing the problem. For any nonperiodic language, repeating this procedure a finite number of times will lead to a branching language.

Lemma 2 ([16]). *For any nonperiodic set of words $L \subseteq \Sigma^+$, there is a branching set of words L' such that $\mathcal{C}(L)$ is rational if and only if $\mathcal{C}(L')$ is rational. Moreover, $\mathcal{C}(L) = L^*$ if and only if $\mathcal{C}(L') = L'^*$.*

We describe in the next result the centralizer of any ternary set. This provides a tool to solve Conjecture 2 and it also gives a very simple proof for Conway's problem in the ternary case, cf. [14].

Theorem 3. *Let F be a ternary set of words over the alphabet Σ.*

(i)   If $1 \in F$, then $\mathcal{C}(F) = \Sigma^*$.
(ii) If $F \subseteq u^+$, for some $u \in \Sigma^+$, then $\mathcal{C}(F) = \rho(u)^*$. Thus, a language X commutes with F if and only if $X = \cup_{i \in I} \rho(u)^i$, for some $I \subseteq \mathbb{N}$.
(iii) If F is a nonperiodic ternary set, $F \subseteq \Sigma^+$, then $\mathcal{C}(F) = F^*$.

Proof. The statements (i) and (ii) follow from Theorem 2 above. For (iii), we can assume by Lemma 2 that F is branching. Thus, let $F = \{u, u', v\}$, where $\mathrm{pref}_1(v) \notin \{\mathrm{pref}_1(u), \mathrm{pref}_1(u')\}$. The following two claims are straightforward to prove.

Claim 1. For any $1 < v' < v$ there is $\alpha \in F$ such that $v'\alpha$ is prefix incomparable with the words of F. Moreover, $v' \notin \mathcal{C}(F)$.

Claim 2. If $1 < v' < v$, then $\mathcal{C}(F) \cap v^+v' = \emptyset$.

Assume now that there is $z \in \mathcal{C}(F) \setminus F^*$. Thus, since $F^*\mathcal{C}(F) \subseteq \mathcal{C}(F)$ and $\text{pref}_1(v) \notin \{\text{pref}_1(u), \text{pref}_1(u')\}$, it follows that $v^*z \subseteq \mathcal{C}(F) \setminus F^*$. Consequently, there is a shortest nonempty word $x \in \Sigma^+$ such that $v^*x \cap (\mathcal{C}(F) \setminus F^*) \neq \emptyset$. In particular, note that $x \notin F^*$.

Let $n \geq 0$ be such that $v^n x \in \mathcal{C}(F) \setminus F^*$. Thus, $v^{n+1}x \in F\mathcal{C}(F) = \mathcal{C}(F)F$ and so, $v^{n+1}x = \alpha\beta$, with $\alpha \in \mathcal{C}(F)$ and $\beta \in F$. Thus, either $\alpha = v^iv'$, $i \leq n$, $v' \leq v$, or $\alpha = v^{n+1}x'$, $x' \leq x$. A simple case analysis based on Claims 1 and 2 shows that this contradicts the choice of x as the shortest word $\delta \in \Sigma^+$ such that $v^*\delta \cap (\mathcal{C}(F) \setminus F^*) \neq \emptyset$. Due to space limitations, we omit here the details.

Consequently, $\mathcal{C}(F) \setminus F^* = \emptyset$, i.e., $\mathcal{C}(F) \subseteq F^*$. Since for all languages L, $L^* \subseteq \mathcal{C}(L)$, it follows that $\mathcal{C}(F) = F^*$. ∎

Theorem 3 implies an affirmative answer for Conway's problem in the ternary case.

Corollary 1. *Conway's problem has an affirmative answer for all ternary sets: for any ternary set F, $\mathcal{C}(F)$ is a rational set.*

3.2   The Commutation with Ternary Sets of Words

In this section, we characterize all sets commuting with a given ternary set $F \subseteq \Sigma^+$. For periodic sets $F \subseteq u^+$, with $u \in \Sigma^+$, a language X commutes with F if and only if $X \subseteq \rho(u)^*$. For nonperiodic ternary sets $F \subseteq \Sigma^+$, we prove here that $XF = FX$ if and only if $X = \cup_{i \in I}F^i$, for some $I \subseteq \mathbb{N}$, a result previously known only for ternary codes, see [13] and [14]. In particular, based on Theorem 3, our proof for the general case turns out to be simpler than that of [13] and [14] for ternary codes, see [20] for a detailed discussion.

The case of codes can be easily solved using Theorem 3 and the following result of [6].

Theorem 4 ([6]). *For any ternary code F, the BTC-property holds for F if and only if $\mathcal{C}(F) = F^*$.*

Using different arguments, we prove next that a similar result holds for all ternary sets. Note that the ternary hypothesis is essential in this result.

Let F be a ternary set of words and X an arbitrary subset of $\mathcal{C}(F)$. We say that a word $x \in X$ satisfies the property $\mathcal{P}_F^X(x)$ if for all $n \in \mathbb{N}$, $x \in F^n$ implies $F^n \not\subseteq X$. Whenever F and X are clearly understood from the context, we simply write $\mathcal{P}(x)$ instead of $\mathcal{P}_F^X(x)$. Note that for any F and for any X, $\mathcal{P}_F^X(1)$ does not hold.

For a finite set of words F, we denote by l_F (L_F, resp.) the length of a shortest (longest, resp.) word in F.

Theorem 5. *Let $F \subseteq \Sigma^+$ be a nonperiodic three word noncode. Then the BTC-property holds for F if and only if $\mathcal{C}(F) = F^*$.*

Proof. If the BTC-property holds for F, then, since $F\mathcal{C}(F) = \mathcal{C}(F)F$, it follows that $F = R^i$ and $\mathcal{C}(F) = \cup_{j \in J} R^j$, for some set of words R and $i \in \mathbb{N}$, $J \subseteq \mathbb{N}$. It is straightforward to prove, based on cardinality arguments, that any nonperiodic ternary set of words is primitive, and so, $i = 1$ and $R = F$. Also, since $\mathcal{C}(F)$ is the largest set commuting with F, it follows that $J = \mathbb{N}$, and thus, $\mathcal{C}(F) = F^*$.

To prove the reverse implication, assume that $\mathcal{C}(F) = F^*$, and let X be a language commuting with F. We prove that in this case, $X = \cup_{i \in I} F^i$, for some $I \subseteq \mathbb{N}$. If $X = \emptyset$, then the claim is trivially true. So, let us assume that $X \neq \emptyset$.

Since F is not a code, there is a nontrivial relation over F:

$$u_1 u_{i_2} \dots u_{i_m} = u_2 u_{i_{m+1}} \dots u_{i_n}, \tag{1}$$

with $u_1 \neq u_2$ and $u_1, u_2, u_{i_k} \in F$, for all $2 \leq k \leq n$. Let u_3 be the third element of F: $F = \{u_1, u_2, u_3\}$.

Since $\mathcal{C}(F)$ includes any set commuting with F, it follows that $X \subseteq \mathcal{C}(F) = F^*$. The following claim is straightforward to prove, based on Graph Lemma.

Claim 1. Let $x \in X$ be such that $\mathcal{P}_F^X(x)$ holds. If $y \in X$, $v \in F$ are such that $u_3 x = yv$, then $\mathcal{P}_F^X(y)$ holds.

Using Claim 1, we can prove the following claim.

Claim 2. If there is $x_1 \in X$ such that $\mathcal{P}_F^X(x_1)$ holds, then $\mathcal{P}_F^X(u_3^q)$ holds, for some positive integer q.

Proof of Claim 2. Since $FX = XF$, there exist for all $n \geq 1$, $v_n \in F$ and $x_{n+1} \in X$ such that

$$u_3 x_n = x_{n+1} v_n. \tag{2}$$

Moreover, by Claim 1, $\mathcal{P}_F^X(x_n)$ holds, for all $n \geq 1$. Since $\mathcal{P}(1)$ never holds, it follows that $x_n \neq 1$, for all $n \geq 1$.

Assume now that $x_n \notin u_3^+$, for all $n \geq 1$. Using the Graph Lemma, it can be easily proved by induction on n that for all $n \geq 1$, $x_n \in u_3^{n-1} F^+$. Consequently,

$$|x_n| \geq |u_3^{n-1}| = (n-1)|u_3|,$$

for all $n \geq 1$. On the other hand, by (2),

$$|x_n| = |x_{n-1}| + |u_3| - |v_{n-1}| \leq |x_{n-1}| + |u_3| - l_F,$$

where $l_F = \min_{u \in F} |u| \geq 1$. Thus, $|x_n| \leq |x_1| + (n-1)(|u_3| - l_F)$. Altogether, we obtain that

$$(n-1)|u_3| \leq |x_n| \leq |x_1| + (n-1)(|u_3| - l_F),$$

for all $n \geq 1$. This further implies that $n \leq 1 + \frac{|x_1|}{l_F}$, for all $n \geq 1$, which is impossible.

Our assumption is thus false: there is $n \geq 1$ such that $x_n \in u_3^+$, i.e., $x_n = u_3^q$, for some positive integer q. Consequently, by Claim 1, $\mathcal{P}(u_3^q)$ holds, proving Claim 2.

Assume now that there is an $x \in X$ such that $\mathcal{P}_F^X(x)$ holds. Then, by Claim 2, there is a positive integer q such that $\mathcal{P}(u_3^q)$ holds. For such a positive integer q, consider arbitrary words $v_1, \dots, v_q \in F$. We prove that $u_3^{q-i}v_1 \dots v_i \in X$, for all $0 \le i \le q$.

Since $\mathcal{P}_F^X(u_3^q)$ holds, necessarily $u_3^q \in X$, proving the claim for $i = 0$. Let now $i \ge 0$ so that $u_3^{q-i}v_1 \dots v_i \in X$, $0 \le i < q$. We prove that $u_3^{q-(i+1)}v_1 \dots v_i v_{i+1} \in X$. Since $XF = FX$, there exist $w \in F$ and $y \in X$ such that

$$u_3^{q-i}v_1 \dots v_i \cdot v_{i+1} = w \cdot y. \tag{3}$$

If $w \ne u_3$, then by Graph Lemma on (1) and (3) we obtain that F is periodic, a contradiction. Thus, $w = u_3$ and so, $y = u_3^{q-(i+1)}v_1 \dots v_i v_{i+1} \in X$.

Consequently, for $i = q$, we obtain that $v_1 \dots v_q \in X$, for all $v_1, \dots, v_q \in F$. Thus, $F^q \subseteq X$, contradicting $\mathcal{P}_F^X(u_3^q)$.

The conclusion is that there is no $x \in X$ such that $\mathcal{P}_F^X(x)$ holds. Equivalently, for any $x \in X$, there is an $m \in \mathbb{N}$ such that $x \in F^m \subseteq X$. In other words, X is of the form $X = \bigcup_{i \in I} F^i$, with $I = \{i \in \mathbb{N} \mid \exists x \in X : x \in F^i \subseteq X\}$. Thus, the BTC-property holds for F. ∎

The following result is a simple consequence of Theorem 2, Theorem 5 and Theorem 3.

Theorem 6. *Let $F \subseteq \Sigma^+$ be a ternary set of words.*

(i) If F is periodic, $F \subseteq u^+$, for some $u \in \Sigma^+$, then a language X commutes with F if and only if $X \subseteq \rho(u)^*$.

(ii) If F is nonperiodic, then a language X commutes with F if and only if $X = \cup_{i \in I} F^i$, for some $I \subseteq \mathbb{N}$.

Note that the cases of ternary and periodic sets are the only known cases of non-codes for which the BTC-property holds. Moreover, the ternary case is the boundary point for the validity of the BTC-property with respect to the cardinality of a finite set, as this property does not hold for languages with at least four words.

4 The Commutation with Codes

The most general result known on Conjecture 1 is that of [11] where it is proved that all ω-codes satisfy the BTC-property. Using the so-called *multiplicity approach*, developed in [11], we give a complete solution for Conjecture 1, proving that all codes satisfy the BTC-property.

In the multiplicity approach, we consider an equation on languages, in this case the commutation equation, and we translate it into the corresponding equation on formal power series. We then solve the problem in terms of formal power series and finally translate the result back into sets of words. The main result used in this respect is Cohn's theorem characterizing the commutation of two formal power series.

Theorem 7 (Cohn's Theorem, [7,8]). *Let K be a commutative field and Σ a finite alphabet. Two formal power series $p, q \in K\langle\langle\Sigma^*\rangle\rangle$ commute if and only if there are some formal power series $r \in K\langle\langle\Sigma^*\rangle\rangle$ and $p', q' \in K[[t]]$ such that $p = p'(r)$ and $q = q'(r)$, for a single variable t.*

Note, however, that in general the commutation of two sets of words does not necessarily imply the commutation of their characteristic formal power series. E.g., the sets of words $X = \{aa, ab, ba, bb, aaa\}$ and $Y = \{a, b, aa, ab, ba, bb, aaa\}$ commute, while their characteristic power series do not.

The following result of [21] is instrumental in our multiplicity approach solution to Conjecture 1.

Lemma 3 ([21]). *Let X be a code and Y a language commuting with X. For any $x \in X$ and $y \in Y$, there exist $k > 0$ and $\alpha \in X^+$ such that $(xy)^k \alpha \in X^+$.*

Recall that for any language X, its centralizer $\mathcal{C}(X)$ is the largest set commuting with X: $X\mathcal{C}(X) = \mathcal{C}(X)X$. Using Lemma 3 we can prove that for any code X, the products $X\mathcal{C}(X)$ and $\mathcal{C}(X)X$ are unambiguous. This implies that the characteristic formal power series of X and $\mathcal{C}(X)$ commute, thus effectively translating the commutation of two languages into the commutation of two formal power series. Based on this result and on Cohn's theorem, we can then characterize the commutation with codes.

Lemma 4. *For any code X, the product $X\mathcal{C}(X)$ is unambiguous.*

Proof. Assume that $X\mathcal{C}(X)$ is ambiguous, i.e., there are $x, y \in X$, $u, v \in \mathcal{C}(X)$ such that $xu = yv$ and $x \neq y$. By Lemma 3, there exists $\alpha \in X^+$ such that $(xu)^k \alpha \in X^+$. Let $z = (xu)^k \alpha$. Then

$$z^\omega = ((xu)^k \alpha)^\omega = (x(ux)^{k-1} u\alpha)^\omega = x((ux)^{k-1} u\alpha x)^\omega = x(wx)^\omega,$$

where $w = (ux)^{k-1} u\alpha \in \mathcal{C}(X)$. As it is easy to see, for any $\delta \in \mathcal{C}(X)$ and any $t \in X$, $(\delta t)^\omega \in X^\omega$ and so, $(wx)^\omega \in X^\omega$. Consequently, $z^\omega \in xX^\omega$.

Analogously, $z^\omega = ((yv)^k \alpha)^\omega \in yX^\omega$ and so, z^ω has two different X-factorizations. Now, a result of [10] states that X is a code if and only if for any $\gamma \in X^+$, γ^ω has exactly one X-factorization. This leads to a contradiction. ∎

By symmetry, it follows from Lemma 3 and Lemma 4 that for any code X, the product $\mathcal{C}(X)X$ is also unambiguous.

The following result gives the exact form of the centralizer in case of codes, the last step in solving Conjecture 1.

Theorem 8. *For any code X, $\mathcal{C}(X) = \rho(X)^*$, where $\rho(X)$ denotes the primitive root of X.*

Proof. By Lemma 4, both products $X\mathcal{C}(X)$ and $\mathcal{C}(X)X$ are unambiguous. Thus, if r_X is the characteristic formal power series of X and $r_{\mathcal{C}(X)}$ that of $\mathcal{C}(X)$, it follows that $r_X r_{\mathcal{C}(X)} = r_{\mathcal{C}(X)} r_X$. By Cohn's theorem, this implies that both r_X and $r_{\mathcal{C}(X)}$ can be expressed as combinations of another series r. If R is the support of r, then we obtain that $X = \cup_{i \in I} R^i$ and $\mathcal{C}(X) = \cup_{j \in J} R^j$, for some $I, J \subseteq \mathbb{N}$. However, X is a code, and so, I is a singleton: $X = R^i$, $i \in N$. It then follows from [11] and [21] that R is a code commuting with X and so, $R = \rho(X)^k$, for some $k \in \mathbb{N} \setminus \{0\}$. Thus, $\mathcal{C}(X) = \cup_{j' \in J'} \rho(X)^{j'}$, $J' \subseteq \mathbb{N}$. Since $\mathcal{C}(X)$ is the largest set commuting with X, it follows then that $\mathcal{C}(X) = \rho(X)^*$. ∎

Corollary 2. *Conway's problem has affirmative answer for all rational codes: if X is a rational code, then both $\rho(X)$ and $\mathcal{C}(X)$ are rational, and $\mathcal{C}(X) = \rho(X)^*$.*

Proof. As proved in [22], see also [5], for any rational language R, if $R = L^n$, for some $L \subseteq \Sigma^*$, then there is a rational language S such that $L \subseteq S$ and $R = S^n$. Thus, there is a rational language S such that $\rho(X) \subseteq S$ and $X = \rho(X)^n = S^n$, for some $n \geq 1$. But then, S is a code and $XS = SX$. By Theorem 1, X and S have the same primitive root: $S = \rho(X)^m$ and so, $\rho(X)^n = \rho(X)^{mn}$, i.e., $m = 1$. Consequently, $\rho(X)$ is rational and so is $\mathcal{C}(X) = \rho(X)^*$.   ∎

Based on Theorem 8, we can solve now Conjecture 1.

Theorem 9. *The BTC-property holds for all codes. Equivalently, for any code X, a language Y commutes with X if and only if there is $I \subseteq \mathbb{N}$ such that $Y = \cup_{i \in I} \rho(X)^i$.*

Proof. Since $XY = YX$, it follows that $Y \subseteq \mathcal{C}(X) = \rho(X)^*$. To prove the claim of the theorem, it is enough to prove that for any $n \geq 0$, if $Y \cap \rho(X)^n \neq \emptyset$, then $\rho(X)^n \subseteq Y$.

Let $u_1, \ldots, u_n \in \rho(X)$ such that $u_1 \ldots u_n \in Y$ and let $\alpha_1, \ldots, \alpha_n$ be arbitrary elements of $\rho(X)$. Let also $X = \rho(X)^k$, $k \geq 1$. Then, since $X^n Y = Y X^n$ and $(\alpha_1 \ldots \alpha_n)^k \in \rho(X)^{nk} = X^n$, it follows that $u_1 \ldots u_n (\alpha_1 \ldots \alpha_n)^k \in X^n Y = \rho(X)^{kn} Y$. Since $Y \subseteq \rho(X)^*$ and $\rho(X)$ is a code, this can only lead to a trivial $\rho(X)$-relation, i.e., $\alpha_1 \ldots \alpha_n \in Y$. Thus, $\rho(X)^n \subseteq Y$, proving the claim.   ∎

5 Conclusions

The commutation of languages turns out to be a very challenging problem in general and it is difficult to even conjecture a possible general characterization. We proved however, that the commutation with a code can be characterized similarly as for words: for any code X, a language L commutes with X if and only if $L = \cup_{i \in I} \rho(X)^i$, for some $I \subseteq \mathbb{N}$, where $\rho(X)$ is the primitive root of X, thus solving an old conjecture of Ratoandromanana [21]. A similar characterization holds also for the commutation with periodic, binary, and - as we proved here - ternary sets of words, but not for languages with at least four words; this solves a conjecture of [13].

The intriguing problem of Conway [9], asking if the centralizer of a rational language is rational, still remains far from being solved. We proved here that the centralizer of any rational code X is rational and in fact, $\mathcal{C}(X) = \rho(X)^*$, an interesting connection between the notions of centralizer and primitive root. Except the "simple" cases of periodic, binary, and ternary languages - solved here with elementary arguments -, nothing else is known on Conway's problem. In fact, it turns out that despite the various techniques we have developed for this problem, see [20], the only cases where we could solve it are those when the centralizer is "trivial", i.e., $\mathcal{C}(X) = \rho(X)^*$ - the proofs however are quite involved in many case, see, e.g., [14] and [21]. We conclude by recalling once again this remarkable problem, as well as some of its possibly simpler variants.

Problem 1 (Conway's Problem). Is it true that for any rational language, its centralizer is rational ?

Problem 2. Is the centralizer of a rational set always: a) recursively enumerable, b) recursive, c) rational ?

Problem 3. Is the centralizer of a finite set always: a) recursively enumerable, b) recursive, c) rational, d) finitely generated ?

References

1. Autebert, J.M., Boasson, L., Latteux, M.: Motifs et bases de langages, *RAIRO Inform. Theor.*, 23(4) (1989) 379–393.
2. Bergman, G.: Centralizers in free associative algebras, *Transactions of the American Mathematical Society* 137 (1969) 327–344.
3. Berstel, J., Perrin, D.: *Theory of Codes*, Academic Press, New York (1985).
4. Choffrut, C., Karhumäki, J.: Combinatorics of Words. In Rozenberg, G., Salomaa, A. (eds.), *Handbook of Formal Languages*, Vol. 1, Springer-Verlag (1997) 329–438.
5. Choffrut, C., Karhumäki, J.: On Fatou properties of rational languages, in Martin-Vide, C., Mitrana, V. (eds.), *Where mathematics, Computer Science, Linguistics and Biology Meet*, Kluwer, Dordrecht (2000).
6. Choffrut, C., Karhumäki, J., Ollinger, N.: The commutation of finite sets: a challenging problem, *Theoret. Comput. Sci.*, 273 (1-2) (2002) 69–79.
7. Cohn, P.M.: Factorization in noncommuting power series rings, *Proc. Cambridge Philos. Soc.* 58 (1962) 452–464.
8. Cohn, P.M.: Centralisateurs dans les corps libres, in Berstel, J. (ed.), *Séries formelles*, Paris, (1978) 45–54.
9. Conway, J.H.: *Regular Algebra and Finite Machines*, Chapman Hall (1971).
10. Devolder, J., Latteux, M., Litovsky, I., Staiger, L.: Codes and infinite words, *Acta Cybernetica* 11 (1994) 241–256.
11. Harju, T., Petre, I.: On commutation and primitive roots of codes, submitted. A preliminary version of this paper has been presented at WORDS 2001, Palermo, Italy.
12. Karhumäki, J.: Challenges of commutation: an advertisement, in *Proc. of FCT 2001*, LNCS 2138, Springer (2001) 15–23.
13. Karhumäki, J., Petre, I.: On the centralizer of a finite set, in *Proc. of ICALP 2000*, LNCS 1853, Springer (2000) 536–546.
14. Karhumäki, J., Petre, I.: Conway's Problem for three-word sets, *Theoret. Comput. Sci.*, 289/1 (2002) 705–725.
15. Karhumäki, J., Petre, I.: Conway's problem and the commutation of languages, *Bulletin of EATCS* 74 (2001) 171–177.
16. Karhumäki, J., Petre, I.: The branching point approach to Conway's problem, LNCS 2300, Springer (2002) 69–76.
17. Lothaire, M.: *Combinatorics on Words* (Addison-Wesley, Reading, MA., (1983).
18. Lothaire, M.: *Algebraic Combinatorics on Words* (Cambridge University Press), (2002).
19. Mateescu, A., Salomaa, A., Yu, S.: On the decomposition of finite languages, TUCS Technical Report 222, http://www.tucs.fi/ (1998).
20. Petre, I.: *Commutation Problems on Sets of Words and Formal Power Series*, PhD Thesis, University of Turku (2002).
21. Ratoandromanana, B.: Codes et motifs, *RAIRO Inform. Theor.*, 23(4) (1989) 425–444.
22. Restivo, A.: Some decision results for recognizable sets in arbitrary monoids, in *Proc. of ICALP 1978*, LNCS 62 Springer (1978) 363–371.
23. Shyr, H.J.: *Free monoids and languages*, Hon Min Book Company, (1991).

On the Confluence of Linear Shallow Term Rewrite Systems

Guillem Godoy*, Ashish Tiwari**, and Rakesh Verma***

[1] Technical University of Catalonia
Jordi Girona 1, Barcelona, Spain
ggodoy@lsi.upc.es
[2] SRI International
Menlo Park, CA
tiwari@csl.sri.com
[3] Computer Science Dept
Univ of Houston, TX
rmverma@cs.uh.edu

Abstract. We show that the confluence of shallow linear term rewrite systems is decidable. The decision procedure is a nontrivial generalization of the polynomial time algorithms for deciding confluence of ground and restricted non-ground term rewrite systems presented in [13,2]. Our algorithm has a polynomial time complexity if the maximum arity of a function symbol in the signature is considered a constant. We also give EXPTIME-hardness proofs for reachability and confluence of shallow term rewrite systems.

1 Introduction

Programming language interpreters, proving equations (e.g. $x^3 = x$ implies the ring is Abelian), abstract data types, program transformation and optimization, and even computation itself (e.g., Turing machine) can all be specified by a set of rules, called a rewrite system. The rules are used to replace ("reduce") subexpressions of given expressions by other expressions (usually equivalent ones in some sense). A fundamental property of a rewrite system is the confluence or Church-Rosser property. Informally, confluence states that if an expression a can be reduced (in zero or more steps) to two different expressions b and c, then there is a common expression d to which b and c can be reduced in zero or more steps. Confluence implies uniqueness of normal ("irreducible") forms and helps to "determinise" their search by avoiding backtracking.

In general, confluence is well-known to be undecidable; however, it is known to be decidable for terminating systems [8] and for the subclass of arbitrary variable-free ("ground") systems [4,11]. Ground systems include as a subclass the tree automata model, which has important computer science applications. The previous decidability

* Partially supported by the Spanish CICYT project MAVERISH ref. TIC2001-2476-C03-01.
** Research supported in part by DARPA under the MoBIES and SEC programs administered by AFRL under contracts F33615-00-C-1700 and F33615-00-C-3043, and NSF CCR-0082560.
*** Research supported in part by NSF grant CCR-9732186.

H. Alt and M. Habib (Eds.): STACS 2003, LNCS 2607, pp. 85–96, 2003.

proofs of confluence for ground systems [4,11] were based on tree-automata techniques and showed that this problem was in EXPTIME, but no nontrivial lower bounds were known. Hence the exact complexity of this problem was open until last year, when a series of papers [7,2,13,6] culminated in a polynomial time algorithm for this problem for shallow and rule-linear systems, which include ground systems as a special case. In a shallow system variables in the rules cannot appear at depth more than one. Shallow systems have been well-studied in other contexts [10,3]. Linearity in [13,6], called rule-linearity here, means each variable can appear at most once in the entire rule. Thus, commutativity ($x+y = y+x$) is a shallow equation but *not* rule-linear, and associativity is neither shallow nor rule-linear.

In this paper, we establish decidability of confluence for shallow systems in which the left-hand side and right-hand side are *independently* linear, i.e., a variable can have two occurrences in a rule—once in each side of the rule. In fact, the decision procedure runs in polynomial time if we assume that the maximum arity of a function symbol in the signature is a constant. The results in this paper subsume the shallow rule-linear systems of [13] as a special case. The algorithm is a nontrivial generalization of the algorithms in [2,13]. We introduce a notion of marked terms and marked rewriting, and then generalize the central concept of top stability in [2]. The conditions to be checked by the algorithm are also generalized and the constructions are more involved. We also prove that the reachability, joinability and confluence problems are all EXPTIME-hard for shallow non-linear systems and all are known to be undecidable for linear non-shallow systems [15], which indicates that the linearity and shallowness assumptions are fairly tight.

1.1 Preliminaries

Let $\mathcal{F}$ be a (finite) set of function symbols with an arity function $arity: \mathcal{F} \to \mathbb{N}$. Function symbols f with $arity(f) = n$, denoted by $f^{(n)}$, are called n-ary symbols (when $n = 1$, one says *unary* and when $n = 2$, *binary*). If $arity(f) = 0$, then f is a *constant symbol*. Let $\mathcal{X}$ be a set of variable symbols. The set of terms over $\mathcal{F}$ and $\mathcal{X}$, denoted by $\mathcal{T}(\mathcal{F}, \mathcal{X})$, is the smallest set containing all constant and variable symbols such that $f(t_1, \ldots, t_n)$ is in $\mathcal{T}(\mathcal{F}, \mathcal{X})$ whenever $f \in \mathcal{F}$, $arity(f) = n$, and $t_1, \ldots, t_n \in \mathcal{T}(\mathcal{F}, \mathcal{X})$. A *position* is a sequence of positive integers. If p is a position and t is a term, then by $t|_p$ we denote the *subterm of t at position p*: we have $t|_\lambda = t$ (where λ denotes the empty sequence) and $f(t_1, \ldots, t_n)|_{i.p} = t_i|_p$ if $1 \leq i \leq n$ (and is undefined if $i > n$). We also write $t[s]_p$ to denote the term obtained by replacing in t the subterm at position p by the term s. For example, if t is $f(a, g(b, h(c)), d)$, then $t|_{2.2.1} = c$, and $t[d]_{2.2} = f(a, g(b, d), d)$. By $|s|$ we denote the *size* (number of symbols) of a term s: we have $|a| = 1$ if a is a constant symbol or a variable, and $|f(t_1, \ldots, t_n)| = 1 + |t_1| + \ldots + |t_n|$. The *depth* of a term s is 0 if s is a variable or a constant, and $1 + max_i depth(s_i)$ if $s = f(s_1, \ldots, s_m)$. Terms with depth 0 are denoted by α, β, with possible subscripts.

If $\to$ is a binary relation on a set S, then $\to^+$ is its transitive closure, $\leftarrow$ is its inverse, and $\to^*$ is its reflexive-transitive closure. Two elements s and t of S are called *joinable* by $\to$, denoted $s \downarrow t$, if there exists a u in S such that $s \to^* u$ and $t \to^* u$. The relation $\to$ is called *confluent* or *Church-Rosser* if the relation $\leftarrow^* \circ \to^*$ is contained in $\to^* \circ \leftarrow^*$, that is, for all s, t_1 and t_2 in S, if $s \to^* t_1$ and $s \to^* t_2$, then $t_1 \downarrow t_2$. An equivalent definition of confluence of $\to$ is that $\leftrightarrow^*$ is contained in $\to^* \circ \leftarrow^*$, that is, all s and t in S such that $s \leftrightarrow^* t$ are joinable.

A *substitution* σ is a mapping from variables to terms. It can be homomorphically extended to a function from terms to terms: using a postfix notation, $t\sigma$ denotes the result of simultaneously replacing in t every $x \in Dom(\sigma)$ by $x\sigma$. Substitutions are sometimes written as finite sets of pairs $\{x_1 \mapsto t_1, \ldots, x_n \mapsto t_n\}$, where each x_i is a variable and each t_i is a term. For example, if σ is $\{x \mapsto f(b,y), y \mapsto a\}$, then $g(x,y)\sigma$ is $g(f(b,y), a)$.

A *rewrite rule* is a pair of terms (l, r), denoted by $l \to r$, with left-hand side (lhs) l and right-hand side (rhs) r. A *term rewrite system* (TRS) R is a finite set of rewrite rules. We say that s rewrites to t in one step at position p (by R), denoted by $s \to_{R,p} t$, if $s|_p = l\sigma$ and $t = s[r\sigma]_p$, for some $l \to r \in R$ and substitution σ. If $p = \lambda$, then the rewrite step is said to be applied *at the topmost position* (at the root) and is denoted by $s \to_R^r t$; it is denoted by $s \to_R^{nr} t$ otherwise. The rewrite relation $\to_R$ induced by R on $\mathcal{T}(\mathcal{F}, \mathcal{X})$ is defined by $s \to_R t$ if $s \to_{R,p} t$ for some position p.

A *(rewrite) derivation or proof* (from s) is a sequence of rewrite steps (starting from s), that is, a sequence $s \to_R s_1 \to_R s_2 \to_R \ldots$. The *size* $|R|$ of a TRS R of the form $\{l_1 \to r_1, \ldots, l_n \to r_n\}$ is $|l_1| + |r_1| + \ldots + |l_n| + |r_n|$.

Definition 1. *A term t is called*

- linear *if no variable occurs more than once in t.*
- shallow *if no variable occurs in t at depth greater than 1, i.e., if $t|_p$ is a variable, then p is a position of length zero or one.*
- flat *if t is a non-constant term of the form $f(s_1, \ldots, s_n)$ where all s_i are variables or constants.*

Definition 2. *Let R be a TRS.*
A term s is reachable *from t by R if $t \to_R^* s$.*
Two terms s and t are equivalent *by R if $s \leftrightarrow_R^* t$.*
Two terms s and t are joinable *by R, denoted by $s \downarrow_R t$, if they are joinable by $\to_R$.*
A term s is R-irreducible *if there is no term t s.t. $s \to_R t$.*
The TRS R is confluent *if the relation $\to_R$ is confluent on $\mathcal{T}(\mathcal{F}, \mathcal{X})$.*

We assume that R is a shallow and linear term rewrite system, that is, if $s \to t$ is a rule in R, then s and t are both linear and shallow terms. Unlike previous results in [13, 6], the terms s and t are allowed to share variables.

2 Confluence of Shallow and Linear Rewrite Systems

Assuming that the maximum arity of a function symbol in $\mathcal{F}$ is bounded by a constant, we show that confluence of shallow and linear term rewrite system R over $\mathcal{F}$ can be decided in polynomial time. The proof of this fact uses suitable generalizations of the techniques in [2,13]. In Section 2.1 we argue that without loss of generality, we can restrict the signature $\mathcal{F}$ to contain exactly one function symbol with nonzero arity. Thereafter, we transform the rewrite system R into a flat rewrite system in Section 2.2. The flat linear term rewrite system is saturated under ordered chaining inference rule in Section 2.3 to construct a rewrite closure, which has several useful properties. The rest of the proof relies on the notion of top-stable and marked top-stable terms (Section 2.4), the ability to compute these sets (Section 2.5), and relating confluence of a saturated flat linear rewrite system to efficiently checkable properties over these sets (Section 2.6).

2.1   Simplifying the Signature

Terms over an arbitrary signature $\mathcal{F}$ can be encoded by terms over a signature $\mathcal{F}'$ containing at most one function symbol with non-zero arity. We may assume that $\mathcal{F}$ contains at least one constant e that does not appear in R.

Proposition 1. *There exists an injective mapping σ from terms over an arbitrary $\mathcal{F}$ to terms over a signature $\mathcal{F}'$ containing exactly one function symbol (with non-zero fixed arity) such that if R' is defined as $\{\sigma(s) \to \sigma(t) : s \to t \in R\}$, then R is confluent if, and only if, R' is confluent.*

Proof. (Sketch) Let m be one plus the maximum arity of any function symbol in $\mathcal{F}$. Define the new signature $\mathcal{F}'$ as

$$\mathcal{F}' = \{h^{(0)} : h^{(l)} \in \mathcal{F}, l > 0\} \cup \{f^{(m)}\} \cup \{c^{(0)} : c^{(0)} \in \mathcal{F}\},$$

where f is a new symbol. Define the map σ as follows: for each $h \in \mathcal{F}$ with arity $l > 0$,

$$\sigma(h(t_1,\ldots,t_l)) = f(\sigma(t_1),\ldots,\sigma(t_l), e,\ldots, e, h)$$

where the number of e's above equals $m-l-1$, and for each $c \in \mathcal{F}$ with arity $0, \sigma(c) = c$. The mapping σ is clearly injective, but not surjective. We can classify terms over $\mathcal{F}'$ into type 1 and type 2 terms (using a simple sorted signature) so that terms of type 1 exactly correspond to $Range(\sigma)$. It is easy to see that there is a bijective correspondence between proofs in R and proofs in R' over terms in $Range(\sigma)$. Combining this observation with a result in [16], which states that proving confluence for arbitrary terms over the signature is equivalent to proving confluence of the well-typed terms according to any many-sorted discipline which is compatible with the rewrite system under consideration, it follows that R is confluent iff R' is confluent.

2.2   Flat Representation

In the transformation described in Section 2.1, the properties of being linear and shallow are preserved. We next flatten the term rewrite system so that the depth of each term is at most one. In particular, given a linear shallow term rewrite system R, it can be transformed so that each rule in R is of the form

$f(\alpha_1,\ldots,\alpha_m) \to c$	(F_c)	$c \to f(\alpha_1,\ldots,\alpha_m)$	(B_c)
$f(\alpha_1,\ldots,\alpha_m) \to x$	(F_x)	$x \to f(\alpha_1,\ldots,\alpha_m)$	(B_x)
$f(\alpha_1,\ldots,\alpha_m) \to f(\beta_1,\ldots,\beta_m)$	(P_f)	$\alpha \to \beta$	(P_c)

where each $\alpha_i, \beta_i, \alpha, \beta$ is a depth 0 term (i.e., either a variable or a constant). Rules of the form F_c and F_x are called *forward* rules and denoted by F, rules of the form B_c

and B_x are called *backward* rules and denoted by B, and rules of the form P_f and P_c are called *permutation* rules and denoted by P. Rules of the form B_x are called *insertion* rules. We call such a rewrite system R a *flat linear* rewrite system.

This transformation is easily done by replacing each non-constant ground term, say s, in R by a new constant, say c, and adding a rule $s \rightarrow c$ or $c \rightarrow s$, depending on whether s occurred on the left- or right-hand side of R.

$$\text{Flatten:} \qquad \frac{u[s] \rightarrow t}{u[c] \rightarrow t, s \rightarrow c} \qquad \frac{t \rightarrow u[s]}{t \rightarrow u[c], c \rightarrow s}$$

where s is a non-constant ground term and c is a new constant.

Exhaustive application of these two rules results in a flat linear shallow rewrite system. This transformation can be done in polynomial time, as the number of applications of the above two rules is bounded by the size of the initial rewrite system R. It is easily seen to preserve confluence, see [2,13] for instance.

2.3 Rewrite Closure

Let $\succ$ order terms based on their size, that is, $s \succ t$ iff $|s| > |t|$. An application of an F-rule results in a smaller term, whereas application of a B-rule gives a bigger term in this ordering.

Definition 3. *A term s is size-irreducible by R if there exists no term t such that $s \rightarrow^*_R t$ and $s \succ t$.*

Definition 4. *A derivation $s \rightarrow^*_R t$ is said to be increasing if for all decompositions $s \rightarrow^*_R s' \rightarrow_{l \rightarrow r, p} t' \rightarrow^*_R t$, there is no step at a prefix position of p in $t' \rightarrow^*_R t$.*

Observe that increasing derivations either have no rewrite step at position λ, or only one at the beginning of the derivation. For simplicity, we eliminate the former case by assuming a dummy rewrite rule $x \rightarrow x$ to be in R, which can always be applied at the λ position in case there is no top step.

A flat linear rewrite system can be saturated under the following ordered chaining inference to give an enlarged flat linear rewrite system with some nice properties.

$$\text{Ordered Chaining:} \qquad \frac{s \rightarrow t \qquad w[u] \rightarrow v}{w[s]\sigma \rightarrow v\sigma} \qquad \frac{s \rightarrow w[t] \qquad u \rightarrow v}{s\sigma \rightarrow w[v]\sigma}$$

where σ is the most general unifier of t and u, neither u nor t is a variable, and $s \not\succ t$ in the first case and $v \not\succ u$ in the second. Note that these restrictions ensure that ordered chaining preserves flatness and shallowness.

Application of ordered chaining preserves confluence. Moreover, if the maximum arity m is a constant, then saturation under ordered chaining can be performed in polynomial time.

Lemma 1. *Let $R = F \cup B \cup P$ be a flat linear rewrite system saturated under the ordered chaining inference rules. If $s \rightarrow^*_R t$, then there is a proof of the form $s \rightarrow^*_F \circ \rightarrow^*_P \circ \rightarrow^*_B t$.*

Lemma 1 can be easily established using proof simplification arguments [1]. Similar proofs have been presented before, but for the special case of ground systems [12] and rule-linear shallow rewrite systems [14]. The generalization to linear shallow case is straightforward and the details are skipped here. The process of saturation, in this context, can be interpreted as asymmetric completion [9].

Lemma 2. *Let R be a flat linear rewrite system saturated under the ordered chaining inference rule above. If s is size-irreducible (or, equivalently F-irreducible) and $s \to_R^* t$, then there is an increasing derivation $s \to_R^* t$.*

Example 1. If $R = \{x + y \to y + x, x \to 0 + x\}$, then the chaining inferences add a new rule $x \to x + 0$ to R. An increasing derivation for $0 + x \to^* (x + 0) + 0$ is $0 + x \to x + 0 \to (x + 0) + 0$.

2.4 Top-Stable Terms, Marked Terms, and Marked Rewriting

In the rest of the paper we assume that R is a flat linear term rewrite system, which is also saturated under the chaining inference rule.

Definition 5. *A term t with depth greater than 0 is said to be* top-stable *if it cannot be reduced to a depth 0 term. A depth 0 term α is* top-stabilizable *if it is equivalent to a top-stable term.*

The following is a simple consequence of Lemma 1.

Lemma 3. *The set $S_0 = \{ f\alpha_1 \ldots \alpha_m : f\alpha_1 \ldots \alpha_m$ is F-irreducible$\}$ is the set of all top-stable flat terms.*

The confluence test relies heavily on the concept of top-stable terms and depth 0 top-stabilizable terms. The basic observation is that if a top-stabilizable constant, say c, occurs at a certain position in a term, say $fct_2 \ldots t_m$, then this term $(fct_2 \ldots t_m)$ is equivalent to a term $t = ft_1t_2 \ldots t_m$ with the property that t rewrites to a depth 0 term via R only if $fxt_2 \ldots t_m$ also does. Here, t_1 is chosen to be top-stable. So, when considering rewrites on $fct_2 \ldots t_m$ or $ft_1 \ldots t_m$, we should treat c and t_1 as variables. This is roughly the intuition behind the following definitions of marked terms and marked rewriting.

Definition 6. *A* marking *M of a term t is a set of leaf positions in t. A term t with a marking M is denoted by (t, M).*

A marked term (s, M) rewrites to (t, N) via marked rewriting *if $s \to_{l \to r \in R, p} t$ for some position $p \notin M$ such that, if $l|_{p_1}$ is a constant then $p.p_1 \notin M$, and the new marking N satisfies: (a) for all q disjoint with p, we have $q \in M$ iff $q \in N$, (b) for all p_1, p_2 and q such that $l|_{p_1}$ and $r|_{p_2}$ are the same variable, $p.p_1.q \in M$ iff $p.p_2.q \in N$, and (c) no more positions are in N.*

A marked flat term $(s = f\alpha_1 \ldots \alpha_m, M)$ is said to be correctly marked *if for all $i \in M$, we have that α_i is top-stabilizable.*

Example 2. The marked term $(0 + x, \{1\})$, denoted as $\underline{0} + x$, cannot be rewritten with the rule $0 + x \to x$, but it can be rewritten with the rule $x + y \to y + x$ to $x + \underline{0}$.

The notions of size-irreducible terms, increasing derivations and top-stable terms can be adapted naturally to marked terms. All the arguments of Lemmas 1 and 2 are also valid for marked rewriting, and we have:

Lemma 4. *If $(s, M) \to_R^* (t, N)$, then there is a derivation of the form $(s, M) \to_F^*$ $\circ \to_P^* \circ \to_B^* (t, N)$. If (s, M) is size-irreducible and $(s, M) \to_R^* (t, N)$, then there is an increasing derivation $(s, M) \to_R^* (t, N)$.*

2.5 The Sets S_∞ and J_∞

The set S_0 of top-stable flat terms can be extended with empty markings to give the new set
$$\{(f\alpha_1 \ldots \alpha_m, \emptyset) : f\alpha_1 \ldots \alpha_m \in S_0\}$$
of marked top-stable terms, which we also denote by S_0. We add new marked flat terms to this set to get the set of all correctly marked top-stable flat terms and top-stabilizable constants (and a variable if some variable is top-stabilizable) using the following fixpoint computation, starting with the new set S_0.

$$S_{j+1} = S_j \cup \{c : c \leftrightarrow^* f\alpha_1 \ldots \alpha_m \text{ for some } (f\alpha_1 \ldots \alpha_m, M) \in S_j\}$$
$$\cup \{(f\alpha_1 \ldots \alpha_m, M) : (f\alpha_1 \ldots \alpha_m, M) \text{ is top-stable and } \forall i \in M : \alpha_i \in S_j\}$$

Note that by Lemma 4, $(f\alpha_1 \ldots \alpha_m, M)$ is top-stable iff it is irreducible by F by marked rewriting.

This iterative procedure of computing larger and larger subsets S_j of the set of all marked flat terms is guaranteed to terminate in a polynomial number of steps. This is because the total number of flat marked terms, up to variable renaming, is polynomial, assuming m is a constant.

Lemma 5. *If S_∞ is the fixpoint of the computation above, then, up to variable renaming, $(f\alpha_1 \ldots \alpha_m, M) \in S_\infty$ iff $(f\alpha_1 \ldots \alpha_m, M)$ is top-stable and correctly marked, and a depth 0 term $c \in S_\infty$ iff c is top-stabilizable.*

Definition 7. *Two marked terms (s, M) and (t, N) are said to be structurally joinable if $(s, M) \to_R^* (s', M')$ and $(t, N) \to_R^* (t', N')$ for some terms s' and t' with the same structure (i.e., $Pos(s') = Pos(t')$, where $Pos(s')$ is the set of all positions in s') and equivalent leaf terms (i.e., for all leaf positions[1] $p \in Pos(s')$, we have that $s'|_p$ and $t'|_p$ are equivalent).*

We use the following fixpoint computation to obtain some structurally joinable pairs of marked terms.

$$J_0 = \{((\alpha, \emptyset), (\beta, M)) : \alpha \leftrightarrow_R^* \beta \text{ and } \alpha, \beta \text{ are depth 0 terms}\}$$
$$J_{j+1} = J_j \cup \{((\alpha, \emptyset), (f\beta_1 \ldots \beta_m, M)) :$$
$$(f\beta_1 \ldots \beta_m, M) \text{ is top-stable,}$$
$$\alpha \to_R^r fa_1 \ldots a_m, (f\beta_1 \ldots \beta_m, M) \to_R^r (fb_1 \ldots b_m, N), \text{ and}$$
$$\forall i \in \{1 \ldots m\} \text{ either } a_i = b_i \text{ or } ((a_i, \emptyset), (b_i, N|_i)) \in J_j\}$$

---

[1] Note that due to Proposition 1, for non-leaf positions $p \in Pos(s')$, $s'|_p = t'|_p = f$.

where $N|_i$ contains the positions p such that $i.p \in N$. Note that the b_i can be considered depth 0 or 1 terms, and that the a_i can be considered depth 0 or satisfying $a_i = b_i$.

Lemma 6. *If J_∞ is the fixpoint of above computation, then it is the set of all structurally joinable pairs of terms of the form $((\alpha, \emptyset), (\beta, M))$ or $((\alpha, \emptyset), (f\beta_1 \ldots \beta_m, M))$, where α, β are depth 0 terms, and $(f\beta_1 \ldots \beta_m, M)$ is a flat top-stable marked term.*

2.6   The Technical Lemma and the Result

Definition 8. *A pair of rules $((l \to r), (l' \to r'))$ is useless if $l = x$ and $l' = y$ for some variables x and y that appear in r and r', respectively, at the same non-root position.*

Two top-stable marked flat terms $(f\alpha_1 \ldots \alpha_m, M)$ and $(f\alpha_1' \ldots \alpha_m', M')$ are first-step joinable if there exist

$$(f\alpha_1 \ldots \alpha_m, M) \to_{(l \to r)\sigma}^r (fs_1 \ldots s_m, N)$$
$$(f\alpha_1' \ldots \alpha_m', M') \to_{(l' \to r')\theta}^r (fs_1' \ldots s_m', N')$$

such that every s_i is equivalent to its corresponding s_i', and $((l \to r), (l' \to r'))$ is not useless.

Note that first-step joinability can be efficiently computed, since it is enough to consider subterms s_i and s_i' of depth 0 or 1: if $r|_i$ is a variable not in l, we can force $r|_i\sigma = r'|_i\theta$ by modifying the substitutions, and the same if $r'|_i$ is a variable not in l'.

The polynomial time test for confluence depends on the following characterization using the sets S_∞ and J_∞ and the notion of first-step joinability.

Lemma 7. *The rewrite system R is confluent if, and only if,*

(c1) Every pair α, β of equivalent depth 0 terms is joinable,
(c2) If $\alpha \leftrightarrow^* f\beta_1 \ldots \beta_m$ and $(f\beta_1 \ldots \beta_m, M) \in S_\infty$ then $(\alpha, \emptyset)$ and $(f\beta_1 \ldots \beta_m, M)$ are structurally joinable, i.e. $((\alpha, \emptyset), (f\beta_1 \ldots \beta_m, M)) \in J_\infty$
(c3) If $(f\alpha_1 \ldots \alpha_m, M) \in S_\infty$ and $(f\beta_1 \ldots \beta_m, N) \in S_\infty$ are such that $f\alpha_1 \ldots \alpha_m \leftrightarrow^* f\beta_1 \ldots \beta_m$, then these two marked terms in S_∞ are first-step joinable.

Proof. (Sketch) A correctly marked flat term $(f\alpha_1 \ldots \alpha_m, M)$ can be *lifted* to a term $fs_1 \ldots s_m$ by replacing the marked α_i's by equivalent top-stable and F-irreducible terms s_i's.

$\Rightarrow$: Suppose R is confluent. Condition (c1) follows from the definition of confluence.

Condition (c2). Suppose $\alpha \leftrightarrow^* f\beta_1 \ldots \beta_m$ and $(f\beta_1 \ldots \beta_m, M) \in S_\infty$. Using Lemma 5, we can lift $(f\beta_1 \ldots \beta_m, M)$ to the term $ft_1 \ldots t_m$. Since $(f\beta_1 \ldots \beta_m, M)$ is top-stable, it follows that $ft_1 \ldots t_m$ is top-stable. Now, α is equivalent to $ft_1 \ldots t_m$ and by confluence they are joinable. Since both are size-irreducible, there are increasing derivations of the form $\alpha \to_R^* u$ and $ft_1 \ldots t_m \to_R^* u$. We can extract an increasing derivation $(f\beta_1 \ldots \beta_m, M) \to_R^* (u' = u[\beta_1']_{p_1} \ldots [\beta_k']_{p_k}, N = \{p_1, \ldots, p_k\})$ from the latter derivation by ignoring all rewrite steps at or below marked positions. Using an auxiliary lemma, we can show that there exists a derivation $\alpha \to_R^* u[\beta_1'']_{p_1} \ldots [\beta_m'']_{p_k}$, such that $\beta_i'' \leftrightarrow_R^* \beta_i'$.

Condition (c3). Let $(f\alpha_1 \ldots \alpha_m, M)$ and $(f\beta_1 \ldots \beta_m, N)$ be marked flat terms in S_∞ such that $f\alpha_1 \ldots \alpha_m \leftrightarrow_R^* f\beta_1 \ldots \beta_m$. Again, using Lemma 5, we can lift $(f\alpha_1 \ldots \alpha_m, M)$ and $(f\beta_1 \ldots \beta_m, N)$ to size-irreducible terms $s = fs_1 \ldots s_m$ and $t = ft_1 \ldots t_m$. By confluence, s and t are joinable, and hence there exist increasing derivations $fs_1 \ldots s_m \to_{(l \to r)\sigma}^r fs'_1 \ldots s'_m \to_R^{*,nr} u$ and $ft_1 \ldots t_m \to_{(l' \to r')\theta}^r ft'_1 \ldots t'_m \to_R^{*,nr} u$. Such a u can be chosen minimally, and consequently the pair $(l \to r, l' \to r')$ is not useless. Clearly, every s'_i is equivalent to the corresponding t'_i.

Now, by suitably modifying the substitutions σ and θ, we can get marked rewrite steps, $(f\alpha_1 \ldots \alpha_m, M) \to_{(l \to r)\sigma'} (fs''_1 \ldots s''_m, M')$ and $(f\beta_1 \ldots \beta_m, N) \to_{(l' \to r')\theta'} (ft''_1 \ldots t''_m, N')$, such that $s''_i \leftrightarrow_R^* s'_i \leftrightarrow_R^* t'_i \leftrightarrow_R^* t''_i$. This shows that the two marked terms $(f\alpha_1 \ldots \alpha_m, M)$ and $(f\beta_1 \ldots \beta_m, N)$ are first-step joinable.

$\Leftarrow$: Suppose conditions (c1), (c2), and (c3) are satisfied, but R is not confluent. Let $\{s, t\}$ be a witness to non-confluence, that is, s and t are equivalent, but not joinable. We compare witnesses by a multiset extension of the ordering $\succ$ defined earlier. First, we note that both s and t can be assumed to be size-irreducible, otherwise we would have a smaller counterexample to confluence.

If $s = fs_1 \ldots s_m$, then each s_i is either top-stable or of depth 0. Similarly, for the term t. Additionally, if the top-stable subterms s_i are equivalent to some depth 0 terms, then the term s can be *projected* onto a correctly marked flat term $(f\alpha_1 \ldots \alpha_m, M)$ where either α_i is the depth 0 term equivalent to s_i and $i \in M$, or $\alpha_i = s_i$. We differentiate the following cases based on the form of s and t:

Case 1. s and t are both depth 0 terms: In this case, Condition (c1) implies that s and t are joinable, a contradiction.

Case 2. s is a depth 0 term α and $t = ft_1 \ldots t_m$: We first claim that each t_i is equivalent to a depth 0 term. If not, then w.l.o.g. let t_1 not be equivalent to any depth 0 term. Then t_1 cannot be "used" in the proof $\alpha \leftrightarrow_R^* ft_1 \ldots t_m$, and hence it can be replaced by a new variable x in this proof to yield a new proof $\alpha' \leftrightarrow_R^* fxt_2 \ldots t_m$. If α' and $fxt_1 \ldots t_m$ are not joinable, then they are a smaller witness to non-confluence, a contradiction. If α' and $fxt_1 \ldots t_m$ are joinable, then $\alpha' = x$, and α and t_1 are equivalent, but not joinable. The pair $\{\alpha, t_1\}$ is a smaller witness to non-confluence, a contradiction again.

Let $(f\beta_1 \ldots \beta_m, M)$ be a projection of t. This marked term is top-stable and correctly marked, and hence by Lemma 5, it is in S_∞. By condition (c2), $(\alpha, \emptyset)$ and $(f\beta_1 \ldots \beta_m, M)$ are structurally joinable, and hence, there exist $(\alpha, \emptyset) \to_R^* (s', \emptyset)$ and $(f\beta_1 \ldots \beta_m, M) \to_R^* (t', M')$ such that $Pos(s') = Pos(t')$, and for every leaf position $p \in Pos(s')$ we have $s'|_p \leftrightarrow_R^* t'|_p$. If $M' = \{p_1 \ldots p_k\}$, then $t = t'[\beta|_{i_1}]_{p_1} \ldots [\beta|_{i_k}]_{p_k}$, for some $i_1 \ldots i_k \subseteq M$.

If we mimic the derivation $f\beta_1 \ldots \beta_m \to_R^* t'$, but now starting from $ft_1 \ldots t_m$, we obtain a derivation of the form $ft_1 \ldots t_m \to_R^* t'' = t'[t_{i_1}]_{p_1} \ldots [t_{i_k}]_{p_k}$. Moreover, each t_{i_j} is equivalent to $t'|_{p_j} = \beta_{i_j}$, and hence, for each leaf position p of s' we have that $s'|_p$ and $t''|_p$ are equivalent, and $size(t''|_p) < size(t)$. Since α and $ft_1 \ldots t_m$ are not joinable, s' and t'' are not joinable, and hence, for some leaf position p of s' we have that $s'|_p$ and $t''|_p$ are not joinable. For such a p, $\{s'|_p, t''|_p\}$ is a smaller witness to non-confluence, a contradiction.

Case 3. $s = fs_1 \ldots s_m$ *and* $t = ft_1 \ldots t_m$: Using arguments similar to the previous case, we can assume that the s_i's and t_i's are equivalent to depth 0 terms. Let $(f\alpha_1 \ldots \alpha_m, M)$ and $(f\beta_1 \ldots \beta_m, N)$ be the projections of s and t. Both these marked terms are top-stable and correctly marked and hence, by Lemma 5, they are in S_∞. By condition (c3), they are first-step joinable, i.e. there exist $(f\alpha_1 \ldots \alpha_m, M) \rightarrow^r_{(l_1 \rightarrow r_1)\sigma}$ $(fs'_1 \ldots s'_m, M')$ and $(f\beta_1 \ldots \beta_m, N) \rightarrow^r_{(l_2 \rightarrow r_2)\theta} ft'_1 \ldots t'_m, N')$ such that every s'_i is equivalent to its corresponding t'_i, and $((l_1 \rightarrow r_1), (l_2 \rightarrow r_2))$ is not useless.

We apply these rewrite steps to the original terms to get $fs_1 \ldots s_m \rightarrow^r_{(l_1 \rightarrow r_1)\sigma'}$ $fs''_1 \ldots s''_m = s''$ and $ft_1 \ldots t_m \rightarrow^r_{(l_2 \rightarrow r_2)\theta'} ft''_1 \ldots t''_m = t''$, by choosing σ' and θ' such that for each $i \in \{1 \ldots m\}$, it is the case that (a) if $r_1|_i$ ($r_2|_i$) is a variable not appearing in l_1 (l_2), then $s''_i = t''_i$, and (b) if not, then either s''_i (t''_i) is a constant or it coincides with one of the s_j (t_j) or it coincides with s (t). But it cannot happen that both s''_i and t''_i coincide with s and t, respectively. This is because the rules $l_1 \rightarrow r_1$ and $l_2 \rightarrow r_2$ are not useless.

By construction, every s''_i is equivalent to its corresponding t''_i. Since s and t are not joinable, s'' and t'' are not joinable, and hence, for some $i \in \{1 \ldots m\}$ we have that s''_i and t''_i are not joinable. This can only happen for case (b) above, and by the previous observation, (s''_i, t''_i) is a smaller witness to non-confluence, a contradiction.

Finally, we are ready to state the main result.

Theorem 1. *Confluence of linear shallow term rewrite systems can be decided in time polynomial in the size of the rewrite system, assuming the maximum arity of any function symbol is bounded by a constant.*

Proof. The input linear shallow rewrite system is transformed into a flat linear rewrite system and then it is saturated under the ordered chaining inference rules. Flattening increases the size $|\mathcal{F}|$ of the signature by a linear factor of the input size. Now, the number of flat linear rewrite rules is bounded by a polynomial in the size $|\mathcal{F}|$ of the signature, and hence these two transformation steps run in polynomial time. Next, the sets S_∞ and J_∞ are computed, again using polynomial time fixpoint computations. Finally, confluence is tested using the characterization given in Lemma 7. The three conditions in Lemma 7 can be tested in polynomial time: (a) Equivalent depth zero terms can be identified because equivalence testing for flat linear rewrite systems can be efficiently done, say using standard completion modulo permutation rules. Joinability of depth zero terms can be tested in polynomial time using simple fixpoint computations, similar to previous work [13]. (b) It is also clear that the conditions (c2) and (c3) can be tested in polynomial time.

Example 3. For the rewrite system R of Example 1, the set S_∞ contains the terms $x + y, 0 + x, x + 0, 0 + 0$, where the positions of 0 are marked. But the pairs $(0, 0 + 0), (x, 0 + x), (x, x + 0)$ are easily seen to be structurally joinable, while $(x + 0, 0 + x)$ is first-step joinable. Hence, this rewrite system is confluent.

3   Relaxing the Restrictions

The reachability, 2-joinability, and confluence problems for shallow term rewrite systems are not known to be decidable. But, we can establish the following lower-bounds.

Theorem 2. *The reachability problem for shallow term rewrite systems is EXPTIME-hard, even when the maximum arity is a constant.*

Proof. We reduce the problem of deciding non-emptiness of language intersection of n bottom-up tree-automata to this problem. The proof is similar to the proof of EXPTIME-hardness of rigid-reachability of shallow terms over ground systems [5]. Let R_1 be the union of the *reversed* transitions of all the n tree-automata. We assume that the tree-automata have disjoint states with accepting states $q_1, q_2, \ldots, q_n$, respectively. Let $R_2 = \{a \to g(q_1, f(q_2, f(q_3, \cdots, f(q_{n-1}, q_n) \cdots))), fxx \to x, gxx \to b\}$, where g, f are two new binary function symbols and a, b are two new constants. Now, a rewrites to b via $R_1 \cup R_2$ iff the intersection of languages accepted by the n automata is nonempty.

For shallow term rewrite systems, EXPTIME-hardness of 2-joinability follows from the hardness of reachability using the reduction in [15]. We next show hardness of deciding confluence of shallow term rewrite systems by modifying the proof of Theorem 2.

Theorem 3. *Deciding confluence of shallow term rewrite systems is EXPTIME-hard, even when the maximum arity is a constant.*

Proof. We add additional rewrite rules to the rewrite system generated in the proof of Theorem 2 to make the system confluent exactly when b is reachable from a. First, we introduce a new constant c in the signature $\mathcal{G}$ of the tree automata and convert all constants d to unary terms $d(c)$. The rules in R_1 are modified to reflect this change. We assume that some ground term can be reached from any tree-automata state q via R_1. Let $R_3 = \{c \to a, \ h(x_1, \ldots, x_{i-1}, b, x_{i+1}, \ldots, x_n) \to b \text{ for all } h \in \mathcal{G}, \ fxb \to b, \ fbx \to b, \ gxb \to b, \ gbx \to b\}$. Now, consider the shallow term rewrite system $R = R_1 \cup R_2 \cup R_3 \cup \{d \to a, d \to b\}$, where R_1 and R_2 are as in proof of Theorem 2 and d is a new constant in the signature. We claim without proof that R is confluent iff the n tree-automata have a non-empty language intersection.

We also note here that reachability, 2-joinability, and confluence problems are undecidable for linear (non-shallow) term rewrite systems [15].

4 Conclusion

In this paper we presented a polynomial time algorithm for deciding confluence of linear shallow term rewrite systems where each variable is allowed at most two occurrences in a rule—one on each side. The time complexity analysis assumes that the maximum arity of a function symbol in the signature is a constant. Our result generalizes those in [2,13]. We also show that the reachability, joinability and confluence problems are all EXPTIME-hard for shallow non-linear systems, and all three are known to be undecidable for linear non-shallow systems, which indicates that our assumptions can not be easily relaxed without considerably losing efficiency. Our technique can be adapted to decide ground confluence of linear shallow term rewrite systems in polynomial time. It is not clear whether our method can give polynomial time algorithms to decide confluence when we have non-fixed arity or for rule-linear rewrite systems (no variable appears twice in the whole rule) and not necessarily shallow, and this is a matter for future work.

Acknowledgments. We would like to thank the reviewers for their helpful comments.

References

1. L. Bachmair. *Canonical Equational Proofs*. Birkhäuser, Boston, 1991.
2. H. Comon, G. Godoy, and R. Nieuwenhuis. The confluence of ground term rewrite systems is decidable in polynomial time. In *42nd Annual IEEE Symposium on Foundations of Computer Science (FOCS)*, Las Vegas, Nevada, USA, 2001.
3. H. Comon, M. Haberstrau, and J.-P. Jouannaud. Syntacticness, cycle-syntacticness, and shallow theories. *Information and Computation*, 111(1):154–191, 1994.
4. M. Dauchet, T. Heuillard, P. Lescanne, and S. Tison. Decidability of the confluence of finite ground term rewrite systems and of other related term rewrite systems. *Information and Computation*, 88(2):187–201, October 1990.
5. H. Ganzinger, F. Jacquemard, and M. Veanes. Rigid reachability: The non-symmetric form of rigid E-unification. *Intl. Journal of Foundations of Computer Science*, 11(1):3–27, 2000.
6. Guillem Godoy, Robert Nieuwenhuis, and Ashish Tiwari. Classes of Term Rewrite Systems with Polynomial Confluence Problems. *ACM Transactions on Computational Logic (TOCL)*, 2002. To appear.
7. A. Hayrapetyan and R.M. Verma. On the complexity of confluence for ground rewrite systems. In *Bar-Ilan International Symposium On The Foundations Of Artificial Intelligence*, 2001. Proceedings on the web at http://www.math.tau.ac.il/~nachumd/bisfai-pgm.html.
8. D. E. Knuth and P. B. Bendix. Simple word problems in universal algebras. In J. Leech, editor, *Computational Problems in Abstract Algebra*, pages 263–297. Pergamon Press, Oxford, 1970.
9. A. Levy and J. Agusti. Bi-rewriting, a term rewriting technique for monotone order relations. In C. Kirchner, editor, *Rewriting Techniques and Applications RTA-93*, pages 17–31, 1993. LNCS 690.
10. R. Nieuwenhuis. Basic paramodulation and decidable theories. In *11th IEEE Symposium on Logic in Computer Science, LICS 1996*, pages 473–482. IEEE Computer Society, 1996.
11. M. Oyamaguchi. The Church-Rosser property for ground term-rewriting systems is decidable. *Theoretical Computer Science*, 49(1):43–79, 1987.
12. A. Tiwari. Rewrite closure for ground and cancellative AC theories. In R. Hariharan and V. Vinay, editors, *Conference on Foundations of Software Technology and Theoretical Computer Science, FST&TCS '2001*, pages 334–346. Springer-Verlag, 2001. LNCS 2245.
13. A. Tiwari. Deciding confluence of certain term rewriting systems in polynomial time. In Gordon Plotkin, editor, *IEEE Symposium on Logic in Computer Science, LICS 2002*, pages 447–456. IEEE Society, 2002.
14. A. Tiwari. *On the combination of equational and rewrite theories induced by certain term rewrite systems*. Menlo Park, CA 94025, 2002. Available at: www.csl.sri.com/~tiwari/combinationER.ps.
15. R. Verma, M. Rusinowitch, and D. Lugiez. Algorithms and reductions for rewriting problems. *Fundamenta Informaticae*, 43(3):257–276, 2001. Also in Proc. of Int'l Conf. on Rewriting Techniques and Applications 1998.
16. H. Zantema. Termination of term rewriting: interpretation and type elimination. *Journal of Symbolic Computation*, 17:23–50, 1994.

Wadge Degrees of ω-Languages of Deterministic Turing Machines

Victor Selivanov*

[1] Universität Siegen
Theoretische Informatik, Fachbereich 6
[2] Novosibirsk Pedagogical University
Chair of Informatics and Discrete Mathematics
vseliv@informatik.uni-siegen.de

Abstract. We describe Wadge degrees of ω-languages recognizable by deterministic Turing machines. In particular, it is shown that the ordinal corresponding to these degrees is ξ^ω where $\xi = \omega_1^{CK}$ is the first non-recursive ordinal known as the Church-Kleene ordinal. This answers a question raised in [Du0?].

Keywords: Wadge degree, hierarchy, reducibility, ω-language, Cantor space, set-theoretic operation.

1 Formulation of Main Result

In [Wag79] K. Wagner described a remarkable hierarchy of regular ω-languages. In [Se94, Se95, Se98] the author gave a new exposition of the Wagner hierarchy relating it to the Wadge hierarchy and the fine hierarchy [Se95a]. In [Du0?] J. Duparc extended the Wagner hierarchy to deterministic context-free ω-languages and formulated a conjecture about Wadge degrees of ω-languages recognized by deterministic Turing machines.

Here we prove the conjecture of J. Duparc and show that Wadge degrees corresponding to deterministic acceptors form an easy structure. This is contrasted by a recent series of papers by O. Finkel showing that for the nondeterministic context-free ω-languages the situation is more complicated.

Our methods come from computability theory and descriptive set theory. We assume some familiarity of the reader with such things as recursive hierarchies, countable and recursive ordinals etc.

Let $\{\Sigma_\alpha^0\}_{\alpha < \omega_1}$, where ω_1 is the first uncountable ordinal, denote the Borel hierarchy of subsets of the Cantor space 2^ω (all results below hold true also for the space $\{0, \ldots, n+1\}^\omega$ for any $n < \omega$ but for notational simplicity we consider only the case $n = 0$) or the Baire space ω^ω. As usual, Π_α^0 denotes the dual class for Σ_α^0 while $\Delta_\alpha^0 = \Sigma_\alpha^0 \cap \Pi_\alpha^0$ — the corresponding ambiguous class. Let $\mathbf{B} = \cup_{\alpha < \omega_1} \Sigma_\alpha^0$ denote the class of all Borel sets.

* Partly supported by the Russian Foundation for Basic Research Grant 00-01-00810.

H. Alt and M. Habib (Eds.): STACS 2003, LNCS 2607, pp. 97–108, 2003.

In [Wad72, Wad84] W. Wadge described the finest possible topological classification of Borel sets by means of a relation $\leq_W$ on subsets of a space $S \in \{2^\omega, \omega^\omega\}$ defined by

$$A \leq_W B \leftrightarrow A = f^{-1}(B),$$

for some continuous function $f : S \to S$. He showed that the structure $(\mathbf{B}; \leq_W)$ is well-founded, proved that for all $A, B \in \mathbf{B}$ either $A \leq_W B$ or $\bar{B} \leq_W A$ (we call structures satisfying these two properties, in which $\bar{B}$ denotes the complement of a set B, *almost well-ordered*), and computed the corresponding (very large) ordinal v. In [VW78, Ste81] it was shown that for any Borel set A which is non-self-dual (i.e., $A \not\leq_W \bar{A}$) exactly one of the principal ideals $\{X | X \leq_W A\}$, $\{X | X \leq_W \bar{A}\}$ has the separation property.

The results cited in the last paragraph give rise to the *Wadge hierarchy of Borel sets* which is, by definition, the sequence $\{\Sigma_\alpha\}_{\alpha < v}$ of all non-self-dual principal ideals of $(\mathbf{B}; \leq_W)$ not having the separation property [Mo80] and satisfying for all $\alpha < \beta < v$ the strict inclusion $\Sigma_\alpha \subset \Delta_\beta$. As usual, we set $\Pi_\alpha = \{\bar{X} | X \in \Sigma_\alpha\}$ and $\Delta_\alpha = \Sigma_\alpha \cap \Pi_\alpha$. Note that the classes

$$\Sigma_\alpha \setminus \Pi_\alpha, \ \Pi_\alpha \setminus \Sigma_\alpha, \ \Delta_{\alpha+1} \setminus (\Sigma_\alpha \cup \Pi_\alpha) \ (\alpha < v),$$

which we call *consituents* of the Wadge hierarchy, are exactly the equivalence classes induced by $\leq_W$ on Borel subsets of the Cantor space. Please distinguish classes of the Wadge hierarchy (denoted without upper index) from corresponding classes of the Borel hierarchy (denoted with upper index 0).

There is a well-known small difference between the Wadge hierachies in the Baire and in the Cantor space with respect to the question for which $\alpha < v$ the class Δ_α has a W-complete set (such sets correspond to the self-dual Wadge degrees). For the Cantor space, these are exactly the successor ordinals $\alpha < v$ while for the Baire space — the successor ordinals and the limit ordinals of countable cofinality [VW78]. This follows easily from the well-known fact that the Cantor space is compact while the Baire space is not.

The Wadge hierarchy on the Cantor space is of interest to the theory of ω-languages since in this theory people also try to classify classes of ω-languages according to their 'complexity'. In [Se94, Se95, Se98] the Wagner hierarchy of regular ω-languages was related to the Wadge hierarchy. The order type of Wadge degrees of regular ω-languages is ω^ω [Wag79]. In [Du0?] the Wadge degrees of ω-languages recognizable by deterministic push-down automata were determined; the corresponding ordinal is $(\omega^\omega)^\omega$. In [Du0?] a conjecture on the structure of Wadge degrees of ω-languages recognizable by deterministic Turing machines was formulated (for the Muller acceptance condition, see [Sta97]) implying that the corresponding ordinal is ξ^ω, where $\xi = \omega_1^{CK}$ is the first non-recursive ordinal known also as the Church-Kleene ordinal.

In this paper we prove the conjecture from [Du0?]. To formulate the corresponding result, define an increasing function $e : \xi^\omega \to \omega_1^\omega$ by

$$e(\xi^n \alpha_n + \cdots + \xi^1 \alpha_1 + \alpha_0) = \omega_1^n \alpha_n + \cdots + \omega_1^1 \alpha_1 + \alpha_0,$$

where $n < \omega$ and $\alpha_i < \xi$. Note that we use some standard notation and facts from ordinal arithmetic (see, for example, [KM67]).

As is well-known, any non-zero ordinal $\alpha < \xi^\omega$ ($\alpha < \omega_1^\omega$) is uniquely representable in the canonical form

$$\alpha = \xi^{n_0}\alpha_0 + \cdots + \xi^{n_k}\alpha_k \quad (\text{resp., } \alpha = \omega_1^{n_0}\alpha_0 + \cdots + \omega_1^{n_k}\alpha_k), \qquad (1)$$

where $k < \omega$, $\omega > n_0 > \cdots > n_k$ and $0 < \alpha_i < \xi$ ($0 < \alpha_i < \omega_1$). The members of the sum (1) will be called *monomials* of the representation. The number n_k will be called *the height* of α.

If we have a similar canonical representation of another non-zero ordinal $\beta < \xi^\omega$ ($\beta < \omega_1^\omega$)

$$\beta = \xi^{m_0}\beta_0 + \cdots + \xi^{m_l}\beta_l \quad (\text{resp., } \beta = \omega_1^{m_0}\beta_0 + \cdots + \omega_1^{m_l}\beta_l),$$

then $\alpha < \beta$ iff the sequence $((n_0, \alpha_0), \ldots, (n_k, \alpha_k))$ is lexicographically less than the sequence $((m_0, \beta_0), \ldots, (m_l, \beta_l))$.

Let DTM_ω denote the class of subsets of the Cantor space recognized by deterministic Turing machines (using the Muller acceptance condition). Our main result is the following

Theorem 1. *(i) For every $\alpha < \xi^\omega$, any of the constituents*

$$\Sigma_{e(\alpha)} \setminus \Pi_{e(\alpha)}, \; \Pi_{e(\alpha)} \setminus \Sigma_{e(\alpha)}, \; \Delta_{e(\alpha+1)} \setminus (\Sigma_{e(\alpha)} \cup \Pi_{e(\alpha)})$$

contains a set from DTM_ω.

(ii) All other constituents of the Wadge hierarchy do not contain sets from DTM_ω.

This result and the obove-mentioned facts on the Wadge hierarchy imply the following

Corollary 1. *The structure $(DTM_\omega; \leq_W)$ is almost well-ordered with the corresponding ordinal ξ^ω.*

2   Set-Theoretic Operations

The first step toward the proof of the main result is to use a result from [Sta97] stating, in our notation, that the class DTM_ω coincides with the boolean closure $bc(\Sigma_2^0)$ of the second level of the arithmetical hierarchy $\{\Sigma_n^0\}_{n<\omega}$ on the Cantor space. Please be careful in distinguishing the levels of the Borel hierarchy (denoted by boldface letters) and the corresponding levels of the arithmetical hierarchy (lightface letters). The result from [Sta97] reduces the problem of this paper to hierarchy theory since it becomes a question on the interplay of (a fragment of) the arithmetical hierarchy (being the effective version of the Borel hierarchy, see e.g. [Mo80]) with the Wadge hierarchy. We will freely use some well-known terminology from computability theory, see e.g. [Ro67].

It remains to describe Wadge degrees of sets in $bc(\Sigma_2^0)$. Note that this last problem makes sense not only for the Cantor space but also for the Baire space. We will get a solution for this case as a consequence of the proof for the Cantor space.

The second step toward the main theorem is to use a close relationship of the Wadge hierarchy to set-theoretic operations established in [Wad84]; a version of this result appeared in [Lo83]. These works describe all levels of the Wadge hirarchy in terms of some countable set-theoretic operations. Let us present a description of an initial segment of the Wadge hierarchy which is (with some notational changes) a particular case of the description in [Lo83].

Let us first define the relevant set-theoretic operations. In definitions below, all sets are subsets of the Cantor or the Baire space. For classes $\mathcal{A}$ and $\mathcal{B}$ of sets, let $\mathcal{A} \cdot \mathcal{B} = \{A \cap B | A \in \mathcal{A}, B \in \mathcal{B}\}$, let $\check{\mathcal{A}} = \{\bar{A} | A \in \mathcal{A}\}$ be the dual class for $\mathcal{A}$ (sometimes it is more convinient to denote the dual class by $co(\mathcal{A})$) and let $\tilde{\mathcal{A}} = \mathcal{A} \cap \check{\mathcal{A}}$ be the corresponding ambiguous class.

Definition 1. *For classes of sets $\mathcal{A}$ and $\mathcal{B}$, let $\mathcal{A} + \mathcal{B}$ denote the class of all symmetric differences $A \triangle B$, where $A \in \mathcal{A}$ and $B \in \mathcal{B}$.*

An ordinal α is called *odd* if $\alpha = 2\beta + 1$, for some ordinal β; the non-odd ordinals are called *even*. For an ordinal α, let $r(\alpha) = 0$ if α is even and $r(\alpha) = 1$, otherwise.

Let us recall the well-known definition of the Hausdorff difference operation.

Definition 2. *(i) For an ordinal α, define an operation D_α sending sequences of sets $\{A_\beta\}_{\beta<\alpha}$ to sets by*

$$D_\alpha(\{A_\beta\}_{\beta<\alpha}) = \bigcup\{A_\beta \setminus \cup_{\gamma<\beta}A_\gamma | \beta < \alpha, r(\beta) \neq r(\alpha)\}.$$

For the sake of brevity, we denote in similar expressions below the set $A_\beta \setminus \cup_{\gamma<\beta}A_\gamma$ by A'_β.

(ii) For an ordinal α and a class of sets $\mathcal{A}$, let $D_\alpha(\mathcal{A})$ be the class of all sets $D_\alpha(\{A_\beta\}_{\beta<\alpha})$, where $A_\beta \in \mathcal{A}$ for all $\beta < \alpha$.

Now define another, more exotic, operation on sets playing a noticible role in the theory of Wadge degrees.

Definition 3. *For classes of sets $\mathcal{A}$, $\mathcal{B}_0$, $\mathcal{B}_1$ and $\mathcal{C}$, let $Bisep(\mathcal{A}, \mathcal{B}_0, \mathcal{B}_1, \mathcal{C})$ be the class of all sets $A_0B_0 \cup A_1B_1 \cup \bar{A}_0\bar{A}_1C$, where XY denotes the intersection of X and Y, $A_0, A_1 \in \mathcal{A}$, $A_0A_1 = \emptyset$, $B_i \in \mathcal{B}_i$ and $C \in \mathcal{C}$.*

For the sake of brevity, we denote the set $Bisep(\Sigma_1^0, \mathcal{A}, co(\mathcal{A}), \mathcal{B})$ also by $\mathcal{A}*\mathcal{B}$.

Let us state some properties of the introduced operations.

Lemma 1. *Let classes $\mathcal{A}, \mathcal{B}, \mathcal{C}$ and their duals be closed under intersections with $\Sigma_1^0 \cup \Pi_1^0$-sets. Then it holds:*

(i) $X \in \mathcal{A}*\mathcal{B}$ iff there are disjoint $U_0, U_1 \in \Sigma_1^0$ such that $XU_0 \in \mathcal{A}, XU_1 \in \check{\mathcal{A}}$ and $X\bar{U}_0\bar{U}_1 \in \mathcal{B}$.

(ii) $co(\mathcal{A} * \mathcal{B}) = \mathcal{A} * co(\mathcal{B})$.

(iii) $\mathcal{A} * (\mathcal{B} * \mathcal{C}) \subseteq (\mathcal{A} * \mathcal{B}) * \mathcal{C}$.

Next we formulate a result describing the initial segment $\{\Sigma_\alpha\}_{\alpha<\omega_1^\omega}$ of the Wadge hierarchy in terms of the introduced operations. The result is a (reformulation of a) particular case of a result from [Wad84, Lo83] providing a similar (quite complicated) description for all levels of the Wadge hierarchy. Our description uses an induction on ordinals and the canonical representation (1) described at the end of the previous section.

Theorem 2. *(i) For $\alpha < \omega_1$, $\Sigma_\alpha = D_\alpha(\Sigma_1^0)$.*
  (ii) For a monomial $\alpha = \omega_1^n(\gamma + 1)$, $0 < n < \omega, \gamma < \omega_1$, $\Sigma_\alpha = \Sigma_\gamma + D_n(\Sigma_2^0)$.
  (iii) For a monomial $\alpha = \omega_1^n \lambda$, $0 < n < \omega, \lambda < \omega_1$, λ a limit ordinal, Σ_α coincides with the class of all sets of the form

$$(\{A'_\beta Y | \beta < \alpha, r(\beta) = 1\}) \cup (\{A'_\beta \bar{Y} | \beta < \alpha, r(\beta) = 0\}),$$

where $\{A_\beta\}_{\beta<\alpha}$ is a sequence of Σ_1^0-sets and $Y \in \Sigma_{\omega_1^n}$.
  (iv) If $\alpha = \beta + \omega_1^n \gamma$, where $0 < n < \omega$, $0 < \gamma < \omega_1$ and β is a non-zero ordinal of height $> n$, then $\Sigma_\alpha = Bisep(\Sigma_1^0, \Sigma_\beta^0, \Pi_\beta^0, \Sigma_{\omega_1^n \gamma})$.
  (v) If $\alpha = \beta + 1 + \gamma$, where $\gamma < \omega_1$ and β is a non-zero ordinal of height > 0, then $\Sigma_\alpha = Bisep(\Sigma_1^0, \Sigma_\beta^0, \Pi_\beta^0, \Sigma_\gamma)$.

Notice that $\Sigma_0 = \{\emptyset\}$, $\Sigma_{\omega_1^n} = D_n(\Sigma_2^0)$ for $0 < n < \omega$, and $\bigcup_{\alpha<\omega_1^\omega} \Sigma_\alpha = bc(\Sigma_2^0)$.

3 Effective Wadge Hierarchy

The third step toward the proof of the main theorem consists in defining an effective analog $\{\mathcal{S}_\alpha\}_{\alpha<\xi^\omega}$ of the sequence $\{\Sigma_\alpha\}_{\alpha<\omega_1^\omega}$. To do this, we turn Theorem 2 into a definition by taking the lightface classes Σ_1^0, Σ_2^0 in place of the boldface ones Σ_1^0, Σ_2^0 and considering recursive well-orderings in place of the countable ordinals.

Recall [Ro67] that a *recursive well-ordering* is a well-ordering of the form $(R; \prec)$ where R is a recursive subset of ω and $\prec$ is a recursive relation on R. Let $r : R \to \{0, 1\}$ be the function induced by the corresponding function on ordinals defined in the last section. As in [Er68], we will consider only the recursive well-orderings such that r is a partial recursive (p.r.) function, and the set of limit elements of $(R; \prec)$ is recursive. Alternatively, one could use the Kleene notation system for recursive ordinals [Ro67].

For a recursive well-ordering $(R; \prec)$ of order type α and a sequence of sets $\{A_x\}_{x\in R}$, let

$$D_\alpha(\{A_x\}_{x\in R}) = \bigcup \{A'_x | x \in R, r(x) \neq r(\alpha)\}, \quad A'_x = A_x \setminus \bigcup_{y \prec x} A_y.$$

The next definition of classes $\mathcal{S}_\alpha$ closely mimicks Theorem 2.

Definition 4. *(i) For $\alpha < \xi$, let $\mathcal{S}_\alpha$ be the class of all sets $D_\alpha(\{A_x\}_{x\in R})$, where $(R; \prec)$ is a recusive well-ordering of order type α and $\{A_x\}_{x\in R}$ is a uniform r.e. sequence.*

(ii) For a monomial $\alpha = \xi^n(\gamma+1)$, $0 < n < \omega, \gamma < \xi$, set $S_\alpha = S_\gamma + D_n(\Sigma_2^0)$.

(iii) For a monomial $\alpha = \xi^n\lambda$, $0 < n < \omega, \lambda < \xi$, λ a limit ordinal, let S_α consist of all sets of the form

$$(\{A'_x Y | x \in R, r(x) = 1\}) \cup (\{A'_x \bar{Y} | x \in R, r(x) = 0\}), \qquad (2)$$

where again $(R; \prec)$ is a recursive well ordering of type λ, $\{A_x\}_{x \in R}$ is an r.e. sequence, and $Y \in S_{\xi^n}$.

(iv) If $\alpha = \beta + \xi^n\gamma$ where $0 < n < \omega$, $0 < \gamma < \xi$ and β is a non-zero ordinal of height $> n$ then set $S_\alpha = S_\beta * S_{\xi^n\gamma}$.

(v) If $\alpha = \beta + 1 + \gamma$ where $\gamma < \xi$ and β is a non-zero ordinal of height > 0 then set $S_\alpha = S_\beta * S_\gamma$.

Let us state an immediate corollary of the last definition and of Theorem 2.

Corollary 2. *For any $\alpha < \xi^\omega$, $S_\alpha \subseteq \Sigma_{e(\alpha)}$.*

Note that Definition 4 resembles the definition of the fine hierarchy studied in [Se95a] which was first defined (for the case of subsets of ω) in [Se83] in terms of some jump operations independently of the work on Wadge degrees. Quite similar to [Se95a] one can check some natural properties of the sequence $\{S_\alpha\}_{\alpha < \xi^\omega}$, e.g.

Lemma 2. *(i) For all $\alpha < \beta < \xi^\omega$, $S_\alpha \subseteq \tilde{S}_\beta$.*

(ii) If $X \in S_\alpha$ and $F : 2^\omega \to 2^\omega$ is recursive then $F^{-1}(X) \in S_\alpha$.

(iii) For $n > 1$, $S_{\xi^n} = \tilde{S}_{\xi^{n-1}} \cdot S_\xi = S_{\xi^{n-1}} + S_\xi$.

(iv) For $n > 1$, $\tilde{S}_{\xi^n} = S_{\xi^{n-1}} + \tilde{S}_\xi$.

(v) If $0 < \alpha < \xi^\omega$ and $e(\alpha)$ is an ordinal of uncountable cofinality then the classes S_α, $\check{S}_\alpha$ and $\tilde{S}_\alpha$ are closed under intersections with Δ_2^0-sets.

(vi) If $X_0, X_1 \in \Sigma_1^0$ and $X_0 Y, X_1 Y \in S_\alpha$ then $(X_0 \cup X_1) Y \in S_\alpha$. The same holds true for the class $\tilde{S}_\alpha$ provided that α is a non-zero ordinal of height > 0.

(vii) For $0 < n < \omega, 0 < \gamma < \xi$, it holds $\tilde{S}_{\xi^n(\gamma+1)} \subseteq S_{\xi^n\gamma} * \tilde{S}_{\xi^n}$.

(viii) If $\alpha = \beta + \xi^n\gamma$, where $0 < n < \omega, 0 < \gamma < \xi$, and β is a non-zero ordinal of height $> n$ then $\tilde{S}_\alpha \subseteq S_\beta * \tilde{S}_{\xi^n\gamma}$.

4   Complete Sets

Now we are in a position to prove the assertion (i) of Theorem 1. We do this by constructing for any $\alpha < \xi^\omega$ a set $C_\alpha \subseteq 2^\omega$ such that $C_\alpha \in S_\alpha$ and any set from $\Sigma_{e(\alpha)}$ is W-reducible to C_α. This really proves the assertion (i) since, by Corollary 2, $C_\alpha \in \Sigma_{e(\alpha)}$ and a fortiori $C_\alpha \in \Sigma_{e(\alpha)} \setminus \Pi_{e(\alpha)}$. For the dual class, the set $\bar{C}_\alpha$ makes the job while for the Δ-level we have

$$C_\alpha \oplus \bar{C}_\alpha \in \Delta_{e(\alpha+1)} \setminus (\Sigma_{e(\alpha)} \cup \Pi_{e(\alpha)}),$$

where $\oplus$ is the join operator on subsets of the Cantor space defined by

$$A \oplus B = \{0^\frown f | f \in A\} \cup \{1^\frown f | f \in B\}.$$

Here $i^\frown f$ is the concatenation of a number i and a function f considered as a sequence. For the construction of the specified sets we need a pair (U_0, U_1) of disjoint Σ_1^0-sets and a set $V \in \Sigma_2^0$ such that:

any pair (X_0, X_1) of disjoint Σ_1^0-sets is W-reducible to (U_0, U_1);

any Σ_2^0-set is W-reducible to V.

These conditions of course imply that U_0 and V are W-complete in Σ_1^0 and Σ_2^0, respectively. For existence of such sets (which can be chosen even as regular ω-languages) see e.g. [Se98].

We will also need canonical computable bijections between sets $2^\omega \times 2^\omega$, $(2^\omega)^\omega$ and the set 2^ω defined by

$$\langle f, g \rangle(2n) = f(n), \quad \langle f, g \rangle(2n+1) = g(n), \quad \langle f_0, f_1, \ldots \rangle \langle m, n \rangle = f_m(n),$$

where $\langle m, n \rangle$ is a computable bijection between $\omega \times \omega$ and ω. As usual, one can also define a computable bijection between $2^\omega \times \cdots \times 2^\omega$ ($n+1$ terms, $n < \omega$) and 2^ω, which is denoted also by $\langle f_0, \ldots, f_n \rangle$.

The following definition of the sets C_α uses the same induction scheme (and the same conditions on ordinals) as Definition 4.

Definition 5. *(i) Let $\alpha < \xi$. For $\alpha = 0$, set $C_0 = \emptyset$. For $0 < \alpha < \omega$, set*

$$C_\alpha = D_\alpha(\{Z_i\}_{i<\alpha}), \text{ where } Z_i = \{\langle f_0, \ldots, f_{\alpha-1} \rangle | f_i \in U_0\}.$$

For $\omega \le \alpha < \xi$, choose a recursive well ordering $(R; \prec)$ of type α and a recursive bijection $p : \omega \to R$ and set

$$C_\alpha = D_\alpha(\{Z_x\}_{x \in R}), \text{ where } Z_{p(i)} = \{\langle f_0, f_1, \ldots \rangle | f_i \in U_0\}.$$

(ii) Let $\alpha = \xi^n(\gamma + 1)$. For $\gamma = 0$, set

$$C_\alpha = D_n(\{Z_i\}_{i<n}), \text{ where } Z_i = \{\langle f_0, \ldots, f_{n-1} \rangle | f_i \in V\}.$$

For $\gamma > 0$, set

$$C_\alpha = X \triangle Y, \text{ where } X = \{\langle f, g \rangle | f \in C_{\xi^n}\}, \quad Y = \{\langle f, g \rangle | g \in C_\gamma\}.$$

(iii) For $\alpha = \xi^n \lambda$, let C_α be of the form (2), where $(R; \prec)$ is a recursive well ordering of type λ, $\{Z_x\}_{x \in R}$ is the sequence defined as in (i) above, and

$$A_x = \{\langle f, g \rangle | f \in Z_x\}, \quad Y = \{\langle f, g \rangle | g \in C_{\xi^n}\}.$$

(iv) For $\alpha = \beta + \xi^n \gamma$, set $C_\alpha = W_0 X_0 \cup W_1 X_1 \cup \bar{W}_0 \bar{W}_1 Y$, where

$$W_i = \{\langle f, g_0, g_1, h \rangle | f \in U_i\}, \quad X_i = \{\langle f, g_0, g_1, h \rangle | g_i \in C_\beta\},$$

and $Y = \{\langle f, g_0, g_1, h \rangle | h \in C_{\xi^n \gamma}\}$.

(v) For $\alpha = \beta + 1 + \gamma$, C_α is defined as in (iv), with γ in place of $\xi^n \gamma$.

The following assertion is easily checked by induction.

Proposition 1. *For every $\alpha < \xi$, $C_\alpha \in S_\alpha$, and any $\Sigma_{e(\alpha)}$-set is W-reducible to C_α.*

5   Effective Hausdorff Theorem

Here we make the fourth step to proving the main theorem by establishing an effective version of the following classical result of F. Hausdorff: a set A is Δ_2^0 iff $A = D_\alpha(\{T_\beta\}_{\beta < \alpha})$, for some $\alpha < \omega_1$ and some sequence $\{T_\beta\}_{\beta < \alpha}$ of open sets. In notation of Section 1 it looks as follows: $\Delta_{\omega_1} = \bigcup_{\alpha < \omega_1} \Sigma_\alpha$.

The effective version of the Hausdoff theorem looks like $\Delta_2^0 = \cup\{S_\alpha | \alpha < \xi\}$. For the case of subsets of ω, the effective version was established in [Er68]; in this case, the equality may be even sharpened to $\Delta_2^0 = \cup\{S_\alpha | \alpha \leq \omega\}$. To the best of my knowledge, for the Cantor (or Baire) space the proof of the corresponding assertion was never published (though it appeared in handwritten notes [Se92] accessible only to a small group of recursion theorists). For this reason, let us sketch here the proof from [Se92]. Note that for the case of Cantor and Baire space the inclusion $\cup\{S_\alpha | \alpha < \gamma\} \subset \Delta_2^0$ is, according to the last section, strict for any $\gamma < \xi$. This is in contrast with the cited result from [Er68].

We need the following connection of Δ_2^0-sets to limiting computations.

Proposition 2. *Let A be a subset of the Cantor (or Baire) space. Then A is Δ_2^0 iff there is a recursive function $G : 2^\omega \times \omega \to \{0,1\}$ such that $A(f) = \lim_{n \to \infty} G(f,n)$.*

PROOF. First note that we identify A with its characteristic function, i.e. $A(f) = 1$ for $f \in A$ and $A(f) = 0$, otherwise. From right to left, the assertion follows from the Tarski-Kuratowski algorithm.

Conversely, let $A \in \Delta_2^0$, then $A = \cap_n B_n$ and $\bar{A} = \cap_n C_n$ for some Σ_1^0-sequences $\{B_n\}, \{C_n\}_{n < \omega}$. For any $n < \omega$, it holds $B_n \cup C_n = 2^\omega$. By Σ_1^0-reduction, there are Δ_1^0-sequences $\{B_n^*\}$ and $\{C_n^*\}$ such that

$$B_n^* \subseteq B_n, \ C_n^* \subseteq C_n, \ B_n^* \cap C_n^* = \emptyset, \ B_n^* \cup C_n^* = 2^\omega.$$

Set $G(f,n) = 1$ for $f \in B_n^*$ and $G(f,n) = 0$ for $f \in C_n^*$. Then the function G has the desired property, completing the proof.

There is a deep and useful connection of the effective difference hierarchy with limiting computations of a special kind which we would like to describe now. Let Φ be a partial function from $S \times \omega$ (where again S is one of $2^\omega, \omega^\omega$) to ω. Relate to Φ and to any recursive well ordering $(R; \prec)$ a partial function $m = m_{a,\Phi}$ from S to R as follows: $m(f)$ is the least element (if any) of $(\{x \in R | \Phi(f,x) \downarrow\}; \prec)$. Note that $m(f) \downarrow$ implies $\Phi(f, m(f)) \downarrow$.

Definition 6. *(i) A function $F : S \to \omega$ is called k-R-computable if there is a p.r. function $\Phi : S \times \omega \to \omega$ (called a k-R-computation of F) such that $F(f) = k$ for $m(f) \uparrow$ and $F(f) = \Phi(f, m(f))$ otherwise.*

(ii) A function $F : S \to \omega$ is called R-computable if $F(f) = \Phi(f, m(f))$ for some p.r. function $\Phi : S \times \omega \to \omega$.

(iii) A set $A \subseteq S$ is called k-R-computable (R-computable) if its characteristic function is k-R-computable (R-computable). Here $k \leq 1$.

Let C_R^k (C_R) denote the class of all k-R-computable (R-computable) functions. Note that any set $A \in C_R^k$ ($k \leq 1$) is k-R-computable by a p.r. function Φ with $rng\Phi \subseteq \{0,1\}$, and similarly for C_R (if Ψ is a p.r. k-R-computation of A then the function Φ defined by

$$\Phi(f,x) = 0 \text{ for } \Psi(f,x) \text{ even and } \Phi(f,x) = 1 \text{ for } \Psi(f,x) \text{ odd,}$$

is also a k-R-computation of A).

If a p.r. function Φ is a k-R-computation of F then any effective stepwise enumerations $\{\Phi^s\}$, $\{R^s\}$ of Φ and of the r.e. set R induce the limiting computations $\{m^s\}$, $\{F^s\}$ of m and F as follows:

$$m^s(f) \text{ is the least element of } (\{x \in R^s | \Phi^s(f,x) \downarrow\}; \prec),$$
$$F^s(f) = k \text{ for } m^s(f) \uparrow \text{ and } F^s(f) = \Phi(f, m^s(f)) \text{ otherwise.}$$

From the well-foundedness of $(R; \prec)$ follows that $m(f) = lim_s m^s(f)$ and $F(f) = lim_s F^s(f)$.

Let us relate the introduced classes of functions to the effective difference hierarchy. Let $\mathcal{S}_R$ denote the class of all sets $D_\alpha(\{A_x\})$, where $\{A_x\}_{x \in R}$ is a r.e. sequence and α is the order type of $(R; \prec)$. Of course, $\mathcal{S}_\alpha$ coincides with with the union of classes $\mathcal{S}_R$, for all recursive well orderings of type α.

Proposition 3. *A set $A \subseteq S$ belongs to $\mathcal{S}_R$ ($\check{\mathcal{S}}_R$, $\tilde{\mathcal{S}}_R$) iff $A \in C_R^0$ (C_R^1, C_R).*

Now we formulate the effective version of the Hausdorff theorem. The proof will appear in the journal version of this paper.

Theorem 3. *The effective difference hierarchy is an exhaustive refinement of the arithmetical hierarchy in the second level, i.e.* $\bigcup_{\alpha < \xi} \mathcal{S}_\alpha = \Delta_2^0$.

For any oracle $h \in 2^\omega$, let $\xi(h)$ be the first ordinal non-recursive in h, and let $\mathcal{S}_\alpha^h$ ($\alpha < \xi(h)$) be the effective difference hierarchy relative to h. The class $\mathcal{S}_\alpha^h$ is defined just as $\mathcal{S}_\alpha$ except this time the well orderings $(R; \prec)$ have to be recursive in h, and the sequences $\{A_x\}_{x \in R}$ r.e. in h. As usual, $\Sigma_n^{0,h}$ denotes the relativization of Σ_n^0 to h, and analogously for other classes of the arithmetical hierarchy. A straightforward relativization of the proof above yields the following

Corollary 3. *For any $h \in 2^\omega$, $\Delta_2^{0,h} = \cup\{\mathcal{S}_\alpha^h | \alpha < \xi(h)\}$.*

Now we extend the result of the previous section to some levels of the effective Wadge hierarchy from Section 3. The effective Hausdorff theorem is a particular case of the following result for $\alpha = \xi$.

Theorem 4. *If $\alpha < \xi^\omega$ and $e(\alpha)$ is a limit ordinal of uncountable cofinality then $\tilde{\mathcal{S}}_\alpha = \cup\{\mathcal{S}_\beta | \beta < \alpha\}$.*

We will need the following straightforward relativization of the preceding theorem.

Corollary 4. *For all $h \in 2^\omega$ and $\alpha < \xi(h)$, if $e(\alpha)$ is a limit ordinal of uncountable cofinality then $\tilde{\mathcal{S}}_\alpha^h = \cup\{\mathcal{S}_\beta^h | \beta < \alpha\}$.*

6   Proof of Main Theorem

The fifth step to the proof of the main theorem are two important facts on the hyperarithmetical sets. The first one states that any ordinal recursive in a hyperarithmetical set is (absolutely) recursive (see e.g. [Ro67, Mo80]). In other words, for any hyperarithmetical oracle $h \in \Delta_1^1$ it holds $\xi(h) = \xi$.

The second fact is related to an effective "hyperarithmetical" version of a theorem in [Lo83] mentioned in Section 2. From definitions in [Lo83] and in Sections 2 and 3 it is not hard to see that the following assertion is a reformulation of a particular case of Theorem 2.4 in [Lo83].

Proposition 4. *For all $\alpha < \xi^\omega$ and $A \in \Delta_1^1 \cap \Sigma_{e(\alpha)}$ there exists a hyperarithmetical oracle $h \in \Delta_1^1$ with $A \in S_\alpha^h$. The same holds true with $\Delta_{e(\alpha)}$ in place of $\Sigma_{e(\alpha)}$ and $\tilde{S}_\alpha^h$ in place of S_α^h.*

Proof of the main theorem. We have to verify the assertion (ii) of the main theorem. Let a set $A \in DTM_\omega$ belong to one of the constituents

$$\Sigma_\alpha \setminus \Pi_\alpha, \ \Pi_\alpha \setminus \Sigma_\alpha, \ \Delta_{\alpha+1} \setminus (\Sigma_\alpha \cup \Pi_\alpha)$$

of the Wadge hierarchy. We consider only the case $A \in \Sigma_\alpha \setminus \Pi_\alpha$ the other two cases being similar.

Since $DTM_\omega \subseteq \bigcup_{\alpha < \omega_1^\omega} \Sigma_\alpha$, it holds $\alpha < \omega_1^\omega$. We have to show that indeed $\alpha = e(\alpha^*)$, for some $\alpha^* < \xi^\omega$. We may of course assume α to be non-zero, hence there is a canonical representation

$$\alpha = \omega_1^{n_0} \alpha_0 + \cdots + \omega_1^{n_k} \alpha_k.$$

It suffices to show that all coefficients $\alpha_0, \ldots, \alpha_k$ are recursive ordinals since then it holds $\alpha = e(\alpha^*)$, where

$$\alpha^* = \xi^{n_0} \alpha_0 + \cdots + \xi^{n_k} \alpha_k.$$

We have $A \in \Delta_\beta, \beta = \omega_1^{n_0+1}$. By Proposition 4, $A \in \tilde{S}_{\xi^{n_0}+1}^h$, for some $h \in \Delta_1^1$. By Corollary 4, $A \in S_{\xi^{n_0}\gamma}^h$ for some $\gamma < \xi(h) = \xi$. By a relativization of Corollary 2, $A \in \Sigma_{\omega_1^{n_0}\gamma}$ for some $\gamma < \xi$. But $A \in \Sigma_\alpha \setminus \Pi_\alpha$, hence $\alpha_0 < \xi$.

If $k = 0$, the proof is over. Otherwise, $A \in \Delta_\beta, \beta = \omega_1^{n_0} \alpha_0 + \omega_1^{n_1+1}$. Arguing as in the last paragraph, we deduce that $A \in \Sigma_\delta, \delta = \omega_1^{n_0} \alpha_0 + \omega_1^{n_1} \gamma$ for some $\gamma < \xi$. But $A \in \Sigma_\alpha \setminus \Pi_\alpha$, hence $\alpha_1 < \xi$. Continuing in this manner, we deduce that really all the ordinals $\alpha_0, \ldots, \alpha_k$ are recursive, completing the proof of the theorem.

Notice that the proof works for any hyperarithmetical sets $A \in \bigcup_{\alpha < \omega_1^\omega} \Sigma_\alpha$. In other words, the Wadge degrees of the hyperarithmetical $bc(\Sigma_2^0)$-sets are the same as the degrees of DTM_ω-sets.

7 Conclusion

For the Baire space, the formulation of the main result looks as follows.

Theorem 5. *(i) For every $\alpha < \xi^\omega$, any of the constituents*

$$\Sigma_{e(\alpha)} \setminus \Pi_{e(\alpha)}, \ \Pi_{e(\alpha)} \setminus \Sigma_{e(\alpha)}, \ \Delta_{e(\alpha+1)} \setminus (\Sigma_{e(\alpha)} \cup \Pi_{e(\alpha)}), \ \Delta_{e(\lambda)} \setminus \bigcup_{\beta < \lambda} \Sigma_{e(\beta)},$$

where $\lambda < \xi^w$ is a limit ordinal such that $e(\lambda)$ is a limit ordinal of countable cofinality, contains a set from $bc(\Sigma_2^0)$.

(ii) All other constituents of the Wadge hierarchy do not contain sets from $bc(\Sigma_2^0)$.

Now we say a couple of words about Lipschitz degrees of $bc(\Sigma_2^0)$-sets. As is well known (see e.g [VW76, An01]), any non-self-dual Wadge degree forms a single Lipschitz degree. Any self-dual Wadge degree splits into an increasing ω-chain of self-dual Lipschitz degrees over the Cantor space, and into an increasing ω_1-chain of self-dual Lipschitz degrees over the Baire space. Accordingly, Lipschitz degrees of DTM_ω-sets are obtained from Wadge degrees of such sets by splitting every self-dual Wadge degrees into an increasing ω-chain of self-dual Lipschitz degrees. For the Baire space, Lipschitz degrees of $bc(\Sigma_2^0)$-sets are obtained from Wadge degrees of such sets by splitting every self-dual Wadge degree into an increasing ξ-chain of self-dual Lipschitz degrees.

It seems that the method of this paper applies also to characterizing Wadge degrees of the arithmetical sets and of the hyperarithmetical sets. At the same time, we do not know any answer to some natural questions similar to those answered above (e.g. whether the effective version of the Hausdorff theorem in the formulation similar to that from Section 5 may be lifted to the higher levels of the hyperarithmetical hierarchy or not).

Acknowledgement. I am grateful to the University of Siegen for supporting my visiting professorship in summer semester 2002 and to Dieter Spreen for the hospitality and for making the stay possible. Thanks are also due to the referees for the careful reading and advices.

References

[An01]   A. Andretta. *Notes on Descriptive Set Theory.* Manuscript, 2001.

[Du0?]   J. Duparc. A hierarchy of deterministic context-free ω-languages. *Theor. Comp. Sci.*, to appear.

[Er68a]   Yu.L. Ershov, On a hierarchy of sets II, *Algebra and Logic*, 7, No 4 (1968), 15–47 (Russian).

[KSW86]   J. Köbler, U. Shöning and K.W. Wagner. The difference and truth-table hierarchies for NP. Dep. of Informatics, Koblenz. Preprint 7 (1986).

[KM67]   K. Kuratowski and A. Mostowski. *Set Theory.* North Holland, Amsterdam, 1967.

[Lo83]   A. Louveau. Some results in the Wadge hierarchy of Borel sets, *Lecture Notes in Math.*, 1019 (1983), 28–55.

[Mos80]   Y.N. Moschovakis. *Descriptive set theory*, North Holland, Amsterdam, 1980.

[Ro67]   H. Rogers jr. *Theory of recursive functions and effective computability.* McGraw-Hill, New York, 1967.

[Se83]   V.L. Selivanov. Hierarchies of hyperarithmetical sets and functions. *Algebra i Logika*, 22, No 6 (1983), 666–692 (English translation: *Algebra and Logic*, 22 (1983), 473–491).

[Se92]   V.L. Selivanov. *Hierarchies, Numerations, Index Sets.* Handritten notes, 1992, 300 pp.

[Se94]   V.L. Selivanov. Fine hierarchy of regular ω-languages. Preprint N 14, 1994, the University of Heidelberg, Chair of Mathematical Logic, 13 pp.

[Se95]   V.L. Selivanov. Fine hierarchy of regular ω-languages. *Lecture Notes in Comp. Sci.*, v. 915. Berlin: Springer, 1995, 277–287.

[Se95a]   V.L. Selivanov. Fine hierarchies and Boolean terms. *J. Symbol. Logic*, 60 (1995), 289–317.

[Se98]   V.L. Selivanov. Fine hierarchy of regular ω-languages. *Theor. Comp. Sci.* 191 (1998), 37–59.

[Sta97]   L. Staiger. ω-languages. In: *Handbook of Formal Languages v. 3*, Springer, Berlin, 1997, 339–387.

[Ste80]   J. Steel. Determinateness and the separation property. *J. Symbol. Logic*, 45 (1980), 143–146.

[Wad72]   W. Wadge. Degrees of complexity of subsets of the Baire space. *Notices A.M.S.* (1972), R-714.

[Wad84]   W. Wadge. *Reducibility and determinateness in the Baire space.* PhD thesis, University of California, Berkeley, 1984.

[Wag79]   K. Wagner. On ω-regular sets. *Inform. and Control*, 43 (1979), 123–177.

[VW76]   Van Wesep. Wadge degrees and descriptive set theory. *Lecture Notes in Math.*, 689 (1978), 151–170.

Faster Deterministic Broadcasting in Ad Hoc Radio Networks*

Dariusz R. Kowalski[1] and Andrzej Pelc[2]

[1] Instytut Informatyki, Uniwersytet Warszawski,
Banacha 2, 02-097 Warszawa, Poland,
darek@mimuw.edu.pl
[2] Département d'informatique, Université du Québec en Outaouais,
Hull, Québec J8X 3X7, Canada,
Andrzej.Pelc@uqo.ca

Abstract. We consider radio networks modeled as directed graphs. In ad hoc radio networks, every node knows only its own label and a linear bound on the size of the network but is unaware of the topology of the network, or even of its own neighborhood. The fastest currently known deterministic broadcasting algorithm working for arbitrary n-node ad hoc radio networks, has running time $\mathcal{O}(n \log^2 n)$. Our main result is a broadcasting algorithm working in time $\mathcal{O}(n \log n \log D)$ for arbitrary n-node ad hoc radio networks of eccentricity D. The best currently known lower bound on broadcasting time in ad hoc radio networks is $\Omega(n \log D)$, hence our algorithm is the first to shrink the gap between bounds on broadcasting time in radio networks of arbitrary eccentricity to a logarithmic factor. We also show a broadcasting algorithm working in time $\mathcal{O}(n \log D)$ for *complete layered* n-node ad hoc radio networks of eccentricity D. The latter complexity is optimal.

1 Introduction

A radio network is a collection of transmitter-receiver stations. It is modeled as a directed graph on the set of these stations, refered to as *nodes*. A directed edge $e = (u, v)$ means that the transmitter of u can reach v. Nodes send messages in synchronous *steps* (time slots). In every step every node acts either as a *transmitter* or as a *receiver*. A node acting as a transmitter sends a message which can potentially reach all of its out-neighbors. A node acting as a receiver in a given step gets a message, if and only if, exactly one of its in-neighbors transmits in this step. The message received in this case is the one that was transmitted. If at least two in-neighbors v and v' of u transmit simultaneously in a given step, none of the messages is received by u in this step. In this case we

---

* This work was done during the first author's stay at the Research Chair in Distributed Computing of the Université du Québec en Outaouais, as a postdoctoral fellow. The work of the second author was supported in part by NSERC grant OGP 0008136 and by the Research Chair in Distributed Computing of the Université du Québec en Outaouais.

H. Alt and M. Habib (Eds.): STACS 2003, LNCS 2607, pp. 109–120, 2003.

say that a *collision* occurred at u. It is assumed that the effect at node u of more than one of its in-neighbors transmitting, is the same as that of no in-neighbor transmitting, i.e., a node cannot distinguish a collision from silence.

The goal of *broadcasting* is to transmit a message from one node of the network, called the *source*, to all other nodes. Remote nodes get the source message via intermediate nodes, along paths in the network. In order to make broadcasting feasible, we assume that there is a directed path from the source to any node of the network. We study one of the most important and widely investigated performance parameters of a broadcasting algorithm, which is the total time, i.e., the number of steps it uses to inform all the nodes of the network.

We consider deterministic distributed broadcasting in ad hoc radio networks. In such networks, a node does not have any *a priori* knowledge of the topology of the network, its maximum degree, its eccentricity, nor even of its immediate neighborhood: the only *a priori* knowledge of a node is its own label and a linear upper bound r on the number of nodes. Labels of all nodes are distinct integers from the interval $[0, ..., r]$. Broadcasting in ad hoc radio networks was investigated, e.g., in [3,5,6,7,9,10,11,12,13,18]. We use the same definition of running time of a broadcasting algorithm working for ad hoc radio networks, as e.g., in [11]. We say that the algorithm works in time t for networks of a given class, if t is the smallest integer such that the algorithm informs all nodes of any network of this class in at most t steps. We do not suppose the possibility of spontaneous transmissions, i.e., only nodes which already got the source message, are allowed to send messages. Of course, since we are only concerned with upper bounds on broadcasting time, all our results remain valid if spontaneous transmissions are allowed. Indeed, a broadcasting algorithm without spontaneous transmissions can be considered in the model with spontaneous transmissions allowed: such transmissions are simply never used by the algorithm. The format of all messages is the same: a node transmits the source message and the current step number.

We denote by n the number of nodes in the network, by r a linear upper bound on n ($r = cn$, for some constant c), by D the eccentricity of the network (the maximum length of a shortest directed path from the source to any other node), and by Δ the maximum in-degree of a node in the network. Among these parameters, only r is known to nodes of the network.

1.1   Related Work

In many papers on broadcasting in radio networks (e.g., [1,2,14,17]), the network is modeled as an undirected graph, which is equivalent to the assumption that the directed graph, which models the network in our scenario, is symmetric. A lot of effort has been devoted to finding good upper and lower bounds on deterministic broadcast time in such radio networks, under the assumption that nodes have full knowledge of the network. In [1] the authors proved the existence of a family of n-node networks of radius 2, for which any broadcast requires time $\Omega(\log^2 n)$, while in [14] it was proved that broadcasting can be done in time

$O(D + \log^5 n)$, for any n-node network of diameter D. (Note that for symmetric networks, diameter is of the order of the eccentricity).

As for broadcasting in ad hoc symmetric radio networks, an $\mathcal{O}(n)$ algorithm assuming spontaneous transmissions was constructed in [9] and a lower bound $\Omega(D \log n)$ was shown in [5], in the case when spontaneous transmissions are precluded.

Deterministic broadcasting in arbitrary directed radio networks was studied, e.g., in [6,7,8,9,10,11,12,13,18]. In [8], a $\mathcal{O}(D \log^2 n)$-time broadcasting algorithm was given for all n-node networks of eccentricity D, assuming that nodes know the topology of the network. Other above cited papers studied broadcasting time in ad hoc directed radio networks. The best known lower bound on this time is $\Omega(n \log D)$, proved in [12]. As for the upper bounds, a series of papers presented increasingly faster algorithms, starting with time $O(n^{11/6})$, in [9], then $O(n^{5/3} \log^{1/3} n)$ in [13], $O(n^{3/2} \sqrt{\log n})$ in [18], $O(n^{3/2})$ in [10], and finally, $O(n \log^2 n)$ in [11], which corresponds to the fastest currently known algorithm, working for ad hoc networks of arbitrary maximum degree. In another approach, broadcasting time is studied for ad hoc radio networks of maximum degree Δ. This work was initiated in [6], where the authors constructed a broadcasting scheme working in time $O(D \frac{\Delta^2}{\log^2 \Delta} \log^2 n)$, for arbitrary n-node networks with eccentricity D and maximum degree Δ. (While the result was stated only for undirected graphs, it is clear that it holds for arbitrary directed graphs, not just symmetric). This result was further investigated, both theoretically and using simulations, in [7,4]. On the other hand, a protocol working in time $O(D\Delta \log^{\log \Delta} n)$ was constructed in [3]. Finally, an $O(D\Delta \log n \log(n/\Delta))$ protocol was described in [12] (for the case when nodes know n but not Δ). If n is also unknown, the algorithm from [12] works in time $O(D\Delta \log^a n \log(n/\Delta))$, for any $a > 1$.

Finally, randomized broadcasting in ad hoc radio networks was studied, e.g., in [2,17]. In [2], the authors give a simple randomized protocol running in expected time $O(D \log n + \log^2 n)$. In [17] it was shown that for any randomized broadcast protocol and parameters D and n, there exists an n-node network of eccentricity D, requiring expected time $\Omega(D \log(n/D))$ to execute this protocol.

1.2   Our Results

The main result of this paper is a deterministic broadcasting algorithm working in time $\mathcal{O}(n \log n \log D)$ for arbitrary n-node ad hoc radio networks of eccentricity D. This improves the best currently known broadcasting time $\mathcal{O}(n \log^2 n)$ from [11], e.g., for networks of eccentricity polylogarithmic in size. Also, for $D\Delta \in \omega(n)$, this improves the upper bound $O(D\Delta \log n \log(n/\Delta))$ from [12]. The best currently known lower bound on broadcasting time in ad hoc radio networks, is $\Omega(n \log D)$ [12], hence our algorithm is the first to shrink the gap between bounds on deterministic broadcasting time for radio networks of arbitrary eccentricity, to a logarithmic factor. Our algorithm is non-constructive, in the same sense as that from [11]. Using the probabilistic method we prove the existence of a combinatorial object, which all nodes use in the execution of the deterministic broadcasting algorithm. (Since we do not count local computations in our time measure, such an object — the same for all nodes — could be found by exhaustive search performed locally by all nodes, without changing our result). It should be noted that a constructive broadcasting protocol working in time $O(n^{1+o(n)})$ was obtained in [15].

We also show a broadcasting algorithm working in time $\mathcal{O}(n \log D)$ for *complete layered* n-node ad hoc radio networks of eccentricity D. The latter complexity is optimal, due to the matching lower bound $\Omega(n \log D)$, proved in [12] for this class of networks, even assuming that nodes know parameters n and D. The best previous upper bound on broadcasting time in complete layered n-node ad hoc radio networks of eccentricity D was $\mathcal{O}(n \log n)$ [11]. Hence we obtain a gain for the same range of values of D as before.

If nodes do not know any upper bound on the size of the network, the upper bound $\mathcal{O}(n \log^2 n)$ from [11] remains valid, using a simple doubling technique, which probes possible values of n. In our case, the doubling technique cannot be used directly, since we deal with two unknown parameters, n and D. However, we can modify our algorithm in this case, obtaining running time $\mathcal{O}(n \log n \log \log n \log D)$, which still beats the time from [11], e.g., for networks of eccentricity polylogarithmic in size. For $D\Delta \in \Omega(n)$, this also improves the upper bound $O(D\Delta \log^a n \log(n/\Delta))$, for any $a > 1$, proved in [12] for the case of unknown n and Δ.

2   The Broadcasting Algorithm

In this section we show a deterministic broadcasting algorithm working in time $\mathcal{O}(n \log n \log D)$ for arbitrary n-node ad hoc radio networks of eccentricity D. Let r be the linear bound on the size of the ad hoc network. Taking $2^{\lceil r \rceil}$ instead of r, we can assume that $\log r$ is a positive integer. The parameter r is known to all nodes. We first show our upper bound under the additional assumption that D is known to all nodes. At the end of this section we show how this assumption can be removed without changing the result. Given a 0-1 matrix $T = [T_i(v)]_{i \leq t; v \leq r}$, where, $t = 3600 \cdot (1 + \log r) \cdot r \log D$, we define the following procedure

Procedure Fast-Broadcasting(T)

After receiving the source message and the current step number $a(1 + \log r) + b$, for some parameters a, b such that $0 \leq b \leq \log r$ and $0 < a(1 + \log r) + b \leq t$, node v waits until step $t_v = (a + 1)(1 + \log r) - 1$.
for $i = t_v + 1, \ldots, t$ do
  Substep A. if $T_i(v) = 1$ **then** v transmits in step i.
  Substep B. if $i \equiv v \mod r$ **then** v transmits in step i.

We now define the following random 0-1 matrix $\hat{T} = [\hat{T}_i(v)]_{i \leq t; v \leq r}$. For all parameters a, b such that $0 \leq b \leq \log r$ and $0 < a(1 + \log r) + b \leq t$, we have $\Pr[T_{a(1+\log r)+b}(v) = 1] = 1/2^b$, and all events $\hat{T}_i(v) = 1$ are independent. The period consisting of steps $a(1 + \log r), \ldots, (a + 1)(1 + \log r) - 1$ is called *stage a*.

Algorithm Fast-Broadcasting is the execution of Procedure Fast-Broadcasting($\hat{T}$), for the above defined random matrix $\hat{T}$.

In what follows, v_0 denotes the source. In order to analyze Algorithm Fast-Broadcasting, we define the following classes of directed graphs. A *path-graph* consists of a directed path $v_0, \ldots, v_k$, where $k \leq D$, possibly with some additional edges $(v_l, v_{l'})$, for $l > l'$, and with some additional nodes v, whose only out-neighbors are among $v_1, \ldots, v_k$. The path $v_0, \ldots, v_k$ is called the *main path*. A *simple-path-graph* consists of a directed path $v_0, \ldots, v_k$, where $k \leq D$, with some additional nodes v, each of which has exactly one out-neighbor and this out-neighbor is among $v_1, \ldots, v_k$. For any path-graph $G = (V, E)$, the graph $\bar{G}$ is the subgraph of G on the same set of nodes, containing the main path and satisfying the condition that for every node v outside the main path, v has exactly one neighbor v_l in $\bar{G}$, where $l = \max\{l' : (v, v_{l'}) \in E\}$. By definition, for every path-graph G, the graph $\bar{G}$ is a simple-path-graph.

In general, path-graphs do not satisfy our assumption that there is a directed path from the source v_0 to any other node. Hence, for path-graphs, we modify our model of broadcasting as follows, and call it the *adversarial wake-up model*. We assume that nodes outside the main path also get the source message, but are woken up by an adaptive adversary in various time-steps not exceeding t. More precisely, the adversary can wake up any node v outside the main path in any step $t_v \leq t$, providing v with the source message and step number t_v. The adversary acts according to some *wake-up pattern* $\mathcal{W}_m$, from the family $\{\mathcal{W}_m\}_{m \leq M}$ of all possible wake-up patterns. A wake-up pattern is a function assigning to every vertex v outside the main path, a step number $t_v \leq t$. For every wake-up pattern $\mathcal{W}_m$ and stage a of Procedure Fast-Broadcasting(T), we define an integer $f_a(m)$ as follows. If node v_k has the source message after stage a, we fix $f_a(m) = k$. Otherwise, $f_a(m) = l$, where v_{l+1} is the last node on the main path which does not have the source message after stage a, but has an in-neighbor having the source message after stage a, assuming that the adversary uses pattern $\mathcal{W}_m$.

We now describe a high-level outline of the proof of our upper bound. First, using the probabilistic method, we show the existence of a matrix T, such that Procedure Fast-Broadcasting(T), applied to any simple-path-graph and to any wake-up pattern, delivers the source message to the last node of the main path, in time $\mathcal{O}(n \log n \log D)$. Next, we show that the same is true for any path-graph. Finally, we show that Procedure Fast-Broadcasting(T) completes broadcasting in all graphs in time $\mathcal{O}(n \log n \log D)$.

Fix an integer a. Suppose that $\hat{T}_i(v)$ are fixed for all $i < (a+1)(1+\log r)$. Our first goal is to show that for a fixed simple-path-graph G, the set $\{m : f_a(m) < f_{a+1}(m) \leq k\}$ contains a constant fraction of values $\{m : f_a(m) < k\}$, for any stage a, with high probability. More precisely, we call stage $a + 1$ *successfull*, if either $\{m : f_a(m) < k\} = \emptyset$ or $|\{m : f_a(m) < f_{a+1}(m) \leq k\}| \geq \frac{1}{36} \cdot |\{m : f_a(m) < k\}|$ in the execution of Algorithm Fast-Broadcasting.

Lemma 1. *With probability at least 0.1, stage $a + 1$ is successfull.*

Proof. The proof will appear in the full version of the paper.

The next lemma implies that the last node of the main path of a simple-path-graph gets the source message by step $\mathcal{O}(n \log n \log D)$ of Algorithm Fast-Broadcasting, with probability at least $1 - (0.45)^{n \log D}$.

Lemma 2. *Fix a simple-path-graph G and a wake-up pattern $\mathcal{W}_m$ in the adversarial wake-up model. By stage $3600 \cdot n \log D$, the number of successful stages is at least $72n \log D$ with probability at least $1 - (0.45)^{n \log D}$. This number of successful stages is sufficient to deliver the source message to node v_k.*

Proof. We consider only substeps of type A. Consider $3600 \cdot n \log D$ consecutive stages, since the beginning of Algorithm Fast-Broadcasting. From Lemma 1, stage a is successful with probability at least 0.1, moreover this probability is at least 0.1 independently of successes in another stages.

The probability, that among $3600n \log D$ consecutive stages at most $72D \log n$ are successful is at most

$$
\sum_{k=0}^{72D \log n} \binom{3600n \log D}{k} \cdot \left(\frac{9}{10}\right)^{3600n \log D - k} \leq
$$

$$
\leq \binom{3600n \log D}{0} \cdot \left(\frac{9}{10}\right)^{3600n \log D} + \binom{3600n \log D}{1} \cdot \left(\frac{9}{10}\right)^{3600n \log D - 1} +
$$

$$
+ \sum_{k=2}^{72D \log n} \frac{(3600n \log D)^{3600n \log D + 1}}{k^k (3600n \log D - k)^{3600n \log D - k}} \cdot \left(\frac{9}{10}\right)^{3600n \log D - k}
$$

$$
\leq 4001n \log D \cdot \left(\frac{9}{10}\right)^{3600n \log D} +
$$

$$
+ \sum_{k=2}^{72D \log n} 3600n \log D \cdot \left(\frac{3600n \log D}{k}\right)^k \left[\frac{9}{10}\left(1 + \frac{k}{3600n \log D - k}\right)\right]^{3600n \log D - k}
$$

$$
\leq 4001n \log D \cdot \left(\frac{9}{10}\right)^{3600n \log D} +
$$

$$
+ 72D \log n \cdot 3600n \log D \cdot \left(\frac{3600n \log D}{72n \log D}\right)^{72n \log D} \left[\frac{9}{10} \cdot \frac{50}{49}\right]^{49 \cdot 72n \log D}
$$

$$
\leq 4001n \log D \cdot \left(\frac{9}{10}\right)^{3600n \log D} + 72D \log n \cdot 3600n \log D \cdot \left(50^{72} \cdot (0.92)^{49 \cdot 72}\right)^{n \log D},
$$

which is at most $(0.45)^{n \log D}$, for sufficiently large n. We used inequalities $\frac{b^b}{e^b} \leq b! \leq \frac{b^{b+1}}{e^b}$, for $b \geq 2$, and the fact that function $\left(\frac{C}{x}\right)^x$ is increasing for $0 < x \leq \frac{C}{e}$, for positive constant C. We also used the inequality $\frac{D}{\log D} \leq \frac{n}{\log n}$.

In order to complete the proof, we show that, if there are at least $72D \log n$ successfull stages by stage a, then $\{m : f_a(m) < k\} = \emptyset$. Let a_x, for $x = 1, \ldots, \log n$, be the first stage after $72D \cdot x$ successfull stages. It is enough to prove, by induction on x, that $|\{m : f_{a_x}(m) < k\}| < n/2^x$. For $x = 1$ this is straightforward. Assume that this inequality is true for x. We show it for $x + 1$. Suppose the contrary: $|\{m : f_{a_{x+1}}(m) < k\}| \geq n/2^{x+1}$. It follows that during each successfull stage b, where $a_x \leq b < a_{x+1}$, at least $\frac{n/2^{x+1}}{36}$ values $f_b(m)$ are smaller than $f_{b+1}(m)$. Since in stage a_x there were fewer than $n/2^x$ such values, and each of them may increase at most D times, we obtain that in stage a_{x+1}, the set $\{m : f_{a_{x+1}}(m) < k\}$ would have fewer than

$$\frac{\frac{n}{2^x} \cdot D}{\frac{n}{36 \cdot 2^{x+1}} \cdot 72D} = 1$$

elements, which contradicts the assumption that $|\{m : f_{a_{x+1}}(m) < k\}| > n/2^{x+1}$.

Lemma 3. *There exists a matrix T of format $r \times 3600r(1 + \log r) \log D$ such that, for any n-node simple-path-graph G of eccentricity at most D, and any wake-up pattern, Procedure Fast-Broadcasting(T) delivers the source message to the last node of the main path of G in $3600n \log D$ stages, for sufficiently large n.*

Proof. First consider D such that $\log D > 20(1+c)$. (Recall that $r = cn$. Knowledge of c is not necessary). In this case we consider only substeps of type A. There are at most

$$\sum_{k=1}^{D} \binom{r}{n} \cdot \binom{n}{k} \cdot k! \cdot k^{n-k} \leq D^{\alpha n}$$

different simple-path-graphs G with n nodes and eccentricity at most D, for some positive constant $\alpha < 1.1$. This is because

$$\binom{r}{n} \cdot \binom{n}{k} \cdot k! \cdot k^{n-k} \leq 2^r \cdot 2^n \cdot k^n = 2^{n(1+c)+n \log k} \leq 2^{n(1+c)+n \log D} < D^{1.05 \cdot n} \,,$$

for any $k \leq D$, and consequently

$$\sum_{k=1}^{D} \binom{r}{n} \cdot \binom{n}{k} \cdot k! \cdot k^{n-k} \leq D \cdot D^{1.05 \cdot n} < D^{1.1 \cdot n}$$

for sufficiently large n ($n > 20$).

Let $\hat{T}$ be the random matrix defined previously. By Lemma 2, the probability that Algorithm Fast-Broadcasting working on $\hat{T}$ delivers the source message to the last node of the main path of G by stage $3600n \log D$, for all simple-path-graphs G and all wake-up patterns $\mathcal{W}_m$, is at least

$$1 - \sum_{G}(0.45)^{n \log D} \geq 1 - D^{\alpha n} \cdot (0.45)^{n \log D} > 1 - D^{1.1n} \cdot (0.45)^{n \log D} > 0 \,,$$

where the sum is taken over all simple-path-graphs G with n nodes chosen from r labels, and of eccentricity at most D. Using the probabilistic method we obtain, that there is a matrix T satisfying the lemma.

Next, consider D such that $\log D \leq 20(1 + c)$. In this case we consider only substeps of type B. Since D is constant, by step $D \cdot r \in \mathcal{O}(n)$, every node of the main path will receive the source message by the round-robin argument. This concludes the proof.

For every wake-up pattern $\mathcal{W}_m$ and step i of Procedure Fast-Broadcasting(T), we define an integer $f_i'(m)$ as follows, by analogy to $f_a(m)$. If node v_k has the source message after step i, we fix $f_i'(m) = k$. Otherwise, $f_i'(m) = l$, where v_{l+1} is the last node on the main path which does not have the source message after step i, but has a neighbor having the source message after step i, assuming that the adversary uses pattern $\mathcal{W}_m$. Obviously $f_a(m) = f_{(a+1)(1+\log r)-1}'(m)$.

Lemma 4. *Fix a path-graph G and a matrix T. For any wake-up pattern $\mathcal{W}_m$ and any step i, values $f_i'(m)$ are the same for G and for $\bar{G}$. Consequently, for any stage a, values $f_a(m)$ are the same for G and for $\bar{G}$.*

Proof. The proof will appear in the full version of the paper.

Theorem 1. *There is a matrix T of format $r \times 3600r(1 + \log r) \log D$ such that for every n-node graph G of eccentricity D, Procedure Fast-Broadcasting(T) performs broadcasting on G in time $\mathcal{O}(n \log n \log D)$.*

Proof. Take the matrix T from Lemma 3. Suppose the contrary: after step $3600n(1+\log r) \log D$, there is a node v without the source message. Consider the subgraph H of G, which contains a shortest directed path $v_0, \ldots, v_k = v$ from the source to node v, with all induced edges between nodes of this path, and all those in-neighbors v' of nodes $v_1, \ldots, v_k$ which received the source message by step $3600n(1 + \log r) \log D$, together with corresponding arcs (v', v_l). By definition, H is a path-graph, and hence $\bar{H}$ is a simple-path-graph. By Lemma 3 we obtain that v received the source message in $\bar{H}$, by step $3600n(1 + \log r) \log D$. (We need to apply Lemma 3 to the wake-up pattern "generated" by Procedure Fast-Broadcasting(T) working on G: every node of $\bar{H}$ outside the main path is woken up in the adversarial wake-up model in the time step in which it gets the source message for the first time when Procedure Fast-Broadcasting(T) is executed on G). From Lemma 4 we obtain, that the same is true in H. Since the considered wake-up pattern is generated by Procedure Fast-Broadcasting(T) working on G, we conclude that v received the source message by step $3600n(1 + \log r) \log D$, when Procedure Fast-Broadcasting(T) is executed on G. This is a contradiction which concludes the proof.

We conclude this section by observing that the assumption that eccentricity D is known to all nodes can be removed without changing our result. It is enough to apply Algorithm Fast-Broadcasting for parameter r and for eccentricities 2^{2^i}, for $i = 1, ..., \lceil \log \log r \rceil$. Broadcasting will be completed after the execution of Algorithm Fast-Broadcasting for $i = \lceil \log \log D \rceil$. Total time will be at most 4 times larger than running time of Algorithm Fast-Broadcasting when D is known.

Observe that using only substeps of type B in Procedure Fast-Broadcasting(T) we can trivially get the estimate $\mathcal{O}(nD)$ on broadcasting time. Hence the upper bound from Theorem 1 can be refined to $\mathcal{O}(n \cdot \min\{\log n \log D, D\})$.

3   Broadcasting with Unknown Bound on Network Size

In this section we show how our estimate of broadcasting time changes if nodes do not know any parameters of the network: neither its eccentricity D nor any upper bound r on the number of nodes. Denote by $AFB(x,y)$ the execution of Algorithm Fast-Broadcasting for the upper bound x on the size of the network and for eccentricity y, running in time $3600x(1 + \log x) \log y$ (it exists by Theorem 1). We construct the following

Algorithm Modified-Fast-Broadcasting

$i := 1$

repeat

$\quad i := i + 1, l := 1$

$\quad$ **while** $2^{2^l} < 2^{i-l}$ **do**

$$AFB(2^{i-l}, 2^{2^l}) \tag{1}$$

$\quad\quad l := l + 1$

$$AFB(2^{i-l}, 2^{i-l}) \tag{2}$$

The following observations hold:

1. For every n-node graph G with parameters r and D, broadcasting on G is completed by the time when algorithm $AFB(2^{\lceil \log r \rceil}, \min\{2^{\lceil \log r \rceil}, 2^{2^{\lceil \log \log D \rceil}}\})$ is executed. This happens for $i = \lceil \log r \rceil + \lceil \log \log D \rceil$ and $l = \lceil \log \log D \rceil$, either in (1) if $\lceil \log r \rceil > 2^{\lceil \log \log D \rceil}$, or in (2) otherwise.

2. $ABF(2^{i-l}, 2^{2^l})$ performs broadcasting in $3600 \cdot 2^{i-l} \cdot (i - l + 1) \cdot 2^l$ steps.

$\quad ABF(2^{i-l}, 2^{i-l})$ performs broadcasting in $3600 \cdot 2^{i-l} \cdot (i - l + 1) \cdot (i - l)$ steps.

3. Fix i. Let l_0 be the largest index l for which $AFB(2^{i-l}, 2^{2^l})$ is executed in (1). The execution of loop "while" lasts at most

$$\sum_{l=1}^{l_0} 3600 \cdot 2^{i-l} \cdot (i - l + 1) \cdot 2^l \leq \sum_{l=1}^{\lceil \log i \rceil} 3600 \cdot 2^{i-l} \cdot (i - l + 1) \cdot 2^l \leq 3600 \cdot \lceil \log i \rceil \cdot 2^i \cdot i$$

steps. The execution of (2) lasts

$$3600 \cdot 2^{i-l_0-1} \cdot (i - l_0 - 1 + 1) \cdot (i - l_0 - 1) \leq 3600 \cdot 2^{i-l_0-1} \cdot (i - l_0 + 1) \cdot 2^{l_0+1}$$

steps. The latter inequality follows from the condition $2^{2^{l_0}} \geq 2^{i-l_0-1}$. We further have

$$3600 \cdot 2^{i-l_0-1} \cdot (i - l_0 + 1) \cdot 2^{l_0+1} = 3600 \cdot 2^{i-l_0}(i - l_0 + 1) \cdot 2^{l_0} \leq 3600 \lceil \log i \rceil \cdot 2^j \cdot i.$$

4. The total number of steps till the execution of "repeat" for $i = i_0$, is at most

$$\sum_{i=2}^{i_0} 2 \cdot 3600 \cdot \lceil \log i \rceil \cdot 2^i \cdot i \leq 7200 \cdot 2^{i_0+1} \cdot i_0 \cdot \lceil \log i_0 \rceil,$$

since (2) lasts at most the same time as the last preceding execution of (1).

5. For any r and D, the total time of Algorithm Modified-Fast-Broadcasting is at most

$$7200 \cdot 2^{\lceil \log r \rceil + \lceil \log \log D \rceil + 1} \cdot (\lceil \log r \rceil + \lceil \log \log D \rceil) \cdot \lceil \log(\lceil \log r \rceil + \lceil \log \log D \rceil) \rceil \,,$$

which is in $\mathcal{O}(r \log r \log \log r \log D) \subseteq \mathcal{O}(n \log n \log \log n \log D)$. This proves the following theorem.

Theorem 2. *Algorithm Modified-Fast-Broadcasting completes broadcasting on any n-node network of eccentricity D, in time $\mathcal{O}(n \log n \log \log n \log D)$, even when nodes do not know any parameters of the network or any bound on its size.*

Similarly as in Section 2, the above upper bound can be refined to $\mathcal{O}(n \cdot \min\{\log n \log \log n \log D, D\})$.

4   Optimal Broadcasting in Complete Layered Networks

In [12] the authors prove a lower bound $\Omega(n \log D)$ on deterministic broadcasting time on any n-node network of eccentricity D. This is done using complete layered networks. All nodes of such networks can be partitioned into layers L_0, L_1,..., L_D where L_0 consists of the source and the set of directed edges is $\{(v, w) : v \in L_i, w \in L_{i+1}, i = 0, 1, ..., D - 1\}$. More precisely, it is shown in [12] that for every deterministic broadcasting algorithm there is a complete layered n-node network of eccentricity D, such that this algorithm requires time $\Omega(n \log D)$ to perform broadcast on this network. This result holds even when n and D are known to all nodes.

In this section we present a deterministic broadcasting algorithm which works on every complete layered n-node network of eccentricity D in time $\mathcal{O}(n \log D)$, and thus it is optimal. Hence, any lower bound sharper than $\Omega(n \log D)$, on broadcasting time in arbitrary radio networks, would have to be established for graphs more complicated than complete layered networks. Our result is also an improvement of the upper bound $\mathcal{O}(n \log n)$, proved in [11] for n-node complete layered networks.

We use the following definition of an (r, k)-selective family. A family $\mathcal{F}$ of subsets of R is called (r, k)-*selective*, for $k \leq r$, if for every subset Z of $\{1, \dots, r\}$, such that $|Z| \leq k$, there is a set $F \in \mathcal{F}$ and element $z \in Z$, such that $Z \cap F = \{z\}$.

Lemma 5. [12] *For every $r \geq 2$ and $k \leq r$, there exists an (r, k)-selective family $\mathcal{F}$ of size $\mathcal{O}(k \log((r + 1)/k))$.*

Let $\mathcal{F}_i$ denote a $(r, 2^i)$-selective family, for $i = 1, \dots, \log r$. By Lemma 5, we can assume, that $f_i = |\mathcal{F}_i| \leq \alpha 2^i \log((r + 1)/2^i)$, for some constant $\alpha > 0$, and for all $i = 1, \dots, \log r$. Let $\mathcal{F}_i = \{F_i(1), \dots, F_i(f_i)\}$.

Algorithm Complete-Layered
    for $i = 1, \dots, \log r$ **do**
      for $j = 1, \dots, f_i$ **do**
        if $v \in F_i(j)$ **then** v transmits
      for $j = 1, \dots, r$ **do**
        if $v = j$ **then** v transmits

Theorem 3. *Algorithm Complete-Layered completes broadcasting in $\mathcal{O}(n \log D)$ time, for any n-node complete layered network of eccentricity D.*

Proof. For $D = 1$ the proof is obvious. Assume $D \geq 2$. Fix an n-node complete layered network G of eccentricity D. Let L_l denote the l-th layer of G, and $d_l = |L_l|$, for $l = 0, \ldots, D$. Let t_l denote the step in which all nodes in L_l received the source message for the first time.

Claim. $t_{l+1} - t_l \leq 4\alpha d_l \log \big(2(r+1)/d_l\big)$, for every $l = 0, \ldots, D-1$.

In step $t_l + 1$ all nodes in L_l start transmitting. After at most

$$\sum_{i=1}^{\lceil \log d_l \rceil} f_i \leq \alpha \sum_{i=1}^{\lceil \log d_l \rceil} 2^i \log((r+1)/2^i)$$

$$\leq \alpha \left[\sum_{i=1}^{\lceil \log d_l \rceil} 2^i \log(r+1) - \sum_{i=1}^{\lceil \log d_l \rceil} i \cdot 2^i \right]$$

$$\leq \alpha \left[2^{\lceil \log d_l \rceil + 1} \log(r+1) - (2^{\lceil \log d_l \rceil + 1} \lceil \log d_l \rceil - 2^{\lceil \log d_l \rceil + 1} + 1) \right]$$

$$\leq \alpha 2^{\lceil \log d_l \rceil + 1} \log \frac{2(r+1)}{d_l}$$

$$\leq 4\alpha d_l \log \frac{2(r+1)}{d_l}$$

steps, all nodes in L_l complete transmissions according to the selective family $\mathcal{F}_{\lceil \log d_l \rceil}$. By definition of $\mathcal{F}_{\lceil \log d_l \rceil}$, there is a step among $t_l + 1, \ldots, t_l + \lfloor 4\alpha d_l \log \frac{2(r+1)}{d_l} \rfloor$ such that exactly one node in L_l transmits in this step. Consequently all nodes in L_{l+1} get the source message by step $t_l + \lfloor 4\alpha d_l \log \frac{2(r+1)}{d_l} \rfloor$. This completes the proof of the Claim.

Since $\sum_{l=0}^{D} d_l = n$ and $t_0 = 0$, we have

$$t_D = \sum_{l=0}^{D-1} (t_{l+1} - t_l) \leq 4\alpha \sum_{l=0}^{D-1} d_l \log \frac{2(r+1)}{d_l}$$

$$= 4\alpha \sum_{l=0}^{D-1} \left(\log \frac{(r+1)^{d_l}}{d_l^{d_l}} + d_l \right) \leq 4\alpha \log \frac{(r+1)^{n-d_D}}{\Pi_{l=0}^{D-1} d_l^{d_l}} + 4\alpha n$$

$$\leq 4\alpha(n - d_D) \log \frac{r+1}{(n-d_D)/D} + 4\alpha n \ .$$

We used the fact, that

$$\Pi_{l=0}^{D-1} d_l^{d_l} \geq \left(\frac{n - d_D}{\sum_{l=0}^{D-1} d_l \cdot \frac{1}{d_l}} \right)^{n-d_D} = \left(\frac{n - d_D}{D} \right)^{n-d_D},$$

which follows from the inequality between geometric and harmonic averages.

Since, for $D \geq 2$, the function $x \cdot \log \frac{D(r+1)}{x}$ is increasing for $x \leq r + 1$, we have

$$4\alpha(n - d_D) \log \frac{r+1}{(n-d_D)/D} + 4\alpha n \leq 4\alpha n \log \frac{r+1}{n/D} + 4\alpha n \in \mathcal{O}(n \log D) \ .$$

References

1. Alon, N., Bar-Noy, A., Linial, N., Peleg, D.: A lower bound for radio broadcast. Journal of Computer and System Sciences 43 (1991) 290–298
2. Bar-Yehuda, R., Goldreich, O., Itai, A.: On the time complexity of broadcast in radio networks: an exponential gap between determinism and randomization. Journal of Computer and System Sciences 45 (1992) 104–126
3. Basagni, S., Bruschi, D., Chlamtac, I.: A mobility-transparent deterministic broadcast mechanism for ad hoc networks. IEEE/ACM Trans. on Networking 7 (1999) 799–807
4. Basagni, S., Myers A.D., Syrotiuk, V.R.: Mobility-independent flooding for realtime multimedia applications in ad hoc networks. Proc. 1999 IEEE Emerging Technologies Symposium on Wireless Communications & Systems, Richardson, TX
5. Bruschi, D., Del Pinto, M.: Lower bounds for the broadcast problem in mobile radio networks. Distributed Computing 10 (1997) 129–135
6. Chlamtac, I., Faragó, A.: Making transmission schedule immune to topology changes in multi-hop packet radio networks. IEEE/ACM Trans. on Networking 2 (1994) 23–29
7. Chlamtac, I., Faragó, A., Zhang, H.: Time-spread multiple access (TSMA) protocols for multihop mobile radio networks. IEEE/ACM Trans. on Networking 5 (1997) 804–812
8. Chlamtac, I., Weinstein, O.: The wave expansion approach to broadcasting in multihop radio networks. IEEE Trans. on Communications 39 (1991) 426–433
9. Chlebus, B., Gąsieniec, L., Gibbons, A., Pelc, A., Rytter, W.: Deterministic broadcasting in unknown radio networks. Distributed Computing 15 (2002) 27–38
10. Chlebus, B., Gąsieniec, L., Östlin, A., Robson, J.M.: Deterministic radio broadcasting. Proc. 27th Int. Coll. on Automata, Languages and Programming (ICALP'2000), LNCS 1853, 717–728
11. Chrobak, M., Gąsieniec, L., Rytter, W.: Fast broadcasting and gossiping in radio networks. Proc. 41st Symposium on Foundations of Computer Science (FOCS'2000), 575–581
12. Clementi, A., Monti, A., Silvestri, R.: Selective families, superimposed codes, and broadcasting on unknown radio networks. Proc. 12th Ann. ACM-SIAM Symposium on Discrete Algorithms (SODA'2001), 709–718
13. De Marco, G., Pelc, A.: Faster broadcasting in unknown radio networks. Information Processing Letters 79 (2001) 53–56
14. Gaber, I., Mansour, Y.: Broadcast in radio networks. Proc. 6th Ann. ACM-SIAM Symposium on Discrete Algorithms (SODA'1995), 577–585
15. Indyk, P.: Explicit constructions of selectors and related combinatorial structures with applications. Proc. 13th Ann. ACM-SIAM Symposium on Discrete Algorithms (SODA'2002), 697–704
16. Kowalski, D., Pelc, A.: Deterministic broadcasting time in radio networks of unknown topology. Proc. 43rd Annual IEEE Symposium on Foundations of Computer Science (FOCS'2002), 63–72.
17. Kushilevitz, E., Mansour, Y.: An $\Omega(D\log(N/D))$ lower bound for broadcast in radio networks. SIAM J. on Computing 27 (1998) 702–712
18. Peleg, D.: Deterministic radio broadcast with no topological knowledge. Manuscript (2000)

Private Computations in Networks: Topology versus Randomness

Andreas Jakoby, Maciej Liśkiewicz[1], and Rüdiger Reischuk

Institut für Theoretische Informatik, Universität zu Lübeck
jakoby/liskiewi/reischuk@informatik.mu-luebeck.de

Abstract. In a distributed network, computing a function privately requires that no participant gains any additional knowledge other than the value of the function. We study this problem for incomplete networks and establish a tradeoff between connectivity properties of the network and the amount of randomness needed. First, a general lower bound on the number of random bits is shown. Next, for every $k \geq 2$ we design a quite efficient (with respect to randomness) protocol for symmetric functions that works in arbitrary k-connected networks. Finally, for directed cycles that compute threshold functions privately almost matching lower and upper bounds for the necessary amount of randmoness are proven.

1 Introduction

Private computation can be defined as follows. Given a collection of players, where each player knows an individual secret, the goal is to compute a specific function that takes these secrets as arguments such that after the computation none of the players knows anything about the secrets of others that cannot be derived from his own secret and the result of the function. An example for such a computation is the secret voting problem. A committee wants to decide on an action based on the number of individual supporting votes of its members. But the ballot should such that after the voting process nobody knows anything about the opinion of the other committee members nor any details about the exact number of yes- and no-votes. The only information available afterwards should be whether the majority has supported or objected the action.

To exchange information the committee members can talk to each other pairwise using secret channels. Depending on the computational power of the players one can distinguish between *cryptographically secure privacy* and *privacy in an information-theoretic sense*. In the first case it is required that no player or set of players can gain any additional information about the inputs within polynomial time (see e.g. [17,18,12,8]), whereas in the second case the computational power of the players is not restricted (see e.g. [4,5]). Hence, the information-theoretic notion of privacy is significantly stronger. It will be used in this paper.

Private computations have been subject of a considerable amount of research. Traditionally, one investigates the number of rounds and random bits as complexity measures for private protocols. Chor and Kushilevitz [7] have studied

---

[1] On leave from Instytut Informatyki, Uniwersytet Wrocławski, Wrocław, Poland.

H. Alt and M. Habib (Eds.): STACS 2003, LNCS 2607, pp. 121–132, 2003.
© Springer-Verlag Berlin Heidelberg 2003

the number of rounds necessary to compute the sum modulo an integer (see also [3,6]). The number of random bits needed to count modulo 2 has been estimated in [15,14]. Gál and Rosén have shown that the parity function cannot be computed by an n-player private protocol using d random bits in $o(\log n/\log d)$ rounds [11]. They have also given an upper bound on the randomness-round-tradeoff for an arbitrary Boolean function depending on its sensitivity. Bounds on the number of rounds needed in the worst-case are established in [1,16].

The number of random bits for private computation of a function f is closely related to its circuit size. Kushilevitz, Ostrovsky, and Rosén [13] have shown that given a complete communication network every n-ary function f with linear circuit size complexity can also be computed by a private protocol using only a constant number of random bits. Their proof yields that every such f can also be computed privately in certain incomplete networks G that posses an independent set of size $m = n/2$. Hence, even networks where many players cannot talk to each other directly remain quite powerful with respect to privacy. In this paper we study more systematically the tradeoff between the amount of randomness and the topology of the communication network.

We assume that the given input bits are distributed among the nodes of the network G. For convenience, let us call the node that gets bit $X[i]$ player P_i. Most previous research has considered the case of complete networks, where each player P_i can talk to every other P_j secretly. However, in reality nodes have bounded degree thus the network has limited connectivity. Franklin and Yung are the first who have studied the role of connectivity in private computations. In [10] they present a protocol for bus networks that can simulate every communication step of a complete network by using a linear number of additional random bits.

In [2] we have investigated private protocols in k-connected networks, in particular, the complete bipartite graph $K_{k,n-k}$ with $n \geq 2k$ consisting of k nodes in one partition and $n-k$ in the other. It is shown that the parity function cannot be computed privately with less than $\frac{n-2}{k-1} - 1$ random bits in $K_{k,n-k}$, and moreover this bound is tight.

In this paper we generalize the result of [2] proving a general trade-off between the size of independent sets in the communication network G, the number of random bits necessary to guarantee privacy, and combinatorial properties of the Boolean function to be computed. We show that for every n-ary function f and every network G with an independent set of size $m \geq n/2$, f cannot be computed by G privately with less than $\frac{s(f)-2}{n-m-1} - 1$ random bits, where $s(f)$ denotes the sensitivity of f (for a definition, see the next section). Furthermore we will generalise the protocol of [2] that computes for arbitrary $k \geq 2$ the parity in k-connected networks, to a protocol for any symmetric Boolean function of small circuit complexity like the *and*, *or*, *majority*, or arbitrary *threshold* functions.

We also investigate the distribution of random players in the communication network if the total number r of random bits used by a protocol is bounded. First, we establish an upper bound on the number of players that may toss coins during a private computation. More specifically, it will be shown that for any private protocol computing the *or* function with r random bits there are at most

$2^{r+1} - 2$ such players. This bound is independent of the underlying communication network and the concrete input. Then we investigate the computation of *threshold* functions in specific networks, namely in directed cycles. It will be shown that these functions cannot be computed privately with a single random player regardless how many random bits he uses. Finally, we prove that any *threshold* function can be computed privately in a directed cycle with 4 random players and provide matching logarithmic upper and lower bounds for the number of random bits.

2   Preliminaries

Let $X = X[1]X[2]\ldots X[n] \in \{0,1\}^n$ denote a binary string of length n, and $\overline{X}$ its bitwise negation: $\overline{X}[i] = \overline{X[i]}$. Throughout the paper, for a set $I \subseteq \{1,..,n\}$ and $\alpha \in \{0,1\}^{|I|}$ we will apply substitutions $X\lceil_{I \leftarrow \alpha}$ to X defined as follows:

$$X\lceil_{I \leftarrow \alpha}[i] := \begin{cases} X[i] \text{ if } i \notin I , \\ \alpha[j] \text{ if } i \text{ is the } j\text{th smallest element in } I. \end{cases}$$

In the special case when we simply want to negate the i-th bit, that means $X\lceil_{\{i\} \leftarrow \overline{X}[i]}$, we will use the shorter notation $X^{\overline{[i]}}$. If $R = (R_1,\ldots,R_n)$ is a vector of binary strings and $\hat{R}_i$ a new string let

$$R\lceil_{\{i\} \leftarrow \hat{R}_i} := (R_1,\ldots,R_{i-1},\hat{R}_i,R_{i+1},\ldots,R_n) .$$

In this paper we will consider n-ary Boolean function $f : \{0,1\}^n \to \{0,1\}$ with a single output bit. For $I \subseteq \{1,..,n\}$ and $\alpha \in \{0,1\}^{|I|}$, we define the partially restricted function $f\lceil_{I \leftarrow \alpha} : \{0,1\}^{n-|I|} \to \{0,1\}$ by fixing the positions in I to the values given by α, i.e. for $X \in \{0,1\}^{n-|I|}$ let $f\lceil_{I \leftarrow \alpha}(X) := f\left((0^n\lceil_{I \leftarrow \alpha})\lceil_{J \leftarrow X}\right)$, where $J = \{1,..,n\} \setminus I$. Furthermore, define $X[I] \in \{0,1\}^{|I|}$ as follows: if i_j is the j-th smallest element in I, then $(X[I])[j] = X[i_j]$.

Definition 1. *A Boolean function f is called **non-degenerated** if for every $i \in \{1,..,n\}$ it holds: $f\lceil_{\{i\} \leftarrow 0} \neq f\lceil_{\{i\} \leftarrow 1}$. The cardinality of the set $S_f(X):= \{i|\ f(X) \neq f(X^{\overline{[i]}})\}$ is called the **sensitivity** of f at X. The sensitivity of f is given by $\mathbf{s}(f) := \max_{X \in \{0,1\}^n} |S_f(X)|$.*

Recall that an undirected graph G is k-connected iff, after deleting an arbitrary subset of at most $k - 1$ nodes, the resulting induced graph remains connected. We will consider n-ary Boolean functions like the *or-, and-, parity-, majority-* or *threshold*-function with threshold b denoted by $\mathtt{OR}_n, \mathtt{AND}_n, \mathtt{XOR}_n, \mathtt{MAJ}_n$ and τ_n^b.

At the beginning each player knows a single bit of the input X. The players can exchange binary messages using point-to-point secure communication links as specified by the communication network $G = (V, E)$. When the computation stops, all players have to know the value $f(X)$. The goal is to compute f such that no player learns anything about the other input bits except for the information he can deduce from his own bit and the result. Such a protocol is called **private**.

Definition 2. [7] *Let C_i be a random variable for the complete sequence of messages exchanged by player P_i with all his neighbours, and $\mathcal{R}_i$ a random variable for the random string used by P_i. A protocol $\mathcal{A}$ for a function f is **private with respect to P_i** if for every pair of input vectors X, Y with $f(X) = f(Y)$ and $X[i] = Y[i]$, and for every possible value C_i of C_i and R_i of $\mathcal{R}_i$,*

$$\Pr[C_i = C_i | \mathcal{R}_i = R_i, X] \ = \ \Pr[C_i = C_i | \mathcal{R}_i = R_i, Y] \ ,$$

where the probability is taken over the random variables $\mathcal{R}_j$ of all other players. $\mathcal{A}$ is private if it is private with respect to each player P_i. A function f can be computed privately in a network G iff G possesses a private protocol for f.

We call a protocol *synchronous* if the communication takes place in rounds and each message consists of a single bit. A protocol is *adaptive* (with respect to coin tosses of the players) if a decision of some player whether to use random bits during the computation or not depends on his specific input and the messages received from others players. If for every player this decision is fixed a priori, we say that the protocol is *non-adaptive*. All protocols given in this paper are synchronous and non-adaptive. However, our lower bounds hold for non-synchronous and adaptive protocols as well.

Definition 3. *For a given private protocol $\mathcal{A}$ and an input vector X define $R_\mathcal{A}(X)$ as the set of all vectors $R = (R_1, \dots, R_n)$, with $R_i \in \{0, 1\}^*$, such that for some execution of $\mathcal{A}$ on X, for every player P_i the value of the random string $\mathcal{R}_i$ is R_i. $\mathcal{A}$ is called **r-random**, if $\max_{X \in \{0,1\}^n} \max_{R \in R_\mathcal{A}(X)} \sum_{R_i \in R} |R_i| \leq r$.*

3   A General Lower Bound for Incomplete Networks

Theorem 1. *For every network G and every n-ary function f with an independent set of size $m \geq n/2$ it holds: f cannot be computed by G privately with less than $\frac{\mathsf{s}(f)-2}{n-m-1} - 1$ random bits.*

Proof: The first part of the proof follows the proof for XOR_n given in [2].

Assume to the contrary that for a given network G with an independent set of size m there exists a private protocol $\mathcal{A}$ that uses $r < \frac{\mathsf{s}(f)-2}{k-1} - 1$ random bits to compute f, where $k := n - m$. Since for $\mathsf{s}(f) < k + 2$ the right expression of the inequality becomes negative it suffices to consider only the case $\mathsf{s}(f) \geq k+2$.

Let $V = \mathcal{U} \cup \mathcal{W}$ be a partition of the nodes V of G such that $\mathcal{W}$ of size m is independent. For an input $X \in \{0,1\}^n$ let $C(X, R) = \langle C_1(X, R), \dots, C_k(X, R) \rangle$ be a complete description of the communication for each of the k players in $\mathcal{U}$ with all his neighbours during the computation of $\mathcal{A}$ on X, when the random strings used by the n players are $R = (R_1, \dots, R_n)$. Since the complementary set $\mathcal{W}$ is independent $C(X, R)$ determines the whole communication of $\mathcal{A}$ for input X and random choices R.

Let Y be an input vector such that f has sensitivity $\mathsf{s}(f)$ at Y. Let $t := \mathsf{s}(f) - k$, and $l_1, l_2, \dots, l_t$ be sensitive positions of Y that belong to players in $\mathcal{W}$. In other words, for every l_i it holds: $f(Y) \neq f(Y^{\overline{[l_i]}})$. Let $\mathcal{X}_Y$ be the set of all inputs $X \in \{0,1\}^n$ that agree with Y on all positions that belong to players in $\mathcal{U}$. $\mathcal{X}_Y$ has size 2^{n-k}. Moreover, choose a bit $b \in \{0,1\}$ such that $\mathcal{X}_{Y,b} := \{X \in \mathcal{X}_Y \mid f(X) = b\}$ has cardinality at least $|\mathcal{X}_Y|/2 = 2^{n-k-1}$.

Now, for $X \in \mathcal{X}_{Y,b}$ and any message pattern C_1 of the first player in $\mathcal{U}$ – let us denote him by $\hat{P}$ in the following – define

$$\mathcal{C}(C_1, X) := \{\langle C_2, \dots, C_k \rangle \mid \exists R \ \ \mathcal{C}(X, R) = \langle C_1, C_2, \dots, C_k \rangle\}.$$

Note that for some message pattern C_1 and input vector X the corresponding set $\mathcal{C}(C_1, X)$ can be empty. However, the privacy of $\mathcal{A}$ implies

Claim 1 $\exists C_1 \ \ \forall X \in \mathcal{X}_{Y,b} \ \ \mathcal{C}(C_1, X) \neq \emptyset$.

Let C_1 be such a message pattern in the following. Next, we approximate the number of different message patterns a player P_i in $\mathcal{U}$ can observe over all choices of input assignments from $\mathcal{X}_{Y,b}$. Let us denote this set by $\mathcal{C}_i$, i.e.

$$\mathcal{C}_i := \bigcup_{X \in \mathcal{X}_{Y,b}} \ \bigcup_R \ \mathcal{C}_i(X, R) .$$

Claim 2 *If $\mathcal{A}$ is r-random then for every player P_i in $\mathcal{U}$ it holds $|\mathcal{C}_i| \leq 2^r$.*

This claim is a slight modification of Lemma 4.10 in [15]. It does not suffice to argue that a private protocol using at most r random bits can see at most 2^r different values for these sequences of bits because in different executions of the protocol these random bits can be used by different players. To show that the above upper bound still holds, one can argue similarly as in [15].

Claim 2 and our general assumption $r < \frac{\mathsf{S}(f)-2}{k-1} - 1 = \frac{\mathsf{S}(f)-k-1}{k-1}$ imply

$$\Big| \bigcup_{X \in \mathcal{X}_{Y,b}} \mathcal{C}(C_1, X) \Big| \ \leq \ |\mathcal{C}_2 \times \dots \times \mathcal{C}_k| \ \leq \ 2^{r(k-1)} \ < \ 2^{\mathsf{S}(f)-k-1} \ = \ 2^{t-1}.$$

Since, by assumption each set $\mathcal{C}(C_1, X)$ is nonempty and $|\mathcal{X}_{Y,b}| \geq 2^{n-k-1}$, by the pigeon hole principle it must hold: $\exists \ C_2, \dots, C_k \ \ \exists \ \mathcal{Z} \subseteq \mathcal{X}_{Y,b} \ \ |\mathcal{Z}| > 2^{n-k-t}$ and $\forall X \in \mathcal{Z} \ \ \langle C_2, \dots, C_k \rangle \in \mathcal{C}(C_1, X)$. Hence, in $\mathcal{Z}$ there exist two strings X and Z such that $X[i] \neq Z[i]$ for some $i \in \{l_1, l_2, \dots, l_t\}$.

Now let R_X and R_Z be random strings such that $\mathcal{C}(X, R_X) = \mathcal{C}(Z, R_Z) = \langle C_1, C_2, \dots, C_k \rangle$, i.e. the players in $\mathcal{U}$ observe the same message pattern in both cases, and thus every player does. This implies that after flipping the i-th bit of X and fixing the random string of each player P_j different from P_i to $R_X[j]$ and for P_i to $R_Z[i]$, the players in $\mathcal{U}$ still have the same message pattern, i.e. $\mathcal{C}(X, R_X) = \mathcal{C}(X^{\overline{[i]}}, R_X \lceil_{\{i\} \leftarrow R_Z[i]})$. This implies in particular for P_i:

$$\mathcal{C}_i(X, R_X) = \mathcal{C}_i(X^{\overline{[i]}}, R_X \lceil_{\{i\} \leftarrow R_Z[i]}). \tag{1}$$

Lemma 1 (Isolated Player). *Let $X \in \mathcal{X}_{Y,b}$ and $R = (R_1, \dots, R_n)$ be an arbitrary vector of random strings used by the players. Moreover, assume that for some player P_i in $\mathcal{W}$ and some random string $\hat{R}_i$ for P_i it holds $\mathcal{C}(X, R) = \mathcal{C}(X^{\overline{[i]}}, R \lceil_{\{i\} \leftarrow \hat{R}_i})$. Then $f(X) = f(X^{\overline{[i]}})$.*

Using Lemma 1 we can conclude from equation (1):

$$f(X) = f(X^{\overline{[i]}}) = b. \tag{2}$$

Since i is one of the sensitive positions of the input Y flipping the i-th bit of Y the function f changes its value. We will only discuss the case $f(Y) = b$. The dual case $f(Y^{\overline{[i]}}) = b$ can be proved in a symmetric way (by equality (2)). Because of the sensitivity at the position i it holds

$$f(Y) = b \quad \text{and} \quad f(Y^{\overline{[i]}}) \neq b. \tag{3}$$

On the other hand, from the privacy of $\mathcal{A}$ with respect to player P_i and by (2) we obtain that there exist random strings $R'_1, \ldots, R'_{i-1}, R'_{i+1}, \ldots, R'_n$ such that $Y[i] = X[i]$ implies $C_i(Y, (R'_1, \ldots, R'_{i-1}, R_X[i], R'_{i+1}, \ldots, R'_n)) = C_i(X, R_X)$. Because of equality (1) the descriptions $C(Y, (R'_1, .., R'_{i-1}, R_X[i], R'_{i+1}, .., R'_n))$ and $C(Y^{\overline{[i]}}, (R'_1, .. R'_{i-1}, R_Z[i], R'_{i+1}, .., R'_n))$ are equal. Note that in both cases P_i exchanges the same messages with the players in $\mathcal{U}$. In case $Y[i] = Z[i]$ one can deduce the same property similarly. Therefore, in both cases the assumptions of Lemma 1 are fulfilled for Y, $Y^{\overline{[i]}}$ and the appropriate random strings. One can conclude that $f(Y) = f(Y^{\overline{[i]}})$, but this contradicts (3). ∎

Corollary 1. *For every $k \leq n/2$ there exists a k-connected network such that no function f can be computed by an $(\frac{\mathsf{s}(f)-2}{k-1} - 2)$-random private protocol. In particular, $\mathtt{OR}_n, \mathtt{AND}_n, \mathtt{XOR}_n$ and $\mathtt{MAJ}_n$ require at least $\frac{n-2}{k-1} - 1$ many random bits.*

This corollary follows immediately from Theorem 1 because the graph $K_{k,n-k}$ is k-connected and has an independent set of size $n - k$.

Corollary 2. *For $m \geq n/2$ and every network with an independent set of size m, the functions $\mathtt{OR}_n, \mathtt{AND}_n, \mathtt{XOR}_n$ and $\mathtt{MAJ}_n$ cannot be computed by a $(\frac{n-2}{n-m-1} - 2)$-random private protocol.*

4   An Almost Optimal Protocol for k-Connected Networks

In this section we show how to compute certain symmetric Boolean functions privately – including the well known ones considered above – in k-connected networks that use an almost optimal number of random bits.

Definition 4. *A symmetric Boolean function f with inputs $X = X[1] \ldots X[n]$ will be called **efficiently computable from counter representation** if given the binary representation of the number of ones in X, that is $N_X := \sum_{i=1}^{n} X[i]$ there exists a circuit of linear size with respect to the length of input N_X, that means of size $O(\log n)$, to compute the function value $f(X)$. Let $\mathcal{E}$ be the set of all Boolean functions that are efficiently computable from counter representation.*

Obviously, functions like $\mathtt{AND}_n, \mathtt{OR}_n, \mathtt{XOR}_n$ or τ_n^b have this property.

Definition 5. *For a graph $G = (V, E)$ let $\Pi_1, \ldots, \Pi_m$ with $\Pi_i = (V_i, E_i)$ for all $i \in \{1, .., m\}$ be a sequence of simple paths covering all nodes of G, i.e. $V = \bigcup_{i \in \{1,..,m\}} V_i$. For a path Π_i let s_i, t_i denote the terminal nodes of Π_i, i.e. s_i and t_i have degree one in Π_i. We call $\Pi_1, \ldots, \Pi_m$ an m-ear cover of G iff for all $i < m$ $s_i, t_i \in \bigcup_{i < j \leq m} V_j$. G is called m-ear coverable iff there exists an m-ear cover of G.*

Egawa, Glas, and Locke [9] have shown, that every k-connected graph G with at least $2d$ vertices and with minimum degree at least d has a cycle of length at least $2d$ through any specified set of k vertices. Using this result and Dirac's Theorem, which ensures the existence of a Hamiltonian cycle in every graph with $n \geq 3$ vertices and minimum degree at least $n/2$, we get the following

Proposition 1. *Let $G = (V, E)$ be k-connected, then for every $\mathcal{V} \subseteq V$ with $|\mathcal{V}| \leq k + 1$ there exists a simple path containing all nodes in $\mathcal{V}$.*

Hence if a graph fulfills the conditions of the Proposition above then it is $\lceil \frac{|V|}{k-1} \rceil$-ear coverable. Let G be an m-ear coverable 2-connected graph, $\Pi_1, \ldots, \Pi_m$ be an m-ear cover of G, and $f \in \mathcal{E}$. The protocol works in two stages. The first stage is based on a protocol for modular addition [7]:

Stage 1. Mark all nodes in G red and for each player P_i set $Z[i] := X[i]$. Then repeat the following step m times for $i = 1, \ldots, m$.

Let P_s, P_t be the first and the last player on path Π_i. We assume that Π_i has at least three red nodes and that P_s and P_t are among the red nodes of Π_i. Then the first player P_s chooses a random number $R_i < 2^{\lceil \log n \rceil}$, computes $(R_i + Z[s]) \bmod 2^{\lceil \log n \rceil}$, and sends the result to the next player in the path. Finally, P_s sets $Z[s] := -R_i \bmod 2^{\lceil \log n \rceil}$.

Each internal player P_j of the path Π_i receives a value w from his predecessor on the path. If P_j is a black player it sends w to his successor. If P_j is a red player it computes $(w + Z[j]) \bmod 2^{\lceil \log n \rceil}$, sends this value to his successor, and changes his colour to black. The last player P_ℓ on the path receives w from his predecessor and computes $Z[\ell] := (Z[\ell] + w) \bmod 2^{\lceil \log n \rceil}$.

From the definition of an m-ear-cover it follows that after this stage is completed, exactly 2 players remain red. W.l.o.g. let us assume that P_1, P_2 are the remaining red players. It is easy to check that $Z[1] + Z[2]$ equals

$$\left(\sum_{i=1}^{n} X[i] + \sum_{j=1}^{m} R_j - \sum_{j=1}^{m} R_j \right) \bmod 2^{\lceil \log n \rceil} = \sum_{i=1}^{n} X[i] = N_X .$$

Stage 2. Now we will use the fact that f has a small circuit if N_X is provided as input. Thus, $f(X)$ can be computed from $(Z[1], Z[2])$ by a circuit C_f' of size $O(\log n)$. To simulate the computation of C_f' we use a protocol similar to the one for directed cycles to be presented in section 6.2. We obtain

Theorem 2. *An arbitrary m-ear coverable 2-connected network can privately compute every function $f \in \mathcal{E}$ using at most $(m + O(1)) \cdot \log n$ random bits.*

Corollary 3. *In every k-connected network $\mathtt{OR}_n, \mathtt{AND}_n, \mathtt{XOR}_n, \mathtt{MAJ}_n$ and all n-ary threshold functions can be computed by a $((\lceil \frac{n}{k-1} \rceil + O(1)) \cdot \log n)$-random private protocol.*

5 Bounding the Number of Random Players

Now let us establish an upper bound on the number of players which can toss coins during a private computation. Let r be the maximum number of random bits used by a protocol $\mathcal{A}$. Recall that for an input $X \in \{0,1\}^n$ by $R(X)$ we denote the set of random strings $(R_1, \ldots, R_n)$ that can occur in a valid computation of $\mathcal{A}$ on X. For each $(R_1, \ldots, R_n) \in R(X)$ it holds $\sum_{i \in \{1,\ldots,n\}} |R_i| \leq r$, i.e. there are at most r players that toss coins. Note that for a fixed input X the protocol $\mathcal{A}$ decides in a deterministic way in which round and which players should make use of their random strings first. Moreover, the decision which group of players toss coins next and in which round, depend only on the results of the initial coin-tossing round, the indices of players who toss coins in the third coin-tossing round depends only on the results of the first and the second coin-tossing rounds and so on. Hence, the number of different players that can toss coins on a valid computation of $\mathcal{A}$ on input X is bounded by $2^r - 1$.

Let $\mathcal{P}_{\mathcal{A}}(X)$ be the set of random players, i.e. the set of those players who toss coins during the computation of $\mathcal{A}$ on X. For two sets $I_1, I_2 \subseteq \{1,..,n\}$ denote by $I_1 \Delta I_2 := (I_1 \cup I_2) \setminus (I_1 \cap I_2)$ the symmetric difference between I_1 and I_2.

Lemma 2. For all $f : \{0,1\}^n \to \{0,1\}$ and for every private protocol $\mathcal{A}$ for f it holds: $\forall X, Y \in \{0,1\}^n \; \forall i \in \mathcal{P}_{\mathcal{A}}(X) \Delta \mathcal{P}_{\mathcal{A}}(Y) \; f(X) = f(Y) \Rightarrow X[i] \neq Y[i]$.

Using this lemma we can show:

Theorem 3. For every r-random private protocol $\mathcal{A}$ computing the OR-function OR_n it holds: $\left| \bigcup_{X \in \{0,1\}^n} \mathcal{P}_{\mathcal{A}}(X) \right| \leq 2^{r+1} - 2$.

Let $\mathcal{A}$ be an r-random private protocol computing τ_n^b. Then for any set $I \subseteq \{1,..,n\}$ with $|I| = b - 1$ it holds $\left| \bigcup_{X \in \{0,1\}^n} \mathcal{P}_{\mathcal{A}}(X \lceil_{I \leftarrow 1^{b-1}}) \right| \leq 2^{r+1} - 2$.

Theorem 4. Let $\mathcal{A}$ be an r-random private protocol computing a Boolean function f. Then there exists a set $I \subseteq \{1,..,n\}$ with $|I| = 2^{r+1} - 2$ and a string $\alpha \in \{0,1\}^{|I|}$ such that $\bigcup_{X \in \{0,1\}^n} \mathcal{P}_{\mathcal{A}}(X \lceil_{I \leftarrow \alpha}) \subseteq I$.

Theorem 5. Let $f(x) := ((\sum_{i \in \{1,..,n\}} x[i]) \bmod b) \circ a$, be a function defined for integers $a \leq n - 1$, $b \geq 2$, and a relation $\circ \in \{=, <, >\}$. Then for every r-random private protocol $\mathcal{A}$ for f it holds $\left| \bigcup_{x \in \{0,1\}^n} \mathcal{P}_{\mathcal{A}}(X) \right| \leq 2^{r+1} + b - 3$.

6 Private Computation in Directed Cycles

In this section we consider private computations in networks with an extremely simple topology – directed cycles. First, we will estimate the number of random players during the computation of a threshold function $\tau_n^b(X)$ (the input X has to contain at least b ones, where b is some number between 1 and n). By duality when negating the input bits we may assume $1 \leq b \leq \lceil \frac{n}{2} \rceil$. Theorem 3 provides an upper bound for this number for arbitrary networks. Below we will prove a nontrivial lower bound on the number of random players. Next, an algorithmic upper bound for the number of random bits for some specific functions will be given. Finally, a sharp lower bound on the number of random bits will be shown.

6.1   One Random Player Is Not Sufficient for Threshold Functions

Assume that for τ_n^b a private protocol $\mathcal{A}$ with a single random player – w.l.o.g this will be P_1 – does exist. For an input W and a fixed random string R of P_1 we denote by $C_i(X, R)$ the complete message pattern between player P_i and P_{i+1}. Now consider the inputs: $X := 0\ 1^{b-1}\ 1\ 0^{n-b-1}$, $Y := 0\ 1^{b-1}\ 1\ 1^{n-b-1}$, and $Z := 0\ 1^{b-1}\ 0\ 1^{n-b-1}$. Note that $\tau_n^b(X) = \tau_n^b(Y) = \tau_n^b(Z) = 1$. Let R be a fixed random string. Then from the privacy of $\mathcal{A}$ it follows that $C_n(X, R) = C_n(Y, R)$ – if not then P_1 gets some additional knowledge about the input. Because P_1 is the only random player and P_i, with $2 \leq i \leq n$, generates the outgoing message pattern deterministically depending on his incoming message pattern and the input bit one can show that $C_{b+1}(X, R) = C_{b+1}(Y, R)$. On the other hand there exists random a string R' such that $C_{b+1}(Y, R) = C_{b+1}(Z, R')$ and $C_n(Y, R) = C_n(Z, R')$. The first equation follows from the privacy of $\mathcal{A}$, the second from the fact that $P_2, \ldots, P_n$ work deterministically. Hence, one can conclude: $C_{b+1}(X, R) = C_{b+1}(Y, R) = C_{b+1}(Z, R')$ and $C_n(X, R) = C_n(Y, R) = C_n(Z, R')$. This means that on input $0\ 1^{b-1}\ 0\ 0^{n-b-1}$ and with random string R' the players $P_1, \ldots, P_{b+1}$ perform the same computation as on input Z with random string R' and that the players $P_{b+2}, \ldots, P_n$ work in the same way as on input X with R. This leads to a contradiction because P_1 with R' does not distinguish between Z and $0\ 1^{b-1}\ 0\ 0^{n-b-1}$.

Theorem 6. *Threshold functions cannot be computed privately in directed cycles if only a single random player is available regardless how many random bits are used.*

Note that this claim is not true for XOR_n. This function can be computed privately with a single random player that even uses just one random bit.

6.2   A Logarithmic Upper Bound for the Number of Random Bits

Now again we will investigate symmetric functions that are efficiently computable from counter representation, a class containing in particular all threshold functions. To compute such a function f privately we will design a protocol that works in two stages. In the first stage player P_1 chooses a random number $R \leq 2^{\lceil \log n \rceil}$ and by sending a message along the cycle the players P_i with $i \leq n-1$ compute successively the values $s_i := (\sum_{j \leq i} X[j]) \bmod 2^{\lceil \log n \rceil}$. This stage stops after the value s_{n-1} is computed. Recall, that f can now be computed by $C_f(s_{n-1} + X[n] - R)$. By assumption, the result of this function can be obtained for input $s_{n-1}, X[n], R$ by a circuit C'_f of size linear in the length of $s_{n-1}, X[n], R$.

To simulate the computation of C'_f we will use a standard technique for the simulation of one gate (see [13]). Such a simulation can be done in the complete graph K_4 by using a constant number of random bits for each gate. Analyzing this simulation one can see that C'_f can be evaluated by an $O(\log n)$-random private protocol using $O(\log n)$ bits of communication. It remains to show how one can simulate this protocol on a directed cycle.

To simulate the transmission of a single bit, we proceed as follows: Whenever a player P_i has to send a bit b to a player $P_j \neq P_{(i \bmod n)+1}$ the receiver P_j generates a random bit R' first and sends it to P_i. P_i computes $b \oplus R'$ and sends this bit to P_j. Finally, P_j computes $b = b \oplus R' \oplus R'$. Note that all other players only see a random bit. Thus, this protocol is private and uses $O(\log n)$ additional random bits. It can be modified such that in case of an undirected cycle only one random player is necessary .

Theorem 7. *Every function $f \in \mathcal{E}$ can be computed privately on a directed cycle with 4 random players using $O(\log n)$ random bits. For an undirected cycle only one random player is necessary.*

Corollary 4. *Threshold functions can be computed privately on a directed cycle with 4 random players using $O(\log n)$ random bits.*

A more general protocol for simulating the transmission of a single bit on a strongly 1-connected and weakly 2-connected network has been presented in [10] by Franklin and Yung. One can obtain our protocol for a directed cycle from that protocol by using only the necessary random bits and combining the messages.

6.3   A Sharp Lower Bound for Threshold Functions

In the following we will show a lower bound on the number of random bits.

Theorem 8. *Threshold functions cannot be computed privately on a directed cycle with less than $\frac{\log n}{2+\varepsilon}$ random bits for any constant $\varepsilon > 0$.*

Below we sketch the proof. For the contrary, let us assume that there exists a private protocol for τ_n^b with $r \leq \frac{\log n}{2+\varepsilon}$ random bits for some constant $\varepsilon > 0$. From Theorem 3 and the pigeon hole principle it follows, that there exists a sequence $P_a, \ldots, P_{a+\sqrt{n}}$ with $b < a \leq n - \sqrt{n}$ such that none of these players uses a random string on any input. W.l.o.g. we assume that $a = n - \sqrt{n}$. Analogously to subsection 6.1 we will restrict the inputs X such that $X[1] = \ldots = X[b-1] = 1$ and $X[b] = \ldots = X[a-1] = 0$. Let $\mathcal{Y}$ be the set of all inputs fulfilling this condition.

Let us now investigate the communication phases of a protocol. We define **phase** t recursively. For phase t let us assume that we can artificially delay all messages on (P_n, P_1) from the beginning of phase t to the end of this phase such that P_1 receives no message from P_n that is sent in this phase. All other edges have no delay.

- Phase 0 starts at the beginning of the computation and ends when no more further messages can be sent by any player. Let $C_i^0(X, R)$ be the message pattern sent by P_i in phase 0.
- Phase t starts directly after the end of phase $t-1$. Assume now every player P_i knows the message pattern sent by his direct predecessor in all previous phases $j < t$. Phase t ends when no more further messages can be sent by any player. Let $C_i^t(X, R)$ be the message pattern sent by P_i in phase t.

For easier notion let $C_i^{\leq t}(X, R)$ be the concatenation of $C_i^1(X, R), \ldots, C_i^t(X, R)$

and call $C_i^t(X, R)$ and $C_i^{\leq t}(X, R)$ **partial message patterns**. Instead of analyzing the complete message pattern between two players at once we will study the partial message pattern phase by phase. Let us call a partial message pattern $C_i^{\leq t}(X, R)$ with $X \in \mathcal{Y}$ **univalent** if for all $Y \in \mathcal{Y}$ with $\tau_b(X) \neq \tau_b(Y)$ and for all random string R' it holds $C_i^{\leq t}(X, R) \neq C_i^{\leq t}(Y, R')$. Otherwise, $C_i^{\leq t}(X, R)$ is called **bivalent**. In a univalent message pattern the final outcome of the protocol basically has been determined and can be observed by P_i and P_{i+1}.

Lemma 3. *Let t be a phase number such that for any random string R and any $X \in \mathcal{Y}$ the partial message pattern $C_{a-1}^{\leq t}(X, R)$ is bivalent. Then for all $Y \in \mathcal{Y}$ there exists a random string R_Y such that $C_{a-1}^{\leq t}(X, R) = C_{a-1}^{\leq t}(Y, R_Y)$. Furthermore, for all $i \in \{a, .., n-1\}$ the message pattern $C_i^{\leq t}(Y, R_Y)$ is bivalent, too.*

Till the end of this section we will show that whenever a partial message pattern $C_{a-1}^{\leq t+1}(X, R)$ becomes univalent the first time, that means $C_{a-1}^{\leq t}(X, R)$ is still bivalent, then player P_a can deduce some information about the input bits. Thus, the protocol would not be private.

Using the upper bound of the number of different message patterns in [15] one can show that the number of different partial message patterns after phase t is at most $2^{r+2} \in o(\sqrt{n})$. On the other hand, the length of the interval $\{a, .., n\}$ is $\sqrt{n}$ Hence, for sufficiently large n the number of players in the interval $P_a, \ldots, P_n$ is larger then the number of different partial message patterns $C_{a-1}^{\leq t}(X, R)$.

Let us now focus on inputs $X, Y \in \mathcal{Y}$ where $X[i] = 0$ and $Y[i] = 1$ for $a \leq i \leq n$. Furthermore, let R_X be a random string. and choose t such that $C_{a-1}^{\leq t}(X, R_X)$ is bivalent. By the pigeon hole principle, there exists a random string R_Y such that the sequences $C_a^{\leq t}(X, R_X), C_{a+1}^{\leq t}(X, R_X), \ldots, C_n^{\leq t}(X, R_X)$ and $C_a^{\leq t}(Y, R_Y), C_{a+1}^{\leq t}(Y, R_Y), \ldots, C_n^{\leq t}(Y, R_Y)$ have at least two partial message patterns $C_j^{\leq t}(X, R_X)$ and $C_k^{\leq t}(X, R_X)$ with $a \leq j < k \leq n$ in common.

Consider the input $Z \in \mathcal{Y}$ with $Z[a] = \ldots = Z[j-1] = 0$, $Z[j] = \ldots = Z[k] = 1$, and $Z[k+1] = \ldots = Z[n] = 0$. Since, the outgoing partial message pattern of each player P_i with $i \in \{a, .., n\}$ only depends on the actual incoming partial message pattern and the input bit of the player, we can conclude for the sequence $C_a^{\leq t}(Z, R_X), C_{a+1}^{\leq t}(Z, R_X), \ldots, C_n^{\leq t}(Z, R_X)$ that $C_i^{\leq t}(Z, R_X) = C_i^{\leq t}(X, R_X)$ for all $i \in \{a, .., j\} \cup \{k, .., n\}$ and $C_i^{\leq t}(Z, R_X) = C_i^{\leq t}(Y, R_Y)$ for all $i \in \{j, .., k\}$. It follows that the inputs X and Z are indistinguishable for the players $P_1, \ldots, P_{a-1}$.

If $C_{a-1}^{\leq t}(X, R_X)$ is bivalent and $C_{a-1}^{\leq t+1}(X, R_X)$ univalent then the protocol does not compute the threshold function correctly. It cannot distinguish between X and Z if the random string turns out to be R_X. Theorem 8 follows directly.

7   Conclusions and Open Problems

In this paper we have investigated the relationship between the topology of a network and the necessary randomness for private computations. One of our main results shows the $\frac{m-1}{n-m-1}$ lower bound for the number of random bits needed to compute some basic Boolean functions like e.g. OR_n, AND_n, XOR_n and MAJ_n in any network with an independent set of size $m \geq n/2$. Hence to compute these functions privately in some specific k-connected networks like the $K_{k,n-k}$ one needs at least $\frac{n-2}{k-1} - 1$ random bits. On the other hand, we have presented an $O(\frac{n}{k-1} \log n)$ random private protocol that computes these functions in arbitrary k-connected network. We leave as an open problem to close the gap.

Finally, we have considered networks with a simple topology, namely directed cycles – the token ring being a popular example – and proved a sharp $\log n$ lower bound on the number of random bits for threshold functions. Does the same bound hold for undirected cycles?

References

1. J. Bar-Ilan, D. Beaver, *Non-Cryptographic Fault-Tolerant Computing in Constant Number of Rounds of Interaction*, Proc. 8. PODC, 1989, 201–209.
2. M. Bläser, A. Jakoby, M. Liśkiewicz, and B. Siebert, *Private Computation – k-connected versus 1-connected Networks*, Proc. 22. CRYPTO, 2002, 194–209.
3. C. Blundo, A. De Santis, G. Persiano, U. Vaccaro, *On the Number of Random Bits in Totally Private Computation*, Proc. 22. ICALP, 1995, 171–182.
4. M. Ben-Or, S. Goldwasser, A. Wigderson, *Completeness Theorem for Non crypto-graphic Fault-tolerant Distributed Computing*, Proc. 20. STOC, 1988, 1–10.
5. D. Chaum, C. Crépeau, I. Damgård, *Multiparty unconditionally secure protocols*, Proc. 20. STOC, 1988, 11–19.
6. B. Chor, M. Geréb-Graus, E. Kushilevitz, *Private Computations Over the Integers*, SIAM J. Computing 24, 1995, 376–386.
7. B. Chor, E. Kushilevitz, *A Communication-Privacy Tradeoff for Modular Addition*, Information Processing Letters 45, 1993, 205–210.
8. R. Canetti, R. Ostrovsky, *Secure Computation with Honest-Looking Parties: What if nobody is truly honest?*, Proc. 31. STOC, 1999, 35–44.
9. Y. Egawa, R. Glas, S.C. Locke, *Cycles and paths through specified vertices in k-connected graphs*, Journal of Combinatorial Theory Series B 52, 1991, 20–29.
10. M. Franklin, M. Yung, *Secure hypergraphs: privacy from partial broadcast (Extended Abstract)*, Proc. 27. STOC, 1995, 36–44.
11. A. Gál, A. Rosén, *A Theorem on Sensitivity and Applications in Private Computation*, Proc. 31. STOC, 1999, 348–357.
12. O. Goldreich, S. Micali, A. Wigderson, *How to Play any Mental Game or a Completeness Theorem for Protocols with Honest Majority*, 28. FOCS, 1987, 218–229.
13. E. Kushilevitz, R. Ostrovsky, A. Rosén, *Characterizing Linear Size Circuits in Terms of Privacy*, Proc. 28. STOC, 1996, 541–550.
14. E. Kushilevitz, Y. Mansour, *Randomness in Private Computations*, SIAM J. Discrete Math 10, 1997, 647–661.
15. E. Kushilevitz, A. Rosén, *A Randomness-Rounds Tradeoff in Private Computation*, SIAM J. Discrete Math 11 , 1998, 61–80.
16. E. Kushilevitz, *Privacy and Communication Complexity*, SIAM J. Discrete Math 5, 1992, 273–284.
17. A. C. Yao, *Protocols for Secure Computations*, Proc. 23. FOCS, 1982, 160–164.
18. A. C. Yao, *How to generate and exchange secrets*, Proc. 27. FOCS, 1986, 162–167.

On Shortest-Path All-Optical Networks without Wavelength Conversion Requirements[★]

Thomas Erlebach and Stamatis Stefanakos

Computer Engineering and Networks Laboratory (TIK)
ETH Zürich, CH-8092 Zürich, Switzerland
{erlebach,stefanak}@tik.ee.ethz.ch

Abstract. In all-optical networks with wavelength-division multiplexing, every connection is routed along a certain path and assigned a wavelength such that no two connections use the same wavelength on the same link. For a given set $\mathcal{P}$ of paths (a routing), let $\chi(\mathcal{P})$ denote the minimum number of wavelengths in a valid wavelength assignment and let $L(P)$ denote the maximum link load. We always have $L(\mathcal{P}) \leq \chi(\mathcal{P})$. Motivated by practical concerns, we consider routings containing only shortest paths. We give a complete characterization of undirected networks for which any set $\mathcal{P}$ of shortest paths admits a wavelength assignment with $L(\mathcal{P})$ wavelengths. These are exactly the networks that do not benefit from the use of (expensive) wavelength converters if shortest-path routing is used. We also give an efficient algorithm for computing a wavelength assignment with $L(\mathcal{P})$ wavelengths in these networks.

1 Introduction

In all-optical networks that employ wavelength-division multiplexing, a connection is established by first choosing a path from the sender to the receiver and then assigning a wavelength for that connection to all the links of the path. The wavelength assignment has to be done so that no two connections that share a link are transmitted through the same wavelength. Since the number of available wavelengths is limited, one is interested in minimizing the number of utilized wavelengths for a given set of connections. One technique that helps cut down the number of necessary wavelengths for operating a network is that of wavelength conversion. A wavelength converter is placed in some node of the network and has the ability of altering the transmitting wavelength of any incoming signal. An interesting algorithmic problem that arises then is that of placing as few converters as possible in suitable positions of the network in order to optimize its capacity. In this paper we focus on networks where connections are always established through shortest paths and we characterize the networks that do not profit from the use of converters.

The network can be naturally modeled by a graph $G = (V, E)$, a connection in the network can be seen as a path on G, and wavelengths can be regarded as colors.

[★] Research partially supported by the Swiss National Science Foundation under Contract No. 21-63563.00 (Project AAPCN) and the EU Thematic Network APPOL II (IST-2001-32007), with funding provided by the Swiss Federal Office for Education and Science (BBW).

H. Alt and M. Habib (Eds.): STACS 2003, LNCS 2607, pp. 133–144, 2003.
© Springer-Verlag Berlin Heidelberg 2003

If the network does not employ wavelength conversion then a wavelength assignment for a set $\mathcal{P}$ of paths is an assignment of a color to each path in $\mathcal{P}$. A valid coloring is one in which no two paths that use the same edge get assigned the same color. For a given set $\mathcal{P}$ of paths (a routing) we denote by $\chi(\mathcal{P})$ the minimum number of colors needed for a valid coloring of $\mathcal{P}$. A trivial lower bound for $\chi(\mathcal{P})$ is the congestion or load $L(\mathcal{P}) = \max_{e \in E} L_e(\mathcal{P})$ of the network, where $L_e(\mathcal{P})$ is the number of paths in $\mathcal{P}$ that use e. If the network has wavelength converters a coloring is an assignment of a color to every edge of each path. In this case, a valid coloring has to satisfy the additional constraint that the color assignments to two consecutive edges of a path can only differ if there is a converter between the two edges.

Obviously, even with the use of converters, we will always need at least $L(\mathcal{P})$ colors for a valid coloring of $\mathcal{P}$. However, the placement of converters in a network can reduce the number of wavelengths needed for a fixed set of connections. It can be the case, for example, that while there is no single wavelength, from the already used ones, available along a path there are different available wavelengths along parts of that path. If there are converters in suitable positions along that path then these different wavelengths can be exploited in order to serve the connection which otherwise would have had to be assigned a new wavelength.

Ideally, by placing converters in some nodes of the network we can guarantee that any routing $\mathcal{P}$ can be accommodated with $L(\mathcal{P})$ wavelengths. This is easily seen to be the case if all nodes of the network are equipped with wavelength converters. However, the cost of these devices is prohibitive for such improvident use. Therefore, the network designer is typically interested in equipping only a small subset of the network nodes with converters, while still achieving the same capacity usage as if there where converters everywhere.

Motivated by this, Wilfong and Winkler [11] introduced the MINIMUM SUFFICIENT SET problem: given a graph $G = (V, E)$, find a sufficient set S for G, i.e., a set $S \subseteq V$ such that any set $\mathcal{P}$ of paths on G can be colored with $L(\mathcal{P})$ colors if we place wavelength converters on the vertices of S; the goal is to minimize the size of S. Wilfong and Winkler [11] proved that MINIMUM SUFFICIENT SET is $\mathcal{N}\mathcal{P}$-hard even for planar bidirected graphs (a bidirected graph is a directed graph where $(u, v) \in E \Rightarrow (v, u) \in E$). Moreover they showed that the only bidirected graphs that admit the empty sufficient set are spiders, i.e., trees with at most one vertex of degree greater than two, and that rings admit a sufficient set of size 1. Finally, they described an efficient way of determining whether a set S is sufficient for a bidirected graph G: one modifies G by "exploding" each node $s \in S$ into degree-of-s-many copies, each of which is made adjacent to one of the old neighbors of s. S is sufficient for G if and only if every component of the graph obtained after this modification is a spider.

Extending this work, Kleinberg and Kumar [7] gave a 2-approximation algorithm for directed graphs and a polynomial time approximation scheme for directed planar graphs using techniques based on the undirected feedback vertex set problem. They also showed that any improvement on the approximation ratio for MINIMUM SUFFICIENT SET on bidirected graphs would lead to a corresponding improvement for vertex cover. The approach of Kleinberg and Kumar can be extended to give a linear time algorithm for MINIMUM SUFFICIENT SET in directed graphs of bounded treewidth [5].

Our Contribution. As described above, the previous work concentrated on the study of MINIMUM SUFFICIENT SET in bidirected or directed graphs. These graphs serve as models for networks with unidirectional links, i.e., networks that support only one-way communication. In this paper, we turn to undirected graphs. Undirected graphs model networks where the physical links are bidirectional or networks with unidirectional links with the additional property that whenever a connection is established in one way, the reverse connection must also be established through the same path and must be transmitted over the same wavelength.

It is easy to see that the only undirected graphs that admit the empty sufficient set are chains. Furthermore, using the technique developed by Wilfong and Winkler, one can show that MINIMUM SUFFICIENT SET is polynomial in undirected graphs: we simply have to place a converter in every node of degree greater than or equal to 3 or in any single node if the graph is a cycle (for more details see [5]). Nevertheless, such placement of converters is not satisfying. For example, in the case where the network is a clique we will need to place a converter in every node; however, it is unlikely that we would need any converter at all in practice since in a clique most connections would be carried over a single link.

In order to capture this real-world scenario we restrict ourselves to shortest-path routings, i.e., we are interested in placing as few converters as possible so that any set $\mathcal{P}$ of shortest paths can be colored with $L(\mathcal{P})$ colors. More formally, we introduce the MINIMUM SP-SUFFICIENT SET problem: given a graph $G = (V, E)$, find an SP-sufficient set S for G, i.e., a set $S \subseteq V$ such that any set $\mathcal{P}$ of shortest paths on G can be colored with $L(\mathcal{P})$ colors if we place wavelength converters on the vertices of S; the goal is to minimize the size of S. We note that this problem is of significant practical importance since shortest-path routing is a common strategy in optical networks, see for example [12]. Moreover, it imposes a weaker condition than MINIMUM SUFFICIENT SET (any sufficient set is an SP-sufficient set) and therefore can help decrease the cost of converters in the design of optical networks.

In this paper we give a complete characterization of the undirected networks that admit the empty SP-sufficient set. We show that the block graph of such networks is a chain and all internal blocks are cliques. The outer blocks are a special case of co-bipartite graphs. If the graph consists of only one block we show that its diameter is less than or equal to 3. For the case of diameter 3 we prove that the graph belongs to a special class of co-bipartite graphs. For the case where the diameter is less than 3 we show that the graph admits the empty SP-sufficient set if and only if its edges can be 2-colored in a certain way. In all cases our proofs provide efficient algorithms for recognizing graphs that admit the empty SP-sufficient set and for optimally coloring shortest-path routings on these graphs. An interesting aspect of our work is that we do not need to consider complicated routings in order to obtain the characterization: all graphs that require a converter admit a witness routing $\mathcal{P}$ with $L(\mathcal{P}) = 2$ and $\chi(\mathcal{P}) = 3$ whose conflict graph is an odd cycle.

Other Related Work. Perhaps the main algorithmic problem that arises in optical networks is that of routing and wavelength assignment, i.e., given a set of connections in a network find a routing and a valid coloring of that routing so that the number of colors is minimized. Most research has focused on the case where the network does not have any conversion capabilities. In that case the problem is known to be $\mathcal{NP}$-hard even for simple topologies like rings and trees [2,11], and in the case of rings, even if the routing is part of the input [6]. Consequently, a lot of effort has been put in designing approximation algorithms for specific network topologies, see e.g., [10,9,3]. There has been another line of research on wavelength converters of bounded degree. A converter of bounded degree does not have full conversion capabilities, i.e., it can transform color i to only a few other colors. We refer the reader to [1] and the references therein.

Outline. The rest of the paper is structured as follows. In the following section we provide the necessary notation and terminology. In Section 3 we characterize graphs that do not need converters and contain cut-vertices while in Section 4 we turn to biconnected graphs. Finally, in Section 5 we summarize our results and discuss future work.

2 Preliminaries

Throughout this paper, a graph $G = (V, E)$ is finite, simple, and undirected unless explicitly stated otherwise. For a graph G we will denote by $V(G), E(G)$ its vertex-set and edge-set respectively. In some cases, in order to simplify notation, we will write $v \in G$ instead of $v \in V(G)$. For $U \subseteq V$, $G[U]$ denotes the graph on U whose edges are the edges of G with both endpoints in U, i.e., $G[U]$ is the subgraph of G *induced* by U. The *distance* $d(u, v)$ in G of two vertices u, v is the number of edges in a shortest $u - v$ path in G (if no such path exists then $d(u, v) := \infty$). The *eccentricity* of a vertex v in G, $\mathrm{ecc}(v) = \max_{u \in V} d(u, v)$, is the maximum distance of v to all other vertices of G. The *diameter* of G, $\mathrm{diam}(G) = \max_{v \in V} \mathrm{ecc}(v)$, is the maximum eccentricity over all vertices of G. The degree, $\deg(v)$, of a vertex v in G is the number of edges incident to v. We denote by $\Delta(G) = \max_{v \in V} \deg(v)$ the maximum degree of G. We denote an induced cycle with k edges and k vertices by C^k and an induced path with k vertices and $k - 1$ edges by P^k. A maximal connected subgraph of G without a cut-vertex is a *block* of G. The *block graph* of G is the bipartite graph on $A \cup B$ and edges aB for $a \in A$ and $B \in \mathcal{B}$ if $a \in B$, where A is the set of cut-vertices of G and $\mathcal{B}$ is the set of blocks of G. The *conflict graph* of a given set of paths is the graph with one vertex for each path and an edge between two vertices if the corresponding paths share an edge.

Obviously, a graph that does not need converters can not contain a configuration that allows us to construct a set $\mathcal{P}$ of shortest paths with $L(\mathcal{P}) < \chi(\mathcal{P})$. The configurations of this type that we will mainly use in this paper are shown in Fig. 1. In the text, in order to exhibit such a configuration we will refer to the vertices that induce the configuration; for example, we say that we have an antenna around $uvzwyx$ (Fig. 1(b)). For the case of the claw we say that we have a claw *at* x. All these configurations allow us to construct a set of shortest paths with load 2 whose conflict graph is an odd cycle and hence require 3 colors for a valid coloring. For example, in the case of the tent we can take $v_1 v_2 u$, $v_2 u v_5$, $u v_5 v_6$, $v_1 v_2 v_3$, $v_2 v_3 v_4$, $v_3 v_4 v_5$, and $v_4 v_5 v_6$. Notice that for some of the configurations shown in Fig. 1, in order to find such a set of paths, we need to be able to find a shortest path of length 3 on them. This is the case for the net, the antenna and the satellite. For the latter for example, we can take the following shortest paths of length 2: uvw, vwx, wxy, yxz, xzv and zvu. In order to have an odd number of paths we need $d(u, x) = 3$ or $d(v, y) = 3$ so that we can exchange two paths of length 2 with one of length 3. Other

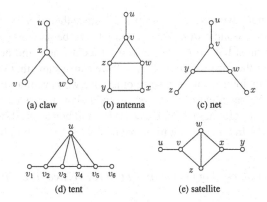

Fig. 1. Some configurations that allow the construction of a set $\mathcal{P}$ of shortest paths with $L(\mathcal{P}) < \chi(\mathcal{P})$.

configurations which allow us to construct a set $\mathcal{P}$ of shortest paths with $L(\mathcal{P}) < \chi(\mathcal{P})$ are induced cycles of length greater than or equal to 5. On an induced odd cycle we can just take all paths of length two. On an induced even cycle the set of all paths of length two has even cardinality and thus we need one path of length three.

Some of the proofs are omitted from this version due to lack of space. All omitted proofs can be found in [4].

3   Graphs with Cut-Vertices

First, assume that $G = (V, E)$ is connected, but not biconnected, and admits the empty SP-sufficient set.

Lemma 1. *The block graph of G is a chain.*

Lemma 2. *Let C be a block of G containing a cut-vertex x. Then the following hold:*

(i)   For all $u \in C$, $d(u,x) \leq 2$.
(ii) For all $u, v \in C$ such that u, v are not cut-vertices, $d(u,v) \leq 2$.

Proof. (i) (sketch) Let $u \in C$ and assume $d(u,x) > 2$. Let $x a_1 \ldots a_k u$, $x b_1 \ldots b_l u$ be two disjoint $x - u$ paths, each of length at least 3 such that $k + l$ is minimum, and let x' be a neighbor of x outside C. Notice that if there are edges $a_i b_{j'}$, $a_{i'} b_j$ with $i < i'$ and $j < j'$ then $i' = i+1$ and $j' = j+1$ since otherwise $k+l$ is not minimum. Also, if there are two such edges then we also have $a_i b_j, a_{i+1} b_{j+1} \in E$ since otherwise we would have a claw. We distinguish cases depending on the edges between a_1, b_2 and a_2, b_1. In all cases we reach a contradiction and the statement follows (case analysis is omitted).

(ii) Let u, v, be two arbitrary vertices in C that are not cut-vertices and assume to the contrary that $d(u,v) > 2$. Let x' be a neighbor of x outside C. By (i) we have that $d(x,u) \leq 2$ and $d(x,v) \leq 2$. We distinguish cases according to the distance of u, v

from x. We can not have $d(x, u) = d(x, v) = 1$ since then $d(u, v) \leq 2$, a contradiction to our assumption. Assume $d(x, v) = 2$ and let xav be an $x - v$ path of length 2. If $d(x, u) = 1$ we must have $au \in E$ because of a claw at x and hence $d(u, v) \leq 2$, a contradiction. For the case where $d(x, u) = 2$ we can assume that there exists an $x - u$ path of length 2 disjoint from xav, since otherwise we have that $d(u, v) \leq 2$. Let xbu be such a path. We have that $ab \in E$ because of a claw at x. This creates a net around $x'xbuav$ with a path of length 3 on it, say $x'xbu$, and hence we should have $au \in E$ or $bv \in E$ or $uv \in E$. In all cases we have that $d(u, v) \leq 2$, a contradiction.   □

Lemma 3. *Every block C of G containing two cut-vertices is a clique.*

Lemma 4. *Let C be a block of G containing exactly one cut-vertex x and let N_1, N_2 be the subgraphs of G induced by the neighborhoods of x in C in distance 1, 2 respectively. The following hold:*

(i) *Both N_1, N_2 are cliques.*
(ii) *There is no C^4 in C.*
(iii) *If there exists a vertex $w \in N_2$ adjacent to two vertices $u, v \in N_1$, then all vertices in N_2 are adjacent to u or to v.*

Proof. For the first part, notice that if N_1 is not a clique we get a claw at x. Consider now two vertices $u, v \in N_2$ and assume $uv \notin E$. If u, v are adjacent to the same vertex of N_1 we have a claw at that vertex. If they are adjacent to different vertices we have a net (since N_1 is a clique) with a path of length 3 on it. We proceed to show the second part. Assume there is a C^4 in C. Since N_1, N_2 are cliques the cycle should contain two vertices, say a, b from N_1 and two vertices, say c, d from N_2. Let $abcd$ be the cycle and let x' be a neighbor of x outside C. We have an antenna around $x'xabcd$ with a path of length 3 on it, a contradiction. For the last part of the lemma assume that there exists a vertex $w \in N_2$ adjacent to two vertices $u, v \in N_1$ and that there exists a vertex $b \in N_2$ adjacent neither to u nor to v. As before, let x' be a neighbor of x outside C. We have a satellite around $bwuvxx'$ with a path of length 3 on it, say $x'xuw$, a contradiction.   □

Now we are able to characterize the graphs that are not biconnected and admit the empty SP-sufficient set:

Theorem 1. *Let G be an undirected graph that is connected but not biconnected. The empty set is SP-sufficient for G if and only if the following hold:*

(i) *The block graph of G is a chain.*
(ii) *Every block of G that contains two cut-vertices is a clique.*
(iii) *Every block C of G that contains only one cut-vertex x does not contain a C^4, N_1^C and N_2^C are cliques and if there exists a vertex $w \in N_2^C$ adjacent to two vertices $u, v \in N_1^C$, then all vertices in N_2^C are adjacent to u or to v, where N_1^C, N_2^C are the subgraphs of G induced by the neighborhoods of x in C in distances 1, 2 respectively. Moreover, no vertex in C is in distance 3 from x.*

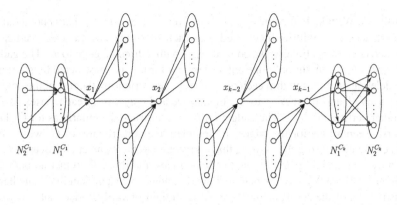

Fig. 2. Illustration of the construction used in the proof of Theorem 1.

Proof. The "only if" part is clear from the previous lemmas. For the "if" part we show how to color a set $\mathcal{P}$ of shortest paths on a given graph G that satisfies the conditions of the statement with $L(\mathcal{P})$ colors. A high level description of our approach follows. We will first modify $\mathcal{P}$ by shortening some paths and discarding some others. We will then construct a directed graph G' and a set $\mathcal{P}'$ of directed paths on G' in 1-1 correspondence with $\mathcal{P}$, with $L(\mathcal{P}') = L(\mathcal{P})$ and with the same conflict graph. We will obtain a coloring for $\mathcal{P}$ with $L(\mathcal{P})$ colors by coloring $\mathcal{P}'$ with $L(\mathcal{P}')$ colors. The coloring of $\mathcal{P}'$ will be done by computing many different local colorings that will be merged to give a single global coloring for $\mathcal{P}'$. A coloring for the initial set of paths will be obtained by coloring greedily the paths that were discarded in the first phase.

Let $C_1, \ldots, C_k$ be the blocks of G ordered so that $|C_i \cap C_{i+1}| = 1$ for all $1 \le i < k$. Let $x_1, \ldots, x_{k-1}$ be the cut-vertices of G ordered so that $C_i \cap C_{i+1} = \{x_i\}$. Let $\mathcal{P}^1 \subseteq \mathcal{P}$ be the set of all paths in $\mathcal{P}$ of length one. We remove these paths from $\mathcal{P}$. Now consider all paths that use an edge in one of $N_2^{C_1}, N_2^{C_k}$. Since we are dealing with shortest paths no such path can use an edge incident to x_1 or x_{k-1}. Therefore, all these paths are of length at most two. We claim that every edge in $N_2^{C_1}, N_2^{C_k}$ will only be used by identical paths. To see this assume that there is an edge uv in, say $N_2^{C_1}$ (the case for C_k is similar), that is used by two paths that are non-identical. Since $N_2^{C_1}$ is a clique and these paths are shortest they must go from u or v to two different vertices of $N_1^{C_1}$, say w, w'. There are two cases: either these paths are wuv and $w'vu$ or wuv and $w'uv$. Since these paths are shortest, in the first case C_1 must contain a C^4, while in the second case $u \in N_2^{C_1}$ is adjacent to both $w, w' \in N_1^{C_1}$ and there exists a vertex in $N_2^{C_1}$, namely v, that is not adjacent to any of w, w'. Both cases contradict the third condition of the statement and hence we obtain that every edge in $N_2^{C_1}, N_2^{C_k}$ will only be used by identical paths of length 2. Hence, we can shorten all such paths so that they do not use any edge in $N_2^{C_1}, N_2^{C_k}$ without modifying the conflict graph of $\mathcal{P}$. If, after this modification, we obtain any paths of length one we remove them from $\mathcal{P}$ and add them to $\mathcal{P}^1$.

Now we proceed to show how we construct G' and the set $\mathcal{P}'$ of directed paths on G' with the properties described above. The construction is illustrated in Fig. 2. For each block C_i of G containing two cut-vertices x_{i-1}, x_i we construct its corresponding gadget as follows. We take two copies v_{in}, v_{out} of every vertex v of C_i that is not a cut-vertex and connect x_{i-1} with a directed edge to each *in* vertex and each *out* vertex with a directed edge to x_i. Finally, we connect x_{i-1} with a directed edge to x_i. The gadget for the blocks that contain only one cut-vertex, say for C_1, is constructed by starting from C_1, deleting all edges within each of $N_1^{C_1}, N_2^{C_1}$ and orienting the rest of the edges from $N_2^{C_1}$ to $N_1^{C_1}$ and from $N_1^{C_1}$ to x_1. Finally we add edges from each vertex of $N_1^{C_1}$ to all other vertices of $N_1^{C_1}$. The gadget for C_k is constructed similarly as for C_1 but with reverse orientations for the edges. To complete the construction of G' we connect all gadgets by identifying the vertices that correspond to the same cut-vertices in G. Since we have removed all paths of length one and have modified $\mathcal{P}$ so that no path uses an edge in $N_2^{C_1}$ and $N_2^{C_k}$, every path in $\mathcal{P}$ has a unique correspondent in G' and hence the construction of the set of paths $\mathcal{P}'$ on G' is straightforward. Notice that every edge in $N_1^{C_1}$ and $N_1^{C_k}$ corresponds to two oppositely directed edges in G'. However, since C_1 and C_k contain no C^4, all paths that use an edge e in $N_1^{C_1}$ or $N_1^{C_k}$ correspond to paths in G' that all use the edge, corresponding to e, in the same direction and hence the conflict graph does not change.

We continue with the coloring of $\mathcal{P}'$. Let $y_1, \ldots, y_l$ be the vertices in G' that correspond to vertices in $N_1^{C_1}$ and let $z_1, \ldots, z_m$ be the vertices in G' that correspond to vertices in $N_1^{C_k}$. Define $V' := \{x_1, \ldots, x_{k-1}, y_1 \ldots, y_l, z_1, \ldots, z_m\}$. For every vertex $v \in V'$, let Q_v^{out} be the set of edges directed out of v and Q_v^{in} be the set of edges directed into v. For $1 \leq i < k$ let $\mathcal{P}_{x_i}'$ be the set of paths that touch x_i (i.e., start at x_i, end at x_i or go over x_i), for $1 \leq i \leq l$ let $\mathcal{P}_{y_i}'$ be the set of paths that go over y_i or start at y_i, and for $1 \leq i \leq m$ let $\mathcal{P}_{z_i}'$ be the set of paths that go over z_i or end at z_i. Notice that $\cup_{v \in V'} \mathcal{P}_v' = \mathcal{P}'$. We show how to color $\mathcal{P}_v'$ for all $v \in V'$ with $L(\mathcal{P}_v')$ colors. Our method is similar to the one given in [3]. We build a bipartite multigraph H_v with one vertex for each edge incident to v. The one part consists of the vertices corresponding to edges in Q_v^{out} and the other consists of the vertices corresponding to edges in Q_v^{in}. Each path in $\mathcal{P}_v'$ that goes over v uses one edge in Q_v^{in}, one in Q_v^{out} and no other edge incident to v. For each such path we add one edge in H_v connecting its *out* vertex to its *in* vertex. For every path that starts or ends at v we add a loop to the vertex of H_v that corresponds to the edge incident to v used by this path. By König's classical result [8], H_v has a proper edge-coloring with $\Delta(H_v)$ colors which we can find in polynomial time. Since two paths in $\mathcal{P}_v'$ intersect in G' if and only if they intersect in an edge incident to v this results in a valid coloring of $\mathcal{P}_v'$. Furthermore, this coloring uses $L(\mathcal{P}_v')$ colors since $\Delta(H_v) = L(\mathcal{P}_v')$ (we assume that a loop contributes 1 to the degree of the vertex).

Let S_v be the coloring obtained for $\mathcal{P}_v'$. We show how we can merge all local colorings S_v into a global coloring S for $\mathcal{P}'$ without increasing the number of colors used. We start by merging S_{x_1} with each of S_v for $v \in \{y_1, \ldots, y_l\}$. The merging for some S_{y_i} is done as follows. The only paths that are in both S_{x_1} and S_{y_i}, and thus might cause conflicts when merging, are the paths that use edge (y_i, x_1). Notice that if a path p colored in S_{y_i} intersects a path q that is colored in S_{x_1} but not in S_{y_i}, then p uses (y_i, x_1) and both p, q use the same outgoing edge incident to x_1, and are both colored with different colors in S_{x_1}. Therefore, to combine the two colorings we maintain S_{x_1} as is and modify S_{y_i}: we permute S_{y_i} (i.e., we rename the colors) so that the paths that use edge (y_i, x_1) get the color they have in S_{x_1}. Since S_{y_i} was a valid coloring, it remains valid after the

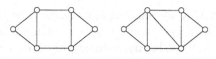

Fig. 3. Examples of graphs of diameter 3 that admit the empty SP-sufficient set.

permutation. Now the two colorings are compatible and can be merged in the obvious way. After the merging, S_{x_1} is the new coloring. After l mergings we have extended S_{x_1} to include the colorings $S_{y_1}, \ldots, S_{y_l}$ without increasing the number of colors used, while maintaining its validity. We repeat the same procedure to merge coloring $S_{x_{k-1}}$ with the colorings $S_{z_1}, \ldots, S_{z_m}$.

Now, we can merge $S_{x_1}, \ldots, S_{x_{k-1}}$ to one global coloring S for $\mathcal{P}'$. To do this we initially set $S = S_{x_1}$ and continue to the processing of S_{x_2}. After processing S_{x_i} we continue to $S_{x_{i+1}}$ and merge the previous global coloring S (which we have obtained by merging colorings $S_{x_1} \ldots S_{x_i}$) with $S_{x_{i+1}}$. The merging of S with $S_{x_{i+1}}$ is done as follows. Consider the paths that use edge (x_i, x_{i+1}) on G'. These paths are the only paths that are in both S and $S_{x_{i+1}}$ and therefore might cause conflicts in the merging. To combine the two colorings we maintain S as is and modify $S_{x_{i+1}}$: we permute $S_{x_{i+1}}$ so that the paths that use edge (x_i, x_{i+1}) get the color they have in S. Since $S_{x_{i+1}}$ was a valid coloring before the modification, it remains valid and now we can merge the two colorings in the obvious way. This way we extend coloring S without increasing the number of colors used in S and $S_{x_{i+1}}$ while maintaining its validity. After $k - 2$ mergings we obtain a global coloring S for $\mathcal{P}'$ that uses $\max_{v \in V'} L(\mathcal{P}'_v) = L(\mathcal{P}')$ colors. We obtain a coloring for the initial set $\mathcal{P}$ with $L(\mathcal{P})$ colors by extending S greedily to the paths in $\mathcal{P}^1$ which we have previously discarded.   □

4   Biconnected Graphs

Now, consider the case where G is biconnected. We assume that G admits the empty SP-sufficient set.

Lemma 5. *For all $u, v \in V$: $d(u, v) \leq 3$.*

Notice that the bound in the statement is tight: there exist graphs that admit the empty SP-sufficient set and have diameter 3. Two such graphs are shown in Fig. 3. On the other hand, if the diameter is at most two (or if we restrict to shortest-path routings of length at most two), then we can efficiently check whether a graph admits the empty SP-sufficient set as the following theorem illustrates.

Theorem 2. *Let $G = (V, E)$ be an undirected graph of diameter at most 2. The empty set is SP-sufficient for G if and only if E can be 2-colored such that every shortest path of length 2 in G uses edges of different colors.*

Proof. For sufficiency we show how to color a set $\mathcal{P}$ of shortest paths on G with $L(\mathcal{P})$ colors. We assume that $\mathcal{P}$ contains only paths of length two since any paths of length one can be colored greedily. We construct a multigraph H on E: for each path uvw in $\mathcal{P}$ we add an edge in H between uv and vw. If the condition holds then H is bipartite and by König's theorem [8] it can be edge-colored with $\Delta(H)$ colors in polynomial time. Since the degree of each vertex of H is equal to the load of the corresponding edge of G this gives a coloring for $\mathcal{P}$ with $L(\mathcal{P})$ colors.

To see that the condition is necessary assume that it does not hold and let $\mathcal{P}$ be the set of all shortest paths of length two on G. We construct a multigraph H on E as before. Since E can not be 2-colored such that every path in $\mathcal{P}$ uses edges of different colors, H is not bipartite and hence contains an odd cycle. The paths corresponding to the edges of the odd cycle form a set of paths $\mathcal{P}'$ with $L(\mathcal{P}') = 2$ and $\chi(\mathcal{P}') = 3$. □

The following theorem settles the case of $\mathrm{diam}(G) = 3$. For $u \in V$, we denote with $N_i(u)$ the subgraph of G induced by the neighborhood of u in distance i. For a vertex $v \in V$ of eccentricity 3 we define $T_G(v)$ to be the bipartite graph with vertex-sets $V_1 = V(N_1(v)) \setminus \{u \in N_1(v) \mid \mathrm{ecc}(u) = 3\}$, $V_2 = V(N_2(v))$ and edge-set $E_T = \{uw \in E \mid u \in V_1, w \in V_2\}$.

Theorem 3. *Let $G = (V, E)$ be an undirected, biconnected graph with $\mathrm{diam}(G) = 3$ and let $v \in V$ be a vertex of eccentricity 3. The empty set is SP-sufficient for G if and only if the following hold:*

(i) $N_1(v)$ and $G[V(N_2(v)) \cup V(N_3(v))]$ are cliques,
(ii) $T_G(v)$ contains no cycle, no P^5, and if it contains a P^4 then it has only one non-trivial connected component.

Proof. The proof for the necessity of the conditions is omitted from this version. For sufficiency we demonstrate how to color a set $\mathcal{P}$ of shortest paths on a graph G with $\mathrm{diam}(G) = 3$ that satisfies (i) and (ii) with $L(\mathcal{P})$ colors. We assume that $\mathcal{P}$ does not contain any paths of length one since these paths can be colored greedily after coloring the longer paths.

Define $Q_1 := \{u \in N_1(v) \mid \mathrm{ecc}(u) = 3\} \cup \{v\}$ and $Q_2 := V(N_3(v))$. Since $\mathrm{diam}(G) = 3$, we have that $Q_1 \cup Q_2 \cup V_1 \cup V_2 = V$, where V_1, V_2 are the vertex-sets of $T_G(v)$, as defined earlier (consider Fig. 4). Notice that by condition (i), $G[Q_1 \cup V_1]$ and $G[V_2 \cup Q_2]$ are cliques. We claim that all vertices of eccentricity 3 are contained in Q_1 and Q_2. By definition V_1 can not contain any vertices of eccentricity 3. Assume that there exists a vertex $u \in V_2$ with $\mathrm{ecc}(u) = 3$. There exists a vertex $x \in V$ with $d(u, x) = 3$. Since $G[V_2 \cup Q_2]$ is a clique we have that $x \notin Q_2$. Also, since $d(u, v) = 2$, u has a neighbor in V_1 and since $G[Q_1 \cup V_1]$ is a clique we have that $x \notin Q_1 \cup V_1$. We reach a contradiction and thus V_2 does not contain any vertices of eccentricity 3.

We will exhibit the coloring algorithm for the case where $T_G(v)$ does not contain a P^4 (the other case is similar and is omitted from this version). Let $C_1, \ldots, C_k$ be the connected components of $T_G(v)$ in some arbitrary ordering. Notice that at least one of $|V(C_i) \cap V_1| = 1$ or $|V(C_i) \cap V_2| = 1$ holds for all $1 \le i \le k$ (i.e., all components of $T_G(v)$ are stars), since no component of $T_G(v)$ contains a P^4 or a P^5. For each $1 \le i \le k$ let l_i be such that $|V(C_i) \cap V_{l_i}| = 1$ and x_i be such that $V(C_i) \cap V_{l_i} = \{x_i\}$. Let p be a path of length 3 in $\mathcal{P}$. Since Q_1, Q_2 contain all vertices of eccentricity 3 in G and both $G[Q_1], G[Q_2]$ are cliques, p goes from a vertex of Q_1 to a vertex of Q_2 and

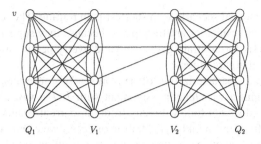

Fig. 4. The structure of biconnected graphs of diameter 3 that do not require converters.

uses two edges incident to some x_i. Moreover, since all paths in $\mathcal{P}$ are shortest paths and $G[Q_1 \cup V_1]$, $G[V_2 \cup Q_2]$ are cliques, a path in $\mathcal{P}$ intersects p in an edge not incident to x_i if and only if it intersects it in an edge incident to x_i. Therefore we can shorten p by restricting it to the edges incident to x_i without modifying the conflict graph of $\mathcal{P}$. We apply the same modification to all paths of length 3 in $\mathcal{P}$. Let $\mathcal{P}^1$ be the set of paths that after this modification have length one. These paths are discarded from $\mathcal{P}$. Now $\mathcal{P}$ contains only paths of length two.

In order to obtain a coloring of $\mathcal{P}$ with $L(\mathcal{P})$ colors it suffices to show that the edges that are used by paths in $\mathcal{P}$ can be 2-colored such that every path in $\mathcal{P}$ uses edges of different colors. Consider first the subgraph of G, $G[V_1 \cup Q_1]$, induced by $V_1 \cup Q_1$. No path in $\mathcal{P}$ uses two edges in this subgraph. Hence, we can color all these edges with one color, say a. The same holds for the subgraph $G[V_2 \cup Q_2]$. Moreover, since $\mathcal{P}$ contains no paths of length 3, no path touches edges from both $G[V_1 \cup Q_1]$, $G[V_2 \cup Q_2]$. Hence, we can color all the edges in $G[V_2 \cup Q_2]$ with color a. We color all remaining edges, i.e., edges in $T_G(v)$, with a second color b. No path uses two edges from $T_G(v)$ and therefore we can obtain a valid coloring of $\mathcal{P}$ with $L(\mathcal{P})$ colors using the reduction to bipartite edge-coloring illustrated in the proof of Theorem 2. We complete the coloring by coloring greedily the paths in $\mathcal{P}^1$ and thus obtain a valid coloring for $\mathcal{P}$ with $L(\mathcal{P})$ colors. $\qquad\square$

Note that Theorem 3 along with Theorem 2 provide a complete characterization of the class of biconnected graphs that admit the empty SP-sufficient set since by Lemma 5 we know that all graphs with diameter greater than 3 need wavelength converters.

5   Conclusion and Future Work

We have given a complete characterization of undirected networks on which any set $\mathcal{P}$ of shortest paths admits a valid wavelength assignment with $L(\mathcal{P})$ wavelengths. These are exactly the networks that do not benefit from the use of wavelength conversion when shortest-path routing is used. It follows from our characterization that this class of networks is efficiently recognizable. We have also given an efficient algorithm for computing a wavelength assignment with $L(\mathcal{P})$ wavelengths for this class of networks. To our knowledge, these are the first theoretical investigations of a wavelength assignment problem in the practical scenario with shortest-path routing. The results should be contrasted with known results for arbitrary paths, because they suggest that the traditional worst-case analysis for arbitrary paths can yield overly pessimistic results concerning wavelength conversion requirements.

We note that a characterization of networks *with* converters that have the same property, i.e., admit a wavelength assignment with $L(\mathcal{P})$ wavelengths for any set $\mathcal{P}$ of shortest paths, does not follow directly from our current results. This is because the approach of Wilfong and Winkler [11] fails in our case: we can no longer "explode" the converters and argue about each component of the resulting graph independently, since such a modification will alter the distance between pairs of vertices.

References

1. V. Auletta, I. Caragiannis, L. Gargano, C. Kaklamanis, and P. Persiano. Sparse and limited wavelength conversion in all-optical tree networks. *Theoretical Computer Science*, 266(1–2):887–934, 2001.
2. T. Erlebach and K. Jansen. The complexity of path coloring and call scheduling. *Theoretical Computer Science*, 255(1–2):33–50, 2001.
3. T. Erlebach, K. Jansen, C. Kaklamanis, M. Mihail, and P. Persiano. Optimal wavelength routing in directed fiber trees. *Theoretical Computer Science*, 221:119–137, 1999.
4. T. Erlebach and S. Stefanakos. On shortest-path all-optical networks without wavelength conversion requirements. TIK-Report 153, ETH Zurich, October 2002. Available electronically at ftp://ftp.tik.ee.ethz.ch/pub/publications/TIK-Report153.pdf.
5. T. Erlebach and S. Stefanakos. Wavelength conversion in networks of bounded treewidth. TIK-Report 132, ETH Zurich, April 2002. Available electronically at ftp://ftp.tik.ee.ethz.ch/pub/publications/TIK-Report132.pdf.
6. M.R. Garey, D.S. Johnson, G.L. Miller, and C.H. Papadimitriou. The complexity of coloring circular arcs and chords. *SIAM Journal on Algebraic and Discrete Methods*, 1(2):216–227, 1980.
7. J.M. Kleinberg and A. Kumar. Wavelength conversion in optical networks. *Journal of Algorithms*, 38(1):25–50, 2001.
8. D. König. Über Graphen und ihre Anwendung auf Determinantentheorie und Mengenlehre. *Math. Ann.*, 77:453–465, 1916.
9. V. Kumar. Approximating circular arc colouring and bandwidth allocation in all-optical ring networks. In *Proceedings of the International Workshop on Approximation Algorithms for Combinatorial Optimization (APPROX '98)*, volume 1444 of *Lecture Notes in Computer Science*, pages 147–158, 1998.
10. Y. Rabani. Path coloring on the mesh. In *Proceedings of the 37th Annual Symposium on Foundations of Computer Science (FOCS'96)*, pages 400–409, 1996.
11. G.T. Wilfong and P. Winkler. Ring routing and wavelength translation. In *Proceedings of the 9th Annual ACM-SIAM Symposium on Discrete Algorithms (SODA'98)*, pages 333–341, 1998.
12. H. Zang, J.P. Jue, and B. Mukherjee. A review of routing and wavelength assignment approaches for wavelength-routed optical WDM networks. *Optical Networks Magazine*, 1(1):47–60, January 2000.

Lattice Reduction by Random Sampling and Birthday Methods

Claus Peter Schnorr

Fachbereiche Mathematik/Biologie-Informatik,
Universität Frankfurt, PSF 111932, D-60054 Frankfurt am Main, Germany.
schnorr@cs.uni-frankfurt.de http://www.mi.informatik.uni-frankfurt.de/

Abstract. We present a novel practical algorithm that given a lattice basis $b_1, ..., b_n$ finds in $O(n^2(\frac{k}{6})^{k/4})$ average time a shorter vector than b_1 provided that b_1 is $(\frac{k}{6})^{n/(2k)}$ times longer than the length of the shortest, nonzero lattice vector. We assume that the given basis $b_1, ..., b_n$ has an orthogonal basis that is typical for worst case lattice bases. The new reduction method samples short lattice vectors in high dimensional sublattices, it advances in sporadic big jumps. It decreases the approximation factor achievable in a given time by known methods to less than its fourth-th root. We further speed up the new method by the simple and the general birthday method.

1 Introduction and Summary

History. The set of all linear combinations with integer coefficients of a set of linearly independent vectors $b_1, ..., b_n \in \mathbf{R}^d$ is a *lattice* of *dimension n*. The problem of finding a shortest, nonzero lattice vector is a landmark problem in complexity theory. This problem is polynomial time for fixed dimension n due to [LLL82] and is NP-hard for varying n [E81, A98, M98]. No efficient algorithm is known to find very short vectors in high dimensional lattices. Improving the known methods has a direct impact on the cryptographic security of many schemes, see [NS00] for a surview.

Approximating the shortest lattice vector to within an *approximation factor* (*apfa* for short) c means to find a nonzero lattice vector with at most c-times the minimal possible length. We consider integer lattices of dimension n in $\mathbf{Z}^n$ with a given lattice basis consisting of integer vectors of Euclidean length $2^{O(n)}$. The LLL-algorithm of LENSTRA, LENSTRA, LOVÁSZ [LLL82] achieves for arbitrary $\varepsilon > 0$ an *apfa* $(\frac{4}{3} + \varepsilon)^{n/2}$ in $O(n^5)$ steps using integers of bit length $O(n^2)$. This algorithm repeatedly constructs short bases in two-dimensional lattices, the two-dimensional problem was already solved by GAUSS. The recent segment LLL-reduction of Koy-Schnorr [KS01a,KS02] achieves the same apfa $(\frac{4}{3} + \varepsilon)^{n/2}$ within $O(n^3 \log n)$ steps.

Finding very short lattice vectors requires additional search beyond LLL-type reduction. The algorithm of KANNAN [K83] finds the shortest lattice vector in time $n^{O(n)}$ by a diligent exhaustive search, see [H85] for an $n^{\frac{n}{2}+o(n)}$ time

H. Alt and M. Habib (Eds.): STACS 2003, LNCS 2607, pp. 145–156, 2003.
© Springer-Verlag Berlin Heidelberg 2003

algorithm. The recent probabilistic sieve algorithm of [AKS01] runs in $2^{O(n)}$ average time and space, but is impractical as the exponent $O(n)$ is about $30\,n$. Schnorr [S87] has generalized the LLL-algorithm in various ways that repeatedly construct short bases of k-dimensional lattices of dimension $k \geq 2$. While $2k$-reduction [S87] runs in $O(n^3 k^{k+o(k)} + n^4)$ time, the stronger BKZ-reduction [S87, SE91] is quite efficient for $k \leq 20$ but lacks a proven time bound. LLL-reduction is the case $k = 1$ of $2k$-reduction.

Our novel method randomly samples short lattice vectors in high dimensional sublattices, and inserts by a global transform short vectors into the lattice basis. It remarkably differs from previous LLL-type reductions that locally transform the lattice basis by reducing small blocks of consecutive basis vectors. The new method applies to lattice bases for which the associated orthogonal basis satisfies two conditions, RA and GSA, defined in Section 2. These conditions are natural for worst case lattice bases and also play in AJTAI's recent worst case analysis [A02] of Schnorr's $2k$-reduction. Our new method is practical and space efficient and outperforms all previous algorithms.

Sampling reduction inserts a short vector found by random sampling into the basis, BKZ–reduces the new basis and iterates the procedure with the resulting basis. We observed sporadic big jumps of progress during BKZ–reduction, jumps that are difficult to analyze. We study the progress of the new method in attacks on the GGH-cryptosysytem [GGH97] where we build on our previous experience. We report in detail, we believe that our findings extend beyond GGH to general applications. We expect that the new algorithms lower the security of lattice based cryptosystems.

Table 1. Theoretic performance of new and previous methods

	time	space/n	apfa
1. sampl. reduction	$n^3(\frac{k}{6})^{k/4}$	1	$(\frac{k}{6})^{n/2k}$
2. simple birthday	$n^3(\frac{4}{3})^{k/3}(\frac{k}{6})^{k/8}$	$(\frac{4}{3})^{k/3}(\frac{k}{6})^{k/8}$	$(\frac{k}{6})^{n/2k}$
3. primal–dual (Koy)	$n^3 k^{k/2+o(k)}$	1	$(\frac{k}{6})^{n/k}$
4. $2k$-reduction [S87]	$n^3 k^{k+o(k)}$	1	$(\frac{k}{3})^{n/k}$

The shown time bounds must be completed by a constant factor and an additive term $O(n^4)$ covering LLL-type reduction. The integer $k, 2 \leq k \leq n/2$, can be freely chosen. The entry c under space/n means that $c+O(n)$ lattice vectors, consisting of $c \cdot n + O(n^2)$ integers, must be stored. The original LLL-algorithm uses integers of bit length $O(n^2)$ required to compute the orthogonal basis in exact integer arithmetic. However, by computing the orthogonal basis in approximate rational arithmetic (floating point arithmetic in practice) LLL-type reduction can be done in $O(n^5)$ arithmetic steps using integers of bit length $O(n)$. The proven analysis of Schnorr [S88] induces diligent steps for error correction, but simple methods of scaling are sufficient in practice [KS01b].

Sampling Reduction repeats the algorithm SHORT of Section 2 $O(n)$-times, see Section 3. SHORT runs in $O(n^2(\frac{k}{6})^{k/4})$ time and decreases with probability $\frac{1}{2}$ an *apfa* greater than $(\frac{k}{6})^{n/2k}$ by a factor $\sqrt{0.99}$. The *apfa* $(\frac{k}{6})^{n/2k}$ is about the

4-th root of the proven *apfa* achievable in the same time by Koy's primal–dual method, the best known fully proven algorithm.

Section 4 combines random sampling with general birthday methods of [W02]. Birthday sampling stores many statistically independent short lattice vectors $\bar{b}_i$ and searches in the spirit of WAGNER's 2^t-list algorithm a vector $b = \sum_{i=1}^{2^t} \bar{b}_i$ that is shorter than b_1. Simple birthday sampling for $t = 1$ further decreases the *apfa* achievable in a given time to its square root, but its practicability hinges on the required space.

Method 4 produces $2k$-reduced bases [S87, Theorem 3.1], and proceeds via shortest lattice vectors in dimension $2k$ while KOY's primal–dual method 3 repeatedly constructs shortest lattice vectors in dimension k [K00]. The *apfa*'s of Methods 3 and 4 assume the realistic bound $\gamma_k \leq k/6$ for $k \geq 24$ for the HERMITE constant γ_k; γ_k is the maximum of $\lambda_1(L)^2(\det(L))^{-2/k}$ for lattices L of dimension k and the length $\lambda_1(L)$ of the shortest, nonzero vector in L.

Notation. An ordered set of linearly independent vectors $b_1, ..., b_n \in \mathbf{Z}^d$ is a *basis* of the integer *lattice* $L = \sum_{i=1}^{n} b_i \mathbf{Z} \subset \mathbf{Z}^d$, consisting of all linear integer combinations of $b_1, ..., b_n$. We write $L = L(b_1, ..., b_n)$. Let $\hat{b}_i$ denote the component of b_i that is orthogonal to $b_1, ..., b_{i-1}$ with respect to the *Euclidean inner product* $\langle x, y \rangle = x^\top y$. The *orthogonal vectors* $\hat{b}_1, ..., \hat{b}_n \in \mathbf{R}^d$ and the *Gram-Schmidt coefficients* $\mu_{j,i}$, $1 \leq i, j \leq n$, associated with the basis $b_1, ..., b_n$ satisfy for $j = 1, ..., n$

$$b_j = \sum_{i=1}^{j} \mu_{j,i}\hat{b}_i, \qquad \mu_{j,j} = 1, \qquad \mu_{j,i} = 0 \text{ for } i > j,$$

$$\mu_{j,i} = \langle b_j, \hat{b}_i \rangle / \langle \hat{b}_i, \hat{b}_i \rangle, \qquad \langle \hat{b}_j, \hat{b}_i \rangle = 0 \text{ for } j \neq i.$$

We let $\pi_i : \mathbf{R}^n \to \mathrm{span}(b_1, ..., b_{i-1})^\perp$ denote the *orthogonal projection*, $\pi_i(b_k) = \sum_{j=i}^{n} \mu_{k,j}\hat{b}_j$, $\pi_i(b_i) = \hat{b}_i$. Let $\|b\| = \langle b, b \rangle^{\frac{1}{2}}$ denote the *Euclidean length* of a vector $b \in \mathbf{R}^d$. Let λ_1 denote the length of the shortest nonzero lattice vector of a given lattice. The *determinant* of lattice $L = L(b_1, ..., b_n)$ is $\det L = \prod_{i=1}^{n} \|\hat{b}_i\|$. For simplicity, let all given lattice bases be bounded so that $\max_i \|b_i\| = 2^{O(n)}$. Our time bounds count arithmetic steps using integers of bit length $O(n)$.

2   Random Sampling of Short Vectors

Let L be a lattice with given basis $b_1, ..., b_n$. As a lattice vector $b = \sum_{j=1}^{n} \mu_j \hat{b}_j$ has length $\|b\|^2 = \sum_{j=1}^{n} \mu_j^2 \|\hat{b}_j\|^2$ the search for short lattice vectors naturally comprises two steps:

1. Decreasing μ_i to $|\mu_i| \leq \frac{1}{2}$ for $i = 1, ..., n$: given $b \in L$ with arbitrary μ_i the vector $b' = b - \mu b_i$ has $\mu'_i = \mu_i - \mu$ and thus $|\mu'_i| \leq \frac{1}{2}$ holds if $|\mu - \mu_i| \leq \frac{1}{2}$.

2. Shortening $\hat{b}_j$, i.e., replacing b_j by a nonzero vector $b \in L(b_j, ..., b_n)$ that minimizes $\|\pi_j(b)\|^2 = \sum_{i=j}^{n} \mu_i^2 \|\hat{b}_i\|^2$ over a suitable subset $S_j \subset L(b_j, ..., b_n)$.

The various reduction algorithms differ by the choice of S_j. The LLL-algorithm uses $S_j = L(b_j, b_{j+1})$, BKZ-reduction [S87, SE91] uses $S_j = L(b_j, ..., b_{j+k-1})$, $2k$-reduction [S87] uses a subset $S_j \subset L(b_j, ..., b_{j+2k-1})$ and HKZ-reduction minimizes over the entire lattice $L(b_j, ..., b_n)$. LLL-type reduction [LLL82, S87, SE91] repeatedly replaces a block $b_j, ..., b_{j+k-1}$ for various j by an equivalent block starting with a vector $b \in L(b_j, \ldots, b_{j+k-1})$ of shorter length $\|\pi_j(b)\| \neq 0$. The vector b is produced by exhaustive enumeration.

The novel *sampling reduction* repeatedly produces via random sampling a nonzero vector $b \in L(b_j, \ldots, b_n)$ with $\|\pi_j(b)\| < \|\widehat{b_j}\|$, and continues with a new basis $b_1, \ldots, b_{j-1}, b, b_j, \ldots, b_{n-1}$. Such b cannot be efficiently produced by exhaustive search, the dimension of $L(b_j, \ldots, b_n)$ is to high. Surprisingly, random sampling in high dimension $n-j+1$ outperforms exhaustive search in low dimension k. Here we introduce random sampling for $j = 1$, the algorithm ESHORT of Section 3 uses a straightforward extension to $j \geq 1$. Random sampling for $j = 1$ searches short vectors $b = \sum_{i=1}^n \mu_i \widehat{b_i} \in L$ with small coefficients $\mu_1, \ldots, \mu_k$. This makes $\|b\|^2 = \sum_{i=1}^n \mu_i^2 \|\widehat{b_i}\|^2$ small. Importantly, the initial vectors $\widehat{b_1}, \ldots, \widehat{b_k}$ are in practice longer than the $\widehat{b_i}$ for $i > k$, so small coefficients $\mu_1, \ldots, \mu_k$ have a bigger impact than small μ_i for $i > k$. We analyse this idea assuming that the lengths $\|\widehat{b_1}\|^2, \ldots, \|\widehat{b_n}\|^2$ are close to a geometric series.

The Sampling Method. Let $1 \leq u < n$ be constant. Given a lattice basis $b_1, \ldots, b_n$ we sample lattice vectors $b = \sum_{i=1}^n t_i b_i = \sum_{i=1}^n \mu_i \widehat{b_i}$ satisfying

$$|\mu_i| \leq \begin{cases} \frac{1}{2} & \text{for } i < n - u \\ 1 & \text{for } n - u \leq i < n \end{cases}, \qquad \mu_n = 1. \tag{1}$$

There are at least 2^u distinct lattice vectors b of this form. The sampling algorithm (SA) below generates a single vector b in time $O(n^2)$. The subsequent algorithm SHORT samples distinct vectors via SA until a vector b is found that is shorter than b_1. The choice of $\mu_n = 1$ implies that $b_1, \ldots, b_{j-1}, b, b_j, \ldots, b_{n-1}$ is a lattice basis.

Sampling Algorithm (SA)

INPUT lattice basis $b_1, \ldots, b_n \in \mathbf{Z}^n$ with coefficients $\mu_{i,j}$.

1. $b := b_n$, $\mu_j := \mu_{n,j}$ for $j = 1, \ldots, n-1$
2. FOR $i = n-1, \ldots, 1$ DO

    Select $\mu \in \mathbf{Z}$ such that $|\mu_i - \mu| \leq \begin{cases} \frac{1}{2} & \text{for } i < n - u \\ 1 & \text{for } i \geq n - u \end{cases}$

    $b := b - \mu\, b_i$, $\mu_j := \mu_j - \mu\, \mu_{i,j}$ for $j = 1, \ldots, i$

OUTPUT b, $\mu_1, \ldots, \mu_n$ satisfying $b = \sum_{i=1}^n \mu_i \widehat{b_i}$ and (1).

The coefficient μ_i is updated $(n - i)$-times. This leads to a nearly uniform distribution of the μ_i, in particular for small i, which is crucial for our method. Note that SA is deterministic, the random sampling is "pseudo-random" in a weak heuristic sense.

Randomness Assumption RA. Let the coefficients μ_i of the vectors $b = \sum_{i=1}^n \mu_i \widehat{b_i}$ sampled by SA be uniformly distributed in $[-\frac{1}{2}, \frac{1}{2}]$ for $i < n - u$ and in $[-1, 1]$ for $n - u \leq i < n$, let the μ_i be statistically independent for distinct i, and let the coefficients μ_i, μ_i' of distinct vectors b, b' sampled by SA be statistically independent.

The Geometric Series Asumption (GSA). Let $\|\widehat{b}_i\|^2/\|b_1\|^2 = q^{i-1}$ for $i = 1, ..., n$ be a geometric series with quotient q, $\frac{3}{4} \leq q < 1$.

The GSA in Practice. In practice the quotients $\|\widehat{b}_i\|^2/\|b_1\|^2$ approximate the q^{i-1} without achieving equality. Importantly, our conclusions under GSA also hold for approximations where $\sum_{i=1}^{n} \mu_i^2 \left(\|\widehat{b}_i\|^2/\|b_1\|^2 - q^{i-1} \right)$ is sufficiently small for random $\mu_i \in_R [-\frac{1}{2}, \frac{1}{2}]$, e.g. smaller than 0.01 for Theorems 1 and 2.

We have tested the GSA for the public GGH-bases according to the cryptosystem of [GGH97]. After either KOY's primal-dual reduction or after BKZ-reduction with block size 20 these bases closely approximate GSA and RA. The GSA-behavior is significantly better after BKZ-reduction than after primal-dual reduction. Lattice bases that are not reduced by an LLL-type reduction usually have bad GSA-behavior.

Under the GSA the values $\log_2(\|b_i\|^2/\|\widehat{b}_i\|^2)$ for $i = 1, ..., n$ are on a straight line. For lattice bases that are BKZ-reduced these values closely approximate a line. Figure 1 shows a GGH-basis generated according to the GGH-cryptosystem [GGH97] after various reductions and a final BKZ-reduction with block size 20.

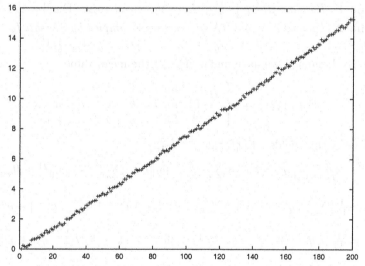

Fig. 1. The values $\log_2(\|b_1\|^2/\|\widehat{b}_i\|^2)$ for $i = 1, ..., 200$ of a BKZ-basis

Worst case bases satisfy the GSA. We show that lattice reduction is harder the better the given basis approximates a geometric series. Lattice bases satisfying the GSA are worst case bases for lattice reduction. Here, let the goal of lattice reduction be to decrease the proven *apfa*, i.e., to decrease $\max_i \|b_1\|/\|\widehat{b}_i\|$ via a new lattice basis. Note that $apfa \leq \max_i \|b_1\|/\|\widehat{b}_i\|$ holds for all bases while GSA implies $apfa \leq q^{(-n+1)/2}$.

We associate with a basis $b_1, ..., b_n$ the quotients $q_i := (\|\widehat{b}_i\|^2/\|b_1\|^2)^{\frac{1}{i-1}}$ for $i = 2, ..., n$, $q := q_n$. As $apfa \leq q^{(-n+1)/2}$ the goal of the reduction is to increase q. Of course our reduction problem gets easier for smaller n and smaller q.

If GSA does not hold we have that $q_i \neq q$ for some i. Select i as to maximize $|q_i - q|$. We transform the given reduction problem into smaller, easier problems with bases that better approximate the GSA.

If $q_i < q$ we reduce the subbasis $b_1, ..., b_i$ by decreasing $\max_i \|b_1\|/\|\widehat{b}_i\|$ via a new lattice basis. If $q_i > q$ we reduce the basis $\pi_i(b_i), ..., \pi_i(b_n)$. The q-value $\overline{q}_i := (\|\widehat{b}_n\|^2/\|\widehat{b}_i\|^2)^{\frac{1}{n-i}}$ of that basis satisfies $q^{n-1} = q_i^{i-1}\overline{q}_i^{n-i}$, thus either $q_i < q$ or $\overline{q}_i < q$.

In either case we solve an easier lattice problem with a smaller q-value and a smaller dimension. Our procedure decreases $|q - q_i|$ providing a basis that better approximates a geometric series. Therefore, lattice bases of the same q-value get harder the better they approximate a geometric series.

Random Sampling Short Vectors. Let $k, u \geq 1$ be constants, $k + u < n$. Consider the event that vectors $b = \sum_{i=1}^{n} \mu_i \widehat{b}_i$ sampled by SA satisfy

$$|\mu_i|^2 \leq \tfrac{1}{4} q^{k-i} \qquad \text{for } i = 1, ..., k. \tag{2}$$

Under RA that event has probability $\prod_{i=1}^{k} q^{(k-i)/2} = q^{\binom{k}{2}/2}$. We study the probability that $\|b\|^2 < \|b_1\|^2$ holds under RA and the conditions (1), (2).

Lemma 1. *Random $\mu_i \in_R [-\tfrac{1}{2}, \tfrac{1}{2}]$ have the mean value* $\mathbf{E}[\mu_i^2] = \tfrac{1}{12}$.

Lemma 2. *Under GSA and RA the vectors b sampled by SA satisfy*
$\mathbf{Pr}\big[\|b\|^2 \|b_1\|^{-2} \leq \tfrac{1}{12}[k\,q^{k-1} + (q^k + 3\,q^{n-u-1})/(1-q)]\big] \geq \tfrac{1}{2}q^{\binom{k}{2}\frac{1}{2}}$.

Proof. By Lemma 1 we have under (1), (2) the mean value

$$\mathbf{E}[\mu_i^2 \,|\, (2)] = \begin{cases} \tfrac{1}{12}q^{k-i} & \text{for } i = 1, \dots, k \\ 1/12 & \text{for } i = k+1, ..., n-u-1 \\ 1/3 & \text{for } i = n-u, \dots, n-1 \end{cases}$$

Under GSA this yields $\mathbf{E}[\|b\|^2 \|b_1\|^{-2} \,|\, (2)]$

$$= \tfrac{1}{12}\Big[\sum_{i=1}^{k} q^{k-i}\|\widehat{b}_i\|^2 + \sum_{i=k+1}^{n-u-1} \|\widehat{b}_i\|^2 + 4\sum_{i=n-u}^{n-1} \|\widehat{b}_i\|^2\Big] + \|\widehat{b}_n\|^2$$

$$= \tfrac{1}{12}\Big[\sum_{i=1}^{k} q^{k-i+i-1} + \sum_{i=k+1}^{n-u-1} q^{i-1} + 4\,q^{n-u-1}\sum_{i=1}^{u} q^{i-1}\Big] + q^{n-1}$$

$$= \tfrac{1}{12}\big[k\,q^{k-1} + [(q^k - q^{n-u-1}) + 4q^{n-u-1}(1 - q^u)]/(1-q)\big] + q^{n-1}$$

$$= \tfrac{1}{12}\big[k\,q^{k-1} + (q^k + 3\,q^{n-u-1} - 4\,q^{n-1})/(1-q)\big] + q^{n-1}.$$

This proves the claim as $4/(q-1) \geq 1$, and (2) holds with probability $q^{\binom{k}{2}/2}$. $\square$

SHORT Algorithm

INPUT lattice basis $b_1, ..., b_n \in \mathbf{Z}^n$ with quotient $q < 1$

Let $u := 1 + \lceil -\binom{k}{2}\tfrac{1}{2}\log_2 q \rceil$ be the minimal integer so that $2^u \geq 2\,q^{-\binom{k}{2}\frac{1}{2}}$.

Sample via SA up to 2^u distinct lattice vectors $b = \sum_{i=1}^{n} \mu_i \widehat{b}_i$ satisfying the inequalities (1) until a vector is found such that $\|b\|^2 < 0.99\,\|b_1\|^2$.

OUTPUT lattice vector b satisfying $\|b\|^2 < 0.99\,\|b_1\|^2$.

Theorem 1. *Given a lattice basis $b_1, ..., b_n$ with quotient $q \leq (\frac{6}{k})^{1/k}$, SHORT runs in $O(n^2 q^{-k^2/4})$ time and finds under GSA and RA with probability $\frac{1}{2}$ for sufficiently large k and n a nonzero lattice vector b so that $\|b\|^2 < 0.99 \|b_1\|^2$.*

Proof. W.l.o.g. let $q^k = \frac{6}{k}$ as the claim holds a fortiori for smaller q. The inequality

$$\tfrac{1}{12}\left[k\, q^{k-1} + (q^k + 3\,q^{n-u-1})/(1-q)\right] \leq \tfrac{1}{2q} + \tfrac{1}{12}\,\tfrac{q^k + 3\,q^{3k}}{1-q} < 0.99.$$

holds for $k = 24$ and $n \geq 3k+u+1$. As $\frac{1}{2q} + \frac{1}{12}\frac{q^k+3q^{3k}}{1-q}$ decreases for $(\frac{6}{k})^{1/k}$ with k the inequality holds for all $k \geq 24$. Hence, the vectors b sampled by SA satisfy $\mathbf{Pr}[\,\|b\|^2\|b_1\|^{-2} < 0.99\,] \geq \frac{1}{2}q^{\binom{k}{2}/2}$ by Lemma 2. As SHORT samples $2\,q^{-\binom{k}{2}/2}$ independent vectors b it finds under RA with probability $1 - e^{-1} > \frac{1}{2}$ some b with $\|b\|^2\|b_1\|^{-2} < 0.99$. □

Remark. **1.** SHORT improves under GSA and RA an *apfa* $(\frac{k}{6})^{n/2k}$ in $O(n^2(\frac{k}{6})^{k/4})$ average time by a factor $\sqrt{0.99}$ for $k \geq 24$ and $n \geq 3k + k\ln k$. **2.** We can replace in Theorems 1 and 2 the constant 6 by an arbitrary large constant δ. This is because $\frac{1}{12}\frac{q^k}{1-q} = \frac{1}{\ln k} + O(k^{-2})$ holds for $q^k = \frac{\delta}{k}$, see [K97, p.107].

Table 2. SHORT performance according to Theorem 1

k	q^k	*apfa*	u	time
48	$8/k$	1.017^n	32	$n^2\,2^{32}$
40	$8/k$	1.020^n	24	$n^2\,2^{24}$
30	$7/k$	1.024^n	17	$n^2\,2^{17}$
24	$6/k$	1.029^n	13	$n^2\,2^{13}$

A comparison with previous methods illustrates the dramatic progress through random sampling: For $k = 24$ $2k$-reduction [S87] achieves *apfa* 1.09^n, Koy's primal-dual reduction achieves *apfa* 1.06^n in $\gg n^3 2^{13}$ time.

Practical Experiments. Consider a basis of dimension 160 consisting of integers of bit length 100, generated according to the GGH-cryptosystem [GGH97]. We reduce this basis in polynomial time by segment–LLL reduction [KS01] and primal–dual segment LLL with segment size 36 [K00]. This takes about 50 minutes and yields a basis with *apfa* about 8.25 and quotient $q \approx 0.946$.

Then a single final enumeration of 2^{12} lattice vectors via SA reduced the length of the shortest found lattice vector by 9%. This took just about one minute. The mean value of the μ_i^2 over the 2^{12} enumerated vectors for $i = 1, ..., 144$ (resp., for $i = 145, ..., 159$) was 0.083344 (resp. 0.3637) while the theoretic mean values under RA is $1/12 = 0.08\overline{3}$ (resp. $1/3 = 0.3\overline{3}$). The discrepancy of the observed mean values of μ_i^2 from the distribution under RA is smaller for small i because the coefficient μ_i gets updated $(n - i)$-times within SA. The initial quotients q_i of primal–dual reduced basis are sligthly larger than

q. This increases the observed mean value of $\|b\|^2/\|b_1\|^2$ for the b produced by SA a bit against the theoretic mean value under RA, GSA.

The observed 9% length reduction via 2^{12} sampled vectors is close to the value predicted under RA and GSA by our refined analysis. All experiments have been done on a 1700 MHz, 512 MB RAM PC using the software packages of NTL 5.1 and GMP 4.0.1.

Refined Analysis of SHORT. The inequalities (2) are sufficient but not necessary to ensure that $\mathbf{E}[\sum_{i=1}^{k} \mu_i^2 \|\widehat{b}_i\|^2] \leq \frac{1}{12} k q^{k-1}$ holds under RA and GSA. SA achieves the $\frac{1}{12} k q^{k-1}$–upper bound with a better probability than $q^{\binom{k}{2}/2}$. In the refined analysis we liberalize the Inequalities (2) by allowing a few larger coefficients $|\mu_i| > \frac{1}{2} q^{(k-i)/2}$ for $1 \leq i < k$ that are balanced by smaller $|\mu_i| < \frac{1}{2} q^{(k-i)/2}$ so that again $\sum_{i=1}^{k} \mu_i^2 \|\widehat{b}_i\|^2 \leq \frac{1}{12} k q^{k-1}$.

3   Sampling Reduction

ESHORT is an extension of SHORT that samples 2^u vectors $b = \sum_{i=1}^{n} \mu_i \widehat{b}_i$ by SA and determines the pair (b, j) for which $\sum_{i=j}^{n} \mu_i^2 \|\widehat{b}_i\|^2 < 0.99 \|\widehat{b}_j\|^2$ holds for the smallest possible $j \leq 10$. (The heuristic bound $j \leq 10$ covers the case that the basis vectors $b_1, ..., b_{10}$ have bad GSA-behaviour, which happens quite often, so that SA cannot succeed for the very first j.)

Sampling Reduction

This algorithm reduces a given basis $b_1, ..., b_n$ under GSA *and* RA.
1.   Search via ESHORT a pair (b, j) so that $\|\pi_j(b)\|^2 < 0.99 \|\widehat{b}_j\|^2$, $j \leq 10$, and terminate if the search fails. Form the new basis $b_1, \ldots, b_{j-1}, b, b_j, \ldots, b_{n-1}$.
2.   BKZ–reduce the new basis $b_1, \ldots, b_{j-1}, b, b_j, \ldots, b_{n-1}$ with block size 20 and go to 1.

Practical Experiments. With the above method C. TOBIAS has reconstructed the secret GGH-basis of the GGH-cryptosystem [GGH97] in dimension $n = 160$ and $n = 180$. This has been done by plain lattice reduction without improving the GGH-lattice by algebraic transforms as has been done by NGUYEN [N99]. The secret GGH–basis was reconstructed in 40 – 80 minutes for dimension $n = 160$ within 4 iterations, and in about 9 hours for $n = 180$ using 20 iterations. This was not possible by previous lattice reduction algorithms. 2^{12} to 2^{17} vectors have been sampled per iteration. BKZ–reduction was done by the BKZ–algorithm of NTL for block size 20. Usually ESHORT succeeds with $j \leq 10$.

Interestingly, BKZ–reduction of the new basis $b_1, \ldots, b_{j-1}, b, b_j, \ldots, b_{n-1}$ triggers sporadic big jumps of progress. Typically the length of the shortest vector is decreased by a factor 1.2 – 1.5 but occasionally by a factor up to 9. Sometimes the shortest lattice vector was found at an early stage. All experiments have been done on a 1700 MHz, 512 MB RAM PC using the software packages of NTL 5.1 and GMP 4.0.1.

What triggers the big jumps during BKZ–*reduction* ? When ESHORT finds a pair (b, j), so that $\sum_{i=j}^{n} \mu_i^2 \|\widehat{b}_i\|^2 < 0.99 \|\widehat{b}_j\|^2$, usually $\mu_j, \ldots, \mu_{j+k-1}$ are particularly small, the subsequent basis $b_1, \ldots, b_{j-1}, b, b_{j+1}, \ldots, b_{n-1}$ has large orthogonal vectors $\widehat{b}_{j+1}, \ldots, \widehat{b}_{j+k-1}$ and badly deviates from the GSA–property in the segment $\widehat{b}_{j+1}, \ldots, \widehat{b}_{j+k-1}$ following b. The lengths $\|\widehat{b}_{j+i}\|$ oscillate heavily up and down with some large values. Bad GSA–behavior of that type obstructs the SHORT algorithm but triggers big jumps of progress within BKZ–reduction because BKZ–reduction closely approximates the GSA. While BKZ–reduction with block size 20 does not improve a GSA–basis, it greatly improves a basis with bad GSA–behavior.

The big jumps of progress markedly differ from the usual BKZ–reduction of LLL–reduced bases. The latter advances in a slow steady progress. It generates intermediate bases where the lengths $\|\widehat{b}_i\|$ gradually decrease in i with only little fluctuation.

4   General Birthday Sampling

The birthday heuristic is a well known method that given a list of m bit integers, drawn uniformly at random, finds a collision of the given integers in $O(2^{m/2})$ average time. The method can easily be extended to find two random k-bit integers having a small difference, less than 2^{k-m}, $O(2^{m/2})$ time. We extend the birthday method from integers to lattice vectors, we extend random sampling to birthday sampling. Let $\mathbf{Q}$ denote the set of rational numbers, let $m, t \geq 1$ be integers.

Wagner's $(2^t, m)$-list algorithm [W02] extends the birthday method to solve the following $(2^t, m)$-*sum problem.* Given 2^t lists $L_1, \ldots, L_{2^t}$ of elements drawn uniformly and independently at random from $\{0, 1\}^m$ find $x_1 \in L_1, \ldots, x_{2^t} \in L_{2^t}$ such that $x_1 \oplus x_2 \oplus \cdots \oplus x_{2^t} = 0$. Wagner's algorithm solves the $(2^t, m)$-sum problem in $O(2^t \, 2^{m/(1+t)})$ average time and space by a tree-like birthday method. The $(4, m)$-list algorithm runs in $O(2^{m/3})$ time and space, and coincides with a previous algorithm of CAMION, PATARIN [CP91]. The *simple* case $t = 1$ is the well known birthday method. Consider the following *small sum problem* (SSP):

$(2^t, m)$-**SSP.** Given 2^t lists $L_1, \ldots, L_{2^t}$ of rational numbers drawn uniformly and independently at random from $[-\frac{1}{2}, \frac{1}{2}] \cap \mathbf{Q}$ find $x_1 \in L_1, \ldots, x_{2^t} \in L_{2^t}$ such that $|\sum_{i=1}^{2^t} x_i| \leq \frac{1}{2} 2^{-m}$.

We extend Wagner's $(2^t, m)$-list algorithm to solve the $(2^t, m)$-SSP. We outline the case $t = 2$ solving the $(4, m)$-SSP in $O(2^{m/3})$ average time. We count for steps additions and comparisons using rational numbers. Let the lists $L_1, \ldots, L_4$ each consist of $\frac{4}{3} 2^{m/3}$ random elements drawn uniformly from $[-\frac{1}{2}, \frac{1}{2}] \cap \mathbf{Q}$. Consider the lists

$$L_1' := \{x_1 + x_2 \mid |x_1 + x_2| \leq \tfrac{1}{2} 2^{-m/3}\}, \quad L_2' := \{x_3 + x_4 \mid |x_3 + x_4| \leq \tfrac{1}{2} 2^{-m/3}\}$$
$$L := \{x_1' + x_2' \mid |x_1' + x_2'| \leq \tfrac{1}{2} 2^{-m}\},$$

where x_i ranges over L_i and x_i' over L_i'. (We also record the source pair (x_1, x_2) of $x_i' = x_1 + x_2 \in L_1'$.) Applying the subsequent Lemma 3 with $\alpha = 2^{-m/3}$ we have that $\mathbf{Pr}[\,|x_1 + x_2| \leq \frac{1}{2} 2^{-m}\,] \geq 2^{-m/3} \frac{3}{4}$. The average size of L_1' (and likewise for L_2') is : $|L_1'| \geq |L_1| \cdot |L_2| \cdot 2^{-m/3} \frac{3}{4} = \frac{4}{3} 2^{m/3}$.

Similarly, we have for $x_i' \in L_i'$ and $\alpha = 2^{-2m/3}$ that $\mathbf{Pr}[\,|x_1' + x_2'| \leq \frac{1}{2} 2^{-m}\,] \geq 2^{-2m/3} \frac{3}{4}$, and thus the average size of L is $|L| \geq |L_1'| \cdot |L_2'| \cdot 2^{-2m/3} \frac{3}{4} = \frac{4}{3}$.

To construct L_1' (and likewise L_2', L) we sort the $x_1 \in L_1$ and the $-x_2 \in -L_2$ according to the numerical values of $x_1, -x_2 \in [-\frac{1}{2}, \frac{1}{2}]$ and we search for close elements $x_1, -x_2$. The sorting and searching is done by bucket sort in $O(2^{m/3})$ average time and space: we partition $[-\frac{1}{2}, \frac{1}{2}]$ into intervals of length $\frac{1}{2} 2^{-m/3}$, we distribute $x_1, -x_2$ to these intervals and we search for pairs $x_1, -x_2$ that fall into the same interval. This solves the $(4, m)$-SSP in $O(\frac{4}{3} 2^{m/3})$ average time and space. More generally the $(2^t, m)$-SSP is solved in $O(2^t \frac{4}{3} 2^{\frac{m}{t+1}})$ average time and space.

Lemma 3. *Let $x_i \in_R [-\frac{1}{2}, \frac{1}{2}]$ for $i = 1, 2$ be uniformly distributed and statistically independent. Then $\mathbf{Pr}[\,|x_1 + x_2| \leq \alpha/2\,] = \alpha\,(1 - \frac{\alpha}{4})$ holds for $0 \leq \alpha \leq 2$.*

Proof. For given $|x_1| \leq \frac{1-\alpha}{2}$ the interval of all $x_2 \in [-\frac{1}{2}, \frac{1}{2}]$ satisfying $|x_1 + x_2| \leq \alpha/2$ has length α. For $|x_1| \geq \frac{1-\alpha}{2}$ the corresponding interval length is $\alpha - y$, where the value $y := |x_1| - \frac{1-\alpha}{2}$ ranges over $[0, \alpha/2]$. Therefore

$$\mathbf{Pr}[\,|x_1 + x_2| \leq \alpha/2\,] = \alpha - 2 \int\limits_{0}^{\alpha/2} y \, dy = \alpha - \tfrac{1}{4} \alpha^2. \qquad \square$$

We extend our method from rational numbers in $[-\frac{1}{2}, \frac{1}{2}]$ to vectors in $\mathbf{Q}^k \cap [-\frac{1}{2}, \frac{1}{2}]^k$. We solve the following *small vector sum problem* $(2^t, m, k)$-VSSP in $O(k\, 2^t\, (\frac{4}{3})^k\, 2^{\frac{m}{t+1}})$ average time and space (in $O(k\, (\frac{4}{3})^{k/2}\, 2^{m/2})$ time for $t = 1$):

$(\mathbf{2^t, m, k})$-**VSSP.** Let an arbitrary partition $m = m_1 + \cdots + m_k$ be given with real numbers $m_1, \ldots, m_k \geq 0$. Given 2^t lists $L_1, \ldots, L_{2^t}$ of rational vectors drawn uniformly and independently from $[-\frac{1}{2}, \frac{1}{2}]^k \cap \mathbf{Q}^k$ find $x_1 \in L_1, \ldots, x_{2^t} \in L_{2^t}$, $x_i = (x_{i,1}, \ldots, x_{i,k}) \in \mathbf{Q}^k$, such that $|\sum_{i=1}^{2^t} x_{i,j}| \leq \frac{1}{2} 2^{-m_j}$ for $j = 1, \ldots, k$.

We extend the $(4, m)$-SSP solution to $(4, m, k)$-VSSP. Let each list L_i consist of $(\frac{4}{3})^k\, 2^{m/3}$ random vectors of $[-\frac{1}{2}, \frac{1}{2}]^k$ for $i = 1, \ldots, 4$. Consider the lists

$$L_1' := \{x_1 + x_2 \mid |x_{1,j} + x_{2,j}| \leq \tfrac{1}{2} 2^{-m_j/3} \text{ for } j = 1, \ldots, k\}$$
$$L_2' := \{x_3 + x_4 \mid |x_{3,j} + x_{4,j}| \leq \tfrac{1}{2} 2^{-m_j/3} \text{ for } j = 1, \ldots, k\}$$
$$L := \{x_1' + x_2' \mid |x_{1,j}' + x_{2,j}'| \leq \tfrac{1}{2} 2^{-m_j} \text{ for } j = 1, \ldots, k\},$$

where x_i ranges over L_i and x_i' ranges over L_i'. Then $|L_2'| = |L_1'| \geq |L_1| \cdot |L_2| \cdot 2^{-m/3} (\frac{3}{4})^k = (\frac{4}{3})^k\, 2^{m/3}$, and $|L| \geq |L_1'| \cdot |L_2'| \cdot 2^{-2m/3} (\frac{3}{4})^k = (\frac{4}{3})^k$ holds for the average list sizes. (For $t = 1$ we only need input lists L_i consisting of $(\frac{4}{3})^{k/2}\, 2^{m/2}$ vectors to succeed with $|L| \geq 1$.)

General Birthday Sampling (GBS). Given a lattice basis $b_1, \ldots, b_n$ and 2^t lists $\bar{L}_1, \ldots, \bar{L}_{2^t}$ of lattice vectors sampled by SA, GBS produces a short vector $b = \sum_{i=1}^{2^t} \bar{b}_i$ with $\bar{b}_i \in \bar{L}_i$ by solving the $(2^t, m, k)$-VSSP for the coefficient vectors $(\bar{\mu}_{1,i}, \ldots, \bar{\mu}_{k,i}) \in \mathbf{Q}^k \cap [-\frac{1}{2}, \frac{1}{2}]^k$ of $\bar{b}_i = \sum_{\ell=1}^n \bar{\mu}_{\ell,i} \widehat{b_\ell} \in \bar{L}_i$ for a suitable m.

Theorem 2. *Given $t \geq 1$ and a lattice basis $b_1, ..., b_n$ with quotient $q \leq (\frac{6}{k})^{1/k}$ GBS finds under GSA and RA, for sufficiently large k and n, a lattice vector $b \neq 0$, $\|b\|^2 < 0.99 \|b_1\|^2$ in $O(n^2 2^t (\frac{4}{3})^{2k/3} q^{-k^2/4(t+1)})$ average time and space.*

Proof. We replace SA in the proof of Theorem 1 by GBS. Initially GBS forms 2^t lists $\bar{L}_i$, each of of $2^{m/(t+1)}$ lattice vectors $\bar{b}_i \in \bar{L}_i$ sampled by SA for $1 \leq i \leq 2^t$. GBS produces a short lattice vector $b = \sum_i \bar{b}_i$ via a solution of the $(2^t, m, k)$-VSSP for the coefficient vectors $(\bar{\mu}_{1,i}, \dots, \bar{\mu}_{k,i}) \in \mathbf{Q}^k \cap [-\frac{1}{2}, \frac{1}{2}]^k$ of $\bar{b}_i = \sum_{j=1}^n \bar{\mu}_{j,i} \hat{b}_j \in \bar{L}_i$. Here let $m := m_1 + \cdots + m_k$ for $m_j := \log_2 q^{(-k+j)/2}$, and thus $m = -\binom{k}{2} \frac{1}{2} \log_2 q$. Under GSA and RA the $(2^t, m, k)$-VSSP solution provides a lattice vector $b = \sum_{j=1}^n \mu_j \hat{b}_j$ such that $|\mu_j| \leq \frac{1}{2} q^{(k-j)/2}$ for $j = 1, \dots, k$ and $\mathbf{E}[\mu_j^2] = \frac{2^t}{12}$ (resp., $\frac{2^t}{3}$) holds for $j = k+1, ..., n-u-1$ (resp., for $j \geq n-u$).

Let k be so large that $q^k \leq \frac{6}{k}$. For $q^k = \frac{6}{k}$, $n \geq 3k+u+1$ we see that

$$\mathbf{E}[\sum_{j=1}^k \mu_j^2 \|\hat{b}_j\|^2 / \|b_1\|^2] \leq \frac{1}{12} \sum_{j=1}^k q^{k-j} q^{j-1} = \frac{k}{12} q^{k-1},$$

$$\mathbf{E}[\sum_{j=k+1}^n \mu_j^2 \|\hat{b}_j\|^2 / \|b_1\|^2] \leq \frac{2^t}{12} \frac{q^k + 3q^{3k}}{1-q},$$

$$\mathbf{E}[\|b\|^2 / \|b_1\|^2] \leq \frac{1}{2q} + \frac{2^t}{12} \frac{q^k + 3q^{3k}}{1-q} < 0.99$$

holds for sufficiently large k, $k \geq e^{2^t(1+o(1))}$.

In this application our $(2^t, m, k)$-VSSP algorithm runs in $k 2^t (\frac{4}{3})^{2k/3} 2^{\frac{m}{t+1}}$ time, and even in $2k (\frac{4}{3})^{k/3} 2^{m/2}$ time for $t = 1$. (We use Lemma 3 with the α-values $q^{i/2}$ for $i = 0, ..., k-1$ and the inequality $\prod_{i=0}^{k-1} (1 - q^{i/2}/4) \geq (\frac{3}{4})^{2k/3}$.) Hence, GBS runs in $O(n^2 2^t (\frac{4}{3})^{2k/3} 2^{\frac{m}{t+1}})$ average time where $m = -\binom{k}{2} \frac{1}{2} \log_2 q$. This yields the claimed time bound. □

Simple GBS for $t = 1$ runs in $O(n^2 (\frac{4}{3})^{k/3} q^{-k^2/8})$ average time. Compared to Theorem 1 it reduces the *apfa* achievable in a given time to its square root, but requires massive space. Iteration of simple GBS via BKZ–reduction achieves in $O(n^3 (\frac{4}{3})^{k/3} q^{-k^2/8})$ average time *apfa* $q^{-n/2}$ for $q \leq (\frac{6}{k})^{1/k}$, $k \geq 60$.

Conclusion. Theorem 1 shows that the new method greatly improves the known algorithms for finding very short lattice vectors. This lowers the security of all lattice based cryptographic schemes. Simple birthday sampling may further decrease that security but its practicability hinges on the required space.

Acknowledgement. I am indepted to C. Tobias for carrying out the practical experiments reported in this paper and for providing Figure 1.

References

[A98]   M. Ajtai, The shortest vector problem in L_2 is NP-hard for randomised reductions. Proc. 30th STOC, pp. 10–19, 1998.

[A02]   M. Ajtai, The worst-case behaviour of Schnorr's algorithm approximating the shortest nonzero vector in a lattice. Preprint 2002.

[AKS01]   M. Ajtai, R. Kumar, and D. Sivakumar, A sieve algorithm for the shortest lattice vector problem. Proc. 33th STOC, 2001.

[CP91]   P. Camion, and J. Patarin, The knapsack hash function proposed at Crypto'89 can be broken.Proc. Eurocrypt'91, LNCS 457, Springer-Verlag, pp. 39–53, 1991.

[E81]   P. van Emde Boas, Another NP-complete partition problem and the complexity of computing short vectors in a lattice. Mathematics Department, University of Amsterdam, TR 81-04, 1981.

[GGH97]   O. Goldreich, S. Goldwasser, and S. Halevi, Public key cryptosystems from lattice reduction problems. Proc. Crypto'97, LNCS 1294, Springer-Verlag, pp. 112–131, 1997.

[H85]   B. Helfrich, Algorithms to construct Minkowski reduced and Hermite reduced bases. Theor. Comp. Sc. 41, pp. 125–139, 1985.

[K83]   R. Kannan, Minkowski's convex body theorem and integer programming. Mathematics of Operations Research 12 pp. 415–440, 1987, Preliminary version in Proc. 13th STOC, 1983.

[K97]   D. E. Knuth, The Art of Computer Programming, Vol 1, Fundamental Algorithms. 3rd Edidtion, Addison-Wesley, Reading, 1997.

[KS02]   H. Koy and C.P. Schnorr, Segment and strong Segment LLL-reduction of lattice bases. Preprint University Frankfurt, 2002, http://www.mi.informatik.uni-frankfurt.de/research/papers.html

[KS01a]   H. Koy and C.P. Schnorr, Segment LLL-reduction of lattice bases. Proc. CaLC 2001, LNCS 2146, Springer-Verlag, pp. 67–80, 2001.

[K00]   H. Koy, Primale/duale Segment-Reduktion von Gitterbasen. Slides of a lecture, Frankfurt, December 2000.

[KS01b]   H. Koy and C.P. Schnorr, Segment LLL-reduction of lattice bases with floating point orthogonalization. Proc. CaLC 2001, LNCS 2146, Springer-Verlag, pp. 81–96, 2001.

[LLL82]   A. K. Lenstra, H. W. Lenstra, and L. Lovász, Factoring polynomials with rational coefficients. Math. Ann. 261, pp. 515–534, 1982.

[M98]   D. Micciancio, The shortest vector in a lattice is NP-hard to approximate to within some constant. Proc. 39th Symp. FOCS, pp. 92–98, 1998, full paper SIAM Journal on Computing, 30 (6), pp. 2008–2035, March 2001.

[N99]   P.Q. Nguyen, Cryptanalysis of the Goldreich-Goldwasser-Halevi Cryptosystem from Crypto'97. Proc. Crypto'99, LNCS 1666, Springer-Verlag, pp. 288–304, 1999.

[NS00]   P.Q. Nguyen and J. Stern, Lattice reduction in cryptology: an update. Proc. ANTS-IV, LNCS 1838, Springer-Verlag, pp. 188–112. full version http://www.di.ens.fr/~{pnguyen,stern}/

[NTL]   V. Shoup, Number Theory Library. http: //www.shoup.net/ntl

[S87]   C.P. Schnorr, A hierarchy of polynomial time lattice basis reduction algorithms. Theor. Comp. Sc. 53, pp. 201–224, 1987.

[S88]   C.P. Schnorr, A more efficient algorithm for lattice reduction. J. of Algor. 9, 47–62, 1988.

[SE91]   C.P. Schnorr and M. Euchner, Lattice Basis Reduction and Solving Subset Sum Problems. Fundamentals of Comput. Theory, Lecture Notes in Comput. Sci., 591, Springer, New York, 1991, pp. 68–85. The complete paper appeared in Math. Programming Studies, 66A, 2, pp. 181–199, 1994.

[W02]   D. Wagner, A Generalized Birthday Problem. Proceedings Crypto'02, LNCS 2442, Springer-Verlag, pp. 288–303, 2002. full version http://www.cs.berkeley.edu/~daw/papers/

On the Ultimate Complexity of Factorials

Qi Cheng

School of Computer Science, the University of Oklahoma,
Norman, OK 73019, USA
qcheng@cs.ou.edu

Abstract. It has long been observed that certain factorization algorithms provide a way to write product of a lot of integers succinctly. In this paper, we study the problem of representing the product of *all* integers from 1 to n ($n!$) by straight-line programs. Formally, we say that a sequence of integers a_n is ultimately $f(n)$-computable, if there exists a nonzero integer sequence m_n such that for any n, $a_n m_n$ can be computed by a straight-line program (using only additions, subtractions and multiplications) of length at most $f(n)$. Shub and Smale [12] showed that if $n!$ is ultimately hard to compute, then algebraic version of $NP \neq P$ is true. Assuming a widely believed number theory conjecture concerning smooth numbers in short interval, a subexponential upper bound ($exp(c\sqrt{\log n \log \log n})$) for the ultimate complexity of $n!$ is proved in this paper, and a random subexponential algorithm constructing such a short straight-line program is presented as well.

Classification of Topics: Computational and structural complexity.

1 Introduction

Computing the factorial function ($n!$) is an important problem in computational complexity research. Due to the size of $n!$, computing it certainly takes exponential time in the Turing model. One can instead study the modular factorial $n! \bmod m$. Given an integer m, the smallest α such that $\gcd(\alpha! \bmod m, m) > 1$ is the smallest prime factor of m. For every $n \geq \alpha$, $\gcd(n! \bmod m, m)$ is greater than 1, hence we can use binary search to find α if we know how to compute $n! \bmod m$ efficiently for any m and n. This shows that the integer factorization problem can be reduced to computing $n! \bmod m$. It is very interesting to compare modular exponentiation with modular factorial. In some sense, the reason that primality testing is easy while factoring is hard is because modular exponentiation is easy but modular factorial is hard. This statement may underestimate the complexity of modular factorial, as it is believed that computing $n! \bmod m$ is much harder than the integer factorization problem. We don't even know whether computing modular factorial is an NP-easy problem or not.

One approach we may take to compute $n! \bmod m$ is to find a short straight-line program for $n!$. This problem relates to the algebraic model of computation [2,1,3]. In the algebraic model, it makes sense to ask whether the factorial problem has a polynomial time algorithm, because in this context, we only count

H. Alt and M. Habib (Eds.): STACS 2003, LNCS 2607, pp. 157–166, 2003.

the number of ring operations used to compute $n!$ in the time complexity, regardless of the size of the operands. Sometimes, algebraic complexity is also called non-scalar complexity. If the function $n!$ has polynomial time algebraic complexity, or equivalently, every integer $n!$ has a short straight-line program uniformly, then by doing modulo m in every step, we obtain a polynomial time algorithm to compute $n! \bmod m$ in the Turing model, which is thought to be unlikely. Throughout this paper, we assume that a straight-line program only contains ring operations. Shamir [11] showed that if division (computing remainder and quotient) is allowed, then $n!$ can be computed by a straight-line program of polynomial length.

The ultimate complexity of a number was first studied in [12] by Shub and Smale. They found a surprising relation between the ultimate complexity of $n!$ and the algebraic version of NP vs. P problem. We say that $n!$ is ultimately hard to compute, if there does not exist a non-zero integer sequence m_n, such that $n!m_n$ can be computed by straight-line programs of length $(\log n)^c$ for an absolute constant c. It was proved in [12]:

> If $n!$ is ultimately hard to compute, then the algebraic version of $NP \neq P$ is true.

Note that in the Turing model, proving that the modular factorial problem is hard does not necessarily imply that $NP \neq P$. There is no corresponding concept of the ultimate complexity in the Turing model.

So far the best algorithm we know computes $n!$ in $O(\sqrt{n} \log^2 n)$ ring operations over $\mathbf{Z}$ [14,4]. No better upper bound has been reported for the ultimate complexity of $n!$. It has long been noticed that certain factorization algorithms provide a way to write product of a lot of primes succinctly. For instance, Lenstra's elliptic curve factorization method [10], performs algebraic computation modulo the integer to be factored, but the operations remain the same for all inputs of certain size. The algebraic computation essentially generates a number with a lot of prime factors, since it factors almost all integers of the size. However, these algorithms do not directly give us a straight-line program to compute a product of $n!$, because

1. Divisions are essential in these algorithms. For instance, in the elliptic curve factorization method, splitting of an integer n happens precisely when taking inverse of a integer modulo n cannot process.
2. More importantly, all the fast factorization algorithms are random in nature. The time complexity of a random algorithm is an *average* measurement. Practically the algorithms should work as expected, but theoretically it is possible that for any integer n, there are bad choices of random bits such that the algorithm will take exponential time to stop. Hence for any choice of random bits of certain length, there are integers which cannot be factored.

In this paper we give a formal proof of a subexponential upper bound for the ultimate complexity of $n!$ under a widely believed number theory conjecture. The essential part of the proof is still based on Lenstra's elliptic curve factorization method. Our result is constructive in the sense that we can construct the straight-line program from n in subexponential time. More precisely, our paper presents a Monte Carlo random algorithm (certainly in the Turing model), given a natural number n as input, output a straight-line program which computes a non-zero multiple of $n!$ with probability better than a constant. The algorithm runs in subexponential time, hence the output straight-line program will have at most a subexponential length. Our result suggests that the ultimate complexity of $n!$ is not as high as the complexity of $n!$. This result also shows that the complexity of certain multiple of $n!$ is much closer to the integer factorization problem than the complexity of $n!$ itself.

It is interesting to note that we don't know whether there exists a subexponential straight line program for the polynomial $(x-1)\cdots(x-n)$, or any polynomial with a lot of distinct integral roots. If we apply the same technique in the paper to construct straight line program, we encounter obstacles from the Uniform Boundedness Theorem [6].

Let $\mathcal{E}$ be an elliptic curve $y^2 = x^3 + ax + b$ with $a,b \in \mathbf{Z}$ and $P_s(x)$ be the univariate s-th division polynomial of $\mathcal{E}$. Given n, $P_n(x)$ can be computed by a straight-line program of length $O(\log n)$ using 1, x, a and b as constants. If x, a and b are integers less than n, then $P_n(x)$ can be calculated by $O(\log n)$ arithmetic operations using 1 as the only constant. Let x be an integer which is not the x-coordinate of a torsion on $\mathcal{E}$, i.e. $P_i(x) \neq 0$ for any positive integer i. For any prime p, we have $p|P_s(x)$ if s is divisible by $|E(\mathbf{F}_p)|$, where E is the reduction of $\mathcal{E}$ at p and x is the x-coordinate of a point on $E(\mathbf{F}_p)$ (x may or may not be an x-coordinate of a point on $\mathcal{E}(\mathbf{Q})$).

If the reduction of $\mathcal{E}$ at a random prime p takes a random number between $p - 2\sqrt{p} + 1$ and $p + 2\sqrt{p} + 1$ as the order over $\mathbf{F}_q$, then with probability greater than 1 over a subexponential function on $\log p$, the reduction curve has a smooth order over $\mathbf{F}_p$. Furthermore, given an elliptic curve $E/\mathbf{F}_p$, a random integer x becomes an x-coordinate of a point on $E(\mathbf{F}_p)$ with a constant probability (about $1/2$). Hence if S is a large smooth number and x is an arbitrary integer, $P_S(x)$ contains a lot of distinct prime factors. In order to get a multiple of $n!$, we only need to collect subexponentially many elliptic curves and evaluate their S-th division polynomials at polynomially many integers. We will show that randomly chosen elliptic curves and integers suffice. The effects of the global torsions will be carefully controlled.

1.1 Main Results

We call a number smooth if all of its prime factors are small. More precisely, a number is said to be y-smooth, if all of its prime factors are less than y. Let

$$\Psi(x,y) = |\{n \leq x, n \text{ is } y - smooth\}|.$$

Throughout this paper, let $L_x(\beta, c)$ denote $e^{c \log^\beta x \log \log^{1-\beta} x}$. A fundamental theorem about $\Psi(x, L_x(1/2, a))$ was proved in [5].

Proposition 1. *For any constant* a, $\Psi(x, L_x(1/2, a)) = x L_x(1/2, -1/(2a) + o(1))$.

It was conjectured that the smooth number in some short interval is as dense as in a large interval. In particular,

Conjecture 1. For any constant $a > 0$,

$$\Psi(p + 1 + 2\sqrt{p}, L_p(1/2, a)) - \Psi(p + 1 - 2\sqrt{p}, L_p(1/2, a)) = \sqrt{p} L_p(1/2, -1/(2a) + o(1)).$$

Though this conjecture has not been proved yet, it is widely believed to be true. See [10,9] for details. In fact, Lenstra's elliptic curve factorization algorithm relies on this conjecture to achieve the subexponential time complexity.

Theorem 1. *Assume that Conjecture 1 is true. There exist absolute constants* c_1 *and* c_2 *such that for any natural number* n, *a non-zero multiple of* $n!$ *can be computed by a straight-line program of length at most* $L_n(1/2, c_1)$. *Furthermore, the straight-line program can be constructed in time* $L_n(1/2, c_2)$ *by a probabilistic Turing machine.*

In this paper, we focus on the theoretical aspects of the problem and do not consider the practical performance. This paper is organized as follows. In Section 2, we define the straight-line program and the ultimate complexity, and prove a lemma about bipartite graphs. In Section 3, we review some facts about elliptic curves. In Section 4, we prove the main theorem. We conclude this paper by a discussion section.

2   Preliminaries

A straight-line program of an integer is a sequence of ring operations, which outputs the integer in the last operation. Formally

Definition 1. *A straight-line program of an integer* m *is a sequence of instructions*

$$z \leftarrow x \alpha y$$

where $\alpha \in \{+, -, *\}$, x, y *are two previously appeared symbols or 1 and* z *is a new symbol, such that after we execute the instructions sequentially, the last symbol will represent the value of* m. *The length of the program is the number of instructions. The length of the shortest straight-line program of* m *is called the straight-line complexity of* m.

An integer n has a straight-line complexity at most $2 \log n$. In some cases, a straight-line program is a very compact description of an integer. It can represent a huge number in small length. For example, the number n^m can be computed using the repeated squaring technique and hence has a straight-line complexity at most $2 \log n + 2 \log m$.

Definition 2. *A sequence of integer a_n is ultimately $f(n)$-computable, if there exists a nonzero integer sequence m_n such that $a_n m_n$ can be computed by a straight-line program of length at most $f(n)$. The smallest $f(n)$ is called the ultimate complexity of a_n.*

In this paper, we study the ultimate complexity of $n!$. First we show that this problem can be reduced to studying the ultimate complexity of the product of primes up to n.

Lemma 1. *Let p_n be the n-th prime number. If the sequence $a_n = p_1 p_2 \cdots p_m$, where p_m is the largest prime less than or equal to n, can be ultimately computed by a straight-line program of length $f(n)$, then $n!$ can be ultimately computed by a straight-line program of length $f(n) + 2 \log n$.*

Proof. This follows from a simple fact that $n! | a_n^n$. Note that the exponent n is minimum possible.

2.1 Lemma about Bipartite Graphs

Given a bipartite graph $G = X \times Y$, we say that a subset $A \subseteq X$ dominates a subset $B \subseteq Y$, if every vertex in B is adjacent to at least one vertex in A.

Lemma 2. *For a simple undirected bipartite graph $X \times Y$, let $m = |X|$ and $n = |Y|$. If every vertex in X has degree greater than $d = \lceil n/r \rceil$, $r > 2$, then there exists a subset $S \subseteq Y$, with cardinality $g = \lceil 2r \log m \rceil$, which dominates X. Moreover if randomly choose a subset of Y with cardinality g, it dominates X with probability greater than $1 - \frac{1}{m}$.*

Proof. From $X \times Y$, we construct a new bipartite graph $X \times \mathcal{Y}$ as follows. $\mathcal{Y}$ is the set of all the subsets of Y with g elements. For any $u \in X$ and $v \in \mathcal{Y}$, u and v are joined by an edge iff in $X \times Y$, u is adjacent to at least one vertex in $v \subseteq Y$.

For every $u \in X$, its degree in $X \times \mathcal{Y}$ is greater than $\binom{n}{g} - \binom{n-d}{g}$. The total number of edges in $X \times \mathcal{Y}$ is thus greater than $m(\binom{n}{g} - \binom{n-d}{g})$. The average degree of elements in $\mathcal{Y}$ is greater than

$$\frac{m(\binom{n}{g} - \binom{n-d}{g})}{\binom{n}{g}} = m(1 - \binom{n-d}{g}/\binom{n}{g}).$$

We have

$$
\begin{aligned}
\binom{n-d}{g} \Big/ \binom{n}{g} &= \frac{(n-d)!/(n-d-g)!}{n!/(n-g)!} \\
&= \frac{(n-d)(n-d-1)\cdots(n-d-g+1)}{n(n-1)\cdots(n-g+1)} \\
&< (1 - \frac{d}{n})^g < (1 - \frac{1}{r})^{2r \log m} \\
&< \frac{1}{m^2}.
\end{aligned}
$$

Suppose that $x|\mathcal{Y}|$ vertices in $\mathcal{Y}$ have degree less than m. The average degree of vertices in $\mathcal{Y}$ is less than $m(1-x)+(m-1)x = m-x$. Hence $m-x > m(1-\frac{1}{m^2})$. This implies that $x < \frac{1}{m}$.

This lemma will be used several times in the paper. First we present a simple consequence of the lemma.

Corollary 1. *Let p be a prime. If we randomly pick $n = \lceil 6 \log p \rceil$ integers $a_1, a_2, \cdots, a_n$ between 2 and p inclusive, then with probability at least $1 - \frac{2 \log p}{p}$, for every prime q, $2 < q \le p$, at least one of integers in $\{a_1, a_2, \cdots, a_n\}$ is a quadratic nonresidue modulo q.*

Proof. For every prime q, $2 < q \le p$, at most $p/3$ of the integers between 2 and p inclusive have prime factor q. For the rest of integers, half of them are quadratic nonresidues modulo q. Hence at least $p/3$ of the integers in the same range are quadratic nonresidues modulo q. By replacing r with 3 in Lemma 2 we get the corollary. Note that there are about $\frac{p}{\log p}$ primes less than p.

3   Elliptic Curves

An elliptic curve is a smooth cubic curve. Let k be a field. If the characteristic of k is not 2 or 3, we may assume that the elliptic curve is given by an equation of the form

$$y^2 = x^3 + ax + b, \qquad a, b \in k.$$

The discriminant of this curve is defined as $-16(4a^3 + 27b^2)$, whose essential part is the discriminant of the polynomial $x^3 + ax + b$. It should be non-zero as the curve is smooth. For detailed information about elliptic curves, we refer to Silverman's book [13].

The set of points on an elliptic curve consists of the solution set of the definition equation plus a point at infinity. These points form an abelian group with the infinity point as the identity. We call a point *torsion* if it has a finite order in the group. The x-coordinates of the torsions of order $n > 3$ are the solutions of $P_n^{\mathcal{E}}(x)$, the n-th division polynomial of E. These polynomials can be computed recursively. Sometimes we omit the superscription $\mathcal{E}$ if there is no confusion about the curve. We can verify that the recursion for $P_n(x)$ does not involve any division. By applying the technique similar to the repeated squaring, we get

Proposition 2. *For any integers $n(> 0)$ and x, the integer $P_n^{\mathcal{E}}(x)$ can be computed by a straight-line program of length $O(\log n + \log(|x| + 1) + \log(|a| + 1) + \log(|b| + 1))$, where $\mathcal{E}$ is the elliptic curve $y^2 = x^3 + ax + b$ with $a, b \in \mathbf{Z}$.*

See [6] for the proof of (a stronger version of) the proposition.

Proposition 3. *Let $\mathcal{E} : y^2 = x^3 + ax + b$ be an elliptic curve defined over $\mathbf{Z}$. Assume that p doesn't divide the discriminant. If x is an integer and*

1. *it is the x-coordinate of a point on* $E(\mathbf{F}_p)$,
2. *the point* $(x, \sqrt{x^3 + ax + b})$ *is not a torsion on* $\mathcal{E}$,

then $P_l(x) \neq 0$ *and* $p | P_l(x)$, *where* l *is any non-zero multiple of* $|E(\mathbf{F}_p)|$.

Let $\mathcal{E} : y^2 = x^3 + ax + b$ be an elliptic curve defined over $\mathbf{Z}$. The torsion points on $\mathcal{E}$ with integral x-coordinates (thus y-coordinates are integers or quadratic algebraic numbers) have order at most 18, as shown in the celebrated Uniform Boundedness Theorem in the quadratic number fields [7,8]. Hence such integers must be the roots of $P_n(x)$ with $n \leq 18$, or of $x^3 + ax + b$. The maximal possible roots of those equations are bounded by the sum of the degrees of the equations, which is an absolute constant. Let B denote this constant. Define

$$R_{\mathcal{E}}(p) = \{x | x \in \mathbf{Z}, 1 \leq x \leq p, (x, \sqrt{x^3 + ax + b}) \text{ is not a torsion on } \mathcal{E}\}.$$

Then $|R_{\mathcal{E}}(p)| \geq p - B$. Given an integer, we can decide whether the integer is in $R_{\mathcal{E}}(p)$ in polynomial time. From Lemma 2, we conclude

Corollary 2. *Let* p *be a prime and* $\mathcal{E} : y^2 = x^3 + ax + b$ *be an elliptic curve defined over* $\mathbf{Z}$ *with* $1 \leq a \leq p-1$ *and* $1 \leq b \leq p-1$. *If* $n = \lceil 6 \log p \rceil$ *integers* $x_1, x_2, \cdots, x_n$ *are randomly chosen from* $R_{\mathcal{E}}(p)$, *then with probability greater than* $1 - \frac{2 \log p}{p}$, *for any prime* q *satisfying* $7B < q \leq p$ *and* $q \nmid 4a^3 + 27b^2$, *one of* x_i *is the x-coordinate of a point on the reduction of* $\mathcal{E}$ *at* q.

Proof. We construct a bipartite graph $P \times Y$ as follows. The set P consists of all the prime numbers from $7B$ to p which are not the prime factors of $4a^3 + 27b^2$. Let $Y = R_{\mathcal{E}}(p)$. For any $q \in P$ and $x \in Y$, draw an edge between q and x iff $x^3 + ax + b$ is a quadratic residue modulo q.

The degree of every element in P is greater than $(\frac{q - 2\sqrt{q}}{2} - B) \times \frac{p}{q} > \frac{q}{3} \times \frac{p}{q} = \frac{p}{3}$ for $q > 7B$. The theorem now follows from Lemma 2.

The j-invariant of the curve $y^2 = x^3 + ax + b$ is defined as $j = 1728 \frac{4a^3}{4a^3 + 27b^2}$. Two elliptic curves with a same j-invariant are isomorphic over the algebraic closed field. For elliptic curves defined over a prime finite field $\mathbf{F}_p$, $p > 3$, two curves with a same j-invariant may not be isomorphic. If $j \neq 0$ or 1728, there are exactly two isomorphic classes which have the same j-invariant, one can be represented by $y^2 = x^3 + kx + k$ and the other by $y^2 = x^3 + c^2 kx + c^3 k$, where $k = \frac{27j}{4(1728-j)}$ and c is a quadratic nonresidue modulo p. There are different number of points over the two classes of curves. There are at most 6 isomorphic classes with $j = 0$, and at most 4 isomorphic classes with $j = 1728$.

We are interested in counting the number of non-isomorphic elliptic curves with the number of points coming from a given set. In [10] the following proposition was proved.

Proposition 4. *There exist two constants* c_1, c_2 *such that if* A *is a set of integers between* $p + 1 - \sqrt{p}$ *and* $p + 1 + \sqrt{p}$, *the number of non-isomorphic classes of elliptic curves defined over* $\mathbf{F}_p$ *whose number of points over* $\mathbf{F}_p$ *are in* A *is*

$$c_1 \sqrt{p}(|A| - 2)/\log p \leq N \leq c_2 \sqrt{p}|A| \log p (\log \log p)^2.$$

4   Proof of the Main Theorem

Our goal is to construct a straight-line program of some multiple of $\alpha_p = 2 \times 3 \times 5 \times \cdots \times p$ in $L_p(1/2, c_1)$ time for some constant c_1. Firstly, we compute a number $S = 2^{e_1} \times 3^{e_2} \cdots \times p_s^{e_s}$, where p_s is the maximal prime less than or equal to $L_p(1/2, 1)$ and for every $1 \leq i \leq s$, $p_i^{e_i}$ is the least p_i-power greater than $p + 1 + 2\sqrt{p}$. Certainly we can compute S in time $L_p(1/2, 2 + o(1))$.

Secondly, we randomly choose $l = \lceil 6 \log p \rceil$ integers $c_1, c_2, \cdots, c_l$ between 2 and p inclusive. The step is successful if for every prime $2 < q \leq p$, at least one of the integers is a quadratic nonresidue mod q. The step succeeds with probability greater than $1 - \frac{2 \log p}{p}$ according to Lemma 1.

Denote by **D** the set of elliptic curve $\{y^2 = x^3 + ax + a | 1 \leq a \leq p\} \cup \{y^2 = x^3 + ac_i^2 x + ac_i^3 | 1 \leq i \leq l, 1 \leq a \leq p\}$. Construct a bipartite graph $X \times \mathbf{D}$ as follows. X consists of all the primes between $7B + 1$ and p inclusive. For any prime $q \in X$ and any elliptic curve $\mathcal{E} \in \mathbf{D}$, connect q and $\mathcal{E}$ by an edge iff the reduction curve E of $\mathcal{E}$ at q is non-singular, and the order of $E(\mathbf{F}_q)$ is $L_p(1/2, 1)$-smooth.

Lemma 3. *The degree of every element in X is greater than $pL_p(1/2, -1/2 + o(1))$ under Conjecture 1.*

Proof. For any prime $7B < q \leq p$, consider the subset of **D**:

$$\mathbf{D}_q = \{y^2 = x^3 + ax + a | 1 \leq a \leq q\} \cup \{y^2 = x^3 + ac_i^2 x + ac_i^3 | 1 \leq i \leq l, 1 \leq a < q\}.$$

The j-invariants of $y^2 = x^3 + ax + a$ and $y^2 = x^3 + ac_i^2 x + ac_i^3$ are $1728 \frac{4a}{4a+27}$. If one of integers in $\{c_1, c_2, \cdots, c_n\}$ is a quadratic nonresidue modulo q, then there exist representations of all the isomorphic classes of elliptic curves over $\mathbf{F}_q$ in $\mathbf{D}_q$, except for the curves with j-invariants 0 or 1728. There are at least $\frac{\sqrt{q}}{L_q(1/2, 1/2 + o(1))}$ $L_q(1/2, 1)$-smooth integers between $q - 2\sqrt{q} + 1$ and $q + 2\sqrt{q} + 1$. Hence there are at least $\sqrt{q} \frac{\sqrt{q}}{L_q(1/2, 1/2 + o(1))} = \frac{q}{L_q(1/2, 1/2 + o(1))}$ curves in $\mathbf{D}_q$ which have $L_q(1/2, 1)$-smooth orders over $\mathbf{F}_q$ according to Proposition 4. In the set **D**, at least $\frac{q}{L_q(1/2, 1/2 + o(1))} \frac{p}{q} > \frac{p}{L_p(1/2, 1/2 + o(1))}$ curves have $L_p(1/2, 1)$-smooth order over $\mathbf{F}_q$. Hence the degree of q in $X \times \mathbf{D}$ is greater than $\frac{p}{L_p(1/2, 1/2 + o(1))}$.

Now we proceed to the third step. We randomly choose $w = \lceil L_p(1/2, 1) \rceil$ curves $\mathcal{E}_1, \cdots, \mathcal{E}_w$ from D. The step is successful if for any prime $7B \leq q \leq p$, q doesn't divide discriminant of at least one of the curves in $\{\mathcal{E}_1, \cdots, \mathcal{E}_w\}$ and the reduction of this curve at q has a $L_p(1/2, 1)$-smooth order over $\mathbf{F}_q$. In the other words, in graph $X \times \mathbf{D}$, $\{\mathcal{E}_1, \cdots, \mathcal{E}_w\} \subseteq \mathbf{D}$ dominates X. Since $L_p(1/2, 1) > 2 \log p L_p(1/2, 1/2 + o(1))$, the step succeeds with probability at least $1 - \frac{2 \log p}{p}$ according to Lemma 2 and Lemma 3.

In the fourth step, for each $1 \leq i \leq w$, we pick $h = \lceil 6 \log p \rceil$ random integers $x_{i,1}, x_{i,2}, \cdots, x_{i,h}$ in $R_{\mathcal{E}_i}(p)$. The i-th sub-step is successful, if for any prime $7B < q \leq p$, at least one integer in $\{x_{i,1}, x_{i,2}, \cdots, x_{i,h}\}$ is the x-coordinate of a $\mathbf{F}_q$-point in the reduction curve of $\mathcal{E}_i$ at q. The successful probability for each sub-step is greater than $1 - \frac{2\log p}{p}$ according to Corollary 2. Hence the successfully probability for this step is greater than $(1 - \frac{2\log p}{p})^w$.

Lemma 4. *All these four steps are successful with probability*

$$(1 - \frac{2\log p}{p})^{L_p(1/2,1/2+o(1))} > 1/3.$$

If all the four steps are successful, then we can get a multiple of α_p by evaluating the S-th division polynomials of $\mathcal{E}_1, \cdots, \mathcal{E}_w$ on $x_{1,1}, x_{1,2}, \cdots, x_{1,h}; \cdots; x_{w,1}, \cdots, x_{w,h}$ respectively and multiplying the results together. Now we are ready to write the straight-line program for a multiple of $2 \times 3 \times 5 \times \cdots \times p$.

1. Start by computing the product of all the primes less than $7B$. Let the result be T_1.
2. Add instructions to compute

$$P_S^{\mathcal{E}_1}(x_{1,1}), \cdots, P_S^{\mathcal{E}_1}(x_{1,h}); \cdots; P_S^{\mathcal{E}_w}(x_{w,1}), \cdots, P_S^{\mathcal{E}_w}(x_{w,h}).$$

3. Add instructions to compute

$$T_2 \leftarrow \prod_{1 \leq i \leq w, 1 \leq k \leq h} P_S^{\mathcal{E}_i}(x_{i,k}).$$

4. Add $T \leftarrow T_1 \times T_2$ into the straight-line program.

Based on the analysis in this paper, it can be verified that the above straight-line program computes α_p ultimately and it has subexponential length.

5   Discussion

The relation between ultimate complexity and integer factorization can be further explored.

Firstly, can we derive a factorization algorithm from a straight-line program for a multiple of $n!$? The only problem here is that the multiple of $n!$, i.e. $n!m_n$, may contain primes greater that n. We must try to restrict the integer m_n such that it only has primes less than n. It seems hard to do this with the algorithm in this paper.

Secondly, is the lower bound of the ultimate complexity of $n!$ also subexponential? Since this problem is closely related to the integer factorization problem, which is believed not having a polynomial time algorithm, we suspect that the answer to this question is positive.

The existence of a short straight-line program for a large number does not imply that we can construct the short straight-line program in reasonable time. Given two integers m, n and a prime p, if m is the generator of $\mathbf{F}_p^*$ and $p \nmid n$, then there exists a short straight-line program for a power of m which is congruent to n modulo p. But we don't know how to construct such a straight-line program from m, n and p, as the problem is equivalent to computing the discrete logarithm problem over $\mathbf{F}_p$. We believe that it may be possible that for some n, $n!$ or a multiple of $n!$ have very short straight-line programs, however constructing the program would be very hard.

References

1. Lenore Blum, Felipe Cucker, Michael Shub, and Steve Smale. *Complexity and Real Computation.* Springer-Verlag, 1997.
2. Lenore Blum, Mike Shub, and Steve Smale. On a theory of computation and complexity over the real numbers: NP-completeness, recursive functions and universal machine. *Bulletin of the American Mathematical Society*, 21(1), 1989.
3. Peter Burgisser. The complexity of factors of multivariate polynomials. In *Proc. 42th IEEE Symp. on Foundations of Comp. Science*, 2001.
4. Peter Burgisser, Michael Clausen, and M. Amin Shokrollahi. *Algebraic Complexity Theory*, volume 315 of *Grundlehren der mathematischen.* Springer-Verlag, 1997.
5. E.R. Canfield, P. Erdos, and C. Pomerance. On a problem of oppenheim concerning "Factorisatio Numerorum". *J of number theory*, pages 1–28, 1983.
6. Qi Cheng. Some remarks on the *L*-conjecture. In *Proc. of the 13th Annual International Symposium on Algorithms and Computation(ISAAC)*, volume 2518 of *Lecture Notes in Computer Science.* Springer-Verlag, 2002.
7. S. Kamienny. Torsion points on elliptic curves and q-coefficients of modular forms. *Inventiones Mathematicae*, 109:221–229, 1992.
8. M. Kenku and F. Momose. Torsion points on elliptic curves defined over quadratic fields. *Nagoya Mathematical Journal*, 109:125–149, 1988.
9. A. Lenstra and H. W. Lenstra Jr. *Handbook of Theoretical Computer Science A*, chapter Algorithms in Number Theory, pages 673–715. Elsevier and MIT Press, 1990.
10. H. W. Lenstra. Factoring integers with elliptic curves. *Annals of Mathematics*, 126:649–673, 1987.
11. A. Shamir. Factoring numbers in $O(\log n)$ arithmetic steps. *Information Processing Letters*, 1:28–31, 1979.
12. M. Shub and S. Smale. On the intractability of Hilbert's nullstellensatz and an algebraic version of "P=NP?". *Duke Math. J.*, 81:47–54, 1995.
13. J.H. Silverman. *The arithmetic of elliptic curves.* Springer-Verlag, 1986.
14. V. Strassen. Einige resultate uber berechnungskomplexitat. *Jber. Deutsch. Math.-Verein*, 78(1):1–8, 1976/77.

On the Effective Jordan Decomposability

Xizhong Zheng[1]*, Robert Rettinger[2], and Burchard von Braunmühl[1]

[1] BTU Cottbus, 03044 Cottbus, Germany
[2] FernUniversität Hagen, 58084 Hagen, Germany

Abstract. The classical Jordan decomposition Theorem says that any real function of bounded variation can be expressed as a difference of two increasing functions. This paper explores the effective version of Jordan decomposition. We give a sufficient and necessary condition for those computable real functions of bounded variation which can be expressed as a difference of two computable increasing functions. Using this condition, we prove further that there is a computable real function which has even a computable modulus of absolute continuity (hence is of bounded variation) but it is not a difference of any two computable increasing functions. The polynomial time version of this result holds too and this gives a negative answer to an open question of Ko in [6].

Topic Classification: Computational Complexity; Computable Analysis

1 Introduction

According to Grzegorczyk [4], a computable real function $f : [0;1] \to \mathbb{R}$ is an effectively uniformly continuous function which maps every computable sequence of real numbers to a computable one. In other words, the computability of real functions is an effectivization of the continuity. Of course, continuity is one of the most important property of a real function. However, there are many problems, especially in applications to physical science, where more precise information than the continuity is required. For example, it is often required to measure how rapidly a real function $f : [0;1] \to \mathbb{R}$ oscillates. Such an oscillatory character of a function can be determined quantitatively by the variation of the function. This quantity turns out to be very useful for problems in physics, engineering, probability theory, and so fourth. Precisely, let $f : [a;b] \to \mathbb{R}$ be a total function on an interval $[a;b]$, the *variation of f over* $[a;b]$, denoted by $V_a^b(f)$, is defined by

$$V_a^b(f) := \sup \sum_{i<n} |f(a_i) - f(a_{i+1})| \qquad (1)$$

where the supremum is taken over all subdivisions $a = a_0 < a_1 < \cdots < a_n = b$ of the interval $[a;b]$. If $V_a^b(f)$ is finite, then we say that f is of *bounded variation*

---

* Corresponding author, email: zheng@informatik.tu-cottbus.de

H. Alt and M. Habib (Eds.): STACS 2003, LNCS 2607, pp. 167–178, 2003.

on $[a; b]$ (BV on $[a; b]$ for short). The class of all BV-functions on $[a; b]$ is denoted by $\mathbb{BV}[a, b]$ or simply $\mathbb{BV}$ if the underlying interval is clear from the context. If $f : [0; 1] \to \mathbb{R}$ is a BV-function, then the function v_f defined by $v_f(x) := V_0^x(f)$ is called the *total variation function* of f. The concept of functions of bounded variation is originated by Camille Jordan in [5]. The most important property of a BV-function is the Jordan decomposition. Namely, any BV-function f can be expressed as a difference $f = g - h$ of two increasing functions g, h. Here, increasing function means always non-strictly increasing, i.e., $g(x) \le g(y)$ for any $x \le y$. If f is a continuous BV-function, then the corresponding g, h can be continuous too. More general properties of BV-functions and their applications are widely discussed in classical mathematics and effective analysis as well as in constructive mathematics ([1,2,3,6,8,10]).

In this paper, we are interested in the computable total functions $f : [a; b] \to \mathbb{R}$ which are of bounded variation (CBV-function, for short) for some computable real numbers $a < b$. According to the *Effective Weierstrass Theorem* (see [4,7,9]), f is a *computable function* if and only if there exists a computable sequence (p_s) of rational polygon functions such that $|f(x) - p_s(x)| \le 2^{-s}$ for any $x \in [a; b]$ and $s \in \mathbb{N}$. Notice that, if a, b are computable real numbers and $f : [a; b] \to \mathbb{R}$ is a CBV-function, then the function $g : [0; 1] \to [0; 1]$ defined by $g(x) := f(a + (b - a)x)/m$ is also a CBV-function such that $V_a^b(f) = V_0^1(g)$, where $m := \max\{f(x) : x \in [a; b]\}$. Therefore, we can restrict ourselves w.l.o.g. to total functions $f : [0; 1] \to [0; 1]$. The class of all total CBV-functions $f : [0; 1] \to [0; 1]$ is denoted by $\mathbb{CBV}[0; 1]$ or simply $\mathbb{CBV}$.

A computable function f is called *effective Jordan decomposable* (EJD, for short) if there are two increasing computable functions f_1, f_2 such that $f = f_1 - f_2$. Not every CBV-function is EJD as shown in [10]. The argument in [10] is based on the following observation that if f is EJD, then its total variation function v_f has a computable modulus of continuity. Therefore, the counterexamples of non-EJD functions given in [10] are not effectively absolutely continuous, where a computable function $f : [0; 1] \to [0; 1]$ is called *effectively absolutely continuous* (EAC for short) if there is a computable function $m : \mathbb{N} \to \mathbb{N}$ (modulus for absolute continuity of f) such that,

$$\sum_{i \le k} |b_i - a_i| \le 2^{-m(n)} \implies \sum_{i \le k} |f(b_i) - f(a_i)| \le 2^{-n}, \tag{2}$$

for any set $\{[a_i; b_i] : i \le k\}$ of non-overlapping subintervals of $[0; 1]$ and for any $n \in \mathbb{N}$. It is well known that any absolutely continuous function is of bounded variation and hence has a Jordan decomposition. Then we can naturally ask, whether every EAC function is also EJD? This question was first asked by Ker-I Ko in [6] for polynomial time computable functions. In [6], Ko has shown at first that there is a polynomial time computable real function f of bounded variation which is not a difference of any two polynomial time computable increasing functions. However he left the following problem open there: whether every polynomial time computable function which has a polynomial modulus of absolutely continuity (PAC function for short) can be expressed as a difference of two polynomial time computable increasing functions? A negative answer to this question will be given in this paper. To this end, we will show a sufficient and necessary condition for EJD first in Section 2. An EAC but not EJD function is constructed in Section 3. In the last Section 4 we construct a PAC function which is not even EJD and this gives a negative answer to Ko's question. Both the proofs in section 3 and 4 apply the criterion for EJD introduced in Section 2.

Let $\langle \cdot, \cdot \rangle : \mathbb{N}^2 \to \mathbb{N}$ be the standard pairing function defined by $\langle n, m \rangle := (n+m)(n+m+1)/2 + n$. $\langle \cdot, \cdot \rangle$ is a 1:1 computable function. For any set A, B, a partial function f with $\text{dom}(f) \subseteq A$ and $\text{range}(f) \subseteq B$ is always denoted by $f :\subseteq A \to B$ while the total functions f from A to B are denoted by $f : A \to B$. Let (M_e) be an effective enumeration of all Turing machines and M_e computes the function $\varphi_e :\subseteq \mathbb{N} \to \mathbb{N}$. Then (φ_e) is an effective enumeration of all partial computable functions $\varphi_e :\subseteq \mathbb{N} \to \mathbb{N}$. Denote by $\varphi_{e,s}$ the function computed by M_e up to step s. Then $(\varphi_{e,s})$ is a uniformly effective approximation of (φ_e) such that $\lim_{s \to \infty} \varphi_{e,s} = \varphi_e$ and the set $\{\langle e, x, y, s \rangle : \varphi_{e,s}(x) = y\}$ is a recursive set. Notice that, if $\varphi_{e,s}(n)$ is defined, then we have $\varphi_{e,t}(n) \downarrow= \varphi_{e,s}(n) = \varphi_e(n)$ for any $t \geq s$, where $\varphi_{e,t}(n) \downarrow$ means that $\varphi_{e,t}(n)$ is defined. The same notations will be also used in this paper for other type of functions like $\varphi_e :\subseteq \mathbb{N} \to \mathbb{Q}$, $\varphi_e :\subseteq \mathbb{Q} \times \mathbb{N} \to \mathbb{Q}$, etc.

2 A Sufficient and Necessary Condition for EJD

Classically, every BV-function is Jordan decomposable. However, as shown in [10], there is a CBV-function which is not EJD. In this section, we show a sufficient and necessary condition for the CBV-functions which are EJD.

Theorem 2.1. *Let* $f \in \mathbb{CBV}$. *Then* f *is EJD iff there is a total computable increasing function* $\varphi : [0; 1] \to \mathbb{R}$ *such that* $V_x^y(f) \leq \varphi(y) - \varphi(x)$, *for any* $x, y \in [0; 1]$ *with* $x \leq y$.

Proof. "$\Rightarrow$". Suppose that $f : [0; 1] \to [0; 1]$ is a CBV-function which has an effective Jordan decomposition $f = f_1 - f_2$ where $f_1, f_2 : [0; 1] \to \mathbb{R}$ are two increasing computable functions. Then, for any $x, y \in [0; 1]$ with $x \leq y$, we have

$$V_x^y(f) = \sup \sum_{i<k} |f(a_i) - f(a_{i+1})|$$

$$\leq \sup \sum_{i<k} (|f_1(a_i) - f_1(a_{i+1})| + |f_2(a_i) - f_2(a_{i+1})|)$$

$$= (f_1(y) - f_1(x)) + (f_2(y) - f_2(x))$$

$$= (f_1(y) + f_2(y)) - (f_1(x) + f_2(x)).$$

where the supremum is taken over all possible subdivisions $x = a_0 < a_1 < \cdots < a_k = y$ of the interval $[x; y]$ and any $k \in \mathbb{N}$. Define the computable function $\varphi : [0; 1] \to \mathbb{R}$ simply by $\varphi(x) := f_1(x) + f_2(x)$ for any $x \in [0; 1]$. Then φ satisfies $V_x^y(f) \leq \varphi(y) - \varphi(x)$, for any $x, y \in [0; 1]$ with $x \leq y$ obviously.

"$\Leftarrow$". Let $f : [0;1] \to [0;1]$ be a CBV-function and φ an increasing computable function such that $V_x^y(f) \leq \varphi(y) - \varphi(x)$ for any $x, y \in [0;1]$ with $x \leq y$. We define two computable functions $f_1, f_2 : [0;1] \to \mathbb{R}$ by $f_1(x) := \varphi(x)$ and $f_2(x) := \varphi(x) - f(x)$ for any $x \in [0;1]$, respectively. Then $f = f_1 - f_2$ holds obviously. It remains to show that f_2 is also an increasing function. This follows immediately from the following inequality. Namely, for any $x, y \in [0;1]$ with $x \leq y$, we have

$$\begin{aligned}
f_2(y) - f_2(x) &= (\varphi(y) - f(y)) - (\varphi(x) - f(x)) \\
&= (\varphi(y) - \varphi(x)) - (f(y) - f(x)) \\
&\geq V_x^y(f) - (f(y) - f(x)) \geq 0.
\end{aligned}$$

Here the last inequality follows from the definition (1) of variation. Thus, $f = f_1 - f_2$ is an effective Jordan decomposition, i.e., f is EJD.

3   EJD and Effectively Absolute Continuity

Using the criterion of EJD of Theorem 2.1, we can construct some non-EJD function by diagonalization against all increasing computable functions φ. Especially, we will construct a computable non-EJD function which has a computable modulus of absolute continuity in this section. Since computable real functions are relatively difficult to manage directly in an effective construction, we use their effective approximations instead. For the increasing computable functions, we can use a special approximation described as follows.

Lemma 3.1. *Let* $[0;1]_\mathbb{Q} := [0;1] \cap \mathbb{Q}$. *For any increasing total computable function* $\varphi : [0;1] \to \mathbb{R}$, *there is a total computable function* $\beta : [0;1]_\mathbb{Q} \times \mathbb{N} \to \mathbb{Q}$ *which satisfies, for all* $n \in \mathbb{N}$ *and* $x, y \in [0;1]_\mathbb{Q}$, *the following conditions.*

$$x < y \Longrightarrow \beta(x, n) < \beta(y, n) \quad and \tag{3}$$

$$|\varphi(x) - \beta(x, n)| \leq 2^{-(n+1)} \;\&\; \beta(x, n) \leq \beta(x, n+1). \tag{4}$$

Lemma 3.1 can be proved easily by effective Weierstrss Theorem (see page 26 of [7]). In this paper, we will call β a *determinator* of φ if conditions (3) and (4) are satisfied.

For any interval $[a;b] \subseteq [0;1]$ and any real number $d \geq 0$, the linear function $f : [a;b] \to \mathbb{R}$ defined by $f(x) := d \cdot (x-a)$ has the variation $V_a^b(f) = d \cdot (b-a)$. If we require a continuous function $f : [0;1] \to \mathbb{R}$ such that $V_a^b(f) = d \cdot (b-a)$ and $f(a) = f(b) = 0$, then f can be defined as a "zigzag" function by $f(x) := d \cdot (x-a)$ if $x \in [a; (b+a)/2]$; $f(x) := d \cdot (b-x)$ if $x \in [(b+a)/2; b]$ and $f(x) := 0$ otherwise. Sometime we require further that $f(x) \leq \delta$ (for some $\delta \geq 0$) and f still have the same variation $d \cdot (b-a)$. In this case, f should have many small "zigzag" in the interval $[a;b]$. Let's call $f : [a;b] \to \mathbb{R}$ a *zigzag function* on $[a;b]$ with *width* 2α (for $2\alpha \leq b - a$) and *height* β if f on $[a; a + 2\alpha]$ is defined by

$$f(x) := \begin{cases} \beta(x-a)/\alpha & \text{if } a \leq x \leq a+\alpha \\ \beta(a+2\alpha-x)/\alpha & \text{if } a+\alpha < x \leq a+2\alpha, \end{cases}$$

and f satisfies condition $f(x) = f(x - 2\alpha)$ for x in $[a + 2\alpha; b]$.

Now we can prove our main result as follows.

Theorem 3.2. *There is a computable total function* $f : [0;1] \to [0;1]$ *which is effectively absolutely continuous but not effectively Jordan decomposable.*

Proof. We will construct a computable sequence (p_s) of rational polygon functions which converges uniformly and effectively to a computable function $f :$ $[0;1] \to [0;1]$ such that f is effectively absolute continuous and f satisfies, for all $e \in \mathbb{N}$, the following requirements.

$$Q_e : [0;1] \subseteq \mathrm{dom}(\psi_e) \Longrightarrow (\exists a, b \in [0;1])(a < b \ \& \ V_a^b(f) > \psi_e(b) - \psi_e(a))$$

where (ψ_e) is an effective enumeration of all partial computable real functions $\psi_e :\subseteq [0;1] \to \mathbb{R}$. Thus, by Theorem 2.1, f is not effectively Jordan decomposable.

If ψ_e is not increasing, then the requirement Q_e can be satisfied trivially. Suppose now that $\psi_e : [0;1] \to \mathbb{R}$ is an increasing total computable functions. We consider the interval $[2^{-(e+1)}; 2^{-e}]$. For such a function ψ_e, there must be a natural number $k_e \geq e$ and two rational numbers $a, b \in [2^{-(e+1)}; 2^{-e}]$ such that

$$a < b \ \& \ (b - a) = 2^{-2k_e} \ \& \ (\psi_e(b) - \psi_e(a))/(b - a) < 2^{k_e}.$$

Define the function $f : [0;1] \to [0;1]$ as a polygon function which connects the points $(0,0)$, $(a,0)$, $((a + b)/2, 2^{-(k_e+1)})$ $(b,0)$ and $(1,0)$. That is, f is a zigzag with width $b - a$ and height $2^{-(k_e+1)}$ on the interval $[a;b]$. Then $V_a^b(f) = 2^{-k_e} > \psi_e(b) - \psi_e(a)$. That is, the requirement Q_e is satisfied. Besides, let $m(e) := \max\{k_i : i \leq e\} + e + 1$. Then m is in fact a modulus of absolute continuity of f.

Unfortunately, the construction described above is not effective, because we cannot calculate the (real number) value $\psi_e(r)$ in finitely many steps even if ψ_e is a total computable real function and r is a rational number. Thus, we cannot guarantee that the functions f and m defined above are computable. However, by Lemma 3.1, ψ_e has always a computable determinator β, if $\psi_e : [0;1] \to \mathbb{R}$ is an increasing computable total function. Therefore, the requirements Q_e can be replaced by the following requirements

$$R_e : \quad \begin{array}{l} [0;1]_\mathbb{Q} \times \mathbb{N} \subseteq \mathrm{dom}(\beta_e) \ \& \ \beta_e \text{ is a determinator of some } \varphi : [0;1] \to \mathbb{R} \\ \Longrightarrow (\exists a, b \in [0;1])(a < b \ \& \ V_a^b(f) > \varphi(b) - \varphi(a)), \end{array}$$

where (β_e) is an effective enumeration of all partial computable functions $\beta_e :\subseteq \mathbb{Q} \times \mathbb{N} \to \mathbb{Q}$. Let $(\beta_{e,s})$ be an effective uniform approximation of (β_e). The strategy to satisfy the requirements R_e is similar to that of Q_e. We reserve the interval $[2^{-(e+1)}; 2^{-e}]$ and all stages $s := \langle e, t \rangle$ of the construction for the requirement R_e for any $e, t \in \mathbb{N}$. At the beginning, let the requirement R_e be in phase I. If, at some stage $s := \langle e, t \rangle$, we can estimate the "speed of increment" of β_e on the interval $[2^{-(e+1)}; 2^{-e}]$ such that $(\beta_{e,s}(2^{-e}, n) - \beta_{e,s}(2^{-(e+1)}, n))/(2^{-e} - 2^{e+1}) \leq 2^{s-1}$ for some $n \leq s$, then denote this s by k_e and put R_e into phase II. If at a later stage $s_1 > s$, we can find some subinterval $[a;b] \subseteq [2^{-(e+1)}; 2^{-e}]$ of length 2^{-2k_e} such that $(\beta_{e,s_1}(b, k_e) - \beta_{e,s_1}(a, k_e))/(b - a) \leq 2^{k_e+2}$, then we define f on this subinterval as a polygon function such that $V_a^b(f) := 2^{-k_e+2}$ and put R_e into phase III. This guarantees that f satisfies the requirement R_e. More precisely, we have the following formal construction.

Stage $s := 0$: For all $e \in \mathbb{N}$, put requirements R_e into phase I and let $k_{e,0}$ be undefined. Define $p_0(x) := 0$ for any $x \in [0; 1]$.

Stage $s := \langle e, t \rangle > 0$ for some $e, t \in \mathbb{N}$. Let $[u_e; v_e] := [2^{-(e+1)}; 2^{-e}]$. We will try to satisfy the requirement R_e at this stage by defining the rational polygon function p_s properly on the interval $[u_e; v_e]$, if R_e is not yet satisfied. We consider the following cases.

Case 1. R_e is in the phase I. The requirement R_e is not yet treated before and $k_{e,s}$ is not defined. If there is a natural number $n \leq s$ such that

$$\beta_{e,s}(u_e, n) \downarrow \ \& \ \beta_{e,s}(v_e, n) \downarrow \ \& \ \beta_{e,s}(v_e, n) - \beta_{e,s}(u_e, n) \geq 2^{-n} \quad (5)$$

$$((\beta_{e,s}(v_e, n) - \beta_{e,s}(u_e, n))/(v_e - u_e)) \leq 2^{s-1}, \quad (6)$$

then let n_e be the minimal such n, define $k_{e,s} := s$ and put the requirement R_e into phase II. Since $k_{e,s}$ will never be changed later, namely, $k_{e,t} = k_{e,s}$ for any $t \geq s$, we denote $k_{e,t}$ simply by k_e for any $t \geq s$. Notice that, for any $i \in \mathbb{N}$, if $k_{i,s}$ is already defined, then $k_{i,s} \leq s$. Besides, by definition of the pairing function, we have in this case also that $e \leq \langle e, t \rangle = s = k_{e,s} = k_e$.

Otherwise, if there is no such n which satisfies both conditions (5) and (6), then go to the next stage.

Case 2. R_e is in phase II. In this case n_e and $k_e := k_{e,s}$ is already defined. Consider the equidistant rational subdivision $u_e = a_0 < a_1 < \cdots < a_{m_e} = v_e$ of the interval $[u_e; v_e]$ such that $a_{i+1} - a_i = 2^{-2k_e}$, for any $i < m_e := 2^{2k_e - (e+1)}$. If the following conditions are satisfied

$$(\forall i \leq m_e)(\beta_{e,s}(a_i, k_e) \downarrow) \text{ and} \quad (7)$$

$$(\exists i < m_e)((\beta_{e,s}(a_{i+1}, k_e) - \beta_{e,s}(a_i, k_e))/(a_{i+1} - a_i) \leq 2^{k_e+1}), \quad (8)$$

then let i_e be the minimal i which satisfies condition (8). Now we define the polygon function p_s such that $p_s(x) := p_{s-1}(x)$ for $x \notin [a_{i_e}; a_{i_e+1}]$ and, on the interval $[a_{i_e}; a_{i_e+1}]$, p_s is a zigzag function with width $2^{-(k_e+s+1)}$ and height 2^{-s}. Finally, put the requirement R_e into the phase III. Notice that, in this case, the number of the zigzags of p_s in the interval $[a_{i_e}; a_{i_e+1}]$ is 2^{s+1-k_e}. Each zigzag has the height 2^{-s}. Therefore $V_{a_{i_e}}^{a_{i_e+1}}(p_s) = 2 \cdot 2^{-s} \cdot 2^{s+1-k_e} = 2^{-k_e+2}$. Furthermore, p_s satisfies the Lipschitz condition that

$$(\forall x, y \in [a_{i_e}; a_{i_e+1}])(|p_s(x) - p_s(y)| \leq 2^{k_e+2} \cdot |x - y|). \quad (9)$$

Otherwise, if no $i < m_e$ satisfies both conditions (7) and (8), then go directly to the next stage.

Case 3. R_e is in phase III. In this case, the requirement R_e is already satisfied. We define $p_s := p_{s-1}$ and go to the next stage.

In all cases, if $k_{i,s-1}$ is defined, then let $k_{i,s} := k_{i,s-1}$ for any $i \in \mathbb{N}$ with $i \neq e$.

This ends the construction. We will show that our construction succeeds by proving the following sublemmas.

Sublemma 3.2.1 *There is a total computable function* $f : [0;1] \to [0;1]$ *such that* $\lim_{s \to \infty} p_s = f$.

Proof. By the construction, (p_s) is a computable sequence of rational polygon functions $p_s : [0;1] \to [0;1]$ which satisfies $|p_s(x) - p_{s+1}(x)| \leq 2^{-s}$ for all $s \in \mathbb{N}$ and $x \in [0;1]$. According to effective Weierstrass Theorem (see e.g. [7]), the limit $f := \lim_{s \to \infty} p_s$ exists and it is a computable real function. $\square$ (sublemma)

Sublemma 3.2.2 *There exist no increasing computable functions* f_1 *and* f_2 *such that* $f = f_1 - f_2$. *Namely,* f *is not effectively Jordan decomposable.*

Proof. By Theorem 2.1, it suffices to show that, for any increasing total computable function $\varphi : [0;1] \to \mathbb{R}$, there are $a, b \in [0;1]$ with $a < b$ such that $V_a^b(f) > \varphi(b) - \varphi(a)$. Let φ be such a function. Then, by Lemma 3.1, there is an $e \in \mathbb{N}$ such that the total computable function $\beta_e : [0;1]_{\mathbb{Q}} \times \mathbb{N} \to \mathbb{Q}$ is a determinator of φ, i.e., it satisfies conditions (3) and (4).

Let's consider the interval $[u_e; v_e] := [2^{-(e+1)}; 2^{-e}] \subseteq [0;1]$. Since φ is an increasing total function on the interval $[0;1]$, we have $\varphi(u_e) < \varphi(v_e)$. Choose an $n \in \mathbb{N}$ such that $\varphi(v_e) - \varphi(u_e) \geq 2^{-n+1}$. Then there is an $s_0 \in \mathbb{N}$ such that both $\beta_{e,s_0}(u_e, n)$ and $\beta_{e,s_0}(v_e, n)$ are defined. Moreover, by (3) and (4), we have

$$\beta_{e,s_0}(v_e, n) - \beta_{e,s_0}(u_e, n) = |\beta_{e,s_0}(v_e, n) - \beta_{e,s_0}(u_e, n)|$$
$$\geq |\varphi(v_e) - \varphi(u_e)| - |\varphi(v_e) - \beta_{e,s_0}(v_e, n)| - |\varphi(u_e) - \beta_{e,s_0}(u_e, n)|$$
$$\geq 2^{-n+1} - 2^{-(n+1)} - 2^{-(n+1)} = 2^{-n}.$$

By the definition of the uniform approximation $(\beta_{e,s})$ at the end of Section 1, we have $\beta_{e,s}(u_e, n) \downarrow= \beta_{e,s_0}(u_e, n)$ and $\beta_{e,s}(v_e, n) \downarrow= \beta_{e,s_0}(v_e, n)$ for any $s \geq s_0$. Therefore, there is an $s \in \mathbb{N}$ such that both conditions (5) and (6) are satisfied.

Suppose that $s_1 = \langle e, t_1 \rangle$ (for some $t_1 \in \mathbb{N}$) is the first stage $s \in \mathbb{N}$ such that both conditions (5) and (6) are satisfied for some $n \in \mathbb{N}$. Then, at stage s_1, we will define $k_e := s_1$ and n_e as the minimal n which satisfies (5) and (6). The requirement R_e is put into phase II at this stage. In this case, we have

$$|\beta_e(v_e, k_e) - \beta_e(u_e, k_e)|$$
$$\leq |\beta_e(v_e, k_e) - \varphi(v_e)| + |\varphi(v_e) - \beta_e(v_e, n_e)| + |\beta_e(v_e, n_e) - \beta_e(u_e, n_e)|$$
$$+ |\beta_e(u_e, n_e) - \varphi(u_e)| + |\varphi(u_e) - \beta_e(u_e, k_e)|$$
$$\leq 2 \cdot 2^{-(k_e+1)} + 2 \cdot 2^{-(n_e+1)} + |\beta_e(v_e, n_e) - \beta_e(u_e, n_e)|$$
$$\leq 2^{-k_e} + 2 \cdot |\beta_e(v_e, n_e) - \beta_e(u_e, n_e)|$$
$$\leq 2^{-k_e} + 2^{k_e} \cdot (v_e - u_e) \leq 2^{(k_e+1)-(e+1)}.$$

That is, $(\beta_e(v_e, k_e) - \beta_e(u_e, k_e))/(v_e - u_e) \leq 2^{k_e+1}$. Therefore, for the equidistant subdivision $u_e = a_0 < a_1 < \cdots < a_{m_e} = v_e$ of the interval $[u_e; v_e]$ with the length $a_{i+1} - a_i = 2^{-2k_e}$ for any $i < m_e := 2^{2k_e - (e+1)}$, there must be a (minimal) $i_e < m_e$ such that $|\beta_e(a_{i_e+1}, k_e) - \beta_e(a_{i_e}, k_e)|/(a_{i_e+1} - a_{i_e}) \leq 2^{k_e+1}$.

Let s_2 be the minimal $s := \langle e, t_2 \rangle > s_1$ (for some $t_2 \in \mathbb{N}$) such that all $\beta_{e,s}(a_i, k_e)$ are defined for $i \leq m_e$, i.e., (7) is satisfied. Then i_e is also the minimal i which satisfies condition (8) for $s := s_2$. By construction, we define a polygon function p_{s_2} at stage s_2 such that $V_{a_{i_e}}^{a_{i_e}+1}(p_{s_2}) = 2^{-k_e+2}$. Moreover, we have the following estimation

$$|\varphi(a_{i_e+1}) - \varphi(a_{i_e})| \leq |\beta_e(a_{i_e+1}, k_e) - \beta_e(a_{i_e}, k_e)|$$
$$+ |\varphi(a_{i_e+1}) - \beta_e(a_{i_e+1}, k_e)| + |\varphi(a_{i_e}) - \beta_e(a_{i_e}, k_e)|$$
$$\leq |\beta_e(a_{i_e+1}, k_e) - \beta_e(a_{i_e}, k_e)| + 2^{-k_e}$$
$$\leq 2^{k_e+1}(a_{i_e+1} - a_{i_e}) + 2^{-k_e}$$
$$= 2^{k_e+1} \cdot 2^{-2k_e} + 2^{-k_e} < 2^{-k_e+2}.$$

On the interval $[a_{i_e}; a_{i_e+1}]$, the function f is equal to the rational polygon function p_{s_2}. Therefore $V_{a_{i_e}}^{a_{i_e}+1}(f) = V_{a_{i_e}}^{a_{i_e}+1}(p_{s_2}) = 2^{-k_e+2} > \varphi_e(a_{i_e+1}) - \varphi_e(a_{i_e})$. Therefore, f is not EJD. $\qquad\square$ (sublemma)

Sublemma 3.2.3 *The function f is effectively absolutely continuous.*
Proof. It suffices to show that the computable function $m : \mathbb{N} \to \mathbb{N}$ defined by $m(s) := 2s + 4$, for any $s \in \mathbb{N}$, is a modulus of absolutely continuity of f, i.e., m satisfies condition (2).

For $n \in \mathbb{N}$ and any set $I := \{[a_i; b_i] : i \leq n_0\}$ (for some $n_0 \in \mathbb{N}$) of non-overlapping subintervals of $[0; 1]$ such that $\sum_{i \leq n_0} |b_i - a_i| \leq 2^{-m(n)} = 2^{-(2n+4)}$. Let

$$A_n := \{[2^{-(e+1)}; 2^{-e}] : e \in \mathbb{N} \; \& \; k_{e,n+1} \text{ is not yet defined}\}$$
$$B_n := \{[2^{-(e+1)}; 2^{-e}] : e \in \mathbb{N} \; \& \; k_{e,n+1} \text{ is defined} \}.$$

Namely, A_n consists of all intervals $[2^{-(e+1)}; 2^{-e}]$ such that R_e is still in phase I at stage $n+1$ and B_n consists of all such intervals corresponding requirement R_e is either in phase II or phase III at stage $n+1$. Notice that, $k_{i,n+1} \leq n+1$ whenever $k_{i,n+1}$ is defined. Therefore, by condition (9), $|f(x) - f(y)| \leq 2^{\max_{i \in \mathbb{N}} k_{i,n+1}+2} \cdot |x - y| \leq 2^{n+3} \cdot |x - y|$, for any $x, y \in \bigcup A_n$. Define

$$I_1 := \{J \subseteq [0; 1] : (\exists J_1 \in I)(\exists J_2 \in A_n)(J = J_1 \cap J_2)\}$$
$$I_2 := \{J \subseteq [0; 1] : (\exists J_1 \in I)(\exists J_2 \in B_n)(J = J_1 \cap J_2)\}.$$

Then we have $\bigcup I = (\bigcup I_1) \cup (\bigcup I_2)$. Let $V_J(f)$ denote the variation of f on J for any set J of non-overlapping subintervals of $[0; 1]$. Namely, if the set $J := \{[a_i; b_i] : i \leq n_1\}$, then $V_J(f) = \sum_{i \leq n_1} V_{a_i}^{b_i}(f)$. Then

$$\sum_{i \leq n_0} |f(b_i) - f(a_i)| \leq V_I(f) = V_{I_1}(f) + V_{I_2}(f)$$

$$\leq V_{A_n}(f) + \mu(I_2) \cdot 2^{n+3}$$

$$\leq \sum_{i > n+1} 2^{-i} + 2^{-m(n)} \cdot 2^{n+3}$$

$$\leq 2^{-(n+1)} + 2^{-(n+1)} = 2^{-n}.$$

where $\mu(I_2)$ is the Lebesgue measure of I_2. Thus, m satisfies condition (2) and is a computable modulus of absolute continuity of f. $\qquad\square$ (sublemma)

Therefore, f is a computable and absolutely continuous function which is not effectively Jordan decomposable. This completes the proof of the Theorem 3.2.

4   Polynomial Time Version of Jordan Decomposition

This section discusses the polynomial time version of the Jordan decomposability. Let's recall first the definition of polynomial time computable real functions. We use the approach of Ko in [6], where the set $\mathbb{D}$ of dyadic rational numbers instead of $\mathbb{Q}$ is used. Namely, let $\mathbb{D}_n := \{m \cdot 2^{-n} : m \in \mathbb{N}\}$ for any $n \in \mathbb{N}$. Then $\mathbb{D} := \cup_{n\in\mathbb{N}}\mathbb{D}_n$. A real function $f : [0;1] \to \mathbb{R}$ is *polynomial time computable* if there is an oracle Turing machine M and a polynomial p such that, for any $n \in \mathbb{N}$, the machine M, with any oracle (x_s) of dyadic rational number sequence such that $x_s \in \mathbb{D}_s$ and $|x - x_s| \leq 2^{-s}$ for any $s \in \mathbb{N}$ and any input n, $M^{(x_s)}(n)$ outputs a dyadic rational number $y \in \mathbb{D}_n$ in $p(n)$ steps such that $|f(x) - y| \leq 2^{-n}$.

A function $f : [a;b] \to \mathbb{R}$ is called a PBV function if it is polynomial time computable and is of bounded variation on $[a;b]$. A PBV function is *polynomial time Jordan decomposable* (PJD for short), if there are two polynomial time computable increasing functions f_1, f_2 such that $f = f_1 - f_2$. Thus, by a similar proof to that of Theorem 2.1, we can show the following result.

Theorem 4.1. *A total function $f : [0;1] \to [0;1]$ is PJD iff there is a polynomial time computable increasing total function $\varphi : [0;1] \to \mathbb{R}$ such that $V_x^y(f) \leq \varphi(y) - \varphi(x)$, for any $x, y \in [0;1]$ with $x \leq y$.*

About the PBV functions, Ko [6] has shown the following results.

Theorem 4.2 (Ko [6]). *Let $f : [0;1] \to [0;1]$ be a PBV function and $v_f(x) := V_0^x(f)$ for any $x \in [0;1]$.*

1. *If f is PJD, then v_f has a polynomial modulus of continuity.*
2. *If v_f is polynomial time computable, then f is PJD.*
3. *There is a PBV function $g : [0;1] \to [0;1]$ which is not PJD.*
4. *There is a PJD function $g : [0;1] \to [0;1]$ such that v_g is not polynomial time computable.*

Ko [6] discusses also the relationship between the PBV functions and PAC functions, where PAC means *polynomial time absolute continuous*. Precisely, a polynomial time computable function $f : [0;1] \to \mathbb{R}$ is called PAC if there is a polynomial $m : \mathbb{N} \to \mathbb{N}$ such that m is a modulus of absolute continuity of f, i.e., m satisfies condition (2). To construct a PBV but non-PJD function f (i.e., Theorem 4.2.3), Ko has constructed a PBV function f such that v_f does not have a polynomial modulus of continuity and apply the result of Theorem 4.2.1. However, this technique does not work for the PAC functions, because, for any PAC function f, v_f does have a polynomial modulus of continuity. Thus, Ko [6] asks the following question: is every PAC function also PJD? Now we will answer this question by applying our criterion for Jordan decomposability.

Notice first that the function constructed in the proof of Theorem 3.2 does have a polynomial function $m(n) := 2n + 4$ of absolute continuity. What is still absent there is only the polynomial time computability. To this end, we show the following technical lemma.

Lemma 4.3. *1. Let $f, g : [0;1] \to [0;1]$ be total functions such that $V_x^y(f) = V_x^y(g)$, for any $x, y \in [0;1]$ and $m : \mathbb{N} \to \mathbb{N}$ any function. Then m is a modulus of absolute continuity of f iff m is a modulus of absolute continuity of g.*

2. Let $(a_s, b_s, d_s)_{s \in \mathbb{N}}$ be a computable sequence of rational triples such that $d_s \neq 0$ and $\{[a_s; b_s] : s \in \mathbb{N}\}$ consists of non-overlapping subintervals of $[0; 1]$. Then there exists a polynomial time computable function $f : [0; 1] \to [0; 1]$ which satisfies, for all $x < y$ in $[0; 1]$ the following condition.

$$V_x^y(f) = \begin{cases} (y - x)d_i & \text{if } a_i \leq x < y \leq b_i \text{ for some } i \in \mathbb{N} \\ 0 & \text{if } [x; y] \cap \bigcup_{i \in \mathbb{N}} [a_i; b_i] = \emptyset. \end{cases} \tag{10}$$

Proof. 1. It follows immediately from the definition of modulus of absolute continuity.

2. Let M be a Turing machine which computes the triple (a_i, b_i, d_i) on input 0^i for any $i \in \mathbb{N}$ and let t_M denote its time complexity. Here, we assume a fixed notation $\nu : \{0, 1\}^* \to \mathbb{Q}$ defined by $\nu(0^n 10^m 10^k) := (n - m)/(k + 1)$ for any $n, m, k \in \mathbb{N}$. Define a function $f : [0; 1] \to [0; 1]$ by

$$f(x) := \begin{cases} g_i(x) & \text{if } x \in [a_i; b_i] \text{ for some } i \in \mathbb{N} \\ 0 & \text{otherwise}, \end{cases} \tag{11}$$

where g_i is the zigzag function on interval $[a_i; b_i]$ with width $2^{-t_M(i)+1}/d_i$ and height $2^{-t_M(i)}$. Notice that, for any $i \in \mathbb{N}$, the polygon function g_i satisfies the Lipschitz condition $(\forall x, y \in [a_i; b_i])(|g_i(x) - g_i(y)| \leq d_i \cdot |x - y|)$.

To show that f is polynomial time computable, we consider the following algorithm: Given $n \in \mathbb{N}$ and a sequence (x_s) of rational numbers such that $|x - x_s| \leq 2^{-s}$ for any $s \in \mathbb{N}$, we simulate the computations $M(0^i)$ for each $i \leq n$ up to n steps. Let $(a_{i_t}, b_{i_t}, d_{i_t})$, $t \leq k_0$, be all triples such that $t_M(i_t) \leq n$. Let

$$d := \max\{d_{i_t} : t \leq k_0\} \quad \text{and}$$

$$c := \min\{|a_{i_t} - b_{i_s}| : t, s \leq k_0 \ \& \ t \neq s \ \& \ a_{i_t} \neq b_{i_s}\}.$$

That is, c is the minimal distance between any two non-connected intervals $[a_{i_t}; b_{i_t}]$ and $[a_{i_s}; b_{i_s}]$ for $t, s \leq k_0$. Find a minimal natural number m such that $2^{-m} \leq \min\{2^{-(n+1)}/d, c/2\}$ and define

$$y := \begin{cases} g_{i_t}(x_m) & \text{if } a_{i_t} \leq x_m \leq b_{i_t} \text{ for some } t \leq k_0 \\ 0 & \text{otherwise}. \end{cases}$$

Notice that, since g_{i_t} is a rational polygon function, $g_{i_t}(x_m)$ is also a rational number which can be computed in polynomial time. Therefore, y can be computed in polynomial time (with respect to n). Moreover, we can show that $|y - f(x)| \leq 2^{-n}$. Let's consider the following cases.

Case 1. $x_m, x \in [a_{i_t}; b_{i_t}]$ for some $t \leq k_0$. In this case we have $|y - f(x)| = |g_{i_t}(x_m) - g_{i_t}(x)| \leq d_{i_t} \cdot |x_m - x| \leq d_{i_t} \cdot 2^{-m} \leq d \cdot 2^{-m} \leq 2^{-n}$.

Case 2. $x_m \in [a_{i_t}; b_{i_t}]$ but $x \notin [a_{i_t}; b_{i_t}]$ for some $t \leq k_0$. In this case we have $x \notin [a_{i_s}; b_{i_s}]$ for any $s \leq k_0$ and $\max\{|a_{i_t} - x_m|, |b_{i_t} - x_m|\} \leq 2^{-m}$. If $x \in [a_i; b_i]$ for some $i \in \mathbb{N}$ such that $t_M(i) > n$, Then $|y - f(x)| \leq |y| + |f(x)| \leq |g_{i_t}(x_m)| + 2^{-t_M(i)} \leq d \cdot 2^{-m} + 2^{-(n+1)} \leq 2^{-n}$. Otherwise, if $x \notin [a_i; b_i]$ for any $i \in \mathbb{N}$, we have $|y - f(x)| \leq |y| \leq |g_{i_t}(x_m)| \leq d \cdot 2^{-m} \leq 2^{-n}$.

Case 3. $x_m \notin [a_{i_t}; b_{i_t}]$ for any $t \leq k_0$. In this case we have $y := 0$. If $x \in [a_{i_t}; b_{i_t}]$ for some $t \leq k_0$, then $\max\{|x - a_{i_t}|, |x - b_{i_t}|\} \leq 2^{-m}$ and hence $|y - f(x)| = |g_{i_t}(x)| \leq d \cdot 2^{-m} \leq 2^{-n}$. If $x \in [a_i; b_i]$ for some $i \in \mathbb{N}$ with $t_M(i) > n$, then $|y - f(x)| = |f(x)| \leq 2^{-t_M(i)} \leq 2^n$. Otherwise, if $x \notin [a_i; b_i]$ for any $i \in \mathbb{N}$, then $f(x) := 0$.

Therefore, f is polynomial time computable. Moreover, it is easy to see that the function f satisfies condition (10).

Now we can give the negative answer to Ko's question. In fact we obtain a stronger result.

Theorem 4.4. *There is a PAC $f : [0; 1] \to [0; 1]$ which is not EJD.*

Proof. According to the proof of Theorem 3.2, there is a computable sequence $(a_s, b_s, d_s)_{s \in \mathbb{N}}$ of rational triples and a computable function $g : [0; 1] \to [0; 1]$ which satisfies the following conditions.

(a) The function $m : \mathbb{N} \to \mathbb{N}$ defined by $m(n) := 2n + 4$ for all $n \in \mathbb{N}$ is a modulus of absolute continuity of g.
(b) $\{[a_s; b_s] : s \in \mathbb{N}\}$ is a set of non-overlapping subintervals of $[0; 1]$.
(c) The function g satisfies condition (10) for g instead of f, and
(d) For any increasing computable function $\varphi : [0; 1] \to \mathbb{R}$, there is an $s \in \mathbb{N}$ such that $V_{a_s}^{a_s}(f) = d_s \cdot (b_s - a_s) > \varphi(b_s) - \varphi(a_s)$.

By Lemma 4.3.2 and the item (b), there is a polynomial time computable function f which satisfies condition (10). By item (c), this implies that $V_x^y(f) = V_x^y(g)$ for any $x, y \in [0; 1]$. Thus m is also a modulus of absolute continuity of f by Lemma 4.3.1 and item (a). That is, f is a PAC function. On the other hand, for any increasing computable function $\varphi : [0; 1] \to \mathbb{R}$, there is an $s \in \mathbb{N}$ such that $V_{a_s}^{b_s}(f) = V_{a_s}^{b_s}(g) > \varphi(b_s) - \varphi(a_s)$. By Theorem 2.1, f is not EJD.

Since any non-EJD function is also non PJD, we have

Corollary 4.5 (Ko [6]). *There is a PAC function which is not PJD.*

References

1. S. K. Berberian. *Fundamentals of Real Analysis*. Universitext. Springer, New York, Berlin, Heidelberg, 1998.
2. D. Bridges. A constructive look at functions of bounded variation. *Bull. London Math. Soc.*, 32(3):316–324, 2000.
3. E. Casas, K. Kunisch, and C. Pola. Some applications of bv functions in optimal control and calculus of variations. *ESAIM: Proceedings*, 4:83–96, 1998.
4. A. Grzegorczyk. On the definitions of computable real continuous functions. *Fundamenta Mathematicae*, 44:61–71, 1957.
5. C. Jordan. *Cours d'analyse de l'Ecole Polytechnique*. 1882–1887.
6. K.-I. Ko. *Complexity Theory of Real Functions*. Progress in Theoretical Computer Science. Birkhäuser, Boston, 1991.
7. M. B. Pour-El and J. I. Richards. *Computability in Analysis and Physics*. Perspectives in Mathematical Logic. Springer, Berlin, 1989.

8. R. Rettinger, X. Zheng, R. Gengler, and B. von Braunmühl. Weakly computable real numbers and total computable real functions. In *Proceedings of COCOON 2001, Guilin, China, August 20-23, 2001*, volume 2108 of *LNCS*, pages 586–595. Springer, 2001.
9. K. Weihrauch. *Computable Analysis, An Introduction.* Springer, Berlin Heidelberg, 2000.
10. X. Zheng, R. Rettinger, and B. von Braunmühl. On the Jordan decomposability for computable functions of bounded variation. Computer Science Reports, BTU Cottbus Report, 03/02, 2002.

Fast Algorithms for Extended Regular Expression Matching and Searching

Lucian Ilie*,**, Baozhen Shan, and Sheng Yu***

Department of Computer Science, University of Western Ontario
N6A 5B7, London, Ontario, CANADA
{ilie,bxshan,syu}@csd.uwo.ca

Abstract. *Extended regular expressions* are an extension of ordinary regular expressions by the operations of intersection and complement. We give new algorithms for extended regular expression *matching* and *searching* which improve significantly the (very old) best upper bound for this problem, due to Hopcroft and Ullman. For an extended regular expression of size m with p intersection and complement operators and an input word of length n our algorithms run in time $\mathcal{O}(mn^2)$ and space $\mathcal{O}(pn^2)$ while the one of Hopcroft and Ullman runs in time $\mathcal{O}(mn^3)$ and space $\mathcal{O}(mn^2)$. Since the matching problem for semiextended regular expressions (only intersection is added) has been very recently shown to be LOGCFL complete, our algorithms are very likely the best one can expect. We also emphasize the importance of the extended regular expressions for software programs currently using ordinary regular expressions and show how the algorithms presented can be improved to run significantly faster in practical applications.

Keywords: extended regular expressions, pattern matching, finite automata, algorithms, complexity

1 Introduction

The importance of regular expressions for applications is well known. They describe lexical tokens for syntactic specifications and textual patterns in text manipulation systems. Regular expressions have become the basis of standard utilities such as scanner generators (Lex), editors (Emacs, vi), or programming languages (Perl, Awk), see [3,5]. The (ordinary) regular expressions, as widely used in applications, have three basic operations: union(+), catenation(·), and iteration(*). They do not include very basic operations like intersections($\cap$) and complement($^-$), though the latter two (especially the complement) would greatly improve the possibilities to write shorter and better expressions. Consider a very simple example: C comments. Say A is the set of all characters that can be introduced from the keyboard. A C comment is any sequence of characters which

---

* Research partially supported by NSERC grant R3143A01.
** corresponding author
*** Research partially supported by NSERC grant OGP0041630.

H. Alt and M. Habib (Eds.): STACS 2003, LNCS 2607, pp. 179–190, 2003.

starts with /* and ends with the first */ encountered. One of the simplest ways to represent a C comment using regular expressions is:

$$/*((A - \{*\}) + **^*(A - \{*,/\}))^**^*/$$

Notice that, though not complicated, this regular expression is by no means obvious. We start with the sequence /*. Then have any number of blocks which either contain no * or consist of one or more *'s followed by a character which is neither * nor /. At the end, we have the sequence */ preceeded by a number of *'s, possibly none. For an average user of, say, Perl, this might be non-trivial. Using just one complement, things become a lot simpler and clearer, as we can express the C comments as

$$/*\overline{A^**/A^*}*/$$

If we increase only a little bit the difficulty by asking for those C comments which contain the word **good** but do not contain the word **bad**, the problem of expressing these comments by regular expressions becomes undoable for most experimented programmers. Of course, not because it is impossible to describe these comments using regular expressions, but simply because of the intricateness of the problem. Using complements and intersections, these comments are, again, easy to express by

$$/*(\overline{A^**/A^*} \cap A^*\text{good}A^* \cap \overline{A^*\text{bad}A^*})*/$$

or the slightly simpler version

$$/*(A^*\text{good}A^* \cap \overline{A^*(*/ + \text{bad})A^*})*/$$

In Perl like notation, assuming we use $^\wedge(\alpha)$ for the complement of α and $\alpha\&\beta$ for the intersection of α and β, we would have:

$$/\backslash*(.*\text{good}.*\&^\wedge(.*(\backslash*/|\text{bad}).*))\backslash*/$$

The advantage of using regular expressions extended with intersection and complement (called hereafter *extended*[1]) is obvious. Extended regular expressions can represent sets of words in a much shorter and clearer way but, on the other hand, that makes some of the related problems a lot harder; see, e.g., [2,8,11,12]. Probably one of the most important reasons that prevented the use of extended regular expressions was the fact that the matching algorithms known are slow. The main goal of this paper is to give efficient algorithms for the matching and searching problems for regular expressions.

Precisely, the matching problem is as follows.

---

[1] The extended regular expressions we investigate in this paper should not to be confused with the extensions of regular expressions by *backreference*, see, e.g., [5]; the latter increase strictly the expressiveness (can represent sets of words which regular expressions cannot) whereas ours do not. The matching problem for those with backreference is NP-complete; see [1].

Problem 1 (Extended regular expression matching) *Given an extended regular expression α and a word w, check whether or not w belongs to the set of words represented by α.*

A closely related problem, which is even more important in applications is the searching problem.

Problem 2 (Extended regular expression searching) *Given an extended regular expression α and a word w, find all subwords of w which belong to the set of words represented by α.*

We assume hereafter that we have an extended regular expression of size m with p complement and intersection operators and an input word of size n. The current (very old) best upper bound for the matching problem is due to Hopcroft and Ullman [7]. Their algorithm uses a straightforward dynamic programming approach and runs in time $\mathcal{O}(mn^3)$ and space $\mathcal{O}(mn^2)$. Two other algorithms with the same running time appeared in [6] and [9]. A somewhat better but very complicated algorithm is given in [16] and runs in time $\mathcal{O}(mn^2 + pn^3)$. Thus, the worst case is actually the same. For the regular expressions extended with intersection only (*semiextended*), [15] claimed an $\mathcal{O}(mn^2)$ algorithm, but according to [16], it does not work in all cases.

In this paper, we give algorithms for the matching and searching problems for extended regular expressions which run in worst case time $\mathcal{O}(mn^2)$ and space $\mathcal{O}(pn^2)$. In practice, our algorithms can be made to run significantly faster.

Theorem 1 *The extended regular expression matching and searching problems can be solved in time $\mathcal{O}(mn^2)$ and space $\mathcal{O}(pn^2)$.*

As shown in [13], the matching problem for semiextended regular expressions is complete in LOGCFL and thus it is very unlikely that a significantly better algorithm than ours exists (even in the case of semiextended regular expressions).

After the main definitions in the next section, we introduce in Section 3 the automata we are going to use in the algorithms; they are ordinary nondeterministic finite automata with ε-transitions used in a suitable way to accommodate complements and intersections. Section 4 contains the matching algorithm; its correctness is proved in Section 5 and its complexity is discussed in Section 6. Section 7 shows how the matching algorithm can be adapted to solve the searching problem. Section 8 is extremely important from practical point of view. For clarity sake, we present in Sections 4 and 7 simplified versions of the algorithms. Section 8 gives improved versions which, even if have the same worst case running time, are much better for practical applications. The last section contains our conclusion about the work and its potential applications for software programs.

2   Extended Regular Expressions

We give in this section the basic definitions we need throughout the paper. For further details, we refer to [7] and [17].

Let A be an alphabet; A^* denotes the set of all words over A; ε denotes the empty word and the length of a word $w \in A^*$ is denoted $|w|$. For $u, w \in A^*$, u is a *subword* of w if $w = xuy$, for some $x, y \in A^*$; if $x = \varepsilon$, then u is a *prefix* of w. For $w = a_1 a_2 \ldots a_n$, $a_i \in A$, we denote $w[i..j] = a_i a_{i+1} \ldots a_j$; if $i > j$, then $w[i..j] = \varepsilon$. A *language* over A is a subset of A^*.

A *regular expression* over A is $\emptyset$, ε, or $a \in A$, or is obtained from these applying the following rules finitely many times: for two regular expressions α and β, the *union*, $(\alpha + \beta)$, the *catenation*, $(\alpha \cdot \beta)$, and the *iteration*, (α^*), are regular expressions. An *extended regular expression* over A is defined as a regular expressions with two more operations: *intersection*, $(\alpha \cap \beta)$, and *complement*, $(\overline{\alpha})$. Some of the parentheses will be omitted by using the precedence rules; in the order of decreasing precedence, we have: $^-$, *, $\cdot$, $\cap$, $+$.

The regular language represented by an extended regular expression α is $L(\alpha)$ and is defined inductively as follows: $L(\emptyset) = \emptyset$, $L(\varepsilon) = \{\varepsilon\}$, $L(a) = \{a\}$, $L(\alpha + \beta) = L(\alpha) \cup L(\beta)$, $L(\alpha \cdot \beta) = L(\alpha) \cdot L(\beta)$, $L(\alpha^*) = L(\alpha)^*$, $L(\alpha \cap \beta) = L(\alpha) \cap L(\beta)$, $L(\overline{\alpha}) = \overline{L(\alpha)} = A^* - L(\alpha)$.

For an extended regular expression α over A, the *size* $|\alpha|$ of α is the number of symbols in α when written in postfix (parentheses are not counted).

A *finite automaton*[2] is a quintuple $M = (Q, A, q_0, \delta, F)$, where Q is the set of states, $q_0 \in Q$ is the initial state, $F \subseteq Q$ is the set of final states, and $\delta \subseteq Q \times (A \cup \{\varepsilon\}) \times Q$ is the transition mapping. The *language* accepted by M is denoted $L(M)$. The *size* of a finite automaton M is $|M| = |Q| + |\delta|$.

3   Finite Automata from Extended Regular Expressions

Our algorithm for extended regular expression matching uses automata which we build in this section. Given an extended regular expression, we construct a finite automaton for it similar to the one of Thompson [14], modified to accommodate complements.

Assume we have a regular expression β and we are interested in the complement of its language, that is, $\overline{L(\beta)}$. Consider an automaton M_β which accepts the language $L(\beta)$; assume also the initial state of M_β has indegree zero and M_β has only one final state which, in addition, has outdegree zero. We add then the states in, fail, and out as in Fig. 1.

Recall first how the algorithm for regular expression matching works; see, e.g., [4]. Given β and M_β, it starts with a set, say $\mathcal{S}$, containing the initial state of M_β. Then, when reading a letter of the input, $\mathcal{S}$ is updated such that it stores the states reachable by the prefix read so far. After reading the whole input, it accepts if and only if the final state of M_β is in $\mathcal{S}$.

To simulate the complement of $L(\beta)$, we start with $\mathcal{S}$ containing the state in and update it to contain the states reachable from the state in. If fail is not in

---

[2] The automaton M is called *deterministic* (DFA) if $\delta : Q \times A \to Q$ is a (partial) function, *nondeterministic* (NFA) if $\delta \subseteq Q \times A \times Q$, and *nondeterministic with ε-transitions* (εNFA) if $\delta \subseteq Q \times (A \cup \{\varepsilon\}) \times Q$. Since we use only εNFAs in this paper, we shall call M simply *finite automaton*.

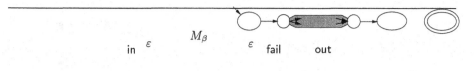

Fig. 1. Handling the complement of an expression β

S, then we add out to S. The idea is that M_β will be part of a bigger automaton and we want to add out whenever we read inside M_β a word which is in $\overline{L(\beta)}$.

We give in Fig. 2 the algorithm to construct the automata which we need for extended regular expression matching later. For simplicity, we eliminate all intersections using DeMorgan's rules. Clearly, this increases the size of the regular expression by a constant factor.

AUTOMATON-BUILDER(α)

- given an extended regular expression α with no $\cap$
- returns a finite automaton denoted $M(\alpha)$

1.   associate indices with the p complements of α: $1, 2, \ldots, p$.
2.   build $M(\alpha)$ inductively (according to the structure of α) as shown in Fig. 3.(i)-
     (vii); in the case (vii), k is the index of the current complement.
3.   **return** $M(\alpha) = (Q, A, q_0, \delta, \{q_F\})$.

Fig. 2. Building finite automata from extended regular expressions

For any k, $1 \leq k \leq p$, we shall refer the part of the automaton $M(\alpha)$ which has in_k as initial state and fail_k as final state, see Fig. 3(vii), as $M_k(\alpha)$; notice that in_k and fail_k are states of $M_k(\alpha)$ while out_k is not. This automaton corresponds to the kth complement of α, let it be $\overline{\beta}$. We denote $L_k(\alpha) = L(\beta)$.

4   Matching

We give in this section our algorithm for extended regular expression matching. We describe first the basic ideas. We combine the ideas of Thompson [14], which uses finite automata, with the one of Hopcroft and Ullman [7], which uses dynamic programming. Assume α is an extended regular expression and $w = a_1 a_2 \ldots a_n$ is an input word. The algorithm of [7] computes, for any subword u of w and any subexpression β of α, whether u is in the language represented by β or not. We shall compute less information. Precisely, we compute, for any subword u of w and any complement k of α, whether u belongs to the language $L_k(\alpha)$ or not. This is done using the idea of Thompson. We consider each automaton $M_k(\alpha)$ and start with a set (queue) S which initially contains only the

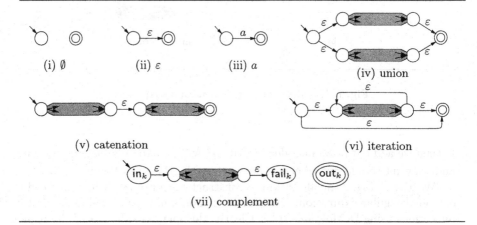

Fig. 3. Inductive construction of the automaton

initial state of $M_k(\alpha)$, that is, in_k. Then, for each letter of u, we update S to contain the states reachable from in_k. The problem appears when a state $in_{k'}$ is added to S because Thompson's algorithm cannot handle complements. The solution is to consider the innermost complements first. This way, by the time when $in_{k'}$ is added to S after reading the letter a_j of u, we know what words $w[j..j']$ are rejected by $M_{k'}(\alpha)$ and we add out_k to S at the appropriate times (after reading $a_{j'}$ from input); just like $w[j..j']$ led $M_k(\alpha)$ from $in_{k'}$ to $out_{k'}$.

The main procedure is EXT-REG-EXP-MATCHER(α, w); see Fig. 4. We first eliminate $\emptyset$'s and $\cap$'s from α. Step 3 is needed for succinctness; we see α as a complement such that we can treat it the same way as the other complements inside α. At step 4 we construct the automaton as described in the previous section. As mentioned above, we need to consider the complements of α such that the innermost come first; the procedure TOPOLOGICAL-SORT(α) sorts the complements $1, 2, \ldots, p+1$ of α such that whenever the complement k contains k', k' comes before k in the ordered sequence $(k_1, k_2, \ldots, k_{p+1})$. We assume that α after step 2 has p complements and so k_{p+1} is the one added at step 3. Using the notation introduced in the previous section, we have $L_{k_{p+1}}(\alpha) = L(\overline{\alpha})$. Recall that $\overline{\alpha}$ is equivalent to the original α (at the beginning of the algorithm).

The main part of the procedure are the steps 6..8 where the values of cont are computed with the meaning:

$$j \in \text{cont}(k, i) \quad \text{iff} \quad w[i+1..j] \notin L_k(\alpha).$$

Put otherwise, when reaching in_k after reading a_i from the input, we can continue from out_k by reading a_{j+1}; just like out_k were reachable by $a_1 a_2 \ldots a_j$ from the initial state q_0 (and by $a_{i+1} \ldots a_j$ from in_k).

The computation of all sets cont is done by CONTINUATION(k, i), given in Fig. 4, which computes $\text{cont}(k, i)$. We start with $\text{cont}(k, i)$ containing all possible

---

EXT-REG-EXP-MATCHER(α, w)

- given an extended regular expression α and a word $w = a_1 a_2 \ldots a_n$
- returns YES if $w \in L(\alpha)$ and NO otherwise

  1.   eliminate $\emptyset$ from α
  2.   eliminate $\cap$ from α
  3. $\alpha \leftarrow \overline{\alpha}$
  4. $M(\alpha) \leftarrow$ AUTOMATON-BUILDER(α)
  5. $(k_1, k_2, \ldots, k_{p+1}) \leftarrow$ TOPOLOGICAL-SORT(α)
  6.   for ℓ from 1 to $p + 1$ do
  7.   for i from 0 to n do
  8.   cont$(k_\ell, i) \leftarrow$ CONTINUATION(k_ℓ, i)
  9.   if $n \notin$ cont$(k_{p+1}, 0)$ then return YES
10.   else return NO

Fig. 4. The matching procedure

elements (step 1). (The bad ones will be removed at step 9 in εCLOSURE.) S is a queue containing the states reachable at any moment by the prefix read so far; initially, it contains only in_k (added at step 5). Then, for all letters after a_i in w, we update S accordingly (steps 7..8). The role of the sets S_j is to keep the states out_k which are to be added to S later at appropriate steps (see steps 3..5 in εCLOSURE).

CONTINUATION(k, i)

  1.   cont$(k, i) \leftarrow [i..n]$
  2. $S \leftarrow$ NEWQUEUE$()$
  3.   for j from i to n do
  4. $S_j \leftarrow \emptyset$
  5.   ENQUEUE(S, in_k)
  6. $S \leftarrow \varepsilon$CLOSURE(S, k, i, i)
  7.   for j from $i + 1$ to n do
  8. $S \leftarrow \varepsilon$CLOSURE$($TRANSIT$(S, a_j), k, i, j)$
  9.   return cont(k, i)

Fig. 5. Computing the continuations

The procedure TRANSIT(S, a), shown in Fig. 6, updates S to contain the states reachable by transitions labelled a. It is very simple because no key state, such as in_k, fail_k, or out_k, can be reached this way.

Our last procedure is εCLOSURE(S, k, i, j); see Fig. 7. It updates S to contain the states reachable by ε-transitions; k means the work is done inside $M_k(\alpha)$, i is the last letter read from the input w before entering $M_k(\alpha)$ and j is the last letter considered so far from the input (if $i = 0$ or $j = 0$, then no letter has been

---

TRANSIT($\mathcal{S}, a$)
1. $\mathcal{T} \leftarrow$ NEWQUEUE()
2.   **while not**(EMPTY($\mathcal{S}$)) **do**
3. $q \leftarrow$ DEQUEUE($\mathcal{S}$)
4.   **for** each $(q, a, p) \in \delta$ **do**
5.   ENQUEUE($\mathcal{T}, p$)
6.   **return** $\mathcal{T}$

Fig. 6. Transitions by a letter

read from w). So, the set $\mathcal{S}$ which is to be updated contains states inside $M_k(\alpha)$ (but not inside inner complements) which were reached from in_k by $a_{i+1} \ldots a_j$. Several special cases are to be considered here. If the current state we consider is $fail_k$ (step 8), then the current prefix $a_{i+1} \ldots a_j$ belongs to $L_k(\alpha)$ and therefore there must be no continuation from out_k by a_{j+1}; j is removed from $\mathrm{cont}(k, i)$ at step 9. (Notice that j might still belong to some other $\mathrm{cont}(k, i')$.) If the current state is some $in_{k'}$ (step 10), the beginning of a complement inside $M_k(\alpha)$, the topological order we did previously tells that all continuations $\mathrm{cont}(k', j')$ have been computed. Using this information, we add the state $out_{k'}$ to the right $\mathcal{S}_{j'}$; later, during the appropriate call of εCLOSURE (steps 3..5), the states in $\mathcal{S}_{j'}$ will be added to $\mathcal{S}$. In the special case $j' = j$, we add $out_{k'}$ directly to $\mathcal{S}$ (and $\mathcal{T}$) at steps 13..14. In the remaining cases (step 15), we simply add the states reachable by ε-transitions (step 17); this is done only if they are not in $\mathcal{S}$ (step 16), in order to avoid considering the same state twice.

5   Correctness

The key result for our correctness proof is the following lemma which states the main idea behind our continuations. Recall the notation $L_k(\alpha)$ introduced after the algorithm for $M(\alpha)$. A rather straightforward technical proof of Lemma 1 is ommitted.

Lemma 1 *For any k, $1 \leq k \leq p + 1$, any i, $0 \leq i \leq n$, and any j, $i \leq j \leq n$, we have $j \in \mathrm{cont}(k, i)$ iff $w[i + 1..j] \notin L_k(\alpha)$.*

Theorem 2 *For any extended regular expression α and any word w, the procedure* EXT-REG-EXP-MATCHER(α, w) *returns* YES *if $w \in L(\alpha)$ and* NO *otherwise.*

Proof. As mentioned before, we have $L_{k_{p+1}}(\alpha) = L(\overline{\alpha})$. Lemma 1 says that $n \in \mathrm{cont}(k_{p+1}, 0)$ iff $w = w[1..n] \notin L_{k_{p+1}}(\alpha)$. Since the original α has been replaced by its complement at step 3 in EXT-REG-EXP-MATCHER, the claim follows. $\square$

εCLOSURE(S, k, i, j)

```
 1.   𝒥 ← NEWQUEUE()
 2.   𝒥 ← S
 3.   while not(EMPTY($S_j$)) do
 4.       q ← DEQUEUE($S_j$)
 5.       ENQUEUE($S, q$); ENQUEUE($𝒥, q$)
 6.   while not(EMPTY($𝒥$)) do
 7.       q ← DEQUEUE($𝒥$)
 8.       if $q = \text{fail}_k$ then
 9.           cont$(k, i) ←$ cont$(k, i) - \{j\}$
10.       else if $(q = \text{in}_{k'}$ and $k' \neq k)$ then
11.           for each $j' \in$ cont$(k', j) - \{j\}$ do
12.               $S_{j'} ← S_{j'} \cup \{\text{out}_{k'}\}$
13.           if $j \in$ cont$(k', j)$ then
14.               ENQUEUE($S, \text{out}_{k'}$); ENQUEUE($𝒥, \text{out}_{k'}$)
15.       else for each $(q, \varepsilon, p) \in \delta$ do
16.               if $p \notin S$ then
17.                   ENQUEUE($S, p$); ENQUEUE($𝒥, p$)
18.   return $S$
```

Fig. 7. Transitions by ε

6   Complexity

We consider first the complexity of the construction of the automaton $M(\alpha)$.

Lemma 2 *For any extended regular expression α of size m, we have*
(i) $|M(\alpha)| = \mathcal{O}(m)$,
(ii) AUTOMATON-BUILDER(α) runs in time and space $\mathcal{O}(m)$.

Proof. The syntax tree of α can be built in time and space $\mathcal{O}(m)$; see, e.g., [3]. Each element of α adds at most three new states and four new transitions to $M(\alpha)$. Moreover, this is done in constant time and space. The claims follow.   □
Next comes the analysis of the complexity of the matching algorithm.

Theorem 3 *For any α and w such that $|\alpha| = m$, $|w| = n$, and α has p complements and intersections, the algorithm EXT-REG-EXP-MATCHER runs in time $\mathcal{O}(mn^2)$ and space $\mathcal{O}(pn^2)$.*

Proof. Consider first the procedure EXT-REG-EXP-MATCHER(α, w). The syntax tree of α can be built in time and space $\mathcal{O}(m)$. Using it, steps 1..3 require time and space $\mathcal{O}(m)$. Also, after step 3 α has size $\mathcal{O}(m)$ and $\mathcal{O}(p)$ complements. (For clarity, we assume at step 5 that α has $p + 1$ complements.) Therefore, by Lemma 2, step 4 requires time and space $\mathcal{O}(m)$ and also $|M(\alpha)| = \mathcal{O}(m)$. The topological sorting at step 5 can be done in linear time and space w.r.t. the size of the graph to be sorted (see, e.g., [10]); this graph is a tree with $p + 1$ nodes and p edges; thus step 5 requires time and space $\mathcal{O}(p)$. At steps 6..8 we have $\mathcal{O}(pn)$ calls to the procedure CONTINUATION.

We shall analyse the time and space required by all calls to CONTINUATION together. Steps 1..6 and 9 require time $\mathcal{O}(pn^2)$. The steps 7 and 8 require $\mathcal{O}(pn^2)$ calls to the two procedures TRANSIT and εCLOSURE. The total time for all these calls will be investigated separately.

For the $\mathcal{O}(pn^2)$ calls to TRANSIT and εCLOSURE, we notice that each state or edge of $M(\alpha)$ will be considered at most n^2 times; it will be considered only in the computation of $\text{cont}(k, i)$, where k is such that $M_k(\alpha)$ is the innermost to contain that state or edge (with the possible exception of in_k which is considered also in the complement immediately containing $M_k(\alpha)$). During each $\text{cont}(k, i)$, each state or edge can be reached at most n times. Therefore, the total time is $\mathcal{O}(mn^2)$.

Consider now the space needed. For all calls to CONTINUATION we need $\mathcal{O}(pn^2)$, since this is the space needed for storing (simultaneously) all values of cont. For all calls to TRANSIT and εCLOSURE, we need $\mathcal{O}(m)$ space since at most one is activated at any given time, so they can share space.   □

Together, Theorems 2 and 3 give a proof for the part of Theorem 1 concerning extended regular expression matching.

7   Searching

The algorithm EXT-REG-EXP-MATCHER(α, w) needs very little change to solve the searching problem. This because the preprocessing part (steps 1..8) computes more than what is needed for matching. Therefore, the searching algorithm, EXT-REG-EXP-SEARCHER(α, w), shown in Fig. 8, should be pretty clear; notice that $[i..n]$ denotes the set $\{i, i+1, \dots, n\}$. Also, its correctness follows from the above. The complexity is the same as for the matching algorithm since the preprocessing part takes most of the time and space; the time and space needed for the (new) steps 9..12 is $\mathcal{O}(n^2)$. Theorem 1 is now completely proved. Very efficient implementations of the two algorithms, matching and searching for extended regular expressions, are discussed in the next section.

8   Very Efficient Implementation

Two important remarks concerning the implementation of the algorithms above are in order. For the purpose of clarity, we gave simplified versions of the algorithms. We show in this section how the two algorithms can be modified such that they become much more efficient in practical applications.

First, and most important, we compute in EXT-REG-EXP-MATCHER(α, w) all continuations $\text{cont}(k_\ell, i)$. The algorithm can be modified such that, in most cases, much fewer continuations are actually computed. The idea is to compute only those continuations which are needed. This can be done as follows. When computing CONTINUATION(k, i), we simulate $L_k(\alpha)$ on a prefix of $w[i+1..n]$. The

---

EXT-REG-EXP-SEARCHER(α, w)

- given an extended regular expression α and a word $w = a_1 a_2 \ldots a_n$
- returns a set $\mathcal{P}$ of pairs such that, for any $(i, j) \in \mathcal{P}$, we have $w[i..j] \in L(\alpha)$

   1.   eliminate $\emptyset$ from α
   2.   eliminate $\cap$ from α
   3. $\alpha \leftarrow \overline{\alpha}$
   4. $M(\alpha) \leftarrow$ AUTOMATON-BUILDER(α)
   5. $(k_1, k_2, \ldots, k_{p+1}) \leftarrow$ TOPOLOGICAL-SORT(α)
   6.   for ℓ from 1 to $p + 1$ do
   7.   for i from 0 to n do
   8.   cont$(k_\ell, i) \leftarrow$ CONTINUATION(k_ℓ, i)
   9.   for i from 0 to n do
 10.   for each $j \in [i..n] -$ cont(k_{p+1}, i) do
 11. $\mathcal{P} \leftarrow \mathcal{P} \cup \{(i+1, j)\}$
 12. return $\mathcal{P}$

Fig. 8. The searching procedure

computation of another continuation is initiated only when an in state is found. That is, we compute a continuation only on request. The details are ommited due to lack of space. The resulting code is much more efficient in practice but longer. Also, it is easy to see that the worst case complexity remains unchanged.

Second, also for clarity sake, we eliminated intersection operators at the cost of increasing the number of complement operators. Intersections can be handled directly, without replacing by complements, in a similar manner. For an intersection $\beta \cap \gamma$ in α, we would construct two automata for β and γ as above, say M_β and M_γ, and then insert some states, say iin, ifail, and iout, with meaning similar to those for complements. The idea is that, once entering iin by some letter a_i, the state iout is added if and only if ifail is simultaneously reached from within M_β and M_γ. When computing the table cont, the essential change is that, when a state such as iin is reached, iout is added to those $S_{j'}$ (see steps 7..9 in εCLOSURE) whenever the two conts corresponding to β and γ contain j' simultaneously. Again, the details are ommited.

9 Conclusion

We investigated the matching and searching problems for extended regular expressions and gave algorithms for them which improve significantly the running times of the best algorithms for these problems. We also explained why our algorithms are most likely essentially optimal. Finally, we gave some ideas on how to implement efficiently the algorithms.

The purpose of our paper is twofold. First we want to give algorithms with very likely the best worst case complexity for the two problems in discussion. Second, and equally important, we want to emphasize the great importance of extended regular expressions for software programs which use only ordinary regular expressions. We explain how our algorithms can be modified to run much faster in practice. The efficient implementations are meant to show that extended regular expressions can efficiently be used as features in software programs. We hope that many software programs, which currently use ordinary regular expressions, will include among their features extended regular expression based tools, such that many users can take advantage of their great expressiveness power.

References

1. Aho, A., Algorithms for finding patterns in strings, in: J. van Leeuwen (ed.), *Handbook of Theoretical Computer Science*, Elsevier, 1990, 256–300.
2. Aho, A., Hopcroft, J., and Ullman, J., *The Design and Analysis of Computer Algorithms*, Addison-Wesley, Reading, MA, 1974.
3. Aho, A., Sethi, R., Ullman, J., *Compilers: Principles, Techniques, and Tools*, Addison-Wesley, MA, 1988.
4. Crochemore, M., Hancart, C., Automata for pattern matching, in: G. Rozenberg, A. Salomaa, eds., *Handbook of Formal Languages, Vol. II*, Springer-Verlag, Berlin, 1997, 399–462.
5. Friedl, J., *Mastering Regular Expressions*, O'Reilly, 1998.
6. Hirst, S.C., A new algorithm for solving membership of extended regular expressions, Rep. 354, Basser Department of Computer Science, Univ. of Sydney, 1989.
7. Hopcroft, J., and Ullman, J., *Introduction to Automata Theory, Languages, and Computation*, Addison-Wesley, Reading, MA, 1979.
8. Hunt, H.B.,III, The equivalence problem for regular expressions with intersection is not polynomial in tape, TR 73-156, Dept. of Computer Science, Cornell Univ., Ithaca, N.Y., 1973.
9. Knight, J., and Myers, E., Super-Pattern Matching, *Algorithmica* **13** (1995), no. 1–2, 211–243.
10. Knuth, D., *The Art of Computer Programming, vol. I*, 3rd edition, Addison-Wesley, Reading, MA, 1997.
11. Meyer, R.E, and Stockmeyer, L., Nonelementary word problems in automata and logic, in: *Proc. AMS Symposium on Complexity of Computation*, 1973.
12. Meyer, R.E, and Stockmeyer, L., Word problems requiring exponential time: preliminary report, in: Proc. of 5th STOC (1973), Assoc. Comput. Mach., New York, 1973, 1–9.
13. Petersen, H., The membership problem for regular expressions with intersection is complete in LOGCFL, in: H. Alt, A. Ferreira, eds., *Proc. of STACS 2002*, Lecture Notes in Comput. Sci. 2285, Springer-Verlag, Berlin, 2002, 513–522.
14. Thompson, K., Regular expression search algorithm, *Comm. ACM* **11** (6) (1968) 419–422.
15. Yamamoto, H., An automata-based recognition algorithm for semi-extended regular expressions, in: M. Nielsen, B. Rovan, eds., *Proc of MFCS 2000*, 699–708, Lecture Notes in Comput. Sci., 1893, Springer-Verlag, Berlin, 2000, 699–708.
16. Yamamoto, H., A New Recognition Algorithm for Extended Regular Expressions, in: P. Eades and T. Takaoka, eds., *Proc of ISAAC 2001*, Lecture Notes in Comput. Sci. 2223, Springer-Verlag Berlin, 2001, 257–267.
17. Yu, S., Regular Languages, in: G. Rozenberg, A. Salomaa, *Handbook of Formal Languages, Vol. I*, Springer-Verlag, Berlin, 1997, 41–110.

Algorithms for Transposition Invariant String Matching
(Extended Abstract)

Veli Mäkinen[1*], Gonzalo Navarro[2**], and Esko Ukkonen[1*]

[1] Department of Computer Science, P.O Box 26 (Teollisuuskatu 23)
FIN-00014 University of Helsinki, Finland.
{vmakinen,ukkonen}@cs.helsinki.fi
[2] Center for Web Research, Department of Computer Science, University of Chile
Blanco Encalada 2120, Santiago, Chile. gnavarro@dcc.uchile.cl

Abstract. Given strings A and B over an alphabet $\Sigma \subseteq \mathbb{U}$, where $\mathbb{U}$ is some numerical universe closed under addition and subtraction, and a distance function $d(A, B)$ that gives the score of the best (partial) matching of A and B, the *transposition invariant distance* is $\min_{t \in \mathbb{U}} \{d(A + t, B)\}$, where $A + t = (a_1 + t)(a_2 + t) \dots (a_m + t)$. We study the problem of computing the transposition invariant distance for various distance (and similarity) functions d, that are different versions of the edit distance. For all these problems we give algorithms whose time complexities are close to the known upper bounds without transposition invariance. In particular, we show how sparse dynamic programming can be used to solve transposition invariant problems.

1 Introduction

Transposition invariant string matching is the problem of matching two strings when all the characters of either of them can be "shifted" by some amount t. By "shifting" we mean that the strings are sequences of numbers and we add or subtract t from each character of one of them.

Interest in transposition invariant string matching problems has recently arisen in the field of music information retrieval (MIR) [2,11,12]. In music analysis and retrieval, one often wants to compare two music pieces to test how similar they are. A reasonable way of modeling music is to consider the pitches and durations of the notes. Often the durations are omitted, too, since it is usually possible to recognize the melody from a sequence of pitches. In this paper, we study distance measures for pitch sequences (of monophonic music) and their computation.

In general, edit distance measures can be used for matching two pitch sequences. There are, however, a couple of properties related to music that should

---

* Supported by the Academy of Finland under grant 22584.
** Supported by Millenium Nucleus Center for Web Research, Grant P01-029-F, Mideplan, Chile.

H. Alt and M. Habib (Eds.): STACS 2003, LNCS 2607, pp. 191–202, 2003.

be taken into account. Transposition invariance is one of those; the same melody is perceived even if the pitch sequence is shifted from one key to another. Another property is the continuity of the alignment; the size of the gaps between matches should be limited, since one long gap in a central part can make a crucial difference in perception. Small gaps should be accepted, e.g. for removing "decorative" notes.

We study how these two properties can be embedded in evaluating the edit distance measures. The summary of our results is given in Section 3.

In the full version of this paper [15] we also study non-gapped measures of similarity under transposition invariance (like Hamming distance), that have other applications besides MIR. Also a more error-tolerant version of edit distances is studied there, where two characters can match if their distance is at most a given constant δ.

2   Definitions

Let Σ be a finite numerical alphabet, which is a subset of some universe $\mathbb{U}$ that is closed under addition and subtraction. Let $A = a_1 a_2 \ldots a_m$ and $B = b_1 b_2 \ldots b_n$ be two *strings* over Σ^*, i.e. the *symbols* (*characters*) a_i, b_j of the two strings are in Σ for all $1 \leq i \leq m, 1 \leq j \leq n$. We will assume w.l.o.g. that $m \leq n$, since the distance measures we consider are symmetric. String A' is a *substring* of A if $A' = A_{i \ldots j} = a_i \ldots a_j$ for some $1 \leq i \leq j \leq m$. String A'' is a *subsequence* of A, denoted by $A'' \sqsubseteq A$, if $A'' = a_{i_1} a_{i_2} \ldots a_{i_{|A''|}}$ for some indexes $1 \leq i_1 < i_2 < \cdots < i_{|A''|} \leq m$.

The following measures can be defined. The length of the *longest common subsequence (LCS)* of A and B is $\mathrm{lcs}(A, B) = \max\{|S| \mid S \sqsubseteq A, S \sqsubseteq B\}$. The *edit distance* [13,18,16] between A and B is the minimum number of edit operations that are needed to convert A into B. Particularly, in the unit cost *Levenshtein distance* d_L the set of edit operations consists of insertions $\epsilon \to b$, deletions $a \to \epsilon$, and substitutions $a \to b$ ($a \neq b$) of one alphabet symbol; here $a, b \in \Sigma$, ϵ is the empty string, and *identity operations* (*matches*) $a \to a$ are not charged. If the substitution operation is forbidden, we get a distance d_{ID}, which is actually a dual problem of evaluating the LCS; it is easy to see that $d_{\mathrm{ID}}(A, B) = m + n - 2 \cdot \mathrm{lcs}(A, B)$. For convenience, we will mainly use the minimization problem d_{ID} (not lcs) in the sequel. If only deletion for symbols of B are allowed, we get a distance d_D.

String A is a *transposed copy* of B (denoted by $A =^t B$) if $B = (a_1 + t)(a_2 + t) \cdots (a_m + t) = A + t$ for some $t \in \mathbb{U}$. Definitions for a transposed substring and a transposed subsequence can be stated similarly. The transposition invariant versions of the above distance measures d_* where $* \in \{L, \mathrm{ID}, D\}$ can now be defined as $d_*^t(A, B) = \min_{t \in \mathbb{U}} d_*(A + t, B)$.

We also define α–limited versions of the edit distance measures, where the distance (gap) between two consecutive matches is limited by a constant $\alpha > 0$, i.e. if $(a_{i'}, b_{j'})$ and (a_i, b_j) are consecutive matches, then $|i - i' - 1| \leq \alpha$ and $|j - j' - 1| \leq \alpha$. We get distances $d_L^{t,\alpha}, d_{\mathrm{ID}}^{t,\alpha}$, and $d_D^{t,\alpha}$.

The *approximate string matching problem*, based on the above distance functions, is to find the minimum distance between A and any substring of B. In this case we call A the *pattern* and denote it $P_{1...m} = p_1 p_2 \cdots p_m$, and call B the *text* and denote it $T_{1...n} = t_1 t_2 \cdots t_n$, and usually assume that $m \ll n$. The *exact string matching problem* is a special case, where all the exact occurrences of P in T are searched. The two string matching problems are also called *search problems*.

In particular, if distance d_D is used in approximate string matching, we obtain a problem known as *episode matching* [14,5]. It can also be stated as follows: Find the shortest substring of the text that contains the pattern as a subsequence.

Our complexity results are different depending on the form of the alphabet Σ. We will distinguish two cases. An *integer* alphabet is any alphabet $\Sigma \subset \mathbb{Z}$. For integer alphabets, $|\Sigma|$ will denote $\max(\Sigma) - \min(\Sigma) + 1$. A *real* alphabet will be any other $\Sigma \subseteq \mathbb{R}$. For any string $A = a_1 \ldots a_m$, we will call $\Sigma_A = \{a_i \mid 1 \leq i \leq m\}$ the alphabet of A.

3   Related Work and Summary of Our Results

The first thing to notice is that the problem of *exact* transposition invariant string matching is extremely easy to solve; one can use the relative encoding of both the pattern $(p'_1 = p_2 - p_1, p'_2 = p_3 - p_2, \ldots)$ and the text $(t'_1 = t_2 - t_1, t'_2 = t_3 - t_2, \ldots)$, and use the whole arsenal of the methods developed for exact string matching. Unfortunately, this relative encoding seems to be of no use when the exact comparison is replaced by an approximate one.

Transposition invariance (as far as we know) was introduced in the string matching context in the work of Lemström and Ukkonen [12]. They proposed transposition invariant longest common subsequence (LCTS) as a measure of similarity between two monophonic music (pitch) sequences. They gave a descriptive nick name for the measure: "Longest Common Hidden Melody". As the alphabet of pitches is some limited integer alphabet $\Sigma \subset \mathbb{Z}$, the transpositions that have to be considered are $\mathbb{T} = \{b - a \mid a, b \in \Sigma\}$. This gives a brute force algorithm for computing the length of the LCTS [12]: Compute $\text{lcs}(A + t, B)$ using $O(mn)$ dynamic programming for each $t \in \mathbb{T}$. The runtime of this algorithm is $O(|\Sigma|mn)$, where typically $|\Sigma| < 256$. In the general case, where Σ could be unlimited, one could instead use the set of transpositions $\mathbb{T}' = \{b - a \mid a \in A, b \in B\}$. This is because some characters must match in any meaningful transposition. The size of $\mathbb{T}'$ could be mn, which gives $O(m^2 n^2)$ worst case time for real alphabets. Thus it is both of practical and theoretical interest to improve this algorithm.

We will also use a brute force approach as described above, but since most transpositions produce sparse instances of the dynamic programming problem, we can use specialized sparse dynamic programming algorithms to get good worst case bounds. Moreover, we show a connection between the resulting sparse dynamic programming problems and semi-static range minimum queries. We obtain simple yet efficient algorithms for the edit distance measures.

For LCS (and thus for d_{ID}) there already exists Hunt-Szymanski [10] type (sparse dynamic programming) algorithms whose time complexities depend on the number r of matching character pairs between the compared strings. The complexity of the Hunt-Szymanski algorithm is $O(r \log m)$ after time $O(n \log m)$ for preprocessing. As we show that the sum of values r over all different transpositions is mn, we get the bound $O(mn \log m)$ for the transposition invariant LCS. This requires constructing the sets of matching character pairs for each transposition; we show that this preprocessing can be done in time $O(mn)$ plus an additive term that depends on the alphabet size. Using improved algorithms for LCS [1,7] yield $O(mn \log \log n)$ bound for the transposition invariant LCS. We improve this to $O(mn \log \log m)$ by giving a new sparse dynamic programming algorithm for LCS. Our algorithm can also be generalized to the case where gaps are limited by a constant α, giving time $O(mn \log(\alpha m))$ for evaluating $d_{\mathrm{ID}}^{t,\alpha}(A, B)$.

Eppstein et al. [7] have proposed sparse dynamic programming algorithms for more complex distance computations such as the Wilbur-Lipman fragment alignment problem [19]. Also the unit cost Levenshtein distance can be evaluated using these techniques [9]. Using this algorithm, the transposition invariant case can be solved in time $O(mn \log \log n)$. However, the algorithm does not generalize to the case of α-limited gaps, and thus we develop an alternative solution that is based on semi-static range minimum queries. This gives us $O(mn \log^2 n \log \log m)$ for evaluating $d_{\mathrm{L}}^{t,\alpha}(A, B)$.

Finally, we give a new sparse dynamic programming algorithm with running time $O(r)$ for episode matching. This further gives us an $O(mn)$ method for the transposition invariant episode matching.

Table 1 compares our results for transposition invariant problems with the known upper bounds for these problems without transposition invariance.

Table 1. Upper bounds for string matching without and with transposition invariance. We have not added, for clarity, the preprocessing times of Lemma 2, which depend on alphabet terms. In some rare cases on real alphabet the preprocessing might take $O(mn \log m)$. The bounds are given for distance evaluation, unless otherwise stated. The corresponding bounds for the transposition invariant search problems are obtained by replacing the factors $\log n$, $\log \log n$ and $\log(\alpha m)$ with $\log m$, $\log \log m$, and $\log m$, respectively.

distance	without transposition invariance	with transposition invariance
exact (searching)	$O(m + n)$	$O(m + n)$
d_{ID}	$O(mn/\log m)$ [4]	$O(mn \log \log m)$
d_{ID}^{α}	$O(\alpha^2 mn)$ (naive DP)	$O(mn \log(\alpha m))$
d_{L}	$O(mn/\log m)$ [4]	$O(mn \log \log n)$
d_{L}^{α}	$O(\alpha^2 mn)$ (naive DP)	$O(mn \log^2 n \log \log m)$
d_{D} (searching)	$O(mn/\log m)$ [5]	$O(mn)$
d_{D}^{α} (searching)	$O(mn)$ [6,3]	$O(mn)$

4   Computation of Transposition Invariant Edit Distances

Let us first review how the edit distances can be computed using dynamic programming [13,18,16]. Let $A = a_1 a_2 \cdots a_m$ and $B = b_1 b_2 \cdots b_n$. One can evaluate an $(m+1) \times (n+1)$ matrix (d_{ij}), $0 \leq i \leq m, 0 \leq j \leq n$, using the recurrence

$$d_{i,j} = \min(d_{i-1,j-1} + w(a_i \to b_j), d_{i-1,j} + w(\epsilon \to a_i), d_{i,j-1} + w(b_j \to \epsilon)), \quad (1)$$

with initialization $d_{i,0} = i$ for $0 \leq i \leq m$ and $d_{0,j} = j$ for $0 \leq j \leq n$. Weight function w gives the costs for the different edit operations. Matrix (d_{ij}) can be evaluated (in some suitable order, like row–by–row or column–by–column) in $O(mn)$ time, each value $d_{i,j}$ giving the weighted edit distance of $A_{1...i}$ and $B_{1...j}$. Especially, value d_{mn} equals the weighted edit distance of A and B.

Algorithms for distances d_{ID}, d_{L}, and d_{D} can now be stated by giving proper weights for the operations. For d_{ID} we have weights $w(a_i \to b_j) = ($ if $a_i = b_j$ then 0 else $\infty)$, and $w(\epsilon \to b_j) = w(a_i \to \epsilon) = 1$. For d_{L}, the above substitution cost ∞ is replaced by 1. For d_{D}, other costs are as in d_{ID}, but $w(a_i \to \epsilon) = \infty$.

The corresponding search problems can be solved by assigning zero to the values in the first row of (d_{ij}) (recall that we identify pattern $P = A$ and text $T = B$). To find the best approximate match, we take $\min_{0 \leq j \leq n} d_{m,j}$.

4.1   The Relation between Sparseness and Transposition Invariance

Let M be the set of matching character pairs between strings A and B, i.e. $M = M(A, B) = \{(i, j) \mid a_i = b_j, 1 \leq i \leq m, 1 \leq j \leq n\}$. Let $r = r(A, B) = |M(A, B)|$. Let us redefine $\mathbb{T}$ to be the set of those transpositions that make some characters match between A and B, that is $\mathbb{T} = \{b_j - a_i \mid 1 \leq i \leq m, 1 \leq j \leq n\}$. One could compute the above edit distances and solve the search problems by running the corresponding recurrences (1) over all pairs $(A+t, B)$, where $t \in \mathbb{T}$. On integer alphabet this takes $O(|\Sigma|mn)$ time, and $O(|\Sigma_A||\Sigma_B|mn)$ time on real alphabet. This kind of a procedure can be significantly speeded up by using suitable "sparse dynamic programming" algorithms, instead.

Lemma 1 *If an algorithm computes a distance $d(A, B)$ in $O(g(r(A, B))f(m, n))$ time, where g is a concave function, then the transposition invariant distance $d^t(A, B) = \min_{t \in \mathbb{T}} d(A+t, B)$ can be computed in $O(g(mn)f(m, n))$ time, by repeating the algorithm for all $t \in \mathbb{T}$.*

Proof. Let $r_t = r(A+t, B)$ be the number of matching character pairs between $A+t$ and B. Then

$$\sum_{t \in \mathbb{T}} g(r_t) f(m, n) = f(m, n) \sum_{t \in \mathbb{T}} g\left(\sum_{i=1}^{m} |\{j \mid a_i + t = b_j, 1 \leq j \leq n\}|\right)$$

$$\leq f(m, n) g\left(\sum_{i=1}^{m} \sum_{t \in \mathbb{T}} |\{j \mid a_i + t = b_j, 1 \leq j \leq n\}|\right) = f(m, n) g(mn). \quad \square$$

The rest of the section devotes to developing algorithms whose running times depend on r.

4.2   Preprocessing

As a first step, we need a way of constructing the match set M sorted in an order that enables sparse evaluation of matrix (d_{ij}). We use *column–by–column order* defined as follows: $j' < j$ or $(j' = j$ and $i' > i)$.[1] The match set corresponding to a transposition t will be called $M_t = \{(i,j) \mid a_i + t = b_j\}$.

Lemma 2 *The match sets $M_t = \{(i,j) \mid a_i + t = b_j\}$, each sorted in column–by–column order, for all transpositions $t \in \mathbb{T}$, can be constructed for an integer alphabet in time $O(|\Sigma| + mn)$, and for a real alphabet in time $O(m \log |\Sigma_A| + n \log |\Sigma_B| + |\Sigma_A||\Sigma_B| \log(\min(|\Sigma_A|, |\Sigma_B|)) + mn)$. Both bounds can be achieved using $O(mn)$ space.*

Proof. (Sketch) For an integer alphabet, $O(|\Sigma| + mn)$ time is obtained by proceeding naively and by using array indexing to get M_t to which each pair (i,j) has to be added.

For a real alphabet, we can construct in $O(m \log |\Sigma_A| + n \log |\Sigma_B|)$ time two binary search trees $\mathcal{T}_A$ and $\mathcal{T}_B$ for characters in A and B, respectively, where each leaf maintains a list of consecutive positions in A (or in B) where that character occurs. Now, consider matching a leaf (character a) of $\mathcal{T}_A$ with a leaf (character b) of $\mathcal{T}_B$. Then, the product of the corresponding position lists gives the part of the match set M_{b-a}, where transposition $t = b - a$ is caused by matching a with b. The same transposition can be caused by several pairs of characters. Thus, when all the $|\Sigma_A||\Sigma_B|$ partial match sets are produced, we can merge the sets that correspond to the same transposition. This can be done in $O(|\Sigma_A||\Sigma_B| \log(\min(|\Sigma_A|, |\Sigma_B|)) + mn)$ time: First, sort the set of partial match sets based on the transposition they belong to by merging $\min(|\Sigma_A|, |\Sigma_B|)$ lists that are already in order (e.g. for each fixed $a \in \Sigma_A$, transpositions $b - a$, $b \in \Sigma_B$, are obtained in increasing order from $\mathcal{T}_B$). This produces a sequence of partial match sets, where the consecutive sets having the same transposition should be merged. Merging can be done efficiently by traversing B and adding a new column j to each match set $M_{b_j - a}$, $a \in \Sigma_A$; each such match set can be found in constant time by maintaining suitable pointers during the sorting phase.   □

The preprocessing can be made more space-efficient with some penalty in the time requirement; the details can be found in [15].

4.3   Computing the Transposition Invariant Longest Common Subsequence

The fastest existing sparse dynamic programming algorithm for LCS is due to Eppstein et al. [7]. Improving an algorithm of Apostolico and Guerra [1], they achieved running time $O(d \log \log \min(d, \frac{mn}{d}))$, where $d \le r$ is the number of dominant matches (see, e.g. [1] for a definition). Using this algorithm with Lemma 1, we get the bound $O(mn \log \log n)$ for the transposition invariant case.

---

[1] Note that our definition differs from the usual column–by–column order in condition $i' > i$. This is to simplify the algorithms later on.

The existing algorithms do not, however, extend to the case of α–limited gaps. We will give here an efficient algorithm for LCS that generalizes to this case. We will also use the same technique for developing an efficient algorithm for the Levenshtein distance with α–limited gaps. Moreover, by replacing the data structure used in this algorithm by a more efficient one, we can achieve $O(r \log \log m)$ complexity, which gives $O(mn \log \log m)$ for the transposition invariant LCS (which is better than the previous bound, since $m \leq n$).

Recall the set of matching character pairs $M = \{(i, j) \mid a_i = b_j\}$. Let $\bar{M} = M \cup \{(0, 0), (m + 1, n + 1)\}$. Using basic properties of the dynamic programming matrix related to d_{ID} one obtains the following sparsity lemma.

Lemma 3 *Distance $d_{\mathrm{ID}}(A, B)$ can be computed by evaluating $d_{i,j}$ for $(i, j) \in \bar{M}$ using the recurrence*

$$d_{i,j} = \min\{d_{i',j'} + i - i' + j - j' - 2 \mid (i', j') \in \bar{M}, i' < i, j' < j\}, \qquad (2)$$

with initialization $d_{0,0} = 0$. Value $d_{m+1,n+1}$ equals $d_{\mathrm{ID}}(A, B)$.

The obvious strategy to use the above lemma is to keep the already computed values $d_{i',j'}$ in a data structure such that their minimum can be retrieved efficiently when computing the value of the next $d_{i,j}$. One difficulty here is that the values stored are not comparable as such since we want the minimum just after $i - i' + j - j' - 2$ is added. This can be solved by storing the *path invariant* values $d_{i',j'} - i' - j'$ instead. Then, after retrieving the minimum value, one can add $i + j - 2$ to get the correct value for $d_{i,j}$. To get the minimum value $d_{i',j'} - i' - j'$ from range $(i', j') \in [-\infty, i) \times [-\infty, j)$, we need a dynamic data structure supporting one-dimensional range queries (our column–by–column traversal order guarantees that all query points are in range $[-\infty, j)$). In addition, the range query should not be output sensitive; it should only report the minimum value, not all the points in the range.

A balanced binary tree can be used as a such data structure. We can use the row number i' as a sort key, and store values $d(i', j') - i' - j'$ in the leaves. Then we can maintain in each internal node the minimum of the values $d(i', j') - i' - j'$ in its subtree. This is summarized in the following well-known lemma.

Lemma 4 *A balanced binary tree $\mathcal{T}$ supports the following operations in $O(\log n)$ amortized time, where n is the amount of elements currently in the tree.*

$p = \mathcal{T}.Insert(k, v)$: *Inserts value v into the tree with key k, and returns a pointer p to the inserted leaf.*

$\mathcal{T}.Delete(p)$: *Deletes the leaf pointed by p.*

$v = \mathcal{T}.Minimum(I)$: *Returns the minimum of values that have key in the one-dimensional range I.*

We are ready to give the algorithm. Initialize a balanced binary tree $\mathcal{T}$ by adding the value of $d_{0,0} - i - j = 0$ with key $i = 0$: $\mathcal{T}.Insert(0, 0)$. Proceed with the match set $\bar{M} \setminus \{(0, 0)\}$ that is sorted in column–by–column order and make the following operations at each pair (i, j):

(1) Take the minimum value from $\mathcal{T}$ whose key is smaller than the current row number i: $d = \mathcal{T}.Minimum([-\infty, i))$. Add $i + j - 2$ to this value: $d \leftarrow d + i + j - 2$.

(2) Add the current value d minus the current row number i and current column number j into $\mathcal{T}$ with the current row number as a key: $\mathcal{T}.Insert(i, d - i - j)$.

Finally, after cell $(m + 1, n + 1)$ has been processed, we have that $d_{\mathrm{ID}}(A, B) = d$.

The above algorithm works correctly: The column–by–column evaluation and the range query restricted by the row number in $\mathcal{T}$ guarantee that the $i' < i$ and $j' < j$ conditions hold. Clearly, the time complexity is $O(r \log r)$.

The queries that we use are semi-infinite. It follows that we can use the result in Lemma 7 (see next section) to obtain $\log \log m$ query time, since our query range is $[0, m]$.

Let us now consider the case with α–limited gaps. The only change we need in our algorithms is to make sure that, in order to compute $d_{i,j}$, we only take into account the matches that are in the range $(i', j') \in [i - \alpha - 1, i) \times [j - \alpha - 1, j)$. What we need to do is to change the range $[-\infty, i)$ into $[i - \alpha - 1, i)$ in $\mathcal{T}$, as well as to delete elements in column $\leq j - \alpha - 1$ after processing elements in column j. The former is easily accomplished by using query $\mathcal{T}.Minimum([i - \alpha - 1, i))$ at phase (1) of the algorithm. The latter needs an auxiliary tree organized by values j, having pointers to the corresponding leaves of the primary tree; then we can use $\mathcal{T}.Delete(p)$ to delete each element p found by a query $[-\infty, j - \alpha - 1]$ from the auxiliary tree. Notice that we can not use the reduction to $\log \log m$ anymore, since the query ranges are no longer semi-infinite. However, since there are at most αm values stored at a time in $\mathcal{T}$, we have $O(\log(\alpha m))$ query time.

By using Lemma 1 and the above algorithms, we get the following result.

Theorem 5 *The transposition invariant distance $d_{\mathrm{ID}}^{\mathrm{t}}(A, B)$ can be computed in $O(mn \log \log m)$ time and $O(mn)$ space. The corresponding search problem can be solved in $O(mn \log \log m)$ time and in $O(m^2)$ space. For the distance $d_{\mathrm{ID}}^{\mathrm{t},\alpha}(A, B)$, the space requirements remain the same, but the time bounds are $O(mn \log(\alpha m))$ for distance computation and $O(mn \log m)$ for searching. The preprocessing bounds of Lemma 2 need to be added to these bounds.*

We refer to [15] for details on how to achieve the above bounds for the search problems.

4.4   Computing the Transposition Invariant Levenshtein Distance

The Levenshtein distance d_{L} has a sparsity property similar to the one given for d_{ID} in Lemma 3. Recall that $\bar{M} = M \cup \{(0,0), (m + 1, n + 1)\}$, where M is the set of matching character pairs. The following lemma is easy to prove.

Lemma 6 *Distance $d_{\mathrm{L}}(A, B)$ can be computed by evaluating $d_{i,j}$ for $(i, j) \in \bar{M}$ using the recurrence*

$$d_{i,j} = \min \begin{cases} \{d_{i',j'} + j - j' - 1 \mid (i', j') \in \bar{M}, i' < i, j' - i' < j - i\} \\ \{d_{i',j'} + i - i' - 1 \mid (i', j') \in \bar{M}, j' < j, j' - i' \geq j - i\} \end{cases} \tag{3}$$

with initialization $d_{0,0} = 0$. Value $d_{m+1,n+1}$ equals $d_{\mathrm{L}}(A, B)$.

This relation is more complex than the one for d_{ID}. In the case of d_{ID} we could store values $d_{i',j'}$ in a comparable format (by storing $d_{i',j'} - i' - j'$ instead) so that the minimum from index range $i' < i, j' < j$ could be retrieved efficiently. For d_{L} there does not seem to be such a comparable format, since the path length from (i', j') to (i, j) may be either $i - i' - 1$ or $j - j' - 1$.

Let us call the two index ranges for i' and j' in the above lemma as the *lower region* $(i' < i, j' - i' < j - i)$ and the *upper region* $(j' < j, j' - i' \geq j - i)$. Our strategy is to maintain separate data structures for both regions. Each value $d_{i',j'}$ will be stored in both structures in such a way that the stored values in each structure are comparable.

Let $\mathcal{L}$ denote the data structure for the lower region and $\mathcal{U}$ the data structure for the upper region. If we store values $d_{i',j'} - j'$ in $\mathcal{L}$, we can take the minimum over those values plus $j - 1$ to get the value of $d_{i,j}$. However, we want this minimum over a subset of values stored in $\mathcal{L}$, i.e. over those $d_{i',j'} - j'$ whose coordinates satisfy $i' < i, j' - i' < j - i$. Similarly, if we store values $d_{i',j'} - i'$ in $\mathcal{U}$, we can take minimum over those values whose coordinates satisfy $j' < j, j' - i' \geq j - i$, plus $i - 1$ to get the value of $d_{i,j}$ (the actual minimum is then the minimum of upper region and the lower region).

What is left to be explained is how the minima of subsets of $\mathcal{L}$ and $\mathcal{U}$ can be obtained. For the upper region, we can use the same structure as for d_{ID}; if we keep values $d_{i',j'} - i'$ in a balanced binary tree $\mathcal{U}$ with keys $j' - i'$, we can make one-dimensional range search to locate the minimum of values $d_{i',j'} - i'$ whose coordinates satisfy $j' - i' \geq j - i$. The column–by–column traversal guarantees that $\mathcal{U}$ only contains values $d_{i',j'} - i'$ for whose coordinates hold $j' < j$. Thus, the upper region can be handled efficiently.

The problem now is the lower region. We could use row–by–row traversal to handle this case efficiently, but then we would have the symmetric problem with the upper region. No traversal order will allow us to limit to one-dimensional range searches in both regions simultaneously; we will need two-dimensional range searches in one of them. Let us consider the two-dimensional range search for the lower region. We would need a query that retrieves the minimum of values $d_{i',j'} - j'$ whose coordinates satisfy $i' < i, j' - i' < j - i$. We make a coordinate transformation to make this triangle region into a rectangle; we map each value $d_{i',j'} - j'$ into an xy-plane to coordinate $i', j' - i'$. What we need in this plane, is a rectangle query $[-\infty, i) \times [-\infty, j - i)$, i.e. a semi-infinite query in both coordinates. We will specify in Lemma 7 an abstract data structure for $\mathcal{L}$ that supports this operation. Such a structure, given in [8], is basically a two-dimensional range tree, such that the secondary trees are replaced by priority queues. The underlying idea is that one-dimensional range minimum queries can be answered using a priority queue. Moreover, these queries can be done in the rank space, and efficient implementations exist for the required operations on an integer range [17].

Lemma 7 ([8]) *There is a data structure $\mathcal{R}$ that, after $O(n \log n)$ time preprocessing, supports the following operations in amortized $O(\log n \log \log n)$ time and $O(n \log n)$ space, where n is the number of elements in the structure:*

$\mathcal{R}.Update(x, y, v)$: *Update value at coordinate* x, y *to* v *(under the condition that the current value must be larger than* v*).*

$v = \mathcal{R}.Minimum(I)$: *Retrieve the minimum of values from range* I*, where* I *is semi-infinite in both coordinates.*

In the one-dimensional case, where the keys are in the integer range $[1, u]$, semi-infinite queries can be answered in $O(\log \log u)$ time.

We are now ready to give a sparse dynamic programming algorithm for the Levenshtein distance. Initialize a balanced binary tree $\mathcal{U}$ for the upper region by adding the value of $d_{0,0} - i = 0$ with key $i = 0$: $\mathcal{U}.Insert(0, 0)$. Initialize a data structure $\mathcal{L}$ for the lower region ($\mathcal{R}$ of Lemma 7) with the triples (i, j, ∞) such that $(i, j) \in \bar{M}$. Update value of $d_{0,0} - j = 0$ with keys $i = 0$ and $j - i = 0$: $\mathcal{L}.Update(0, 0, 0)$. Proceed with the match set $\bar{M} \setminus \{(0, 0)\}$ that is sorted in column–by–column order and make the following operations at each pair (i, j):

(1) Take the minimum value from $\mathcal{U}$ whose key is larger or equal to the current diagonal $j - i$: $d' = \mathcal{U}.Minimum([j - i, \infty])$. Add $i - 1$ to this value: $d' \leftarrow d' + i - 1$.
(2) Take the minimum value from $\mathcal{L}$ inside the rectangle $[-\infty, i) \times [-\infty, j - i)$: $d'' = \mathcal{L}.Minimum([-\infty, i) \times [-\infty, j - i))$. Add $j - 1$ to this value: $d'' \leftarrow d'' + j - 1$.
(3) Choose the minimum of d' and d'' as the current value $d = d_{i,j}$.
(4) Add the current value d minus i into $\mathcal{U}$ with key $j - i$: $\mathcal{U}.Insert(j - i, d - i)$.
(5) Add the current value d minus j into $\mathcal{L}$ with keys i and $j - i$: $\mathcal{L}.Update(i, j - i, d - j)$.

Finally, after cell $(m + 1, n + 1)$ has been processed, we have that $d_{\mathrm{L}}(A, B) = d$.

The correctness of the algorithm should be clear from the above discussion. The time complexity is $O(r \log r \log \log r)$ (r elements are inserted and updated in the lower region structure, and r times it is queried). The space usage is $O(r \log r)$. We can reduce the time complexity to $O(r \log r \log \log m)$ since the $\log \log n$ factor in Lemma 7 is actually $\log \log u$, where $1 \ldots u$ is the the range of values added to the secondary structure. We can implement the structure in Lemma 7 so that $u = m$.

Using this algorithm, the transposition invariant distance can be evaluated in $O(mn \log n \log \log m)$ time. This is, by a $\log n$ factor, worse than what can be achieved by using an $O(r \log \log \min(r, mn/r))$ time algorithm of Eppstein et al. [7].

However, the advantage of our range query approach is that we can now easily evaluate the distance with α–limited gaps. Consider the lower region. We need the minimum over the values whose coordinates (i', j') satisfy $i' \in [i - \alpha - 1, i)$, $j' \in [j - \alpha - 1, j)$, and $j' - i' \in [-\infty, j - i)$. We map each $d_{i',j'} - j'$ into three-dimensional coordinate $(i', j', j' - i')$. The data structure of Lemma 7 can be generalized to answer three-dimensional range minimum queries when the ranges are semi-infinite in each coordinate [8]. We can use query $\mathcal{R}.Minimum([-\infty, i) \times [j - \alpha - 1, \infty] \times [-\infty, j - i))$ when computing the value of $d_{i,j}$ from the lower region, since $i' \geq i - \alpha - 1$ when $j' - i' < j - i$, and column–by–column order guarantees that $j' < j$. The upper region case is now symmetric and can be handled similarly. The data structure $\mathcal{R}$ can be implemented so that we get overall time complexity $O(r \log^2 r \log \log m)$.

Using Lemma 1 with the above algorithms (and with the algorithm of Eppstein et al.), we obtain the following result for the transposition invariant case.

Theorem 8 *Transposition invariant Levenshtein distance $d_{\mathrm{L}}^{\mathrm{t}}(A, B)$ can be computed in $O(mn \log \log n)$ time and in $O(mn)$ space. The corresponding search problem can be solved in $O(mn \log \log m)$ time and $O(m^2)$ space. For the distance $d_{\mathrm{L}}^{\mathrm{t},\alpha}(A, B)$ the time requirements are $O(mn \log^2 n \log \log m)$ and $O(mn \log^2 m \log \log m)$, and space requirements $O(mn \log^2 n)$ and $O(m^2 \log^2 m)$, respectively, for distance computation and for searching. The preprocessing bounds of Lemma 2 need to be added to these bounds.*

Again, we refer to [15] for details on how to achieve the above bounds for the search problems.

4.5   Transposition Invariant Episode Matching

Finally we look at the episode matching problem and the $d_{\mathrm{D}}^{\mathrm{t}}$ distance, which has a simple sparse dynamic programming solution. Recall that $\bar{M} = M \cup \{(0,0), (m+1, n+1)\}$, where M is the set of matching character pairs. The following lemma for d_{D} is easy to prove.

Lemma 9 *Distance $d_{\mathrm{D}}(A, B)$ can be computed by evaluating $d_{i,j}$ for $(i, j) \in \bar{M}$ using the recurrence*

$$d_{i,j} = \min\{d_{i-1,j'} + j - j' - 1 \mid (i-1, j') \in \bar{M}\}, \qquad (4)$$

with initialization $d_{0,0} = 0$. Value $d_{m+1,n+1}$ equals $d_{\mathrm{D}}(A, B)$.

Consider an algorithm that traverses the match set $\bar{M}$ in the column–by–column order. We will maintain for each row a value $c(i)$ that gives the largest $j' < j$ such that $a_i = b_{j'}$, and a value $d(i) = d_{i,j'}$. First, initialize these values to ∞, except that $c(0) = 0$ and $d(0) = 0$. Let $(i, j) \in \bar{M}$ be the current pair whose value we need to evaluate. Then $d = d_{i,j} = d(i-1) + j - c(i-1) - 1$. We can now update the values of the current row: $c(i) = j$ and $d(i) = d$. This takes overall $O(|\bar{M}|) = O(|M|)$ time. (Preprocessing time $O(n \log |\Sigma_A| + |M|)$ for constructing M needs to be added to this.)

The above algorithm generalizes to the search problem and to the episode matching problem by implicitly initializing values $c(0) = j - 1$ and $d(0) = 0$ for the values in the first row. The problem of α–limited gaps can also be handled easily; we simply avoid updating $d(i)$ as defined when $j - c(i-1) - 1 > \alpha$. In this case we set $d(i) = \infty$. Using Lemma 1 we get:

Theorem 10 *The transposition invariant episode matching problem can be solved in $O(mn)$ time. The same bound applies in the case of α–limited gaps. The preprocessing bounds of Lemma 2 need to be added to these bounds.*

References

1. A. Apostolico and C. Guerra. The longest common subsequence problems revisited. *Algorithmica* 2:315–336, 1987.
2. T. Crawford, C.S. Iliopoulos, and R. Raman. String matching techniques for musical similarity and melodic recognition. *Computing in Musicology* 11:71–100, 1998.
3. M. Crochemore, C. Iliopoulos, C. Makris, W. Rytter, A. Tsakalidis, and K. Tsichlas. Approximate string matching with gaps. *Nordic Journal of Computing* 9(1):54–65, Spring 2002.
4. M. Crochemore, G. Landau, and M. Ziv-Ukelson. A sub-quadratic sequence alignment algorithm for unrestricted cost matrices. In *Proc. SODA'2002*, pp. 679–688. ACM-SIAM, 2002.
5. G. Das, R. Fleischer, L. Gasieniec, D. Gunopulos, and J. Kärkkäinen. Episode matching. In *Proc. CPM'97*, LNCS 1264, Springer, pp. 12–27, 1997.
6. M.J. Dovey. A technique for "regular expression" style searching in polyphonic music. In *Proc. ISMIR 2001*, pp. 179–185, October 2001.
7. D. Eppstein, Z. Galil, R. Giancarlo, and G. F. Italiano. Sparse dynamic programming I: linear cost functions. *J. of the ACM* 39(3):519–545, July 1992.
8. H. N. Gabow, J. L. Bentley, and R. E. Tarjan. Scaling and related techniques for geometry problems. *Proc. STOC'84*, pp. 135–143, 1984.
9. Z. Galil and K. Park. Dynamic programming with convexity, concavity and sparsity. *Theoretical Computer Science* 92:49–76, 1992.
10. J. W. Hunt and T. G. Szymanski. A fast algorithm for computing longest common subsequences. *Commun. ACM*, 20(5):350–353, May 1977.
11. K. Lemström and J. Tarhio. Searching monophonic patterns within polyphonic sources. In *Proc. RIAO 2000* , pp. 1261–1279 (vol 2), 2000.
12. K. Lemström and E. Ukkonen. Including interval encoding into edit distance based music comparison and retrieval. In *Proc. AISB 2000*, pp. 53–60, 2000.
13. V. Levenshtein. Binary codes capable of correcting deletions, insertions and reversals. *Soviet Physics Doklady* 6:707–710, 1966.
14. H. Mannila and H. Toivonen, and A. I. Verkamo. Discovering frequent episodes in sequences. In *Proc. KDD'95*, AAAI Press, pp. 210–215, 1995.
15. V. Mäkinen, G. Navarro, and E. Ukkonen. Algorithms for Transposition Invariant String Matching. *TR/DCC-2002-5*, Dept. of CS, Univ. Chile, July 2002, "ftp://ftp.dcc.uchile.cl/pub/users/gnavarro/ti_matching.ps.gz"
16. P. Sellers. The theory and computation of evolutionary distances: Pattern recognition. *J. of Algorithms*, 1(4):359–373, 1980.
17. P. van Emde Boas. Preserving order in a forest in less than logarithmic time and linear space. *Inf. Proc. Letters* 6(3):80–82, 1977.
18. R. Wagner and M. Fisher. The string-to-string correction problem. *J. of the ACM* 21(1):168–173, 1974.
19. W. J. Wilbur and D. J. Lipman. The contect-dependent comparison of biological sequence. *SIAM J. Appl. Math.* 44(3):557–567, 1984.

On the Complexity of Finding a Local Maximum of Functions on Discrete Planar Subsets

Anton Mityagin

Department of Computer Science, Weizmann Institute of Science
Rehovot, 76100, Israel
`mityagin@weizmann.ac.il`

Abstract. We study how many values of an unknown integer-valued function f one needs to know in order to find a local maximum of f. We consider functions defined on finite subsets of discrete plane. We prove upper bounds for functions defined on rectangles and present lower bounds for functions defined on arbitrary domains in terms of the size of the domain and the size of its border.

Keywords: computational complexity, decision trees, local maximum search.

1 Introduction

Finding a local maximum point of a function is one of the classical problems in Computer Science. Many methods for finding a local maximum point were developed as well as many facts about properties of such points were established. For a fixed domain we are interested how many values of a given function one has to query in the worst case.

Let Ω be a finite subset of a discrete plane $\mathbb{Z} \times \mathbb{Z}$, $a = (x_0, y_0) \in \Omega$. Let $U(a)$ denote an 8-*neighborhood* of a:

$$U(a) = \{(x, y) \in \mathbb{Z} \times \mathbb{Z} : |x - x_0| \leq 1, |y - y_0| \leq 1\}.$$

A function $f : \Omega \to \mathbb{Z}$ *has a local maximum* in a point $a \in \Omega$, or $a \in \Omega$ is a *local maximum point* of a function f if $f(b) \leq f(a)$ for all $b \in U(a) \cap \Omega$. If Ω is non-empty then every function $f : \Omega \to \mathbb{Z}$ has at least one such point. For example, a point of absolute maximum of f in Ω is a local maximum point.

As a computational model we consider *decision tree algorithms*. A decision tree algorithm A on Ω is a rooted tree whose nodes are labelled by elements of Ω and for each internal node all the outgoing edges are labelled by elements of $\mathbb{Z}$ (forming a one-to-one correspondence between outgoing edges and $\mathbb{Z}$). Given an input function $f : \Omega \to \mathbb{Z}$ the result of A on f, $A(f)$, is defined as follows. Starting at the root, we evaluate the function f on the label v of the current node and move along the edge labelled by $f(v)$ until we come to a leaf of the tree. The label of that leaf is the result of A on f. The labels of nodes we have passed are called queries of A and the sequence of all queries is denoted by $s(A, f)$.

H. Alt and M. Habib (Eds.): STACS 2003, LNCS 2607, pp. 203–211, 2003.

As a complexity measure $q(A)$ of A we consider the height of A, that is, the maximum number of queries over all input functions f: $q(A) = \max_f |s(A, f)|$. We say that a decision tree algorithm A on Ω finds a local maximum point if $A(f)$ is a local maximum point of f for every input function $f : \Omega \to \mathbb{Z}$. Let $q(\Omega)$ be the minimum $q(A)$ over all decision tree algorithms A on Ω.

Decision trees with height as complexity measure were studied as a computational model both for language recognition and for searching problems, see [1], [8], [5], [10], [7] and [3]. A. Rastsvetaev and L. Beklemishev [6] established matching lower and upper bounds for discrete intervals in $\mathbb{Z}$ in a decision tree model with M parallel queries, generalizing well-known algorithm from early '50s that was analyzed by Kiefer in [9]. Lower bounds of $q(\Omega)$ were used by L. Beklemishev [4] to show the independence of the schema of induction for decidable predicates $I\Delta_1$ from the set of all true arithmetical Π_2-sentences.

2 Upper Bounds

In this section we establish upper bounds of $q(R)$ for rectangles, presenting an algorithm to find a local maximum point of functions on those domains.

Definition 1. *Let $R_{n,m}(a, b)$ denote a $n \times m$ rectangle with left-down corner in the point (a, b):*

$$R_{n,m}(a, b) = \{(x, y) \in \mathbb{Z} \times \mathbb{Z} : a \leq x < a + n, b \leq y < b + m\}.$$

Let $R_{n,m}$ stand for $R_{n,m}(1, 1)$.

Theorem 1. *For every $m \leq n$ we have*

$$q(R_{n,m}) \leq m\lfloor \log_2 \frac{n}{m} \rfloor + 2m + \frac{n}{2^{\lfloor \log_2 \frac{n}{m} \rfloor}} + 6\log_2(mn).$$

Proof. We design a recursive program P that takes as input a sub-rectangle R of $R_{n,m}$. If $R \neq R_{n,m}$ it takes also a point $x \in R$ as an extra input. The program P finds a local maximum of f on $R_{n,m}$ provided that $f(x)$ is greater than or equal to all values of f on the boundary of R. The boundary of R is defined as the set of all $c \in (U(d) \cap R_{n,m}) \setminus R$ for $d \in R$.

Program P on inputs $R = R_{k,l}(a, b)$ and $x \in R$ works as follows (we assume that $k \geq l$; the other case is entirely similar):

1. If the rectangle R consists of only one point then return that point. It is a local maximum point of f on $R_{n,m}$: either $R = R_{n,m}, m = n = 1$ and the statement is trivial or the only point of R is an extra input x.
2. Otherwise divide R by a vertical middle cut into three parts: $A = R_{\lfloor k/2 \rfloor, l}(a, b)$ (the left part), $B = R_{1,l}(a + \lfloor k/2 \rfloor, b)$ (the cut), $C = R_{k - \lfloor k/2 \rfloor - 1, l}(a + \lfloor k/2 \rfloor + 1, b)$ (the right part).
3. Ask for values of f in all points of the cut and find a point c in the cut with maximal value.

4. If $R \neq R_{n,m}$ and $f(c) < f(x)$, run the program recursively on that of parts containing x with x as the extra input—$f(x)$ is greater than the values of f on the boundaries of both left and right parts.
5. Otherwise (when $R = R_{n,m}$ or $f(c) \geq f(x)$) ask for values of f in all points of $U(c) \cap R$. If $f(c) \geq f(d)$ for all $d \in U(c) \cap R$, return c. Note that c is a local maximum of f in $R_{k,l}$.
6. Otherwise $f(d) > f(c)$ for some $d \in U(c) \cap R$. In this case d belongs either to the left part or to the right part and $f(d)$ is greater than the value of f on boundaries of both parts. Run the program recursively on that of parts containing d with d as the extra input.

Let us estimate the number of queries made by the program. Given a rectangle $k \times l$ ($k \geq l$) it asks at most $l+6$ queries and proceeds to a rectangle of size at most $k/2 \times l$. Therefore the total number of recursive calls is at most $\log_2(mn)$. Assume first that $R_{n,m}$ is balanced, that is, $m \leq n < 2m$. In this case vertical and horizontal cuts of R alternate and the total number of queries is bounded by

$$m + n/2 + m/2 + n/4 + m/4 + \cdots + 6 \log_2(mn) \leq 2m + n + 6 \log_2(mn).$$

If $n \geq 2m$ we make first $\lfloor \log_2 \frac{n}{m} \rfloor$ vertical cuts obtaining a balanced rectangle of size $(n/2^{\lfloor \log_2 \frac{n}{m} \rfloor}) \times m$. The total number of queires in this case is bounded by

$$m \lfloor \log_2 \frac{n}{m} \rfloor + 2m + \frac{n}{2^{\lfloor \log_2 \frac{n}{m} \rfloor}} + 6 \log_2(mn). \square$$

For particular cases we obtain the following corollary.

Corollary 1. *For every $m \leq n \leq 2m$ we have $q(R_{n,m}) \leq 2m + n + O(\log_2(m))$. For every n we have $q(R_{n,1}) = O(\log_2(n))$.*

The bound $q(R_{n,1}) = O(\log_2(n))$ has the same order of magnitude as the exact bound proved in [6]: $q(T_{n,1}) = \log_\alpha(n) + O(1)$, where $\alpha = \frac{1+\sqrt{5}}{2}$ is the golden ratio.

3   Reduction to a Coloring Problem

A sequence $c_1, \ldots, c_n \in \Omega$ is called *a path from c_1 to c_n* if $c_{k+1} \in U(c_k)$ for all $k = 1, \ldots, n-1$. Points $a, b \in \Omega$ are *connected* if there is a path from a to b. The set Ω is called *connected* if every two its points are connected.

We will consider later only connected Ω, as the complexity of finding a local maximum point in Ω is equal to the minimum complexity of finding a local maximum point in its connected components.

Searching for a local maximum point of an unknown function on Ω is closely related to coloring Ω using the following rules. We are allowed to color any point of Ω in red until the non-colored part of Ω is not connected. Once the non-colored part Ω' of Ω is not connected, all the connected components of Ω'

except the largest one get colored in blue and we may continue coloring in red until the non-colored part gets not connected for the second time, and so on. It turns out that using these rules we can get Ω colored so that at most $q(\Omega)$ points are colored in red. Using geometric methods we will obtain lower bounds for the number of red of points necessary to color Ω, which are lower bounds for $q(\Omega)$ too.

Let us state this differently. Define a notion of *a coloring sequence* $s = \{s_1, \ldots, s_k\}$ *for* Ω by induction on k. If $k = 0$, that is, s is empty then s is a coloring sequence for the empty Ω only. Otherwise let G be the largest connected component of $\Omega \setminus \{s_1\}$ (if there are more than one largest components, let G be any of them). Then $s = \{s_1, \ldots, s_k\}$ is a coloring sequence for Ω if $\{s_2, \ldots, s_k\}$ is a coloring sequence for G.

For connected $\Omega \subset \mathbb{Z} \times \mathbb{Z}$ let the complexity of coloring Ω, $c(\Omega)$, be the minimal length of a coloring sequence for Ω.

Theorem 2. *If $\Omega \subset \mathbb{Z} \times \mathbb{Z}$ is connected and consists of more then 1 point then $q(\Omega) \geq c(\Omega)$.*

Proof. We will prove by induction on $|\Omega| > 1$ the following statement: For every algorithm A and for every number C there is a function f on Ω whose values are greater than C and such that either the sequence of queries made by A on f, $s(A, f)$, is a coloring sequence for Ω, or the result of A on f, $A(f)$, is incorrect.

The base of induction, when $|\Omega| = 2$, is trivial.

Given Ω with $|\Omega| > 2$ and C we have to construct a function f on Ω. Run A. If A outputs a result without making any queries we can easily define f so that $A(f)$ is incorrect. Otherwise let s_1 be the first query of A to f. Let G be, as in the definition of a coloring sequence, the largest connected component of $\Omega \setminus \{s_1\}$. Let $H_1, \ldots, H_n$ be all the other connected components of $\Omega \setminus \{s_1\}$. Consider two cases.

Case 1: $n = 0$, that is, $\Omega \setminus \{s_1\}$ is connected. Then $|G| > 1$. Let $f(s_1) = C+1$. Apply the induction hypothesis to G, $C + 1$ and the algorithm A' equal to the subtree of A rooted at that node where A goes after obtaining the answer $f(s_1) = C + 1$. Let g be the function on G existing by the induction hypothesis. Define $f(v) = g(v)$ for all $v \in G$. We have to prove that $s(A, f)$ is a coloring sequence for Ω provided that $A(f)$ is a local maximum of f on Ω. Assume that $A(f)$ is a local maximum of f on Ω. Obviously s_1 is not a local maximum of f on Ω. Thus $A(f)$ is a local maximum of g on G and $s(A', g)$ is a coloring sequence for G. Hence $s(A, f) = \{s_1\} \cup s(A', g)$ is a coloring sequence for Ω.

Case 2: $n > 0$. Define the value of f on $H_1 \cup \cdots \cup H_n$ as follows. Fix $j \leq n$. Since Ω is connected but $\Omega \setminus \{s_1\}$ is not, every point v in H_j is connected to s_1 by a path P lying in H_j (except the last point s_1 of P). Indeed, connect v to a point $w \in G$ by a path in Ω; this path passes through s_1; take the beginning of the path till the first occurence of s_1. Call the *rank of* v the minimum length of such path. Let d be the maximum rank of points in $H_1 \cup \cdots \cup H_n$. For $v \in H_j$ define $f(v)$ be equal to $d + C + 1$ minus the rank of v (thus $C < f(v) \leq d + C$). Then apply the induction hypothesis to A', $d + C + 1$ and $G \cup \{s_1\}$. Here A' is the decision tree algorithm on $G \cup \{s_1\}$ that is obtained from A by answering $f(v)$ to all queries v in $H_1 \cup \cdots \cup H_n$. Let g be the function on $G \cup \{s_1\}$ existing by induction hypothesis and let $f(v) = g(v)$ for all $v \in G \cup \{s_1\}$.

Let us prove that f has no local maximums outside $G \cup \{s_1\}$. For any point v in H_j consider the shortest path in H_j connecting v to s_1. The values of f increase along this path hence v is not a local maximum point.

Assume that $A(f)$ is a local maximum of f on Ω. Then $A(f)$ is a local maximum of g on $G \cup \{s_1\}$, consequently, $s(A', g)$ is a coloring sequence for $G \cup \{s_1\}$. Hence the sequence $\{s_1\} \cup s(A', g)$ is a coloring sequence for Ω. This implies easily that $s(A, f)$ is also a coloring sequence for Ω.

4   Lower Bounds

We start with a simple lower bound of $q(R_{n,m})$. Later we will prove by a more involved arguments a stronger bound.

Theorem 3. *For all m, n we have $q(R_{n,m}) \geq \min(m, n, \frac{1}{2}\max(m, n))$.*

Proof. By Theorem 2 it suffices to prove that $c(R_{n,m}) \geq \min(m, n, \frac{1}{2}\max(m, n))$. Without loss of generality assume that $m \leq n$. Assume that the rectangle $R_{n,m}$ can be colored so that $M < \min(\frac{1}{2}n, m)$ points are red. Fix any such coloring. Then at least $\frac{1}{2}n$ columns of the rectangle and at least 1 row contain only blue points. Let Λ denote the union of all points in those rows and that column. Obviously Λ is connected, as it is a union of some rows and a column. The number of points in Λ is greater than $\frac{1}{2}nm = |R_{n,m}|/2$. Hence at each step of the coloring Λ is included in the largest connected component—a contradiction.

Corollary 2. *For rectangles $R_{m,n}$ with $m = \Omega(n), n \to \infty$ theorems 4.1 and 2.1 provide asymptotically linear upper and lower bounds $q(R_{m,n}) = \Theta(n), n \to \infty$.*

Define the boundary of $C \subset \mathbb{Z}^2$, $B(C)$, as the set of all $c \in U(d) \setminus C$ for $d \in C$.

Lemma 1. *Let W be a finite subset of $\mathbb{Z}^2$. Let $C_1, C_2, \ldots, C_n$ be all the connected components of $\mathbb{Z}^2 \setminus W$. Then*

$$\sum_{k=1}^{n} |B(C_k)| \leq 2|W| + 2n - 4. \qquad (1)$$

Proof. Turn W into a planar graph G: connect by straight line segments all the pairs of points of the form (i, j), $(i, j + 1)$ and of the form (i, j), $(i + 1, j)$. Consider on the plain $\mathbb{R}^2$ the usual topology. The resulting set of segments divides the plain $\mathbb{R}^2$ into connected components. Call them *faces* (to distingush from $C_1, C_2, \ldots, C_n$). It is easy to see that for every $k \leq n$ there is a face $\tilde{C}_k$ such that $C_k = \tilde{C}_k \cap \mathbb{Z}^2$ and the boundary of $\tilde{C}_k$ (with respect to the topology on $\mathbb{R}^2$) is equal to the subgraph G_k of G induced by $B(C_k)$. Fix k and consider a connected component H of the graph G_k. If H is a tree then remove all the points of H from W. The removed points do not belong to the boundary of C_i for $i \neq k$. The removal therefore does not decrease the number of connected components. The left hand side of (1) will decrease by $|H|$ while the right hand side by $2|H|$. Thus it suffices to prove the inequality (1) for resulting set W. In other words, we may assume that each connected component H of G_k has a cycle. In every connected graph that has a cycle, the number of nodes is less than or equal to the number of edges. Therefore we may replace, in the equation (1), $|B(C_k)|$ by the number of edges in G_k.

Note that there might be faces different from $\tilde{C}_1, \ldots, \tilde{C}_n$. Those faces are squares of size 1×1 (see Fig. 1).

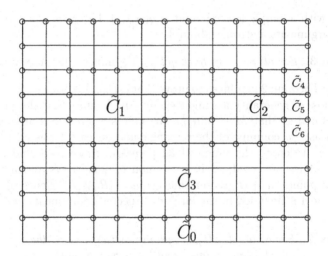

Fig. 1.

Let $\tilde{C}_{n+1}, \ldots, \tilde{C}_{n+l}$ be such faces and $G_{n+1}, \ldots, G_{n+l}$ the corresponding subgraphs of G. Each of the subgraphs $G_{n+1}, \ldots, G_{n+l}$ has 4 edges. Add $4l$ both to the left hand side and to the right hand side of (1). Now the left hand side of (1) does not exceed 2 times the number of edges of G and the right hand side of (1) is equal to $2(|W| + n - 2) + 4l \geq 2(|W| + n + l - 2)$ and the inequality holds by Euler's theorem for planar graphs.

The next theorem strengthens Theorem 3 and generalizes it to the case of arbitrary subsets of discrete plane.

Theorem 4. *Let Ω be a connected subset of $\mathbb{Z} \times \mathbb{Z}$ then*

$$q(\Omega) + 2(1 + \sqrt{2})\sqrt{q(\Omega)} \geq 2(1 + \sqrt{2})\sqrt{|\Omega|} - B/2,$$

where $B = 2|B(\Omega)| - |B(\mathbb{Z}^2 \setminus (\Omega \cup B(\Omega)))|$ $(|B(\Omega)| \leq B \leq 2|B(\Omega)|)$

Proof. By Theorem 3.1 it suffices to prove the statement of the theorem for $c(\Omega)$ instead of $q(\Omega)$. Let $s = \{s_1, \ldots, s_m\}$ be a shortest coloring sequence for Ω. Let $C_1, \ldots, C_n$ stand for connected components of $\Omega \setminus \{s_1, \ldots, s_m\}$ and $C_0, C_{n+1}, \ldots, C_r$ for connected components of $\mathbb{Z}^2 \setminus (\Omega \cup B(\Omega))$, where C_0 is the unbounded one. Applying Lemma 1 to the set $\{s_1, \ldots, s_m\} \cup B(\Omega)$ we obtain

$$\sum_{k=0}^{r} |B(C_k)| \leq 2(m + |B(\Omega)|) + 2(r+1) - 4.$$

For each k $|B(C_k)| \geq 4$, so

$$|B(C_0)| + \sum_{k=1}^{n} |B(C_k)| \leq 2(m + |B(\Omega)|) + 2n.$$

$B(C_0) \subset B(\mathbb{Z}^2 \setminus (\Omega \cup B(\Omega)))$ so we have $B \geq 2|B(\Omega)| - |B(C_0)|$ and therefore

$$\sum_{k=1}^{n} |B(C_k)| \leq 2m + 2n + B.$$

According to [2], for every finite connected set $C \subset \mathbb{Z}^2$ we have $4\sqrt{|C|} + 4 \leq |B(C)|$ (the equality holds when C is a square). Therefore we have

$$4 \sum_{k=1}^{n} \sqrt{|C_k|} \leq B + 2m. \tag{2}$$

Let us prove by induction on n that

$$\sum_{k=1}^{n} \sqrt{|C_k|} \geq (\sqrt{2} + 1)(\sqrt{|\Omega|} - \sqrt{m}). \tag{3}$$

For $n = 0$ we have $|\Omega| = m$ and the inequality turnes into the trivial equality.

Assume that $n > 0$. Obviously, $|C_1| \leq |\Omega|/2$. The sequence $s_1, \ldots, s_m$ is a coloring sequence for $\Omega \setminus C_1$ too. By induction hypothesis we have $\sum_{k=2}^{n} \sqrt{|C_k|} \geq (\sqrt{2} + 1)(\sqrt{|\Omega| - |C_1|} - \sqrt{m})$. Therefore it suffices to prove that

$$\sqrt{|C_1|} + (\sqrt{2} + 1)\sqrt{|\Omega| - |C_1|} \geq (\sqrt{2} + 1)\sqrt{|\Omega|}.$$

For $|C_1| = 0$ and for $|C_1| = |\Omega|/2$ this is true. By concavity of $\sqrt{x}$ function this is true also for all $0 \leq |C_1| \leq |\Omega|/2$.

Combining inequalities (2) and (3) we obtain

$$4(\sqrt{2} + 1)(\sqrt{|\Omega|} - \sqrt{m}) \leq B + 2m$$

and

$$2(\sqrt{2} + 1)\sqrt{|\Omega|} - B/2 \leq m + 2(\sqrt{2} + 1)\sqrt{m}. \qquad \Box$$

Corollary 3. *For the square $R_{n,n}$ the theorem gives the lower bound $q(R_{n,n}) \geq 2\sqrt{2}n + O(\sqrt{n})$ while the upper bound provided by Theorem 1 is $q(R_{n,n}) \leq 3n + O(\log(n))$.*

5 Finding a Local Maximum for Other Neighborhoods

Consider the 4-*neighborhood* instead of 8-*neighborhood*:

$$U'(x_0, y_0) = \{(x,y) \in \mathbb{Z} \times \mathbb{Z} : |x - x_0| + |y - y_0| \le 1\}.$$

The notions of a local maximum point, $q'(\Omega)$, $c'(\Omega)$ and $B'(\Omega)$ are similarly defined for the 4-neighborhood topology. In this case we have $|B'(\Omega)| \ge 2\sqrt{|\Omega|} + 4$ for every connected Ω (the equality holds for a rhombus). Repeating the arguments from the proof of Theorem 4 we obtain

Theorem 5. *Let Ω be a connected subset of $\mathbb{Z} \times \mathbb{Z}$ (in 4-neighborhood topology). Let $B' = 2|B'(\Omega)| - |B'(\mathbb{Z}^2 \setminus (\Omega \cup B'(\Omega)))|$. Then*

$$q'(\Omega) + (1 + \sqrt{2})\sqrt{q'(\Omega)} \ge (1 + \sqrt{2})\sqrt{|\Omega|} - B'/2.$$

In place of Theorem 1 we obtain

Theorem 6. *Let $Rh_{n,m}$ stand for a rhombus of size $n \times m$ where $1 \le m \le n$. Then*
$$q'(Rh_{n,m}) \le m\lfloor \log_2 \frac{n}{m} \rfloor + 2m + \frac{n}{2^{\lfloor \log_2 \frac{n}{m} \rfloor}} + 6\log_2(mn).$$

One can consider the similar problem for functions on subsets of n-dimensional discrete space. All our theorems generalize naturally to this case.

Acknowledgements. The author wishes to thank all participants of the Kolmogorov seminar in Moscow State University and especially Alexander Shen and Alexander Rastsvetaev for helpful comments and criticism. The author is very grateful to Nikolai Vereshchagin for lots of advices concerning both contents and presentation of the material.

References

[1] R. Impagliazzo, M. Naor. Decision Trees and Downward Closures, *Third Annual Conference on Structure in Complexity Theory*, pp. 29–38, 1988.

[2] D.Wang, P.Wang. Discrete isoperimetric problems. *SIAM J. Appl.Math.* vol. 32, pp 860–870, 1977.

[3] M. Blum and R. Impagliazzo. General oracle and oracle classes. *28th Annual IEEE Symposium on Foundation of Computer Science*, pp 118–126, 1987.

[4] L. Beklemishev. On the induction schema for decidable predicates. Department of Philosophy, Utrecht University, *Logic Group Preprint series*, 2000.

[5] M. Santha. On the Monte Carlo boolean decision tree complexity of read-once formulae. In *6th Annual Conference on Structure in Complexity Theory*, pp 180–187, 1991.

[6] A. Rastsvetaev and L. Beklemishev. On the query complexity of finding a local maximum point. Department of Philosophy, Utrecht University, *Logic Group Preprint Series* 206, 2000. To appear in *Information Processing Letters*.

[7] G. Tardos. Query complexity or why is it difficult to separate $NP^A \cap Co\text{-}NP^A$ from P^A by a random oracle. *Combinatorica*, 9:385–392, 1990.

[8]   L. Lovasz, M. Naor, I. Newman, A. Wigderson. Search problems in the decision tree model. *32th FOCS*, pp 576–585, 1991.
[9]   J. Kiefer. Sequential minimax search for a maximum. *Proc. Amer. Math. Soc.*, 4, pp 502–506, 1953.
[10] P. Crescenzi, R. Silvestri. Sperner's lemma and Robust Machines. *Eighth annual structure in complexity theory conference*, pp 194–199, 1993.

Some Results on Derandomization

Harry Buhrman[1], Lance Fortnow[2], and A. Pavan[3]

[1] CWI, Amsterdam, The Netherlands. buhrman@cwi.nl.
[2] NEC Laboratories America, Princeton, NJ, USA. fortnow@nec-labs.com.
[3] Iowa State University, Ames, Iowa. pavan@cs.iastate.edu.

Abstract. We show several results about derandomization including

1. If NP is easy on average then efficient pseudorandom generators exist and P = BPP.
2. If NP is easy on average then given an NP machine M we can easily on average find accepting computations of $M(x)$ when it accepts.
3. For any A in EXP, if $NEXP^A$ is in $P^A/poly$ then $NEXP^A = EXP^A$.
4. If A is Σ_k^p-complete and $NEXP^A$ is in $P^A/poly$ then $NEXP^A = EXP = MA^A$.

1 Introduction

The past several years have seen several exciting results in the area of derandomization (e.g. [14,15,17,20,11]). These results have given strong evidence that we can often eliminate randomness from computation. These papers use hardness results to get derandomization and exhibit many exciting applications of these methods.

Can the collapse of complexity classes cause derandomization to occur? If P = NP then one can easily show that efficient pseudorandom generators exist. We weaken this assumption to show that even if NP is just easy on average then pseudorandom number generators exist and P = BPP.

We use this result to study relations between distributional search problems and decision problems. These relations are well understood in the in the context of worst-case complexity. For example, every NP search problem is reducible to a decision problem in NP, this implies that if P = NP, then accepting computations of NP machines can be computed in polynomial time. We also know that if P = NP, then all optimization problems are solvable in polynomial time, and indeed the entire polynomial-time hierarchy is in P. We do not have analogous results in average-case complexity. Say a class $\mathcal{C}$ is easy on average if for every language L in $\mathcal{C}$ and every polynomial-time computable distribution μ, L can be decided in average-polynomial time with respect to μ. Ideally, one would like to show if NP is easy on average, then the entire polynomial-time hierarchy is easy on average. However, this question is open. The conventional methods used in worst-case complexity that exploit the self reducibility of SAT do not seem to carry over to the average-case world. There are some partial results in this direction. Ben-David *et al* showed that every distributional NP search problem

H. Alt and M. Habib (Eds.): STACS 2003, LNCS 2607, pp. 212–222, 2003.

is random truth-table reducible to a distributional decision problem [6]. Schuler and Watanabe [22] showed that P^{NP} is easy on average, if every language in NP is solvable in average-polynomial time with respect to every P^{NP}-samplable distribution. Note that the hypothesis here is much stronger than what we would like have, i.e., NP is easy on average.

Here we show that if NP is easy on average, then all search problems in NP are also easy on average. The proof uses the earlier mentioned derandomization result. Thus we obtain an a result that is analogous to the result in worst-case complexity. Moreover we extend the results of Ben-David et. al. [6] and show that if NP is easy on average then one can find in exponential time a witness to any NE predicate.

Impagliazzo, Kabanets and Wigderson [11] use derandomization techniques to show that if all languages in nondeterministic exponential time (NEXP) have polynomial-size circuits then NEXP = MA where MA is the class of languages with Merlin-Arthur protocols. Their result gives the first proof that if NEXP is in P/poly then NEXP = EXP.

Can one extend their result to questions like what happens if $NEXP^{SAT}$ is in P^{SAT}/poly? One cannot apply the Impagliazzo-Kabanets-Wigderson result directly as their proof does not relativize.

For any A in EXP, we get $NEXP^A$ is in P^A/poly implies $NEXP^A = EXP^A$.

If A is Σ_k^p-complete for some k (for example $A = SAT$) we get that if $NEXP^A$ is in P^A/poly then $NEXP^A = MA^A = EXP$ where the final EXP is unrelativized. As a corollary we get that $NEXP^{\Sigma_k^p}$ is in $P^{\Sigma_k^p}$/poly for at most one k.

We combine techniques from derandomization, average-case complexity, interactive proof systems, diagonalization and the structure of the polynomial-time and exponential-time hierarchies to prove our results.

2   Preliminaries

We assume a standard background in basic computational complexity.

We let $\Sigma = \{0,1\}$ and A^n represent $A \cap \Sigma^n$. For strings x and y of the same length n we let $x \cdot y$ be the dot product of x and y, viewing x and y as n-dimensional vectors over GF[2].

A tally set is a subset of 1^*.

We define the classes $E = \text{DTIME}(2^{O(n)})$, $EXP = \text{DTIME}(2^{n^{O(1)}})$ and $EE = \text{DTIME}(2^{2^{O(n)}})$. The classes NE, NEXP and NEE are the nondeterministic versions of E, EXP and EE respectively.

A language A is in $\text{SIZE}(s(n))$ if there is a constant c such that for every n there is a circuit of size $c\,s(n)$ that computes A^n.

A language A is in io$-[\mathcal{C}]$ for some class $\mathcal{C}$ if there is a language B in $\mathcal{C}$ such that for an infinite number of n, $A^n = B^n$.

The class Σ_k^p represents the kth-level of the polynomial-time hierarchy with $\Sigma_1^p = NP$. The classes Σ_k^{EXP} represent the kth-level of the exponential-time hierarchy. For $k \geq 1$, $\Sigma_k^{EXP} = NEXP^{\Sigma_{k-1}^p}$.

A theorem *relativizes* if for all A, the theorem holds if *all* machines involved have access to A. Most theorems in computational complexity relativize.

2.1 Average-Case Complexity

Even if P $\neq$ NP, we still could have that NP is easy for all practical purposes, i.e. given any *reasonable* distribution, NP is easy on average on that distribution.

To formalize this problem, Levin [19] developed a notion of average-polynomial time. Given a language L and a distribution μ on Σ^*, we say L is polynomial-time on average with respect to μ if there exists a Turing machine M computing L and an $\epsilon > 0$ such that

$$\sum_{x \in \Sigma^*} \frac{(T_M(x))^\epsilon}{|x|} \mu'(x) < \infty$$

Here $\mu(x)$ represents the probability that a chosen string is lexicographically at most x, $\mu'(x) = \mu(x) - \mu(x-1)$ is the probability that the chosen string is x and $T_M(x)$ is the running time of machine M on input x. We denote the set of such (L, μ) by average$-$P.

Blass and Gurevich [7] extended Levin's definition to randomized average-polynomial time. We use a slightly different and simpler definition for our purposes. We say L is in randomized average-polynomial time with respect to μ, if here exists a probabilistic Turing machine M that decides L and an $\epsilon > 0$ such that

$$\sum_{\{x,R\} \in \Sigma^* \times R_x} \frac{T_M^\epsilon(x)}{|x|} \mu'(x) 2^{-|R|} < \infty,$$

where R_x is the set of random strings that M uses on x.

Levin defines the class dist$-$NP which contains pairs of (L, μ) where L is in NP and μ is polynomial-time computable. We say NP is easy on average if dist$-$NP is in average$-$P. It remains open, even in relativized worlds, whether this conjecture is truly different than P $=$ NP or if this conjecture implies the collapse of the polynomial-time hierarchy.

We recommend the survey of Wang [24] for a full background on average-case complexity.

2.2 Arthur-Merlin Games

Babai [2,5] developed Arthur-Merlin and Merlin-Arthur games. In this model an arbitrarily powerful Merlin is trying to convince an untrusting probabilistic polynomial-time Arthur that a string x is in some language L. For x in L Merlin should succeed in causing Arthur to accept with high probability. For x not in L, no variation of Merlin should be able to convince Arthur to accept with more than some small probability.

The class MA consists of languages with protocols where Merlin sends a message followed by Arthur's probabilistic computation. The class AM consists of protocols where Arthur first sends random coins to Merlin who responds with a message that Arthur now deterministically decides whether to accept. Babai shows that MA is contained in AM and this result relativizes.

The classes MAEXP and AMEXP are the same as MA and AM except we allow Arthur to run in time $2^{n^{O(1)}}$.

2.3   If EXP Has Small Circuits

Informally a *probabilistically checkable proof system* is a proof for an accepting nondeterministic computation that can be verified with high confidence using only a few probabilistically-chosen queries to the proof. Probabilistically checkable proof systems were developed by Fortnow, Rompel and Sipser [10] and given their name by Arora and Safra [1].

Babai, Fortnow and Lund [4] show the following result for EXP.

Theorem 1 (BFL). *Every language in* EXP *has an (exponentially-large) probabilistically checkable proof system whose answers are computable in* EXP. *The probabilistic checking algorithm runs in polynomial time with random access to the proof.*

Nisan observed that one can use this formulation to get a consequence of EXP having small circuits.

Theorem 2 (BFL-Nisan). *If* EXP *is in* P/poly *then* EXP = MA.

Theorems 1 and 2 do not relativize though we can get a limited relativization for Theorem 2.

Theorem 3. *For any A, if* EXP *is in* P^A/poly *then* EXP $\subseteq$ MA^A.

Proof: Let L be in EXP and x be in L. Merlin gives Arthur the circuit C such that C^A computes the probabilistic checkable proof for x in L. Arthur then uses C making queries to A as necessary to verify that x is in L. If $x \notin L$ then every proof will make Arthur reject with high probability and hence also the proof described by the C^A circuit. ∎

3   Average-Case Complexity and Derandomization

In this section we will show that the assumption that NP is easy on average gives us derandomization.

Theorem 4. *If* NP *is easy on average then pseudorandom generators exist and* P = BPP.

We need several results from the areas of derandomization and average-case complexity. First Impagliazzo and Wigderson [14] show that for full derandomization it suffices if E requires large circuits.

Theorem 5 (Impagliazzo-Wigderson). *Suppose there exists a language L in* E *such that for some $\epsilon > 0$ and all but a finite number of n, computing L on strings of length n cannot be done by circuits of size $2^{\epsilon n}$. We then have that efficient pseudorandom generators exist and* P = BPP.

Suppose the assumption needed for Theorem 5 fails, i.e. for every $\epsilon > 0$, E has circuits of size $2^{\epsilon n}$. We would like some weak version of Theorem 2 but Theorem 1 is too weak to apply to this case. We instead use the following result due to Babai, Fortnow, Levin and Szegedy [3] and Polishchuk and Spielman [21].

Theorem 6 (BFLS-PS). *For any ϵ, $0 < \epsilon \leq 1$, and given the computational path of a nondeterministic machine using time $t(n)$, there exists a probabilistically checkable proof of this computation computable in time $t^{1+\epsilon}(n)$ and probabilistically verifiable in time $\log^{O(1/\epsilon)} t(n)$.*

From Theorem 6 we can get the following lemma.

Lemma 1. *Suppose that for all ϵ and A in E and for infinitely many lengths n, A^n has circuits of size $2^{\epsilon n}$. For every $\epsilon > 0$ and A in E, there exists a Merlin-Arthur protocol where Arthur uses time $2^{\epsilon n}$ for for infinitely many n, if $|x| = n$ then*

- *If x is in A then Arthur will accept with probability at least $2/3$.*
- *If x is not in A, no matter what is Merlin's message, A will accept with probability at most $1/3$.*

Proof: Fix K a linear-time complete set for E. Fix an input x. If x is in K then there is a computation path for this of length $2^{O(n)}$. By Theorem 6 there is a probabilistically checkable proof of this computation computable in time $2^{O(n)}$. By assumption, for $\epsilon > 0$ and infinitely many n, this proof can be described by a circuit of size $2^{\epsilon n}$. Merlin just sends over this circuit and Arthur probabilistically verifies this proof in time polynomial in n with random access to the proof which he can simulate by evaluating the circuit in time $2^{\epsilon n}$.   ∎

We also need some results on average-case complexity due to Ben-David, Chor, Goldreich and Luby [6] and Köbler and Schuler [18].

Theorem 7 (BCGL). *If NP is easy on average then E = NE*

The idea of the proof uses a μ' that puts enough measure on the tally strings such that any algorithm that is polynomial on μ average has to be correct on all the tally sets. Hence the tally sets in NP are in P which implies that E = NE.

Theorem 8 (Köbler-Schuler). *If NP is easy on average then MA = NP.*

To see that this is true consider the following set in coNP.

$$A = \{x \mid x \text{ is the truth table of a boolean } f \text{ of } > \epsilon n \text{ circuit complexity}\}$$

Note that $f : \{0,1\}^{\log n} \to \{0,1\}$. Most functions f have circuit complexity at least ϵn. Hence a constant fraction of the strings of length n are in A. Now consider the uniform distribution that is clearly polynomial-time computable. Hence our assumption yields that A is polynomial time on average with respect to the uniform distribution. Let's say that this is witnessed by some machine M. This implies that there must be some $x_0 \in A$ on which M runs in polynomial time. So given this x_0 one can check in polynomial time that it belongs to A and thus has high circuit complexity. Using Theorem 5, given such a function with circuit complexity at least ϵn, one can derandomize MA to NP. So first the NP machine guesses x_0 and checks that x_0 is indeed in A, by running the machine M for polynomial many steps. Next if M halts and accepts x_0 the NP machine uses x_0 as a hard function to derandomize the MA protocol.

Combining Lemma 1 and the techniques of Theorem 8 we get the following.

Lemma 2. *If for all $\epsilon > 0$, E infinitely often has circuits of size $2^{\epsilon n}$ then there is a constant c such that for all A in E and $\epsilon > 0$ there is a language L in* NTIME($2^{\epsilon n}$) *such that for infinitely many n, $A^n = L^n$.*

We are now ready to prove the main theorem of this section.

Proof of Theorem 4: Suppose that there exists a language A in E and an ϵ such that for almost every n, A^n has circuit complexity at least $2^{\epsilon n}$, then by Theorem 5 we have that pseudorandom generators exist and we are done. We will show that indeed there exists such a hard language in E. By contradiction assume that for every $\epsilon > 0$ and set $A \in$ E there exist infinitely many n such that A^n has circuits of size $2^{\epsilon n}$. For some d to be chosen later, let A be a set in DTIME(2^{dn}) such that for every Turing machine M using time bounded by $2^{(d-1)n}$, and for all but a finite number of n, M fails to compute A correctly on some input of length n. One can construct such an A by simple diagonalization.

By Lemma 2, we have that there is a language B in NTIME(2^n) such that $B^n = A^n$ for infinitely many n. Next by Theorem 7 we have B in DTIME(2^{en}), where e is a fixed constant that witnesses that NTIME(2^n) $\subseteq$ DTIME(2^{en}). Choosing $d = e + 1$ gives a contradiction with the construction of A. ∎

Now we show some interesting applications of the previous theorem.

Theorem 9. *Let T be a tally set accepted by an NP machine M. If* NP *is easy on average then there exists a polynomial time algorithm such that for every $0^n \in T$ outputs an accepting computation of $M(0^n)$.*

Proof: The idea is to first use the assumption that NP is easy on average to show a probabilistic algorithm that with high probability outputs an accepting computation of $M(0^n)$ for 0^n in T. Theorem 4 shows that we can derandomize that algorithm.

Suppose M has computation paths encoded as strings of length n^k. Consider the NP language L consisting of tuples $\langle 0^n, i, j, r_1, \ldots, r_{n^k} \rangle$ such that there exists a computation path p of M such that

1. p is an accepting computation path of $M(0^n)$,
2. $p \cdot r_\ell = 1$ for all ℓ, $1 \le \ell \le j$ and
3. The ith bit of p is one.

Valiant and Vazirani [23] show that if $M(0^n)$ accepts and the r_ℓ's are chosen at random then with probability at least one-fourth there is a j, $0 \le j \le n^k$ such that exactly one path p fulfills conditions 1 and 2 above.

Consider the polynomial-time computable distribution μ with

$$\mu'(\langle 0^n, i, j, r_1, \ldots, r_{n^k} \rangle) = \frac{1}{n^{2+4k} 2^{n^{2k}}}$$

for $1 \le i \le n^k$ and $0 \le j \le n^k$.

Since NP is easy on average then L is easy with distribution μ. This implies that there is a polynomial-time algorithm that computes the correct answers to L on all n, i and j and most choices of the r_ℓ.

Our probabilistic algorithm on input 0^n works by choosing several possibilities for the r_ℓ and then if our polynomial-time algorithm computes L for all i

and j, tries for all j to read off a path using i. If there is only one path fulfilling conditions 1 and 2 we will succeed in finding a path. The analysis above shows that we will succeed with high probability. ∎

Theorem 9 now yields an extension to Theorem 7.

Corollary 1. *If* NP *is easy on average then for every* NE *predicate one can find a witness in time* $2^{O(n)}$.

This is indeed a strengthening of Theorem 7 since Impagliazzo and Tardos [13] constructed a relativized world where E = NE but there exists a NE predicate for which no witnesses can be found in time $2^{O(n)}$

We can generalize theorem 9 by allowing the inputs to also be chosen according to some distribution giving the following result.

Theorem 10. *If* NP *is easy on average then for any* NP *machine* N *and any polynomial-time computable distribution* μ *there is a function* f *computable in average polynomial-time according to* μ *such that* $f(x)$ *is an accepting path of* N *if* $N(x)$ *accepts.*

Impagliazzo and Levin [12] showed that for every NP search problem (R, μ) with a p-samplable distribution, there exists a search problem (R', ν), where ν is the uniform distribution, such that (R, μ) is randomly reducible to (R', ν). Schuler and Watanabe [22], by slightly modifying there proof, showed that if NP is easy on average, then for every language L in NP and for every p-samplable distribution μ, the distributional problem (L, μ) can be solved in randomized average-polynomial time. Moreover, the machine that solves (L, μ) never errs. We obtain the following improvement.

Theorem 11. *If* NP *is easy on average, then every distributional* NP *problem with a* p-samplable *distribution can be solved in average-polynomial time.*

4   NEXP and Polynomial-Size Circuits

Impagliazzo, Kabanets and Wigderson [11] use derandomization techniques to prove the following consequence of NEXP having small circuits.

Theorem 12 (IKW). *If* NEXP *is in* P/poly *then* NEXP = EXP = MA.

van Melkebeek [11] shows that the converse also holds.

Theorem 12 does not relativize: Buhrman, Fortnow and Thierauf [8] exhibit an oracle relative to which MAEXP is in P/poly. If Theorem 12 applied in this relativized world then we would have by padding that NEE is in MAEXP but by Kannan [16] even EE does not have polynomial-size circuits in any relativized world.

We show in this section that one can get some weak relativizations of Theorem 12.

Theorem 13. *For any* A *in* EXP, *if* NEXP^A *is in* P^A/poly *then* NEXP^A = EXP^A.

We can do better if A is complete for some level of the polynomial-time hierarchy.

Theorem 14. *Let A be complete for Σ_k^p for any $k \geq 0$. If NEXP^A is in $\mathrm{P}^A/\mathrm{poly}$ then $\mathrm{NEXP}^A = \mathrm{MA}^A = \mathrm{EXP}$.*

Theorem 14 has the following interesting corollary.

Corollary 2. *There is at most one k such that $\mathrm{NEXP}^{\Sigma_k^p}$ is in $\mathrm{P}^{\Sigma_k^p}/\mathrm{poly}$.*

Proof: Suppose that we had a $j < k$ such that the statement in Corollary 2 holds for both j and k. By Theorem 14 we have that $\mathrm{NEXP}^{\Sigma_k^p} = \mathrm{EXP} = \mathrm{NEXP}^{\Sigma_j^p}$ in $\mathrm{P}^{\Sigma_j^p}/\mathrm{poly}$. Kannan [16] gives a relativizable proof that Σ_2^{EXP} is not in P/poly. Relativizing this to Σ_j^p shows that $\Sigma_{j+2}^{\mathrm{EXP}}$ is not in $\mathrm{P}^{\Sigma_j^p}/\mathrm{poly}$. But $\mathrm{NEXP}^{\Sigma_k^p} = \Sigma_{k+1}^{\mathrm{EXP}}$ contains $\Sigma_{j+2}^{\mathrm{EXP}}$ contradicting the fact that $\mathrm{NEXP}^{\Sigma_k^p}$ is in $\mathrm{P}^{\Sigma_j^p}/\mathrm{poly}$. ∎

To prove Theorem 14 we first observe that the techniques of Impagliazzo, Kabanets and Wigderson do relativize to show

Lemma 3. *For all A, if NEXP^A is in $\mathrm{P}^A/\mathrm{poly}$ and EXP^A is in AM^A then $\mathrm{NEXP}^A = \mathrm{EXP}^A$.*

Theorem 12 follows immediately from Lemma 3 and Theorem 2.

Buhrman and Homer [9] give a relativizing proof of the following result.

Theorem 15 (Buhrman-Homer). *If $\mathrm{EXP}^{\mathrm{NP}}$ is in $\mathrm{EXP}/\mathrm{poly}$ then $\mathrm{EXP}^{\mathrm{NP}} = \mathrm{EXP}$.*

We show the following generalization.

Theorem 16. *For any $k \geq 0$, if $\mathrm{EXP}^{\Sigma_k^p}$ is in $\mathrm{EXP}/\mathrm{poly}$ then $\mathrm{EXP}^{\Sigma_k^p} = \mathrm{EXP}$.*

Proof: By induction in k. The case $k = 0$ is trivial. For $k > 0$, we have $\Sigma_k^p = \mathrm{NP}^{\Sigma_{k-1}^p}$.

Relativizing Theorem 15 to Σ_{k-1}^p gives

$$\mathrm{EXP}^{\Sigma_k^p} \subseteq \mathrm{EXP}^{\Sigma_{k-1}^p}/\mathrm{poly} \Rightarrow \mathrm{EXP}^{\Sigma_k^p} = \mathrm{EXP}^{\Sigma_{k-1}^p}. \tag{1}$$

If $\mathrm{EXP}^{\Sigma_k^p}$ is in $\mathrm{EXP}/\mathrm{poly}$ then $\mathrm{EXP}^{\Sigma_{k-1}^p}$ is in $\mathrm{EXP}/\mathrm{poly}$ so by induction $\mathrm{EXP}^{\Sigma_{k-1}^p} = \mathrm{EXP}$. Plugging this into Equation (1) gives us Theorem 16. ∎

Proof of Theorem 14: Suppose $\mathrm{NEXP}^{\Sigma_k^p}$ is in $\mathrm{P}^{\Sigma_k^p}/\mathrm{poly}$. Since $\mathrm{P}^{\Sigma_k^p}$ is in EXP and $\mathrm{EXP}^{\Sigma_k^p} \subseteq \mathrm{NEXP}^{\Sigma_k^p}$, by Theorem 16 we have $\mathrm{EXP}^{\Sigma_k^p} = \mathrm{EXP}$. We also have EXP in $\mathrm{P}^{\Sigma_k^p}/\mathrm{poly}$ so by Theorem 3 we have EXP in $\mathrm{MA}^{\Sigma_k^p}$ and thus $\mathrm{EXP}^{\Sigma_k^p}$ in $\mathrm{MA}^{\Sigma_k^p}$. By Lemma 3 with A a complete language for Σ_k^p, we have $\mathrm{NEXP}^{\Sigma_k^p} = \mathrm{EXP}^{\Sigma_k^p}$ and thus $\mathrm{NEXP}^{\Sigma_k^p} = \mathrm{EXP} = \mathrm{MA}^{\Sigma_k^p}$. ∎

To prove Theorem 13 we need the following relativizable results due to Impagliazzo, Kabanets and Wigderson [11].

Theorem 17 (IKW, Theorem 6). *If* NEXP $\neq$ EXP *then for every* $\epsilon > 0$,

$$\text{AM} \subseteq \text{io}-[\text{NTIME}(2^{n^\epsilon})/n^\epsilon].$$

Theorem 18 (IKW, Claim 11). *If* NEXP *is in* P/poly *then there is a universal constant* d_0 *such that*

$$\text{NTIME}(2^n)/n \subseteq \text{SIZE}(n^{d_0})$$

for all sufficiently large n.

Proof of Theorem 13: Suppose we have an A in EXP such that NEXP^A is in P^A/poly and NEXP$^A \neq$ EXP^A. Since Theorem 17 relativizes we have that

$$\text{AM}^A \subseteq \text{io}-[\text{NTIME}^A(2^{n^\epsilon})/n^\epsilon].$$

We also have EXP in P^A/poly so by Theorem 3 we have EXP in MA^A which is contained in AM^A and thus io$-[\text{NTIME}^A(2^{n^\epsilon})/n^\epsilon]$.

Since Theorem 18 relativizes we have

$$\text{NTIME}^A(2^n)/n \subseteq \text{SIZE}^A(n^{d_0})$$

for some fixed d_0 and sufficiently large n.

Combining we get

$$\text{EXP} \subseteq \text{io}-[\text{SIZE}^A(n^{d_0})]. \tag{2}$$

Since A is a fixed language in EXP, a straightforward diagonalization argument shows that Equation (2) is false. ∎

5   Further Research

Perhaps one can use Theorem 4 to get other consequences from NP being easy on average with the ultimate goal to show that this assumption is equivalent to P = NP.

It would be nice to unify the results in Section 4 to show that for any A in EXP if NEXP^A is in P^A/poly then NEXP^A = MA^A = EXP.

Also one should look for other ways to bring in various theoretical techniques to prove other new and interesting results on derandomization.

Acknowledgments. Thanks to Peter Bro Miltersen, Russell Impagliazzo, Dieter van Melkebeek, Valentine Kabanets and Osamu Watanabe for helpful discussions.

References

1. S. Arora and S. Safra. Probabilistic checking of proofs: A new characterization of NP. *Journal of the ACM*, 45(1):70–122, January 1998.
2. L. Babai. Trading group theory for randomness. In *Proceedings of the 17th ACM Symposium on the Theory of Computing*, pages 421–429. ACM, New York, 1985.
3. L. Babai, L. Fortnow, L. Levin, and M. Szegedy. Checking computations in poly-logarithmic time. In *Proceedings of the 23rd ACM Symposium on the Theory of Computing*, pages 21–31. ACM, New York, 1991.
4. L. Babai, L. Fortnow, and C. Lund. Non-deterministic exponential time has two-prover interactive protocols. *Computational Complexity*, 1(1):3–40, 1991.
5. L. Babai and S. Moran. Arthur-Merlin games: a randomized proof system, and a hierarchy of complexity classes. *Journal of Computer and System Sciences*, 36(2):254–276, 1988.
6. S. Ben-David, B. Chor, O. Goldreich, and M. Luby. On the theory of average case complexity. *Journal of Computer and System Sciences*, 44:193–219, 1992.
7. A. Blass and Y. Gurevich. Randomizing reductions of search problems. *SIAM Journal of Computing*, 22:949–975, 1993.
8. H. Buhrman, L. Fortnow, and T. Thierauf. Nonrelativizing separations. In *Proceedings of the 13th IEEE Conference on Computational Complexity*, pages 8–12. IEEE, New York, 1998.
9. H. Buhrman and S. Homer. Superpolynomial circuits, almost sparse oracles and the exponential hierarchy. In *Proceedings of the 12th Conference on the Foundations of Software Technology and Theoretical Computer Science*, volume 652 of *Lecture Notes in Computer Science*, pages 116–127. Springer, Berlin, Germany, 1992.
10. L. Fortnow, J. Rompel, and M. Sipser. On the power of multi-prover interactive protocols. *Theoretical Computer Science A*, 134:545–557, 1994.
11. R. Impagliazzo, V. Kabanets, and A. Wigderson. In search of an easy witness: Exponential versus probabilistic time. In *Proceedings of the 16th IEEE Conference on Computational Complexity*, pages 2–12. IEEE, New York, 2001.
12. R. Impagliazzo and L. Levin. No better ways to generate hard NP instances than picking uniformly at random. In *Proceedings of the 31st Annual Symposium on Foundations of Computer Science*, pages 812–821. IEEE Computer Society Press, 1990.
13. R. Impagliazzo and G. Tardos. Decision versus search problems in super-polynomial time. In *Proceedings of the 30th IEEE Symposium on Foundations of Computer Science*, pages 222–227. IEEE, New York, 1989.
14. R. Impagliazzo and A. Wigderson. P = BPP if E requires exponential circuits: Derandomizing the XOR lemma. In *Proceedings of the 29th ACM Symposium on the Theory of Computing*, pages 220–229. ACM, New York, 1997.
15. R. Impagliazzo and A. Wigderson. Randomness vs. time: Derandomization under a uniform assumption. *Journal of Computer and System Sciences*, 63(4):672–688, December 2001.
16. R. Kannan. Circuit-size lower bounds and non-reducibility to sparse sets. *Information and Control*, 55:40–56, 1982.
17. A. Klivans and D. van Melkebeek. Graph nonisomorhism has subexponential size proofs unless the polynomial-time hierarchy collapses. In *Proceedings of the 31st ACM Symposium on the Theory of Computing*, pages 659–667. ACM, New York, 1999.

18. J. Köbler and R. Schuler. Average-case intractability vs. worst-case intractability. In *The 23rd International Symposium on Mathematical Foundations of Computer Science*, volume 1450 of *Lecture Notes in Computer Science*, pages 493–502. Springer, 1998.

19. L. Levin. Average case complete problems. *SIAM Journal on Computing*, 15:285–286, 1986.

20. P. Miltersen and V. Vinodchandran. Derandomizing Arthur-Merlin games using hitting sets. In *Proceedings of the 40th IEEE Symposium on Foundations of Computer Science*, pages 71–80. IEEE, New York, 1999.

21. A. Polishchuk and D. Spielman. Nearly-linear size holographic proofs. In *Proceedings of the 26th ACM Symposium on the Theory of Computing*, pages 194–203. ACM, New York, 1994.

22. R. Schuler and O. Watanabe. Towards average-case complexity analysis of NP optimization problems. In *Proceedings of the tenth Annual Conference in complexity theory*, pages 148–161. IEEE Computer Society, 1995.

23. L. Valiant and V. Vazirani. NP is as easy as detecting unique solutions. *Theoretical Computer Science*, 47:85–93, 1986.

24. J. Wang. Average-case computational complexity theory. In L. Hemaspaandra and A. Selman, editors, *Complexity Theory Retrospective II*, pages 295–328. Springer, 1997.

On the Representation of Boolean Predicates of the Diffie-Hellman Function

Eike Kiltz

Lehrstuhl Mathematik & Informatik, Fakultät für Mathematik,
Ruhr-Universität Bochum, 44780 Bochum, Germany.
kiltz@lmi.ruhr-uni-bochum.de, http://www.ruhr-uni-bochum.de/lmi/kiltz/

Abstract. In this work we give a non-trivial upper bound on the spectral norm of various Boolean predicates of the Diffie-Hellman function. For instance, we consider every individual bit and arbitrary unbiased intervals. Combining the bound with recent results from complexity theory we can *rule out* the possibility that such a Boolean function can be represented by *simple functions* like depth-2 threshold circuits with a small number of gates.

1 Introduction

Recently, Forster [6] could prove that every arrangement of linear halfspaces that represents a ±1 matrix with small spectral norm must be of high dimension. As a striking consequence from the work of [3] one gets that every attempt to achieve a representation of this matrix by means of *simple functions* is doomed to fail. For example, a matrix with small spectral norm cannot be represented by *depth-2-threshold circuits* with sub-exponential number of threshold gates, where the weights of the top layer are unbounded and the weights of the bottom layer are polynomially bounded.

In this work we present a result on circuit lower bounds for specific Boolean functions. More precisely we show that the matrix representing the binary labels of a Boolean predicate of the Diffie-Hellman function has a small spectral norm. This result holds for various Boolean predicates b that are not too much *biased* towards -1 or $+1$.

It is widely believed that the Diffie-Hellman function itself is hard to compute (computational Diffie-Hellman assumption) or even hard to decide (Diffie-Hellman indistinguishability assumption, see [1]). So circuit lower bounds of this kind are not a great surprise. On the other hand it would have a dramatic impact on modern cryptography if such a simple representation does exist. This observation was the motivation of various research papers that are closely related to our work. In [2,10,15,16,17,18], lower bounds on several complexity measures of the Diffie-Hellman function, the related Squaring Exponent function and the discrete logarithm are given. It is shown, for instance, that any polynomial representation of the Diffie-Hellman functions, even for small subsets of their input, must

H. Alt and M. Habib (Eds.): STACS 2003, LNCS 2607, pp. 223–233, 2003.

inevitably have many non-zero coefficients and, thus, high degree. In contrast to the mentioned work we show similar unpredictability results already hold for various (explicitly given) Boolean predicates of the Diffie-Hellman function. We mention that our paper technically completely differs from the work cited above.

The main technical contribution of our paper is to give a non-trivial upper bound on the spectral norm of the matrix $A(b)$ representing the binary labels of the Diffie-Hellman function which is given by the mapping

$$(g^x, g^y) \mapsto b(g^{xy}).$$

This bound will only depend on the Boolean predicate b. The main tool to archive this bound are exponential sums.

Although it is not hard to show that for *almost all* Boolean functions the spectral norm is small [4], in general it seems hard to give *specific* Boolean functions with a small spectral norm. As far as we know, no non-trivial upper bound on a spectral norm of a cryptographically relevant function was known until now.

The proof methods build on recent work of Shparlinski [14] and Shaltiel [13]. Shparlinski gives a non-trivial upper bound for two related measures, the discrepancy and the $\| \cdot \|_\infty$ norm of the Fourier coefficients. These bounds hold for a specific Boolean predicate of the Diffie-Hellman function, the least significant bit. Shparlinski left it as an open problem to extend his techniques to the case of every bit. Shaltiel showed in his paper that the discrepancy and the spectral norm of a ± 1 matrix are related. Combining the two results we immediately get a bound on the spectral norm of least significant bit of the Diffie-Hellman function and thus solving Question 13.19 of [15].
Our contribution is to extend the techniques to general Boolean predicates and to improve on the bound on the spectral norm that is directly implied by the results of Shparlinski and Shaltiel.

We start in Section 2 by giving some basic definitions and review some known results. In Section 3, we formalize our main results and mention complexity theoretic implications by means of the least significant bit. In Section 4, we give a general upper bound on the spectral norm of arbitrary ± 1 matrices that may be of independent interest. In Section 5 we prove our main results. Finally, in Section 6, we discuss some extensions and limitations of our techniques.

2   Preliminaries

We first give the basic definitions of the terms that are used throughout this paper. Let M be a $p - 1 \times p - 1$ matrix with entries in $\{-1, 1\}$.

Let p be an odd prime with $2^n < p < 2^{n+1}$. For $x \in \mathbb{Z}_p^*$, $0 \leq k \leq n$, $\text{bit}_k(x) \in \{-1, 1\}$ denotes the kth bit of the binary representation of x, i.e. $\text{bit}_k(x) = 2(\lfloor x/2^k \rfloor \bmod 2) - 1$. In particular, we put $\text{lsb}(x) = \text{bit}_0(x)$.

For an integer p we define the exponential function

$$\mathbf{e}(z) = \exp(2\pi i z/p).$$

Note that for any integer z, $\mathbf{e}(z)$ has length one.

Let $b : \mathbb{Z}_p^* \to \{-1, +1\}$ be a Boolean predicate. We consider the $p - 1 \times p - 1$ matrix $A(b)$ defined via the following mapping $A(b) : \mathbb{Z}_p^* \times \mathbb{Z}_p^* \to \{-1, 1\}$ given by

$$A(b)_{g^x, g^y} := b(g^{xy} \bmod p), \quad x, y \in \mathbb{Z}_p^*,$$

where g is a generator of $\mathbb{Z}_p^*$.

For a vector x we use $||x||_2$ to denote the Euclidean norm. The *spectral* or *operator norm* $||M||_2$ of M is given by:

$$||M||_2 = \max_{\substack{x \in \mathbb{R}^{p-1} \\ ||x||_2 = 1}} ||Mx||_2 = \max_{\substack{x, y \in \mathbb{R}^{p-1} \\ ||x||_2, ||y||_2 = 1}} |x^t M y|.$$

Trivial bounds for ± 1 matrices are $\sqrt{p-1} \le ||M||_2 \le p - 1$.

We recall the definition of *communication complexity*. Let there be two parties, one (Alice) knows a value x and the other (Bob) knows a value y where one party has no information about the others value. The common goal is to create a communication protocol P between Alice and Bob where at least one party at the end is able to compute a public Boolean function $f(x, y)$. The largest number of bits exchanged by such a protocol P, taken over all possible inputs x and y is called the communication complexity of P. The smallest possible value taken over all possible protocols P is called the communication complexity of the function f.

In this paper we consider two different types of communication complexity. First, for the *deterministic communication complexity*, $CC(f)$, we require the protocol to always compute the correct value $f(x, y)$. See [9].

Second, for the *probabilistic communication complexity with unbounded error*, $PCC(f)$, we require the protocol to compute for all possible inputs x and y the correct $f(x, y)$ with probability greater than half, where the probability is taken over all random coin flips of the protocol P. See [11] for a formal definition.

Definition 1. *A d-dimensional linear arrangement representing a matrix $A \in \{-1, +1\}^{p-1 \times p-1}$ is given by collections of vectors (u_x) and (v_y) from $\mathbb{R}^d$ such that $\text{sign}\langle u_x, v_y \rangle = A_{x,y}$ for all $x, y \in \mathbb{Z}_p^*$.*

Lemma 2 (Forster). *There is no d-dimensional linear arrangement representing $A \in \{-1, +1\}^{p-1 \times p-1}$ unless $d \ge p/||A||_2$.*

The lemma of Forster implies that A is hard to represent (or to compute) in a broad sense. The most striking conclusion was drawn by Forster himself by combining Lemma 2 with a well-known relation between linear arrangements and probabilistic communication complexity (which is due to Paturi and Simon [11]).

Corollary 3. $PCC(A) \ge n - \log ||A||_2$.

The next conclusion is from [3].

Corollary 4. *Consider depth-2 threshold circuits in which the top gate is a linear threshold gate with unrestricted weights and the bottom level has s linear threshold gates using integer weights of absolute value at most W. A circuit of this kind cannot compute A (as a function in $x = x_1, \ldots, x_{n+1}, y = y_1, \ldots y_{n+1}$) unless $s = \Omega\left(\frac{p}{nW\|A\|_2}\right)$.*

3   The Results

We state our main results about the spectral norm of the matrix $A(b)_{g^x, g^y} = b(g^{xy})$ in the Theorems 5 and 6. Note that $A(b)$ implicitly also depends on n, the prime p and the generator g.

Theorem 5 (Main Theorem). *Let $b : \mathbb{Z}_p^* \to \{-1, 1\}$ be a Boolean predicate, let $H_+(b) = \{x \in \mathbb{Z}_p^* : b(x) = 1\}$ and $H_-(b) = \{x \in \mathbb{Z}_p^* : b(x) = -1\}$. We define the following two bounds depending on b:*

$$C_1(b) = |2|H_+(b)| - p|, \quad C_2(b) = \frac{1}{p} \sum_{a=1}^{p-1} \left| \sum_{z \in H_+(b)} e(az) - \sum_{z \in H_-(b)} e(az) \right|.$$

Then the bound

$$\|A(b)\|_2 \le C_1(b) + 2^{\frac{23}{24}n + o(n)} C_2(b)$$

holds.

The proof of this theorem will be given in Section 5. Note that $C_1(b) = |2|H_+(b)| - p| = \|H_+(b)| - |H_-(b)|\| = |\sum_{z \in \mathbb{Z}_p^*} b(z)|$ reflects how much *biased* towards $+1$ or -1 the predicate b is. The bound gets trivial whenever one of $|H_+(b)|$ or $|H_-(b)|$ is of the order $o(p)$. In the sequel we are interested in Boolean predicates leading to a bound of $C_2(b) = O(\log p)$.

For some *special* Boolean predicates we can give non-trivial bounds on $C_1(b)$ and $C_2(b)$ that lead to a non-trivial bound on $\|A\|_2$.

We call a Boolean predicate $b : \mathbb{Z}_p^* \to \{-1, 1\}$ *semilinear of length k* if there are integers M_i, K_i, L_i such that for all inputs x

$$b(x) = 1 \quad \Longleftrightarrow \quad x \in \bigcup_{i=1\ldots k} H_i, \quad H_i = \{M_i z + K_i \bmod p, \quad 1 \le z \le L_i\},$$

where the sets H_i are pairwise disjoint. For instance, the lsb and the predicate with $b(x) = 1$ iff x is in a fixed interval are semilinear functions of length 1.

Theorem 6. *1. Let $p > 2$ be a prime. If b_ε is semilinear of length $k = 2^{o(n)}$ and $\|H_+(b_\varepsilon)| - p/2| \le p^{23/24+\varepsilon}/2$ for a constant $\varepsilon > 0$, then*

$$\|A(b_\varepsilon)\|_2 \le 2^{(\frac{23}{24}+\varepsilon)n + o(n)}.$$

2. For any $0 \le k = o(n)$, we have

$$\|A(\text{bit}_k)\|_2 \le 2^{\frac{23}{24}n + o(n)}.$$

3. Let $(p_n)_{n \in \mathbb{N}}$ be a sequence of n-bit primes $(2^n \leq p_n \leq 2^{n+1} - 1)$ such that we have $p_n = 2^n + 2^{o(n)}$. Then, for any $0 \leq k < n$,

$$||A(\text{bit}_k)||_2 \leq 2^{\frac{23}{24}n + o(n)}.$$

To prove Theorem 6 we have to bound $C_1(b)$ and $C_2(b)$ for the predicates b of Theorem 6. This is done by exploiting some facts about exponential sums that can be looked up in [8]. Though not very difficult it is quite technical and can be looked up in the full version [7] of this paper.

EXAMPLE APPLICATIONS. We quickly discuss the complexity theoretic implications of the main theorem by means of a specific Boolean predicate, the *least significant bit*, lsb. Let $A(\text{lsb})_{g^x, g^y} = \text{lsb}(g^{xy})$ be the matrix representing the least significant bit of the Diffie-Hellman function. By Theorem 6 we get that $||A(\text{lsb})||_2 \leq 2^{\frac{23}{24}n + o(n)}$. So by the results from Section 2 we get:

- Let Ψ be a depth-2 threshold circuit in which the top gate is a linear threshold gate with unrestricted weights and the bottom level has s linear threshold gates using polynomially bounded integer weights. Then Ψ cannot represent $A(\text{lsb})$ unless

$$s \geq 2^{\frac{n}{24} + o(n)}.$$

- The deterministic (probabilistic) communication complexity of $A(\text{lsb})$ is bounded by

$$CC(A(\text{lsb})) \geq n/16 + o(n), \quad PCC(A(\text{lsb})) \geq n/24 + o(n).$$

More complexity theoretic results about the representation of $A(b)$ by polynomials over the reals and by Boolean decision trees are mentioned in the full version of this paper [7].

4   A Bound on the Spectral Norm for Any ± 1 Matrix

In this section we show that the spectral norm of an arbitrary ± 1 matrix $A(b) = (a_{ij})$ essentially only depends on sums of the form $|\sum x_i a_{ij} y_j|$ over subrows and subcolumns of $A(b)$. The summation is done over all indices (i, j), where x_i and y_i are not simultaneously small.

Lemma 7. Let $A = (a_{ij})$ be a matrix from $\{-1, +1\}^{p-1 \times p-1}$ and let $0 \leq \delta < p^{-1/2}$. Then there are vectors x and y with $||x||_2 = ||y||_2 = 1$ such that

$$||A||_2 \leq \frac{1}{1 - \delta^2 p} \left| x_{\leq \delta}^t A y_{> \delta} + x_{> \delta}^t A y \right|,$$

where $x_{\leq \delta}$ is the vector obtained from x by keeping all entries satisfying $|x_i| \leq \delta$ and setting the remaining entries to zero. The vectors $x_{> \delta}, y_{> \delta}$ and $y_{\leq \delta}$ are defined analogously.

Proof. Let x and y be vectors such that $||x||_2 = ||y||_2 = 1$ and $||A||_2 = |x^t A y|$. Since the Euclidean length of the vectors $\frac{1}{\sqrt{p\delta}} x_{\leq \delta}$ and $\frac{1}{\sqrt{p\delta}} y_{\leq \delta}$ is at most 1, we get

$$|x^t_{\leq \delta} A y_{\leq \delta}| \leq \delta^2 p ||A||_2. \tag{1}$$

Note that, by construction, $x = x_{\leq \delta} + x_{> \delta}$ and $y = y_{\leq \delta} + y_{> \delta}$. Thus,

$$||A||_2 \leq |x^t_{\leq \delta} A y_{\leq \delta}| + |x^t_{\leq \delta} A y_{> \delta} + x^t_{> \delta} A y|.$$

Applying (1) we get

$$||A||_2 \leq |x^t_{\leq \delta} A y_{\leq \delta}| + |x^t_{\leq \delta} A y_{> \delta} + x^t_{> \delta} A y|$$
$$\leq \delta^2 p ||A||_2 + |x^t_{\leq \delta} A y_{> \delta} + x^t_{> \delta} A y|,$$

which yields the lemma.

5   A Bound on $||A(b)||_2$

As already mentioned in the introduction, by combining the results from [14] and [13] we immediately get the bound

$$||A(\text{lsb})||_2 \leq 2^{\frac{71}{72} n + o(n)}.$$

In this section we improve this result by a factor 3 in the exponent to

$$||A(\text{lsb})||_2 \leq 2^{\frac{23}{24} n + o(n)}.$$

Furthermore we generalize it to general (unbiased) Boolean predicates.
Define the matrix $A'(b)$ as $A'(b)_{x,y} = b(g^{xy})$. Since g is a generator of $\mathbb{Z}_p^*$, the functions $x \mapsto g^x, y \mapsto g^y$ both define permutations on $\mathbb{Z}_p^*$. It is therefore clear that $||A(b)||_2 = ||A'(b)||_2$. In the sequel of this section we will therefore concentrate on the matrix $A'(b)$ rather than $A(b)$. We discuss the drawback of this observation in Section 6.

We quickly recall the well known fact, see Theorem 5.2 of Chapter 1 of [12], that for any integer $m \geq 2$, the number of integer divisors $\tau(m)$ of m satisfies

$$\tau(m) \leq 2^{(1+o(1))\frac{\ln m}{\ln \ln m}}. \tag{2}$$

For the exponential function $\mathbf{e}(\cdot)$, we have the following identity, see [8]:

Lemma 8 (Identity). *For every integer x and $p \geq 2$,*

$$\frac{1}{p} \sum_{a=0}^{p-1} \mathbf{e}(xa) = \begin{cases} 0 &:& \text{if } x \neq 0 \pmod{p} \\ 1 &:& \text{if } x = 0 \pmod{p}. \end{cases}$$

The next lemma is taken from [14] and will be used in our proofs. We recall that $A \ll B$ as well as $B \gg A$ is equivalent to $A = O(B)$.

Lemma 9 (Shparlinski [14]). *Let* $\mathcal{I} \subseteq \mathbb{Z}_p^*$ *and* $I(d) = \{i \in \mathcal{I} : \gcd(i, p-1) = d\}$. *Then the bound*

$$\sum_{y,z \in \mathcal{I}(d)} \left| \sum_{x=1}^{p-1} \mathbf{e}(a(g^{xy} - g^{xz})) \right|^4 \ll |\mathcal{I}(d)| p^{14/3} d^{1/3}$$

holds.

The next lemma shows how to bound sums of the form $|\sum x_i a_{ij} y_j|$ for $a_{ij} = b(g^{ij})$ and not too small x_i in the sense of Lemma 7. Techniques from [14] are extended to bound a weighted form of the exponential sum $T_a = \sum_{i \in \mathcal{I}} \sum_{j \in \mathcal{J}} x_i y_j \mathbf{e}(ag^{ij})$.

Lemma 10. *Let* $0 < \delta(n) < 1$, $\|x\|_2, \|y\|_2 \leq 1$, $\mathcal{I} = \{i \in \mathbb{Z}_p^* : x_i > \delta\}$ *and* $\mathcal{J} \subseteq \mathbb{Z}_p^*$. *Let* $b : \mathbb{Z}_p^* \to \{-1, 1\}$ *be a Boolean predicate and* $H_+(b) = \{x \in \mathbb{Z}_p^* : b(x) = 1\}$. *Then*

$$S = \left| \sum_{i \in \mathcal{I}} \sum_{j \in \mathcal{J}} x_i y_j b(g^{ij}) \right| \ll C_1(b) + p^{5/8} \delta^{-2/3} \tau(p) C_2(b).$$

Proof. Put

$$S_a = \left| \frac{1}{p} \sum_{i \in \mathcal{I}} \sum_{j \in \mathcal{J}} x_i y_j \left(\sum_{z \in H_+(b)} \mathbf{e}(a(g^{ij} - z)) - \sum_{z \notin H_+(b)} \mathbf{e}(a(g^{ij} - z)) \right) \right|$$

and $T_a = \sum_{i \in \mathcal{I}} \sum_{j \in \mathcal{J}} x_i y_j \mathbf{e}(ag^{ij})$. From Lemma 8 we expand S to

$$S = \left| \frac{1}{p} \sum_{a=0}^{p-1} \sum_{i \in \mathcal{I}} \sum_{j \in \mathcal{J}} x_i y_j \left(\sum_{z \in H_+(b)} \mathbf{e}(a(g^{ij} - z)) - \sum_{z \notin H_+(b)} \mathbf{e}(a(g^{ij} - z)) \right) \right|.$$

Splitting the outer sum into $a = 0$ and $a \neq 0$ we can continue as follows:

$$\leq |S_0| + \left| \sum_{a=1}^{p-1} S_a \right|$$

$$\leq |S_0| + \max_{1 \leq a \leq p-1} \{|T_a|\} \frac{1}{p} \sum_{a=1}^{p-1} \left| \sum_{z \in H_+(b)} \mathbf{e}(az)) - \sum_{z \notin H_+(b)} \mathbf{e}(az) \right|$$

$$= |S_0| + \max_{1 \leq a \leq p-1} \{|T_a|\} C_2(b).$$

We now prove the desired bounds on $|T_a|$ for $1 \leq a \leq p-1$ and on $|S_0|$. Define $\mathcal{I}(d) := \{i \in \mathcal{I} : \gcd(i, p-1) = d\}$. For each divisor d of $p-1$, we consider the "d-slice" of T_a,

$$\sigma(d) = \sum_{i \in \mathcal{I}(d)} \sum_{j \in \mathcal{J}} x_i y_j \mathbf{e}(ag^{ij}),$$

such that

$$|T_a| = \left| \sum_{d|p-1} \sigma(d) \right| \le \tau(p-1) \max_{d|p-1} \{|\sigma(d)|\}.$$

We now present a bound on $|\sigma(d)|$. Using Cauchy inequality and the fact that $|w|^2 = w\bar{w}$ for complex w we get

$$|\sigma(d)|^2 \le \sum_{j \in \mathcal{J}} |y_j|^2 \sum_{j \in \mathcal{J}} \left| \sum_{i \in \mathcal{I}(d)} x_i \mathbf{e}(ag^{ij}) \right|^2 \le \sum_{j=1}^{p-1} |y_j|^2 \sum_{j=1}^{p-1} \left| \sum_{i \in \mathcal{I}(d)} x_i \mathbf{e}(ag^{ij}) \right|^2$$

$$\le \sum_{j=1}^{p-1} \left| \sum_{i \in \mathcal{I}(d)} x_i \mathbf{e}(ag^{ij}) \right|^2 = \sum_{k,l \in \mathcal{I}(d)} x_k x_l \sum_{j=1}^{p-1} \mathbf{e}(a(g^{kj} - g^{lj})).$$

Note that $|\mathcal{I}| \le 1/\delta^2$. By Hölder inequality with exponents $3/2$ and 3 we get

$$\sum_{i \in \mathcal{I}} |x_i|^{4/3} \le \left(\sum_{i \in \mathcal{I}} |x_i|^2 \right)^{2/3} \cdot \left(\sum_{i \in \mathcal{I}} 1^3 \right)^{1/3} \le \delta^{-2/3}. \tag{3}$$

Again by applying Hölder inequality with exponents $4/3$ and 4 to $|\sigma(d)|^2$ we have

$$|\sigma(d)|^8 \le \left(\sum_{k,l \in \mathcal{I}(d)} |x_k x_l|^{4/3} \right)^3 \sum_{k,l \in \mathcal{I}(d)} \left| \sum_{j=1}^{p-1} \mathbf{e}(a(g^{kj} - g^{lj})) \right|^4$$

$$\le \delta^{-4} \sum_{k,l \in \mathcal{I}(d)} \left| \sum_{j=1}^{p-1} \mathbf{e}(a(g^{kj} - g^{lj})) \right|^4 \ll \delta^{-4} |\mathcal{I}(d)| d^{1/3} p^{14/3},$$

where the second inequality is obtained from (3) and the last inequality comes from Lemma 9. Since $|\mathcal{I}(d)| \le |\mathcal{I}| \le \delta^{-2}$ and $|\mathcal{I}(d)| \le p/d$, we obtain

$$|\mathcal{I}(d)| = |\mathcal{I}(d)|^{2/3} |\mathcal{I}(d)|^{1/3} \le \delta^{-4/3} \left(\frac{p}{d} \right)^{1/3}.$$

Thus, $|\sigma(d)^8| \ll \delta^{-4} \delta^{-4/3} p^5$ and therefore for any $d|p-1$,

$$|\sigma(d)| \ll \delta^{-2/3} p^{5/8}.$$

This implies the bound on $|T_a|$. It leaves to bound $|S_0|$. Note that

$$\left| \sum_{z \in H_+(b)} \mathbf{e}(0z) - \sum_{z \notin H_+(b)} \mathbf{e}(0z) \right| = C_1(b).$$

Thus,

$$|S_0| = \left| \frac{1}{p} \sum_{i \in \mathcal{I}} \sum_{j \in \mathcal{J}} x_i y_j \left(\sum_{z \in H_+(b)} e(0z) - \sum_{z \notin H_+(b)} e(0z) \right) \right|$$

$$\leq \frac{C_1(b)}{p} \sum_{i \in \mathcal{I}} |x_i| \sum_{j \in \mathcal{J}} |y_j| \leq \frac{C_1(b)}{p} \sum_{i=1}^{p-1} |x_i| \sum_{j=1}^{p-1} |y_j| \leq \frac{C_1(b)}{p} p \leq C_1(b).$$

Now we are ready to prove our main theorem.

Proof (of Theorem 5). Set $\delta = 1/\sqrt{2p}$. By Lemma 7 we get

$$||A'(b)||_2 \leq 2 \left| x_{\leq \delta}^t A y_{> \delta} + x_{> \delta}^t A y \right|.$$

Bounding $|x_{\leq \delta}^t A y_{> \delta}|$ and $|x_{> \delta}^t A y|$ by Lemma 10 and $\tau(p-1)$ by equation (2) we get

$$||A'(b)||_2 \ll C_1(b) + p^{5/8} \delta^{-2/3} \tau(p-1) C_2(b) \ll C_1(b) + 2^{\frac{23}{24}n + o(n)} C_2(b).$$

6 Remarks

FAULTY REPRESENTATIONS. Using techniques from [5] our methods can be extended to handle faulty representations of the matrix $A(b)$. Consider the Matrix $\tilde{A}(b)$ that coincides with $A(b)$ in all but t entries, i.e.

$$A(b)_{g^x, g^y} = \tilde{A}(b)_{g^x, g^y}, \quad \forall (x, y) \in T,$$

where $T \subseteq \mathbb{Z}_p^* \times \mathbb{Z}_p^*$ is a set of cardinality $|T| = p^2 - t$. Then the perturbation matrix $P(b) = \tilde{A}(b) - A(b)$ has at most t non-zero entries in $+2, -2$ and hence, as observed in Lemma 3 of [5], the spectral norm is bounded by $||P(b)||_2 \leq 2\sqrt{t}$. So

$$||\tilde{A}(b)||_2 = ||A(b) + P(b)||_2 \leq ||A(b)||_2 + ||P(b)||_2 \leq ||A(b)||_2 + 2\sqrt{t}.$$

Let Ψ be a depth-2-threshold circuit with s threshold gates, where the weights of the top layer are unbounded and the weights of the bottom layer are polynomially bounded. Then, as a consequence of our observation, Ψ cannot coincide with $A(\text{lsb})$ in all but $2^{o(n)}$ entries unless $s \geq 2^{\frac{n}{24} + o(n)}$.

FOURIER COEFFICIENTS. It is not hard to show that for ± 1 valued Boolean functions $f : \{0, 1\}^n \times \{0, 1\}^n \rightarrow \{-1, +1\}$ the inequalities

$$L_\infty(f) \leq ||A_f||_2 / 2^n, \quad L_1(f) \leq 2^n ||A_f||_2$$

hold, where $L_\infty(f)$ denotes the maximum and $L_1(f)$ the sum of the absolute Fourier coefficients of the function f. $||A_f||_2$ is the spectral norm of the matrix $A_{x,y} = f(x, y)$. So the bounds obtained in this work can be applied to get bounds on the L_1 and L_∞ norms of the function $f(g^x, g^y) = b(g^{xy})$. Note that such a result for the special case of the least significant bit was already (directly) obtained by Shparlinski [14].

LIMITATIONS. One might ask the question, if for *every* (possibly biased) Boolean predicate, the bound on $C_2(b)$ is of the order $O(\log p)$. Unfortunately this is not the case since for the predicate $b(x) = 1$ iff x is a quadratic residue modulo p, we can show by means of Gaussian Sums [8] that $C_2(b) = \Theta(p^{1/2})$. In this case the bound from Theorem 5 gets trivial. We must leave it as an open question if the spectral norm can be bounded for every biased predicate.

A different view of our result is that in a restricted model of computation (when computations are restricted to depth-2 threshold circuits with polynomially many gates), computing various Boolean predicates of the Diffie-Hellman function is hard. Unfortunately this restricted model does not properly separate between easy and hard functions: As observed in Section 5, the spectral norm of $A(b)$ and $A'(b)$ are equal. So the unpredictability results of Section 3 do also hold for $A'(b)$. Therefore in this restricted model of computation the predicate $b(g^{xy})$ cannot be efficiently computed even though x and y are given as input.

Acknowledgment. The author is grateful to Tanja Lange, Igor Shparlinski and Hans-Ulrich Simon for useful discussions and helpful ideas. I am also thankful to anonymous referees for pointing out some relevant previous work and helpful comments.

References

1. D. Boneh. The decision Diffie-Hellman problem. *Lecture Notes in Computer Science*, 1423:48–63, 1998.
2. D. Coppersmith and I. Shparlinski. On polynomial approximation of the Discrete Logarithm and the Diffie-Hellman mapping. *Journal of Cryptology*, 13(3):339–360, March 2000.
3. Forster, Krause, Lokam, Mubarakzjanov, Schmitt, and Simon. Relations between communication complexity, linear arrangements, and computational complexity. *FSTTCS: Foundations of Software Technology and Theoretical Computer Science*, 21, 2001.
4. J. Forster. *personal communication*, 2002.
5. J. Forster and H. U. Simon. On the smallest possible dimension and the largest possible margin of linear arrangements representing given concept classes. In *Proceedings of the 13th International Conference on Algorithmic Learning Theory*, pages 128–138, 2002.
6. Juergen Forster. A linear lower bound on the unbounded error probabilistic communication complexity. In *Proceedings of the Sithteenth Annual Conference on Computational Complexity*, pages 100–106. IEEE Computer Society, 2001.
7. E. Kiltz. On the representation of boolean predicates of the diffie-hellman function. *Manuscript*, 2003.

8. N. M. Korobov. *Exponential Sums and their Applications*. Kluwer Academic Publishers, 1992.
9. E. Kushilevitz and N. Nisan. *Communication Complexity*. Cambridge University Press, Cambridge, 1997.
10. T. Lange and A. Winterhof. Polynomial interpolation of the elliptic curve and XTR discrete logarithm. In *Proceedings of the 8th Annual International Computing and Combinatorics Conference (COCOON'02)*, pages 137–143, 2002.
11. R. Paturi and J. Simon. Probabilistic communication complexity. *Journal of Computer and System Sciences*, 33(1):106–123, aug 1986.
12. K. Prachar. *Primzahlverteilung*. Springer-Verlag, Berlin, 1957.
13. R. Shaltiel. Towards proving strong direct product theorems. In *Proceedings of the Sithteenth Annual Conference on Computational Complexity*, pages 107–119. IEEE Computer Society, 2001.
14. I. Shparlinski. Communication complexity and fourier coefficients of the Diffie-Hellman key. *Proc. the 4th Latin American Theoretical Informatics Conf., LNCS 1776*, pages 259–268, 2000.
15. I. E. Shparlinski. *Number Theoretic Methods in Cryptography*. Birkhäuser Verlag, 1999.
16. I. E. Shparlinski. *Cryptographic Application of Analytic Number Theory*. Birkhäuser Verlag, 2002.
17. A. Winterhof. A note on the interpolation of the Diffie-Hellman mapping. In *Bulletin of the Australian Mathematical Society*, volume 64, pages 475–477, 2001.
18. A. Winterhof. Polynomial interpolation of the discrete logarithm. *Designs, Codes and Cryptography*, 25:63–72, 2002.

Quantum Circuits with Unbounded Fan-out

Peter Høyer[1,*] and Robert Špalek[2,**]

[1] Dept. of Comp. Sci., Univ. of Calgary, AB, Canada. hoyer@cpsc.ucalgary.ca
[2] Centrum voor Wiskunde en Informatica, Amsterdam, The Netherlands.
Robert.Spalek@cwi.nl

Abstract. We demonstrate that the unbounded fan-out gate is very powerful. Constant-depth polynomial-size quantum circuits with bounded fan-in and unbounded fan-out over a fixed basis (denoted by QNC_f^0) can approximate with polynomially small error the following gates: parity, mod[q], And, Or, majority, threshold[t], exact[q], and counting. Classically, we need logarithmic depth even if we can use unbounded fan-in gates. If we allow arbitrary one-qubit gates instead of a fixed basis, then these circuits can also be made exact in log-star depth. Sorting, arithmetical operations, phase estimation, and the quantum Fourier transform can also be approximated in constant depth.

1   Introduction

In this paper, we study the power of shallow quantum circuits. Long quantum computations encounter various problems with decoherence, hence we want to speed them up as much as possible. We can exploit two types of parallelism:

1. Gates on different qubits can be applied at the same time.
2. Commuting gates can be applied on the same qubits at the same time.

The first possibility is straightforward. There are clues that also the second possibility might be physically feasible: ion-trap [3] and bulk-spin NMR [5]. If two quantum gates commute, so do their Hamiltonians and thus we can apply their joint operation by simply performing both evolutions at the same time.

We define an *unbounded fan-out gate* as a sequence of controlled-not gates sharing one control qubit. This gate is universal for all commuting gates: We show that the parallelisation method of [10,6] can apply general commuting gates in parallel using just the fan-out gate and one-qubit gates.

Classically, the main classes computed by polynomial-size, $(\log^k n)$-depth circuits with unbounded fan-out are:

- NC^k: bounded fan-in gates,
- AC^k: unbounded fan-in gates,
- TC^k: unbounded threshold gates,
- $AC^k[q]$: unbounded fan-in and mod[q] gates, and $ACC^k = \bigcup_q AC^k[q]$.

---

* Supported by the Alberta Ingenuity Fund and the Pacific Institute for the Mathematical Sciences.
** Work conducted in part while at Vrije Universiteit, Amsterdam. Partially supported by EU fifth framework project QAIP, IST-1999-11234 and RESQ, IST-2001-37559.

It is known that TC^k is strictly more powerful than ACC^k [11], and that $AC^k[q] \neq AC^k[q']$ for powers of distinct primes [14].

The main quantum circuit classes corresponding to the classical classes are QNC^k, QAC^k, QTC^k, and $QACC^k$. We use subscript 'f' to indicate circuits where we allow the fan-out gate (e.g. QNC^k_f). In contrast to the classical case, allowing $\mathrm{mod}[q]$ gates with different moduli always leads to the same quantum classes: $QACC^k = QAC^k[q]$ for every q [6]. Furthermore, parity is equivalent to unbounded fan-out, hence $QAC^k_f = QAC^k[2] = QACC^k$.

In this paper, we show that even threshold gates can be approximated with fan-out and single qubit gates in constant depth. This implies that the bounded-error versions of the classes are equal: $B\text{-}QNC^k_f = B\text{-}QAC^k_f = B\text{-}QTC^k_f$.

We first construct a circuit for the exact$[q]$ gate (which outputs 1 if the input is of Hamming weight q, and 0 otherwise) and then use it for all other gates. The exact$[q]$ gate can be approximated in constant depth thanks to the parallelisation method. Furthermore, we show how to achieve exact computation at the cost of log-star depth.

Sorting and several arithmetical problems including addition and multiplication of n integers are computed by constant-depth threshold circuits [13], hence they are in $B\text{-}QNC^0_f$. By optimising the methods of [4] to use the fan-out gate, we also put quantum phase estimation and the quantum Fourier transform in $B\text{-}QNC^0_f$. By results of [12,4], polynomial-time bounded-error algorithms with oracle $B\text{-}QNC^0_f$ can factorise numbers and compute discrete logarithms. Thus, if $B\text{-}QNC^0_f$ can be simulated by a BPP machine, then factorisation can be done in polynomial time by bounded-error Turing machines.

2   Quantum Circuits with Unbounded Fan-out

Quantum circuits resemble classical reversible circuits. A quantum circuit is a sequence of quantum gates ordered into *layers*. The gates are consecutively applied in accordance with the order of the layers. Gates in one layer can be applied in parallel. The *depth* of a circuit is the number of layers and the *size* is the number of gates. A circuit can solve problems of a fixed size, so we define *families* of circuits containing one circuit for every input size. We consider only *uniform* families, whose description can be generated by a log-space Turing machine.

A *quantum gate* is a unitary operator applied on some subset of qubits. We usually use gates from a fixed *universal basis* (Hadamard gate, rotation by an irrational multiple of π, and the controlled-not gate) that can approximate any quantum gate with good precision [1]. The qubits are divided into 2 groups: *Input/output* qubits contain the description of the input at the beginning and they are measured in the computational basis at the end. *Ancilla qubits* are initialised to $|0\rangle$ at the beginning and the circuits usually clean them at the end, so that the output qubits are in a pure state and the ancillas could be reused.

Since unitary evolution is reversible, every operation can be undone. Running the computation backward is called *uncomputation* and is often used for cleaning ancilla qubits.

2.1 Definition of Quantum Gates

Quantum circuits cannot use a naive quantum fan-out gate mapping every superposition $|\phi\rangle|0\rangle \ldots |0\rangle$ to $|\phi\rangle \ldots |\phi\rangle$ due to the no-cloning theorem [16]. Such a gate is not linear, let alone unitary. Instead, our fan-out gate copies only classical bits and the effect on superpositions is determined by linearity. It acts as a controlled-not-not-$\ldots$-not gate, i.e. it is an unbounded sequence of controlled-not gates sharing one control qubit. Parity is a natural counterpart of fan-out. It is an unbounded sequence of controlled-not gates sharing one target qubit.

Definition 1. *The fan-out gate maps* $|x\rangle|y_1\rangle \ldots |y_n\rangle \to |x\rangle|y_1 \oplus x\rangle \ldots |y_n \oplus x\rangle$, *where* $x \oplus y = (x+y) \bmod 2$. *The parity gate maps* $|x_1\rangle \ldots |x_n\rangle|y\rangle \to |x_1\rangle \ldots |x_n\rangle$ $|y \oplus (x_1 \oplus \ldots \oplus x_n)\rangle$.

Example 1. As used in [6], parity and fan-out can simulate each other in constant depth. Recall the Hadamard gate $H = \frac{1}{\sqrt{2}}\begin{pmatrix} 1 & 1 \\ 1 & -1 \end{pmatrix}$ and that $H^2 = I$. If a controlled-not gate is preceded and succeeded by Hadamard gates on both qubits, it just turns around. Since parity is a sequence of controlled-not gates, we can turn around all of them in parallel. The circuit is shown in the following figure:

In this paper, we investigate the circuit complexity of among others these gates:

Definition 2. *Let* $x = x_1 \ldots x_n$ *and let* $|x|$ *denote the Hamming weight of* x. *The following* $(n+1)$-*qubit gates map* $|x\rangle|y\rangle \to |x\rangle|y \oplus g(x)\rangle$, *where* $g(x) = 1$ *iff*

$|x| > 0$: Or,	$|x| \geq \frac{n}{2}$: majority,	$|x| = q$: exact[q],
$|x| = n$: And (Toffoli),	$|x| \geq q$: threshold[q],	$|x| \bmod q = 0$: mod[q].

The counting gate is any gate that maps $|x\rangle|0^m\rangle \to |x\rangle|\,|x|\,\rangle$ *for* $m = \lceil \log(n+1) \rceil$.

2.2 Quantum Circuit Classes

Definition 3. $\mathrm{QNC_f}(d(n))$ *contains operators computed exactly (i.e. without error) by uniform families of quantum circuits with fan-out of depth* $\mathrm{O}(d(n))$, *polynomial-size, and over a fixed basis.* $\mathrm{QNC_f^k} = \mathrm{QNC_f}(\log^k n)$. R-$\mathrm{QNC_f^k}$ *contains operators approximated with one-sided, and* B-$\mathrm{QNC_f^k}$ *with two-sided, polynomially small error.*

Remark 1. Every s-qubit quantum gate can be decomposed into a sequence of one-qubit and controlled-not gates of length $O(s^3 4^s)$ [2]. Hence it does not matter whether we allow one-qubit or fixed-size gates in the basis. All our circuits below are over a fixed basis, unless explicitly mentioned otherwise. Some of our circuits need arbitrary one-qubit gates to be exact.

We do not distinguish between a quantum operator computed by a quantum circuit, a classical function induced by that operator by a measurement, and a language decided by that function. All of them are denoted by QNC_f^k.

3 Parallelisation Method

In this section, we describe a general parallelisation method for achieving very shallow circuits. Furthermore, we apply it on the rotation by Hamming weight and the rotation by value and show how to compute them in constant depth.

3.1 General Method

The unbounded fan-out gate is universal for commuting gates in the following sense: Using fan-out, gates can be applied on the same qubits at the same time whenever (1) they commute, and (2) we know the basis in which they all are diagonal, and (3) we can efficiently change into the basis. The method reduces the depth, however it costs more ancilla qubits.

Lemma 1. [8, Theorem 1.3.19] *For every set of pairwise commuting unitary gates, there exists an orthogonal basis in which all the gates are diagonal.*

Theorem 1. [10,6] *Let $\{U_i\}_{i=1}^n$ be pairwise commuting gates on k qubits. Gate U_i is controlled by $|x_i\rangle$. Let T be a gate changing the basis according to Lemma 1. There exists a quantum circuit with fan-out computing $U = \prod_{i=1}^n U_i^{x_i}$ having depth $\max_{i=1}^n \operatorname{depth}(U_i) + 4 \cdot \operatorname{depth}(T) + 2$, size $\sum_{i=1}^n \operatorname{size}(U_i) + (2n+2) \cdot \operatorname{size}(T) + 2$, and using $(n-1)k$ ancillas.*

Proof. Consider a circuit that applies all U_i sequentially. Put $TT^\dagger = I$ between U_i and U_{i+1}. Take $V_i = T^\dagger U_i T$ as new gates. They are diagonal in the computational basis, hence they just impose some phase shifts. The circuit follows:

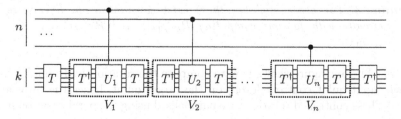

Multiple phase shifts on entangled states multiply, so can be
applied in parallel. We use fan-
out gates twice: first to create n
entangled copies of target qubits
and then to destroy the entan-
glement. The final circuit with
the desired parameters follows:

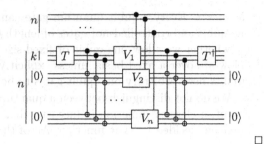

□

Example 2. As used in [6], it is simple to prove that $\text{mod}[q] \in \text{QNC}_f^0$: Each
input qubit controls one increment modulo q on a counter initialised to 0. At
the end, we obtain $|x| \bmod q$. The modular increments commute and thus can
be parallelised. Since q is fixed, changing the basis and the increment can both
be done in constant depth.

3.2   Rotation by Hamming Weight and Value

In this paper, we often use a *rotation by Hamming weight* $R_z\,(\varphi|x|)$ and a *rotation
by value* $R_z\,(\varphi x)$, where $R_z\,(\alpha)$ is one-qubit rotation around z-axis by angle α:
$R_z\,(\alpha) = |0\rangle\langle 0| + e^{i\alpha}|1\rangle\langle 1|$. They both can be computed in constant depth.

First of all, it is convenient to use controlled one-qubit gates as basic elements. The following lemma shows that they can be simulated by one-qubit and
controlled-not gates.

Lemma 2. [2, Lemma 5.1] *For every one-
qubit gate U, there exist one-qubit gates
A, B, C and rotation $P = R_z\,(\alpha)$ such that
the controlled gate U is computed by the following constant-depth circuit:*

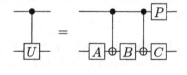

Remark 2. If a qubit controls more one-qubit gates, then we can still use this
method. The controlled-not gate is just replaced by the fan-out gate and the
rotations P are multiplied.

Lemma 3. *For every angle φ, there exist constant-depth, linear-size quan-
tum circuits with fan-out computing $R_z\,(\varphi|x|)$ and $R_z\,(\varphi x)$ on input $x =
x_{n-1} \ldots x_1 x_0$.*

Proof. The left figure shows how to compute the rotation by Hamming weight:
Each input qubit controls $R_z\,(\varphi)$ on the target qubit, hence the total angle is
$\varphi|x|$. These controlled rotations are parallelised using the parallelisation method.

The right figure shows the rotation by value. It is similar to the rotation by Hamming weight, only the input qubit $|x_j\rangle$ controls $R_z\left(\varphi 2^j\right)$, hence the total angle is $\varphi \sum_{j=0}^{n-1} 2^j x_j = \varphi x$.

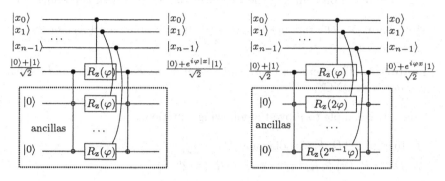

$\square$

Remark 3. The construction uses rotations $R_z\left(\varphi\right)$ for arbitrary $\varphi \in \mathbb{R}$. However, we are only allowed to use a fixed set of one-qubit gates. It is easy to see that every rotation can be approximated with polynomially small error by $R_z\left(\theta q\right) = \left(R_z\left(\theta\right)\right)^q$, where $\sin\theta = \frac{3}{5}$ and q is a polynomially large integer [1]. These q rotations commute, so can be applied in parallel and the depth is preserved.

4   Approximate Circuits

In this section, we present very shallow approximate circuits for all gates from Definition 2.

4.1   Quantum Fourier Transform

QFT is a very powerful tool used in several quantum algorithms, e.g. factorisation of integers [12]. In this section, we use it for a different purpose: parallelisation of increment gates.

Definition 4. *The quantum Fourier transform (QFT) performs the Fourier transform on the quantum amplitudes of the state, i.e. it maps*

$$F_n : |x\rangle \to |\psi_n^x\rangle = \frac{1}{2^{n/2}} \sum_{y=0}^{2^n-1} e^{2\pi i x y / 2^n} |y\rangle. \tag{1}$$

Shor has shown in [12] how to compute QFT in quadratic depth, quadratic size, and without ancillas. The depth has further been improved to linear. Cleve and Watrous have shown in [4] that QFT can be approximated with error ε in depth $O\left(\log n + \log\log\frac{1}{\varepsilon}\right)$ and size $O\left(n \log \frac{n}{\varepsilon}\right)$. Furthermore, they have shown that logarithmic depth is necessary (in the model without fan-out).

4.2   Circuits of Double-Logarithmic Depth

Circuits in this sub-section are not optimal, however they give a good insight. They are based on counting the weight of the input, which is parallelised.

Definition 5. *The increment gate maps* $\text{Incr}_n : |x\rangle \to |(x+1) \bmod 2^n\rangle$.

Lemma 4. *The increment gate is diagonal in the Fourier basis and its diagonal version is in* QNC^0.

Proof. It is simple to prove the following equations:

1. $\text{Incr}_n = F_n^\dagger D_n F_n$ for diagonal $D_n = \sum_{x=0}^{2^n-1} e^{2\pi i x/2^n} |x\rangle\langle x|$,
2. $D_n = R_z(\pi) \otimes R_z(\pi/2) \otimes \ldots \otimes R_z(\pi/2^{n-1})$.   □

Remark 4. Classically, $\text{Incr} \in \text{NC}^1$. The circuits for QFT mentioned above also have logarithmic depth and they are only approximate. However, quantum circuits of this type can be parallelised, which we cannot do with classical circuits.

Furthermore, the addition of a fixed integer q is as hard as the increment: by Lemma 4, $\text{Incr}^q = F^\dagger D^q F$ and $(R_z(\varphi))^q = R_z(\varphi q)$, hence the diagonal version of the addition of q is also in QNC^0.

Theorem 2. *Using fan-out, the counting gate can be approximated with error ε in depth $\text{O}\big(\log\log n + \log\log \frac{1}{\varepsilon}\big)$ and size $\text{O}\big((n + \log \frac{1}{\varepsilon})\log n\big)$.*

Proof. Compute the Hamming weight of the input: Each input qubit controls one increment on an m-qubit counter initialised to 0, where $m = \lceil \log(n+1) \rceil$. The increments Incr_m are parallelised, so we apply the quantum Fourier transform F_m twice and the n constant-depth controlled D_m gates in parallel.   □

Remark 5. Other gates are computed from the counting gate by standard methods: threshold[t] can be computed as the most significant qubit of the counter if we align it to a power of 2 by adding fixed integer $2^m - t$.

4.3   Constant-Depth Circuits

Rotations by Hamming weight computed for many elementary angles in parallel can be used for approximating the Or and exact[q] gates in constant depth.

Define one-qubit state $|\mu_\varphi^w\rangle = (H \cdot R_z(\varphi w) \cdot H) |0\rangle = \frac{1+e^{i\varphi w}}{2}|0\rangle + \frac{1-e^{i\varphi w}}{2}|1\rangle$. By Lemma 3, $|\mu_\varphi^{|x|}\rangle$ can be computed in constant-depth and linear-size.

Theorem 3. *Or $\in$ R-QNC$_f^0$, i.e. it can be approximated with one-sided polynomially small error in constant-depth.*

Proof. Let $m = a \cdot n$, where a will be chosen later. For all $k \in \{0, 1, \dots, m-1\}$, compute in parallel $|y_k\rangle = |\mu_{\varphi_k}^{|x|}\rangle$ for angle $\varphi_k = \frac{2\pi}{m}k$. If $|y_k\rangle$ is measured in the computational basis, the expected value is

$$E[Y_k] = \left|\frac{1 - e^{i\varphi_k|x|}}{2}\right|^2 = \left|e^{-i\varphi_k|x|}\right| \cdot \frac{\left|e^{i\varphi_k|x|} + e^{-i\varphi_k|x|} - 2\right|}{4} = \frac{1 - \cos(\varphi_k|x|)}{2}.$$

If all these m qubits $|y\rangle$ are measured, the expected Hamming weight is

$$E[|Y|] = E\left[\sum_{k=0}^{m-1} Y_k\right] = \frac{m}{2} - \frac{1}{2}\sum_{k=0}^{m-1}\cos\left(\frac{2\pi k}{m}|x|\right) = \begin{cases} 0 & \text{if } |x| = 0, \\ \frac{m}{2} & \text{if } |x| \neq 0. \end{cases}$$

The qubits $|y\rangle$ are actually not measured, but their Hamming weight $|y|$ controls another rotation on a new ancilla qubit $|z\rangle$. So compute $|z\rangle = |\mu_{2\pi/m}^{|y|}\rangle$.

Let Z be the outcome after $|z\rangle$ is measured. If $|y| = 0$, then $Z = 0$ with certainty. If $\left||y| - \frac{m}{2}\right| \leq \frac{m}{\sqrt{n}}$, then

$$P[Z = 0] = \left|\frac{1 + e^{i\frac{2\pi}{m}|y|}}{2}\right|^2 = \frac{1 + \cos\left(\frac{2\pi}{m}|y|\right)}{2} \leq \frac{1 - \cos\frac{2\pi}{\sqrt{n}}}{2} = O\left(\frac{1}{n}\right).$$

Assume that $|x| \neq 0$. Since $0 \leq Y_k \leq 1$, we can use Hoeffding's Lemma 5 below and obtain $P\left[||Y| - \frac{m}{2}| \geq \varepsilon m\right] \leq \frac{1}{2^{\varepsilon^2 m}}$. Fix $a = \log n$ and $\varepsilon = \frac{1}{\sqrt{n}}$. Now,

$$P\left[||y| - \frac{m}{2}| \geq \frac{m}{\sqrt{n}}\right] \leq \frac{1}{2^{m/n}} = \frac{1}{2^a} = \frac{1}{n}.$$ Hence $P[Z = 0] = \begin{cases} 1 & \text{if } |x| = 0, \\ O\left(\frac{1}{n}\right) & \text{if } |x| \neq 0. \end{cases}$

The circuit has constant depth and size $O(mn) = O(n^2 \log n)$. It is outlined in the following figure. The figure is slightly simplified: unimportant qubits and uncomputation of ancillas are omitted.

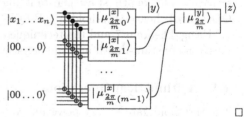

Lemma 5 (Hoeffding). [7] *If* $Y_1, \dots, Y_m$ *are independent random variables bounded by* $a_k \leq Y_k \leq b_k$, *then, for all* $\varepsilon > 0$,

$$P\left[|S - E[S]| \geq \varepsilon m\right] \leq 2\exp\frac{-2m^2\varepsilon^2}{\sum_{k=1}^{m}(b_k - a_k)^2}, \quad \text{where } S = \sum_{i=k}^{m} Y_k.$$

Remark 6. Since the outcome is a classical bit, we can save it and clean all ancillas by uncomputation. It remains to prove that the intermediate qubits $|y\rangle$ need not be measured, in order to be able to uncompute them.

We have only proved that the output qubit is a good approximation of the logical Or, if $|y\rangle$ is immediately measured. By the principle of deferred measurement, we can use controlled quantum operations and measure $|y\rangle$ at the end. However, the outcome is a classical bit hardly entangled with $|y\rangle$, hence it does not matter whether $|y\rangle$ is measured.

Remark 7. If we need smaller error $\frac{1}{n^c}$, we create c copies and compute exact Or of them by a binary tree of Or gates. The tree has depth $\log c = O(1)$. Using Theorem 6, the size can be reduced to $O\left(dn \log^{(d)} n\right)$ and the depth is $O(d)$.

Theorem 4. *exact[q]* $\in$ R-QNC$_\mathrm{f}^0$.

Proof. Slight modification of the circuit for Or: As outlined in the figure, by adding rotation $R_z(-\varphi q)$ to the rotation by Hamming weight in the first layer, we obtain $|\mu_\varphi^{|x|-q}\rangle$ instead of $|\mu_\varphi^{|x|}\rangle$. The second layer stays the same. If the output qubit $|z\rangle$ is measured, then

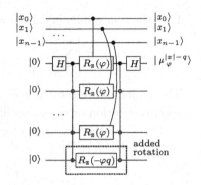

$$P[Z=0] = \begin{cases} 1 & \text{if } |x| = q, \\ O\!\left(\frac{1}{n}\right) & \text{if } |x| \neq q. \end{cases}$$

We obtain an approximation of the exact[q] gate with one-sided polynomially small error.   □

Remark 8. Other gates are computed from the exact[q] gate by standard methods: threshold[t] can be computed as the parity of exact[t], exact[t + 1], ..., exact[n]. The depth stays constant and the size is just n-times bigger, i.e. $O\!\left(n^3 \log n\right)$, hence threshold[t] $\in$ B-QNC$_\mathrm{f}^0$.

Using Theorem 7 and the technique of Theorem 2, threshold[t] can be computed in constant-depth and smaller size $O(n \log n)$.

4.4   Arithmetical Operations

The threshold gate is very powerful, so the fan-out gate is powerful too:

Theorem 5. *The following functions are in* B-QNC$_\mathrm{f}^0$*: addition and multiplication of n integers, division of two integers, and sorting of n integers.*

Proof. By [13], these functions are computed by constant-depth, polynomial-size threshold circuits. The depths are really small, from 2 to 5. A threshold circuit is built of weighted threshold gates. It is simple to prove that also the weighted threshold gate (with polynomially large integer weights) is in B-QNC$_\mathrm{f}^0$.   □

5   Exact Circuits

In the previous section, we have shown how to approximate the exact[q] gate in constant depth. In this section, we show how to compute it exactly in log-star depth. The circuits in this section need arbitrary one-qubit gates instead of a fixed basis, otherwise they would not be exact.

Theorem 6. *Or on n qubits can be reduced exactly to Or on $m = \lceil \log(n+1) \rceil$ qubits in constant-depth and size $O(n \log n)$.*

Proof. For $k \in \{1, 2, \ldots, m\}$, compute in parallel $|y_k\rangle = |\mu_{\varphi_k}^{|x|}\rangle$ for angle $\varphi_k = \frac{2\pi}{2^k}$:

- If $|x| = 0$, then $|y_k\rangle = |0\rangle$ for each k.
- If $|x| \neq 0$, take unique decomposition $x = 2^a(2b+1)$ where $a, b \in \mathbb{N}_0$. Then

$$\langle 1|y_{a+1}\rangle = \frac{1 - e^{i\varphi_{a+1}|x|}}{2} = \frac{1 - e^{i\pi(2b+1)}}{2} = \frac{1 - e^{i\pi}}{2} = 1.$$

It follows that $|x| = 0 \iff |y| = 0$. Hence the original problem is exactly reduced to a problem of logarithmic size. $\square$

Remark 9. If all input qubits are zero, then also all output qubits are zero. Otherwise the output qubits are in a general superposition such that the amplitude of the zero state is 0.

Corollary 1. *exact[q] $\in$ QAC$_f^0$.*

Proof. Using the same method as in Theorem 4, also the exact[q] gate can be reduced to Or, which is in QAC$_f^0$. $\square$

Corollary 2. *exact[q] $\in$ QNC$_f$($\log^* n$).*

Proof. Repeat the exact reduction $\log^* n$ times until the input size ≤ 2. Compute and save the outcome and clean ancillas by uncomputation. $\square$

6 Quantum Fourier Transform and Phase Estimation

6.1 Constant-Depth QFT

We show that the approximate circuit for QFT from [4] can be compressed to constant depth, if we use the fan-out gate. Recall Definition 4 of QFT.

Theorem 7. *QFT $\in$ B-QNC$_f^0$.*

Proof. The operator $F_n : |x\rangle \to |\psi_n^x\rangle$ can be computed by composing:

1. Fourier state construction (QFS): $|x\rangle|0\rangle \ldots |0\rangle \quad \to |x\rangle|\psi_n^x\rangle|0\rangle \ldots |0\rangle$
2. Copying Fourier state: $|x\rangle|\psi_n^x\rangle|0\rangle \ldots |0\rangle \to |x\rangle|\psi_n^x\rangle \ldots |\psi_n^x\rangle$
3. Uncomputing phase estimation (QFP): $|\psi_n^x\rangle \ldots |\psi_n^x\rangle|x\rangle \to |\psi_n^x\rangle \ldots |\psi_n^x\rangle|0\rangle$
4. Uncopying Fourier state: $|\psi_n^x\rangle \ldots |\psi_n^x\rangle|0\rangle \to |\psi_n^x\rangle|0\rangle \ldots |0\rangle$

The following lemmas show that each of these four operators is in B-QNC$_f^0$. $\square$

Lemma 6. *QFS $\in$ QNC$_f^0$.*

Proof. QFS maps $|x\rangle|0\rangle \to |x\rangle|\psi_n^x\rangle$. Define $|\rho_r\rangle = \frac{|0\rangle + e^{2\pi r i}|1\rangle}{\sqrt{2}}$. It is simple to prove that $|\psi_n^x\rangle = |\rho_{x/2^1}\rangle|\rho_{x/2^2}\rangle \ldots |\rho_{x/2^n}\rangle$. The n qubits $|\rho_{x/2^k}\rangle$ can be computed from x in parallel. Computation of $|\rho_{x/2^k}\rangle = R_z\left(\frac{2\pi}{2^k}x\right)\frac{|0\rangle + |1\rangle}{\sqrt{2}}$ is done by the rotation by value (Lemma 3) in constant depth and linear size.   □

Definition 6. *Let $x_1, \ldots, x_m$ be n-bit integers. The reversible addition-gate maps* $\text{add}_n^m : |x_1\rangle \ldots |x_m\rangle \to |x_1\rangle \ldots |x_{m-1}\rangle|y\rangle$, *where* $y = \left(\sum_{i=1}^m x_i\right) \bmod 2^n$.

Lemma 7. $\text{add}_n^m \in \text{B-QNC}_f^0$.

Proof. By Theorem 5, $y = \left(\sum_{i=1}^m x_i\right) \bmod 2^n$ can be approximated in constant depth and polynomial size. The result is, however, stored into ancilla qubits. Uncompute $x_m = \left(y - \sum_{i=1}^{m-1} x_i\right) \bmod 2^n$ in the same way (subtraction is as hard as addition).   □

Lemma 8. *Copying Fourier state is in* B-QNC$_f^0$.

Proof. Take the reversible addition-gate: $(\text{add}_n^2)|y\rangle|x\rangle = |y\rangle|(x+y) \bmod 2^n\rangle$. It is simple to prove that $(\text{add}_n^2)^{-1}|\psi_n^y\rangle|\psi_n^x\rangle = |\psi_n^{x+y}\rangle|\psi_n^x\rangle$. Hence $(\text{add}_n^2)^{-1}|\psi_n^0\rangle|\psi_n^x\rangle = |\psi_n^x\rangle|\psi_n^x\rangle$. The state $|\psi_n^0\rangle = H^{\otimes n}|0^n\rangle$ is easy to prepare in constant depth.

By the same arguments, $(\text{add}_n^m)^{-1}|\psi_n^0\rangle \ldots |\psi_n^0\rangle|\psi_n^x\rangle = |\psi_n^x\rangle \ldots |\psi_n^x\rangle|\psi_n^x\rangle$.   □

Lemma 9. $QFP \in \text{B-QNC}_f^0$.

Proof. QFP maps $|\psi_n^x\rangle \ldots |\psi_n^x\rangle|0\rangle \to |\psi_n^x\rangle \ldots |\psi_n^x\rangle|x\rangle$. By Cleve and Watrous [4, sub-section 3.3], we can compute x with probability $\geq 1 - \varepsilon$ from $O\left(\log \frac{n}{\varepsilon}\right)$ copies of $|\psi_n^x\rangle$ in depth $O\left(\log n + \log\log \frac{1}{\varepsilon}\right)$ and size $O\left(n \log \frac{n}{\varepsilon}\right)$. Use $\varepsilon = \frac{1}{poly(n)}$. It is easy to see that their circuit can have constant depth, if we use fan-out, parity, And, Or, majority gate. All these gates are in B-QNC$_f^0$.   □

6.2   Quantum Phase Estimation

The method of computing QFT can be also used for phase estimation:

Theorem 8. *Given a gate* $S_x : |y\rangle|\phi\rangle \to |y\rangle R_z\left(\frac{2\pi x}{2^n}y\right)|\phi\rangle$ *for basis states* $|y\rangle$, *where* $x \in \mathbb{Z}_{2^n}$ *is unknown, we can determine x with probability $\geq 1 - \varepsilon$ in constant depth, size $O\left(n \log \frac{n}{\varepsilon}\right)$, and using the S_x gate $O\left(n \log \frac{n}{\varepsilon}\right)$ times.*

Proof. Compute $|\rho_{x/2^k}\rangle = R_z\left(\frac{2\pi x}{2^k}\right)\frac{|0\rangle + |1\rangle}{\sqrt{2}} = R_z\left(\frac{2\pi x}{2^n}2^{n-k}\right)\frac{|0\rangle + |1\rangle}{\sqrt{2}}$, which is the result of one application of $S_x\left(|2^{n-k}\rangle\frac{|0\rangle + |1\rangle}{\sqrt{2}}\right)$. Apply QFP on $O\left(\log \frac{n}{\varepsilon}\right)$ copies of $|\psi_n^x\rangle = |\rho_{x/2^1}\rangle|\rho_{x/2^2}\rangle \ldots |\rho_{x/2^n}\rangle$.   □

7   Concluding Remarks

7.1   Relations of Quantum Circuit Classes

We have shown that $\text{B-QNC}_f^0 = \text{B-QAC}_f^0 = \text{B-QACC}^0 = \text{B-QTC}_f^0$ (Theorem 4). If we allow arbitrary one-qubit gates, then also $\text{QAC}_f^0 = \text{QTC}_f^0 \subseteq \text{QNC}_f(\log^* n)$ (Corollaries 1 and 2). Several open problems of [6] have thus been solved.

Only little is known about classes that do not include the fan-out gate. For example, we do not know whether $\text{TC}^0 \subseteq \text{QTC}^0$, we only know that $\text{TC}^0 \subseteq \text{QTC}_f^0$. It is simple to prove that parity is in TC^0: take Or of exact[1], exact[3], exact[5], ..., and compute exact[k] from threshold[k] and threshold[k + 1]. However, this method needs fan-out to copy the input bits.

7.2   Randomised versus Quantum Depth

We compare depths of randomised classical circuits and quantum circuits, both with bounded fan-in and unbounded parity and fan-out. Quantum upper bounds are proved in this paper. Classical lower bounds can be proved by Yao's principle and the polynomial method (with polynomials modulo 2).

Gate	Randomised	Quantum
Or and threshold[t] exactly	$\Theta(\log n)$	$O(\log^* n)$
mod[q] exactly	$\Theta(\log n)$	$\Theta(1)$
Or with error $\frac{1}{n}$	$\Theta(\log\log n)$	$\Theta(1)$
threshold[t] with error $\frac{1}{n}$	$\Omega(\log\log n)$	$\Theta(1)$

7.3   Upper Bounds for B-QNC$_f^0$

Shor's original factoring algorithm uses modular exponentiation and the quantum Fourier transform followed by a polynomial-time deterministic algorithm. The modular exponentiation a^x can be replaced by multiplication of some subset of numbers $a, a^2, a^4, \ldots, a^{2^{n-1}}$ [4]. Numbers a^{2^k} are precomputed classically.

Since both multiplication of n numbers (Theorem 5) and QFT (Theorem 7) are in B-QNC$_f^0$, there is a polynomial-time bounded-error algorithm with oracle B-QNC$_f^0$ factoring numbers, i.e. factoring $\in \text{RP}[\text{B-QNC}_f^0]$. If B-QNC$_f^0 \subseteq \text{BPP}$, then factoring $\in \text{RP}[\text{BPP}] \subseteq \text{BPP}[\text{BPP}] = \text{BPP}$.

Acknowledgements. We would like to thank Harry Buhrman, Hartmut Klauck, and Hein Röhrig at CWI in Amsterdam, and Fred Green at Clark University in Worcester for plenty helpful discussions, and Ronald de Wolf at CWI for help with writing the paper. We are grateful to Schloss Dagstuhl, Germany, for providing an excellent environment, where part of this work was carried out.

References

1. L. M. Adleman, J. DeMarrais, and M. A. Huang. Quantum computability. *SIAM Journal on Computing*, 26(5):1524–1540, 1997.
2. A. Barenco, C. Bennett, R. Cleve, D. P. DiVincenzo, N. Margolus, P. Shor, T. Sleator, J. A. Smolin, and H. Weinfurter. Elementary gates for quantum computation. *Physical Review A*, 52:3457–3467, 1995. quant-ph/9503016.
3. J. I. Cirac and P. Zoller. Quantum computations with cold trapped ions. *Phys. Rev. Lett.*, 74:4091–4094, 1995.
4. R. Cleve and J. Watrous. Fast parallel circuits for the quantum Fourier transform. In *Proc. of the 41st IEEE Symp. on Foundations of Computer Science*, pages 526–536, 2000.
5. N. Gershenfeld and I. Chuang. Bulk spin resonance quantum computation. *Science*, 275:350–356, 1997. http://citeseer.nj.nec.com/gershenfeld97bulk.html.
6. F. Green, S. Homer, C. Moore, and C. Pollett. Counting, fanout, and the complexity of quantum ACC. *Quantum Information and Computation*, 2(1):35–65, 2002. quant-ph/0106017.
7. W. Hoeffding. Probability inequalities for sums of bounded random variables. *J. Amer. Statist. Assoc.*, 58:13–30, 1963.
8. R. A. Horn and C. R. Johnson. *Matrix Analysis*. Cambridge University Press, 1985.
9. C. Moore. Quantum circuits: Fanout, parity, and counting. quant-ph/9903046, 1999.
10. C. Moore and M. Nilsson. Parallel quantum computation and quantum codes. *SIAM Journal on Computing*, 31(3):799–815, 2002. quant-ph/9808027.
11. A. A. Razborov. Lower bounds for the size of circuits of bounded depth with basis $\{\&, \oplus\}$. *Math. Notes Acad. Sci. USSR*, 41(4):333–338, 1987.
12. P. W. Shor. Algorithms for quantum computation: discrete logarithms and factoring. In *Proc. of the 35th Annual Symp. on FOCS*, pages 124–134, Los Alamitos, CA, 1994. IEEE Press. http://citeseer.nj.nec.com/14533.html.
13. K.-Y. Siu, J. Bruck, T. Kailath, and T. Hofmeister. Depth efficient neural networks for division and related problems. *IEEE Transactions on Information Theory*, 39(3):946–956, 1993.
14. R. Smolensky. Algebraic methods in the theory of lower bounds for Boolean circuit complexity. In *Proc. 19th Annual ACM Symposium on Theory of Computing*, pages 77–82, 1987.
15. R. Špalek. Quantum circuits with unbounded fan-out. Master's thesis, Faculty of Sciences, Vrije Universiteit, Amsterdam, 2002. http://www.ucw.cz/~robert/qncwf/. Shorter version and improved results in quant-ph/0208043.
16. W. K. Wootters and W. H. Zurek. A single quantum cannot be clone. *Nature*, 299:802–803, 1982.

Analysis of the Harmonic Algorithm for Three Servers

Marek Chrobak[1] and Jiří Sgall[2]

[1] Department of Computer Science, University of California, Riverside, CA 92521.
marek@cs.ucr.edu,
[2] Mathematical Inst., AS CR, Žitná 25, CZ-11567 Praha 1, Czech Republic.
sgall@math.cas.cz, http://www.math.cas.cz/sgall,

Abstract. HARMONIC is a randomized algorithm for the k-server problem that, at each step, given a request point r, chooses the server to be moved to r with probability inversely proportional to the distance to r. For general k, it is known that the competitive ratio of HARMONIC is at least $\frac{1}{2}k(k+1)$, while the best upper bound on this ratio is exponential in k. It has been conjectured that HARMONIC is $\frac{1}{2}k(k+1)$-competitive for all k. This conjecture has been proven in a number of special cases, including $k = 2$ and for the so-called lazy adversary.

In this paper we provide further evidence for this conjecture, by proving that HARMONIC is 6-competitive for $k = 3$. Our approach is based on the random walk techniques and their relationship to the electrical network theory. We propose a new potential function Φ and reduce the proof of the validity of Φ to several inequalities involving hitting costs. Then we show that these inequalities hold for $k = 3$.

1 Introduction

In the *k-server problem* we are given a metric space M populated by k mobile servers. At each time step a request $r \in M$ is specified, and we need to move one server to r to "satisfy" the request. The decision as to which server to move needs to be made *online*, without the knowledge about future requests. The cost of this move is the distance traveled by the server that moves to r.

An online algorithm **A** for the k-server problem is called *R-competitive* if its total cost on any request sequence ϱ is at most R times the optimal (offline) cost on this sequence, plus possibly an additive constant independent of ϱ. The same definition applies to randomized algorithms as well, we just need to replace our cost by our *expected* cost on ϱ. The *competitive ratio* of **A** is the smallest R for which **A** is R-competitive.

Manasse, McGeoch and Sleator [13] gave a 2-competitive algorithm for 2 servers and they proved that no deterministic online algorithm for k servers can be better than k-competitive. They also formulated the still-open *k-server conjecture* that a k-competitive algorithm exists for each k. The best known upper bound on the competitive ratio for $k \geq 3$ is $2k - 1$, achieved by the Work

H. Alt and M. Habib (Eds.): STACS 2003, LNCS 2607, pp. 247–259, 2003.
© Springer-Verlag Berlin Heidelberg 2003

Function Algorithm, as proven by Koutsoupias and Papadimitriou [12]. We refer the reader to [4,7] for more information on the k-server problem.

HARMONIC is a randomized k-server algorithm proposed by Raghavan and Snir [15]. If the distance between server j and the request point r is d_{jr}, the algorithm moves j to r with probability proportional to its inverted distance to r, that is $d_{jr}^{-1}/(\sum_i d_{ir}^{-1})$.

As shown in [15], the competitive ratio of HARMONIC is at least $\frac{1}{2}k(k+1)$. This is also conjectured to be the exact competitive ratio of HARMONIC (see [3,4]). Although this conjecture remains open in general, there is a number of results that support this conjecture in some special cases. For $k = 2$, Raghavan and Snir [15] showed that the competitive ratio of HARMONIC is at most 6. This case was later settled by Chrobak and Larmore [6], who proved that for two servers HARMONIC is 3-competitive, thus matching the lower bound. Another, simpler proof for $k = 2$ was later given by Chrobak and Sgall [8]. For $k = 3$, Berman et $al.$ [3] proved that HARMONIC's ratio is finite (the value of the upper bound in [3] was astronomical). The best currently known upper bound for arbitrary k, by Bartal and Grove [2,4], is $O(2^k \log k)$. In particular, for $k = 3$, the upper bound formula from [4] gives the value 25, while the lower bound is 6.

The competitive analysis of HARMONIC is considered one of the main open problems in the area of on-line algorithms. There are several reasons for this. First, unlike most other k-server algorithms studied in the literature, HARMONIC is very natural and simple. It is also memoryless, that is, the choice of the server at each step depends only on the current configuration and the location of the request, not on the past requests. It can be trivially implemented in time $O(k)$ per step, independently of n (the number of requests). The only other algorithm known to have a constant competitive ratio is the Work Function Algorithm [12] which, in general, requires exponential space and time per step to implement (roughly $O(n^k)$). The past research on HARMONIC [15,5,1] revealed some close connections to the analysis of random walks and electrical networks, which are worth investigating for their own interest. Finally, it is reasonable to expect that insights gained through the analysis of HARMONIC will help with resolving the question whether there exists a memoryless k-competitive randomized algorithm for k-servers in arbitrary metric spaces (see [10].)

In this paper we prove that HARMONIC is 6-competitive for $k = 3$. This matches the lower bound, and it provides further evidence for the conjectured $\frac{1}{2}k(k+1)$-competitiveness of HARMONIC. Our approach is based on the random walk techniques and their relationship to the electrical network theory. The first part of the proof is valid for any k. First, in Section 2, we propose a candidate potential function Φ, and we show that proving $\frac{1}{2}k(k+1)$-competitiveness of HARMONIC can be reduced to proving four key inequalities involving hitting costs of random walks. In Section 3 we develop a framework for calculating and manipulating the hitting costs in terms of certain rational functions of the distances. Next, in Section 4, we apply this framework to one of the key inequalities, reducing it to a polynomial inequality in the distances. This reduction is valid for general k, while the verification of the resulting inequality in the second part of the proof is for $k = 3$ only. In Section 5 we prove another key inequality for arbitrary k. The second part of the proof, Section 6, involves deriving the remaining three key inequalities in their polynomial form for $k = 3$, implying that HARMONIC is 6-competitive for 3 servers. The details of these somewhat tedious calculations are omitted here, see the technical report [9].

We note that all these upper bounds on HARMONIC's competitive ratio hold even against the so-called *adaptive online adversary* that can choose requests adaptively, based on the information on the locations of HARMONIC's servers.

2   A General Framework of the Proof

Throughout the paper, M is an arbitrary but fixed metric space. We will analyze the performance of HARMONIC in M. By d_{uv} we denote the distance between points $u, v \in M$, and by $c_{uv} = 1/d_{uv}$ we denote the inverse of the distance, also called the *conductance* of edge (u, v).

In our calculations we often deal with sets from which we remove one element or add a new element. For simplicity of notation, for $t \notin G$ and $x \in G$, we will write $G + t = G \cup \{t\}$, $G - x = G - \{x\}$, and $G + t - x = (G \cup \{t\}) - \{x\}$.

Harmonic random walks. Let G be a k-point subspace of M. Let $s \in G$ and $t \in M - G$ be two nodes called the *start node* and *target node*. The *harmonic random st-walk in G* is defined as follows: We start at node s. At each step, if the current location is $u \neq t$, move from u to another vertex $v \in G + t$ with probability inversely proportional to the distance between u and v. In terms of conductances, this probability is $c_{uv}/\sum_{x \in G+t-u} c_{ux}$. The *cost* of move from u to v is d_{uv}. The random walk terminates when $u = t$.

By H_{st} we denote the expected total cost of a harmonic st-walk in G, and we call it the *hitting cost*. (We omit G in this notation, as it is always fixed or understood from context.)

For a given target node t, the values H_{ut}, for $u \in G$, are related to each other by the following system of linear equations:

$$\left(\textstyle\sum_{v \in G+t-u} c_{uv}\right) \cdot H_{ut} = k + \textstyle\sum_{v \in G-u} c_{uv} H_{vt}. \tag{1}$$

System (1) has a finite and unique solution that can be expressed as a rational function in the distances (see, for example, [11]). We discuss this solution in more detail in Section 3.

Doubled points. We also consider a limit case of harmonic random st-walks in G when the target node t is at distance 0 from some node $v \in G$. More formally, let t be a "copy" of v, that is, for all $u \in G - v$ we have $c_{ut} = c_{uv}$ and the value of c_{vt} is arbitrary. We define $\widehat{H}_{uv} = \lim_{c_{vt} \to \infty} H_{ut}$. Intuitively, for v infinitely close to t, this is the same a as a hitting cost of the harmonic random walk in $G - v$ with two target nodes v and t, where t is a copy of v, so that the walk starts at s and terminates if we reach either v or t (and thus c_{vt} is irrelevant).

Key inequalities. As before, let G be a k-point subspace of M, $s, v \in G$ and $t \in M - G$. We now formulate a number of inequalities involving hitting costs H_{st} and $\widehat{H}_{sv}$. Later in this section we show that these inequalities imply $\frac{1}{2}k(k+1)$-competitiveness of HARMONIC. In the following sections, we give a proof of inequality (D), and we prove the remaining three inequalities for $k = 3$. These key inequalities are:

$$\sum_{v \in G + t - s} \left| \frac{\partial H_{st}}{\partial d_{sv}} \right| \leq \frac{1}{2}k(k+1) \tag{A}$$

$$H_{st} - H_{vt} \leq \widehat{H}_{sv} \tag{B}$$

$$(c_{vt} - c_{vs})\widehat{H}_{sv} \leq c_{vt}H_{st} \tag{C}$$

$$(c_{vs} - c_{vt})H_{vt} \leq c_{vs}H_{st} \tag{D}$$

Inequality (A) has a natural interpretation. The *edge stretch* of a random walk from s to t is the ratio between the expected cost of this walk and d_{st}. As shown in [1], the edge stretch of a harmonic random st-walk is at most $\frac{1}{2}k(k+1)$. Inequality (A) extends this result, since it says that if the position of s changes by some distance $\epsilon > 0$, then the hitting cost changes at most by $\frac{1}{2}k(k+1)\epsilon$.

In inequality (B), notice that it becomes an equality for $t = v$. So inequality (B) addresses the following question: where should we place the target point t so that it is much more expensive to reach from s than from v? The assertion of (B) is that it should be at point v.

In inequality (D), consider the special case when s is in the interval between v and t, i.e. $d_{vs} + d_{st} = d_{vt}$. Then (D) asserts that $H_{vt}/d_{vt} \leq H_{st}/d_{st}$. This agrees with the intuition that the stretch factor increases when we move two endpoints towards each other. As we show in Section 5, this inequality is indeed true for all k. Inequality (C) has a similar interpretation when t is between s and v. (However, counter-intuitively, we can show that, although true for $k = 3$, this inequality does not hold for large values of k.)

The potential function. We use the harmonic random walks introduced above to define a potential function for the analysis of HARMONIC. By $\mathcal{A} = \{1, 2, \ldots, k\}$ and $\mathcal{H} = \{1', \ldots, k'\}$ we denote the configurations of the adversary's k servers and HARMONIC's k servers, respectively. To simplify notation, we will often use the same notation for the server (j or j') and its position in M. Without loss of generality, we assume that $|\mathcal{A}| = |\mathcal{H}| = k$, that is, at any step, the adversary's servers and the HARMONIC's servers are each on k distinct points. Of course, some of HARMONIC's servers may coincide with the adversary's servers.

The random walks we now consider take place in the subspace $G = \mathcal{A}$, with the target node $t \in \mathcal{H}$. In terms of HARMONIC, H_{it} can be interpreted as follows. Suppose that HARMONIC's configuration is $\mathcal{H} = \mathcal{A} + t - i$. Then H_{it} is equal to the expected cost of HARMONIC if the adversary only makes *lazy* requests, that is requests on points in $\mathcal{A}$ (without moving its servers). This can be seen by noting that the harmonic random walk from i to t exactly describes the moves of the "hole", i.e., the point covered by the adversary but not by HARMONIC.

Consider the weighted, complete bipartite graph $\mathcal{B} \subseteq \mathcal{A} \times \mathcal{H}$ in which the weight of each edge (i, j') is $H_{ij'}$. In the degenerate case when $j' \in \mathcal{A}$, this weight becomes the cost of a harmonic random walk with a doubled target point j' and thus, by definition, $H_{ij'} = \widehat{H}_{ij'}$. In particular for $i = j'$, we have $H_{ij'} = 0$. The potential of configuration $(\mathcal{A}, \mathcal{H})$ is defined as

$$\Phi(\mathcal{A}, \mathcal{H}) = \text{minimum weight of a perfect matching in } \mathcal{B}.$$

Theorem 2.1. *For any $k \geq 2$, if inequalities* (A)–(D) *are true, then* HARMONIC *is $\frac{1}{2}k(k+1)$-competitive against an adaptive on-line adversary.*

Proof. We view the process as follows: At each time step, the adversary issues a request specified by a point $r \in M$. Since the adversary is adaptive, he can choose r depending on the current position of HARMONIC's servers. Once the request is issued, first the adversary serves it with one of its servers and then HARMONIC chooses randomly which server to move to r.

Consider an arbitrary, but fixed, configuration $(\mathcal{A}, \mathcal{H})$. It suffices to prove

$$\Delta cost_H + \Delta\Phi \leq \tfrac{1}{2}k(k+1)\Delta cost_{adv}, \tag{2}$$

where $\Delta cost_H$, $\Delta\Phi$, and $\Delta cost_{adv}$, denote, respectively, the expected cost of HARMONIC, the expected potential change, and the adversary cost, in the current step. By definition, $\Phi \geq 0$, and the theorem follows from (2) by routine amortization (see, for example, [4]).

In our proof, we divide each step into two sub-steps: the adversary move and the HARMONIC's move, and we prove (2) separately for each sub-step. Without loss of generality, we assume that in the matching which realizes the minimum in $\Phi(\mathcal{A}, \mathcal{H})$ before each sub-step considered, the adversary server j is matched to HARMONIC's server j', for each $j = 1, \ldots, k$, and thus $\Phi = \sum_{j=1}^{k} H_{jj'}$.

The adversary move. We can assume that in this sub-step the adversary moves just one server, say server 1. (Otherwise, we can move the servers one by one and analyze each move separately). Denoting by z the distance between the old and new location of server 1, we have $\Delta cost_{adv} = z$. The weight of the initial minimum matching changes by $\Delta H_{11'}$, where $\Delta H_{11'}$ denotes the change of the hitting cost $H_{11'}$. The potential after the move cannot be greater than the new weight of this initial matching, so $\Delta\Phi \leq |\Delta H_{11'}|$.

We think about this move as a continuous change of the position of server 1 along some line. Then we can view $H_{11'} = H_{11'}(z)$ as composed of two mappings: $z \to (d_{12}, d_{13}, \ldots, d_{1k}, d_{11'}) \to H_{11'}$. Since $|\partial d_{1i}/\partial z| \leq 1$ for all i, we have

$$|\partial H_{11'}/\partial z| \leq |\partial H_{11'}/\partial d_{11'}| + \sum_{i=2}^{k}|\partial H_{11'}/\partial d_{1i}| \leq \tfrac{1}{2}k(k+1),$$

using (A) with $s = 1$ and $t = 1'$. Therefore $\Delta\Phi \leq |\Delta H_{11'}| \leq \tfrac{1}{2}k(k+1)\Delta cost_{adv}$, and, since $\Delta cost_H = 0$, we conclude that (2) holds in the adversary move.

HARMONIC's move. Without loss of generality, we assume that the adversary makes a request on server 1. We first claim that, for each $j = 2, \ldots, k$,

$$(c_{1j'} - c_{1j})(H_{j1'} - H_{11'}) = (c_{1j} - c_{1j'})(H_{11'} - H_{j1'}) \leq c_{1j'}H_{jj'}. \tag{3}$$

We distinguish two cases. If $c_{1j} \leq c_{1j'}$ then, using (B) with $v = 1$, $s = j$ and $t = 1'$ and (C) with $v = 1$, $s = j$, $t = j'$, we have

$$(c_{1j'} - c_{1j})(H_{j1'} - H_{11'}) \leq (c_{1j'} - c_{1j})\widehat{H}_{j1} \leq c_{1j'}H_{jj'}.$$

Otherwise $c_{1j} \geq c_{1j'}$. The minimality of the matching implies $H_{11'} + H_{jj'} \leq H_{1j'} + H_{j1'}$. Then

$$(c_{1j} - c_{1j'})(H_{11'} - H_{j1'}) \leq (c_{1j} - c_{1j'})(H_{1j'} - H_{jj'})$$
$$= c_{1j'}H_{jj'} + [(c_{1j} - c_{1j'})H_{1j'} - c_{1j}H_{jj'}] \leq c_{1j'}H_{jj'},$$

where the last inequality follows from inequality (D) with $v = 1$, $s = j$ and $t = j'$. This completes the proof of (3).

Now we prove (2). For $u = 1$ and $t = 1'$, we can rewrite (1) as

$$k - c_{11'}H_{11'} = \sum_{j=2}^{k} c_{1j}(H_{11'} - H_{j1'}) \tag{4}$$

Let $\tilde{c} = \sum_{i=1}^{k} c_{1i'}$. The probability that HARMONIC's server j' moves to 1 is $c_{1j'}/\tilde{c}$, and the cost of moving j' is $d_{1j'} = 1/c_{1j'}$. Thus $\Delta cost_H = k/\tilde{c}$. If $1'$ moves to 1, the potential decreases by $H_{11'}$. If $j' \neq 1'$ moves to 1, the potential increases by at most $H_{j1'} - H_{11'} - H_{jj'}$, since after the move we can match $1'$ with j and j' with 1. So

$$\tilde{c} \cdot (\Delta cost_H + \Delta\Phi) \leq k + [-c_{11'}H_{11'} + \sum_{j=2}^{k} c_{1j'}(H_{j1'} - H_{11'} - H_{jj'})]$$
$$= \sum_{j=2}^{k} [c_{1j}(H_{11'} - H_{j1'}) + c_{1j'}(H_{j1'} - H_{11'} - H_{jj'})]$$
$$= \sum_{j=2}^{k} [(c_{1j} - c_{1j'})(H_{11'} - H_{j1'}) - c_{1j'}H_{jj'}] \leq 0,$$

where the first equality follows from (4) and the last inequality from (3). Since $\Delta cost_{adv} = 0$, (2) holds when HARMONIC moves. This completes the proof. □

In configurations where exactly one of HARMONIC's servers is not covered by adversary servers, only a single edge in the minimum matching has non-zero cost, and Φ is equal to the cost of HARMONIC when the adversary makes only lazy requests (i.e., does not move his server). This is clearly a lower bound on a potential function, and in [1] it was shown that this potential is sufficient against the lazy adversary. Our potential generalizes this idea to all configurations.

3   Computing Hitting Costs

In this section we describe formulas for the hitting costs H_{it} as rational expressions in the conductances. We also introduce notation useful for their manipulation in subsequent analysis. Throughout this section, we assume that $G = \{1, 2, \ldots, k\}$ and $t \in M - G$. We treat $G + t$ as a complete graph with weights c_{ij} assigned to its edges (unordered pairs or nodes).

Admittance matrix. Let $\bar{c}_i = \sum_{j=1}^{k} c_{ij}$, for all i (where $c_{ii} = 0$ for all i). Define the *admittance matrix* C as

$$C = \begin{bmatrix} \bar{c}_1 + c_{1t} & -c_{12} & -c_{13} \cdots & -c_{1k} \\ -c_{21} & \bar{c}_2 + c_{2t} & -c_{23} \cdots & -c_{2k} \\ \vdots & & \ddots & \vdots \\ -c_{k1} & -c_{k2} & -c_{k3} \cdots & \bar{c}_k + c_{kt} \end{bmatrix}$$

Let $\overline{C}_j$ denote a matrix obtained from C by replacing column j by vector $\mathbf{1} = (1, 1, \ldots, 1)^T$. By C_{ij} we denote C with row i and column j removed. We have $\det(\overline{C}_j) = \sum_{i=1}^{k}(-1)^{i+j}\det(C_{ij})$. The equations for hitting costs (1) can be rewritten as $C\mathbf{H}_t = k\mathbf{1}$, where $\mathbf{H}_t = (H_{1t}, \ldots, H_{kt})^T$. Thus

$$H_{jt} = k \cdot \frac{\det(\overline{C}_j)}{\det(C)}. \tag{5}$$

ℓ-Trees and conductance products. An *ℓ-tree* is defined to be a forest with vertex set $G+t$ and with exactly ℓ connected components (trees). Any ℓ-tree has $k+1-\ell$ edges. If S is an ℓ-tree, then by r_{jS} we denote the size of the component of j in S.

For any set S of edges of $G+t$, let $\xi_S = \prod_{(i,j)\in S} c_{ij}$. The theorem below is a classical result from the theory of electrical networks (see [14,16]). It characterizes determinants and cofactors of admittance matrices in terms of certain sums of conductance products ξ_S.

Theorem 3.1. (a) $\det(C) = \sum_T \xi_T$, *where the sum is taken over all spanning trees T of $G+t$.*
(b) $\det(C_{ij}) = (-1)^{i+j}\sum_S \xi_S$, *where the sum is taken over all spanning 2-trees S of $G+t$ in which i, j are in one component and t in the other.*

Corollary 3.2. $\det(\overline{C}_j) = \sum_S r_{jS}\xi_S$, *where the sum is taken over all spanning 2-trees S of $G+t$ in which j and t are in different components.*

Proof. Applying Theorem 3.1.b, we have

$$\det(\overline{C}_j) = \sum_{i=1}^{k}(-1)^{i+j}\det(C_{ij}) = \sum_{i=1}^{k}\sum_{S_i}\xi_{S_i},$$

where, for each fixed i, the second summation is over all spanning 2-trees S_i of $G+t$ in which i, j are in one component and t in the other. Thus each ξ_S is counted r_{jS} times, and the corollary follows. □

Using Theorem 3.1 and Corollary 3.2, we can express the hitting costs H_{it} and $\widehat{H}_{ij}$ in terms of ℓ-tree conductance products.

Corollary 3.3. (a) H_{it} *can be written as*

$$H_{it} = k \cdot \frac{\sum_S r_{iS}\xi_S}{\sum_T \xi_T}, \tag{6}$$

*where T ranges over all spanning trees of G + t and S ranges over all spanning
2-trees S of G + t in which i and t are in different components.*

(b) *For i ≠ j, $\widehat{H}_{ij}$ can be written as*

$$\widehat{H}_{ij} = k \cdot \frac{\sum_Y r_{iY} 2^{\deg_Y(j)} \xi_Y}{\sum_Z 2^{\deg_Z(j)} \xi_Z}, \tag{7}$$

*where Z ranges over all spanning trees of G, Y ranges over all spanning 2-trees
of G that separate i from j, and $\deg_Q(u)$ denotes the degree of u in a graph Q.*

Proof. Part (a) follows immediately from the formula (5) for hitting costs, Theorem 3.1.a, and Corollary 3.2.

It remains to prove (b). The proof is by letting t converge to j and applying (a). Let t be a copy of j and $c_{jt} \to \infty$. Then, using notation from (6), we have

$$H_{it} = k \cdot \frac{\sum_S r_{iS} \xi_S}{\sum_T \xi_T} \to k \cdot \frac{\sum_{S \ni (j,t)} r_{iS} \xi_{S-(j,t)}}{\sum_{T \ni (j,t)} \xi_{T-(j,t)}} = k \cdot \frac{\sum_P r_{iP} \xi_P}{\sum_R \xi_R}$$

where P ranges over all spanning 2-trees of G + t that separate i, j from t, and R ranges over all spanning 2-trees of G + t that separate j from t.

For each such R, by contracting j with t we get a spanning tree Z of G with $\xi_Z = \xi_R$. Further, each Z can be obtained from such a contraction in exactly $2^{\deg_Z(j)}$ ways. So $\sum_R \xi_R = \sum_Z 2^{\deg_Z(j)} \xi_Z$. By the same argument, $\sum_P r_{iP} \xi_P = \sum_Y r_{iY} 2^{\deg_Y(j)} \xi_Y$, where Y is as in (7). This completes the proof of part (b). □

$\mathcal{X}$-Expressions. We use the symbol $\mathcal{X}\langle \cdot \rangle$ to denote sums of terms ξ_S or $r_{jS} \xi_S$ over various sets S specified by appropriate parameters of $\mathcal{X}\langle \cdot \rangle$. The argument of $\mathcal{X}\langle \cdot \rangle$ is a list of nodes separated into $\ell \geq 0$ fields by symbols | or ‖. The sum in $\mathcal{X}\langle \cdot \rangle$ is taken over all spanning ℓ-trees of G + t in which the nodes in different fields are in different components and the nodes in the same field are in the same component. The separator ‖ appears at most once; and if present, each term ξ_S is multiplied by the total number of vertices in the components corresponding to the fields before ‖. For example:

$\mathcal{X}\langle \rangle = \sum_S \xi_S$, where the sum is taken over all spanning trees S of G + t.

$\mathcal{X}\langle 1|2 \rangle = \sum_S \xi_S$, where the sum is taken over all spanning 2-trees S of G + t in which 1 and 2 are in different components.

$\mathcal{X}\langle 1, 2\|3 \rangle = \sum_S r_{1S} \xi_S$, where the sum is taken over all spanning 2-trees S of G + t in which 1, 2 are in the same component and 3 in the other component.

$\mathcal{X}\langle 1|2\|3 \rangle = \sum_S (r_{1S} + r_{2S}) \xi_S$, where the sum is taken over all spanning 3-trees S of G + t in which 1, 2 and 3 are in different components.

In terms of $\mathcal{X}$-expressions, Theorem 3.1 and Corollaries 3.2 and 3.3 imply $\det(C) = \mathcal{X}\langle \rangle$, $\det(\overline{C}_j) = \mathcal{X}\langle j\|t \rangle$, and

$$H_{it} = k \cdot \frac{\mathcal{X}\langle i\|t \rangle}{\mathcal{X}\langle \rangle} \tag{8}$$

Partitioning $\mathcal{X}$-expressions. We now introduce two useful identities for $\mathcal{X}$-expressions. Let $i \neq j$ be two arbitrary nodes in $G + t$. Then:

$$\mathcal{X}\langle\rangle = c_{ij}\mathcal{X}\langle i|j\rangle + \sum_{\ell \neq i,j} c_{i\ell}\mathcal{X}\langle i|j,\ell\rangle \tag{9}$$

$$(k-1)\mathcal{X}\langle\rangle = \sum_{\ell \neq i} c_{i\ell}\mathcal{X}\langle \ell\|i\rangle \tag{10}$$

The equation (9) is obtained by grouping all spanning trees according to the first node $\ell \neq i$ on the path from i to j.

Equation (10) can be derived as follows. The LHS(10) is the sum of all terms $(k-1)\xi_T$, where T is a spanning tree of $G + t$. Given a spanning tree T, we think of T as being rooted at i. For each neighbor ℓ of i, let k_ℓ be the size of the subtree rooted at ℓ. Then $(k-1)\xi_T = \sum_{(i,\ell)\in T} k_\ell c_{i\ell}\xi_{T-(i,\ell)}$. Thus LHS(10) is the sum of the above expressions over all trees T. By exchanging the order of summations and grouping the terms according to $c_{i\ell}$, we obtain the RHS(10).

These two identities play crucial role in our proofs. They allow us to manipulate $\mathcal{X}$-expressions by breaking them into smaller expressions. Both (9) and (10) can be extended to more complex $\mathcal{X}$-expressions and, in fact, we will usually use these more general forms later in the paper. We refer to these two methods (9) and (10) of partitioning $\mathcal{X}$-expressions as *cut-partitioning* and *neighbor-partitioning*, respectively.

4   An Auxiliary Inequality

In this section we focus on inequality (A). We show that, for any k, inequality (A) can be reduced to a statement about certain tree products. The proof uses Theorem 2.b from [1]. We restate it here in our notation.

Lemma 4.1. [1, Theorem 2.b] *Let $s \in G$ and $t \notin G$. Then*

$$2c_{st}\mathcal{X}\langle s\|t\rangle \leq (k+1)\mathcal{X}\langle\rangle. \tag{11}$$

Lemma 4.2. *Let $s \in G$ and $t \notin G$. Suppose that for every $v \in G - s$,*

$$|\mathcal{X}\langle s|v\|t\rangle\mathcal{X}\langle\rangle - \mathcal{X}\langle s|v\rangle\mathcal{X}\langle s\|t\rangle| \leq \tfrac{1}{2}(k+1)\mathcal{X}\langle s|v,t\rangle\mathcal{X}\langle s|v\rangle. \tag{12}$$

Then inequality (A) holds.

Proof. By (8) we have $H_{st} = k \cdot \mathcal{X}\langle s\|t\rangle/\mathcal{X}\langle\rangle$. Thus (A) can be written as

$$\sum_{v \in G+t-s}\left|\frac{\partial\mathcal{X}\langle s\|t\rangle}{\partial d_{sv}}\mathcal{X}\langle\rangle - \frac{\partial\mathcal{X}\langle\rangle}{\partial d_{sv}}\mathcal{X}\langle s\|t\rangle\right| \leq \tfrac{1}{2}(k+1)(\mathcal{X}\langle\rangle)^2. \tag{13}$$

Since $\partial c_{sv}/\partial d_{sv} = -c_{sv}^2$, for any set of edges Q we have $\partial\xi_Q/\partial d_{sv} = -c_{sv}^2\xi_{Q-(s,v)}$ if $(s,v) \in Q$, and equals 0 otherwise, and thus

$$\frac{\partial\mathcal{X}\langle\rangle}{\partial d_{sv}} = -c_{sv}^2\mathcal{X}\langle s|v\rangle,$$

$$\frac{\partial\mathcal{X}\langle s\|t\rangle}{\partial d_{sv}} = -c_{sv}^2\mathcal{X}\langle s|v\|t\rangle \quad \text{for } v \neq t, \text{ and} \qquad \frac{\partial\mathcal{X}\langle s\|t\rangle}{\partial d_{st}} = 0.$$

Substituting into (13) we obtain

$$\text{LHS}(13) = c_{st}^2 \mathcal{X}\langle s|t\rangle \mathcal{X}\langle s\|t\rangle + \sum_{v\in G-s} c_{sv}^2 |\mathcal{X}\langle s|v\|t\rangle \mathcal{X}\langle\rangle - \mathcal{X}\langle s|v\rangle \mathcal{X}\langle s\|t\rangle|$$
$$\leq \tfrac{1}{2}(k+1)c_{st}\mathcal{X}\langle s|t\rangle \mathcal{X}\langle\rangle + \sum_{v\in G-s} \tfrac{1}{2}(k+1)c_{sv}^2 \mathcal{X}\langle s|v,t\rangle \mathcal{X}\langle s|v\rangle$$
$$\leq \tfrac{1}{2}(k+1)\mathcal{X}\langle\rangle \left[c_{st}\mathcal{X}\langle s|t\rangle + \sum_{v\in G-s} c_{sv}\mathcal{X}\langle s|v,t\rangle \right] = \text{RHS}(13).$$

The first inequality follows from (12) and from Lemma 4.1. The second inequality follows from $c_{sv}\mathcal{X}\langle s|v\rangle \leq \mathcal{X}\langle\rangle$. In the last step we used the equality $\mathcal{X}\langle\rangle = c_{st}\mathcal{X}\langle s|t\rangle + \sum_{v\in G-s} c_{sv}\mathcal{X}\langle s|v,t\rangle$, which is the cut-partitioning identity (9). □

5 Proof of Inequality (D)

In this section we prove (D), that is, $(c_{vs} - c_{vt})H_{vt} \leq c_{vs}H_{st}$. By applying the formula (8), multiplying by $\mathcal{X}\langle\rangle/k$, and applying the equation $\mathcal{X}\langle v|t\rangle - \mathcal{X}\langle s\|t\rangle = \mathcal{X}\langle v\|s,t\rangle - \mathcal{X}\langle s\|v,t\rangle$, we reduce (D) to:

$$c_{vs}(\mathcal{X}\langle v\|s,t\rangle - \mathcal{X}\langle s\|v,t\rangle) \leq c_{vt}\mathcal{X}\langle v\|t\rangle. \tag{14}$$

To prove (14), we prove the following, slightly stronger lemma.

Lemma 5.1. *Let* $s,v \in G$ *and* $t \in M - G$. *Then*

$$c_{vs}|\mathcal{X}\langle v\|s,t\rangle - \mathcal{X}\langle s\|v,t\rangle| \leq c_{vt}\mathcal{X}\langle v\|t,s\rangle \tag{15}$$

Proof. The proof is by induction on k, the size of G. In the base step, for $k = 2$, inequality (15) reduces to the triangle inequality $c_{vs}|c_{st} - c_{vt}| \leq c_{vt}c_{st}$.

Suppose now that $k \geq 3$ and that the lemma holds for all $k' < k$. First, we claim that the following identities hold; the sums are over $\bar{s} \in G - v - s$:

$$\mathcal{X}\langle v\|s,t\rangle = c_{st}\mathcal{X}\langle v\|t|s\rangle + \sum_{\bar{s}} c_{s\bar{s}}\mathcal{X}\langle v\|\bar{s},t|s\rangle \tag{16}$$
$$\mathcal{X}\langle s\|v,t\rangle = \mathcal{X}\langle v,t|s\rangle + \sum_{\bar{s}} c_{s\bar{s}}\mathcal{X}\langle \bar{s}\|v,t|s\rangle \tag{17}$$
$$c_{vt}\mathcal{X}\langle v\|t|s\rangle + \sum_{\bar{s}} c_{v\bar{s}}\mathcal{X}\langle v\|\bar{s},t|s\rangle = \mathcal{X}\langle v,t|s\rangle + \sum_{\bar{s}} c_{v\bar{s}}\mathcal{X}\langle \bar{s}\|v,t|s\rangle \tag{18}$$

The equality (16) is obtained using the cut-partitioning formula (9), applied to expression $\mathcal{X}\langle v\|s,t\rangle$, with $\bar{s}$ denoting the first node on the path from v to s. Similarly, the equality (17) is obtained using the neighbor-partitioning formula (10), applied to expression $\mathcal{X}\langle v\|s,t\rangle$, with $\bar{s}$ ranging over the neighbors of s.

Equation (18) is derived using cut- and neighbor-partitioning, but this time the expression we partition is not expressible using the $\mathcal{X}$-notation. For a 2-tree S that separates v,t from s, denote by q_S the size of the component of v in a 3-tree obtained from S by removing the first edge on the path from v to t. We obtain (18) by partitioning the expression $\sum_S q_S \xi_S$ (where S ranges over all 2-trees separating v,t from s) in two different ways: the LHS(18) is obtained using cut-partitioning, and RHS(18) is obtained using neighbor-partitioning.

We derive (15) in several steps; the sums are over $\bar{s} \in G - v - s$:

LHS(15)

$$= |c_{vs}c_{st}\mathcal{X}\langle v\|t|s\rangle + \sum_{\bar{s}} c_{vs}c_{s\bar{s}}(\mathcal{X}\langle v\|\bar{s},t|s\rangle - \mathcal{X}\langle\bar{s}\|v,t|s\rangle) - c_{vs}\mathcal{X}\langle v,t|s\rangle| \quad (19)$$

$$\leq |c_{vs}c_{vt}\mathcal{X}\langle v\|t|s\rangle + \sum_{\bar{s}} c_{vs}c_{v\bar{s}}(\mathcal{X}\langle v\|\bar{s},t|s\rangle - \mathcal{X}\langle\bar{s}\|v,t|s\rangle) - c_{vs}\mathcal{X}\langle v,t|s\rangle| \quad (20)$$

$$+ c_{vt}c_{st}\mathcal{X}\langle v\|t|s\rangle + \sum_{\bar{s}} c_{v\bar{s}}c_{s\bar{s}}|\mathcal{X}\langle v\|\bar{s},t|s\rangle - \mathcal{X}\langle\bar{s}\|v,t|s\rangle|$$

$$\leq c_{vt}c_{st}\mathcal{X}\langle v\|t|s\rangle + \sum_{\bar{s}} c_{s\bar{s}}c_{vt}\mathcal{X}\langle v\|t,\bar{s}|s\rangle \ = \ \text{RHS(15)} \quad (21)$$

We now explain the steps in the above derivation. The equality (19) is obtained by substituting (16) and (17) into LHS(15).

In step (20) we use triangle inequalities. For the first term, we apply $|c_{vs}c_{st} - c_{vs}c_{vt}| \leq c_{vt}c_{st}$ as follows: this triangle inequality implies that $|\alpha c_{vs}c_{st} + \beta| \leq |\alpha c_{vs}c_{vt} + \beta| + |\alpha|c_{vt}c_{st}$ for any reals α and β. We apply this form with $\alpha = \mathcal{X}\langle v\|t|s\rangle$ and with β equal to the sum of the remaining terms within the absolute value. Similarly, for each term in the sum, one at a time, we use $|c_{vs}c_{s\bar{s}} - c_{vs}c_{v\bar{s}}| \leq c_{v\bar{s}}c_{s\bar{s}}$ as follows: this triangle inequality implies that $|\alpha c_{vs}c_{v\bar{s}} + \beta| \leq |\alpha c_{vs}c_{v\bar{s}} + \beta| + |\alpha|c_{v\bar{s}}c_{s\bar{s}}$ for any reals α and β. We apply this form with $\alpha = \mathcal{X}\langle v\|\bar{s},t|s\rangle - \mathcal{X}\langle\bar{s}\|v,t|s\rangle$ and β equal to the sum of the remaining terms within the absolute value.

In step (21), all terms within the absolute value in the first line cancel, by identity (18). Then we use the inductive assumption on each element of the sum in the second line of (20), with the component of s removed. For every fixed tree in the component of s, both terms contain the same factor of ξ_S and the remainder is just the left-hand side of LHS(15) for a smaller set of vertices (note that we removed at least one node, namely s).

Finally, the last equality in (21) is obtained by applying the cut-partitioning method (9). This completes the proof of the lemma. □

6   Analysis of HARMONIC for $k = 3$

The framework developed above can be used to prove that HARMONIC is 6-competitive for 3 servers.

The inequalities (A)–(C) for $k = 3$ can be verified by somewhat tedious calculations. Inequality (12) does not require any triangle inequality, and thus can be easily verified in Maple or Mathematica. Other inequalities can also be verified in this manner, but this process is complicated by the fact that the choice of the triangle inequalities and the order of their application is not always easy to determine. We omit the details and refer the reader to the technical report [9].

As shown in Section 5, (D) holds for all k. Therefore, from Theorem 2.1, we obtain the main result of this paper:

Theorem 6.1. *For $k = 3$, HARMONIC is 6-competitive.*

7   Final Comments

We proved that HARMONIC is 6-competitive for 3 servers. It is easy to verify
that inequalities (A)–(C) also hold for $k = 2$, giving yet another proof that
HARMONIC is 3-competitive for $k = 2$. We believe that inequalities (A) and (B)
are valid for arbitrary k, and that our potential can be applied to show that
HARMONIC is $\frac{1}{2}k(k + 1)$-competitive in other cases, and quite possibly for the
general case as well. However, we can show that inequality (C) fails for large k,
and thus even if Φ works in general, the proof will require a different approach.

The inequalities (A) and (B) represent interesting properties of harmonic
random walks and they are of independent interest. They also may be applicable to obtain other results on HARMONIC. For example, inequality (B) can be
thought of as a triangle inequality for harmonic random walks. Proving inequality (A) alone would show that HARMONIC is $\frac{1}{2}k(k + 1)$ competitive against a
semi-lazy adversary, which is like the lazy adversary with the additional freedom
of moving the server that is not covered by our servers.

Acknowledgements. We are grateful to anonymous referees for helpful
comments. Partially supported by cooperative research grant KONTAKT-
ME476/CCR-9988360-001 from MŠMT ČR and NSF. First author supported
by NSF grants CCR-9988360 and CCR-0208856. Second author partially supported by Institute for Computer Science, Prague (project L00A056 of MŠMT
ČR), by grant 201/01/1195 of GA ČR and by grant A1019901 of GA AV ČR.

References

1. Y. Bartal, M. Chrobak, J. Noga, and P. Raghavan. More on random walks, electrical networks, and the harmonic k-server algorithm. *Information Processing Letters*,
   84:271–276, 2002.
2. Y. Bartal and E. Grove. The harmonic k-server algorithm is competitive. *Journal
   of the ACM*, 47(1):1–15, 2000.
3. P. Berman, H. Karloff, and G. Tardos. A competitive algorithm for three servers.
   In *Proc. 1st Symp. on Discrete Algorithms*, pages 280–290, 1990.
4. A. Borodin and R. El-Yaniv. *Online Computation and Competitive Analysis*. Cambridge University Press, 1998.
5. A. K. Chandra, P. Raghavan, W. L. Ruzzo, R. Smolensky, and P. Tiwari. The electrical resistance of a graph captures its commute and cover times. *Computational
   Complexity*, 6:312–340, 1997.
6. M. Chrobak and L. L. Larmore. HARMONIC is three-competitive for two servers.
   Theoretical Computer Science, 98:339–346, 1992.
7. M. Chrobak and L. L. Larmore. Metrical task systems, the server problem, and
   the work function algorithm. In *Online Algorithms: State of the Art*, pages 74–94.
   Springer-Verlag, 1998.
8. M. Chrobak and J. Sgall. A simple analysis of the harmonic algorithm for two
   servers. *Information Processing Letters*, 75:75–77, 2000.
9. M. Chrobak and J. Sgall. Analysis of the harmonic algorithm for three servers.
   Technical Report ITI Series 2002-102, Charles University, Prague, 2002.
   http://iti.mff.cuni.cz/series/index.html

10. D. Coppersmith, P. G. Doyle, P. Raghavan, and M. Snir. Random walks on weighted graphs and applications to on-line algorithms. *Journal of the ACM*, 40:421–453, 1993.
11. P. G. Doyle and J. L. Snell. *Random Walks and Electrical Networks.* Mathematical Association of America, 1984.
12. E. Koutsoupias and C. Papadimitriou. On the k-server conjecture. *Journal of the ACM*, 42:971–983, 1995.
13. M. Manasse, L. A. McGeoch, and D. Sleator. Competitive algorithms for server problems. *Journal of Algorithms*, 11:208–230, 1990.
14. W. Percival. Solution of passive electrical networks by means of mathematical trees. *J. Inst. Elect. Engrs.*, 100:143–150, 1953.
15. P. Raghavan and M. Snir. Memory versus randomization in on-line algorithms. *IBM Journal on Research and Development*, 38, 1994.
16. S. Seshu and M. B. Reed. *Linear Graphs and Electrical Networks.* Addison-Wesley, 1961.

Non-clairvoyant Scheduling for Minimizing Mean Slowdown

N. Bansal[1], K. Dhamdhere[1], J. Könemann[2], and A. Sinha[2]

[1] School of Computer Science, Carnegie Mellon University, Pittsburgh, PA 15213, USA,
{nikhil,kedar}@cs.cmu.edu
[2] Graduate School of Industrial Administration, Carnegie Mellon University, Pittsburgh, PA 15213, USA,
jochen@cmu.edu
asinha@andrew.cmu.edu

Abstract. We consider the problem of scheduling jobs online non-clairvoyantly, that is, when job sizes are not known. Our focus is on minimizing mean *slowdown*, defined as the ratio of flow time to the size of the job. We use *resource augmentation* in terms of allowing a faster processor to the online algorithm to make up for its lack of knowledge of job sizes.

Our main result is an $O(1)$-speed $O(\log^2 B)$-competitive algorithm for minimizing mean slowdown non-clairvoyantly, when B is the ratio between the largest and smallest job sizes. On the other hand, we show that any $O(1)$-speed algorithm, deterministic or randomized, is at least $\Omega(\log B)$ competitive.

The motivation for bounded job sizes is supported by an $\Omega(n)$ lower bound for arbitrary job sizes, where n is the number of jobs. Furthermore, a lower bound of $\Omega(B)$ justifies the need for resource augmentation even with bounded job sizes. For the static case, i.e. when all jobs arrive at time 0, we give an $O(\log B)$ competitive algorithm which does not use resource augmentation and a matching $\Omega(\log B)$ lower bound on the competitiveness.

1 Introduction

1.1 Motivation

While scheduling algorithms in general have received a lot of interest in the past, most algorithms assume that the input is completely known. However, there are several situations where the scheduler has to schedule jobs without knowing the sizes of the jobs. This lack of knowledge, known as *non-clairvoyance*, is a significant impediment in the scheduler's task, as one might expect. The study of non-clairvoyant scheduling algorithms was initiated by Motwani *et al.* [12].

We consider the problem of minimizing the total slowdown (also called stretch) non-clairvoyantly on a single processor with preemptions. This was posed as an important open problem by Becchetti *et al.* [3]. Slowdown or stretch of a job was first considered by Bender *et al.* [6]. It is the ratio of the flow time of a job to its size. Slowdown as a metric has received much attention lately [9,15,

H. Alt and M. Habib (Eds.): STACS 2003, LNCS 2607, pp. 260–270, 2003.

13,5,4,1,7], since it captures the notion of "fairness". Note that a low slowdown implies that jobs are delayed in proportion to their size, hence smaller jobs are delayed less and large jobs are delayed proportionately more. Muthukrishnan *et al.* [13] first studied mean slowdown, and showed that the shortest remaining processing time (SRPT) algorithm achieves a competitive ratio of 2 for the single machine and 14 for the multiple machine case. Note that SRPT requires the knowledge of job sizes and hence cannot be used in the non-clairvoyant setting. Similarly, the various extensions and improvements [8,1,4,7] to the problem have all been in the clairvoyant setting. Since clairvoyant scheduling does not accurately model many systems, there is significant interest in the non-clairvoyant version of slowdown.

As expected, non-clairvoyant algorithms usually have very pessimistic bounds (See for example [12]). A major advance in the study of non-clairvoyant algorithms was made by Kalyanasundaram and Pruhs [10], who proposed the model of *resource augmentation* where the online algorithm is compared against an adversary that has a slower processor. This provides a nice framework for comparing the performance of algorithms where the traditional competitive analysis gives a very pessimistic guarantee for all algorithms. Our analysis makes use of resource augmentation model.

1.2   Model

We consider a single machine scenario, where jobs arrive dynamically over time. The size of a job i, denoted by p_i, is its total service requirement. However, p_i is not known by the scheduler at any time before the job completes, in particular p_i is not known when job i arrives. Obviously, p_i becomes known when the job completes. The flow time of a job is the difference of the times when the job completes (c_i) and the time when it arrives, i.e. its release date (r_i). The slowdown of a job is the ratio of its flow time to its size. We are interested in minimizing the mean (or equivalently total) slowdown.

Traditionally, an online algorithm is said to be c-competitive if the worst case ratio (over all possible inputs) of the performance of the online algorithm is no more than c times the performance of the optimum offline adversary. In the resource augmentation model [10], we say that an algorithm is $s - speed$, $c - competitive$ if it uses an s times faster processor than the optimum algorithm and produces a solution that has competitive ratio no more than c against the optimal algorithm with no speedup.

1.3   Results

1. *General Case:* Our main contribution is an algorithm for minimizing mean slowdown non-clairvoyantly which is $O(\log^2 B)$-competitive with an $O(1)$-speedup, where B is the ratio of the maximum to the minimum job size. We also note that our algorithm does not need to know B and is fully online. We will use the existence of B only in our analysis.
2. *Need for Resource Augmentation:* It is not hard to see that in the absence of an upper bound on job sizes, no algorithm for minimizing mean slowdown

262   N. Bansal et al.

can be $\Omega(n)$ competitive. Surprisingly, it turns out that no algorithm (deterministic or randomized) can be $o(n/k)$-competitive even with a k-speedup. In particular, we need at least $\Omega(n)$ speedup to be competitive.

3. The above lower bounds require instances where the range of job sizes varies exponentially in n. In a more realistic scenario, when job size ratios are bounded by B, we show that in the absence of speedup, any algorithm (deterministic or randomized) is $\Omega(B)$ competitive. Moreover, we show that even with a factor k speedup any algorithm (deterministic or randomized) is at least $\Omega(\log B/k)$ competitive. Note that the performance our algorithm matches the lower bound upto a $\log B$ factor in the competitive ratio.

4. *Static Case:* When all the requests arrive at time 0, we settle the question exactly. We give an $O(\log B)$ competitive algorithm that does not use any speed up and a matching $\Omega(\log B)$ lower bound on the competitive ratio of any deterministic or randomized algorithm.

We also stress that our algorithm for the general case is fast and easy to implement, which is a useful advantage when it comes to putting it to work in real systems, say a web server. In fact, it can be implemented in $O(n)$ time.

1.4   Relations to Other Problems

Minimizing total slowdown can be thought of as a special case of minimizing total weighted flow time, where the weight of a job is inversely proportional to its size. However, what makes the problem interesting and considerably harder in a non-clairvoyant setting is that the sizes (hence weights) are not known. Hence not only does the online scheduler have no idea of job sizes, it also has no idea as to which job is more important (has a higher weight).

An interesting aspect of slowdown is that neither resource augmentation nor randomization seems to help in the general case. This is in sharp contrast to weighted flow time where a $(1 + \epsilon)$ speed gives a $(1 + \frac{1}{\epsilon})$ competitive algorithm [2]. Similarly, for flow time there is a lower bound of $n^{1/3}$ for any deterministic algorithm, whereas using randomization, algorithms which are $O(\log n)$ competitive can be obtained [11,3]. In contrast, even with randomization and a k speed-up the slowdown problem is at least $\Omega(n/k)$ competitive.

In terms of proof techniques, all previous techniques using resource augmentation [10,5,2] relied on *local competitiveness*[1]. For non-clairvoyant slowdown, proving local competitiveness is unlikely to work: it could easily be the case that the optimal clairvoyant algorithm has a few jobs of size B, while the non-clairvoyant online algorithm is left with a few jobs of size 1, thus being extremely uncompetitive locally.

Our main idea is to define a new metric, which we call the *inverse work* metric. We use this metric to connect the performance of the online algorithm to the offline algorithm. We also use resource augmentation in a novel way which can potentially find uses elsewhere.

---

[1] Intuitively, an algorithm is locally competitive if the cost incurred in objective in each unit step is within a constant factor of that incurred by optimal algorithm. Local competitiveness is both necessary and sufficient for global competitiveness [5] when there is no speedup.

2   Scheduling for Mean Slowdown

We first present an $O(1)$-speed, $O(\log^2 B)$-competitive algorithm for minimizing mean slowdown non-clairvoyantly. The comparison is made against the best possible clairvoyant scheduling algorithm. Here B is an upper bound on the ratio between the biggest and smallest job size. We will show later that we cannot design significantly better algorithms unless we have bounded job sizes, and resource augmentation.

Let J be a scheduling instance on n jobs with processing times $p_1, \ldots, p_n$ and release times $r_1, \ldots, r_n$. For a scheduling algorithm A, we let $S(A, s, J)$ denote the slowdown when A is provided with an s speed processor. Let $Opt_S(s, J)$ denote the slowdown of the algorithm which has optimal slowdown on the instance J using an s speed processor.

Let J' be a scheduling instance on the same jobs as J but with job sizes $p'_i = 2^{\lceil \log_2 p_i \rceil}$ for all $1 \le i \le n$. It is easy to see that

$$Opt_S(2s, J') \le Opt_S(s, J) \tag{1}$$

Imagine an algorithm with a 2-speed processor that simply mimics the optimum algorithm for the speed 1 processor. It can possibly sit idle when the speed 1 processor is still working on a job which the speed two processor has already finished. And hence slowdown on rounded-up instance of optimal algorithm will be as good as the slowdown of optimal algorithm with a speed 1 processor. In fact, with a speed-up 2, we can assume that job sizes are of the form $2^i - 1$.

Secondly, using a speed up of 2 we also know that *Shortest Job First* (SJF) is a 2-competitive algorithm for slowdown. This follows from a more general result due to Becchetti *et al.* [5]. They show that for the problem of minimizing weighted flow time, the greedy algorithm *Highest Density First* (HDF) is a $(1+\epsilon)$-speed, $(1 + 1/\epsilon)$ competitive algorithm. The HDF algorithm works on the job with the highest weight to size ratio at any time. Since slowdown is a special case of weighted flow time with weights equal to the reciprocal of size, we get that,

$$S(SJF, 2s, J) \le 2Opt_S(s, J) \tag{2}$$

2.1   Algorithm Description

The input consists of jobs arriving over a period of time. We assume that the smallest job has size 1. More than one job may arrive at a given instant of time. We invoke Equation 1 to assume that all jobs have sizes which are of the form $2^i - 1$.

We use the *Multi-Level Feedback* (MLF) algorithm [14] as our scheduling algorithm. MLF scheduling algorithm is used by Unix operating system. In MLF, jobs are queued in different queues according to the amount of work done on them. Queue i holds jobs which have between $2^i - 1$ to $2^{i+1} - 1$ work done on them. MLF always works on the lowest queue Q_l, and within each queue MLF processes jobs in *first come first serve* manner. If a job in Q_l has received 2^l amount of work but hasn't yet finished, it is promoted to the next queue, Q_{l+1}. Moreover, MLF will be given a constant speed-up, and we will compute this constant later.

2.2   Analysis

We define a new metric called the "inverse work" metric as an upper bound for slowdown and denote it by W^{-1}. Consider a scheduling instance J, and an algorithm A operating on it. At time step t, if $w_j(t)$ is the amount of work done on job j and $2^i - 1 \leq w_j(t) < 2^{i+1} - 1$, then define $j_t = 2^{i+1} - 1$. Then the total inverse work under A is defined as $\sum_{j \in J} \int_{r_j}^{c_j} \frac{1}{j_t}$, where r_j and c_j denote the arrival and completion times of job j respectively.

As usual, we will denote the total inverse work under algorithm A on instance J, using an s speed processor, by $W^{-1}(A, s, J)$. Similarly the optimal algorithm for W^{-1} on instance J and which uses an s speed processor will be denoted by $Opt_{W^{-1}}(s, J)$.

The main intuition of analysis is the following: Note that, any non-clairvoyant algorithm must spend some processing time probing for jobs of small sizes. However, this can cause it to postpone a small job while working on a large job. A clairvoyant algorithm on the other hand, has the advantage that it can postpone working on a large job for a long time. The inverse work metric takes care of essentially that, it captures the advantage of "probing". A second nice property of this metric is that a good schedule for slowdown can be converted into a good schedule for inverse work, with some speed up. This two ideas form the crux of our analysis. The details of the argument are given below.

Lemma 1. *For any algorithm A, any input instance J and any $s > 0$, we have* $S(A, s, J) \leq W^{-1}(A, s, J)$.

Proof. Work done on a job is always less or equal to size of the job. Also, the job sizes are powers of 2 minus 1, we have $j_t \leq p_j$ at all times before the job is completed. Hence, the inequality follows from the definitions of slowdown and inverse work.

Given an instance J, we define $J(m)$ to be the instance on the same jobs as J, but with jobs sizes about 2^{-m} times that of J. Moreover, if the size of a job is less than 1, we consider it as 0. Formally, if $r_m(i)$ and $p_m(i)$ denote the release time and size of job i in $J(m)$, then $r_m(i) = r_i$ and $p_m(i) = \lfloor 2^{-m}(p_i + 1) \rfloor$. Let $J' = J(1) \cup J(2) \cup \ldots \cup J(\log B)$. Thus, a job of size $2^i - 1$ in J is replaced by jobs of size $1, 2, 4, \ldots, 2^{i-1}$ in J'.

Lemma 2. *For an instance J, let J' be as defined above. Then for any $s > 0$,*

$$W^{-1}(MLF, s, J) \leq S(SJF, s, J')$$

Proof. Observe that running MLF on J exactly corresponds to running SJF on J'.

We now compare $W^{-1}(MLF, 1, J)$ with $S(SJF, 1, J')$. Let j_t be the work done by MLF on job $j \in J$ at some time t. If $j_t \leq 1$, then the contribution of j to W^{-1} is 1, and the contribution of the corresponding jobs in J' to the slowdown is at least 1. If $j_t > 1$, let $i > 0$ be such that $2^i - 1 \leq j_t < 2^{i+1} - 1$. Then the contribution of this job to $W^{-1}(MLF, s, J)$ at time t is $1/2^{i+1} - 1$. Moreover, $j_t < 2^{i+1} - 1$ means that, in J', SJF has at most finished the chunks $1, 2, \ldots, 2^{i-1}$. So the contribution of the rest of the chunks to $S(SJF, s, J')$ is at least $1/2^i$. Thus $W^{-1}(MLF, s, J) \leq S(SJF, s, J')$.

We now give a couple of useful lemmas whose proofs are fairly obvious. These will play a crucial role in our subsequent analysis.

Lemma 3. *Let $f_{SJF}(s,j)$ denote the flow time of job j under SJF with a s-speed processor. Then under SJF with an $s \geq 1$ speed processor, we have $f_{SJF}(s,j) \leq (1/s)f_{SJF}(1,j)$.*

Proof. Let $w(x,1,j)$ denote the work done on job j, after x units of time since it arrived, under SJF using a 1 speed processor. Similarly, let $w(x,s,j)$ denote the work done on job j, after x units of time since it arrived, under SJF using an s speed processor. We will show a stronger invariant that for all jobs j and all times t, $w((t-r_j)/s,s,j) \geq w(t-r_j,1,j)$. Notice that this stronger invariant trivially implies the result of the lemma.

Consider some instance where this condition is violated. Let j be the job and t be the earliest time for which $w((t-r_j)/s,s,j) < w(t-r_j,1,j)$. Clearly, the speed s processor (SJF(s)) is not working on j at time t, due to minimality of t. Thus, SJF(s) is working on some other smaller job j'. Since SJF(1) is not working on j', it has already finished j' by some time $t' < t$. However, this means that $w((t'-r_{j'}),s,j') < w(t'-r_{j'},1,j')$, which contradicts the minimality of t.

Lemma 4. *Let J be an instance, and $J(m)$ be defined as above. Then, if $x \geq 2^{-m}$,*

$$S(SJF,xs,J(m)) \leq \frac{1}{x}S(SJF,s,J)$$

Proof. It is easy to see that the flow time of every job $j \in J$ under SJF using an s speed processor is at least that of the corresponding job $j(m) \in J(m)$, when we run SJF on $J(m)$ using a speed $2^{-m}s$ processor.

Now, by Lemma 3, it is easy to see that if we run SJF on $J(m)$ using a speed xs processor, then the flow time of each job is $1/x$ times smaller than that if we run SJF on $J(m)$ using an $2^{-m}s$ speed processor, which is the same as that if we run SJF on J using a speed s processor.

Since the sizes of jobs in $J(m)$ are 2^{-m} times smaller than in J, the slowdown of $j(m) \in J(m)$ is $\frac{1}{x}$ times of $j \in J$.

Lemma 5. *For an instance J and J' defined as above,*

$$S(SJF,4s,J') \leq (\log^2 B)S(SJF,s,J)$$

Proof. By Lemma 4, if we run SJF on $J(i)$ with a speed $x_i \geq 1/2^i$ processor, then we can guarantee the slowdown in $J(i)$ to be at most $1/x_i$ times the slowdown in the schedule produced by SJF on J.

So for each $J(i)$, we use a processor of speed $x_i \geq 1/2^i$, which guarantees us that slowdown on $J(i)$ is within factor $1/x_i$ of $S(SJF,1,J)$. We run all these processors simultaneously.

Here, the total speedup needed is $\sum_{i=1}^{\log B} x_i$ and the slowdown for J' (denote by $S(J')$) is related to the slowdown of J by $S(J') \leq S(J)\sum_{i=1}^{\log B} 1/x_i$.

Choosing $x_1 = 1, x_2 = 1/2, \ldots, x_{\log\log B} = 1/\log B$, and $x_i = 1/\log B$ for $i > \log\log B$ ensures that $\sum x_i \leq 3$ and $\sum 1/x_i \leq \log^2 B$. This proves that $S(SJF,4s,J') \leq (\log^2 B)S(SJF,s,J)$.

Theorem 1. *There is an 16-speed, $2(\log^2 B)$-competitive algorithm for minimizing mean slowdown non-clairvoyantly.*

Proof. Consider an instance J where job sizes are integers and powers of 2 minus 1. Then we have following set of inequalities:

By Lemma 1, we have $S(MLF, 8s, J) \leq W^{-1}(MLF, 8s, J)$.

By Lemma 2, we have $W^{-1}(MLF, 8s, J) \leq S(SJF, 8s, J')$.

By Lemma 5, we have $S(SJF, 8s, J') \leq (\log^2 B)S(SJF, 2s, J)$.

Lastly, by Equation 2, we have $S(SJF, 2s, J) \leq 2Opt_S(s, J)$.

Putting these together we get:

$$S(MLF, 8s, J) \leq 2(\log^2 B)Opt_S(s, J)$$

Finally, with another speed up of 2, we can get rid of the assumption that job sizes are powers of 2 minus 1(using Equation 1). Thus the result follows.

In the above, we rounded the job sizes to powers of 2, which is also the base of the logarithm in $\log B$. We use 2 as the base of the exponent in the analysis for ease of exposition. Choosing the base as $1 + \epsilon$ and a more careful analysis gives a $(1 + \epsilon)$ speed, $(1 + 1/\epsilon)^4 \log^2 B$ competitive algorithm, for any $\epsilon > 0$. The proofs of these are deferred to the full version.

3   Lower Bounds

We now give lower bounds which motivate our algorithm in the previous section, in particular the need for bounded job sizes and the need for resource augmentation. Moreover, all of the lower bounds also hold if we allow the algorithm to be randomized.

3.1   Bounded Job Sizes without Resource Augmentation

Without resource augmentation the performance of any non-clairvoyant algorithm is really bad, even when job sizes are bounded. We show that any non-clairvoyant algorithm *without speedup*, deterministic or randomized is at least $\Omega(B)$ competitive.

Theorem 2. *Any non-clairvoyant algorithm, deterministic or randomized is at least $\Omega(B)$ competitive for mean slowdown, where B is the ratio of job sizes.*

Proof. Consider an instance where nB jobs of size 1, and n jobs of size B, arrive at time $t = 0$. At time $t = nB$, the adversary gives a stream of m jobs of size 1 every unit of time.

The optimum algorithm finishes all jobs of size 1 by nB and continues to work on the stream of size 1 jobs. The slowdown incurred due to jobs of size B is at most $n(2nB + m)/B$ and due to the jobs of size 1 is $nB + m$. The deterministic non-clairvoyant algorithm, on the other hand, can be made to have at least n jobs of size 1 remaining at time $t = nB$. Thus, it incurs a tot al slowdown of at least nm. Choosing $n > B$ and $m > nB$, it is easy to see that the slowdown is at least $\Omega(B)$.

For the randomized case, we use Yao's Lemma, and observe that by time $t = nB$, the algorithm will have at least $\frac{B}{B+1}n$ jobs of size 1 remaining in expectation. Thus the result follows.

3.2   Lower Bound with Bounded Size and Resource Augmentation

We now consider lower bounds when resource augmentation is allowed. We first consider a static (all jobs arrive at time $t = 0$) scheduling instance, and give an $\Omega(\log B)$ lower bound without resource augmentation. While this in itself is weaker than Theorem 2, the scheduling instance being static implies that even resource augmentation by k times can only help by a factor of k.

Lemma 6. *No deterministic or randomized algorithm can have performance ratio better than $\Omega(\log B)$ for a static scheduling instance, where B is the ratio of the largest to the smallest job size.*

Proof. We consider an instance with $\log B$ jobs $j_0, \ldots, j_{\log B}$ such that job j_i has size 2^i.

We first look at how SRPT behaves on this problem instance. The total slowdown for SRPT is $\sum_{i=0}^{\log B} \frac{1}{2^i} \left(\sum_{j=0}^{i} 2^j \right) = O(\log B)$. This basically follows from the fact that SRPT has to finish jobs $j_1, \ldots, j_{i-1}$ before it finishes j_i by the definition of our instance.

Now we show that for any non-clairvoyant deterministic algorithm A, the adversary can force the total slowdown to be $\Omega(\log^2 B)$. The rough idea is as follows: We order the jobs of our instance such that for all $0 \le i \le \log B$, A spends at least 2^i work on jobs $j_{i+1}, \ldots, j_{\log B}$ before it finishes job j_i. In this case, the theorem follows because the total slowdown of A on the given instance is

$$\sum_{i=1}^{\log B} \frac{1}{2^i} \left(\log(B) - i + 1 \right) 2^i = \Omega(\log^2(B)).$$

It remains to show that we can order the jobs in such a way. Since A is a deterministic algorithm that does not use the size of incoming jobs, we can determine the order in which jobs receive a total of at least 2^i work by A. We let j_i be the $(\log(B) - i + 1)^{th}$ job that receives 2^i work for all $0 \le i \le \log B$. It is clear that this yields the claimed ordering.

The example can be randomized to prove that even a randomized algorithm has mean slowdown no better than $\Omega(\log B)$. The idea is to assume that the instance is a random permutation of the jobs $j_0, j_1, \ldots, j_{\log B}$. Then to finish job j_i, the scheduler has to spend at least 2^i work on at least half of $j_{i+1}, \ldots, j_{\log B}$ (in expectation). Thus its expected slowdown is $\frac{1}{2}(\log B - i + 1)$ and the total slowdown is $\Omega(\log^2 B)$. We now use Yao's Minimax Lemma to obtain the result.

As the input instance in Lemma 6 is static, a k-speed processor can at most improve all the flow times by a factor of k. Hence the mean slowdown can go down by the same factor. This gives us the following theorem.

Theorem 3. *Any k-speed deterministic or randomized, non-clairvoyant algorithm has an $\Omega(\log B/k)$ competitive ratio for minimizing mean slowdown, in the static case (and hence in the online case).*

3.3 Scheduling with General Job Sizes

The previous result also implies the following lower bound when job sizes could be arbitrarily large. In particular, we can choose the job sizes to be $1, 2, \ldots 2^n$ which gives us the following theorem.

Theorem 4. *Any k-speed non-clairvoyant algorithm, either deterministic or randomized, has $\Omega(n/k)$ performance ratio for minimizing mean slowdown.*

Theorems 2 and 4 show that in the absence of resource augmentation and bounded job sizes achieving any reasonable guarantee on the competitive ratio is impossible. This motivates our model in Section 2, where we assume bounded job sizes and use resource augmentation.

4   Static Scheduling

Static scheduling is usually substantially easier than the usual dynamic scheduling, and the same is the case here. We do not need resource augmentation here, and we show that the *Round Robin* (RR) algorithm is $O(\log B)$ competitive, hence matching the lower bound shown in the previous section.

4.1   Optimal Clairvoyant Algorithm

Note that it follows from a simple interchange argument that the optimal clairvoyant algorithm for minimizing mean slowdown in the static case is exactly SRPT or equivalently SJF.

Lemma 7. *For any scheduling instance with n jobs all arriving at time 0, SRPT has $\Omega(n/\log B)$ mean slowdown.*

Proof. We approximate the total slowdown of SRPT as follows: for a job of size x, we require SRPT to work only $2^{\lfloor \lg x \rfloor}$ amount in order to finish the job and we divide the flow time by $2^{\lceil \lg x \rceil}$ to get slowdown. Thus, we can round down all the job sizes to a power of 2 and have the new total slowdown within a factor of 2 of the total slowdown of original instance.

Now we have $x_0, x_1, \ldots, x_{\log B}$ jobs of sizes $1, 2, \ldots, B$ respectively. We also have $\sum x_i = n$. Contribution to the total slowdown by jobs of size 2^i is at least $(1/2)x_i^2$. It is easy to see that $\sum(1/2)x_i^2 = \Omega(n^2/\log B)$. Thus the mean slowdown for SRPT is $\Omega(n/\log B)$.

4.2   Competitiveness of Round-Robin

The RR algorithm simply time shares the processor equally among all jobs. Thus, if there at m jobs at some time, each job gets a $1/m$ fraction of the processor.

Lemma 8. *For a scheduling instance with n jobs, RR has a $O(n)$ mean slowdown.*

Proof. Since each job shares the processor equally with at most n jobs, it can have a slowdown of at most n.

Combining the results of Lemmas 7 and 8, we get the following result:

Theorem 5. *For static scheduling with bounded job sizes, the Round Robin algorithm is $O(\log B)$-competitive for minimizing mean slowdown.*

5   Open Questions

The only "gap" in our paper is the discrepancy between the upper and lower bounds for the main problem of non-clairvoyant scheduling to minimize mean slowdown. It would be interesting to close this.

Acknowledgments. The authors would like to thank Avrim Blum, Moses Charikar, Kirk Pruhs and R. Ravi for useful discussions.

References

1. B. Awerbuch, Y. Azar, S. Leonardi, and O. Regev. Minimizing the flow time without migration. In *ACM Symposium on Theory of Computing*, pages 198–205, 1999.
2. N. Bansal and K. Dhamdhere. Minimizing weighted flow time. In *14th Annual ACM-SIAM Symposium on Discrete Algorithms*, 2003.
3. L. Becchetti and S. Leonardi. Non-clairvoyant scheduling to minimize the average flow time on single and parallel machines. In *ACM Symposium on Theory of Computing (STOC)*, pages 94–103, 2001.
4. L. Becchetti, S. Leonardi, and S. Muthukrishnan. Scheduling to minimize average stretch without migration. In *Symposium on Discrete Algorithms*, pages 548–557, 2000.
5. Luca Becchetti, Stefano Leonardi, Alberto Marchetti-Spaccamela, and Kirk R. Pruhs. Online weighted flow time and deadline scheduling. *Lecture Notes in Computer Science*, 2129:36–47, 2001.
6. M. Bender, S. Chakrabarti, and S. Muthukrishnan. Flow and stretch metrics for scheduling continuous job streams. In *ACM-SIAM Symposium on Discrete Algorithms (SODA)*, pages 270–279, 1998.
7. M. Bender, S. Muthukrishnan, and R. Rajaraman. Improved algorithms for stretch scheduling. In *13th Annual ACM-SIAM Symposium on Discrete Algorithms*, 2002.
8. C. Chekuri, S. Khanna, and A. Zhu. Algorithms for weighted flow time. In *ACM Symposium on Theory of Computing (STOC)*, 2001.

9. M. Crovella, R. Frangioso, and M. Harchol-Balter. Connection scheduling in web servers. In *USENIX Symposium on Internet Technologies and Systems*, 1999.
10. B. Kalyanasundaram and K. Pruhs. Speed is as powerful as clairvoyance. *Journal of the ACM*, 47(4):617–643, 2000.
11. Bala Kalyanasundaram and Kirk Pruhs. Minimizing flow time nonclairvoyantly. In *IEEE Symposium on Foundations of Computer Science*, pages 345–352, 1997.
12. R. Motwani, S. Phillips, and E. Torng. Nonclairvoyant scheduling. *Theoretical Computer Science*, 130(1):17–47, 1994.
13. S. Muthukrishnan, R. Rajaraman, A. Shaheen, and J. Gehrke. Online scheduling to minimize average stretch. In *IEEE Symposium on Foundations of Computer Science (FOCS)*, pages 433–442, 1999.
14. A. Silberschatz and P. Galvin. *Operating System Concepts*. Addison-Wesley, 5th Edition, 1998.
15. H. Zhu, B. Smith, and T. Yang. Scheduling optimization for resource-intensive web requests on server clusters. In *ACM Symposium on Parallel Algorithms and Architectures*, pages 13–22, 1999.

Space Efficient Hash Tables with Worst Case Constant Access Time[*]

Dimitris Fotakis[1], Rasmus Pagh[2,**], Peter Sanders[1], and Paul Spirakis[3,* * *]

[1] Max-Planck-Institut für Informatik, Saarbrücken, Germany.
{fotakis,sanders}@mpi-sb.mpg.de
[2] IT University of Copenhagen, Denmark. pagh@itu.dk
[3] Computer Technology Institute (CTI), Greece. spirakis@cti.gr

Abstract. We generalize Cuckoo Hashing [16] to *d-ary Cuckoo Hashing* and show how this yields a simple hash table data structure that stores n elements in $(1 + \epsilon) n$ memory cells, for any constant $\epsilon > 0$. Assuming uniform hashing, accessing or deleting table entries takes at most $d = O(\ln \frac{1}{\epsilon})$ probes and the expected amortized insertion time is constant. This is the first dictionary that has worst case constant access time and expected constant update time, works with $(1 + \epsilon) n$ space, and supports satellite information. Experiments indicate that $d = 4$ choices suffice for $\epsilon \approx 0.03$. We also describe a hash table data structure using explicit constant time hash functions, using at most $d = O(\ln^2 \frac{1}{\epsilon})$ probes in the worst case.

A corollary is an expected linear time algorithm for finding maximum cardinality matchings in a rather natural model of sparse random bipartite graphs.

1 Introduction

The efficiency of many programs crucially depends on hash table data structures, because they support constant expected access time. We also know hash table data structures that support *worst case* constant access time for quite some time [8,7]. Such worst case guarantees are relevant for real time systems and parallel algorithms where delays of a single processor could make all the others wait. A particularly fast and simple hash table with worst case constant access time is *Cuckoo Hashing* [16]: Each element is mapped to two tables t_1 and t_2 of size $(1 + \epsilon) n$ using two hash functions h_1 and h_2, for any $\epsilon > 0$. A factor above two in space expansion is sufficient to ensure with high probability (henceforth "whp."[1]) that each element e can be stored either in $t_1[h_1(e)]$ or $t_2[h_2(e)]$. The

---

[*] This work was partially supported by DFG grant SA 933/1-1 and the Future and Emerging Technologies programme of the EU under contract number IST-1999-14186 (ALCOM-FT).

[**] The present work was initiated while this author was at BRICS, Aarhus University, Denmark.

[* * *] Part of this work was done while the author was at MPII.

[1] In this paper "whp." will mean "with probability $1 - O(1/n)$".

H. Alt and M. Habib (Eds.): STACS 2003, LNCS 2607, pp. 271–282, 2003.

main trick is that insertion moves elements to different table entries to make room for the new element.

To our best knowledge, all previously known hash tables with worst case constant access time and sublinear insertion time share the drawback of a factor at least two in memory blowup. In contrast, hash tables with only expected constant access time that are based on open addressing can work with memory consumption $(1+\epsilon)\, n$. In the following, ϵ stands for an arbitrary positive constant.

The main contribution of this paper is a hash table data structure with worst case constant access time and memory consumption only $(1 + \epsilon)\, n$. The access time is $O(\ln \frac{1}{\epsilon})$ which is in some sense optimal, and the expected insertion time is also constant. The proposed algorithm is a rather straightforward generalization of Cuckoo Hashing to d-ary Cuckoo Hashing: Each element is stored at the position dictated by one out of d hash functions. In our analysis, insertion is performed by breadth first search (BFS) in the space of possible ways to make room for a new element. In order to ensure that the space used for bookkeeping in the BFS is negligible, we limit the number of nodes that can be searched to $o(n)$, and perform a rehash if this BFS does not find a way of accommodating the elements. For practical implementation, a random walk can be used. Unfortunately, the analysis that works for the original (binary) Cuckoo Hashing and $\log n$-wise independent hash functions [16] breaks down for $d \geq 3$. Therefore we develop new approaches. Section 2 constitutes the main part of the paper and gives an analysis of the simple algorithm outlined above for the case that hash functions are truly random and that no element is deleted and later inserted again.[2]

Section 3 complements this analysis by experiments that indicate that Cuckoo Hashing is even better in practical situations. For example, at $d = 4$, we can achieve 97 % space utilization and at 90 % space utilization, insertion requires only about 20 memory probes on the average, i.e., only about a factor two more than uniform hashing.

In Section 4 we present *Filter Hashing*, an alternative to d-ary Cuckoo Hashing that uses *explicit* constant time evaluable hash functions (polynomial hash functions of degree $O(\ln \frac{1}{\epsilon})$). It has the same performance as d-ary Cuckoo Hashing except that it uses $d = O(\ln^2 \frac{1}{\epsilon})$ probes for an access in the worst case.

A novel feature of both d-ary Cuckoo Hashing (in the variant presented in Section 3) and Filter Hashing is that we use hash tables having size only a fraction of the number of elements hashed to them. This means that high space utilization is ensured, even though there is only one possible location for an element in each table. Traditional hashing schemes use large hash tables where good space utilization is achieved by having many possible locations for each element.

---

[2] For theoretical purposes, the restriction on deletions is easily overcome by just *marking* deleted elements, and only removing them when periodically rebuilding the hash table with new hash functions.

1.1 Related Work

Space efficient dictionaries. A *dictionary* is a data structure that stores a set of elements, and associates some piece of information with each element. Given an element, a dictionary can look up whether it is in the set, and if so, return its associated information. Usually elements come from some universe of bounded size. If the universe has size m, the information theoretical lower bound on the number of bits needed to represent a set of n elements (without associated information) is $B = n \log(em/n) - \Theta(n^2/m) - O(\log n)$. This is roughly $n \log n$ bits less than, say, a sorted list of elements. If $\log m$ is large compared to $\log n$, using n words of $\log m$ bits is close to optimal.

A number of papers have given data structures for storing sets in near-optimal space, while supporting efficient lookups of elements, and other operations. Cleary [4] showed how to implement a variant of linear probing in space $(1+\epsilon)B + O(n)$ bits, under the assumption that a truly random permutation on the key space is available. The expected average time for lookups and insertions is $O(1/\epsilon^2)$, as in ordinary linear probing. A space usage of $B + o(n) + O(\log \log m)$ bits was obtained in [15] for the *static* case. Both these data structures support associated information using essentially optimal additional space.

Other works have focused on dictionaries *without* associated information. Brodnik and Munro [3] achieve space $O(B)$ in a dictionary that has worst case constant lookup time and amortized expected constant time for insertions and deletions. The space usage was recently improved to $B + o(B)$ bits by Raman and Rao [17]. Since these data structures are not based on hash tables, it is not clear that they extend to support associated information. In fact, Raman and Rao mention this extension as a goal of future research.

Our generalization of Cuckoo Hashing uses a hash table with $(1+\epsilon)n$ entries of $\log m$ bits. As we use a hash table, it is trivial to store associated information along with elements. The time analysis depends on the hash functions used being truly random. For many practical hash functions, the space usage can be decreased to $(1+\epsilon)B + O(n)$ bits using *quotienting* (as in [4,15]). Thus, our scheme can be seen as an improvement of the result of Cleary to worst case lookup bounds (even having a better dependence on ϵ than his average case bounds). However, there remains a gap between our experimental results for insertion time and our theoretical upper bound, which does not beat Cleary's.

Open addressing schemes. Cuckoo Hashing falls into the class of open addressing schemes, as it places keys in a hash table according to a sequence of hash functions. The worst case $O(\ln(1/\epsilon))$ bound on lookup time matches the *average* case bound of classical open addressing schemes like double hashing. Yao [22] showed that this bound is the best possible among all open addressing schemes that do not move elements around in the table. A number of hashing schemes move elements around in order to improve or remove the dependence on ϵ in the average lookup time [1,9,11,12,18].

The *worst case* retrieval cost of the classical open addressing schemes is $\Omega(\log n)$. Bounding the worst case retrieval cost in open addressing schemes was investigated by Rivest [18], who gave a polynomial time algorithm for arranging keys so as to minimize the worst case lookup time. However, no bound was shown on the expected worst case lookup time achieved. Rivest also considered the dynamic case, but the proposed insertion algorithm was only shown to be expected constant time for low load factors (in particular, nothing was shown for $\epsilon \leq 1$).

Matchings in random graphs. Our analysis uses ideas from two seemingly unrelated areas that are connected to Cuckoo Hashing by the fact that all three problems can be understood as finding matchings in some kind of random bipartite graphs.

The proof that space consumption is low is similar in structure to the result in [21,20] that two hash functions suffice to map n elements (disk blocks) to D places (disks) such that no disk gets more than $\lceil n/D \rceil +1$ blocks. The proof details are quite different however. In particular, we derive an analytic expression for the relation between ϵ and d. Similar calculations may help to develop an analytical relation that explains for which values of n and D the "+1" in $\lceil n/D \rceil +1$ can be dropped. In [20] this relation was only tabulated for small values of n/D.

The analysis of insertion time uses expansion properties of random bipartite graphs. Motwani [14] uses expansion properties to show that the algorithm by Hopcroft and Karp [10] finds perfect matchings in random bipartite graphs with $m > n \ln n$ edges in time $O(m \log n / \log \log n)$. He shows an $O(m \log n / \log d)$ bound for the *d-out* model of random bipartite graphs, where all nodes are constrained to have degree at least $d \geq 4$. Our analysis of insertion can be understood as an analysis of a simple incremental algorithm for finding perfect matchings in a random bipartite graph where n nodes on the left side are constrained to have constant degree d whereas there are $(1 + \epsilon)\, n$ nodes on the right side without a constraint on the degree. We feel that this is a more natural model for sparse graphs than the *d-out* model because there seem to be many applications where there is an asymmetry between the two node sets and where it is unrealistic to assume a lower bound on the degree of a right node (e.g., [21,19,20]). Under these conditions we get *linear* run time even for very sparse graphs using a very simple algorithm that has the additional advantage to allow incremental addition of nodes. The main new ingredient in our analysis is that besides expansion properties, we also prove *shrinking properties* of nodes not reached by a BFS. An aspect that makes our proof more difficult than the case in [14] is that our graphs have weaker expansion properties because they are less dense (or less regular for the *d-out* model).

2   *d*-Ary Cuckoo Hashing

A natural way to define and analyze *d*-ary Cuckoo Hashing is through matchings in asymmetric bipartite graphs. In particular, the elements can be thought of as the left vertices and the memory cells can be thought of as the right vertices of a bipartite graph $B(L, R, E)$. For *d*-ary Cuckoo Hashing, the number of right vertices is $(1 + \epsilon)$ times the number of left vertices. An edge connecting a left vertex to a right vertex indicates that the corresponding element can be stored at the corresponding memory cell. For d-ary Cuckoo Hashing, the edge set of the bipartite graph B is a random set determined by letting each left vertex select exactly d neighbors randomly and independently (with replacement) from the set of right vertices. Any one-to-one assignment of elements/left vertices to memory cells/right vertices forms a matching in B. Since every element is stored in some cell, this matching is L-perfect, i.e. it covers all the left vertices.

In the following, we only consider bipartite graphs resulting from d-ary Cuckoo Hashing, i.e. graphs $B(L, R, E)$ where $|L| = n$, $|R| = (1 + \epsilon) n$, and each left vertex has exactly d neighbors selected randomly and independently (with replacement) from R. We also assume that the left vertices arrive (along with their d random choices/edges) one-by-one in an arbitrary order and the insertion algorithm incrementally maintains an L-perfect matching in B.

Having fixed an L-perfect matching M, we can think of B as a directed graph, where all the edges of E are directed from left to right, except for the edges of M, which are directed from right to left. Hence, the matching M simply consists of the edges directed from right to left. In addition, the set of *free vertices* $F \subseteq R$ simply consists of the right vertices with no outgoing edges.

When a new left vertex v arrives, all its edges are considered as outgoing (i.e. directed from left to right), since v is not currently matched. Then, any directed path from v to F is an augmenting path for M, because if we reverse the directions of all the edges along such a path, we obtain a new matching M' which also covers v. The insertion algorithm we analyze always augments M along a shortest directed path from v to F. Such a path can be found by the equivalent of a Breadth First Search (BFS) in the directed version of B, which is implicitly represented by the d hash functions and the storage table. To ensure space efficiency, we restrict the number of vertices the BFS can visit to $o(n)$.

2.1   Existence of an L-Perfect Matching

We start by showing that for appropriately large values of d, d-ary Cuckoo Hashing leads to bipartite graphs that contain an L-perfect matching whp.

Lemma 1. *Given a constant $\epsilon \in (0, 1)$, for any integer $d \geq 2(1 + \epsilon) \ln(\frac{e}{\epsilon})$, the bipartite graph $B(L, R, E)$ contains a perfect matching with probability at least $1 - O(n^{4-2d})$.*

Proof Sketch. We apply Hall's Theorem and show that any subset of left vertices X has at least $|X|$ neighbors with probability at least $1 - O(n^{4-2d})$. □

By applying our analysis for particular values of ϵ, we obtain that if $\epsilon \geq 0.57$ and $d = 3$, if $\epsilon \geq 0.19$ and $d = 4$, and if $\epsilon \geq 0.078$ and $d = 5$, B contains an L-perfect matching whp. The experiments in Section 3 indicate that even smaller values of ϵ are possible.

In addition, we can show that this bound for d is essentially best possible.

Lemma 2. *If $d < (1 + \epsilon) \ln(1/\epsilon)$ then $B(L, R, E)$ does not contain a perfect matching whp.*

Proof Sketch. If $d < (1 + \epsilon) \ln(1/\epsilon)$, there are more than ϵn isolated right vertices whp. □

2.2   The Average Insertion Time of d-Ary Cuckoo Hashing

To avoid any dependencies among the random choices of a newly arrived left vertex and the current matching, we restrict our attention to the case where a left vertex that has been deleted from the hash table cannot be reinserted. In addition, we assume that the bipartite graph contains an L-perfect matching, which happens whp. according to Lemma 1.

Theorem 1. *For any positive $\epsilon \leq 1/5$ and $d \geq 5 + 3\ln(1/\epsilon)$, the incremental algorithm that augments along a shortest augmenting path needs $(1/\epsilon)^{O(\log d)}$ expected time per left vertex/element to maintain an L-perfect matching in B.*

The proof of Theorem 1 consists of three parts. We first prove that the number of vertices having a directed path to the set of free vertices F of length at most λ grows exponentially with λ, whp., until almost half of the vertices have been reached. We call this the *expansion property*. We next prove that for the remaining right vertices, the number of right vertices having no path to F of length at most λ decreases exponentially with λ, whp. We call this the *shrinking property*. The proofs of both the expansion property and the shrinking property are based on the fact that for appropriate choices of d, d-ary Cuckoo Hashing results in bipartite graphs that are good expanders, whp. Finally, we put the expansion property and the shrinking property together to show that the expected insertion time per element is constant. The same argument implies that the number of vertices visited by the BFS is $o(n)$ whp.

Proof of Theorem 1. In the proof of Theorem 1, we are interested in bounding the distance (respecting the edge directions) of matched right vertices from F, because a shortest directed path from a matched right vertex u to F can be used for the insertion of a newly arrived left vertex which has an edge incident to u.

We measure the distance of a vertex v from F by only accounting for the number of left to right edges (*free edges* for short), or, equivalently, the number of left vertices appearing in a shortest path (respecting the edge directions) from v to F. We sometimes refer to this distance as the *augmentation distance* of v. Notice that the augmentation distance depends on the current matching M. We use the augmentation distance of a vertex v to bound the complexity of searching for a shortest directed path from v to F.

The Expansion Property. We first prove that if d is chosen appropriately large, for any small constant δ, any set of right vertices Y that is not too close to size $n/2$ has at least $(1 + \delta)|Y|$ neighbors in L whp. (cf. Lemma 3). This implies that the number of right vertices at augmentation distance at most λ is at least $\epsilon(1 + (1 + \delta)^\lambda)n$, as long as λ is so small that this number does not exceed $n/2$ (cf. Lemma 4).

Lemma 3. *Given a constant $\epsilon \in (0, 1/4)$, let δ be any positive constant not exceeding $\frac{4(1-4\epsilon)}{1+4\epsilon}$, and let d be any integer such that $d \geq 3 + 2\delta + 2\epsilon(1 + \delta) + (2 + \delta)\epsilon \ln\left(\frac{1+\epsilon}{\epsilon}\right) / \ln(1 + \epsilon)$. Then, any set of right vertices Y of cardinality $\epsilon n \leq |Y| \leq \frac{n}{2(1+\delta)}$ has at least $(1 + \delta)|Y|$ neighbors with probability $1 - 2^{-\Omega(n)}$.*

Proof Sketch. We show that any subset of left vertices X, $\frac{n}{2} \leq |X| \leq (1 - (1 + \delta)\epsilon)n$, has at least $(1 + \epsilon)n - \frac{n-|X|}{1+\delta}$ neighbors with probability $1 - 2^{-\Omega(n)}$, and that any such bipartite graph satisfies the conclusion of the lemma.   □

The following lemma, which can be proven by induction on λ, concludes the proof of the expansion property.

Lemma 4. *Given a constant* $\epsilon \in (0, 1/4)$, *let* δ *be any positive constant not exceeding* $\frac{4(1-4\epsilon)}{1+4\epsilon}$, *and let* $B(L, R, E)$ *be any bipartite graph satisfying the conclusion of Lemma 3. Then, for any integer* λ, $1 \leq \lambda \leq \log_{(1+\delta)}\left(\frac{1}{2\epsilon}\right)$, *the number of right vertices at augmentation distance at most* λ *is at least* $\epsilon(1 + (1 + \delta)^{\lambda})n$.

For $\epsilon \leq 1/5$ we can take $\delta = 1/3$. Then, using the fact that $\ln(1+\epsilon) > \epsilon - \epsilon^2/2$, the requirement of Lemma 3 on d can be seen to be satisfied if $d \geq 5 + 3\ln(1/\epsilon)$.

The Shrinking Property. By the expansion property, there is a set consisting of nearly half of the right vertices that have augmentation distance smaller than some number λ^*. The second thing we show is that for any constant $\gamma > 0$, any set of left vertices X, $|X| \leq \frac{n}{2(1+\gamma)}$, has at least $(1+\gamma)|X|$ neighbors in R whp. This implies that the number of right vertices being at augmentation distance greater than $\lambda^* + \lambda$ decreases exponentially with λ (cf. Lemma 6).

Lemma 5. *Let* γ *be any positive constant, and let* $d \geq (1 + \log e)(2 + \gamma) + \log(1 + \gamma)$ *be an integer. Then, any set of right vertices* Y, $|Y| \geq (1/2 + \epsilon)n$, *has at least* $n - \frac{(1+\epsilon)n - |Y|}{(1+\gamma)}$ *neighbors, with probability* $1 - O(n^{3+\gamma-d})$.

Proof Sketch. We show that any subset of left vertices X, $|X| \leq \frac{n}{2(1+\gamma)}$, has at least $(1+\gamma)|X|$ neighbors with probability $1 - O(n^{3+\gamma-d})$, and that any bipartite graph with this property satisfies the conclusion of the lemma.   □

The following lemma can be proven by induction on λ.

Lemma 6. *Given a constant* $\epsilon \in (0, 1/4)$, *let* δ *be any positive constant not exceeding* $\frac{4(1-4\epsilon)}{1+4\epsilon}$ *and let* γ *be any positive constant. In addition, let* $B(L, R, E)$ *be any bipartite graph satisfying the conclusions of both Lemma 3 and Lemma 5 and let* $\lambda^* = \left\lceil \log_{(1+\delta)}\left(\frac{1}{2\epsilon}\right) \right\rceil$. *Then, for any integer* $\lambda \geq 0$, *the number of right vertices at augmentation distance greater than* $\lambda + \lambda^*$ *is at most* $\frac{n}{2(1+\gamma)^{\lambda}}$.

Bounding the Average Insertion Time. We can now put everything together to bound the average insertion time of d-ary Cuckoo Hashing, thus concluding the proof of Theorem 1.

Let T_v be the random variable denoting the time required by the insertion algorithm to add a newly arrived left vertex v to the current matching. In addition, let Y_λ denote the set of right vertices at augmentation distance at most λ. To bound v's expected insertion time, we use the assumption that the current matching and the sets Y_λ do not depend on the random choices of v.

At first we assume that the bipartite graph B satisfies the conclusions of both Lemma 3 and Lemma 5. Also, assume for now that no rehash is carried out if there are too many nodes in the BFS.

If at least one of the d neighbors of v is at augmentation distance at most λ, an augmenting path starting at v can be found in time $O(d^{\lambda+1})$. Therefore, for any integer $\lambda \geq 0$, with probability at least $1 - (1 - \frac{|Y_\lambda|}{(1+\epsilon)n})^d$, T_v is $O(d^{\lambda+1})$. Hence, the expectation of T_v can be bounded by

$$\mathbb{E}[T_v] = \sum_{t=1}^{\infty} \Pr[T_v \geq t]$$

$$\leq O(d) + \sum_{\lambda=0}^{\infty} O(d^{\lambda+2}) \left(1 - \frac{|Y_\lambda|}{(1+\epsilon)n} \right)^d$$

$$\leq O(d^{\lambda^*+2}) + \frac{d^{\lambda^*+2}}{2^d} \sum_{\lambda=0}^{\infty} \left(\frac{d}{(1+\gamma)^d} \right)^{\lambda},$$

where the last inequality holds because for any $\lambda \geq 0$, $1 - \frac{|Y_{\lambda^*+\lambda}|}{(1+\epsilon)n} \leq \frac{1}{2(1+\gamma)^\lambda}$, by Lemma 6. For $\gamma = 0.55$ and any $d \geq 6$, $\frac{d}{(1+\gamma)^d} < \frac{1}{2}$ and the expectation of T_v can be bounded by $d^{O(\lambda^*)}$.

On the other hand, for such γ and d, the bipartite graph B does not have the desired properties (i.e. it violates the conclusions of Lemma 3 or Lemma 5) with probability $O(n^{-2})$. Since T_v is always bounded by $O(n)$, this low probability event has a negligible contribution to the expectation of T_v. Similarly, the expected contribution of a rehash due to too many nodes in the BFS can be shown to be negligible. $\square$

3   Experiments

Our theoretical analysis is not tight with respect to the constant factors and lower order terms in the relation between the worst case number of probes d and the waste of space ϵn. The analysis is even less accurate with respect to the insertion time. Since these quantities are important to judge how practical d-ary Cuckoo Hashing might be, we designed an experiment that can partially fill this gap. We decided to focus on a variant that looks promising in practice: We use d separate tables of size $(1 + \epsilon)n/d$ because then it is not necessary to reevaluate the hash function that led to the old position of an element to be moved. Insertion uses a random walk, i.e., an element to be allocated randomly picks one of its d choices even if the space is occupied. In the latter case, the displaced element randomly picks one of its $d-1$ remaining choices, etc., until a free table entry is found. The random walk insertion saves us some bookkeeping that would be needed for insertion by BFS. Figure 1 shows the average number of probes needed for insertion as a function of the space utilization $1/(1+\epsilon)$ for $d \in \{2, 3, 4, 5\}$. Since $1/\epsilon$ is a lower bound, the y-axis is scaled by ϵ. We see that all schemes are close to the insertion time $1/\epsilon$ for small utilization and grow quickly as they approach a capacity threshold that depends on d. Increasing d strictly decreases expected insertion time so that we get clear trade-off between worst case access time guarantees and average insertion time.

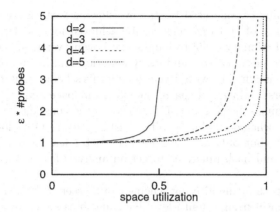

Fig. 1. Scaled average number of memory probes for insertion into a d-ary Cuckoo Hash table with 100 000 entries as a function of the memory utilization $n/10^5$ ($\epsilon = 1 - n/10^5$). Starting from $n = 1000 \cdot k$ ($k \in \{1, \ldots, 100\}$), a random element is removed and a new random element is inserted. This is repeated 1000 times for each of 100 independent runs. Hash functions are full lookup tables filled with random elements generated using [13]. The curves stop when any insertion fails after 1000 probes.

The maximum space utilization approaches one quickly as d is incremented. The observed thresholds were at 49 % for $d = 2$, 91 % at $d = 3$, 97 % at $d = 4$, and 99 % at $d = 5$.

4 Filter Hashing

In this section, we describe and analyze *Filter Hashing*, a simple hashing scheme with worst case constant lookup time, that can be used in combination with essentially any other hashing scheme to improve the space efficiency of the latter. More precisely, Filter Hashing space efficiently stores almost all elements of a set. The remaining elements can then be stored using a less space efficient hashing scheme, e.g., [5].

To explain Filter Hashing, we again switch to the terminology of bipartite graphs. For a parameter γ, $0 < \gamma < 1$ we split the right vertices into $d = \Theta(\ln^2(1/\gamma))$ parts, called *layers*, of total size at most n. Each left vertex is associated with exactly one neighbor in each of the d layers, using hash functions as described below. A newly arrived vertex is always matched to an unmatched neighbor in the layer with the *smallest possible* number. The name filter hashing comes from the analogy of a particle (hash table element / left vertex) passing through a cascade of d filters (layers). If all the neighbors in the d layers have been matched, the vertex is not stored, i.e., it is left to the hashing scheme handling such "overflowing" vertices. We will show that this happens to at most γn elements whp.

If the hashing scheme used for the overflowing vertices uses linear space, a total space usage of $(1 + \epsilon)n$ cells can be achieved for $\gamma = \Omega(\epsilon)$. For example, if we use the dictionary of [5] to handle overflowing vertices, the space used for overflowing vertices is $O(\gamma n)$, and every insertion and lookup of an overflowing vertex takes constant time whp. Even though this scheme exhibits relatively high constant factors in time and space, the effect on space and average time of the combined hashing scheme is small if we choose the constant γ to be small.

A hashing scheme similar to filter hashing, using $O(\log \log n)$ layers, was proposed in [2], but only analyzed for load factor less than $1/2$. Here, we use stronger tools and hash functions to get an analysis for load factors arbitrarily close to 1.

What happens in the filtering scheme can be seen as letting the left vertices decide their mates using a multi-level balls and bins scenario, until the number of unmatched left vertices becomes small enough. The scheme gives a trade-off between the number of layers and the fraction γ of overflowing vertices.

We proceed to describe precisely the bipartite graph $B(L, R, E)$ used for the scheme, where $|L| = |R| = n$. We partition R into d layers R_i, $i = 1 \ldots d$, where $d = \lceil \ln^2(4/\gamma) \rceil$ and $|R_i| = \left\lfloor \frac{n}{\ln(4/\gamma)} \left(1 - \frac{1}{\ln(4/\gamma)}\right)^{i-1} \right\rfloor$. Suppose that $L \subseteq \{1, \ldots, m\}$ for some integer m, or, equivalently, that we have some way of mapping each vertex to a unique integer in $\{1, \ldots, m\}$. The edges connecting a vertex $v \in L$ to R_i, for $i = 1, \ldots, d$, are given by function values on v of the hash functions

$$h_i(x) = (\sum_{j=0}^{t} a_{ij} x^j \bmod p) \bmod |R_i| \tag{1}$$

where $t = 12 \lceil \ln(4/\gamma) + 1 \rceil$, $p > mn$ is a prime number and the a_{ij} are randomly and independently chosen from $\{0, \ldots, p - 1\}$.

For n larger than a suitable constant (depending on d), the total size $\sum_{i=1}^{d} |R_i|$ of the d layers is in the range

$$\left[\sum_{i=1}^{d} \frac{n}{\ln(4/\gamma)} \left(1 - \frac{1}{\ln(4/\gamma)}\right)^{i-1} - d \; ; \; \sum_{i=1}^{\infty} \frac{n}{\ln(4/\gamma)} \left(1 - \frac{1}{\ln(4/\gamma)}\right)^{i-1} \right]$$

$$= \left[n\left(1 - \left(1 - \frac{1}{\ln(4/\gamma)}\right)^{d}\right) - d \; ; \; n \right] \subseteq \left[(1 - \frac{\gamma}{2})n \; ; \; n\right]$$

From the description of filter hashing, it is straightforward that the worst case insertion time and the worst case access time are at most d. In the following, we prove that at most γn left vertices overflow whp., and that the average time for a successful search is $O(\ln(1/\gamma))$. Both these results are implied by the following lemma.

Lemma 7. *For any constant γ, $0 < \gamma < 1$, for $d = \lceil \ln^2(4/\gamma) \rceil$ and n larger than a suitable constant, the number of left vertices matched to vertices in R_i is at least $(1 - \gamma/2)|R_i|$ for $i = 1, \ldots, d$ with probability $1 - O\left(\left(\frac{1}{\gamma}\right)^{O(\log \log(\frac{1}{\gamma}))} \frac{1}{n}\right)$.*

Proof Sketch. We use tools from [6] to prove that each of the layers has at least a fraction $(1 - \gamma/2)$ of its vertices matched to left vertices with probability $1 - O((\frac{1}{\gamma})^{O(\log\log(\frac{1}{\gamma}))}\frac{1}{n})$. As there are $O(\ln^2(1/\gamma))$ layers, the probability that this happens for all layers is also $1 - O((\frac{1}{\gamma})^{O(\log\log(\frac{1}{\gamma}))}\frac{1}{n})$.   □

To conclude, there are at most $\frac{\gamma}{2}n$ of the n right side vertices that are not part of $R_1, \ldots, R_d$, and with probability $1 - O(\frac{1}{n})$ there are at most $\frac{\gamma}{2}n$ vertices in the layers that are not matched. Thus, with probability $1 - O(\frac{1}{n})$ no more than γn vertices overflow.

The expected average time for a successful search can be bounded as follows. The number of elements with search time $i \leq d$ is at most $|R_i|$, and the probability that a random left vertex overflows is at most $\gamma + O(\frac{1}{n})$, i.e., the expected total search time for all elements is bounded by:

$$(\gamma + O(\tfrac{1}{n}))\,nd + \sum_{i=1}^{d}|R_i|i \leq (\gamma + O(\tfrac{1}{n}))\left\lceil \ln^2(\tfrac{4}{\gamma}) \right\rceil n + \frac{n}{\ln(\frac{4}{\gamma})}\sum_{i=0}^{\infty}\left(1 - \frac{1}{\ln(\frac{4}{\gamma})}\right)^i i$$

$$= O(n\ln(\tfrac{4}{\gamma})) \ .$$

The expected time to perform a rehash in case too many elements overflow is $O(\ln(1/\gamma)\,n)$. Since the probability that this happens for any particular insertion is $O((\frac{1}{\gamma})^{O(\log\log(\frac{1}{\gamma}))}\frac{1}{n})$, the expected cost of rehashing for each insertion is $(\frac{1}{\gamma})^{O(\log\log(\frac{1}{\gamma}))}$. Rehashes caused by the total number of elements (including those marked deleted) exceeding n have a cost of $O(\ln(\frac{1}{\gamma})/\gamma)$ per insertion and deletion, which is negligible.

5   Conclusions and Open Problems

From a practical point of view, d-ary Cuckoo Hashing seems a very advantageous approach to space efficient hash tables with worst case constant access time. Both worst case access time and average insertion time are very good. It also seems that one could make *average* access time quite small. A wide spectrum of algorithms could be tried out from maintaining an optimal placement of elements (via minimum weight bipartite matching) to simple and fast heuristics.

Theoretically, many open questions remain. Can we work with (practical) hash functions that can be evaluated in constant time? What are tight (high probability) bounds for the insertion time?

Filter hashing is inferior in practice to d-ary Cuckoo Hashing but it might have specialized applications. For example, it could be used as a *lossy* hash table with worst case constant *insertion* time. This might make sense in real time applications where delays are not acceptable whereas losing some entries might be tolerable, e.g., for gathering statistic information on the system. In this context, it would be theoretically and practically interesting to give performance guarantees for simpler hash functions.

Acknowledgement. The second author would like to thank Martin Dietzfelbinger for discussions on polynomial hash functions.

References

1. R. P. Brent. Reducing the retrieval time of scatter storage techniques. *Communications of the ACM*, 16(2):105–109, 1973.
2. A. Z. Broder and A. R. Karlin. Multilevel adaptive hashing. In *Proc. 1st Annual ACM-SIAM Symposium on Discrete Algorithms*, pages 43–53. ACM Press, 2000.
3. A. Brodnik and J. I. Munro. Membership in constant time and almost-minimum space. *SIAM J. Comput.*, 28(5):1627–1640, 1999.
4. J. G. Cleary. Compact hash tables using bidirectional linear probing. *IEEE Transactions on Computers*, C-33(9):828–834, September 1984.
5. M. Dietzfelbinger, J. Gil, Y. Matias, and N. Pippenger. Polynomial hash functions are reliable (extended abstract). In *Proc. 19th International Colloquium on Automata, Languages and Programming*, volume 623 of *LNCS*, pages 235–246. Springer-Verlag, 1992.
6. M. Dietzfelbinger and T. Hagerup. Simple Minimal Perfect Hashing in Less Space In *Proc. 9th European Symposium on Algorithms*, volume 2161 of *LNCS*, pages 109–120. Springer-Verlag, 2001.
7. M. Dietzfelbinger, A. Karlin, K. Mehlhorn, F. Meyer auf der Heide, H. Rohnert, and R. E. Tarjan. Dynamic perfect hashing: Upper and lower bounds. *SIAM J. Comput.*, 23(4):738–761, 1994.
8. M. L. Fredman, J. Komlós, and E. Szemerédi. Storing a sparse table with $O(1)$ worst case access time. *J. ACM*, 31(3):538–544, 1984.
9. G. H. Gonnet and J. I. Munro. Efficient ordering of hash tables. *SIAM J. Comput.*, 8(3):463–478, 1979.
10. J. E. Hopcroft and R. M. Karp. An $O(n^{5/2})$ algorithm for maximum matchings in bipartite graphs. *SIAM J. Comput.*, 2:225–231, 1973.
11. J. A. T. Maddison. Fast lookup in hash tables with direct rehashing. *The Computer Journal*, 23(2):188–189, May 1980.
12. E. G. Mallach. Scatter storage techniques: A uniform viewpoint and a method for reducing retrieval times. *The Computer Journal*, 20(2):137–140, May 1977.
13. M. Matsumoto and T. Nishimura. Mersenne twister: A 623-dimensionally equidistributed uniform pseudo-random number generator. *ACMTMCS: ACM Transactions on Modeling and Computer Simulation*, 8:3–30, 1998.
14. R. Motwani. Average-case analysis of algorithms for matchings and related problems. *J. ACM*, 41(6):1329–1356, November 1994.
15. R. Pagh. Low redundancy in static dictionaries with constant query time. *SIAM J. Comput.*, 31(2):353–363, 2001.
16. R. Pagh and F. F. Rodler. Cuckoo hashing. In *Proc. 9th European Symposium on Algorithms*, volume 2161 of *LNCS*, pages 121–133. Springer-Verlag, 2001.
17. R. Raman and S. Srinivasa Rao. Dynamic dictionaries and trees in near-minimum space. Manuscript, 2002.
18. R. L. Rivest. Optimal arrangement of keys in a hash table. *J. ACM*, 25(2):200–209, 1978.
19. P. Sanders. Asynchronous scheduling of redundant disk arrays. In *12th ACM Symposium on Parallel Algorithms and Architectures*, pages 89–98, 2000.
20. P. Sanders. Reconciling simplicity and realism in parallel disk models. In *12th ACM-SIAM Symposium on Discrete Algorithms*, pages 67–76, 2001.
21. P. Sanders, S. Egner, and J. Korst. Fast concurrent access to parallel disks. In *11th ACM-SIAM Symposium on Discrete Algorithms*, pages 849–858, 2000.
22. A. Yao. Uniform hashing is optimal. *J. ACM*, 32(3):687–693, 1985.

Randomized Jumplists: A Jump-and-Walk Dictionary Data Structure*

Hervé Brönnimann, Frédéric Cazals, and Marianne Durand

[1] Polytechnic University, CIS, Six Metrotech, Brooklyn NY 11201, USA;
hbr@poly.edu
[2] INRIA Sophia-Antipolis, Projet Prisme, F-06902 Sophia-Antipolis, France;
Frederic.Cazals@sophia.inria.fr
[3] INRIA Rocquencourt, Projet Algo, F-78153 Le Chesnay, France;
Marianne.Durand@inria.fr

Abstract. This paper presents a data structure providing the usual dictionary operations, i.e. CONTAINS, INSERT, DELETE. This data structure named *Jumplist* is a linked list whose nodes are endowed with an additional pointer, the so-called jump pointer. Algorithms on jumplists are based on the *jump-and-walk* strategy: whenever possible use to the jump pointer to speed up the search, and walk along the list otherwise. The main features of jumplists are the following. They perform within a constant factor of binary search trees. Randomization makes their dynamic maintenance easy. Jumplists are a compact data structure since they provide rank-based operations and forward iterators at a cost of three pointers/integers per node. Jumplists are trivially built in linear time from sorted linked lists.

Keywords. Dictionary data structures. Searching and sorting. Randomization. Asymptotic analysis.

1 Introduction

Dictionaries, Binary Search Trees (BST) and alternatives. Dictionaries and related data structures have a long standing history in theoretical computer science. These data structures were originally designed so as to organize pieces of information and provide efficient storage and access functions. But in addition to the standard CONTAINS, INSERT and DELETE operations, several other functionalities were soon felt necessary.

In order to accommodate divide-and-conquer algorithms, split and merge operations are mandated. For priority queues, access to the minimum and/or maximum must be supported. For applications involving order statistics, rank-based operations must be provided. Additionally, the data structure may be requested to incorporate knowledge about the data processed —e.g. random or sorted keys.

---

* Extended abstract. Due to space limitations, all proofs have been omitted. Refer to [1] for the full paper.

H. Alt and M. Habib (Eds.): STACS 2003, LNCS 2607, pp. 283–294, 2003.
© Springer-Verlag Berlin Heidelberg 2003

This variety of constraints lead to the development of a large number of data structures. The very first ones were randomly grown BST [15,14] as well as deterministic balanced BST [26,14,7]. Since then, solutions more geared to provably good amortized or randomized performance were proposed. Splay trees [24], treaps [5], skip lists [18] and more recently randomized BST [16] fall into this category. An important feature of the randomized data structures —in particular randomized BST and skip lists— is their ease of implementation. Skip lists are particularly illustrative since the deterministic versions are substantially more difficult to code [17].

This paper presents a data structure providing the usual dictionary operations, i.e. CONTAINS, INSERT and DELETE. This data structure named *jumplist* is an ordered list whose nodes are endowed with an additional pointer, the so-called jump pointer. Algorithms on jumplists are based on the *jump-and-walk* strategy: whenever possible use to the jump pointer to speed-up the search, and walk along the list otherwise.

Like skip lists, we use jump pointers to speed-up searches. Similarly to skip lists too, the profile of the data structure does not depend on the ordering of the keys processed. Instead, a jumplist depends upon random tosses independent from the keys processed. Unlike skip lists, however, jumplists do not have a hierarchy of jump pointers. In particular, every node contains exactly two pointers, so the storage required is linear —as opposed to expected linear storage for skip lists. The data structures closest to jumplists in addition to skip lists are randomized BST. Similarly to skip lists or jumplists, the profile of randomized BST is independent from the sequence of keys processed.

The main features of jumplists are the following. Their performance are within a constant factor of optimal BST. Randomization makes their dynamic maintenance very easy. Jumplists are a compact data structure since they provide rank-based operations and forward iterators at a cost of three pointers/integers per node. Jumplists are trivially built in linear time from sorted linked lists.

Overview. This paper is organized as follows. The jumplist data structure and its search algorithms are described in Section 2. Section 3 analyzes the expected performance of the data structure. Dynamic maintenance of jumplists is presented in Section 4.

2   Jumplist: Data Structure and Searching Algorithms

2.1   The Data Structure

The jumplist is stored as a singly or doubly connected list, with next[x] pointing forward and prev[x] backward. The reverse pointers are not needed, except for backward traversal. If bidirectional traversal is not supported, they can be omitted from the presentation. The list is circularly connected, and the successor (resp. predecessor) of the header is the first (resp. last) element of the list, or the header in either case if the list is empty. See Figure 1. Following standard implementation technique, the list always has a node header[L] which contains no value, called its *header*, and which comes before all the other nodes. This facilitates insertions in a singly-linked list.

To each node is associated a *key*, and we assume that the list is sorted with respect to the keys. It is convenient to treat the header as the first node of the jumplist and give it a key of $-\infty$, especially for expressing the invariants and in the proofs. In this way, there is always a node with key less than k, for any k. We will be careful to state our algorithms such that the header key is never referenced. We may introduce for each node x an interval [key[x], key[next[x]]] which corresponds to the keys that are *not* present in the jumplist. Thus if n represents the number of nodes, there are $n - 1$ keys. When all $n - 1$ keys are distinct, there are n intervals defined by the keys and by the header.

We denote by $x \prec y$ the relation induced by the order of the list (beginning at the first element and ending at the header). If all nodes have distinct keys, this relation is the same as that induced by the keys: $x \prec y$ iff key[x] < key[y], and $x \preceq y$ iff key[x] $\leq$ key[y]. When some keys are identical, it is inefficient to test whether $x \prec y$ when key[x] = key[y] (one must basically traverse the list).

Traversal of jumplists, unlike BST, is extremely simple: simply follow the list pointers.

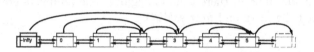

Fig. 1. A possible jumplist data structure over 6 elements. The last node (in dashed lines) is identical to the first one on this representation.

In addition to the list pointers, each node also has a pointer jump[x] which points to a successor of x in the list. We refer to the pair $(x, \text{jump}[x])$ as an *arch*, and to the arch starting at header[L] as the *fundamental arch*. As mentioned earlier, the jump pointers have to satisfy $x \prec \text{jump}[x]$ for every node x not equal to the last node, as well as the non-crossing condition: for any pair of nodes x, y, we cannot have $x \prec y \prec \text{jump}[x] \prec \text{jump}[y]$. Thus if $x \prec y$, then either $y \prec \text{jump}[y] \prec \text{jump}[x]$ (*strictly nested*), jump[x] $\preceq y \prec \text{jump}[y]$ (*semi-disjoint*), or $y \prec \text{jump}[y] = \text{jump}[x]$ if jump[y] = next[y] (*exceptional pointer*). In order to close the loop, we also require that the jump pointer of the last element points back to the header. This last pointer is also called exceptional.

Remark. Exceptional pointers are necessary because otherwise we could not set the jump pointers of $y = \text{prev}[\text{jump}[x]]$, for nodes x such that next[x] $\prec$ jump[x]. The value jump[y] = next[y] is the only one that does not break a non-strictly nested condition. We could also get rid of exceptional pointers by putting jump[x] = NIL in that case. This actually complicates the algorithms which have to guard against null jump pointers. Since it will be clear during the discussion of the search algorithm that these exceptional jump pointers are never followed anyway, it only hurts to set them to NIL. Therefore, we choose to set them to next[x] in the data structure, so that the jump pointers are never NIL. This also has the advantage that the search can be started from anywhere, not just from the header, without changing the search algorithm, because in that case the exceptional jump pointers are followed automatically.

Remark. The list could be singly or doubly linked, although the version we present here is singly linked and is sufficient to express all of our algorithms. Moreover, the predecessor can be searched efficiently in a singly-linked jumplist, unlike singly-linked lists. All in all, we therefore have the following invariants:

I1 (LIST) L is a singly- or doubly-connected list, with header[L] a special node whose key is $-\infty$, key[x] $\leq$ key[next[x]] for every node $x \neq$ header[L].

I2 (Jump forward) $x \prec$ jump[x] for every node, except the last node y for which jump[y] = next[y] = header[L].

I3 (Non-crossing jumps) for any two nodes $x \prec y$, either $x \prec y \prec$ jump[y] $\prec$ jump[x], or jump[x] $\preceq y$, or jump[y] = jump[x] = next[y].

As shown in [1] it is possible to verify the invariant in $O(n)$ time.

Let C be a jumplist node and let J and N be the nodes pointed by its jump and next pointers. The jumplist rooted at C will be denoted by the triple (C, J, N) or just by C if there is no ambiguity. The jumplists pointed by J and N are called the *jump* and *next sublists*.

2.2 Randomized Jumplists

Following the randomized BST of Martinez and Roura [16], we define a randomized jumplist as a jumplist in which the jump pointer of the header takes any value in the list, and the jump and next sublists recursively have the randomized property.

In order to construct randomized jumplists, we need to augment the nodes with a size information. Each jumplist node is endowed with two fields jsize and nsize corresponding to the sizes of the two sublists. Thus jsize[C] = size(J) and nsize[C] = size(N), and for a jumplist rooted at C, we have

$$\text{size}(C) = 1 + \text{jsize}[C] + \text{nsize}[C].$$

Remark. The size of the jumplist therefore counts the header. If only the number of keys stored in the jumplist is desired, then subtract one.

2.3 Searching

As already pointed out, the basic search algorithm is *jump-and-walk*: follow the jump pointers if you can, otherwise continue with the linear search. Note that if two arches $(x, \text{jump}[x])$ and $(y, \text{jump}[y])$ crossed, i.e. $x \prec y \prec \text{jump}[x] \prec \text{jump}[y]$, the second one would never be followed, since to reach y the key would be less than that of jump$[x]$, and by transitivity less than jump$[y]$. Note also that in order to be useful, arches should neither be too long (or they would never be used) nor too short (or else they would not speed up the search).

Several searching strategies are available depending on which pointers are tested first. Two search algorithms are presented on Figure 2. Our search algorithm is JUMPLIST-FIND-LAST-LESS-THAN-OR-EQUAL, and it returns the node y after which a key k should be inserted to preserve the list ordering. (Note that if there are many keys equal to k, this will return the last such key; inserting after that key preserves the list ordering as well, and the equal keys will be stored in the the order of their insertions). With this, testing if a key is present is easy: simply check whether key$[y] = k$.

To evaluate both strategies, let us compute the average and worst-case costs of accessing —at random with uniform probability— one of the n keys of a jumplist. If the jump and next pointers are tested in this order, accessing all the elements of the jump and next sublists, as well as the header respectively requires $n - i + 1$, $2(i - 2)$ and 3 comparisons (there is an extra comparison for testing if key$[y] = k$), if the jump pointer points to the ith element. This leads to an average of $(n + i)/n$ and a worst-case of 3. If the next and jump pointers are tested in this order, accessing any element always requires exactly 2 comparisons. We shall resort to the first solution which has a better average case.

JUMPLIST-FIND-LAST-LESS-THAN-OR-EQUAL(k)	JUMPLIST-FIND-LAST-LESS-THAN-OR-EQUAL(k)
1: $y \leftarrow$ header$[L]$	1: $y \leftarrow$ header$[L]$
2: ◁ *Current node, never more than* k	2: ◁ *Current node, never more than* k
3: **while** next$[y] \neq$ header$[L]$ **do**	3: **while** next$[y] \neq$ header$[L]$ **do**
4: **if** key$[$jump$[y]] \leq k$ **then**	4: **if** key$[$next$[y]] \leq k$ **then**
5: $y \leftarrow$ jump$[y]$	5: **if** key$[$jump$[y]] \leq k$ **then**
6: **else if** key$[$next$[y]] \leq k$ **then**	6: $y \leftarrow$ jump$[y]$
7: $y \leftarrow$ next$[y]$	7: **else**
8: **else**	8: $y \leftarrow$ next$[y]$
9: **return** y	9: **else**
10: **return** y	10: **return** y
	11: **return** y

Fig. 2. Two implementation of the search algorithm (left) with $1 + \frac{i}{n}$ comparisons per node on average. (right) with exactly 2 comparisons per node.

2.4 Finding Predecessor

If we change the inequalities in the algorithm JUMPLIST-FIND-LAST-LESS-THAN-OR-EQUAL to strict inequalities, then we obtain an algorithm JUMPLIST-FIND-LAST-LESS-THAN which can be used to find the predecessor of a node x with key k in the same time as a search. Thus, unlike linked lists, jumplists allow to trade off storage (one prev pointer) for predecessor access (constant vs. logarithmic time). One optimization for predecessor-finding is that, as soon as a node y such that $jump[y] = x$ is found, no more comparisons are needed: simply follow the jump pointers until $next[y] = x$.

2.5 Correspondence with Binary Search Trees

From a structural point of view and if one forgets the labels of the nodes of a jumplist —i.e. the keys, there is a straightforward bijection between a jumplist and a binary tree: the jump and next sublists respectively correspond to the left and right subtrees. The jump sublist however is never empty, so that the number of unlabeled jumplists of a given size is expected to be less than the n-th Catalan number.

If one compares the key stored into a jumplist node with those of the jump and next sublists, a jumplist is organized as a heap-ordered binary tree and not a BST —the root points to two larger keys. Stated differently, the keys are stored not in the binary search tree order, but in preorder. But this bijection does not help much for insertion and deletion algorithms since there is no equivalent of rotation in jumplists. Rotations as performed for BST are actually impossible since the head node must contain the smallest value.

As is known in the folklore, BST can be searched with at most one comparison per node, at the cost of one extra comparison on the average at the end of the search (simply remember the last potential equal node along the path—for which the strict comparison fails, and test it when reaching a leaf). Unfortunately, the same trick cannot work for jumplists: consider a jumplist storing the keys $[1..n]$ for which $jump[i] = n+1-i$ for each $i < (n+1)/2$. For the keys in $[1..(n+1)/2]$, no jump pointer will be followed, so storing the last node during the search for which a jump pointer failed will not disambiguate between these keys, and the comparisons with the next pointers seem necessary in that case.

3 Expected Performances of Randomized Jumplists

3.1 Internal Path Length, Expected Number of Comparisons, Jumplists Profiles

We start the analysis of jumplists with the Internal Path Length (IPL) statistic. Similarly to BST, the IPL is defined as the sum of the depths of the internal nodes of the structure, the depth of a node being the number of pointers traversed from the header of the list. The analysis uses the so-called level polynomial and its associated bivariate generating function.

Definition 1. *Let $s_{n,k}$ denote the expected number of nodes at depth k from the root in a randomized jumplist of size n. The level polynomial is defined by $S_n(u) = \sum_{k \geq 0} s_{n,k}\, u^k$. The associated bivariate generating function is defined by $S(z, u) = \sum_{n \geq 0} S_n(u) z^n$.*

The expected value of the IPL is given by $S'_n(1)$, and can easily be extracted from $S(z, u)$. The bivariate generating function can also be used to study the distribution of the nodes' depths.

Internal Path Length. Let γ stand for the Euler constant and Ei denote the exponential integral function [2]. The following shows that the leading term of IPL for jumplists matches that of randomized BST [25,15]:

Theorem 1. *The expected internal path length of a jumplist of size n is asymptotically equivalent to*

$$2n \ln n + n(-3 - 2Ei(1,1) + e^{-1}) + 2\ln n - 2Ei(1,1) + e^{-1} + 3 + o(1).$$

Profile of a jumplist. Theorem 1 shows that the expected depth of a node is $2 \ln n$. We can actually be more precise and exhibit the corresponding distribution. This distribution is Gaussian, and its variance matches that of BST [15]:

Theorem 2. *The random variable X_n defined by $P(X_n = k) = s_{n,k}/S_n(1)$ is asymptotically Gaussian, with average $2 \ln n$ and variance $2 \ln n$.*

Moreover the depth of the kth node is shown to be equivalent to $2 \log n$ when k, $n - k$, and n are proportional.

Expected number of comparisons along a search. The previous analysis investigates the number of pointers traversed from the root to the nodes of a jumplist. But the search cost of a key requires a more careful analysis since, according to the search algorithm JUMPLIST-FIND-LAST-LESS-THAN-OR-EQUAL of Figure 2(left), accessing the nodes of the jump and next sublists requires one and two comparisons, while accessing the root requires three comparisons. This accounts for the following

Definition 2. *Consider a jumplist C with n nodes, i.e. $n - 1$ used-defined keys together with the sentinel. The internal search length (external search length) ISL(C) (ESL(C)) is defined as the total number of comparisons performed by all possible successful (unsuccessful) searches, i.e. needed to access the $n - 1$ user-defined keys (n intervals).*

Since the only difference between a successful and an unsuccessful search is that the former cannot reach the sentinel i.e. the node whose key if $-\infty$, and that accessing the first node of a jumplist requires three comparisons —recall that the jump and next pointers are checked first, we have:

Proposition 1. *The external and internal search lengths satisfy* ESL = ISL + 3.

Interestingly, ESL and ISL are within a constant as opposed to BST where one has $EPL = IPL + 2n$ with n the number of internal nodes. Using a variation of the analysis performed for the internal path length we get:

Theorem 3. *The expected number of comparisons ESL_n performed when accessing all the nodes of a jumplist of size n is asymptotically equivalent to*

$$\mathrm{ESL}_n \sim 3n \ln n + n(-3 - 3Ei(1,1) + 3e^{-1}) + 3 \ln n - 3Ei(1,1) + 3e^{-1} + 9/2 + o(1).$$

It is not hard to see that algorithm JUMPLIST-FIND-LAST-LESS-THAN of Section 2.4 has an expected cost of $\mathrm{ESL}_n + n + o(n)$, since in addition to the search, it follows all the jump pointers of the next sublist to arrive at the predecessor. The number of comparisons is the same, however, with the optimization mentioned at the end of Section 2.4.

4   Insertion and Deletion Algorithms

So far, we have shown that the performance of randomized jumplists are as good as those of randomized BST. We now show how to maintain the randomized property upon insertions and deletions. This maintenance will make the performance of jumplists identical for random keys or sorted keys —a property similar to that of randomized Binary Search Trees [16].

4.1   Creating a Jumplist from a Sorted Linked List

Constructing a jumplist from a list is very simple: we only have to choose the jump pointer of the header, and recursively build the next and jump sublists. We use a recursive function REBALANCEINTERVAL(L, x, n), which creates a randomized jumplist for the n elements of L starting at x, next$[x]$, ... , $z = \text{next}^{n-1}[x]$. The element $y = \text{next}^n[x]$ acts as a sentinel (the last element z jumps to it, but jump$[y]$ is not set). It is this element y which is returned. Hence:

JUMPLISTFROMLIST(L)
  1: $y \leftarrow \text{header}[L]$
  2: $n \leftarrow 1 + \text{nsize}[y] + \text{jsize}[y]$
  3: REBALANCEINTERVAL(y, n)
REBALANCEINTERVAL(x, n)
  1: **while** $n > 1$ **do**
  2:    $m \leftarrow$ random number in $[2, n]$
  3:    $y \leftarrow \text{next}[x]$
  4:    jump$[x] \leftarrow$
  5:    REBALANCEINTERVAL($y, m - 2$)
  6:    $x \leftarrow \text{jump}[x]$
  7:    $n \leftarrow n - m + 1$
  8: **return** x

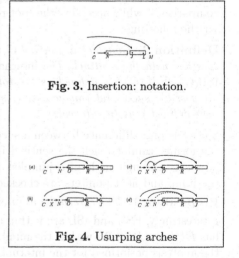

Fig. 3. Insertion: notation.

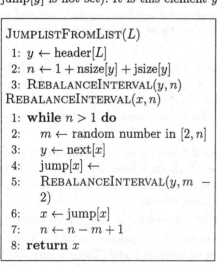

Fig. 4. Usurping arches

4.2   Maintaining the Randomness Property upon an Insertion

Suppose as depicted on Figure 3 that we aim at inserting the key x into the jumplist (C, J, N), and let X be the jumplist node to be allocated in order to accommodate x. The general pattern of the insertion algorithm is the search algorithm of Figure 2(left) since we need to figure out the position of x. But on the other hand we have to maintain the randomness property. We now describe algorithm JUMPLISTINSERT. Refer to Figure 3 for the notation, and to [1] for the pseudo-code.

When inserted into the list rooted at C, the node X containing x becomes a candidate as the endpoint of the fundamental arch starting at C. Assume that jsize[C] + nsize[C] = $n - 1$. Since x is inserted into (C, J, N), X has to be made the fundamental arch of $C \cup X$ with probability $1/n$. (Notice that if C is the end of the list, this probability is one so that we create a length one arch between the last item of the list and X, and exit.) With probability $1 - 1/n$ the fundamental arch of C does not change. If x is more than the keys of J or N, we recursively insert into the jump or next sublist. If not, x has to be inserted right after C, that is X becomes the successor of C and the randomness property must be restored for the jumplist rooted at X. To summarize, the recursive algorithm JUMPLISTINSERT stops when one of the following events occur:

- Case 1: x is inserted in the list rooted at C, and $[C, X]$ becomes the new fundamental arch. The randomness property of the list rooted at C, that is $C \cup X$, has to be restored.
- Case 2: x is inserted right after C. The randomness property of the list rooted at X that is $X \cup N$ has to be restored.

Following this discussion, we have the following

Proposition 2. *Algorithm* JUMPLISTINSERT *maintains the randomness property of the jumplist under insertion of any key.*

We proceed with the complexity of algorithm JUMPLISTINSERT. The reorganization to be performed in cases 1 and 2 consists of maintaining the randomness property of a jumplist whose root and size are known. This can be done using the algorithm JUMPLISTFROMLIST of section 4.1. Alas, algorithm JUMPLISTFROMLIST has linear complexity and the following observation shows that applying JUMPLISTFROMLIST to cases 1 and 2 is not optimal.

Observation 4 *Let C be a randomized jumplist of size n and suppose that x has to be inserted right after C. The expected number N_n of keys involved in the restoration of the randomness property of $C \cup X$ satisfies $N_n \sim n/2$.*

4.3   Insertion Algorithm

As just observed, inserting a key after the header of the list may require restoring the randomness property over a linear number of terms. We show that this can be done at a logarithmic cost. The intuition is that instead of computing from scratch a new randomized jumplist on $X \cup N$, and since by induction N is a randomized jumplist, one can reuse the arches of N in order to maintain randomness. More precisely, we shall make X usurp N and proceed recursively.

Algorithm USURPARCHES runs as follows. Due to the lack of space, we also omit the pseudo-code.

As depicted on Figure 4(a,b), assume that x is inserted after C and that prior to this insertion C has a successor N which is the root of the jumplist (N, O, R). Let n be the size of N. With probability $1/n$ we just create the length one arch $[X, N]$ —Figure 4(c). If not and with probability $1 - 1/n$, the arch of X has to be chosen as any of the nodes in the jumplist rooted at N. But since by induction hypothesis (N, O, R) is a randomized jumplist, X can usurp the arch of N —Figure 4(d). The next sublist of X then has to be re-organized. We do so recursively by having N usurp its successor.

Following the previous discussion we have the following

Proposition 3. *Algorithm* USURPARCHES *maintains the randomness property of the jumplist.*

Analyzing the complexity of algorithm USURPARCHES requires counting the number of jump pointers updates along the process. We have:

Proposition 4. *Let X be a jumplist node whose next sublist (N, O, R) has size n. The expected number S_n of jump pointers updates during the recursive usurping strategy starting at X satisfies $S_n \sim \ln n$.*

We are now ready to state the main result of this section:

Theorem 5. *Algorithm* JUMPLISTINSERT *using algorithm* JUMPLISTFROM-LIST *for Case 1 and algorithm* USURPARCHES *for Case 2 returns a randomized jumplist. Moreover, its complexity is $O(\log n)$.*

4.4 Deletion Algorithm

We show in this section that removing a key from a jumplist is exactly the symmetric operation of the insertion. Assume key x has to be removed from a jumplist C. To begin with, the node containing x has to be located. To do so, we search for x as usual following the pattern of the search algorithm of Figure 2(left). To be more precise, we actually seek the node C such that key[jump[C]] $= x$ or key[next[C]] $= x$. The actual removal of the node X containing x distinguishes between the following two situations.

Removing X with key[jump[C]] $= x$. First, X is removed from the linked list. If we assume a singly connected linked list is used, the removal of X requires the knowledge of its predecessor, but this can be obtained by using the algorithm JUMPLIST-PREDECESSOR(X) of section 2. Second, the system of arches starting at X needs to be recomputed. To do so, we again use the function JUMPLISTFROMLIST.

Removing X with key[next[C]] $= x$. Here too, we first remove X and second maintain the randomness property of the next sublist of C. The former operation is trivial. For the second one, first observe that if we have a length one arch, i.e. jump[X] = next[X], the situation is trivial too. Consider the situation where this is not the case. To create a random arch rooted at N, we recursively unwind the usurping operation described for the insertion algorithm. The operations performed are exactly the opposite of those described for the usurping algorithm, and the expected complexity is also logarithmic.

5 Conclusion

In this paper, we have presented a data structure called jumplist, which is inspired by skip lists and by randomized binary search trees, and which shares many of their properties.

There are a few advantages to randomized jumplists over randomized BST and treaps, the main one being the low storage used (if only forward traversal is needed; note that BST require parent pointers in order to provide the successor operation), and that the traversal is very simple (simply follow the underlying list, in $O(1)$ worst-case time per element). Moreover, it is conceivable to avoid storing the sublist sizes, but the insertion is more involved and does not have the randomness property. It is an open challenge to design deterministic jumplists. Determinism would likely make the data structure faster, and can probably be achieved by weight balancing (since the sublist sizes are known). Yet we have not carried it to its conclusion. Another challenge is to identify a splay operation on jumplists (in the manner of splay trees) in order to provide good amortized performance.

The main advantage to jumplists over skip lists is the size requirement: the jumplist stores exactly one jump pointer per node, whereas this number is not constant for skip lists (although the total expected storage remains linear).

Thus, jumplists provide an alternative to the classical dictionary data structures, and like skip lists, they have the potential to extend for higher-dimensional search structures in computational geometry [9]. Anecdotally, this is the reason we started to investigate jumplists. We plan to continue our research in this direction.

Acknowledgments. Philippe Flajolet, Marc Glisse and Bruno Salvy are acknowledged for discussions on the topic.

References

1. Hervé Brönnimann, Frédéric Cazals and Marianne Durand. Randomized jumplists: A jump-and-walk dictionary data structure. Research Report RR-xxxx, INRIA, 2003.
2. M. Abramowitz and I. A. Stegun. *Handbook of Mathematical Functions.* Dover, 1973. A reprint of the tenth National Bureau of Standards edition, 1964.

3. N. Amenta, S. Choi, T. K. Dey, and N. Leekha. A simple algorithm for homeomorphic surface reconstruction. In *Proc. 16th Annu. ACM Sympos. Comput. Geom.*, pages 213–222, 2000.
4. Nina Amenta and Marshall Bern. Surface reconstruction by Voronoi filtering. *Discrete Comput. Geom.*, 22(4):481–504, 1999.
5. C. Aragon and R. Seidel. Randomized search trees. In *Proc. 30th Annu. IEEE Sympos. Found. Comput. Sci.*, pages 540–545, 1989.
6. Jean-Daniel Boissonnat and Frédéric Cazals. Smooth surface reconstruction via natural neighbour interpolation of distance functions. In *Proc. 16th Annu. ACM Sympos. Comput. Geom.*, pages 223–232, 2000.
7. T. H. Cormen, C. E. Leiserson, and R. L. Rivest. *Introduction to Algorithms*. MIT Press, Cambridge, MA, 1990.
8. Mark de Berg, Marc van Kreveld, Mark Overmars, and Otfried Schwarzkopf. *Computational Geometry: Algorithms and Applications*. Springer-Verlag, Berlin, Germany, 2nd edition, 2000.
9. Olivier Devillers. Improved incremental randomized Delaunay triangulation. In *Proc. 14th Annu. ACM Sympos. Comput. Geom.*, pages 106–115, 1998.
10. P. Flajolet and A. M. Odlyzko. Singularity analysis of generating functions. *SIAM J. Disc. Math.*, 3(2):216–240, 1990.
11. G. H. Gonnet and R. Baeza-Yates. *Handbook of Algorithms and Data Structures*. Addison-Wesley, 1991.
12. D. H. Greene and D. E. Knuth. *Mathematics for the analysis of algorithms*. Birkhäuser, Boston, 1982.
13. H.-K. Hwang. *Théorèmes limites pour les structures combinatoires et les fonctions arithmetiques*. PhD thesis, École Polytechnique, Palaiseau, France, December 1994.
14. Donald E. Knuth. *The Art of Computer Programming, Vol 3., @nd Edition.* Addison-Wesley, 1998.
15. H.M. Mahmoud. *Evolution of random search trees*. Wiley, 1992.
16. C. Martínez and S. Roura. Randomized binary search trees. *J. Assoc. Comput. Mach.*, 45(2), 1998.
17. J.I. Munro, T. Papadakis, and R. Sedgewick. Deterministic skip lists. In *SODA*, Orlando, Florida, United States, 1992.
18. W. Pugh. Skip lists: a probabilistic alternative to balanced trees. *Commun. ACM*, 33(6):668–676, 1990.
19. S. Roura. An improved master theorem for divide-and-conquer recurrences. *J. Assoc. Comput. Mach.*, 48(2), 2001.
20. R. Sedgewick. *Algorithms in C++, Parts 1-4: Fundamentals, Data Structure, Sorting, Searching*. Addison-Wesley, third edition, 1998.
21. R. Sedgewick and P. Flajolet. *An Introduction to the Analysis of algorithms*. Addison-Wesley, 1996.
22. R. Sedgewick and P. Flajolet. Analytic combinatorics—symbolic combinatorics. To appear, 2002.
23. R. Seidel and C. R. Aragon. Randomized search trees. *Algorithmica*, 16:464–497, 1996.
24. D. D. Sleator and R. E. Tarjan. Self-adjusting binary search trees. *J. ACM*, 32(3):652–686, 1985.
25. J.S. Vitter and P. Flajolet. Average-case analysis of algorithms and data structures. In J. van Leeuwen, editor, *Algorithms and Complexity*, volume A of *Handbook of Theoretical Computer Science*, pages 432–524. Elsevier, Amsterdam, 1990.
26. N. Wirth. *Algorithms + Data Structures = Programs*. Prentice-Hall, 1975.

Complexity Theoretical Results on Nondeterministic Graph-Driven Read-Once Branching Programs

(Extended Abstract)

Beate Bollig*

FB Informatik, LS2, Univ. Dortmund,
44221 Dortmund, Germany
bollig@ls2.cs.uni-dortmund.de

Abstract. Branching programs are a well-established computation and representation model for boolean functions, especially read-once branching programs (BP1s) have been studied intensively. Recently two restricted nondeterministic (parity) BP1 models, called well-structured nondeterministic (parity) graph-driven BP1s and nondeterministic (parity) graph-driven BP1s, have been investigated. The consistency test for a BP-model M is the test whether a given BP is really a BP of model M. Here it is shown that the complexity of the consistency test is a difference between the two nondeterministic (parity) graph-driven models. Moreover, a new lower bound technique for nondeterministic graph-driven BP1s is presented which is applied in order to answer in the affirmative the open question whether the model of nondeterministic graph-driven BP1s is a proper restriction of nondeterministic BP1s (with respect to polynomial size). Furthermore, a function f in n^2 variables is exhibited such that both the function f and its negation $\neg f$ can be computed by Σ_p^3-circuits, f and $\neg f$ have simple nondeterministic BP1s of small size but f requires nondeterministic graph-driven BP1s of size $2^{\Omega(n)}$. This answers an open question stated by Jukna, Razborov, Savický, and Wegener (1999).

1 Introduction

1.1 Branching Programs and Circuits

Branching programs (BPs), sometimes called binary decision diagrams (BDDs), belong to the most important nonuniform models of computation. (For a history of results on branching programs see, e.g., the monograph of Wegener [19]).

Definition 1. *A* branching program (BP) *or* binary decision diagram (BDD) *on the variable set $X_n = \{x_1, \ldots, x_n\}$ is a directed acyclic graph with one source and two sinks labeled by the constants 0 and 1. Each non-sink node (or decision*

---

* Supported in part by DFG We 1066/9.

H. Alt and M. Habib (Eds.): STACS 2003, LNCS 2607, pp. 295–306, 2003.

node) is labeled by a boolean variable and has two outgoing edges, one labeled by 0 and the other by 1. A nondeterministic branching program ($\vee$-BP *for short) is a branching program with some additional unlabeled nodes, called nondeterministic nodes, which have out-degree 2.*

An input $a \in \{0,1\}^n$ *activates all edges consistent with* a, *i.e., the edges labeled by* a_i *which leave nodes labeled by* x_i *and all unlabeled edges. A* computation path *for an input* a *in a BP is a path of edges activated by the input* a *that leads from the source to a sink. A computation path for an input* a *that leads to the 1-sink is called* accepting path *for* a.

Let B_n *denote the set of all boolean functions* $f : \{0,1\}^n \rightarrow \{0,1\}$. *The (nondeterministic) BP* G *represents a function* $f \in B_n$ *for which* $f(a) = 1$ *iff there exists an accepting path for the input* a. *A* parity branching program *(or* $\oplus$-BP *for short) is syntactically a nondeterministic branching program but instead of the usual existential nondeterminism the parity acceptance mode is used. An input* a *is accepted iff the number of its accepting paths is odd.*

A BP is called read-once *(or BP1 for short), if on each path each variable is tested at most once.*

The size *of a branching program* G *is the number of its nodes and is denoted by* $|G|$. *The* branching program size *of a boolean function* f *is the size of the smallest BP representing* f. *The* length *of a branching program is the maximum length of a path.*

In order to learn more about the power of branching programs, various restricted models have been investigated intensively and several interesting restricted types of BPs could be analyzed quite successfully (for the latest breakthrough for semantic super-linear length BPs see [1],[2], and [3], where using a subtle combinatorial reasoning super-polynomial lower bounds were obtained). Besides this complexity theoretical viewpoint people have used branching programs in applications. Bryant [8] introduced ordered binary decision diagrams (OBDDs) which are up to now the most popular representation for formal circuit verification.

Definition 2. *An* OBDD *is a branching program with a* variable order *given by a permutation* π *on the variable set. On each path from the source to the sinks, the variables at the nodes have to appear in the order prescribed by* π *(where some variables may be left out).*

Since several important and also quite simple functions have exponential OBDD size, more general representations with good algorithmic behavior are necessary. Generalizing the concept of variable order to graph order Gergov and Meinel [9,10] and Sieling and Wegener [17] have proposed how deterministic read-once branching programs can be used for verification.

Definition 3. *A* graph order *is a (deterministic) branching program with a single sink, where on each path from the source to the sink all variables appear exactly once. Let* $\omega \in \{\vee, \oplus\}$. *An* ω-nondeterministic graph-driven BP1 *is an* ω-nondeterministic BP1 G *for which there exists a graph order* G_0 *with the* *following property: If for an input a, a variable x_i appears on a computation path of a in G before the variable x_j, then x_i also appears on the unique computation path of a in G_0 before x_j.*

An ω-nondeterministic graph-driven BP1 G is well-structured iff there exists a graph order G_0 and a mapping α from the node set of G to the node set of G_0 such that for every node v in G the node $\alpha(v)$ is labeled with the same variable as v, and such that if a computation path of an input a passes through v, then the computation path of a in G_0 passes through $\alpha(v)$.

In the following if nothing else is mentioned nondeterministic graph-driven BP1s mean $\vee$-nondeterministic graph-driven BP1s.

Any OBDD is well-structured since there exists exactly one x_i-node in any variable order for each variable x_i. In [4] it has been shown that even restricted well-structured nondeterministic graph-driven BP1s, called tree-driven nondeterministic BP1s, are a proper generalization of nondeterministic OBDDs. The result can easily be extended to the parity case. Well-structured graph-driven BP1s according to a fixed graph order G_0 can easily be obtained in the following way. Since we are interested in lower bounds, we may assume that each graph order does not contain identical subgraphs. We start by a complete decision tree which is ordered according to G_0, afterwards we merge all identical subgraphs. Finally, all nodes which have the same 0-and 1-successor are deleted.

The difference between the two graph-driven models is the following one. For graph-driven BP1s G according to a graph order G_0 it is possible that a node v with label x_i is reached on the computation paths for two inputs a and b in G whereas the nodes with label x_i on the computation paths for the inputs a and b in G_0 are different. This is not allowed in the well-structured case. Figure 1 shows an example of a graph order G_0, a well-structured ω-nondeterministic G_0-driven BP1 G_1 and an ω-nondeterministic G_0-driven BP1 G_2.

The concept of graph-driven branching programs has turned out to be also useful in other settings, see e.g. [14] and [18]. Gergov and Meinel [9] were the first ones who suggested parity graph-driven BP1s as a data structure for boolean functions. Another reason for investigating parity graph-driven BP1s is that until now exponential lower bounds on the size of parity read-once branching programs for explicitly defined boolean functions are unknown. One step towards the proof of such bounds might be to investigate BP models *inbetween* deterministic and parity BP1s. Nondeterministic and parity graph-driven BP1s have been investigated more intensively in [4], [7], [5], and [6].

Besides BPs or BDDs boolean circuits are one of the standard representations for boolean functions.

Definition 4. Σ_p^d, Π_p^d *are the classes of functions that can be computed by polynomial size depth-d circuits over the de Morgan basis $\{\wedge, \vee, \neg\}$ (negations are allowed only at the input variables and do not contribute to the depth) that have $\vee$ (respectively, $\wedge$) as output gate.*

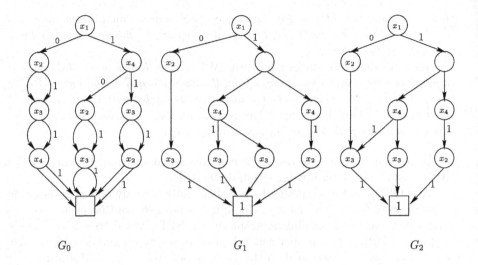

Fig. 1. *A graph order G_0, a (well-structured) ω-nondeterministic G_0-driven BP1 G_2 (G_1) representing the function $f(x_1, x_2, x_3, x_4) = \overline{x_1}\ \overline{x_2}x_3 \vee x_1x_2x_4 \vee x_1x_3x_4 \vee x_1\overline{x_3}\ \overline{x_4}$. Missing edges are leading to the 0-sink.*

1.2   The Results

We investigate the question whether well-structured ω-nondeterministic graph-driven BP1s, $\omega \in \{\vee, \oplus\}$, are proper restrictions of ω-nondeterministic graph-driven BP1s in Section 2. We prove that a difference between the two models is the complexity of the *consistency test*. The consistency test for a BP-model M is the test whether a given BP is really a BP of model M. A rule of thumb that can be obtained by comparing several variants of BPs is that variants with a larger class of functions with small-size representations usually have less efficient algorithms. In Section 2, we prove the surprising result that this rule is not true for nondeterministic (parity) BP1s and the consistency test. Although there exist polynomial time algorithms for the consistency test for nondeterministic (parity) BP1s and well-structured ω-nondeterministic graph-driven BP1s the consistency test for ω-nondeterministic graph-driven BP1s is coNP-complete.

In Section 3, we describe a new lower bound technique for nondeterministic graph-driven BP1s which generalizes a method presented in [6]. Applying this method we prove that $\vee$-nondeterministic graph-driven BP1s are in fact significantly more restricted than nondeterministic BP1s.

The following general question has been widely studied for various computational models:

Suppose that both a computational problem f and its complement $\neg f$ posses an efficient nondeterministic computation in some model. Does this imply that f can also be computed efficiently and deterministically in the same model?

This question is often called the P *versus* NP∩co-NP *question* and for boolean (nonuniform) complexity, polynomial size instead of polynomial time is investigated. In [13] an explicitly defined boolean function is presented that is in Σ_p^3 and in NP∩co-NP for read-once branching programs but has exponential deterministic read-once branching program size. Another function belongs to the smaller class $\Sigma_p^3 \cap \Pi_p^3$, but the separation is only quasipolynomial. Jukna, Razborov, Savický, and Wegener [13] asked whether the class $\Sigma_p^3 \cap \Pi_p^3$ contains a function separating NP∩co-NP from quasipolynomial size in the context of read-once branching programs. In Section 4, we answer the question in the affirmative. Moreover, we prove that not only the deterministic BP1 size but even the ∨-nondeterministic graph-driven BP1s size of the selected function is exponential.

2   The Consistency Test

The consistency test for nondeterministic (parity) BP1s is simple. Let G be the given BP. Let $label(u)$ be the variable which is the label of the node u if u is a decision node and the empty set if u is a nondeterministic node. According to a topological order of the nodes, we compute for each decision node w the set X_w of variables tested on some path from the source to w excluding the label of w. If $label(w) \in X_w$, G cannot be a nondeterministic BP1. Otherwise the consistency test succeeds.

For well-structured ω-nondeterministic graph-driven BP1s, $\omega \in \{\vee, \oplus\}$, we can use an observation of Brosenne, Homeister, and Waack [7] which is a slight generalization of a result for the deterministic case described in [17]. Let G be a well-structured ω-nondeterministic graph-driven BP1 according to the graph order G_0 and let α be the function which maps the nodes from G to the nodes from G_0. A c-successor, $c \in \{0,1\}$, of a node v in G is a node which can be reached via the c-edge of v and unlabeled edges. The crucial observation is the following one. If w is one of the c-successors of v in G, then all paths to the sink in G_0 which leave $\alpha(v)$ via the c-edge pass through $\alpha(w)$. Due to the lack of space we have to omit the polynomial time algorithm.

For unrestricted ω-nondeterministic graph-driven BP1s the situation is more difficult.

Theorem 1. *The problem to decide for a given branching program G whether G is an ω-nondeterministic graph-driven BP1, $\omega \in \{\vee, \oplus\}$, is co-NP-complete.*

Proof. We prove that the inconsistency test for ω-nondeterministic graph-driven BP1s is NP-complete. The problem is contained in NP since we can guess two computation paths for an input a with two contradictory order of the variables.

Now we present a polynomial time reduction from 3-SAT. Let (X, C) be an instance of 3-SAT where $X = \{x_1, \dots, x_n\}$ is the set of variables and $C = \{c_1, \dots, c_m\}$ is the set of clauses. W.l.o.g. we assume that each clause consists of three different literals and each variable occurs in each clause at most once. Furthermore, we assume for each clause $c_i = x_{i_1}^{b_{i,1}} \vee x_{i_2}^{b_{i,2}} \vee x_{i_3}^{b_{i,3}}$, $1 \leq i \leq m$ and $b_{i,j} \in \{0,1\}$ for $1 \leq j \leq 3$, that $i_1 < i_2 < i_3$.

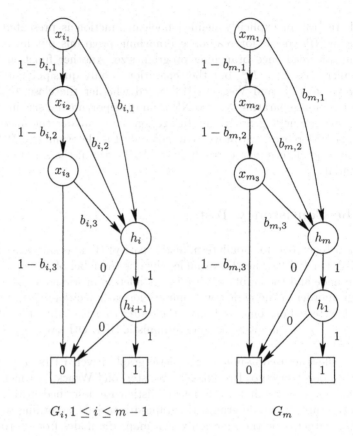

$$G_i, 1 \leq i \leq m-1 \qquad\qquad G_m$$

Fig. 2. *The components G_i, $1 \leq i \leq m-1$, and G_m in the co-NP-completeness proof.*

First, we introduce m new variables $h_1, \ldots, h_m$. Then we construct for each clause c_i the component G_i, $1 \leq i \leq m$, shown in Figure 2. The constructed branching program G represents the disjunction of all these components. The upper part of G looks like a switch and consists of $m-1$ nondeterministic nodes (see Figure 3).

We have to prove that the resulting branching program G is not an ω-nondeterministic graph-driven BP1 iff the instance (X, C) for 3-SAT is satisfiable.

$\Leftarrow$ We prove that there exists an input with at least two computation paths with different variable order. Let $a = \{0, 1\}^n$ be a satisfying assignment for (X, C). Let a' be the assignment which consists of the assignment a to the x-variables and $h_i = 1$, $1 \leq i \leq m$. There exist m accepting paths for a', on each of these accepting paths two h-variables are tested. The corresponding order of the h-variables are $h_i \to h_{i+1}$, $1 \leq i \leq m-1$, which means that the variable h_i has to be tested before the variable h_{i+1} and h_m has to be tested before h_1. Contradiction.

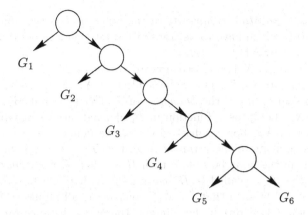

Fig. 3. *The structure of the upper part of G, if $m = 6$.*

$\Rightarrow$ We construct a graph order G_0 such that G is a ω-nondeterministic graph-driven BP1 if (X, C) is unsatisfiable. Since (X, C) is unsatisfiable there exists for each assignment a of the x-variables a first clause which is unsatisfied by a. Our graph order starts with a complete binary tree of size $2^n - 1$ on the x-variables, where the x-variables are ordered according to $x_1, \ldots, x_n$. If c_i is the first clause that is unsatisfied by the assignment a of the x-variables the path that corresponds to a in G_0 continues with the order $h_{i+1}, \ldots, h_m, h_1, \ldots, h_i$. $\qquad\Box$

3   A Lower Bound Method for $\vee$-Nondeterministic Graph-Driven BP1s

The theory of communication complexity is a powerful tool for proving lower bounds on the size of restricted nondeterministic oblivious BPs (a BP is called *oblivious* if the nodes can be partitioned into levels such that edges point only from lower to higher levels and all decision nodes of one level are labeled by the same variable). (See, e.g., [11] and [16] for the theory of communication complexity.) In [6] it has been shown how this tool can be used for proving large lower bounds on the size of well-structured nondeterministic and parity graph-driven BP1s.

First, we restate the notation described in [6]. Consider a Boolean function $f \in B_n$ which is defined on the variables in $X_n = \{x_1, \ldots, x_n\}$, and let $\Pi = (\mathcal{X}_A, \mathcal{X}_B)$ be a partition of X_n. Assume that Alice has access only to the input variables in $\mathcal{X}_A$ and Bob has access only to the input variables in $\mathcal{X}_B$. In a one-way communication protocol, upon a given input x, Alice is allowed to send a single message (depending on the input variables in $\mathcal{X}_A$) to Bob who must then be able to compute the answer $f(x)$. In a nondeterministic communication protocol Alice is allowed to *guess* a message. The function value is one if there exists at least one guess such that Bob accepts the input. The *nondeterministic one-way communication complexity* of the function f is the number of bits of communication which have to be transmitted by such a protocol that computes f. It is denoted by $\mathrm{ND}^{A \to B}(f, \Pi)$.

A *filter* of a set X is a closed upward subset of 2^X (i.e. if $S \in \mathcal{F}$, then all supersets of S are in $\mathcal{F}$). Let $\mathcal{F}$ be a filter of $X_n = \{x_1, \dots, x_n\}$. A subset $B \subseteq X_n$ is said to be in the *boundary* of $\mathcal{F}$ if $B \notin \mathcal{F}$ but $B \cup \{x_i\} \in \mathcal{F}$ for some $x_i \in X_n$. Let f be a function in B_n defined on the variables in X_n and $\mathcal{F}$ be a filter of X_n. For a subset $Z \subseteq X_n$, we denote by $\mathcal{A}(Z)$ the set of all possible assignments to the variables in Z. Let $\Pi = (X_A, X_B)$ be a partition of X_n. If X_B is in the boundary of $\mathcal{F}$, then Π is called $\mathcal{F}$-*partition* of X_n. Finally, a function $f' \in B_n$ is called (ϵ, Π)-*close* to f, if there exists a set $R \subseteq \mathcal{A}(X_A)$ with $|R| \geq \epsilon \cdot 2^{|X_A|}$, such that f and f' coincide on all inputs in $R \times \mathcal{A}(X_B)$.

A function f is *d-rare* if the minimal Hamming distance for two arbitrary inputs $a, b \in f^{-1}(1)$ is at least d.

Theorem 2. *Let $f \in B_n$ be a 2-rare function and $\mathcal{F}$ be a filter on X_n, $0 < \epsilon \leq 1$, and $\ell \in \mathbb{N}$. Let $\Pi = (X_A, X_B)$ be an $\mathcal{F}$-partitioning of X_n and $f_{|X_A}$ be a subfunction of f, where we have replaced the X_A-variables by constants. If for every $\mathcal{F}$-partition Π of X_n and for every subfunction $f_{|X_A}$ it holds $f_{|X_A} \neq 0$, and for every function f' which is (ϵ, Π)-close to f it is $\mathrm{ND}^{A \to B}(f', \Pi) > \ell$, then any $\vee$-nondeterministic graph-driven BP1 representing f has a size of at least $\min\{2^\ell, \epsilon^{-1}\}$.*

The function $n/2$-RC_n is defined on $n \times n$ boolean matrices X on the variables $X_n = \{x_{1,1}, \dots, x_{n,n}\}$. (For the ease of readability we assume that n is an even number.) Its function value is 1 if and only if there exist exactly $n/2$ ones in each row or exactly $n/2$ ones in each column.

Theorem 3. *The function $n/2$-RC_n can be represented by nondeterministic BP1s with one nondeterministic node in size $O(n^3)$ but its $\vee$-nondeterministic graph-driven BP1 size is $\Omega(2^{n/4})$.*

Sketch of proof.

We present some ideas for the proof of the lower bound. First, we have to define an appropriate filter $\mathcal{F}$ on the variables $x_{1,1}, \dots, x_{n,n}$. A set T is in the filter $\mathcal{F}$ if T contains all variables from $n/2 + 1$ arbitrary rows and $n/2 + 1$ arbitrary columns. If $\Pi = (X_A, X_B)$ is an $\mathcal{F}$-partition, then by definition $X_B \notin \mathcal{F}$ and there exists a variable $x_{i,j}$ such that $X_B \cup \{x_{i,j}\} \in \mathcal{F}$. Hence, X_A contains exactly $n/2$ variables from different rows and at most $n/2$ variables from different columns or vice versa. Since every maxterm of $n/2$-RC_n contains at least $n/2+1$ literals from different rows and $n/2 + 1$ literals from different columns, $n/2$-$\mathrm{RC}_{n|X_a} \neq 0$. Furthermore, the function $n/2$-RC_n is 2-rare since each 1-input consists of exactly $n/2 \cdot n$ 1-entries.

Let $\epsilon = 2^{-n/4}$. It remains to prove that for every $\mathcal{F}$-partition Π of X_n and for every function f' which is (ϵ, Π)-close to $n/2$-RC_n it is $\mathrm{ND}^{A \to B}(f', \Pi) > n/4$. For this we use some of the ideas presented in [6]. $\qquad \square$

4   The Separation Result for Circuits and ∨-Nondeterministic Graph-Driven BP1s

The function f_n is defined on $n \times n$ boolean matrices X on the variable set $X_n = \{x_{1,1}, \ldots, x_{n,n}\}$ and outputs 1 iff the matrix X contains exactly one 1-entry in each row or exactly $n - 1$ columns with exactly one 1-entry and $n - 1$ 1-entries altogether. (For the ease of readability we assume w.l.o.g. that n is an even number.) The lower bound technique described in Section 3 cannot be applied directly since f_n is not 2-rare. Therefore, we have to add some arguments. First, we need some further notation.

Definition 5. *Let f be a boolean function defined on the variables in $X_n = \{x_1, \ldots, x_n\}$. A set $A(f) = \{(\alpha_1, \beta_1), (\alpha_2, \beta_2), \ldots, (\alpha_k, \beta_k)\}$, $\alpha_i \in \{0,1\}^{n'}$ and $\beta_i \in \{0,1\}^{n-n'}$, is called a strong 1-fooling set if*

i) $f(\alpha_i, \beta_i) = 1$ for all $i \in \{1, \ldots, k\}$, and
ii) $i, j \in \{1, \ldots, k\}$ and $i \neq j$ implies that $f(\alpha_i, \beta_j) = 0$ and $f(\alpha_j, \beta_i) = 0$.

For a subset $Z \subseteq X_n$ we denote by $\mathcal{A}(Z)$ the set of all possible assignments to the variables in Z. Let G_0 be a graph order and v a node in G_0. Let X_v be the set of variables tested on a path from the source to v (excluding the variable which is the label of v) and $\mathcal{A}_v \subseteq \mathcal{A}(X_v)$ a set of partial assignments which lead in G_0 from the source to v. Using well-known facts from communication complexity the following is easy to prove. Let $v_1, \ldots, v_k$ be nodes in G_0, where $X_{v_1} = \ldots = X_{v_k}$, $X_v := X_{v_1}$, and $\mathcal{A}_v := \cup_{1 \le j \le k} \mathcal{A}_{v_j}$. If there exists a strong 1-fooling set $A(f) = \{(\alpha_1, \beta_1), (\alpha_2, \beta_2), \ldots, (\alpha_{|\mathcal{A}_v|}, \beta_{|\mathcal{A}_v|})\}$, $\alpha_i \in \mathcal{A}_v$, and $\beta_i \in \mathcal{A}(X_n \setminus X_v)$, then any ω-nondeterministic graph-driven BP1, $\omega \in \{\vee, \oplus\}$, representing f according to G_0 has a size of at least $|\mathcal{A}_v|$.

Now our proof idea is the following one. Let G be a nondeterministic graph-driven BP1 representing f_n and let G_0 be a graph order such that G is ordered according to G_0. We choose a large set of subpaths in G_0 and define V as the set of nodes $v_1, \ldots, v_l$ which are reached by at least one of the chosen subpaths. If there is large number of subpaths which lead to the nodes $v_{i_1}, \ldots, v_{i_k}$ in V, where $X_{v_{i_1}} = \ldots = X_{v_{i_k}}$, then we can construct a large strong 1-fooling set. Otherwise, there are many nodes $v_{j_1}, \ldots, v_{j_m}$ in V, where $X_{v_{j_l}} \neq X_{v_{j_{l'}}}$, $1 \le l < l' \le m$. Let S_w be the union of all variables tested on a path from w to a sink (including the label of w). Let $x_{i',j'}$ be a variable that is not contained in $X_{v_{j_1}} \cup \ldots \cup X_{v_{j_m}}$. If for each set $X_{v_{j_i}}$ there exists a node w in G such that $S_w = X_n \setminus \{X_{v_{j_i}} \cup \{x_{i',j'}\}\}$ or $S_w = X_n \setminus X_{v_{j_i}}$, we can conclude that the size of G is at least m.

Theorem 4. *The function f_n is in $\Sigma_p^3 \cap \Pi_p^3$ and can be represented by nondeterministic BP1s with one nondeterministic node in linear size and the function $\neg f_n$ has size $O(n^4)$ for nondeterministic OBDDs but the $\vee$-nondeterministic graph-driven BP1 size for f_n is $2^{\Omega(n)}$.*

Due to the lack of space we have to omit the proof of the upper bounds.

Proof of the lower bound.

Jukna (1989) and Krause, Meinel, and Waack (1991) have presented exponential lower bounds on the size of nondeterministic BP1s representing the function $PERM_n$, the test whether a boolean matrix contains exactly one 1-entry in each row and in each column. Here for the choice of the considered subpaths some of their ideas are used.

Let G be a nondeterministic graph-driven BP1 representing f_n and G_0 be a graph order such that G is ordered according to G_0. We consider all paths in G_0 that correspond to permutation matrices which means that there exist n variables $x_{i_1,j_1}, \ldots, x_{i_n,j_n}$ which are set to 1, where $i_l \neq i_{l'}$ and $j_l \neq j_{l'}$ if $l \neq l'$, and all other variables are set to 0. The number of these paths is $n!$. Next, we define a cut through all these paths after exactly $n/2 + 1$ variables are set to 1. For each assignment a corresponding to a chosen path to the cut there exist $(n/2 - 1)!$ different possibilities to complete a to a permutation matrix. Therefore, there exist $\frac{n!}{(n/2-1)!}$ different paths from the source to the cut. Let R_p (C_p) be the set of indices i for which a variable $x_{i,\cdot}$ $(x_{\cdot,i})$ is set to 1 on p. If $(n/2 + 1)$ rows and columns have been chosen, there are $(n/2 + 1)!$ possibilities to map the indices of the rows to the indices of the columns. Therefore, there is a set P of different paths, $|P| \geq \binom{n}{n/2+1}$, such that for two different paths p and p' it is $R_p \neq R_{p'}$ or $C_p \neq C_{p'}$. Using the pigeonhole principle we can conclude that there exists a variable $x_{i,j}$ such that for at least $|P|/n^2$ paths p in P the variable $x_{i,j}$ is the last variable tested 1 on p. Let $P' \subseteq P$ be the set of these paths. The set V consists of all nodes v on a path from P' labeled by $x_{i,j}$. Now we consider all subpaths from the paths in P' to a node $v \in V$. Let P'' be the set of these paths. Obviously, $|P''| = |P'|$. By the definition of P and the fact that $x_{i,j}$ is set to 1 on all paths in P', we know that $R_p \neq R_{p'}$ or $C_p \neq C_{p'}$ for two different paths $p, p' \in P''$. For the application of our lower bound method it is important that all paths in P'' do not contain a variable $x_{i,\cdot}$ of the ith row or a variable $x_{\cdot,j}$ of the jth column that is set to 1.

In the following let a_p be the corresponding (partial) assignment of the variables tested on a path p. Let v_p be a node that is reached by a path $p \in P''$. We distinguish two cases.

1) There are at least $|P''|^{2/3}$ paths in P'' such that for any two of them $X_{v_p} = X_{v_{p'}}$ or

2) there are at least $|P''|^{1/3}$ paths in P'' such that for any two of them p, p', where $p \neq p'$, $X_{v_p} \neq X_{v_{p'}}$.

- We consider the first case. One of the following properties is true. There are at least $|P''|^{1/3}$ paths in P'' such that for any two of them $R_p \neq R_{p'}$ or at least $|P''|^{1/3}$ paths such that $C_p \neq C_{p'}$ for two different paths $p, p' \in P''$.
  - If there are at least $|P''|^{1/3}$ paths in P'' such that for any two of them $R_p \neq R_{p'}$, let $\mathcal{A}_v$ be the set of the partial assignments which correspond to these paths. The variable $x_{i,j}$ is the label of the nodes reached by these paths. We choose for each $a_p \in \mathcal{A}_v$ a partial assignment a_p^R of the variables in $X_n \setminus X_{v_p}$ such that (a_p, a_p^R) corresponds to a permutation

matrix and $x_{i,j} = 1$. By the definition of P'' such an assignment a_p^R exists. The function value $f(a_p, a_p^R)$ is 1 since in each row there exists exactly one 1-entry. For $a_{p'} \in \mathcal{A}_v$, let $a_{p'}^R$ be a partial assignment such that $(a_{p'}, a_{p'}^R)$ corresponds to a permutation matrix and $x_{i,j} = 1$. For $(a_{p'}, a_p^R)$, $p' \neq p$, there exists at least one row without a variable set to 1. Since the number of ones in X is n, $f(a_{p'}, a_p^R) = 0$. With the same arguments $f(a_p, a_{p'}^R) = 0$.

- If there are at least $|P''|^{1/3}$ paths in P'' such that for any two of them $C_p \neq C_{p'}$, let $\mathcal{A}_v$ be the set of the partial assignments which correspond to these paths. Again $x_{i,j}$ is the label of the nodes reached by these paths. We choose for each a_p a partial assignment a_p^C which resembles a_p^R but with the exception that $x_{i,j}$ is set to 0. Since there are exactly $n - 1$ columns with one 1-entry and the number of 1-entries is $n - 1$ altogether, the function value $f(a_p, a_p^C)$ is 1. For $(a_{p'}, a_p^C)$, $p' \in P''$ and $p' \neq p$, there exist at least two columns without a variable set to 1. Since the number of 1-entries in X is $n - 1$, we can conclude $f(a_{p'}, a_p^C) = 0$. With the same arguments $f(a_p, a_{p'}^C) = 0$.

Altogether, we have proved that there exists a strong 1-fooling set of size at least $|P''|^{1/3}$.

- Now we consider the case that for at least $|P''|^{1/3}$ paths in P'' for any two of them $X_{v_p} \neq X_{v_{p'}}$, $p \neq p'$. For each path p we consider a partial assignments a_p^C of the variables in $X_n \setminus X_{v_p}$. As a first step we consider a partial assignment a_p^R such that (a_p, a_p^R) corresponds to a permutation matrix and $x_{i,j} = 1$. By the definition of P'' such an assignment a_p^R exists. Now the partial assignment a_p^C resembles a_p^R but with the exception that $x_{i,j}$ is set to 0. Obviously, $f(a_p, a_p^C) = 1$. For each of the chosen assignments, we consider one accepting path in G. Each of these accepting paths has length of at least $n^2 - 1$ and the variable $x_{i,j}$ is the only variable that can be left out. We define a cut in G through all these accepting paths after exactly $n/2$ variables are set to 1. The variable $x_{i,j}$ cannot be tested on any of these paths, because $x_{i,j}$ is set to 0 and after the test of $x_{i,j}$ only $n/2 - 1$ variables are set to 1. Let W be the set of nodes reached for one of these accepting paths. If p is one of the chosen paths in P'' and v_p is the node in G_0 that is reached by p, then there exist a node $w \in W$ such that $X_w = X_{v_p}$ and $S_w = X_n \setminus X_{v_p}$ or $S_w = X_n \setminus (X_{v_p} \cup \{x_{i,j}\})$. By assumption the sets X_{v_p} for the chosen paths in P'' are all different, therefore, there are at least $|P''|^{1/3}$ nodes in W.

Summarizing we can conclude that the $\vee$-nondeterministic graph-driven BP1 complexity is at least $|P''|^{1/3} \geq (|P|/n^2)^{1/3} = \left(\binom{n}{n/2+1} / n^2 \right)^{1/3} = 2^{\Omega(n)}.$ $\square$

Acknowledgement. The author would like to thank Martin Sauerhoff, Ingo Wegener, and Philipp Woelfel for proofreading and fruitful discussions on the subject of the paper.

References

1. Ajtai, M. (1999). A non-linear time lower bound for boolean branching programs. Proc. of 40th FOCS, 60–70.
2. Beame, P., Saks, M., Sun, X., and Vee, E. (2000). Super-linear time-space tradeoff lower bounds for randomized computation. Proc. of 41st FOCS, 169–179.
3. Beame, P. and Vee, E. (2002). Time-space trade-offs, multiparty communication complexity, and nearest neighbor problems. Proc. of 34th STOC, 688–697.
4. Bollig, B. (2001). Restricted nondeterministic read-once branching programs and an exponential lower bound for integer multiplication. RAIRO Theoretical Informatics and Applications, 35:149–162.
5. Bollig, B., Waack, St., and Woelfel, P. (2002). Parity graph-driven read-once branching programs and an exponential lower bound for integer multiplication. Proc. of 2nd IFIP International Conference on Theoretical Computer Science, 83-94.
6. Bollig, B. and Woelfel, P. (2002). A lower bound technique for nondeterministic graph-driven read-once branching programs and its applications. Proc. of MFCS 2002, 131-142.
7. Brosenne, H., Homeister, M., and Waack, St. (2001). Graph-driven free parity BDDs: algorithms and lower bounds. Proc. of MFCS, 212–223.
8. Bryant, R. E. (1986). Graph-based algorithms for boolean function manipulation. IEEE Trans. on Computers 35, 677–691.
9. Gergov, J. and Meinel, C. (1993). Frontiers of feasible and probabilistic feasible boolean manipulation with branching programs. Proc. of STACS, LNCS 665, 576–585.
10. Gergov, J. and Meinel, C. (1994). Efficient boolean manipulation with OBDDs can be extended to FBDDs. IEEE Trans. on Computers 43, 1197–1209.
11. Hromkovič, J. (1997). *Communication Complexity and Parallel Computing*. Springer.
12. Jukna, S. (1989). The effect of null-chains on the complexity of contact schemes. Proc. of FST, LNCS 380, 246–256.
13. Jukna, S., Razborov, A., Savický, P., and Wegener, I. (1999). On P versus NP∩co-NP for decision trees and read-once branching programs. Computational Complexity 8, 357–370.
14. Krause, M. (2002). BDD-based cryptanalysis of keystream generators. Proc. of EUROCRYT, 222-237.
15. Krause, M., Meinel, C., and Waack, St. (1991). Separating the eraser Turing machine classes L_e, NL_e, co-NL_e and P_e. Theoretical Computer Science 86, 267-275.
16. Kushilevitz, E. and Nisan, N. (1997). *Communication Complexity*. Cambridge University Press.
17. Sieling, D. and Wegener, I. (1995). Graph driven BDDs - a new data structure for boolean functions. Theoretical Computer Science 141, 283–310.
18. Sieling, D. and Wegener, I. (2001). A comparison of free BDDs and transformed BDDs. Formal Methods in System Design 19, 223–236.
19. Wegener, I. (2000). *Branching Programs and Binary Decision Diagrams - Theory and Applications*. SIAM Monographs on Discrete Mathematics and Applications.

Randomness versus Nondeterminism for Read-Once and Read-k Branching Programs

Martin Sauerhoff[*]

Universität Dortmund, FB Informatik, LS 2, 44221 Dortmund, Germany
sauerhof@ls2.cs.uni-dortmund.de

Abstract. Recent breakthroughs have lead to strong methods for proving lower bounds on the size of branching programs (BPs) with quite weak restrictions. Nevertheless, lower bounds for the randomized and nondeterministic variants of the established BP models still offer many challenges. Here, the knowledge on the randomized case is extended as follows:

(i) The so-far open problem of proving that randomization with arbitrary bounded error can be weaker than nondeterminism for *read-once BPs* is solved in the following strong sense: It is shown that the so-called "weighted sum function" requires strongly exponential size for randomized read-once BPs with error bounded by any constant smaller than $1/2$, while both the function and its complement have polynomial size for nondeterministic read-once BPs.

(ii) For randomized *read-k BPs*, an exponential lower bound for a natural, graph-theoretical function that is easy to compute nondeterministically is presented. This is the first such bound for the boolean BP model. The function $cl_{3,n}$ deciding whether an n-vertex graph contains a triangle is obviously easy for nondeterministic read-once BPs while its complement is known to require strongly exponential size in this model. It is proved here that the function still requires size $2^{\Omega(k^{-2}2^{-4k}\cdot\sqrt{n})}$ for randomized read-k BPs with error at most $2^{-c2^{2k}}$ for some positive constant c.

1 Introduction

Branching programs (BPs) are a well-established model for studying the space complexity of sequential, nonuniform algorithms. For a thorough introduction, we refer to the monograph of Wegener [24].

Definition 1. *A (deterministic) branching program (BP) on the variable set* $X = \{x_1, \ldots, x_n\}$ *is a directed acyclic graph with one source and two sinks. The sinks are labeled by the constants 0 and 1, resp. Each interior node is labeled by a variable* x_i *and has two outgoing edges carrying labels 0 and 1, resp. The BP computes a function* $f\colon \{0,1\}^n \to \{0,1\}$ *defined on* X. *For an input* $a = (a_1, \ldots, a_n) \in \{0,1\}^n$, $f(a)$ *is equal to the label of the sink reached by the*

---

[*] Supported by DFG grant We 1066/9.

H. Alt and M. Habib (Eds.): STACS 2003, LNCS 2607, pp. 307–318, 2003.
© Springer-Verlag Berlin Heidelberg 2003

computation path for a, *which is the path from the source to a sink obtained by following the edge labeled by a_i for nodes labeled by x_i. The* size $|G|$ *of a branching program is the number of its nodes. The* length (*or* time) *is the maximum number of edges on a computation path.*

Nondeterministic and randomized BPs are defined by introducing additional *nondeterministic nodes* or *randomized nodes*. When reaching such nodes during a computation, the successor on the computation path is nondeterministically guessed or determined by flipping a fair coin, resp. Furthermore, different modes of acceptance with (bounded or unbounded) one-sided and two-sided error probability are defined as usual. Nondeterministic and randomized variants of restricted models of BPs are obtained by applying the respective restrictions to the nodes labeled by variables on each path or on each computation path.

By proving superpolynomial lower bounds for explicitly defined functions in less and less restricted variants of the general model of BPs, one hopes to ultimately be able to separate complexity classes such as L and P. Although this goal has not been attained so far, impressive progress has been made along this line of research during the last years, culminating in the recent bounds for length-restricted BPs that can also be regarded as time-space tradeoffs for general BPs [10,6,2,3,7,8]. Apart from this, branching programs are also an interesting model of computation for studying the relationships between determinism, nondeterminism, and randomness in the space-bounded scenario. So far, we have a more or less complete picture of the power of the basic modes of computation only for *oblivious BPs*, which are BPs where the sequence of variables appearing on each of the paths has to be consistent with one fixed variable sequence [1, 20,12]. For the more general, non-oblivious models our knowledge still is scant, though. There are essentially only two quite restricted classes of functions for which randomized lower bounds could be proved so far for the usual boolean model of BPs: (i) Functions based on inner products and quadratic forms with respect to matrices with strong rigidity properties [10,6,2,3,7,8,9]. These functions yield the best known randomized lower bounds, but are also difficult in the nondeterministic case (for some models, the latter is only a conjecture). (ii) Certain artificial matrix test functions [17,19,23], which have been used to prove the previously best separation results between randomness and nondeterminism.

We discuss the bounds of the second type in more detail. It has been shown in [17] that randomized read-once BPs with two-sided error bounded by a constant smaller than $27/128$ (improved to $1/3$ in [19]) can require exponential size for functions that are easy for the nondeterministic variant. Thathachar [23] has extended this result to *(syntactic) read-k BPs*, which are BPs where on each path each variable may appear at most k times. For his function, he has obtained an exponential lower bound for randomized read-k BPs with two-sided error bounded by $(1/3) \cdot 2^{-5^{k+1}}$, while the same function has polynomial size for deterministic read-$(k+1)$ BPs and nondeterministic read-once BPs. The bound on the error can be relaxed to $(1/3)^k$ by applying ideas from [19].

All these results have the drawback that the lower bound for the randomized case does not work for arbitrary bounded error. This cannot be mended as usual by probability amplification because of the read-restrictions for the considered models. By known upper bounds, it is in fact clear that none of the functions from the mentioned papers is hard for arbitrary bounded error. Here this situation is remedied for the case of read-once BPs (see the next section).

The power of the different modes of computation has also been investigated in the recent breakthrough papers of Ajtai [2] and Beame, Saks, Sun, and Vee [7] on lower bounds for BPs with restricted length. Apart from other results, these papers contain exponential lower bounds on the size of deterministic resp. randomized *multiway BPs* with length $O(n\sqrt{\log n/\log\log n})$, n the input size, for the variant of the element distinctness function where the variables take values in a large, nonboolean domain. The complement of this function is easily computable by the nondeterministic variant of the model. But so far, no separation results between randomized and nondeterministic variants of the more general models of BPs are known for the usually considered boolean case.

2   Our Results

We consider the following function due to Savický and Žák [21]. For a natural number n, let $p(n)$ be the smallest prime greater than n. The function WS_n ("weighted sum") is defined on $x = (x_1, \ldots, x_n) \in \{0,1\}^n$. Let $s = s_n(x) = \left(\sum_{i=1}^{n} i x_i\right) \bmod p(n)$. Then $\mathrm{WS}_n(x) = x_s$ if $1 \le s \le n$, and (say) $\mathrm{WS}_n(x) = x_1$ otherwise.

Using that $p(n) \le 2n$ by the prime number theorem, it is easy to see that WS_n and the complement $\neg\,\mathrm{WS}_n$ have nondeterministic read-once BPs of size $O(n^3)$. One even gets *oblivious* read-once BPs (better known as OBDDs, ordered binary decision diagrams), where the variables appear in a fixed order along each path. Savický and Žák [21] have shown that, on the other hand, deterministic read-once BPs for WS_n require size $2^{n-O(\sqrt{n})}$, which nearly matches the asymptotically largest possible size of $O(2^n/n)$. Finally, Ablayev [1] has proved a lower bound of the same order for the randomized variant of OBDDs. These facts do not imply hardness for randomized read-once BPs, though: It is known that the so-called "matrix storage access function" that is also easy for nondeterministic and co-nondeterministic OBDDs and hard for deterministic read-once BPs and randomized OBDDs has randomized read-once BPs of polynomial size, even with zero error [20]. The function WS_n turns out to have a different behavior with respect to randomized read-once BPs:

Theorem 1. *Each randomized read-once branching program for* WS_n *with two-sided error bounded by an arbitrary constant smaller than* $1/2$ *has strongly exponential size* $2^{\Omega(n)}$.

Due to the upper bounds for nondeterministic read-once BPs, it is clear that there are large combinatorial rectangles, i.e., sets of inputs of the form $R = A \times B$, A and B sets of assignments to disjoint sets of variables, on which the function WS_n is the constant 0 or 1. This implies that the usual proof methods using upper bounds on the size of combinatorial rectangles fail for this function.

We introduce a new approach for proving Theorem 1 that can be regarded as a randomized variant of the (deterministic) method of Simon and Szegedy [22].

Functions deciding whether a graph contains a clique (i. e., a complete subgraph) of a given size build a class of combinatorially important functions that has been studied for a long time in the literature on boolean functions. Our second result is for the triangle (or 3-clique) function $cl_{3,n}$ defined on $\binom{n}{2}$ boolean variables representing the edges of an n-vertex graph and checking whether this graph has a triangle.

It is easy to see that $cl_{3,n}$ has nondeterministic read-once BPs of polynomial size. On the other hand, it has been shown by Duriš et al. [11] that the complement $\neg cl_{3,n}$ requires strongly exponential size $2^{\Omega(n^2)}$ for nondeterministic read-once BPs. Jukna and Schnitger [13] have proved related results on the so-called multipartition communication complexity of this function. Furthermore, they have established that the function $\neg cl_{4,n}$ detecting the absence of 4-cliques in n-vertex graphs requires exponential size for nondeterministic read-k BPs. Their bound is even strongly exponential for constant k.

We introduce a simple subfunction of $\neg cl_{3,n}$ that nevertheless turns out to be difficult for read-k BPs with $k > 1$. An exponential lower bound for this subfunction and thus for $\neg cl_{3,n}$ can already be shown by a straightforward application of the arguments from [13]. Here we consider the randomized case for which additional ideas are required.

Theorem 2. *There is a constant $c > 0$ such that for any $\varepsilon < 2^{-c2^{2k}}$, each randomized read-$k$ BP for $cl_{3,n}$ with two-sided error at most ε requires size $2^{\Omega(k^{-2}2^{-4k}\cdot\sqrt{n})}$.*

For the proof of the theorem, we follow the usual approach of proving a small upper bound on the size of appropriately generalized combinatorial rectangles on which the considered function is nearly constant. The main ingredient is a carefully chosen input distribution that allows us to apply a variant of a rectangle size lower bound for the set disjointness function due to Babai, Frankl, and Simon [5].

Overview on the rest of the paper. In the next section, we describe the methods for proving lower bounds on the size of randomized read-once and read-k BPs used here. In Section 4 and 5, we prove Theorem 1 and Theorem 2, resp.

3   Proof Methods

Read-once BPs. We introduce some notation first. Let X be an n-element set that we regard as a set of *variables*. Let 2^X denote the set of all *assignments to X*, i.e., mappings from X to $\{0,1\}$ that are identified with boolean vectors from $\{0,1\}^n$ as usual. Given a partition $\Pi = (X_1, X_2)$ of X, i.e., $X = X_1 \cup X_2$ and $X_1 \cap X_2 = \emptyset$, a *(combinatorial) rectangle with respect to the partition Π* is a set R such that there are sets of assignments $A \subseteq 2^{X_1}$ and $B \subseteq 2^{X_2}$ with

$R = A \times B$. We call A and B the *first* and *second part* of the rectangle, resp. If $|X_1| = \ell$, for $\ell \in \{1, \ldots, n-1\}$, we call R an ℓ-*rectangle*. If $B = 2^{X_2}$, we call R a *one-way rectangle*.

Let f be a boolean function on X. A one-way rectangle $R = A \times 2^{X_2}$, $A \subseteq 2^{X_1}$, is called f-*uniform*, if for each $x, x' \in A$ and $y \in 2^{X_2}$, $f(x,y) = f(x',y)$. Let $R(f, \ell)$ be the minimum number of ℓ-rectangles in a partition of the input space of f into f-uniform one-way ℓ-rectangles. By examining the proof methods of Borodin, Razborov, and Smolensky [10] and Okol'nishnikova [16] for deterministic and nondeterministic read-k BPs, it easy to see that the same arguments also yield lower bounds on the size of read-once BPs in terms of one-way rectangles.

Proposition 1 ([16,10]). *Let f be a boolean function on n variables and let G be a deterministic read-once BP for f. Let $\ell \in \{1, \ldots, n-1\}$. Then $|G| \geq 2n \cdot R(f, \ell)$.*

In our terminology, the proof method for deterministic read-once BPs due to Simon and Szegedy [22] (in its most simple form) provides a lower bound on $R(f, \ell)$ by upper bounding the measure of one-way rectangles under the uniform distribution: If $|R|/2^n \leq \beta$ for each one-way ℓ-rectangle R that is f-uniform, then obviously $R(f, \ell) \geq 1/\beta$. We now extend these ideas to the randomized case. As usual, we prove lower bounds for read-once BPs computing an approximation of the considered function and then obtain a lower bound for randomized read-once BPs with the same error probability via averaging. Call a function g an ε-*approximation* of f if it differs from f on at most an ε-fraction of all inputs. Let $R_\varepsilon(f, \ell)$ be the minimum of $R(g, \ell)$ taken over all ε-approximations g of f. As a corollary of the above proposition, we get:

Proposition 2. *Let f be a boolean function on n variables and let G be a deterministic read-once BP computing an ε-approximation of f. Let $\ell \in \{1, \ldots, n-1\}$. Then $|G| \geq 2n \cdot R_\varepsilon(f, \ell)$.*

It remains to discuss how lower bounds on $R_\varepsilon(f, \ell)$ can be proved. Let R be a one-way rectangle that is g-uniform. We say that g *approximates f uniformly on R with error ε*, if for each x in the first part of R the fraction of y in the second part of R for which $g(x,y) \neq f(x,y)$ is bounded above by ε. The following lemma, together with Proposition 2, allows us to prove lower bounds for approximating read-once BPs.

Lemma 1. *Let $0 \leq \varepsilon < \varepsilon' < 1/2$. Let β be an upper bound on $|R|/2^n$ for all one-way ℓ-rectangles R that are g-uniform for a function g that approximates f uniformly on R with error at most ε'. Then $R_\varepsilon(f, \ell) \geq (1 - \varepsilon/\varepsilon')/\beta$.*

Proof. Let $k = R_\varepsilon(f, \ell)$. Let g be an ε-approximation of f, and let $R_1, \ldots, R_k$ form a partition of the input space of f into one-way ℓ-rectangles that are all g-uniform. We show that there is a one-way ℓ-rectangle R with $|R|/2^n \geq (1 - \varepsilon/\varepsilon')/k$, which is g-uniform, and which is such that g approximates f uniformly on R with error at most ε'.

Let A be the set of all partial inputs in the first parts of the rectangles $R_1, \ldots, R_k$. Then $|A| = 2^\ell$. For each $x \in A$, let $X_1(x)$ with $|X_1(x)| = \ell$ be its respective variable set (belonging to the first part of its rectangle). Let $\varepsilon(x)$ be the fraction of inputs from $\{x\} \times 2^{X - X_1(x)}$ for which g differs from f. Due to the definitions, the sets $\{x\} \times 2^{X - X_1(x)}$, $x \in A$, partition the whole input space. Hence, by the law of total probability, $\sum_{x \in A} \varepsilon(x) \, 2^{-\ell} \le \varepsilon$. Let $A' = \{x \in A \mid \varepsilon(x) \le \varepsilon'\}$. By Markov's inequality, $|A'| \ge (1 - \varepsilon/\varepsilon')|A|$. By averaging, there is a set $A'' \subseteq A'$ of size $|A''| \ge |A'|/k$ such that all inputs from A'' belong to the same rectangle. Furthermore, there is a fixed set X_1 of size ℓ such that $X_1(x) = X_1$ for all $x \in A''$. Now it is obvious that the set $R = A'' \times 2^{X - X_1}$ of size $|A''| \cdot 2^{n-\ell} \ge (1 - \varepsilon/\varepsilon') \, 2^n / k$ is a one-way rectangle with the desired properties. $\qquad\square$

Read-k BPs. Again, we start by turning a given randomized read-k BP into a deterministic read-k BP computing an approximation. We use the following fact that can be proved by arguments similar to those of Newman [15] showing that public coin communication protocols can simulate private coin protocols with small overhead.

Lemma 2 ([18]). *Let $0 \le \varepsilon < \varepsilon' < 1/2$. Let G be a randomized read-k BP that represents the boolean function f on n variables with error ε. Then there are $m = \mathrm{poly}(n)/(\varepsilon' - \varepsilon)$ deterministic read-k BPs $G_1, \ldots, G_m$ representing functions $g_1, \ldots, g_m$ such that $|G_i| \le |G|$ for $i = 1, \ldots, m$ and for all inputs x, $(1/m) \sum_{i=1}^{m} (g_i(x) \ne f(x)) \le \varepsilon'$.*

In particular, the lemma ensures that there is a read-k BP representing an ε'-approximation of f that is no larger than the original randomized read-k BP.

Let X be a set of n variables. Call a pair of sets $X_1, X_2 \subseteq X$ a λ-*balanced cover* of X, where $0 < \lambda \le 1/2$, if $X = X_1 \cup X_2$ and $|X_i - X_{3-i}| \ge \lambda|X|$ for $i = 1, 2$. Call a set of inputs $R \subseteq 2^X$ a *(generalized combinatorial) rectangle with respect to Γ* (Γ-*rectangle* for short), $\Gamma = (X_1, X_2)$ a cover of X, if the characteristic function of R can be written as $R = R_1 \wedge R_2$, where for $i = 1, 2$ the function R_i only depends on the variables in X_i. Finally, call R a c-*colored rectangle* for a boolean function f, if $R \subseteq f^{-1}(c)$, $c \in \{0, 1\}$.

We decompose read-k BPs using the following theorem obtained by combining the proof methods of Okol'nishnikova [16] and Borodin, Razborov, and Smolensky [10] with a combinatorial lemma on balanced covers due to Thathachar [23] (see also [6,7,13] for similar results).

Theorem 3 ([16,10,23]). *Let k be a natural number, $8k^2 2^k \le r \le n$, and $\lambda = 2^{-k-1}$. Then each deterministic read-k BP G for a boolean function f on n variables defines a partition of the input space of f into at most $(2|G|)^r$ 0- and 1-colored rectangles with respect to λ-balanced covers of the variables.*

Call a collection of rectangles obtained by this theorem a *rectangle representation of f*. The theorem also yields an upper bound on the number of rectangles in a rectangle representation of an approximation for f. On the other hand, we can lower bound the number of such rectangles by the following lemma obtained again by averaging arguments analogous to Lemma 1.

Lemma 3. *Let $\varepsilon, \varepsilon'$ be constants with $0 \le \varepsilon < \varepsilon' < 1/2$. Let f be a boolean function and let g be an ε-approximation of f with respect to a distribution μ on* the inputs. Let $R_1, \ldots, R_k$ be the 1-colored rectangles in a rectangle representation of g. Then there is a rectangle R_i with $\mu(R_i) \geq \left(\mu\left(f^{-1}(1)\right) - \varepsilon/\varepsilon'\right)/k$ and $\mu\left(R_i \cap f^{-1}(0)\right)/\mu(R_i) \leq \varepsilon'$.

4 Proof of Theorem 1

Our aim is to apply the method described in Section 3 to WS_n. We collect some technical lemmas before we do this. Recall that $p(n)$ is the smallest prime greater than n.

Lemma 4. Let $q = q(n)$ be a sequence of primes and let $n \leq q - 1$ and $n = \Omega\left(q^{2/3+\delta}\right)$ for any constant $\delta > 0$. Let $a_1, \ldots, a_n, b \in \mathbb{Z}_q^* = \mathbb{Z}_q - \{0\}$ where the $a_1, \ldots, a_n$ are pairwise different. Then for $(x_1, \ldots, x_n) \in \{0,1\}^n$ chosen uniformly at random, $\left|\Pr\{a_1 x_1 + \cdots + a_n x_n \equiv b \bmod q\} - 1/q\right| = 2^{-\Omega\left(q^{3\delta}\right)}$.

Lemma 5. $p(n) = n + O(\log n)$.

The proof of the first lemma, which we have to omit due to lack of space, is based on a suitable formula for the exact number of boolean solutions of the equation $a_1 x_1 + \cdots + a_n x_n \equiv b \bmod \mathbb{Z}_q$ (see, e. g, [4]). The second lemma follows from the prime number theorem.

Since we want to prove an upper bound on the size of one-way rectangles on which the considered function is well approximated, it is natural to try to extend a suitable non-approximability result for one-way communication protocols. We consider the well-known "index function" IND_n, defined on $x = (x_1, \ldots, x_n) \in X = \{0,1\}^n$ and $y \in Y = \{1, \ldots, n\}$ by $\mathrm{IND}_n(x, y) = x_y$. Kremer, Nisan, and Ron [14] have shown that one-way communication protocols which approximate this function with two-sided error bounded by $\varepsilon < 1/8$ have linear complexity. Using the ideas from their paper, it is easy to get the following extended result:

Lemma 6. Let $\varepsilon, \varepsilon^*$ be constants with $0 \leq \varepsilon < \varepsilon^* < 1/2$. Let $R = A \times Y$ be a one-way rectangle that is g-uniform for a function g that differs from IND_n on at most an ε-fraction of R. Then $|R|/(n \cdot 2^n) \leq 2^{-(1-H(\varepsilon^*))n-\Omega(\log n)}$, where $H(x) = -(x \log x + (1-x) \log(1-x))$.

We now obtain the following upper bound on the size of one-way rectangles for WS_n:

Lemma 7. Let n be a natural number and $p = p(n)$, where $p(n)$ is the smallest prime greater than n. Let $\ell = n - \Theta\left(p^{2/3+\delta}\right)$, where $\delta > 0$ is an arbitrarily small constant. Let $\varepsilon, \varepsilon^*$ be arbitrary constants with $0 \leq \varepsilon < \varepsilon^* < 1/2$. Then for each one-way ℓ-rectangle R that is g-uniform for a function g approximating WS_n uniformly on R with error at most ε, $|R|/2^n \leq 2^{-(1-H(\varepsilon^*))n+o(n)}$.

Before turning to the proof, we use this to derive our first main result.

Proof of Theorem 1. Let ε be any constant smaller than $1/2$. Choose $\varepsilon^* = (\varepsilon + 1/2)/2 = \varepsilon/2 + 1/4$ in the above lemma. Together with Lemma 1 and Proposition 2, we get that each read-once BP G that approximates WS_n with error at most ε has size at least $2^{(1-H(\varepsilon/2+1/4))n-o(n)}$. Via averaging, this also gives Theorem 1. $\square$

Proof of Lemma 7. Identify the variables of WS_n with their index set $X = \{1,\ldots,n\}$. Let R and g be as in the lemma. Let (X_1, X_2) be a partition of X with $|X_1| = \ell$ and let $R = A_R \times 2^{X_2}$ with $A_R \subseteq 2^{X_1}$. For $0 \le i \le p-1$, let $A_{R,i}$ be the set of all partial assignments $x \in A_R$ with $\sum_{j \in X_1} j x_j \equiv i \bmod p$. Then there is an i with $|A_{R,i}| \ge |A_R|/p$. Fix such an i for the rest of the proof and let $A = A_{R,i}$. Since g approximates WS_n uniformly on R with error at most ε, g differs from WS_n on at most an ε-fraction of all inputs in $A \times 2^{X_2}$.

For $0 \le j \le p-1$, define B_j as the set of all assignments x to X_2 with $\sum_{r \in X_2} r x_r \equiv j \bmod p$. Let $X_1 = \{j_1,\ldots,j_\ell\}$, where $j_1 < \cdots < j_\ell$ and $\ell = n - \Theta(p^{2/3+\delta})$ by assumption. Due to Lemma 5, we know that $\ell \ge n - o(n)$ and $\ell/p \ge 1 - o(1)$. Define $B = B_{(j_1-i) \bmod p} \cup \cdots \cup B_{(j_\ell-i) \bmod p}$. By Lemma 4, $|B_j| \ge (2^{|X_2|}/p) \cdot (1-o(1))$ for all j. Thus $|B|/2^{|X_2|} \ge (\ell/p) \cdot (1-o(1)) \ge 1-o(1)$. Since g differs from WS_n on at most an ε-fraction of $A \times 2^{X_2}$, it is incorrect only for at most a fraction of $\varepsilon' \le \varepsilon \cdot 2^{|X_2|}/|B| \le \varepsilon \cdot (1+o(1))$ of all inputs in $A \times B$.

Since all B_j are non-empty (for sufficiently large input size) and almost of the same size by Lemma 4, we can apply the law of total probability and Markov's inequality to find $b_1 \in B_{(j_1-i) \bmod p}, \ldots, b_\ell \in B_{(j_\ell-i) \bmod p}$ such that g differs from WS_n on at most an ε''-fraction of all inputs in $R' = A \times \{b_1,\ldots,b_\ell\}$ with $\varepsilon'' \le \varepsilon' \cdot (1+o(1)) \le \varepsilon \cdot (1+o(1))$.

Now we are ready to apply the result for the index function. Let $R'' = A \times \{1,\ldots,\ell\}$. Define h on inputs $x \in \{0,1\}^\ell$ and $y \in \{1,\ldots,\ell\}$ by $h(x,y) = g(x, b_y)$. Then the rectangle R'' is h-uniform since R' is g-uniform. Since g differs at most on an ε''-fraction of R' from WS_n, h differs from IND_ℓ on at most an ε''-fraction of R''. Furthermore, $\varepsilon < \varepsilon^*$ and $\varepsilon'' \le \varepsilon \cdot (1 + o(1))$, which implies $\varepsilon'' < \varepsilon^*$ for sufficiently large input size. By Lemma 6, $|A|/2^\ell = |R''|/(\ell \cdot 2^\ell) \le 2^{-(1-H(\varepsilon^*))\ell - \Omega(\log \ell)}$. Now $|A| \ge |A_R|/p$ and $|R| = |A_R| \cdot 2^{n-\ell}$. Thus, $|R|/2^n = |A_R|/2^\ell \le p|A|/2^\ell \le p \cdot 2^{-(1-H(\varepsilon^*))\ell - \Omega(\log \ell)}$. Since $p \le n + o(n)$ and $\ell \ge n - o(n)$, this bound is of the desired size. □

5 Proof of Theorem 2

We now prove the lower bound for the triangle function $cl_{3,n}$. Here it is more convenient to consider the complement of this function, the *triangle-freeness function* $\neg cl_{3,n}$. To apply the method from Section 3, we prove that each rectangle containing a large fraction of inputs that encode triangle-free graphs is necessarily small. One of the main ingredients of the proof is a similar bound on the size of rectangles for the *(set) disjointness function*, which is a variant of a well-known result due to Babai, Frankl, and Simon [5].

The disjointness function $DISJ_n$ is defined by $DISJ_n(x,y) = 1$ if the boolean vectors $x = (x_1,\ldots,x_n)$ and $y = (y_1,\ldots,y_n)$ represent disjoint subsets of $\{1,\ldots,n\}$ and 0 otherwise. Babai, Frankl, and Simon have considered the measure of usual combinatorial rectangles on which this function is almost the constant 1 for input sets of fixed size $\sqrt{n}$ chosen uniformly at random. Since we will need a product distribution on the single input bits later on, we work with the following approximation of their distribution: Define μ on the inputs of $DISJ_n$

by setting each bit to 1 with probability $1/\sqrt{n}$ and doing this independently for different bits. Using the fact that μ still produces sets of size approximately $\sqrt{n}$ with high probability, the proof of Babai, Frankl, and Simon can be adapted to yield the following result (the proof is omitted due to the space constraints).

Lemma 8. *Let* $\Pi = (X_1, X_2)$ *with* $X_1 = \{x_1, \ldots, x_n\}$ *and* $X_2 = \{y_1, \ldots, y_n\}$. *For any constant* ε *with* $0 \le \varepsilon \le 0.1$, *each* Π-*rectangle* R *with* $\mu(R \cap \mathrm{DISJ}_n^{-1}(0))/\mu(R) \le \varepsilon$ *satisfies* $\mu(R) = 2^{-\Omega(\sqrt{n})}$.

Instead of the simple partition of the input variables in the above result, we have to cope with the more general covers of the variables arising from the decomposition of read-k BPs. The idea how this is done is captured in the following lemma.

Lemma 9. *Let* μ *be a product distribution on* $\{0,1\}^{2n}$ *and* $1 \le s \le n$. *Let* $\Gamma = (X_1, X_2)$ *be a cover of the input variables* $X = \{x_i, y_i \mid 1 \le i \le n\}$ *of* DISJ_n *with* $X_1' = \{x_1, \ldots, x_s\} \subseteq X_1 - X_2$ *and* $X_2' = \{y_1, \ldots, y_s\} \subseteq X_2 - X_1$. *Let* R *be a* Γ-*rectangle with* $\mu(R \cap \mathrm{DISJ}_n^{-1}(0))/\mu(R) \le \varepsilon$, *where* $0 \le \varepsilon < 1/2$. *Then for each* ε' *with* $\varepsilon < \varepsilon' < 1/2$ *there is a* Π-*rectangle* R', $\Pi = (X_1', X_2')$, *such that* *(i)* $\mu(R' \cap \mathrm{DISJ}_s^{-1}(0))/\mu(R') \le \varepsilon'$; *and (ii)* $\mu(R') \ge (1 - \varepsilon)(1 - \varepsilon/\varepsilon')\mu(R)$.

Proof. Let A be the set of all inputs of DISJ_n where for each $i = s+1, \ldots, n$ at least one of the variables x_i and y_i is set to 0. Let $S = R \cap A$. Since $\overline{A} \subseteq \mathrm{DISJ}_n^{-1}(0)$, we have $\varepsilon \ge \mu(R \cap \mathrm{DISJ}_n^{-1}(0))/\mu(R) = \left(\mu(S \cap \mathrm{DISJ}_n^{-1}(0)) + \mu(R \cap \overline{A})\right) / \left(\mu(S) + \mu(R \cap \overline{A})\right) \ge \mu(S \cap \mathrm{DISJ}_n^{-1}(0))/\mu(S)$. Furthermore, $\mu(R \cap \overline{A}) \le \mu(R \cap \mathrm{DISJ}_n^{-1}(0)) \le \varepsilon\,\mu(R)$ and thus $\mu(S) = \mu(R \cap A) \ge (1 - \varepsilon)\,\mu(R)$.

Let $X' = X_1' \cup X_2'$ and let B be the projection of A to the variables in $X - X'$. For an assignment $b \in B$, let S_b be the set of assignments in S that agree with b on $X - X'$. Observe that S is the disjoint union of all S_b with $b \in B$. Let $S_b = S_b' \times \{b\}$, where S_b' is a set of assignments to X'. Let B' be the set of all $b \in B$ with $\mu(S_b \cap \mathrm{DISJ}_n^{-1}(0))/\mu(S_b) \le \varepsilon'$. By Markov's inequality, $\sum_{b \in B'} \mu(S_b) \ge (1 - \varepsilon/\varepsilon')\mu(S)$. Since μ is a product distribution, $\mu(S_b) = \mu(S_b')\mu(b)$, and we get $\sum_{b \in B'} \mu(S_b') \cdot \mu(b)/\mu(B') \ge (1 - \varepsilon/\varepsilon')\mu(S)/\mu(B') \ge (1 - \varepsilon/\varepsilon')\mu(S)$. Hence, there is a $b \in B'$ with $\mu(S_b') \ge (1 - \varepsilon/\varepsilon')\mu(S) \ge (1 - \varepsilon/\varepsilon')(1 - \varepsilon)\mu(R)$. Due to the definition of the disjointness function and the fact that μ is a product distribution, $\mu(S_b \cap \mathrm{DISJ}_n^{-1}(0))/\mu(S_b) = \mu(S_b' \cap \mathrm{DISJ}_s^{-1}(0))/\mu(S_b')$. Altogether, $R' = S_b'$ is a Π-rectangle with the properties claimed in the lemma. $\square$

The second main step in the proof of Theorem 2 is to carefully choose a hard subfunction of the triangle-freeness function $\neg \mathrm{cl}_{3,n}$.

Duriš *et al.* [11] have proved that there is a subfunction of $\neg \mathrm{cl}_{3,n}$ that requires strongly exponential size for nondeterministic read-once BPs. Their subfunction cannot be used to apply the method from Section 3 for the read-k case, though, as argued in [13]. It turns out that an even further restricted subfunction works, but we have to pay with a weaker lower bound for the additional restrictions. We consider a subfunction of the triangle-freeness function $\neg \mathrm{cl}_{3,n+1}$ that we apply to graphs whose vertices are partitioned into sets $U = \{0\}$ and $V = \{1, \ldots, n\}$.

For a set of edges $E \subseteq V \times V$ that will be chosen below, we fix all variables belonging to $V \times V$ by setting $x_{ij} = 1$ if $\{i, j\} \in E$ and $x_{ij} = 0$ otherwise. This gives us a subfunction Δ_E of $\neg \mathrm{cl}_{3,n+1}$ that only depends on the variables belonging to $U \times V$.

A triangle formed by two edges in $U \times V$ connecting the vertex 0 to two vertices $v, w \in V$ and the edge $\{v, w\} \in E$ is called a *test* for E. Call the edges in $U \times V$ *free*. Two tests are said to *collide*, if a triangle can be formed by picking one free edge from each test and an edge from E. In particular, tests collide if they share a free edge. Given a cover Γ of $U \times V$, we call a test *split by* Γ if its free edges belong to different parts of the cover. The following lemma allows us to choose E such that for any not too large collection of balanced covers, there are many split tests for each cover in the collection. This will be sufficient for proving the desired lower bound.

Lemma 10. *Let $n^{-1/6+\delta} \le \lambda \le 1/2$ for some constant $\delta > 0$. Then there is a set of edges $E \subseteq V \times V$ and a constant $a > 0$ such that for all λ-balanced covers $\Gamma_1, \ldots, \Gamma_p$ of $U \times V$ with $p \le 2^{a\lambda^6 n}$ there is a collision-free set T of tests for E, $|T| = \Omega(\lambda^4 n)$, such that for each $i = 1, \ldots, p$ there is a subset $T_i \subseteq T$ of $\Omega(\lambda^2 |T|)$ tests that are split by Γ_i.*

This lemma is proved by a straightforward combination of ideas from [11] and [13] (the proof is omitted due to lack of space). Now we are ready to prove our second main result.

Proof of Theorem 2. Choose E according to Lemma 10. We consider the subfunction Δ_E of $\neg \mathrm{cl}_{3,n+1}$ obtained by fixing all variables belonging to $V \times V$ accordingly. Let G be a randomized read-k BP for Δ_E with error at most ε, $0 \le \varepsilon < 2^{-c2^{2k}}$ for some constant $c > 0$ chosen below.

Lemma 2 yields deterministic read-k BPs $G_1, \ldots, G_m$ and functions $g_1, \ldots, g_m$ with $m = \mathrm{poly}(n)$ such that each g_i approximates Δ_E on a fraction of the inputs. Set $r = 8k^2 2^k$ and $\lambda = 2^{-k-1}$. By Theorem 3, each read-k BP G_i defines a partition of the input space into at most $(2|G_i|)^r \le (2|G|)^r$ rectangles that are 0- or 1-colored rectangles for g_i and defined according to λ-balanced covers of the variables of Δ_E. Hence, there are $p \le m \cdot (2|G|)^r = O(\mathrm{poly}(n)|G|^r)$ such covers of the input variables altogether in all rectangles. Suppose that $p \le 2^{a\lambda^6 n}$, where $a > 0$ is the constant from Lemma 10 (otherwise, we are done), and apply the lemma to get the sets T and $T_1, \ldots, T_p$ of tests for E. Let $t = |T| = \Omega(\lambda^4 n)$ and $|T_i| = s = b\lambda^2 t$ for each i, where $b > 0$ is a suitable constant.

Define the distribution μ on the inputs of Δ_E as follows. Set all variables not belonging to edges in tests of T to 0. For each test in T and each of its two free edges, set the respective variable to 1 with probability $p = 1/\sqrt{s}$, and do this independently for all different variables. By the usual averaging argument, we obtain an $i \in \{1, \ldots, m\}$ such that g_i differs from Δ_E on at most an ε-fraction of all inputs with respect to μ. Let $R_1, \ldots, R_\ell$, $\ell \le (2|G|)^r$, be the 1-colored rectangles in the partition of the input space induced by G_i.

Let ε' be a constant chosen such that $\varepsilon < \varepsilon' < 1/2$. By Lemma 3, we get a rectangle $R = R_j$ with $\mu(R_j \cap \Delta_E^{-1}(0))/\mu(R_j) \le \varepsilon'$ and $\mu(R_j) \ge \rho =$

$\left(\mu\left(\Delta_E^{-1}(1)\right) - \varepsilon/\varepsilon'\right)/\ell$. Using the definitions, it is easy to prove that $\mu\left(\Delta_E^{-1}(1)\right) \geq e^{-2/(b\lambda^2)}$. We choose $\varepsilon' = \varepsilon/\left((1-\delta)\,\mu\left(\Delta_E^{-1}(1)\right)\right)$ for some small constant $\delta > 0$. Then $\mu\left(\Delta_E^{-1}(1)\right) - \varepsilon/\varepsilon' = \delta \cdot \mu\left(\Delta_E^{-1}(1)\right)$ is bounded below by a positive constant and thus $\rho \geq c'/\ell$ for some constant $c' > 0$.

Due to the definition of μ, we may assume that the rectangle R obtained above only contains inputs for Δ_E where all variables outside of T are set to 0 without changing the bounds on the fraction of non-accepted inputs in R and on $\mu(R)$. Then Δ_E is equal to 1 for an input in R if and only if any of the tests in T has both of its free edges set to 1. Hence, Δ_E decides whether the input bits corresponding to the free edges encode disjoint sets. Let T' with $|T'| = s$ be the set of split tests according to Lemma 10 for the cover Γ of the input variables used by R. Let $(x_1, y_1), \ldots, (x_s, y_s)$ be the pairs of variables belonging to the free edges of the tests in T'. Then R can be identified with a rectangle as described in the hypothesis of Lemma 9.

Let $\varepsilon'' = \varepsilon'/(1 - \delta')$ for some small constant $\delta' > 0$ such that $\varepsilon' < \varepsilon'' < 1/2$. By Lemma 9, we get a rectangle R' with respect to the usual partition $\Pi = (\{x_1, \ldots, x_s\}, \{y_1, \ldots, y_s\})$ of the variables of DISJ_s with $\mu\left(R' \cap \mathrm{DISJ}_s^{-1}(0)\right)/\mu(R') \leq \varepsilon''$ and $\mu(R') \geq (1-\varepsilon')\left(1 - \varepsilon'/\varepsilon''\right)\rho = (1-\varepsilon')\,\delta'\rho = \Omega(\rho)$. Recall that $\rho \geq c'/\ell$ for some constant $c' > 0$. If we start with ε small enough such $\varepsilon'' \leq 0.1$, then $\mu(R') = 2^{-\Omega(\sqrt{s})}$ by Lemma 8. Thus, $2^{-\Omega(\sqrt{s})} \geq \mu(R') \geq c''/\ell$ for some constant $c'' > 0$ and $\ell = (2|G|)^r$, which implies $|G| = 2^{\Omega(\sqrt{s}/r)} = 2^{\Omega((\lambda^3/r)\sqrt{n})}$. Substituting $\lambda = 2^{-k-1}$ and $r = 8k^2 2^k$ gives the desired lower bound, $|G| = 2^{\Omega(k^{-2}2^{-4k}\cdot\sqrt{n})}$.

Finally, it remains to choose an appropriate upper bound for ε. We need $\varepsilon'' \leq 0.1$. By the above definitions, $\varepsilon'' = \varepsilon'/(1-\delta')$ and $\varepsilon' = \varepsilon/\left((1-\delta)\mu\left(\Delta_E^{-1}(1)\right)\right)$. Hence, $\varepsilon'' \leq 0.1$ is equivalent to $\varepsilon \leq 0.1 \cdot (1-\delta) \cdot (1-\delta') \cdot \mu\left(\Delta_E^{-1}(1)\right) \leq 2^{-c2^{2k}}$ for an appropriate constant $c > 0$. □

Acknowledgment. Thanks to Ingo Wegener for proofreading and to Stasys Jukna and Ingo Wegener for helpful discussions.

References

1. F. Ablayev. Randomization and nondeterminism are incomparable for polynomial ordered binary decision diagrams. In *Proc. of ICALP, LNCS 1256*, 195–202. Springer, 1997.
2. M. Ajtai. Determinism versus non-determinism for linear time RAMs with memory restrictions. In *Proc. of 31st STOC*, 632–641, 1999.
3. M. Ajtai. A non-linear time lower bound for boolean branching programs. In *Proc. of 40th FOCS*, 60–70, 1999.
4. L. Babai. The Fourier transform and equations over finite abelian groups. Lecture Notes, version 1.2, Dec. 1989.
5. L. Babai, P. Frankl, and J. Simon. Complexity classes in communication complexity theory. In *Proc. of 27th FOCS*, 337–347, 1986.

6. P. Beame, T. S. Jayram, and M. Saks. Time-space tradeoffs for branching programs. *Journal of Computer and System Sciences*, 63(4):542–572, 2001.
7. P. Beame, M. Saks, X. Sun, and E. Vee. Super-linear time-space tradeoff lower bounds for randomized computation. In *Proc. of 41st FOCS*, 169–179, 2000.
8. P. Beame and E. Vee. Time-space tradeoffs, multiparty communication complexity, and nearest neighbor problems. In *Proc. of 34th STOC*, 688–697, 2002.
9. B. Bollig, M. Sauerhoff, I. Wegener. On the nonapproximability of boolean functions by OBDDs and read-k-times branching programs. *Information and Computation*, 178:263–278, 2002.
10. A. Borodin, A. A. Razborov, and R. Smolensky. On lower bounds for read-k-times branching programs. *Computational Complexity*, 3:1–18, 1993.
11. P. Duriš, J. Hromkovič, S. Jukna, M. Sauerhoff, and G. Schnitger. On multipartition communication complexity. In *Proc. of 18th STACS, LNCS 2010*, 206–217. Springer, 2001.
12. J. Hromkovič and M. Sauerhoff. Tradeoffs between nondeterminism and complexity for communication protocols and branching programs. In *Proc. of 17th STACS, LNCS 1770*, 145–156. Springer, 2000.
13. S. Jukna and G. Schnitger. On multi-partition communication complexity of triangle-freeness. *Combinatorics, Probability & Computing* 11(6):549–569, 2002.
14. I. Kremer, N. Nisan, and D. Ron. On randomized one-round communication complexity. *Computational Complexity*, 8(1):21–49, 1999.
15. I. Newman. Private vs. common random bits in communication complexity. *Information Processing Letters*, 39(2):67–71, 1991.
16. E. A. Okol'nishnikova. On lower bounds for branching programs. *Siberian Advances in Mathematics*, 3(1):152–166, 1993.
17. M. Sauerhoff. Lower bounds for randomized read-k-times branching programs. In *Proc. of 15th STACS, LNCS 1373*, 105–115. Springer, 1998.
18. M. Sauerhoff. *Complexity Theoretical Results for Randomized Branching Programs*. PhD thesis, Univ. Dortmund. Shaker, Aachen, 1999.
19. M. Sauerhoff. Approximation of boolean functions by combinatorial rectangles. *ECCC*, Technical Report 58, 2000. To appear in *Theoretical Computer Science*.
20. M. Sauerhoff. On the size of randomized OBDDs and read-once branching programs for k-stable functions. *Computational Complexity*, 10:155–178, 2001.
21. P. Savický and S. Žák. A read-once lower bound and a $(1, +k)$-hierarchy for branching programs. *Theoretical Computer Science*, 238(1-2):347–362, 2000.
22. J. Simon and M. Szegedy. A new lower bound theorem for read-only-once branching programs and its applications. In J.-J. Cai, editor, *Advances in Computational Complexity Theory, DIMACS Series in Discrete Mathematics and Theoretical Computer Science 13*, 183–193. American Mathematical Society, 1993.
23. J. Thathachar. On separating the read-k-times branching program hierarchy. In *Proc. of 30th STOC*, 653–662, 1998.
24. I. Wegener. *Branching Programs and Binary Decision Diagrams—Theory and Applications*. Monographs on Discrete and Applied Mathematics. SIAM, Philadelphia, PA, 2000.

Branch-Width, Parse Trees, and Monadic Second-Order Logic for Matroids

(Extended Abstract)

Petr Hliněný*

Institute of Mathematics and Comp. Science (IMI SAV), Matej Bel University
Severná ulica 5, 974 00 Banská Bystrica, Slovakia
and
Institute for Theoretical Comp. Science (ITI MFF)**, Charles University
Malostranské nám. 25, 118 00 Praha 1, Czech Republic
hlineny@member.ams.org

Abstract. We introduce "matroid parse trees" which, using only a limited amount of information, can build up all matroids of bounded branch-width representable over a finite field. We prove that if $\mathcal{M}$ is a family of matroids described by a sentence in the second-order monadic logic of matroids, then the parse trees of bounded-width representable members of $\mathcal{M}$ can be recognized by a finite tree automaton. Since the cycle matroids of graphs are representable over any finite field, our result directly extends the well-known "MS_2-theorem" for graphs of bounded tree-width by Courcelle and others. This work has algorithmic applications in matroid or coding theories.

Keywords: representable matroid, branch-width, monadic second-order logic, fixed-parameter complexity.
Classification: parametrized complexity, and logic in computer science.
(Math subjects 05B35, 68R05, 03D05.)

1   Introduction

We assume that the reader is familiar with basic concepts of graph theory, for example Diestel [8]. In the past decade, the notion of a *tree-width* of graphs [20, 3] attracted plenty of attention, both from graph-theoretical and computational points of view. This attention followed the pioneer work of Robertson and Seymour on the Graph Minor Project [19], and results of various researchers using tree-width in parametrized complexity.

The theory of parametrized complexity provides a background for analysis of difficult algorithmic problems which is finer than classical complexity theory.

---

* This work is based on an original research that the author carried out at the Victoria University of Wellington in New Zealand, supported by a Marsden Fund research grant to Geoff Whittle.
** ITI is supported by Ministry of Education of Czech Republic as project LN00A056.

H. Alt and M. Habib (Eds.): STACS 2003, LNCS 2607, pp. 319–330, 2003.

For an introduction, we suggest [9]. Briefly saying, a problem is "fixed-parameter tractable" if there is an algorithm having its running time with the (possible) super-polynomial part separated in terms of some "parameter", which is supposed to be small even for large input in practice. Successful practical applications of this concept are known, for example, in computational biology or in database theory: Imagine a query of a small size k to a large database of size $n \gg k$; then an $O(2^k \cdot n)$ parametrized algorithm may be better in practice than, say, an $O(n^k)$ algorithm, or even than an $O((kn)^c)$ polynomial algorithm. We are, in particular, interested in studying problems that are parametrized by a tree-like structure of the input.

The notion of a *branch-width* is closely related to that of a tree-width [20], but a branch-decomposition does not refer to vertices and so branch-width directly generalizes from graphs to matroids. We postpone formal definitions till the next sections. Branch-width has recently shown to be a very interesting structural matroid parameter, too. Besides other results, we mention well-quasi-ordering of matroids of bounded branch-width over finite fields [11].

We show in this paper that the branch-width and branch-decompositions of representable matroids also have interesting computation-theoretical aspects. Namely we prove a main result analogous to so called "MS_2-theorem" by Courcelle [5] (also [2] and [4]), for matroids represented by matrices over a finite field $\mathbb{F}$ (Theorem 4.3): If $\mathcal{M}$ is a family of matroids described by a sentence in the second-order monadic logic of matroids, then the "parse trees" of bounded-branch-width $\mathbb{F}$-represented members of $\mathcal{M}$ are recognizable by a finite tree automaton. This result covers, among other applications, the cycle matroids of graphs. Our proof follows the main ideas of Abrahamson–Fellows' [1] automata-theoretical approach to Courcelle's theorem.

The results here are formulated in the language of matroid theory since it is natural and convenient, and since it shows the close relations of this research to well-known graph structural and computational concepts. Our work could be, as well, viewed as results about matrices, point configurations, or about linear codes over a finite field $\mathbb{F}$. The key to the subject is the notion of parse trees for bounded-width $\mathbb{F}$-represented matroids, defined in Section 3. We propose these parse trees as a powerful tool for handling matroids or point configurations of bounded branch-width in general.

Our main result, the above mentioned Theorem 4.3, has important consequences mainly in the field of parametrized complexity. A related result of ([15] or Theorem 5.1) presents an algorithm that constructs a bounded-width parse tree of an $\mathbb{F}$-represented matroid M in cubic time if the branch-width of M is bounded by a constant. Hence we prove that matroid properties expressible in the monadic second-order logic are fixed-parameter tractable for $\mathbb{F}$-represented matroids of bounded branch-width (Corollary 5.2). The applications, in addition to extensions of some "classical" graph problems like hamiltonicity for bounded tree-width, include testing all minor-closed properties of matroids for bounded branch-width ([16] or Theorem 5.4). Another indirect application via ([12] or Theorem 5.5) gives algorithms for computing the Tutte polynomial of a matroid, the critical index, and the Hamming weight or the weight enumerator of a linear code. Read more in Section 5.

Our research involves areas of theoretical computer science, structural matroid theory, and also of logic. In order to make the paper accessible to a wide audience of computer scientists, we provide sufficient introductory definitions for all of these areas. We refer to [14] for full proofs of our results.

2   Basics of Matroids

We refer to Oxley [18] for basic matroid terminology. A *matroid* is a pair $M = (E, \mathcal{B})$ where $E = E(M)$ is the ground set of M (elements of M), and $\mathcal{B} \subseteq 2^E$ is a nonempty collection of *bases* of M. Moreover, matroid bases satisfy the "exchange axiom"; if $B_1, B_2 \in \mathcal{B}$ and $x \in B_1 - B_2$, then there is $y \in B_2 - B_1$ such that $(B_1 - \{x\}) \cup \{y\} \in \mathcal{B}$. Subsets of bases are called *independent sets*, and the remaining sets are *dependent*. The *rank function* $\mathrm{r}_M : 2^E \to \mathbb{N}$ tells the maximal cardinality of an independent subset of a set in M.

If G is a graph, then its *cycle matroid* on the ground set $E(G)$ is denoted by $M(G)$. The bases of $M(G)$ are the spanning forests of G, and the minimal dependent sets of $M(G)$ are the circuits of G. In fact, a lot of matroid terminology is inherited from graphs. Another typical example of a matroid is a finite set of vectors with usual linear dependency.

The *dual* matroid M^* of M is defined on the same ground set E, and the bases of M^* are the set-complements of the bases of M. An element e of M is called a *loop* (a *coloop*), if $\{e\}$ is dependent in M (in M^*). The matroid $M \setminus e$ obtained by *deleting* a non-coloop element e is defined as $(E - \{e\}, \mathcal{B}^-)$ where $\mathcal{B}^- = \{B : B \in \mathcal{B}, e \notin B\}$. The matroid M/e obtained by *contracting* a non-loop element e is defined using duality $M/e = (M^* \setminus e)^*$. (This corresponds to contracting an edge in a graph.) A *minor* of a matroid is obtained by a sequence of deletions and contractions of elements.

Branch-Decomposition

The connectivity function λ_M of a matroid M is defined for all $A \subseteq E$ by
$$\lambda_M(A) = \mathrm{r}_M(A) + \mathrm{r}_M(E - A) - \mathrm{r}(M) + 1.$$
Here $\mathrm{r}(M) = \mathrm{r}_M(E)$. Notice that $\lambda_M(A) = \lambda_M(E - A)$. A *sub-cubic tree* is a tree in which all vertices have degree at most three. Let $\ell(T)$ denote the set of leaves of a tree T.

Let M be a matroid on the ground set $E = E(M)$. A *branch-decomposition* of M is a pair (T, τ) where T is a sub-cubic tree, and τ is an injection of E into $\ell(T)$, called *labelling*. Let e be an edge of T, and T_1, T_2 be the connected components of $T - e$. We say the e *displays* the partition (A, B) of E where $A = \tau^{-1}(\ell(T_1))$, $B = \tau^{-1}(\ell(T_2))$. The *width* of an edge e in T is $\omega_T(e) = \lambda_M(A) = \lambda_M(B)$. The width of the branch-decomposition (T, τ) is maximum of the widths of all edges of T, and the *branch-width* of M is the minimal width over all branch-decompositions of M. If T has no edge, then we take its width as 0.

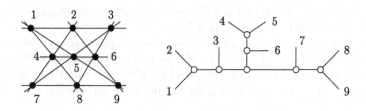

Fig. 1. An example of a width-3 branch-decomposition of the Pappus matroid.

An example of a branch-decomposition is presented in Fig. 1. We remark that the branch-width of a graph is defined analogously, using connectivity function λ_G where $\lambda_G(F)$ is the number of vertices incident both with F and $E(G) - F$. Clearly, the branch-width of a graph G is never smaller than the branch-width of its cycle matroid $M(G)$. It is still an open conjecture that these numbers are actually equal. On the other hand, branch-width is within a constant factor of tree-width in graphs [20].

Represented Matroids

We now turn our attention to matroids represented over a fixed finite field $\mathbb{F}$. This is a crucial part of our introductory definitions. A *representation* of a matroid M is a matrix $\boldsymbol{A}$ whose columns correspond to the elements of M, and maximal linearly independent subsets of columns form the bases of M. We denote by $M(\boldsymbol{A})$ the matroid represented by a matrix $\boldsymbol{A}$. Let $\boldsymbol{I}_k$ (shortly $\boldsymbol{I}$) denote the $k \times k$ unit matrix. Typically, a matroid representation is given in the so called standard form $\boldsymbol{A} = [\boldsymbol{I} \mid \boldsymbol{A}']$ with a displayed unit submatrix.

We denote by $PG(n, \mathbb{F})$ the *projective geometry (space)* obtained from the vector space $\mathbb{F}^{n+1}$. For a set $X \subseteq PG(n, \mathbb{F})$, we denote by $\langle X \rangle$ the span (affine closure) of X in the space. The (projective) rank $r(X)$ of X is the maximal cardinality of a linearly independent subset of X. A projective transformation is a mapping between two projective spaces over $\mathbb{F}$ that is induced by a linear transformation between the underlying vector spaces. Clearly, the matroid $M(\boldsymbol{A})$ represented by a matrix $\boldsymbol{A}$ is unchanged when columns are scaled by non-zero elements of $\mathbb{F}$. Hence we may view a loopless matroid representation $M(\boldsymbol{A})$ as a multiset of points in the projective space $PG(n, \mathbb{F})$ where n is the rank of $M(\boldsymbol{A})$.

Definition. We call a finite multiset of points in a projective space over $\mathbb{F}$ a *point configuration*; and we represent a loop in a point configuration by the empty subspace $\emptyset$. Two point configurations P_1, P_2 in projective spaces over $\mathbb{F}$ are *equivalent* if there is a non-singular projective transformation between the projective spaces that maps P_1 onto P_2 bijectively. We define an $\mathbb{F}$-*represented matroid* to be an equivalence class of point configurations over $\mathbb{F}$.

Standard matroidal terms are inherited from matroids to represented matroids. Obviously, all point configurations in one equivalence class belong to the same isomorphism class of matroids, but the converse is not true in general. When we want to deal with an $\mathbb{F}$-represented matroid, we actually pick an arbitrary point configuration from the equivalence class.

Width of a Matrix

The previous theory allows us to define a notion of a matrix "width" that is invariant under standard matrix row operations, or, in other words, invariant under the projective equivalence of point configurations: The *branch-width* of a matrix A over $\mathbb{F}$ is the branch-width of the matroid $M([I \mid A])$. Hence the matrix branch-width of A is not changed when standard row operations are applied to $[I \mid A]$ (i.e. when pivoting in A), or when A is transposed; and so it is a robust measure of a "complexity" of A.

Moreover, $\mathbb{F}$-represented matroids are in a one-to-one correspondence with linear codes over $\mathbb{F}$ since a projective equivalence of point configurations coincides with the standard equivalence of linear codes. Thus we may define a branch-width of a linear code C as the branch-width of the generator matrix of the code C.

Let us mention that some authors use another matrix "width" parameter defined as follows. For a matrix $A = [a_{i,j}]_{i,j=1}^{n}$, let G_A be the graph on the vertex set $\{1, \ldots, n\}$ and the edge set consisting of all $\{i, j\}$ such that $a_{i,j} \neq 0$ or $a_{j,i} \neq 0$. The tree-width of the matrix A is given by the tree-width of the graph G_A. However, this notion of tree-width for a matrix is not robust in the above sense — applying a row operation to a matrix A may dramatically change the tree-width of G_A, while the corresponding vector configuration is still the same. That is why we think that tree-width of the graph G_A is not a good measure of a "complexity" of the matrix A. Moreover, such matrices of bounded tree-width are very sparse; and they may be viewed, in fact, just as edge-coloured graphs.

Let J_n be the $n \times n$ matrix with all entries 1. Look at the following example of a matrix $D = J_n - I_n$: The graph G_D defined by this matrix is a clique, and so it has tree-width $n - 1$. On the other hand, the (3-connected) matroid $M([I_n \mid D])$ is a so called spike, and $M([I_n \mid D])$ has branch-width 3. More similar examples may be given easily.

3   Parse Trees for Matroids

In this section we introduce our basic formal tool — the parse trees for represented matroids of bounded branch-width. We use this tool to link the matroids with formal languages and automata. Loosely speaking, a parse tree shows how to "build up" the matroid along the tree using only fixed amount of information at each tree node. We are inspired by analogous parse trees known for graphs of bounded tree-width, see [1] and [9, Section 6.4].

We refer the reader to Hopcroft-Ullmann [17] for a basic introduction to automata theory. A *rooted ordered sub-binary tree* is such that each of its vertices has at most two sons that are ordered as "left" and "right". (If there is only one son, then it may be either left or right.) Let Σ be a finite alphabet. We denote by Σ^{**} the class of rooted ordered sub-binary trees with vertices labelled by symbols from Σ.

A deterministic finite leaf-to-root *tree automaton* is $\mathcal{A} = (K, \Sigma, \delta_t, q_0, F)$, where a set of states K, an alphabet Σ, an initial state q_0, and accepting states F are like in a classical automaton. The transition function δ_t is defined as a mapping from $K \times K \times \Sigma$ to K. Let the function $eval_\mathcal{A}$ for $\mathcal{A}$ be defined recursively by $eval_\mathcal{A}(\emptyset) = q_0$, and $eval_\mathcal{A}(T) = \delta_t(eval_\mathcal{A}(T_l), eval_\mathcal{A}(T_r), a)$ for $T \in \Sigma^{**}$, where T_l, T_r is the left and right, respectively, subtree of the root of T, and where a is the root symbol. A tree T is accepted by $\mathcal{A}$ if $eval_\mathcal{A}(T) \in F$. A tree language $\mathcal{L} \subseteq \Sigma^{**}$ is *finite state* if it is accepted by a finite tree automaton.

Boundaried Matroids

All matroids throughout this section are $\mathbb{F}$-represented for some fixed finite field $\mathbb{F}$. Hence, for simplicity, if we say a "(represented) matroid", then we mean an $\mathbb{F}$-represented matroid. If we speak about a projective space, we mean a projective geometry over the field $\mathbb{F}$. Let $[s, t]$ denote the set $\{s, s+1, \ldots, t\}$.

Definition. A pair $\bar{N} = (N, \delta)$ is called a *t-boundaried (represented) matroid* if the following holds: t is a non-negative integer, N is a represented matroid, and $\delta : [1, t] \to E(N)$ is an injective mapping such that $\delta([1, t])$ is independent in N.

We call $J(\bar{N}) = E(N) - \delta([1, t])$ the *internal elements* of $\bar{N}$, elements of $\delta([1, t])$ the *boundary points* of $\bar{N}$, and t the *boundary rank* of $\bar{N}$. In particular, the boundary points are not loops. We denote by $\partial(\bar{N})$ the boundary subspace spanned by $\delta([1, t])$. We say that that the boundaried matroid $\bar{N}$ is *based on* the represented matroid $N \setminus \delta([1, t])$ which is the restriction of N to $J(\bar{N})$. The notion of a t-boundaried represented matroid is similar to "rooted configurations" defined in [11], but it is more flexible. The basic operation we will use is the *boundary sum* described in the next definition.

Definition. Let $\bar{N}_1 = (N_1, \delta_1)$, $\bar{N}_2 = (N_2, \delta_2)$ be two t-boundaried represented matroids. We denote by $\bar{N}_1 \bar{\oplus} \bar{N}_2 = N$ the represented matroid defined as follows: Let Ψ_1, Ψ_2 be projective spaces such that the intersection $\Psi_1 \cap \Psi_2$ has rank exactly t. Suppose that, for $i = 1, 2$, $P_i \subset \Psi_i$ is a point configuration representing N_i, such that $P_1 \cap P_2 = \delta_1([1, t]) = \delta_2([1, t])$, and $\delta_2(j) = \delta_1(j)$ for $j \in [1, t]$. Then N is the matroid represented by $(P_1 \cup P_2) - \delta_1([1, t])$.

Informally, the boundary sum $\bar{N}_1 \bar{\oplus} \bar{N}_2 = N$ on the ground set $E(N) = J(\bar{N}_1) \dot{\cup} J(\bar{N}_2)$ is obtained by gluing the representations of N_1 and N_2 on a common subspace (the boundary) of rank t, so that the boundary points of both are identified in order and then deleted. Keep in mind that a point configuration is a multiset. It is a matter of elementary linear algebra to verify that the boundary sum is well defined with respect to equivalence of point configurations.

We write "$\leq t$-boundaried" to mean t'-boundaried for some $0 \leq t' \leq t$. We now define a composition operator (over the field $\mathbb{F}$) which will be used to generate large boundaried matroids from smaller pieces.

Definition. A $\leq t$-*boundaried composition* operator is defined as a quadruple $\odot = (R, \gamma_1, \gamma_2, \gamma_3)$, where R is a represented matroid, $\gamma_i : [1, t_i] \to E(R)$ is an injective mapping for $i = 1, 2, 3$ and some fixed $0 \leq t_i \leq t$, each $\gamma_i([1, t_i])$ is an independent set in R, and $(\gamma_i([1, t_i]) : i = 1, 2, 3)$ is a partition of $E(R)$.

The $\leq t$-boundaried composition operator $\odot$ is a binary operator applied to a t_1-boundaried represented matroid $\bar{N}_1 = (N_1, \delta_1)$ and to a t_2-boundaried represented matroid $\bar{N}_2 = (N_2, \delta_2)$. The result of the composition is a t_3-boundaried represented matroid $\bar{N} = (N, \gamma_3)$, written as $\bar{N} = \bar{N}_1 \odot \bar{N}_2$, where a matroid N is defined using boundaried sums: $N' = \bar{N}_1 \bar{\oplus} (R, \gamma_1)$, $N = (N', \gamma_2) \bar{\oplus} \bar{N}_2$.

Speaking informally, a boundaried composition operator is a bounded-rank configuration with three boundaries distinguished by $\gamma_1, \gamma_2, \gamma_3$, and with no other internal points. The meaning of a composition $\bar{N} = \bar{N}_1 \odot \bar{N}_2$ is that, for $i = 1, 2$, we glue the represented matroid N_i to R, matching $\delta_i([1, t_i])$ with $\gamma_i([1, t_i])$ in order. The result is a t_3-boundaried matroid $\bar{N}$ with boundary $\gamma_3([1, t_3])$. One may shortly write the result as $\bar{N} = \left((\bar{N}_1 \bar{\oplus} (R, \gamma_1), \gamma_2) \bar{\oplus} \bar{N}_2, \gamma_3 \right)$.

Parse Trees

Next we present the main outcome of this section — that it is possible to generate all represented matroids of branch-width at most $t + 1$ using $\leq t$-boundaried composition operators on parse trees, and then to handle them by tree automata. The finiteness of the field $\mathbb{F}$ is cruical here.

Let $\bar{\Omega}_t$ denote the *empty* t-boundaried matroid (Ω, δ_0) where $t \geq 0$ and $\delta_0([1, t]) = E(\Omega)$ (t will often be implicit in the context). If $\bar{N} = (N, \delta)$ is an arbitrary t-boundaried matroid, then $\bar{N} \bar{\oplus} \bar{\Omega}_t$ is actually the restriction of $\bar{N}$ to $E(N) - \delta([1, t])$. Let $\bar{\Upsilon}$ denote the *single-element* 1-boundaried matroid (Υ, δ_1) where $E(\Upsilon) = \{x, x'\}$ are two parallel elements, and $\delta_1(1) = x'$. Let $\bar{\Upsilon}_0$ denote the *loop* 0-boundaried matroid (Υ_0, δ_0) where $E(\Upsilon_0) = \{z\}$ is a loop, and δ_0 is an empty mapping. Let $\mathcal{R}_t^{\mathbb{F}}$ denote the finite set of all $\leq t$-boundaried composition operators over the field $\mathbb{F}$. We set $\Pi_t = \mathcal{R}_t^{\mathbb{F}} \cup \{\bar{\Upsilon}, \bar{\Upsilon}_0\}$.

Let $T \in \Pi_t^{**}$ be a rooted ordered sub-binary tree with vertices labelled by the alphabet Π_t. Considering a vertex v of T; we set $\varrho(v) = 1$ if v is labelled by $\bar{\Upsilon}$, $\varrho(v) = 0$ if v is labelled by $\bar{\Upsilon}_0$, and $\varrho(v) = t_3(\odot)$ if v is labelled by $\odot$. We call T a $\leq t$-*boundaried parse tree* if the following are true:

- only leaves of T are labelled by $\bar{\Upsilon}$ or $\bar{\Upsilon}_0$;
- if a vertex v of T labelled by a composition operator $\odot$ has no left (no right) son, then $t_1(\odot) = 0$ ($t_2(\odot) = 0$);
- if a vertex v of T labelled by $\odot$ has left son u_1 (right son u_2), then $t_1(\odot) = \varrho(u_1)$ ($t_2(\odot) = \varrho(u_2)$).

Informally, the boundary ranks of composition operators and/or single-element terminals must "agree" across each edge. Notice that $\bar{\Upsilon}$ or $\bar{\Upsilon}_0$ are the only labels from Π_t that "create" elements of the resulting represented matroid $M(T)$ in the next definition. See an illustration example in Fig. 2.

Definition. Let T be a $\leq t$-boundaried parse tree. The $\leq t$-boundaried represented *matroid* $\bar{M}(T)$ *parsed by* T is recursively defined as follows:

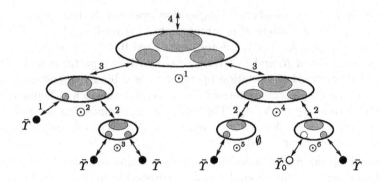

Fig. 2. An example of a boundaried parse tree. The ovals represent composition operators, with shaded parts for the boundaries and edge-numbers for the boundary ranks. (E.g. $\odot^4 = (R^4, \gamma_1^4, \gamma_2^4, \gamma_3^4)$ where $\gamma_1^4, \gamma_2^4 : [1,2] \to E(R^4)$, $\gamma_3^4 : [1,3] \to E(R^4)$.)

- if T is an empty tree, then $\bar{M}(T) = \bar{\Omega}_0$;
- if T has one vertex labelled by $\bar{\Upsilon}$ (by $\bar{\Upsilon}_0$), then $\bar{M}(T) = \bar{\Upsilon}$ ($= \bar{\Upsilon}_0$);
- if the root r of T is labelled $\odot_r$, and r has a left subtree T_1 and a right subtree T_2 (possibly empty trees), then $\bar{M}(T) = \bar{M}(T_1) \odot_r \bar{M}(T_2)$.

The composition is well defined according to the parse-tree description in the previous paragraph. The represented matroid parsed by T is $M(T) = \bar{M}(T) \,\bar{\oplus}\, \bar{\Omega}$.

Theorem 3.1. *A $\mathbb{F}$-represented matroid M has branch-width at most $t + 1$ if and only if M is parsed by some $\leq t$-boundaried parse tree.*

Let $\mathcal{B}_t$ be the set of all $\mathbb{F}$-represented matroids that have branch-width at most t. Let $\mathcal{T}_t \subset \Pi_{t-1}^{**}$ be the language of all $\leq(t-1)$-boundaried parse trees over the alphabet Π_{t-1}, and let $\bar{\mathcal{B}}_t$ be the set of all $\leq(t-1)$-boundaried matroids parsed by the trees from $\mathcal{T}_t$. We know from Theorem 3.1 that $N \in \mathcal{B}_t$ if and only if $N = \bar{N}' \,\bar{\oplus}\, \bar{\Omega}$ for some $\bar{N}' \in \bar{\mathcal{B}}_t$. However, notice that not all $\leq(t-1)$-boundaried matroids based on members of $\mathcal{B}_t$ necessarily belong to $\bar{\mathcal{B}}_t$.

It is easy to see that $\mathcal{T}_t$ itself is finite state. Suppose that $\mathcal{M}$ is a set of represented matroids. We say that $\mathcal{M}$ is *t-finite state* if the collection of all parse trees parsing the members of $\mathcal{M} \cap \mathcal{B}_t$ is finite state. We define an equivalence $\approx_{\mathcal{M}}$ for $\bar{N}_1, \bar{N}_2 \in \bar{\mathcal{B}}_t$ as follows: $\bar{N}_1 \approx_{\mathcal{M}} \bar{N}_2$ if and only if $\bar{N}_1 \,\bar{\oplus}\, \bar{M} \in \mathcal{M} \iff \bar{N}_2 \,\bar{\oplus}\, \bar{M} \in \mathcal{M}$ for all $\bar{M} \in \bar{\mathcal{B}}_t$. Finally, we state an analogue of the Myhill-Nerode Theorem for matroid parse trees.

Theorem 3.2. *Let $t \geq 1$ and $\mathbb{F}$ be a finite field. A set of $\mathbb{F}$-represented matroids $\mathcal{M}$ is t-finite state if and only if the equivalence $\approx_{\mathcal{M}}$ has finite index over $\bar{\mathcal{B}}_t$.*

Remark. The property "$\approx_{\mathcal{M}}$ has finite index" over $\bar{\mathcal{B}}_t$ is also known as "*t-cutset regularity*". We add an informal remark to this interesting concept since its formal definition may sound confusing. The true meaning of $\approx_{\mathcal{M}}$ having finite index is that, regardless of a choice of $\bar{N} \in \bar{\mathcal{B}}_t$, only a bounded amount of information concerning membership in $\mathcal{M}$ may "cross" the boundary of $\bar{N}$.

4   Monadic Second-Order Logic

Our is inspired by the well-known "MS_2-theorem" (Theorem 4.1) for graphs. The syntax of the (extended) *second-order monadic logic MS_2 of graphs* has variables for vertices, edges, and their sets, the quantifiers $\forall, \exists$ applicable to these variables, the logical connectives $\wedge, \vee, \neg$, and the next binary relations:

1. $=$, the equality for vertices, edges, and their sets,
2. $v \in W$, where v is a vertex and W is a vertex set variables,
3. $e \in F$, where e is an edge and F is an edge set variables,
4. $\mathrm{inc}(v, e)$, where v is a vertex variable and e is an edge variable, and the relation tells whether v is incident with e.

Parse trees for graphs of bounded tree-width had been defined before in [1], and they are analogous to our parse trees for matroids of bounded branch-width in Section 3. Let $\mathcal{G}$ be a graph family. Analogously to the previous section, we say that $\mathcal{G}$ is *t-finite state* if the collection of parse trees corresponding to the tree-width-t members of $\mathcal{G}$ is finite state. The MS_2-theorem (in a tree-automata formulation) follows.

Theorem 4.1. (Courcelle [5,6,7]) *Let $t \geq 1$. If $\mathcal{G}$ is a family of graphs described by a sentence in the second-order monadic logic MS_2, then $\mathcal{G}$ is t-finite state for graphs of tree-width bounded from above by t.*

Similar results were independently obtained also by Arnborg, Lagergren, and Seese [2], and later by Borie, Parker, and Tovey [4].

As noted earlier, the concept of branch-width is very close to that of tree-width in graphs. If we want to extend Theorem 4.1 to matroids, we have to define a similar logic for matroids. The *monadic second-order logic of matroids MS_M* is defined as follows: The syntax of MS_M includes variables for matroid elements and element sets, the quantifiers $\forall, \exists$ applicable to these variables, the logical connectives $\wedge, \vee, \neg$, and the next predicates:

1. $=$, the equality for elements and their sets,
2. $e \in F$, where e is an element and F is an element set variables,
3. $\mathrm{indep}(F)$, where F is an element set variable, and the predicate tells whether F is independent in the matroid.

Recall that, for $M = M(G)$ being the cycle matroid of a graph G, a set $F \subseteq E(M)$ is independent if and only if F contains no circuits in G. Let $G \uplus H$ denote the graph obtained from disjoint copies of G and H by adding all edges between them. To show the close relation of the MS_M logic of matroids to graphs, we present the next result.

Theorem 4.2. *Let G be a loopless graph, and let M be the cycle matroid of the graph $G \uplus K_3$. Then any sentence about G in the MS_2 logic can be expressed as a sentence about M in the MS_M logic.*

The main result of our paper is formulated in the following theorem:

Theorem 4.3. *Let $t \geq 1$, and let $\mathbb{F}$ be a finite field. If $\mathfrak{M}$ is a set of represented matroids over $\mathbb{F}$ described by a sentence in the second-order monadic logic MS_M of matroids, then $\mathfrak{M}$ is t-finite state for matroids of branch-width bounded by t.*

Sketch of proof. (See [14].) According to Theorem 3.2, it is enough to prove that the equivalence $\approx_\mathfrak{M}$ over $\bar{\mathcal{B}}_t$, defined as above, has finite index. We extend $\approx_\mathfrak{M}$ to an equivalence $\approx_\phi$ over "equipped" boundaried matroids, which is defined analogously for second-order monadic formulas ϕ (like those describing $\mathfrak{M}$) that are not necessarily closed. Then we proceed by induction on the length of ϕ. The claim is straightforward for the atomic formulas; and the inductive step uses similar arguments as the graphic proof in [1]. ∎

If G is a graph of tree-width t, then the graph $G' = G \uplus K_3$ has tree-width at most $t + 3$, and so branch-width at most $t + 4$ by [20]. Hence the branch-width of the cycle matroid $M = M(G')$ of G' is at most $t + 4$ as well. The cycle matroids of graphs are representable over any finite field, as already noted above. Therefore Courcelle's Theorem 4.1 for (loopless) graphs follows from Theorem 4.3 and Theorem 4.2.

Remark. We note in passing, that the proofs of Theorems 3.2 and 4.3 are constructive in the following sense: If $\mathfrak{M}$ is a set of represented matroids given by an MS_M formula ϕ, then we can, inductively on ϕ, construct the equivalence classes of $\approx_\mathfrak{M}$. These equivalence classes then, essentially, provide the states of the constructed finite tree automaton in Theorem 3.2. Hence there is an algorithm that computes this tree automaton for given $\mathbb{F}$, ϕ, and t.

5   Concluding Remarks

Notice an interesting complexity aspect of our parse trees — to describe a n-element rank-r matroid of bounded branch-width representable over a finite field $\mathbb{F}$, it is enough to give only linear amount $O(n)$ of information in the boundaried parse tree, regardless of r. On the other hand, to describe the same matroid by the whole matrix representation, one has to provide all $r \cdot n$ entries of the matrix. Even when the matrix is given in the standard form $\mathbf{A} = [\mathbf{I}_r \mid \mathbf{A}']$, the input size is $r \cdot (n - r)$, which typically means order of $\Omega(n^2)$.

To use Theorem 4.3 in a computation, we first have to construct a parse tree for the given represented matroid of bounded branch-width.

Theorem 5.1. (PH [15]) *Let $\mathbb{F}$ be a finite field, and let $t \geq 1$. There is an algorithm $\mathcal{P}$ that inputs $\mathbb{F}$, t, and an $r \times n$ matrix $\mathbf{A} \in \mathbb{F}^{r \times n}$ of rank r such that the branch-width of the matroid $M(\mathbf{A})$ is at most t; and $\mathcal{P}$ computes a $(\leq 3t)$-boundaried spanning parse tree T such that the represented matroid parsed by T is $M(\mathbf{A})$. If $\mathbb{F}$ and t are fixed, then $\mathcal{P}$ computes in time $O(n^3)$.*

We refer to [9] for the definition of basic terms of parametrized complexity. Using Theorem 5.1, one may easily derive the following corollary of Theorem 4.3.

Corollary 5.2. *Let $t \geq 1$, let $\mathbb{F}$ be a finite field, and let ϕ be a sentence in the second-order monadic logic MS_M of matroids. Assume that an input matrix $\boldsymbol{A} \in \mathbb{F}^{r \times n}$ of rank $r \leq n$ is such that the branch-width of the matroid $M(\boldsymbol{A})$ is at most t. The question whether ϕ is true for the matroid $M(\boldsymbol{A})$ is uniformly fixed-parameter tractable with respect to the parameter $(\mathbb{F}, t, \phi)$. If $\mathbb{F}$, t, and ϕ are fixed, then the answer can be computed from $\boldsymbol{A}$ in time $O(n^3)$.*

On the other hand, we show in [13] that there exists a sentence ψ in the MS_M logic of matroids with the following property: Given a square matrix $\boldsymbol{B} \in \mathbb{N}^{n \times n}$ of branch-width 3 over the integers, it is NP-hard to decide whether ψ is true for $M([\boldsymbol{I}_n \mid \boldsymbol{B}])$. (The entries of $\boldsymbol{B}$ have length $O(n)$.) This hardness result, in contrast to Corollary 5.2, clearly shows that it is absolutely necessary to consider matroids represented over finite fields in our work.

Unfortunately, the algorithm in Theorem 5.1 does not necessarily produce the optimal branch-decomposition / parse tree. By [10], the excluded minors for the class of matroids of branch-width at most k have size at most $(6^{k+1} - 1)/5$. For every matroid N there is an MS_M formula ψ_N testing for an N-minor in the given matroid, and so one may construct an MS_M formula ϕ_k such that $\phi_k(M)$ is true iff M has branch-width at most k. Details can be found in [15].

Corollary 5.3. *Let $\mathbb{F}$ be a finite field, and let $t \geq 1$. There is an algorithm that, given a rank-r matrix $\boldsymbol{A} \in \mathbb{F}^{r \times n}$ such that the branch-width of the matroid $M(\boldsymbol{A})$ is at most t, finds the exact branch-width of $M(\boldsymbol{A})$ in time $O(n^3)$.*

Moreover, the main result of [11] implies that, for any minor-closed matroid family $\mathcal{M}$ and any k, there are only finitely many $\mathbb{F}$-representable excluded minors for $\mathcal{M}$ of branch-width at most k. (This result is non-constructive.) Hence the membership in $\mathcal{M}$ can be expressed by an MS_M sentence with respect to $\mathbb{F}$ and k, and we may conclude:

Theorem 5.4. (PH [16]) *Let $\mathbb{F}$ be a finite field, let t be an integer constant, and let $\mathcal{M}$ be a minor-closed matroid family. If $\boldsymbol{A} \in \mathbb{F}^{r \times n}$ is a given rank-r matrix such that branch-width of the matroid $M(\boldsymbol{A})$ is at most t, then one can decide whether the matroid $M(\boldsymbol{A})$ belongs to $\mathcal{M}$ in time $O(n^3)$.*

More similar algorithmic applications can be found in [16]. Besides applications based directly on Theorem 4.3, we may use the machinery of matroid parse trees from Section 3 for solving other problems. For example, we provide a straightforward recursive formula and an algorithm for computing the Tutte polynomial of a represented matroid in [12].

Theorem 5.5. (PH [12]) *Let $\mathbb{F}$ be a finite field, and let t be an integer constant. If $\boldsymbol{A} \in \mathbb{F}^{r \times n}$ is a given rank-r matrix such that branch-width of the matroid $M(\boldsymbol{A})$ is at most t, then the Tutte polynomial $T(M(\boldsymbol{A}); x, y)$ of $M(\boldsymbol{A})$ can be computed in time $O(n^6 \log n \log \log n)$.*

As we have already noted in Section 2, a linear code over a field $\mathbb{F}$ could be viewed as a matroid represented by the generator matrix. Then the Hamming weight of a code equals to the smallest cocircuit size in the corresponding matroid, and the weight enumerator is an evaluation of the Tutte polynomial. Thus our Theorem 5.5 shows that the Hamming weight and the weight enumerator can be efficiently computed for a linear code of bounded branch-width.

References

1. K.A. Abrahamson, M.R. Fellows, *Finite Automata, Bounded Treewidth, and Well-Quasiordering*, In: Graph Structure Theory, Contemporary Mathematics 147, American Mathematical Society (1993), 539–564.
2. S. Arnborg, J. Lagergren, D. Seese, *Problems easy for Tree-decomposible Graphs (extended abstract)*, Proc. 15th Colloq. Automata, Languages and Programming, Lecture Notes in Computer Science 317, Springer-Verlag (1988), 38–51.
3. H.L. Bodlaender, *A Tourist Guide through Treewidth*, Acta Cybernetica 11 (1993), 1–21.
4. R.B. Borie, R.G. Parker, C.A. Tovey, *Automatic generation of linear-time algorithms from predicate calculus descriptions of problems on recursively constructed graph families*, Algorithmica 7 (1992), 555–582.
5. B. Courcelle, *Recognizability and Second-Order Definability for Sets of Finite Graphs*, technical report I-8634, Universite de Bordeaux, 1987.
6. B. Courcelle, *Graph Rewriting: an Algebraic and Logic Approach*, In: Handbook of Theoretical Computer Science Vol. B, Chap. 5, North-Holland 1990.
7. B. Courcelle, *The Monadic Second-Order Logic of Graphs I. Recognizable sets of Finite Graphs* Information and Computation 85 (1990), 12–75.
8. R. Diestel, Graph theory, Graduate Texts in Mathematics 173, Springer-Verlag, New York 1997, 2000.
9. R.G. Downey, M.R. Fellows, Parametrized Complexity, Springer-Verlag New York, 1999, ISBN 0-387-94833-X.
10. J.F. Geelen, A.H.M. Gerards, N. Robertson, G.P. Whittle, *On the Excluded Minors for the Matroids of Branch-Width k*, manuscript, 2002.
11. J.F. Geelen, A.H.M. Gerards, G.P. Whittle, *Branch-Width and Well-Quasi-Ordering in Matroids and Graphs*, J. Combin. Theory Ser. B 84 (2002), 270–290.
12. P. Hliněný, *The Tutte Polynomial for Matroids of Bounded Branch-Width*, submitted, 2002.
13. P. Hliněný, *It is Hard to Recognize Free Spikes*, submitted, 2002.
14. P. Hliněný, *Branch-Width, Parse Trees, and Monadic Second-Order Logic for Matroids*, submitted, 2002.
15. P. Hliněný, *A Parametrized Algorithm for Matroid Branch-Width*, submitted, 2002.
16. P. Hliněný, *Branch-Width and Parametrized Algorithms for Representable Matroids*, in preparation, 2002.
17. J. Hopcroft, J. Ullmann, Introduction to Automata Theory, Adisson-Wesley 1979.
18. J.G. Oxley, Matroid Theory, Oxford University Press, 1992,1997, ISBN 0-19-853563-5.
19. N. Robertson, P.D. Seymour, *Graph Minors – A Survey*, Surveys in Combinatorics, Cambridge Univ. Press 1985, 153–171.
20. N. Robertson, P.D. Seymour, *Graph Minors X. Obstructions to Tree-Decomposition*, J. Combin. Theory Ser. B 52 (1991), 153–190.

Algebraic Characterizations of Small Classes of Boolean Functions

Ricard Gavaldà[1*] and Denis Thérien[2**]

[1] Department of Software (LSI), Universitat Politècnica de Catalunya. Jordi Girona Salgado 1–3, E-08034 Barcelona, Spain. gavalda@lsi.upc.es.
[2] School of Computer Science, McGill University. 3480 University, Montréal, Québec H3A 2A7 Canada. denis@cs.mcgill.ca.

Abstract. Programs over semigroups are a well-studied model of computation for boolean functions. It has been used successfully to characterize, in algebraic terms, classes of problems that can, or cannot, be solved within certain resources. The interest of the approach is that the algebraic complexity of the semigroups required to capture a class should be a good indication of its expressive (or computing) power.
In this paper we derive algebraic characterizations for some "small" classes of boolean functions, all of which have depth-3 AC^0 circuits, namely k-term DNF, k-DNF, k-decision lists, decision trees of bounded rank, and DNF. The interest of such classes, and the reason for this investigation, is that they have been intensely studied in computational learning theory.

1 Introduction

Programs over semigroups are a well established model of computation for boolean functions. One of the reasons why the model is so appealing is that the program itself is a very simple device and essentially all the computation is left to the semigroup. Hence, the algebraic complexity of the semigroups required to compute a function should reflect the complexity of that function. More in general, the program model provides a way to connect complexity classes to classes of algebraic structures, hence to obtain a deeper insight on both.

Many complexity classes, originally defined in terms of small-depth circuits, have been characterized as those problems solved by programs over particular classes of semigroups. Some early examples are the classes NC^1, CC^0, and AC^0 [Bar89,BT88,BST90]. In the context of communication complexity, functions computable by two players in $O(1)$ communication have also been characterized by programs over semigroups [Sze93].

Here we deal with small complexity classes, all of which lie within AC_3^0 (=depth 3 AC^0), and mostly subclasses of DNF or CNF formulas. Limited as

---

* Partially supported by the IST Programme of the EU under contract number IST-1999-14186 (ALCOM-FT), by the Spanish Government TIC2001-1577-C03-02 (LOGFAC), by CIRIT 1997SGR-00366, and TIC2000-1970-CE.
** Supported by NSERC, FCAR, and the von Humboldt Foundation.

H. Alt and M. Habib (Eds.): STACS 2003, LNCS 2607, pp. 331–342, 2003.

these classes are, this range of expressibility is the interesting one in some contexts. One of them is the communication complexity mentioned before, where **DA** (a subclass of AC_2^0) is the borderline between "easy" and "hard" communication complexity for star-free regular sets.

Another such context is computational learning theory: a large part of the literature in this area deals with learning functions well within AC_2^0 or AC_3^0. Since in this paper we do not address learning problems directly, we refer the reader to textbooks [KV94] or to surveys listed in the learning theory server (www.learningtheory.org) for definitions and results on learning theory. Let us say only that its goal is to investigate mathematical models of the process of learning, and the algorithmic resources necesssary or sufficient to learn a given class of functions. Two of the most popular models of learning are Valiant's PAC learning model [Val84], in which algorithms receive a set of randomly chosen examples, and Angluin's query-based model of exact learning [Ang87,Ang88], in which algorithms can ask Evaluation queries, or Equivalence queries, or both.

Research in learning theory has produced a large body of results stating that a specific class of boolean functions is or is not learnable in a specific learning model. Effort has concentrated on classes inspired by empirical machine learning research, since these usually represent a good trade-off between naturality (e.g., they express knowledge in a way understandable by human beings), expressive power, and feasible inference algorithms. Among these classes, many are rule-like formalisms in propositional logic, and, in particular, subclasses of DNF (Disjunctive Normal Form) and CNF (Conjunctive Normal Form) formulas.

Learning DNF and CNF formulas efficiently (say, in polynomial time) in a reasonably general model of learning has been an open problem since the seminal works in the field [Val84,Ang88]. The only polynomial-time result is Jackson's algorithm [Jac97] for PAC learning DNF with Membership queries; unfortunately, the algorithm assumes that random examples provided to the PAC algorithm come from the uniform distribution, which is considered an impractical requirement. Despite the intense effort, the problem remains wide open and there is not even a clear consensus about the direction (DNF learnable or DNF nonlearnable) in which it will eventually be resolved.

Naturally, progress on the question has been attempted by considering subclasses of DNF or CNF that can be proved learnable. The list of results on learnable subclasses of DNF or CNF is too numerous to list here. We mention only those to which our results apply: k-term DNF, k-DNF, decision trees, and the CDNF class (see Section 2.1 for definitions). Contrary to DNF, these classes are known to be learnable in different protocols [Val84,Ang88,Bsh95]. A related class, not included in DNF, is that of decision lists (see Section 2.1); k-decision lists, where each node contains a term of length k, is also known to be learnable [Riv87,Sim95,CB95].

In this paper we give algebraic characterizations of some such classes of boolean functions, included or close to DNF formulas. An easy example is the following result:

"A boolean function can be (nonuniformly) represented as a constant-size boolean combination of terms if and only if it is computable by a nonuniform family of programs over a monoid in the variety called $\mathbf{J_1}$."

We obtain the analogous results for k-DNF formulas, decision trees of bounded rank, and k-decision lists. The corresponding semigroup classes are $\mathbf{J_1} * \mathbf{D}$, $\mathbf{DA}$, and $\mathbf{DA} * \mathbf{D}$, where $\mathbf{J_1}$ and $\mathbf{DA}$ are monoid classes, $\mathbf{D}$ is a semigroup class, and $*$ denotes wreath product. Finally, we point out that existing results give a similar algebraic characterization for the classes of DNF and CNF formulas.

Some points are worth noting about these results: First, varieties $\mathbf{J_1}$, $\mathbf{J}$, $\mathbf{DA}$, and $\mathbf{D}$ are not ad-hoc defined for this paper; they rather constitute central and very natural examples that were among the first to arise in investigations on algebraic theory of automata [Eil76]. Second, results about terms and k-DNF are simple corollaries of known results but those about decision trees of bounded rank and k-decision lists involve new insights in important classes of boolean functions and algebras. Third, some of our results give as a byproduct proof that provably different classes of semigroups in fact have the same computing power in the program sense; most notably, the monoid class $\mathbf{DA}$ is no more powerful than its strict subclass $\mathbf{R}$ of R-trivial monoids. Finally, it is worth noting that, by known results, "polynomial size" can be imposed on both sides of the "iff" without restricting at all the classes of boolean functions captured.

Figure 1 shows most classes considered in this paper and their inclusion relationships.

2 Definitions

2.1 Boolean Functions

We denote by $\mathcal{B}_n$ the set of boolean functions $\{0,1\}^n \to \{0,1\}$.

A DNF formula is an OR of terms, i.e., ANDs of literals. A CNF formula is an AND of clauses, i.e., ORs of literals. The *size* of a DNF (CNF) is its number of terms (clauses). By CDNF we mean the set of boolean functions having both polynomial-size CNF and polynomial-size DNF formulas (note that the name CDNF has been used with related, but different, meanings).

A *decision tree* is a rooted binary tree whose internal nodes are labelled with variables, the left (resp., right) outgoing edge of each internal node has label 0 (resp., 1), and each leaf is labelled either 0 or 1. The size of a decision tree is its number of nodes, internal or leaves. A tree is called *reduced* if each variable appears at most once along each path from the root to a leaf.

A decision tree using variables $\{x_1, \ldots, x_n\}$ computes a function in $\mathcal{B}_n$ in the obvious way: each assignment to the variables defines a path from the root to a leaf by following, at each internal node, the edge labelled with the value assigned to the variable at the node; thus, the path branches left on a node labelled x_i if the ith input bit is false, and branches right otherwise. The result on that assignment is the label of the leaf reached in this way.

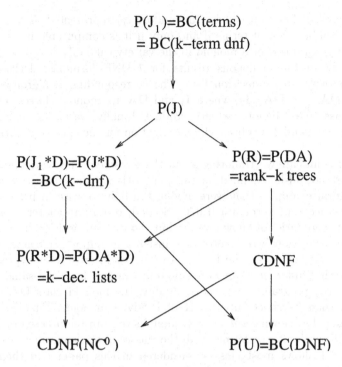

Fig. 1. Summary of classes. Arrows denote inclusions, BC denotes Boolean closure, and P(V) functions computed by programs over variety V.

A *decision list* is an ordered list of pairs (or "nodes") (t_1, b_1), $(t_2, b_2), \ldots, (t_m, b_m)$, where each t_i is a term, each b_i is a bit, and t_m is the constant true term. A list computes a boolean function in the following way: on a given assignment a, let i be the first index such that $t_i(a)$ is true; then b_i is the output of the list on a.

A k-DNF is a DNF having only terms of length at most k. A k-CNF is a CNF having only clauses of length at most k. A k-decision list is one whose nodes only contain terms of length k.

Rather than individual boolean functions, often we consider families of boolean functions, i.e., sequences $\{f_n\}_{n\geq 0}$. A size measure for boolean functions is extended to families of them in the standard way. Equally standard, for a family of boolean functions f, a statement such as "f is accepted by k-DNF formulas" means that each f_n is accepted by a k-DNF formula; these k-DNF formulas may be nonuniform, i.e., there may be no algorithmic way to produce them from n.

2.2 Semigroups, Monoids, and Programs

A *semigroup* is a pair $(S, \cdot)$ where S is a set and $\cdot$, the product over S, is a binary associative operation on S. Usually we use just S to denote the semigroup. In this paper we consider only finite semigroups. When $\cdot$ has an identity element, S is called a *monoid*.

Monoids are classically used as language recognizers in the following way: a language $L \subseteq A^*$ is *recognized* by S iff there exist a morphism $\phi : A^* \to S$ and a subset F of S such that $L = \phi^{-1}(F)$. Similarly, semigroups can be used to recognize subsets of A^+, i.e. the empty word is no longer taken into consideration.

A monoid is *aperiodic* if it has no submonoid which is a non-trivial group. Equivalently, a monoid M is aperiodic if there exists an integer t such that, for all $m \in M$, we have $m^t = m^{t+1}$.

A variety is a class of finite monoids closed under direct product, morphic image and submonoid. A similar definition can be given for semigroups. In this paper we consider some subvarieties of the variety of aperiodic monoids known as **DA**, **R**, **J**, and $\mathbf{J_1}$. We note that $\mathbf{J_1} \subseteq \mathbf{J} \subseteq \mathbf{R} \subseteq \mathbf{DA}$.

The variety **DA** can be defined in a good number of ways; see [TT01]. For example, monoid M is in **DA** iff $MeM = MfM$ and $e = e^2$ imply $f = f^2$. In this paper we will use the following characterization in language terms.

Fix an alphabet A. An unambiguous concatenation is a language of the form $A_0^* a_0 A_1^* a_1 A_2^* \ldots A_{k-1}^* a_k A_k^*$ where $A_0, \ldots, A_k \subseteq A$ and $a_1, \ldots, a_k \in A$, with the property that every string in the language admits at most one factorization with respect to that regular expression.

Theorem 1. *[Sch76] A monoid M is in **DA** iff every language recognized by M is a disjoint union of unambiguous concatenations.*

The variety **R** is a subvariety of **DA**. In parallel with the definition of **DA** that we gave before, a monoid M is in **R** iff $eM = fM$ implies $e = f$. In language terms, **R** is characterized as in the previous theorem, with the extra requirement that all unambiguous concatenations $A_0^* a_0 A_1^* a_1 A_2^* \ldots A_{k-1}^* a_k A_k^*$ that are involved can be chosen such that $a_i \notin A_{i-1}$.

A monoid M is in the variety **J** iff $MeM = MfM$ implies $e = f$. In parallel to Theorem 1, M is in **J** iff every language recognized by M is a boolean combination of languages of the form $A^* a_1 A^* a_2 \ldots a_k A^*$. That is, membership of a string in L is completely determined by the set of its subwords up to some fixed length (letters of a subword need not appear consecutively in the string). When the value of k is required to be 1 in this theorem, membership to L depends only on the *letters* appearing in the word. The corresponding variety of monoids is called $\mathbf{J_1}$, and is exactly the variety of idempotent, commutative monoids.

We use a basic family of semigroups **D** to move out of monoid varieties. A semigroup S is in **D** iff, for some k, it satisfies the equation $xy^k = y^k$. In language terms, L is recognized by S iff membership to L is completely determined by the last k consecutive letters of a word, for some constant k. It is clear from the equation (taking y to be the identity) that **D** contains no monoid except the trivial one.

Two semigroups S and T, either of which can be a monoid, may be combined together using the wreath product operation. The new semigroup $S \circ T$ has underlying set $S^T \times T$ and its operation is given by the rule $(f_1, t_1) \cdot (f_2, t_2) =$

$(f, t_1 t_2)$, where $f(t) = f_1(t) f_2(t t_1)$. This operation is totally natural from an automata-theoretic point of view as it corresponds to connecting automata in series. We extend this operation to varieties by setting $\mathbf{V} * \mathbf{W}$ to be the variety generated by all semigroups of the form $S \circ T, S \in \mathbf{V}, T \in \mathbf{W}$.

A *program* over semigroup S with domain D is a list of *instructions* together with an accepting set $Acc \subseteq S$. An instruction is a pair (i, f), where $i \in \{1, \ldots, n\}$ and f is a function $D \mapsto S$. The program computes a function from D^n to $\{0, 1\}$ as follows: Each instruction (i, f) is interpreted as reading the value of variable x_i and emitting element $f(x_i)$. All these elements are multiplied in S, producing some $s \in S$. The output of the program is 1 if s is in Acc, and 0 otherwise. In this paper we deal only with programs computing boolean functions, i.e., where $D = \{0, 1\}$; thus, we usually write an instruction (i, f) as a triple $(i, f(0), f(1)) \in \{1, \ldots, n\} \times S \times S$.

Most classes of semigroups of interest to us admit a characterization in language-theoretic terms. It will be convenient to use these characterizations directly in programs, hence we give an alternative (but equivalent in power) definition of boolean programs.

Let S be a semigroup and $L \subseteq A^*$ be a language that can be recognized by S. A program P over S is a program whose instructions are of the form (i, f) and f maps $\{0, 1\}$ to A^*, together with L as accepting language. $P(x)$ is the string in A^* formed by concatenation of the strings emitted by all instructions on input x. The program accepts iff $P(x)$ is in language L.

3 k-Term DNF, k-DNF, k-Decision Lists

The most natural subclasses one can consider on DNF formulas is restricting the number or length of their terms. Indeed, k-term DNF and k-DNF were among the first classes to be studied in computational learning theory, and shown to be learnable in time $O(n^k)$ in both the PAC model [Val84] and from Equivalence queries [Ang88]. It has been shown later that even $O(\log n)$-term DNF are learnable in polynomial time [BR95]. Decision lists were introduced in [Riv87], and shown to be PAC-learnable when they contain only terms of length k; k-decision lists are also known to be polynomial-time learnable from Equivalence queries [CB95,Sim95].

Note that a k-term DNF can be rewritten into a k-CNF by distributivity, and that both k-CNF and k-DNF can be rewritten as k-decision lists. It is a bit more complex to show that rank-k decision trees (to be defined in the Section 4) can be rewritten as k-decision lists [Blu92].

The proofs of the results in this section are omitted in this version. Their structure is similar to those in Section 4. Theorems 2 and 3 are not hard to derive from results in [BT88] and [MPT00].

Theorem 2. *A family of boolean functions is accepted by programs over a fixed monoid in $\mathbf{J_1}$ iff it is a fixed boolean combination of functions each of which is accepted by a 1-DNF formula, i.e., a clause.*

Note that boolean combinations of terms coincide with boolean combinations of k-term DNFs, so the previous theorem applies to k-term DNFs too.

Theorem 3. *A family of boolean functions is accepted by programs over a fixed semigroup in* $\mathbf{J_1} * \mathbf{D}$ *iff it is accepted by programs over a fixed semigroup in* $\mathbf{J} * \mathbf{D}$ *iff it is a fixed boolean combination of functions each of which is accepted by k-DNF formulas, for some k.*

Theorem 4. *A family of boolean functions is accepted by programs over a fixed semigroup in* $\mathbf{R} * \mathbf{D}$ *iff it is accepted by programs over a fixed semigroup in* $\mathbf{DA} * \mathbf{D}$ *iff it is accepted by k-decision lists, for some k.*

Observe that for terms and k-DNF we require boolean combinations in the theorem, as these arise naturally in characterizations using the standard notion of programs over semigroups. This is irrelevant for k-decision lists as any boolean combination of k-decision lists has a k'-decision list for some k' depending on k and the size of the boolean combination only.

4 DA and Decision Trees

This section contains the most important results in this paper, namely, the equivalence of constant-rank decision trees and programs over $\mathbf{DA}$, and that $\mathbf{DA}$ is no more powerful than its subclass $\mathbf{R}$ of left-to-right unambiguous monoids.

As we argue next, these results are interesting both as the first circuit-like characterization of a fundamental algebraic object, $\mathbf{DA}$, and as the characterization of a subclass of trees studied in learning theory.

The rank of a binary tree, first defined in [EH89], has proven to be an interesting measure of the complexity of decision trees. It is, in essence, the depth of the largest complete tree that can be embedded in the tree; we give an equivalent but more convenient definition.

Definition 1. *[EH89] The* rank *of a binary tree is defined as follows: 1) The rank of a leaf is 0; 2) if the two children of a tree T have ranks r_L and r_R, then the rank of T is $r_L + 1$ if $r_L = r_R$, and is $\max\{r_L, r_R\}$ if $r_L \neq r_R$.*
For a boolean function f, $rank(f)$ is the rank of the decision tree computing f with smallest rank.

It is known [EH89] that trees with rank r have size at most $O(n^r)$.

In [EH89] an algorithm for PAC-learning decision trees of rank r in time $n^{O(r)}$ is developed. Hence, the algorithm runs in polynomial time precisely on trees of constant rank. Learnability also follows from that of its superclass, k-decision lists [Riv87,Sim95,CB95]. In fact, the whole class of decision trees is known to be learnable in polynomial time, although that algorithm uses both Membership and Equivalence queries and outputs the result as a DNF formula [Bsh95].

Tree rank seems to be relevant even in empirical machine learning. For example, it has been argued that, when inducing a decision tree, controlling the growth of tree rank may be more effective than than controlling tree size [Elo99].

The class **DA** is a very robust one in the sense that it admits several, quite different, characterizations, and recently has proved relevant in several contexts. For some algorithmic problems on monoids, it is known that feasibility depends essentially on membership to **DA**. For example, the membership problem is known to be PSPACE-complete for any aperiodic monoid outside of **DA** [BMT92]. Also, the word problem for an aperiodic monoid can be resolved in $O(1)$ communication complexity (in the $O(1)$-player setting) iff the monoid belongs to **DA** [RTT98]. In [GTT01], it was shown that expressions over **DA** are learnable from Evaluation queries, and it was conjectured that it was the largest aperiodic class with this property. See [TT01] for a recent survey on known results involving **DA**.

These results indicate that **DA** captures an interesting type of computation. Up to this day, however, there was no characterization of the boolean functions computed by programs over **DA**. In order to prove our characterization, we develop yet another characterization of **DA**.

Definition 2. *The language classes DA_1, DA_2, $\ldots$, are defined as follows:*

- *L is in DA_1 iff $L = A^*$ for some alphabet A.*
- *$L \subseteq A^*$ is in DA_k iff*
  - *either $L \in DA_{k-1}$, or*
  - *there are languages $L_1 \in DA_i$, $L_2 \in DA_j$, with $i + j = k - 1$, and a letter $a \in A$ such that $L = L_1 \cdot a \cdot L_2$ and either $A^*aA^* \cap L_1 = \emptyset$ or $A^*aA^* \cap L_2 = \emptyset$.*

Lemma 1. *A monoid M is in **DA** iff there is some k such that every language recognizable by M is a boolean combination of languages in DA_k.*

The proof is omitted in this version. Because of this property, we overload the name DA_k to denote also a class of monoids, namely, those recognizing only boolean combinations of languages in the language class DA_k.

Theorem 5. *If $f \in \mathcal{B}_n$ is computed by a program over a monoid M in **DA**, then f is computed by a decision tree of fixed rank r (hence, size $n^{O(r)}$), where r depends on M only.*

Proof. Since M is in **DA**, there is some fixed k such that every language recognized by M is a boolean combination of languages in DA_k. A program over M can thus be thought as a program emitting outputs in A^* and using such a language L as its accepting set.

We will assume first that L is a language in DA_k, and show by induction on k that f is computed by a decision tree of rank k. Then we will deal with the case that L is a boolean combination of languages in DA_k.

Let P be a program computing f with accepting language L, We introduce the following simplifications to describe our trees. Let ℓ be the length of the longest string emitted by an instruction in P. By splitting each instruction into at most ℓ consecutive instructions, and adding a neutral letter to A and L, we can assume that each instruction in P always emits a letter, so $P(x)$ always has the same length as P. This transformation increases program length by a constant ($|M|$, at most).

Let $x_1, \ldots, x_n$ be the input variables to P, and m be the length of P. We describe the claimed decision trees using as variables a set of symbols $\{y_{j,B} \mid j = 1, \ldots, m, B \subseteq A\}$. The intended meaning of symbol $y_{j,B}$ is "the jth letter in $P(x)$ is in B". This predicate depends exclusively on the value of variable x_i read by the j-th instruction in P, hence it is logically equivalent to one of $true$, $false$, x_i, or $\bar{x}_i$ as these are the only functions on x_i. So each symbol $y_{j,B}$ in fact stands for with one of these four expressions, which (after simplification) will give a valid decision tree over $x_1, \ldots, x_n$. Also for simplicity, we write $y_{j,a}$ instead of $y_{j,\{a\}}$.

We will use the following lemma on ranks, which is easily proved by induction:

Lemma 2. *For any f_1, f_2 we have 1) $rank(not(f_1)) = rank(f_1)$; 2) $rank(f_1 \cup f_2) \leq rank(f_1) + rank(f_2)$.*

Assume first that L is in DA_k. We prove the claim by induction on k.

Base case. For $k = 1$, we have $L = B^\star$ for some $B \subseteq A^\star$. A decision tree equivalent to P has to check whether $P(x)$ is in $B^\star$, namely, $y_{j,B}$ for each j. Therefore, a tree equivalent to P consists of a long branch checking in sequence $y_{1,B}, \ldots, y_{m,B}$, and accepting iff all checks are true. This tree has rank 1.

Induction step. For $k \geq 1$, assume $L = L_1 a L_2$ for $L_1 \in DA_i$ and $L_2 \in DA_j$ ($i + j = k - 1$), and (w.l.o.g) that $A^\star a A^\star \cap L_1 = \emptyset$.

Intuitively, the decision tree for f looks for the first position in P containing an a, then recursively checks that the prefix of P up that point is in L_1 and that the suffix of P after that position is in L_2. Recursion reaches depth k, and searching for the first a is a linear tree, so we obtain a tree of rank k.

We now describe the tree in more detail. Fix an index $p \leq m$. Let P_p be a program that accepts iff the prefix of $P(x)$ of length p is in L_1. Let S_p be a program that accepts iff the suffix of $P(x)$ of length $m - p$ is in L_2. Program P_p is easily obtained from the first p instructions in P, and similarly for S_p.

Languages L_1 and L_2 are in DA_i and DA_j, respectively. By induction, for each P_p there is an equivalent tree of rank i, and for each S_p there is an equivalent tree of rank j. By Lemma 2, there is a tree of rank $k - 1$ accepting the AND of the functions computed by P_p and S_p, call it T_p.

A tree T for L is defined in terms of $T_1, \ldots, T_m$. More precisely, it consists of a branch checking variables $y_{1,a}, \ldots, y_{m,a}$ in order. (In case $A^\star a A^\star \cap L_2 = \emptyset$, use the reverse order). The 0-edge of the node labelled $y_{j,a}$ ($1 \leq j < m$) leads to the next node in the branch. The 0-edge of the node labelled $y_{m,a}$ is a 0-leaf, indicating "no a letter has been found". The 1-edge of the node labelled $y_{j,a}$ leads to a copy of tree T_p.

It is easy to show inductively that this tree has rank k given that all T_p trees have rank $k - 1$. This completes the induction argument.

When, instead, L is a boolean combination of languages in DA_k, the previous argument for languages in DA_k together with Lemma 2 yields constant rank trees for L.   ∎

Now we state the converse direction: decision trees of fixed rank are simulated by **DA** monoids.

Theorem 6. *Let $f \in \mathcal{B}_n$ be computed by a decision tree of rank k and size m. Then f is computed by a program of length $O(m)$ over a monoid M_k in* **DA**, *depending on k but not on f or m.*

Proof. Let $A_0 = \{yes, no\}$, and $A_k = A_{k-1} \cup \{e_k, i_k, f_k, l_k, l'_k, r_k, r'_k\}$. We will construct a program φ where each instruction emits a letter in the alphabet A_k. We will then define a **DA**-recognized language Y_k and argue that for any input $x \in \{0,1\}^n$, $\varphi(x)$ is in Y_k iff $f(x) = 1$.

The construction is by recurrence on the structure of the tree.

Base case. Let v be a node of height 0, i.e., a leaf. To this node we associate the constant instruction *yes* if v is an accepting leaf, and *no* if v is a rejecting leaf. We also define the accepting language for level 0 to be $Y_0 = \{yes\} \subseteq A_0^\star$.

Induction step. Let v be a node of height > 0; assume the node is querying variable x_i and let T_L and T_R be the left and right subtrees of tree T rooted at v. We distinguish two cases:

Case 1. $rank(T_L) = rank(T_R)$, say rank $k - 1$. Hence $rank(T) = k$. By induction there exist programs φ_L and φ_R (both outputting strings in $A_{k-1}^\star$) and a language $Y_{k-1} \subseteq A_{k-1}^\star$ such that $T_L(x) = 1$ iff $\varphi_L(x) \in Y_{k-1}$ and $T_R(x) = 1$ iff $\varphi_R(x) \in Y_{k-1}$. We associate to T the program $l'_k \varphi_L (x_i, l_k, r_k) \varphi_R r'_k$.

Case 2. $rank(T_L) \neq rank(T_R)$. Hence $rank(T) = \max(rank(T_L), rank(T_R))$. Let us assume that $rank(T) = rank(T_L) > rank(T_R)$; the other case is symmetric. We associate to T the program $(x_i, e_k, i_k) \varphi_R f_k \varphi_L$. (Note that the program "of smaller rank" φ_R comes before the program of "higher rank" φ_L: if T_L has smaller rank than T_R, we could use $(x_i, i_k, e_k) \varphi_L f_k \varphi_R$.)

In this way we get a program φ associated to the decision tree computing f. The length of the program is $O(m)$, in fact it can be made exactly m if we allow coalescing the constant instructions in the neighboring ones.

We then define $B_k, C_k, Y_k \subseteq A_k^\star$ as

$$B_k = A_k - \{l'_k, r_k, i_k\}$$
$$C_k = A_k - \{r_k, i_k\}$$
$$Y_k = Y_{k-1} \cup B_k^\star i_k Y_{k-1} f_k A_k^\star \cup B_k^\star l'_k Y_{k-1} l_k A_{k-1}^\star \cup C_k^\star r_k Y_{k-1} r'_k A_{k-1}^\star.$$

By induction we can assume that Y_{k-1} is **DA**-recognizable. By inspection of their alphabets, we can see that the other three languages forming Y_k are unambiguous concatenations, disjoint among themselves, and disjoint from Y_{k-1}. Therefore, Y_k is **DA**-recognizable.

It remains to show that $f(x) = 1$ iff $\varphi(x) \in Y_k$. There are two kinds of paths to consider for a given input x.

Case 1. The path on x reaches a node v where both subtrees have rank $k - 1$. Suppose v is labeled x_i and that the *ith* bit of the input is 0. The program constructed produces an output of the form $z_0 l'_k w l_k z_1$, with no i_k in z_0 as there is no previous branching towards smaller-rank subtrees. Observe that in a (sub)tree of rank k there can be at most one node where the rank of both subtrees has rank $k-1$, hence there can be at most one occurrence of a pair (l'_k, l_k). Therefore, the decision tree accepts x iff the left subtree of v accepts x iff, by induction, w is in Y_{k-1}. If the ith bit of the input is 1, the situation is completely analogous using r_k and r'_k.

Case 2. Otherwise the path on x branches s times, $s \geq 0$, to a subtree of rank k and reaches a node v where the left subtree T_L has rank k, the right subtree T_R has rank less than k (or vice versa), the node v is labeled x_i, and the ith bit of the input is 1 (or 0). The program on x produces a string of the form $e_k w_1 f_k \ldots e_k w_s f_k i_k w f_k z$, where $w_1, \ldots w_s$ are in A^*_{k-1}, hence do not contain i_k. The decision tree accepts x iff T_R accepts x iff, by induction, w is in Y_{k-1}. The proof is complete.   ∎

In fact, the proof of this theorem yields an accepting language which is unambiguous left to right. This means that monoid M_k is in the subvariety **R**, and therefore proves that programs over **DA** are equivalent in power to programs over **R**, up to polynomial differences in length.

Theorems 5 and 6 and the preceding comment give the following corollary.

Corollary 1. *A family of boolean functions is accepted by programs over a fixed monoid in* **DA** *iff it is accepted by programs over a fixed monoid in* **R** *iff it is accepted by decision trees of bounded rank.*

To conclude with our results, we note the following fact on DNF that is easily derived from known results. It involves the monoid we call U, discussed in [Thé89], which is the syntactic monoid of the language $(a+ba)^*(\epsilon+b)$, checking that no two consecutive bs appear in a word. It has 6 elements, namely, 1, $0 = bb$, a, b, ab, and ba. It was shown in [Thé89], that every boolean function f can be computed by a program over U whose length is polynomial in the size of the smallest DNF formula for f, hence its name. The converse direction can be derived from the results in [MPT00].

Putting these two facts together we have:

Theorem 7. *For any size $s(n)$, a family of boolean functions is accepted by a fixed boolean combination of DNF formulas of size $s(n)^{O(1)}$ iff it is computed by programs of length $s(n)^{O(1)}$ over a monoid in the variety generated by U.*

References

[Ang87]   D. Angluin. Learning regular sets from queries and counterexamples. *Information and Computation*, 75:87–106, 1987.

[Ang88]   D. Angluin. Queries and concept learning. *Machine Learning*, 2:319–342, 1988.

[Bar89]   D.A. Barrington. Bounded-width polynomial-size branching programs recognize exactly those languages in NC^1. *Journal of Computer and System Sciences*, 38:150–164, 1989.

[Blu92]   A.L. Blum. Rank- r decision trees are a subclass of r-decision lists. *Information Processing Letters*, 42:183–185, 1992.

[BMT92]   M. Beaudry, P. McKenzie, and D. Thérien. The membership problem in aperiodic transformation monoids. *Journal of the ACM*, 39(3):599–616, 1992.

[BR95]   A. Blum and S. Rudich. Fast learning of k-term dnf formulas with queries. *Journal of Computer and System Sciences*, 51:367–373, 1995.

[Bsh95]   N.H. Bshouty. Exact learning boolean functions via the monotone theory. *Information and Computation*, 123:146–153, 1995.

[BST90]   D.A. Mix Barrington, H. Straubing, and D. Thérien. Non-uniform automata over groups. *Information and Computation*, 89:109–132, 1990.

[BT88]   D.A. Mix Barrington and D. Thérien. Finite monoids and the fine structure of NC^1. *Journal of the ACM*, 35:941–952, 1988.

[CB95]   J. Castro and J.L. Balcázar. Simple PAC learning of simple decision lists. In *Algorithmic Learning Theory, 6th International Workshop, ALT '95, Proceedings*, volume 997, pages 239–248. Springer, 1995.

[EH89]   A. Ehrenfeucht and D. Haussler. Learning decision trees from random examples. *Information and Computation*, 82:231–246, 1989.

[Eil76]   S. Eilenberg. *Automata, Languages, and Machines*, volume B. Academic Press, 1976.

[Elo99]   T. Elomaa. The biases of decision tree pruning strategies. In *Advances in Intelligent Data Analysis: Proc. 3rd Intl. Symp.*, pages 63–74, 1999.

[GTT01]   R. Gavaldà, D. Thérien, and P. Tesson. Learning expressions and programs over monoids. In *Technical Report R01–38, Department LSI, UPC*, pages –, 2001.

[Jac97]   J.C. Jackson. An efficient membership-query algorithm for learning dnf. *Journal of Computer and System Sciences*, 53:414–440, 1997.

[KV94]   M.J. Kearns and U.V. Vazirani. *An Introduction to Computational Learning Theory*. The MIT Press, 1994.

[MPT00]   A. Maciel, P. Péladeau, and D. Thérien. Programs over semigroups of dot-depth one. *Theoretical Computer Science*, 245:135–148, 2000.

[Riv87]   Ronald L. Rivest. Learning decision lists. *Machine Learning*, 2(3):229–246, 1987.

[RTT98]   J.-F. Raymond, P. Tesson, and D. Thérien. An algebraic approach to communication complexity. In *Proc. ICALP'98, Springer-Verlag LNCS*, volume 1443, pages 29–40, 1998.

[Sch76]   M.P. Schützenberger. Sur le produit de concaténation non ambigu. *Semigroup Forum*, 13:47–75, 1976.

[Sim95]   H.U. Simon. Learning decision lists and trees with equivalence queries. In *Proceedings of the 2nd European Conference on Computational Learning Theory (EuroCOLT'95)*, pages 322–336, 1995.

[Sze93]   M. Szegedy. Functions with bounded symmetric communication complexity, programs over commutative monoids, and acc. *Journal of Computer and System Sciences*, 47:405–423, 1993.

[Thé89]   D. Thérien. Programs over aperiodic monoids. *Theoretical Computer Science*, 64(3):271–280, 29 1989.

[TT01]   D. Thérien and P. Tesson. Diamonds are forever: the DA variety. In *submitted*, 2001.

[Val84]   L.G. Valiant. A theory of the learnable. *Communications of the ACM*, 27:1134–1142, 1984.

On the Difficulty of Some Shortest Path Problems

John Hershberger[1], Subhash Suri[2], and Amit Bhosle[3]

[1] Mentor Graphics Corp., 8005 SW Boeckman Road, Wilsonville, OR 97070.
john_hershberger@mentor.com
[2] University of California, Santa Barbara, CA 93106. suri@cs.ucsb.edu.[†]
[3] University of California, Santa Barbara, CA 93106. bhosle@cs.ucsb.edu.[†]

Abstract. We prove super-linear lower bounds for some shortest path problems in directed graphs, where no such bounds were previously known. The central problem in our study is the *replacement paths* problem: Given a directed graph G with non-negative edge weights, and a shortest path $P = \{e_1, e_2, \ldots, e_p\}$ between two nodes s and t, compute the shortest path distances from s to t in each of the p graphs obtained from G by deleting one of the edges e_i. We show that the replacement paths problem requires $\Omega(m\sqrt{n})$ time in the worst case whenever $m = O(n\sqrt{n})$. Our construction also implies a similar lower bound for the k shortest paths problem for a broad class of algorithms that includes all known algorithms for the problem. To put our lower bound in perspective, we note that both these problems (replacement paths and k shortest paths) can be solved in near linear time for *undirected* graphs.

1 Introduction

Some shortest path problems seem to have resisted efficient algorithms for *directed* graphs, while their *undirected* counterparts have been solved to optimality or near optimality. One notorious example is the problem of determining the k shortest simple paths between a pair of nodes. Specifically, given a directed graph G with non-negative edge weights, a positive integer k, and two vertices s and t, the problem asks for the k shortest paths from s to t in increasing order of length, where the paths are required to be *simple* (loop free); if the paths are allowed to be *non-simple*, then an optimal algorithm is known for both directed and undirected graphs [7]. The best algorithm known for computing the k shortest simple paths in a directed graph is due to Yen [24,25]. The worst-case time complexity of his algorithm, using modern data structures, is $O(kn(m + n \log n))$; in other words, it requires $\Theta(n)$ single-source shortest path computations for each of the k shortest paths that are produced. In the past thirty years, there have been several "practical" improvements to Yen's algorithm, but none has succeeded in improving the worst-case asymptotic complexity of the problem. (See the references [6,10,18,19,23] for implementations and modifications of Yen's algorithm.)

[†] Supported in part by NSF grants IIS-0121562 and CCR-9901958.

H. Alt and M. Habib (Eds.): STACS 2003, LNCS 2607, pp. 343–354, 2003.
© Springer-Verlag Berlin Heidelberg 2003

By contrast, if the graph G is undirected, then one can compute the k shortest simple paths in time $O(k(m + n \log n))$ [12,14].

A second problem with similar behavior is the *replacement paths* problem. Given a directed graph G with non-negative edge weights, and a pair of nodes s, t, let $P = (e_1, e_2, \ldots, e_p)$ denote the sequence of edges in a shortest path from s to t in G. The replacement paths problem is to compute the shortest path from s to t, in each of the graphs $G \setminus e_i$, that is, G with edge e_i removed, for $i = 1, 2, \ldots, p$. (Another variant of the replacement paths problem arises when *nodes* of the path P are removed one at a time. These two problems have equivalent computational complexity.) The replacement paths problem is motivated both by routing applications, in which a new route is needed in response to individual link failures [8], and by computational mechanism design, in which the Vickrey-Clarke-Groves scheme is used to elicit true link costs in a distributed but self-interested communication setting, such as the Internet [21]. In the VCG payment scheme, the bonus to a link agent e_i equals $d(s, t; G \setminus e_i) - d(s, t; G|_{e_i=0})$, where the former is the replacement path length and the latter is the shortest path distance from s to t in the graph G with the cost of e_i set to zero. Computing all these payments is equivalent to solving the replacement paths problem for s, t.

A naïve algorithm for the replacement paths problem runs in $O(n(m + n \log n))$ time, executing the single-source shortest path algorithm up to n times, but no better algorithm is known. By contrast, if G is *undirected*, then the problem can be solved in $O(m + n \log n)$ time [11]; that is, roughly one single-source shortest path computation suffices! (Our FOCS 2001 paper [11] erroneously claims to work for both directed and undirected graphs; however, there is a flaw that invalidates the algorithm for directed graphs.) In the case of undirected graphs, the same bound is also achieved by Nardelli, Proietti, and Widmayer [20], who solve the *most vital node* problem with an algorithm that also solves the replacement paths problem. The work of Nardelli, Proietti, and Widmayer is in turn based on earlier work by Malik, Mittal and Gupta [17], Ball, Golden, and Vohra [4] and Bar-Noy, Khuller, and Schieber [5]. All of these algorithms are restricted to undirected graphs, and again the directed version of the problem seems to have eluded efficient solutions.

Main Results

Our main result is to prove a *lower bound* on the replacement paths problem in directed graphs, thereby explaining our lack of success in finding an efficient algorithm. Our lower bound applies to the *path comparison model* proposed by Karger, Koller, and Phillips [13]. In this model, an algorithm learns about shortest paths by comparing the lengths of two different paths. A path-comparison based algorithm can perform all the standard operations in unit time, but its access to edge weights is only through the comparison of two paths. Most of the known shortest path algorithms, including those of Dijkstra, Bellman-Ford, Floyd, Spira, Frieze-Grimmet, as well as the hidden paths algorithm of Karger-Koller-Phillips [13] fit into the path comparison model. On the other hand, the $o(n^3)$ time all-pairs shortest paths algorithms of Fredman [9] does not fit the path comparison model—it compares sums of weights of edges that do not necessarily form paths. The matrix-multiplication based algorithms by Alon, Galil, Margalit [1] and Zwick [26,27] also fall outside the path comparison model; however, these algorithm assume the weights lie in a small integer range.

Despite its obvious appeal and power, we discovered that the path comparison model also has an unfortunate weakness: adding extra edges to a graph, even ones that obviously do not affect the solution to the problem being solved, can invalidate lower bounds. We discuss this limitation in Section 3.2. Because of this limitation, we must restrict our lower bound to algorithms that do not compare certain paths that cannot belong to any solution. All presently known algorithms for the replacement paths problem satisfy this restriction.

We show that the replacement paths problem requires $\Omega(m\sqrt{n})$ time in the worst case for a directed graph G that has n vertices and m edges, with $m = O(n\sqrt{n})$. (Another way of phrasing the limitation on m is to say that the lower bound is $\Omega(\min(n^2, m\sqrt{n}))$.) Our lower bound construction also has implications for the k simple shortest paths problem. All known algorithms for this problem compute the best candidate path for each possible branch point off previously chosen paths. Our lower bound construction can be adapted to show that computing this best candidate path requires $\Omega(m\sqrt{n})$ time and, therefore, all k shortest path algorithms of this genre are subject to our lower bound, even for $k = 2$. Our construction can be used to establish nontrivial lower bounds on some related shortest problems as well, such as the replacement paths problem in which nodes of the s–t path are removed one at a time, and the replacement paths problem with the ⟨length⟩ × ⟨hop count⟩ metric. To the best of our knowledge, these are the first nontrivial lower bounds for these problems.

2   A Lower Bound Construction

Our main theorem establishes a lower bound on the replacement paths problem by tying the computation of replacement paths to computing shortest path distances between pairs of vertices in another graph. We define the following natural problem:

k-Pairs Shortest Paths Problem: Given a directed graph G, with non-negative edge weights, and k source-destination pairs (s_i, t_i), for $i = 1, 2, \ldots, k$, compute the shortest path distance for each pair (s_i, t_i). (The sources and destinations need not be unique.)

Our lower bound relates the n pair distances in an instance of the n-pairs shortest paths problem to the replacement path distances in a "wrapper" graph. We describe this general construction below.

We start with a directed path $P = (v_0, v_1, \ldots, v_n)$, with $\|v_i, v_{i+1}\| = 0$, for $i = 0, 1, \ldots, n - 1$, where $\|v_i, v_{i+1}\|$ is the length (weight) of the edge (v_i, v_{i+1}). Set $s = v_0$ and $t = v_n$ to be the source and destination pair for our replacement paths problem. Let H be an arbitrary directed graph with n nodes, m edges, and non-negative edge weights. We choose n source-destination pairs (s_i, t_i) in

H. Let w be a weight at least as large as the maximum weight of any simple path in H; we can crudely bound w by n times the largest edge weight in H. We choose $W = 10w$. We create edges between the nodes of the path P and the graph H as follows.

We map the n edges of P bijectively onto the n source-destination pairs (s_j, t_j) in H. Suppose the edge $(v_i, v_{i+1}) \in P$ is mapped to the pair (s_j, t_j). We create a directed edge (v_i, s_j) with

$$\|v_i, s_j\| = (n - i)W.$$

Similarly, we create a directed edge (t_j, v_{i+1}) with

$$\|t_j, v_{i+1}\| = (i + 1)W.$$

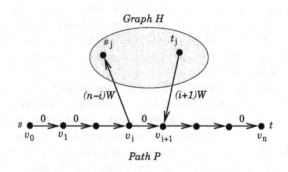

Fig. 1. The main lower bound construction.

Figure 1 illustrates the construction. Note that the index of the pair (s_j, t_j) plays no rôle in the cost of these edges, but only influences which nodes of the path P the edges are joined to. We call the entire graph G; it has $O(n)$ vertices and $O(m)$ edges. Observe that P is the shortest path from s to t in G. The following lemma states a crucial property of this graph.

Lemma 1. *Suppose the edge $(v_i, v_{i+1}) \in P$ is mapped to the source-destination pair (s_j, t_j) in H. Then the shortest path from s to t in the graph $G \setminus (v_i, v_{i+1})$ has cost*

$$(n + 1)W + d_H(s_j, t_j),$$

where $d_H(s_j, t_j)$ denotes the shortest path distance between s_j and t_j in the subgraph H.

Proof. We first note that the replacement path that starts at s, follows P up to node v_i, uses edge (v_i, s_j), takes the shortest path from s_j to t_j in H, returns to v_{i+1} using the edge (t_j, v_{i+1}), and finally traces the path P from v_{i+1} to t has length

$$(n - i)W + d_H(s_j, t_j) + (i + 1)W = (n + 1)W + d_H(s_j, t_j).$$

We next show that any other replacement path candidate is longer. Since the cost of edges connecting a vertex in P to a vertex of H is at least W, which is substantially larger than any subpath in P or any path in H, an optimal replacement path makes exactly one trip to H. Let us consider a path that follows P up to a node v_a, for $a \leq i$, and then returns to a node v_b, for $b \geq i+1$, where either $a \neq i$ or $b \neq i+1$. Supposing $a \neq i$, then this path has length at least $(n-a)W + (i+1)W \geq (n+2)W$. Since $W = 10w$, it follows that

$$(n+2)W > (n+1)W + d_H(s_j, t_j).$$

Similarly, any path for which $b \neq i+1$ is also longer. This completes the proof.

In order to apply this construction to the replacement paths problem, we need a lower bound on the k-pairs shortest paths problem. While such a lower bound is easily derived by modifying the construction of Karger, Koller, and Phillips [13], a subtle point about the *robustness* of path comparison lower bounds must be made.

Definition 1. *A path comparison model lower bound for a graph problem is path addition insensitive if it holds even if the graph is modified by adding edges and vertices that do not affect the solution to the problem.*

Because Lemma 1 shows that computing the replacement paths for the graph G also computes the n-pairs shortest path distances for H, we have the following black-box reduction theorem.

Theorem 1. *If there is a directed graph H with n nodes, m edges, and n source-destination pairs (s_i, t_i) such that any algorithm to compute the shortest path distances between all the (s_i, t_i) pairs must take $f(n, m)$ time, and this lower bound is path addition insensitive, then there is a lower bound of $\Omega(f(n, m))$ time for the replacement paths problem on a graph with $O(n)$ nodes and $O(m)$ edges.*

3   The Path Comparison Model and Lower Bounds

We begin with a lower bound on the k-pairs shortest paths problem. Our argument uses a small modification of the construction of Karger, Koller, and Phillips [13]. Unfortunately, we then show in Section 3.2 that the Karger-Koller-Phillips lower bound is not path addition insensitive. As a result, the class of algorithms to which our lower bound result applies is slightly restricted, though it still includes all known algorithms for the replacement paths problem.

3.1   A Lower Bound for the k-Pairs Shortest Paths Problem

Our k-pairs shortest paths lower bound is based on the lower bound construction of Karger, Koller, and Phillips [13]. The graph is a directed tripartite graph, with vertices x_a, y_b, and z_c, where the indices a, b, c range from 0 to $n-1$. The edges are of the form (x_a, y_b) or (y_b, z_c). The construction of [13] is for the all-pairs

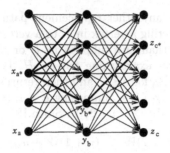

Fig. 2. The lower bound construction of Karger, Koller and Phillips [13]. The edges whose weights are modified are shown as thick lines.

shortest paths problem, where the goal is to compute the shortest path distances $d(x_a, z_c)$, for all $0 \leq a, c \leq n - 1$. See Figure 2 for an illustration. The lower bound argument depends on the fact that there is a choice of edge weights so that if the algorithm does not consider some x-z path, say $(x_{a^*}, y_{b^*}, z_{c^*})$, then an adversary can modify the weights in such a way that

1. The path $(x_{a^*}, y_{b^*}, z_{c^*})$ becomes the shortest path between x_{a^*} and z_{c^*}, and
2. The relative order of all the remaining paths remains unchanged.

Thus an algorithm is forced to consider all $\Theta(n^3)$ paths from x-nodes to z-nodes, giving a lower bound of $\Omega(n^3)$ path comparisons for the special case of $m = \Theta(n^2)$. In order to extend this construction to any value of m, we simply reduce the number of vertices in the middle column to m/n, which gives $\Theta(mn)$ triples of the type (x_a, y_b, z_c). The specific weights used by Karger, Koller, and Phillips have the following form:

$$\|x_a, y_b\| = [1, \ 0, \ a, \ 0, \ b, \ 0, \ 0]_{n+1} \qquad (1)$$
$$\|y_b, z_c\| = [0, \ 1, \ 0, \ c, \ 0, \ -b, \ 0]_{n+1} \qquad (2)$$

using numbers in base $n + 1$. That is, $[\phi_r, \ldots, \phi_0]_\alpha$ is defined as $[\phi_r, \ldots, \phi_0]_\alpha = \sum_{i=0}^{r} \phi_i \alpha^i$. The negative numbers do not cause any problems because the graph is a directed acyclic graph; they are introduced only to simplify the presentation. In order to make $(x_{a^*}, y_{b^*}, z_{c^*})$ the new shortest path, the weights of all the edges (x_{a^*}, y_b), for $b \leq b^*$, and the edge (y_{b^*}, z_{c^*}) are reduced as follows:

$$\|x_{a^*}, y_b\|' = [1, \ 0, \ a^*, \ 0, \ 0, \ b, \ b^*]_{n+1}, \text{ for all } b \leq b^* \qquad (3)$$
$$\|y_{b^*}, z_{c^*}\|' = [0, \ 1, \ 0, \ c^*, \ 0, \ -b^*, \ -n]_{n+1} \qquad (4)$$

It is now easy to check that using the new weights, the path $(x_{a^*}, y_{b^*}, z_{c^*})$ becomes the shortest path between x_{a^*} and z_{c^*}, while all other pairs of paths retain their relative ordering.

An easy modification of this construction leads to an $\Omega(\min(nk, m\sqrt{k}))$ lower bound for the k-pairs shortest paths problem. We use $\sqrt{k}$ vertices in the first and the third column, and $\min(n, m/\sqrt{k})$ vertices in the middle column. That is, for vertices x_a and z_c, the indices range from 0 to $\lceil \sqrt{k} \rceil$, and, for the y_b vertices, the indices range from 0 to $\min(n, \lceil m/\sqrt{k} \rceil)$. Thus, as long as $m = O(n\sqrt{k})$, then the total number of vertices in the graph is $O(n)$, the number of edges is $\Theta(m)$, and there are $\Omega(m\sqrt{k})$ triples of the form (x_a, y_b, z_c). A path-comparison based algorithm must examine all the triples to compute all x_a, z_c distances correctly.

Theorem 2. *For any n, k, and m, with $1 \le k \le n^2$ and $m = O(n\sqrt{k})$, there exists a directed graph H with n nodes, m non-negatively weighted edges, and k source-destination pairs (s_i, t_i) such that any path-comparison based algorithm must spend $\Omega(m\sqrt{k})$ time computing the shortest path distances for the k pairs (s_i, t_i).*

3.2 Limitations of Path-Comparison Lower Bounds

The path comparison lower bound of Karger et al. [13] is the only nontrivial lower bound known for the all-pairs shortest paths problem. Since most shortest path algorithms fit into the path comparison model, such a lower bound has considerable intellectual value. Unfortunately, the lower bound is weaker than it appears at first glance. In particular, adding extra nodes and edges to the graph, even ones that clearly have no effect on the shortest paths under investigation, may invalidate the lower bound argument. The argument in [13] rests on the fact that any three-vertex path $(x_{a^*}, y_{b^*}, z_{c^*})$ can be made the shortest path linking x_{a^*} to z_{c^*} without affecting the relative order of the other paths. We show below that this crucial property is invalidated by the addition of some superfluous nodes and edges, which otherwise have no effect on the shortest path problem.

Consider the strongly-connected graph H' obtained from the directed acyclic graph H of Figure 2 by connecting the third column of vertices back to the first column by two fans of zero-weight edges and a single looping-back edge with weight W, where W is ten times the weight of the heaviest path in H. See Figure 3. Then for any pair (x_a, z_c), the shortest path joining them is identical

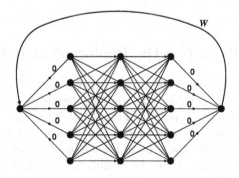

Fig. 3. The lower bound breaks down when we add a useless loop.

to the shortest path in the original graph H. Furthermore, it is far from obvious how the existence of the additional edges in H' might make it easier to find the shortest path for any (x_a, z_c). Nevertheless, the additional edges invalidate Karger, Koller, and Phillips's lower bound by creating paths whose relative order *does* change when the graph is re-weighted to make $(x_{a^*}, y_{b^*}, z_{c^*})$ the shortest path from x_{a^*} to z_{c^*}. For example, suppose $3 < b^* < n-2$ and $a^* \neq 0$. Consider the following two paths:

$$\pi_1 = (x_{a^*}, y_3, z_0, \ldots \text{loop} \ldots, x_0, y_{n-2}, z_1)$$
$$\pi_2 = (x_{a^*}, y_0, z_0, \ldots \text{loop} \ldots, x_0, y_{n-1}, z_1)$$

In the unmodified graph, we have $\|\pi_1\| > \|\pi_2\|$, because the path weights are

$$\|\pi_1\| = [2, 2, a^*, 1, n+1, -(n+1), 0]_{n+1} + W$$
$$\|\pi_2\| = [2, 2, a^*, 1, n-1, -(n-1), 0]_{n+1} + W.$$

After the edge weights are modified to make $(x_{a^*}, y_{b^*}, z_{c^*})$ the shortest path from x_{a^*} to z_{c^*}, the two path weights become

$$\|\pi_1\| = [2, 2, a^*, 1, n-2, -(n-2), 3]_{n+1} + W$$
$$\|\pi_2\| = [2, 2, a^*, 1, n-1, -(n-1), 0]_{n+1} + W,$$

and now we have $\|\pi_1\| < \|\pi_2\|$. Thus, the relative order of $\|\pi_1\|$ and $\|\pi_2\|$ changes when the weights are modified, even though neither π_1 nor π_2 contains either edge (x_{a^*}, y_{b^*}) or (y_{b^*}, z_{c^*}). Thus, the lower bound of Karger, Koller, and Phillips is not path addition insensitive.

Another potential weakness of the path comparison model is that it allows *any* two paths to be compared at unit cost, without regard to the number of edges in the paths; that is, the computation is assumed to be free. But just as the unbounded word-size RAM model can be (theoretically) abused to sort numbers in $O(n)$ time by arithmetic on long numbers [15,22], path comparison lower bounds can be invalidated by the creation of long paths, which may encode all relevant paths as subpaths. In response to these shortcomings of the path comparison model, we limit the scope of our lower bound somewhat in the following section.

4   A Lower Bound on the Replacement Paths Problem

We now return to the replacement paths lower bound. We will use the construction described above, with $k = n$, as the subgraph H. If the Karger-Koller-Phillips construction were path addition insensitive, this would immediately give an $\Omega(m\sqrt{n})$ lower bound on the replacement paths problem, by Theorem 1. However, because the k-pairs lower bound is not path addition insensitive, we must show that when the weights are modified to make a particular triple $(x_{a^*}, y_{b^*}, z_{c^*})$ the shortest path between x_{a^*} and z_{c^*}, then all relevant path orders are preserved. Since the graph H is supplemented with the path P and the edges connecting P and H, many new paths are introduced in the graph G whose orders are not considered in the lower bound argument of Section 3.1. In fact, if paths visit H more than once, then path orders may *not* be preserved, as in the example of Section 3.2.

Definition 2. *Given a path π from s to t, a* detour *is any subpath of π that begins and ends on P but contains no edges of P.*

We focus on algorithms that examine paths with at most one detour. This is reasonable, because of the following fact.

Lemma 2. *Given a directed graph G, and an instance of the replacement paths problem with s–t shortest path $P = (e_1, e_2, \ldots, e_p)$, suppose P_i is the replacement path when e_i is deleted, for $i = 1, 2, \ldots, p$. Then P_i cannot have more than one detour.*

Proof. A detour that leaves P at some node x and returns to P on the subpath between s and x creates a cycle and can be shortened. (Recall that edges of G have non-negative costs.) If a path has two forward-going detours, at most one of them bypasses e; the other one can be shortcut by following P. That is, by subpath optimality, the subpath of P connecting the detour endpoints is no longer than the detour.

Definition 3. *A* single-detour path-comparison based algorithm *for the replacement paths problem considers only subpaths of paths with at most one detour.*

All the algorithms for the replacement paths problem that we know are single-detour algorithms. This is because, as Lemma 2 shows, a multiple-detour path is guaranteed not to be (a subpath of) a shortest replacement path for any edge on P. More generally, all the shortest path algorithms that we know exploit local optimality: suppose π_1 and π_2 are two paths between u and v, the algorithm has determined that $\|\pi_1\| < \|\pi_2\|$, and both paths satisfy any other criteria imposed by the problem. Then any path the algorithm later constructs that includes a u–v subpath will not use π_2. Our focus on single-detour paths essentially disallows such fruitless comparisons.

Lemma 3. *Consider a single-detour path-comparison based algorithm that solves the replacement paths problem for the directed graph G, and suppose the algorithm does not consider a triple $(x_{a^*}, y_{b^*}, z_{c^*})$ in the subgraph H. If we modify the weights of some of the edges of the subgraph H as in Eqs. 3 and 4, then this triple becomes a subpath of some replacement path, and all other subpaths of G considered by the algorithm maintain their relative order.*

Proof. Each path can be divided into three parts: edges of P, edges of H, and edges connecting P and H. The weight from edges of P is zero; the weight from edges of H is at most $w = W/10$; and the total weight of the connector edges is some multiple of W. (The bound on the weight of the H edges follows because the algorithm examines only single-detour paths.) If the two paths have different weights from their connector edges, then their relative order is determined by these edges only—the modification to H changes any path length by less than

$w = W/10$. If the two paths have equal connector edge weights, then their relative order is determined by the edges of H they contain, and the Karger-Koller-Phillips argument shows that their relative order is unchanged (since H is visited at most once by a single-detour path).

Combining the results of Theorem 2 and Lemma 3, we get our main result.

Theorem 3. *For any n and m, with $m = O(n\sqrt{n})$, there exists a directed graph G with n nodes, m non-negatively weighted edges, and two specified vertices s and t such that any single-detour path-comparison based algorithm that solves the replacement paths problem for s, t in G must spend $\Omega(m\sqrt{n})$ time.*

Our lower bound applies to the node version of the replacement paths problem as well: we can convert our lower bound construction into a node-based problem by adding an extra node in the middle of each edge of P. Finding replacement paths for these nodes would solve our edge replacement problem, and hence both problems are subject to the same lower bound.

5   Upper Bounds

The best upper bound known for the replacement paths problem is $O(n(m + n \log n))$. Thus, there remains a substantial gap between our lower bound and the upper bounds. However, for the specific problem instances constructed for our lower bounds, we can solve the n-pairs shortest paths subproblem in $O(\sqrt{n}(m + n \log n))$ time—we run a single-source shortest path algorithm from each x_a node. Once all the $O(n)$ x–z distances in the subgraph H have been computed, we show that the replacement paths (for the shortest path metric) can be computed in additional $O(n\sqrt{n}\log n)$ time.

We present our upper bound for a slight generalization of the lower bound construction. Suppose that every detour from the path P passes through at least one of a set of $O(\sqrt{n})$ pre-specified *gateway* vertices not on P. We compute the distances from each gateway to all the vertices of P in the graph $G \setminus P$, that is, using only paths disjoint from P until they reach P. Likewise, we compute the distances from all the vertices of P to the gateways, using paths in $G \setminus P$. (This is done by finding shortest paths *from* the gateways in a graph obtained from $G \setminus P$ by reversing all the edges.) These path computations take $O(\sqrt{n}(m + n \log n))$ time. In additional $O(n\sqrt{n})$ time, we can compute, for each gateway g and each edge $e \in P$, the length of the shortest path from s to g that leaves P before e, and the length of the shortest path from g to t that first touches P after e. The sum of these distances is the length of the shortest replacement path for e that passes through g. Minimizing over all gateways g gives the overall shortest replacement path for e. (If the graph also happens to have some single-edge detours, which necessarily do not pass through any gateway, we can find the shortest replacement paths that use these single-edge detours in $O(m + n \log n)$ time [11].)

6   Related Problems and Concluding Remarks

We have shown that the replacement paths problem in a directed graph with $m = O(n\sqrt{n})$ edges has a lower bound of $\Omega(m\sqrt{n})$ for any single-detour path-comparison based algorithm. Our construction can be used to establish nontrivial lower bounds on some related shortest problems as well, such as the k shortest paths, replacement paths with $\langle$length$\rangle \times \langle$hop count$\rangle$ metric, and replacement shortest path trees. We discuss the k shortest paths problem briefly.

The best algorithm for k simple shortest paths in a directed graph dates back to Yen [24]. The worst-case complexity of his algorithm, using modern data structures, is $O(kn(m + n \log n))$. In essence, his algorithm performs $\Theta(n)$ single-source shortest path computations for each of the k shortest paths. By contrast, the undirected version of the same problem can be solved in time $O(k(m + n \log n))$ [12,14]. The underlying idea of Yen's algorithm, as well as those of Lawler [16], Katoh, Ibaraki, and Mine [14], and the various heuristic improvements to Yen's algorithm [6,10,12,18,19,23] is the following simple observation: the ith shortest path must differ from each of the first $(i-1)$ shortest paths in at least one edge. Thus, these algorithms generate new candidate paths by computing the shortest candidate paths for each possible branch point off previously chosen paths. The lower bound construction of our paper implies an $\Omega(m\sqrt{n})$ lower bound for this computation even if $k = 2$.

With minor modifications in the construction, we can show an $\Omega(m\sqrt{n})$ lower bound for the replacement paths problem with the $\langle$length$\rangle \times \langle$hop count$\rangle$ metric, and an $\Omega(nm)$ lower bound for the replacement *shortest path trees* problem. The $\langle$length$\rangle \times \langle$hop count$\rangle$ metric has potential applications in frugal path mechanisms [2,3], and the replacement shortest path tree is a natural generalization of the replacement shortest path. Due to limited space, we omit further details.

Finally, a worthwhile distinction to make is between directed acyclic graphs (DAGs) and general directed graphs. The lower bound of Karger, Koller, and Phillips [13] holds even for DAGs. On the other hand, both the replacement paths problem and the k shortest paths problems can be solved much more efficiently for DAGs. In particular, our undirected graph algorithms work for DAGs, so the replacement paths problem can be solved in $O(m + n \log n)$ time for DAGs [11]; the k shortest paths problem can be solved in $O(k + m + n \log n)$ time by a method not based on replacement paths [7].

References

1. N. Alon, Z. Galil, and O. Margalit. On the exponent of the all pairs shortest path problem. In *32nd Symposium on Foundations of Computer Science*, 569–575, 1991.
2. A. Archer. Private communication, 2001.
3. A. Archer and E. Tardos. Frugal path mechanisms. In *Proc. 13th Annual ACM-SIAM Symposium on Discrete Algorithms*, pages 991–999, 2002.
4. M. O. Ball, B. L. Golden, and R. V. Vohra. Finding the most vital arcs in a network. *Oper. Res. Letters*, 8:73–76, 1989.

5. A. Bar-Noy, S. Khuller, and B. Schieber. The complexity of finding most vital arcs and nodes. Technical Report CS-TR-3539, Institute for Advanced Studies, University of Maryland, College Park, MD, 1995.
6. A. Brander and M. Sinclair. A comparative study of K-shortest path algorithms. In *Proc. of 11th UK Performance Engineering Workshop*, pages 370–379, 1995.
7. D. Eppstein. Finding the k shortest paths. *SIAM J. Computing*, 28(2):652–673, 1998.
8. B. Fortz and M. Thorup. Internet traffic engineering by optimizing OSPF weights. In *INFOCOM*, pages 519–528, 2000.
9. M. Fredman. New bounds on the complexity of the shortest path problem. *SIAM J. of Computing*, 5:83–89, 1976.
10. E. Hadjiconstantinou and N. Christofides. An efficient implementation of an algorithm for finding K shortest simple paths. *Networks*, 34(2):88–101, September 1999.
11. J. Hershberger and S. Suri. Vickrey prices and shortest paths: What is an edge worth? In *Proceedings of the 42nd Annual IEEE Symposium on Foundations of Computer Science*, pages 252–259, 2001.
12. J. Hershberger, M. Maxel, and S. Suri. Finding the k Shortest Simple Paths: A New Algorithm and its Implementation. To appear in *Proc. of ALENEX*, 2003.
13. D. R. Karger, D. Koller, and S. J. Phillips. Finding the hidden path: Time bounds for all-pairs shortest paths. *SIAM J. Comput.*, 22:1199–1217, 1993.
14. N. Katoh, T. Ibaraki, and H. Mine. An efficient algorithm for k shortest simple paths. *Networks*, 12:411–427, 1982.
15. D. Kirkpatrick and S. Reisch. Upper bounds for sorting integers on random access machines. *Theoretical Computer Science*, 28(3):263–276, 1984.
16. E. L. Lawler. A procedure for computing the K best solutions to discrete optimization problems and its application to the shortest path problem. *Management Science*, pages 401–405, 1972.
17. K. Malik, A. K. Mittal, and S. K. Gupta. The k most vital arcs in the shortest path problem. *Oper. Res. Letters*, 8:223–227, 1989.
18. E. Martins and M. Pascoal. A new implementation of Yen's ranking loopless paths algorithm. Submited for publication, Universidade de Coimbra, Portugal, 2000.
19. E. Martins, M. Pascoal, and J. Santos. A new algorithm for ranking loopless paths. Technical report, Universidade de Coimbra, Portugal, 1997.
20. E. Nardelli, G. Proietti, and P. Widmayer. Finding the most vital node of a shortest path. In *Proc. COCOON*, 2001.
21. N. Nisan and A. Ronen. Algorithmic mechanism design. In *Proc. 31st Annu. ACM Sympos. Theory Comput.*, 1999.
22. W. J. Paul and J. Simon. Decision trees and random access machines. In *Proc. International Symp. on Logic and Algorithmic*, pages 331–340, 1980.
23. A. Perko. Implementation of algorithms for K shortest loopless paths. *Networks*, 16:149–160, 1986.
24. J. Y. Yen. Finding the K shortest loopless paths in a network. *Management Science*, 17:712–716, 1971.
25. J. Y. Yen. Another algorithm for finding the K shortest loopless network paths. In *Proc. of 41st Mtg. Operations Research Society of America*, volume 20, 1972.
26. U. Zwick. All Pairs Shortest Paths in Weighted Directed Graphs—exact and almost exact algorithms. In *Proc. IEEE Symposium on Foundations of Computer Science*, 310–319, 1998.
27. U. Zwick. All Pairs Shortest Paths using Bridging Sets and Rectangular Matrix Multiplication. In *Electronic Colloquium on Computational Complexity*, 2000.

Representing Graph Metrics with Fewest Edges

T. Feder*, A. Meyerson **, R. Motwani***, L. O'Callaghan[†], and
R. Panigrahy[‡]

Carnegie-Mellon University and Stanford University

Abstract. We are given a graph with edge weights, that represents the
metric on the vertices in which the distance between two vertices is the
total weight of the lowest-weight path between them. Consider the problem of representing this metric using as few edges as possible, provided
that new "steiner" vertices (and edges incident on them) can be added.
The compression factor achieved is the ratio k between the number of
edges in the original graph and the number of edges in the compressed
graph. We obtain approximation algorithms for unit weight graphs that
replace cliques with stars in cases where the cliques so compressed are
disjoint, or when only a constant number of the cliques compressed meet
at any vertex. We also show that the general unit weight problem is essentially as hard to approximate as graph coloring and maximum clique.

1 Introduction

Suppose we are given a finite metric space, represented as a graph $G = (V, E)$ on
n nodes, with positive edge weights $l(e)$. We wish to find a graph $G' = (V', E')$,
where $V \subseteq V'$, such that $|E'|$ is substantially smaller than E while ensuring
that the metric is preserved exactly (i.e., pairwise distances for the vertices in
V remain the same). The compression achieved by an algorithm is the ratio
$k = |E|/|E'|$. If V' is constrained to be exactly V (i.e., if we are not allowed
to add any vertices), we can find the edge-minimal graph in polynomial time.
If we are allowed to add new vertices, the problem becomes more complex. We
give polynomial-time algorithms which find graphs that are approximately edge-
minimal.

---

* Email: tomas@theory.stanford.edu
** 4110 Wean Hall, Department of Computer Science, Carnegie-Mellon University,
   Pittsburgh, PA 15213. Research supported by NSF Grant CCR-0122581 and ARO
   Grants DAAG-55-98-1-0170 and DAAG-55-97-1-0221. Email: adam@cs.cmu.edu
*** Department of Computer Science, Stanford University, Stanford, CA 94305. Rese-
   arch supported by NSF Grant IIS-0118173, an Okawa Foundation Research Grant,
   and Veritas. Email: rajeev@cs.stanford.edu.
† Department of Computer Science, Stanford University, Stanford, CA 94305. Rese-
   arch supported by an NSF Graduate Fellowship, an ARCS Fellowship, and NSF
   Grants IIS-0118173, IIS-9811904, and EIA-0137761. Email: loc@cs.stanford.edu
‡ Cisco Systems. Email: rinap@cisco.com

H. Alt and M. Habib (Eds.): STACS 2003, LNCS 2607, pp. 355–366, 2003.
© Springer-Verlag Berlin Heidelberg 2003

Main Techniques. The key tool in our algorithms is the following. Consider the case in which we are given a collection of weighted graphs H_i called compressions, where distinct H_i may only share vertices in G. The candidate weighted graphs G' for the compression problem described above are obtained by selecting some of the compressions H_i, taking their union, and adding some edges from G itself. We show that if some such G' achieves a compression ratio k, then we can find one such G' achieving compression ratio at least $k/\log k$.

Suppose we are given a graph G in that has unit edge weights. We show that we achieve the best compression by replacing cliques by stars — the vertices of each replaced clique become the leaves of a star whose center is a new vertex, and each edge has weight $1/2$. In general, it is hard to select an appropriate collection of stars H_i to which to apply the compression algorithm. We show that this can be done if G satisfies certain degree constraints. We also study the case where G is weighted and sparse, and the H_i are trees.

Summary of Results. The unit weight problem varies in hardness depending on whether we consider very special optima and solutions or more general ones. We summarize the results that we have obtained. At one end of the spectrum, when we consider the compression of a single clique, we obtain constant factor positive and negative approximation results. At the other end of the spectrum, where we are considering the compression of an arbitrary number of cliques that intersect arbitrarily, we cannot hope to obtain approximation algorithms, since the problem is essentially as hard to approximate as graph coloring and maximum clique. Three intermediate levels exhibit intermediate hardness in approximation: With respect to an optimum that compresses arbitrarily many disjoint cliques, the approximation factor achieved is the logarithm of the optimum compression; for the compression of two cliques that are not necessarily disjoint, the approximation factor achieved is the square root of the optimum compression; and with respect to an optimum that compresses arbitrarily many cliques that are not necessarily disjoint but where each clique compressed meets only a constant number of the other cliques compressed, the approximation factor achieved exceeds the square root of the optimum compression by a logarithm. Thus five levels of generality in the problem give five levels of hardness of approximation.

We describe the results in more detail now. In the unit weight case, we look for algorithms that perform well relative to an optimal solution in which the cliques corresponding to the H_i share no vertices. For the problem of finding a single clique, we give a linear-time 2-approximate algorithm, and show that the problem is as hard to approximate as vertex cover, hence hard to approximate within $7/6$ by the result of Håstad [15] and within 1.3606 by the result of Dinur and Safra [8]. More generally, if some r disjoint cliques achieve compression ratio k, then we can find n cliques such that some r of them achieve compression ratio at least $k/3$. The earlier algorithm applied to these n cliques achieves compression at least $k/(3 \log k)$.

The problem where the cliques for the optimal choice of H_i may share vertices is harder to approximate. If compressing two cliques achieves ratio k, then we can find two cliques that give compression $\sqrt{k}/4$. We look for algorithms that perform well relative to an optimal solution in which each clique compressed meets at most r of the other cliques compressed; call such a solution an r-*sparse* solution. If an r-sparse solution with r constant achieves compression ratio k, then we find a solution achieving compression at least $\sqrt{k}/c\log k$, where c depends on r.

We then consider a related problem, where G is bipartite, and the stars H_i must have leaves forming a complete bipartite subgraph of G. Feder and Motwani [10] considered this problem for dense graphs and achieve compression factor $\log n/\log(n^2/m)$ on graphs with n vertices and m edges. The case of a single star H_i again has a 2-approximation algorithm, this time based on parametric flows. Here also, if disjoint stars achieve compression k, then we find a collection of stars containing a subset achieving ratio $k/3$, and we can then find a subset achieving ratio $k/(3\log k)$. With respect to an optimal r-sparse solution with r constant, we also get an algorithm with compression at least $\sqrt{k}/(c\log k)$ in the bipartite case.

Finally, we show that the unit weight compression problem is hard to approximate within a factor $n^{\frac{1}{2}-\epsilon}$ on instances with optimum compression at most $n^{\frac{1}{2}}$ for any constant $\epsilon > 0$ unless NP=ZPP, and for $\epsilon = 1/(\log n)^{\gamma}$ for some constant $\gamma > 0$ unless NP $\subseteq$ ZPTIME$(2^{(\log n)^{O(1)}})$.

Related Work. In 1964, Hakimi and Yau [14] defined the *optimal realization* of a distance matrix — a graph that preserves shortest-path distances while minimizing the *total sum of edge weights*. They give a solution in the case where the optimal realization is a tree. Since then, there has been substantial work on this problem [4,9,17,18], and Althofer [1] has established the NP-hardness of finding the optimal realization if the matrix entries are integral. Chung et al. [6] give an algorithm to find a graph that is approximately optimal in the above sense, and in which the shortest-path distances are no shorter than in the given distance matrix.

Under the model of Arya et al. [2], Das et al. [7], and Rao and Smith [19], the vertices are points in Euclidean space, and the goal is to find *spanners* — subgraphs of the complete Euclidean graph that approximately preserve shortest path distances and have approximately minimal total weight (i.e., the sum of weights of edges in the subgraph). In a similar vein, Gupta [13] shows that some vertices can be removed with only constant distortion to distances on the remaining vertex set. Other related work includes that of Bartal [3], who introduces the idea of probabilistically approximating metric spaces by distributions over sets of other metric spaces, and of Charikar et al. [5], who derandomize this algorithm.

2   Generic Compression Algorithm

A *weighted graph* is a graph G with a positive weight on each edge. The distance $d(x,y)$ between two vertices x,y in G is the minimum over all paths p from x to y in G of the sum of the weights on p. If for three distinct vertices x,y,z we have $d(x,y) = d(x,z) + d(z,y)$, then the edge (x,y) is redundant and can be removed from the graph. If G is initially a complete graph on n vertices, and G' is obtained from G by removing redundant edges so that G' has only m edges, then we can obtain G' from G in $O(n(m + n\log n))$ time.

A *compression* for G is a weighted graph H such that $V = V(G) \cap V(H)$ is nonempty, with $d_H(x, y) \geq d_G(x, y)$ for all x, y in V, and such that H does not have an edge (x, y) with both x, y in V. We say that G' is the weighted graph obtained by applying compression H to G if G' is obtained from $G \cup H$ by removing all edges (x, y) in G with both x, y in V such that $d_H(x, y) = d_G(x, y)$. Let C be a set of compressions for G. We can obtain a weighted graph G' by successively applying each of the compressions in C, starting with G. We say that C compresses m to m/k if $|E(G)| = m$ and $|E(G')| = m/k$. We also say that C has *compression factor* k.

Suppose we are given a set C of candidate compressions, and suppose that some subset of C has compression factor k. Theorem 1 establishes that we can find a subset of C with compression factor at least $k/(1 + \log k)$. The next step is to determine which weighted graphs H should be used as compressions. We focus on the *unit weight* case, where every edge of G has weight 1. Theorem 2 shows that we can assume without loss of generality that H is a star with edges of weight $1/2$ whose leaves form a clique in G.

Theorem 1 *Assume G has arbitrary weights, and let C be a given set of candidate compressions. Suppose that some subset of C compresses m to m/k. Then we can find in polynomial time a subset of C that compresses m to at most $\frac{m}{k}(1 + \log k)$.*

Proof Sketch: Each compression H_i we apply replaces $\lambda_i p_i$ edges with p_i new edges, for some $\lambda_i > 1$. [1] Let r be the number of edges from G that we do not replace in this way. Then the original graph G has $r + \sum_i \lambda_i p_i = m$ edges, while the graph G' obtained by applying the compressions H_i has $r + \sum_i p_i = m'$ edges.

The algorithm is greedy: Repeatedly select the compression H_i with $\lambda_i > 1$ largest. Define the compression factor of an edge e in G to be the value $s(e) = 1/\lambda_i$ when the algorithm uses a compression H_i to replace $\lambda_i p_i$ edges including e with just p_i edges; let $s(e) = 1$ otherwise. In the end, the number of edges in G' will be $\sum_{e \in E} s(e)$.

When $s > m/k$ edges remain that have not been removed by applying a compression, since the optimal solution compresses them to at most m/k edges, the compression factor for the edges replaced when the next H_i is applied by the algorithm is at most $m/(ks)$. Therefore $\sum_e s(e) \leq \frac{m}{k} + \sum_{\frac{m}{k} < s \leq m} \frac{m}{ks} = \frac{m}{k}(1 + H_m - H_{\frac{m}{k}}) \leq \frac{m}{k}(1 + \log k)$. □

The following theorem implies that the question, "Can we reduce the number of edges by p by adding a new vertex?" is NP-complete.

Theorem 2 *In the unit weight case, one can assume without loss of generality that each compression H used is a star with edges of weight $1/2$ whose leaves form a clique in G.*

Proof Sketch: Let H be a compression for G. Suppose H has an edge (x_0, y) with x_0 in G and $d_H(x_0, y) < 1/2$. Then there is no vertex $x \neq x_0$ in $V(G) \cap V(H)$

---

[1] Note that if we use two compressions H_i and H_j that would both replace a common edge e, and we apply H_i first, then we will credit only H_i for the replacement of e.

such that $d_H(x, y) \leq 1/2$; otherwise we would have $d_G(x_0, x) \leq d_H(x_0, x) < 1$. Consequently, if $d_H(x', y) + d_H(x'', y) = 1$ for some x', x'' in G, such that the edge (x', x'') can be removed from G, then one of x', x'' must be x_0. We can obtain a smaller H' by removing y and its incident edges (y, y') from H and adding edges (x_0, y') for each such $y' \neq x_0$, with $d_{H'}(x_0, y') = d_H(x_0, y) + d_H(y, y')$.

Therefore, we can assume that if H has an edge (x, y) with x in G, then $d_H(x, y) \geq 1/2$. An edge (x', x'') in G can thus only be removed if x' and x'' have a common neighbor y in H ($y \notin V$) with $d_H(x, y) = d_H(x', y) = 1/2$. That is, the compression H is a union of stars with edges of weight $1/2$ whose leaves form a clique in G.   □

3   Compression and the Disjoint Optimum

We continue to assume our graph G has unit weights. We have seen that in this case we can assume that each compression corresponds to a clique in G. It remains to determine which cliques should be chosen for compression. We consider here a comparision with a compression that compresses either a single clique or disjoint cliques.

Theorem 3 *In the unit weight case, compression by selecting a single clique has a 2-approximation algorithm that runs in $O(m)$ time.*

Proof Sketch: We consider the unit weight case with a single additional vertex, and give a 2-approximation algorithm. If the maximum clique has size k, then the optimal compression is from m to $m + k - \binom{k}{2} = \alpha\binom{k}{2}$. With no compression, we have $m \leq (\alpha + 1)\binom{k}{2}$ edges, for an approximation factor of $(\alpha + 1)/\alpha = 1 + 1/\alpha$, giving the result for $\alpha \geq 1$. We now focus on the case $\alpha < 1$.

We repeatedly remove vertices of degree at most $k - 2$, until every vertex has degree at least $k - 1$. There are $m - \binom{k}{2} = \alpha\binom{k}{2} - k$ edges not in the clique of size k. If there are v vertices not in the clique, then the number of edges not in the clique is at least $(k - 1)v/2$, which gives $v \leq \alpha(k - 1)$. The number of edges not in the clique is also at least $(k - 1)v - \binom{v}{2}$. The inequality $(k - 1)v - \binom{v}{2} \leq \alpha\binom{k}{2} - k$ yields $v \leq (k - 1) + \frac{1}{2} - \sqrt{((1 - \alpha)(k - 1)^2 + (3 - \alpha)(k - 1) + 9/4)}$, which implies $v \leq (1 - \sqrt{1 - \alpha})(k - 1)$. The v vertices form a vertex cover in the complement graph. Since vertex cover has a 2-approximation algorithm by means of a maximal matching, we can obtain a vertex cover with at most $2v$ vertices, and the vertices not in the vertex cover give a clique in the original graph with $l \geq k - v$ vertices. Compressing this clique, the resulting number of edges is $m + l - \binom{l}{2} = (\alpha + 1)\binom{k}{2} - k + l - \binom{l}{2} \leq (\alpha + 1)\binom{k}{2} - \binom{l}{2} \leq 2\alpha\binom{k}{2}$. This last inequality follows from the equivalent inequality $\binom{l}{2} \geq (1 - \alpha)\binom{k}{2}$, since $l - 1 \geq \sqrt{1 - \alpha}(k - 1)$ and $l \geq \sqrt{1 - \alpha}(k)$.

The algorithm is as follows:

1. Find a sequence of graphs G_t, where $G_1 = G$, and if v_t is a vertex of minimum degree d_t in G_t, then $G_{t+1} = G_t \setminus \{v_t\}$ is obtained by removing v_t and its incident edges from G_t.

2. Find a maximal matching M_t in the complement graph $\overline{G}_t$ for each t. A maximal matching M_t in $\overline{G}_t$ can be obtained from a maximal matching M_{t+1} in $\overline{G}_{t+1}$ by letting $M_t = M_{t+1} \cup \{(v_t, u)\}$ if v_t has a neighbor u in $\overline{G}_t$ such that u is not in an edge of M_{t+1}; otherwise $M_t = M_{t+1}$.

3. The vertices in G_t not incident to an edge of M_t form a clique Q_t — compress the largest such clique Q_t. □

Theorem 4 *If compression by selecting a single clique has an α-approximation algorithm with $\alpha < 2$, then vertex cover has an $(\alpha + \epsilon)$-approximation algorithm for all $\epsilon > 0$.*

Proof Sketch: Let G be an instance of vertex cover with n vertices and minimum vertex cover of size b. We can assume $b > 1/\epsilon$, since otherwise the minimum vertex cover can be found by considering all subsets of size b. Let G' be the graph on N vertices obtained by adding at least n/ϵ vertices to G, with no new edges.

Consider the complement graph $\overline{G}'$ as an instance of the single clique problem. The maximum clique in $\overline{G}'$ has size $N - b$, and after compressing this clique we have $\mathrm{OPT} \leq N(b+1)$ edges in the compressed graph.

Use the α-approximation algorithm to find a solution that compresses a clique of size $N - a$, thus giving a vertex cover of size a in the original graph. We have $\alpha\mathrm{OPT} \geq \mathrm{SOL} \geq (N - a - 1)(a - b) + \mathrm{OPT}$, implying that $(N - a - 1)(a - b) \leq (\alpha - 1)\mathrm{OPT} \leq (\alpha - 1)N(b+1)$. This gives $(N - a)a \leq \alpha N(b+1)$, implying that $a \leq \frac{1 + \frac{1}{b}}{1 - \frac{a}{N}}\alpha b \leq \frac{1+\epsilon}{1-\epsilon}\alpha b \leq (1 + 3\epsilon)\alpha b$. We have an $\alpha + \epsilon'$ approximation if we let $\epsilon = \epsilon'/(3\alpha)$. □

Theorem 5 *In the unit weight case, suppose r disjoint cliques compress m to m/k. Then we can find n cliques such that some r of them compress m to at most $3m/k$.*

Proof Sketch: Consider the r disjoint cliques Q_i of size q_i. Let $d_i + q_i - 1$ be the minimum degree of a vertex in Q_i. Then there are at least $d_i q_i$ edges coming out of Q_i so $\frac{m}{k} \geq \sum_i \frac{d_i q_i}{2}$. If $d_i \geq q_i$, then not compressing Q_i costs at most $\binom{q_i}{2} \leq \frac{d_i q_i}{2}$ extra edges. Suppose next $d_i < q_i$. Let v_i be a vertex of degree $d_i + q_i - 1$ in Q_i. Let G_i be the graph induced by v_i and its neighbors. The complement graph $\overline{G}_i$ has a vertex cover of size d_i consisting of the d_i vertices not in Q_i. We can find a maximal matching M_i on $\overline{G}_i$. The matching M_i will have at most d_i edges, and involve at most d_i vertices in Q_i. The vertices of G_i not incident to an edge of M_i give a clique R_i that has at least $q_i - d_i$ vertices in Q_i. Compressing R_i leaves at most $d_i(q_i - 1)$ edges of Q_i not compressed. Furthermore R_i has at most d_i more vertices than Q_i, so the total extra cost is at most $d_i(q_i - 1) + d_i \leq d_i q_i$ extra edges. The total extra cost for the entire graph is therefore at most $\sum_i d_i q_i \leq 2m/k$ extra edges. □

The next result follows immediately from Theorems 1 and 5.

Theorem 6 *In the unit weight case, suppose r disjoint cliques compress m to m/k. Then we can find a compression from m to at most $3m/k(1 + \log(k/3))$.*

4   Compression and the NonDisjoint Optimum

It is more difficult to obtain algorithms that perform as well with respect to the optimum which can compress cliques that are not necessarily disjoint.

Theorem 7 *In the unit weight case, suppose two (not necessarily disjoint) cliques compress m to m/k. Then we can find two cliques that compress m to at most $4m/\sqrt{k}$.*

Proof Sketch: Suppose the two cliques Q_1 and Q_2 have a vertices in common, and are of size $a + b_1$ and $a + b_2$ respectively, with $b_2 \leq b_1$. Let r be the number of edges not in the two cliques, and write $r = d_1 b_1 = d_2 b_2$.

Suppose first $d_2 \leq b_2$. Then $d_1 \leq b_1$ as well. For $i = 1, 2$, some vertex v_i out of the b_i vertices in clique Q_i but not in the other clique has at most d_i edges incident to v_i and not in Q_i.

As in the proof of Theorem 5, we can find a clique R_i contained in the graph induced by v_i and its neighbors, so that when we compress it, the extra number of edges is at most $d_i(a + b_i)$. The total number of edges after both R_1 and R_2 are compressed is thus at most $\frac{m}{k} + d_1(a + b_1) + d_2(a + b_2) \leq 3\frac{m}{k} + 2d_2 a$ with $(d_2 a)^2 = d_2^2 a^2 \leq r(2m) \leq \frac{2m^2}{k}$, so that $d_2 a \leq \frac{\sqrt{2}m}{\sqrt{k}}$. Suppose next $d_2 > b_2$. If we only compress Q_1, the total number of edges resulting is at most $\frac{m}{k} + b_2(a + b_2) \leq 2\frac{m}{k} + b_2 a$ with $(b_2 a)^2 = b_2^2 a^2 \leq r(2m) \leq \frac{2m^2}{k}$, so that $b_2 a \leq \frac{\sqrt{2}m}{\sqrt{k}}$. We can find a 2-approximation to compressing Q_1 by Theorem 3.

In both cases, the bound is at most $4\frac{m}{k} + \frac{2\sqrt{2}m}{\sqrt{k}} = (2\sqrt{2} + \frac{4}{\sqrt{k}})\frac{m}{\sqrt{k}}$.

If $k \leq 16$, then the m original edges give a $4m/\sqrt{k}$ bound; if $k \geq 16$ then the above bound is at most $(2\sqrt{2} + 1)m/\sqrt{k}$ and the result follows.   □

Consider a solution involving some l compressed cliques H_i, such that each H_i intersects at most r other H_i, for constant r. Suppose this solution has compression factor k. We define *sectors* so that two vertices are in the same sector if and only if the set of H_i to which they belong is same. The number of sectors within a clique H_i is at most s for some $s \leq 2^r$. We define an *associated graph* whose vertices are the sectors S_j; two sectors are adjacent if they belong to the same clique H_i for some i. The max-degree in the associated graph is $d \leq rs$.

Theorem 8 *There is a constant c such that we can find a collection of at most $n^{\lfloor d/2 \rfloor + 1}$ cliques containing a subcollection of at most ls^2 cliques that achieve compression factor at least $\sqrt{k}/(cd^4)$.*

Proof Sketch: Consider two adjacent sectors S_1 and S_2, and allow $S_1 = S_2$. Consider the sector S_3 adjacent to both S_1 and S_2 that has the largest number v of vertices in it; we allow S_3 to be S_1 or S_2 as well. Suppose the number of edges joining S_1 to S_2 is at most $v^2/\sqrt{k}$. We charge these edges to the $v^2/2$ edges of S_3. The sector S_3 will be so charged at most d^2 times, so the total charge is at most $md^2/\sqrt{k}$.

Conversely, suppose the number of edges joining S_1 to S_2 is at least $v^2/\sqrt{k}$. Then both S_1 and S_2 have at least $v/\sqrt{k}$ vertices. Let Q be a maximal clique in the associated graph containing S_1, S_2, S_3, such that each sector in Q has at least $v/\sqrt{k}$ vertices. For each sector S_i in Q, let u_i be the vertex in S_i that has the smallest number t_i of edges incident to it going to vertices in sectors S_j such that S_i and S_j are not adjacent. Let H be the induced subgraph whose vertices are all the vertices w that are either equal to some u_i or adjacent to all u_i.

We can find a single clique that gives a 2-approximation in H as in Theorem 5. The bound on the number of edges of Q not compressed plus the number of additional edges in the compression is fq, where f is the number of vertices in H that are not in Q, and q is the size of Q. We bound f and q. Clearly $q \leq dv$.

There are at most d sectors adjacent to all sectors in Q but not in Q, and each such sector has at most $v/\sqrt{k}$ vertices, for a total of $dv/\sqrt{k}$ vertices. Multiplying this quantity by q gives at most $d^2 v^2/\sqrt{k}$ edges, which can be charged again to the $v^2/2$ edges of S_3. The sector S_3 will be so charged at most d^2 times, so the total charge is at most $md^4/\sqrt{k}$.

The remaining t vertices have at least one neighbor u_i such that their edge to u_i is not compressed in the optimum. Thus, $t \leq dt_i$ for some d; multiplying this quantity by q gives at most $d^2 v t_i$ edges, which can be charged to the $\frac{v}{\sqrt{k}} t_i$ edges not compressed coming out of S_i. Again S_i will be charged at most d^2 times, for a total charge of $d^4 \sqrt{k}$ per edge not compressed in the optimum. Since the number of edges not compressed in the optimum is at most $\frac{m}{k}$, the total charge is at most $\frac{md^4}{\sqrt{k}}$.

Finally, each vertex is involved in at most d^2 cliques Q, giving at most $nd^2 \leq \frac{md^2}{k}$ new edges.

The algorithm is thus as follows: For each choice of at most d vertices u_i forming a clique, find a single clique in the graph of the common neighbors of the u_i as in Theorem 5. We can reduce the number of chosen vertices to $\lfloor d/2 \rfloor + 1$ as follows. Either Q has at most this many sectors, or the number of sectors not in Q adjacent to a chosen sector in Q is at most $\lceil d/2 \rceil - 1$. We will need to choose at most $\lceil d/2 \rceil - 1$ extra sectors in Q to rule out the neighbor sectors that are not common neighbors of all sectors in Q, for a total of $\lceil d/2 \rceil \leq \lfloor d/2 \rfloor + 1$ chosen vertices. □

Theorem 9 follows immediately from Theorems 1 and 8.

Theorem 9 *We can find a collection of cliques achieving compression factor* $\sqrt{k}/(cd^4 \log k)$.

In general, it is not possible to find all the cliques and apply the generic compression algorithm. However, this can be done if all vertices have degree $O(\log n)$, or for slightly smaller cliques in a graph of a slightly larger degree. We consider again the sequence of graphs G_t, where $G_1 = G$, and if v_t is a vertex of minimum degree d_t in G_t, then $G_{t+1} = G_t \setminus \{v_t\}$ is obtained by removing v_t and its incident edges from G_t. Every clique in G is a clique containing v_t in G_t for some t.

Theorem 10 *If v_t has degree $d_t \leq f_t \log n$ in G_t, then we can find the polynomially many cliques containing v_t in G_t of size $O(\log n / \log f_t)$. These are all the cliques if d_t is $O(\log n)$.*

In the weighted case, for a vertex v and a constant c, we denote by $N^c(v)$ the set of vertices joined by a path with at most c edges to v in G. A *good compression* H has all its vertices from G inside $N^c(v)$ for some v.

Theorem 11 *In the weighted case, there are polynomially many good compressions by trees of size $O(\log n / \log\log n)$ in a graph of maximum degree $O(\log^d n)$, and these can be found in poly-time.*

5 Bipartite Compression

Consider the following situation. We have identified two cliques R_1, R_2 to be compressed, and every additional clique Q that we may compress has vertices in either R_1 or R_2. We may assume $V(R_1) \cap V(R_2) = \emptyset$ and $V(R_1) \cup V(R_2) = V(G)$. Then $G' = G \setminus (R_1 \cup R_2)$ is a bipartite graph $G' = (V(R_1), V(R_2), E)$. Compressing a clique Q in G corresponds to compressing a complete bipartite subgraph of G', which we refer to as a *bi-clique*.

We consider here the case where G' has a collection of r bi-cliques sharing no vertices giving a compression factor k, and obtain three results analogous to the three results in Section 3. We consider then the optimum where every bi-clique compressed meets at most a constant number of the other bi-cliques compressed, and has compression k. We obtain results analogous to those in Section 4.

Theorem 12 *Compression by a single bi-clique has a 2-approximation algorithm.*

Proof Sketch: Suppose the optimal bi-clique Q has q_1 vertices in R_1 and q_2 vertices in R_2. The optimal compression is thus from m to $m + q_1 + q_2 - q_1 q_2$. Let v_1 be a vertex in $Q \cap R_1$ of minimum degree $q_2 + d_2$, and let v_2 be a vertex in $Q \cap R_2$ of minimum degree $q_1 + d_1$. The number of edges not compressed incident to Q is at least $s = q_1 d_2 + q_2 d_1$. Let $\hat{G}$ be the subgraph induced by the $q_1 + d_1$ neighbors of v_2 and the $q_2 + d_2$ neighbors of v_1.

We consider first the case where $q_1 = q_2 = q$. Find a maximal matching M in the bipartite complement of $\hat{G}$. The veritces not in M form a bi-clique T in $\hat{G}$. Note that M has at most d_2 vertices in $Q \cap R_1$ and at most d_1 vertices in $Q \cap R_2$. Thus the number of edges in Q but not in T is at most $q d_2 + q d_1 = s$. This gives a 2-approximation when we compress T instead of Q.

When q_1 and q_2 are not necessarily equal, we can define $\hat{G}'$ obtained from $\hat{G}$ by making q_2 copies of each vertex in $\hat{G} \cap R_1$ and q_1 copies of each vertex in $\hat{G} \cap R_2$. Now Q gives a bi-clique Q' in G' with $q_1 q_2 = q_2 q_1 = q$ vertices in each side.

A maximal matching M' in the complement of $\hat{G}'$ has at most $d_2 q_1$ vertices in $Q' \cap R_1'$ and at most $d_1 q_2$ vertices in $Q' \cap R_2'$. The vertices not in M' form a bi-clique T' in $\hat{G}'$. The number of edges in Q' but not in T' is at most $q d_2 q_1 + q d_1 q_2 = qs$. We can add vertices to T' until the vertices not in T' form a minimal vertex cover in the complement of $\hat{G}'$. Then T' will have either all or none of the q_2 copies of a vertex in R_1, and either all or none of the q_1 copies of a vertex in R_2, so T' corresponds to a bi-clique T in $\hat{G}$, and the number of edges in Q but not in T is at most s.

We may thus try all possible pairs of values q_1, q_2. Alternatively, we can find minimum instead of minimal vertex covers in the complement of $\hat{G}$. For a parameter $0 < \lambda < 1$, assign weight λ to the vertices in R_1 and weight $1 - \lambda$ to the vertices in R_2. We can find a collection of at most $1 + \min(|\hat{G} \cap R_1|, |\hat{G} \cap R_2|)$ weighted minimum vertex covers over all values of λ, by a parametric flow [12]. One of these weighted minimum vertex covers will correspond to $q_1/q_2 = (1 - \lambda)/\lambda$. □

Theorem 13 *Suppose r disjoint bi-cliques compress m to m/k. Then we can find at most n^3 bi-cliques such that some r of them compress m to at most $3m/k$.*

Theorem 14 *Suppose r disjoint bi-cliques compress m to m/k. Then we can find a compression from m to at most $3\frac{m}{k}(1 + \log \frac{k}{3})$.*

Consider a solution involving some l compressed bi-cliques H_i, such that each H_i intersects at most r other H_i, for constant r. Suppose this solution has compression factor k. We define *sectors* so that two vertices in the same R_p ($p = 1, 2$) are in the same sector if and only if the set of H_i to which they belong is the same. The number of sectors within a bi-clique H_i and in the same R_p is at most s for some $s \leq 2^r$. We define an *associated graph* whose vertices are the sectors S_j, where two sectors are adjacent if they belong to the same clique H_i for some i and they are in different R_p. The max-degree in the associated graph is $d \leq rs$.

Theorem 15 *There is a constant c such that we can find a collection of at most n^{d+2} bi-cliques containing a subcollection of at most ls^2 bi-cliques that achieve compression factor at least $\sqrt{k}/(cd^4)$.*

Theorem 16 follows from Theorem 15 by the algorithm of Theorem 1.

Theorem 16 *We can find a collection of bi-cliques achieving compression factor $\frac{\sqrt{k}}{cd^4 \log k}$.*

6   Hardness of Approximation

We establish the following result and its corollary.

Theorem 17 *Finding an $(\frac{r}{4+\log_p n})$-approximation for the unit weight compression problem, on instances with n^2 vertices and optimum compression factor at most n, is as hard as finding an independent set of size n/rp in a p-colorable graph with n vertices, where r, p may depend on n.*

Proof Sketch: Let G be a p-colorable graph where we wish to find an independent set of size n/rp. We assume $n = pq$, where p and q are prime numbers. We define a graph H with n^2 vertices of the form (x, y, z), where $0 \leq x < n$, $0 \leq y < p$, and $0 \leq z < q$. We view x, y, z as integers modulo n, p, q respectively.

The graph H has all the edges between two vertices (x, y, z) and (x', y', z') such that $x \neq x'$ and $y \neq y'$. The number of such edges is $M_1 = n(n - 1)p(p - 1)q^2/2$. In addition, H has all the edges between two vertices (x, y, z) and (x', y, z) such that $x \neq x'$ and (x, x') is not an edge in G. The number of such edges is at most $M_2 = n(n - 1)pq/2$.

We exhibit a compression of H, using the fact that G is p-colorable. We shall not compress the M_2 edges, although these edges may belong to compressed cliques. Note that $M_2 \leq \frac{1}{(p-1)q}M_1$.

Let R be the clique consisting of the n vertices $(x, y, 0)$ such that vertex x in G has color y in the p-coloring. For $0 \leq i < p$, $1 < j < p$, and $0 \leq k, l < q$, let R_{ijkl} be the clique consisting of the n vertices $(x, i + jy, k + ly)$ such that $(x, y, 0)$ is in R. These $p(p - 1)q^2$ cliques compress all the edges between two vertices (x, y, z) and (x', y', z') with $x \neq x'$ and $y \neq y'$ and such that x and x' have different colors in the p-coloring, and introduce $np(p - 1)q^2 \leq \frac{2}{n-1}M_1$ new edges.

It remains to compress the edges between two vertices (x, y, z) and (x', y', z') with $x \neq x'$ and $y \neq y'$ and such that x and x' have the same color d in the p-coloring. Let s_d be the number of vertices of color d, so that $\sum_{0 \leq d < p} s_d = n$. There exist $\lceil \log_p s_d \rceil$ p-colorings of the s_d vertices such that every pair of distinct vertices among the s_d vertices gets different colors in at least one such p-coloring. Using the previous argument, we find $p(p - 1)q^2 \lceil \log_p s_d \rceil$ cliques of size s_d for the vertices (x, y, z) with x of color d. The number of new edges introduced is $\sum_d s_d p(p - 1)q^2 \lceil \log_p s_d \rceil \leq np(p - 1)q^2 \lceil \log_p n \rceil \leq \frac{2\lceil \log_p n \rceil}{n-1}M_1$. Thus, we have a compression factor k such that $\frac{1}{k} \leq \frac{1}{(p-1)q} + \frac{2}{n-1} + \frac{2\lceil \log_p n \rceil}{n-1} \leq \frac{6 + 2\log_p n}{n-1}$. Suppose we can find a compression of H with compression factor l. Then the compression includes a clique with s vertices in H such that $(s - 1)/2 \geq l$. This clique gives is a p-coloring of s vertices of G, and hence an independent set of size at least s/p in G. If this independent set is of size smaller than n/rp, then $s \leq n/p$ and $l \leq n/2r \leq (\frac{4 + \log_p n}{r})k$. $\quad\square$

Feige and Kilian [11] prove that chromatic number is hard to approximate within factor $n^{1-\epsilon}$ for any constant $\epsilon > 0$, unless NP=ZPP. This implies that it is also hard to find an independent set of size $\frac{n^\epsilon}{p}$ for any constant $\epsilon > 0$, where p is the chromatic number of the graph; otherwise we could repeatedly select large independent sets and get a coloring with $pn^{1-\epsilon} \log n$ colors. Since the graphs in the preceding theorem have size n^2, setting $r = n^{1-\epsilon}$ gives the following.

Theorem 18 *The unit weight compression problem is inapproximable in polynomial time within factor $n^{\frac{1}{2}-\epsilon}$ on instances with optimum compression factor at most $n^{\frac{1}{2}}$ for any constant $\epsilon > 0$, unless NP=ZPP.*

Khot [16] improves the chromatic number result to $\epsilon = \frac{1}{(\log n)^\gamma}$ for a constant $\gamma > 0$, if it is not the case that NP $\subseteq$ ZPTIME$(2^{(\log n)^{O(1)}})$. This can be carried over to the above theorem as well.

References

1. I. Althofer. "On optimal realizations of finite metric spaces by graphs." *Discrete Comp. Geom* 3, 1988.
2. S. Arya, G. Das, D. M. Mount, J. S. Salowe, and M. H. M. Smid. "Euclidean spanners: short, thin, and lanky." In *Proc. STOC*, 1995.
3. Y. Bartal. "Probabilistic approximation of metric spaces and its algorithmic applications." In *Proc FOCS*, 1996.
4. F. Boesch. "Properties of the distance matrix of a tree." *Quart. Appl. Math.* 26 (1968-69), 607–609.
5. M. Charikar, C. Chekuri, A. Goel, S. Guha, and S. Plotkin. "Approximating a finite metric by a small number of tree metrics." In *Proc. FOCS*, 1998.
6. F. Chung, M. Garrett, R. Graham, and D. Shallcross. "Distance realization problems with applications to internet tomography." Preprint, http://www.math.ucsd.edu/~fan.
7. G. Das, G. Narasimhan, and J. Salowe. "A new way to weigh malnourished Euclidean graphs." In *Proc. SODA*, 1995.
8. I. Dinur and M. Safra. Personal communication.
9. A. W. M. Dress. "Trees, tight extensions of metric spaces, and the cohomological dimension of certain groups." Advances in Mathematics 53 (1984), 321–402.
10. T. Feder and R. Motwani. "Clique compressions, graph partitions and speeding-up algorithms." *JCSS* 51 (1995), 261–272.
11. U. Feige and J. Kilian. "Zero-knowledge and chromatic number." In *Proc. Annual Conf. on Comp. Complex.* (1996).
12. G. Gallo, M. D. Grigoriadis, and R. E. Tarjan. "A fast parametric maximum flow algorithm and applications." *SICOMP 18* (1989) 30–55.
13. A. Gupta. "Steiner points in tree metrics don't (really) help." In *Proc. 12th SODA* 2001, pp 220-227.
14. S. L. Hakimi and S. S. Yau. "Distance matrix of a graph and its realizability." *Quart. Appl. Math.* 22 (1964), 305–317.
15. J. Håstad. "Some optimal inapproximability results." In *Proc STOC* (1997) 1–10.
16. S. Khot. "Improved inapproximability results for max clique, chromatic number and approximate graph coloring." In *Proc FOCS* (2001).
17. J. Nieminen. "Realizing the distance matrix of a graph." *Elektron. Informationsverarbeit. Kybernetik* 12(1-2):1976, 29–31.
18. J. Pereira. "An algorithm and its role in the study of optimal graph realizations of distance matrices." *Discrete Math.* 79(3):1990, 299–312.
19. S. B. Rao and W. D. Smith. "Improved approximation schemes for geometrical graphs via spanners and banyans." In *Proc. STOC* (1998), 540–550.

Computing Shortest Paths with Uncertainty

T. Feder *, R. Motwani **, L. O'Callaghan * * *, C. Olston [†], and
R. Panigrahy [‡]

Stanford University

Abstract. We consider the problem of estimating the length of a short-est path in a DAG whose edge lengths are known only approximately but can be determined exactly at a cost. Initially, each edge e is known only to lie within an interval $[l_e, h_e]$; the estimation algorithm can pay c_e to find the exact length of e. In particular, we study the problem of finding the cheapest set of edges such that, if exactly these edges are queried, the length of the shortest path will be known, within an additive $\kappa > 0$ that is given as an input parameter. We study both the general problem and several special cases, and obtain both easiness and hardness approximation results.

1 Introduction

Consider a weighted DAG G with a single source s of in-degree zero and a single sink t of out-degree zero, whose exact edge lengths are not known with certainty. Assume that for every edge e of G the length of e is only known to lie in an interval $[l_e, h_e]$. The length of a path in G can be computed with uncertainty and represented as an interval containing the exact length. An interval containing the length of the shortest path between two nodes in G can similarly be computed.

Suppose that for every edge e, the exact length of e can be found at query cost c_e. An optimization problem that arises naturally is: *Given a DAG G, a source s, a sink t, a set P of s-t paths, edge-length intervals $[l_{e_1}, h_{e_1}], \ldots, [l_{e_m}, h_{e_m}]$, edge query costs $c_{e_1} \ldots c_{e_m}$, and precision parameter κ, find a minimum-cost set of edges $E' = \{e_{i_1}, \ldots, e_{i_k}\}$ such that if exactly the edges in E' are queried, an interval of width at most κ can be identified, that is guaranteed to contain the length of the shortest $p \in P$.* P may be given explicitly, it may be specified

* Email: `tomas@theory.stanford.edu`
** Department of Computer Science, Stanford University, Stanford, CA 94305. Research supported by NSF Grant IIS-0118173, an Okawa Foundation Research Grant, and Veritas. Email: `rajeev@cs.stanford.edu`.
* * * Department of Computer Science, Stanford University, Stanford, CA 94305. Research supported by an NSF Graduate Fellowship, an ARCS Fellowship, and NSF Grants IIS-0118173, IIS-9811904, and EIA-0137761. Email: `loc@cs.stanford.edu`
[†] Department of Computer Science, Stanford University, Stanford, CA 94305. Research supported by an NSF Graduate Research Fellowship, and by NSF Grants IIS-9817799, IIS-9811947, and IIS-0118173. Email: `olston@cs.stanford.edu`
[‡] Cisco Systems. Email: `rinap@cisco.com`

H. Alt and M. Habib (Eds.): STACS 2003, LNCS 2607, pp. 367–378, 2003.
© Springer-Verlag Berlin Heidelberg 2003

implicitly, or it may not be specified at all, in which case P can be assumed to be the set of all s-t paths in G. It turns out that an actual s-t path, of length within κ of the shortest s-t path, will be obtained as well. Note that there are two natural versions of this problem. In the *online* version, the sequence of queries is chosen adaptively — each query is answered before the next one is chosen. In the *offline* version, the entire set of queries must be specified completely before the answers are provided, and it must be *guaranteed* that the length of the shortest path can be pinned down as desired, regardless of the results of the queries. In this paper we consider the offline formulation of the problem.

The problem of finding a cheap set of edge queries is neither in NP nor co-NP unless NP = co-NP; however, it is in the class Σ_2. We therefore study the hardness of the problem under various types of restrictions. The special case of zero-error is solvable in polynomial time if the set of paths is given explicitly or has a particular type of implicit description. If P is given explicitly, then the number of paths we consider is clearly polynomial in the size of the input, and the zero-error problem has a polynomial-time solution; we also show that if P admits a recursive description as defined later, which includes the case of series-parallel graphs, it can be analyzed in polynomial time. If the set of paths is unrestricted, however, for all $\delta > 0$ the zero-error problem is hard to approximate within $n^{1-\delta}$, even if the error *and* cost requirements are relaxed substantially. In order to obtain polynomial algorithms or reasonable approximation bounds, we must therefore consider suitable restrictions on the structure of instances of the problem. In many cases, under such restrictions we obtain matching upper and lower bounds.

We consider the case in which κ may be greater than zero, and examine different types of restrictions on the path structure of the graph. The first such restriction is the unique-upper-length requirement — that all paths in the graph have the same upper bound on length. This restriction alone is not enough to make the problem tractable. With certain assumptions on κ, the length intervals, and the edge structure, however, the problem can be solved in polynomial time or with small approximation factors. We consider restrictions on κ, restrictions on the edge structure of the graph, and restrictions on the number and types of nontrivial edges (i.e., edges whose lengths are not known exactly) on the paths under consideration.

1.1 Motivation and Related Work

Our problem is motivated by the work of Olston and Widom [11] on query processing over replicated databases, where local cached copies of databases are used to support quick processing of queries at client sites. There is a master copy of the data where all the updates to the database are maintained. The frequency of updates makes it infeasible to maintain exact consistency between the cached copies and the master copy, and the data values in the cache are likely to become stale and drift from the master values. However, the cached copies store for each data value an interval that is guaranteed to contain the master value. Systems considerations sometimes make it desirable to perform all queries to the master copy en masse; in this case, the offline formulation is much more relevant.

In some cases, the data appears as a graph with edges whose lengths are updated over time, and queries request a short path between two nodes. It is desirable to find a path whose length is within κ of that of the shortest, where the value of κ is specified along with the query. This problem has applications in areas such as network monitoring and computerized route selection for cars. Here, edge lengths may change rapidly, and excessive queries will result in high communication costs.

Another class of queries, aggregation with uncertainty, has been studied in the context of the interval caching framework. Olston and Widom [11] consider aggregation functions such as *sum*, *min*, and *max* for the offline formulation. The shortest-path problem we consider in this paper is a strict and common generalization of all three of these aggregation functions. Feder *et al.* [4] consider both the online and the offline formulations of the selection problem, with *median* as a special case. Finally, Khanna and Tan [8] extend some of the results for the selection and sum problems, focusing on other precision parameter formulations. In the model of Papadimitriou *et al.* [12,3] (similar to the above-described online model), a robot tries to learn the map of a Euclidean region, keeping its travel distance competitive with the cheapest proof to the map. Karasan *et al.* [7] and Montemanni and Gambardella [10] assume a digraph with source and sink and, for each edge, an interval containing the length; they study how to find a path with lowest worst-case length (resp. worst-case competitive ratio) [1].

2 The Zero-Error Case

We consider first the case with no error, i.e., $\kappa = 0$. Consider two paths p and q from s to t in P. Let L be the sum of l_e over edges e in p but not in q. Let H be the sum of h_e over edges e in q but not in p. Say that p *dominates* q if $L > H$, or if $L = H$ and p and q do not have the same nontrivial edges (edges e with $l_e \neq h_e$).

Proposition 1 *A choice of edges e to query guarantees zero error if and only if it queries all the nontrivial edges in each path $p \in P$ that does not dominate any path $q \in P$. (The zero-error problem is thus in co-NP.)*

Proof Sketch: Suppose a nontrivial edge e is in a path p that does not dominate any path q. Then for every path q that does not have the same nontrivial edges as p we have $L \leq H + \delta$ for some $\delta > 0$; we can choose δ smaller than the length of the interval for e. Therefore, if we query all edges f other than e, and obtain the answer l_f for f in p, and the answer h_f for f not in p, then the resulting interval will contain at least $[L_0, L_0 + \delta]$, where L_0 is the resulting length of the path p when we choose l_e for e as well.

---

[1] Here, the competitive ratio of an s-t path, given some assignment of lengths to edges, is its total length, divided by that of the shortest path.

If p dominates some q, then we can ignore p because for all possible answers to the queries, the length of p is at least the length of q, and the domination relation is acyclic.   □

Theorem 1 *If the collection P is given explicitly, then the zero-error problem can be solved in polynomial time.*

For the proof, observe that by Proposition 1, it suffices to test for each path p in P whether p dominates some q in P, and query all edges in p if it does not.

We consider next the following *implicit description* of P for G. Either (1) G consists of a single edge $e = (s, t)$, and P has the single path given by e; or, (2) We are given an implicit description of P' for G' with source s and sink t, where G' contains an edge $e = (s', t')$, and we are also given an explicit description of P'' for G'' with source s' and sink t'. The graph G is obtained by taking G' and replacing e with G''; the paths P are obtained by taking the paths P' and replacing the occurrences of e in these paths with each of the paths in P''.

Proposition 2 *In the implicit description of P for G, suppose that each explicit description of P'' for G'' used contains all the paths in G''. Then P contains all the paths in G. In particular, if G is a series-parallel graph, then each such G'' consists of just two edges (either in series or in parallel), giving a polynomial description for all paths in G.*

For the proof, note that all paths in G correspond to paths in G' which may go through the special edge e, and if they do, to a choice of path in G''. The result follows by induction. A series-parallel graph can be reduced to a single edge by repeatedly replacing G'' consisting of two edges in series or two edges in parallel with a single edge.

Theorem 2 *If the collection of paths P for G is implicitly presented as described earlier, then the zero-error problem can be solved in polynomial time.*

Proof Sketch: Suppose p in P dominates some q in P. We may choose q so that p and q have the same edges from s to some s', are disjoint from s' to some t', and have the same edges from t' to t.

Simplify the graph G by repeatedly replacing the components G'' with single edges, until we obtain a component G'' containing both s' and t', for all choices of s' and t'. In this component G'', consider all pairs of paths p'' and q'' from P''; some such pair corresponds to p and q.

We can then determine the values L and H for p'' and q''. To determine H, choose h_f for all edges f in q'' not in p''. If such an edge f corresponds to a subcomponent, find the shortest path in that subcomponent corresponding to taking h_g for each edge g in the subcomponent. To determine L, choose l_f for all edges f in p'' not in q''. If such an edge f corresponds to a subcomponent, find the shortest path in that subcomponent corresponding to taking l_g for each edge g in the subcomponent; here, however, we only consider paths in the subcomponent that do not dominate other paths, which we can assume have been precomputed by the same algorithm. We also only consider paths going through the edge e being tested for membership in a path p that does not dominate any q.   □

Theorem 3 *If P consists of all the paths in the given G, and each $[l_e, h_e]$ is either $[0, 0]$ or $[0, 1]$, the zero-error problem is co-NP-complete; in the unit-cost case it is hard to approximate within $n^{1-\delta}$, for any constant $\delta > 0$. In fact, it is hard to distinguish the case where there is a zero-error solution of cost w from the case where there is no solution of error at most $\kappa = n^{\delta_1}$ and cost at most wn^{δ_2} for all constants $\delta_1, \delta_2 > 0$ satisfying $\delta_1 + \delta_2 < 1$.*

Proof Sketch: We show that testing whether an edge e belongs to some path p that does not dominate any path q is NP-complete. The reduction is from 3-colorability.

We will construct a DAG as follows. We have a DAG S with source s and sink s', where all edges have interval $[0, 1]$, and a DAG T with source t' and sink t, where all edges have interval $[0, 1]$. The edge e goes from s' to t'. There is also a set E of edges with interval $[0, 0]$ from vertices in S to vertices in T. The required path p consists then of a path p_S in S from s to s' and a path p_T in T from t' to t. Such a path does not dominate any other path if and only if no edge in E joins a vertex in p_S and a vertex in p_T.

Let G be the instance of 3-colorability, with vertices $v_1, \ldots, v_n$ and edges $e_1, \ldots, e_m$. The DAG S, in addition to s and s', has vertices $(v_i, 1)$, $(v_i, 2)$ and $(v_i, 3)$ for each v_i in G. The edges in S go from each (v_i, j) to each (v_{i+1}, j'), plus edges from s to each (v_1, j) and from (v_n, j) to s'. The path p_S thus chooses a (v_i, j) for each v_i, i.e., assigns color j to v_i.

The DAG T, in addition to t' and t, has vertices $(e_l, 1, 1)$, $(e_l, 2, 1)$, $(e_l, 3, 1)$, $(e_l, 1, 2)$, $(e_l, 2, 2)$, $(e_l, 3, 2)$ for each e_l in G. The DAG T has a path from t' to t going through all $(e_l, j, 1)$; the $(e_l, j, 2)$ are connected similarly to the $(e_l, j, 1)$. Thus the path p_T chooses a k in (e_l, j, k) for each e_l, j.

If G has an edge $e_l = (v_i, v_{i'})$, Then E has the edges $((v_i, j), (e_l, j, 1))$ and $((v_{i'}, j), (e_l, j, 2))$. Thus p_T will be able to choose a k in (e_l, j, k) for e_l, j if and only if v_i and $v_{i'}$ are not both assigned color j. Thus the choice of paths p_S, p_T corresponds to a 3-coloring of G.

This gives NP-completeness. The hardness of approximation follows from the fact that the special edge e can be given an arbitrarily large cost, much larger than the sum of all the other costs. In the unit cost case, we can use a large number of parallel edges for e to achieve the same effect as a large cost, and insert extra edges of length $[0, 0]$ if parallel edges are not allowed. Thus, with unit costs, we have m_0 edges other than e, and $m_0^c \approx n$ edges corresponding to e, giving an approximation hardness of $m_0^{c-1} = n^{1-\delta}$.

The stronger hardness of approximation follows by replacing each edge with length $[0, 1]$ with a path of r edges of length $[0, 1]$. Then we have $m_0 r$ edges other than e, and $m_0^c r \approx n$ edges corresponding to e. Setting $r = n^{\delta_1}$, $m_0^{c-1} = n^{\delta_2}$, and $m_0 = n^{1-\delta_1-\delta_2}$ gives the result.   □

3   The Unique-Upper-Length Case

We now allow error $\kappa \geq 0$. For each edge e with interval $[l_e, h_e]$ we assume both l_e and h_e are integer. This implies that only integer κ is interesting. The unique-upper-length case is the case where there is an integer H such that for each path p from s to t, $\sum_{e \in p} h_e = H$.

Proposition 3 *In the unique-upper-length case, a choice of queried edges guarantees error κ if and only if it guarantees error κ for each path from s to t in P. (The unique-upper-length problem is thus in NP.)*

We assume next that P consists of all paths from s to t. Let $K \leq H$ denote the total error when no interval is queried.

Theorem 4 *The unique-upper-length case can be solved in polynomial time for $\kappa = 0, 1, K - 1, K$.*

Proof Sketch: Error $\kappa = K$ requires no queries, while $\kappa = 0$ requires querying all nontrivial intervals (by Proposition 3).

Suppose $\kappa = K - 1$. We must query at least one edge in each path p such that the sum of the lengths of the intervals $[l_e, h_e]$ over edges e in p equals K. Choose the l_e values for all edges e, and compute for each vertex v the length l_v of the shortest path from s to v. Then $K = H - l_t$. The sum of the lengths of the intervals $[l_e, h_e]$ over edges $e = (v, v')$ in p equals K if and only if $l_v + l_e = l_{v'}$ for all such e. Therefore, if we remove all edges e with $l_v + l_e > l_{v'}$, then we must query at least one edge for each path p from s to t in the resulting graph. This is the same as computing a minimum cost cut between s and t, which can be obtained by a maximum flow computation.

Suppose next $\kappa = 1$. Then all edges e with interval $[l_e, h_e]$ such that $h_e - l_e > 1$ must be queried, and all but at most one of the edges e with interval $[l_e, h_e]$ with $h_e - l_e = 1$ on a path p must be queried. Define an associated graph G' whose vertices are the edges e with $h_e - l_e = 1$, with an edge from e to e' in G' if e precedes e' in some path p. The graph G' is transitively closed, and paths p correspond to cliques in G'; we must therefore query all but at most one e in each such clique, i.e., the queried edges must form a vertex cover in G'.

Suppose e has incoming edges from A and outgoing edges to B in G'. Since G' is transitively closed, it also has all edges from A to B, and thus a vertex cover must choose all of A or all of B. Replace e with two vertices e_1, e_2, so that e_1 has the incoming edges from A and e_2 has the outgoing edges to B. Then a vertex cover will only choose at most one of e_1, e_2, since it chooses all of A or all of B. Furthermore, a vertex cover choosing e corresponds to a vertex cover choosing one of e_1, e_2, and vice versa. If we apply this transformation for each e, then each vertex e_1 has only incoming edges and each vertex e_2 has only outgoing edges. Therefore the graph is a bipartite graph, with vertices e_1 in one side and vertices e_2 in the other side. A minimum cost vertex cover in a bipartite graph can be obtained by a maximum flow computation.   □

Theorem 5 *The unique-upper-length case is NP-complete, and hard to approximate within $1+\delta$ for some $\delta > 0$, if $2 \le \kappa \le K-2 \le H-2$. This includes the case $K = H = 4$ and $\kappa = 2$, even if: (1) all $[l_e, h_e]$ intervals are $[0,0], [0,1]$, or $[0,2]$, with at most four nontrivial intervals per path (which has a 2-approximation algorithm); or, (2) all $[l_e, h_e]$ intervals are $[0,1], [0,2]$ or $[1,1]$, with at most three nontrivial intervals per path (which has a 1.5-approximation algorithm).*

Proof Sketch: Suppose $K = H = 4$ and $\kappa = 2$. We do a reduction from vertex cover for a graph G. In fact, we consider vertex cover for a graph G' obtained from G by replacing each edge (v, v') in G with a path (v, x, y, v') of length 3 in G'. The optimal vertex covers are related by $\text{opt}(G') = \text{opt}(G) + |E(G)|$.

The DAG has a vertex v for each v in G, and a corresponding edge (s, v) with interval $[0,1]$. For each edge (v, v') in G, the DAG has two extra vertices a, b, an edge (v, a) with interval $[0,2]$ corresponding to x, an edge (a, t) with interval $[0,1]$ corresponding to y, an edge (v', b) with interval $[0,1]$ of very large cost that will not be queried, and an edge (b, a) with interval $[0,1]$ of zero cost that will be queried for Case (1) or with interval $[1,1]$ for Case (2).

For the paths (s, v, a, t), we must either query (v, a) corresponding to x or query both (s, v) and (a, t) corresponding to v and y respectively. This corresponds to covering the two edges (v, x) and (x, y). For the paths (s, v', b, a, t) we must query either (s, v') or (a, t) corresponding to v' and y respectively. This corresponds to covering the edge (y, v'), completing the reduction.

If only unit costs are allowed, then for Case (2) use a large number of parallel paths (v', b, a) from v' to a to simulate a large cost. For Case (1), use parallel edges to simulate cost, and add edges with interval $[0,0]$ if parallel edges are not allowed.

This proves NP-completeness. For the hardness of approximation, take G of maximum degree d for some constant $d \ge 3$, for which vertex cover is known to be hard to approximate within $1 + \delta$ for some $\delta > 0$ [2]. We know that the optima r, r' for G, G' are related by $r' = r + m$, where $m = |E(G)|$. Furthermore $r \ge \frac{m}{d}$. Therefore $r = \frac{r}{d+1} + \frac{rd}{d+1} \ge \frac{r}{d+1} + \frac{m}{d+1} = \frac{r'}{d+1}$, so an excess of δr for the vertex cover obtained in G corresponds to an excess of $\frac{\delta}{d+1} r'$ for the vertex cover obtained in G', giving hardness within $1 + \frac{\delta}{d+1}$ for G' and hence for the corresponding DAG with source s and sink t.

To obtain the result for any κ, K, H with $2 \le \kappa \le K - 2 \le H - 2$, insert right after s a path with $H - K$ edges with interval $[1,1]$, and $K - 4$ edges with interval $[0,1]$ of which $\kappa - 2$ have very large cost and will not be queried, and $K - \kappa - 2$ have zero cost and will be queried. Again large different costs can be simulated with different numbers of parallel paths of two edges in the unit cost case.

The approximation upper bounds follow from Corollary 1 in Section 4. $\square$

Theorem 6 *The unique-upper-length case has a $(K - \kappa)$-approximation algorithm; and for all integers d has an $(\frac{\kappa}{d} + d)$-approximation algorithm, in particular a $(2\sqrt{\kappa} + 1)$-approximation algorithm. We can get a 2-approximation with respect to the optimum cost for a given κ if we allow replacing the error κ by 2κ as well.*

374   T. Feder et al.

In the case where all intervals $[l_e, h_e]$ satisfy $h_e - l_e \leq 1$, we get an $H_{K-\kappa} \leq 1 + \log(K - \kappa)$ approximation algorithm, where $H_v = \sum_{1 \leq i \leq v} \frac{1}{v} \leq 1 + \log v$; and for all integer d an $(\frac{\kappa}{d} + 1 + H_{d-1})$-approximation algorithm, in particular a $(3 + \log \kappa)$-approximation algorithm.

Proof Sketch: We give the $(K - \kappa)$-approximation algorithm. Let $\mu = K - \kappa$, and let w be the cost of the optimal solution r_0. Suppose we have found a solution r that reduces the error by $\lambda \leq \mu - 1$. We shall find extra edges of total cost at most w which when combined with the solution r reduce the error by at least $\lambda + 1$. Applying this at most μ time yields the result.

For each edge e in r, replace its interval $[l_e, h_e]$ with $[h_e, h_e]$. The extra edges must reduce the error in this new graph by at least 1, and an optimal solution for this problem can be obtained from Theorem 4. This optimal solution has cost at most w because the edges in r_0 and not in r provide such a solution and have cost at most w.

We give the $(\frac{\kappa}{d} + d)$-approximation algorithm. We consider the linear programming relaxation of the problem. We introduce a variable $0 \leq x_e \leq h_e - l_e$ for each edge e. The idea is that $x_e = 0$ if e is queried and $x_e = h_e - l_e$ if e is not queried. We want the longest path using the x_e values to be at most κ. This can be expressed by introducing a variable y_v for each vertex, and adding the conditions $y_s = 0$, $y_t \leq \kappa$, and $y_{v'} \geq y_v + x_e$ for each edge $e = (v, v')$. The objective function to be minimized is $\sum_e w_e (1 - \frac{x_e}{h_e - l_e})$. Let w be the value of the optimum. If we query all edges for which $1 - \frac{x_e}{h_e - l_e} \geq \frac{d}{\kappa + d}$, we incur cost at most $\frac{\kappa + d}{d} w$. For each path p from s to t we have $\sum_{e \in p} x_e \leq \kappa$. For each edge e in p not queried, we have $1 - \frac{x_e}{h_e - l_e} < \frac{d}{\kappa + d}$, or $h_e - l_e < \frac{\kappa + d}{\kappa} x_e$. Therefore the sum of the lengths $h_e - l_e$ of intervals over edges e in p not queried is at most $\kappa + d - 1$. So we must reduce the error further by $d - 1$. After replacing the intervals for edges e previously queried with $[h_e, h_e]$, this will cost at most $(d - 1)w$ by the previous result. The total cost is thus at most $(\frac{\kappa + d}{d} + d - 1)w = (\frac{\kappa}{d} + d)w$, as required.

The remark on doubling both the cost of the optimum and the error allowed follows from setting $d = \kappa$ in the linear programming rounding.

Suppose now all intervals $[l_e, r_e]$ satisfy $r_e - l_e \leq 1$. We give the $H_{K-\kappa}$-approximation algorithm. Let $\mu = K - \kappa$, and let w be the cost of the optimal solution r_0. Suppose we have found a solution r that reduces the error by $\lambda \leq \mu - 1$. We shall find extra edges of total cost at most $\frac{w}{\nu}$ for $\nu = \mu - \lambda$ which when combined with the solution r reduce the error by at least $\lambda + 1$. Applying this at most μ times yields the result.

For each edge e in r, replace its interval $[l_e, h_e]$ with $[h_e, h_e]$. The extra edges must reduce the error in this new graph by at least 1, and an optimal solution for this problem can be obtained from Theorem 4. We must show that this optimal solution costs at most $\frac{w}{\nu}$. The edges in r_0 and not in r reduce the error in this graph by at least ν: call these edges r_1. Assign to each edge e the value l_e, and compute for each vertex v the length l_v of a shortest path from s to v. A solution must query at least one nontrivial edge in each path p such that every nontrivial edge $e = (v, v')$ in p has $l_{v'} - l_v = l_e$, that is, in every path p after we remove all edges e that have $l_{v'} - l_v < l_e$. Now every path p has at least ν edges in r_1, so we can divide the edges in r_1 into ν sets such that each set is a cut: the ith set contains all the edges that occur at the earliest as the ith edge from r_1 in some path p. Thus, some such set constitutes a cut with cost at most $\frac{w}{\nu}$. The optimal solution from Theorem 4 then has cost at most $\frac{w}{\nu}$ as well.

The $(\frac{\kappa}{d} + 1 + H_{d-1})$-approximation algorithm follows as before. We incur cost $\frac{\kappa + d}{d} w$ to reduce the error to at most $\kappa + d - 1$ by linear programming, and we then incur cost $H_{d-1} w$ to reduce the error to κ by the algorithm just described. $\square$

Theorem 7 *The unique-upper-length case on series-parallel graphs has a $(1+\delta)$-approximation algorithm for all $\delta > 0$, and an exact polynomial time algorithm if κ is bounded by a polynomial.*

Proof Sketch: We compute an optimal solution on G for all $\kappa' \leq \kappa$. If G consists of two graphs G_1, G_2 in parallel, then we must have a solution for κ' on each of G_1, G_2. If G consists of two graphs G_1, G_2 in series, then we must have a solution for κ_1 on G_1 and a solution for κ_2 on G_2 with $\kappa_1 + \kappa_2 = \kappa'$. We can try all such decompositions of κ'. The result follows by induction on the structure of G, when κ is bounded by a polynomial.

For larger κ, we introduce an approximation factor of $1 + \frac{\delta}{n}$, and do a binary search for possible values of κ', stopping each branch of the binary search when the solutions for two consecutive values of κ' differ by a factor of at most $1 + \frac{\delta}{n}$. Thus, a polynomial number of solutions will be maintained. When the solutions for G_1, G_2 are combined to obtain solutions for G, the number of solutions obtained for G may be larger, but can be reduced again by incurring another factor of $1 + \frac{\delta}{n}$. Since graphs are combined at most n times, the total factor incurred is at most $(1 + \frac{\delta}{n})^n \approx 1 + \delta$. $\square$

Note that the special case where the series-parallel graph is a single path is the same as knapsack [11] and thus NP-complete.

The case where the collection of paths P is given explicitly and does not contain all paths is harder.

Theorem 8 *For a given collection of paths P for G, with intervals $[l_e, h_e]$ given by $[0, 0]$, $[0, 1]$ or $[1, 1]$, the unique-upper-length case is as hard to approximate as $(\kappa + 1)$-hypergraph vertex cover even if $\kappa = K - 1$. With arbitrary intervals $[l_e, h_e]$, the problem has an $(\kappa + 1)$-approximation algorithm.*

Proof Sketch: Let R be an instance of $(\kappa+1)$-hypergraph vertex cover consisting of n vertices, and hyperedges each having $\kappa + 1$ vertices. Construct a DAG consisting of a path of length n, where each edge is replaced by two parallel edges with intervals $[0, 1]$ and $[1, 1]$. (Use an additional edge with interval $[0, 0]$ if parallel edges are not allowed.) Each edge on the path of length n corresponds to a vertex of the hypergraph. We choose a path for each hyperedge, selecting the edges $[0, 1]$ corresponding to the vertices in the hyperedge, otherwise selecting the edges $[1, 1]$. A solution must query at least one edge in each such path, and thus corresponds to a vertex cover in the hypergraph.

The $(\kappa + 1)$-approximation algorithm is obtained again by the linear programming relaxation. We introduce a variable $0 \leq x_e \leq h_e - l_e$ for each edge e. The idea is that $x_e = 0$ if e is queried and $x_e = h_e - l_e$ if e is not queried. Write $\sum_{e \in p} x_e \leq \kappa$ for each path p in P. The objective function to be minimized is $\sum_e w_e(1 - \frac{x_e}{h_e - l_e})$. Let w be the value of the optimum. If we query all edges for which $1 - \frac{x_e}{h_e - l_e} \geq \frac{1}{\kappa+1}$, we incur cost at most $(\kappa+1)w$. For each edge e in p not queried we have $1 - \frac{x_e}{h_e - l_e} < \frac{1}{\kappa+1}$, or $h_e - l_e < \frac{\kappa+1}{\kappa} x_e$. Therefore the sum of the lengths $h_e - l_e$ of intervals over edges e in p not queried is at most κ.   □

4   The General Case

We now consider the case of arbitrary $\kappa \geq 0$ and arbitrary intervals $[l_e, h_e]$.

Proposition 4 *A choice of queried edges guarantees error κ if and only if for each path p in P there is a path q in P (possibly $q = p$) such that $H - L + x \leq \kappa$. Here H is the sum of h_e over edges e in q and not in p, L is the sum of l_e over edges e in p and not in q, and x is the sum of the interval lengths $h_e - l_e$ over edges e in both p and q that are not queried. (The general case is thus in Σ_2.)*

Proof Sketch: Suppose there is a path p such that $H - L + x > \kappa$ for each path q. Answer the queries to edges e in p by l_e, and answer the queries to all other edges f by h_e. Let L_0 be the minimum possible length on p. Then for every path q, the maximum possible length is at least $L_0 + (H - L) + x > L_0 + \kappa$.

Suppose for every path p there is a path q such that $H - L + x \leq \kappa$. After the queries, the minimum possible length of a shortest path is the minimum possible length L_0 for some p. The maximum possible length for the corresponding q is at most $L_0 + (H - L) + x \leq L_0 + \kappa$.   □

Theorem 9 *If each path in P contains at most r nontrivial edges, then the problem is as easy and as hard to approximate as r-hypergraph vertex cover, for $r = O(\log n)$; in particular, it has an r-approximation algorithm.*

Proof Sketch: The hardness was shown in Theorem 8 with $r = \kappa + 1$. For the easiness, consider each path p. For a choice of edges X to be queried in p, we can determine whether $H - L + x > \kappa$ for each path q as in Proposition 4. If this is the case, then at least some nontrivial edge in p not in X must be queried, giving a set of at most r nontrivial edges to be covered, and thus defining a corresponding hyperedge of size at most r. The number of choices X to be considered for p is at most 2^r, thus polynomial in n.   □

Theorem 10 *If the nontrivial edges of G have r colors, so that the nontrivial edges in each path p from P have distinct colors, then the problem is as easy and as hard to approximate as r-partite hypergraph vertex cover, for $r = O(\log n)$; in particular, it has an $\frac{r}{2}$-approximation algorithm [9,5].*

Proof Sketch: Let R be an instance of r-partite hypergraph vertex cover, with r parts R_i. Construct a DAG consisting of a path of length r, where the ith edge is replaced by $|R_i|$ parallel edges with intervals $[0, 1]$ of color i. Each hyperedge in R thus corresponds to a path in the DAG since it selects an element from R_i for each i. Setting $\kappa = r - 1$ forces each hyperedge to be covered, proving the hardness.

The easiness is as in Theorem 9. The hyperedges correspond to the nontrivial edges in p not in X for some X, so they all have elements of different colors, i.e., elements in different parts R_i of they r-partite hypergraph. □

Corollary 1 *If P consists of all paths in G, and each such path has at most r nontrivial edges, then the problem has an $\frac{r}{2}$-approximation algorithm, if r is constant. In particular, the case $r = 2$ can be solved exactly. (The case $r = 3$ is NP-hard to approximate within $1 + \delta$ for some $\delta > 0$.)*

Proof Sketch: Assign to each nontrivial edge e a color i which is the maximum i such that e occurs as the ith nontrivial edge in a path p. There are at most r colors, and all the nontrivial edges in each path p have different colors. The choice of at most r nontrivial edges for a path p can be assumed to determine p, since we can choose the remaining trivial edges so as to minimize the length of the path. Thus the number of paths to be considered is $O(m^r)$.

The hardness for $r = 3$ was established in Theorem 5. □

A *bi-tree* is a series-parallel graph consisting of a tree S with root s, edges oriented down from the root, a tree T with root t, edges oriented up to the root, where the leaves of S coincide with the leaves of T.

Theorem 11 *Bi-trees have a $(1+\delta)$-approximation algorithm for all $\delta > 0$, and an exact polynomial time algorithm if κ is bounded by a polynomial.*

Proof Sketch: Consider the optimal solution. Consider a series-parallel sub-graph G' of the given bi-tree which is given by a subtree rooted at some s' of the tree rooted at s, and a subtree rooted at some t' of the tree rooted at t. The paths p going through G' such that the corresponding q satisfying Proposition 4 does not go through G' are determined by a lower bound L_0 on the sum of l_e over the edges e of p in G'. Thus, there are at most n different scenarios for which paths p going through G' must have their corresponding q going through G' as well. If the set of such p is nonempty, then the sum x_0 of the $h_e - l_e$ for edges from s to s' or from t' to t not queried must satisfy $x_0 \leq \kappa$, for a total of κn scenarios for each such G'.

Each such G' is obtained by combining two G_1', G_2' in parallel, or by taking a G_1' having some s_1', t_1' as source and sink, and adding either an edge (s', s_1') or an edge (t_1', t'). The optimal solution for each scenario for G can be obtained from the optimal solution for the scenarios for G_1' for this last case or for the scenarios for both G_1', G_2' in the previous case, because given x_0 it is easy to determine which paths q in G_1' take care of paths p in G_2' and vice versa.

For large κ, we proceed as in Theorem 7 and introduce at each step an approximation factor of $1 + \frac{\delta}{n}$, so that the total factor incurred is at most $(1 + \frac{\delta}{n})^n \approx 1 + \delta$. ☐

Note that the special case where the bi-tree is a single path is the same as knapsack [11] and thus NP-complete.

Theorem 12 *If P consists of all paths in G, and G is made up of components G_i in series, where each G_i has an α-approximation algorithm, then G has an $\alpha(1 + \delta)$-approximation algorithm for all $\delta > 0$, and an α-approximation algorithm if κ is bounded by a polynomial.*

Proof Sketch: A solution must combine κ_i for each G_i that add up to at most κ. We proceed inductively for the possible values of partial sums $\kappa_1 + \cdots + \kappa_i$, and combine solutions accordingly.

For large κ, we proceed again as in Theorem 7 and introduce at each step an approximation factor of $1 + \frac{\delta}{n}$, so that the total factor incurred is at most $(1 + \frac{\delta}{n})^n \approx 1 + \delta$. ☐

References

1. D. Aingworth, C. Chekuri, P. Indyk, and R. Motwani. "Fast estimation of diameter and shortest paths (without matrix multiplication)." *SIAM Journal on Computing* 28(1999):1167–1181.
2. P. Berman and M. Karpinski. "On some tighter inapproximability results." DIMACS Technical Report 99–23 (1999).
3. X. Deng, T. Kameda, and C. Papadimitriou. "How to learn an unknown environment." To appear in the *Journal of the ACM*.
4. T. Feder, R. Motwani, R. Panigrahy, C. Olston, and J. Widom. "Computing the median with uncertainty." In *Proceedings of the 32nd Annual ACM Symposium on Theory of Computing*, 2000, pages 602–607.
5. Z. Füredi. "Matchings and covers in hypergraphs." *Graphs and Combinatorics* 4(1988):115–206.
6. O.H. Ibarra and C.E. Kim. "Fast approximation algorithms for the knapsack and sum of subsets problems." *Journal of the ACM* 22(1975):463–468.
7. O. Karasan, M. Pinar, and H. Yaman. "The robust shortest path problem with interval data." Manuscript, August 2001.
8. S. Khanna and W. Tan. "On computing function with uncertainty." In *Proceedings of the ACM SIGACT-SIGMOD-SIGART Symposium on Principles of Database Systems*, 2001, pages 171–182.
9. L. Lovász. "On minimax theorems of combinatorics." Doctoral Thesis, Mathematikai Lapok 26(1975):209–264. (Hungarian)
10. R. Montemanni and L. M. Gambardella. "An algorithm for the relative robust shortest path problem with interval data." Tech. Report IDSIA-05-02, 2002.
11. C. Olston and J. Widom. "Offering a precision-performance tradeoff for aggregation queries over replicated data." In *Proceedings of the 26th International Conference on Very Large Data Bases*, 2000, pages 144–155.
12. C.H. Papadimitriou and M. Yannakakis. "Shortest paths without a map." In *Proceedings of the 16th International Colloquium on Automata, Languages, and Programming*, Lecture Notes in Computer Science 372(1989):610–620.

Solving Order Constraints in Logarithmic Space

Andrei Krokhin[1][*] and Benoit Larose[2][**]

[1] Department of Computer Science, University of Warwick
Coventry, CV4 7AL, UK
andrei.krokhin@dcs.warwick.ac.uk
[2] Champlain Regional College,
900 Riverside Drive, St-Lambert, Québec, Canada J4P 3P2, and
Department of Mathematics and Statistics, Concordia University
1455 de Maisonneuve West, Montréal, Québec, Canada H3G 1M8
larose@mathstat.concordia.ca

Abstract. We combine methods of order theory, finite model theory, and universal algebra to study, within the constraint satisfaction framework, the complexity of some well-known combinatorial problems connected with a finite poset. We identify some conditions on a poset which guarantee solvability of the problems in (deterministic, symmetric, or non-deterministic) logarithmic space. On the example of order constraints we study how a certain algebraic invariance property is related to solvability of a constraint satisfaction problem in non-deterministic logarithmic space.

1 Introduction

A wide range of combinatorial search problems encountered in artificial intelligence and computer science can be naturally expressed as 'constraint satisfaction problems' [7,35], in which the aim is to find an assignment of values to a given set of variables subject to specified constraints. For example, the standard propositional satisfiability problem [37,41] may be viewed as a constraint satisfaction problem where the variables must be assigned Boolean values, and the constraints are specified by clauses. Further examples include graph colorability, clique, and Hamiltonian circuit problems, conjunctive-query containment, and many others (see [20,26]). One advantage of considering a common framework for all of these diverse problems is that it makes it possible to obtain generic structural results concerning the computational complexity of constraint satisfaction problems that can be applied in many different areas such as machine vision, belief maintenance, database theory, temporal reasoning, type reconstruction, graph theory, and scheduling (see, e.g., [12,20,26,27,35,42]).

The general constraint satisfaction problem (CSP) is **NP**-complete. Therefore, starting with the seminal paper by Schaefer [41], much work has been done

---

* Partially supported by the UK EPSRC grant GR/R29598
** Partially supported by NSERC Canada

H. Alt and M. Habib (Eds.): STACS 2003, LNCS 2607, pp. 379–390, 2003.
© Springer-Verlag Berlin Heidelberg 2003

on identifying restrictions on the problem that guarantee lower complexity (see, e.g., [3,4,5,7,8,9,12,13,14,20,21,22,23,24,26]).

The constraint satisfaction problem has several equivalent definitions (see, e.g., [5,7,14,20,26]). For the purposes of this paper, we define it, as in [8,12,20, 26], as the homomorphism problem: given two relational structures, $\mathcal{A}$ and $\mathcal{B}$, the question is whether there is a homomorphism from $\mathcal{A}$ to $\mathcal{B}$. One distinguishes uniform and non-uniform CSPs depending on whether both $\mathcal{A}$ and $\mathcal{B}$ are parts of the input or not (see, e.g., [26]). The case when the choice of $\mathcal{A}$ is restricted has been studied in connection with database theory (see, e.g., [13,14,16,26]). In this paper, as in [8,12,20], we will consider non-uniform constraint satisfaction problems CSP($\mathcal{B}$) where the structure $\mathcal{B}$ is fixed.

We will concentrate on the much understudied case of constraint problems in logarithmic space complexity classes. The class **NL** of problems solvable in non-deterministic logarithmic space has received much attention in complexity theory. It is known that **NL** $\subseteq$ **NC**, and so problems in **NL** are highly parallelizable.

Despite the large amount of tractable constraint satisfaction problems identified so far, to the best of our knowledge, only two families of concrete problems CSP($\mathcal{B}$) are known to be in **NL**. The first family consists of restricted versions of the Boolean satisfiability problem: the so-called bijunctive constraints [41], including the 2-SATISFIABILITY problem, and implicative Hitting-Set Bounded constraints [7,8]. The second family consists of implicational constraints [24] which are a generalization of bijunctive constraints to non-Boolean problems, and which, in turn, have been slightly generalized in [8]. Dalmau introduced 'bounded path duality', a general sufficient condition for a CSP($\mathcal{B}$) to be in **NL**, which is shown to be satisfied by all problems from the two families [8]. However, this condition is not easy to apply; in general, it is not known to be decidable.

Our main motivation in this paper is to clarify the relation between solvability of a CSP in non-deterministic logarithmic space and the algebraic condition of invariance under a near-unanimity operation that is also satisfied by all concrete problems above. It is known that many important properties of CSPs can be captured by algebraic invariance conditions (see, e.g., [9,12,20]), and that the condition mentioned above exactly corresponds to the following property: for some l, every instance of CSP($\mathcal{B}$) has precisely the same solutions as the system of all its subinstances of size l [12,21]. It is unknown whether this property is sufficient and/or necessary for a CSP($\mathcal{B}$) to be in **NL**. We answer this question for problems connected with posets by showing that, for such problems, the condition is sufficient but not necessary (and therefore it is not necessary in general).

In order theory, there is a rich tradition of linking combinatorial and algebraic properties of posets (see, e.g., [6,28,29,30,39,45,46]). We use such results in order to get combinatorial properties of posets that can be described in first-order logic with the transitive closure operator. As is known from finite model theory, sentences in this logic can be evaluated in non-deterministic logarithmic space. We use the poset retraction problem as a medium for applying finite model theory results to the considered CSPs. The notion of retraction plays an important role in order theory (see, e.g., [11,28,39]), and this problem has also been studied in computer science along with the problem of satisfiability of inequalities in a poset and the extendibility problem (see, e.g., [2,12,29,38]). As intermediate results that are of independent interest, we describe two classes of posets for which the retraction problem is solvable in deterministic or (complete for) symmetric logarithmic space.

The complexity class **SL** (symmetric log-space) appears in the literature less often than **L** and **NL**, so we say a few words about it here. It is the class of problems solvable by symmetric non-deterministic Turing machines in logarithmic space [31], and $\mathbf{L} \subseteq \mathbf{SL} \subseteq \mathbf{NL}$. This class can be characterized in a number of ways (see, e.g., [1,15,36]) and is known to be closed under complementation [36]. It contains such important problems as the UNDIRECTED st-CONNECTIVITY problem and the GRAPH 2-COLORABILITY problem [1,36,40]; it was shown in [25] that the GRAPH ISOMORPHISM problem restricted to colored graphs with color multiplicities 2 and 3 is **SL**-complete. It is known (see, e.g., [15]) that sentences of $\mathbf{FO} + \mathbf{STC}$, first-order logic with the symmetric transitive (reflexive) closure operator, can be evaluated in symmetric logarithmic space.

Proofs of all results in this paper are omitted due to space constraints.

2   Constraint Satisfaction Problems

A *vocabulary* τ is a finite set of relation symbols or predicates. Every relation symbol $R \in \tau$ has an *arity* $ar(R) \geq 0$. A τ-structure $\mathcal{A}$ consists of a set A, called the *universe* of $\mathcal{A}$, and a relation $R^{\mathcal{A}}$ of arity $ar(R)$ for every $R \in \tau$. All structures considered in this paper are *finite*.

A *homomorphism* from a τ-structure $\mathcal{A}$ to a τ-structure $\mathcal{B}$ is a mapping $h : A \to B$ such that $(a_1, \ldots, a_{ar(R)}) \in R^{\mathcal{A}}$ implies $(h(a_1), \ldots, h(a_{ar(R)})) \in R^{\mathcal{B}}$ for every $R \in \tau$. If there exists a homomorphism from $\mathcal{A}$ to $\mathcal{B}$, we write $\mathcal{A} \to \mathcal{B}$.

We consider the problem CSP($\mathcal{B}$) where only structure $\mathcal{A}$ is the input, and the question is whether $\mathcal{A} \to \mathcal{B}$. Of course, one can view CSP($\mathcal{B}$) as the class of all τ-structures $\mathcal{A}$ such that $\mathcal{A} \to \mathcal{B}$. So we have a class of problems parameterized by finite structures $\mathcal{B}$, and the ultimate goal of our research is to classify the complexity of such problems. The classic problems of this type are various versions of GRAPH COLORABILITY and SATISFIABILITY of logical formulas.

Example 1 (H-coloring). If $\mathcal{B}$ is an undirected irreflexive graph H then CSP($\mathcal{B}$) is the GRAPH H-COLORING problem. This problem is **SL**-complete if H is bipartite [36,40], and **NP**-complete otherwise [17]. If H is a complete graph K_k then CSP($\mathcal{B}$) is the GRAPH k-COLORABILITY problem.

Example 2 (2-SAT). Let $\mathcal{B}$ be the stucture with universe $\{0,1\}$ and all at most binary relations over $\{0,1\}$. Then CSP($\mathcal{B}$) is exactly the 2-SATISFIABILITY problem, it is **NL**-complete [37].

Example 3 (NAE-SAT). Let $\mathcal{B}$ be the stucture with universe $\{0,1\}$ and one ternary relation $R = \{(a,b,c) \mid \{a,b,c\} = \{0,1\}\}$. Then $\mathrm{CSP}(\mathcal{B})$ is exactly the NOT-ALL-EQUAL SATISFIABILITY problem as defined in [41], it is **NP**-complete.

The most significant progress in classifying the complexity of $\mathrm{CSP}(\mathcal{B})$ has been made via methods of finite model theory [12,26] and methods of universal algebra [3,4,5,20,21,22,23]. In [8,9], both approaches are present.

The finite model theory approach aims at defining $\mathrm{CSP}(\mathcal{B})$ or its complement in various logics. Most of the known results in this direction make extensive use of the logic programming language Datalog.

A *Datalog program* over a vocabulary τ is a finite collection of rules of the form

$$t_0 : -t_1, \ldots, t_m$$

where each t_i is an atomic formula $R(v_1, \ldots, v_l)$. The predicates occuring in the heads of the rules are the *intensional database* predicates (IDBs), while all others are *extensional database* predicates (EDBs) and must belong to τ. One of the IDBs is designated as the goal predicate. A Datalog program is a recursive specification of the IDBs with semantics obtained via least fixed-points (see [43]).

For $0 \le j \le k$, (j,k)-Datalog is the collection of Datalog programs with at most k variables per rule and at most j variables per rule head. A Datalog program is *linear* if every rule has at most one IDB in its body.

In [12], tractability of many constraint satisfaction problems was explained in the following way: if $\neg\mathrm{CSP}(\mathcal{B})$, the complement of $\mathrm{CSP}(\mathcal{B})$, is definable in (j,k)-Datalog for some j,k then $\mathrm{CSP}(\mathcal{B})$ is solvable in polynomial time. Dalmau [8] introduced "bounded path duality" for $\mathcal{B}$, a general sufficient condition for $\mathrm{CSP}(\mathcal{B})$ to be in **NL**. This condition is characterized in seven equivalent ways, one of them being definability of $\neg\mathrm{CSP}(\mathcal{B})$ in linear Datalog. However, only definability by linear $(1,k)$-Datalog programs is known to be decidable (for any fixed k) [8]. It is noted in that paper that all known concrete structures $\mathcal{B}$ with $\mathrm{CSP}(\mathcal{B})$ in **NL** have bounded path duality.

The algebraic approach to $\mathrm{CSP}(\mathcal{B})$ uses the notion of a polymorphism.

Definition 1 *Let R be an m-ary relation on B, and f an n-ary operation on B. Then f is said to be a* polymorphism *of R (or R is* invariant *under f) if, for every $(b_{11}, \ldots, b_{m1}), \ldots, (b_{1n}, \ldots, b_{mn}) \in R$, we have*

$$f\begin{pmatrix} b_{11} & b_{12} & \cdots & b_{1n} \\ b_{21} & b_{22} & \cdots & b_{2n} \\ \vdots & \vdots & & \vdots \\ b_{m1} & b_{m2} & \cdots & b_{mn} \end{pmatrix} = \begin{pmatrix} f(b_{11}, b_{12}, \ldots, b_{1n}) \\ f(b_{21}, b_{22}, \ldots, b_{2n}) \\ \vdots \\ f(b_{m1}, b_{m2}, \ldots, b_{mn}) \end{pmatrix} \in R.$$

An operation on B is said to be a polymorphism of a τ-structure $\mathcal{B}$ if it is a polymorphism of every $R^{\mathcal{B}}$, $R \in \tau$.

Example 4. 1) It is easy to check that both permutations on $\{0,1\}$ are polymorphisms of the relation R from Example 3, while the binary operation min is not.

2) The polymorphisms of a poset are simply the monotone operations on it.

3) The *dual discriminator* on B is defined by

$$\mu_B(x,y,z) = \begin{cases} y & \text{if } y = z, \\ x & \text{otherwise.} \end{cases}$$

One can verify that $\mu_{\{0,1\}}$ is a polymorphism of the structure $\mathcal{B}$ from Example 2.

Let $\mathrm{Pol}(\mathcal{B})$ denote the set of all polymorphisms of $\mathcal{B}$. This set determines the complexity of $\mathrm{CSP}(\mathcal{B})$, as the following result shows.

Theorem 1 ([20]) *Let $\mathcal{B}_1$ and $\mathcal{B}_2$ be structures with the same universe B. If $\mathrm{Pol}(\mathcal{B}_1) \subseteq \mathrm{Pol}(\mathcal{B}_2)$ then $\mathrm{CSP}(\mathcal{B}_2)$ is polynomial-time reducible to $\mathrm{CSP}(\mathcal{B}_1)$.*

Moreover, if the equality relation $=_B$ on B can be expressed by using predicates of $\mathcal{B}_1$, conjunction, and existential quantification, then the above reduction is logarithmic-space.

Examples of classifying the complexity of $\mathrm{CSP}(\mathcal{B})$ by particular types of polymorphisms can be found in [3,4,5,8,9,20,21,22,23]. It follows from [22,24] that if μ_B is a polymorphism of $\mathcal{B}$ then $\mathrm{CSP}(\mathcal{B})$ is in **NL**. This result is generalized in [8] to give a slightly more general form of ternary polymorphism guaranteeing that $\mathrm{CSP}(\mathcal{B})$ is in **NL**.

Definition 2 *Let f be an n-ary operation. Then f is said to be idempotent if $f(x,\ldots,x) = x$ for all x, and it is said to be Taylor if, in addition, it satisfies n identities of the form*

$$f(x_{i1},\ldots,x_{in}) = f(y_{i1},\ldots,y_{in}), \quad i = 1,\ldots,n$$

where $x_{ij}, y_{ij} \in \{x,y\}$ for all i,j and $x_{ii} \neq y_{ii}$ for $i = 1,\ldots,n$.

For example, it is easy to check that any binary idempotent commutative operation and the dual discriminator are Taylor operations. The following result links hard problems with the absence of Taylor polymorphisms.

Theorem 2 ([29]) *Let $\mathcal{B}$ be a structure such that $\mathrm{Pol}(\mathcal{B})$ consists of idempotent operations. If $\mathcal{B}$ has no Taylor polymorphism then $\mathrm{CSP}(\mathcal{B})$ is **NP**-complete.*

3   Poset-Related Problems

In this section we introduce the particular type of problems we work with in this paper, the poset retraction problem and the order constraint satisfaction problem. Here and in the following the universes of $\mathcal{A}, \mathcal{B}, \mathcal{P}, \mathcal{Q}$ etc. are denoted by A, B, P, Q etc., respectively.

Let $\mathcal{A}$, $\mathcal{B}$ be structures such that $B \subseteq A$. Recall that a *retraction* from $\mathcal{A}$ onto $\mathcal{B}$ is a homomorphism $h : \mathcal{A} \to \mathcal{B}$ that fixes every element of $\mathcal{B}$, that is $h(b) = b$ for every $b \in B$.

Fix a poset $\mathcal{P}$. An instance of the *poset retraction problem* PoRet($\mathcal{P}$) is a poset $\mathcal{Q}$ such that the partial order of $\mathcal{P}$ is contained in the partial order of $\mathcal{Q}$, and the question is whether there is a retraction from $\mathcal{Q}$ onto $\mathcal{P}$. Again, we can view PoRet($\mathcal{P}$) as the class of all posets $\mathcal{Q}$ with positive answer to the above question, and then $\neg$PoRet($\mathcal{P}$) is the class of all instances $\mathcal{Q}$ with the negative answer. This problem was studied in [2,12,38]. For example, it was proved in [12] that, for every structure $\mathcal{B}$, there exists a poset $\mathcal{P}$ such that CSP($\mathcal{B}$) is polynomial-time equivalent to PoRet($\mathcal{P}$).

In [38], a poset $\mathcal{P}$ is called *TC-feasible* if $\neg$PoRet($\mathcal{P}$) is definable (in the class of all instances of PoRet($\mathcal{P}$)) by a sentence in a fragment of $\mathbf{FO} + \mathbf{TC}$, first-order logic with the transitive closure operator, over the vocabulary τ_1 that contains one binary predicate R interpreted as partial order and the constants c_p, $p \in P$, always interpreted as the elements of $\mathcal{P}$. The fragment is defined by the condition that negation and universal quantification are disallowed. Since any sentence in $\mathbf{FO} + \mathbf{TC}$ is verifiable in non-deterministic logarithmic space [15] and $\mathbf{NL} = \mathbf{CoNL}$ [19], TC-feasibility of a poset $\mathcal{P}$ implies that PoRet($\mathcal{P}$) is in $\mathbf{NL}$. Note that even if $\neg$PoRet($\mathcal{P}$) is definable by a sentence in full $\mathbf{FO} + \mathbf{TC}$ then, of course, PoRet($\mathcal{P}$) is still in $\mathbf{NL}$.

Let $P = \{p_1, \ldots, p_n\}$ and let τ_2 be a vocabulary containing one binary predicate R and n unary predicates $S_1, \ldots, S_n$. We denote by $\mathcal{P}_P$ the τ_2-structure with universe P, $R^{\mathcal{P}_P}$ being the partial order $\leq$ of $\mathcal{P}$, and $S_i^{\mathcal{P}_P} = \{p_i\}$. We call problems of the form CSP($\mathcal{P}_P$) *order constraint satisfaction problems*. Note that if $\mathcal{B}$ is simply a poset then CSP($\mathcal{B}$) is trivial because every mapping sending all elements of an instance to a fixed element of $\mathcal{B}$ is a homomorphism. The main difference between CSP($\mathcal{P}_P$) and PoRet($\mathcal{P}$) is that an instance of CSP($\mathcal{P}_P$) is not necessarily a poset, but an arbitrary structure over τ_2.

The following theorem will provide us with a way of obtaining order constraint satisfaction problems in $\mathbf{NL}$.

Theorem 3 *Let $\mathcal{P}$ be a poset. If $\neg$PoRet($\mathcal{P}$) is definable (within the class of all posets containing $\mathcal{P}$) by a τ_1-sentence in $\mathbf{FO} + \mathbf{TC}$ then CSP($\mathcal{P}_P$) is in $\mathbf{NL}$.*

By analysing the (proof of the) above theorem one can show that the problems PoRet($\mathcal{P}$) and CSP($\mathcal{P}_P$) are polynomial-time equivalent. Invoking the result from [12] mentioned in this section, we get the following statement.

Corollary 1 *For every structure $\mathcal{B}$, there exists a poset $\mathcal{P}$ such that CSP($\mathcal{B}$) and CSP($\mathcal{P}_P$) are polynomial-time equivalent.*

It is not known whether the equivalence can be made logarithmic-space, and so the corollary cannot now help us in studying CSPs in $\mathbf{NL}$. We believe that Corollary 1 is especially interesting in view of Theorem 1 because polymorphisms of a poset may be easier to analyse. In particular, note that all polymorphisms of $\mathcal{P}_P$ are idempotent, and so, according to Theorem 2, in order to classify the complexity of CSP($\mathcal{B}$) up to polynomial-time reductions, it suffices to consider posets with a Taylor polymorphism.

Other poset-related problems studied in the literature are the satisfiability of inequalities, denoted $\mathcal{P}$-SAT, and the extendibility problem Ext($\mathcal{P}$). An instance of $\mathcal{P}$-SAT consists of a system of inequalities involving constants from $\mathcal{P}$ and variables, and the question is whether this system is satisfiable in $\mathcal{P}$. The problem $\mathcal{P}$-SAT plays an important role in type reconstruction (see, e.g., [2,18,38]), it was shown to be polynomial-time equivalent to PoRet($\mathcal{P}$) for the same $\mathcal{P}$ [38]. An instance of Ext($\mathcal{P}$) consists of a poset $\mathcal{Q}$ and a partial map f from $\mathcal{Q}$ to $\mathcal{P}$, and the question is whether f extends to a homomorphism from $\mathcal{Q}$ to $\mathcal{P}$. The problem Ext($\mathcal{P}$) was studied in [29], where it was shown to be polynomial-time equivalent to CSP($\mathcal{P}_P$), so all four poset-related problems are polynomial-time equivalent.

It is easy to see that PoRet($\mathcal{P}$) is the restriction of Ext($\mathcal{P}$) to instances where Q contains P and $f = \mathrm{id}_P$ is the identity function on P, while the problem Ext($\mathcal{P}$) can be viewed as the restriction of CSP($\mathcal{P}_P$) to instances where the binary relation is a partial order. Of course, if CSP($\mathcal{P}_P$) is in **NL** then so are the other three problems.

4   Near-Unanimity Polymorphisms

In this section we prove that posets with a polymorphism of a certain form, called *near-unanimity* operation, give rise to order constraint satisfaction problems in **NL**.

Definition 3 *A* near-unanimity *(NU)* *operation is an l-ary (l $\geq$ 3) operation satisfying*

$$f(y, x, \ldots, x) = f(x, y, x, \ldots, x) = \cdots = f(x, \ldots, x, y) = x$$

for all x, y.

Near-unanimity operations have attracted much attention in order theory and universal algebra (see, e.g., [10,28,29,30,39,45,46]). For example, the posets having a ternary NU polymorphism (known as a *majority* operation) are precisely retracts of direct products of fences [39], where a fence is a poset on $\{a_0, \ldots, a_k\}$ such that $a_0 < a_1 > a_2 < \ldots a_k$ or $a_0 > a_1 < a_2 > \ldots a_k$, and there are no other comparabilities. Posets with an NU polymorphism (of some arity) are characterized in a number of ways in [28,30].

It was proved in [12,21] (using different terminology) that if a structure $\mathcal{B}$ has an NU polymorphism then CSP($\mathcal{B}$) is solvable in polynomial time. In [12], such problems are said to have *bounded strict width*, while in [21] they are shown to be related to the so-called *strong consistency*, a notion from artificial intelligence. Moreover, it was shown in these papers that the presence of an $(l+1)$-ary NU polymorphism f is equivalent to the l-Helly property for $\mathcal{B}$ and to definability of $\neg$CSP($\mathcal{B}$) by an (l, k)-Datalog program with a special property. Not going into formal definitions here, the intuition behind these properties is that every relation invariant under f is decomposable into its l-fold projections, and so a mapping from $\mathcal{A}$ to $\mathcal{B}$ that is not a homomorphism can be shown to be such using only at most l-element subsets of $\mathcal{A}$.

Interestingly, up until now all known concrete structures $\mathcal{B}$ with CSP($\mathcal{B}$) in **NL** have an NU polymorphism. It is mentioned in [8] that all two-element structures $\mathcal{B}$ with an NU polymorphism (which precisely correspond to the first family mentioned in the introduction) have bounded path duality, and so the correponding problems CSP($\mathcal{B}$) belong to **NL**. All known concrete structures $\mathcal{B}$ that have at least three elements and such that CSP($\mathcal{B}$) in **NL** have a ternary NU polymorphism derived from dual discriminators (see Example 4) [8], these are the problems from the second family.

Our next result shows that every order constraint satisfaction problem with a near-unanimity polymorphism is in **NL**. Note that by [10], for every $l \geq 3$, there exists a poset that has an l-ary NU polymorphism, but no such polymorphism of smaller arity. Moreover, it was proved in [28] that it can be decided in polynomial time whether a poset has an NU polymorphism, while in [12] this was shown to be true for any fixed arity l for general structures.

In the following we deal mostly with posets, and so we use symbol $\leq$ (rather than R) to denote partial order. Recall that a poset is called *connected* if its comparability graph is connected.

An important tool used in the proof of the next theorem is the notion of a poset *zigzag* introduced in [45]. Intuitively, a $\mathcal{P}$-zigzag is a minimal obstruction for the extendibility problem Ext($\mathcal{P}$). More formally, a poset $\mathcal{X} = (X, \leq^{\mathcal{X}})$ is said to be contained in a poset $\mathcal{Q} = (Q, \leq^{\mathcal{Q}})$ if $\leq^{\mathcal{X}} \subseteq \leq^{\mathcal{Q}}$, and it is said to be properly contained if $\mathcal{X} \neq \mathcal{Q}$. A $\mathcal{P}$-zigzag is a pair $(\mathcal{Q}, f)$ such that f is a partial mapping from $\mathcal{Q}$ to $\mathcal{P}$ that cannot be extended to a full homomorphism, but, for every poset $\mathcal{X}$ properly contained in $\mathcal{Q}$, the mapping $f|_X$ is extendible to a full homomorphism from $\mathcal{X}$ to $\mathcal{P}$. The key fact in the proof of the following theorem is that a connected poset has a near-unanimity polymorphism if and only if the number of its zigzags is finite [30].

Theorem 4 *Let $\mathcal{P}$ be a poset with an NU polymorphism. Then* PoRet($\mathcal{P}$) *is in* **SL** *and* CSP($\mathcal{P}_P$) *is in* **NL**. *If, in addition, $\mathcal{P}$ is connected then* PoRet($\mathcal{P}$) *is in* **L**.

Theorem 4 generalizes a result from [38] where it was shown that if poset $\mathcal{P}$ has the 2-Helly property then PoRet($\mathcal{P}$) is in **NL**. Moreover, together with Theorem 2 it completely covers the classification of complexity of PoRet($\mathcal{P}$) for bipartite posets given in [38].

5   Series-Parallel Posets

In this section we exhibit the first concrete examples of structures without NU polymorphism but with constraint satisfaction problem solvable in non-deterministic logarithmic space. We deal with the poset-related problems for series-parallel posets. These posets have been studied in computer science because they play an important role in concurrency (see, e.g., [32,33,34]).

Recall that a linear sum of two posets $\mathcal{P}_1$ and $\mathcal{P}_2$ is a poset $\mathcal{P}_1 + \mathcal{P}_2$ with the universe $P_1 \cup P_2$ and partial order $\leq^{\mathcal{P}_1} \cup \leq^{\mathcal{P}_2} \cup \{(p_1, p_2) \mid p_1 \in P_1, p_2 \in P_2\}$.

Definition 4 *A poset is called* series-parallel *if it can be constructed from singletons by using disjoint union and linear sum.*

Let **k** denote a k-antichain (that is, disjoint union of k singletons). A 4-*crown* is a poset isomorphic to $\mathbf{2} + \mathbf{2}$. The *N-poset* can be described as $\mathbf{2} + \mathbf{2}$ with one comparability missing (its Hasse diagram looks like the letter "N"). Series-parallel posets can be characterized as *N*-free posets, that is, posets not containing the *N*-poset as a subposet [44].

Denote by $\mathbb{A}$ the class of all series-parallel posets $\mathcal{P}$ with the following property: if $\{a, a', b, b'\}$ is a 4-crown in $\mathcal{P}$, a and a' being the bottom elements, then at least one of the following conditions holds:

1. there is $e \in P$ such that a, a', e, b, b' form a subposet in $\mathcal{P}$ isomorphic to $\mathbf{2} + \mathbf{1} + \mathbf{2}$;
2. $\inf_{\mathcal{P}}(a, a')$ exists;
3. $\sup_{\mathcal{P}}(b, b')$ exists.

Recall that $\inf_{\mathcal{P}}$ and $\sup_{\mathcal{P}}$ denote the greatest common lower bound and the least common upper bound in $\mathcal{P}$, respectively.

Theorem 5 *Let $\mathcal{P}$ be a series-parallel poset. If $\mathcal{P} \in \mathbb{A}$ then* PoRet$(\mathcal{P})$ *is in* **SL** *and* CSP$(\mathcal{P}_P)$ *is in* **NL**. *Otherwise both problems are* **NP**-*complete (via polynomial-time reductions).*

The class $\mathbb{A}$ can be characterized by means of "forbidden retracts".

Lemma 1 *A series-parallel poset $\mathcal{P}$ belongs to $\mathbb{A}$ if and only if it has no retraction onto one of $\mathbf{1} + \mathbf{2} + \mathbf{2}, \mathbf{2} + \mathbf{2} + \mathbf{2} + \mathbf{1}, \mathbf{1} + \mathbf{2} + \mathbf{2} + \mathbf{2} + \mathbf{1}, \mathbf{2} + \mathbf{2}$, and $\mathbf{2} + \mathbf{2} + \mathbf{2}$,.*

It was proved in [46] that a series-parallel poset $\mathcal{P}$ has an NU polymorphism if and only if it has no retraction onto one of $\mathbf{2} + \mathbf{2}, \mathbf{1} + \mathbf{2} + \mathbf{2}, \mathbf{2} + \mathbf{2} + \mathbf{1}$, and $\mathbf{1} + \mathbf{2} + \mathbf{2} + \mathbf{1}$. Combining this result with Theorem 5 and Lemma 1 we can get, in particular, the following.

Corollary 2 *If $\mathcal{P}$ is one of $\mathbf{1} + \mathbf{2} + \mathbf{2}, \mathbf{2} + \mathbf{2} + \mathbf{1}$, and $\mathbf{1} + \mathbf{2} + \mathbf{2} + \mathbf{2} + \mathbf{1}$ then* PoRet$(\mathcal{P})$ *is in* **SL** *and* CSP$(\mathcal{P}_P)$ *is in* **NL** *but $\mathcal{P}_P$ has no NU polymorphism.*

It should be noted here that TC-feasibility of posets $\mathbf{2} + \mathbf{2} + \mathbf{1}$ and $\mathbf{1} + \mathbf{2} + \mathbf{2}$ was proved in [38].

In fact, we can say more about the complexity of PoRet$(\mathcal{P})$ for the posets mentioned in Theorem 5.

Proposition 1 *If $\mathcal{P} \in \mathbb{A}$ and $\mathcal{P}$ has no NU polymorphisms then* PoRet$(\mathcal{P})$ *is* **SL**-*complete (under many-one logarithmic space reductions).*

6   Conclusion

It is an open problem whether the presence of an NU polymorphism leads to solvabilty of a CSP($\mathcal{B}$) in non-deterministic logarithmic space for general structures. In this paper we solved the problem positively in the case of order constraints and also proved that this algebraic condition is not necessary. Hence, if the general problem above has a positive answer, then the algebraic condition characterizing CSPs in **NL** (if it exists) is of a more general form than an NU polymorphism.

Motivated by the problem, Dalmau asked [8] whether structures with an NU polymorphism have bounded path duality. At present, we do not know whether the structures $\mathcal{P}_P$ from the two classes described in this paper have bounded path duality. It follows from results of [9,29] that if $\mathcal{P}_P$ has an NU polymorphism then $\neg CSP(\mathcal{P}_P)$ is definable in $(1,2)$-Datalog, and the same can be shown for structures $\mathcal{P}_P$ with $\mathcal{P} \in \mathbb{A}$. There may be some way of using special properties of posets for transforming such $(1,2)$-Datalog programs into linear Datalog programs.

It is easy to see that the posets from our two classes are TC-feasible (see definition in Section 3). It was shown in [38] that the class of TC-feasible posets is closed under isomorphism, dual posets, disjoint union, direct products, and retractions. Using these operations one can easily construct TC-feasible posets lying outside the two classes. For example, direct product of $2 + 2 + 1$ and its dual poset neither has an NU polymorphism nor is series-parallel. In this way one can find further order constraint satisfaction problems in **NL**. Theorem 1 provides another way of obtaining further CSPs in **NL** because, clearly, the equality relation $=_P$ can be expressed in $\mathcal{P}_P$.

Acknowledgment. Part of this research was done while the first author was visiting Concordia University, Montreal. Financial help of NSERC Canada is gratefully acknowledged.

References

1. M. Alvarez and R. Greenlaw. A compendium of problems complete for symmetric logarithmic space. *Computational Complexity*, 9(2):123–145, 2000.
2. M. Benke. Some complexity bounds for subtype inequalities. *Theoretical Computer Science*, 212:3–27, 1999.
3. A.A. Bulatov. A dichotomy theorem for constraints on a three-element set. In *Proceedings 43rd IEEE Symposium on Foundations of Computer Science, FOCS'02*, pages 649–658, 2002.
4. A.A. Bulatov, A.A. Krokhin, and P.G. Jeavons. Constraint satisfaction problems and finite algebras. In *Proceedings 27th International Colloquium on Automata, Languages and Programming, ICALP'00*, volume 1853 of *Lecture Notes in Computer Science*, pages 272–282. Springer-Verlag, 2000.
5. A.A. Bulatov, A.A. Krokhin, and P.G. Jeavons. The complexity of maximal constraint languages. In *Proceedings 33rd ACM Symposium on Theory of Computing, STOC'01*, pages 667–674, 2001.

6. E. Corominas. Sur les ensembles ordonnés projectif et la propriétré du point fixe. *C. R. Acad. Sci. Paris Serie I Math.*, 311:199–204, 1990.

7. N. Creignou, S. Khanna, and M. Sudan. *Complexity Classifications of Boolean Constraint Satisfaction Problems*, volume 7 of *SIAM Monographs on Discrete Mathematics and Applications*. 2001.

8. V. Dalmau. Constraint satisfaction problems in non-deterministic logarithmic space. In *Proceedings 29th International Colloquium on Automata, Languages and Programming, ICALP'02*, volume 2380 of *Lecture Notes in Computer Science*, pages 414–425. Springer-Verlag, 2002.

9. V. Dalmau and J. Pearson. Set functions and width 1 problems. In *Proceedings 5th International Conference on Constraint Programming, CP'99*, volume 1713 of *Lecture Notes in Computer Science*, pages 159–173. Springer-Verlag, 1999.

10. J. Demetrovics, L. Hannák, and L. Rónyai. Near-unanimity functions of partial orders. In *Proceedings 14th International Symposium on Multiple-Valued Logic, ISMVL'84*, pages 52–56, 1984.

11. D. Duffus and I. Rival. A structure theory for ordered sets. *Discrete Mathematics*, 35:53–118, 1981.

12. T. Feder and M.Y. Vardi. The computational structure of monotone monadic SNP and constraint satisfaction: A study through Datalog and group theory. *SIAM Journal of Computing*, 28:57–104, 1998.

13. G. Gottlob, L. Leone, and F. Scarcello. A comparison of structural CSP decomposition methods. *Artificial Intelligence*, 124:243–282, 2000.

14. G. Gottlob, L. Leone, and F. Scarcello. Hypertree decomposition and tractable queries. *Journal of Computer and System Sciences*, 64(3):579–627, 2002.

15. E. Grädel. Capturing complexity classes by fragments of second-order logic. *Theoretical Computer Science*, 101:35–57, 1992.

16. M. Grohe, T. Schwentick, and L. Segoufin. When is the evaluation of conjunctive queries tractable? In *Proceedings 33rd ACM Symposium on Theory of Computing, STOC'01*, pages 657–666, 2001.

17. P. Hell and J. Nešetřil. On the complexity of H-coloring. *Journal of Combinatorial Theory, Ser.B*, 48:92–110, 1990.

18. M. Hoang and J.C. Mitchell. Lower bounds on type inference with subtypes. In *Proceedings 22nd ACM Symposium on Principles of Programming Languages, POPL'95*, pages 176–185, 1995.

19. N. Immerman. Nondeterministic space is closed under complementation. *SIAM Journal on Computing*, 17:935–939, 1988.

20. P.G. Jeavons. On the algebraic structure of combinatorial problems. *Theoretical Computer Science*, 200:185–204, 1998.

21. P.G. Jeavons, D.A. Cohen, and M.C. Cooper. Constraints, consistency and closure. *Artificial Intelligence*, 101(1-2):251–265, 1998.

22. P.G. Jeavons, D.A. Cohen, and M. Gyssens. A unifying framework for tractable constraints. In *Proceedings 1st International Conference on Constraint Programming, CP'95*, volume 976 of *Lecture Notes in Computer Science*, pages 276–291. Springer-Verlag, 1995.

23. P.G. Jeavons, D.A. Cohen, and M. Gyssens. Closure properties of constraints. *Journal of the ACM*, 44:527–548, 1997.

24. L. Kirousis. Fast parallel constraint satisfaction. *Artificial Intelligence*, 64:147–160, 1993.

25. J. Köbler and J. Torán. The complexity of graph isomorphism for colored graphs with color classes of size 2 and 3. In *Proceedings 19th Symposium on Theoretical Aspects of Computer Science, STACS'02*, pages 121–132, 2002.

26. Ph.G. Kolaitis and M.Y. Vardi. Conjunctive-query containment and constraint satisfaction. *Journal of Computer and System Sciences*, 61:302–332, 2000.
27. V. Kumar. Algorithms for constraint satisfaction problems: A survey. *AI Magazine*, 13(1):32–44, 1992.
28. G. Kun and Cs. Szabó. Order varieties and monotone retractions of finite posets. *Order*, 18:79–88, 2001.
29. B. Larose and L. Zádori. The complexity of the extendibility problem for finite posets. manuscript, obtainable from
   http://cicma.mathstat.concordia.ca/faculty/larose/.
30. B. Larose and L. Zádori. Algebraic properties and dismantlability of finite posets. *Discrete Mathematics*, 163:89–99, 1997.
31. H. Lewis and C. Papadimitriou. Symmetric space bounded computation. *Theoretical Computer Science*, 19:161–188, 1982.
32. K. Lodaya and P. Weil. Series-parallel posets: algebra, automata and languages. In *Proceedings 15th Symposium on Theoretical Aspects of Computer Science, STACS'98*, pages 555–565, 1998.
33. R.H. Möhring. Computationally tractable classes of ordered sets. In *Algorithms and Order (Ottawa, 1987)*, pages 105–193. Kluwer, Dordrecht, 1989.
34. R.H. Möhring and M.W. Schäffter. Scheduling series-parallel orders subject to 0/1-communication delays. *Parallel Computing*, 25:23–40, 1999.
35. U. Montanari. Networks of constraints: Fundamental properties and applications to picture processing. *Information Sciences*, 7:95–132, 1974.
36. N. Nisan and A. Ta-Shma. Symmetric logspace is closed under complementation. *Chicago Journal on Theoretical Computer Science*, 1, 1995. (electronic).
37. C.H. Papadimitriou. *Computational Complexity*. Addison-Wesley, 1994.
38. V. Pratt and J. Tiuryn. Satisfiabilty of inequalities in a poset. *Fundamenta Informaticae*, 28:165–182, 1996.
39. R.W. Quackenbush, I. Rival, and I.G. Rosenberg. Clones, order varieties, near-unaminity functions and holes. *Order*, 7:239–248, 1990.
40. J.H. Reif. Symmetric complementation. In *Proceedings 14th ACM Symposium on Theory of Computing, STOC'82*, pages 201–214, 1982.
41. T.J. Schaefer. The complexity of satisfiability problems. In *Proceedings 10th ACM Symposium on Theory of Computing, STOC'78*, pages 216–226, 1978.
42. E. Tsang. *Foundations of Constraint Satisfaction*. Academic Press, London, 1993.
43. J.D. Ullman. *Principles of Database and Knowledge-Base Systems*, volume 1 & 2. Computer Science Press, 1989.
44. J. Valdes, R.E. Tarjan, and E.L. Lawler. The recognition of series-parallel digraphs. *SIAM Journal on Computing*, 11:298–313, 1982.
45. L. Zádori. Posets, near-unanimity functions and zigzags. *Bulletin of Australian Mathematical Society*, 47:79–93, 1993.
46. L. Zádori. Series parallel posets with nonfinitely generated clones. *Order*, 10:305–316, 1993.

The Inversion Problem for Computable Linear Operators

Vasco Brattka[*]

Theoretische Informatik I, Informatikzentrum
FernUniversität, 58084 Hagen, Germany
vasco.brattka@fernuni-hagen.de

Abstract. Given a program of a linear bounded and bijective operator T, does there exist a program for the inverse operator T^{-1}? And if this is the case, does there exist a general algorithm to transfer a program of T into a program of T^{-1}? This is the inversion problem for computable linear operators on Banach spaces in its non-uniform and uniform formulation, respectively. We study this problem from the point of view of computable analysis which is the Turing machine based theory of computability on Euclidean space and other topological spaces. Using a computable version of Banach's Inverse Mapping Theorem we can answer the first question positively. Hence, the non-uniform version of the inversion problem is solvable, while a topological argument shows that the uniform version is not. Thus, we are in the striking situation that any computable linear operator has a computable inverse while there exists no general algorithmic procedure to transfer a program of the operator into a program of its inverse. As a consequence, the computable version of Banach's Inverse Mapping Theorem is a powerful tool which can be used to produce highly non-constructive existence proofs of algorithms. We apply this method to prove that a certain initial value problem admits a computable solution.

Keywords: computable analysis, linear operators, inversion problem.

1 Introduction

Given two Banach spaces X, Y and a linear bounded and bijective operator $T : X \to Y$, Banach's Inverse Mapping Theorem (see e.g. [11]) guarantees that the inverse $T^{-1} : Y \to X$ is a linear bounded operator as well. Hence, it is reasonable to ask for computable versions of this fact, i.e. do the implications

(1) T computable $\Longrightarrow T^{-1}$ computable (**non-uniform inversion problem**),
(2) $T \mapsto T^{-1}$ computable (**uniform inversion problem**)

hold? Of course, both questions have to be specified carefully. In particular, the computability notion in the uniform case should reflect the fact that algorithms of T are transfered into algorithms of T^{-1}.

---

[*] Work partially supported by DFG Grant BR 1807/4-1

H. Alt and M. Habib (Eds.): STACS 2003, LNCS 2607, pp. 391–402, 2003.
© Springer-Verlag Berlin Heidelberg 2003

Such meaningful computability notions are provided by computable analysis, which is the Turing machine based theory of computability on real numbers and other topological spaces. Pioneering work on this theory has been presented by Turing [23], Banach and Mazur [2], Lacombe [18] and Grzegorczyk [12]. Recent monographs have been published by Pour-El and Richards [20], Ko [16] and Weihrauch [25]. Certain aspects of computable functional analysis have already been studied by several authors, see for instance [1,19,10,24,27,28,26].

From the computational point of view Banach's Inverse Mapping Theorem is interesting, since its classical proof relies on the Baire Category Theorem and therefore it counts as "non-constructive" (see [6] for a discussion of computable versions of the Baire Category Theorem). However, a "non-constructive" application of the Baire Category Theorem suffices in order to prove that the non-uniform inversion problem (1) is solvable; but at the same time the non-constructiveness is the reason why the uniform inversion problem (2) is not solvable.

We close the introduction with a short survey of the organisation of this paper. In the following Section 2 we will present some preliminaries from computable analysis. In Section 3 we discuss computable metric spaces, computable Banach spaces and effective open subsets. In Section 4 we investigate computable versions of the Open Mapping Theorem and based on these results we study Banach's Inverse Mapping Theorem in Section 5. Finally, in Section 6 we apply the computable version of this theorem in order to prove that a certain initial value problem admits a computable solution. In this extended abstract most proofs are omitted; they can be found in [5].

2   Preliminaries from Computable Analysis

In this section we briefly summarize some notions from computable analysis. For details the reader is refered to [25]. The basic idea of the representation based approach to computable analysis is to represent infinite objects like real numbers, functions or sets, by infinite strings over some alphabet Σ (which should at least contain the symbols 0 and 1). Thus, a *representation* of a set X is a surjective mapping $\delta :\subseteq \Sigma^\omega \to X$ and in this situation we will call (X, δ) a *represented space*. Here Σ^ω denotes the set of infinite sequences over Σ and the inclusion symbol is used to indicate that the mapping might be partial. If we have two represented spaces, then we can define the notion of a computable function.

Definition 1 (Computable function). Let (X, δ) and (Y, δ') be represented spaces. A function $f :\subseteq X \to Y$ is called (δ, δ')–*computable*, if there exists some computable function $F :\subseteq \Sigma^\omega \to \Sigma^\omega$ such that $\delta'F(p) = f\delta(p)$ for all $p \in \mathrm{dom}(f\delta)$.

Of course, we have to define computability of functions $F :\subseteq \Sigma^\omega \to \Sigma^\omega$ to make this definition complete, but this can be done via Turing machines: F is computable if there exists some Turing machine, which computes infinitely long and transforms each sequence p, written on the input tape, into the corresponding sequence $F(p)$, written on the one-way output tape. Later on, we will also need computable multi-valued operations $f :\subseteq X \rightrightarrows Y$, which are defined analogously to computable functions by substituting $\delta' F(p) \in f\delta(p)$ for the equation in Definition 1 above. If the represented spaces are fixed or clear from the context, then we will simply call a function or operation f *computable*.

For the comparison of representations it will be useful to have the notion of *reducibility* of representations. If δ, δ' are both representations of a set X, then δ is called *reducible* to δ', $\delta \leq \delta'$ in symbols, if there exists a computable function $F :\subseteq \Sigma^\omega \to \Sigma^\omega$ such that $\delta(p) = \delta' F(p)$ for all $p \in \mathrm{dom}(\delta)$. Obviously, $\delta \leq \delta'$ holds, if and only if the identity $\mathrm{id} : X \to X$ is (δ, δ')–computable. Moreover, δ and δ' are called *equivalent*, $\delta \equiv \delta'$ in symbols, if $\delta \leq \delta'$ and $\delta' \leq \delta$.

Analogously to the notion of computability we can define the notion of (δ, δ')–*continuity* for single- and multi-valued operations, by substituting a continuous function $F :\subseteq \Sigma^\omega \to \Sigma^\omega$ for the computable function F in the definitions above. On Σ^ω we use the *Cantor topology*, which is simply the product topology of the discrete topology on Σ. The corresponding reducibility will be called *continuous reducibility* and we will use the symbols $\leq_t$ and $\equiv_t$ in this case. Again we will simply say that the corresponding function is *continuous*, if the representations are fixed or clear from the context. If not mentioned otherwise, we will always assume that a represented space is endowed with the final topology induced by its representation.

This will lead to no confusion with the ordinary topological notion of continuity, as long as we are dealing with *admissible* representations. A representation δ of a topological space X is called *admissible*, if δ is maximal among all continuous representations δ' of X, i.e. if $\delta' \leq_t \delta$ holds for all continuous representations δ' of X. If δ, δ' are admissible representations of topological spaces X, Y, then a function $f :\subseteq X \to Y$ is (δ, δ')–continuous, if and only if it is sequentially continuous, cf. [21,7].

Given a represented space (X, δ), we will occasionally use the notions of a *computable sequence* and a *computable point*. A *computable sequence* is a computable function $f : \mathbb{N} \to X$, where we assume that $\mathbb{N} = \{0, 1, 2, ...\}$ is represented by $\delta_\mathbb{N}(1^n 0^\omega) := n$ and a point $x \in X$ is called *computable*, if there is a constant computable sequence with value x.

Given two represented spaces (X, δ) and (Y, δ'), there is a canonical representation $[\delta, \delta']$ of $X \times Y$ and a representation $[\delta \to \delta']$ of certain functions $f : X \to Y$. If δ, δ' are *admissible* representations of sequential topological spaces, then $[\delta \to \delta']$ is actually a representation of the set $\mathcal{C}(X, Y)$ of continuous functions $f : X \to Y$. If $Y = \mathbb{R}$, then we write for short $\mathcal{C}(X) := \mathcal{C}(X, \mathbb{R})$. The function space representation can be characterized by the fact that it admits evaluation and type conversion.

Proposition 2 (Evaluation and type conversion). *Let (X, δ) and (Y, δ') be admissibly represented sequential topological spaces and let (Z, δ'') be a represented space. Then:*

(1) **(Evaluation)** ev : $\mathcal{C}(X,Y) \times X \to Y, (f,x) \mapsto f(x)$ *is* $([[\delta \to \delta'],\delta],\delta')$-*computable,*

(2) **(Type conversion)** $f : Z \times X \to Y$, *is* $([\delta'',\delta],\delta')$-*computable, if and only if the function* $\check{f} : Z \to \mathcal{C}(X,Y)$, *defined by* $\check{f}(z)(x) := f(z,x)$ *is* $(\delta'',[\delta \to \delta'])$-*computable.*

The proof of this proposition is based on a version of the smn– and utm–Theorem, see [25,21]. If (X,δ), (Y,δ') are admissibly represented sequential topological spaces, then in the following we will always assume that $\mathcal{C}(X,Y)$ is represented by $[\delta \to \delta']$. It follows by evaluation and type conversion that the computable points in $(\mathcal{C}(X,Y),[\delta \to \delta'])$ are just the (δ,δ')-computable functions $f : X \to Y$. Since evaluation and type conversion are even characteristic properties of the function space representation $[\delta \to \delta']$, we can conclude that this representation actually reflects the properties of programs. That is, a name p of a function $f = [\delta \to \delta'](p)$ can be considered as a "program" of f since it just contains sufficiently much information in order to evaluate f. This corresponds to the well-known fact that the compact-open topology is the appropriate topology for programs [22] and actually, if (X,δ), (Y,δ') are admissibly represented separable Banach spaces, one obtains the compact open topology as final topology of $[\delta \to \delta']$ (see [21]).

If (X,δ) is a represented space, then we will always assume that the set of sequences $X^\mathbb{N}$ is represented by $\delta^\mathbb{N} := [\delta_\mathbb{N} \to \delta]$. The computable points in $(X^\mathbb{N},\delta^\mathbb{N})$ are just the computable sequences in (X,δ). Moreover, we assume that X^n is always represented by δ^n, which can be defined inductively by $\delta^1 := \delta$ and $\delta^{n+1} := [\delta^n,\delta]$.

3   Computable Metric and Banach Spaces

In this section we will briefly discuss computable metric spaces and computable Banach spaces. The notion of a computable Banach space will be the central notion for all following results. Computable metric spaces have been used in the literature at least since Lacombe [18]. Pour-El and Richards have introduced a closely related axiomatic characterization of sequential computability structures for Banach spaces [20] which has been extended to metric spaces by Mori, Tsujii, and Yasugi [27].

We mention that we will denote the *open balls* of a metric space (X,d) by $B(x,\varepsilon) := \{y \in X : d(x,y) < \varepsilon\}$ for all $x \in X$, $\varepsilon > 0$ and correspondingly *closed balls* by $\overline{B}(x,\varepsilon) := \{y \in X : d(x,y) \le \varepsilon\}$. Occasionally, we denote complements of sets $A \subseteq X$ by $A^c := X \setminus A$.

Definition 3 (Computable metric space).   A tuple (X,d,α) is called a *computable metric space,* if

(1) $d : X \times X \to \mathbb{R}$ is a metric on X,
(2) $\alpha : \mathbb{N} \to X$ is a sequence which is dense in X,
(3) $d \circ (\alpha \times \alpha) : \mathbb{N}^2 \to \mathbb{R}$ is a computable (double) sequence in $\mathbb{R}$.

Here, we tacitly assume that the reader is familiar with the notion of a computable sequence of reals, but we will come back to that point below. Occasionally, we will say for short that X is a *computable metric space*. Obviously, a computable metric space is especially separable. Given a computable metric space (X, d, α), its *Cauchy representation* $\delta_X :\subseteq \Sigma^\omega \to X$ can be defined by

$$\delta_X(01^{n_0+1}01^{n_1+1}01^{n_2+1}...) := \lim_{i\to\infty} \alpha(n_i)$$

for all n_i such that $(\alpha(n_i))_{i\in\mathbb{N}}$ converges and $d(\alpha(n_i), \alpha(n_j)) \le 2^{-i}$ for all $j > i$ (and undefined for all other input sequences). In the following we tacitly assume that computable metric spaces are represented by their Cauchy representations. If X is a computable metric space, then it is easy to see that $d : X \times X \to \mathbb{R}$ is computable [7]. All Cauchy representations are admissible with respect to the corresponding metric topology.

An important computable metric space is $(\mathbb{R}, d_\mathbb{R}, \alpha_\mathbb{R})$ with the Euclidean metric $d_\mathbb{R}(x, y) := |x - y|$ and some standard numbering of the rational numbers $\mathbb{Q}$, as $\alpha_\mathbb{R}\langle i, j, k \rangle := (i - j)/(k + 1)$. Here, $\langle i, j \rangle := 1/2(i + j)(i + j + 1) + j$ denotes *Cantor pairs* and this definition is extended inductively to finite tuples. Similarly, we can define $\langle p, q \rangle \in \Sigma^\omega$ for sequences $p, q \in \Sigma^\omega$. For short we will occasionally write $\overline{k} := \alpha_\mathbb{R}(k)$. In the following we assume that $\mathbb{R}$ is endowed with the Cauchy representation $\delta_\mathbb{R}$ induced by the computable metric space given above. This representation of $\mathbb{R}$ can also be defined, if $(\mathbb{R}, d_\mathbb{R}, \alpha_\mathbb{R})$ just fulfills (1) and (2) of the definition above and this leads to a definition of computable real number sequences without circularity. Computationally, we do not have to distinguish the complex numbers $\mathbb{C}$ from $\mathbb{R}^2$. We will use the notation $\mathbb{F}$ for a field which always might be replaced by both, $\mathbb{R}$ or $\mathbb{C}$. Correspondingly, we use the notation $(\mathbb{F}, d_\mathbb{F}, \alpha_\mathbb{F})$ for a computable metric space which might be replaced by both computable metric spaces $(\mathbb{R}, d_\mathbb{R}, \alpha_\mathbb{R})$ and $(\mathbb{C}, d_\mathbb{C}, \alpha_\mathbb{C})$ (defined analogously). We will also use the notation $Q_\mathbb{F} = \text{range}(\alpha_\mathbb{F})$, i.e. $Q_\mathbb{R} = \mathbb{Q}$ and $Q_\mathbb{C} = \mathbb{Q}[i]$.

For the definition of a computable Banach space it is helpful to have the notion of a computable vector space which we will define next.

Definition 4 (Computable vector space). A represented space (X, δ) is called a *computable vector space* (over $\mathbb{F}$), if $(X, +, \cdot, 0)$ is a vector space over $\mathbb{F}$ such that the following conditions hold:

(1) $+ : X \times X \to X, (x, y) \mapsto x + y$ is computable,
(2) $\cdot : \mathbb{F} \times X \to X, (a, x) \mapsto a \cdot x$ is computable,
(3) $0 \in X$ is a computable point.

If (X, δ) is a computable vector space over $\mathbb{F}$, then $(\mathbb{F}, \delta_\mathbb{F})$, (X^n, δ^n) and $(X^\mathbb{N}, \delta^\mathbb{N})$ are computable vector spaces over $\mathbb{F}$. If, additionally, (X, δ), (Y, δ') are admissibly represented second countable T_0–spaces, then the function space $(\mathcal{C}(Y, X), [\delta' \to \delta])$ is a computable vector space over $\mathbb{F}$. Here we tacitly assume that the vector space operations on product, sequence and function spaces are defined componentwise. The proof for the function space is a straightforward application of evaluation and type conversion. The central definition for the present investigation will be the notion of a computable Banach space.

Definition 5 (Computable normed space). A tuple $(X, \|\ \|, e)$ is called a *computable normed space*, if

(1) $\|\ \| : X \to \mathbb{R}$ is a norm on X,
(2) $e : \mathbb{N} \to X$ is a *fundamental sequence*, i.e. its linear span is dense in X,
(3) (X, d, α_e) with $d(x, y) := \|x - y\|$ and $\alpha_e \langle k, \langle n_0, ..., n_k \rangle \rangle := \sum_{i=0}^{k} \alpha_{\mathbb{F}}(n_i) e_i$, is a computable metric space with Cauchy representation δ_X,
(4) (X, δ_X) is a computable vector space over $\mathbb{F}$.

If in the situation of the definition the underlying space $(X, \|\ \|)$ is even a Banach space, i.e. if (X, d) is a complete metric space, then $(X, \|\ \|, e)$ is called a *computable Banach space*. If the norm and the fundamental sequence are clear from the context or locally irrelevant, we will say for short that X is a *computable normed space* or a *computable Banach space*. We will always assume that computable normed spaces are represented by their Cauchy representations, which are admissible with respect to the norm topology. If X is a computable normed space, then $\|\ \| : X \to \mathbb{R}$ is a computable function. Of course, all computable Banach spaces are separable. In the following proposition some computable Banach spaces are defined.

Proposition 6 (Computable Banach spaces). *Let $p \in \mathbb{R}$ be a computable real number with $1 \le p < \infty$ and let $a < b$ be computable real numbers. The following spaces are computable Banach spaces over $\mathbb{F}$.*

(1) $(\mathbb{F}^n, \|\ \|_\infty, e)$ *with*
- $\|(x_1, x_2, ..., x_n)\|_\infty := \max\limits_{k=1,...,n} |x_k|$,
- $e_i = e(i) = (e_{i1}, e_{i2}, ..., e_{in})$ *with* $e_{ik} := \begin{cases} 1 \ if \ i = k \\ 0 \ else \end{cases}$.

(2) $(\ell_p, \|\ \|_p, e)$ *with*
- $\ell_p := \{x \in \mathbb{F}^{\mathbb{N}} : \|x\|_p < \infty\}$,
- $\|(x_k)_{k\in\mathbb{N}}\|_p := \sqrt[p]{\sum\limits_{k=0}^{\infty} |x_k|^p}$,
- $e_i = e(i) = (e_{ik})_{k\in\mathbb{N}}$ *with* $e_{ik} := \begin{cases} 1 \ if \ i = k \\ 0 \ else \end{cases}$.

(3) $(\mathcal{C}^{(n)}[a, b], \|\ \|_{(n)}, e)$ *with*
- $\mathcal{C}^{(n)}[a, b] := \{f : [a, b] \to \mathbb{R} : f \ n\text{--}times \ continuously \ differentiable\}$,
- $\|f\|_{(n)} := \sum\limits_{i=0}^{n} \max\limits_{t\in[a,b]} |f^{(i)}(t)|$,
- $e_i(t) = e(i)(t) = t^i$.

We leave it to the reader to check that these spaces are actually computable Banach spaces. If not stated differently, then we will assume that $(\mathbb{F}^n, \|\ \|)$ is endowed with the maximum norm $\|\ \|_\infty$. It is known that the Cauchy representation $\delta_{\mathcal{C}[a,b]}$ of $\mathcal{C}^{(0)}[a, b] = \mathcal{C}([a, b], \mathbb{R})$ is equivalent to $[\delta_{[a,b]} \to \delta_{\mathbb{R}}]$, where $\delta_{[a,b]}$ denotes the restriction of $\delta_{\mathbb{R}}$ to $[a, b]$ (cf. Lemma 6.1.10 in [25]). In the following we will occasionally utilize the sequence spaces ℓ_p to construct counterexamples.

Since we will study the Open Mapping Theorem in the next section, we have to compute with open sets. Therefore we need representations of the hyperspace $\mathcal{O}(X)$ of open subsets of X. Such representations have been studied in the Euclidean case in [9,25] and for the metric case in [8].

Definition 7 (Hyperspace of open subsets). Let (X, d, α) be a computable metric space. We endow the hyperspace $\mathcal{O}(X) := \{U \subseteq X : U \text{ open}\}$ with the representation $\delta_{\mathcal{O}(X)}$, defined by $\delta_{\mathcal{O}(X)}(p) := \bigcup_{i=0}^{\infty} B(\alpha(n_i), \overline{k_i})$ for all sequences $p = 01^{\langle n_0, k_0 \rangle + 1} 01^{\langle n_1, k_1 \rangle + 1} 01^{\langle n_2, k_2 \rangle + 1} \dots$ with $n_i, k_i \in \mathbb{N}$.

Those open subsets $U \subseteq X$ which are computable points in $\mathcal{O}(X)$ are called *r.e. open*. We close this section with a helpful proposition which states that we can represent open subsets by preimages of continuous functions. This is a direct consequence of results in [8] and an effective version of the statement that open subsets of metric spaces coincide with the functional open subsets.

Proposition 8 (Functional open subsets). *Let X be a computable metric space. The map $Z : \mathcal{C}(X) \to \mathcal{O}(X), f \mapsto X \setminus f^{-1}\{0\}$ is computable and admits a computable right-inverse $\mathcal{O}(X) \rightrightarrows \mathcal{C}(X)$.*

4   The Open Mapping Theorem

In this section we will study the effective content of the Open Mapping Theorem, which we formulate first. The classical proof of this theorem can be found in [11] or other textbooks on functional analysis.

Theorem 9 (Open Mapping Theorem). *Let X, Y be Banach spaces. If $T : X \to Y$ is a linear surjective and bounded operator, then T is open, i.e. $T(U) \subseteq Y$ is open for any open $U \subseteq X$.*

Whenever $T : X \to Y$ is an open operator, we can associate the function

$$\mathcal{O}(T) : \mathcal{O}(X) \to \mathcal{O}(Y), U \mapsto T(U)$$

with it. Now we can ask for three different computable versions of the Open Mapping Theorem. If $T : X \to Y$ is a linear computable and surjective operator, does the following hold true:

(1) $U \subseteq X$ r.e. open $\Longrightarrow T(U) \subseteq Y$ r.e. open?
(2) $\mathcal{O}(T) : \mathcal{O}(X) \to \mathcal{O}(Y), U \mapsto T(U)$ is computable?
(3) $T \mapsto \mathcal{O}(T)$ is computable?

Since any computable function maps computable inputs to computable outputs, we can conclude (3)$\Longrightarrow$(2)$\Longrightarrow$(1). In the following we will see that questions (1) and (2) can be answered in the affirmative, while question (3) has to be answered in the negative. The key tool for the positive results will be Theorem 10 on effective openness. It states that $T : X \to Y$ is computable, if and only if $\mathcal{O}(T) : \mathcal{O}(X) \to \mathcal{O}(Y)$ is computable, provided that T is a linear bounded and and open operator and X, Y are computable normed spaces.

Theorem 10 (Effective openness). *Let X, Y be computable normed spaces and let $T : X \to Y$ be a linear and bounded operator. Then the following conditions are equivalent:*

(1) $T : X \to Y$ is open and computable,
(2) $\mathcal{O}(T) : \mathcal{O}(X) \to \mathcal{O}(Y), U \mapsto T(U)$ is well-defined and computable.

The proof of "(1)$\Longrightarrow$(2)" can be performed in two steps. First one uses the fact that T is open and linear in order to prove that given $x \in X$ and an open subset $U \subseteq X$, we can effectively find some radius $r > 0$ such that $B(Tx, r) \subseteq T(U)$. This is the non-uniform part of the proof since it is based on the fact that for any open T there exists some radius $r' > 0$ with $B(0, r') \subseteq T(B(0, 1))$ (such an r' always exists, but it cannot be effectively determined from T). In the second step, computability of T is exploited in order to prove that $\mathcal{O}(T)$ is computable. Using the previous theorem we can directly conclude a computable version of the Open Mapping Theorem as a corollary of the classical Open Mapping Theorem.

Theorem 11 (Computable Open Mapping Theorem). *Let X, Y be computable Banach spaces and let $T : X \to Y$ be a linear computable operator. If T is surjective, then T is open and $\mathcal{O}(T) : \mathcal{O}(X) \to \mathcal{O}(Y)$ is computable. Especially, $T(U) \subseteq Y$ is r.e. open for any r.e. open set $U \subseteq X$.*

This version of the Open Mapping Theorem leaves open the question whether the map $T \mapsto \mathcal{O}(T)$ itself is computable. This question is answered negatively by the following example.

Example 12. The mapping $T \mapsto \mathcal{O}(T)$, defined for linear, bounded and bijective operators $T : \ell_2 \to \ell_2$ with $\|T\| = 1$, is not $([\delta_{\ell_2} \to \delta_{\ell_2}], \delta_{\mathcal{O}(\ell_2)})$–continuous.

Although the mapping $T \mapsto \mathcal{O}(T)$ is discontinuous, we know by Theorem 10 that $\mathcal{O}(T) : \mathcal{O}(\ell_2) \to \mathcal{O}(\ell_2)$ is computable whenever $T : \ell_2 \to \ell_2$ is computable. On the one hand, this guarantees that $T \mapsto \mathcal{O}(T)$ is not too discontinuous [4]. On the other hand, we have to use sequences to construct a computable counterexample for the uniform version of the Open Mapping Theorem. The proof is based on appropriately defined diagonal matrices.

Proposition 13. *There exists a computable sequence $(T_n)_{n \in \mathbb{N}}$ in $\mathcal{C}(\ell_2, \ell_2)$ of linear computable and bijective operators $T_n : \ell_2 \to \ell_2$ such that $(T_n B(0, 1))_{n \in \mathbb{N}}$ is a sequence of r.e. open subsets of ℓ_2 which is not computable in $\mathcal{O}(\ell_2)$.*

5   Banach's Inverse Mapping Theorem

In this section we want to study computable versions of Banach's Inverse Mapping Theorem. Again we start with a formulation of the classical theorem.

Theorem 14 (Banach's Inverse Mapping Theorem). *Let X, Y be Banach spaces and let $T : X \to Y$ be a linear bounded operator. If T is bijective, then $T^{-1} : Y \to X$ is bounded.*

Similarly as in case of the Open Mapping Theorem we have two canonical candidates for an effective version of this theorem: the non-uniform version (1) and the uniform version (2), as formulated in the Introduction. Again we will see that the non-uniform version admits a solution while the uniform version does not. Analogously as we have used Theorem 10 on effective openness to prove the computable Open Mapping Theorem 11, we will use Theorem 15 on effective continuity to prove the computable version of Banach's Inverse Mapping Theorem. We sketch the first part of the proof which is based on Proposition 8.

Theorem 15 (Effective continuity). *Let X, Y be computable metric spaces and let $T : X \to Y$ be a function. Then the following conditions are equivalent:*

(1) $T : X \to Y$ is computable,
(2) $\mathcal{O}(T^{-1}) : \mathcal{O}(Y) \to \mathcal{O}(X), V \mapsto T^{-1}(V)$ is well-defined and computable.

Proof. "(1)$\Longrightarrow$(2)" If $T : X \to Y$ is computable, then it is continuous and hence $\mathcal{O}(T^{-1})$ is well-defined. Given a function $f : Y \to \mathbb{R}$ such that $V = Y \setminus f^{-1}\{0\}$, we obtain $T^{-1}(V) = T^{-1}(Y \setminus f^{-1}\{0\}) = X \setminus (fT)^{-1}\{0\}$. Using Proposition 8 and the fact that composition $\circ : \mathcal{C}(Y, \mathbb{R}) \times \mathcal{C}(X, Y) \to \mathcal{C}(X, \mathbb{R}), (f, T) \mapsto f \circ T$ is computable (which can be proved by evaluation and type conversion), we obtain that $\mathcal{O}(T^{-1})$ is computable. □

Now we note the fact that for Banach spaces X, Y and bijective linear operators $T : X \to Y$, the operation $\mathcal{O}(T^{-1})$, associated with T according to the previous theorem, is the same as the operation $\mathcal{O}(S)$, associated with $S = T^{-1}$ according to Theorem 10. Thus, we can directly conclude the following computable version of the Inverse Mapping Theorem as a corollary of Theorem 15 and Theorem 10.

Theorem 16 (Computable Inverse Mapping Theorem). *Let X, Y be computable Banach spaces and let $T : X \to Y$ be a linear computable operator. If T is bijective, then $T^{-1} : Y \to X$ is computable too.*

In contrast to Theorem 10 on effective openness, we can formulate a uniform version of the Theorem 15 on effective continuity: $\omega : T \mapsto \mathcal{O}(T^{-1})$, defined for all continuous $T : X \to Y$, is $([\delta_X \to \delta_Y], [\delta_{\mathcal{O}(Y)} \to \delta_{\mathcal{O}(X)}])$–computable and its inverse ω^{-1} is computable in the corresponding sense too. Using this positive result, applied to T^{-1}, we can transfer our negative results on the Open Mapping Theorem to the Inverse Mapping Theorem. As a corollary of this fact and Example 12 we obtain the following result.

Example 17. The inversion map $T \mapsto T^{-1}$, defined for linear bounded and bijective operators $T : \ell_2 \to \ell_2$ with $||T|| = 1$, is not continuous.

This holds with respect to $([\delta_{\ell_2} \to \delta_{\ell_2}], [\delta_{\ell_2} \to \delta_{\ell_2}])$–continuity, that is with respect to the compact open topology. Correspondingly, we can construct a computable counterexample for the uniform version of the Inverse Mapping Theorem. As a corollary of Proposition 13 we obtain the following counterexample.

Corollary 18. *There exists a computable sequence* $(T_n)_{n \in \mathbb{N}}$ *in* $\mathcal{C}(\ell_2, \ell_2)$ *of linear computable and bijective operators* $T_n : \ell_2 \to \ell_2$, *such that* $(T_n^{-1})_{n \in \mathbb{N}}$ *is a sequence of computable operators* $T_n^{-1} : \ell_2 \to \ell_2$ *which is not computable in* $\mathcal{C}(\ell_2, \ell_2)$.

We can even assume that $\|T_n\| = 1$ for all $n \in \mathbb{N}$. It is worth mentioning that in case of the previous corollary there is no computable function which is completely above the sequence $(\|T_n^{-1}\|)_{n \in \mathbb{N}}$. We close this section with an application of the Computable Inverse Mapping Theorem 16 which shows that any two comparable computable complete norms are computably equivalent.

Theorem 19. *Let* $(X, \| \ \|)$, $(X, \| \ \|')$ *be computable Banach spaces and let* δ, δ' *be the corresponding Cauchy representations of* X. *If* $\delta \leq \delta'$ *then* $\delta \equiv \delta'$.

Proof. If $\delta \leq \delta'$, then the identity id : $(X, \| \ \|) \to (X, \| \ \|')$ is (δ, δ')–computable. Moreover, the identity is obviously linear and bijective. Thus, the inverse identity id^{-1} : $(X, \| \ \|') \to (X, \| \ \|)$ is (δ', δ)–computable by the Computable Inverse Mapping Theorem. Consequently, $\delta' \leq \delta$. $\square$

6   An Initial Value Problem

In this section we will discuss an application of the computable version of Banach's Inverse Mapping Theorem to the initial value problem of ordinary linear differential equations. Consider the linear differential equation with initial values

$$\sum_{i=0}^{n} f_i(t) x^{(i)}(t) = y(t) \text{ with } x^{(j)}(0) = a_j \text{ for } j = 0, ..., n-1. \tag{1}$$

Here, $x, y : [0, 1] \to \mathbb{R}$ are functions, $f_i : [0, 1] \to \mathbb{R}$ are coefficient functions with $f_n \neq 0$ and $a_0, ..., a_{n-1} \in \mathbb{R}$ are initial values. It is known that for each $y \in \mathcal{C}[0, 1]$ and all values $a_0, ..., a_{n-1}$ there is exactly one solution $x \in \mathcal{C}^{(n)}[0, 1]$ of this equation, see for instance [13]. Given f_i, a_i and y, can we effectively find this solution? The positive answer to this question can easily be deduced from the computable Inverse Mapping Theorem 16.

Theorem 20 (Initial Value Problem). *Let* $n \geq 1$ *be a natural number and let* $f_0, ..., f_n : [0, 1] \to \mathbb{R}$ *be computable functions with* $f_n \neq 0$. *The solution operator* $L : \mathcal{C}[0, 1] \times \mathbb{R}^n \to \mathcal{C}^{(n)}[0, 1]$ *which maps each tuple* $(y, a_0, ..., a_{n-1}) \in \mathcal{C}[0, 1] \times \mathbb{R}^n$ *to the unique function* $x = L(y, a_0, ..., a_{n-1})$ *which fulfills Equation (1), is computable.*

Proof. $L^{-1} : \mathcal{C}^{(n)}[0, 1] \to \mathcal{C}[0, 1] \times \mathbb{R}^n, x \mapsto \left(\sum_{i=0}^{n} f_i x^{(i)}, x^{(0)}(0), ..., x^{(n-1)}(0) \right)$ is obviously linear. Using the evaluation and type conversion property and the fact that the i–th differentiation operator $\mathcal{C}^{(n)}[0, 1] \to \mathcal{C}[0, 1], x \mapsto x^{(i)}$ is computable for $i \leq n$, one can easily prove that L^{-1} is computable. By the computable Inverse Mapping Theorem 16 it follows that L is computable too. $\square$

We obtain the following immediate corollary on computability of solutions of ordinary linear differential equations.

Corollary 21. *Let $n \geq 1$ and let $y, f_0, ..., f_n : [0,1] \to \mathbb{R}$ be computable functions and let $a_0, ..., a_{n-1} \in \mathbb{R}$ be computable real numbers. Then the unique function $x \in C^{(n)}[0,1]$ which fulfills Equation (1) is a computable point in $C^{(n)}[0,1]$. Especially, $x^{(0)}, ..., x^{(n)} : [0,1] \to \mathbb{R}$ are computable functions.*

7 Conclusion

We have investigated the non-uniform and the uniform version of the inversion problem for computable linear operators. The computable version of Banach's Inverse Mapping Theorem, Theorem 16, shows that the non-uniform version is solvable while Example 17 proves by a topological argument and Corollary 18 proves by a computability argument that the uniform version is not solvable (this is because any computable operation is continuous and maps computable sequences to computable sequences, respectively).

The negative result corresponds to what is known in constructive analysis [3]. However, our positive results on Banach's Inverse Mapping Theorem cannot be deduced from known positive results in constructive analysis [14,15] since these theorems have stronger assumptions (such as "effective bijectivity").

In a certain sense, the negative result seems to be in contrast with the so-called Banach's Inversion Stability Theorem [17]. However, this theorem states that $T \mapsto T^{-1}$ is continuous with respect to the *operator norm topology* on the function space. In the infinite-dimensional case this topology is different from the compact open topology and it is only the latter which reflects the meaning of "programs". Additionally, the operator norm topology is not separable in these cases and hence it is not obvious how to handle it computationally [5]. Moreover, it is worth mentioning that in case of finite-dimensional spaces X, Y, the uniform version of the inversion problem becomes solvable as well [5]. In this case $T \mapsto T^{-1}$ is computable (and continuous with respect to the compact open topology).

Altogether we are in the somewhat surprising situation that in the general case of infinite-dimensional Banach spaces X, Y any computable linear operator $T : X \to Y$ has a computable inverse T^{-1} while there is no general algorithmic procedure to transfer programs of T into programs of T^{-1}. As we have demonstrated with Theorem 20, this leads to highly non-effective existence proofs of algorithms: the proof of Theorem 20 shows that there exist an algorithm which solves the corresponding initial value problem without a single hint how such an algorithm could look like. Nevertheless, this is a meaningful insight, since only in case of existence the search for a concrete algorithm is promising (of course, in the special case of the initial value problem such concrete algorithms are known).

References

1. O. Aberth, *Computable Analysis*, McGraw-Hill, New York 1980.
2. S. Banach and S. Mazur, Sur les fonctions calculables, *Ann. Soc. Pol. de Math.* **16** (1937) 223.

3. E. Bishop and D. S. Bridges, *Constructive Analysis*, Springer, Berlin 1985.
4. V. Brattka, Computable invariance, *Theoret. Comp. Sci.* **210** (1999) 3–20.
5. V. Brattka, Computability of Banach space principles, Informatik Berichte 286, FernUniversität Hagen, Fachbereich Informatik, Hagen, June 2001.
6. V. Brattka, Computable versions of Baire's category theorem, in: J. Sgall, A. Pultr, and P. Kolman (eds.), *Mathematical Foundations of Computer Science 2001*, vol. 2136 of *Lect. Not. Comp. Sci.*, Springer, Berlin 2001, 224–235.
7. V. Brattka, Computability over topological structures, in: S. B. Cooper and S. Goncharov (eds.), *Computability and Models*, Kluwer Academic Publishers, Dordrecht (to appear), 93–136.
8. V. Brattka and G. Presser, Computability on subsets of metric spaces, *Theoret. Comp. Sci.* (accepted for publication).
9. V. Brattka and K. Weihrauch, Computability on subsets of Euclidean space I: Closed and compact subsets, *Theoret. Comp. Sci.* **219** (1999) 65–93.
10. X. Ge and A. Nerode, Effective content of the calculus of variations I: semicontinuity and the chattering lemma, *Ann. Pure Appl. Logic* **78** (1996) 127–146.
11. C. Goffman and G. Pedrick, *First Course in Functional Analysis*, Prentince-Hall, Englewood Cliffs 1965.
12. A. Grzegorczyk, On the definitions of computable real continuous functions, *Fund. Math.* **44** (1957) 61–71.
13. H. Heuser, *Functional analysis*, John Wiley & Sons, Chichester 1982.
14. H. Ishihara, A constructive version of Banach's inverse mapping theorem, *New Zealand J. Math.* **23** (1994) 71–75.
15. H. Ishihara, Sequential continuity of linear mappings in constructive mathematics, *J. Univ. Comp. Sci.* **3** (1997) 1250–1254.
16. K.-I. Ko, *Complexity Theory of Real Functions*, Birkhäuser, Boston 1991.
17. S. Kutateladze, *Fundamentals of Functional Analysis*, Kluwer Academic Publishers, Dordrecht 1996.
18. D. Lacombe, Quelques procédés de définition en topologie récursive, in: A. Heyting (ed.), *Constructivity in mathematics*, North-Holland, Amsterdam 1959, 129–158.
19. G. Metakides, A. Nerode, and R. Shore, Recursive limits on the Hahn-Banach theorem, in: M. Rosenblatt (ed.), *Errett Bishop: Reflections on Him and His Research*, vol. 39 of *Contemporary Mathematics*, AMS, Providence 1985, 85–91.
20. M. B. Pour-El and J. I. Richards, *Computability in Analysis and Physics*, Springer, Berlin 1989.
21. M. Schröder, Extended admissibility, *Theoret. Comp. Sci.* **284** (2002) 519–538.
22. M. Smyth, Topology, in: S. Abramsky, D. Gabbay, and T. Maibaum (eds.), *Handbook of Logic in Computer Science, Vol. 1*, Clarendon Press, Oxford 1992, 641–761.
23. A. M. Turing, On computable numbers, with an application to the "Entscheidungsproblem", *Proc. London Math. Soc.* **42** (1936) 230–265.
24. M. Washihara, Computability and tempered distributions, *Mathematica Japonica* **50** (1999) 1–7.
25. K. Weihrauch, *Computable Analysis*, Springer, Berlin 2000.
26. K. Weihrauch and N. Zhong, Is wave propagation computable or can wave computers beat the Turing machine?, *Proc. London Math. Soc.* **85** (2002) 312–332.
27. M. Yasugi, T. Mori, and Y. Tsujii, Effective properties of sets and functions in metric spaces with computability structure, *Theoret. Comp. Sci.* **219** (1999) 467–486.
28. N. Zhong, Computability structure of the Sobolev spaces and its applications, *Theoret. Comp. Sci.* **219** (1999) 487–510.

Algebras of Minimal Rank over Arbitrary Fields

Markus Bläser

Institut für Theoretische Informatik, Universität zu Lübeck
Wallstr. 40, 23560 Lübeck, Germany
blaeser@tcs.uni-luebeck.de

Abstract. Let $R(A)$ denote the rank (also called bilinear complexity) of a finite dimensional associative algebra A. A fundamental lower bound for $R(A)$ is the so-called Alder–Strassen bound $R(A) \geq 2 \dim A - t$, where t is the number of maximal twosided ideals of A. The class of algebras for which the Alder–Strassen bound is sharp, the so-called algebras of minimal rank, has received a wide attention in algebraic complexity theory.

As the main contribution of this work, we characterize all algebras of minimal rank over arbitrary fields. This finally solves an open problem in algebraic complexity theory, see for instance [12, Sect. 12, Problem 4] or [6, Problem 17.5].

1 Introduction

One of the most important problems in algebraic complexity theory is the question about the costs of multiplication, say of matrices, triangular matrices, or polynomials (modulo a fixed polynomial). To be more specific, let A be a finite dimensional associative k-algebra with identity 1. By fixing a basis of A, say $v_1, \ldots, v_N$, we can define a set of bilinear forms corresponding to the multiplication in A. If $v_\mu v_\nu = \sum_{\kappa=1}^{N} \alpha_{\mu,\nu}^{(\kappa)} v_\kappa$ for $1 \leq \mu, \nu \leq N$ with *structural constants* $\alpha_{\mu,\nu}^{(\kappa)} \in k$, then these constants and the equation

$$\left(\sum_{\mu=1}^{N} X_\mu v_\mu \right) \left(\sum_{\nu=1}^{N} Y_\nu v_\nu \right) = \sum_{\kappa=1}^{N} b_\kappa(X, Y) v_\kappa$$

define the desired bilinear forms $b_1, \ldots, b_N$. The *rank* (also called *bilinear complexity*) of $b_1, \ldots, b_N$ is the smallest number of essential bilinear multiplications necessary and sufficient to compute $b_1, \ldots, b_N$ from the indeterminates $X_1, \ldots, X_N$ and $Y_1, \ldots, Y_N$. More precisely, the bilinear complexity of $b_1, \ldots, b_N$ is the smallest number r of products $p_\rho = u_\rho(X) \cdot v_\rho(Y)$ with linear forms u_ρ and v_ρ in the X_i and Y_j, respectively, such that $b_1, \ldots, b_N$ are contained in the linear span of $p_1, \ldots, p_r$. From this definition, it is obvious that the bilinear complexity of $b_1, \ldots, b_N$ is independent of the choice of $v_1, \ldots, v_N$, thus we may speak about the bilinear complexity of (the multiplication in) A. For a modern introduction to this topic and to algebraic complexity theory in general, we recommend [6].

H. Alt and M. Habib (Eds.): STACS 2003, LNCS 2607, pp. 403–414, 2003.

A fundamental lower bound for the rank of an associative algebra A is the so-called Alder–Strassen bound [1]. It states that the rank of A is bounded from below by twice the dimension of A minus the number of twosided ideals in A. This bound is sharp in the sense that there are algebras for which equality holds. Since then, a lot of effort has been spent on characterizing these algebras in terms of their algebraic structure. There has been some success for certain classes of algebras, like division algebras, commutative algebras, etc., but up to now, the general case is still unsolved. As the main contribution of the present work, we determine *all* algebras of minimal rank over *arbitrary* fields, thus solving this problem completely.

Model of computation. In the remainder of this work, we use a coordinate-free definition of rank, which is more appropriate when dealing with algebras of minimal rank, see [6, Chap. 14]. For a vector space V, V^* denotes the dual space of V, that is, the vector space of all linear forms on V.

Definition 1. *Let k be a field, U, V, and W finite dimensional vector spaces over k, and $\phi : U \times V \to W$ be a bilinear map.*

1. *A sequence $\beta = (f_1, g_1, w_1, \ldots, f_r, g_r, w_r)$ such that $f_\rho \in U^*$, $g_\rho \in V^*$, and $w_\rho \in W$ is called a bilinear computation of length r for ϕ if*

$$\phi(u, v) = \sum_{\rho=1}^{r} f_\rho(u) g_\rho(v) w_\rho \quad \text{for all } u \in U, v \in V.$$

2. *The length of a shortest bilinear computation for ϕ is called the bilinear complexity or the rank of ϕ and is denoted by $R(\phi)$.*

3. *If A is a finite dimensional associative k-algebra with identity, then the rank of A is defined as the rank of the multiplication map of A, which is a bilinear map $A \times A \to A$. The rank of A is denoted by $R(A)$.*

Closely related to the bilinear complexity is the *multiplicative complexity*. Here we allow that f_ρ and g_ρ are linear forms on $U \times V$ in the above definition. The multiplicative complexity is clearly a lower bound for the bilinear complexity and it is easy to show that twice the multiplicative complexity is an upper bound for the bilinear complexity, see e.g. [6, Chap. 14].

Structure of associative algebras. Before we discuss the results obtained in this paper, we briefly review some facts about the structure of associative algebras. For detailed informations, the reader is referred to [7]. Throughout this work, the term "algebra" means a finite dimensional associative k-algebra with identity 1.

The first building block of associative algebras are *division algebras*. An algebra D is a division algebra, if all nonzero elements of D are invertible, i.e., for all $d \in D \setminus \{0\}$ there is a $c \in D$ such that $cd = dc = 1$. If D is commutative, too, then D is an *extension field* of k. Such an extension field is called simply generated, if there is an $a \in D$ such that $D = k(a)$. In other words, $D \cong k[X]/(p)$ for some irreducible polynomial $p \in k[X]$. If k is perfect, then each extension field is simply generated by the primitive element theorem. A prominent division algebra that is not commutative are the Hamiltonian quaternions, which form a four-dimensional division algebra over $\mathbb{R}$.

From division algebras, one can construct *simple algebras*: let I be a vector space of some algebra A. I is called a *two-sided ideal* if $A \cdot I \cdot A \subseteq I$ (pointwise multiplication). In other words, I is closed under left and right multiplication with elements from A. Now an algebra A is called simple, if A has only two twosided ideals, the zero ideal $\{0\}$ and A itself. It is a fundamental theorem that every simple algebra A is isomorphic to a matrix algebra over a division algebra D, i.e., $A \cong D^{n \times n}$ for some integer $n \geq 1$.

The next larger class are *semisimple algebras*. A twosided ideal is called maximal, if it is not contained in any other proper twosided ideal. The *radical* of an algebra A, denoted by $\operatorname{rad} A$, is the intersection of all maximal twosided ideals, which is again a twosided ideal. The radical is nilpotent, that means, there is an integer N (for instance $N = \dim A$) such that $(\operatorname{rad} A)^N = \{0\}$. An algebra A is called semisimple if $\operatorname{rad} A = \{0\}$. Wedderburn's structure theorem states that an algebra is semisimple if and only if it is isomorphic to a direct product of simple algebras $A_1 \times \cdots \times A_t$ (with componentwise addition and multiplication). As mentioned above, each such A_τ is isomorphic to a matrix algebra over a division algebra D_τ, i.e., $A_\tau \cong D_\tau^{n_\tau \times n_\tau}$. The number t in this decomposition equals the number of maximal twosided of A.

The structure of arbitrary algebras (with $\operatorname{rad} A \neq \{0\}$) in general is complicated (and not understood well). One important fact is that for any algebra A, the quotient algebra $A/\operatorname{rad} A$ is semisimple and has the same number of maximal twosided ideals as A. An algebra A is *basic*, if $A/\operatorname{rad} A$ is a direct product of division algebras, that is, we have $n_\tau = 1$ for all τ in the decomposition of the semisimple algebra $A/\operatorname{rad} A$

Previous results. Using the notations from Definition 1, the Alder–Strassen bound [1] reads as

$$R(A) \geq 2 \dim A - t,$$

where t is the number of maximal twosided ideals in A. Algebras for which this bound is sharp are called *algebras of minimal rank*. They have received a wide attention in algebraic complexity theory. One prominent algebra of minimal rank is $k^{2 \times 2}$, the algebra of 2×2–matrices [11]. We have $R(k^{2 \times 2}) = 7$. It has been a longstanding open problem whether $k^{3 \times 3}$ is of minimal rank or not, see [6, Problem 17.1]. This question motivated the concept of algebras of minimal rank. The idea was that if one could characterize all algebras of minimal rank in terms of their algebraic structure, then one simply has to verify whether $k^{3 \times 3}$ has this structure or not. Meanwhile, we know that $k^{3 \times 3}$ is not of minimal rank [2]. Nevertheless, the characterization of the algebras of minimal rank is an interesting and important topic in algebraic complexity theory on its own, see e.g. [12, Sect. 12, Problem 4] or [6, Problem 17.5].

Let us review which characterization results have already been obtained: De Groote [8] determined all division algebras D of minimal rank. Over infinite fields, these are all simply generated extension fields of k. If k is finite, then D has minimal rank if in addition $\#k \geq 2 \dim D - 2$, the latter result follows from the classification of the algorithm variety of polynomial multiplication modulo some irreducible polynomial by Winograd [13]. De Groote and Heintz [9] characterize all commutative algebras of minimal rank over infinite fields. Next Büchi and Clausen [5] classify all local algebras of minimal rank over infinite fields. Then Heintz and Morgenstern [10] determine all basic algebras over algebraically closed fields. Thereafter, all semisimple algebras of minimal rank over arbitrary fields and all algebras of minimal rank over algebraically closed fields have been characterized [3]: first, semisimple algebras of minimal rank are isomorphic to a finite direct product of division algebras of minimal rank (as considered by de Groote and Winograd) and of copies of $k^{2\times2}$. Second, algebras of minimal rank over algebraically closed fields are isomorphic to a direct product of copies $k^{2\times2}$ and a basic algebra B of minimal rank (as characterized by Heintz and Morgenstern). In [4], all algebras of minimal rank over perfect fields have finally been characterized. The result proven there states that an algebra over a perfect field k has minimal rank if and only if

$$A \cong C_1 \times \cdots \times C_s \times k^{2\times2} \times \cdots \times k^{2\times2} \times B \tag{1}$$

where $C_1, \ldots, C_s$ are local algebras of minimal rank with $\dim(C_\sigma / \operatorname{rad} C_\sigma) \geq 2$ (as determined by Büchi and Clausen) and $\#k \geq 2 \dim C_\sigma - 2$, and B is an algebra of minimal rank such that $B / \operatorname{rad} B \cong k^r$ for some r. In the following, we call algebras with $B / \operatorname{rad} B \cong k^r$ *superbasic*. The proof in [4] then continues by determining the superbasic algebras of minimal rank. This proof is even valid over arbitrary fields: a superbasic algebra B has minimal rank if and only if there exist $w_1, \ldots, w_m \in \operatorname{rad} B$ with $w_i w_j = 0$ for $i \neq j$ such that

$$\operatorname{rad} B = \mathsf{L}_B + B w_1 B + \cdots + B w_m B = \mathsf{R}_B + B w_1 B + \cdots + B w_m B$$

and $\#k \geq 2N(B) - 2$. Here L_B and R_B denote the left and right annihilator of $\operatorname{rad} B$, that is, the set of all $x \in \operatorname{rad} B$ with $x(\operatorname{rad} B) = \{0\}$ or $(\operatorname{rad} B)x = \{0\}$, respectively. $N(B)$ is the largest natural number s such that $(\operatorname{rad} B)^s \neq \{0\}$.

New results. As our main result, we characterize all algebras of minimal rank over arbitrary fields (Theorem 1). After more than two decades, this answers a major open problem in algebraic complexity theory completely.

It turns out that the structure of the algebras of minimal rank over arbitrary fields is exactly the same as over perfect fields, as stated above in (1). A key tool in the proofs in [4] was a subalgebra B of A with $A / \operatorname{rad} A \cong B$ and $B \oplus \operatorname{rad} A$. Over perfect fields, such an algebra is guaranteed to exist by the Wedderburn–Malcev Theorem [7, Thm. 6.2.1]. (In fact, only the existence of B is needed in [4] and not the perfectness of the underlying field. Thus the results in [4] can be directly transferred to algebras such that $A / \operatorname{rad} A$ is separable.) One is perhaps tempted to conjecture that if such an algebra B does not exists, then the structure of A is "too irregular" and A is not an algebra of minimal rank. This conjecture is however wrong: let $k = GF(2)(t)$ for some indeterminate t and consider $A = k[X]/(X^4 - t^2)$ for some indeterminate X. We have $(X^4 - t^2) = (X^2 - t)^2$. Furthermore, $X^2 - t$ is irreducible, since k does not contain a square root of t. Hence $A \cong k[X]/(p^\ell)$ for some irreducible polynomial p and some $\ell \geq 1$. It is however well known that such algebras have minimal rank (over large enough fields), see [6, Sect. 17.4].

Astonishingly, the solution of the characterization problem over arbitrary fields required more new algebraic insights than novel lower bound techniques. Particularly, we show that the phenomenon of the non-existence of the above mentioned algebra B is in some sense "local". More precisely, we show that there always exists a subalgebra B of A with $B/\operatorname{rad} B \cong A/\operatorname{rad} A$ that has certain properties (Section 2). With this algebra, many of the techniques used in [3] and [4] can be transferred to the case of an arbitrary field. Still, some new lower bounds are needed (Section 3).

Organization of the paper. In Section 2, we derive some structural properties of associative algebras, which will enable us to deal with the difficulty that over non-perfect fields, a subalgebra B with $A/\operatorname{rad} A \cong B$ and $B \oplus \operatorname{rad} A$ need not exist. In Section 3, we prove our characterization results for the algebras of minimal rank over arbitrary fields. In the course of the proof of this main result, we rely on some lower bound techniques due to Alder and Strassen [1]. An overview of these techniques can be found in [3, Sect. 3.1]. For convenience, we will refer to [3, Sect. 3.1] instead of the original work of Alder and Strassen.

2   Structural Insights

In the proof of the characterization results in [4], the existence of a subalgebra B of A with $B \oplus \operatorname{rad} A = A$ and $B \cong A/\operatorname{rad} A$ plays a crucial role. Over perfect fields, such an algebra always exists by the Wedderburn–Malcev Theorem, see [7, Thm. 6.2.1]. Over non-perfect fields, such an algebra need not exist, see [7, Ch. 6, Ex. 4] for an example.

In the present section, we show that this phenomenon is in some sense "local". This enables us to transfer many of the techniques from [4] to the case of non-perfect fields. Instead of the above algebra B, we here use a lifting of the idempotents in $A/\operatorname{rad} A$. Let $\bar{\ } : A \to A/\operatorname{rad} A$ denote the canonical projection defined by $a \mapsto \bar{a} = a + \operatorname{rad} A$. Let $A_1 \oplus \cdots \oplus A_t$ be the decomposition of $A/\operatorname{rad} A$ into simple factors. We here write the decomposition of $A/\operatorname{rad} A$ in an additive way, that is, we consider the A_τ as subspaces of $A/\operatorname{rad} A$. This is done to simplify notations, mainly to write A_τ instead of $\{0\} \times \cdots \times \{0\} \times A_\tau \times \{0\} \times \cdots \times \{0\}$, which would be the corresponding subspace of $A_1 \times \cdots \times A_t$. Let e_τ denote the identity of A_τ for $1 \leq \tau \leq t$. Then $e_1 + \cdots + e_t$ is a *decomposition of the identity* of $A/\operatorname{rad} A$, that is,

1. $1 = e_1 + \cdots + e_t$,
2. the e_τ are idempotents, i.e., $e_\tau^2 = e_\tau$ for all τ, and
3. the e_τ annihilate each other, i.e., $e_\sigma e_\tau = 0$ for $\sigma \neq \tau$.

Now the crucial point is that this decomposition can be lifted to A: by [7, Cor. 3.3.9], there is a corresponding decomposition of the identity $1 = f_1 + \cdots + f_t$ in A such that

$$e_\tau = \bar{f}_\tau = f_\tau + \operatorname{rad} A \qquad \text{for all } \tau. \tag{2}$$

We remark that the decomposition $e_1 + \cdots + e_t$ is a *central* decomposition of the identity in $A/\operatorname{rad} A$, i.e., $e_\tau a = a e_\tau$ holds for all τ and for all $a \in A/\operatorname{rad} A$.

The next lemma shows that $B := f_1 A f_1 + \cdots + f_t A f_t$ is a subalgebra of A such that $B/\operatorname{rad} B \cong A_1 \oplus \cdots \oplus A_t$. Note that the algebra B always exists, particularly, its existence does not depend on the perfectness of the ground field. The important point is that we have found an algebra with $B/\operatorname{rad} B \cong A_1 \oplus \cdots \oplus A_t = A/\operatorname{rad} A$ for which $B \cap f_i A f_j = \{0\}$ holds for all $i \neq j$. (This is what we meant by "local".) In Section 3.2, we use this property to transfer many of the results from [4] for the case of a perfect field directly to the general case.

Lemma 1. *Let A be an algebra over an arbitrary field and $A_1 \oplus \cdots \oplus A_t$ be its decomposition into simple algebras. Let $f_1 + \cdots + f_t$ be a decomposition of the identity as in (2). Then the following holds:*

1. *$B := f_1 A f_1 + f_2 A f_2 + \cdots + f_t A f_t$ is a subalgebra of A,*
2. *$B_\tau := f_\tau A f_\tau$ is a subalgebra of B for all τ,*
3. *$B_\sigma \cdot B_\tau = \{0\}$ for all $\sigma \neq \tau$,*
4. *$\operatorname{rad} B_\tau = f_\tau (\operatorname{rad} A) f_\tau$ for all τ, and*
5. *$B_\tau / \operatorname{rad} B_\tau \cong A_\tau$ for all τ.*

Proof. 1. B is clearly closed under addition and scalar multiplication. It remains to show that it is also closed under multiplication. This follows from the fact that the f_τ annihilate each other: we have

$$(f_1 a_1 f_1 + \cdots + f_t a_t f_t)(f_1 b_1 f_1 + \cdots + f_t b_t f_t)$$
$$= f_1 a_1 f_1 b_1 f_1 + \cdots + f_t a_t f_t b_t f_t \in B, \quad (3)$$

   since $f_\sigma f_\tau = 0$ for $\sigma \neq \tau$.
2. B_τ is obviously closed under addition, scalar multiplication, and multiplication. Furthermore, f_τ is the identity of B_τ.
3. We have $f_\sigma a f_\sigma \cdot f_\tau a f_\tau = 0$ for all $f_\sigma a f_\sigma \in B_\sigma$ and $f_\tau b f_\tau \in B_\tau$, as $f_\sigma \cdot f_\tau = 0$.
4. Since $e_1 + \cdots + e_t$ is a central decomposition, this follows from [7, Thm. 3.5.3].
5. We have $A_\tau \cong e_\tau (A/\operatorname{rad} A) e_\tau$ where $e_1 + \cdots + e_t$ is a decomposition of the identity of $A/\operatorname{rad} A$ as in (2). On the other hand, $B_\tau = f_\tau A f_\tau / f_\tau (\operatorname{rad} A) f_\tau$ by the second and fourth statement. We define a mapping $i : A_\tau \to B_\tau$ by $e_\tau (a + \operatorname{rad} A) e_\tau \mapsto f_\tau a f_\tau + f_\tau (\operatorname{rad} A) f_\tau$. We claim that i is an isomorphism. First, i is well defined: consider elements a and b fulfilling $e_\tau (a + \operatorname{rad} A) e_\tau = e_\tau (b + \operatorname{rad} A) e_\tau$. Then by (2),

$$f_\tau a f_\tau + \operatorname{rad} A = (f_\tau + \operatorname{rad} A)(a + \operatorname{rad} A)(f_\tau + \operatorname{rad} A)$$
$$= (f_\tau + \operatorname{rad} A)(b + \operatorname{rad} A)(f_\tau + \operatorname{rad} A)$$
$$= f_\tau b f_\tau + \operatorname{rad} A. \tag{4}$$

Since $(f_j A f_\ell) \cap (f_{j'} A f_{\ell'}) = \{0\}$ for $j \neq j'$ or $\ell \neq \ell'$ (see for instance [7, Eq. 1.7.1]), (4) implies

$$i(a) = f_\tau a f_\tau + f_\tau (\operatorname{rad} A) f_\tau = f_\tau b f_\tau + f_\tau (\operatorname{rad} A) f_\tau = i(b).$$

Thus i is well defined.

Obviously, $i(\alpha x) = \alpha i(x)$ and $i(x + y) = i(x) + i(y)$ for all $x, y \in A_\tau$. For $x = e_\tau (a + \operatorname{rad} A) e_\tau$ and $y = e_\tau (b + \operatorname{rad} A) e_\tau$ in B_τ, we have

$$xy = e_\tau (a + \operatorname{rad} A) e_\tau \cdot e_\tau (b + \operatorname{rad} A) e_\tau = e_\tau (a f_\tau b + \operatorname{rad} A) e_\tau. \qquad (5)$$

Hence,

$$\begin{aligned} i(xy) &= f_\tau a f_\tau b f_\tau + f_\tau (\operatorname{rad} A) f_\tau \\ &= (f_\tau a f_\tau + f_\tau (\operatorname{rad} A) f_\tau)(f_\tau b f_\tau + f_\tau (\operatorname{rad} A) f_\tau) \\ &= i(x) i(y). \end{aligned}$$

Finally, since $f_\tau a f_\tau \in f_\tau (\operatorname{rad} A) f_\tau$ if and only if $e_\tau (a + \operatorname{rad} A) e_\tau \subseteq \operatorname{rad} A$, i is also bijective.   □

3   Main Result

Throughout this section, k denotes an infinite field. This is no restriction, since finite fields are perfect and for perfect fields, the characterization results are proven in [4]. A denotes a k-algebra of minimal rank and $A_1 \oplus \cdots \oplus A_t$ denotes the decomposition of $A/\operatorname{rad} A$ into simple algebras. Again we write the decomposition in an additive way.

By [3, Cor. 3.4 & Lem. 3.6(i)], $A/\operatorname{rad} A$ is of minimal rank. By [3, Thm. 2], either $A_\tau \cong k$ or $A_\tau \cong k^{2 \times 2}$ or A_τ is a proper extension field of k for all τ.

Let e_τ be the identity of A_τ and let $f_1 + \cdots + f_t$ be a decomposition of the identity of A as in (2). Consider the *Peirce decomposition* (see [7, p. 26]) of A with respect to $f_1, \ldots, f_t$:

$$A = \bigoplus_{1 \le \sigma, \tau \le t} f_\sigma A f_\tau \qquad \text{and} \qquad \operatorname{rad} A = \bigoplus_{1 \le \sigma, \tau \le t} f_\sigma (\operatorname{rad} A) f_\tau.$$

Since $f_1, \ldots, f_t$ are idempotents that annihilate each other, the pairwise intersections of the $f_\sigma A f_\tau$ equal in fact $\{0\}$.

3.1   A First Decomposition Step

Assume that say A_1 is either an extension field of dimension at least two or isomorphic to $k^{2 \times 2}$. Moreover, assume that $f_1(\operatorname{rad} A) f_j = f_j(\operatorname{rad} A) f_1 = \{0\}$ for all $j \ge 2$. Then we may decompose A according to the subsequent lemma.

Lemma 2. *With the above assumptions,*

$$A \cong f_1 A f_1 \times (f_2 + \cdots + f_t) A (f_2 + \cdots + f_t).$$

Proof. Let $f' := f_2 + \cdots + f_t$. First note that $f_1 A f_1$ is an algebra by Lemma 1(2). Along the same lines, it is easy to see that $f'Af'$ is an algebra.

Since $e_\tau = f_\tau + \mathrm{rad}\, A$ is the identity of A_τ for all τ, e_τ is contained in the center of $A/\mathrm{rad}\, A$ (i.e., $e_\tau a = a e_\tau$ for all $a \in A/\mathrm{rad}\, A$). Thus by [7, Thm. 3.5.3], we have $f_\sigma A f_\tau = f_\sigma (\mathrm{rad}\, A) f_\tau$ for all $\sigma \neq \tau$. In particular, $f_1 A f_j = f_j A f_1 = \{0\}$ for all $j \geq 2$.

Let $a \in A$ be arbitrary. We can write $a = a_1 + a'$ with unique $a_1 \in f_1 A f_1$ and $a' \in f'Af'$, since $f_1 A f_1 \oplus f'Af' = A$. We define a mapping $A \to f_1 A f_1 \times f'Af'$ by $a \mapsto (a_1, a')$. Exploiting the facts that $f_1 A f' = f'A f_1 = \{0\}$ as well as $f_1 f' = f' f_1 = 0$, it is easy to verify that this is an isomorphism of algebras. $\square$

Next we show that if A has minimal rank, both $f_1 A f_1$ and $f'Af'$ have minimal rank. This follows directly from the work of Alder and Strassen.

Lemma 3. *Let B_1 and B_2 be two algebras. The algebra $B = B_1 \times B_2$ has minimal rank if and only if both B_1 and B_2 have minimal rank.*

Proof. Assume that B has minimal rank. Note that $\mathrm{rad}\, B = \mathrm{rad}\, B_1 \times \mathrm{rad}\, B_2$. By [3, Cor. 3.4 & Lem. 3.6(i)], $R(B) \geq 2 \dim \mathrm{rad}\, B_1 + R((B_1/\mathrm{rad}\, B_1) \times B_2)$. If we now apply [3, Lem. 3.6(ii)], we obtain $R(B) \geq 2 \dim B_1 - 1 + R(B_2)$. Thus B_2 has to have minimal rank. B_1 is treated completely alike.

The other direction is trivial. $\square$

By Lemma 3, the algebra $f_1 A f_1$ has to have minimal rank. By Lemma 1(5), we obtain $(f_1 A f_1)/\mathrm{rad}(f_1 A f_1) = e_1(A/\mathrm{rad}\, A)e_1 \cong A_1$. By our assumption at the beginning of this subsection, A_1 either is an extension field of dimension at least two or isomorphic to $k^{2 \times 2}$. In the first case, $f_1 A f_1$ is a local algebra with $\dim f_1 A f_1/(\mathrm{rad}\, f_1 A f_1) \geq 2$. Such algebras have been characterized by Büchi and Clausen [5]. They are isomorphic to $k[X]/(p(X)^m)$ for some irreducible polynomial p with $\deg p \geq 2$ and some $m \geq 1$. The second case is settled by the following lemma. It turns out that in this case, we necessarily have $f_1 A f_1 \cong k^{2 \times 2}$.

Lemma 4. *Let C be an algebra fulfilling $C/\mathrm{rad}\, C \cong k^{2 \times 2}$. Then $R(C) \geq \frac{5}{2} \dim C - 4$. In particular, if $\mathrm{rad}\, C \neq \{0\}$, then C is not of minimal rank.*

Proof. Let $\beta = (f_1, g_1, w_1, \ldots, f_r, g_r, w_r)$ be a bilinear computation for C. We may assume w.l.o.g. that $f_1, \ldots, f_N$ is a basis of C^*, where $N = \dim C$. Let $u_1, \ldots, u_N$ be its dual basis. Furthermore, assume that $\bar{u}_{N-3}, \ldots, \bar{u}_N$ is a basis of $C/\mathrm{rad}\, C \cong k^{2 \times 2}$, where $\bar{} : C \to C/\mathrm{rad}\, C$ denotes the canonical projection. Choose $\alpha_1, \ldots, \alpha_4$ such that $1 = \alpha_1 \bar{u}_{N-3} + \cdots + \alpha_4 \bar{u}_N$ (in $C/\mathrm{rad}\, C$). By Nakayama's Lemma [7, Lem. 3.1.4], $\alpha_1 u_{N-3} + \cdots + \alpha_4 u_N$ is invertible in C. Exploiting a technique called sandwiching (see e.g. [3, Sect. 3.3]), we can achieve that $1 = \alpha_1 u_{N-3} + \cdots + \alpha_4 u_N$ in C. Choose $\xi_1, \ldots, \xi_4$ and $\eta_1, \ldots, \eta_4$ such that for $x = \xi_1 u_{N-3} + \cdots + \xi_4 u_N$ and $y = \eta_1 u_{N-3} + \cdots + \eta_4 u_N$,

$$\bar{x} = \begin{pmatrix} 1 & 0 \\ 0 & 0 \end{pmatrix} \quad \text{and} \quad \bar{y} = \begin{pmatrix} 0 & 1 \\ 1 & 0 \end{pmatrix}.$$

An easy calculation shows that $\bar{x}\bar{y} - \bar{y}\bar{x}$ is invertible in $C/\operatorname{rad}C$. Once more by Nakayama's Lemma, $xy - yx$ is invertible in C. Now by [3, Lem. 3.8] (with $m = N - 4$),

$$r \geq \tfrac{5}{2}\dim C - 4. \tag{6}$$

Since $C/\operatorname{rad}C \cong k^{2\times2}$, the algebra $C/\operatorname{rad}C$ is separable, that is, for any extension field K of k, $(C/\operatorname{rad}C)\otimes K \cong K^{2\times2}$ is semisimple. By the Wedderburn–Malcev Theorem [7, Thm. 6.2.1], C contains a subalgebra $C' \cong k^{2\times2}$ with $C' \oplus \operatorname{rad}C = C$. Hence $\operatorname{rad}C$ is a $k^{2\times2}$-bimodule. By [3, Fact 8.5], $\operatorname{rad}C \neq \{0\}$ implies $\dim\operatorname{rad}C \geq 4$. Now we can rewrite (6) as

$$r \geq \tfrac{5}{2}\dim C - 4 = 2\dim C + \underbrace{\tfrac{1}{2}(4 + \dim\operatorname{rad}C) - 4}_{\geq 0 \text{ if } \operatorname{rad}C \neq \{0\}}.$$

This proves the second statement. $\qquad\square$

By [7, Thm. 3.5.3], $\operatorname{rad}(f'Af') = f'(\operatorname{rad}A)f'$ holds. From this, it follows that $(f'Af')/\operatorname{rad}(f'Af') = e'(A/\operatorname{rad}A)e' = A_2 \oplus \cdots \oplus A_t$ as in the proof of Lemma 1(5), where $e' = e_2 + \cdots + e_t$. Hence we can proceed recursively with $f'Af'$. The following lemma summarizes the above considerations.

Lemma 5. *Let A be an algebra of minimal rank over an arbitrary field k. Then*

$$A \cong C_1 \times \cdots \times C_s \times \underbrace{k^{2\times2} \times \cdots \times k^{2\times2}}_{u \text{ times}} \times B,$$

where

1. *the C_σ are local algebras of minimal rank with $\dim C_\sigma/\operatorname{rad}C_\sigma \geq 2$ and*
2. *B is an algebra of minimal rank.*

Furthermore, if $B_1 \oplus \cdots \oplus B_r$ is a decomposition of $B/\operatorname{rad}B$ into simple factors, e_ρ denotes the identity of B_ρ for $1 \leq \rho \leq r$, and $f_1 + \cdots + f_r$ is a decomposition of the identity of B as in (2), then for all B_ρ such that B_ρ is either isomorphic $k^{2\times2}$ or a proper extension field of k, there is a $j_\rho \neq \rho$ such that $f_\rho(\operatorname{rad}A)f_{j_\rho} \neq \{0\}$ or $f_{j_\rho}(\operatorname{rad}A)f_\rho \neq \{0\}$. Above, s and u may be zero and the factor B is optional. $\qquad\square$

3.2 B Is Superbasic

The aim of the present section is to show that the algebra B in Lemma 5 is always superbasic, that is, all the B_ρ in the decomposition of $B/\operatorname{rad}B$ into simple algebras are isomorphic to k. To do so, we construct under the assumption that one of the factors B_ρ is not isomorphic to k, some algebras that have to have minimal rank because B has minimal rank. Then we prove that none of the constructed algebras has minimal rank, obtaining a contradiction.

The next lemma shows that it suffices to consider the problem modulo $(\operatorname{rad}B)^2$. The lemma holds for arbitrary algebras C: let $\tilde{\cdot} : C \to C/(\operatorname{rad}C)^2$ denote the canonical projection defined by $a \mapsto \tilde{a} = a + (\operatorname{rad}C)^2$.

Lemma 6. *Let C be an algebra and $C_1 \oplus \cdots \oplus C_r$ be the decomposition of $C/\operatorname{rad} C$ into simple factors. Let e_ρ be the unit element of C_ρ for all ρ and let $f_1 + \cdots + f_r$ be a decomposition of the identity as in (2). For all indices i and j with $i \neq j$ such that $f_i(\operatorname{rad} C)f_j \neq \{0\}$, there is an index j' with $i \neq j'$ such that $\tilde{f}_i(\operatorname{rad} \tilde{C})\tilde{f}_{j'} \neq \{0\}$. In the same way, there is an index i' with $i' \neq j$ such that $\tilde{f}_{i'}(\operatorname{rad} \tilde{C})\tilde{f}_j \neq \{0\}$.*

Proof. Let x such that $f_i x f_j \neq 0$ and $f_i x f_j \in f_i(\operatorname{rad} C)f_j$. Since $f_i C f_j = f_i(\operatorname{rad} C)f_j$ by [7, Thm. 3.5.3], we may assume $x \in \operatorname{rad} C$. Let $n \geq 1$ be the unique number such that $x \in (\operatorname{rad} C)^n \setminus (\operatorname{rad} C)^{n+1}$. By definition, there are $x_1, \ldots, x_n \in \operatorname{rad} C \setminus (\operatorname{rad} C)^2$ such that $x = x_1 \cdots x_n$. Since $1 = f_1 + \cdots + f_r$,

$$f_i x_1 (f_1 + \cdots + f_r) x_2 (f_1 + \cdots + f_r) \cdots (f_1 + \cdots + f_r) x_n f_j = f_i x f_j \neq 0.$$

Thus there are indices $h_1, \ldots, h_{n-1}$ such that

$$f_i x_1 f_{h_1} x_2 f_{h_2} \cdots f_{h_{n-1}} x_n f_j \neq 0.$$

If all h_ν equal i, then $f_{h_{n-1}} x_n f_j = f_i x_n f_j \neq 0$ and we set $j' = j$. Since $f_i x_n f_{j'} \in \operatorname{rad} C \setminus (\operatorname{rad} C)^2$, $\tilde{f}_i \tilde{x}_n \tilde{f}_{j'} \neq 0$, hence we are done. Otherwise let ν_0 be the smallest index such that $h_{\nu_0} \neq i$. Now we set $j' = h_{\nu_0}$ and can conclude in the same way as before that $\tilde{f}_i(\operatorname{rad} \tilde{C})\tilde{f}_{j'} \neq \{0\}$.

The index i' is constructed completely alike. $\qquad\square$

Assume that w.l.o.g. B_1 is either isomorphic to $k^{2\times 2}$ or a proper extension field of k. We first construct two algebras which have to have minimal rank, since B has minimal rank. Next we prove that these algebras cannot have minimal rank if B_1 is isomorphic to $k^{2\times 2}$ or a proper extension field of k. Hence $B_1 \cong k$ must hold. By symmetry, this has to hold for all $B_1, \ldots, B_r$. It follows that B is superbasic.

We decompose $\operatorname{rad} B$ as

$$\operatorname{rad} B = \bigoplus_{1 \leq \rho, \eta \leq r} f_\rho(\operatorname{rad} B)f_\eta. \tag{7}$$

By Lemma 5, there is an index $j \geq 2$ such that $f_1(\operatorname{rad} B)f_j \neq \{0\}$. The case where $f_j(\operatorname{rad} B)f_1 \neq \{0\}$ is treated symmetrically. By Lemma 6, there is an $j' \geq 2$ such that $\tilde{f}_1(\operatorname{rad} \tilde{B})\tilde{f}_{j'} \neq \{0\}$. W.l.o.g. $j' = 2$.

Consider $C := f_1 B f_1 + f_2 B f_2$. As in Lemma 1(1), it is easy to see that C is a subalgebra of B. As in Lemma 1(4), $\operatorname{rad} C = f_1(\operatorname{rad} B)f_1 + f_2(\operatorname{rad} B)f_2$. Finally $C/\operatorname{rad} C \cong B_1 \oplus B_2$. Since the pairwise intersection of the $f_\rho(\operatorname{rad} B)f_\eta$ in (7) are the nullspace, we also have $C/\operatorname{rad} B \cong B_1 \oplus B_2$.

Let

$$I = \bigoplus_{(\rho,\eta) \neq (1,2)} f_\rho(\operatorname{rad} B)f_\eta.$$

From $(\operatorname{rad} B)^2 \subseteq I$, it follows that I is a twosided ideal of B (see also [3, p. 105]). Since B is of minimal rank and $I \subseteq \operatorname{rad} B$, B/I is also of minimal rank by [3,

Cor. 3.4 & Lem. 3.6(i)]. By [7, Cor. 3.1.14], we have $(\operatorname{rad} B/I) = (\operatorname{rad} B)/I \cong f_1(\operatorname{rad} B)f_2/I \neq \{0\}$. Let $\hat{\cdot} : B \to B/I$ denote the canonical projection defined by $a \mapsto \hat{a} = a + I$. Since $I \subseteq \operatorname{rad} B$, $\hat{f}_1, \ldots, \hat{f}_r$ are idempotent elements that annihilate each other, because this also holds for $f_1, \ldots, f_r$. Let $\hat{f} = \hat{f}_3 + \cdots + \hat{f}_r$. Then for all $a \in \hat{f}(B/I)\hat{f}$, we have $a(\operatorname{rad} B/I) = (\operatorname{rad} B/I)a = \{0\}$, since the $\hat{f}_\rho$ annihilate each other. Thus by [3, Lem. 8.3]

$$B/I \cong (\hat{f}_1 + \hat{f}_2)(B/I)(\hat{f}_1 + \hat{f}_2) \times \hat{f}(B/I)\hat{f}.$$

By Lemma 3, B/I is of minimal rank only if $D := (\hat{f}_1 + \hat{f}_2)(B/I)(\hat{f}_1 + \hat{f}_2)$ is of minimal rank. The next two paragraphs show that this cannot be the case. Hence we have obtained the following lemma.

Lemma 7. *The algebra B in Lemma 5 is superbasic.* $\square$

Before we prove the two statements mentioned above, we first collect some structural properties of D. Because $C := f_1 B f_1 + f_2 B f_2$ is a subalgebra of $(f_1 + f_2)B(f_1 + f_2)$, $\hat{C}$ is a subalgebra of D. This subalgebra $\hat{C}$ is isomorphic to $B_1 \oplus B_2$, since $\operatorname{rad} C = f_1(\operatorname{rad} B)f_1 + f_2(\operatorname{rad} B)f_2$. Hence D has a subalgebra that is isomorphic to $B_1 \oplus B_2$. The radical of D is $\operatorname{rad} D \cong f_1(\operatorname{rad} B)f_2/I \neq \{0\}$.

Case 1: B_1 is isomorphic to $k^{2\times 2}$. The radical $\operatorname{rad} D$ is a nonzero B_1-left module and a nonzero B_2-right module. Moreover, D is isomorphic to $B_1 \times B_2 \times \operatorname{rad} D$ equipped with the multiplication law $(a, b, c)(a', b', c') = (aa', bb', ac' + cb')$. Now [3, Lem. 8.7] shows that C is not of minimal rank if $B_2 \cong k$. If $B_2 \cong k^{2\times 2}$, then this follows from [3, Lem. 8.8]. Finally, the case where B_2 is a proper extension field of k is settled by [4, Lem. 20].

Case 2: B_1 is a proper extension field. First note that we may assume that B_2 is also an extension field of k. If B_2 was isomorphic to $k^{2\times 2}$, then we could exchange the roles of B_1 and B_2. But this case has already been treated above. Now [4, Lem. 19] shows that C cannot have minimal rank. This completes the proof of Lemma 7.

3.3 The Final Proof

By Lemma 5, A is isomorphic to

$$A \cong C_1 \times \cdots \times C_s \times \underbrace{k^{2\times 2} \times \cdots \times k^{2\times 2}}_{u \text{ times}} \times B,$$

where the C_σ are local algebras of minimal rank with $\dim C_\sigma / \operatorname{rad} C_\sigma \geq 2$. By Lemma 7, B is a superbasic algebra of minimal rank. By exploiting the characterization results in [4] for superbasic algebras of minimal rank (which were proven there for arbitrary fields), we obtain the desired characterization of algebras of minimal rank over arbitrary fields.

414   M. Bläser

Theorem 1. *An algebra A over an arbitrary field k is an algebra of minimal rank if and only if*

$$A \cong C_1 \times \cdots \times C_s \times \underbrace{k^{2\times2} \times \cdots \times k^{2\times2}}_{u \ times} \times B \tag{8}$$

where $C_1, \ldots, C_s$ are local algebras of minimal rank with $\dim(C_\sigma/\operatorname{rad} C_\sigma) \geq 2$, i.e., $C_\sigma \cong k[X]/(p_\sigma(X)^{d_\sigma})$ for some irreducible polynomial p_σ with $\deg p_\sigma \geq 2$, $d_\sigma \geq 1$, and $\#k \geq 2 \dim C_\sigma - 2$, and B is a superbasic algebra of minimal rank, that is, there exist $w_1, \ldots, w_m \in \operatorname{rad} B$ with $w_i w_j = 0$ for $i \neq j$ such that

$$\operatorname{rad} B = \mathsf{L}_B + Bw_1B + \cdots + Bw_mB = \mathsf{R}_B + Bw_1B + \cdots + Bw_mB$$

and $\#k \geq 2N(B) - 2$. Any of the integers s, u, or m may be zero and the factor B in (8) is optional.

The results for finite fields follow also from [4], because finite fields are perfect.

References

1. A. Alder and V. Strassen. On the algorithmic complexity of associative algebras. *Theoret. Comput. Sci.*, 15:201–211, 1981.
2. Markus Bläser. Lower bounds for the multiplicative complexity of matrix multiplication. *Comput. Complexity*, 8:203–226, 1999.
3. Markus Bläser. Lower bounds for the bilinear complexity of associative algebras. *Comput. Complexity*, 9:73–112, 2000.
4. Markus Bläser. Algebras of minimal rank over perfect fields. In *Proc. 17th Ann. IEEE Computational Complexity Conf. (CCC)*, pages 113–122, 2002.
5. Werner Büchi and Michael Clausen. On a class of primary algebras of minimal rank. *Lin. Alg. Appl.*, 69:249–268, 1985.
6. Peter Bürgisser, Michael Clausen, and M. Amin Shokrollahi. *Algebraic Complexity Theory*. Springer, 1997.
7. Yurij A. Drozd and Vladimir V. Kirichenko. *Finite Dimensional Algebras*. Springer, 1994.
8. Hans F. de Groote. Characterization of division algebras of minimal rank and the structure of their algorithm varieties. *SIAM J. Comput.*, 12:101–117, 1983.
9. Hans F. de Groote and Joos Heintz. Commutative algebras of minimal rank. *Lin. Alg. Appl.*, 55:37–68, 1983.
10. Joos Heintz and Jacques Morgenstern. On associative algebras of minimal rank. In *Proc. 2nd Applied Algebra and Error Correcting Codes Conf. (AAECC)*, Lecture Notes in Comput. Sci. 228, pages 1–24. Springer, 1986.
11. Volker Strassen. Gaussian elimination is not optimal. *Num. Math.*, 13:354–356, 1969.
12. Volker Strassen. Algebraic complexity theory. In J. van Leeuwen, editor, *Handbook of Theoretical Computer Science Vol. A*, pages 634–672. Elsevier Science Publishers B.V., 1990.
13. Shmuel Winograd. On multiplication in algebraic extension fields. *Theoret. Comput. Sci.*, 8:359–377, 1979.

Evolutionary Algorithms and the Maximum Matching Problem

Oliver Giel* and Ingo Wegener*

FB Informatik, LS 2, Univ. Dortmund, 44221 Dortmund, Germany
{giel, wegener}@ls2.cs.uni-dortmund.de

Abstract. Randomized search heuristics like evolutionary algorithms are mostly applied to problems whose structure is not completely known but also to combinatorial optimization problems. Practitioners report surprising successes but almost no results with theoretically well-founded analyses exist. Such an analysis is started in this paper for a fundamental evolutionary algorithm and the well-known maximum matching problem. It is proven that the evolutionary algorithm is a polynomial-time randomized approximation scheme (PRAS) for this optimization problem, although the algorithm does not employ the idea of augmenting paths. Moreover, for very simple graphs it is proved that the expected optimization time of the algorithm is polynomially bounded and bipartite graphs are constructed where this time grows exponentially.

1 Introduction

The design and analysis of problem-specific algorithms for combinatorial optimization problems is a well-studied subject. It is accepted that randomization is a powerful concept for theoretically and practically efficient problem-specific algorithms. Randomized search heuristics like random local search, tabu search, simulated annealing, and variants of evolutionary algorithms can be combined with problem-specific modules. The subject of this paper are general and not problem-specific search heuristics. Practitioners report surprisingly good results which they have obtained with such search heuristics. Nevertheless, one cannot doubt that problem-specific algorithms outperform general search heuristics – if they exist. So the area of applications of general search heuristics is limited to situations where good problem-specific algorithms are not known. This may happen if one quickly needs an algorithm for some subproblem in a large project and there are not enough resources (time, money, or experts) available to develop an efficient problem-specific algorithm. In many real-life applications, especially in engineering disciplines, there is no possibility to design a problem-specific algorithm. E. g., we may have the rough draft of a machine but we still have to choose between certain alternatives to obtain an explicit description of the machine. If we have m binary decisions to take, the search space (the space of all

---

* Supported in part by the Deutsche Forschungsgemeinschaft (DFG) as part of the Collaborative Research Center "Computational Intelligence" (SFB 531).

H. Alt and M. Habib (Eds.): STACS 2003, LNCS 2607, pp. 415–426, 2003.

possible solutions) equals $\{0,1\}^m$. Then there exists a function $f\colon \{0,1\}^m \to \mathbb{R}$ such that $f(a)$ measures the quality of the machine if the vector of alternatives $a = (a_1, \ldots, a_m)$ is chosen. However, often no closed form of f is known and we obtain $f(a)$ only by an experiment (or its simulation).

We conclude that general randomized search heuristics have applications and that their analysis is necessary to understand, improve, and teach them.

It is not possible to analyze algorithms on "unknown" functions f. However, one can improve the knowledge on a search heuristic by

– analyzing its behavior on some classes of functions,
– analyzing its behavior on some well-known combinatorial problems,
– constructing example functions showing special properties of the heuristic.

Such results have been obtained recently for evolutionary algorithms. Evolutionary algorithms have been analyzed on unimodal functions [2], linear functions [3], quadratic polynomials [16], and monotone polynomials [15]. Among other properties the effect of crossover has been studied [9,10]. A first step to study evolutionary algorithms on combinatorial problems has been made by Scharnow, Tinnefeld and Wegener [13] who studied sorting as minimization of unsortedness of a sequence and the shortest path problem. These problems allow improvements by local steps. Here, we investigate one of the best-known combinatorial optimization problems in P, namely the maximum matching problem.

We work with the following model of the problem. For graphs with n vertices and m edges, we have to decide for each edge whether we choose it. The search space is $\{0,1\}^m$ and a search point $a = (a_1, \ldots, a_m)$ describes the choice of all edges e_i where $a_i = 1$. The function f to be optimized has the value $a_1 + \cdots + a_m$ (the number of edges) for all a describing matchings, i.e., edge sets where no two edges share a vertex. For all non-matchings a, the so-called fitness value $f(a)$ is $-c$ where the collision number c is the number of edge pairs e_i and e_j that are chosen by a and share a vertex. This definition is crucial. If we chose $f(a) = 0$ for all non-matchings a then our algorithm (and many other randomized search heuristics) would not find any matching in polynomial time, e.g., for the complete graph.

The maximum matching problem has the following nice properties:

– There is a well-known optimization strategy by Hopcroft and Karp [8] which is based on non-local changes along augmenting paths,
– there are graphs where the Hamming distance between a second-best search point a and the only optimal search point is as large as possible (see Sect. 3), namely m,
– for each non-maximum matching a, there is a sequence $a^0 = a, a^1, \ldots, a^\ell$ such that $f(a^0) = f(a^1) = \cdots = f(a^{\ell-1}) < f(a^\ell)$, Hamming distances $H(a^i, a^{i+1}) \leq 2$, and $\ell \leq \lceil n/2 \rceil$,
– and Sasaki and Hajek [12] have investigated simulated annealing on it.

Simulated annealing only explores Hamming neighbors and, therefore, has to accept worse matchings from time to time. Evolutionary algorithms frequently consider new search points with larger Hamming distance to their current search point. We investigate a simple mutation-based evolutionary algorithm (EA) with population size one. Our conjecture is that larger populations and crossover do not help. The basic EA consists of an initialization step and an infinite loop. Special mutation operators will be introduced in the next paragraph.

Initialization: Choose $a \in \{0,1\}^m$ according to the uniform distribution.

Loop: Create a' from a by mutation and replace a by a' iff $f(a') \geq f(a)$.

In applications, we need a stopping criterion but typically we never know whether a is optimal. Hence, we are interested in X, the minimum t such that we obtain an optimal a in step t. This random variable X is called the optimization time of the algorithm. The standard mutation operator decides for each bit a_i of a independently whether it should be flipped (replaced by $1 - a_i$). The flipping probability equals $1/m$ implying that the expected number of flipping bits equals one. This algorithm is called (1+1) EA. We can compute $f(a)$ for the first a in time $O(m)$ and all successive a' in expected time $O(1)$ each (see [5]). Hence, $\mathrm{E}(X)$ is an approximative measure of the runtime. Since we have seen that steps with at most two flipping bits suffice to find an improvement, we also investigate the *local* (1+1) EA; in each step with probability $1/2$ a randomly chosen bit a_i flips and with probability $1/2$ a randomly chosen pair a_i and a_j flips. Sometimes it is easier to understand some ideas when discussing the local (1+1) EA. However, only the (1+1) EA is a general randomized search heuristic optimizing eventually each function $f : \{0,1\}^m \to \mathbb{R}$. In particular, the (1+1) EA (and also its local variant) does not employ the idea of augmenting paths and it is interesting to investigate whether it nevertheless randomly finds augmenting paths. Such a result would be a hint that evolutionary algorithms may implicitly use an optimization technique without knowing it. Again we stress that our aim is the investigation of evolutionary algorithms and we definitely do not hope to improve the best known maximum matching algorithms [1,11,14]. Here, we mention that our model of the matching problem allows a polynomial-time algorithm even if the graph is not given explicitly and the algorithm only sees f-values (see [5]).

In Sect. 2, we show that the considered EAs always find matchings easily. It is proved that the EAs are polynomial-time randomized approximation schemes (PRAS) for optimization problems. This is a fundamental result, since approximation is the true aim of heuristics. In Sect. 3, we describe how the EAs work efficiently on paths and, in Sect. 4, we describe graphs where the EAs have an exponential expected optimization time. In most cases, only the main ideas of the proofs are described and complete proofs can be found in [5].

2   Evolutionary Algorithms Are PRAS

For many graphs it is very likely that the initial search point is a non-matching. However, the (local) (1+1) EA finds matchings quickly. Only steps decreasing the collision number are accepted. It can be shown that there are always at least $\sqrt{c}$ 1-bit mutations which decrease c. This leads (see [5]) to the following result.

Lemma 1. *The (local) (1+1) EA discovers a matching in expected time $O(m^2)$.*

Now we are prepared to prove that the (local) (1+1) EA efficiently finds at least almost optimal matchings.

Theorem 1. *For $\varepsilon > 0$, the (local) (1+1) EA finds a $(1 + \varepsilon)$-approximation of a maximum matching in expected time $O(m^{2\lceil 1/\varepsilon \rceil})$.*

Proof. By Lemma 1, we can assume that the EA has found a matching M. If M is not optimal, there exists an augmenting path $e_{i(1)}, \ldots, e_{i(\ell)}$, where ℓ is odd, $e_{i(j)} \notin M$ for j odd, $e_{i(j)} \in M$ for j even, and no edge in M meets the first or last vertex of the path. The (1+1) EA improves M by flipping exactly the edges of the augmenting path. This happens with probability $\Omega(m^{-\ell})$. The local (1+1) EA improves M by $\lfloor \ell/2 \rfloor$ 2-bit mutations shortening the augmenting path from left or right and a final 1-bit mutation changing the free edge of the resulting augmenting path of length one into a matching edge. The probability that this happens within the next $\lfloor \ell/2 \rfloor + 1$ steps is bounded below by $\Omega((m^{-2})^{\lfloor \ell/2 \rfloor} \cdot m^{-1}) = \Omega(m^{-\ell})$. If we can ensure that there always exists an augmenting path whose length is at most $\ell = 2\lceil \varepsilon^{-1} \rceil - 1$, the expected time to improve the matching is bounded by $O(m^\ell)$ for the (1+1) EA and $O(\ell \cdot m^\ell)$ for the local (1+1) EA. For ε a constant, $O(\ell \cdot m^\ell) = O(m^\ell)$. In fact, the bound $O(m^\ell)$ for the local (1+1) EA holds for arbitrary $\varepsilon > 0$ (see [5]). Hence, for both EAs, the expected overall time is $O(m^2) + O(m) \cdot O(m^\ell) = O(m^{2\lceil \varepsilon^{-1} \rceil})$.

We can apply the known theory on the maximum matching problem to prove that bad matchings imply short augmenting paths. Let M^* be an arbitrary but fixed maximum matching. We assume $|M^*| > (1+\varepsilon)|M|$, i.e., the (1+1) EA has not yet produced a $(1 + \varepsilon)$-approximation. Furthermore, let $|M| \geq 1$; otherwise there exists a path of length $1 \leq \ell$. Consider the graph $G' = (V, E')$ with edge set $E' = M \oplus M^*$, where $\oplus$ denotes the symmetric difference. G' consists of paths and cycles, forming the components of G'. All cycles and all paths of even length consist of the same number of M-edges as M^*-edges, whereas paths of odd length have a surplus of one M^*-edge or one M-edge. That means, all paths of odd length starting with an M^*-edge also end with an M^*-edge and are augmenting paths relative to M. Let $k := |M^*| - |M|$. Then $|M^*| > (1 + \varepsilon)|M|$ implies $k/|M| > \varepsilon$. There exist at least k disjoint paths of the last kind and at least one of them has no more than $\lfloor |M|/k \rfloor \leq \lfloor \varepsilon^{-1} \rfloor$ M-edges. In fact, if ε^{-1} is an integer, then $|M|/k < \varepsilon^{-1}$ implies $\lfloor |M|/k \rfloor < \lfloor \varepsilon^{-1} \rfloor$. Thus the path has at most $\lceil \varepsilon^{-1} \rceil - 1$ M-edges and its total length is at most $\ell = 2\lceil \varepsilon^{-1} \rceil - 1$. □

The next corollary is an easy application of Markov's inequality.

Corollary 1. *According to Theorem 1, let $p_\varepsilon(m)$ be a polynomial in m and an upper bound on the expected number of fitness evaluations for the (local) EA to find a $(1 + \varepsilon)$-approximation. The (local) (1+1) EA with an efficient implementation of the mutation operator and the fitness function that halts after $4p_\varepsilon(m)$ fitness evaluations is a PRAS for the maximum matching problem, i.e., it finds a $(1 + \varepsilon)$-optimal solution with probability at least $3/4$.*

3   Paths

Here, we prove that the (local) (1+1) EA finds maximum matchings for graphs consisting of a path of m edges in expected polynomial time. Among all graphs, these graphs allow the maximum length m for an augmenting path if m is odd. We prepare our analysis by describing the matchings on a fitness level distinct from the level of all maximum matchings. During the exploration of a fitness level, the number of disjoint augmenting paths is unchanged; otherwise the matching size would change, too. However, individual augmenting paths may vanish and new augmenting paths are created at the same time. Figure 1 depicts such a mutation. Solid lines indicate matching edges, dashed lines indicate free edges; the path's names after the mutation step are chosen arbitrarily. The shortest augmenting paths are edges with two exposed endpoints. We term these edges *selectable*, e.g., A' is a path consisting of a single selectable edge.

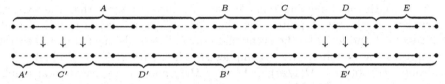

Fig. 1. Mutation step.

With Lemma 1, we can assume that the (local) (1+1) EA has arrived at a matching after expected time $O(m^2)$ and there are at most $\lceil m/2 \rceil$ fitness levels left to climb. At each point of time during the exploration of a fitness level, we focus an augmenting path P but consider only relevant steps. A *relevant* step alters P and produces a new string accepted by the EA. Furthermore, we distinguish two situations. In *Situation 1*, the current matching is not maximal, i.e., there exists some selectable edge e. The current matching can be improved by flipping exactly the right bit in the next step. We choose $P = \{e\}$ and for both EAs the probability that the next step is relevant (event R) is $\Theta(1/m)$. In *Situation 2*, the matching is maximal and, therefore, cannot be improved by a 1-bit flip. Shortest augmenting paths have length at least three. For all choices of P, the probability of a relevant step is $\Theta(1/m^2)$: It is lower bounded by the probability that only a specific pair of edges at one end of P flips and upper bounded by the probability that at least one of at most four edge pairs at both ends of P flip (only for the (1+1) EA there are some more possibilities where at least three edges in line flip). Clearly, both EAs have a not very small probability to leave the current fitness level in Situation 1, whereas for the (1+1) EA it is much harder to leave the level in Situation 2 and for the local (1+1) EA even impossible. The EAs enter a fitness level in either situation and may shift from one situation to the other several times until they finally leave the level. We name such a mutation step *improving*. As we have seen, at any time, the probability of a relevant step is at least $\Omega(1/m^2)$. Hence, the expected number of steps per relevant step is at most $O(m^2)$. If an expected number of T relevant steps is necessary to reach some target then the expected total number of steps is $\sum_{0 \le t < \infty} E(\#\text{steps} \mid T = t) \cdot \text{Prob}(T = t) \le \sum_{0 \le t < \infty} O(m^2) \cdot t \cdot \text{Prob}(T =$

$t) = O(m^2) \cdot E(T)$. We use this property in the following way to show that it takes expected time $O(m^4)$ to find a maximum matching on a path with m edges. The size of the maximum matching equals $\lceil m/2 \rceil$. If the current matching size is $\lceil m/2 \rceil - i$, there exist i disjoint augmenting path; one of length at most $\ell := m/i$. If an expected number of $O(\ell^2)$ relevant steps are sufficient to improve the matching by one edge then $\sum_{1 \leq i \leq \lceil m/2 \rceil} O((m/i)^2) = O(m^2)$ relevant steps are sufficient for the optimum.

As a beginning, we consider the local (1+1) EA and demonstrate that central ideas of our proofs are easy to capture. Our analysis of the (1+1) EA only takes advantage of mutation steps where at most two bits flip, too. All other mutation steps only complicate the analysis and lengthen proofs considerably.

Theorem 2. *For a path of m edges, the expected runtime of the local (1+1) EA is $O(m^4)$.*

Proof. With our foregoing remarks we only have to show that the expected number of relevant steps to leave the current fitness level is $O(\ell^2)$. Consider Situation 1 and let A be the event that only e flips in the next step and thereby improves the matching, i.e., A implies R. Then $\text{Prob}_R(A) := \text{Prob}(A \mid R) = \text{Prob}(A)/\text{Prob}(R) = \Omega(1/m)/O(1/m) = \Omega(1)$ and the expected total number of relevant steps the EA spends in Situation 1 is $O(1)$. Let B be the event that the next step is not improving and leads to Situation 2; again B implies R. We want to bound $\text{Prob}_R(B) := \text{Prob}(B \mid R) = \text{Prob}(B)/\text{Prob}(R)$ from above. Only a mixed 2-bit flip can preserve the matching size. By definition, the selectable edge e has no neighbor in the matching. Hence, one of at most two neighbored pairs next to e has to flip. Thus $\text{Prob}_R(B) = O(1/m^2)/\Omega(1/m) = O(1/m)$. As A and B are disjoint events, the conditional probability to improve the matching when leaving Situation 1 in a relevant step is $\text{Prob}_R(A \mid A \cup B) = \text{Prob}_R(A)/(\text{Prob}_R(A) + \text{Prob}_R(B)) = 1 - \Omega(1/m)$. Thus the expected number of times the EA leaves Situation 1 is at most $1 + O(1/m)$. Consequently, the expected number of times the EA leaves Situation 2 is bounded by $1 + O(1/m) = O(1)$, too. Now it suffices to show that the expected number of relevant steps to leave Situation 2 is $O(\ell^2)$. To this end, the rest of this proof shows for some constants c and $\alpha > 0$, the probability to leave Situation 2 within $c\ell^2$ relevant steps is bounded below by α. Since the proof will be independent of the initial string when the EA enters Situation 2, it implies the $O(\ell^2)$ bound for leaving Situation 2. Having this, the expected number of relevant steps to leave the level is dominated by the product of the expected number of steps to leave Situation 2 and the number of times to leave Situation 2. Since in our analysis both numbers are upper bounded by independent random variables, expectations multiply to $O(\ell^2)$.

In Situation 2, there are no selectable edges. Straightforward considerations show that only pairs of neighbored edges located at one end of an alternating path can flip in an accepted step. Consider a phase of $c\ell^2$ relevant steps, possibly finished prematurely when an augmenting path of length one is created. Within the phase, augmenting paths have minimum length three. We focus attention to an augmenting path P whose initial length is at most ℓ and estimate the probability that P shrinks to length one and finishes the phase. If another augmenting path accomplishes this before P does, so much the better. We call a relevant step a *success* if it shortens P (by two edges). Since P can always shrink by two edges at both ends but sometimes cannot grow at both ends, the probability of a success is lower bounded by $1/2$ in each relevant step. As the initial length of P is at most ℓ, $\ell/4$ more successes than the expected value guarantee that P shrinks to length one in that time. We want to estimate the probability of at least $(1/2)c\ell^2 + (1/4)\ell$ successes within $c\ell^2$ relevant steps. With c chosen sufficiently large, $N := c\ell^2$ and b a constant, the probability of exactly k successful steps is

$$\binom{N}{k}\left(\tfrac{1}{2}\right)^k\left(1-\tfrac{1}{2}\right)^{N-k} \le \binom{N}{N/2}2^{-N} \le \frac{\sqrt{3\pi}e^{-N}N^{N+1/2}2^{-N}}{\left(\sqrt{2\pi}e^{-N/2}(N/2)^{N/2+1/2}\right)^2} = bN^{-1/2} \le \tfrac{1}{2\ell}.$$

The probability of less than $(1/2)c\ell^2 + (1/4)\ell$ successes is bounded above by

$$\text{Prob(less than } (1/2)c\ell^2 \text{ successes)} + \sum_{k=(1/2)c\ell^2}^{(1/2)c\ell^2+(1/4)\ell-1} \tfrac{1}{2\ell} \le \tfrac{1}{2} + \tfrac{\ell}{4}\cdot\tfrac{1}{2\ell} = \tfrac{5}{8}. \qquad \square$$

Theorem 3. *For a path of m edges, the (1+1) EA's expected runtime is $O(m^4)$.*

Proof. As in the previous proof, we only have to show that the expected number of relevant steps to leave a level is $O(\ell^2)$. In Situation 1, the probability that a relevant step is improving again is $\text{Prob}_R(A) = \Omega(1)$. A necessary condition to move to Situation 2 in a relevant step is that e or at least one of at most two neighbors of e is turned into a matching edge. Thus $\text{Prob}_R(B) = O(1/m^2)/\Omega(1/m) = O(1/m)$. As before, the expected total number of relevant steps in Situation 1 is $O(1)$ and the expected number of times the (1+1) EA leaves Situation 2 is at most $1 + O(1/m)$. It suffices to show that $c\ell^2$ relevant steps succeed in leaving Situation 2 with a probability $\alpha = \Omega(1)$.

In Situation 2, we ignore improving steps; they may take place and only shorten the time to leave the fitness level. Again we focus on an augmenting path P whose initial length is at most ℓ and consider a phase of $c\ell^2$ relevant steps. The phase is finished prematurely when P or another augmenting path shrinks to length one. The (1+1) EA allows mutation steps where the path's length $|P|$ changes by more than two edges or P vanishes completely as depicted in Fig. 1. The following properties ensure that $|P|$ never changes by more than two edges (implying none or two) in any step of the phase. Let x and y be the vertices at the current endpoints of P. Furthermore, let $E_x = \{\{u,v\} \in E \mid \text{dist}(x,u) \le 3\}$ be the set of edges where one endpoint has at most distance three to x, analogously for y (Fig. 2). The first property is that no step is accepted

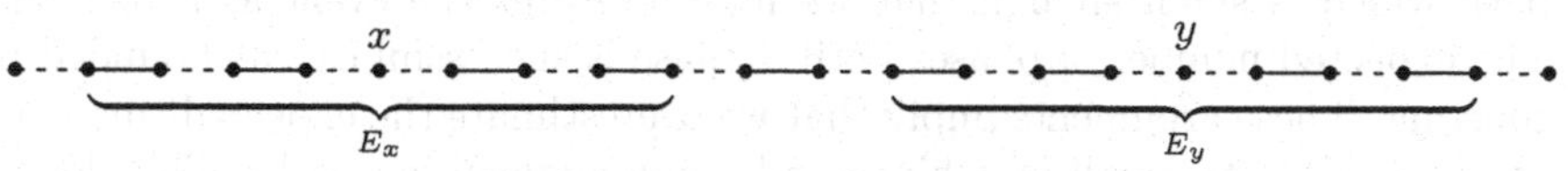

Fig. 2. Environments E_x and E_y.

and flips more than three edges in $E_x \cup E_y$. The second property is that no step flips three or more edges in line on the extended path $P' := P \cup E_x \cup E_y$ and is accepted. We call steps that respect both properties or finish the phase *clean.* Obviously, we only have to ensure that all $c\ell^2$ relevant steps are clean and call this event C. In [5] we show that, given a step is relevant, it is clean with a probability $1 - O(1/m^2)$. Then for a certain constant d and $m^2 \geq 2d$, $\text{Prob}(C) \geq (1 - d/m^2)^{c\ell^2} \geq (1 - d/m^2)^{cm^2} \geq e^{-2cd} = \Omega(1)$ holds.

In the proof for the local (1+1) EA we have already seen, for initial path length at most ℓ, $c\ell^2$ relevant steps produce $\ell/4$ more successes than the expected value and succeed in decreasing the path length to one with probability at least 3/8 if c is sufficiently large and a relevant step shortens the path with probability at least 1/2. We call the event, given $c\ell^2$ relevant and clean steps, these steps succeed in decreasing the path to length one, event S. Then the success probability of a phase is at least $\text{Prob}(C \cap S) = \text{Prob}(C) \cdot \text{Prob}(S \mid C)$. Given that a relevant step in Situation 2 is clean implies that it either finishes a phase or it flips one and only one pair of neighbored edges at one end of P (and perhaps some more edges not affecting P). In the latter case, the probability to flip a pair of edges shortening the path is at least 1/2 and the probability to lengthen the path is at most 1/2, since there are at least two shortening pairs and sometimes only one lengthening pair. Thus $\text{Prob}(S \mid C) \geq 3/8$. $\square$

We discuss the results of Theorem 2 and 3. Paths are difficult since augmenting paths tend to be rather long in the final stages of optimization. The (1+1) EA can cope with this difficulty. Paths are easy since there are not many possibilities to lengthen an augmenting path. The time bound $O(m^4) = O(n^4)$ is huge but can be explained by the characteristics of general (and somehow blind) search. If we consider a step relevant if it alters any augmenting path, there are many irrelevant steps, including steps which are rejected. In the case of $O(1)$ augmenting paths and no selectable edge a step is relevant only with a probability of $\Theta(1/m^2)$. The expected number of relevant steps is bounded by only $O(m^2) = O(n^2)$. Indeed, the search on the level of second-best matchings is already responsible for this. Since lengthenings and shortenings of the augmenting path have almost always the same probability for the local (1+1) EA, we are in a situation of fair coin tosses and have to wait for the first point of time where we have $\Theta(m)$ more heads than tails and this takes $\Theta(m^2)$ coin tosses with large probability. This implies that our bounds are tight if we have one augmenting path of length $\Theta(m)$. The situation is more difficult for the (1+1) EA. It is likely that from time to time several simultaneously flipping bits change the scenario drastically. We have no real control of these events. However, by focussing on one augmenting path we can ignore these events for the other paths and can prove a bound on the probability of a bad event for the selected augmenting path which is small enough that we may interpret the event as a bad phase. The expected number of phases until a phase is successful can be bounded by a constant. These arguments imply that we overestimate the expected time on the fitness levels with small matchings and many augmenting paths. This does not matter since the last improvement has an expected time of $\Theta(m^4)$ if we start with an augmenting path of length $\Theta(m)$.

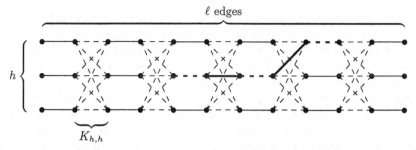

Fig. 3. The graph $G_{h,\ell}$ and an augmenting path.

4   Example with Exponential Time

After having seen that the (local) (1+1) EA computes maximum matchings on very simple graphs efficiently, we present a class of bipartite graphs where both EAs have an exponential expected optimization time. The graph $G_{h,\ell}$ is defined on $n = h \cdot (\ell + 1)$ nodes where $\ell \geq 3$ is odd. To describe the graphs we consider the nodes as grid points (i,j), $1 \leq i \leq h$, $0 \leq j \leq \ell$. The nodes $(\cdot, j)$ belong to the jth column. Between column j, j even, and $j + 1$ there are the edges $\{(i,j),(i,j+1)\}$ and between column j, j odd, and $j + 1$ we have all edges of a complete bipartite graph (see Fig. 3). Sasaki and Hajek [12] have proved that simulated annealing has an exponential expected optimization time on these graphs for $h = \ell$. Our result is the following one.

Theorem 4. *The (local) (1+1) EA has an exponential expected optimization time $2^{\Omega(\ell)}$ on $G_{h,\ell}$ if $h \geq 3$.*

Here we describe the main ideas of the proof. The full paper [5] contains complete proofs including the more difficult case of the general (1+1) EA. It is interesting that our result holds also in the case of constant $h = 3$ where the degree of the graph is bounded by $d = 4$. Hence, the (local) (1+1) EA is not successful on graphs of constant degree. Observe that we obtain a path if $h = 1$. Theorem 4 even holds for $h = 2$ but then special situations have to be investigated more carefully. It is easy to see that the unique perfect matching contains all edges $\{(i,j),(i,j+1)\}$, j even. We are mostly interested in the situation where the algorithm has found an almost optimal matching of size $n/2 - 1$. Then it is also easy to see that there exists a unique augmenting path. Its length ℓ' is an odd number bounded by ℓ. It contains nodes $(i_0, j), (i_1, j + 1), \ldots, (i_{\ell'}, j + \ell')$, i.e., it runs from left to right possibly changing the level (see Fig. 3). The main observation is that an accepted 2-bit flip can shorten or lengthen the augmenting path at either endpoint. However, at each endpoint (if not in column 0 or ℓ) there are h possibilities to lengthen the augmenting path and only one possibility to shorten it. From a more global point of view, we may consider "semi-augmenting" paths, i.e., alternating paths starting at a free node which cannot be lengthened to an augmenting path. The number of semi-augmenting paths is exponential (if ℓ' is not close to ℓ). The (local) (1+1) EA searches more or less locally and cannot distinguish immediately between semi-augmenting paths and augmenting paths. Our conjecture is that the presence of exponentially many semi-augmenting paths and only polynomially many augmenting paths at many points of time prevents the (local) (1+1) EA from being efficient. This also explains why paths are easy and why trees should be easy for the (local) (1+1) EA. The proof of Theorem 4 follows this intuition.

Step 1. The first search point is with overwhelming probability a non-matching and the first matching found is not perfect with a probability of at least $\Omega(1/h)$. The conjecture is that the probability of interest is close to 1, at least $1 - o(1)$. However, the proven result is strong enough to prove the theorem.

Step 2. Under the assumption that the (local) (1+1) EA finds at first a non-perfect matching, the probability of finding a matching of size $n/2 - 1$ before finding a perfect matching is $1 - O(1/m)$.

Step 3. Assume that the (local) (1+1) EA has found a matching of size $n/2 - 1$. Let ℓ_1 be the length of the unique augmenting path with respect to the matching. We conjecture that ℓ_1 is with probability $1 - o(1)$ large, i.e., at least ℓ^ε for some $\varepsilon > 0$. As a substitute of this unproven conjecture we use the following arguments. Before constructing the perfect matching the (local) (1+1) EA searches on the plateau of matchings of size $n/2 - 1$, since no other search point is accepted. In order to investigate the search on this plateau, we describe the random search point in step k by X_k and its value by x_k. Moreover, let L_k be the random length of the augmenting path of X_k and ℓ_k the corresponding value of L_k. The search stops iff $\ell_k = 0$. Let T be this stopping time. Before that, ℓ_k is an odd number bounded above by ℓ.

Claim 1. If $\ell_1 = 1$, the probability that $\ell_k \geq 3$ for some $k < T$ is $\Omega(h/m)$ for the (local) (1+1) EA.

Claim 2. If $\ell_1 \geq 3$ the probability that $\ell_k = \ell$ for some $k < T$ is $\Omega(1)$ for the (local) (1+1) EA.

To discuss the main ideas, we consider here the local (1+1) EA. Let $b_k = (\ell_k - 1)/2$. The local (1+1) EA can change the b-value by at most 1 as long as its value is positive. Obviously, $b_1 \geq 1$ and pessimistically $b_1 = 1$. We are interested in the probability of reaching the maximal b-value $\ell/2$ before the value 0. There are two 2-bit flips decreasing the b-value by 1 and there are at least h (one endpoint of the augmenting path can be in column 0 or ℓ) 2-bit flips increasing the b-value. Hence, the next step changing the b-value leads to $b - 1$ with a probability of at most $2/(h + 2)$ and leads to $b + 1$ with a probability of at least $h/(h + 2)$. If $h \geq 3$ (for $h = 2$ one can prove the same result discussing the situation of augmenting paths ending in column 0 or ℓ more carefully), we have an unfair game and can apply the result of the gambler's ruin problem. Alice owns A \$ and Bob B \$. They play a coin-tossing game with a probability of $p \neq 1/2$ that Alice wins a round in this game, i.e., Bob pays a dollar to Alice. Let $t := (1 - p)/p$. Then Alice wins, i.e., she has $(A + B)$ \$ before being ruined, with a probability of $(1 - t^A)/(1 - t^{A+B}) = 1 - t^A(1 - t^A)/(1 - t^{A+B})$ (e.g., [4]). We consider a game with $p = h/(h + 2)$, $A = 1$, and $B = (\ell - 1)/2 - 1$. Then the probability that $\ell_k = \ell$ for some $k < T$ is larger than the probability that Alice wins. The last probability equals $(1 - t)/(1 - t^{(\ell-1)/2+1})$. Since

$t = 2/h \leq 2/3$, this probability is at least $1/3$ and for general h it is at least $1 - O(1/h)$. Moreover, if ℓ_1 is not too small, this probability is even close to 1.

Step 4. In order to prove the theorem, we can now assume that we start with an augmenting path of length ℓ. We discuss here again the local (1+1) EA and use the notation with the b-values. We start with a b-value of $(\ell-1)/2$. In order to reach the value 0 we have to reach $(\ell-1)/4$ first. Using the arguments from above the probability to reach then $(\ell-1)/2$ before 0 is at least $1 - t^{(\ell-1)/4}(1-t^{(\ell-1)/4})/(1-t^{(\ell-1)/2})$. Since $t = 2/h \leq 2/3$, this probability is $1 - (2/h)^{\Theta(\ell)}$ and the probability of reaching 0 before $(\ell-1)/2$ can be bounded by $(2/h)^{\Theta(\ell)} = 2^{-\Omega(\ell)}$. Finally, the expected number of such phases is bounded below by $2^{\Omega(\ell)}$. In order to obtain a similar result for the (1+1) EA we have to analyze the steps of many flipping bits much more carefully. Here we describe the proof technique. The assumption is that $L_1 = \ell$. For the (1+1) EA it is still easy to prove that $E(L_{t+1} - L_t)$ is positive for each fixed search point with an L_t-value of less than ℓ. There are steps decreasing the length of the augmenting path significantly. In order to prove that the success probability of the (1+1) EA within an exponential (in ℓ) number of steps is exponentially small we have to take special care of these deceasing steps, although they have small probability. It is not enough to argue about $E(L_{t+1} - L_t)$. In order to obtain a result for a large number of steps by considering only $L_{t+1} - L_t$ we use the trick of the Chernoff bound method and investigate $E(e^{\lambda(L_{t+1}-L_t)})$ for some appropriate $\lambda < 0$. This implies that the steps where $-(L_{t+1} - L_t)$ is large have a much larger influence than in the expectation of $L_{t+1} - L_t$. If $E(e^{\lambda(L_{t+1}-L_t)}) \leq 1 - \alpha(m)$ and $\alpha(m) > 0$ is not too small, this can be interpreted as a strong drift of increasing the length of the augmenting path. This leads by a drift theorem [6,7] to the result described above which is even stronger than an exponential lower bound on the expected optimization time.

We have seen that our results are not difficult in the case of the local (1+1) EA. Only the general (1+1) EA can escape eventually from each local optimum. The probability of flipping many bits in one step is essential and makes the analysis difficult. The result of Theorem 4 has the drawback of stating a lower bound on the expected optimization time and not a an exponential lower bound which holds with a probability exponentially close to 1. This offers the chance that a multistart strategy may have a polynomially bounded expected optimization time. However, many of our results hold already with a probability exponentially close to 1. The missing link is that we obtain a non-perfect matching with $n/2 - O(1)$ edges and one augmenting path of length $\Omega(\ell^\varepsilon)$ with large probability. It is already difficult to derive properties of the first matching created by the algorithms.

Conclusions

Evolutionary algorithms without problem-specific modules are analyzed for the maximum matching problem. The results show how heuristics can "use" algorithmic ideas not known to the designer of the algorithm. Moreover, this is one of the first results where an EA is analyzed on a well-known combinatorial problem.

References

1. N. Blum. A simplified realization of the Hopcroft-Karp approach to maximum matching in general graphs. Technical report, Universität Bonn, 1999.
2. S. Droste, T. Jansen, and I. Wegener. On the optimization of unimodal functions with the (1+1) evolutionary algorithm. In *Proc. of the 5th Conf. on Parallel Problem Solving from Nature (PPSN V)*, LNCS 1498, pages 13–22, 1998.
3. S. Droste, T. Jansen, and I. Wegener. On the analysis of the (1+1) evolutionary algorithm. *Theoretical Computer Science*, 276:51–82, 2002.
4. W. Feller. *An Introduction to Probability Theory and Its Applications*. Wiley, 1971.
5. O. Giel and I. Wegener. Evolutionary algorithms and the maximum matching problem. Technical Report CI 142/02, Universität Dortmund, SFB 531, 2002. http://ls2-www.cs.uni-dortmund.de/~wegener/papers/Matchings.ps.
6. B. Hajek. Hitting-time and occupation-time bounds implied by drift analysis with applications. *Advances in Applied Probability*, 14:502–525, 1982.
7. J. He and X. Yao. Drift analysis and average time complexity of evolutionary algorithms. *Artificial Intelligence*, 127:57–85, 2001.
8. J. E. Hopcroft and R. M. Karp. An $n^{5/2}$ algorithm for maximum matchings in bipartite graphs. *SIAM Journal on Computing*, 2(4):225–231, 1973.
9. T. Jansen and I. Wegener. Real royal road functions – where crossover provably is essential. In *Proc. of the 3rd Genetic and Evolutionary Computation Conference (GECCO 2001)*, pages 375–382, 2001.
10. T. Jansen and I. Wegener. The analysis of evolutionary algorithms – a proof that crossover really can help. *Algorithmica*, 34:47–66, 2002.
11. S. Micali and V. V. Vazirani. An $O(\sqrt{|V|} \cdot |E|)$ algorithm for finding maximum matching in general graphs. In *Proc. 21st Annual Symp. on Foundations of Computer Science (FOCS)*, pages 17–27, 1980.
12. G. H. Sasaki and B. Hajek. The time complexity of maximum matching by simulated annealing. *Journal of the ACM*, 35:387–403, 1988.
13. J. Scharnow, K. Tinnefeld, and I. Wegener. Fitness landscapes based on sorting and shortest paths problems. In *Proc. of the 7th Conf. on Parallel Problem Solving from Nature (PPSN VII)*, LNCS 2439, pages 54–63, 2002.
14. V. V. Vazirani. A theory of alternating paths and blossoms for proving correctness of the $O(\sqrt{V}E)$ maximum matching algorithm. *Combinatorica*, 14(1):71–109, 1994.
15. I. Wegener. Theoretical aspects of evolutionary algorithms. In *Proc. of the 28th Internat. Colloq. on Automata, Languages and Programming (ICALP)*, LNCS 2076, pages 64–78, 2001.
16. I. Wegener and C. Witt. On the analysis of a simple evolutionary algorithm on quadratic pseudo-boolean functions. *Journal of Discrete Algorithms*, 2002. To appear.

Alternative Algorithms for Counting All Matchings in Graphs

Piotr Sankowski

Institute of Informatics
Warsaw University, ul. Banacha 2, 02-097 Warsaw
sank@mimuw.edu.pl

Abstract. We present two new methods for counting all matchings in a graph. Both methods are alternatives to methods based on the Markov Chains and both are unbiased. The first one is a generalization of a Godman-Godsil estimator. We show that it works in time $O(1.0878^n \epsilon^{-2})$ for general graphs. For dense graphs (every vertex is connected with at least $(\frac{1}{2} + \alpha)n$ other vertices) it works in time $O(n^{4+(6\ln 6)/\alpha} \epsilon^{-2})$, where n is the number of vertices of a given graph and $0 < \epsilon < 1$ is an expected relative error. We also analyze the efficiency of the presented algorithm applied for random graphs. The second method uses importance sampling. This method works in exponential time but it can easily be enriched in some heuristics leading to very efficient algorithms in practice. Experiments show that our methods give better estimates than the Markow Chain approach.

1 Introduction

Algorithms for counting matchings find applications in physics where this problem appears in statistical mechanics and is called the monomer-dimer problem. In this problem one considers an undirected graph $G = (V, E)$. The vertices of the graph can be covered with monomers (molecules occupying one vertex) and dimers (molecules occupying two adjacent vertices). Usually G is a regular grid in two or three dimensions. Three dimensional monomer-dimer problem is used to model liquids containing one-atom molecules and two-atom molecules. The two dimensional version models absorption of two atom molecules on a cristalic surface. For a more complete description see [HL72]. All physical properties of the system can be derived from its partition function. Generally speaking, the partition function is the number of states of the system. As far as the monomer-dimer system is concerned, it is exactly the number of all matchings. The problems of counting perfect matchings and all matchings are #P-complete, therefor it is reasonable to seek approximation algorithms.

A Fully Polynomial Randomized Approximation Scheme (FPRAS) for counting all matchings was proposed by Jerrum and Sinclair [JS89]. It works in time $O(n^4 m (\ln n)^3 \epsilon^{-2})$, n is the number of vertices, m is the number of edges in a given graph, and ϵ is a relative error the algorithm makes with the probability $\geq \frac{3}{4}$. Van den Berg and Brouver showed [BB] that the Jerrum and Sincliar's

H. Alt and M. Habib (Eds.): STACS 2003, LNCS 2607, pp. 427–438, 2003.
© Springer-Verlag Berlin Heidelberg 2003

algorithm applied to lattices works in time $O(n^3(\ln n)^4 \epsilon^{-2})$. It seemed that practical application of this algorithm could have been possible. We implemented the algorithm in accordance with the prescription given in [JSMCMC] and in [BB]. Due to a very high constant in the time complexity function, the algorithm happened to be practically useless. It required about 5 minutes for a graph with 4 vertices using Athlon 1.2Ghz computer.

The existence of a polynomial time algorithm for counting perfect matchings was shown by Jerrum, Sinclair and Vigoda [JSV2000]. Their algorithm is only a theoretical result and should be considered as a hint that the efficient solutions for the problem exists and only have to be found. The algorithm is based on the Monte Carlo Markov Chain method. The literature contains two alternative approximation methods for counting perfect matchings.

The number of perfect matching in a graph given by the adjacency matrix is equal to the permanent (definition in chapter 2) of this matrix. Similarity between the permanent and the determinant suggests in a natural way a usage of the determinant for approximation of the permanent. This method was discovered by Godsil and Gutman [GG] and than analyzed and improved by Karmarkar, Karp, Lipton, Lovasz and Luby [KKLLL]. They suggested a complex version of the algorithm. Frieze and Jerrum [FJ95] analyzed the algorithm in case of dense and random graphs. The further improvement was made by Barvinok [AB99] who proposed a quaternion version of the algorithm and extended it to matrix algebras [AB00]. In 2002, Chien, Rasmussen and Sinclair [CRS02] extended this estimator to Clifford algebras.

The other group of approximation algorithms for cunting perfect matchings is based on an approach called importance sampling. Rasmussen [LER92] showed a very simple algorithm working in polynomial time for almost all matrices. Beichl and Sullivan [BS99] applied the algorithm, with some heuristics added, for computing two-dimensional and three-dimensional dimer constants.

In this paper, for the first time we present how the above methods can be applied to the problem of counting all matchings in general graphs. We show a determinant approximation method for counting all matchings. This approximation scheme works in time $O(n^3 1.0878^n \epsilon^{-2})$. For dense graphs (every vertex is connected with at least $(\frac{1}{2} + \alpha)n$ other vertices) the running time is $O(n^{4+(6\ln 6)/\alpha} \epsilon^{-2})$. The algorithms also works in expected polynomial time for random graphs. It seems that this method may lead to FPRAS for all matrices (see [CRS02]). We also apply importance sampling to the problem of counting all matchings and get an estimator working in linear time depending on the size of a given graph. It gives a randomized approximation scheme working in time $O((n+m)n \binom{|E|}{n/2} \epsilon^{-2})$. It can be easily enriched with some heuristics giving efficient algorithms. Both algorithms described in this paper are unbiased and can be extended for approximating the partition function of monomer-dimer systems. Experiments show that in practice both algorithms give better approximation than the known algorithms based on the Monte Carlo Markov Chain approach.

2   Basics

Let Γ_n be the set of all permutations of $\{1, \ldots, n\}$. For even n let Π_n be the set of all permutations such that $p \in \Pi_n : p_1 < p_3 < p_5 < \cdots < p_{n-1}$ and $p_1 < p_2, p_3 < p_4, \ldots, p_{n-1} < p_n$. Such permutations are equivalent to partitions of a n-element set into unordered pairs.

Definition 1. *The permanent of a matrix A of size $n \times n$, is defined as*

$$Per(A) = \sum_{p \in \Gamma_n} \prod_{i=1}^{n} a_{i,p_i}. \tag{1}$$

Permanent $Per(A)$ is equal to the number of perfect matchings in the bipartite graph $n \times n$ for that A is the adjacency matrix. In this matrix, A_{ij} means that the edge between vertex i in one group of vertices and vertex j in the other group exists.

Definition 2. *The hafnian of a matrix A of size $n \times n$, n even, is defined as*

$$Haf(A) = \sum_{p \in \Pi_n} \prod_{i=1}^{\frac{n}{2}} a_{p_{2i-1}, p_{2i}}. \tag{2}$$

Hafnian $Haf(A)$ is equal to the number of perfect matchings of a graph with the adjacency matrix A.

Let Ω denote a set and let $f : \Omega \to R$ be a function which values we want to compute. A *randomized approximation scheme* for f is a randomized algorithm that takes $x \in \Omega$ and $\epsilon > 0$ as an input and returns a number Y (the value of a random variable) such that

$$P((1 - \epsilon)f(x) \leq Y \leq (1 + \epsilon)f(x)) \geq \frac{3}{4}. \tag{3}$$

We say that a randomized approximation scheme is fully polynomial if it works in time polynomially dependent on the size of input data x and ϵ^{-1}. Algorithms that approximate numerical values are called estimators. An estimator is unbiased if its result is a random variable with expected value equal to the value being approximated.

Our randomized approximation schemes are constructed in the following way. First, we find an unbiased estimator. Then, in order to get a randomized approximation scheme, we run the estimator a number of times and return the average of results given by the estimator. Let the result of the estimator be a value of the random variable Y. Consider the following algorithm:

- run the estimator $4Var(Y)E(Y)^{-2}\epsilon^{-2}$ times;
- return the average of obtained values.

Since the different runs of the estimator are independent the relative variance of the random variable Z returned by the algorithm satisfies the following

$$\frac{Var(Z)}{(E(Z))^2} = \frac{Var(Y)}{4Var(Y)E(Y)^{-2}\epsilon^{-2}E(Y)^2} = \frac{\epsilon^2}{4}. \tag{4}$$

Applying Chebyshev's inequality, we get the following

$$Pr((1-\epsilon)E(Y) \le Z \le (1+\epsilon)E(Y)) \ge \frac{3}{4}. \tag{5}$$

Notice, that in order to construct an approximation scheme from a given estimator, it is sufficient to know the quotient of the variance and the square of the expected value of the estimator. The working time of this randomized approximation scheme is $O(Var(y)E(y)^{-2}\epsilon^{-2})$ multiplied by the running time of the estimator.

3 Application of Determinant to Counting All Matchings

As stated in the introduction the idea of application of the determinant to approximate the number of perfect matchings was deeply investigated by several authors. This method uses the similarity between the determinant and the permanent as well as the fact that the determinant can be computed in time $O(n^3)$. As noticed by Barvinok [AB99] the determinant can be used for approximating hafnian, in other words, for approximating the number of perfect matchings in general graphs. We show that a similar method can be used for approximating the number of all matchings in general graphs. On the contrary, it seems that the permanent approximation method cannot be so easily modified.

3.1 Hafnian Approximation Method

Let Δ denote the set of all skew symmetric matrices $n \times n$ with all non-diagonal elements equal to 1 or -1. We have $|\Delta| = 2^{\frac{n(n-1)}{2}}$. For matrices A and D of size $n \times n$, let $A(D)$ be a matrix $n \times n$ which elements are products of elements from A and D, i.e. $(A(D))_{ij} = A_{ij}D_{ij}$. Similarly, let $(A^{(r)})_{ij}$ dentote $(A_{ij})^r$, where r is an arbitrary real number.

Theorem 1 (Barvinok 99). *For an arbitrary $n \times n$ symmetric matrix A of non-negative real numbers we have*

$$Haf(A) = \frac{1}{|\Delta|} \sum_{D \in \Delta} Det(A^{(\frac{1}{2})}(D)). \tag{6}$$

In other words, $Haf(A)$ is the average of $Det(A^{(\frac{1}{2})}(D))$ over all matrices $D \in \Delta$. In order to approximate $Haf(A)$, one can compute the average of $Det(A^{\frac{1}{2}}(D))$ over k random matrices D from Δ. The following theorem gives bound on the error of this method. From now we will only consider zero-one matrices. In such a case we have $A(D) = A^{\frac{1}{2}}(D)$.

Theorem 2. *For random matrix $D \in \Delta$ the variance of $Det(A(D))$ is bounded by $3^{\frac{n}{4}}(Haf(A))^2$.*

This proof and other proofs omitted can be found in the full version of the paper.

Corollary 1. *The estimator from Theorem 1 leads to the approximation scheme for computing the number of perfect matchings and working in time $O(\epsilon^{-2}n^3 3^{\frac{n}{4}})$.*

3.2 Approximating the Number of All Matchings

The following theorem allows us to construct an approximation algorithm for computing the number of all matchings in an arbitrary graph.

Theorem 3. *If A is the adjacency matrix of a graph G, I is the identity matrix, then the number of all matchings in G equals*

$$S(A) = \frac{1}{|\Delta|} \sum_{D \in \Delta} Det(A(D) + I). \tag{7}$$

Proof. Let M be a matrix of size $n \times n$. By $M^{0,k}$ we denote the matrix M with M_{kk} set to 0. Let $M^{1,k}$ be the matrix obtained from M by setting to 0 all elements in row k and column k except M_{kk}. The determinant of M can be expressed as the sum of two determinants $Det(M) = Det(M^{1,k}) + Det(M^{0,k})$. Using this decomposition to the right hand side of (7) as many times as possible we get

$$\sum_{I \in \{0,1\}^n} \frac{1}{|\Delta|} \sum_{D \in \Delta} Det((\ldots((A(D) + I)^{I_1,1})^{I_2,2} \ldots)^{I_n,n}) \tag{8}$$

Let as denote by κ a matrix operation, that for all k removes row k and column k if the only nonzero element in them is 1 on diagonal. After applying κ to a matrix the determinant does not change. Notice that in case of the matrix in (8) we get a matrix having only zeros on diagonal. Now, we may apply the Theorem 1 to each element of the sum separately and get the following

$$\sum_{I \in \{0,1\}^n} Haf(\kappa((\ldots((A(D) + I)^{I_1,1})^{I_2,2} \ldots)^{I_n,n})). \tag{9}$$

We may interpret each element of the sum as the number of perfect matchings in some subgraph of the graph G. Now we see that each element of the sum gives the number of perfect matchings in some subgraph of the graph. The number of all matchings is equal to the number of perfect matchings in all subgraphs of the graph and the theorem follows.   □

Theorem 4. *The variance of $Det(A(D) + I)$ over $D \in \Delta$ is bounded by*

$$1.4^{\frac{n}{4}}(S(A))^2 \approx 1.0877573^n(S(A))^2.$$

Corollary 2. *The estimator from Theorem 3 yields a randomized approxima-tion scheme for computing the number of all matchings that works in time* $O(\epsilon^{-2}n^3 1.0877573^n)$.

The following theorem is an adaptation of Theorem 8 from [FJ95].

Theorem 5. *Let* $\alpha > 0$ *be a constant and let* G *be an* n-*vertex graph with the minimal degree* $\delta(G) \geq (\frac{1}{2} + \alpha)n$. *Then the variance of* $Det(A(D) + I)$ *over* $D \in \Delta$ *is bounded by* $O(n^{1+(6\ln 6)/\alpha})$.

Corollary 3. *The estimator from Theorem 3 yields a randomized approxima-tion sheme for counting all matchings that works in time* $O(n^{4+(6\ln 6)/\alpha})$ *for dense graphs.*

Our algorithm can be easily modified to the complex and quaternionic version but it does not bring anything new to the topic. Moreover, Corollary 6 and 7 from [FJ95] can be directly applied to the sum of determinants in Theorem 3 in the complex version of the algorithm. As the result we have

Corollary 4. *The estimator from Theorem 3 in complex version yields a ran-domized approximation scheme for counting all matchings that works in polyno-mial time for random graphs in the model* $B(n, p = \frac{1}{2})$.

The above method can also be used for approximating the partition function of the monomer-dimer system. The partition function is given by the following formula

$$Z(\lambda) = \sum_{k=0}^{n/2} m_k \lambda^k, \tag{10}$$

m_k is the number of matchings of cardinality k and λ is a positive real parameter. The following theorem can be shown exactly in the same way as Theorem 3.

Theorem 6. *If* A *is an adjacency matrix describing a monomer-dimer system then the partition function of the system is equal to*

$$Z(\lambda) = \frac{1}{|\Delta|\lambda^{n/2}} \sum_{D \in \Delta} Det(A(D) + \frac{1}{\sqrt{\lambda}}I). \tag{11}$$

4   Application of Importance Sampling for Counting All Matchings

4.1   Importance Sampling

Importance sampling is based on a direct sampling and does not require generat-ing samples with the uniform distribution. Let us suppose, we want to compute the size of a finite set Ω. If we are able to generate elements of Ω randomly with some known distribution π and in such a way that all elements are generated with non-zero probability then the size of Ω is given by a simple formula

$$|\Omega| = \sum_{X \in \Omega} 1 = \sum_{X \in \Omega} \pi(X) \frac{1}{\pi(X)}. \tag{12}$$

In other words, the size of Ω is given by the average of $1/\pi(X)$ over the distribution π. Notice, that this equation is correct even if $\pi(\Omega) < 1$. Number $1 - \pi(\Omega)$ should be interpreted as the probability that the algorithm does not generate a sample from Ω. In such a case the algorithm fails. The variance of the method is given by the formula

$$Var(\pi) = \sum_{X \in \Omega} \pi(X) \frac{1}{\pi(X)^2} - |\Omega|^2 = \sum_{X \in \Omega} \frac{1}{\pi(X)} - |\Omega|^2 \tag{13}$$

We see that the variance is small if the distribution π is close to the uniform distribution.

4.2   Estimator for the Number of Perfect Matchings

By using this method, Rasmussen proposed the following algorithm for counting perfect matchings in bipartite graphs [LER92]. Let A_{ij} be elements of the adjacency matrix of a given graph. The variable x is an approximated value of the number of perfect matchings. We modified the algorithm of Rasmussen by applying the idea of Beichl and Sullivan [BS99] and instead of uniform distribution we use an arbitrary distribution.

Algorithm 4.1 Rasmussen's Estimator

$x := 1$
for $i := 1$ to n **do**
  $W := \{j : A_{ij} = 1\}$
  if $W = \emptyset$ **then**
    $x := 0$
  else
    choose index j from W with some distribution ρ_i (choose an edge (i, j))
    for $k := 1$ to n **do**
      $A_{kj} := 0$
    end for
    $x := x/\rho_i(j)$
  end if
end for

Distributions ρ_i may be different in every step and in every run of the algorithm and may depend on previous choices. The distributions just have to guarantee that every matching will be generated with a nonzero probability. Value $1/x$ is the probability $\pi(X)$ of the chosen matching $X - \pi(X)$ is the product of

$\rho_i(j)$'s because the choices are made independently. Even if the distributions ρ_i are uniform the resulting distribution π does not have to be uniform.

Theorem 7. *The variance of the Rasmussen's estimator for uniform distributions is bounded by $n! per(A)^2$.*

The proof of the theorem can be found in [LER92]. The theorem follows from (13) as well, because the probability of generating any matching is always greater then $1/n!$.

Corollary 5. *The Rassmussen's estimator yields a randomized approximation scheme for counting the number of perfect matchings that works in time $O((m + n)n!\epsilon^{-2})$.*

Since we prefer graphs rather than matrices, we will rewrite this algorithm in the other form. $G = (V, E)$ is an input graph.

Algorithm 4.2 Estimator for the Number of Perfect Matchings I

$x := 1$
while $V \neq \emptyset$ **do**
    choose any vertex $v \in V$
    $W := \{w : (v, w) \in E\}$
    if $W = \emptyset$ **then**
        $x := 0$
    else
        choose w from W with the uniform distribution
        remove w and v from G
        $x := |W|x$
    end if
end while

As we can see, it works for all graphs, not only bipartite ones. If degrees of vertices in the graph are bounded by d then the variance is bounded by $d^{\frac{n}{2}} per(A)^2$ because the size of set W will be always smaller than d. Using this property, we may add a simple heuristic to our algorithm. In every step of the algorithm, we choose a vertex with the smallest degree. It turns out that this approach works very well for cubic grids. We notice that the degree of each chosen vertex is at most 3, so the variance in this case is bounded by $3^{\frac{n}{2}} per(A)^2$. This algorithm allowed to estimate three dimensional dimer constant for cubic lattice ($\lambda_3 = 0.446 \pm 0.001$) in about 15 hours on a personal computer. This result should be compared with the result of Beichl and Sullivan [BS99] ($\lambda_3 = 0.4466 \pm 0.0006$) which, as far as the author knows, was obtained using massive computations. It should be noted that the error does not come from the analysis of the algorithm but from a statistical analysis of the results obtained in this computer experiment.

4.3   Estimator for the Number of All Matchings

Let us return to the problem of counting all matchings. In order to see how to use importance sampling to solve this problem let us modify Algorithm 4.2 again in order to work directly with edges instead of vertices.

Algorithm 4.3 Estimator for the Number of Perfect Matchings II

$x := 1$
$t := 1$
while $E \neq \emptyset$ **do**
    $x := |E| * x/t$
    choose (v, w) from E with uniform distribution
    remove vertices w and v and all edges incident with w or v from G
    $t := t + 1$
end while
if $V \neq \emptyset$ **then**
    x := 0
end if

Just like before, we generate a perfect matching with some distribution and the same time compute the probability of getting this matching. The factor $1/t$ comes from the fact that in step t of the algorithm we generate a matching of cardinality t from some matching of cardinality $t - 1$ and every matching of cardinality t may be obtained from t different matchings of cardinality $t - 1$. In order to get an estimator for the number of all matchings we modify the Algorithm 4.3 getting the Algorithm 4.4.

Algorithm 4.4 Estimator for the Number of All Matchings

$wtg := 1$
$t := 1$
$x := 1$
while $E \neq \emptyset$ **do**
    $wtg := |E| * wtg/t$
    $x := x + wtg$
    choose (v, w) from E with uniform distribution
    remove vertices w and v and all edges incident with w or v from G
    $t := t + 1$
end while
if $V \neq \emptyset$ **then**
    x := 0
end if

One can prove the following bounds on the variance.

Theorem 8. *The variance of the Estimator 4.3 is bounded by*

$$\frac{Var(x)}{(E(x))^2} \leq \binom{|E|}{n/2} \tag{14}$$

Proof. Observe that x is multiplied by at most $|E|/1$ in the first step of the algorithm, by at most $(|E|-1)/2$ in the second step and by at most $(|E|-i)/i$ in i'th step. □

Corollary 6. *The Estimator 4.3 yields a randomized approximation scheme for the number of perfect matchings that works in time* $O((n+m)\binom{|E|}{n/2}\epsilon^{-2})$.

Theorem 9. *The variance of the Estimator 4.4 is bounded by*

$$\frac{Var(x)}{(E(x))^2} \leq \frac{n}{2}\binom{|E|}{n/2}. \tag{15}$$

Proof. Denote by x_i the value that is added to x in i'th step. Then

$$\frac{Var(x)}{(E(x))^2} = \frac{Var(\sum_{i=0}^{\frac{n}{2}} x_i)}{(E(\sum_{i=0}^{\frac{n}{2}} x_i))^2} \leq \frac{\frac{n}{2}\sum_{i=0}^{\frac{n}{2}} Var(x_i)}{(\sum_{i=0}^{\frac{n}{2}} E(x_i))^2} \leq \frac{\frac{n}{2}\sum_{i=0}^{\frac{n}{2}}\binom{|E|}{i}E(x_i)^2}{(\sum_{i=0}^{\frac{n}{2}} E(x_i))^2} \leq \tag{16}$$

$$\leq \frac{\frac{n}{2}\binom{|E|}{n/2}\sum_{i=0}^{\frac{n}{2}} E(x_i)^2}{(\sum_{i=0}^{\frac{n}{2}} E(x_i))^2} \leq \frac{n}{2}\binom{|E|}{n/2} \tag{17}$$

□

Corollary 7. *The Estimator 4.4 yields a randomized approximation scheme for the number of all matchings that works in time* $O((n+m)n\binom{|E|}{n/2}\epsilon^{-2})$.

These bounds are not satisfactory because they do not lead to polynomial time algorithms. But the strength of the method is its simplicity and that it can be enriched with some heuristics. Using this algorithm we tried to estimate monomer-dimmer constant for a cubic lattice but in this case statistical analysis does not allow to get reliable error estimates. The Estimator 4.2 can be also used to construct an estimator for computing the number of all matchings. If vertices are chosen randomly then it generates every matching with nonzero probability. It seems however that it leads to a slower algorithm. The extension of the Algorithm 4.4 for computing the value of partition function at point λ is simple. We only have to multiply wtg by λ in each step of the algorithm.

5   Experiments

We implemented the algorithm of Jerrum and Sinclair exactly following the prescription given in [JSMCMC]. This algorithm worked about 5 minutes on Athlon 1.2GHz for a graph with 4 vertices and with the relative error 0.1. We see that the algorithm in this form cannot be used in practice. However, it can be applied when we lower the certainty of obtaining precise results. In this case it is sufficient to use smaller number of samples. In order to compare our algorithms with the Jerrum and Sinclair's algorithm we chose the number of samples so that the algorithms always worked about 10 seconds. We made some series of tests on some families of graphs, i.e. two and three dimensional grids, some types of sparse graphs and random dense graphs. For all types of graphs we have got very similar results. Here we present only the results for two dimensional rectangular grids.

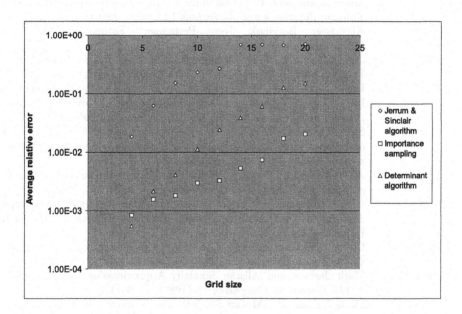

In the algorithm of Jerrum and Sinclair, the bound on mixing time of the chain given in [BB] was used. For computing the determinant Gauss elimination for sparse matrices in list representation was used. We see that both methods presented in this work give more precise results than the Jerrum and Sinclair's algorithm.

6   Summary

In this paper, we presented two unbiased estimators for computing the number of all matchings. The estimators can be also applied to computing the partition function of the monomer-dimer system. Both algorithms give better estimates than the Monte Carlo Markov Chain approach. We presented that the determinant approach works in polynomial time for dense and random graphs. We conjecture that the importance sampling approach works in polynomial time for random graphs as the Rasmussen estimator does. We conclude that if the idea presented in [CRS02] leads to a polynomial time algorithm for the permanent then it will also lead to polynomial time algorithms for counting all matchings and for computing the partition function of the monomer-dimer system. Additionally we confirmed the estimation of the three dimensional dimer constant given in [BS99].

References

[AB99]   Alexander Barvinok, Polynomial time algorithms to approximate permanents and mixed discriminants within a simply exponential factor, *Random Structures and Algorithms* **14** (1999), 29–61.

[AB00]   Alexander Barvinok, New Permanent Estimators Via Non-Commutative Determinants, preprint.

[BB]   J. van den Berg and R. Brouwer, Random Sampling for the Monomer-Dimer Model on a Lattice, 1999 CWI Research Report.

[BS99]   Isabel Beichl and Francis Sullivan, Approximating the Permanent via Importance Sampling with Application to Dimer Covering Problem, *Journal of computational Physics* **149**, 1, February 1999.

[CRS02]   Steve Chien, Lars Rasmussen and Alistair Sinclair, Clifford Algebras and Approximating the Permanent, *Proceedings of the 34th ACM Symposium on Theory of Computing*, 2001, 712–721.

[FJ95]   A. Frieze and M. Jerrum, An analysis of a Monte Carlo algorithm for estimating the permanent , *Combinatorica*, **15** (1995), 67–83.

[GG]   C.D. Godsil and I. Gutman, On the matching polynomial of a graph, *Algebraic Methods in Graph Theory, Vol. I, II, (Szeged, 1978)*, North-Holland, Amsterdam - New York, 1981, 241–249.

[HL72]   Ole J. Heilmann, Elliott H.Lieb, Theory of Monomer-Dimer Systems, *Communications in mathematical Physics* **25** (1972), 190–232.

[JS89]   Mark Jerrum and Alistair Sinclair, Approximating the permanent, *SIAM Journal on Computing* **18** (1989), 1149–1178.

[JSMCMC]   Mark Jerrum and Alistair Sinclair, The Markow Chain Monte Carlo method: an approach to approximate counting and integration, in *Approximation Algorithms for NP-hard Problems* (Dorit Hochbaum, ed.), PWS 1996.

[JSV2000]   Mark Jerrum, Alistair Sinclair and Eric Vigoda, A polynomial-time approximation algorithm for the permanent of a matrix with non-negative entries, *Electronic Colloquium on Computational Complexity*, Report No. 79 (2000).

[KKLLL]   N. Karmarkar, R. Karp, R. Lipton, L. Lovász and M. Luby, A Monte Carlo algorithm for estimating the permanent, *SIAM Journal on Computing* **22** (1993), 284–293.

[LER92]   Lars Eilstrup Rasmussen, Approximating the Permanent: a Simple Approach, *Random Structures Algorithms* **5** (1994), 349–361.

Strong Stability in the Hospitals/Residents Problem

Robert W. Irving, David F. Manlove*, and Sandy Scott

Computing Science Department, University of Glasgow, Glasgow G12 8QQ, Scotland
{rwi,davidm,sas}@dcs.gla.ac.uk. *Fax:* +44 141 330 4913.

Abstract. We study a version of the well-known Hospitals/Residents problem in which participants' preferences may involve ties or other forms of indifference. In this context, we investigate the concept of *strong stability*, arguing that this may be the most appropriate and desirable form of stability in many practical situations. When the indifference is in the form of ties, we describe an $O(a^2)$ algorithm to find a strongly stable matching, if one exists, where a is the number of mutually acceptable resident-hospital pairs. We also show a lower bound in this case in terms of the complexity of determining whether a bipartite graph contains a perfect matching. By way of contrast, we prove that it becomes NP-complete to determine whether a strongly stable matching exists if the preferences are allowed to be arbitrary partial orders.

Keywords: stable matching problem; strong stability; hospitals/residents problem; polynomial-time algorithm; lower bound; NP-completeness.

1 Introduction

The Hospitals/Residents problem [3] is a many-to-one extension of the classical Stable Marriage problem (SM), so-called because of its widespread application to matching schemes that match graduating medical students (residents) to hospital posts. In particular the National Resident Matching Program (NRMP) in the USA [14], the Canadian Resident Matching Service [1], and the Scottish PRHO Allocations (SPA) matching scheme [6] all make use of algorithms that solve variants of this problem.

An instance of the classical Hospitals/Residents problem (HR) involves two sets, a set R of *residents* and a set H of *hospitals*. Each resident in R seeks to be *assigned* to exactly one hospital, and each hospital $h \in H$ has a specified number p_h of posts, referred to as its *quota*. Each resident ranks a subset of H in strict order of preference, and each hospital ranks, again in strict order, those residents who have ranked it. These are the *preference lists* for the instance. Note that preference lists are *consistent* in the sense that a resident r appears on a hospital h's list if and only if h appears on r's list. Consistency of preference lists will

---

* Supported by award NUF-NAL-02 from the Nuffield Foundation and grant GR/R84597/01 from the Engineering and Physical Sciences Research Council.

H. Alt and M. Habib (Eds.): STACS 2003, LNCS 2607, pp. 439–450, 2003.
© Springer-Verlag Berlin Heidelberg 2003

be assumed throughout. A resident-hospital pair (r, h) are *mutually acceptable* if they are each on the other's preference list, and we denote by A (a subset of $R \times H$) the set of mutually acceptable pairs, with $|A| = a$.

A *matching* M is a subset of A such that $|\{h : (r, h) \in M\}| \leq 1$ for all r and $|\{r : (r, h) \in M\}| \leq p_h$ for all h. For a matching M, we denote by $M(r)$ the hospital to which r is assigned in M (this is null if r is unassigned), and by $M(h)$ the set of residents assigned to h in M. For a pair $(r, h) \in A$, we define $h \prec_r M(r)$ to mean that either r is unassigned in M, or r prefers h to $M(r)$. Likewise, we define $r \prec_h M(h)$ to mean that either $|M(h)| < p_h$, or h prefers r to at least one of the members of $M(h)$.

A matching M in an instance of HR is *stable* if there is no pair $(r, h) \in A \backslash M$, such that $h \prec_r M(r)$ and $r \prec_h M(h)$. If such a pair (r, h) exists it is said to be a *blocking pair* for the matching M, or to *block* M. The existence of a blocking pair potentially undermines the matching, since both members of the pair could improve their situation by becoming matched to each other.

An instance of HR can be solved by an extension of the Gale/Shapley (GS) algorithm for SM [2,3]. SM can be viewed as a restriction of HR in which each hospital has quota 1 (and the residents and hospitals are re-named men and women). As with SM, a stable matching exists for every instance of HR, and all stable matchings for a given instance have the same size [3]. The extension of the GS algorithm finds one such matching in $O(a)$ time [3].

Recent pressure from student bodies associated with the NRMP has ensured that the extended version of the GS algorithm that is employed by the scheme is now *resident-oriented*, meaning that it produces the *resident-optimal* stable matching for a given instance of HR [13]. This is the unique stable matching M_0 in which every resident assigned in M_0 is assigned the best hospital that he/she could obtain in any stable matching, and any resident unassigned in M_0 is unassigned in any stable matching.

In this paper we consider generalisations of HR in which preferences involve some form of indifference. This is highly relevant for practical matching schemes — for example, a popular hospital may be unable or unwilling to produce a strict ranking over all of its many applicants.

The most natural form of indifference involves ties. A set R' of k residents forms a *tie* of length k in the preference list of hospital h if h does not prefer r_i to r_j for any $r_i, r_j \in R'$ (i.e. h is *indifferent* between r_i and r_j), while for any other resident r who is acceptable to h, either h prefers r to all of the residents in R', or h prefers all of the residents in R' to r. A tie on a resident's list is defined similarly. For convenience in what follows, we consider an untied entry in a preference list as a tie of length 1.

We denote by HRT the version of the problem in which preference lists can include arbitrary ties, and by HRP the version in which each preference 'list' can be an arbitrary partial order. This latter version allows for more complex forms of indifference, generalising the case of lists with ties. For a given practical application, a variety of external factors could contribute to a given preference structure, yielding a more complex form of indifference represented by an arbitrary partial order. Given an instance I of HRT or HRP, a *derived* instance of HR is any instance of HR obtained from I by resolving the indifference (breaking all of the ties or extending each partial order to a total order).

These extensions of the original problem force a re-evaluation of the concept of a blocking pair. We could view a pair (r, h) to be a blocking pair if, by coming together (a) both parties would be better off, or (b) neither party would be worse off, or (c) one party would be better off and the other no worse off. These three possibilities give rise to the notions of weak stability, super-stability, and strong stability, respectively, first considered by Irving [5] in the context of the Stable Marriage problem. We now formally define these three forms of stability.

A matching M in an instance of HRT or HRP is *weakly stable* if there is no pair $(r, h) \in A \setminus M$, such that $h \prec_r M(r)$ and $r \prec_h M(h)$.

A weakly stable matching exists for every instance of HRP, and can be found by forming a derived instance of HR, and applying the extended GS algorithm. It turns out that, in contrast to HR, weakly stable matchings for an instance of HRT may have different sizes, and it is notable that the problem of finding the largest weakly stable matching, and various other problems involving weak stability, are NP-hard [9,12].

To define super-stability and strong stability we need to extend our notation. For a given matching M and pair $(r, h) \in A$, we define $h \lhd_r M(r)$ to mean that r is unassigned in M, or that r prefers h to $M(r)$, or is indifferent between them. Likewise, $r \lhd_h M(h)$ means that $|M(h)| < p_h$, or h prefers r to at least one member of $M(h)$, or is indifferent between r and at least one member of $M(h)$.

A matching M is *super-stable* if there is no pair $(r, h) \in A \setminus M$, such that $h \lhd_r M(r)$ and $r \lhd_h M(h)$.

By contrast with weak stability, it is trivial to show that there are instances of HRT and HRP for which no super-stable matching exists. However, there is an $O(a)$ algorithm to determine whether an instance of HRT admits a super-stable matching, and to find one if it does [7]. With some straightforward modifications, this algorithm is also applicable in the more general context of HRP.

A matching M is *strongly stable* if there is no pair $(r, h) \in A \setminus M$, such that either (i) $h \prec_r M(r)$ and $r \lhd_h M(h)$; or (ii)$h \lhd_r M(r)$ and $r \prec_h M(h)$. If $(r, h) \in M$ for some strongly stable matching M we say that (r, h) is a *strongly stable pair*.

Again, it is easy to construct an instance of HRT that does not admit a strongly stable matching (see [5] for further details). Clearly, as is implied by the terminology, a super-stable matching is strongly stable, and a strongly stable matching is weakly stable.

There is a sense in which strong stability can be viewed as the most appropriate criterion for a practical matching scheme when there is indifference in the preference lists, and that in cases where a strongly stable matching exists, it should be chosen instead of a matching that is merely weakly stable. Consider a weakly stable matching M for an instance of HRT or HRP, and suppose that $h \prec_r M(r)$ while h is indifferent between r and at least one member r' of $M(h)$. Such a pair (r, h) would not constitute a blocking pair for weak stability (if

$|M(h)| = p_h$). However, r might have such an overriding preference for h over $M(r)$ that he is prepared to engage in persuasion, even bribery, in the hope that h will reject r' and accept r instead. Hospital h, being indifferent between r and r' may yield to such persuasion, and, of course, a similar situation could arise with the roles reversed. However, the matching cannot be potentially undermined in this way if it is strongly stable. On the other hand, insisting on super-stability seems unnecessarily restrictive (and is less likely to be attainable).

Hence, strong stability avoids the possibility of a matching being undermined by persuasion or bribery, and is therefore a desirable property in cases where it can be achieved.

In this paper we present an $O(a^2)$ algorithm for finding a strongly stable matching, if one exists, given an instance of HRT, thus solving an open problem described in [7]. Our algorithm is resident-oriented in that it finds a strongly stable matching with similar optimality properties to those of the resident-optimal stable matching in HR, as mentioned above. This algorithm is a non-trivial extension of the strong stability algorithms for the stable marriage problem due to Manlove [10] and Irving [5]. We also prove that the complexity of any algorithm for HRT under strong stability has the same lower bound as applies to the problem of determining if a bipartite graph has a perfect matching. By contrast, we show that the problem of deciding whether a given instance of HRP admits a strongly stable matching is NP-complete.

The remainder of this paper is structured as follows. In Section 2 we present the polynomial-time algorithm that finds a strongly stable matching in an instance of HRT, when one exists. In Section 3 we establish the complexity of the algorithm to be $O(a^2)$. The lower bound for the problem of finding a strongly stable matching in an instance of HRT is given in Section 4, whilst Section 5 contains the NP-completeness result for HRP under strong stability. Finally, Section 6 presents our conclusions and some open problems.

2   An Algorithm for Strong Stability in HRT

In this section we describe our algorithm for finding a strongly stable matching, if one exists, given an instance of HRT, and prove its correctness. Before doing so, we present some definitions relating to the algorithm.

A hospital h such that $|\{r : (r, h) \in M\}| = p_h$ is said to be *full* in the matching M. During the execution of the algorithm residents become *provisionally assigned* to hospitals, and it is possible for a hospital to be provisionally assigned a number of residents that exceeds its quota. At any stage, a hospital is said to be *over-subscribed, under-subscribed* or *fully-subscribed* according as it is provisionally assigned a number of residents greater than, less than, or equal to, its quota. We describe a hospital as *replete* if at any time during the execution of the algorithm it has been over-subscribed or fully subscribed.

The algorithm proceeds by deleting from the preference lists pairs that cannot be strongly stable. By the *deletion* of a pair (r, h), we mean the removal of r and h from each other's lists, and, if r is provisionally assigned to h, the breaking of this provisional assignment. By the *head* and *tail* of a preference list at a given point we mean the first and last ties respectively on that list (recalling that a tie can be of length 1). We say that a resident r is *dominated* in a hospital h's list if h prefers to r at least p_h residents who are provisionally assigned to it.

A resident r who is provisionally assigned to a hospital h is said to be *bound* to h if h is not over-subscribed or r is not in h's tail (or both). The *provisional assignment graph* G has a vertex for each resident and each hospital, with (r, h) forming an edge if resident r is provisionally assigned to hospital h. A *feasible matching* in a provisional assignment graph is a matching M such that, if r is bound to one or more hospitals, then r is matched with one of these hospitals in M, and subject to this restriction, M has maximum possible cardinality.

A *reduced assignment graph* G_R is formed from a provisional assignment graph as follows. For each resident r, for any hospital h such that r is bound to h, we delete the edge (r, h) from the graph, and we reduce the quota of h by one; furthermore, we remove *all* other edges incident to r. Each isolated resident vertex is then removed from the graph. Finally, if the quota of any hospital h is reduced to 0, or h becomes an isolated vertex, then h is removed from the graph. For each surviving h we denote by p'_h the revised quota.

Given a set Z of residents in G_R, define $\mathcal{N}(Z)$, the *neighbourhood* of Z, to be the set of hospital vertices adjacent in G_R to a resident vertex in Z. The *deficiency* of Z is defined by $\delta(Z) = |Z| - \sum_{h \in \mathcal{N}(Z)} p'_h$. It is not hard to show that, if Z_1 and Z_2 are maximally deficient, then so also is $Z_1 \cap Z_2$, so there is a unique minimal set with maximum deficiency. This is the *critical set*.

The algorithm, displayed in Figure 1, begins by assigning each resident to be free (i.e. not assigned to any hospital), and each hospital h to be non-replete. The iterative stage of the algorithm involves each free resident in turn being provisionally assigned to the hospital(s) at the head of his list. If by gaining a new provisional assignee a hospital h becomes fully or over-subscribed then it is set to be replete, and each pair (r, h), such that r is dominated in h's list, is deleted. This continues until every resident is provisionally assigned to one or more hospitals or has an empty list. We then find the reduced assignment graph G_R and the critical set Z of residents. As we will see later, no hospital in $\mathcal{N}(Z)$ can be assigned a resident from those in its tail in any strongly stable matching, so all such pairs are deleted. The iterative step is then reactivated, and this entire process continues until Z is empty, which must happen eventually, since if Z is found to be non-empty, then at least one pair is subsequently deleted from the preference lists.

Let M be any feasible matching in the final provisional assignment graph G. Then M is a strongly stable matching unless either (a) some replete hospital h is not full in M, or (b) some non-replete hospital h has a number of assignees in M less than its degree in G; in cases (a) and (b) no strongly stable matching exists.

The correctness of Algorithm HRT-strong, and an optimality property of any strongly stable matching that it finds, are established by means of three

    assign each resident to be free;
    assign each hospital to be non-replete;
    repeat {
        while some resident r is free and has a non-empty list
            for each hospital h at the head of r's list {
                provisionally assign r to h;
                if h is fully-subscribed or over-subscribed {
                    set h to be replete;
                    for each resident r' dominated on h's list
                        delete the pair (r', h); } }
        form the reduced assignment graph;
        find the critical set Z of residents;
        for each hospital $h \in \mathcal{N}(Z)$
            for each resident r in the tail of h's list
                delete the pair (r, h);
    } **until** $Z == \emptyset$;
    let G be the final provisional assignment graph;
    let M be a feasible matching in G;
    if (some replete hospital is not full in M) **or**
    (some non-replete hospital has fewer assignees in M than its degree in G)
        no strongly stable matching exists;
    else
        output the strongly stable matching specified by M;

Fig. 1. Algorithm HRT-strong

key lemmas. Here we give sketch proofs of the three lemmas. The full proofs, together with some auxiliary lemmas, appear in [8].

Lemma 1. *A matching output by Algorithm HRT-strong is strongly stable.*

Proof (Sketch). We suppose that a pair (r, h) blocks a matching M output by Algorithm HRT-strong, and show that (r, h) cannot have been deleted. Hence r must be bound to both h and $M(r)$ in G, the final provisional assignment graph. But we also show that, if any resident is bound to two hospitals in G then the algorithm reports that no strongly stable matching exists, a contradiction. □

Lemma 2. *No strongly stable pair is ever deleted during an execution of Algorithm HRT-strong.*

Proof (Sketch). We let (r, h) be the first strongly stable pair deleted during some execution of the algorithm, and let M be a strongly stable matching containing (r, h). We consider both points in the algorithm at which (r, h) could be deleted. If (r, h) is deleted because r becomes dominated in h's list, we show that one of the residents provisionally assigned to h at this point must form a blocking pair for M with h. If, on the other hand, (r, h) is deleted because r is in h's tail at a point when h is provisionally assigned a resident from the critical set Z, then we show that there is some resident $r' \in Z$ and some hospital $h' \in \mathcal{N}(Z)$, with r' provisionally assigned to h' at this point, such that r' is not assigned in M to a hospital from the head of his current list, and h' is assigned in M at least one resident from the tail of its current list. It follows that (r', h') blocks M.   □

Lemma 3. *Let M be a feasible matching in the final provisional assignment graph G. If (a) some non-replete hospital h has fewer assignees in M than provisional assignees in G, or (b) some replete hospital h is not full in M, then no strongly stable matching exists.*

Proof (Sketch). We suppose that condition (a) or (b) is satisfied, so that some hospital h must satisfy $|M(h)| < \min(d_G(h), p_h)$, where $d_G(h)$ denotes the degree of vertex h in G (i.e. the number of residents provisionally assigned to h). We also suppose that there is a strongly stable matching, M', for the instance. We show that $|M'| \leq |M|$, so that some hospital h' must also satisfy $|M(h')| < \min(d_G(h'), p_{h'})$. Clearly h' then forms a blocking pair for M' with one of the residents provisionally assigned to it in G.   □

Lemmas 1 and 3 prove the correctness of Algorithm HRT-strong. Further, Lemma 2 shows that there is an optimality property for each assigned resident in any strongly stable matching output by the algorithm. To be precise, we have proved:

Theorem 1. *For a given instance of HRT, Algorithm HRT-strong determines whether or not a strongly stable matching exists. If such a matching does exist, all possible executions of the algorithm find one in which every assigned resident is assigned as favourable a hospital as in any strongly stable matching, and any unassigned resident is unassigned in every strongly stable matching.*

3   Implementation and Analysis of Algorithm HRT-Strong

For the implementation and analysis of Algorithm HRT-strong, we require to describe the efficient construction of maximum cardinality matchings and critical sets in a context somewhat more general than that of simple bipartite graphs.

Consider a *capacitated* bipartite graph $G = (V, E)$, with bipartition $V = R \cup H$, in which each vertex $h \in H$ has a positive integer *capacity* c_h. In this context, a *matching* is a subset M of E such that $|\{h : \{r, h\} \in M\}| \leq 1$ for all $r \in R$, and $|\{r : \{r, h\} \in M\}| \leq c_h$ for all $h \in H$.

For any vertex x, a vertex joined to x by an edge of M is called a *mate* of x. A vertex $r \in R$ with no mate, or a vertex $h \in H$ with fewer than c_h mates, is said to be *exposed*. An *alternating path* in G relative to M is any simple path in which edges are alternately in, and not in, M. An *augmenting* path is an alternating path both of whose end points are exposed. It is immediate that an augmenting path is of odd length, with one end point in R and the other in H.

The following lemmas may be established by straightforward extension of the corresponding results for one-to-one bipartite matching.

Lemma 4. *Let P be the set of edges on an augmenting path relative to a matching M in a capacitated bipartite graph G. Then $M' = M \oplus P$ is a matching of cardinality $|M| + 1$ in G.*

Lemma 5. *A matching M in a capacitated bipartite graph has maximum cardinality if and only if there is no augmenting path relative to M in G.*

The process of replacing M by $M' = M \oplus P$ is called *augmenting M* along path P.

With these lemmas, we can extend to the context of capacitated bipartite graphs the classical augmenting path algorithm for a maximum cardinality matching. The algorithm starts with an arbitrary matching – say the empty matching – and repeatedly augments the matching until there is no augmenting path. The search for an augmenting path relative to M is organised as a restricted breadth-first search in which only edges of M are followed from vertices in H and only edges not in M are followed from vertices in R, to ensure alternation. The number of iterations is $O(\min(|R|, \sum c_h))$, and each search can be completed in $O(|R| + |H| + |E|)$ time.

During the breadth-first search, we record the parent in the BFS spanning tree of each vertex. This enables us to accomplish the augmentation in time $O(|R| + |H| + |E|)$, observing that, for each vertex $h \in H$, the set of mates can be updated in constant time by representing the set as, say, a doubly linked list, and storing a pointer into this list from any child node in the BFS spanning tree.

Hence, overall, the augmenting path algorithm in a capacitated bipartite graph can be implemented to run in $O((\min(|R|, \sum c_h))((|R| + |H| + |E|))$ time.

The following lemma (whose proof appears in [8]) points the way to finding the critical set.

Lemma 6. *Given a maximum cardinality matching M in the capacitated bipartite graph G_R, the critical set Z consists of the set U of unmatched residents together with the set U' of residents reachable from a vertex in U via an alternating path.*

During each iteration of the repeat-until loop of Algorithm HRT-strong we need to form the reduced assignment graph, which takes $O(a)$ time, then search for a maximum cardinality matching in the bipartite graph G_R. This allows us to use Lemma 6 to find the critical set. The key to the analysis of Algorithm HRT-strong, as with Algorithm STRONG in [5], is bounding the total amount of work done in finding the maximum cardinality matchings.

It is clear that work done other than in finding the maximum cardinality matchings and critical sets is bounded by a constant times the number of deleted pairs, and so is $O(a)$.

Suppose that Algorithm HRT-strong finds a maximum cardinality matching M_i in the reduced assignment graph G_R at the ith iteration. Suppose also that, during the ith iteration, x_i pairs are deleted because they involve residents in the critical set Z, or residents tied with them in the list of a hospital in $\mathcal{N}(Z)$. Suppose further that in the $(i+1)$th iteration, y_i pairs are deleted during the proposal sequence. Note that any edge in G_R at the ith iteration which is not one of these $x_i + y_i$ deleted pairs must be in G_R at the $(i+1)$th iteration, since a resident can only become bound to a hospital when he becomes provisionally assigned to it. In particular at least $|M_i| - x_i - y_i$ pairs of M_i remain in G_R at the $(i+1)$th iteration. Hence, in that iteration, we can start from these pairs and find a maximum cardinality matching in $O(min(na, (x_i + y_i + z_i)a))$ time, where n is the number of residents and z_i is the number of edges in G_R at the $(i + 1)$th iteration which were not in G_R at the ith iteration.

Let s denote the number of iterations carried out, let $S = \{1, 2, \ldots, s\}$, and let $S' = S \setminus \{s\}$. Let $T \subseteq S'$ denote those indices i such that $min(na, (x_i + y_i + z_i)a) = na$, and let $t = |T|$. Then the algorithm has time complexity $O(min(n, p)a + tna + a \sum_{i \in S' \setminus T}(x_i + y_i + z_i))$, where p is the total number of posts, and the first term is for the first iteration. But $\sum_{i \in S'}(x_i + y_i) \leq a$ and $\sum_{i \in S'} z_i \leq a$ (since each of these summations is bounded by the total number of deletions and proposals, respectively), and since $x_i + y_i + z_i \geq n$ for each $i \in T$, it follows that

$$tn + \sum_{i \in S' \setminus T}(x_i + y_i + z_i) \leq \sum_{i \in S'}(x_i + y_i + z_i) \leq 2a. \text{ Thus } \sum_{i \in S' \setminus T}(x_i + y_i + z_i) \leq 2a - tn.$$

Hence the overall complexity of Algorithm HRT-strong is $O(min(n, p)a + tna + a(2a - tn)) = O(a^2)$.

4   A Lower Bound for Finding a Strongly Stable Matching

To establish the lower bound of this section, we let STRONGLY STABLE MATCHING IN HRT be the problem of deciding whether a given instance of HRT admits a strongly stable matching.

Let n denote the number of participants in an instance of HRT. We show that, for any function f on n, where $f(n) = \Omega(n^2)$, the existence of an $O(f(n))$ algorithm for STRONGLY STABLE MATCHING IN HRT would imply the existence of an $O(f(n))$ algorithm for PERFECT MATCHING IN BIPARTITE GRAPHS (the problem of deciding whether a given bipartite graph admits a perfect matching).

The result is established by the following simple reduction from PERFECT MATCHING IN BIPARTITE GRAPHS to STRONGLY STABLE MATCHING IN HRT.

Let $G = (V, E)$ be a bipartite graph with bipartition $V = R \cup H$. Let $R = \{r_1, \ldots, r_n\}$ and $H = \{h_1, \ldots, h_n\}$, and, without loss of generality, assume that G contains no isolated vertices. Also, for each i $(1 \leq i \leq n)$, let P_i denote the set of vertices in H adjacent to r_i.

We form an instance I of HRT as follows. Let $p_i = 1$ for all i. Form a preference list for each participant in I as follows:

$$r_i : (P_i) \; (H \setminus P_i) \qquad\qquad h_i : (R) \qquad\qquad (1 \leq i \leq n)$$

In a given participant's preference list (S) denotes all members of the set S listed as a tie in the position where the symbol occurs.

It is straightfoward to verify that G admits a perfect matching if and only if I admits a strongly stable matching. Clearly the reduction may be carried out in $O(n^2)$ time. Hence, for any function f on n, where $f(n) = \Omega(n^2)$, an $O(f(n))$ algorithm for STRONGLY STABLE MATCHING IN HRT would solve PERFECT MATCHING IN BIPARTITE GRAPHS in $O(f(n))$ time. The current best algorithm for PERFECT MATCHING IN BIPARTITE GRAPHS has complexity $O(\sqrt{n}m)$ [4], where m is the number of edges in G.

5   NP-Completeness of Strong Stability in HRP

In this section we establish NP-completeness of STRONGLY STABLE MATCHING IN SMP, which is the problem of deciding whether a given instance of SMP admits a strongly stable matching. Here, SMP denotes the variant of SM in which each person's preferences over the members of the opposite sex are represented as an arbitrary partial order (hereafter this preference structure is referred to as a *preference poset*). Clearly SMP is a special case of HRP in which $A = R \times H$ and each hospital has quota 1. It therefore follows immediately that the problem of deciding whether a given instance of HRP admits a strongly stable matching is also NP-complete.

To prove our result we give a reduction from the following problem:

Name: RESTRICTED SAT.
Instance: Boolean formula B in CNF, where each variable v occurs in exactly two clauses of B as literal v, and in exactly two clauses of B as literal $\overline{v}$.
Question: Is B satisfiable?

RESTRICTED SAT is NP-complete (see [8] for further details). We now state and prove the main result of this section.

Theorem 2. STRONGLY STABLE MATCHING IN SMP *is NP-complete.*

Proof. Clearly STRONGLY STABLE MATCHING IN SMP is in NP. To show NP-hardness, we give a polynomial reduction from RESTRICTED SAT, which is NP-complete as mentioned above. Let B be a Boolean formula in CNF, given as an instance of this, in which $X = \{x_1, x_2, \ldots, x_n\}$ is the set of variables and $C = \{c_1, c_2, \ldots, c_m\}$ is the set of clauses. For each i $(1 \leq i \leq n)$ and for each r $(1 \leq r \leq 2)$, let $c(x_i^r)$ (respectively $c(\overline{x}_i^r)$) denote the clause corresponding to the rth occurrence of literal x_i (respectively $\overline{x}_i$).

We now construct an instance I of SMP, as follows. Let $U = X^1 \cup X^2 \cup \overline{X}^1 \cup \overline{X}^2 \cup Z$ be the set of men in I, and let $W = Y^1 \cup Y^2 \cup \overline{Y}^1 \cup \overline{Y}^2 \cup C$ be the set of women in I, where

$$X^r = \{x_i^r : 1 \leq i \leq n\}\ (1 \leq r \leq 2), \quad Y^r = \{y_i^r : 1 \leq i \leq n\}\ (1 \leq r \leq 2),$$
$$\overline{X}^r = \{\overline{x}_i^r : 1 \leq i \leq n\}\ (1 \leq r \leq 2), \quad \overline{Y}^r = \{\overline{y}_i^r : 1 \leq i \leq n\}\ (1 \leq r \leq 2),$$
$$Z = \{z_i : 1 \leq i \leq m\}, \quad C = \{c_i : 1 \leq i \leq m\}.$$

Clearly $|U| = |W| = 4n + m$. Now, for each person p in I, we formulate $\prec_p^*$, the preference poset of p. In order to define $\prec_p^*$, we will construct a relation $\prec_p$, where $q \prec_p r$ implies that p prefers q to r. We then obtain the partial order $\prec_p^*$ by taking the transitive closure of $\prec_p$. Note that p is indifferent between q and r if and only if q, r are incomparable in $\prec_p^*$ (i.e. neither $q \prec_p^* r$ nor $r \prec_p^* q$ holds). For each person q we will also define a subset $P(q)$ of members of the opposite sex; if $r \in P(q)$ we say that r is *proper* for q.

 – *Preference poset of* x_i^r $(1 \leq i \leq n, 1 \leq r \leq 2)$: $\overline{y}_i^1 \prec_{x_i^r} c(x_i^r)$, $\overline{y}_i^2 \prec_{x_i^r} c(x_i^r)$, $y_i^r \prec_{x_i^r} p$, for every $p \in W \backslash P(x_i^r)$, where $P(x_i^r) = \{c(x_i^r), y_i^r, \overline{y}_i^1, \overline{y}_i^2\}$.

- *Preference poset of* $\overline{x}_i^r$ *($1 \le i \le n$, $1 \le r \le 2$):* $y_i^1 \prec_{\overline{x}_i^r} c(\overline{x}_i^r)$, $y_i^2 \prec_{\overline{x}_i^r} c(\overline{x}_i^r)$, $\overline{y}_i^r \prec_{\overline{x}_i^r} p$, *for every* $p \in W \backslash P(\overline{x}_i^r)$, *where* $P(\overline{x}_i^r) = \{c(\overline{x}_i^r), \overline{y}_i^r, y_i^1, y_i^2\}$.
- *Preference poset of* z_i *($1 \le i \le m$):* $y \prec_{z_i} p$, *for every* $y \in P(z_i)$ *and for every* $p \in W \backslash P(z_i)$, *where* $P(z_i) = Y^1 \cup Y^2 \cup \overline{Y}^1 \cup \overline{Y}^2$.
- *Preference poset of* y_i^r *($1 \le i \le n$, $1 \le r \le 2$):* $x_i^r \prec_{y_i^r} \overline{x}_i^1$, $x_i^r \prec_{y_i^r} \overline{x}_i^2$. *Let* $P(y_i^r) = \{x_i^r, \overline{x}_i^1, \overline{x}_i^2\} \cup Z$.
- *Preference poset of* $\overline{y}_i^r$ *($1 \le i \le n$, $1 \le r \le 2$):* $\overline{x}_i^r \prec_{\overline{y}_i^r} x_i^1$, $\overline{x}_i^r \prec_{\overline{y}_i^r} x_i^2$. *Let* $P(\overline{y}_i^r) = \{\overline{x}_i^r, x_i^1, x_i^2\} \cup Z$.
- *Preference poset of* c_i *($1 \le i \le m$):* $\prec_{c_i} = \emptyset$. *Let* $P(c_i)$ *contain those members of* $X^1 \cup X^2 \cup \overline{X}^1 \cup \overline{X}^2$ *corresponding to the literal-occurrences in clause* c_i.

It is easy to verify that, for any two people q, r of the opposite sex, r is proper for q if and only if q is proper for r.

Now suppose that B admits a satisfying truth assignment f. We form a matching M in I as follows. For each clause c_i in B ($1 \le i \le m$), pick any literal-occurrence $x \in X^1 \cup X^2 \cup \overline{X}^1 \cup \overline{X}^2$ corresponding to a true literal in c_i, and add (x, c_i) to M. For any x_i^r left unmatched ($1 \le i \le n$, $1 \le r \le 2$), add (x_i^r, y_i^r) to M. Similarly, for any $\overline{x}_i^r$ left unmatched ($1 \le i \le n$, $1 \le r \le 2$), add $(\overline{x}_i^r, \overline{y}_i^r)$ to M. Finally, there remain m members of $Y^1 \cup Y^2 \cup \overline{Y}^1 \cup \overline{Y}^2$ that are as yet unmatched. Add to M a perfect matching between these women and the men in Z. It is straightforward to verify that M is strongly stable in I.

Conversely suppose that I admits a strongly stable matching M. Then it is not difficult to see that $(m, w) \in M$ implies that w is proper for m and vice versa. Also, for each i ($1 \le i \le n$), c_i is matched in M to some man $x \in X^1 \cup X^2 \cup \overline{X}^1 \cup \overline{X}^2$ corresponding to an occurrence of a literal in clause c_i of B. Suppose that $x = x_i^r$ for some i ($1 \le i \le n$) and r ($1 \le r \le 2$) (the argument is similar if $x = \overline{x}_i^r$). Then by the strong stability of M, $(\overline{x}_i^1, \overline{y}_i^1) \in M$ and $(\overline{x}_i^2, \overline{y}_i^2) \in M$. Thus we may form a truth assignment f for B as follows: if $x = x_i^r$ then set variable x_i to have value T, otherwise if $x = \overline{x}_i^r$ then set variable x_i to have value F. Any remaining variable whose truth value has not yet been assigned can be set to T. Clearly f is a satisfying truth assignment for B.   □

6   Conclusion and Open Questions

In this paper we have described a polynomial-time algorithm for the problem of finding a strongly stable matching, if one exists, given an instance of HRT. By constrast we have shown that the corresponding existence question becomes NP-complete for HRP. However, much remains to be investigated, and the following questions are particularly noteworthy:

1. For a given instance I of HRT that admits a strongly stable matching M, it is possible that I can admit weakly stable matchings of sizes $> |M|$ and $< |M|$ (see [8] for an example). However all strongly stable matchings for a given instance of HRT have the same cardinality [8]. In this paper, the robustness

of a strongly stable matching (against situations of persuasion or bribery) has been our primary motivation for studying HRT under strong stability. Nevertheless, further consideration should be given to the relative sizes of weakly stable matchings compared to the size of strongly stable matchings, given an instance of HRT.

2. The current algorithm for strong stability in HRT is resident-oriented. However, for super-stability in HRT there are both resident-oriented and hospital-oriented algorithms [7]. The problem of describing a hospital-oriented algorithm for HRT under strong stability remains open.

3. For a given instance of the stable marriage problem with ties, it is known that the set of strongly stable matchings forms a distributive lattice, when the set is partitioned by a suitable equivalence relation [11]. It remains open to characterise any similar structure in the set of strongly stable matchings for a given instance of HRT.

References

1. Canadian Resident Matching Service. How the matching algorithm works. Web document available at http://www.carms.ca/matching/algorith.htm.
2. D. Gale and L.S. Shapley. College admissions and the stability of marriage. *American Mathematical Monthly*, 69:9–15, 1962.
3. D. Gusfield and R.W. Irving. *The Stable Marriage Problem: Structure and Algorithms*. MIT Press, 1989.
4. J.E. Hopcroft and R.M. Karp. A $n^{5/2}$ algorithm for maximum matchings in bipartite graphs. *SIAM Journal on Computing*, 2:225–231, 1973.
5. R.W. Irving. Stable marriage and indifference. *Discrete Applied. Mathematics*, 48:261–272, 1994.
6. R.W. Irving. Matching medical students to pairs of hospitals: a new variation on a well-known theme. In *Proceedings of ESA '98*, volume 1461 of *Lecture Notes in Computer Science*, pages 381–392. Springer-Verlag, 1998.
7. R.W. Irving, D.F. Manlove, and S. Scott. The Hospitals/Residents problem with Ties. In *Proceedings of SWAT 2000*, volume 1851 of *Lecture Notes in Computer Science*, pages 259–271. Springer-Verlag, 2000.
8. R.W. Irving, D.F. Manlove and S. Scott. Strong Stability in the Hospitals/Residents Problem. Technical Report TR-2002-123, University of Glasgow, Computing Science Department, 2002.
9. K. Iwama, D. Manlove, S. Miyazaki, and Y. Morita. Stable marriage with incomplete lists and ties. In *Proceedings of ICALP '99*, volume 1644 of *Lecture Notes in Computer Science*, pages 443–452. Springer-Verlag, 1999.
10. D.F. Manlove. Stable marriage with ties and unacceptable partners. Technical Report TR-1999-29, University of Glasgow, Computing Science Department, 1999.
11. D.F. Manlove. The structure of stable marriage with indifference. *Discrete Applied Mathematics*, 122:167–181, 2002.
12. D.F. Manlove, R.W. Irving, K. Iwama, S. Miyazaki, and Y. Morita. Hard variants of stable marriage. *Theoretical Computer Science*, 276:261–279, 2002.
13. M. Mukerjee. Medical mismatch. *Scientific American*, 276(6):40-41, 1997.
14. A.E. Roth. The evolution of the labor market for medical interns and residents: a case study in game theory. *Journal of Political Economy*, 92(6):991–1016, 1984.

The Inference Problem for Propositional Circumscription of Affine Formulas Is coNP-Complete

Arnaud Durand[1] and Miki Hermann[2]

[1] LACL Paris 12 and LAMSADE Paris 9 (CNRS UMR 7024), Dept. of Computer Science, Université Paris 12, 94010 Créteil, France. durand@univ-paris12.fr
[2] LIX (CNRS, UMR 7650), École Polytechnique, 91128 Palaiseau cedex, France. hermann@lix.polytechnique.fr

Abstract. We prove that the inference problem of propositional circumscription for affine formulas is coNP-complete, settling this way a longstanding open question in the complexity of nonmonotonic reasoning. We also show that the considered problem becomes polynomial-time decidable if only a single literal has to be inferred from an affine formula.

1 Introduction and Summary of Results

Various formalisms of nonmonotonic reasoning have been investigated during the last twenty-five years. Circumscription, introduced by McCarthy [McC80], is a well-developed formalism of common-sense reasoning extensively studied by the artificial intelligence community. It has a simple and clear semantics, and benefits from high expressive power, that makes it suitable for modeling many problems involving nonmonotonic reasoning. The key idea behind circumscription is that we are interested only in the *minimal models* of formulas, since they are the ones with as few "exceptions" as possible, and embody therefore common sense. Moreover, propositional circumscription inference has been shown by Gelfond *et al.* [GPP89] to coincide with reasoning under the extended closed world assumption, which is one of the main formalisms for reasoning with incomplete information. In the context of Boolean logic, circumscription amounts to the study of models of Boolean formulas that are *minimal* with respect to the *pointwise partial order* on models.

Several algorithmic problems have been studied in connection with propositional circumscription: among them the *model checking* and the *inference* problems. Given a propositional formula φ and a truth assignment s, the model checking problem asks whether s is a minimal model of φ. Given two propositional formulas φ and ψ, the inference problem asks whether ψ is true in every minimal model of φ. Cadoli proved in [Cad92] the model checking problem to be coNP-complete, whereas Kirousis and Kolaitis settled in [KK01a] the question of the dichotomy theorem for this problem. The inference problem was proved $\Pi_2 P$-complete by Eiter and Gottlob in [EG93]. Cadoli and Lenzerini proved in [CL94] that the inference problem becomes coNP-complete if φ is a

H. Alt and M. Habib (Eds.): STACS 2003, LNCS 2607, pp. 451–462, 2003.

Krom or a dual Horn formula. See also [CMM01] for an exhaustive overview of existing complexity results in nonmonotonic reasoning and circumscription. The complexity of the inference problem for affine formulas remained open for ten years. It was known that the problem is in coNP, but there was no proved coNP-hardness lower bound.

This paper is a partial result of our effort to find an output-polynomial algorithm for enumerating the minimal models of affine formulas, an open problem stated in [KSS00]. Following the result of Berlekamp *et al.* [BMvT78], it is clear that we cannot develop an output-polynomial algorithm for this enumeration problem by producing consecutive minimal models of the affine system with increasing Hamming weight, unless P = NP. Another natural approach consists of producing partial assignments to the variables that are extended to minimal models afterwards. However, as our result indicates, this new approach does not lead to an output-polynomial algorithm either, unless the same collapse occurs.

We settle in this paper the complexity of the inference problem for the propositional circumscription of affine formulas, proving that the problem is coNP-complete. First, we prove a new criterion for determining whether a given partial solution of an affine system can be extended to a minimal one. This criterion, which is interesting on its own, is then extensively used in the subsequent coNP-hardness proof of the inference problem for affine formulas. More precisely, we prove the NP-hardness of the problem, given a partial solution s of an affine system S, whether it can be extended to a minimal solution $\bar{s}$. To our knowledge, this proof uses a new approach combining matroid theory, combinatorics, and computational complexity techniques. The inference problem for affine circumscription is then the dual problem to minimal extension, what proves the former to be coNP-complete. Finally, we prove that the restriction of the affine inference problem with ψ being a single literal is decidable in polynomial time.

2   Preliminaries

Let $s = (s_1, \ldots, s_n)$ and $s' = (s'_1, \ldots, s'_n)$ be two Boolean vectors from $\{0, 1\}^n$. We write $s < s'$ to denote that $s \neq s'$ and $s_i \leq s'_i$ holds for every $i \leq n$. Let $\varphi(x_1, \ldots, x_n)$ be a Boolean formula having $x_1, \ldots, x_n$ as its variables and let $s \in \{0, 1\}^n$ be a truth assignment. We say that s is a *minimal model* of φ if s is a satisfying truth assignment of φ and there is no satisfying truth assignment s' of φ that satisfies the relation $s' < s$. This relation is called the *pointwise partial order* on models.

Let $\varphi(x_1, \ldots, x_n)$ be a propositional formula in conjunctive normal form. We say that $\varphi(x)$ is *Horn* if φ has at most one positive literal per clause, *dual Horn* if φ has at most one negative literal per clause, *Krom* if φ has at most two literals per clause, and *affine* if φ is a conjunction of clauses of the type $x_1 \oplus \cdots \oplus x_n = 0$ or $x_1 \oplus \cdots \oplus x_n = 1$, where $\oplus$ is the exclusive-or logical connective, what is equivalent to an affine system of equations $S : Ax = b$ over $\mathbb{Z}_2$.

Let φ and ψ be two propositional formulas in conjunctive normal form. We say that ψ follows from φ in propositional circumscription, denoted by $\varphi \models_{\min} \psi$, if ψ is true in every minimal model of φ. Since ψ is a conjunction $c_1 \wedge \cdots \wedge c_k$ of clauses c_i, then $\varphi \models_{\min} \psi$ if and only if $\varphi \models_{\min} c_i$ for each i. Hence we can restrict ourselves to consider only a single clause instead of a formula ψ at the right-hand side of the propositional inference problem $\varphi \models_{\min} c$. We can further restrict the clause c to one containing only negative literals $c = \neg u_1 \vee \cdots \vee \neg u_n$, as it was showed in [KK01b].

If x and y are two vectors, we denote by $z = xy$ the vector obtained by concatenation of x and y. Let $S \colon Az = b$ be a $k \times n$ affine system of equations over $\mathbb{Z}_2$. Without loss of generality, we assume that the system S is in standard form, i.e., that the matrix A has the form $(I\ B)$, where I is the $k \times k$ identity matrix and B is an arbitrary $k \times (n - k)$ matrix of full column rank. For convenience, we denote by x the variables from z associated with I and by y the ones associated with B. Hence, we consider affine systems of the form $S \colon (I\ B)(xy) = b$.

If A is a $k \times n$ matrix, we denote by $A(i, j)$ the element of A positioned at row i and column j. The vector forming the row i of the matrix A is denoted by $A(i, -)$, whereas the column vector j of A is denoted by $A(-, j)$. Let $I \subseteq \{1, \ldots, k\}$ and $J \subseteq \{1, \ldots, n\}$ be two index sets. Then $A(I, -)$ denotes the submatrix of A restricted to the rows I. Similarly, $A(-, J)$ is then the submatrix of A restricted to the columns J, whereas $A(I, J)$ stands for the submatrix of A restricted to the rows I and columns J. There are also two matrices with a special notation: the $k \times k$ identity matrix I_k and the $k \times n$ all-zero matrix O_k^n.

For a $k \times n$ affine system $S \colon Az = b$ over $\mathbb{Z}_2$, an index set $J = \{j_1, \ldots, j_m\} \subseteq \{1, \ldots, n\}$ of cardinality $|J| = m$, and a Boolean vector $v = (v_1, \ldots, v_m)$ of length m, we denote by $S[J/v]$ the new system $S' \colon A'z' = b'$ formed by replacing each variable z_{j_i} by the value v_i. We also denote by $one(v) = \{i \mid v_i = 1\}$ and $zero(v) = \{i \mid v_i = 0\}$ the positions in the vector v assigned to the values 1 and 0, respectively. The Hamming weight $wt(v)$ of a vector v is equal to the cardinality of the set $one(v)$, i.e., $wt(v) = |one(v)|$.

Each affine system $S \colon Az = b'$ can be transformed to the standard form $(I\ B)(xy) = b$ by means of Gaussian elimination in polynomial time. without changing the ordering of solutions. Indeed, a row permutation or addition does not change the solutions of S. A column permutation permutes the variables and therefore also the positions in each solution uniformly. However, for each column permutation π and a couple of solutions s, s', the relation $s < s'$ holds if and only if $\pi(s) < \pi(s')$. This allows us to consider affine systems in the form $S \colon (I\ B)(xy) = b$ without loss of generality.

Suppose that s is a variable assignment for the variables y, i.e., for each $y_i \in y$ there exists a value $s(y_i) \in \mathbb{Z}_2$. The vector s is a *partial assignment* for variables $z = xy$. An *extension* of the vector s is a variable assignment $\bar{s}$ for each variable from z, i.e., for each $z_i \in z$ there exists a value $\bar{s}(z_i) \in \mathbb{Z}_2$, such that $s(y_i) = \bar{s}(y_i)$ for each y_i. If s is a variable assignment for the variables y in the affine system $S \colon (I\ B)(xy) = b$ then the extension $\bar{s}$ to a solution of the system S is *unique*. If the variables y in the system $S \colon (I\ B)(xy) = b$ have been assigned, then the values for the variables x are already determined. In connection with the previous notions we define the following two **index sets**

$$eq(s) = \{i \mid (Bs)_i = b_i\} \qquad \text{and} \qquad neq(s) = \{i \mid (Bs)_i \neq b_i\},$$

where $b = (b_1, \ldots, b_k)$ and $(Bs)_i$ means the i-th position of the vector obtained after multiplication of the matrix B by the vector s. The set $eq(s)$ (resp. $neq(s)$) is the subset of row indices i for which the unique extension $\bar{s}$ satisfies the equality $\bar{s}(x_i) = 0$ (resp. $\bar{s}(x_i) = 1$). It is clear that $eq(s) \cap neq(s) = \emptyset$ and $eq(s) \cup neq(s) = \{1, \ldots, k\}$ hold for each s.

3   A New Criterion for Affine Minimality

There exists a straightforward method to determine in polynomial time whether a solution s is minimal for an affine system S over $\mathbb{Z}_2$. However, this method is unsuitable for testing whether a partial solution s can be extended to a minimal solution $\bar{s}$ of S. We propose here a completely new method well-suited to decide whether an extension $\bar{s}$ is a minimal solution of S.

Proposition 1. *Let* $S\colon (I\ B)(xy) = b$ *be an affine* $k \times n$ *system over* $\mathbb{Z}_2$ *and let* s *be a Boolean vector of length* $n - k$. *The extension* $\bar{s}$ *is a minimal solution of* S *if and only if* $B(eq(s), one(s))$ *is a matrix of column rank* $wt(s)$, *i.e., all its columns are linearly independent.*

Proof. Suppose that $\bar{s}$ is minimal and the matrix $B(eq(s), one(s))$ has the column rank smaller than $wt(s)$. This means that the columns of $B(eq(s), one(s))$ are linearly dependent, therefore there exists a subset $J \subseteq one(s)$, such that $\sum_{j \in J} B(eq(s), j) = \mathbf{0}$ holds. Let t be a Boolean vector satisfying the condition $one(t) = one(s) \smallsetminus J$. The columns of the matrix $B(eq(s), one(s))$ can be partitioned into two sets: those in J and those in $one(t)$. Knowing that the columns in J add up to the zero vector $\mathbf{0}$, we derive the following equality.

$$\sum_{j \in one(s)} B(eq(s), j) = \sum_{j \in one(t)} B(eq(s), j) + \sum_{j \in J} B(eq(s), j) = \sum_{j \in one(t)} B(eq(s), j)$$

The vector t is smaller than s in the pointwise order. We will show that also the extensions $\bar{s}$ and $\bar{t}$ satisfy the relation $\bar{t} < \bar{s}$. For each row $i \in eq(s)$, the coefficients $B(i, j)$ sum up to the value b_i, i.e., that $\sum_{j \in one(s)} B(i, j) = \sum_{j \in one(t)} B(i, j) = b_i$. Recall that each variable in the vector x occurs in the system S exactly once, because of the associated identity matrix I_k. Since already the assignments s and t to the variables y sum up to the value b_i, this determines the value of the variable x_i in the extensions $\bar{s}$ and $\bar{t}$ to be $\bar{s}(x_i) = \bar{t}(x_i) = 0$ for each row $i \in eq(s)$. In the same spirit, the assignment s to the variables y sums up to the value $1 - b_i$ for each row $i \in neq(s)$, what determines the value of the variable x_i in the extension $\bar{s}$ to be $\bar{s}(x_i) = 1$. Therefore we have $\bar{t}(x_i) \leq \bar{s}(x_i) = 1$ for each row $i \in neq(s)$. This shows that $\bar{t}$ is a solution of S smaller than $\bar{s}$, what contradicts our assumption that $\bar{s}$ is minimal.

Conversely, suppose that the matrix $B(eq(s), one(s))$ has the column rank $wt(s)$ but $\bar{s}$ is not minimal. The latter condition implies that there exists a variable assignment t, such that the extension $\bar{t}$ is a solution of S satisfying the relation $\bar{t} < \bar{s}$. Let $J = one(\bar{s}) \smallsetminus one(\bar{t})$ be the set of positions on which the extensions $\bar{s}$ and $\bar{t}$ differ. Both extensions $\bar{s}$ and $\bar{t}$ are solutions of S, therefore we have $(I \ B)\bar{s} + (I \ B)\bar{t} = \sum_{j \in J}(I \ B)(-,j) = \mathbf{0}$. The index set J can be partitioned into two disjoint sets J_1 containing the positions smaller or equal to k, that are associated with the identity matrix I, and the set J_2 containing the positions greater than k, that are associated with the matrix B. Hence the inclusion $J_2 \subseteq one(s)$ holds. The columns of the identity matrix I are linearly independent, therefore the set J_2 must be nonempty in order to get the above sum equal to $\mathbf{0}$. The partition of J implies the equality $\sum_{j \in J_1} I(-,j) + \sum_{j \in J_2} B(-,j) = \mathbf{0}$. The restriction of this equality to the rows in $eq(s)$ yields $\sum_{j \in J_1} I(eq(s),j) + \sum_{j \in J_2} B(eq(s),j) = \mathbf{0}$. The vector $\bar{s}$ is a solution of S and for each row $i \in eq(s)$ we have $\bar{s}(x_i) = 0$, since already the values $s(y_j)$ with $j \in J_2$ sum up to b_i. This implies together with the previous equation that $i \notin J_1$, since $i \leq k$ holds, and for all indices $j \in J_1$ the column $I(eq(s),j)$ is the all-zero vector. This yields the equality $\sum_{j \in J_1} I(eq(s),j) = \mathbf{0}$, what implies the final equality $\sum_{j \in J_2} B(eq(s),j) = \mathbf{0}$. Since J_2 is a subset of the columns $one(s)$, this contradicts the fact that the matrix $B(eq(s), one(s))$ has the column rank $wt(s)$. $\qquad\square$

4   Extension and Inference Problems

In this paper we will be interested in the complexity of the inference problem of propositional circumscription with affine formulas. Since affine propositional formulas are equivalent to affine systems $S\colon Az = b$ over $\mathbb{Z}_2$, this problem can be formulated as follows.

Problem: AFFINF
Input: An affine system $S\colon Az = b$ over $\mathbb{Z}_2$ with a Boolean $k \times n$ matrix A, a Boolean vector b of length k, a variable vector $z = (z_1, \ldots, z_n)$, and a negative clause $c = \neg u_1 \vee \cdots \vee \neg u_m$, where $u_i \in z$ holds for each i.
Question: Does $S \models_{\min} c$ hold?

Another interesting problem, closely related to the previous one, is the problem of extending a Boolean vector to a minimal solution of an affine system.

Problem: MINEXT
Input: An affine system $S\colon Az = b$ over $\mathbb{Z}_2$ with a Boolean $k \times n$ matrix A, a Boolean vector b of length k, a variable vector $z = (z_1, \ldots, z_n)$, and a partial assignment s for the variables y, where $z = xy$.
Question: Can s be extended to a vector $\bar{s}$, such that $\bar{s}$ is a minimal solution of the system S?

The minimal extension problem appears naturally within algorithms enumerating minimal solutions. For any given class of propositional formulas, when the corresponding minimal extension problem is polynomial-time decidable, then there exists an algorithm that enumerates each consecutive pair of minimal solutions with polynomial delay.

To derive the lower bound of the complexity of the latter problem, we need to consider the following well-known NP-complete problem.

Problem: POSITIVE 1-IN-3 SAT

Input: A propositional formula φ in conjunctive normal form with three positive literals per clause.
Question: Is there a truth assignment to the variable of φ, such that exactly one literal is assigned to *true* and the two others are assigned to *false* in every clause?

Theorem 2. MINEXT *is* NP-*complete even if the partial assignment* s *contains no 0.*

Proof. Membership of MINEXT in NP is obvious. For the lower bound, we construct a polynomial reduction from the problem POSITIVE 1-IN-3 SAT.

Let $\varphi(x_1, \ldots, x_n)$ be a propositional formula in conjunctive normal form $c_1 \wedge \cdots \wedge c_m$ with the clauses $c_i = x_i^1 \vee x_i^2 \vee x_i^3$. We construct an affine system $S\colon (I\ B)(zxy) = b$, where I is the $(4m + n) \times (4m + n)$ identity matrix, z, x, and y are variable vectors of respective lengths $4m + n$, n, and $3m$, and B is a special $(4m+n) \times (3m+n)$ matrix encoding the formula φ. We also construct a partial assignment s and show that the formula φ has a model satisfying exactly one variable per clause if and only if s can be extended to a minimal solution of S.

The matrix B is composed from six blocks as follows

$$\begin{pmatrix} B_1^1 & B_1^2 \\ B_2^1 & B_2^2 \\ B_3^1 & B_3^2 \end{pmatrix}$$

The matrix B_1^1 of size $m \times n$ is the clause-variable incidence matrix of the formula φ, i.e., $B_1^1(i, j) = 1$ holds if and only if $x_j \in c_i$. The matrix B_1^2 of size $m \times 3m$ is the identity matrix I_m with each column tripled, i.e., it verifies the conditions $B_1^2(i, 3(i - 1) + 1) = B_1^2(i, 3(i-1) + 2) = B_1^2(i, 3i) = 1$ for all i and $B_1^2(i, j) = 0$ otherwise. The matrix B_2^1 of size $3m \times n$ encodes the polynomials $x_i^1 + x_i^2$, $x_i^2 + x_i^3$, and $x_i^3 + x_i^1$ over $\mathbb{Z}_2$ for each clause $c_i = x_i^1 \vee x_i^2 \vee x_i^3$. This encoding is done for each $i = 1, \ldots, m$ in three consecutive rows. Hence, we have $B_2^1(3i, i_1) = B_2^1(3i, i_2) = 1$, $B_2^1(3i + 1, i_2) = B_2^1(3i + 1, i_3) = 1$, and $B_2^1(3i + 2, i_3) = B_2^1(3i + 2, i_1) = 1$, where i_j is the position of the variable x_i^j in the vector $x = (x_1, \ldots, x_n)$. Otherwise we have $B_2^1(3i + q, j) = 0$ for $q = 0, 1, 2$ and $j \neq i_1, i_2, i_3$. In another words, the rows $B_2^1(3i, -)$, $B_2^1(3i + 1, -)$, and $B_2^1(3i + 2, -)$ are the incidence vectors of the polynomials $x_i^1 + x_i^2$, $x_i^2 + x_i^3$, and $x_i^3 + x_i^1$, respectively. The matrix B_2^2 of size $3m \times 3m$ is the identity matrix I_{3m}. The matrix B_3^1 of size $n \times n$ is the identity matrix I_n, whereas the matrix B_3^2 of size $n \times 3m$ is the all-zero matrix O_n^{3m}. Note that due to the blocks B_2^2 and B_3^1, that are identity matrices, as well as the block B_3^2 that is an all-zero matrix, the matrix B has the column rank $n + 3m$. Denote by B_1 the submatrix of B restricted to the first m rows, i.e., $B_1 = B(\{1, \ldots, m\}, -)$. Analogously, we define $B_2 = B(\{m+1, \ldots, 4m\}, -)$ and $B_3 = B(\{4m+1, \ldots, 4m+n\}, -)$. In the same spirit, we denote by $B^1 = B(-, \{1, \ldots, n\})$ and $B^2 = B(-, \{n+1, \ldots, n+3m\})$ the left and the right part of the columns, respectively, of the matrix B.

The vector b of length $4m + n$ in the system S is a concatenation of three vectors b_1, b_2, and b_3, where b_1 is the all-zero vector of length m, b_2 is the all-zero vector of length $3m$, and b_3 is the all-one vector of length m. The parts b_i of the vector b correspond to the row blocks B_i of the matrix B for $i = 1, 2, 3$. Figure 1 describes the constructed matrix B and vector b.

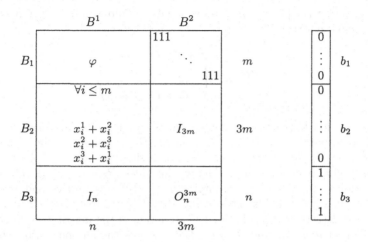

Fig. 1. Matrix B and the associated vector b

Finally, we set the vector s of size $3m$ to be equal to 1 in each coordinate, i.e., $s(y_i) = 1$ for each $i = 1, \ldots, 3m$ and the Hamming weight of s is $wt(s) = 3m$.

Let v be a model of the formula φ satisfying exactly one literal per clause. We will prove that when we append the all-one vector s to v, forming the vector $t = vs$, then the extension $\bar{t}$ is a minimal solution of S. Let us study the set $eq(t)$. Since every clause $c_i = x_i^1 \vee x_i^2 \vee x_i^3$ of φ is satisfied, the sum of literal values is equal to $v(x_i^1) + v(x_i^2) + v(x_i^3) = 1$. Moreover, for each $j = 1, \ldots, m$ we have $s(x_j) = 1$, therefore all m rows of B_1 belong to $eq(t)$. Exactly two of the polynomials $x_i^1 + x_i^2$, $x_i^2 + x_i^3$, and $x_i^3 + x_i^1$ are evaluated to 1 for each clause c_i and for each $j = 1, \ldots, 3m$ we have $s(x_j) = 1$, what implies that exactly $2m$ rows from B_2 belong to the set $eq(t)$. The row i of B_1 and the rows $3(i-1)+1$, $3(i-1)+2$, and $3i$ of B_2 correspond to the clause c_i. Form the corresponding row index set $I(i) = \{i, \, m+3(i-1)+1, \, m+3(i-1)+2, \, m+3i\}$ for a given i. Consider the restriction of the block B^2 to the rows $I(i)$. This restriction $B^2(I(i), -)$ will have plenty of all-zero columns. Keep only the columns containing at least one value 1. These columns will be $3(i-1)+1$, $3(i-1)+2$, and $3i$. Form the corresponding column index set $J(i) = \{n+3(i-1)+1, \, n+3(i-1)+2, \, n+3i\}$ for a given i. The restriction of B to the rows $I(i)$ and columns $J(i)$ is the matrix

$$B(I(i), J(i)) = \begin{pmatrix} 1\ 1\ 1 \\ 1\ 0\ 0 \\ 0\ 1\ 0 \\ 0\ 0\ 1 \end{pmatrix} = B^*(i).$$

Note that the first row of $B^*(i)$ and exactly two out of the three last rows of $B^*(i)$ are also represented in the set $eq(t)$. If we delete one of the last three rows of $B^*(i)$, the resulting square matrix will remain non-singular. Note that the column index sets $J(i)$ are pairwise disjoint and that their union equals the index set $J^* = \{n+1, \ldots, n+3m\}$. Since $B(-, J^*) = B^2$ holds, we easily see that the restriction $B^2(\{1, \ldots, 4m\}, -)$ is equal, modulo a suitable row permutation, to the block matrix

$$B^2_{1+2} = \begin{pmatrix} B^*(1) & O & O \\ O & \ddots & O \\ O & O & B^*(m) \end{pmatrix}.$$

The restriction $B^2(eq(t), -)$ deletes from B^2_{1+2} one of the last three rows of each block corresponding to $B^*(i)$. The matrix B^2_{1+2} is non-singular, what implies that the restriction $B^2(eq(t), -)$ is also non-singular, since $B^*(i)$ with one row deleted remains non-singular. Finally, the block B_3 contributes $wt(v)$ rows to $eq(t)$. Hence, the set $eq(t)$ contains $3m + wt(v)$ row indices and the equality $wt(t) = 3m + wt(v)$ holds. This means that $B(eq(t), one(t))$ is a square matrix. Note that $B(eq(t), one(t))$ is the concatenation of the matrices $B(eq(t), one(v))$ and $B(eq(t), one(s))$, since $t = vs$. Because s is the all-one vector, the matrix $B(eq(t), one(s))$ is equal to $B^2(eq(t), -)$. Notice that $B(eq(t) \cap \{4m+1, \ldots, 4m+n\}, one(v))$ (i.e. the restriction of $B^1(eq(t), one(v))$ to rows of B^1_3) is once more an identity matrix, what makes the block $B^1(eq(t), one(v)) = B(eq(t), one(v))$ non-singular. Finally, the block B^2_3 is an all-zero matrix, therefore the concatenation of matrices $B(eq(t), one(v))B(eq(t), one(s)) = B(eq(t), one(t))$ is non-singular, what means that its columns are linearly independent. According to Proposition 1, the extension $\bar{t}$ is a minimal solution of S, hence s can be extended to a minimal solution of the system S.

Conversely, suppose that s can be extended to a minimal solution of S. Then there exists a partial assignment v to the variables x, forming with s the concatenation $t = vs$, such that $\bar{t}$ is minimal and $wt(t) = 3m + wt(v)$ holds. Note that independently from the choice of the values $v(x^1_i)$, $v(x^2_i)$, and $v(x^3_i)$, at most two of the polynomials $x^1_i + x^2_i$, and $x^2_i + x^3_i$, and $x^3_i + x^1_i$ evaluate to 1. Hence, at most $2m$ rows of B_2 are evaluated to 0 by the assignment t.

Let us analyze the row indices of B that belong to $eq(t)$. The block B_2 contributes always at most $2m$ elements and the block B_3 contributes exactly $wt(v)$ elements to $eq(t)$. Suppose that not all indices of B_1 belong to $eq(t)$. In this case, the block B_1 contributes at most $m-1$ elements to $eq(t)$. This implies that the cardinality of the set $eq(t)$ is smaller or equal than $3m - 1 + wt(v)$ and $B(eq(t), one(t))$ is a $(3m - 1 + wt(v)) \times (3m + wt(v))$ matrix. In this case the column rank of the matrix $B(eq(t), one(t))$ is smaller than $3m + wt(v)$, i.e., the columns are linearly dependent. Following Proposition 1, the extension $\bar{t}$ cannot be minimal. Hence, all m row indices of B_1 must belong to $eq(t)$.

Since all m rows of B_1 belong to $eq(t)$ and $s(y_j) = 1$ holds for each j, the structure of B^1_1, encoding the clauses $c_i = x^1_i \vee x^2_i \vee x^3_i$ of φ, implies that the equality $t(x^1_i) + t(x^2_i) + t(x^3_i) = v(x^1_i) + v(x^2_i) + v(x^3_i) = 1$ holds over $\mathbb{Z}_2$ for each i. There are two cases to analyze: (1) either $v(x^1_i) = v(x^2_i) = v(x^3_i) = 1$ or

(2) exactly one of the values $v(x_i^1)$, $v(x_i^2)$, $v(x_i^3)$ is equal to 1 and the two others are equal to 0. Suppose that there exists an i such that Case 1 is satisfied. Then the maximal number of row indices in $eq(t)$ contributed by B_2 is $2(m-1)$. This is because the equalities $v(x_i^1) + v(x_i^2) = v(x_i^2) + v(x_i^3) = v(x_i^3) + v(x_i^1) = 0$ hold over $\mathbb{Z}_2$. The cardinality of $eq(t)$ is then bounded by $3m-2+wt(v)$, what implies once more that the columns of $B(eq(t), one(t))$ are linearly dependent and this leads to the same contradiction, implying that the extension $\bar{t}$ is not minimal, as in the previous paragraph. Case 2 presents a valid 1-in-3 assignment to φ.   □

Theorem 3. *The problem* AFFINF *is* coNP-*complete.*

Proof. The problem AFFINF is the dual of the problem MINEXT. Note that, given a formula φ and a clause $c = \neg u_1 \vee \cdots \vee \neg u_k$, the condition $\varphi \models_{\min} \neg u_1 \vee \cdots \vee \neg u_k$ holds if and only if there is no minimal model m of φ that satisfies $m(u_1) = \cdots = m(u_k) = 1$. The latter is true if and only if the partial assignment s with $s(u_1) = \cdots = s(u_k) = 1$ cannot be extended to a minimal model of φ, or equivalently, to a minimal solution of the affine system S corresponding to φ.   □

5   Decompositions and Polynomial-Time Decidable Cases

Eiter and Gottlob proved in [EG93] that the inference problem $\varphi \models_{\min} c$ for propositional circumscription remains $\Pi_2 P$-complete even if the clause c consists of a single negative literal $\neg u$. However, it is not guaranteed that the complexity remains the same for one-literal clauses c for the usual subclasses of propositional formulas. Concerning the considered inference problem, Cadoli and Lenzerini proved in [CL94] that for dual Horn formulas it remains coNP-complete but for Krom formulas it becomes polynomial-time decidable for a clause c consisting of a single negative literal. It is a natural question to ask what happens in the case of affine formulas in the presence of a single literal. In the rest of the section we will focus on the restrictions AFFINF$_1$ and MINEXT$_1$ of the respective problems AFFINF and MINEXT to a single negative literal clause $c = \neg u$.

To be able to investigate the complexity of MINEXT$_1$ and AFFINF$_1$, we need to define a neighborhood and a congruence closure on the columns.

Definition 4. *Let B be a $k \times n$ matrix over $\mathbb{Z}_2$ and let $j \in \{1, \ldots, n\}$ be a column index. The* **p-*neighborhood*** $N_p(j)$ *of the column j in B, for $p = 0, 1, \ldots, n$, is defined inductively by*

$$N_0(j) = \{j\},$$
$$N_{p+1}(j) = \{m \mid (\forall q)[(q \leq p) \rightarrow (m \notin N_q(j))] \wedge$$
$$(\exists \ell)(\exists i)[(\ell \in N_p(j)) \wedge (B(i, \ell) = B(i, m) = 1)]\}.$$

The **connected component** $CC(j)$ *of the column j in B is the union of the p-neighborhoods for all p, i.e., $CC(j) = \bigcup_{p=0}^{n} N_p(j)$.*

Speaking in terms of hypergraphs and matroids, where B is interpreted as the vertex-hyperedge incidence matrix, the p-neighborhood $N_p(j)$ is the set of vertices reachable from the vertex j by a path of length p. The vertex ℓ belongs to $N_p(j)$ if and only if the shortest path from j to ℓ in B has the length p. The connected component $CC(j)$ is the set of all reachable vertices from j.

Example 5. Consider the following following affine system $S \colon (I\ B)(xy) = b$, where I, B and b are represented by the successive blocks of the following matrix.

$$(I \mid B \mid b) = \begin{pmatrix} 1\,0\,0\,0\,0\,0 & 0 & 1 & \boxed{1} & \boxed{1} & 0 & 0 \\ 0\,1\,0\,0\,0\,0 & 1 & 0 & 0 & 0 & 0 & 0 \\ 0\,0\,1\,0\,0\,0 & 1 & 1 & 1 & 0 & 0 & 0 \\ 0\,0\,0\,1\,0\,0 & \boxed{1} & 0 & \boxed{1} & 0 & 0 & 0 \\ 0\,0\,0\,0\,1\,0 & 0 & 0 & 0 & \boxed{1} & 1 & 1 \\ 0\,0\,0\,0\,0\,1 & 0 & 0 & 0 & 0 & 1 & 1 \end{pmatrix}$$

Take $j = 7$ and compute the p-neighborhood from vertex 7 in the matrix B for each $p = 0, 1, \ldots, 6$. We obtain $N_0(7) = \{7\}$, $N_1(7) = \{8, 9\}$, $N_2(7) = \{10\}$, $N_3(7) = \{11\}$, and $N_4(7) = N_5(7) = N_6(7) = \emptyset$. The connected component of the vertex 7 is $CC(7) = \{7, 8, 9, 10, 11\}$.

When computing the connected component for all columns of a given matrix B, we may get two or more disjoint sets of vertices. In this case we say that the matrix B is *decomposable*. The following lemma shows that we can compute the problems MINEXT and AFFINF by connected components without increasing the complexity.

Lemma 6. *Let $S \colon (I\ B)(xy) = b$ be an affine system over $\mathbb{Z}_2$. Suppose that the matrix B can be decomposed, up to a permutation of rows and columns, into the components*

$$\begin{pmatrix} B_1 & O \\ O & B_2 \end{pmatrix}$$

where B_1 is a $k_1 \times n_1$ matrix and B_2 is a $k_2 \times n_2$ matrix. Let b_1 and b_2 be two vectors of respective size n_1 and n_2, such that $b = b_1 b_2$. Then the set of minimal solutions of S is equal, up to a permutation, to the Cartesian product $M_1 \times M_2$ of the sets of minimal solutions M_1 and M_2 of the systems $S_1 \colon (I\ B_1)(x'y') = b_1$ and $S_2 \colon (I\ B_2)(x''y'') = b_2$, respectively, where $x = x'x''$ and $y = y'y''$.

The proof of the following theorem shows that finding a minimal extension $\bar{s}$ of a Boolean vector s with $wt(s) = 1$ can be done by finding a shortest path in a connected component of the matrix B from a given column to an inhomogeneous equation in the system S.

Theorem 7. MINEXT$_1$ *and* AFFINF$_1$ *are decidable in polynomial time.*

Proof. (*Hint*) Suppose without loss of generality that S is a $k \times n$ system of the form $S \colon (I\ B)(xy) = b$ and that the variable assigned by s is y_1. This can be achieved through a suitable permutation of rows and columns. We also suppose that the matrix B is indecomposable. Otherwise, we could apply the method described in this proof to one of the subsystems S_1 or S_2 separately, following Lemma 6. Since B is indecomposable, the connected component of the first column is $CC(1) = \{1, \ldots, n\}$, i.e., there are no unreachable columns.

The following condition holds for extensions of vectors with weight 1 to minimal solutions: There exists a minimal solution $\bar{s}$ with $\bar{s}(y_1) = 1$ if and only if $b \neq \mathbf{0}$.

If $b = \mathbf{0}$ then the system S is homogeneous and the all-zero assignment for xy is the unique minimal solution of S, what contradicts the existence of a minimal solution $\bar{s}$ with $\bar{s}(y_1) = 1$.

Conversely, suppose that $b \neq \mathbf{0}$. We construct a partial assignment s for the variables y with $s(y_1) = 1$, such that $\bar{s}$ is minimal. We must find the first inhomogeneous equation reachable from y_1. Since $b \neq \mathbf{0}$, there exists a shortest path through $p+1$ hyperedges $j_0 = 1, j_1, \dots, j_p$ of the hypergraph corresponding to the matrix B, such that the following conditions hold: (1) each hyperedge j_q, $q \leq p$, is reachanble from j_0 since each pair of consecutive hyperedges j_q and j_{q+1} has a common vertex, (2) the existence of a vertex i in a hyperedge j_q, where $q < p$, implies $b_i = 0$, and (3) there exists a vertex i in the last hyperedge j_p, such that $b_i = 1$. Define the partial assignment s for the variables y by $s(y_{j_q}) = 1$ for each $q \leq p$ and set $s(y_j) = 0$ otherwise. This assignment corresponds to the shortest hyperpath starting from a vertex of the hyperedge j_0 and finishing in a vertex i of the hyperedge j_p, such that $b_i = 1$. It is easy to see that $\bar{s}$ is a minimal solution of S corresponding to the shortest hyperpath. Each vertex i_q, except the last one, occurs twice in the shortest hyperpath, what allows us to have $b_{i_q} = 0$. The last vertex i_p appears only once, what implies $b_{i_p} = 1$. The variables x are all set equal to 0. Both a shortest hyperpath and the connected component can be computed in polynomial time, therefore both problems MINEXT$_1$ and AFFINF$_1$ are polynomial-time decidable. $\qquad\square$

Example 8 (Example 5 continued). Start with the column $j_0 = 7$ and compute a shortest path reaching an inhomogeneous equation. There is a shortest path from the column 7 through the columns $j_0 = 7$, $j_1 = 9$, $j_2 = 10$, reaching the inhomogeneous row 5. The path $B(4,7) \to B(4,9) \to B(1,9) \to B(1,10) \to B(5,10)$ is indicated in the matrix by boxed values. Hence, we computed the partial assignment $s = (1,0,1,1,0)$ for the variables y and the extension $\bar{s} = (0,1,0,0,0,1,1,0,1,1,0)$ is a minimal solution of the system S.

6   Conclusion

We proved that the inference problem of propositional circumscription for affine formulas is coNP-complete. It also shows that reasoning under the extended

formula φ	clause inference c	literal inference c
CNF	Π_2P-complete [EG93]	Π_2P-complete [EG93]
Horn	in P	in P
dual Horn	coNP-complete [CL94]	coNP-complete [CL94]
Krom	coNP-complete [CL94]	in P [CL94]
affine	coNP-complete [Theorem 3]	in P [Theorem 7]

Fig. 2. Complexity of the inference problem of propositional circumscription

closed world assumption is intractable for affine formulas. In fact, the exact complexity of affine inference was an open problem since the beginning of the 1990s when several researchers started to investigate the propositional circumscription from algorithmic point of view. We also proved that the inference problem for affine formulas becomes polynomial-time decidable when only a single literal has to be inferred. The complexity classification of the inference problem of propositional circumscription for the usual classes of formulas is presented in Figure 2.

References

[BMvT78] E. R. Berlekamp, R. J. McEliece, and H. C. A. van Tilborg. On the inherent intractability of certain coding problems. *IEEE Trans. on Inf. Theory*, IT-24(3):384–386, 1978.

[Cad92] M. Cadoli. The complexity of model checking for circumscriptive formulae. *Inf. Proc. Letters*, 44(3):113–118, 1992.

[CL94] M. Cadoli and M. Lenzerini. The complexity of propositional closed world reasoning and circumscription. *JCSS*, 48(2):255–310, 1994.

[CMM01] S. Coste-Marquis and P. Marquis. Knowledge compilation for closed world reasoning and circumscription. *J. Logic and Comp.*, 11(4):579–607, 2001.

[EG93] T. Eiter and G. Gottlob. Propositional circumscription and extended closed-world reasoning are Π_2^p-complete. *TCS*, 114(2):231–245, 1993.

[GPP89] M. Gelfond, H. Przymusinska, and T. C. Przymusinski. On the relationship between circumscription and negation as failure. *Artificial Intelligence*, 38(1):75–94, 1989.

[KK01a] L. M. Kirousis and P. G. Kolaitis. The complexity of minimal satisfiability problems. In A. Ferreira and H. Reichel, (eds), *Proc. 18th STACS, Dresden (Germany)*, LNCS 2010, pp 407–418. Springer, 2001.

[KK01b] L. M. Kirousis and P. G. Kolaitis. A dichotomy in the complexity of propositional circumscription. In *Proc. 16th LICS, Boston (MA)*, pp 71–80. 2001.

[KSS00] D. J. Kavvadias, M. Sideri, and E. C. Stavropoulos. Generating all maximal models of a boolean expression. *Inf. Proc. Letters*, 74(3-4):157–162, 2000.

[McC80] J. McCarthy. Circumscription — A form of non-monotonic reasoning. *Artificial Intelligence*, 13(1-2):27–39, 1980.

Decidable Theories of Cayley-Graphs[*]

Dietrich Kuske[1] and Markus Lohrey[2]

[1] Institut für Algebra, Technische Universität Dresden
D-01062 Dresden, Germany kuske@math.tu-dresden.de
[2] Universität Stuttgart, Institut für Informatik,
Breitwiesenstr. 20-22, D-70565 Stuttgart, Germany
lohrey@informatik.uni-stuttgart.de

Abstract. We prove that a connected graph of bounded degree with only finitely many orbits has a decidable MSO-theory if and only if it is context-free. This implies that a group is context-free if and only if its Cayley-graph has a decidable MSO-theory. On the other hand, the first-order theory of the Cayley-graph of a group is decidable if and only if the group has a decidable word problem. For Cayley-graphs of monoids we prove the following closure properties. The class of monoids whose Cayley-graphs have decidable MSO-theories is closed under free products. The class of monoids whose Cayley-graphs have decidable first-order theories is closed under general graph products. For the latter result on first-order theories we introduce a new unfolding construction, the factorized unfolding, that generalizes the tree-like structures considered by Walukiewicz. We show and use that it preserves the decidability of the first-order theory.
Most of the proofs are omitted in this paper, they can be found in the full version [17].

1 Introduction

The starting point of our consideration was a result by Muller and Schupp [21] showing that the Cayley-graph of any context-free group has a decidable monadic second-order theory (MSO-theory). The questions we asked ourselves were: is there a larger class of groups with this property? Can one show similar results for first-order theories (FO-theories) of Cayley-graphs? Are there analogous connections in monoid theory? Similarly to Muller and Schupp's work, this led to the investigation of graph classes with decidable theories that now forms a large part of the paper at hand. Due to potential applications for the verification of infinite state systems, recently such graph classes have received increasing interest, see [28] for an overview.

Courcelle showed that the class of graphs of tree-width at most b has a decidable MSO-theory (for any $b \in \mathbb{N}$) [5]. A partial converse was proved by Seese [24]

---

[*] This work was done while the first author worked at University of Leicester, parts of it were done while the second author was on leave at IRISA, Campus de Beaulieu, 35042 Rennes Cedex, France and supported by the INRIA cooperative research action FISC.

H. Alt and M. Habib (Eds.): STACS 2003, LNCS 2607, pp. 463–474, 2003.

(in conjunction with another result by Courcelle [6]) showing that any class of graphs of bounded degree whose MSO-theory is decidable is of bounded tree-width. On the other hand, there are even trees with an undecidable FO-theory. We therefore restrict attention to connected graphs of bounded degree whose automorphism group has only finitely many orbits. If such a graph G has finite tree-width, then it is context-free (Theorem 3.1). Our proof of this fact is based on the construction of a tree decomposition with quite strong combinatorial properties, using techniques from the theory of groups acting on graphs [10]. By another result of Muller and Schupp [21], G has a decidable MSO-theory.

Using this general result on graphs, we can show that Muller and Schupp's result on Cayley-graphs of context-free groups is optimal: any finitely generated group whose Cayley-graph has a decidable MSO-theory is context-free (Corollary 4.1). A similar result will be also shown for first-order logic: the FO-theory of the Cayley-graph of a group is decidable if and only if the word problem of the group is decidable (Proposition 4.2). One implication is simple since one can express by a first-order sentence that a given word labels a cycle in the Cayley-graph. The other implication follows from Gaifman's locality theorem for first-order logic [14] which allows to restrict quantifications over elements of the Cayley-graph to certain spheres around the unit.

These results for groups do not carry over to monoids, e.g., there is a monoid with a decidable word problem whose Cayley-graph has an undecidable FO-theory (Proposition 6.1). On the other hand, we are able to prove some closure properties of the classes of monoids whose Cayley-graphs have decidable theories. Using a theorem of Walukiewicz [31] (the original statement goes back to work by Stupp [27], Shelah [26], Muchnik, and Semenov, see [31] for an account) on MSO-theories of unfoldings, we prove that the class of finitely generated monoids whose Cayley-graphs have decidable MSO-theories is closed under free products (Theorem 6.3(2)). Moreover, we show that the class of finitely generated monoids whose Cayley-graphs have decidable FO-theories is closed under graph products (Theorem 6.3(1)) which is a well-known construction in mathematics, see e.g. [15,30]; it generalizes both, the free and the direct product of monoids. In order to show this closure property, we introduce the notion of a factorized unfolding in Section 5, which is also of independent interest (see the discussion in Section 5): Walukiewicz's unfolding of a structure $\mathcal{A}$ consists of the set of words over the set of elements of $\mathcal{A}$. This set of words is equipped with the natural tree structure. Hence the successors of any node of the tree can be identified with the elements of $\mathcal{A}$ and can therefore naturally be endowed with the structure of $\mathcal{A}$. Basically, a factorized unfolding is the quotient of this structure with respect to Mazurkiewicz's trace equivalence (in fact, it is a generalization of this quotient). We show that the FO-theory of a factorized unfolding can be reduced to the FO-theory of the underlying structure (Theorem 5.7). The proof of this result uses techniques of Ferrante and Rackoff [13] and a thorough analysis of factorized unfoldings using ideas from the theory of Mazurkiewicz traces [8]. From this result on factorized unfoldings, we obtain the closure under graph products similarly to the closure under free products.

Our results on FO-theories of Cayley-graphs should be also compared with the classical results about FO-theories of monoids: the FO-theory of a monoid $\mathcal{M}$ contains all true first-order statements about $\mathcal{M}$ that are built over the signature containing the monoid operation and all monoid elements as constants. Thus the FO-theory of the Cayley-graph of $\mathcal{M}$ can be seen as a fragment of the whole FO-theory of $\mathcal{M}$ in the sense that only equations of the form $xa = y$, with x and y variables and $a \in \mathcal{M}$ are allowed. In this context we should mention the classical results of Makanin, stating that the existential FO-theory of a free monoid [18] or free group [19] is decidable, see [7] for a more detailed overview.

2 Preliminaries

Let $\mathcal{A} = (A, (R_i)_{i \in K})$ be a relational structure with carrier set A and relations R_i of arbitrary arity. *First-order logic* (FO) and *monadic second-order logic* (MSO) over the structure $\mathcal{A}$ are defined as usual. The *FO-theory* (resp. *MSO-theory*) of $\mathcal{A}$ is denoted by $\mathrm{FOTh}(\mathcal{A})$ (resp. $\mathrm{MSOTh}(\mathcal{A})$).

A Σ-*labeled directed graph* (briefly *graph*) is a relational structure $G = (V, (E_a)_{a \in \Sigma})$, where Σ is a finite set of labels, and $E_a \subseteq V \times V$ is the set of all a-labeled edges. The undirected graph that results from G by forgetting all labels and the direction of edges is denoted by $\mathrm{undir}(G)$. We say that G is *connected* if $\mathrm{undir}(G)$ is connected. We say that G has *bounded degree*, if for some constant $c \in \mathbb{N}$, every node of G is incident with at most c edges in $\mathrm{undir}(G)$. The *diameter* of $U \subseteq V$ in G is the maximal distance in $\mathrm{undir}(G)$ between two nodes $u, v \in U$ (which might be ∞).

In Section 3 we will consider graphs of bounded tree-width. We will omit the formal definition of tree-width (see e.g. [9]) since we are mainly interested in the stronger notion of *strong tree-width*. A *strong tree decomposition* of an undirected graph $G = (V, E)$ is a partition $P = \{V_i \mid i \in K\}$ of V such that the quotient graph $G/_P = (P, E/_P)$, where $E/_P = \{(V_i, V_j) \in P \times P \mid V_i \times V_j \cap E \neq \emptyset\}$, is a forest, i.e., acyclic [23]. The *width of* P is the supremum of the cardinalities $|V_i|$, $i \in K$. If there exists a strong tree decomposition P of G of width at most b then G has *strong tree-width at most* b.

3 Graphs with a Decidable MSO-Theory

In [21], Muller & Schupp gave a graph-theoretical characterization of the transition graphs of pushdown automata, which are also called *context-free graphs*. Moreover, in [21] it is shown that the MSO-theory of any context-free graph is decidable. In this section, we outline a proof of the converse implication for graphs with a high degree of symmetry. More precisely, we consider graphs with only *finitely many orbits*. Here the orbits of a graph $G = (V, (E_a)_{a \in \Sigma})$ are the equivalence classes with respect to the equivalence $\sim$ defined as follows: $u \sim v$ for $u, v \in V$ if and only if there exists an automorphism f of G with $f(u) = v$.

Theorem 3.1. *Let* $G = (V, (E_a)_{a \in \Sigma})$ *be a connected graph of bounded degree with only finitely many orbits. Then* $\mathrm{MSOTh}(G)$ *is decidable if and only if* $\mathrm{undir}(G)$ *has finite tree-width if and only if* G *is context-free.*

Proof (sketch). Assume that $\mathrm{MSOTh}(G)$ is decidable. Notice that MSO only allows quantification over sets of nodes, whereas quantification over sets of edges is not possible. On the other hand, for graphs of bounded degree, Courcelle [6] has shown that the extension of MSO by quantification over sets of edges, which is known as MSO_2, can be defined within MSO. Thus the MSO_2-theory of G is decidable. A result of Seese [24] implies that $\mathrm{undir}(G)$ has finite tree-width.

Thus, assume that $H = \mathrm{undir}(G)$ has tree-width at most b for some $b \in \mathbb{N}$. Then also any finite subgraph of H has tree-width at most b. Since the degree of H is bounded by some constant d, the same holds for its finite subgraphs. Hence, by a result from [3], any finite subgraph of H has strong tree-width at most $c = (9b + 7)d(d + 1)$. From these strong tree decompositions of the finite subgraphs of H, one can construct a strong tree decomposition P of H of width at most c as follows. Since H is connected and of bounded degree, H must be countable. Thus we can take an ω-sequence $(G_i)_{i \in \mathbb{N}}$ of finite subgraphs of H whose limit is H. From the non-empty set of all strong-tree decompositions of width at most c of the graphs G_i, $i \in \mathbb{N}$, we construct a finitely branching tree as follows. Put an edge between a strong tree decomposition P_i (of width at most c) of G_i and a strong tree decomposition P_{i+1} (of width at most c) of G_{i+1} if P_i results from P_{i+1} by restriction to the nodes of G_i. By König's Lemma, this tree contains an infinite path. Taking the limit along this path results in a strong tree decomposition P of H of width at most c.

By splitting some of the partition classes of P, we can refine P into a strong tree decomposition Q of width at most c with the following property: for all edges (V_1, V_2) of the quotient graph H/Q, removing all edges between V_1 and V_2 (note that there are at most c^2 such edges) splits H into exactly two connected components. In the terminology of [10,29], the set of edges connecting V_1 and V_2 is called a *tight* c^2-*cut* of H. By [10, Paragraph 2.5] (see also [29, Prop. 4.1] for a simplified proof), every edge of H is contained in only finitely many tight c^2-cuts. From this fact and the assumption that G, and hence also H, has only finitely many orbits, one can deduce that the diameter of every partition class in Q is bounded by some fixed constant $\gamma \in \mathbb{N}$. Using this, one can show that the graph G can be $(2\gamma + 1)$-triangulated [20], this step is similar to the proof of [2, Thm. 8]. Then essentially the same argument that was given in the proof of [21, Thm. 2.9] for a vertex-transitive graph (i.e., a graph that has only one orbit) shows that G is context-free.

The remaining implication "G context-free $\Rightarrow$ $\mathrm{MSOTh}(G)$ decidable" was shown in [21]. □

Remark 3.2. In [25] it was shown that if G is context-free then also $G/_\sim$ is context-free. Thus, a natural generalization of the previous theorem could be the following: Let G be a connected graph of bounded degree such that the quotient graph $G/_\sim$ is context-free with finitely many orbits. Then G has a decidable MSO-theory if and only if G is context-free. But this is false: take $\mathbb{Z}$ together with the successor relation and add to every number $m = \frac{1}{2}n(n+1)$ $(n \in \mathbb{Z})$ a copy m' together with the edge (m, m'), whereas for every other number m we add two copies m' and m'' together with the edges (m, m') and (m, m''). The resulting graph is not context-free, but it has a decidable MSO-theory [11] (see also [21]) and $G/\sim$ is context-free with just two orbits.

4 Cayley-Graphs of Groups

Let $\mathcal{G}$ be a group generated by the finite set Γ. Its *Cayley graph* $\mathbb{C}(\mathcal{G}, \Gamma)$ has as vertices the elements of $\mathcal{G}$ and as a-labeled edges the pairs (x, xa) for $x \in \mathcal{G}$ and $a \in \Gamma$. The word problem of $\mathcal{G}$ wrt. Γ is the set of words over $\Gamma \cup \{a^{-1} \mid a \in \Gamma\}$ that represent the identity of $\mathcal{G}$. It is well known that the decidability of the word problem does not depend on the chosen generating set; henceforth we will speak of *the* word problem regardless of the generators. The group $\mathcal{G}$ is called *context-free* if its word problem is a context free language [1,20]. By [21] this is equivalent to saying that $\mathbb{C}(\mathcal{G}, \Gamma)$ is a context-free graph. The automorphism group of any Cayley-graph acts transitively on the vertices (i.e., has just one orbit). Furthermore, Cayley-graphs are always connected. If the group is finitely generated, then moreover its Cayley-graph has bounded degree. Thus, from Theorem 3.1, we get the following (the implication "$\Rightarrow$" is due to Muller & Schupp)

Corollary 4.1. *Let $\mathcal{G}$ be a group finitely generated by Γ. Then $\mathcal{G}$ is context-free if and only if* $\mathrm{MSOTh}(\mathbb{C}(\mathcal{G}, \Gamma))$ *is decidable.*

For FO-theories we obtain

Proposition 4.2. *Let $\mathcal{G}$ be a group finitely generated by Γ. Then the following are equivalent:*

(1) The word problem of $\mathcal{G}$ is decidable.
(2) $\mathrm{FOTh}(\mathbb{C}(\mathcal{G}, \Gamma))$ *is decidable.*
(3) The existential FO-theory of the Cayley-graph $\mathbb{C}(\mathcal{G}, \Gamma)$ is decidable.

Proof (sketch). The implication (2) $\Rightarrow$ (3) is trivial. The implication (3) $\Rightarrow$ (1) is easily shown since a word over $\Gamma \cup \{a^{-1} \mid a \in \Gamma\}$ represents the identity of $\mathcal{G}$ if and only if it labels some cycle in the Cayley-graph, an existential property expressible in first-order logic. The remaining implication is shown using Gaifman's theorem [14]: since the automorphism group of $\mathbb{C}(\mathcal{G}, \Gamma)$ acts transitively on the vertices, it implies that it suffices to decide first-order properties of spheres in $\mathbb{C}(\mathcal{G}, \Gamma)$ around the identity of $\mathcal{G}$. But these spheres are finite and effectively computable since the word problem is decidable. □

In the complete version of this extended abstract [17], we prove that every FO-sentence is equivalent in $\mathbb{C}(\mathcal{G}, \Gamma)$ to the same sentence but with all quantifiers restricted to spheres around the unit of at most exponential diameter. This proof uses techniques developed by Ferrante and Rackoff [13]. In addition to the above result, it provides a tight relationship between the word problem of $\mathcal{G}$ and $\mathrm{FOTh}(\mathbb{C}(\mathcal{G}, \Gamma))$ in terms of complexity: the space complexity of $\mathrm{FOTh}(\mathbb{C}(\mathcal{G}, \Gamma))$ is bounded exponentially in the space complexity of the word problem of $\mathcal{G}$ [17].

5   Factorized Unfoldings

In [31], Walukiewicz proved that the MSO-theory of the tree-like unfolding of a relational structure can be reduced to the MSO-theory of the underlying structure. The origin of this result goes back to [26,27]. Tree-like unfoldings are defined as follows:

Definition 5.1. *Let* $\mathcal{A} = (A, (R_i)_{1 \leq i \leq n})$ *be a relational structure where the relation* R_i *has arity* p_i*. On the set of finite words* A^**, we define the following relations:*

$$\widehat{R_i} = \{(ua_1, ua_2, \ldots, ua_{p_i}) \mid u \in A^*, (a_1, a_2, \ldots, a_{p_i}) \in R_i\}$$
$$\mathrm{suc} = \{(u, ua) \mid u \in A^*, a \in A\}$$
$$\mathrm{cl} = \{(ua, uaa) \mid u \in A^*, a \in A\}$$

Then the relational structure $\widehat{\mathcal{A}} = (A^*, (\widehat{R_i})_{1 \leq i \leq n}, \mathrm{suc}, \mathrm{cl})$ *is called the* tree-like unfolding *of* $\mathcal{A}$*.*

Theorem 5.2 (cf. [31]). *Let* $\mathcal{A}$ *be a relational structure. Then* $\mathrm{MSOTh}(\widehat{\mathcal{A}})$ *can be reduced to* $\mathrm{MSOTh}(\mathcal{A})$*.*

We will in particular use the immediate consequence that $\mathrm{MSOTh}(\widehat{\mathcal{A}})$ is decidable whenever the MSO-theory of $\mathcal{A}$ is decidable. The main result of this section is a FO-analogue of the above result (Theorem 5.7).

The relations of the tree-like unfoldings are instances of a more general construction that will be crucial for our notion of factorized unfoldings. Let $\varphi(x_1, x_2, \ldots, x_n)$ be a first-order formula over the signature of $\mathcal{A}$ with n free variables. For a word $w = a_1 a_2 \cdots a_n \in A^*$ of length n we write $\mathcal{A} \models \varphi(w)$ if $\mathcal{A} \models \varphi(a_1, a_2, \ldots, a_n)$. An n-ary relation R over A^* is k-*suffix definable* in $\mathcal{A}$ if there are $k_1, \ldots, k_n \leq k$ ($k_i = 0$ is allowed) and a first-order formula φ over the signature of $\mathcal{A}$ with $\sum_{i=1}^{n} k_i$ free variables such that

$$R = \{(uu_1, uu_2, \ldots, uu_n) \mid u, u_i \in A^*, |u_i| = k_i, \mathcal{A} \models \varphi(u_1 u_2 \cdots u_n)\}.$$

Obviously, all relations of $\widehat{\mathcal{A}}$ are 2-suffix definable in $\mathcal{A}$. On the other hand, there exist 2-suffix definable relations such that adding them to $\widehat{\mathcal{A}}$ makes Theorem 5.2 fail. To see this, let

$$\mathrm{eq} = \{(ua, uba) \mid u \in A^*, a, b \in A\},$$

which is 2-suffix definable in $\mathcal{A}$. Define the prefix order $\preceq$ on A^* by $\preceq = \{(u, uv) \mid u, v \in A^*\}$, it is the reflexive transitive closure of the relation suc from $\widehat{\mathcal{A}}$, thus it is MSO-definable in $\widehat{\mathcal{A}}$. Let $A = \mathbb{N} \cup \{a, b\}$ be the set of natural numbers together with two additional elements. On A we define the predicates $S = \{(n, n+1) \mid n \in \mathbb{N}\}$, $U_a = \{a\}$, and $U_b = \{b\}$. Then the structure $\mathcal{A} = (A, S, U_a, U_b)$ has a decidable MSO-theory. We consider the structure $\mathcal{B} = (A^*, \widehat{S}, \widehat{U_a}, \widehat{U_b}, \mathrm{suc})$, which is a reduct of the tree-like unfolding of $\mathcal{A}$. Using FO-logic over $(\mathcal{B}, \mathrm{eq}, \preceq)$, we can express that a given 2-counter machine terminates. Thus we obtain

Proposition 5.3. $\text{FOTh}(\mathcal{B}, \text{eq}, \preceq)$ *is undecidable.*

In particular, the MSO-theory of $(\mathcal{B}, \text{eq})$ is undecidable. Thus, the presence of the relation eq makes Walukiewicz's result fail.

Recall that the underlying set of the tree-like unfolding of a structure $\mathcal{A}$ is the set of words over the carrier set of $\mathcal{A}$. In factorized unfoldings that we introduce next, this underlying set consists of equivalence classes of words wrt. Mazurkiewicz's trace equivalence:

A (not necessarily finite) set A together with an irreflexive and symmetric relation $I \subseteq A \times A$ is called *independence alphabet*, the relation I is the *independence relation*. With any such independence alphabet, we associate the least congruence $\equiv_I$ on A^* identifying ab and ba for $(a, b) \in I$. The quotient $\mathbb{M}(A, I) = A^*/{\equiv_I}$ is the *free partially commutative* or *(Mazurkiewicz) trace monoid* generated by (A, I). The trace that is represented by the word $w \in A^*$ is denoted by $[w]_I$. Note that for $I = \emptyset$, the trace monoid $\mathbb{M}(A, I)$ is isomorphic to the free monoid A^*. In the other extreme, i.e., if $I = (A \times A) \setminus \{(a, a) \mid a \in A\}$, we have $\mathbb{M}(A, I) \cong \mathbb{N}^A$, i.e., the trace monoid is free commutative generated by A. For a trace $t \in \mathbb{M}(A, I)$, we let $\min(t) = \{a \in A \mid \exists s \in A^* : t = [as]_I\}$ the set of minimal symbols of t. The set $\max(t)$ of maximal symbols of t is defined analogously. For an n-ary relation R over A^*, we define its I-quotient

$$R/_I = \{([u_1]_I, \dots, [u_n]_I) \mid (u_1, \dots, u_n) \in R\}.$$

Definition 5.4. *Let $\mathcal{A}$ be a relational structure with carrier set A. Let furthermore*

- *$I \subseteq A \times A$ be an independence relation which is first-order definable in $\mathcal{A}$,*
- *$\eta : \mathbb{M}(A, I) \to S$ be a monoid morphism into some finite monoid S such that $\eta^{-1}(s) \cap A$ is first-order definable in $\mathcal{A}$ for all $s \in S$.*
- *R_i be a k_i-suffix definable relation in $\mathcal{A}$ for $1 \le i \le n$.*

Then the structure $\mathcal{B} = (\mathbb{M}(A, I), (\eta^{-1}(s))_{s \in S}, (R_i/_I)_{1 \le i \le n})$ is a factorized unfolding of $\mathcal{A}$.

Note that in contrast to the tree-like unfolding there are many different factorized unfoldings of $\mathcal{A}$. The notion of a factorized unfolding is a proper generalization of the tree-like unfolding even in case $I = \emptyset$: by Proposition 5.3, the relation eq cannot be defined in the tree-like unfolding, but since it is 2-suffix definable it may occur in a factorized unfolding. On the other hand, if $I = \emptyset$, then, since $\eta^{-1}(s) \cap A$ is first-order definable in $\mathcal{A}$, the set $\eta^{-1}(s) \subseteq \mathbb{M}(A, I) = A^*$ is MSO-definable in the tree-like unfolding of $\mathcal{A}$. Since Walukiewicz was interested in the MSO-theory of his unfolding, the relations $\eta^{-1}(s)$ are "effectively present" in $\widehat{\mathcal{A}}$.

The structure $(\mathcal{B}, \text{eq}, \preceq)$ from Proposition 5.3 has an undecidable FO-theory. Thus, allowing the relation $\preceq/_I$ in factorized unfoldings would make the main result of this section (Theorem 5.7) fail. In Theorem 5.7, we will also assume that there are only finitely many different sets $I(a) = \{b \in A \mid (a, b) \in I\}$, which roughly speaking means that traces from $\mathbb{M}(A, I)$ have only "bounded parallelism". The reason is again that otherwise the result would fail:

Proposition 5.5. *There exists an infinite structure $\mathcal{A}$ and a factorized unfolding $\mathcal{B}$ of $\mathcal{A}$ such that* FOTh($\mathcal{A}$) *is decidable but* FOTh($\mathcal{B}$) *is undecidable.*

Proof (sketch). Let $(V, E) = K_{\aleph_0}$ be a countable complete graph, $A = V \dot\cup E$, and $R \subseteq (V \times E)$ be the incidence relation. Furthermore, $I = (A \times A) \setminus \text{id}_A$. Then we think of a trace $t \in \mathbb{M}(A, I)$ as representing the subgraph $\max(t) \subseteq A$ of $K_{\aleph_0}$. This allows to reduce the FO-theory of all finite graphs to the FO-theory of the factorized unfolding $(\mathbb{M}(A, I), \text{cl}/_I, R/_I)$ of $\mathcal{A}$. The former theory is undecidable by a result of Trakhtenbrot. $\square$

In Proposition 5.3 and 5.5 we used infinite structures $\mathcal{A}$. Infinity is needed as the following shows:

Proposition 5.6. *Let $\mathcal{A}$ be a finite structure and $\mathcal{B}$ be a factorized unfolding of $\mathcal{A}$. Then $(\mathcal{B}, \preceq/_I)$ is an automatic structure [16]; hence its FO-theory is decidable.*

Proof (sketch). The underlying set of the structure $\mathcal{B}$ is the set of traces $\mathbb{M}(A, I)$. For these traces, several normal forms are known [8], here we use the Foata normal form. Since $\mathcal{A}$ is finite, all the relations in $(\mathcal{B}, \preceq/_I)$ (more precisely: their Foata normal form incarnations) are synchronized rational relations. $\square$

Now we finally formulate the main result of this section:

Theorem 5.7. *Let $\mathcal{A}$ be a relational structure and consider a factorized unfolding $\mathcal{B} = (\mathbb{M}(A, I), (\eta^{-1}(s))_{s \in S}, (R_i/_I)_{1 \leq i \leq n})$ of $\mathcal{A}$ where $\{I(a) \mid a \in A\} \subseteq 2^A$ is finite. Then* FOTh($\mathcal{B}$) *can be reduced to* FOTh($\mathcal{A}$).

Proof (sketch). For a trace $t \in \mathbb{M}(A, I)$, let $|t|$ be the length of any word representing t. We will write $\exists x \leq n : \psi$ as an abbreviation for $\exists x : |x| \leq n \wedge \psi$, i.e., $\exists x \leq n$ restricts quantification to traces of length at most n. In order to use techniques similar to those developed by Ferrante and Rackoff [13], one then defines a computable function $H : \mathbb{N} \times \mathbb{N} \to \mathbb{N}$ with the following property:

Let $\varphi = Q_1 x_2 Q_2 x_2 \ldots Q_d x_d \psi$ be a formula in prenex normal form over the signature of $\mathcal{B}$, where $Q_i \in \{\forall, \exists\}$. Then $\mathcal{B} \models \varphi$ if and only if

$$\mathcal{B} \models Q_1 x_1 \leq H(1, d)\, Q_2 x_2 \leq H(2, d) \ldots Q_d x_d \leq H(d, d) : \psi \qquad (1)$$

In order to be able to define H, the assumption that there are only finitely many sets $I(a)$ is crucial.

At this point we have restricted all quantifications to traces of bounded length. Now a variable x that ranges over traces of length n can be replaced by a sequence of first-order variables $y_1 \cdots y_n$ ranging over $\mathcal{A}$. Since I is FO-definable in $\mathcal{A}$, we can express in FO-logic over $\mathcal{A}$ that two such sequences represent the same trace. Since also $\eta^{-1}(s)$ is first-order definable in $\mathcal{A}$ for every $s \in S$ and all other relations in $\mathcal{B}$ result from k-suffix definable relations, it follows that (1) can be translated into an equivalent first-order statement about $\mathcal{A}$. $\square$

Remark 5.8. The function $H : \mathbb{N} \times \mathbb{N} \to \mathbb{N}$ refered to in the above proof satisfies $H(i, d) \leq H(i + 1, d)$ and $H(d, d) \in 2^{O(d)}$ (values for $H(i, d)$ with $i > d$ are not used in the proof). This allows to show that this procedure transforms a formula φ over the signature of $\mathcal{B}$ into a formula of size $2^{2^{O(|\varphi|)}}$ over the signature of $\mathcal{A}$.

6   Cayley-Graphs of Monoids

The Cayley-graph $\mathbb{C}(\mathcal{M}, \Gamma)$ of a monoid $\mathcal{M}$ wrt. some finite set of generators Γ can be defined analogously to that of a group. It will turn out to be convenient to consider the *rooted Cayley-graph* $(\mathbb{C}(\mathcal{M}, \Gamma), 1)$ that in addition contains a constant 1 for the unit element of the monoid $\mathcal{M}$.

It is easily checked that the implications $(2) \Rightarrow (3) \Rightarrow (1)$ from Proposition 4.2 carry over to monoids, but the situation for the remaining implication is different. The following proposition follows from [22, Thm. 2.4].

Proposition 6.1. *There is a finitely presented monoid $\mathcal{M}$ with a decidable word problem such that $\mathbb{C}(\mathcal{M}, \Gamma)$ has an undecidable existential FO-theory.*

On the decidability side let us mention that Cayley-graphs of automatic monoids [4] have decidable FO-theories since they are automatic structures [16].

In the sequel, we will prove closure properties of classes of monoids with decidable theories. Using simple MSO-interpretations it is easy to see that the class of finitely generated monoids, whose Cayley-graphs have decidable MSO-theories, is closed under finitely generated submonoids and Rees-quotients w.r.t. rational ideals. Moreover, if $\mathrm{MSOTh}(\mathbb{C}(\mathcal{M}, \Gamma), 1)$ is decidable and S is a finite monoid then also $\mathrm{MSOTh}(\mathbb{C}(\mathcal{M} \times S, \Gamma \cup S), 1)$ is easily seen to be decidable.

Now, we consider graph products of monoids [15] which generalize both, the direct and the free product. In order to define it, let (Σ, J) be some finite independence alphabet and let $\mathcal{M}_\sigma$ be a monoid for $\sigma \in \Sigma$. Then the *graph product* $\prod_{(\Sigma, J)} \mathcal{M}_\sigma$ is the quotient of the free product of the monoids $\mathcal{M}_\sigma$ subject to the relations $ab = ba$ for $a \in \mathcal{M}_\sigma$, $b \in \mathcal{M}_\tau$ and $(\sigma, \tau) \in J$. If $J = \emptyset$, then there are no such relations, i.e., the graph product equals the free product $*_{\sigma \in \Sigma} \mathcal{M}_\sigma$. If, in the other extreme, $J = (\Sigma \times \Sigma) \setminus \{(\sigma, \sigma) \mid \sigma \in \Sigma\}$, then the graph product equals the direct product $\prod_{\sigma \in \Sigma} \mathcal{M}_\sigma$. For the subsequent discussions, fix some finite independence alphabet (Σ, J) and for every $\sigma \in \Sigma$ a monoid $\mathcal{M}_\sigma = (\mathcal{M}_\sigma, \circ_\sigma, 1_\sigma)$, which is generated by the finite set Γ_σ. Furthermore, let $\mathcal{M} = (\mathcal{M}, \circ, 1) = \prod_{(\Sigma, J)} \mathcal{M}_\sigma$ be the graph product of these monoids wrt. (Σ, J). This monoid is generated by the finite set $\Gamma = \bigcup_{\sigma \in \Sigma} \Gamma_\sigma$.

We will prove decidability results for the theories of the rooted Cayley-graph $(\mathbb{C}(\mathcal{M}, \Gamma), 1)$ using Theorems 5.2 and 5.7. In these applications, the underlying structure $\mathcal{A}$ will always be the disjoint union of the rooted Cayley-graphs $(\mathbb{C}(\mathcal{M}_\sigma, \Gamma_\sigma), 1_\sigma)$. Hence the carrier set A of the structure $\mathcal{A}$ is the disjoint union of the monoids $\mathcal{M}_\sigma$. It has binary edge-relations $E_a = \{(x, x \circ_\sigma a) \mid x \in \mathcal{M}_\sigma\} \subseteq \mathcal{M}_\sigma \times \mathcal{M}_\sigma$ for all $\sigma \in \Sigma$ and $a \in \Gamma_\sigma$, as well as unary relations $A_\sigma \subseteq A$ comprising all elements of the monoid $\mathcal{M}_\sigma$, and unary relations $U_\sigma = \{1_\sigma\}$. We now define a factorized unfolding $\mathcal{B}$ of this disjoint union $\mathcal{A}$: the independence relation

$$I = \bigcup_{(\sigma, \tau) \in J} \mathcal{M}_\sigma \times \mathcal{M}_\tau$$

is FO-definable in $\mathcal{A}$ using the unary predicates A_σ. Since Σ is finite, there are only finitely many sets $I(a)$ for $a \in A$. The relations $\widehat{E_a}, \widehat{U_\sigma}, \mathrm{suc}$, and $\mathrm{suc}_a =$

$\{(x, xa) \mid x \in A^*\}$, where $\sigma \in \Sigma$ and $a \in \Gamma \subseteq A$, are 1-suffix definable in $\mathcal{A}$ (note that every $a \in \Gamma$ is FO-definable in $\mathcal{A}$). We define the monoid morphism η in such a way that we are able to interpret the rooted Cayley-graph $(\mathbb{C}(\mathcal{M}, \Gamma), 1)$ in the factorized unfolding $\mathcal{B}$. In particular, elements of the graph product $\mathcal{M}$ will be represented by traces over (A, I). To this aim, the following paragraph defines the mapping η as follows:

For $a \in A$, let $\mu(a) \in \Sigma$ be the unique index with $a \in \mathcal{M}_{\mu(a)}$ and define $\mu(t) = \{\mu(a) \mid a \text{ occurs in } t\}$ for $t \in \mathbb{M}(A, I)$. Then set $\eta(t) = \perp$ if there is $\sigma \in \Sigma$ such that 1_σ is a factor of the trace t, or if there are $a, b \in \mathcal{M}_\sigma$ such that the trace ab is a factor of the trace t. If this is not the case, let $\eta(t) = (\mu(\min(t)), \mu(t), \mu(\max(t)))$. Thus, η is a mapping from $\mathbb{M}(A, I)$ into some finite set S. Then the kernel $\{(s, t) \in \mathbb{M}(A, I) \times \mathbb{M}(A, I) \mid \eta(s) = \eta(t)\}$ of η is a monoid congruence. In other words, the set S can be endowed with a monoid structure such that η is actually a monoid morphism into some finite monoid.

Now we have collected all the ingredients for our factorized unfolding of $\mathcal{A}$:

$$\mathcal{B} = (\mathbb{M}(A, I), (\eta^{-1}(s))_{s \in S}, (\widehat{E_a}/I)_{a \in \Gamma}, (\widehat{U_\sigma}/I)_{\sigma \in \Sigma}, \text{suc}/I, (\text{suc}_a/I)_{a \in \Gamma})$$

is a factorized unfolding of $\mathcal{A}$. Note that it does not contain the relation eq/I. Therefore, in case $J = \emptyset$ (i.e., $I = \emptyset$) $\mathcal{B}$ is MSO-definable in the tree-like unfolding $\widehat{\mathcal{A}}$, which will allow to apply Theorem 5.2. A major step towards a proof of Theorems 6.3 is

Lemma 6.2. *There is a first-order interpretation of the rooted Cayley-graph $(\mathbb{C}(\mathcal{M}, \Gamma), 1)$ in the factorized unfolding $\mathcal{B}$ of $\mathcal{A}$.*

Proof (sketch). The elements of the graph product $\mathcal{M}$ can be identified with those traces t that satisfy $\eta(t) \neq \perp$ (in the terminology of [30], they are Γ-equivalence classes of words of the form $S(u)$). In order to define the edges of the rooted Cayley-graph $(\mathbb{C}(\mathcal{M}, \Gamma), 1)$ within $\mathcal{B}$, let us take $s, t \in \mathbb{M}(A, I)$ with $\eta(s) \neq \perp \neq \eta(t)$ and let $\sigma \in \Sigma$, $a \in \Gamma_\sigma$. Then one can show that $s \circ a = t$ (here we view s and t as elements of $\mathcal{M}$) if and only if the following holds in $\mathbb{M}(A, I)$:

- $sa = t$, or
- there is $b \in \mathcal{M}_\sigma$ such that $b \circ_\sigma a = 1_\sigma$ and $tb = s$, or
- there is $b \in \mathcal{M}_\sigma$ and $u \in \mathbb{M}(A, I)$ with $s = ub$, $b \circ_\sigma a \neq 1_\sigma$, and $t = u (b \circ_\sigma a)$.

All these properties can be easily expressed in first-order logic over $\mathcal{B}$.   □

Now we can show the main result of this section:

Theorem 6.3. *Let $\mathcal{M} = \prod_{(\Sigma, J)} \mathcal{M}_\sigma$, where (Σ, J) is a finite independence alphabet and $\mathcal{M}_\sigma$ is a monoid finitely generated by Γ_σ ($\sigma \in \Sigma$). Let $\Gamma = \bigcup_{\sigma \in \Sigma} \Gamma_\sigma$.*

(1) *If $\text{FOTh}(\mathbb{C}(\mathcal{M}_\sigma, \Gamma_\sigma), 1_\sigma)$ is decidable for all $\sigma \in \Sigma$, then $\text{FOTh}(\mathbb{C}(\mathcal{M}, \Gamma), 1)$ is decidable as well.*

(2) *If $J = \emptyset$ and $\text{MSOTh}(\mathbb{C}(\mathcal{M}_\sigma, \Gamma_\sigma), 1_\sigma)$ is decidable for all $\sigma \in \Sigma$, then $\text{MSOTh}(\mathbb{C}(\mathcal{M}, \Gamma), 1)$ is decidable as well.*

Proof. First assume that FOTh($\mathbb{C}(\mathcal{M}_\sigma, \Gamma_\sigma), 1_\sigma$) is decidable for all $\sigma \in \Sigma$. Lemma 6.2 implies that we can reduce FOTh($\mathbb{C}(\mathcal{M}, \Gamma), 1$) to the FO-theory of the factorized unfolding $\mathcal{B}$, which is decidable by Theorem 5.7 since FOTh($\mathcal{A}$) is decidable by [12]. The second statement on MSO-theories follows similarly by refering to [31] and [26] instead of Theorem 5.7 and [12], respectively.   □

Statement (2) from Theorem 6.3 does not generalize to graph products.

Proposition 6.4. *Let* (Σ, J), $\mathcal{M}_\sigma$, Γ_σ, $\mathcal{M}$, *and* Γ *as in Theorem 6.3 with* $\mathcal{M}_\sigma$ *non-trivial. Assume furthermore that* MSOTh($\mathbb{C}(\mathcal{M}, \Gamma), 1$) *is decidable. Then*

(a) (Σ, J) *does not contain an induced cycle of length 4 (also called C4),*
(b) if $(\sigma, \tau) \in J$ *and* $\mathcal{M}_\sigma$ *is infinite, then* $\mathcal{M}_\tau$ *is finite,*
(c) if $(\sigma, \sigma_1), (\sigma, \sigma_2) \in J$, $\sigma_1 \neq \sigma_2$, *and* $\mathcal{M}_\sigma$ *is infinite, then* $(\sigma_1, \sigma_2) \in J$,
(d) MSOTh($\mathbb{C}(\mathcal{M}_\sigma, \Gamma_\sigma), 1_\sigma$) *is decidable for every* $\sigma \in \Sigma$.

Proof (sketch). Condition (a), (b), and (c) hold, since otherwise $\mathcal{M}$ contains a direct product of two infinite monoids and thus ($\mathbb{C}(\mathcal{M}, \Gamma), 1$) contains an infinite grid. In order to show (d), one defines ($\mathbb{C}(\mathcal{M}_\sigma, \Gamma_\sigma), 1_\sigma$) in ($\mathbb{C}(\mathcal{M}, \Gamma), 1$).   □

It remains open whether the four conditions in Proposition 6.4 characterize graph products, whose corresponding Cayley-graphs have decidable MSO-theories.

References

1. A. V. Anisimov. Group languages. *Kibernetika*, 4:18–24, 1971. In Russian; English translation in: *Cybernetics 4*, 594–601, 1973.
2. A. Blumensath. Prefix-recognizable graphs and monadic second-order logic. Technical Report 2001-06, RWTH Aachen, Department of Computer Science, 2001.
3. H. L. Bodlaender. A note on domino treewidth. *Discrete Mathematics & Theoretical Computer Science*, 3(4):141–150, 1999.
4. C. M. Campbell, E. F. Robertson, N. Ruškuc, and R. M. Thomas. Automatic semigroups. *Theoretical Computer Science*, 250(1-2):365–391, 2001.
5. B. Courcelle. The monadic second-order logic of graphs, II: Infinite graphs of bounded width. *Mathematical Systems Theory*, 21:187–221, 1989.
6. B. Courcelle. The monadic second-order logic of graphs VI: On several representations of graphs by relational structures. *Discrete Applied Mathematics*, 54:117–149, 1994.
7. V. Diekert. Makanin's algorithm. In M. Lothaire, editor, *Algebraic Combinatorics on Words*, pages 342–390. Cambridge University Press, 2001.
8. V. Diekert and G. Rozenberg, editors. *The Book of Traces*. World Scientific, 1995.
9. R. Diestel. *Graph Theory, Second Edition*. Springer, 2000.
10. M. J. Dunwoody. Cutting up graphs. *Combinatorica*, 2(1):15–23, 1981.
11. C. C. Elgot and M. O. Rabin. Decidability and undecidability of extensions of second (first) order theory of (generalized) successor. *Journal of Symbolic Logic*, 31(2):169–181, 1966.
12. S. Feferman and R. L. Vaught. The first order properties of products of algebraic systems. *Fundamenta Mathematicae*, 47:57–103, 1959.
13. J. Ferrante and C. Rackoff. *The Computational Complexity of Logical Theories*, number 718 of *Lecture Notes in Mathematics*. Springer, 1979.

14. H. Gaifman. On local and nonlocal properties. In J. Stern, editor, *Logic Colloquium '81*, pages 105–135. North Holland, 1982.

15. E. R. Green. *Graph Products of Groups*. PhD thesis, The University of Leeds, 1990.

16. B. Khoussainov and A. Nerode. Automatic presentations of structures. In *LCC: International Workshop on Logic and Computational Complexity*, number 960 in Lecture Notes in Computer Science, pages 367–392, 1994.

17. D. Kuske and M. Lohrey. Decidable theories of graphs, factorized unfoldings and cayley-graphs. Technical Report 2002/37, University of Leicester, MCS, 2002.

18. G. S. Makanin. The problem of solvability of equations in a free semigroup. *Math. Sbornik*, 103:147–236, 1977. In Russian; English translation in: *Math. USSR Sbornik 32, 1977*.

19. G. S. Makanin. Equations in a free group. *Izv. Akad. Nauk SSR*, Ser. Math. 46:1199–1273, 1983. In Russian; English translation in *Math. USSR Izvestija 21, 1983*.

20. D. E. Muller and P. E. Schupp. Groups, the theory of ends, and context-free languages. *Journal of Computer and System Sciences*, 26:295–310, 1983.

21. D. E. Muller and P. E. Schupp. The theory of ends, pushdown automata, and second-order logic. *Theoretical Computer Science*, 37(1):51–75, 1985.

22. P. Narendran and F. Otto. Some results on equational unification. In M. E. Stickel, editor, *Proceedings of the 10th International Conference on Automated Deduction (CADE 90), Kaiserslautern (Germany)*, number 449 in Lecture Notes in Computer Science, pages 276–291. Springer, 1990.

23. D. Seese. Tree-partite graphs and the complexity of algorithms. In L. Budach, editor, *Proceedings of Fundamentals of Computation Theory (FCT'85), Cottbus (GDR)*, number 199 in Lecture Notes in Computer Science, pages 412–421, 1985.

24. D. Seese. The structure of models of decidable monadic theories of graphs. *Annals of Pure and Applied Logic*, 53:169–195, 1991.

25. G. Sénizergues. Semi-groups acting on context-free graphs. In F. M. auf der Heide and B. Monien, editors, *Proceedings of the 28th International Colloquium on Automata, Languages and Programming (ICALP 96), Paderborn (Germany)*, number 1099 in Lecture Notes in Computer Science, pages 206–218. Springer, 1996.

26. S. Shelah. The monadic theory of order. *Annals of Mathematics, II. Series*, 102:379–419, 1975.

27. J. Stupp. The lattice-model is recursive in the original model. The Hebrew University, Jerusalem, 1975.

28. W. Thomas. A short introduction to infinite automata. In W. Kuich, G. Rozenberg, and A. Salomaa, editors, *Proceedings of the 5th International Conference on Developments in Language Theory (DLT 2001), Vienna (Austria)*, number 2295 in Lecture Notes in Computer Science, pages 130–144. Springer, 2001.

29. C. Thomassen and W. Woess. Vertex-transitive graphs and accessibility. *Journal of Combinatorial Theory, Series B*, 58:248–268, 1993.

30. A. Veloso da Costa. Graph products of monoids. *Semigroup Forum*, 63(2):247–277, 2001.

31. I. Walukiewicz. Monadic second-order logic on tree-like structures. *Theoretical Computer Science*, 275(1–2):311–346, 2002.

The Complexity of Resolution with Generalized Symmetry Rules

Stefan Szeider*

Department of Computer Science, University of Toronto,
M5S 3G4 Toronto, Ontario, Canada
szeider@cs.toronto.edu

Abstract. We generalize Krishnamurthy's well-studied symmetry rule for resolution systems by considering homomorphisms instead of symmetries; symmetries are injective maps of literals which preserve complements and clauses; homomorphisms arise from symmetries by releasing the constraint of being injective.
We prove that the use of homomorphisms yields a strictly more powerful system than the use of symmetries by exhibiting an infinite sequence of sets of clauses for which the consideration of global homomorphisms allows exponentially shorter proofs than the consideration of local symmetries. It is known that local symmetries give rise to a strictly more powerful system than global symmetries; we prove a similar result for local and global homomorphisms. Finally, we pinpoint an exponential lower bound for the resolution system enhanced by the local homomorphism rule.

1 Introduction

Informal proofs often contain the phrase "...without loss of generality, we assume that..." indicating that it suffices to consider one of several symmetric cases. Krishnamurthy [8] made this informal feature available for the resolution system; he introduced a global symmetry rule (exploiting symmetries of the refuted CNF formula) and a local symmetry rule (exploiting symmetries of those clauses of the refuted CNF formula which are actually used at a certain stage of the derivation). — Similar rules have been formulated for cut-free Gentzen systems by Arai [1,3].

In the quoted paper, Krishnamurthy observes that the resolution system, equipped with the global symmetry rule, permits short proofs (i.e., proofs of polynomial length) of several combinatorial principles, including the pigeon hole formulas; however, it is well known that the pigeon hole formulas require resolution proofs of exponential length ([7]; see also [5]). A formal proof of this separation (resolution from resolution + global symmetry) can be found in [12]. Moreover, Arai and Urquhart [4] showed that for resolution systems the local symmetry rule attains an exponential speedup over the global symmetry rule.

---

* Supported by the Austrian Science Fund (FWF) Project J2111.

H. Alt and M. Habib (Eds.): STACS 2003, LNCS 2607, pp. 475–486, 2003.

Random formulas contain almost no nontrivial global symmetries, but it is expected that random formulas contain a lot of local symmetries [12].

The symmetries of CNF formulas considered by Krishnamurthy are special cases of *CNF homomorphisms*, introduced in [10]. A homomorphism from a CNF formula F to a CNF formula G is a map φ from the literals of F to the literals of G which preserves complements and clauses; i.e., $\varphi(\overline{x}) = \overline{\varphi(x)}$ for all literals x of F, and $\{ \varphi(x) : x \in C \} \in G$ for all clauses $C \in F$ — symmetries are nothing but injective homomorphisms (see Section 3 for a more detailed definition). Allowing homomorphisms instead of symmetries in the formulation of the global and local symmetry rule gives raise to more general rules which we term *global* and *local homomorphism rule*, respectively.

Separation results. We show that the consideration of homomorphisms gives an exponential speedup over symmetries. We provide a sequence of formulas for which even *global* homomorphisms outperform *local* symmetries (Section 5).

Furthermore, in Section 6 we exhibit a sequence of formulas for which proofs using local homomorphisms are exponentially shorter than shortest proofs using global homomorphisms (a similar result is shown in [4] for symmetries). Fig. 1 gives an overview of our results on the relative efficiency of the considered systems in terms of *p-simulation* (system A p-simulates system B if refutations of system B can be transformed in polynomial time into refutations of system A, cf. [11]).

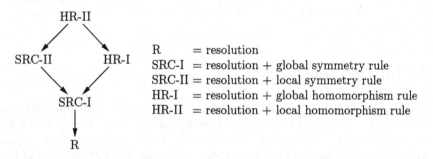

R   = resolution
SRC-I   = resolution + global symmetry rule
SRC-II = resolution + local symmetry rule
HR-I   = resolution + global homomorphism rule
HR-II   = resolution + local homomorphism rule

Fig. 1. Proof system map. $A \to B$ indicates that system A p-simulates system B, but B cannot p-simulate A.

Lower bounds. The exponential lower bound for resolution + local symmetry rule established in [4] does not extend to the more general homomorphism system: to prevent any symmetries, it suffices to modify formulas which are hard for resolution (e.g., pigeon hole formulas) so that all clauses have different width (besides some unit clauses). This can be achieved by adding "dummy variables" to clauses and by providing unit clauses which contain the negations of the dummy variables. However, since widths of clauses may decrease under homomorphisms, such approach is not applicable for homomorphisms.

We achieve an exponential lower bound for the local homomorphism rule by a "link construction," which transforms any formula F which is hard for resolution into a formula F° which is hard for resolution + local homomorphism rule. The trick is to take a new variable for every literal occurrence of F, and to interconnect the obtained clauses by certain sets of clauses ("links") which cannot be mapped to F° by a non-trivial homomorphism. This construction is presented in Section 7.

2   Definitions and Preliminaries

We consider propositional formulas in conjunctive norm form (CNF) represented as sets of clauses: We assume an infinite set var of (propositional) *variables*. A *literal* ℓ is a variable x or a negated variable $\neg x$; we write lit $:= \{\, x, \neg x : x \in \mathsf{var}\,\}$. For a literal ℓ we put $\overline{\ell} := \neg x$ if $\ell = x$, and $\overline{\ell} := x$ if $\ell = \neg x$. For a set of literals C we put $\overline{C} := \{\,\overline{\ell} : \ell \in C\,\}$. A set of literals is *tautological* if $C \cap \overline{C} \neq \emptyset$. A finite non-tautological set of literals is a *clause*; a finite set of clauses is a *formula*. The *length* of a formula F is given by its cardinality $|F|$, and its *size* by $\|F\| := \sum_{C \in F} |C|$. Note that always $|F| \leq \|F\| + 1$. A formula F *mentions* a variable x if F contains a clause C such that $x \in C \cup \overline{C}$; $\mathsf{var}(F)$ denotes the set of variables mentioned by F. Similarly we put $\mathsf{lit}(F) := \mathsf{var}(F) \cup \overline{\mathsf{var}(F)}$. A literal ℓ is a *pure literal* of a formula F if some clauses of F contain ℓ but no clause contains $\overline{\ell}$.

A formula F is *satisfiable* if there is a map $t : \mathsf{var}(F) \to \{0, 1\}$ such that every clause of F contains either a variable x with $t(x) = 1$ or a literal $\neg x$ with $t(x) = 0$. A formula is *minimally unsatisfiable* if it is unsatisfiable but every proper subset is satisfiable.

If $C_1 \cap \overline{C_2} = \{\ell\}$ for clauses C_1, C_2 and a literal ℓ, then the *resolution rule* allows the derivation of the clause $D = (C_1 \cup C_2) \setminus \{\ell, \overline{\ell}\}$; D is the *resolvent* of C_1 and C_2, and we say that D is obtained by *resolving on* ℓ. Let F be a formula and C a clause. A sequence $S = C_1, \ldots, C_k$ of clauses is a *resolution derivation* of C_k *from* F if for each $i \in \{1, \ldots, k\}$ at least one of the following holds.

1. $C_i \in F$ ("C_i is an axiom");
2. C_i is a resolvent of C_j and $C_{j'}$ for some $1 \leq j < j' < i$ ("C_i is obtained by resolution");
3. $C_i \supseteq C_j$ for some $1 \leq j < i$ ("C_i is obtained by weakening").

We write $|S| := k$ and call k the *length* of S. If C_k is the empty clause, then S is a *resolution refutation* of F.

It is well known that resolution is a complete proof system for unsatisfiable formulas; i.e., a formula F is unsatisfiable if and only if there exists a resolution refutation of it.

The *resolution complexity* $\mathsf{Comp_R}(F)$ of an unsatisfiable formula F is the length of a shortest resolution refutation of F (for satisfiable formulas we put $\mathsf{Comp_R}(F) := \infty$). Here, R stands for the resolution system, and we will use similar notations for other proof systems considered in the sequel.

We call a resolution derivation *weakening-free* if no clause is obtained by weakening. It is well known that weakening is inessential for the length of resolution refutations.

If a formula F contains a unit clause $\{\ell\}$, then we can reduce F to a formula F' by removing $\{\ell\}$ from F and $\bar{\ell}$ from all other clauses. We say that F *can be reduced to F^* by unit resolution* if F^* can be obtained from F by multiple applications of this reduction. Evidently, F is satisfiable if and only if F^* is satisfiable. The following can be shown easily.

Lemma 1 *Let F and F^* be formulas such that F can be reduced to F^* by unit resolution. Then* $\mathsf{Comp_R}(F^*) \leq \mathsf{Comp_R}(F)$.

The *pigeon hole formulas* PH_n, $n = 1, 2, \ldots$ encode the fact that $n+1$ pigeons do not fit into n holes if each hole can hold at most one pigeon (i.e., Dirichlet's Box Principle); formally, we take variables $x_{i,j}$, $1 \leq i \leq n+1$ and $1 \leq j \leq n$ (with the intended meaning 'pigeon i sits in hole j') and put

$$\mathrm{PH}_n := \{\, \{x_{i,1}, \ldots, x_{i,n}\} : 1 \leq i \leq n+1 \,\} \cup$$
$$\{\, \{\neg x_{i,j}, \neg x_{i',j}\} : 1 \leq j \leq n,\ 1 \leq i < i' \leq n+1 \,\}.$$

Since PH_n contains $n+1$ clauses of width n and $n\binom{n+1}{2}$ clauses of width 2, we have $|\mathrm{PH}_n| = (n^3 + n^2)/2 + n + 1 = \mathcal{O}(n^3)$, and $\|\mathrm{PH}_n\| = n^3 + 2n^2 + n = \mathcal{O}(n^3)$. Furthermore, the following can be verified easily.

Lemma 2 PH_n *is minimally unsatisfiable for every $n \geq 1$.*

Note that the weaker "onto" variant of the pigeon hole formula is not minimally unsatisfiable.

The following seminal result on the length of resolution refutations is due to Haken [7]; see also [5] for a simpler proof. This result is the basis for our separation and lower bound results.

Theorem 1 *Shortest resolution refutations of PH_n have length $2^{\Omega(n)}$.*

3   Homomorphisms

Consider a finite set $L \subseteq \mathsf{lit}$ of literals. A map $\rho : L \to \mathsf{lit}$ is a *renaming* if for every pair $\ell, \bar{\ell} \in L$ we have $\overline{\rho(\ell)} = \rho(\bar{\ell})$ (note that in our setting, renamings are not necessarily injective). For a subset $C \subseteq L$ we put $\rho(C) := \{\, \rho(\ell) : \ell \in C \,\}$, and for a formula F with $\mathsf{lit}(F) \subseteq L$ we put $\rho(F) := \{\, \rho(C) : C \in F \,\}$. Since for a clause C, $\rho(C)$ may be tautological, we define $\rho_{\mathsf{cls}}(F)$ as the set of all non-tautological $\rho(C)$ with $C \in F$.

Lemma 3 *Let F be a formula, $S = C_1, \ldots, C_k$ a resolution derivation from F, and $\rho : \mathsf{lit}(F) \to \mathsf{lit}$ a renaming. If $\rho(C_k)$ is a clause, then $\rho(C_1), \ldots, \rho(C_k)$ contains a subsequence which is a resolution derivation of $\rho(C_k)$ from $\rho_{\mathsf{cls}}(F)$.*

Proof. By induction on the length of S.   □

Let F_1, F_2 be formulas and $\varphi : \mathsf{lit}(F_1) \to \mathsf{lit}(F_2)$ a renaming. We call φ a *homomorphism from F_1 to F_2* if $\varphi(F_1) \subseteq F_2$ (thus, for every $C \in F_1$, $\varphi(C)$ is a clause and belongs to F_2). The set of all homomorphisms from F_1 to F_2 is denoted by $\mathsf{Hom}(F_1, F_2)$. A homomorphism $\varphi \in \mathsf{Hom}(F_1, F_2)$ is a *monomorphism* if the map $\varphi : \mathsf{lit}(F_1) \rightarrow \mathsf{lit}(F_2)$ is injective. Homomorphisms from a formula to itself are called *endomorphisms*; an endomorphism φ of F is called *automorphism* (or *symmetry*) if $\varphi(F) = F$; otherwise it is a *proper endomorphism*. We denote by id_F the automorphism of F which maps every literal of F to itself. Finally, we call a homomorphism $\varphi \in \mathsf{Hom}(F_1, F_2)$ *positive* if $\varphi(\mathsf{var}(F_1)) \subseteq \mathsf{var}(F_2)$; i.e., literals are mapped to literals of the same polarity.

We state some direct consequences of Lemma 3.

Lemma 4 *Let F_1 and F_2 be formulas.*

1. *If $C_1, \ldots, C_k$ is a resolution derivation from F_1, $\varphi \in \mathsf{Hom}(F_1, F_2)$, and $\varphi(C_k)$ is a clause, then $\varphi(C_1), \ldots, \varphi(C_k)$ contains a subsequence which is a resolution derivation of $\varphi(C_k)$ from F_2.*
2. *If $\mathsf{Hom}(F_1, F_2) \neq \emptyset$, then $\mathsf{Comp_R}(F_2) \leq \mathsf{Comp_R}(F_1)$.*
3. *If $\mathsf{Hom}(F_1, F_2) \neq \emptyset$ and F_1 is unsatisfiable, then F_2 is unsatisfiable.*
4. *Let φ be an endomorphism of F_1. Then F_1 is satisfiable if and only if $\varphi(F_1)$ is satisfiable.*

Parts 3 and 4 of the previous lemma have short semantic proofs as well, see [10]. In view of part 4 we can reduce a formula F by endomorphisms until we end up with a subset F' of F for which every endomorphism is an automorphism. We call such F' a *core* of F, and we call F a core if it is a core of itself. In general, there may be different ways of reducing F by endomorphisms, and we may end up with different cores. However, in [10] it is shown that all cores of a formula are isomorphic; thus, in a certain sense, this reduction is confluent. In the quoted paper it is also shown that recognition of cores (i.e., formulas without proper endomorphisms) is a co-NP-complete problem. The following is a direct consequence of the last part of Lemma 4.

Lemma 5 *Minimally unsatisfiable formulas are cores.*

4   The Homomorphism Rule

Consider a derivation S from a formula F and a subsequence S' of S which is a derivation of a clause C from a subset $F' \subseteq F$. If there is a homomorphism $\varphi \in \mathsf{Hom}(F', F)$ such that $\varphi(C)$ is non-tautological, then the *local homomorphism rule* allows the derivation of $\varphi(C)$. We call the restricted form of this rule which can only be applied if $F' = F$ the *global homomorphism rule*. The systems **HR-I** and **HR-II** arise from the resolution system by addition of the global and local homomorphism rule, respectively.

Lemma 6 *The homomorphism rule is sound; i.e., formulas having an HR-II refutation are unsatisfiable.*

Proof. Let F be a formula, $S = C_1, \ldots, C_k$ an HR-II refutation of F, and $n(S)$ the number of applications of the homomorphism rule. We show by induction on $n(S)$ that S can be transformed into a resolution refutation S' of F.

If $n(S) = 0$, then this holds vacuously. Assume $n(S) > 0$ and choose $i \in \{1, \ldots, k\}$ minimal such that C_i is obtained from some C_j, $1 \leq j < i$, using the homomorphism rule. Thus, there is some $F' \subseteq F$ and a homomorphism $\varphi \in \mathsf{Hom}(F', F)$ such that $\varphi(C_j) = C_i$, and $C_1, \ldots, C_j$ contains a subsequence S' which is a derivation of C_j from F'. By the choice of i, S' is a resolution derivation. Applying Lemma 4(1), we conclude that $\varphi(C_1), \ldots, \varphi(C_j)$ contains a subsequence S'' which is a resolution derivation of $\varphi(C_j)$ from $\varphi(F') \subseteq F$. By juxtaposition of S'' and S we thus get an HR-II refutation S^*; since $n(S'') = 0$, and since we replaced one application of the homomorphism rule by a weakening, we have $n(S^*) = n(S) - 1$. By induction hypothesis, S^* can be transformed into a resolution refutation of F. $\square$

The proof of Lemma 6 gives a reason for considering HR-II refutations as *succinct representations of resolution refutations*. Note that the transformation defined in this proof may cause an exponential growth of refutation length (this is the case for the formulas constructed in Section 7).

Krishnamurthy's systems of *symmetric resolution* **SR-λ** and **SRC-λ**, $\lambda \in \{\mathrm{I}, \mathrm{II}\}$, arise as special cases of HR-λ: In SRC-λ, applications of the homomorphism rule are restricted to cases where φ is a monomorphism (for $\lambda = \mathrm{I}$ this means that φ is an automorphism of the refuted formula); SR-λ arises from SRC-λ by considering only positive monomorphisms (variables are mapped to variables). In the context of SR-λ and SRC-λ we refer to the homomorphism rule as the *symmetry rule*. Note that weakening is inessential for all these systems.

Borrowing a notion from category theory, we call a formula *rigid* if it has no automorphism except the identity map (cf. [9]). Since an SRC-I refutation of a rigid formula is nothing but a resolution refutation, we have the following.

Lemma 7 *If a formula F is rigid, then* $\mathsf{Comp}_{\mathrm{SRC\text{-}I}}(F) = \mathsf{Comp}_{\mathrm{R}}(F)$.

We say that F is *locally rigid* if for every integer $n \geq 2$ there is at most one clause $C \in F$ with $|C| = n$. The next result is due to Arai and Urquhart [4].

Lemma 8 *If F is locally rigid, then* $\mathsf{Comp}_{\mathrm{R}}(F) = \mathsf{Comp}_{\mathrm{SRC\text{-}II}}(F)$.

5   Separating HR-I from SRC-II

For this section, F denotes some arbitrary but fixed unsatisfiable formula and $S = C_1, \ldots, C_k$ a weakening-free SRC-I refutation of F. Let $h(1) < \cdots < h(n)$ be the indexes $h(i) \in \{1, \ldots, k\}$ such that $C_{h(i)}$ is obtained by the symmetry rule; let $\alpha_{h(i)}$ denote the automorphism used to obtain $C_{h(i)}$.

We construct a formula $F^\times$ as follows. For each $i = 1, \ldots, n$ we take a variable-disjoint copy F_i of F, using a new variable $\langle x, i \rangle$ for each $x \in \mathrm{var}(F)$. To unify notation, we write $\langle x, 0 \rangle := x$ and $F_0 := F$. By disjoint union we obtain the formula $F^\times := \bigcup_{i=0}^n F_i$, and we observe that $\|F^\times\| \leq |S| \cdot \|F\|$. We call an endomorphism ψ of $F^\times$ *increasing* if for every variable $\langle x, j \rangle$ we have $\psi(\langle x, j \rangle) \in \{\langle y, j' \rangle, \neg\langle y, j' \rangle\}$ with $j \leq j'$.

Lemma 9 *There is a weakening-free HR-I refutation $S^\times$ of $F^\times$ such that $|S^\times| \leq |S|^2$ and all endomorphisms used in $S^\times$ are proper and increasing.*

Proof. (Sketch.) The above SRC-I refutation $S = C_1, \ldots, C_k$ is clearly an HR-I refutation of $F^\times$ (the symmetries of F extend to symmetries of $F^\times$ by mapping clauses in $F^\times \setminus F$ to themselves). Consider $C_{h(1)}$, the first clause in S which is obtained by the symmetry rule, say $C_{h(1)} = \alpha_{h(1)}(C_j)$ for some $j < h(1)$. We obtain $S' := C_1, \ldots, C_{h(1)-1}, \pi(C_1), \ldots, \pi(C_k)$ where π is the endomorphism of $F^\times$ defined by $\pi(\langle x, 0 \rangle) = \langle x, 1 \rangle$. Now S' is an HR-I refutation of $F^\times$, and $\pi(C_{h(1)})$ can be obtained from C_j by the proper and increasing endomorphism $\alpha' := \pi \circ \alpha$. By multiple application of this construction we can replace successively all symmetries by proper and increasing endomorphisms. $\square$

Next we modify $F^\times$ so that it becomes a locally rigid formula $F^\sharp$, deploying a similar construction as used by Arai and Urquhart [4]. Let $E_1, \ldots E_m$ be a sequence of all the clauses of $F^\times$ such that for any $E_j \in F_i$ and $E_{j'} \in F_{i'}$ we have

$$i > i' \text{ implies } j < j';$$
$$i = i' \text{ and } |E_j| < |E_{j'}| \text{ implies } j < j'.$$

For each clause E_j we take new variables $y_{j,1}, \ldots, y_{j,j}$, and we define the formula

$$Q_j := \{E_j \cup \{y_{j,1}, \ldots, y_{j,j}\}, \{\neg y_{j,1}\}, \ldots, \{\neg y_{j,j}\}\}.$$

Finally we define $F^\sharp := \bigcup_{j=1}^m Q_j$.

We observe that $\|F^\sharp\| \leq \|F^\times\| + 2|F^\times|^2$, and we state a direct consequence of the above definitions.

Lemma 10 *$F^\sharp$ is locally rigid and can be reduced to $F^\times$ by unit resolution.*

Every increasing endomorphism $\varphi^\times$ of $F^\times$ can be extended to an endomorphism $\varphi^\sharp$ of $F^\sharp$ as follows. Since for every pair of clauses $E_j, E_{j'} \in F^\times$, $\varphi^\times(E_j) = E_{j'}$ implies $j \leq j'$, we can put

$$\varphi^\sharp(y_{j,i}) := y_{j',\min(i,j')}.$$

Hence $\varphi^\sharp(Q_j) = Q_{j'}$ follows. In view of this construction, it is not difficult to see that an HR-I refutation of $F^\times$ can be transformed into an HR-I refutation of $F^\sharp$ with a polynomial increase of length; in particular, we have the following estimation.

Lemma 11 $\mathsf{Comp}_{\text{HR-I}}(F^\sharp) \leq |S^\times| + |F^\sharp|$.

The following lemma is due to Urquhart [12], see also Krishnamurthy [8]. (In [12], the lemma is formulated for certain formulas PHC_n with $\text{PH}_n \subseteq \text{PHC}_n$; its proof, however, does not rely on the clauses in $\text{PHC}_n \setminus \text{PH}_n$.)

Lemma 12 *There are SR-I refutations of length $(3n+1)n/2$ for the pigeon hole formulas PH_n.*

Theorem 2 *There is an infinite sequence of formulas F_n, $n = 1, 2, \ldots$ such that the size of F_n is $\mathcal{O}(n^{10})$, F_n has HR-I refutations of length $\mathcal{O}(n^{10})$, but shortest SRC-II refutations have length $2^{\Omega(n)}$.*

Proof. By Lemma 12, pigeon hole formulas PH_n have SRC-I refutations S_n of length $\mathcal{O}(n^2)$. We apply the above constructions and consider $\mathrm{PH}_n^\times$, $\mathrm{PH}_n^\sharp$ and the corresponding HR-I refutations $S_n^\times$, $S_n^\sharp$, respectively. We put $F_n := \mathrm{PH}_n^\sharp$. Lemmas 9, 12, and 11 yield $|S_n^\sharp| = \mathcal{O}(n^{10})$. Clearly putting $\varphi(\langle x, i \rangle) := x$ defines a homomorphism from $F^\times$ to F. Thus, by Lemmas 4(2), 1, 10, and 8, respectively, we have

$$\mathsf{Comp}_{\mathrm{R}}(\mathrm{PH}_n) \leq \mathsf{Comp}_{\mathrm{R}}(\mathrm{PH}_n^\times) \leq \mathsf{Comp}_{\mathrm{R}}(\mathrm{PH}_n^\sharp) = \mathsf{Comp}_{\mathrm{SRC\text{-}II}}(\mathrm{PH}_n^\sharp).$$

The result now follows from Theorem 1.   □

Corollary 1 *SR-II (and so SR-I) p-simulates neither HR-I nor HR-II.*

6   Separating HR-I from HR-II

In [4] it is shown that SR-II has an exponential speed up over SR-I. We show an analogous result for HR-II and HR-I, using a similar construction.

Consider the pigeon hole formula $\mathrm{PH}_n = \{E_1, \ldots, E_t\}$. For each clause E_j we take new variables $y_{j,1}, \ldots, y_{j,j}$, and we define

$$Q_j := \{E_j \cup \{y_{j,1}\}, \{\neg y_{j,1}, y_{j,2}\}, \ldots, \{\neg y_{j,j-1}, y_{j,j}\}, \{\neg y_{j,j}\}\},$$

and put $\mathrm{PH}_n^\sim := \bigcup_{j=1}^t Q_j$. Note that $\|\mathrm{PH}_n^\sim\| \leq \|\mathrm{PH}_n\| + 2|\mathrm{PH}_n|^2 = \mathcal{O}(n^6)$.

Theorem 3 *There is an infinite sequence of formulas F_n, $n = 1, 2, \ldots$ such that the size of F_n is $\mathcal{O}(n^6)$, F_n has SR-II refutations (and so HR-II refutations) of length $\mathcal{O}(n^6)$, but shortest HR-I refutations have length $2^{\Omega(n)}$.*

Proof. By means of Lemma 1 we conclude from Theorem 1 that

$$\mathsf{Comp}_{\mathrm{R}}(\mathrm{PH}_n^\sim) = 2^{\Omega(n)}.$$

Since PH_n is minimally unsatisfiable, so is $\mathrm{PH}_n^\sim$; thus $\mathrm{PH}_n^\sim$ is a core by Lemma 5. It is not difficult to show that $\mathrm{PH}_n^\sim$ is rigid. Thus every HR-I refutation of $\mathrm{PH}_n^\sim$ is nothing but a resolution refutation, and we get

$$\mathsf{Comp}_{\mathrm{HR\text{-}I}}(\mathrm{PH}_n^\sim) = \mathsf{Comp}_{\mathrm{R}}(\mathrm{PH}_n^\sim) = 2^{\Omega(n)}.$$

By a straightforward construction, an SR-I refutation of PH_n can be transformed into an SR-II refutation of $\mathrm{PH}_n^\sim$ adding less than $2|\mathrm{PH}_n|^2$ steps of unit resolution. Hence, the Theorem follows by Lemma 12.   □

Corollary 2 *HR-I p-simulates neither SR-II nor HR-II.*

In view of Corollary 1 we also have the following.

Corollary 3 *HR-I and SR-II are incomparable in terms of p-simulation.*

7   An Exponential Lower Bound for HR-II

In this section, F denotes some arbitrarily chosen formula without unit clauses. We assume a fixed order $E_1, \ldots, E_m$ of the clauses of F, and a fixed order of the literals in each clause, so that we can write

$$F = \{\{\ell_1, \ldots, \ell_{i_1}\}, \{\ell_{i_1+1}, \ldots, \ell_{i_2}\}, \ldots, \{\ell_{i_{m-1}+1}, \ldots, \ell_s\}\}; \; s = \|F\|.$$

From F we construct a formula F° as follows. For every $j \in \{1, \ldots, s\}$ we take new variables $y_{j,1}, \ldots, y_{j,j+2}$ and z_j. We define the formula

$$L_j := \{\{\neg y_{j,1}, y_{j,2}, z_j\}, \{\neg y_{j,2}, y_{j,3}, z_j\}, \ldots, \{\neg y_{j,j+1}, y_{j,j+2}, z_j\}, \{\neg y_{j,j+2}, \ell_j\}\}$$

and put

$$L_j' := L_j \cup \{\{\neg z_j\}\}.$$

We call L_j a *link*. Furthermore, we define for every clause $E_i = \{\ell_j, \ldots, \ell_{j+|E_i|}\}$ a corresponding clause

$$E_i^\circ := \{y_{j,1}, \ldots, y_{j+|E_i|,1}\}.$$

Finally, we put the above definitions together and obtain the formula

$$F^\circ := \{E_1^\circ, \ldots, E_m^\circ\} \cup \bigcup_{j=1}^{s} L_j'.$$

The size of L_j' is less than $3(s+2)$, thus $\|F^\circ\| \le 3s^2 + 7s$.

We will refer to clauses E_i° as *main clauses*, and to clauses in L_j' as *link clauses*. For a subset $F' \subseteq F^\circ$ we define its *body* $b(F')$ to be the subset of F' consisting (i) of all main clauses $E_i^\circ \in F'$ for which $L_j \subseteq F'$ holds for at least one $y_{j,1} \in E_i^\circ$, and (ii) of all link clauses of F' which belong to some L_j' with $L_j \subseteq F'$. Roughly speaking, we obtain $b(F')$ from F' by removing all incomplete links and all main clauses which are not adjacent to some complete link. For a subset $F' \subseteq F^\circ$, homomorphisms $\varphi \in \mathsf{Hom}(F', F)$ are "almost" the identity map on $b(F')$.

The proof of the next lemma is elementary but lengthy, since several cases must be distinguished; we omit the proof due to space constraints.

Lemma 13 *Every $\varphi \in \mathsf{Hom}(L_j, F^\circ)$ is a monomorphism, $1 \le j \le s$.*

Corollary 4 $\mathsf{Hom}(L_j, F^\circ) = \{id_{L_j}\}$ *for every $j \in \{1, \ldots, s\}$.*

Lemma 14 *For any $F' \subseteq F^\circ$ and $\varphi \in \mathsf{Hom}(F', F^\circ)$ we have the following.*

1. $\varphi(\ell) = \ell$ *for all literals ℓ with $\ell, \bar{\ell} \in \bigcup_{C \in b(F')} C$;*
2. $\varphi(C) = C$ *for all $C \in b(F')$;*

3. *if there is a weakening-free resolution derivation of a clause D from $b(F')$, then $\varphi(D) = D$.*

Proof. The first part follows from Corollary 4. To show the second part, choose a clause $C \in b(F')$ arbitrarily. If C is a link clause, then $\varphi(C) = C$ follows from the first part; hence assume that C is a main clause. By definition of $b(F')$, C contains at least one literal ℓ such that $\bar{\ell}$ belongs to some link clause of $b(F')$; consequently $\varphi(\ell) = \ell$. Since main clauses are mutually disjoint, we conclude $\varphi(C) = C$; thus part 2 follows. Part 3 follows from the first two parts by induction on the length of the resolution derivation.   □

We take a new variable z and define a renaming $\rho : \mathrm{lit}(F^\circ) \to \mathrm{lit}(F) \cup \{z, \neg z\}$ by setting

$$
\begin{aligned}
\rho(y_{j,i}) &:= \ell_j & (j = 1, \ldots, s;\ i = 1, \ldots, j+2), \\
\rho(\ell_j) &:= \ell_j & (j = 1, \ldots, s), \\
\rho(z_j) &:= z & (j = 1, \ldots, s).
\end{aligned}
$$

Consequently, for link clauses C we have either $\rho(C) = \{\neg z\}$ or $\rho(C) \supseteq \{\ell_j, \bar{\ell}_j\}$ for some $j \in \{1, \ldots, s\}$; for main clauses C° we have $\rho(C^\circ) = C$. Hence $\rho_{\mathsf{cls}}(F^\circ)$ as defined in Section 3 is nothing but $F \cup \{\{\neg z\}\}$, and $\neg z$ is a pure literal of $\rho_{\mathsf{cls}}(F^\circ)$.

Lemma 15 *Let $S = C_1, \ldots, C_n$ be a resolution derivation from $F' \subseteq F^\circ$. If $\rho(C_n)$ is non-tautological, then either some subsequence S' of $\rho(C_1), \ldots, \rho(C_n)$ is a resolution derivation of $\rho(C_n)$ from $\rho_{\mathsf{cls}}(b(F'))$, or there is some $D \in F'$ with $\rho(D) \subseteq \rho(C_n)$.*

Proof. We assume, w.l.o.g., that no proper subsequence of S is a resolution derivation of C_n from F'. Hence, if ℓ is a literal of some axiom of S, and if no clause of S is obtained by resolving on ℓ, then the last clause of S contains ℓ as well. By Lemma 3 some subsequence S' of $\rho(C_1), \ldots, \rho(C_n)$ is a resolution derivation of $\rho(C_n)$ from $\rho_{\mathsf{cls}}(F')$. Assume that S' is not a resolution derivation of $\rho(C_n)$ from $\rho_{\mathsf{cls}}(b(F'))$. That is, some axiom D' of S' belongs to $\rho_{\mathsf{cls}}(F') \setminus \rho_{\mathsf{cls}}(b(F'))$. Consequently, there is an axiom $D \in F' \setminus b(F')$ of S with $\rho(D) = D'$. We will show that $D' = \rho(D) \subseteq \rho(C_n)$.

First assume that D is a main clause; consequently $D' \in F$. Thus, for some $j \in \{1, \ldots, s\}$,

$$
D = \{y_{j,1}, \ldots, y_{j+|D|,1}\} \text{ and } D' = \{\ell_j, \ldots, \ell_{j+|D|}\}.
$$

Consider any $j' \in \{j, \ldots, j + |D|\}$. Since $D \notin b(F')$, $L_{j'} \not\subseteq F'$ by definition of $b(F')$. Then, however, some $y_{j',i'}$, $i' \in \{1, \ldots, j'+2\}$, is a pure literal of F'. Since S is assumed to be minimal, $y_{j',i'} \in C_n$, and so $\rho(y_{j',i'}) = \ell_{j'} \in \rho(C_n)$ follows. Whence $\rho(D) \subseteq \rho(C_n)$.

Second, assume that D is a link clause. Since $\rho(D)$ is non-tautological, $D = \{\neg z_j\}$ for some $j \in \{1, \ldots, s\}$; hence $D' = \{\neg z\}$. However, since $\neg z$ is a pure literal of F', we conclude as in the previous case that $\neg z \in \rho(C_n)$. Thus again $\rho(D) \subseteq \rho(C_n)$.   □

Lemma 16 $\mathsf{Comp_R}(F) \leq \mathsf{Comp_{HR\text{-}II}}(F^\circ) + |F|$.

Proof. (Sketch.) Let $S = C_1, \ldots, C_n$ be a weakening-free HR-II resolution refutation of F°, and let C_i be the first clause which is obtained from some clause C_j, $j < i$, by the homomorphism rule, say $\varphi \in \mathsf{Hom}(F', F^\circ)$ and $C_i = \varphi(C_j)$. If $\rho(C_j)$ is non-tautological, then it follows from Lemma 15 that either some subsequence of $\rho(C_1), \ldots, \rho(C_j)$ is a resolution derivation of $\rho(C_j)$ from $\rho_{\mathsf{cls}}(b(F'))$, or $\rho(E_k) \subseteq \rho(C_j)$ for some $k \in \{1, \ldots, m\}$ (recall that $F = \{E_1, \ldots, E_m\}$). In the first case, Lemma 14 yields $\varphi(b(F')) = b(F')$ and $C_i = \varphi(C_j)$; thus $\rho(C_i) = \rho(C_j)$. In the second case we can obtain $\rho(C_i)$ by weakening from E_k. By multiple applications of this argument, we can find a subsequence of $E_1, \ldots, E_m, \rho(C_1), \ldots, \rho(C_n)$ which is a resolution refutation of F. □

Theorem 4 *There is an infinite sequence of unsatisfiable formulas F_n, $n = 1, 2, \ldots$ such that the size of F_n is $\mathcal{O}(n^6)$, and shortest HR-II refutations of F_n have length $2^{\Omega(n)}$.*

Proof. Again we use the pigeon hole formulas and put $F_n = \mathrm{PH}_{n+1}^\circ$ (we avoid PH_1 since it contains unit clauses). By construction, we have $\|\mathrm{PH}_{n+1}^\circ\| \leq \mathcal{O}(\|\mathrm{PH}_{n+1}\|^2) = \mathcal{O}(n^6)$. The theorem follows by Lemma 16 and Theorem 1. □

Corollary 5 *SR-II cannot p-simulate HR-I or HR-II; SR-I cannot p-simulate HR-I or HR-II.*

8   Discussion and Further Generalizations

The Achilles' heel of HR-II appears to be the fact that the local homomorphism rule cannot take advantage of structural properties of the input formula if these properties are slightly "disguised;" that is, if the properties are not explicitly present in the input formula, but can be made explicit by a simple preprocessing using resolution. We used this observation for showing the exponential lower bound for HR-II: though pigeon hole formulas PH_n have short HR-II refutations, disguised as PH_n° they require HR-II refutations of exponential length.

Other proof systems like *cutting plane proofs (CP)* and *simple combinatorial reasoning (SCR)* (see [6] and [2], respectively), which also allow short refutations of the pigeon hole formulas, are more robust with respect to such disguise. This was observed in [4], where it is shown that SRC-II cannot p-simulate CP or SCR (CP cannot p-simulate SR-I neither). Using a similar argument, it can be shown that HR-II cannot p-simulate CP or SCR. Thus we conclude that HR-II and CP are incomparable in terms of p-simulation.

However, the described flaw of HR-II can be fixed; inspection of the soundness proof (Lemma 6) yields that we can generalize the local homomorphism rule as follows, without loosing soundness.

Consider a derivation $S = C_1, \ldots, C_k$ from F and a subsequence S' of S which is a derivation of a clause C from some formula F'. If there is a homomorphism $\varphi \in \mathsf{Hom}(F', \{C_1, \ldots, C_k\})$ such that $\varphi(C)$ is non-tautological, then the *dynamic homomorphism rule* allows the derivation of $\varphi(C)$.

Note that we have released two constraints of the local homomorphisms rule: F' is not necessarily a subset of the input formula F, and φ is not necessarily a homomorphism from F' to F (but a homomorphism from F' to the set of clauses appearing in S). Let SR-III, SRC-III, and HR-III denote the proof systems arising from the respective systems using the dynamic homomorphism rule. The formulas which are used to show exponential lower bounds for the global and local systems (see [4,12] and Theorems 3 and 4 of the present paper) have evidently refutations of polynomial length even in the weakest dynamic system SR-III.

The complexities of SR-III, SRC-III, and HR-III, and their relations to CP and SCR remain as interesting open problems (it seems to be feasible to defeat SR-III by formulas obtained from the pigeon hole formulas by suitable flipping of polarities of literals).

References

1. N. H. Arai. Tractability of cut-free Gentzen type propositional calculus with permutation inference. *Theoretical Computer Science*, 170(1-2):129–144, 1996.
2. N. H. Arai. No feasible monotone interpolation for simple combinatorial reasoning. *Theoretical Computer Science*, 238(1-2):477–482, 2000.
3. N. H. Arai. Tractability of cut-free Gentzen-type propositional calculus with permutation inference. II. *Theoretical Computer Science*, 243(1-2):185–197, 2000.
4. N. H. Arai and A. Urquhart. Local symmetries in propositional logic. In R. Dyckhoff, editor, *Automated Reasoning with Analytic Tableaux and Related Methods (Proc. TABLEAUX 2000)*, volume 1847 of *Lecture Notes in Computer Science*, pages 40–51. Springer Verlag, 2000.
5. E. Ben-Sasson and A. Wigderson. Short proofs are narrow—resolution made simple. *Journal of the ACM*, 48(2):149–169, 2001.
6. W. Cook, C. R. Coullard, and G. Turán. On the complexity of cutting-plane proofs. *Discrete Applied Mathematics*, 18(1):25–38, 1987.
7. A. Haken. The intractability of resolution. *Theoretical Computer Science*, 39:297–308, 1985.
8. B. Krishnamurthy. Short proofs for tricky formulas. *Acta Informatica*, 22:253–275, 1985.
9. A. Pultr and V. Trnková. *Combinatorial, Algebraic and Topological Representations of Groups, Semigroups and Categories*. North-Holland Publishing Co., Amsterdam, 1980.
10. S. Szeider. Homomorphisms of conjunctive normal forms. To appear in *Discrete Applied Mathematics*.
11. A. Urquhart. The complexity of propositional proofs. *The Bulletin of Symbolic Logic*, 1(4):425–467, 1995.
12. A. Urquhart. The symmetry rule in propositional logic. *Discrete Applied Mathematics*, 96/97:177–193, 1999.

Colouring Random Graphs in Expected Polynomial Time

Amin Coja-Oghlan and Anusch Taraz

Humboldt-Universität zu Berlin, Institut für Informatik,
Unter den Linden 6, 10099 Berlin, Germany
{coja,taraz}@informatik.hu-berlin.de

Abstract. We investigate the problem of colouring random graphs $G \in G(n, p)$ in polynomial expected time. For the case $p < 1.01/n$, we present an algorithm that finds an optimal colouring in linear expected time. For sufficiently large values of p, we give algorithms which approximate the chromatic number within a factor of $O(\sqrt{np})$. As a by-product, we obtain an $O(\sqrt{np}/\ln(np))$-approximation algorithm for the independence number which runs in polynomial expected time provided $p \gg \ln^6 n/n$.

1 Introduction and Results

The problem of determining the minimum number of colours needed to colour a graph G – denoted by the $\chi(G)$, the chromatic number of G – was proven to be NP-hard by Karp [19]. In fact, already deciding whether a given graph is 3-colourable is NP-complete. These results have been completed in the last decade by non–approximability theorems. Feige and Kilian [9] showed that, unless $coRP = NP$, no polynomial time algorithm with approximation ratio less than $n^{1-\varepsilon}$ exists.

However, these hardness results are deeply rooted in the worst–case paradigm and thus the question arises whether algorithms can be designed and analyzed that perform well on *average* or *random* instances. The binomial model for such a *random graph*, usually denoted by $G(n, p)$, is defined as follows. In a graph with vertex set $[n] = \{1, \ldots, n\}$ every possible edge is present with probability p independently of all others. Here $p = p(n)$ can be a function in n. In the particular case of $G(n, \frac{1}{2})$ this is exactly the uniform distribution on the set of all graphs with vertex set $[n]$. Admittedly, the $G(n, p)$-model may not be appropriate in every setting, but it is definitely the standard model of a random graph. We shall only mention those results on random graphs here which are important for our purposes and refer the interested reader to [4,16,13] for general background information and to the survey [24] for a comprehensive overview on algorithmic random graph colouring.

We say that $G(n, p)$ has a certain property A *almost surely* or *with high probability*, if the probability that $G \in G(n, p)$ has A tends to 1 as $n \to \infty$.

H. Alt and M. Habib (Eds.): STACS 2003, LNCS 2607, pp. 487–498, 2003.
© Springer-Verlag Berlin Heidelberg 2003

Let $b = 1/(1-p)$. Bollobás and Łuczak (cf. [4,16]) proved that almost surely a random graph $G \in G(n,p)$ satisfies

$$\chi(G) \sim \begin{cases} \frac{n}{2\log_b n} & \text{for constant } p \\ \frac{np}{2\ln np} & \text{for } \frac{C}{n} < p = o(1) \end{cases} \qquad \alpha(G) \sim \begin{cases} 2\log_b n & \text{for constant } p \\ \frac{2\ln np}{p} & \text{for } \frac{C}{n} < p = o(1), \end{cases} \quad (1)$$

where C is a sufficiently large constant. For even smaller values of p, and constant k, the situation is as follows. For any number $k \geq 2$, there are constants c_k^- and c_k^+ such that for $p \leq (1-\varepsilon)c_k^-/n$ (and $p \geq (1+\varepsilon)c_k^+/n$ respectively), $G(n,p)$ almost surely is (respectively is not) k-colourable. In fact, it is conjectured that one can actually choose $c_k^- = c_k^+$ [1].

These structural issues are accompanied by the obvious algorithmic question: are there good algorithms to compute a (near-) optimal colouring of $G \in G(n,p)$? The greedy colouring algorithm achieves good results in this setting. The algorithm considers the vertices in an arbitrary order and assigns to each vertex the smallest possible colour. Grimmett and McDiarmid [14] proved that almost surely the number of colours used for $G \in G(n, \frac{1}{2})$ is asymptotic to $n/\log_2 n$. A slight improvement over the greedy algorithm using randomization has been obtained by Krivelevich and Sudakov [25]. Still, to date the best known approximation ratio is asymptotically 2.

It is worth emphasizing that for inputs from $G(n,p)$ the greedy algorithm *always* has polynomial running time (in fact, linear) and achieves a 2-approximation *with high probability*. Nevertheless, the algorithm itself does not find a certificate for a lower bound on the chromatic number of the input graph G, and thus it cannot guarantee a good approximation ratio. Indeed, there are input instances for which the approximation ratio achieved by the greedy algorithm is quite bad (i.e. close to n), even for almost all permutations of the vertices [26]. This motivates the following question by Karp [20]:

Is there an algorithm that for inputs from $G(n,p)$ has *expected* polynomial running time and *always* uses the minimum number of colours?   (2)

Here the expected running time of an algorithm $\mathcal{A}$ is defined as $\sum_G R_{\mathcal{A}}(G)P(G)$, where the sum runs over all graph G with vertex set $[n]$, $R_{\mathcal{A}}(G)$ is the running time needed by $\mathcal{A}$ for input G, and $P(G)$ denotes the probability that G is chosen according to the distribution $G(n,p)$.

The obvious approach to design an algorithm that meets the requirements of (2) is as follows. Typical inputs will have certain structural properties which enable us to efficiently find an optimal solution. In the exceptional cases, we have to invest more time, but this doesn't affect the expected running time much as it will happen only with small probability. However, the crucial point of this approach is *how to decide efficiently whether the input is typical or not*.

Our first result answers question (2) in the affirmative for the case of small edge probabilities p.

Theorem 1. *For $p \leq 1.01/n$ there exists a colouring algorithm which finds an optimal colouring and, when applied to $G(n,p)$, has linear expected running time.*

We remark that the constant 1.01 is certainly not best possible. It merely demonstrates that the algorithm can find optimal colourings even after $G(n,p)$ has passed the *phase transition* where it suddenly becomes much more complex. Variants of question (2) have been addressed by several authors (see e.g. [5,7,12, 29]). They, too, give exact algorithms with expected polynomial running time, but over a different probability distribution than $G(n,p)$.

A new line of research has recently been initiated by Krivelevich and Vu [22], relaxing question (2) in the following way. Is there an algorithm, that for inputs from $G(n,p)$ has *expected* polynomial running time and *always* approximation ratio r? Here r may be a constant or a function in n (and p).

Krivelevich and Vu prove in [22] that for $n^{-1/2+\varepsilon} \leq p \leq 0.99$, there exists a colouring algorithm with approximation ratio $O(\sqrt{np}/\ln n)$ and expected polynomial running time over $G(n,p)$. Moreover they ask [22,24] whether it is possible to find similarly good approximation algorithms for smaller values of p. Our next two results give positive answers to this question.

Theorem 2. *For* $\ln(n)^6/n \ll p \leq 3/4$ *there exists a colouring algorithm with approximation ratio* $O(\sqrt{np})$ *and polynomial expected running time over* $G(n,p)$.

For smaller values of p, where the above theorem does not hold, we can prove the following.

Theorem 3. *Let* $p \geq 1/n$. *Then there exists a colouring algorithm with approximation ratio* $\frac{e^2}{3}np$ *and linear expected running time over* $G(n,p)$.

Observe that in the range where p is too small for Theorem 2 to apply, Theorem 3 is almost as good: here both approximation ratios are of order $poly(\ln n)$.

Instead of considering algorithms that find (near-) optimal colourings for a given graph, one also considers the problem of deciding, for some integer k, whether the input graph is k-colourable or not. Krivelevich [23] has shown that for every fixed $k \geq 3$ there exists a constant $C = C(k)$ so that k-colourability can be decided in expected polynomial time over $G(n,p)$, provided that $p(n) \geq C/n$. Furthermore, he asks [24] whether it is possible to find such algorithms (i) for the case where $k = k(n)$ is a function growing in n and (ii) for any value of the edge probability $p = p(n)$. It turns out that the methods we have developed for our colouring algorithms contribute to answering the above questions.

Our techniques also capture a more general *semirandom* situation. Semirandom graph problems have been studied e.g. in [10,5]. In contrast to $G(n,p)$, semirandom settings allow for an adversary to change the outcome of the random experiment to a certain degree. In this way they capture a larger class of input distributions and therefore require rather robust algorithmic techniques. Here we introduce two semirandom models for the decision version of k-colouring, $G(n,p)^+$ and $G(n,p)^-$, and give algorithms for deciding k-colourability. In the model $G(n,p)^+$, instances are created in two steps as follows:

1. First, a random graph $G_0 = G(n,p)$ is chosen.
2. Then, an adversary may *add* edges to G_0, in order to produce the input instance G.

Note that the adversary can change the vertex degrees, the independence number, the chromatic number and even the spectrum of the random graph G_0. Instances of the semirandom model $G(n,p)^-$ are created similarly, but the adversary is only allowed to *remove* edges instead of adding edges. Given $G_0 = G(n,p)$, let $\mathcal{I}(G_0)^+$ denote the set of all *instances over* G_0, i.e. the set of all graphs G that can be produced by adding edges to G_0. We define $\mathcal{I}(G_0)^-$ similarly. An algorithm $\mathcal{A}$ is said to run in *expected polynomial time applied to the semirandom model* $G(n,p)^+$ if there is a constant l such that for any map I that to each graph G_0 associates an element $I(G_0) \in \mathcal{I}(G_0)^+$, we have

$$\sum_{G_0} R_{\mathcal{A}}(I(G_0))P(G_0) = O(n^l),$$

where $R_{\mathcal{A}}(I(G_0))$ denotes the running time of $\mathcal{A}$ on input $I(G_0)$. We define a similar notion for the model $G(n,p)^-$.

Theorem 4. *For any $k = k(n)$ and $p \le k/(10n)$ there exists an algorithm which decides whether a graph G is k-colourable, and, when applied to $G(n,p)^-$, has linear expected running time.*

Theorem 5. *Let $k = k(n)$ and $p = p(n) \ge \max\{\ln(n)^7/n, 200k^2/n\}$. There exists an algorithm which decides whether a graph G is k-colourable, and, when applied to $G(n,p)^+$, has polynomial expected running time.*

Observe that Theorem 4 deals with the range where $G(n,p)$ is almost surely k-colourable, while Theorem 5 concerns graphs which almost surely are not k-colourable. The threshold for k-colourability has order $k \ln k/n$, which lies in the remaining gap for $\Omega(k/n) \le p \le \tilde{O}(k^2/n)$.

As an immediate by-product from our work on graph colouring, we obtain the following approximation algorithm for the independent set problem.

Theorem 6. *For $\ln(n)^6/n \ll p \le 3/4$ there exists an algorithm that approximates the independence number with approximation ratio $O(\sqrt{np}/\ln(np))$ and has polynomial expected running time over $G(n,p)$.*

The remainder of the paper is organized as follows. After introducing the necessary technical terminology, we shall first prove Theorem 3 in Section 2, as this is the simplest algorithm. Section 3 deals with Theorem 1. The algorithms for Theorems 2 and 6 can be found in Section 4. Finally, in Section 5 we give the algorithms for Theorems 4 and 5.

The Euler constant will be denoted by $e = \exp(1)$. For a graph $G = (V, E)$ and a subset $S \subseteq V$ we let $G[S]$ be the subgraph of G induced by S. We write $H \subseteq G$ if H is a (weak) subgraph of G. Denote by $|G| = |V|$ the order of G. For a vertex v in G we let $d(v) = d_G(v)$ be the degree of v in G. Denote by $\delta(G)$ the minimal degree in G. In this paper we disregard rounding issues since these do not affect our arguments.

2   A Simple $O(np)$-Approximation Algorithm

Consider an integer $k \geq 2$. The *k-core* of a graph G is the unique subgraph $H \subseteq G$ of maximum cardinality such that $\delta(H) \geq k$. The following observation is elementary but important for our algorithms: if G has no k-core, then a k-colouring of G can be found in linear time. Indeed, consecutively push vertices of degree at most $k-1$ from the graph onto a stack until the graph is empty. Then, while putting the vertices back into the graph in reverse order, assign to each vertex the smallest colour possible, which is at most k since the vertex is (at that point) connected to at most $k-1$ other vertices. This procedure is known as the smallest–last heuristic [2]. Observe that the order in which we remove vertices from the graph does not matter, as long as they have degree at most $k-1$ at the time of removal.

Pittel, Spencer and Wormald [28] computed the constants γ_k which determine the precise threshold $p = \gamma_k/n$ for the appearance of the k-core in $G(n,p)$. To give an example, $\gamma_3 \approx 3.35$. By the preceding discussion, it is clear that this marks a lower bound for the threshold of k-colourability. It was an open question of Bollobás whether the appearance of the k-core in $G(n,p)$ coincides with the threshold of k-colourability. A few years ago, Achlioptas and Molloy [1] answered this in the negative by showing that $G(n,p)$ remains 3-colourable for at least as long as $p \leq 3.84/n$.

Let $k \geq 3$. We emphasize that even though almost surely $G(n,p)$ does not contain a k-core of size, say, $\ln^{10}(n)$ at $p = 1/n$, the probability for this event is not exponentially small in n, and hence such an event is not sufficiently unlikely to allow us to find the optimal colouring by brute force *on the whole graph*. Instead, the following simple algorithm will only need to deal with the core.

Algorithm 7. $\texttt{CoreColour}(G, k)$
Input: A graph $G = (V, E)$ and an integer k.
Output: A colouring of G.

1. Let $G' := G$, set stack $S := \emptyset$.
2. While in G' there exists a vertex v with $d_{G'}(v) \leq k - 1$ do
   push v onto S, $G' := G' - v$.
3. Colour G' exactly by Lawler's algorithm [27] in time $O((1 + \sqrt[3]{3})^{n'})$, where $n' = |G'|$.
4. Extend this to a colouring of G: while $S \neq \emptyset$ do
   pick the top vertex v from S, assign to v the least possible colour.

Proposition 8. *For any graph G and any integer $k \geq 2$, $\texttt{CoreColour}(G, k)$ finds a colouring of G with at most $\max\{\chi(G), k\}$ colours.*

Proof. Step 3 uses $\chi(G') \leq \chi(G)$ colours. Let v be an arbitrary vertex. Since $d_{G'}(v) \leq k - 1$ when v was removed, Step 4 will never need a colour greater than k, since v has at most $k-1$ coloured neighbours at the moment it is coloured.   □

Observe next that at the beginning of step 3, the vertices in G' make up the k-core. Therefore, in order to examine the expected running time of CoreColour, we need an upper bound on the probability that G contains a k-core of a certain size.

Lemma 9. *Let $p \geq 1/n$ and consider an integer $k \geq e^2np$. Then*

$$P(G(n,p) \text{ contains a } k\text{-core on } \nu \text{ vertices}) < e^{-\nu}$$

Proof. The probability that there exists a k-core on ν vertices is bounded from above by the probability that there exists a weak subgraph H on ν vertices with $\mu = k\nu/2$ edges. Let $N = \binom{\nu}{2}$. Then the expected number of such subgraphs is given by

$$\binom{n}{\nu}\binom{N}{\mu}p^\mu \leq \left(\frac{en}{\nu}\right)^\nu \left(\frac{e\nu^2p}{2\mu}\right)^\mu \leq \left[\frac{en}{\nu}\left(\frac{e\nu p}{k}\right)^{k/2}\right]^\nu \leq \left[\left(\frac{\nu}{en}\right)^{\frac{k}{2}-1}\right]^\nu \leq e^{-\nu},$$

where the last step used $\nu \leq n$ and $k \geq e^2np \geq 4$ because of $p \geq 1/n$. $\square$

Proof of Theorem 3. We can easily check whether $\chi(G) \leq 2$, and, if so, output an optimal colouring. So assume that $\chi(G) > 2$. Set $k := e^2np$ and apply CoreColour$(G(n,p), k)$.

Proposition 8 shows that the algorithm has approximation ratio at most $k/3 = (e^2/3)np$ as claimed. Its running time is linear for step 2 and step 4. As $1 + \sqrt[3]{3} < 2.443 < e$, Lemma 9 implies that the expected running time of step 3 can be bounded by $\sum_{\nu=0}^n (1 + \sqrt[3]{3})^\nu \cdot P(|G'| = \nu) = O(n)$. $\square$

3   Finding an Optimal Colouring

This section contains a sketch of the proof of Theorem 1. The underlying idea is basically the same as in Section 2, but we need to refine both the algorithm and the computations in order to achieve an approximation ratio 1.

We will employ an exact algorithm by Beigel and Eppstein [3] which decides in time $O(1.3447^n)$ whether a given graph is 3-colourable or not. Suppose that it answers this question in the affirmative, then it is obviously not hard to actually find such a 3-colouring in $O(n^2 1.3447^n)$ steps, simply by trying to add edges to the graph as long as the graph remains 3-colourable. Thus in the end we have a complete 3-partite graph and can read off the colouring.

The basic idea of our algorithm is as follows: having checked for 2-colourability, we peel off vertices until we get to the 3-core. For the 3-core we merely decide whether it is 3-colourable because applying the Beigel–Eppstein *decision* algorithm is much cheaper than *finding* an optimal colouring via Lawler's algorithm. If yes, we are done. If not, then we know that $\chi(G) \geq 4$ and continue shaving off vertices until we get to the 4-core, which will most probably be substantially smaller than the 3-core. Now we optimally colour the 4-core using Lawler's algorithm.

Algorithm 10. ExactColour(G)
Input: A graph $G = (V, E)$
Output: A colouring of G.

1. If $\chi(G) \leq 2$ then find an optimal colouring of G and stop.
2. Let $G' := G$, set stack $S := \emptyset$.
   While there exists a vertex v with $d_{G'}(v) \leq 2$ do
      push v onto S, let $G' := G' - v$.
3. Check whether $\chi(G') \leq 3$ using Beigel and Eppstein's algorithm.
4.   If yes then find a 3-colouring of G' in time $O(n'^2 1.3447^{n'})$, where $n' = |G'|$.
5.   If no then run CoreColour($G', 4$).
6. Extend the obtained colouring of G' to a colouring of G as follows:
   while $S \neq \emptyset$ do
      pick the top vertex v from S, assign to v the least possible colour.

By similar arguments as in Section 2, it is clear that the algorithm produces an optimal colouring.

In order to show that the expected running time remains indeed linear for p as large as $1.01/n$, one needs to sharpen the bound given in Lemma 9. Apart from obvious room for improvements by using more careful technical estimates, most of the gain is won by only counting those subgraphs which have indeed minimum degree k and not just $k\nu/2$ edges.

4 $O(\sqrt{np})$-Approximation via the ϑ-Function

Throughout this section we shall assume that $p \gg \ln(n)^6/n$. Let $\omega = np$ and $G \in G(n, p)$ be a random graph. On input $(G, 10np)$, the algorithm CoreColour presented in Section 2 produces a colouring of G using at most $\max\{\chi(G), 10np\}$ colours. Thus, CoreColour immediately gives an $O(np)$-approximation algorithm for colouring $G(n, p)$ within expected polynomial time. In order to improve the approximation ratio $O(np)$, it is crucial to develop techniques to compute good lower bounds on the chromatic number of the input graph, since we need to distinguish graphs with a chromatic number near the "typical value" (1) from "pathological cases" efficiently. To this end, we shall use the following trivial inequality. For any graph G of order n we have $\chi(G) \geq n/\alpha(G)$. Thus, instead of lower bounding the chromatic number $\chi(G)$, we may also compute an upper bound on the independence number $\alpha(G)$.

The same approach has been used in [22], where the independence number of $G(n, p)$ is bounded from above as follows. Given a random graph $G \in G(n, p)$ with vertex set $V = \{1, \ldots, n\}$, let the matrix $M = (m_{ij})_{i,j \in V}$ have entries

$$m_{ij} = \begin{cases} 1 & \text{if } \{i, j\} \notin E(G) \text{ or } i = j \\ \frac{p-1}{p} & \text{otherwise.} \end{cases}$$

Then the largest eigenvalue $\lambda_1(M)$ provides an upper bound on $\alpha(G)$. In order to base an approximation algorithm with polynomial expected running time on computing the bound $n/\lambda_1(M) \leq \chi(G)$, it is necessary to estimate the probability that $\lambda_1(M)$ is "large". To this end, it is shown in [22] that

$$P(\lambda_1(M) \geq 4(n/p)^{1/2}) \leq 2^{-\omega/8}. \tag{3}$$

Note that the exponent on the right hand side depends on ω (and thus both on p and n). Nonetheless, in the case $\omega \geq n^{1/2+\varepsilon}$, $\varepsilon > 0$, the concentration result (3) suffices to obtain an $O(\sqrt{np}/\ln n)$-approximation algorithm for both the chromatic number and the independence number of $G(n, p)$. The proof of (3) is based on Talagrand's inequality (cf. [16]).

However, in the case $p \ll n^{-1/2}$, (3) seems not to be sharp enough in order to construct an algorithm with expected polynomial running time and approximation ratio $\tilde{O}(np)^{1/2}$. Therefore, we shall use the *Lovász number* $\vartheta(G)$ as an upper bound on $\alpha(G)$ instead of the eigenvalue $\lambda_1(M)$; it is well-known that $\alpha(G) \leq \vartheta(G)$ for any graph G. For a thorough introduction to the Lovász number the reader is referred to [15,21]. The Lovász number $\vartheta(G)$ can be computed within polynomial time, using the ellipsoid algorithm [15].

What is the relation between $\vartheta(G)$ and the eigenvalue $\lambda_1(M)$? Let us call a real symmetric matrix $A = (a_{vw})_{v,w \in V}$ *feasible for* G if $a_{vv} = 1$ for all v and $a_{uv} = 1$ for any pair u, v of non-adjacent vertices of G. Then we have $\vartheta(G) = \min\{\lambda_1(A)|\ A$ is feasible for $G\}$, where as above $\lambda_1(A)$ denotes the maximum eigenvalue of A. Observing that the matrix M defined above is feasible, we obtain

$$\alpha(G) \leq \vartheta(G) \leq \lambda_1(M). \tag{4}$$

Thus, our colouring algorithm uses $n/\vartheta(G)$ as a lower bound on $\chi(G)$. Consequently, we need to estimate the expectation of $\vartheta(G(n, p))$.

Lemma 11. *With high probability,* $\vartheta(G(n, p)) \leq 3(n/p)^{1/2}$.

The proof of Lemma 11 is based on (4) and on the fact that M is a random symmetric matrix as considered in [11], except for the fact that the entries m_{ij} may be unbounded. The analysis of our colouring algorithm makes use of the following concentration result on the Lovász number of random graphs [6].

Lemma 12. *Let* μ *be a median of* $\vartheta(G(n, p))$. *Let* $\xi \geq \max\{10, \sqrt{\mu}\}$. *Then*

$$P(\vartheta(G(n, p)) \geq \mu + \xi) \leq 30 \exp\left(-\frac{\xi^2}{5\mu + 10\xi}\right).$$

Consequently, if $a \geq 10\sqrt{n/p}$, *then*

$$P(\vartheta(G(n, p)) \geq a) \leq \exp(-a/30). \tag{5}$$

The proof of Lemma 12 is based on Talagrand's inequality. Note that, contrary to the estimate (3), the exponent on the right hand side of (5) does not increase as p decreases. The following algorithm for colouring $G(n, p)$ resembles the colouring algorithm proposed in [22]. The main difference is that we use the algorithm CoreColour instead of the greedy algorithm in step 1, and that we bound the chromatic number via the Lovász number.

Algorithm 13. ApproxColour(G)
Input: A graph $G = (V, E)$.
Output: A colouring of G.

1. Run CoreColour($G, 10np$). Let $\mathcal{C}$ be the resulting colouring of G.
2. Compute $\vartheta(G)$. If $\vartheta(G) \leq 10(n/p)^{1/2}$, then output $\mathcal{C}$ and terminate.
3. Check whether there exists a subset S of V, $|S| = 25\ln(np)/p$, such that $|V \setminus (S \cup N(S))| > 10(n/p)^{1/2}$, by enumerating all subsets of V of cardinality $25\ln(np)/p$. If no such set exists, then output $\mathcal{C}$ and terminate.
4. Check whether in G there is an independent set of size $10(n/p)^{1/2}$ by enumerating all subsets of V of cardinality $10(n/p)^{1/2}$. If this is not the case, then output $\mathcal{C}$ and terminate.
5. Colour G exactly by Lawler's algorithm [27] in time $O(n^3(1 + \sqrt[3]{3})^n)$.

It is straightforward to verify that ApproxColour achieves the approximation ratio stated in Theorem 2. The proof that the expected running time over $G(n,p)$ is polynomial is based on (5).

In [17] it is observed that with high probability $\vartheta(G(n,p)) = \Omega(n/p)^{1/2}$. Since the chromatic number $\chi(G(n,p))$ satisfies (1) with high probability, we cannot hope for an approximation ratio considerably better than $O(np)^{1/2}$ using $n/\vartheta(G(n,p))$ as a lower bound on $\chi(G(n,p))$.

Let us conclude this section noting that the techniques behind ApproxColour immediately lead to an approximation algorithm for the maximum independent set problem in $G(n,p)$. In order to obtain an algorithm ApproxMIS that satisfies the requirements of Theorem 6, we adapt ApproxColour as follows. Instead of CoreColour, we use the greedy (colouring) algorithm and keep the largest colour class it produces. Then we estimate the independence number of the input graph as ApproxColour does. Finally, instead of using an exact graph colouring algorithm in step 5, we use an exact algorithm for the maximum independent set problem.

Lemma 14. *The probability that the largest colour class produced by the greedy colouring algorithm is of size* $< \ln(np)/(2p)$ *is at most* 2^{-n}.

The proof uses a similar argument as given already in [22] for the case that $p \geq n^{\varepsilon - 1/2}$.

Algorithm 15. ApproxMIS(G)
Input: A graph $G = (V, E)$.
Output: An independent set of G.

1. Run the greedy algorithm for graph colouring on input G. Let I be the largest resulting colour class. If $|I| < \ln(np)/(2p)$, then go to 5.
2. Compute $\vartheta(G)$. If $\vartheta(G) \leq 10(n/p)^{1/2}$, then output I and terminate.
3. Check whether there exists a subset S of V, $|S| = \frac{25\ln(np)}{p}$, such that $|V \setminus (S \cup N(S))| > 10(n/p)^{1/2}$. If no such set exists, then output I and terminate.
4. Check whether in G there is an independent set of size $10\sqrt{n/p}$. If this is not the case, then output I and terminate.
5. Enumerate all subsets of V and output a maximum independent set.

5 Deciding k-Colourability

In this final section we shall apply the methods established so far to the following problem. Given a semirandom graph $G = G(n,p)^+$ or $G = G(n,p)^-$, we are to decide within polynomial expected time whether G is $k = k(n)$-colourable. First, let us see what the algorithm CoreColour contributes to the problem of deciding k-colourability in the model $G(n,p)^-$.

Algorithm 16. CoreDecide(G)
Input: A graph $G = (V, E)$.

1. Run CoreColour($G, 10np$). Let C be the resulting colouring of G.
2. If the number of colours used in C is at most k, output "G is k-colourable".
   Otherwise, output "G is not k-colourable".

Let $G_0 = G(n,p)$ be a random graph, $p \leq k/(10n)$, and let $G \in \mathcal{I}(G_0)^-$ be the instance chosen by the adversary. Because the adversary can only remove edges, the $10np$-core of G is contained in the $10np$-core of G_0. Thus, the argument given in the proof of Lemma 9 shows that the expected running time of CoreDecide is linear. Moreover, it is easy to see that the answer produced by CoreDecide is always correct. For if CoreDecide states that G is k-colourable, then CoreColour must have found a proper k-colouring of G. Conversely, by Proposition 8 we know that the number of colours that CoreColour uses is at most $\max\{\chi(G), 10np\}$. Thus, by our choice of p, if $\chi(G) \leq k$, then CoreColour finds a k-colouring of G.

Let us now assume that $np \geq \max\{200k^2, \ln(n)^7\}$. In this range we almost surely have $\chi(G(n,p)) > k$. Applying the techniques behind ApproxColour to the problem of deciding k-colourability of the semirandom graph $G(n,p)^+$, we obtain the following algorithm.

Algorithm 17. Decide(G)
Input: A graph $G = (V, E)$.

1. Compute $\vartheta(G)$. If $\vartheta(G) \leq 10(n/p)^{1/2}$, then terminate with output "G is not k-colourable".
2. Check whether there exists a subset S of V, $|S| = 25\ln(np)/p$, such that $|V \setminus (S \cup N(S))| > 10(n/p)^{1/2}$. If no such set exists, then output "G is not k-colourable" and terminate.
3. Check whether in G there is an independent set of size $10(n/p)^{1/2}$. If this is not the case, then output "G is not k-colourable" and terminate.
4. Compute the chromatic number of G exactly by Lawler's algorithm [27] in time $O(e^n)$ and answer correctly.

In order to prove that Decide has polynomial expected running time applied to $G(n,p)^+$, we consider a random graph $G_0 \in G(n,p)$ and an instance $G \in \mathcal{I}(G_0)^+$. Because G_0 is a subgraph of G, we have $\vartheta(G) \leq \vartheta(G_0)$, by the monotonicity of the ϑ-function (cf. [21]). Moreover, if G admits a set S as in step 2 of Decide, then a fortiori G_0 admits such a set S. Finally, $\alpha(G) \leq \alpha(G_0)$.

Thus, the arguments that we used to prove that the expected running time of `ApproxColour` applied to $G(n,p)$ is polynomial immediately yield that the expected running time of `Decide` applied to the semirandom model $G(n,p)^+$ is polynomial.

We finally claim that `Decide`(G) always answers correctly. For assume that `Decide`(G) asserts that G is k-colourable. Then step 4 of `Decide`(G) must have found a proper k-colouring of G; hence, G is k-colourable. On the other hand, assume that `Decide`(G) asserts that G is not k-colourable. Then, `Decide`(G) has found out that

$$\alpha(G) \le 10(n/p)^{1/2} + 25\ln(np)/p \le 11(n/p)^{1/2}.$$

Thus,

$$\chi(G) \ge \frac{n}{\alpha(G)} \ge \frac{n}{11(n/p)^{1/2}} = \frac{(np)^{1/2}}{11} > k,$$

as claimed.

6   Conclusion

No serious attempts have been made to optimize the constants involved. For example, an easy way to slightly enlarge the range of p for which the algorithm `ExactColour` has expected polynomial running time would be to use Eppstein's recent algorithm [8] for finding an optimal colouring instead of Lawler's – with $O(2.415^n)$ running time instead of $O(2.443^n)$. However it is of course clear that our general approach must fail at the latest when $p = 3.84/n$, as here the 3-core appears almost surely and has linear size.

References

1. Achlioptas, D., Molloy, M.: The analysis of a list-coloring algorithm on a random graph, Proc. 38th. IEEE Symp. Found. of Comp. Sci. (1997) 204-212
2. Beck, L.L., Matula, D.W.: Smallest-last ordering and clustering and graph coloring algorithms, J. ACM **30** (1983) 417–427
3. Beigel, R., Eppstein, D.: 3-coloring in time $O(1.3446^n)$: a no-MIS algorithm, Proc. 36th. IEEE Symp. Found. of Comp. Sci. (1995) 444-453
4. Bollobás, B.: Random graphs, 2nd edition, Cambridge University Press 2001
5. Coja-Oghlan, A.: Finding sparse induced subgraphs of semirandom graphs. Proc. 6th. Int. Workshop Randomization and Approximation Techniques in Comp. Sci. (2002) 139–148
6. Coja-Oghlan, A.: Finding large independent sets in expected polynomial time. To appear in Proc. STACS 2003
7. Dyer, M., Frieze, A.: The solution of some NP-hard problems in polynomial expected time, J. Algorithms **10** (1989) 451–489
8. Eppstein, D.: Small maximal independent sets and faster exact graph coloring. To appear in J. Graph Algorithms and Applications

9. Feige, U., Kilian, J.: Zero knowledge and the chromatic number. Proc. 11. IEEE Conf. Comput. Complexity (1996) 278–287
10. Feige, U., Kilian, J.: Heuristics for semirandom graph problems. J. Comput. and System Sci. **63** (2001) 639–671
11. Füredi, Z., Komloś, J.: The eigenvalues of random symmetric matrices, Combinatorica **1** (1981) 233–241
12. Fürer, M., Subramanian, C.R., Veni Madhavan, C.E.: Coloring random graphs in polynomial expected time. Algorithms and Comput. (Hong Kong 1993), Springer LNCS 762, 31–37
13. Frieze, A., McDiarmid, C.: Algorithmic theory of random graphs. Random Structures and Algorithms **10** (1997) 5–42
14. Grimmett, G., McDiarmid, C.: On colouring random graphs. Math. Proc. Cam. Phil. Soc **77** (1975) 313–324
15. Grötschel, M., Lovász, L., Schrijver, A.: Geometric algorithms and combinatorial optimization. Springer 1988
16. Janson, S., Łuczak, T., Ruciński, A.: Random Graphs. Wiley 2000
17. Juhász, F.: The asymptotic behaviour of Lovász ϑ function for random graphs, Combinatorica **2** (1982) 269–280
18. Karger, D., Motwani, R., Sudan, M.: Approximate graph coloring by semidefinite programming. Proc. of the 35th. IEEE Symp. on Foundations of Computer Science (1994) 2–13
19. Karp, R.: Reducibility among combinatorial problems. In: Complexity of computer computations. Plenum Press (1972) 85–103.
20. Karp, R.: The probabilistic analysis of combinatorial optimization algorithms. Proc. Int. Congress of Mathematicians (1984) 1601–1609.
21. Knuth, D.: The sandwich theorem, Electron. J. Combin. **1** (1994)
22. Krivelevich, M., Vu, V.H.: Approximating the independence number and the chromatic number in expected polynomial time. J. of Combinatorial Optimization **6** (2002) 143–155
23. Krivelevich, M.: Deciding k-colorability in expected polynomial time, Information Processing Letters **81** (2002) 1–6
24. Krivelevich, M.: Coloring random graphs – an algorithmic perspective, Proc. 2nd Coll. on Mathematics and Computer Science, B. Chauvin et al. Eds., Birkhauser, Basel (2002) 175–195.
25. Krivelevich, M., Sudakov, B.: Coloring random graphs. Informat. Proc. Letters **67** (1998) 71–74
26. Kučera, L.: The greedy coloring is a bad probabilistic algorithm. J. Algorithms **12** (1991) 674–684
27. Lawler, E.L.: A note on the complexity of the chromatic number problem, Information Processing Letters **5** (1976) 66–67
28. Pittel, B., Spencer, J., Wormald, N.: Sudden emergence of a giant k-core in a random graph. JCTB **67** (1996) 111–151
29. Prömel, H.J., Steger, A.: Coloring clique-free graphs in polynomial expected time, Random Str. Alg. **3** (1992) 275–302

An Information-Theoretic Upper Bound of Planar Graphs Using Triangulation

(Extended Abstract)

Nicolas Bonichon, Cyril Gavoille, and Nicolas Hanusse

LaBRI, Université Bordeaux I, France.
{bonichon,gavoille,hanusse}@labri.fr

Abstract. We propose a new linear time algorithm to represent a planar graph. Based on a specific triangulation of the graph, our coding takes on average 5.03 bits per node, and 3.37 bits per node if the graph is maximal. We derive from this representation that the number of unlabeled planar graphs with n nodes is at most $2^{\alpha n + O(\log n)}$, where $\alpha \approx 5.007$. The current lower bound is $2^{\beta n + \Theta(\log n)}$ for $\beta \approx 4.71$. We also show that almost all unlabeled and almost all labeled n-node planar graphs have at least $1.70n$ edges and at most $2.54n$ edges.

1 Introduction

How many information can contain a simple planar graph with n nodes? The question is highly related to the number of planar graphs. Counting the number of (non-isomorphic) planar graphs with n nodes is a well-known long-standing unsolved graph-enumeration problem (cf. [25]). There is no known close formula, neither asymptotic nor even an asymptotic on the logarithm of this number. Any asymptotic on the logarithm would give a bound on the number of independent random bits needed to generate randomly and uniformly a planar graph (but not necessary in polynomial time).

Random combinatorial object generation is an important activity regarding average case complexity analysis of algorithms, and testing algorithms on typical instances. Unlike random graphs (the Erdös-Rényi graph Model), still little is known about random planar graphs. This is mainly due to the fact that adding an edge in a planar graph highly depends on the location of all previous edges. Random planar maps, i.e., plane embeddings of planar graphs, have been investigated more successfully. Schaeffer [32], and then Banderier et al. [2] have showed how to generate in polynomial time several planar map families, e.g. 3-connected planar maps. Unfortunately, this generating gives quite few information about random planar graphs because there are many ways to embed a planar graph into the plane. On the positive side, some families of planar graphs support efficient random generation: Trees [1], maximal outerplanar graphs [3, 15], and more recently labeled and unlabeled outerplanar graphs [5].

Besides the combinatorial aspect and random generation, a lot of attention is given in Computer Science to *efficiently* represent discrete objects. Efficiently

H. Alt and M. Habib (Eds.): STACS 2003, LNCS 2607, pp. 499–510, 2003.

means that the representation is succinct, i.e., the storage of these objects uses few bits, and that the time to compute such representation is polynomial in their size. Fast manipulation of the so encoded objects and easy access to a part of the code are also desirable properties. At least two scopes of applications of high interests are concerned with planar graph representation: Computer Graphics [21, 22,31] and Networking [16,17,26,34].

1.1 Related Works

Succinct representation of n-node m-edge planar graphs has a long history. Turán [35] pioneered a $4m$ bit encoding, that has been improved later by Keeler and Westbrook [20] to $3.58m$. Munro and Raman [28] then proposed a $2m + 8n$ bit encoding based of the 4-page embedding of planar graphs (see [37]). In a series of articles, Lu et al. [10,12] refined the coding to $4m/3+5n$ thanks to orderly spanning trees, a generalization of Schnyder's trees [33]. Independently, codings have been proposed for triangulations, where $m = 3n - 6$. A $4n$ bit encoding has been obtained by several authors [7,12,31], interestingly with rather different techniques. Then improved by the Rossignac's Edgebreaker [22], guaranteeing $3.67n$ bits for triangulations, moreover computable in $O(n)$ time. Actually, He, Kao and Lu [19] showed that, in $O(n \log n)$ time, a space optimal encoding for triangulations and for unlabeled planar graphs can be achieved. So, a $O(n \log n)$ time and a $3.24n$ bit encoder for triangulations exists. For that, they use a recursive separator decomposition of the graph, and an exponential coding algorithm for the very end components of sub-logarithmic size. However, the time complexity hidden in the big-O notation could be of limited use in practice. To implement the encoder, one needs, for instance, to implement planar isomorphism and Lipton-Tarjan planar separator [24]. Recently the time complexity has been improved to $O(n)$ for planar graphs by Lu [27]. Although the length of the coding is optimal, the approach of [19,27] does not give any explicit bound of the number of bits use in the representation.

If we are interested only in the Information-Theoretic bound of planar graphs or in statistical properties of planar graphs (what a random planar graph looks like: number of edges, connectivity, etc.), other tools can be used. Denise et al. [13] specified a Markov chain on the space of all labeled planar graphs whose limit distribution is the uniform distribution. Their experiments show that random planar graphs have approximately $2n$ edges, are connected but not 2-connected. Although the Markov chain converges to the uniform distribution, it is not proved whether this Markov chain becomes sufficiently close to the uniform distribution after a polynomial number of steps. It is however proved that almost all labeled planar graphs have at least $1.5n$ edges, and that the number $p(n)$ of unlabeled planar graphs satisfies that $\frac{1}{n} \log_2 p(n)$ tends to a constant γ such that $\log_2(256/27) \leqslant \gamma \leqslant \log_2(256/27) + 3$. The bounds on γ easily derive from Tutte's formula [36]: Triangulations are planar graphs, and every planar graph is a subgraph of a triangulation, thus having 2^{3n-6} possible subsets of edges. There are also no more than $n!2^{\gamma n+o(n)}$ labeled planar graphs as there are at most $n!$ ways to label the nodes of a graph.

Osthus, Prömel, and Taraz [29] investigated triangulations containing any planar graph, and they showed that there is no more than $n!2^{5.22n+o(n)}$ labeled planar graphs. They also showed that almost all labeled planar graphs have at most $2.56n$ edges. A lower bound of $13n/7 \approx 1.85n$ has been obtained by Gerke and McDiarmid [18], improving the $1.5n$ lower bound of [13]. They also derived that almost all unlabeled planar graphs have at most $2.69n$ edges. Note that there is no evidence that labeled and unlabeled planar graphs have the same growing rate (up to the $n!$ term), unlike general graphs. So upper bounds on labeled planar graphs do not transfer to upper bounds on unlabeled planar graphs, but the reverse is true.

Using generating function techniques, Bender, Gao and Wormald [4] proved that the number of labeled 2-connected planar graphs tends to $n!2^{4.71n+O(\log n)}$. Note that enumeration of simple planar maps gives an upper bound on planar graphs whose algebraic generating function has been given by Liu [23]. Their number is asymptotic to $2^{5.098n+O(\log n)}$ (cf. [9]), providing an upper bound for unlabeled and labeled planar graphs.

1.2 Our Results

In this paper we show an upper bound of $2^{5.007n+O(\log n)}$ on $p(n)$, the number of unlabeled planar graphs with n nodes. This result implies that the number of labeled planar graphs is no more than $n!2^{5.007n+O(\log n)}$, an improvement upon the bounds of [29] and of [9,23].

Since our upper bound can be parameterized with the number of edges, and using the lower bound of [4], we are able to show that almost all unlabeled graphs have at least $1.70n$ edges and at most $2.54n$ edges, setting a new lower bound and improving the $2.69n$ upper bound of [29]. Moreover the result holds for labeled planar graphs, connected labeled, and connected unlabeled planar graphs. Thus our bounds also slightly improve the previous $2.56n$ upper bound [29] for labeled planar graphs, the $13n/7$ lower bound of [18] remaining better in the labeled case.

Beside the fundamental aspect of planar graph enumeration, our technique is based on an *explicit* representation, quite easily and linear time constructible. In addition we give a simple linear time $3.37n$ bit coding of triangulations, and a $5.03n$ bit coding for planar graphs. We also show[1] that our coding never exceeds $2.90m$ bits, the worst-case is attained for $m = 1.39n$ edges. Our triangulation representation already improves the Edgebreaker compressor [22], and there is no doubt that our explicit construction for planar graphs can be used for Networking, in particular to improve the result of [26].

1.3 Outline of the Paper

Let us sketch our technique (due to space limitation the proofs moved to [8]). A natural approach to represent an n-node planar graph G is to consider a triangulation of G, i.e., a supergraph S of G such that S is planar, has n nodes

---

[1] This result appears in the full version.

and $3n-6$ edges. Then, G can be obtained by coding S and a set M_S of edges such that $E(S) \setminus M_S = E(G)$. This way of representing a planar graph is suggested by the $(\log_2(256/27) + 3)n = 6.24n$ bit upper bound of [13] given above. To obtain a representation more compact than $6.24n$ bits we need to carefully construct S. In particular, crucial steps are the way we embed G into the plane, and the way we triangulate its faces. We show that, if G is connected, the supergraph S of G, called hereafter *super-triangulation* defined in Section 2, has the property that given S and a node $v \in S$ only, one can perform in a unique manner a traversal of S and find a spanning tree T rooted in v such that T is contained in G. So, given the super-triangulation S of G, the edges of M_S can be described among the possible edges of $S \setminus T$ only, i.e., with at most $2n$ bits. This already provides a $(\log_2(256/27) + 2)n = 5.24n$ bit upper bound. Observe that the case G not connected can be easily transformed (in linear time) into a new connected graph $\tilde{G}$, e.g. by merging all the connected components of G into a single node (see Section 4 for more details).

Actually, the $5.24n$ bound discussed above does not lead to an explicit representation because it is based on a $3.24n$ bit theoretical coding of triangulation, and no such explicit representation is known. To overcome this problem, we represent the super-triangulation S (that is a specific triangulation) by a triple of trees forming a very specific *realizer*, that is a partition of the edges into three particular trees (T_0, T_1, T_2) also called Schnyder's trees [33]. And, we show how to uniquely recover the tree partition from a triangulation. Among the properties of S, we have that T_0 is contained in G (assumed to be connected from the transformation). Moreover every edge (u, v) of S, where u is the parent of v in T_1 and where u is an inner node in T_2, must belong to G. This significantly saves bits in the coding of M_S since many edges of G can be guessed from S. An extra property is that two nodes belonging to the same branch of T_2 have the same parent in T_1 (a branch is a maximal set of related nodes obtained in a clockwise depth-first search of the tree, and such that a node belongs to only one branch at the time, see Section 3). This latter property simplifies a lot the representation of S. Knowing T_2, T_1 does not need to be fully represented. Only one relevant edge per branch of T_2 suffices. As any tree of a realizer can be deduced from the two others, the representation of S can be compacted in a very efficient way, storing for instance T_2 and the relevant edges of T_1.

Finally, in Section 3, the explicit representation of G is done with 8 binary strings of different density (namely the ratio between the number ones it contains and its length): 7 for representing S (5 for T_2 and 2 for the relevant edges of T_1), and 1 for M_S. We compact each string with a variant of the Pagh's compressor [30]. This allows to reach an optimal entropy coding, i.e., with $\log_2 \binom{n}{k} + o(n)$ bits for an n-bit string of k ones[2]. Taken in parameter the number of branches of T_2 (or equivalently its number of leaves), an entropy analysis shows that $3.37n$ bits suffice to represent S, and $5.03n$ bits suffice for G. The final $5.007n$ bound is reached with a more sophisticated encoding of S.

---

[2] The original compressor runs in expected linear time. We give in this paper a simpler guaranteed linear time construction with asymptotically the same performances.

We propose a rather simple algorithm to construct a super-triangulation. It works in two steps. After a specific plane embedding of the graph, an intermediate object is constructed: the *well-orderly tree*. It consists in a specific spanning tree of the graph that mainly corresponds to one of the tree of the super-triangulation. Then, faces of the embedding are specifically triangulated to obtain the super-triangulation with all its properties.

2   Embedding and Triangulating Algorithms

A *plane embedding of a graph*, or shortly a *plane graph*, is a mapping of each node to a point of the plane and of each edge to the continuous curve joining the two ends of this edge such that edges do not crossing except, possibly, on a common extremity. A graph that has a plane embedding is a planar graph.

In this paper we deal with simple (no loops and no multi-edges) and undirected graphs. If we cut the plane along the edges, the remainder falls into connected regions of the plane, called *faces*. Each plane graph has a unique unbounded face, called the *outerface*. The *boundary* of a face is the set of incident edges. The *interior* edges are the edges non incident to the boundary of the outerface, similarly for interior nodes.

A *triangulation* is a plane embedding of a maximal planar graph that is a planar graph with n nodes and $3n - 6$ edges. There is only one way to embed in the plane (up to a continuous transformation) a maximal planar graph whose three nodes are chosen to clockwise lie on the outerface.

2.1   Well-Orderly Tree, Realizer, and Super-Triangulation

Let T be a rooted spanning tree of a plane graph H. Two nodes are *unrelated* if neither of them is an ancestor of the other in T. An edge of H is unrelated if its endpoints are unrelated.

We introduce *well-orderly trees*, a special case of *orderly spanning trees* of Chiang, Lin and Lu in [10], referred as simply orderly trees in the next. Let $v_1, \ldots, v_n$ be the clockwise preordering of the nodes in T. Recall that a node v_i is *orderly* in H w.r.t. T if the incident edges of v_i in H form the following four blocks (possibly empty) in clockwise order around v_i:

- $B_P(v_i)$: the edge incident to the parent of v_i;
- $B_<(v_i)$: unrelated edges incident to nodes v_j with $j < i$;
- $B_C(v_i)$: edges incident to the children of v_i; and
- $B_>(v_i)$: unrelated edges incident to nodes v_j with $j > i$.

A node v_i is *well-orderly* in H w.r.t. T if it is orderly, and if:

- the clockwise first edge $(v_i, v_j) \in B_>(v_i)$, if it exists, verifies that the parent of v_j is an ancestor of v_i.

T is a *well-orderly tree* of H if all the nodes of T are well-orderly in H, and if the root of T belongs to the boundary of the outerface of H (similarly for simply orderly tree). Note that an orderly tree (simply or well-orderly) is necessarily a spanning tree. Observe also that the incident edges in H of a node of T are either in T or unrelated.

A *realizer* of a triangulation is a partition of its interior edges in three sets T_0, T_1, T_2 of directed edges such that for each interior node v it holds:

- the clockwise order of the edges incident with v is: leaving in T_0, entering in T_1, leaving in T_2, entering in T_0, leaving in T_1 and entering in T_2;
- there is exactly one leaving edge incident with v in T_0, T_1, and in T_2.

Cyclic permutations of a realizer are not in general the only distinct realizers of a given triangulation. Fig. 1 depicts two realizers for a same triangulation. Actually, the number of n-node realizers is asymptotically $2^{4n+O(\log n)}$ (cf. [7]), whereas the number of triangulations is only $(256/27)^{n+O(\log n)}$ (cf. [36]).

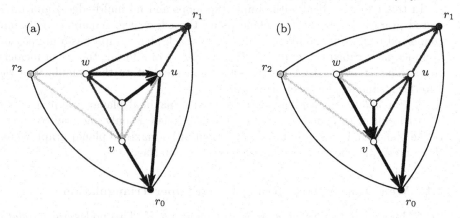

Fig. 1. Two realizers for a triangulation. The tree $\overline{T}_0$ rooted in r_0 (the tree with directed blue bold edges augmented with the edges (r_0, r_1) and (r_0, r_2)) is well-orderly in (b), and simply orderly in (a) (the node v is not well-orderly).

Schnyder showed in [33] that each set T_i of a realizer induces a tree rooted in one node of the outerface. Moreover, he described a linear time algorithm to compute such trees. Hereafter, if $R = (T_0, T_1, T_2)$ is a realizer, then for notational convenience R also denotes the underlying triangulation.

For each tree T_i of a realizer, we denote by $\overline{T}_i$ the tree composed of T_i augmented with the two edges of the outerface incident to the root of T_i. A node of a rooted tree is *inner* if it is neither the root, nor a leaf. For every non-root node $u \in T_i$, we denote by $p_i(u)$ the parent of u in T_i.

Definition 1. *A realizer* $S = (T_0, T_1, T_2)$ *is a* super-triangulation *of a graph* G *if:*

1. $V(S) = V(G)$ *and* $E(G) \subseteq E(S)$;
2. $E(T_0) \subseteq E(G)$;
3. $\overline{T}_0$ *is a well-orderly tree of* S; *and*
4. *for every inner node* v *of* T_2, $(v, p_1(v)) \in E(G)$.

Theorem 1. *Every connected planar graph with at least three nodes has a super-triangulation, computable in linear time.*

Intuitively, a super-triangulation of a graph G is a specific triangulation of the faces of a specific plane embedding of G. One of the interesting properties of super-triangulations is the following one:

Theorem 2. *For every 3-connected planar graph G and every triple of nodes r_0, r_1, r_2 of a face of G, there exists only one super-triangulation (T_0, T_1, T_2) of G such that T_i has root r_i, for every $i \in \{0, 1, 2\}$.*

In particular, triangulations have exactly one super-triangulation, up to a cyclic permutation of the roots on the outerface. This property will play an important role in Section 4 for our upper bound on the number of planar graphs. Fig. 1(b) is a super-triangulation, whereas Fig. 1(a) is not because the tree $\overline{T}_0$ cannot be well-orderly.

2.2   Computing a Super-Triangulation

Theorem 1 and Theorem 2 follow from the next three lemmas. A plane graph H is a *well-orderly embedding rooted in v* if H has a well-orderly tree of root v.

Lemma 1. *Every well-orderly embedding rooted in some node v has a unique well-orderly tree of root v.*

Lemma 2. *Let G be a connected planar graph, and let v be any node of G. Then G has a well-orderly embedding of root v. Moreover, the well-orderly tree and the well-orderly embedding can be computed in linear time.*

Lemma 3. *Let T be the well-orderly tree of H rooted in some node r_0, and assume that T has at least two leaves. Let r_2 and r_1 be the clockwise first and last leaves of T respectively. Then, there is a unique super-triangulation (T_0, T_1, T_2) of the underlying graph of H, preserving the embedding H, and such that each T_i has root r_i. Moreover, $T_0 = T \setminus \{r_1, r_2\}$ and the super-triangulation is computable in linear time.*

3   Encoding a Planar Graph with a Super-Triangulation

3.1   Properties of Super-Triangulations

A *cw-triangle*, is a triple a nodes (u, v, w) of a realizer such that $p_2(u) = v$, $p_1(v) = w$, and $p_0(w) = u$. In the realizer depicted in Fig. 1(a), (u, v, w) forms a cw-triangle, whereas the realizer of Fig. 1(b) has no cw-triangle.

Let $v_1, \ldots, v_n$ be the clockwise preordering of the nodes of a tree T. The subsequence $v_i, \ldots, v_j$ is a *branch* of T if it is a chain (i.e., v_t is the parent of v_{t+1} for every $i \leqslant t < j$), and if $j - i$ is maximal.

The tree $\overline{T}_0$ of a realizer (T_0, T_1, T_2) has the *branch property* if for all nodes v_j and $v_i = p_0(v_j)$, either $p_2(v_j) = p_2(v_i)$, or $v_k = p_2(v_j)$ with $i < k < j$ (that is

$p_2(v_j)$ is a descendant of v_i clockwise before v_j in $\overline{T}_0$). An important feature of the branch property is that all the nodes of a given branch of T_0 (maybe except the root of T_0) must have the same parent in T_2. Indeed, v_j and $v_i = p_0(v_j)$ belong to the same branch implies that $j = i + 1$, and thus, because there is no index k such that $i < k < j$, $p_2(v_j) = p_2(v_i)$ must hold.

Property 1. Let $S = (T_0, T_1, T_2)$ be any realizer. The following statements are equivalent:

1. S is a super-triangulation for some graph G.
2. S has no cw-triangle.
3. $\overline{T}_i$ is well-orderly in S, for every $i \in \{0, 1, 2\}$.
4. $\overline{T}_i$ has the branch property in S, for every $i \in \{0, 1, 2\}$.

Property 2. A tree of a realizer is uniquely determined by given the two others. Moreover, given the embedding of the two trees, it takes a linear time to construct the third one.

3.2   Representation of Planar Graphs with Binary Strings

Along this section, we consider $S = (T_0, T_1, T_2)$ be any super-triangulation of G, a connected planar graph with n nodes and m edges. To show how to use S to efficiently represent G, we define two sets of edges. The *relevant-edge* set is $R_S := \{(v, p_1(v)) \mid v \text{ leaf of } T_2\}$. The *remaining-edge* set is the set M_S of edges of G that are neither in T_0, nor defined by Rule 4 of Def. 1. More formally $M_S := E(G) \setminus (E(T_0) \cup \{(v, p_1(v)) \mid v \text{ inner node of } T_2\})$.

Theorem 3. *Let $S = (T_0, T_1, T_2)$ be any super-triangulation of a graph G.*

1. *Given T_2 and R_S, one can determine S in linear time.*
2. *Given S and M_S, one can determine G in linear time.*

Let $v_1, \ldots, v_n$ be the clockwise preordering of the nodes of a tree T. A leaf v_i of T is a *bud* if the parent of v_i and of v_{i+1} is v_{i-1}. Informally, a bud is a leaf that is clockwise a first child and having at least a sibling. Buds allow a compact encoding of the relevant-edge leaving the leaf v_j located just after a bud v_i. Indeed, as we will see in more details in Lemma 5 how to save bits, if $T = T_0$ has the branch property, then $p_2(v_j) \in \{v_i, p_2(v_{i-1})\}$.

Consider the Eulerian tour $e_1, \ldots, e_{2(n-1)}$ of edges of T obtained clockwise around the outerface, and starting from the root. The *Eulerian string* of T is the binary string such that the ith bit is 1 if and only if $e_i = (u, v)$ with u parent of v. Observe that T has a bud for every 1101-sequence of the Eulerian string.

Let S be a binary string. We denote by $\#S$ the number of binary strings having the same length and the same number of ones than S. More precisely, if S is of length x and has y ones, then we set $\#S := \binom{x}{y}$. Along this section, we assume that T_2 has $n - 2$ nodes (i.e., $\overline{T}_2$ has n nodes), has ℓ leaves and b buds.

Lemma 4. *Knowing S and m, the set M_S can be coded by a binary strings C_1 such that: $\#C_1 = \binom{n+\ell}{3n-m-6}$. Moreover, knowing S and m, coding and decoding M_S take a linear time.*

Lemma 5. *Knowing T_2, the set R_S can be coded by two binary strings B_1, B_2, and by an integer $t \in [0, b]$, such that: $\#B_1 = \binom{b}{t}$, and $\#B_2 = \binom{n-t+\ell-b-3}{\ell-b-1}$. Moreover, knowing T_2, coding and decoding R_S take a linear time.*

Lemma 6. *Knowing n, ℓ, b, the tree T_2 can be coded by five binary strings $A_1, \ldots, A_5$, and by some integers $p \in [1, \ell]$ and $w \in [b, n - \ell]$, such that: $\#A_1 = \binom{\ell-1}{p-1}$, $\#A_2 = \binom{p}{b}$, $\#A_3 = \binom{w}{b-1}$, $\#A_4 = \binom{n-\ell-w+p-b-3}{p-b-1}$, and $\#A_5 = \binom{n-\ell-2}{p-b-1}$. Moreover, knowing n, ℓ, b, coding and decoding T_2 take a linear time.*

Lemma 7. *Any binary string S of length n can be coded into a binary string of length $\log_2(\#S) + o(n)$. Moreover, knowing n, coding and decoding S can be done in linear time, assuming a RAM model of computation on $\Omega(\log n)$ bit words.*

4   Entropy Analysis

Theorem 4. *The number $p(n)$ of unlabeled planar graphs on n nodes satisfies, for every n large enough: $\beta n - \Theta(\log n) \leqslant \log_2 p(n) \leqslant \alpha n + O(\log n)$ with the constants $\alpha \approx 5.007$ and $\beta \approx 4.710$.*

The lower bound in Theorem 4 results from the number $g(n)$ of labeled 2-connected planar graphs. Clearly, $p(n) \geqslant g(n)/n!$. From [4] we have $g(n) \sim \Theta(n^{-7/2}) c^{-n} n!$, where $c \approx 0.03819$. It follows that for n large enough $\log_2 p(n) \geqslant \log_2(g(n)/n!) = \beta n - \Theta(\log n)$, with $\beta = -\log_2 c \approx 4.71066$.

Let us show the upper bound. Let $q(n)$ denote the number of unlabeled connected planar graphs on n nodes. To rely $p(n)$ and $q(n)$, we represent every planar graph G with $k \geqslant 1$ connected components by a triple $(k, \tilde{G}, v)$, where $\tilde{G}$ and v are defined as follows. Let v_i be any non cut-vertex of the ith component of G. (Recall that a *cut-vertex* of a graph is a node whose removal strictly increases the number of connected components. A leaf of a tree being not a cut-vertex, it is clear that every connected graph has a non cut-vertex). We merge all the connected components of G by identifying all the v_i's into a single node v. Clearly $\tilde{G}$ is planar and connected. One can obtained G from $(k, \tilde{G}, v)$ by splitting v in $\tilde{G}$. All the $k' \leqslant k$ connected components obtained by this way are included in G (there is no risk to disconnect a single connected component of G as v_i's are not cut-vertices of each connected component of G). To fully recover G, we may add $k - k'$ isolated nodes because $\tilde{G}$. The number of nodes of $\tilde{G}$ is $n - k + 1$. From this representation, it turns out that $\log_2 p(n) \leqslant \log_2 q(n) + O(\log n)$. So to prove Theorem 4, it remains to prove that, for every n large enough, $\log_2 q(n) \leqslant \alpha n + O(\log n)$, for some constant $\alpha \approx 5.007$.

From our representation, a connected planar graph G can be represented by 8 compressed binary strings: $A_1, \ldots, A_5$, B_1, B_2, and C_1. Let us sketch the entropy analysis (the full analysis can be founded in [8]). Let $\lambda := \ell/n$, and let

$$f(\lambda) := \max_{b,t,w,p} \left\{ \frac{1}{n} \log_2 (\#A_1 \cdots \#A_5 \cdot \#B_1 \cdot \#B_2) \right\}$$

where b, t, w, p are some integers ranging in $[0, n]$ and defined by lemmas 4, 5, and 6. As $\#C_1 = \binom{n+\ell}{3n-m-6}$, $\frac{1}{n} \log_2(\#C_1) \leqslant 1 + \lambda$. So, we have an explicit representation of G with $\max_{0 \leqslant \lambda \leqslant 1} \{f(\lambda) + 1 + \lambda\} n + O(\log n)$ bits.

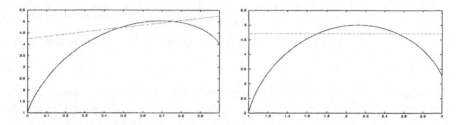

Fig. 2. The behavior of $f(\lambda) + 1 + \lambda$, and of $h(m/n)$ vs. 4.71.

The maximum can be read on the graph of the function $f(\lambda) + 1 + \lambda$ depicted on Fig. 2. Note that if G is a triangulation, then $\log_2(\#C_1) = 0$ (as $m = 3n - 6$). The bit count of our triangulation representation is given by the maximum of $f(\lambda)n + O(\log n)$ (see [8] for details). Therefore, we have:

Theorem 5. *Triangulations and connected planar graphs on n nodes can be represented by a binary string of length at most $3.37n$ and $5.03n$ respectively. Moreover, coding and decoding can be done in linear time.*

To prove the $5.007n$ upper bound, we need refinements. Observe that triangulations have only one super-triangulation (Theorem 2). Therefore, we have an alternative to represent a super-triangulation: either we use our representation (given by Theorem 3) thanks to the pair (T_2, R_S) (this can be done with $f(\lambda)n + O(\log n)$ bits), or we can use a theoretical optimal encoding of the underlying triangulation of S with $\log_2(256/27)n$ bits. Recover the super-triangulation from the triangulation can be done in a unique way by Theorem 2. So, the entropy of a planar graph is no more that $\max_{\lambda} \{\min \{f(\lambda), \log_2(256/27)\}\} n \leqslant 5.007n$. This value can be read on the left side of Fig. 2. The coding is represented by taking the minimum between $f(\lambda) + 1 + \lambda$ and the line defined by: $\lambda \mapsto \log_2(256/27) + 1 + \lambda \approx 4.24 + \lambda$.

Theorem 6. *Almost all unlabeled and almost all labeled planar graphs on n nodes have at least $1.70n$ edges and at most $2.54n$ edges. Moreover, the result holds also for unlabeled connected and labeled connected planar graphs.*

Proof. (sketch). Our representation can be parameterized with the number of edges. The length of the coding is no more than $h(m/n)n + O(\log n)$ bits, where

$$h(m/n) := \max_{0 \leqslant \lambda \leqslant 1} \left\{ \min \{f(\lambda), \log_2(256/27)\} + (1 + \lambda) \cdot H\left(\frac{3 - m/n}{1 + \lambda}\right) \right\}$$

and where $\mathrm{H}(x) = \lim_n \frac{1}{n} \log_2 \binom{n}{xn}$ is the Entropy Function. Using a reduction from arbitrary planar graphs to connected planar graphs (labeled or unlabeled), we can apply our upper bound. Combined with the $4.71n$ bit lower bound of [4], we derive two numbers $\mu_1 = 1.707$ and $\mu_2 = 2.536$ such that our representation is below 4.71. See on the right side of Fig. 2.   □

5   Conclusion

An interesting problem left open in this paper is to know if the approach consisting in representing a planar graph by a suitable triangulation and a set of edges to delete would lead to an optimal information-theoretic bound.

We think that our upper bound could be improved, still using a triangulation approach, because we have proposed a $3.37n$ bit encoding for triangulations, whereas $3.24n$ is possible. An hypothetic gain of $0.13n$ bits should be possible, leading to a $(5.00 - 0.13)n = 4.87n$ bit upper bound for general planar graphs. Independently, if the number of leaves ℓ of T_2 in a random triangulation is close to the value we have obtained here, that is $\ell \approx 0.62n$, the remaining-edge set M_S could be stored with $n + \ell = 1.62n$ bits (we need n bits to describe the edges of T_2 that are in G, and of ℓ bits for the edges of T_1 that are in G but that are leaving a leaf of T_2). This would give a $(3.24 + 1.62)n = 4.84n$ bit random planar graph encoding. By the Occam's razor principle, a conjecture must be simple, and since we have the numerical coincidence that is $4.85 \approx 5\mathrm{H}(2/5)$, we propose:

Conjecture 1. For every n large enough, $p(n) \leqslant \binom{5n}{2n}$.

References

1. L. Alonso, J. L. Rémy, and R. Schott. A linear-time algorithm for the generation of trees. *Algorithmica*, 17(2):162–182, 1997.
2. C. Banderier, P. Flajolet, G. Schaeffer, and M. Soria. Planar maps and airy phenomena. In 27th *ICALP*, vol. 1853 of LNCS, pp. 388–402. Springer, Jul. 2000.
3. E. Barcucci, A. del Lungo, and E. Pergola. Random generation of trees and other combinatorial objects. *Theoretical Computer Science*, 218(2):219–232, 1999.
4. E.A. Bender, Z. Gao, and N.C. Wormald. The number of labeled 2-connected planar graphs, 1999. Manuscript.
5. M. Bodirsky and M. Kang. Generating random outerplanar graphs. In *ALICE '03*.
6. B. Bollobás. *Extremal Graph Theory*. Academic Press, New York, 1978.
7. N. Bonichon. A bijection between realizers of maximal plane graphs and pairs of non-crossing Dyck paths. In *FPSAC*, Jul. 2002.
8. N. Bonichon, C. Gavoille, and N. Hanusse. An information upper bound of planar graphs using triangulation. TR #1279-02, LaBRI, Univ. of Bordeaux, Sep. 2002.
9. M. Bousquet-Mélou, 2002. Private communication.
10. Y.-T. Chiang, C.-C. Lin, and H.-I Lu. Orderly spanning trees with applications to graph encoding and graph drawing. In 12th *SODA*, pp. 506–515, Jan. 2001.
11. N. Chiba, T. Nishizeki, S. Abe, and T. Ozawa. A linear algorithm for embedding planar graphs using pq-trees. *J. of Comp. and Sys. Sci.*, 30(1):54–76, 1985.

12. R.C.-N. Chuang, A. Garg, X. He, M.-Y. Kao, and H.-I Lu. Compact encodings of planar graphs via canonical orderings and multiple parentheses. In 25th ICALP, vol. 1443 of LNCS, pp. 118–129. Springer, Jul. 1998.
13. A. Denise, M. Vasconcellos, and D.J.A. Welsh. The random planar graph. Congressus Numerantium, 113:61–79, 1996.
14. R. Diestel. Graph Theory, vol. 173 of Graduate Texts in Math. Springer, 2000.
15. P. Epstein and J.-R. Sack. Generating triangulations at random. ACM Trans. Model. and Comput. Simul., 4:267–278, 1994.
16. G.N. Frederickson and R. Janardan. Efficient message routing in planar networks. SIAM J. on Computing, 18(4):843–857, August 1989.
17. C. Gavoille and N. Hanusse. Compact routing tables for graphs of bounded genus. In 26th ICALP, vol. 1644 of LNCS, pp. 351–360. Springer, Jul. 1999.
18. S. Gerke and C. McDiarmid. Adding edges to planar graphs, 2001. Preprint.
19. X. He, M.-Y. Kao, and H.-I Lu. A fast general methodology for information-theoretically optimal encodings of graphs. SIAM J. on Comp., 30:838–846, 2000.
20. K. Keeler and J. Westbrook. Short encodings of planar graphs and maps. Discr. Appl. Math., 58:239–252, 1995.
21. A. Khodakovsky, P. Alliez, M. Desbrun, and P. Schröder. Near-optimal connectivity encoding of 2-manifold polygon meshes. Graphical Models, 2002. To appear.
22. D. King and J. Rossignac. Guaranteed 3.67V bit encoding of planar triangle graphs. In 11th CCCG, pp. 146–149, Aug. 1999.
23. Y. Liu. Enumeration of simple planar maps. Util. Math., 34:97–104, 1988.
24. R.J. Lipton and R.E. Tarjan. A separator theorem for planar graphs. SIAM J. on Applied Mathematics, 36(2):177–189, Apr. 1979.
25. V.A. Liskovets and T.R. Walsh. Ten steps to counting planar graphs. Congressus Numerantium, 60:269–277, 1987.
26. H.-I Lu. Improved compact routing tables for planar networks via orderly spanning trees. In 8th COCOON, vol. 2387 of LNCS, pp. 57-66. Springer, Aug. 2002.
27. H.-I Lu. Linear-time compression of bounded-genus graphs into information-theoretically optimal number of bits. In 13th SODA, pp. 223–224, Jan. 2002.
28. J.I. Munro and V. Raman. Succinct representation of balanced parentheses, static trees and planar graphs. In 38th FOCS, pp. 118–126. IEEE Press, 1997.
29. D. Osthus, H.J. Prömel, and A. Taraz. On random planar graphs, the number of planar graphs and their triangulations. Submitted for publication, 2002.
30. R. Pagh. Low redundancy in static dictionaries with constant query time. SIAM J. on Computing, 31(2):353–363, 2001.
31. J. Rossignac. Edgebreaker: Connectivity compression for triangle meshes. IEEE Transactions on Visualization and Computer Graphics, 5(1):47–61, 1999.
32. G. Schaeffer. Random sampling of large planar maps and convex polyhedra. In 31st STOC, pp. 760–769, May 1999.
33. W. Schnyder. Embedding planar graphs on the grid. In SODA '90, pp. 138–148.
34. M. Thorup. Compact oracles for reachability and approximate distances in planar digraphs. In 42th FOCS. IEEE Computer Society Press, Oct. 2001.
35. G. Turán. Succinct representations of graphs. Discr. Appl. Math., 8:289–294, 1984.
36. W.T. Tutte. A census of planar triangulations. Can. J. of Math., 14:21–38, 1962.
37. M. Yannakakis. Embedding planar graphs in four pages. J. of Comp. and Sys. Sci., 38:36–67, 1989.

Finding Large Independent Sets in Polynomial Expected Time

Amin Coja-Oghlan[*]

Humboldt-Universität zu Berlin, Institut für Informatik,
Unter den Linden 6, 10099 Berlin, Germany
coja@informatik.hu-berlin.de

Abstract. We consider instances of the maximum independent set problem that are constructed according to the following semirandom model. First, let $G_{n,p}$ be a random graph, and let S be a set consisting of k vertices, chosen uniformly at random. Then, let G_0 be the graph obtained by deleting all edges connecting two vertices in S. Adding to G_0 further edges that do not connect two vertices in S, an adversary completes the instance $G = G_{n,p,k}^*$. We propose an algorithm that in the case $k \geq C(n/p)^{1/2}$ on input G within polynomial expected time finds an independent set of size $\geq k$.

1 Introduction and Results

An *independent set* in a graph $G = (V, E)$ is a set S of vertices of G such that no two vertices that belong to S are adjacent. The *independence number* $\alpha(G)$ is the size of a largest independent set. Already in 1972 Karp proved that the *maximum independent set problem*, i.e. to find, on input G, a maximum independent set, is NP-hard [19]. Indeed, a non-approximability result due to Håstad [14] shows that it is hard to approximate $\alpha(G)$ within a factor $n^{1-\varepsilon}$, for any $\varepsilon > 0$. Consequently, since we cannot hope for polynomial time algorithms that compute reasonable approximate solutions in the worst case, it is natural to ask for algorithms that perform well on *average* instances, as proposed by Karp as early as in 1976 [20].

But how can we model "average" instances? One could take the *binomial model* $G_{n,p}$ of a random graph on n vertices for a first answer. Given a parameter $p = p(n)$, $0 < p < 1$, the random graph $G_{n,p}$ is obtained by including each possible edge with probability p independently of all others. For example, in the case $p = 1/2$ the probability distribution $G_{n,p}$ is simply the uniform distribution on the class of all graphs of order n. Since the seminal work of Erdős and Rényi, $G_{n,p}$ has been the standard model of a random graph. Therefore, both the study of the combinatorial structure and the algorithmic theory of $G_{n,p}$ have become important and fruitful fields of investigation [5,15,11]. In fact, it is known [15]

---

[*] Research supported by the Deutsche Forschungsgemeinschaft (grant DFG FOR 413/1-1)

H. Alt and M. Habib (Eds.): STACS 2003, LNCS 2607, pp. 511–522, 2003.

that the independence number of the random graph $G = G_{n,p}$ almost surely (i.e. with probability tending to 1 as $n \to \infty$) satisfies

$$\alpha(G) \sim 2\log_2(n) \text{ for } p = 1/2, \ \alpha(G) \sim 2\ln(np)/p \text{ for } 1/n \ll p \ll 1. \quad (1)$$

Concerning the algorithmic aspect, in the case of $G_{n,1/2}$ a simple greedy heuristic (that merely computes a *maximal* independent set) almost surely finds an independent set of size $\approx \log_2(n)$ and hence approximates the independence number within a factor of 2. Remarkably, no polynomial time algorithm is known that almost surely finds an independent set of size $\geq (1 + \varepsilon)\log_2(n)$.

Although the greedy heuristic performs well almost surely, it suffers a serious disadvantage: The greedy heuristic does not compute an upper bound on the independence number, and hence cannot distinguish between such random graphs $G = G_{n,p}$ with a "low" $\alpha(G)$, cf. (1), and "exceptional" inputs with $\alpha(G)$ much larger. Indeed, it is easy to figure out graphs G for which the ratio between $\alpha(G)$ and the output of the greedy heuristic is close to n.

In order to guarantee an approximation ratio better than $n^{1-\varepsilon}$, Krivelevich and Vu [23] consider an algorithm that applied to $G_{n,p}$ has a polynomial *expected* running time. Note that by the hardness result [14], we cannot hope for algorithms with a good approximation ratio that are always fast. In the case $p \geq n^{\varepsilon-1/2}$, $\varepsilon > 0$, the algorithm given in [23] approximates $\alpha(G)$ within a factor $O(\sqrt{np}/\ln(np))$, for any input graph G. (Very recently, an algorithm with similar performance for all $p \geq n^{\varepsilon-1}$ has been proposed by Coja-Oghlan and Taraz [8].) Both [23,8] combine the greedy heuristic with a technique to compute an upper bound on the independence number.

More precisely, as a first step, [23] uses a greedy heuristic to find an independent set S of a size similar to (1). Then, the algorithm computes (in polynomial time) an upper bound λ on the independence number. If this upper bound is $\leq 4(n/p)^{1/2}$, then the output is just S, $\#S$ being within the desired approximation ratio. However, in the case that $\lambda > 4(n/p)^{1/2}$, the running time will be superpolynomial. Still, the expected running time of the algorithm remains polynomial, because the probability that $\lambda > 4(n/p)^{1/2}$ is extremely small.

Semirandom models. However, there are several reasons why $G_{n,p}$ may fail to provide an appropriate model of the "average" instances we are confronted with. First, since all maximal independent sets of $G_{n,p}$ are of size (1) up to a constant factor, $G_{n,p}$ cannot model such instances that actually contain a very large independent set that we are to find. Indeed, in the case $\alpha(G) > 4(n/p)^{1/2}$ the algorithm [23] has an exponential running time. Moreover, $G_{n,p}$ enjoys several well-studied structural properties (cf. [5,15]) that the "average" instances we have in mind may not have. For example, almost surely all vertex degrees of $G_{n,p}$ are roughly np. Hence, it would be desireable to study a model that captures the case of "average" instances containing some large independent set, and can be used to describe instances that lack some typical properties of $G_{n,p}$.

In this paper, we study a *semirandom model* for the maximum independent set problem that meets the above requirements. The first semirandom models, concerning the k-coloring problem, were studied by Blum and Spencer [4]. Instances of semirandom graph problems consist of a random share and a worst case part constructed by an adversary. Hence, semirandom graphs intermediate between the worst-case and the $G_{n,p}$ model. A semirandom model for the maximum independent set problem has been proposed by Feige and Krauthgamer [10]: Let $V = \{1, \ldots, n\}$, and let $c > 0$ be a constant. First, a set $S \subset V$, $\#S = c\sqrt{n}$, is chosen uniformly at random. Then, every edge $\{v, w\}$, $v \in V$, $w \in V \setminus S$, is included into the graph G_0 with probability $p = 1/2$ independently. Thus, G_0 is a random graph $G_{n,1/2}$ with a planted independent set of size $c\sqrt{n}$. Finally, the adversary may add to G_0 further edges $\{v, w\}$, $v \in V$, $w \in V \setminus S$, thereby completing the instance G. Since S is an independent set of G, $\alpha(G) \geq c\sqrt{n}$. We emphasize the fact that the adversary's decisions are *not* random decisions. Obviously, the adversary can change the vertex degrees, the eigenvalues of the adjacency matrix etc. The algorithm studied by Feige and Krauthgamer [10] always runs in polynomial time and with high probability finds the hidden independent set S and certifies its optimality.

In this paper, we shall study two semirandom models. Given an edge probability $p = p(n)$, $0 < p < 1$, and a number $k = k(n)$, instances of our first semirandom model $G^*_{n,p,k}$ are made up as follows.

1. A set $S \subset V$ consisting of k vertices is chosen uniformly at random.
2. The random graph $G_0 = G_{n,p,k}$ is obtained by including every edge $\{v, w\}$, $v \in V$, $w \in V \setminus S$, with probability p independently of all other edges.
3. The adversary may add to G_0 edges $\{v, w\}$, $v \in V$, $w \in V \setminus S$, thereby completing the instance $G = G^*_{n,p,k}$.

Clearly, S is an independent set in G, whence $\alpha(G) \geq k$. Note that $G^*_{n,1/2,\Omega(\sqrt{n})}$ coincides with the model treated in [10].

Instances of our second semirandom model $G^*_{n,p}$ are constructed as follows. First, choose a random graph $G_0 = G_{n,p}$. Then the adversary may add to G_0 arbitrary edges, thereby completing the instance $G = G^*_{n,p}$. Thus, the model $G^*_{n,p}$ describes instances of the maximum independent set problem that do not contain a very large independent set.

For each $G_0 = G_{n,p,k}$, $\mathcal{I}(G_0)$ denotes the set of all graphs that can be obtained from G_0 according to 3. above. Let $R_A(G)$ denote the running time of an algorithm A on input G. We say that *applied to* $G^*_{n,p,k}$, A *runs within expected polynomial time* if there is a constant $l > 0$ such that for any map I that to each $G_0 = G_{n,p,k}$ assigns an element of $\mathcal{I}(G_0)$ we have $\sum_{G_0 \in G_{n,p,k}} R_A(I(G_0))P(G_0) = O(n^l)$. The main result of this paper is the following theorem.

Theorem 1. *Suppose that* $1/2 \geq p \gg \ln(n)^6/n$ *and that* $k \geq C\sqrt{n/p}$ *for some sufficiently large constant* C. *There is an algorithm* **Find** *that satisfies the following properties.*

1. *Both applied to* $G^*_{n,p,k}$ *and to* $G^*_{n,p}$ **Find** *runs in polynomial expected time.*
2. *If* $\alpha(G) \geq k$, **Find**(G) *outputs an independent set of size* $\geq k$.
3. *If* $\alpha(G) < k$, **Find**(G) *outputs* $\emptyset$.

Thus, **Find** improves over the algorithm studied in [10] in the two following respects.

- Applied to $G^*_{n,p,k}$, Find *always* succeeds in finding an independent set of size $\geq k$, and has polynomial expected running time.
- The running time of Find is polynomial even when applied to $G^*_{n,p}$. Hence, Find can distinguish between $G^*_{n,p}$ and $G^*_{n,p,k}$ efficiently.

However, the algorithm [10] with high probability certifies the optimality of its output, which Find does not.

Moreover, Find complements the algorithms studied by Krivelevich and Vu [23] and by Coja-Oghlan and Taraz [8] as follows. First, Find can exhibit a large independent set efficiently, if there is any. In contrast, the running time of the algorithms studied in [23,8] will be exponential if the input instance contains a large independent set. Further, Find can handle *semirandom* instances, hence is applicable to a wider class of input distributions than just $G_{n,p}$.

Similar arguments as in [9] can be used to show the following hardness result.

Theorem 2. *Suppose that* $2\ln(n)/n \leq p \leq n^{-1/100}$. *Furthermore, let* $n^{1/200} \leq k \leq (1-\varepsilon)\ln(n)/p$, $\varepsilon > 0$. *Then, there is no polynomial time algorithm that applied to* $G^*_{n,p,k}$ *with high probability finds an independent set of size* $\geq k$, *unless* NP $\subset$ BPP.

Techniques and related work. The first to study the planted independent set model $G_{n,1/2,k}$ were Jerrum [16] and Kučera [22]. Kučera observed that in the case $k \geq c\sqrt{n\ln(n)}$ for a certain $c > 0$, one can recover the planted independent set by simply picking the k vertices of least degree. However, this approach fails if $k = o(\sqrt{n\ln(n)})$. Alon, Krivelevich and Sudakov presented an algorithm based on spectral techniques that in the model $G_{n,1/2,k}$, $k = \Omega(\sqrt{n})$, with high probability finds the hidden independent set [2]. No polynomial time algorithm is known that can handle the case $k = o(\sqrt{n})$. Though, it is easy to recover a planted independent set S of size $k = o(\sqrt{n})$ in time $n^{c\ln(n)}$: Enumerate all subsets T of size $c\ln(n)$ and consider the set of all non-neighbors of T; if $T \subset S$, then with high probability the non-neighborhood of T is precisely S.

The semirandom model $G^*_{n,p,k}$, $p = 1/2$, $k = \Omega(\sqrt{n})$, first was considered by Feige and Krauthgamer [10]. Their algorithm relies on the fact that in the case $k \geq c\sqrt{n}$ with high probability $\vartheta(G^*_{n,p,k}) = k$, where ϑ denotes the Lovász number of a graph [21] (note that trivially $\vartheta(G^*_{n,p,k}) \geq k$). Since it is not hard to extend this result to smaller values of p, say $p \gg n^{\varepsilon-1}$, $k \geq C(n/p)^{1/2}$, one obtains an algorithm that recovers the hidden independent set with high probability. Moreover, in the case $p = 1/2$, $k = \Omega(\sqrt{n})$, the method immediately yields an algorithm with polynomial expected running time, because the hidden independent set can most probably be recovered in time $n^{O(\ln(n))}$. However, if $p \leq n^{-1/2}$, then this approach (even in combination with the large deviation result [3] on the eigenvalues of random symmetric matrices) seems not to lead to an algorithm that runs in polynomial expected time.

Feige and Kilian [9] studied the problem of finding with high probability an independent set of size $\Omega(n)$ hidden in a semirandom graph. Their ideas can be extended to get algorithms with polynomial expected running time [7,

6]. The algorithms [9,7,6] are based on semidefinite programming techniques developed in [1,18]. However, in the case of an independent set of size $o(n)$, these semidefinite programming techniques seem not to be applicable.

The spectral techniques used in [2] seem to fail in the semirandom model $G^*_{n,p,k}$. Further, even in the case $p = 1/2$, $k \geq c\sqrt{n \ln(n)}$, the degree trick [22] does not work. Moreover, it is clear that the adversary can easily prevent the simple greedy heuristic from finding large independent sets. Thus, one may conclude that the semirandom models $G^*_{n,p,k}$ and $G^*_{n,p}$ require comparatively robust algorithmic techniques.

The main technical result of this paper is a new large deviation result on the Lovász number of the random graph $G_{n,p}$, and may be of independent interest.

Theorem 3. *Suppose that $p \leq 0.99$. Let m be a median of $\vartheta(G_{n,p})$.*

1. *Let $\xi \geq \max\{10, m^{1/2}\}$. Then $P(\vartheta(G_{n,p}) \geq m + \xi) \leq 30 \exp\left(-\frac{\xi^2}{5m+10\xi}\right)$.*

2. *Let $\xi > 10$. Then $P(\vartheta(G_{n,p}) \leq m - \xi) \leq 3 \exp\left(-\frac{\xi^2}{10m}\right)$.*

In combination with similar flow techniques as used in [7,6], Thm. 3 enables us to design an algorithm as asserted in Thm. 1. Remarkably, due to Thm. 3, we can avoid deriving an upper bound on the probable value of $\vartheta(G^*_{n,p,k})$, which is the main technical difficulty in [10].

2   The Large Deviation Result

In this section, we sketch the proof of Thm. 3. Let us briefly recall the definition of the Lovász number $\vartheta(G)$ (cf. [21,13] for thorough treatments). Let $G = (V, E)$ be a graph, $V = \{1, \ldots, n\}$. Let d be a positive integer. Following [21], an *orthogonal labeling* of G is a tuple $(v_1, \ldots, v_n)$ of vectors $v_i \in \mathbf{R}^d$ such that for any two vertices $i, j \in V$, $i \neq j$, with $\{i, j\} \notin E$ we have $(v_i | v_j) = 0$, where $(\cdot | \cdot)$ denotes the scalar product of vectors. We define the *cost* of a d-dimensional vector $a = (a_1, \ldots, a_d)$ by letting $c(a) = a_1^2/(a|a)$ if $a \neq 0$, and $c(0) = 0$. Let $\mathrm{TH}(G)$ be the set of all vectors $(x_1, \ldots, x_n) \in \mathbf{R}^n$, $x_i \geq 0$ for all i, that satisfy the following condition: For any orthogonal labeling $(v_1, \ldots, v_n)$ of G we have $\sum_{i=1}^n x_i c(v_i) \leq 1$. Then $\mathrm{TH}(G)$ is a compact convex set, and $\mathrm{TH}(G) \subset [0; 1]^n$. The *Lovász number* of G is $\vartheta(G) = \max\{(1|x)| \ x \in \mathrm{TH}(G)\}$, where 1 denotes the n-dimensional vector with all entries equal to 1, i.e. $(1|x) = \sum_{i=1}^n x_i$. It is not hard to prove that $\alpha(G) \leq \vartheta(G) \leq \chi(\bar{G})$. In order to bound the probability that the Lovász number $\vartheta(G_{n,p})$ is far from its median, we shall apply the following version of *Talagrand's inequality* (cf. [15, p. 44]).

Theorem 4. *Let $\Lambda_1, \ldots, \Lambda_N$ be probability spaces. Let $\Lambda = \Lambda_1 \times \cdots \times \Lambda_N$. Let $A, B \subset \Lambda$ be measurable sets such that for some $t \geq 0$ the following condition is satisfied: For every $b \in B$ there is $\alpha = (\alpha_1, \ldots, \alpha_N) \in \mathbf{R}^N \setminus \{0\}$ such that for all $a \in A$ we have $\sum_{a_i \neq b_i} \alpha_i \geq t \left(\sum_{i=1}^N \alpha_i^2\right)^{1/2}$, where a_i (respectively b_i) denotes the ith coordinate of a (respectively b). Then $P(A)P(B) \leq \exp(-t^2/4)$.*

The main ingredient to the proof of Thm. 3 is the following lemma.

Lemma 5. *Let m be a median of $\vartheta(G_{n,p})$. Let $\vartheta_0 \geq 0$ be any number, and let $\xi \geq 10$. Then $P(m + \xi \leq \vartheta(G_{n,p}) \leq \vartheta_0) \leq 2\exp(-\xi^2/(5\vartheta_0))$.*

Proof. For $i \geq 2$, let Λ_i denote the first $i - 1$ entries of the ith row of the adjacency matrix of $G_{n,p}$. Then $\Lambda_2, \ldots, \Lambda_n$ are independent random variables, and Λ_i determines to which of the $i - 1$ vertices $1, \ldots, i-1$ vertex i is adjacent. Therefore, we can identify $G_{n,p}$ with the product space $\Lambda_2 \times \cdots \times \Lambda_n$. Let $\pi_i : G_{n,p} \to \Lambda_i$ be the ith projection. Let $A = \{G \in G_{n,p}| \ \vartheta(G) \leq m\}$. Then $P(A) \geq 1/2$, because m is a median.

Let $H \in G_{n,p}$ be such that $m + \xi \leq \vartheta(H) \leq \vartheta_0$. Then there is $x \in \mathrm{TH}(H)$ satisfying $(1|x) = \vartheta(H)$. Put $\alpha = (x_2, \ldots, x_n)$. Let $G \in A$, and put $y_i = x_i$, if $\pi_i(G) = \pi_i(H)$, and $y_i = 0$ otherwise, $2 \leq i \leq n$. We claim that $y = (0, y_2, \ldots, y_n) \in \mathrm{TH}(G)$. Thus, let $(v_1, \ldots, v_n)$ be any orthogonal labeling of G, i.e. if $i, j \in \{1, \ldots, n\}$ are non-adjacent in G, then $(v_i|v_j) = 0$. Put $w_i = v_i$, if $\pi_i(G) = \pi_i(H)$, and $w_i = 0$ otherwise. If $i, j \in \{1, \ldots, n\}$ are non-adjacent in H, $i < j$, then we either have $\pi_j(G) = \pi_j(H)$ or $w_j = 0$. In the first case, i and j are non-adjacent in G, whence $(w_i|w_j) = (v_i|v_j) = 0$. If $w_j = 0$, then obviously $(w_i|w_j) = 0$. Hence $(w_1 = 0, w_2, \ldots, w_n)$ is an orthogonal labeling of H. Therefore, $1 \geq \sum_{i=1}^{n} c(w_i)x_i = \sum_{i=1}^{n} c(v_i)y_i$, whence $y \in \mathrm{TH}(G)$.

Since $y \in \mathrm{TH}(G)$, we have $(1|y) \leq \vartheta(G) \leq m$. On the other hand, $(1|x) = \vartheta(H) \geq m + \xi$. Therefore,

$$\xi \leq (1|x - y) = x_1 + \sum_{i=2}^{n} x_i - y_i \leq 1 + \sum_{\pi_i(G) \neq \pi_i(H)} x_i.$$

Furthermore, we have $\sum_{i=1}^{n} x_i^2 \leq \sum_{i=1}^{n} x_i = (1|x) = \vartheta(H) \leq \vartheta_0$. Thus, $\left(\sum_{i=1}^{n} x_i^2\right)^{1/2} \leq \vartheta_0^{1/2}$. Put $t = (\xi - 1)/\sqrt{\vartheta_0}$. Then, for any $G \in A$ we have $\sum_{\pi_i(G) \neq \pi_i(H)} x_i \geq t \left(\sum_{i=1}^{n} x_i^2\right)^{1/2}$. Thus, Thm. 4 entails

$$P(A)P(m + \xi \leq \vartheta(G_{n,p}) \leq \vartheta_0) \leq \exp(-t^2/4) \leq \exp\left(-\frac{\xi^2}{5\vartheta_0}\right).$$

Finally, our assertion follows from the fact that $P(A) \geq 1/2$.   □

Now, the bound on the upper tail follows from Lemma 5 by estimating the geometric sum

$$P(m + \xi \leq \vartheta) \leq \sum_{l=1}^{\infty} P(m + l\xi \leq \vartheta \leq m + (l+1)\xi) \leq 2\sum_{l=1}^{\infty} \exp\left(-\frac{l\xi^2}{5(m + 2\xi)}\right),$$

thereby proving the first half of Thm. 3. Since for the analysis of our algorithm Find we do not need the bound on the lower tail, we skip the proof.

3   The Algorithm Find

Let us fix a parameter $k \geq C(n/p)^{1/2}$, where C denotes a sufficiently large constant. Furthermore, assume that $\ln(n)^6/n \ll p \leq 1/2$. Before we come down

Algorithm 6. Find(G)
Input: A graph $G = (V, E)$.
Output: A subset of V.

1. Let $I' = $ Filter(G). If $\#I' > 2k$, then let $\tilde{S} = $ Exact(G) and terminate with output $\tilde{S}$. If $I' = \emptyset$, then terminate with output $\emptyset$.
2. Let $\eta_{\max}$ be the largest number $\leq k/2$ satisfying

$$\left(\frac{2k}{\eta_{\max}} \right)^{10} \leq \exp(k/\ln\ln(n)).$$

   For $\eta = 0, \ldots, \eta_{\max}$ do:
3.   Let $I''' = $ Search(G, I', η). If $I''' \neq \emptyset$, then terminate with output I'''.
4. Let $\tilde{S} = $ Exact(G). Output $\tilde{S}$.

Fig. 1. The Algorithm Find.

Algorithm 7. Filter(G)
Input: A graph $G = (V, E)$.
Output: A subset I' of V.

1. Compute a vector $x \in$ TH(G) such that $(1|x) = \vartheta(G)$. If $\vartheta(G) < k$, then return with output $\emptyset$.
2. Let $I = \{v \in V | \ x_v \geq 1/2\}$.
3. Compute a maximal matching M in $G[I]$. Return with output $I' = I \setminus V(M)$.

Fig. 2. The Algorithm Filter.

Algorithm 8. Exact(G)
Input: A graph $G = (V, E)$.
Output: A subset of V.

1. Enumerate all sets $\tilde{S} \subset V$, $\#\tilde{S} = k/\ln(n)^2$. For each such $\tilde{S}$ enumerate all $Y \subset V$, $\#Y \leq k/\ln(n)^2$. If $V \setminus (N(\tilde{S}) \cup Y)$ is an independent set of cardinality $\geq k$, then return with output $V \setminus (N(\tilde{S}) \cup Y)$. If $\#V \setminus N(\tilde{S}) < k$ for all $\tilde{S}$, then return with output $\emptyset$.
2. Enumerate all subsets $\tilde{S} \subset V$ of cardinality k. For each $\tilde{S}$, check whether $\tilde{S}$ is an independent set. In this case, return with output $\tilde{S}$.
3. Return with output $\emptyset$.

Fig. 3. The Algorithm Exact.

to the details, we shall sketch how the algorithm Find(G) (see Fig. 1) proceeds. Let us assume that the input graph is $G = G^*_{n,p,k}$. As a first step, Find(G) runs the subroutine Filter, which tries to compute a set I' that contains a large share of the hidden independent set S but only few vertices in $V \setminus S$. If Filter succeeds, then Find applies the subroutine Search to I' in order to recover the entire set S (or another independent set of size $\geq k$). The subroutine Search (Fig. 4) relies on a certain expansion property that the random share

Algorithm 9. Search(G, I', η)
Input: A graph $G = (V, E)$, a set $I' \subset V$, a number $\eta \in \{0, 1, 2, \ldots\}$.
Output: A subset of V.

1.   For all $D' \subset I'$, $\#D' \leq \eta$, do
2.   Construct the following network N:

   - The vertices of N are s, t, s_v for $v \in I' \setminus D'$, and t_w for $w \in V$.
   - The arcs of N are (s, s_v) for $v \in I' \setminus D'$, (t_w, t) for $w \in V$, and (s_v, t_w) for $\{v, w\} \in E$, $v \in I' \setminus D'$.
   - The capacity c is given by $c(s, s_v) = \lceil \frac{50n}{k} \rceil$ and $c(s_v, t_w) = c(t_w, t) = 1$.

   Compute a maximum integer flow f in N and put

   $$L = \{v \in I' \setminus D' \mid f(s, s_v) = c(s, s_v)\}.$$

   Let $I'' = I' \setminus (L \cup D')$.
3.   Let $\tilde{V} = V \setminus N(I'')$. If $\#\tilde{V} > 3k/2$, then iterate 1 (i.e. try the next D').
4.   Enumerate all subsets $Y \subset \tilde{V}$ of cardinality $\#Y \leq 6\eta$. Check for each such Y whether $I'' \cup Y$ is an independent set of cardinality at least k. In this case, return with output $I'' \cup Y$.
5.   For any $D \subset \tilde{V}$, $\#D \leq \eta$, do
6.   Let $I''' = I''$. For $\tau = 0, 1, \ldots, \lceil \log_2(n) \rceil$ do
7.   Compute $V' = V \setminus (N(I''') \cup D)$. Construct the following network N.

   - The vertices of N are s, t, s_v for $v \in V' \setminus I'''$, and t_w for $w \in V'$.
   - The arcs of N are (s, s_v) for $v \in V' \setminus I'''$, (t_w, t) for $w \in V'$, and (s_v, t_w) if $\{v, w\} \in E$.
   - The capacity c of N is given by $c(s, s_v) = 6$ and $c(s_v, t_w) = c(t_w, t) = 1$.

   Compute a maximum integer flow f in N. Put

   $$L = \{v \in V' \setminus I''' \mid f(s, s_v) = c(s, s_v)\}$$

   and $I''' = V' \setminus L$.
8.   If $\#I''' \geq k$ and I''' is an independent set, then return with output I'''.
9.   Return $\emptyset$.

Fig. 4. The Algorithm Search.

G_0 contained in the input instance G with high probability has. Since Find is supposed to work properly on any input, Find uses the parameter η in order to take care of instances that violate the expansion property. Finally, if Search does not succeed for any value of η, then Find calls the subroutine Exact that will always find an independent set of size $\geq k$. The crucial point in the analysis of Find is to show that with extremely high probability the output I' of Filter indeed contains most vertices of S but only few vertices not in S. The proof is based on our large deviation result Thm. 3.

The subroutine Filter. Since the computation in the first step of Filter can be implemented in polynomial time using the ellipsoid algorithm [13], Filter runs in polynomial time. We need the following estimate on the Lovász number of random graphs (cf. [17]).

Lemma 10. *With high probability we have $\vartheta(G_{n,p}) \leq 3(n/p)^{1/2}$.*

Note that the Lovász number of $G^*_{n,p,k}$ is much larger, since $\vartheta(G^*_{n,p,k}) \geq k$. Further, we make use of the *monotonicity* of the Lovász number, cf. [21]: If $H_1 = (V, E_1)$ is a subgraph of $H_2 = (V, E_2)$, then $\vartheta(H_2) \leq \vartheta(H_1)$. Suppose that the input graph is $G = G^*_{n,p,k}$. Let $G_0 = G_{n,p,k}$ denote the random share contained in the input instance G, and let S be the hidden independent set.

Lemma 11. *If $C \geq 2000$, then with probability at least $1 - \exp(-\Omega(k))$ the set I computed in step 2 of Filter enjoys the following properties: $\#I \setminus S < k/100$ and $\#I \cap S > 99k/100$.*

Sketch of proof. We claim that in the case that either $\#I \setminus S \geq k/100$ or $\#I \cap S \leq 99k/100$, we have $\vartheta(G_0 \setminus S) \geq k/200$. Since $G_0 - S$ is a random graph $G_{n-k,p}$, Lemma 10 in combination with Thm. 3 then yields our assertion. Thus, assume that $\#I \setminus S \geq k/100$. Then $\sum_{v \in V \setminus S} x_v \geq \sum_{v \in I \setminus S} x_v \geq k/200$. Because $(x_v)_{v \in V \setminus S} \in \mathrm{TH}(G - S)$, by the monotonicity of ϑ we have $\vartheta(G_0 - S) \geq \vartheta(G - S) \geq k/200$. In the case $\#I \cap S \leq 99k/100$, a similar argument works.   □

Let M denote the maximal matching computed by step 3 of Filter. Suppose that $\#I \setminus S \leq k/100$ and $\#I \cap S \geq 99k/100$. Because S is an independent set, every edge $\{v, w\} \in M$ is incident with a vertex in $I \setminus S$. Therefore, $\#V(M) \leq 2\#I \setminus S \leq k/50$, where $V(M)$ is the set of all vertices that are incident with an edge in M. As a consequence, $\#I' \geq (1 - 3/100)k$ and $\#I' \cap S \geq 49k/50$. Since M is maximal, I' is an independent set.

Lemma 12. *With probability $\geq 1 - \exp(-\Omega(k))$ the following holds. If $T \subset V \setminus S$, $\#T \geq 1/p$, and $U \subset S$, $\#U \geq (1 - 1/25)k$, then there is a T-U-edge.*

Proposition 13. *Let $k \geq C(n/p)^{1/2}$ for some large constant $C > 0$. Then with probability $\geq 1 - \exp(-\Omega(k))$ the output I' of Filter on input $G^*_{n,p,k}$ enjoys the following properties: $\#I' \cap S \geq (1 - 1/50)k$ and $\#I' \setminus S \leq 1/p$.*

Now assume that the input graph is $G = G^*_{n,p}$. By monotonicity, Lemma 10 gives an upper bound on the median of $\vartheta(G)$. Thus, the probability that $\vartheta(G) \geq k$, is $\exp(-\Omega(k))$. Hence, on input $G = G^*_{n,p}$, with extremely high probability the output of Filter is $\emptyset$.

The subroutine Search. Let $G = G^*_{n,p,k}$ be the input graph, and let S be the independent set hidden in G. The subroutine $\mathrm{Search}(G, I', \eta)$ proceeds in two phases. In the first phase (step 1–2), Search attempts to purify I', i.e. to identify and to remove $I' \setminus S$, thereby obtaining I''. In the second phase (step 3–8), the aim is to enlarge $I'' \subset S$ several times, each time adding to the current set $I''' \subset S$ at least half of the remaining vertices in $S \setminus I'''$. Thus, after at most $\log_2(n)$ steps, we have $I''' = S$. Since a similar approach has been used in [9, 7,6], we only give a brief sketch of the analysis of Search, which relies on the following expansion property of $G_{n,p,k}$.

Lemma 14. *Let $\gamma > 0$ be fixed. Further, let $d = 50n/k$, and $\eta \in \{0, 1, \ldots, k/2\}$. Then, with probability $\geq 1 - o\left(\dfrac{2k}{\eta}\right)^{-\gamma}$ the graph G_0 has the following property: For any set $T \subset V \setminus S$, $\#T \leq k/(2d)$, there exists a set $D \subset T$, $\#D \leq \eta$, such that $T \setminus D$ admits a complete d-fold matching to S.*

We summarize the analysis of the first phase of Search as follows.

Proposition 15. *Suppose that the set I' satisfies $\#I' \setminus S \leq 1/p$, $\#I' \cap S \geq (1-1/50)k$. Further, assume that the expansion property in Lemma 14 is satisfied with defect $\leq \eta$. Then there exists a set $D' \subset I'$, $\#D' \leq \eta$, such that the set I'' computed in step 2 of $\mathrm{Search}(G, I', \eta)$ satisfies $I'' \subset S$ and $\#I'' \geq (1-1/25)\#S$.*

As for the second phase of Search, we have the following result.

Proposition 16. *Suppose that the set I'' satisfies $\#I'' \geq (1 - 1/25)k$, $I'' \subset S$. Further, assume that the property in Lemma 14 is satisfied with defect $\leq \eta$, and that the property in Lemma 12 is satisfied. Then Search outputs an independent set of size $\geq k$.*

Note that, up to polynomial factors, the running time of $\mathrm{Search}(G, I', \eta)$ is

$$\leq \binom{2k}{\eta}\left(\binom{3k/2}{6\eta} + \binom{3k/2}{\eta}\right) = O\left(\frac{2k}{\eta}\right)^7, \tag{2}$$

provided $\#I' \leq 2k$.

The subroutine Exact. Let us assume that the input graph is $G = G^*_{n,p,k}$. Then $\mathrm{Exact}(G)$ always finds an independent set of size $\geq k$. Moreover, a simple probabilistic argument shows that with probability $\geq 1 - \exp(-2k\ln n)$ step 1 of $\mathrm{Exact}(G)$ already succeeds. Thus, the probability that Exact runs step 2 is $\leq \exp(-2k\ln(n))$, whence the expected time spent in step 2 is $o(1)$. Now assume that the input graph is $G = G^*_{n,p}$. Clearly, in the case $\alpha(G) \geq k$, Exact will output an independent set of size $\geq k$. On the other hand, if Exact runs step 2, then there is a set $\tilde{S}$, $\#\tilde{S} \geq k/\ln(n)^2$, such that $V \setminus N(\tilde{S}) \geq k$. The probability of this event is $\leq \exp(-2k\ln(n))$, whence also on input $G^*_{n,p}$ the expected time spent in step 2 of Exact is $o(1)$.

The running time of Find. In order to show that the expected running time of Find applied to $G^*_{n,p,k}$ or to $G^*_{n,p}$ is polynomial, let us summarize our results on the running time of the subroutines.

- Filter(G) always runs in polynomial time.
- The running time of Search(G, I', η) is bounded by (2).
- The running time of the first step of Exact(G) is $\ll \exp(2k/\ln(n))$. The expected running time of the second step of Exact is negligible.

By Thm. 3 and Lemma 10, the probability that step 1 of Find branches to Exact is $\exp(-\Omega(k))$. Hence, the expected time spent executing step 1 of Find is polynomial on input $G^*_{n,p,k}$ or $G^*_{n,p}$. Moreover, if the input is $G^*_{n,p}$, then by Thm. 3 once more, the probability that step 1 of Find does not terminate with output $\emptyset$ is $\exp(-\Omega(k))$. Since the choice of $\eta_{\max}$ ensures that the time Find spends executing step 2 and 3 is $\leq \exp(o(k))$, we conclude that on input $G^*_{n,p}$ the expected running time of Find is polynomial.

Lemmas 14 and 12 in combination with our analysis of Search above, ensure that the expected time spent executing Search is polynomial on input $G^*_{n,p,k}$. Observe that Search does succeed in finding an independent set of size $\geq k$, unless one of the following events occurs:

- Filter fails, i.e. we have $\#I' \cap S < (1 - 1/50)k$ or $\#I' \setminus S > 1/p$.
- The property stated in Lemma 12 is violated.
- The expansion property in Lemma 14 is violated for some $\eta > \eta_{\max}$.

Since the probability of each of the above events is $\ll \exp(-2k/\ln(n))$, the expected running time of Find applied to $G^*_{n,p,k}$ is polynomial. It is straightforward to check that the output of Find always is an independent set of size $\geq k$ if the input graph contains any, and that the output is $\emptyset$ otherwise.

4   Concluding Remarks

Though Thm. 2 gives a lower bound on the size of independent sets that can be recovered within polynomial expected time, there remains a gap between this lower bound and the upper bound provided by Find. Therefore, it is a challenging problem to either construct an algorithm that beats the upper bound provided by Find, or to prove a better lower bound.

References

1. Alon, N., Kahale, N.: Approximating the independence number via the ϑ-function. Math. Programming **80** (1998) 253–264.
2. Alon, N., Krivelevich, M., Sudakov, B.: Finding a large hidden clique in a random graph. Random Structures & Algorithms **13** (1998) 457–466
3. Alon, N., Krivelevich, M., Vu, V.H.: On the concentration of the eigenvalues of random symmetric matrices. to appear in Israel J. of Math.
4. Blum, A., Spencer, J.: Coloring random and semirandom k-colorable graphs. J. of Algorithms **19(2)** (1995) 203–234

5. Bollobás, B.: Random graphs, 2nd edition. Cambridge University Press (2001)
6. Coja-Oghlan, A.: Finding sparse induced subgraphs of semirandom graphs. Proc. 6. Int. Workshop RANDOM (2002) 139–148
7. Coja-Oghlan, A.: Coloring k-colorable semirandom graphs in polynomial expected time via semidefinite programming, Proc. 27th Int. Symp. on Math. Found. of Comp. Sci. (2002) 201–211
8. Coja-Oghlan, A., Taraz, A.: Colouring random graphs in expected polynomial time. To appear in STACS 2003.
9. Feige, U., Kilian, J.: Heuristics for semirandom graph problems. J. Comput. and System Sci. **63** (2001) 639–671
10. Feige, U., Krauthgamer, J.: Finding and certifying a large hidden clique in a semi-random graph. Random Structures & Algorithms **16** (2000) 195–208
11. Frieze, A., McDiarmid, C.: Algorithmic theory of random graphs. Random Structures & Algorithms **10** (1997) 5–42
12. Füredi, Z., Komlós, J.: The eigenvalues of random symmetric matrices, Combinatorica **1** (1981) 233–241
13. Grötschel, M., Lovász, L., Schrijver, A.: Geometric algorithms and combinatorial optimization. Springer (1988)
14. Håstad, J.: Clique is hard to approximate within $n^{1-\varepsilon}$. Proc. 37th Annual Symp. on Foundations of Computer Science (1996) 627–636
15. Janson, S., Łuczak, T., Ruciński, A.: Random Graphs. Wiley (2000)
16. Jerrum, M.: Large cliques elude the metropolis process. Random Structures & Algorithms **3** (1992) 347–359
17. Juhász, F.: The asymptotic behaviour of Lovász ϑ function for random graphs. Combinatorica **2** (1982) 269–280
18. Karger, D., Motwani, R., Sudan, M.: Approximate graph coloring by semidefinite programming. J. Assoc. Comput. Mach. **45** (1998) 246–265
19. Karp, R.: Reducibility among combinatorial problems. Miller, R.E., Thatcher, J.W. (eds.): Complexity of Computer Computations. Plenum Press (1972) 85–103
20. Karp, R.: Probabilistic analysis of some combinatorial search problems. Traub, J.F. (ed.): Algorithms and complexity: New Directions and Recent Results. Academic Press (1976) 1–19
21. Knuth, D.: The sandwich theorem, Electron. J. Combin. **1** (1994)
22. Kučera, L.: Expected complexity of graph partitioning problems. Discrete Applied Math. **57** (1995) 193–212
23. Krivelevich, M., Vu, V.H.: Approximating the independence number and the chromatic number in expected polynomial time. J. of Combinatorial Optimization **6** (2002) 143–155

Distributed Soft Path Coloring

Peter Damaschke

Chalmers University, Computing Sciences, 41296 Göteborg, Sweden
ptr@cs.chalmers.se

Abstract. In a soft coloring of a graph, a few adjacent vertices may receive the same color. We study soft coloring in the distributed model where vertices are processing units and edges are communication links. We aim at reducing coloring conflicts as quickly as possible over time by recoloring. We propose a randomized algorithm for 2-coloring the path with optimal decrease rate. Conflicts can be reduced exponentially faster if extra colors are allowed. We generalize the results to a broader class of locally checkable labeling problems on enhanced paths. A single result for grid coloring is also presented.

Keywords: distributed algorithms, coloring, locality, randomization

1 Problem Description

Soft coloring means to color the vertices of a graph such that only for a few conflict edges, both endvertices receive the same color. This relaxation is meaningful if coloring must be done quickly in a distributed fashion: Vertices are processing units in a network, equipped with local memory, and edges are communication links. Every vertex knows only its own color and can learn the colors of its neighbors, and it can recolor itself. It can distinguish between incident edges (i.e. assign identifiers to them). For ease of presentation we suppose that all actions are done in synchronized rounds, although our results do not rely on this. Time t is measured by the number of rounds, starting at $t = 1$. In regularly structured graphs we do not assume a global sense of direction (for a formal treatment see e.g. [7]). For example, in a path, there is no global agreement on what is left and right.

Distributed coloring has applications in resource sharing in networks, frequency assignment etc., see e.g. [6,8,10]. A specific motivation is described in [6]. There, colors are time slots in a cyclic schedule, and adjacent units share some device and cannot work at the same time. However a few conflicts are acceptable. If the network is permanently changing, any proper coloring can be easily destroyed anyway, and a small number of colors might be more important than total avoidance of conflicts. Even if the topology is fixed and known to all units (e.g. a grid), a new coloring from scratch (reset) may be required under some circumstances. Nevertheless the remaining conflicts should be reduced over time. Other labeling problems with local constraints are of similar nature.

In [6], some oblivious randomized local recoloring rules have been experimentally studied for grids. However a rigorous analysis has not been given. Parallel

H. Alt and M. Habib (Eds.): STACS 2003, LNCS 2607, pp. 523–534, 2003.
© Springer-Verlag Berlin Heidelberg 2003

graph coloring by symmetry breaking is well studied in [2,9,3], but in a stronger model where processors have IDs. Edge coloring and other labeling problems in the distributed framework have been addressed e.g. in [1,13,17]. Likewise, [14, 16] studied labeling problems that can be solved locally, i.e. in $O(1)$ time. Local mending of conflicts [11,12] means to recover a legal global state. However the goal in these papers is to reach a proper result, not a fast decrease of conflicts as a function of the time passed by. Although these issues are certainly related, they are not identical: 2-coloring a path is completely trivial from the viewpoint of time complexity, but obtaining the smallest possible conflict density at *any* time is more subtle.

Our contributions: This is apparently the first analysis of distributed soft coloring and other locally checkable labeling problems from the viewpoint of conflict density (number of conflicts divided by number of edges) as a function of time t. The main ideas can be already developed on such a trivial graph as the path of n vertices. The rules proposed in [6] do not yield the fastest possible descent of conflicts if we use the minimum number of colors, whereas the rule we propose is provably optimal in that case, but still simple. Conflict density behaves as $O(1/t)$. (Density bounds refer to times $t < n$. Later, conflicts can be reduced to 0 anyway.) For $k > 2$ colors we get $O(a^{t^2})$ with some constant $a < 1$. A similar result is easy to obtain in any graph if $k > \Delta$ (the maximum degree), but there is room for improvements: We show that a grid ($\Delta = 4$) can be 4-colored that fast. Later we extend the results to other labeling problems on path-like graphs. We define skew-symmetric labelings on the path which naturally captures problems that can be solved without sense of direction. It includes e.g. coloring of powers of paths. Finally we mention various open problems.

Due to space limitations we give the proofs in an informal way.

2   Two Colors on a Path

Let G be a path with vertices $v_1, \ldots, v_n$ and edges $v_i v_{i+1}$, for $1 \leq i < n$. At any moment, the conflict edges of the current 2-coloring subdivide G into properly colored segments. The best conflict density we can expect after t rounds is $\Theta(1/t)$, because the vertices have to communicate with each other to agree on one of the two proper colorings, but until time t any vertex can have received messages only from vertices within radius t.

If $v_i v_{i+1}$ is a conflict edge, to move the conflict to the left (right) means to recolor v_i (v_{i+1}). It is quite obvious that two colliding conflicts annihilate each other. A moving conflict is also resolved if it arrives at an end of the path. This suggests the idea of a distributed soft 2-coloring algorithm: We try to let many conflicts collide as soon as possible. A first attempt is Brownian motion: Every conflict moves to the left/right with probability $1/2$, and a new decision is made in every round. However, due to well-known results for random walks (see e.g. [15]), the expected time to reach a vertex at distance k grows as k^2, thus we have to wait too long for collisions. A contrary and more purposeful approach is the following algorithm.

RANDOMDIRECTION: Every conflict decides upon a direction at random but then it sticks to this initial direction until it is resolved.

Note that this rule does not require sense of direction. Now a conflict at initial distance k will be encountered after $k/2$ steps if it moves in the opposite direction. In the following we analyze RANDOMDIRECTION. Infinite processes where moving particles annihilate or coalesce have been analyzed in [5,4].

Consider a conflict moving to the left/right as a right/left bracket. Since all conflicts move at the same speed, the conflicts that annihilate each other are exactly the matching pairs of left and right brackets, irrespective of their distances. (Note that matching pairs in arbitrary sequences of brackets are well-defined.) This observation will allow a purely combinatorial analysis. Conflicts/brackets without mate are not resolved by collisions, instead they disappear at an end of the path. We say that they *escape*. A sequence no bracket escapes from is called *balanced*. These are exactly the bracket strings in algebraic expressions (Dyck language). As a well-known fact, the Catalan number $\binom{2k}{k}/(k+1)$ is the number of balanced strings of k pairs of brackets.

Let $Esc(n)$ be the expected number of escaped brackets in a random string of length n. According to RANDOMDIRECTION, characters are, independently, left or right brackets with probability $1/2$. Let $p(k)$ be the probability of a left/right bracket to find a mate among the next k brackets to the left/right. By linearity of expectation we have

Lemma 1.

$$Esc(n) = n - \frac{1}{2}\sum_{k=1}^{n}(p(k-1) + p(n-k)) = \sum_{k=0}^{n-1}(1 - p(k)).$$

In order to compute the $p(k)$, observe that two brackets in a sequence are mates iff they "look at each other" and the enclosed sequence is balanced. For $k > 0$ this gives

$$p(k) = \frac{1}{2}\sum_{i=0}^{\lfloor (k-1)/2 \rfloor} \frac{\binom{2i}{i}}{(i+1)4^i}.$$

The number of "open" brackets in increasing prefixes of a random sequence is a random walk on the line. Since it eventually reaches any point with probability 1, we have $\lim_{k\to\infty} p(k) = 1$. The asymptotic behaviour can be figured out by applying Stirling's formula to the Catalan numbers. We omit this routine calculation. The result is:

Lemma 2. $1 - p(k) \sim \sqrt{2/\pi}\sqrt{1/k}$.

From the asymptotics for $p(k)$ it follows immediately

Theorem 1. $Esc(n) \sim \sqrt{8/\pi}\sqrt{n}$.

Since $1 - p(k)$ is monotone decreasing in k, Esc is subadditive.

Now we study the expected conflict density after t rounds. Let c denote the initial conflict density, $c = 1$ at worst. To get an upper bound, we split the path into segments of length t. Brackets that find a mate within their own segments are annihilated before time t. Some escaped brackets might also collide with others escaped from other segments before time t, but we ignore these collisions. Still, the expected number of brackets after time t is bounded by $\sum_i Esc(n_i)$ where n_i is the initial number of brackets in the ith segment. Since Esc is subadditive, the worst case is that brackets are evenly spread over the path. This shows:

Theorem 2. *The expected conflict density of* RANDOMDIRECTION *is at most* $Esc(ct)/t \approx \sqrt{8/\pi}\sqrt{c/t}$ *for large enough t. Hence the expected density decreases as $O(1/\sqrt{t})$.*

However this is not the fastest possible decrease of conflicts. Another disadvantage of RANDOMDIRECTION shows up if conflicts are concentrated in a small subpath. Then a square-root fraction of conflicts escapes and is never resolved until they reach the ends. Thus, after a vehement annihilation phase, conflicts are not further reduced for a long time.

Now we propose an algorithm somewhere between Brownian motion and RANDOMDIRECTION which overcomes this weakness and is optimal, subject to some constant factor.

RANDOMDIRECTIONDOUBLEDTIMES (RDDT): At each time $t = 2^i$ (i integer), every conflict decides upon a direction at random and then it sticks to this direction until time $2t$.

Theorem 3. *In RDDT, the conflict density is $O(1/t)$, most of the time.*

Proof. At time t, let a be the average number of conflicts in subpaths of length t, in some partition of the path. As shown before, the expected new a at time $2t$ is $O(\sqrt{a})$. By Markov's inequality, the probability that the actual new a exceeds some constant multiple of $\sqrt{a}$ can be made smaller than some other constant. Hence, on the logarithmic time axis, $\log\log a$ decreases by 1 with some constant positive probability, and it can only slightly increase (as a is at most doubled) else. Thus $\log\log a$ follows a biased random walk. By well-known properties of random walks, it remains constantly bounded most of the time.   □

3   More Colors

Soft coloring a path with $k \geq 3$ colors is dramatically faster. Vertices need no longer agree upon one of the few proper global colorings, instead they can resolve conflicts locally. This intuition suggests an algorithm like

CONSERVATIVERANDOMRECOLORING (CRR): Every vertex is either active or ultimate. Initially all vertices are active. Ultimate vertices do not further change their colors (except in one case mentioned below).
In every round, do the following for every active inner vertex.
(1) If both neighbors have colors different to yours, declare yourself as ultimate.
(2) If exactly one neighbor is ultimate then take a color, different from the color of your ultimate neighbor, at random, and declare yourself as ultimate. Maybe

two adjacent vertices become ultimate in this way, but have the same color. In this case recolor one of them at random in the next step.
(3) If both neighbors are active then assign a random color to yourself.

We highlight that CRR requires only $O(1)$ memory per vertex and $O(1)$ computations per round.

Theorem 4. *When CRR is applied, the expected conflict density at time t is $O(a^{t^2})$ where $a = 2/k$.*

Proof. An active vertex becomes ultimate "for internal reasons", i.e. by applying (3) followed by (1), with probability at least $(k-2)/k = 1 - a$, and "by external force", i.e. by rule (2), for sure within the next i steps, if there is already an ultimate vertex at distance i. Hence, a conflict can survive the first t steps only if the internal case did not apply to any vertex at distance $i \le t$ at any time $j \le t - i$. Since these are t^2 vertex-time pairs, this happens with probability at most a^{t^2}. By linearity of expectation, this is also the expected conflict density. □

CRR can be generalized to graphs of maximum vertex degree Δ and $k > \Delta$ colors: A variant of CRR which we call SIMPLECRR declares a vertex ulimate if its color differs from all colors of neighbors, and recolors an active vertex by a random color not used by any ultimate neighbor.

Theorem 5. *When SIMPLECRR is applied to graphs with maximum degree Δ and $k > \Delta$ colors, the expected conflict density at time t is $O(a^t)$, for some $a < 1$ depending on Δ and k only.*

Proof. Observe the following precoloring extension property: Every proper coloring of a subset of vertices can be extended to all vertices. SIMPLE CRR ensures proper precoloring of the ultimate vertices at any time, and a positive probability for every active vertex to become ultimate in the next step. This immediately implies the assertion. □

The idea of Theorem 5 does not work in general if $k \le \Delta$. However, in many cases we might still achieve exponential decrease. We illustrate this by 4-coloring a grid graph. Vertices are points in the plane with integer coordinates, and every vertex has at most four neighbors, one step to the North, East, South and West, respectively.

Theorem 6. *Grid graphs can be 4-colored such that the expected conflict density at time t is $O(a^t)$ for some constant $a < 1$.*

Proof. We start with SIMPLECRR. A vertex v remains active only if its neighbors are ultimate and have four different colors. The picture shows such a vertex v, symbolized by 0. We enhance SIMPLECRR by an additional procedure that repairs such cases. We can obviously resolve the conflict involving v if we can recolor one of v's neighbors. This is impossible only if, for each of these four vertices, its three neighbors other than v are in turn ultimate and have three different colors. By symmetry we can w.l.o.g. assign color 3 to the vertex to the Northeast of v. Then the above condition forces the coloring of all eight vertices at distance 2 from v shown in the picture. Still, we can resolve the conflict if we can exchange the colors of two adjacent vertices among the eight vertices surrounding v. This exchange involves a corner of this 3×3 square of vertices. Exchange is impossible only if the color brought to that corner is already used by a neighbor of this corner. Straightforward tracking shows that eight further colors in the picture are enforced by this condition. But now we can e.g. recolor the vertex to the North of v by 4, and then its northern neighbor can use either 2 or 3.

Finally, if several nearby conflicts have to be solved, these processes might interfere. But one can randomly assign priorities to the conflicts and solve a conflict only if it has currently the highest priority in a region of bounded radius. This delays the process for each conflict only by a constant number of steps and does not affect the exponential bound.   □

$$- 1 \ 4 \ 1 -$$
$$4 \ 2 \ 1 \ 3 \ 2$$
$$3 \ 4 \ 0 \ 2 \ 1$$
$$4 \ 1 \ 3 \ 4 \ 2$$
$$- 3 \ 2 \ 3 -$$

4   Color Reduction

Theorem 4 implies that a path G of n vertices is properly colored in $O(\sqrt{\log n})$ expected steps. Note that [2] gave a deterministic $O(\log^* n)$ algorithm for 3-coloring a path, however under the stronger assumption that G has already an initial coloring (with many colors) or that vertices have unique IDs. Instead of given IDs, vertices may use homemade random bit strings, such that adjacent vertices get different strings with high probability. However this still requires unbounded local memory, unlike CRR.

In this section we propose a way to use larger (but still constant) local memory to improve the speed of conflict resolution if $k > 2$.

Suppose that colors are numbered, and that a local color reduction rule as in [2,13,14], restricted to K colors, is implemented or hard-wired in our processors, i.e. vertices of G. Such rules take a proper k-coloring and compute a new proper coloring with $k' < k$ colors in constant time. One can achieve $k' = O(\log \log k)$ [14]. (The rule in [14] does not need a sense of direction, as it exploits only the - unordered - set of colors of neighbors.) Hence the decrease of k is rapid in the beginning, and for the last steps (down to three colors on the path) one may use an old observation from [2]: $k > 3$ colors can be reduced to $k - 1$ colors in one round. To this end, each vertex with a color larger than all colors in its neighborhood recolors itself, using the smallest currently available color. Hence no adjacent vertices change their colors simultaneuosly and produce conflicts, and color k disappears in one round. Now we sketch the use of such schemes for our problem.

Theorem 7. *Suppose that processors can store integers up to K and a fast reduction rule restricted to K colors. Then, at time t, we can reach conflict density $O(2/K^{(t-d)^2})$, where the delay d is $O(\log^* K)$.*

Proof. Basically we apply CCR, but instead of k colors we use $K > k$ auxiliary colors $1, 2, 3, \ldots, K$. Since base a in Theorem 4 is now $2/K$, the conflict density in this extended coloring decreases as $O((2/K)^{t^2})$.

To obtain a k-coloring at any time, let the true color of a vertex be its auxiliary color if it is less than k, and k otherwise. This k-coloring has conflicts also between neighbored vertices with auxiliary colors larger than $k - 1$, but every such conflict is resolved after d rounds. (The bound on delay d comes from the aforementioned results for color reduction.) Hence, if we shift the time axis by d units, we get a bound on the actual conflict density. $\quad\square$

5 More Edges and More Labelings

Consider paths with additional edges, for example, the $(d - 1)$th power of the path, which has edges between vertices of distance less than d in the original path. As a motivation, note that in frequency assignment in cellular networks, the same frequency may not be used in nearby cells because of possible interference. However we will, straightaway, study a much more general problem:

Given a path $G = (V, E)$, a set of states S, and a digraph $T = (S, F)$ called the transition graph. We allow loops and opposite arcs in T. Now we describe the task: Establish a labeling $\lambda : V \to S$ such that, for any adjacent $u, v \in V$, their labels are adjacent, too: $\lambda(u)\lambda(v) \in F$ or $\lambda(v)\lambda(u) \in F$. In an obvious sense, λ induces an edge labeling $E \to F$ and an orientation of edges in G. (If there exist arcs from $\lambda(u)$ to $\lambda(v)$ and back, choose one.) Our second demand is that these edge orientations in G have to respect one of the two global orientations of G. If, with every arc in T, the opposite arc is also in T, we may consider T as an undirected graph, and the global orientation demand is vacuously satisfied.

A (directed!) path in T is defined as usual, however we allow arbitrarily many occurences of arcs in a path. Suppose that T admits at least one infinite path, otherwise our problem becomes trivial. Now the problem can be rephrased: Map path G onto some path in T.

Coloring the chordless path is in fact a special case: Labels are colors, and T is the complete but loop-free undirected graph. Coloring of translation-invariant enhanced paths with bandwidth d (w.l.o.g. odd) can also be described by a transition graph T whose states are sequences of d colors. More precisely: Let the state of any v contain the colors of d consecutive vertices with v in the center, ordered from the left to the right. There is an arc $st \in F$ if the last $d - 1$ colors in s equal the first $d - 1$ colors in t. We encode the coloring constraints already in the state set: Only sequences respecting the constraints are in S. These transition graphs $T = (S, F)$ are no longer undirected, and the global orientation requirement must be fulfilled. It is evident that one can state any locally checkable labeling problem, such as maximal independent set, minimal dominating set, etc., in such a way.

Now we want to label G softly: Every edge in E not mapped to an arc in F is called a conflict. Fix some global orientation of G. The key observation is: If u, v are vertices of G, with v standing e positions to the right of u, and if some path from $\lambda(u)$ to $\lambda(v)$ has exactly e arcs, then we can make the subpath of G

from u to v conflict-free by relabeling this subpath only. For any two states s, t, let $L(s, t)$ be the set of all $e \geq 0$ such that some path from s to t has exactly e arcs.

Lemma 3. *For a strongly connected finite directed graph T, there exist positive integers m and c such that, for any two vertices s, t, the members of $L(s, t)$ greater than m form an arithmetic sequence $m + h(s, t) + ic$, with $0 \leq h(s, t) < c$ and $i = 0, 1, 2, 3, \ldots$. In particular, c is the greatest common divisor of cycle lengths in T.*

We omit the straightforward proof. Connected undirected graphs can be classified in two types, since they always have cycles of length 2 (an edge forth and back). These types are: bipartite graphs ($c = 2$), and all other graphs ($c = 1$).

Back to our labeling problem, if T is bipartite, vertices of G must agree upon one of the two assignments of partite sets of T to the color classes of G. If T is not bipartite then, for some m, T has paths of suitable length between the current states of any two vertices of T of distance at least m. Thus one can locally resolve conflicts by relabeling subpaths of length m, thereby protecting properly labeled subpaths. Sense of direction is not required, since paths in T are undirected. Now it is not hard to adapt the previous algorithms to any undirected T.

Since T is fixed, we will assume that a distributed algorithm, known to all vertices of G, can identify the states of T by names. Hence it is able choose states from some convenient subgraph of T only.

Theorem 8. *For any fixed undirected transition graph T, there exist distributed labeling algorithms for the path, such that conflicts decrease as $O(1/t)$ if T is bipartite, and as $O(a^{t^2})$ for some constant $a < 1$, otherwise. (In the following we refer to these cases as "slow" and "fast".)*

Proof. If T is bipartite, we simply choose two adjacent states of T as labels, which is nothing but 2-coloring.

Otherwise, T has an odd cycle. Since a distributed algorithm can choose it in advance, we can w.l.o.g. consider connected T. Apply CRR, generalized in an obvious way. The only facts needed in the proof of Theorem 4 are that, in a path of conflict edges, any inner conflict is resolved with some constant positive probability, and the conflict path is shortened at the ends for sure, in each step. This remains true for random relabeling, just because T is fixed and contains some edges.

It remains to resolve isolated conflict edges in G. Let m be the constant from Lemma 3 for T. Every vertex involved in a conflict can explore its neighborhood of radius m in $O(1)$ time. If the next conflicts are more than m edges away, relabel a subpath of length m including the conflict. However there can be chains of conflicts at distance less than m. Apply CRR once more to these chains, to select conflict edges of pairwise distance larger than m and resolve them simultaneously. (One may e.g. construct a 3-coloring of the chain with labels 0,1,2 and then always resolve conflicts which have a larger label than both neighbors in the chain.) It follows that also the density of isolated conflicts decreases as fast as in Theorem 4. □

If T is directed then vertices of G must, in general, also agree upon some global orientation, thus only a slow algorithm would apply, regardless of c. However an algorithm may restrict the choice of states to some undirected and non-bipartite subgraph of T, if one exists. Then we get fast labeling.

6 Skew-Symmetric Labelings

There is an interesting extension of the undirected case. A directed graph $T = (S, F)$ is said to be skew-symmetric if there exists a mapping $s \mapsto s'$ on S, such that $s'' = s$ for every vertex (note that $s' = s$ is also allowed), and for every arc $st \in F$, contraposition $t's'$ is also in F.

The skew-symmetric case appears naturally, for example, in coloring enhanced paths. Recall that any state $s = \lambda(v)$ is a sequence of d different colors, with v as the central vertex. Then s' is simply this sequence reversed.

Given that the vertices of G lack sense of direction, a vertex does not know whether s or s' is used. Nevertheless, $v_i, v_{i+1}, v_{i+2}, \ldots, v_{i+j}$ can, by communication in $O(j)$ time, coordinate their knowledge in the following sense: They can make their unordered pairs of states $\{s_i, s'_i\}$ ordered pairs (s_i, s'_i), such that both $s_i, s_{i+1}, s_{i+2}, \ldots, s_{i+j}$ and $s'_{i+j}, \ldots, s'_{i+2}, s'_{i+1}, s'_i$ are paths in T. So the decision at one end determines the decision at the opposite end. The point is that, for skew-symmetric T, vertices at distance j in G can locally resolve conflicts in between, whenever T has a path of length exactly j between the current states (connecting the correct members of both pairs of states). As in Theorem 8, this yields, for the characteristic c from Lemma 3:

Theorem 9. *For any fixed skew-symmetric, strongly connected transition graph T with $c = 1$, there exists a fast distributed labeling algorithm for the path.*

As an example consider coloring of the $(d-1)$th power of a path. First of all, note that the vertices can figure out their neighbors in the original path in $O(1)$ time: Neighbors in the path are characterized by sharing $2d - 2$ neighbors in G.

The only possible colorings with $k = d$ colors are periodic repetitions of an arbitrary but fixed permutation of colors, and T consists of disjoint directed cycles of length d, such that every $L(s, t)$ is within a residue class modulo m. Therefore only the slow algorithm applies. For $k > d$ however, T satisfies the condition that enables fast distributed coloring:

Lemma 4. *For $k > d$, the transition graph for coloring the $(d-1)$th power of a path contains directed paths of arbitrary lengths above some m depending on d.*

Proof. Clearly, it suffices to prove the all-paths-lengths property of T for $k = d + 1$ only. First let d be odd. We defined the state of v as the sequence of colors of d consecutive vertices with v in the center. States may be considered as directed cycles of d fields painted with d distinct colors, with a distinguished cutpoint between two neighbored fields. Colored cycles (with cutpoint) obtained by rotation are equivalent. Then, walking an arc in T corresponds to the following operation: Move the cutpoint one field forward. Either keep the color of the passed field or replace it with the color currently not in the cycle. Clearly, one can reach every state: To add the next desired color to an initial sequence, move the cutpoint around the cycle, evict the desired color, and re-insert it in the target field. (If the desired color is already in the next field, we may do a lap of honour.) Hence T is strongly connected. Moreover, above some threshold depending on d, there exist paths of arbitrary length, since one can initially fix an arbitrary position in the cycle as cutpoint in the representation of the destination state, resulting in a shift of any desired length.

For even d the argument is slightly more complicated. Define the state of v as the sequence of colors of $d + 1$ consecutive vertices with v in the center. Now the fields incident to the cutpoint may have the same color (such that one color is absent) or different colors. To reach a desired state we can perform almost the same operation as above, but as long as some color is absent, we must, with each move of the cutpoint, paint the passed field with the next color in the cycle. Colors are evicted and re-inserted in an obvious way. □

Theorem 10. *For $k > d$, there exists a distributed algorithm that colors the path such that each color appears only once among any d consecutive vertices, and conflicts decrease as $O(a^{t^2})$ after time t.*

Proof. Label every vertex of G by a state from T. Apply Theorem 9 to the labeling problem. By Lemma 4 we have $c = 1$. At every moment, define the color of each vertex v to be the central color in $\lambda(v)$ (which is an odd sequence).

It remains to show is that conflicts in the resulting coloring are linearly bounded by the number of labeling conflicts. This in turn is proved if any properly labeled subpath (of length exceeding some threshold) is also properly colored. However this follows from the definition of T:

Consider two vertices u, v of distance less than d within such a subpath. There is a subpath of d' vertices ($d' = d$ if d is odd, and $d' = d + 1$ if d is even) including both u and v. Let w be its central vertex. If we walk from w to v edge by edge, the first $d' - 1$ colors of the next state are always equal to the last $d' - 1$ colors of the previous state. By induction, all vertices on the way have the colors "claimed" by $\lambda(w)$. We argue similarly for the path from w to u. It follows that the colors of u and v are correctly displayed in a single state, $\lambda(w)$. Since states in T never assign the same color to vertices of distance $d - 1$ or less, u and v have got different colors. □

7   Open Questions

First of all, can RDDT be generalized to regular graphs with more dimensions, such as enhanced grids, or to even more general graphs $G = (V, E)$, and what assumptions about prior local knowledge must be made? Is there a clear dichotomy (fast and slow case), as for path coloring?

Our results may be further extended to graphs $G = (V, E)$ of bounded bandwidth or pathwidth. The idea is to encode, in the states and transitions, the proper labelings of all possible isomorphism classes of local subgraphs (of constant size). Then one can work on the transition graph as above, and always construct the actual labeling from the state and the actual local subgraph. However it remains to elaborate on the details for specific problems and graph classes.

If the underlying total order is not given as an extra information, it is also a non-trivial problem to sort the vertices locally.

What is the best hidden constant in the $O(1/t)$ result for 2-coloring the path? A disadvantage of RDDT is that vertices need a local clock and must hold $O(\log t)$ bits at time t. This may be reduced to $O(\log \log t)$ by randomized approximate counting. However, is there an algorithm where vertices need only constant local memory? We outline a preliminary idea: Neighbored conflicts negotiate, by probe and handshake messages, disjoint pairs and then recolor the enclosed segments. One may first try to align every conflict to the nearest other conflict. In chains of segments whose lengths form long monotone sequences, conflicts that have been given the brush-off change their favourite partners with some constant probability in the next pairing attempt. The problem is to analyze the speed of reducing the conflict density.

For $k > 2$ we conjecture that CRR is the optimal $O(1)$ memory algorithm. There seems to be no better way to break a monochrome path by local rules, as there always remain monochrome subpaths of exponentially distributed lengths. (Interestingly, a so-called weak coloring tolerating isolated conflicts cannot be established locally in graphs with even-degree vertices [16].)

Can we develop CRR-based algorithms for $k > \Delta$ colors such that conflicts decrease as $a^{f(t)}$ with f superlinear? A stronger bound may hold already for SIMPLECRR: Neighbors of ultimate vertices are more likely to become ultimate, since they must only hit colors not used by their active neighbors. However the resulting acceleration depends on the topology.

What can be done in general in cases where $k \leq \Delta$? Can the precoloring extension property be relaxed, to allow some local patches?

Acknowledgements. I would like to thank several colleagues at the CS Department for interesting discussions, and especially Olle Häggström and Jeff Steif (Maths Department, Chalmers) for some pointers to the stochastic process literature. Finally, my master's student Martin Antonsson contributed some ideas to Section 7.

References

1. B. Awerbuch, A. Goldberg, M. Luby, S. Plotkin: Network decomposition and locality in distributed computation, *30th IEEE FOCS'89*
2. R. Cole, U. Vishkin: Deterministic coin tossing with applications to optimal parallel list ranking, *Info. and Control* 70 (1986), 32–53
3. G. De Marco, A. Pelc: Fast distributed graph coloring with $O(\Delta)$ colors, *12th Symp. on Discrete Algorithms SODA'2001*
4. M.S. Ermakov: Exact probabilities and asymptotics for the one-dimensional coalescing ideal gas, *Stochastic Processes and their Applications* 71 (1997), 275–284
5. R. Fisch: Clustering in the one-dimensional three-color cyclic cellular automaton, *Ann. Probab.* 20 (1992), 1528–1548
6. S. Fitzpatrick, L. Meertens: An experimental assessment of a stochastic, anytime, decentralized, soft colourer for sparse graphs, *1st Symp. on Stochastic Algorithms, Foundations and Applications SAGA'2001*, Berlin, *LNCS* 2264, 49–64

7. P. Flocchini, B. Mans, N. Santoro: Sense of direction: Definitions, properties and classes, *Networks* 32 (1998), 165–180

8. N. Garg, M. Papatriantafilou, P. Tsigas: Distributed list coloring: How to dynamically allocate frequencies to mobile base stations, *Wireless Networks* 8 (2002), 49–60

9. A. Goldberg, S. Plotkin, G. Shannon: Parallel symmetry-breaking in sparse graphs, *19th Symp. on Theory of Computing STOC'87*, 315–324

10. J. Janssen, L. Naranayan: Approximation algorithms for channel asignment with constraints, *10th Int. Symp. on Algorithms and Computation ISAAC'99*, 327–336

11. S. Kutten, D. Peleg: Fault-local distributed mending, *J. Algorithms* 30 (1999), 144–165

12. S. Kutten, D. Peleg: Tight fault locality, *SIAM J. Computing* 30 (2000), 247–268

13. N. Linial: Locality in distributed graph algorithms, *SIAM J. Computing* 21 (1992), 193–201

14. A. Mayer, M. Naor, L. Stockmeyer: Local computations in static and dynamic graphs, *3rd Israel. Symp. on Theory of Computing and Systems* 1995, 268–278

15. R. Motwani, P. Raghavan: *Randomized Algorithms*, Cambridge Univ. Press 1995

16. M. Naor, L. Stockmeyer: What can be computed locally? *SIAM J. Computing* 24 (1995), 1259–1277

17. A. Panconesi, A. Srinivasan: Randomized distributed edge coloring via an extension of the Chernoff-Hoeffding bounds, *SIAM J. Computing* 26 (1997), 350–368

Competing Provers Yield Improved Karp–Lipton Collapse Results[*]

Jin-Yi Cai[1], Venkatesan T. Chakaravarthy[1], Lane A. Hemaspaandra[2], and Mitsunori Ogihara[2]

[1] Comp. Sci. Dept., Univ. of Wisc., Madison, WI 53706. {jyc,venkat}@cs.wisc.edu
[2] Dept. of Comp. Sci., Univ. of Rochester, Rochester, NY 14627. {lane,ogihara}@cs.rochester.edu

Abstract. Via competing provers, we show that if a language A is self-reducible and has polynomial-size circuits then $S_2^A = S_2$. Building on this, we strengthen the Kämper–AFK Theorem, namely, we prove that if $NP \subseteq (NP \cap coNP)/poly$ then the polynomial hierarchy collapses to $S_2^{NP \cap coNP}$. We also strengthen Yap's Theorem, namely, we prove that if $NP \subseteq coNP/poly$ then the polynomial hierarchy collapses to S_2^{NP}. Under the same assumptions, the best previously known collapses were to ZPP^{NP} and $ZPP^{NP^{NP}}$ respectively ([20,6], building on [18,1,17,30]). It is known that $S_2 \subseteq ZPP^{NP}$ [8]. That result and its relativized version show that our new collapses indeed improve the previously known results. Since the Kämper–AFK Theorem and Yap's Theorem are used in the literature as bridges in a variety of results—ranging from the study of unique solutions to issues of approximation—our results implicitly strengthen all those results.

1 Proving Collapses via Competing Provers

The symmetric alternation class S_2 was introduced by Canetti [10] and Russell and Sundaram [23]. In one model that captures this notion, we have two all-powerful competing provers, the Yes-Prover and the No-prover, and a polynomial-time verifier. Given an input string x, the Yes-prover and the No-prover attempt to convince the verifier of $x \in L$ and of $x \notin L$, respectively. To do so, they provide proofs (i.e., bitstrings) y and z, respectively. Then the verifier simply checks whether y is a correct (in whatever sense of "correct" that the verifier happens to enforce) one for $x \in L$ and whether z is a correct one for $x \notin L$, and votes in favor of one of the provers. We require that if $x \in L$ then the Yes-prover has an irrefutable proof y that can withstand any challenge z from the No-prover; and if $x \notin L$ then the No-prover has an irrefutable proof z that can withstand any challenge y from the Yes-prover. Languages with such a proof system are said to be in the class S_2. We define the class formally in Section 3.

---

[*] Supported in part by NIH grants RO1-AG18231 and P30-AG18254, and NSF grants CCR-9322513, INT-9726724, CCR-9701911, INT-9815095, DUE-9980943, EIA-0080124, CCR-0196197, and EIA-0205061.

H. Alt and M. Habib (Eds.): STACS 2003, LNCS 2607, pp. 535–546, 2003.

When we allow the verifier to have access to an oracle A, we obtain the relativized class S_2^A. Our main result gives a partial characterization for sets that are not useful as oracles to the verifier:

THEOREM: If A is self-reducible and has polynomial-size circuits then $S_2^A = S_2$.

We note that similar results are known for NP^{NP} [3] and ZPP^{NP} [20]. The above result is useful in obtaining a number of conditional collapse results. For example, we can show that if NP has polynomial-size circuits then the polynomial hierarchy, PH, collapses to S_2. This follows from the fact that SAT is a self-reducible complete problem for NP. Though this result is already known (see [8]), our result provides a general method to obtain conditional collapses for other classes. We can apply the theorem to other complexity classes with a set of self-reducible languages that are "collectively" complete for the class (e.g., UP, FewP, NP, Σ_k^p, $\oplus$P). Moreover, by using a relativized version of the above theorem, we can obtain collapses for the first time to $S_2^{NP \cap coNP}$ (under assumptions weaker than used to obtain collapses to S_2). For example, we will show that (i) if $NP \subseteq$ $(NP \cap coNP)/poly$ then PH collapses to $S_2^{NP \cap coNP}$; (ii) if $NP \subseteq coNP/poly$ then PH collapses to S_2^{NP}. Previously the best known collapse results under the same assumptions were to ZPP^{NP} [20] and $ZPP^{NP^{NP}}$. Since $S_2 \subseteq ZPP^{NP}$ [8] (and because this result relativizes), we see that the new collapses are indeed improvements. In Section 2, we discuss the motivation behind the theorem in more detail.

We introduce and use the novel technique of a "dynamic contest" to prove the above theorem. The first hurdle is to show that if A is self-reducible and has polynomial-size circuits then $A \in S_2$. Upon input x, suppose each prover provides a circuit of appropriate size to the verifier. Of course, the honest prover can provide the correct circuit. And by simulating the circuit with x as input, the verifier can determine the membership of x in A correctly. But the issue is that the polynomial-time verifier needs to first find out which one of the two circuits is the correct one! The idea is to simulate the circuits on a sequence of successively smaller strings. The strings are chosen dynamically, using the self-reducing algorithm of A and the outputs of the circuits on earlier strings in the sequence. Using this idea of a dynamic contest between the circuits, we show how the verifier can always choose a correct circuit (if at least one of the two is correct). Then the honest prover can provide the correct circuit and win the vote, irrespective of the circuit provided by the other prover. We then extend this to even prove that $S_2^A = S_2$.

2   Background and Motivation

Karp and Lipton [18] proved that if $NP \subseteq P/poly$ (equivalently, if some sparse set is Turing-hard for NP) then $NP^{NP} = PH$. Köbler and Watanabe [20] (see also Bshouty et al. [6]) strengthened this result by showing that if $NP \subseteq P/poly$ then $ZPP^{NP} = PH$. It is known [2] that there are relativized worlds in which

NP $\subseteq$ P/poly does not imply the collapse of the boolean hierarchy (and so certainly does not imply P = NP).

The just-mentioned Köbler–Watanabe result has itself been further strengthened, via the combination of two results of independent importance: First, Sengupta (see [8]) noted that an alternate proof of the Karp–Lipton theorem by Hopcroft [16] shows that NP $\subseteq$ P/poly implies S_2 = PH, where S_2 is the symmetric alternation class of Canetti [10] and Russell and Sundaram [23]. Second, Cai [8] proved that $S_2 \subseteq \text{ZPP}^{\text{NP}}$, showing that the Hopcroft–Sengupta collapse of PH to S_2 is at least as deep a collapse as that of Bshouty et al. and Köbler–Watanabe. Currently this is the strongest form of the Karp–Lipton Theorem.

The Karp–Lipton result and the Köbler–Watanabe result have been generalized to "lowness" results. Regarding the former, we have for example the lowness result of Balcázar, Book, and Schöning [3] that every Turing self-reducible set A in P/poly is low for NP^{NP}, i.e., $\text{NP}^{\text{NP}^A} = \text{NP}^{\text{NP}}$. Regarding the latter, Köbler and Watanabe [20] themselves proved that every Turing self-reducible set A in (NP $\cap$ coNP)/poly is low for ZPP^{NP}, i.e., $\text{ZPP}^{\text{NP}^A} = \text{ZPP}^{\text{NP}}$.

There are at least two reasons why such transitions from conditional collapse results to lowness results are valuable. The first reason is aesthetic and philosophical. A result of the form "NP $\subseteq$ P/poly $\Longrightarrow$ NP^{NP} = PH" is, in the eyes of many, probably merely stating that false implies false. That is, if the hypothesis does not hold, the result yields nothing. In contrast, consider for example the related lowness result: Every Turing self-reducible set in P/poly is low for NP^{NP}. This result not only has the practical merit of (as can be shown) implying the former result, but also proves, unconditionally, that a class of sets (the class of all Turing self-reducible sets in P/poly) exhibits a simplicity property (namely, giving absolutely no additional power to NP^{NP} when used as free information).

A second reason why making the transition from conditional collapse results to lowness results can be valuable is directly utilitarian. Lowness results are generally a more broadly applicable tool in obtaining collapse consequences for a wide variety of classes. In practice, lowness results in our settings will often apply crisply and directly to all classes having Turing self-reducible complete sets, and even to all classes $\mathcal{C}$ having a set of self-reducible languages that are "collectively" complete for the class. Among natural classes having such properties are UP, FewP, NP, coUP, coFewP, coNP, Σ_k^p, Π_k^p, $\oplus$P, and PSPACE. Even if it is possible to prove these results for each class separately, it is much more desirable to have a single proof.

Thus making a transition from conditional collapse results to lowness results, as has been done already for the Karp–Lipton Σ_2^p result and Köbler–Watanabe ZPP^{NP} result, is of interest. The present paper essentially achieves this transition for the Hopcroft–Sengupta S_2 result.

We say "essentially" since S_2 presents strong barriers to the transition— barriers that are not present, even by analogy, for the NP^{NP} and ZPP^{NP} cases. The source of the barrier is deeply ingrained in the nature of S_2. While NP^{NP} and ZPP^{NP} both have as their "upper level" an unfettered access to existential quantification, S_2 by its very definition possesses a quite subtly constrained quantification structure. For the case of P/poly, this problem does not affect us, and we are able to establish the following lowness result: *All Turing self-reducible sets in* P/poly *are low for* S_2. However, for the case of $(NP \cap coNP)/poly$ the restrictive structure of S_2 is remarkably hostile to obtaining pure lowness results. Nonetheless, we obtain the following lowness-like result that we show is useful in a broad range of new, strongest-known collapses: *For all Turing self-reducible sets* A *in* $(NP \cap coNP)/poly$ *and all sets* B *that are Turing reducible to* A *(or even* $\leq_T^{rs}$-*reducible to* A*),* $S_2^B \subseteq S_2^{NP \cap coNP}$.

In showing that the above lowness and lowness-like results do yield conditional collapse consequences, it will be important to know that Cai's $S_2 \subseteq ZPP^{NP}$ result relativizes, and we note that it does. We also establish (as Theorem 8) that the Hopcroft–Sengupta result relativizes flexibly.

So, putting this all together, we establish a collection of lowness results and tools that allow, for a very broad range of classes, strong uniform-class consequences to be read off from assumed containments in nonuniform classes. The most central of these results is that we strengthen the Kämper–AFK Theorem (which says that $NP \subseteq (NP \cap coNP)/poly \implies PH = NP^{NP}$), relative to both its just-mentioned original version [1,17] and the strengthened version due to Köbler–Watanabe [20]. In particular, we prove that $NP \subseteq (NP \cap coNP)/poly \implies PH = S_2^{NP \cap coNP}$. Another central result we strengthen is Yap's Theorem. Yap's Theorem [30] states that if $NP \subseteq coNP/poly$ (or, equivalently, if $coNP \subseteq NP/poly$) then $PH = \Sigma_3^p$. Köbler and Watanabe strengthened Yap's Theorem by showing that if $NP \subseteq coNP/poly$ then $PH = ZPP^{\Sigma_2^p}$ [20]. We further strengthen Yap's Theorem, via showing that if $NP \subseteq coNP/poly$ then $PH = S_2^{NP}$. (Note: $S_2^{NP} \subseteq ZPP^{\Sigma_2^p} \subseteq \Sigma_3^p$.)

This paper explores the relationship between $(NP \cap coNP)/poly$ and classes up to and including PSPACE. Regarding the relationship between $(NP \cap coNP)/poly$ and classes beyond PSPACE, we commend to the reader the work of Variyam [27,28], who shows for example that an exponential-time analog of AM is not contained in $(NP \cap coNP)/poly$.

The paper is organized as follows. Section 3 presents definitions. Section 4 presents our main lowness theorems about S_2. In Section 5, our lowness theorems yield collapse results for many complexity classes.

Due to space limitations, some results and discussion, a section on open problems, and all but one proof are omitted here. We urge the interested reader to see the full version of this paper [9].

3 Definitions

In this section we present the required definitions and notations. Throughout this paper all polynomials are without negative coefficients so they are monotonically nondecreasing and for all $n \geq 0$ their values at n are nonnegative. Throughout this paper, $(\exists^m y)$ will denote $(\exists y : |y| = m)$, and $(\forall^m y)$ will denote $(\forall y : |y| = m)$.

We now define the symmetric alternation class S_2.

Definition 1 ([10,23]). *A language L is in S_2 if there exists a polynomial-time computable 3-argument boolean predicate P and a polynomial p such that, for all x,*

1. *$x \in L \iff (\exists^{p(|x|)} y)(\forall^{p(|x|)} z)[P(x,y,z) = 1]$, and*
2. *$x \notin L \iff (\exists^{p(|x|)} z)(\forall^{p(|x|)} y)[P(x,y,z) = 0]$.*

Since relativizing S_2 will be important in this paper, we generalize S_2 in a flexible way that allows us to rigorously specify what we mean by relativized S_2, and that also potentially itself opens the door to the study of S_2-like notions applied to a wide variety of classes of predicates.

Definition 2. *Let $\mathcal{C}$ be any complexity class. We define $S_2[\mathcal{C}]$ to be the class of all sets L such that there exists a 3-argument boolean predicate $P \in \mathcal{C}$ and a polynomial q such that*

1. *$x \in L \iff (\exists^{q(|x|)} y)(\forall^{q(|x|)} z)[P(x,y,z) = 1]$, and*
2. *$x \notin L \iff (\exists^{q(|x|)} z)(\forall^{q(|x|)} y)[P(x,y,z) = 0]$.*

We now define our relativizations of S_2 that give the P-time predicate of S_2 access to an oracle.

Definition 3. *1. For each set A, S_2^A denotes $S_2[\mathrm{P}^A]$.*
2. For each class $\mathcal{C}$, $S_2^{\mathcal{C}}$ denotes $\bigcup_{A \in \mathcal{C}} S_2^A$.

Definition 4 (see [21,11]). *We say that a nondeterministic polynomial-time Turing machine (NPTM) M is strong with respect to (an oracle) B, if for all inputs x, $M^B(x)$ satisfies: (i) each computation path halts in one of the states accept, reject or "?", (ii) if there is an accepting path there are no rejecting paths, and (iii) if there is a rejecting path there are no accepting paths, and (iv) at least one path accepts or rejects. M is said to be robustly strong if M is strong with respect to every oracle B. The language defined by M^B is the set of all strings x such that $M^B(x)$ has at least one accepting path.*

Proposition 1 ([21]). *A language A is in $\mathrm{NP}^B \cap \mathrm{coNP}^B$ if and only if there is a NPTM M such that M is strong with respect to B and $A = L(M^B)$. In particular, a set A belongs to $\mathrm{NP} \cap \mathrm{coNP}$ if and only if there is a NPTM M strong with respect to $\emptyset$ such that $A = L(M)$.*

Definition 5 ([11]). *We say that $B \leq_T^{rs} A$, if there is a robustly strong NPTM M such that $B = L(M^A)$.*

For each a and b such that $\leq_b^a$ is a defined reduction, and for each class $\mathcal{C}$, $R_a^b(\mathcal{C})$ will denote $\{B \mid (\exists A \in \mathcal{C})[B \leq_a^b A]\}$. In the following definition and in invocations of it, $\langle \cdot, \cdot \rangle$ denotes some fixed, standard pairing function (having the standard nice properties such as injectivity, surjectivity, polynomial-time computability, and polynomial-time invertibility).

Definition 6 ([18]). *Let C be a complexity class. A language L is said to be in C/poly if there exist a language $L' \in C$, a function s, and a polynomial p for which the following conditions hold:*

1. For all $n \geq 0$, $s(1^n)$ is a string bounded in length by $p(n)$.
2. For all x, $x \in L \iff \langle x, s(1^{|x|}) \rangle \in L'$.

Definition 7 (see the survey [19]). *For each class C for which relativization is well-defined and for each set A, we say that A is low for C if $C^A = C$. For a class D, we say that D is low for C exactly if each set in D is low for C.*

Self-reducibility is a central, widely used concept in complexity theory, and will be important in this paper. (We note that every claim/theorem made in this paper for Definition 8's length-based notion of self-reducibility also holds under a "nice-p-order"-based definition of Turing self-reducibility.)

Definition 8 (see, e.g., [4]). *A set B is said to be Turing self-reducible if there is a (clocked, deterministic) polynomial-time Turing machine M that has the following two properties: (1) $B = L(M^B)$, and (2) on each input string x, regardless of the answers it receives to oracle queries, M never queries any string of length greater than or equal to $|x|$.*

Definition 9. *1. For sets B and C, we say that B is Turing self-reducible with respect to C if there is a (clocked, deterministic) polynomial-time Turing machine M such that the following properties hold.*
  a) $B = L(M^{B,C})$, where in this model M has both an oracle tape for querying B and a separate oracle tape for querying C.
  b) On each input string x, regardless of the answers it receives to previous oracle queries from either of its oracles, M never queries on its oracle tape corresponding to B any string of length greater than or equal to $|x|$.
2. For a set B and a class C, we say that B is Turing self-reducible with respect to C if there is a set $C \in C$ such that B is Turing self-reducible with respect to C.

All complexity classes (e.g., NP, coNP, ZPP, Mod_kP, PSPACE, etc.) have their standard definitions (see [15]). For clarity, we mention explicitly that, as is standard, Mod_kP is the class of all sets A such that for some nondeterministic polynomial-time machine M, and each x, it holds that $x \in A$ if and only if the number of accepting paths of $M(x)$ is not congruent to 0 mod k.

4 Lowness Results for S_2

In this section, we present some lowness results about S_2 and $S_2^{\text{NP}\cap\text{coNP}}$. We also give a lowness transference lemma. The results in Section 5 stating conditional collapses for many complexity classes are based on these underpinnings.

Theorem 1. *If $A \in$ P/poly and A is Turing self-reducible then A is low for S_2, i.e., $S_2^A = S_2$.*

Proof. Let A be as in the hypothesis of the theorem. Let L be an arbitrary language in S_2^A. There exist a 3-argument predicate $B \in P^A$ and a polynomial p such that, for all x, (i) $x \in L \iff (\exists^{p(|x|)} y)(\forall^{p(|x|)} z)[B(x, y, z) = 1]$, and (ii) $x \notin L \iff (\exists^{p(|x|)} z)(\forall^{p(|x|)} y)[B(x, y, z) = 0]$. Since $B \in P^A$ there exists a polynomial-time oracle Turing machine M_0 that decides B with oracle A. Let q be a polynomial such that, for all x, y, and z satisfying $|y| = |z| = p(|x|)$, all query strings of M_0 on input $\langle x, y, z \rangle$ have length at most $q(|x|)$ irrespective of its oracle. Since $A \in$ P/poly, there exist a language in P and an advice function s witnessing that $A \in$ P/poly. Let S be the function such that, for all x, $S(x) = s(1^0) \# s(1) \# \cdots \# s(1^{|x|})$, where $\#$ is a delimiter. Then there exists a polynomial r such that S is polynomially length-bounded by r and there exists a polynomial-time machine M_{adv} such that for all x and n, $n \geq |x|$, M_{adv} on $\langle x, S(1^n) \rangle$ correctly decides whether $x \in A$. (This is what is sometimes called in the literature "strong advice"—advice that works not just at one length but also on all strings up to that given length, see [5].) Since A is Turing self-reducible, there exists a polynomial-time oracle Turing machine M_{sr} such that M_{sr}, given A as its oracle, correctly decides A, and such that, for all x, irrespective of the oracle, every query string (if any) of M_{sr} on input x has length strictly less than $|x|$.

We first define a deterministic polynomial-time machine called the *A-simulator*, which takes as its input a triple $\langle w, s_1, s_2 \rangle$ and outputs either $M_{\text{adv}}(\langle w, s_1 \rangle)$ or $M_{\text{adv}}(\langle w, s_2 \rangle)$. On input $\langle w, s_1, s_2 \rangle$, the A-simulator first computes $M_{\text{adv}}(\langle w, s_1 \rangle)$ and $M_{\text{adv}}(\langle w, s_2 \rangle)$. If they agree on their outcome, the A-simulator outputs that outcome and halts. Otherwise, it sets a variable α to w and simulates $M_{\text{sr}}(\alpha)$ answering its queries β by running both $M_{\text{adv}}(\langle \beta, s_1 \rangle)$ and $M_{\text{adv}}(\langle \beta, s_2 \rangle)$ as long as they agree. If for some query β, $M_{\text{adv}}(\langle \beta, s_1 \rangle)$ and $M_{\text{adv}}(\langle \beta, s_2 \rangle)$ disagree, then the A-simulator sets α to β and starts over the above procedure. Since M_{sr} is a self-reduction, the queries are always shorter than the input, so the length of α becomes smaller on each iteration. Thus, if the "otherwise" case above was reached, then there is eventually a point at which α satisfies (i) $M_{\text{adv}}(\langle \alpha, s_1 \rangle)$ and $M_{\text{adv}}(\langle \alpha, s_2 \rangle)$ disagree and (ii) for all queries β of M_{sr} on input α, $M_{\text{adv}}(\langle \beta, s_1 \rangle)$ and $M_{\text{adv}}(\langle \beta, s_2 \rangle)$ agree. For such an α, there is exactly one $t \in \{s_1, s_2\}$ such that $M_{\text{adv}}(\langle \alpha, t \rangle)$ agrees with $M_{\text{sr}}(\alpha)$ when all the queries are answered by M_{adv} with t as the advice string (note that, on these particular queries made by M_{sr}, using s_1 as the advice string and using s_2 as the advice string produce the exact same answers). The A-simulator finds which of s_1 and s_2 is this t and outputs the value of $M_{\text{adv}}(\langle w, t \rangle)$. Since the length of α decreases each iteration and both M_{adv} and M_{sr} are polynomial time-bounded, the A-simulator runs in polynomial time.

We show that L is in S_2 by developing its verification scheme with the Yes- and No-provers as discussed in the paragraph after Definition 1. Let x be an input. The Yes-prover's certificate Y and the No-prover's certificate Z are of the form $\langle y, s_1 \rangle$ and $\langle z, s_2 \rangle$, respectively, where $|y| = |z| = p(|x|)$ and $|s_1|, |s_2| \leq$

$r(q(|x|))$. The verifier attempts to evaluate $B(x, y, z)$ using s_1 and s_2. To do this, the verifier simulates M_0 on input $\langle x, y, z \rangle$. When M_0 makes a query, say w, the verifier computes the answer from the oracle by running the A-simulator on input $\langle w, s_1, s_2 \rangle$. The verification process clearly runs in polynomial time.

Suppose $x \in L$. Take the string y to be such that for all z satisfying $|z| = p(|x|)$ it holds that $B(x, y, z) = 1$, and take s_1 to be $S(1^{q(|x|)})$. With s_1 as the advice string, for all strings w, $|w| \le q(|x|)$, M_{adv} correctly decides whether $w \in A$, and thus M_{sr} correctly decides whether $w \in A$ with M_{adv} acting as the oracle. This implies that, for all z and s_2, the A-simulator on input $\langle w, s_1, s_2 \rangle$ outputs $M_{\mathrm{adv}}(\langle w, s_1 \rangle)$, which is the membership of w in A. Thus, the verifier correctly evaluates $B(x, y, z)$. So, y is an irrefutable certificate. By symmetry, if $x \notin L$, the No-prover can provide an irrefutable certificate for $x \notin L$. $\square$

A careful examination shows that Theorem 1 can be relativized as follows.

Theorem 2. *Let A and B be sets such that A is self-reducible with respect to B. If $A \in \mathrm{P}^B/\mathrm{poly}$ then $\mathrm{S}_2^A \subseteq \mathrm{S}_2^B$.*

We also can prove the following.

Theorem 3. *Let A be a Turing self-reducible (or even self-reducible with respect to $\mathrm{NP} \cap \mathrm{coNP}$) set in $(\mathrm{NP} \cap \mathrm{coNP})/\mathrm{poly}$. For each set B, if $B \le_T^{rs} A$ then $\mathrm{S}_2^B \subseteq \mathrm{S}_2^{\mathrm{NP} \cap \mathrm{coNP}}$.*

We end this section with a lowness transference lemma. We need a relativized version of Cai's result $\mathrm{S}_2 \subseteq \mathrm{ZPP}^{\mathrm{NP}}$ [8]; the fact below holds via adapting the proof of Cai into a relativized setting.

Fact 1. *For each A, $\mathrm{S}_2^A \subseteq \mathrm{ZPP}^{\mathrm{NP}^A}$.*

Corollary 1. $\mathrm{S}_2^{\mathrm{NP} \cap \mathrm{coNP}} \subseteq \mathrm{ZPP}^{\mathrm{NP}}$.

We now state our lowness transference lemma.

Lemma 1. *If $\mathrm{S}_2^A \subseteq \mathrm{S}_2^{\mathrm{NP} \cap \mathrm{coNP}}$ then A is low for $\mathrm{ZPP}^{\mathrm{NP}}$ and A is low for $\mathrm{NP}^{\mathrm{NP}}$. In particular, if A is low for S_2 then A is low for $\mathrm{ZPP}^{\mathrm{NP}}$ and A is low for $\mathrm{NP}^{\mathrm{NP}}$.*

5   Applications of S_2 Lowness Theorems

We use our lowness theorems proven in Section 4 to get collapse results for many complexity classes. The following theorems are useful in proving those results. (For proofs, please see our full version [9].) We mention in passing that Buhrman and Fortnow [7] have constructed an oracle separating $\Sigma_2^p \cap \Pi_2^p$ from S_2.

Proposition 2. $\mathrm{NP} \subseteq \mathrm{S}_2 \subseteq \mathrm{S}_2^{\mathrm{NP} \cap \mathrm{coNP}} \subseteq \Sigma_2^p$ *and* $\Pi_2^p \subseteq \mathrm{S}_2^{\mathrm{NP}} \subseteq \mathrm{PH} \subseteq \mathrm{PSPACE}$.

Theorem 4. *Let $\mathcal{C}$ be any complexity class that has a self-reducible Turing-complete set. Then $\mathcal{C} \subseteq \mathrm{P}/\mathrm{poly}$ implies $\mathrm{S}_2^{\mathcal{C}} = \mathrm{S}_2$. Furthermore, if $\mathrm{S}_2 \subseteq \mathcal{C}$, then $\mathcal{C} \subseteq \mathrm{P}/\mathrm{poly}$ implies $\mathcal{C} = \mathrm{S}_2$.*

Theorem 5. *Let $\mathcal{C}$ be any complexity class that has a self-reducible Turing-complete set. Then $\mathcal{C} \subseteq (\mathrm{NP} \cap \mathrm{coNP})/\mathrm{poly}$ implies $\mathrm{S}_2^{\mathcal{C}} \subseteq \mathrm{S}_2^{\mathrm{NP} \cap \mathrm{coNP}}$.*

5.1   Collapse Consequences of NP Having Small Circuits

Karp and Lipton showed that if $\text{NP} \subseteq \text{P/poly}$ then $\text{PH} = \Sigma_2^p \cap \Pi_2^p$ [18]. Köbler and Watanabe strengthened their result to show that if $\text{NP} \subseteq (\text{NP} \cap \text{coNP})/\text{poly}$ then $\text{PH} = \text{ZPP}^{\text{NP}}$ [20]. The following shows further strengthenings of this result.

Theorem 6. *1. (Hopcroft and Sengupta, see [8])* $\text{NP} \subseteq \text{P/poly} \Longrightarrow \text{PH} = S_2$.
2. $\text{NP} \subseteq (\text{NP} \cap \text{coNP})/\text{poly} \Longrightarrow \text{PH} = S_2^{\text{NP} \cap \text{coNP}}$.

An important open question that we commend to the reader is whether one can strengthen Theorem 6 by proving that $\text{NP} \in (\text{NP} \cap \text{coNP})/\text{poly}$ implies $\text{PH} = S_2$.

Theorem 6 has consequences in the study of reducing solutions of NP functions. Hemaspaandra et al. [14] prove that if NP has unique solutions (see that paper for a full definition) then the polynomial hierarchy collapses to ZPP^{NP}. Their actual proof involves showing that if NP has unique solutions then $\text{NP} \subseteq (\text{NP} \cap \text{coNP})/\text{poly}$. Thus, in light of part 2 of Theorem 6, one has the following stronger collapse consequence from the assumption that NP has unique solutions.

Theorem 7. *If* NP *has unique solutions then* $\text{PH} = S_2^{\text{NP} \cap \text{coNP}}$.

We now turn to the following result, which shows that part 1 of Theorem 6 can be relativized in a particularly flexible fashion (the natural relativization would simply be the $A = B$ case).

Theorem 8. *For any two sets A and B, if* $\text{NP}^A \subseteq \text{P}^B/\text{poly}$ *and* $A \in \text{P}^B$ *then* $\text{NP}^{\text{NP}^A} \subseteq S_2^B$.

We now immediately apply Theorem 8 to improve Yap's Theorem. Recall, as mentioned in the introduction, that Yap's Theorem [30] is $\text{NP} \subseteq \text{coNP/poly} \Longrightarrow \text{PH} = \Sigma_3^p$ and that the best previously known strengthening of it is the result of Köbler and Watanabe [20] that $\text{NP} \subseteq \text{coNP/poly} \Longrightarrow \text{PH} = \text{ZPP}^{\Sigma_2^p}$.

Corollary 2. *For each $k \geq 1$,* $\Sigma_k^p \subseteq \Pi_k^p/\text{poly} \Longrightarrow \text{PH} \subseteq S_2^{\Sigma_k^p}$. *In particular,* $\text{NP} \subseteq \text{coNP/poly} \Longrightarrow \text{PH} \subseteq S_2^{\text{NP}}$.

Note that the final sentence of the statement of Corollary 2, in one fell swoop, strengthens all known results that are based on Yap's Theorem.

Part 2 of Theorem 6 improved the collapse consequences that follow from $\text{NP} \subseteq (\text{NP} \cap \text{coNP})/\text{poly}$. In Section 5.2 we will similarly improve such collapse consequences for the cases of various other classes, such as UP and FewP. In each case, the conclusions involve S_2-related classes. We will not extensively discuss what happens when Mod-based and larger classes are contained in $(\text{NP} \cap \text{coNP})/\text{poly}$. The reason is that the assumption that $(\text{NP} \cap \text{coNP})/\text{poly}$ contains such powerful classes (classes that, basically due to Toda's Theorem, in the presence of the BP operator subsume the polynomial hierarchy [25,26]) is so sweeping as to immediately imply collapses that are even deeper than S_2-related collapses. In particular, it is known that:

Theorem 9 (see [20,22]).

1. *If* $PP \subseteq P/poly$, *then* $PH^{\#P} = MA$.
2. *If* $PSPACE \subseteq P/poly$, *then* $PSPACE = MA$.

And with regard to $Mod_k P$, we observe that a related collapse holds under the assumption that $Mod_k P$ is in $P/poly$.

Theorem 10. *For each integer* $k \geq 2$, *it holds that* $Mod_k P \subseteq P/poly$ *implies* $NP^{Mod_k P} \subseteq MA$.

The above collapse is stronger than the collapse found in the literature, i.e., Köbler and Watanabe's [20] result that for each integer $k \geq 2$ it holds that $Mod_k P \subseteq P/poly$ implies $Mod_k P \subseteq MA$. Since $MA \subseteq S_2$ [23], the collapses of Theorems 10 and 9 are indeed very severe ones (obtained from very severe assumptions). Though one must be careful in doing so, one can from the proof idea of the preceding theorem show collapse results from containments in $(NP \cap coNP)/poly$. In particular, we have the following theorem which, since $MA \subseteq S_2$ relativizes, reflects an even deeper collapse than would a collapse to $S_2^{NP \cap coNP}$.

Theorem 11. *1. If* $PP \subseteq (NP \cap coNP)/poly$, *then* $PH^{\#P} = MA^{NP \cap coNP}$.
2. If $PSPACE \subseteq (NP \cap coNP)/poly$, *then* $PSPACE = MA^{NP \cap coNP}$.
3. For each integer $k \geq 2$, *it holds that* $Mod_k P \subseteq (NP \cap coNP)/poly$ *implies* $NP^{Mod_k P} \subseteq MA^{NP \cap coNP}$.

Note that in Theorem 11, we are not relativizing Theorems 9 and 10, but rather we (to prove them) argue regarding the power of the provers.

Notwithstanding the paragraph immediately preceding Theorem 11, we do mention one collapse to an S_2-related class that is an application of Theorem 5, and that—though involving modulo-based classes—does not seem to follow from such MA-based collapses as part 3 of Theorem 11.

Theorem 12. *For each* $k \geq 2$, $Mod_k P \subseteq (NP \cap coNP)/poly \implies Mod_k P^{PH} = S_2^{NP \cap coNP}$.

5.2   Results in the Absence of Self-Reducible Complete Sets

Theorem 1 connects nonuniform containments with uniform collapse consequences for classes having self-reducible complete sets. But there are some complexity classes for which self-reducible complete sets are not known. In this section, we handle some such classes.

We first show that Theorem 1 can even be applied to some classes $\mathcal{C}$ that potentially lack complete sets, but that do satisfy $\mathcal{C} \subseteq R_m^p(\mathcal{C} \cap$ Turing-self-reducible), i.e., classes whose Turing self-reducible sets are known to be, in some sense, "collectively" complete for the class. For example, neither UP nor FewP is known to have complete sets, and indeed each is known to lack complete sets in relativized worlds [12,13]. Nonetheless we have the following claim.

Theorem 13. *Let C be any member of this list:* UP, coUP, FewP, coFewP. *If $C \subseteq P/poly$ then C is low for each of S_2, ZPP^{NP}, and NP^{NP}.*

We can also obtain the following theorem.

Theorem 14.   *1.* $UP \subseteq (NP \cap coNP)/poly \Longrightarrow S_2^{UP} \subseteq S_2^{NP \cap coNP}$.
2. $FewP \subseteq (NP \cap coNP)/poly \Longrightarrow S_2^{FewP} \subseteq S_2^{NP \cap coNP}$.

The final complexity class that we consider is $C_=P$ [29,24]. Our goal is to prove a complexity class collapse from the assumption that $C_=P \subseteq (NP \cap coNP)/poly$. However, $C_=P$ is not known to have self-reducible complete sets, and so Theorem 1 cannot be used directly. Nonetheless:

Theorem 15. *If* $C_=P \subseteq P/poly$ *then* $PH^{\#P} = PH^{C_=P} = S_2^{C_=P} = S_2 = MA$.

Theorem 16. *If* $C_=P \subseteq (NP \cap coNP)/poly$ *then* $PH^{\#P} = MA^{NP \cap coNP}$.

To conclude the paper, we remind the reader of what we consider the most interesting open issue: Can one prove not merely Theorem 6 but even that $NP \in (NP \cap coNP)/poly$ implies $PH = S_2$?

Acknowledgments We thank Edith Hemaspaandra for helpful discussions and comments, Vinodchandran Variyam for exchanging papers with us, and Alina Beygelzimer and Mayur Thakur for proofreading an earlier draft.

References

1. M. Abadi, J. Feigenbaum, and J. Kilian. On hiding information from an oracle. *Journal of Computer and System Sciences*, 39:21–50, 1989.
2. V. Arvind, Y. Han, L. Hemachandra, J. Köbler, A. Lozano, M. Mundhenk, M. Ogiwara, U. Schöning, R. Silvestri, and T. Thierauf. Reductions to sets of low information content. In K. Ambos-Spies, S. Homer, and U. Schöning, editors, *Complexity Theory*, pages 1–45. Cambridge University Press, 1993.
3. J. Balcázar, R. Book, and U. Schöning. The polynomial-time hierarchy and sparse oracles. *Journal of the ACM*, 33(3):603–617, 1986.
4. J. Balcázar, J. Díaz, and J. Gabarró. *Structural Complexity I*. EATCS Texts in Theoretical Computer Science. Springer-Verlag, 2nd edition, 1995.
5. J. Balcázar and U. Schöning. Logarithmic advice classes. *Theoretical Computer Science*, 99(2):279–290, 1992.
6. N. Bshouty, R. Cleve, S. Kannan, R. Gavaldà, and C. Tamon. Oracles and queries that are sufficient for exact learning. In *Proceedings of the 17th ACM Conference on Computational Learning Theory*, pages 130–139, 1994. JCSS 52(3):421-433 (1996).
7. H. Buhrman and L. Fortnow, Sept. 2001. Personal communication.
8. J. Cai. $S_2^p \subseteq ZPP^{NP}$. In *Proceedings of the 42nd IEEE Symposium on Foundations of Computer Science*, pages 620–629. IEEE Computer Society Press, Oct. 2001.
9. J. Cai, V. Chakaravarthy, L. Hemaspaandra, and M. Ogihara. Some Karp–Lipton-type theorems based on S_2. Technical Report TR-759, Department of Computer Science, Univ. of Rochester, Rochester, NY, Sept. 2001. Revised, Nov. 2002. Available (as the TR-759 entry) at http://www.cs.rochester.edu/trs/theory-trs.html.

10. R. Canetti. More on BPP and the polynomial-time hierarchy. *Information Processing Letters*, 57(5):237–241, 1996.
11. R. Gavaldà and J. Balcázar. Strong and robustly strong polynomial time reducibilities to sparse sets. *Theoretical Computer Science*, 88(1):1–14, 1991.
12. J. Hartmanis and L. Hemachandra. Complexity classes without machines: On complete languages for UP. *Theoretical Computer Science*, 58(1–3):129–142, 1988.
13. L. Hemaspaandra, S. Jain, and N. Vereshchagin. Banishing robust Turing completeness. *International Journal of Foundations of Computer Science*, 4(3):245–265, 1993.
14. L. Hemaspaandra, A. Naik, M. Ogihara, and A. Selman. Computing solutions uniquely collapses the polynomial hierarchy. *SIAM Journal on Computing*, 25(4):697–708, 1996.
15. L. Hemaspaandra and M. Ogihara. *The Complexity Theory Companion*. Springer-Verlag, 2002.
16. J. Hopcroft. Recent directions in algorithmic research. In *Proceedings 5th GI Conference on Theoretical Computer Science*, pages 123–134. Springer-Verlag *Lecture Notes in Computer Science #104*, 1981.
17. J. Kämper. Non-uniform proof systems: A new framework to describe non-uniform and probabilistic complexity classes. *Theoretical Computer Science*, 85(2):305–331, 1991.
18. R. Karp and R. Lipton. Some connections between nonuniform and uniform complexity classes. In *Proceedings of the 12th ACM Symposium on Theory of Computing*, pages 302–309. ACM Press, Apr. 1980. An extended version has also appeared as: Turing machines that take advice, *L'Enseignement Mathématique*, 2nd series, 28, 1982, pages 191–209.
19. J. Köbler. On the structure of low sets. In *Proceedings of the 10th Structure in Complexity Theory Conference*, pages 246–261. IEEE Computer Society Press, June 1995.
20. J. Köbler and O. Watanabe. New collapse consequences of NP having small circuits. *SIAM Journal on Computing*, 28(1):311–324, 1998.
21. T. Long. Strong nondeterministic polynomial-time reducibilities. *Theoretical Computer Science*, 21:1–25, 1982.
22. C. Lund, L. Fortnow, H. Karloff, and N. Nisan. Algebraic methods for interactive proof systems. *Journal of the ACM*, 39(4):859–868, 1992.
23. A. Russell and R. Sundaram. Symmetric alternation captures BPP. *Computational Complexity*, 7(2):152–162, 1998.
24. J. Simon. *On Some Central Problems in Computational Complexity*. PhD thesis, Cornell University, Ithaca, N.Y., Jan. 1975. Available as Cornell Department of Computer Science Technical Report TR75-224.
25. S. Toda. PP is as hard as the polynomial hierarchy. *SIAM Journal on Computing*, 20(5):865–877, 1991.
26. S. Toda and M. Ogihara. Counting classes are at least as hard as the polynomial-time hierarchy. *SIAM Journal on Computing*, 21(2):316–328, 1992.
27. V. Variyam. A note on NP ∩ coNP/poly. Technical Report RS-00-19, BRICS, Aarhus, Denmark, Aug. 2000. Note: The author uses NP ∩ coNP/poly to denote what in the present paper is denoted (NP ∩ coNP)/poly.
28. V. Variyam. $AM_{exp} \not\subseteq$ (NP ∩ coNP)/poly. Manuscript, Oct. 2002.
29. K. Wagner. The complexity of combinatorial problems with succinct input representations. *Acta Informatica*, 23(3):325–356, 1986.
30. C. Yap. Some consequences of non-uniform conditions on uniform classes. *Theoretical Computer Science*, 26(3):287–300, 1983.

One Bit of Advice

Harry Buhrman[1], Richard Chang[2*], and Lance Fortnow[3]

[1] CWI & University of Amsterdam. Address: CWI, INS4, P.O. Box 94709, Amsterdam, The Netherlands. buhrman@cwi.nl.
[2] Department of Computer Science and Electrical Engineering, University of Maryland Baltimore County, 1000 Hilltop Circle, Baltimore, MD 21250, USA. chang@umbc.edu.
[3] NEC Laboratories America, 4 Independence Way, Princeton, NJ 08540, USA. fortnow@nec-labs.com

Abstract. The results in this paper show that coNP is contained in NP with 1 bit of advice (denoted NP/1) if and only if the Polynomial Hierarchy (PH) collapses to D^P, the second level of the Boolean Hierarchy (BH). Previous work showed that $BH \subseteq D^P \implies coNP \subseteq NP/poly$. The stronger assumption that $PH \subseteq D^P$ in the new result allows the length of the advice function to be reduced to a single bit and also makes the converse true. The one-bit case can be generalized to any constant k:

$$PH \subseteq BH_{2^k} \iff coNP \subseteq NP/k$$

where BH_{2^k} denotes the 2^k-th level of BH and NP/k denotes the class NP with k-bit advice functions.

1 Introduction

The results in this paper are motivated in part by the search for a total upward collapse of the Polynomial Hierarchy (PH) under the assumption that one query to NP is just as powerful as two queries — i.e., the assumption that $P^{NP[1]} = P_{tt}^{NP[2]}$. Kadin was first to show that if $P^{NP[1]} = P_{tt}^{NP[2]}$ then the PH collapses to Σ_3^P. Chang and Kadin improved the collapse of PH to the Boolean Hierarchy over Σ_2^P [12]. This was further improved by Beigel, Chang and Ogihara to a class just above Σ_2^P [3]. Most recently Fortnow, Pavan and Sengupta, building on Buhrman and Fortnow [4], pushed the collapse below the Σ_2^P level and showed that $P^{NP[1]} = P_{tt}^{NP[2]}$ implies that $PH \subseteq S_2^P$ [13]. Separately, Chang and Kadin noted that $P^{NP[1]} = P_{tt}^{NP[2]}$ implies that $P^{NP[O(\log n)]} \subseteq P^{NP[1]}$ [11]. This was further improved by Buhrman and Fortnow to $P^{NP} \subseteq P^{NP[1]}$ [4]. Since $S_2^P \subseteq ZPP^{NP}$ [7], we have the following situation:

$$P^{NP[1]} = P_{tt}^{NP[2]} \implies PH \subseteq ZPP^{NP}.$$

$$P^{NP[1]} = P_{tt}^{NP[2]} \implies P^{NP} \subseteq P^{NP[1]}.$$

---

* Supported in part by the University of Maryland Institute for Advanced Computer Studies.

H. Alt and M. Habib (Eds.): STACS 2003, LNCS 2607, pp. 547–558, 2003.
© Springer-Verlag Berlin Heidelberg 2003

This is almost a complete upward collapse of PH down to $P^{NP[1]}$ except for the "gap" between P^{NP} and ZPP^{NP}. Closing this gap might be done with a proof that $P^{NP[1]} = P_{tt}^{NP[2]} \implies ZPP^{NP} \subseteq P^{NP}$. However, the possibility remains for less direct approaches.

The question we ask in this paper is: under what conditions could we get a total collapse of PH below $P_{tt}^{NP[2]}$? We show that

$$PH \subseteq D^P \iff coNP \subseteq NP/1.$$

Here, the NP/1 is NP with one bit of advice and the class D^P consists of those languages that can be expressed as the difference of two NP languages. Note that $P^{NP[1]} \subseteq D^P \subseteq P_{tt}^{NP[2]}$ and that the proofs of most of the upward collapse results involving $P^{NP[1]}$ and $P_{tt}^{NP[2]}$ actually start with the argument that $P^{NP[1]} = P_{tt}^{NP[2]}$ implies that D^P is closed under complementation. Previous results have shown that under the weaker assumption that the Boolean Hierarchy collapses to D^P, coNP $\subseteq$ NP/poly [16]. In contrast, the results in this paper make the stronger assumption that PH collapses to D^P. The stronger assumption allows us to reduce the advice to just one bit and also allows us to prove the converse. Our results also generalize to the k-bit case. We are able to show that:

$$PH \subseteq BH_{2^k} \iff coNP \subseteq NP/k.$$

2   Preliminaries

We use the standard definition and notation for complexity classes with advice (a.k.a., non-uniform complexity) [17]. An important consideration in the definition below is that the advice function depends only on the length of x and not on x itself.

Definition 1. Let L be a language and $f : \mathbb{N} \to \{0,1\}^*$ be an *advice function*. Then, we define $L/f = \{x \mid \langle x, f(|x|) \rangle \in L\}$. For a complexity class $\mathcal{C}$ and a class of functions $\mathcal{F}$, $\mathcal{C}/\mathcal{F} = \{L/f \mid L \in \mathcal{C}, f \in \mathcal{F}\}$. Thus, NP/poly denotes the class of languages recognized by NP machines with advice functions f where $|f(n)|$ is bounded by a polynomial in n. For this paper, we will consider the classes NP/1 and NP/k where the NP machines have, respectively, one-bit and k-bit advice functions (i.e., $|f(n)| = 1$ and $|f(n)| = k$).

The Boolean Hierarchy is a generalization of the class D^P defined by Papadimitriou and Yannakakis [20]. For constant k, the kth level of the Boolean Hierarchy can be defined simply as nested differences of NP languages [5,6].

Definition 2. Starting with $BH_1 = NP$ and $BL_1 = SAT$, we define the kth level of the Boolean Hierarchy and its complete languages by:

$$BH_{k+1} = \{ L_1 - L_2 \mid L_1 \in NP \text{ and } L_2 \in BH_k \}$$
$$coBH_k = \{ L \mid \overline{L} \in BH_k \}$$

$$BL_{2k} = \{ \langle x_1, \ldots, x_{2k} \rangle \mid \langle x_1, \ldots, x_{2k-1} \rangle \in BL_{2k-1} \text{ and } x_{2k} \in \overline{\text{SAT}} \}$$

$$BL_{2k+1} = \{ \langle x_1, \ldots, x_{2k+1} \rangle \mid \langle x_1, \ldots, x_{2k} \rangle \in BL_{2k} \text{ or } x_{2k+1} \in \text{SAT} \}$$

$$\text{coBL}_k = \{ \langle x_1, \ldots, x_k \rangle \mid \langle x_1, \ldots, x_k \rangle \notin BL_k \}.$$

Thus, BL_k is $\leq^P_m$-complete for BH_k [5,6]. We let $BH = \bigcup_{k=1}^\infty BH_k$. Also, for historical convention, we let $D^P = BH_2$, co-$D^P = \text{coBH}_2$, $\text{SAT} \wedge \overline{\text{SAT}} = BL_2$ and $\overline{\text{SAT}} \vee \text{SAT} = \text{coBL}_2$.

The complexity of the Boolean Hierarchy is closely related to the complexity of the bounded query classes which we now define.

Definition 3. Let $q(n)$ be a polynomial-time computable function. We use $P^{\text{SAT}[q(n)]}$ to denote the set of languages recognized by deterministic polynomial-time Turing machines which on inputs of length n ask at most $q(n)$ queries to SAT, the canonical $\leq^P_m$-complete language for NP. When the queries are made in parallel, we use the notation $P^{\text{SAT}[q(n)]}_{\text{tt}}$. We will use P^{SAT} and $P^{\text{SAT}}_{\text{tt}}$ when the machines are allowed polynomial many queries.

The connection between the Boolean Hierarchy and bounded queries to SAT is rich and varied. We ask the reader to consult the literature for a full accounting [1,2,3,4,5,6,8,9,10,12,14,15,16,18,21,22,23]. For this paper, we make use of the following facts about the Boolean Hierarchy and bounded queries to SAT.

$$P^{\text{SAT}[k-1]}_{\text{tt}} \subseteq BH_k \cap \text{coBH}_k \subseteq BH_k \cup \text{coBH}_k \subseteq P^{\text{SAT}[k]}_{\text{tt}} \quad [2,18].$$

$$P^{\text{SAT}[k]} = P^{\text{SAT}[2^k-1]}_{\text{tt}} \quad [2,22,23].$$

$$BH_k = \text{coBH}_k \implies BH = BH_k \quad [5,6].$$

$$BH_k = \text{coBH}_k \implies \overline{\text{SAT}} \in \text{NP/poly} \quad [16].$$

3   Proof of Main Theorem

In this section we will prove the main result, the 1-bit case. The proof for the general case is deferred to the next section. We prove the main result in two parts, one for each direction of the if and only if.

Theorem 1. $\text{coNP} \subseteq \text{NP}/1 \implies \text{PH} \subseteq D^P$.

Proof: We prove this direction in two steps:

$$\text{coNP} \subseteq \text{NP}/1 \implies \Sigma^P_2 \subseteq P^{\text{SAT}} \tag{1}$$

$$\text{coNP} \subseteq \text{NP}/1 \implies P^{\text{SAT}} \subseteq D^P. \tag{2}$$

To prove (1), let U be a $\leq^P_m$-complete language for Σ^P_2 with the usual padding properties and which can be written as:

$$U = \{ \langle x, y \rangle \mid (\exists^P y')(\forall^P z)[y' \leq y \wedge R(x, y', z)] \}$$

for some polynomial-time computable relation R. Since coNP $\in$ NP/1 by assumption, we can construct an NP/1 machine N_U that recognizes U using standard oracle replacement techniques. We only need to note that by padding, all the oracles queries we replace have the same length. Thus, only one bit of advice is needed for the entire computation.

Next, we construct a P^{SAT} machine D_U which recognizes U without any advice bits. On input $\langle x, y \rangle$, D_U looks for y'_{max}, the largest $y' \le y$ such that $(\forall z)[R(x, y', z)]$. D_U finds y'_{max} using binary search and queries to N_U, which can be answered by SAT if D_U had the advice bit for N_U. (The same advice bit can be used for all the queries to N_U during one binary search.) Since D_U does not have the advice bit, it simply tries both 0 and 1. Let y'_0 and y'_1 be the two values produced by the two trials. Then

$$\langle x, y \rangle \in U \iff (\forall^P z)[R(x, y'_0, z)] \vee (\forall^P z)[R(x, y'_1, z)]. \tag{3}$$

D_u can verify the right hand side of (3) with its SAT oracle. Thus, $\Sigma_2^P \subseteq P^{SAT}$ and we have established (1).

To prove that (2) also holds, we show that coNP $\subseteq$ NP/1 implies that LexMaxSat, defined below, is in D^P.

LexMaxSat =
    { φ | the lexically largest satisfying assignment of φ ends with 1 }.

Since LexMaxSat is $\le_m^P$-complete for P^{SAT} [19], we have $P^{SAT} \subseteq D^P$. Note that $\Sigma_2^P \subseteq P^{SAT} \implies PH \subseteq P^{SAT}$. Thus by (1) we have PH $\subseteq D^P$.

Using the assumption that coNP $\subseteq$ NP/1, we can construct an NP/1 machine N_{LMS} that given φ outputs α_{max}, the lexically largest satisfying assignment for φ. When N_{LMS} is given the correct advice bit, all of its computation paths that produce an output (henceforth, the output paths) will output α_{max}. If N_{LMS} has the wrong advice bit, it might output different values on different output paths or it might not have any output paths. We program N_{LMS} to explicitly check that every string it outputs is at least a satisfying assignment of φ. This will be useful below when we need to consider the behavior of N_{LMS} given the wrong advice.

We define two NP languages A_1 and A_2 and claim that LexMaxSat = $A_1 - A_2$. First, we let

$A_1 = \{ \varphi \mid N_{LMS}(\varphi, 0)$ or $N_{LMS}(\varphi, 1)$ outputs a value that ends with 1 $\}$.

Recall that in our notation $N_{LMS}(\varphi, 0)$ and $N_{LMS}(\varphi, 1)$ represents the computations of N_{LMS} given advice bit 0 and 1, respectively.

The language A_2 is defined by an NP machine N_{A_2}. On input φ, N_{A_2} looks for a computation path of $N_{LMS}(\varphi, 0)$ and a computation path of $N_{LMS}(\varphi, 1)$ that output different satisfying assignments for φ. Call these assignments α_1 and α_2 and w.o.l.o.g. assume that $\alpha_1 < \alpha_2$. $N_{A_2}(\varphi)$ accepts if α_1 ends with a 1 and α_2 ends with a 0.

Clearly, A_1 and A_2 are NP languages. To see that LexMaxSat = $A_1 - A_2$, first suppose that $\varphi \in$ LexMaxSat. Since one of $N_{LMS}(\varphi, 0)$ and $N_{LMS}(\varphi, 1)$

has the correct advice bit, one of them must output $\alpha_{\max}$. Since $\alpha_{\max}$ ends with 1, $\varphi \in A_1$. On the other hand, φ cannot be in A_2 by maximality of $\alpha_{\max}$. Thus, $\varphi \in \text{LexMaxSat} \Longrightarrow \varphi \in A_1 - A_2$.

Conversely, suppose that $\varphi \notin \text{LexMaxSat}$. Then, the largest satisfying assignment ends with a 0. So, the computation with the correct advice bit will never output a value ending with a 1. Thus, $\varphi \in A_1$ only in the case that the computation with the wrong advice bit outputs a value $\alpha < \alpha_{\max}$ and α ends with a 1. However, in this case, φ is also in A_2. Thus, $\varphi \notin \text{LexMaxSat} \Longrightarrow \varphi \notin A_1 - A_2$. $\qquad \square$

In the next theorem, we show that $\text{PH} \subseteq \text{D}^P \Longrightarrow \text{coNP} \subseteq \text{NP}/1$ using the hard/easy argument which was used to show that $\text{D}^P = \text{co-D}^P$ implies a collapse of PH [16]. Suppose that $\text{D}^P = \text{co-D}^P$. Then $\text{SAT} \wedge \overline{\text{SAT}} \leq^P_m \overline{\text{SAT}} \vee \text{SAT}$ via some polynomial-time reduction h. Using the reduction h, we define a *hard string*:

Definition 4. Suppose $\text{SAT} \wedge \overline{\text{SAT}} \leq^P_m \overline{\text{SAT}} \vee \text{SAT}$ via some polynomial-time reduction h. Then, a string H is called a *hard string for length* n, if $|H| = n$, $H \in \overline{\text{SAT}}$ and for all x, $|x| = n$, $\langle x, H \rangle \overset{h}{\longmapsto} \langle G_1, G_2 \rangle$ with $G_2 \notin \text{SAT}$. If $F \in \overline{\text{SAT}}$, $|F| = n$ and F is not a hard string for length n, then we say that F is an *easy string*.

Suppose that we were given a hard string H for length n. Then the NP procedure below accepts a formula F of length n if and only if $F \in \overline{\text{SAT}}$.

PROCEDURE Hard(F): Compute $h(F, H) = \langle G_1, G_2 \rangle$. Guess a truth assignment α to the variables in G_1. Accept if α satisfies G_1.

On the other hand, if there are no hard strings for length n — i.e., all formulas in $\overline{\text{SAT}}^{=n}$ are easy — we also have an NP procedure for $\overline{\text{SAT}}^{=n}$.

PROCEDURE Easy(F): Guess a string x with $|x| = |F|$. Compute $h(x, F) = \langle G_1, G_2 \rangle$. Guess a truth assignment α to the variables of G_2. Accept if α satisfies G_2.

The correctness of Procedures Hard and Easy follows directly from the definitions of $\text{SAT} \wedge \overline{\text{SAT}}$, $\overline{\text{SAT}} \vee \text{SAT}$ and hard strings [16]. Since a polynomial advice function can provide an NP machine with a hard string for each length n or with the advice that all strings in $\overline{\text{SAT}}^{=n}$ are easy, $\text{D}^P = \text{co-D}^P \Longrightarrow \text{coNP} \subseteq \text{NP}/\text{poly}$. For this paper, we want to show that $\text{PH} \subseteq \text{D}^P \Longrightarrow \text{coNP} \subseteq \text{NP}/1$ which is both a stronger hypothesis and a strong consequence. Hence, we need to exploit the assumption that $\text{PH} \subseteq \text{D}^P$.

Theorem 2. $\text{PH} \subseteq \text{D}^P \Longrightarrow \text{coNP} \subseteq \text{NP}/1$.

Proof: Suppose that $\text{PH} \subseteq \text{D}^P$. Then, $\text{D}^P = \text{co-D}^P$ via some $\leq^P_m$-reduction h. Now, fix a length n and consider only inputs strings φ of length n. Our goal is to find a hard string for length n or determine that there are no hard strings for length n. Then we can use Procedure Hard or Easy to accept if and only if $\varphi \in \overline{\text{SAT}}$.

Note that the lexically smallest hard string for length n can be found by a $\text{P}^{\text{NP}^{\text{NP}}}$ machine, because the set of hard strings is in coNP. Since $\text{PH} \subseteq \text{D}^\text{P}$, the language HARDBITS defined below is also in D^P.

$$\text{HARDBITS} = \{\langle 1^n, 0 \rangle \mid \text{there are no hard strings for length } n\}$$
$$\cup \{\langle 1^n, i \rangle \mid \text{the } i\text{th bit of the lexically smallest hard string}$$
$$\text{for length } n \text{ is } 1\}.$$

Since $\text{D}^\text{P} \subseteq \text{P}_{\text{tt}}^{\text{SAT}[2]}$, HARDBITS is recognized by some $\text{P}_{\text{tt}}^{\text{SAT}[2]}$ machine M_{HB}. Now, consider the following $n+1$ computations of M_{HB}: $M_{\text{HB}}(1^n, 0)$, $M_{\text{HB}}(1^n, 1)$, $M_{\text{HB}}(1^n, 2)$, ... $M_{\text{HB}}(1^n, n)$. If we are given the accept/reject results of all $n+1$ computations, then we can recover the lexically smallest hard string for length n or conclude that there are no hard strings for length n. Let W be the set of oracle queries made to SAT in these $n + 1$ computations. Without loss of generality we assume that the queries have the same length m. In the remainder of the proof we construct an NP/1 machine that can determine the satisfiability of the formulas in W. The one-bit of advice for our NP/1 computation is 0 if all the strings in $W \cap \overline{\text{SAT}}$ are easy and 1 if W contains at least one hard string for length m. Note that the set W depends only on $|\varphi|$ and not on φ itself, so the one bit of advice is indeed the same for all inputs of length n. Our NP/1 computation is divided into two cases, depending on the advice. Putting the two cases together gives us coNP $\subseteq$ NP/1.

Case 1: all strings in $W \cap \overline{\text{SAT}}$ are easy. We construct an NP machine N_e that accepts if and only if the original input φ of length n is unsatisfiable. N_e first constructs the set W by simulating M_{HB}. Then, for each string w in W, N_e either guesses a satisfying assignment for w or uses Procedure Easy to verify that w is unsatisfiable. The only computation branches of N_e that survive this step are the ones that have correctly guessed the satisfiability of each $w \in W$.

Next, N_e simulates each of the $n + 1$ computations of M_{HB} for HARDBITS. Since N_e has the answers to each oracle query, the simulations can be completed and N_e can reconstruct the lexically smallest hard string for length n or determine that there are no hard strings for length n. Then N_e uses either Procedure Hard or Easy and accepts the original input φ if and only if $\varphi \in \overline{\text{SAT}}$.

Case 2: W contains at least one hard string. Our advantage in this case is that we can look for a hard string for length m just among the $\leq 2n + 2$ strings in W instead of all 2^m strings of length m. We construct an NP machine N_h which nondeterministically places each string $w \in W$ into three sets: W_{SAT}, W_{easy} and W_{hard}. The intention is that W_{SAT} has all the strings in $W \cap \text{SAT}$, W_{easy} has all the easy strings in $W \cap \overline{\text{SAT}}$ and W_{hard} has the remaining strings, the hard strings in $W \cap \overline{\text{SAT}}$. As in the previous case, N_h can verify that the strings in W_{SAT} are satisfiable and that the strings in W_{easy} are easy. However, N_h will not be able to verify that the strings in W_{hard} are unsatisfiable and hard. Fortunately, we only need to know that the strings in W_{hard} are unsatisfiable. That would be enough to simulate the $n+1$ computations of M_{HB} for HARDBITS.

To check that $W_{\text{hard}} \subseteq \overline{\text{SAT}}$, N_h takes each string x in W_{hard}, assumes for the moment that x is indeed hard and uses x to check that every string w in W_{hard} is unsatisfiable. That is, for each pair of strings w, x in W_{hard}, N_h computes $h(w, x) = \langle G_1, G_2 \rangle$ and guesses a satisfying assignment for G_1. If N_h succeeds for every pair (w, x) then we say that W_{hard} has been verified.

Now, suppose that some computation branch of N_h has verified W_{SAT}, W_{easy} and W_{hard}. Since W contains at least one hard string z, N_h must have placed z in W_{hard}. Then z would have been used to test that every $w \in W_{\text{hard}}$ is indeed unsatisfiable. Since z really is a hard string, we are assured that $W_{\text{hard}} \subseteq \overline{\text{SAT}}$. Thus, we can claim that $W_{\text{SAT}} = W \cap \text{SAT}$. Furthermore, some computation branch of N_h guessed W_{SAT}, W_{easy} and W_{hard} to be exactly the satisfiable, easy and hard strings in W. Therefore, at least one computation branch of N_h has verified its W_{SAT}, W_{easy} and W_{hard} and has determined the satisfiability of every string in W.

As in the previous case, since N_h knows the answer to every oracle query in the $n+1$ computations of M_{HB} for HARDBITS, N_h can recover the lexically smallest hard string for length n and use it to nondeterministically recognize the unsatisfiability of the original input φ. $\square$

4   Generalizations

In this section we generalize the main theorem and show that $\text{PH} \subseteq \text{BH}_{2^k} \iff \text{coNP} \subseteq \text{NP}/k$. Recall that BH_{2^k} is the 2^kth level of the Boolean Hierarchy and that $\text{D}^{\text{P}} = \text{BH}_2$, so the preceding theorems are special cases of the ones in this section. As before, we prove each direction separately:

Theorem 3. $\text{coNP} \subseteq \text{NP}/k \implies \text{PH} \subseteq \text{BH}_{2^k}$.

Proof: The first step of the proof of Theorem 1 showed that $\text{coNP} \subseteq \text{NP}/1 \implies \text{PH} \subseteq \text{P}^{\text{SAT}}$. This step generalizes to k bits of advice in a straightforward manner. The P^{SAT} machine D_U for the Σ_2^{P}-complete language U simply has to try all 2^k possible advice strings. D_U on input $\langle x, y \rangle$ obtains 2^k candidates $y_1', \ldots, y_{2^k}'$ for y_{max}'. For each y_i', it checks whether $(\forall^{\text{P}} z)[R(x, y_i', z)]$ using its NP oracle. Then, $\langle x, y \rangle \in U$ if and only if $(\forall^{\text{P}} z)[R(x, y_i', z)]$ for some i, $1 \le i \le 2^k$.

Next, we show that LEXMAXSAT $\in \text{BH}_{2^k}$ which completes the proof, since LEXMAXSAT is $\le_m^{\text{P}}$-complete for P^{SAT}. As in Theorem 1, we use an NP/k machine N_{LMS} which, given the right advice, outputs the largest satisfying assignment α_{max} of its input formula φ. We will consider 2^k computation trees of N_{LMS} on input φ denoted $N_{\text{LMS}}(\varphi, 0^k), \ldots, N_{\text{LMS}}(\varphi, 1^k)$ (one for each k-bit advice string).

Given the correct advice, N_{LMS} will output α_{max} on all of its output paths. Given the wrong advice, N_{LMS} might output one or more incorrect values or have no output paths. Recall that we had previously rigged N_{LMS} so that it only outputs satisfying assignments of φ even when it is given the wrong advice. Our objective is to construct 2^k NP languages $A_1, \ldots A_{2^k}$ such that

$$\varphi \in \text{LexMaxSat} \iff \varphi \in A_1 - (A_2 - (\cdots - A_{2^k}) \cdots).$$

We use the mind-change technique which was used to show that $P^{\text{SAT}[k]} = P^{\text{SAT}[2^k-1]}_{\text{tt}}$ [2,22,23]. We construct 2^k NP machines $N_{A_1}, \ldots, N_{A_{2^k}}$. On input φ, N_{A_i} does the following:

PROCEDURE $A_i(\varphi)$
1. Guess i different advice strings $\sigma_1, \ldots, \sigma_i \in \{0,1\}^k$ in any possible order.
2. For each j, $1 \le j \le i$, guess a computation path of $N_{\text{LMS}}(\varphi, \sigma_j)$ that produces an output. Call this output string α_j.
3. Verify that $\alpha_1 < \alpha_2 < \cdots < \alpha_i$ in lexical ordering.
4. Verify that the last bit of α_1 is a 1 and that for each j, $1 \le j < i$, the last bit of α_j is different from the last bit of α_{j+1}.
5. Accept if Steps 2, 3 and 4 succeed.

When N_{A_i} accepts, we think of $\alpha_1 < \alpha_2 < \cdots \alpha_i$ as a sequence of mind changes. Now suppose that the longest sequence of mind changes is $\alpha_1 < \alpha_2 < \cdots < \alpha_\ell$. Then, the last bit of α_ℓ must be the same as the last bit of $\alpha_{\max}$. This is true if $\ell = 2^k$, since then all advice strings in $\{0,1\}^k$ were used in Step 2, including the correct advice which produces $\alpha_{\max}$. Since every α_j is satisfiable, $\alpha_{\max}$ must equal α_ℓ. If $\ell \ne 2^k$ then appending $\alpha_{\max}$ to the end of the sequence would create a longer sequence of mind changes contradicting the maximality of ℓ. Appending $\alpha_{\max}$ would be possible since N_{LMS} outputs $\alpha_{\max}$ given the correct advice.

Let A_i be the set of formulas φ accepted by N_{A_i} and let $A = A_1 - (A_2 - (\cdots - A_{2^k}) \cdots)$. We claim that $\varphi \in \text{LexMaxSat}$ if and only if $\varphi \in A$. Let ℓ be the largest i such that $\varphi \in A_i$ or 0 if no such i exists. Note that $\varphi \in A$ if and only if ℓ is odd because $A_1 \supseteq A_2 \supseteq \cdots \supseteq A_{2^k}$. Now suppose that $\varphi \in \text{LexMaxSat}$. Then, the last bit of $\alpha_{\max}$ is 1, so $\varphi \in A_1$ and $\ell \ge 1$. Let $\alpha_1 < \alpha_2 < \cdots < \alpha_\ell$ be a mind change sequence found by N_{A_ℓ} on input φ. As we argued above, the last bit of α_ℓ must be the same as the last bit of $\alpha_{\max}$ which is a 1. Since α_1 ends with 1 and the α_j's must alternate between ending with 1 and ending with 0, ℓ must be odd. Thus, $\varphi \in A$.

Conversely, suppose that $\varphi \in A$. Then, ℓ must be odd. Again, looking at the mind change sequence, $\alpha_1 < \alpha_2 < \cdots < \alpha_\ell$, we conclude that α_ℓ must end with 1 and thus $\alpha_{\max}$ must end with 1. Therefore, $\varphi \in \text{LexMaxSat}$. □

In the next proof, we extend the hard/easy argument in the previous section to show that $PH \subseteq BH_{2^k} \implies coNP \subseteq NP/k$. A key element of the proof is the generalization of a hard string to a hard sequence [12].

Definition 5. For $\boldsymbol{y} = \langle y_1, \ldots, y_s \rangle$, let $\boldsymbol{y}^R = \langle y_s, \ldots, y_1 \rangle$ be the reversal of the sequence. Let π_i and $\pi_{i,j}$ be the projection functions such that $\pi_i(\boldsymbol{y}) = y_i$ and $\pi_{i,j}(\boldsymbol{y}) = \langle y_i, \ldots, y_j \rangle$.

Definition 6. Suppose that $BL_r \leq_m^P coBL_r$ via a polynomial-time reduction h. For $0 \leq s \leq r - 1$, let $\ell = r - s$. Then, $\boldsymbol{x} = \langle x_1, \ldots, x_s \rangle$ is a *hard sequence for length n with respect to h*, if the following hold:

1. for each i, $1 \leq i \leq s$, $x_i \in \{0,1\}^n$.
2. for each i, $1 \leq i \leq s$, $x_i \in \overline{\text{SAT}}$.
3. for all $u_1, \ldots, u_\ell \in \{0,1\}^n$, let $\boldsymbol{u} = \langle u_1, \ldots, u_\ell \rangle$ and let $\langle v_1, \ldots, v_s \rangle = \pi_{\ell+1,r}(h(\boldsymbol{u}, \boldsymbol{x}^R))$. Then, $v_i \in \overline{\text{SAT}}$, for all $1 \leq i \leq s$.

We refer to s as the *order* of the hard sequence $\boldsymbol{x}$ and for notational convenience, we define the empty sequence to be a hard sequence of order 0. Furthermore, given a hard sequence $\boldsymbol{x}$, we say that a string w is *easy* with respect to $\boldsymbol{x}$ if $|w| = n$ and there exists $u_1, \ldots, u_{\ell-1} \in \{0,1\}^n$ such that $\pi_\ell(h(u_1, \ldots, u_{\ell-1}, w, \boldsymbol{x}^R)) \in$ SAT. We say that a hard sequence $\boldsymbol{x}$ is a *maximal hard sequence*, if for all $w \in \{0,1\}^n$, $\langle x_1, \ldots, x_s, w \rangle$ is not a hard sequence. A *maximum* hard sequence is a hard sequence with maximum order among all the hard sequences for length n.

As with hard strings, a hard sequence $\boldsymbol{x}$ allows an NP machine to verify that a string w is unsatisfiable when w is easy with respect to $\boldsymbol{x}$. It follows directly from the definitions of BL_r and $coBL_r$, that if $\boldsymbol{x}$ is a hard sequence and $\pi_\ell(h(u_1, \ldots, u_{\ell-1}, w, \boldsymbol{x}^R)) \in$ SAT for any $u_1, \ldots, u_{\ell-1} \in \{0,1\}^n$, then w must be unsatisfiable [12]. Since every string of length n in $\overline{\text{SAT}}$ must be easy with respect to a maximum hard sequence for length n, finding a maximum hard sequence will allow us to recognize $\overline{\text{SAT}}^{=n}$ with an NP machine.

Theorem 4. $PH \subseteq BH_{2^k} \implies coNP \subseteq NP/k$.

Proof: Suppose that $PH \subseteq BH_{2^k}$. Then, $BL_{2^k} \leq_m^P coBL_{2^k}$ via some polynomial-time reduction h. Fix a length n and consider only input strings φ of length n. Our goal is to find a maximum hard sequence $\boldsymbol{x}$ for length n using an NP machine with k bits of advice. Since φ must be easy with with respect to $\boldsymbol{x}$, we get an NP procedure that accepts if and only if $\varphi \in \overline{\text{SAT}}$.

Since the set of hard sequences is in coNP, a $P^{NP^{NP}}$ machine can use binary search to find the lexically smallest maximum hard sequence for length n. Moreover, since $PH \subseteq BH_{2^k}$, the language HARDBITS defined below can be recognized by a $P_{tt}^{SAT[2^k]}$ machine M_{HB}.

$$\text{HARDBITS} = \{\langle 1^n, 0 \rangle \mid \text{the maximum hard sequence for length } n \text{ has order } 0\}$$
$$\cup \{\langle 1^n, i \rangle \mid \text{the } i\text{th bit of the lexically smallest maximum}$$
$$\text{hard sequence for length } n \text{ is } 1\}.$$

Running M_{HB} on $2^k n + 1$ input strings will allow us to recover a maximum hard sequence for length n. As before, we assume that all the queries made by M_{HB} in these computations have a fixed length m. Let W be the set of these length m queries. There are at most $2^{2k} n + 2^k$ strings in W.

Let $\mathrm{HARD}(m, W)$ be the set of hard sequences for length m where every component of the hard sequence is a string from W. Since the number of all possible sequences $< 2^k(2^{2k}n + 2^k)^{2^k}$, $|\mathrm{HARD}(m, W)|$ is polynomially bounded. Furthermore, $\mathrm{HARD}(m, W)$ depends only on $|\varphi|$ and not on φ itself. Thus, we can define a k-bit advice function that provides the maximum order of the hard sequences in $\mathrm{HARD}(m, W)$. Call this value z.

We construct an NP$/k$ machine N for $\overline{\mathrm{SAT}}$ as follows. On input x and given advice z, N first guesses two sets W_{SAT} and H. The set W_{SAT} is a subset of W. If N guesses W_{SAT} correctly, then W_{SAT} would be exactly $W \cap \mathrm{SAT}$. The set H is a set of sequences with $\leq z$ components where each component is a string in W. One correct guess for H is the set $\mathrm{HARD}(m, W)$. There may be other correct guesses for H.

N verifies W_{SAT} and H as follows. For each $w \in W_{\mathrm{SAT}}$, N guesses a satisfying assignment for w. It remains possible that some $w \in W - W_{\mathrm{SAT}}$ is satisfiable. Next, we try to verify that each sequence $\boldsymbol{y} = \langle y_1, \ldots, y_s \rangle \in H$ is a hard sequence. First, each y_i must be an element of $W - W_{\mathrm{SAT}}$, since the components of a hard sequence must be unsatisfiable. Also, for each $\boldsymbol{y} = \langle y_1, \ldots, y_s \rangle \in H$ and each $w \in W - W_{\mathrm{SAT}}$, if $\langle \boldsymbol{y}, w \rangle \notin H$, then w should be easy with respect to $\boldsymbol{y}$. This can be confirmed using the following NP procedure:

PROCEDURE EasyTest($\langle y_1, \ldots, y_s \rangle, w$)
1. Let $\ell = 2^k - s$.
2. Guess a sequence $u_1, \ldots, u_{\ell-1} \in \{0, 1\}^m$.
3. Compute the formula $G = \pi_\ell(h(u_1, \ldots, u_{\ell-1}, w, \boldsymbol{y}^R))$.
4. Guess a satisfying assignment for G.

Clearly, if $W_{\mathrm{SAT}} = W \cap \mathrm{SAT}$ and $H = \mathrm{HARD}(m, W)$, then every verification step will succeed. We claim that if W_{SAT} and H pass every verification step, then $W_{\mathrm{SAT}} = W \cap \mathrm{SAT}$. (Note: we do not claim that H must also equal $\mathrm{HARD}(m, W)$.)

Suppose that H passes every verification step. Let $\boldsymbol{y} = \langle y_1, \ldots, y_s \rangle$ be any hard sequence from $\mathrm{HARD}(m, W)$. We claim that $\boldsymbol{y}$ must be in H. Suppose not. W.o.l.o.g. we can assume that the empty sequence is in H. Thus there exists i, $0 \leq i < s$, such that $\langle y_1, \ldots, y_i \rangle \in H$ but $\langle y_1, \ldots, y_{i+1} \rangle \notin H$. Then, y_{i+1} should be easy with respect to the hard sequence $\langle y_1, \ldots, y_i \rangle$. This will prompt N to run the EasyTest procedure on $\langle y_1, \ldots, y_i \rangle$ and y_{i+1}. However, $\langle y_1, \ldots, y_{i+1} \rangle$ is in reality a hard sequence, so EasyTest($\langle y_1, \ldots, y_i \rangle, y_{i+1}$) will fail. Thus, H would not have passed every verification, which is a contradiction. Therefore, $\mathrm{HARD}(m, W) \subseteq H$.

Next, we claim that $W_{\mathrm{SAT}} = W \cap \mathrm{SAT}$. Fix a string $w \in W - W_{\mathrm{SAT}}$ and let $\boldsymbol{x}$ be a hard sequence in H of order z (the maximum order given by the advice function). We know that such a hard sequence exists since z was given by the advice function and we have just shown that $\mathrm{HARD}(m, W) \subseteq H$. Since $\langle \boldsymbol{x}, w \rangle \notin H$, N must have succeeded in the procedure call EasyTest($\boldsymbol{x}, w$). Then, there exists $u_1, \ldots, u_{\ell-1} \in \{0, 1\}^m$ such that $\pi_\ell(h(u_1, \ldots, u_{\ell-1}, w, \boldsymbol{x}^R)) \in \mathrm{SAT}$, where $\ell = 2^k - z$. By the definitions of BL_{2^k} and coBL_{2^k}, this is enough to imply that $w \in \overline{\mathrm{SAT}}$. Thus, every string $w \in W - W_{\mathrm{SAT}}$ must be unsatisfiable. Since every string in W_{SAT} was already confirmed to be satisfiable, it follows that $W_{\mathrm{SAT}} = W \cap \mathrm{SAT}$.

Finally, some computation path of N will guess the correct W_{SAT} and a correct H which passes every verification step. On such a path, N knows the elements of $W \cap \mathrm{SAT}$. Thus, N can carry out the simulations of M_{HB} and recover the lexically smallest hard sequence for length n. Using this hard sequence, N can then accept the original input φ if and only if $\varphi \in \overline{\mathrm{SAT}}$. Therefore, $\mathrm{coNP} \subseteq \mathrm{NP}/k$. □

5 Discussion

The results in this paper show a tight connection between the number of bits of advice that an NP machine needs to recognize $\overline{\mathrm{SAT}}$ and the collapse of the Polynomial Hierarchy. On a technical level, this connection is borne out by the mind change technique and the hard/easy argument. We need exactly k bits to encode the order of the maximum hard sequence given a $\leq_{\mathrm{m}}^{\mathrm{P}}$-reduction h from BL_{2^k} to coBL_{2^k} and 2^k mind changes is exactly what we need to recognize a Σ_2^{P} language assuming $\mathrm{coNP} \in \mathrm{NP}/k$. In comparison, Chang and Kadin showed that if $\mathrm{BH} \subseteq \mathrm{BH}_{2^k}$ then an $\mathrm{NP}^{\mathrm{NP}}$ machine could recognize a Σ_3^{P}-complete language, if it is given the order of the maximum hard sequence as advice [12, Lemma 4.4]. Since this advice can also be encoded in k bits, this previous result showed that $\mathrm{BH} \subseteq \mathrm{BH}_{2^k} \implies \mathrm{PH} \subseteq \mathrm{NP}^{\mathrm{NP}}/k$.

Our new results are obtained not only by strengthening the hypothesis to $\mathrm{PH} \subseteq \mathrm{BH}_{2^k}$ but also through improvements in the hard/easy argument. The technique used by Chang and Kadin required an existential search for a hard sequence (hence requiring an $\mathrm{NP}^{\mathrm{NP}}$ machine). The current technique involves a search for the hard string or hard sequence in a polynomial sized domain. This technique was first introduced by Hemaspaandra, Hemaspaandra and Hempel [14] and further refined by Buhrman and Fortnow [4] and by Chang [9].

One direction of our results holds true when we consider non-constant advice length. It is fairly easy to extend Theorem 3 to show that $\mathrm{coNP} \subseteq \mathrm{NP}/\log \implies \mathrm{PH} \subseteq \mathrm{P}^{\mathrm{NP}}$. However, the techniques used in Theorem 4 assumes a constant number of queries and cannot be used to show the converse. It remains an open question whether $\mathrm{coNP} \subseteq \mathrm{NP}/\log \iff \mathrm{PH} \subseteq \mathrm{P}^{\mathrm{NP}}$.

References

1. A. Amir, R. Beigel, and W. I. Gasarch. Some connections between bounded query classes and non-uniform complexity. In *Proceedings of the 5th Structure in Complexity Theory Conference*, pages 232–243, 1990.
2. R. Beigel. Bounded queries to SAT and the Boolean hierarchy. *Theoretical Computer Science*, 84(2):199–223, July 1991.

3. R. Beigel, R. Chang, and M. Ogiwara. A relationship between difference hierarchies and relativized polynomial hierarchies. *Mathematical Systems Theory*, 26(3):293–310, July 1993.
4. H. Buhrman and L. Fortnow. Two queries. *Journal of Computer and System Sciences*, 59(2):182–194, 1999.
5. J. Cai, T. Gundermann, J. Hartmanis, L. Hemachandra, V. Sewelson, K. Wagner, and G. Wechsung. The Boolean hierarchy I: Structural properties. *SIAM Journal on Computing*, 17(6):1232–1252, December 1988.
6. J. Cai, T. Gundermann, J. Hartmanis, L. Hemachandra, V. Sewelson, K. Wagner, and G. Wechsung. The Boolean hierarchy II: Applications. *SIAM Journal on Computing*, 18(1):95–111, February 1989.
7. Jin-Yi Cai. $S_2^p \subseteq ZPP^{NP}$. In *Proceedings of the IEEE Symposium on Foundations of Computer Science*, pages 620–629. IEEE Computer Society, October 2001.
8. R. Chang. On the structure of bounded queries to arbitrary NP sets. *SIAM Journal on Computing*, 21(4):743–754, August 1992.
9. R. Chang. Bounded queries, approximations and the Boolean hierarchy. *Information and Computation*, 169(2):129–159, September 2001.
10. R. Chang, W. I. Gasarch, and C. Lund. On bounded queries and approximation. *SIAM Journal on Computing*, 26(1):188–209, February 1997.
11. R. Chang and J. Kadin. On computing Boolean connectives of characteristic functions. *Mathematical Systems Theory*, 28(3):173–198, May/June 1995.
12. R. Chang and J. Kadin. The Boolean hierarchy and the polynomial hierarchy: A closer connection. *SIAM Journal on Computing*, 25(2):340–354, April 1996.
13. Lance Fortnow, Aduri Pavan, and Samik Sengupta. Proving SAT does not have small circuits with an application to the two queries problem. Technical Report 2002-L014N, NEC Laboratories America Technical Note, 2002.
14. E. Hemaspaandra, L. A. Hemaspaandra, and H. Hempel. Downward collapse within the polynomial hierarchy. *SIAM Journal on Computing*, 28(2):383–393, April 1999.
15. A. Hoene and A. Nickelsen. Counting, selecting, sorting by query-bounded machines. In *Proceedings of the 10th Symposium on Theoretical Aspects of Computer Science*, volume 665 of *Lecture Notes in Computer Science*. Springer-Verlag, 1993.
16. J. Kadin. The polynomial time hierarchy collapses if the Boolean hierarchy collapses. *SIAM Journal on Computing*, 17(6):1263–1282, December 1988.
17. R. Karp and R. Lipton. Turing machines that take advice. *L'Enseignement Mathématique*, 28:191–209, 1982.
18. J. Köbler, U. Schöning, and K. Wagner. The difference and truth-table hierarchies for NP. *RAIRO Theoretical Informatics and Applications*, 21:419–435, 1987.
19. M. W. Krentel. The complexity of optimization problems. *Journal of Computer and System Sciences*, 36(3):490–509, 1988.
20. C. Papadimitriou and M. Yannakakis. The complexity of facets (and some facets of complexity). *Journal of Computer and System Sciences*, 28(2):244–259, April 1984.
21. K. Wagner. Bounded query computations. In *Proceedings of the 3rd Structure in Complexity Theory Conference*, pages 260–277, June 1988.
22. K. Wagner. Bounded query classes. *SIAM Journal on Computing*, 19:833–846, 1990.
23. K. Wagner and G. Wechsung. On the Boolean closure of NP. In *Proceedings of the 1985 International Conference on Fundamentals of Computation Theory*, volume 199 of *Lecture Notes in Computer Science*, pages 485–493. Springer-Verlag, 1985.

Strong Reductions and Immunity for Exponential Time

Marcus Schaefer[1] and Frank Stephan[2]*

[1] School of CTI, DePaul University, 243 South Wabash Avenue, Chicago, Illinois 60602, USA, schaefer@cs.depaul.edu.
[2] Mathematisches Institut, Universität Heidelberg, Im Neuenheimer Feld 294, 69120 Heidelberg, Germany, EU, fstephan@math.uni-heidelberg.de.

Abstract. This paper investigates the relation between immunity and hardness in exponential time. The idea that these concepts are related originated in computability theory where it led to Post's program. It has been continued successfully in complexity theory [10,14,20]. We study three notions of immunity for exponential time. An infinite set A is called
- **EXP**-immune, if it does not contain an infinite subset in **EXP**;
- **EXP**-hyperimmune, if for every infinite sparse set $B \in$ **EXP** and every polynomial p there is an $x \in B$ such that $\{y \in B : p^{-1}(|x|) \leq |y| \leq p(|x|)\}$ is disjoint from A;
- **EXP**-avoiding, if the intersection $A \cap B$ is finite for every sparse set $B \in$ **EXP**.

EXP-avoiding sets are always **EXP**-hyperimmune and **EXP**-hyperimmune sets are always **EXP**-immune but not vice versa. We analyze with respect to which polynomial-time reducibilities these sets can be hard for **EXP**. **EXP**-immune sets cannot be conjunctively hard for **EXP** although they can be hard for **EXP** with respect to disjunctive and parity-reducibility. **EXP**-hyperimmunes sets cannot be hard for **EXP** with respect to any of these three reducibilities. There is a relativized world in which there is an **EXP**-avoiding set which is hard with respect to positive truth-table reducibility. Furthermore, in every relativized world there is some **EXP**-avoiding set which is Turing-hard for **EXP**.

Keywords. Computational and structural complexity, hardness for exponential time, polynomial time reducibilities.

1 From Post's Program to Complexity Theory

Concepts of immunity have a long tradition in computability theory beginning with the famous paper of Post [18] which introduced simple sets and showed that they are not hard in the sense that the halting problem cannot be many-one reduced to a simple set or its complement. In fact, no set without an infinite

* This work was written while F. Stephan was visiting DePaul University. He was supported by the Deutsche Forschungsgemeinschaft (DFG) Heisenberg grant Ste 967/1-1 and DePaul University.

H. Alt and M. Habib (Eds.): STACS 2003, LNCS 2607, pp. 559–570, 2003.
© Springer-Verlag Berlin Heidelberg 2003

computable subset can be many-one hard for the halting problem. These sets are called *immune* and the present paper extends the study of resource bounded versions of this notion. Post also considered the more restrictive notions of hyperimmune and hypersimple sets which are not truth-table hard for the halting problem. Post's program to find a Turing-incomplete set defined by abstract properties more restrictive than being simple was completed in the 1970s when Dëgtev and Marchenkov proved that η-maximal semirecursive simple sets exist for a suitable positive equivalence-relation η and that such sets neither have the Turing-degree of the empty set nor of the halting problem, but are intermediate [17, Section III.5]. Furthermore, Harrington and Soare [12] found a condition which is definable in the lattice defined by set-inclusion of the recursively enumerable sets and which enforces that these sets have an intermediate Turing degree.

Within complexity theory, Berman and Hartmanis [9] started the search for structural properties which imply that a set is not hard for the classes **NP**, **PSPACE** or **EXP**. Berman [8], for example, showed that **NEXP**-complete sets are not **EXP**-immune. This result was strengthened by Tran [21] who showed that **NEXP**-complete sets, and their complements, are not **P**-immune. As a matter of fact, 1-tt and conjunctively complete sets for **NEXP** are even **P**-levelable, and therefore not **P**-immune [20]. Hartmanis, Li and Yesha [14] studied whether **NP**-simple sets can be complete for **NP** where a set A is **NP**-simple iff $A \in$ **NP** and A is co-infinite and no infinite set $B \in$ **NP** is disjoint to A. They showed that an **NP**-simple set cannot be many-one hard, unless every problem in **NP** can be decided in subexponential time. Agrawal communicated to the first author, that under the assumption that **P** and **NP** are not equal, no **NP**-simple set A can be complete for **NP** with respect to honest bounded truth-table reducibility. Schaefer and Fenner [20] showed that no **NP**-hyperimmune set is hard for **NP** with respect to honest Turing reducibility if $\mathbf{P} \neq \mathbf{UP}$, that is, there is a set $A \in \mathbf{NP} \setminus \mathbf{P}$ which has unique witnesses, in the sense that we can write $A = \{x : (\exists y \in \{0,1\}^{p(|x|)})[(x,y) \in B]\}$ where p is a polynomial, $B \in \mathbf{P}$ and there is, for every x, at most one $y \in \{0,1\}^{p(|x|)}$ with $(x,y) \in B$. Furthermore, under various stronger assumptions similar results were shown with respect to arbitrary, not necessarily honest polynomial time reducibilities. For example, **NP**-simple sets cannot be bounded truth-table hard for **NP**, unless **SUBEXP** $\subseteq$ **UP** $\cap$ **coUP**.

The class **EXP** is much more well-behaved than **NP**. It is different from **P** in all relativized worlds and therefore contains difficult sets. Furthermore, one can build sets in **EXP** by doing a polynomially length-bounded search within other sets in **EXP**. For example, one can construct for any set in **EXP** a sparse but infinite subset which is still in **EXP** (a set is *sparse*, if it contains at most polynomially many strings at every length). The study of immunity notions for **EXP** often yields results like this which are true for all relativized worlds and do not depend on any unproven assumptions such as the non-collapse of the polynomial hierarchy or some **NP**-complete problem not being computable in subexponential time.

In the present paper, we investigate three immunity-notions for **EXP**. Namely, an infinite set A is

- **EXP**-immune, if it does not contain an infinite subset in **EXP**;
- **EXP**-hyperimmune, if for every infinite sparse set $B \in$ **EXP** and every polynomial p there is an $x \in B$ such that $\{y \in B : p^{-1}(|x|) \leq |y| \leq p(|x|)\}$ is disjoint from A;
- **EXP**-avoiding, if the intersection $A \cap B$ is finite for every sparse set $B \in$ **EXP**.

Every **EXP**-avoiding set is **EXP**-hyperimmune and every **EXP**-hyperimmune set is **EXP**-immune. Note that the condition of B being sparse is necessary in the definition of **EXP**-avoiding sets, since every infinite set has an infinite intersection with a set in **P** namely $\{0,1\}^*$.

In this paper we investigate whether sets immune in one of the senses above can be hard for **EXP** for different types of reducibilities. This continues similar research by Schaefer and Fenner [20] on **NP**-simple, **NEXP**-simple and **NP**-immune sets.

2   Basic Definitions and Theorems

The complexity classes **P**, **Q** and **EXP** denote the sets of languages for which there is a constant c such that their characteristic function is computable with time bound n^c, $n^{\log^c(n)}$ and 2^{n^c}, respectively, where n is the length of the corresponding input (viewed as a binary string; actually n has to be set to 3 if the string has length 0, 1 or 2). **NP** is the class corresponding to **P** which permits non-deterministic computations, that is, $A \in$ **NP** iff there is a function M and a constant c such that, for all $x \in A$, some computation of M with input x halts in time n^c and, for all $x \notin A$, no computation with input x halts, whatever time the computation needs. Note that non-determinism permits M to have different computations on the same input, this is an essential part of the definition of **NP**. A set is *sparse*, if for some polynomial p we have $|A \cap \{0,1\}^n| \leq p(n)$ for all n.

The following definition of an **EXP**-immune set is analogous to those of **P**-immune, **NP**-immune and **PSPACE**-immune sets found in the literature [6].

Definition 2.1. *A set A is **EXP**-immune iff A is infinite and it does not contain an infinite subset in **EXP**.*

We observed earlier that every infinite set in **EXP** has an infinite sparse subset in **EXP**, it follows that a set A is **EXP**-immune iff A does not have an infinite sparse subset in **EXP**. This property is strengthened in the following definition of **EXP**-avoiding sets.

Definition 2.2. *A set A is **EXP**-avoiding iff A is infinite and the intersection of A with any sparse set in **EXP** is finite.*

The next definition is obtained by adapting the notion of **NP**-hyperimmune from Schaefer and Fenner [20] to exponential time. In the definition we use the notation "$f\{x\}$" to indicate that the output of f is not a string but a set of strings which, of course, could be coded as a string again.

Definition 2.3. *Call a (partial) function f an* **EXP**-*array, if f is computable in* **EXP**, *it has infinite domain and there is a polynomial p such that the cardinality of $f\{x\}$ is at most $p(|x|)$ for all x and $p^{-1}(|x|) \le |y| \le p(|x|)$ for all $y \in f\{x\}$, where $p^{-1}(n) = \min\{m : p(m) \ge n\}$. (The final condition assures that f is honest with respect to every element it outputs.)*

A set A is **EXP**-*hyperimmune iff for all* **EXP**-*arrays f there is an x in the domain of f such that $f\{x\}$ and A are disjoint.*

The following theorem shows that the definition of **EXP**-hyperimmunity given in Definition 2.3 is equivalent to the one given in the introduction.

Theorem 2.4. *A set A is* **EXP**-*hyperimmune iff for every infinite* **EXP**-*sparse set B and every polynomial p, there is an $x \in B$ such that $\{y \in B : p^{-1}(|x|) \le |y| \le p(|x|)\}$ is disjoint from A. In particular, every* **EXP**-*avoiding set is* **EXP**-*hyperimmune.*

A reducibility is an algorithm to compute a set A relative to a set B. B is often called an *oracle*. In this paper we only consider polynomial time algorithms for reducibilities. Different types of reducibilities are obtained by restricting the algorithm, and its access to B; we include a partial list.

In the following, M denotes the machine computing the reduction and f computes the set of elements queried by M. Since the function f is denoting the set of strings queried and not a single string, we write $f\{x\}$ instead of $f(x)$ for the output of f on input x.

Definition 2.5. *A is* Turing *reducible to B iff there is a polynomial p such that some Turing machine M computes $A(x)$ in time $p(|x|)$ with queries to B. We write $A(x) = M^B(x)$. Due to time constraints, the cardinality of $f\{x\}$ is at most $p(|x|)$ and every $y \in f\{x\}$ satisfies $|y| \le p(|x|)$. If furthermore all x and $y \in f\{x\}$ satisfy $|y| \ge p^{-1}(|x|)$ then the reduction is called* honest.

A is truth-table reducible (tt-reducible) to B iff M, f can be chosen such that the set $f\{x\}$ of queries does not depend on the oracle. That is, $f\{x\}$ can be computed before any query is evaluated.

The following refinements of tt-reductions depend only on the cardinality of $f\{x\} \cap B$. A is conjunctive reducible (c-reducible) iff there are M, f such that $x \in A \Leftrightarrow f\{x\} \subseteq B$. A is disjunctive reducible (d-reducible) to B iff there are M, f such that $x \in A \Leftrightarrow f\{x\}$ intersects B. A is parity-reducible (or linear reducible) to B iff there are M, f such that $x \in A \Leftrightarrow f\{x\} \cap B$ has odd cardinality. A is many-one reducible (m-reducible) to B iff there is a polynomial-time computable function h such that $A(x) = B(h(x))$.

A is positive reducible (ptt-reducible) to B iff there is a tt-reduction M such that $(\forall x)\,[M^C(x) \le M^D(x)]$ whenever $C \subseteq D$.

Furthermore, a $g(n)$-r-reducibility is a reducibility where on input of length n one can ask at most $g(n)$ questions. For example, a $\log(n)$-tt-reduction requires that the cardinality of the set $f\{x\}$ is always at most $\log(|x|)$.

In the following, reducibilities are also used to define a generalization of the notion of classes using advice. The best-known related concept is the class **P/poly** which is the class of all sets that can be Turing-reduced to polynomial-sized advice.

Definition 2.6. *We call a set E in* **EXP** *compressible via r-reducibility, if there is an r-reduction M and for infinitely many lengths $n > 0$ there is a set $A_n \subseteq \{0,1\}^{<n}$, called the* advice *such that M r-reduces E to A_n on the domain $\{0,1\}^n$. Otherwise E is called* incompressible via r-reducibility.

We say **EXP** *is* compressible via r-reducibility *iff every set E in* **EXP** *is compressible via r-reducibility, and* incompressible via r-reducibility, *otherwise.*

Theorem 2.7. **EXP** *is incompressible via any of the following reducibilities: conjunctive reducibility, disjunctive reducibility, parity-reducibility. This result relativizes.*

Note that **EXP** $\subseteq$ **P/poly** iff there is a tally Turing-complete set for **EXP**. Wilson [23] constructed a relativized world in which **EXP** $\subseteq$ **P/poly**.

Hence, we cannot expect to improve the statement of the theorem to Turing reductions without making further assumptions.

Proof. Recall that the reducibilities above compute on input x some set $f\{x\}$ such that $x \in E$ iff $f\{x\}$ intersects A_n in case of disjunctive reducibility, $f\{x\}$ is a subset of A_n in the case of conjunctive reducibility, and $f\{x\}$ has an odd number of elements in common with A_n in the case of parity-reducibility.

Disjunctive Reducibility. Given a reducibility $f = \varphi_e$ which — without loss of generality — is computable in time 2^e on input of length e one defines a partial function g from $\{0,1\}^e$ to $\{0,1\}^{<e}$ which on input x outputs some string z iff this z (and perhaps some other ones) is in the set $f\{x\}$ but not in any set $f\{y\}$ with $y \in \{0,1\}^e \setminus \{x\}$. Therefore, whenever $g(x)$ and $g(y)$ are defined, then $g(x) \neq g(y)$. As there are 2^e strings of length e but only $2^e - 1$ strings of strictly shorter length, there is a string $x \in \{0,1\}^e$ such that $g(x)$ is undefined. For each e, let x_e be the lexicographically first string in $\{0,1\}^e$ where g is undefined and let $E = \{x_0, x_1, \ldots\}$ be the set of all these x_e.

To see that E is in **EXP** note that g is based on the function $f = \varphi_e$ which is computable in time 2^e. We initialize an array of length 2^e with entry 0 for every $x \in \{0,1\}^e$. For every z we can check in time 4^e whether $z \in \varphi_e\{x\}$ for exactly one $x \in \{0,1\}^e$ and if so, set the corresponding entry to 1. Repeating this for all z allows us to compute x_e as the lexicographic first string in $\{0,1\}^e$ whose entry is still 0 in at most 8^e steps.

For every e, the function $f = \varphi_e$ does not compute $E \cap \{0,1\}^e$ using any advice A_e, because $x_e \in E$ implies that A_e intersects $\varphi_e\{x_e\}$ at some element z and, since $g(x_e)$ is undefined, there is some $y \in \{0,1\}^e \setminus \{x_e\}$ with $z \in \varphi_e\{y\}$. It would follow that $y \in E$ which contradicts x_e being the only element of E of length e.

Conjunctive Reducibility. The proof is analogous to the one for disjunctive reducibility. We can take the complement $\overline{E}$ of the previously constructed set E and use the fact that φ_e reduces E disjunctively to advice A_e iff φ_e reduces $\overline{E}$ conjunctively to $\overline{A_e}$.

Parity-Reducibility. For the case of parity-reducibility we use the fact that every $f = \varphi_e$ defines a linear mapping from the $(2^n - 1)$-dimensional Boolean vector space of the subsets of $\{0,1\}^{<n}$ into the vector space of the subsets of

$\{0,1\}^n$ for each n. As the space of all characteristic functions on $\{0,1\}^n$ is 2^n-dimensional, there is some possible characteristic function on $\{0,1\}^n$ which is not in the linear closure of the images $\varphi_e^{-1}(\{z\})$ where $|z| < n$. Hence, for every length $n = e$, we can determine in exponential time the characteristic function on all strings of length e in such a way that it does not coincide with any possible image $\varphi_e^{-1}(A_e)$ for any $A_e \subseteq \{0,1\}^{<n}$. Thus we get a set $E \in \mathbf{EXP}$ which is incompressible via parity-reducibility. $\square$

3 Immunity and Hardness for EXP

The goal of this section is to investigate for which reducibilities r there are **EXP**-immune, **EXP**-hyperimmune, and **EXP**-avoiding sets which are r-hard for **EXP**.

Theorem 3.1. *No c-hard set for* **EXP** *is* **EXP**-*immune.*

Proof. From Theorem 2.7 it follows that there is a set $\{x_0, x_1, \dots\} \in \mathbf{EXP}$ whose complement E is incompressible via conjunctive reducibility. Assume by way of contradiction that there is a conjunctive reduction f from E to some **EXP**-immune set A and let $F = \{x : f\{x\}$ contains some string z with $|z| \geq |x|\}$. If $E \cap F$ is infinite then the set

$$U = \{z : (\exists x \in E \cap F)\,[|x| \leq |z| \wedge z \in f\{x\}]\}$$

is also infinite and it is a subset of A, since $f\{x\} \subseteq A$ for every $x \in E$. This contradicts A being **EXP**-immune. So $E \cap F$ is finite and we can modify f to obtain the following function g also computable in polynomial time. Without loss of generality let $\lambda \in A$. Now g is defined by taking the first of the following cases which applies.

$$g\{x\} = \begin{cases} f\{x\} & \text{if } f\{x\} \subseteq \{0,1\}^{<|x|}; \\ \{\lambda\} & \text{if } x \in E \cap F; \\ \emptyset & \text{otherwise.} \end{cases}$$

This g would then witness that E can be compressed using a conjunctive reduction, contradicting the choice of E. Thus E cannot be c-reduced to an **EXP**-immune set. $\square$

Buhrman [10] showed that there are d-complete sets in **EXP** which are **P**-immune. This result also holds for higher levels.

Fact 3.2 (Buhrman [10]). *Let* **DEXP** *denote the class of all sets which are computable in double exponential time, that is, in time* $2^{2^{p(n)}}$ *for some polynomial* p. *For a given* **DEXP**-*complete set* A *one can construct an* **EXP**-*immune set* $B \in \mathbf{DEXP}$ *such that* $x \in A \Leftrightarrow x0 \in B \vee x1 \in B$. *$B$ is clearly d-complete for* **DEXP** *and d-hard for* **EXP**.

In contrast to the existence of **EXP**-immune sets which are d-hard for **EXP**, the next result shows that d-hard sets for **EXP** cannot be **EXP**-hyperimmune.

Theorem 3.3. *No d-hard set for* **EXP** *is* **EXP**-*hyperimmune.*

Proof. Let $E = \{x_0, x_1, \ldots\}$ be the language constructed in Theorem 2.7 which is incompressible with regard to disjunctive reductions. Assume that $E \leq_d A$ for an hyperimmune set A via a reduction g, that is, $x \in E$ iff $g\{x\} \cap A \neq \emptyset$. By the padding-lemma there is an infinite polynomial-time computable set U of indices of g. For every $e \in U$ we can compute for every $z \in f_e\{x_e\}$ with $|z| < |x_e|$ one input $y \in \{0, 1\}^e \setminus \{x_e\}$ such that $z \in f_e\{y\}$ (this is possible by the definition of x_e). This gives a polynomial-sized subset F_e of $\{0, 1\}^e$. We define

$$h\{x_e\} = \{z : (\exists y \in F_e \cup \{x_e\})\,[|z| \geq e \wedge z \in f_e\{y\}]\}$$

where $h\{y\}$ is undefined whenever $y \notin E$ or $|y| \notin U$. As U is infinite, h has an infinite domain and there is an e such that $h\{x_e\} \cap A = \emptyset$. Now f_e disjunctively reduces the elements of $F_e \cup \{x_e\}$ to $A^{<e}$, which contradicts the construction in Theorem 2.7. $\square$

Theorem 3.3 leads to the question whether **EXP**-hyperimmune sets can be parity-hard for **EXP**.

Theorem 3.4. *No* **EXP**-*hyperimmune set is hard for* **EXP** *with respect to parity-reducibility.*

The corresponding question for general truth-table reductions is open. Buhrman [11] proved the related result that no **NEXP**-simple set is btt-hard for **EXP** where a **NEXP**-simple set is a co-infinite set $A \in$ **NEXP** such that no infinite set in **NEXP** is disjoint to A. Later, Schaefer [19] showed that no $\alpha \log(n)$-tt-hard set for **EXP** is **EXP**-hyperimmune, where α can be any constant; the proof of this result implied Buhrman's result. It also showed that **NEXP**-simple sets cannot be btt-hard for **EXP**. Buhrman actually showed a stronger result for tt-reductions that comes closer to the general case.

Fact 3.5 (Buhrman [11]). **EXP**-*hyperimmune sets are not* n^α-*tt-hard for* **EXP** *where* $\alpha < \frac{1}{3}$.

This also implies that **EXP**-hyperimmune sets are not $\alpha \log n$-Turing hard for **EXP** where $\alpha < \frac{1}{3}$. This bound is almost optimal as the next result shows that there is even an **EXP**-avoiding set which is $\log^2(n)$-Turing-hard for **EXP**.

Theorem 3.6. *There is an* **EXP**-*avoiding set which is* $\log^2(n)$-*Turing-hard for* **EXP**.

It is unknown what happens in the case of a $\log(n)$-Turing reduction, or a general truth-table reduction. The next result shows that there is at least a relativized world where there is a truth-table hard set for **EXP** which is **EXP**-avoiding. Remember that there is a relativized world in which **EXP** $\subseteq$ **P/poly** [23].

Theorem 3.7. *In the relativized world where* **EXP** $\subseteq$ **P/poly**, *there is a* **EXP**-*avoiding and thus* **EXP**-*hyperimmune set which is hard for* **EXP** *with respect to positive truth-table reducibility.*

4   Immunity and Related Concepts

In this section, we investigate to which extent immunity notions are compatible with other well-known complexity theoretic properties such as randomness, genericity, approximability, and simplicity.

4.1   Randomness and Genericity

Within this section, the notions of **EXP**-Immunity are compared to the notions of resource-bounded randomness and genericity. These notions are effective variants of concepts from measure theory and Baire category. These notions need to be more restrictive than their classical counterparts in order to be meaningful on complexity classes as these classes are countable (except in the case of non-uniform classes like **P/poly**). In a classical sense countable classes are always small; that is, they have measure 0, and are meager. Lutz [16] introduced the following notion of measure 0 and random sets with respect to quasi-polynomial time computations. Quasi-polynomial means time $n^{\log^c(n)}$ for some constant c. This class corresponds to **EXP** if one uses inputs of exponential size (as is done in the case of functionals).

Definition 4.1 (Lutz [16]). *Let a_x be the x-th string with respect to the length-lexicographically ordered list $\lambda, 0, 1, 00, 01, 10, 11, 000, 001, \ldots$ of all strings. That is, $a_0 = \lambda$, $a_1 = 0$, $a_2 = 1$, $a_3 = 00$, and so on. Furthermore, call f a **Q**-functional iff the domain of f is the set of prefixes of characteristic functions, and, for arbitrary sets A, the value $f(A(a_0)A(a_1)\ldots A_{a_x})$ is computed in quasi-polynomial time, that is, in time $x^{\log^c(x)}$ for some constant c.*

*A **Q**-functional f is a **Q**-martingale iff the values of f are (codes for) positive rational numbers and f satisfies,*

$$f(B(a_0)\ldots B(a_x)) \geq \tfrac{1}{2} \cdot (f(B(a_0)\ldots B(a_x)0) + f(B(a_0)\ldots B(a_x)1)),$$

*for every set B, and every x. A **Q**-martingale f succeeds on a set A iff for every rational number r, there is an x such that $f(A(a_0)A(a_1)\ldots A(a_x)) > r$.*

*A class has **Q**-measure 0 iff there is a **Q**-martingale which succeeds on every set in the class. A set A is called **Q**-random iff no **Q**-martingale succeeds on the set A.*

Note that a class may fail to have **Q**-measure 0 although it does not contain **Q**-random sets. The most prominent example for such a class is the class **EXP** itself as on the one hand **EXP** does not have **Q**-measure 0 while on the other hand no set in **EXP** is **Q**-random [4,15,16]. The next proposition shows that **Q**-random sets cannot be **EXP**-hyperimmune. The proof uses a standard technique which can also be applied to show that general generic sets (as defined below) are not random [1].

Proposition 4.2. *The class of all **EXP**-hyperimmune sets has **Q**-measure 0. In particular, no **EXP**-hyperimmune set is **Q**-random.*

Ambos-Spies, Fleischhack and Huwig [2] introduced a notion of genericity which is compatible with the notion of randomness in the sense that every random set is generic but not vice versa. Lutz [15] transferred the original general definition from Computability theory to complexity theory.

Definition 4.3 (Ambos-Spies, Fleischhack and Huwig [2]; Lutz [15]). *A set A is* **Q**-*generic iff for every* **Q**-*functional* f,

- *either* $f(A(a_0)A(a_1)\ldots A(a_x)) \notin \{0,1\}$ *for almost all* x
- *or* $f(A(a_0)A(a_1)\ldots A(a_x)) = A(a_{x+1})$ *for infinitely many* x.

The condition $f(A(a_0)A(a_1)\ldots A(a_x)) \notin \{0,1\}$ *permits* f *not to make a prediction. If* f *makes infinitely many predictions on* A, *then the second case must pertain.*

A set A *is* general **Q**-*generic, if the functional is either almost always undefined, or it infinitely often predicts the next quasi-polynomially many values and one of these predictions is met by* A. *Predicting quasi-polynomially many values means that* f *predicts* $A(a_{x+1})$ *up to* $A(a_{q(x)})$ *where* $q(x) = 2^{\log^c(x)}$ *for some constant* c.

On the one hand, Ambos-Spies [1] showed that no general **Q**-generic set is **Q**-random. On the other hand, every **Q**-random set is still **Q**-generic [3,4] so that these two notions of genericity are different. It follows from the definition that every **Q**-generic set is **EXP**-hyperimmune. Since any **EXP**-avoiding set A contains only finitely many strings from the set $\{0\}^*$, one can easily show that A is not **Q**-generic by considering a functional which predicts every element of $\{0\}^* - A$ to be in A: none of these predictions is correct.

Fact 4.4. *Every general* **Q**-*generic set is* **EXP**-*hyperimmune. Therefore, no general* **Q**-*generic set is hard for* **EXP** *with respect to d-reducibility, c-reducibility or parity-reducibility. Some but not all* **Q**-*generic sets are* **EXP**-*hyperimmune. No* **EXP**-*avoiding set is* **Q**-*generic.*

Theorem 4.5 proves that in the case of truth-table and Turing reducibilities, it depends on the relativized world whether general **Q**-generic sets can be hard for **EXP** or not.

Theorem 4.5. *In every relativized world where* **EXP** $\subseteq$ **P/poly** *there is a positive truth-table hard set for* **EXP** *which is general* **Q**-*generic.*

There is a relativized world in which there is no Turing hard set for **EXP** *which is general* **Q**-*generic.*

4.2 Approximability

We can ask how immunity notions for exponential time relate to the notions of approximability of sets as defined by Beigel, Kummer and Stephan [7] where a set is called *approximable* iff there is a **P**-function f and a constant k such that for every input $x_1, x_2, \ldots, x_k$ the function f computes in polynomial time k bits $y_1, y_2, \ldots, y_k$ such that one of these bits coincides with the characteristic function of A: $y_l = A(x_l)$ for some $l \in \{1, 2, \ldots, k\}$. We consider two special cases of approximability: a set $\{b_0, b_1, \ldots\}$ is **P**-*retraceable* iff there is a **P**-function f with $f(a_x) = a_y$ for some $y \le x$ and all natural numbers x and $f(b_{n+1}) = b_n$ for all n. A **P**-*semirecursive set* is a set B where one can compute from any finite set D of strings in time polynomial in the sum of their lengths an input string a member $x \in D$ such that either $x \in B$ or $D \subseteq \overline{B}$. Note that **P**-retraceable and **P**-semirecursive sets are both approximable with the constant k having the value 2.

Theorem 4.6. *An* **EXP**-*avoiding set can be* **P**-*retraceable but it is never* **P**-*semirecursive.*

Note that the second result also holds with "**EXP**-semirecursive" in place of "**P**-semirecursive".

The second result cannot be strengthened to **EXP**-hyperimmune sets. Dekker constructed a set A which is hypersimple and semirecursive in the computability theoretic sense [17, Theorem II.6.16 and Theorem III.3.13]. An easy modification of the construction makes the set A **P**-semirecursive. The complement $\overline{A}$ of A is then also **P**-semirecursive. Furthermore, $\overline{A}$ is **EXP**-hyperimmune as $\overline{A}$ is already hyperimmune in the sense of computability theory.

4.3   Simplicity

NEXP-simple sets are those which are in **NEXP** and have an **NEXP**-immune complement. Such sets can be defined in a very general way.

Definition 4.7 (Balcázar, Diaz, Gabarró [6]). *Let* **CC** *be a class of sets. A set A is called* **CC**-*immune, if it is infinite and does not contain any infinite subset in* **CC**. *A set A is called* **CC**-*simple, if A is infinite, $A \in$ **CC** and the complement $\overline{A}$ is* **CC**-*immune.*

Simplicity has been studied at many levels, ranging from computability [17,18, 19] to complexity theory [5,13,20,22]. A major open problem is the existence of simple sets under some natural assumptions (maybe involving measure theory). Easy padding arguments give us the following relations between complexity-theoretic variants of simple sets. If there is a **NEXP**-simple set, then there is a **NE**-simple set (namely $\{x : x[1 \ldots |x|^{1/k}] \in L\}$ where $L \in$ **NTIME**(2^{n^k}) is **NEXP**-simple). If there is an **NE**-simple set, then there is an **NP**-simple set in **P**/1 ($\{x : |x| = y, y \in L\}$, where L is **NE**-simple).

For any complexity class **CC** that allows us to run Ladner's delayed diagonalization technique we can show that any **CC**-simple degree bounds a **CC**-simple degree which is Turing-incomplete for **CC**. For example, any **NP**-simple degree bounds a degree Turing-incomplete for **NP**, which contains an **NP**-simple set (the same is true for **NEXP**). To prove this, we use Ladner's construction to construct a set $A \in P$ such that $S \cup A$ is Turing-incomplete for **CC**, and remains **CC**-simple. Of course $S \cup A \leq_T S$. It might be worth to try to exploit these ideas in order to attack the following open problem.

Open Question 4.8. *Is there a relativized world in which* **NEXP**-*simple sets exist but none of them is Turing hard for* **NEXP**.

5   Conclusion

The central topic of the paper is the question, for which reducibilities can **EXP**-immune, **EXP**-hyperimmune and **EXP**-avoiding sets be hard for **EXP**. With respect to many-one and conjunctive reducibilities, Theorem 3.1 shows that no **EXP**-immune set can be hard for **EXP**, which implies the same result for the more restrictive notions of **EXP**-hyperimmune and **EXP**-avoiding sets. With respect to disjunctive reducibility, Buhrman [10] constructed an **EXP**-immune set which is hard for **EXP**, while Theorem 3.3 states that there are no **EXP**-hyperimmune and thus also no **EXP**-avoiding sets which are hard for **EXP**. These results were obtained by using Theorem 2.7 which states that **EXP**-complete sets cannot be compressed conjunctively or disjunctively. Theorem 3.4 shows that there are no **EXP**-hyperimmune and thus also no **EXP**-avoiding sets which are parity-hard for **EXP**.

In the case of positive truth-table reducibility one has, on the one hand, a relativized world with an **EXP**-hyperimmune, even **EXP**-avoiding set which is ptt-hard for **EXP** while, on the other hand, Buhrman's result (Fact 3.5) implies that, for any $\alpha < \frac{1}{3}$ and in any relativized world there is no **EXP**-hyperimmune set which is hard for **EXP** with respect to truth-table reductions asking n^α many queries. Theorem 3.6 states that there is an **EXP**-avoiding Turing-hard set for **EXP** with respect to reductions using $\log^2(n)$ queries while as an immediate consequence of Fact 3.5 we know that no **EXP**-hyperimmune set is Turing-hard for **EXP** with respect to reductions using $\alpha \log(n)$ queries where $\alpha < \frac{1}{3}$. Finally, Theorem 4.5 shows that general **Q**-generic sets can be positive truth-table hard for **EXP** in those relativized worlds where **EXP** $\subseteq$ **P/poly** while there are, by Theorem 4.5, other relativized worlds where general **Q**-generic sets are even not Turing-hard for **EXP**.

Acknowledgments. We would like to thank Stephen Fenner for discussions and Wolfgang Merkle for important comments.

References

1. Klaus Ambos-Spies. Resource-bounded genericity. In S. B. Cooper et al., editor, *Computability, Enumerability, Unsolvability*, volume 224 of *London Mathematical Society Lecture Notes Series*, pages 1–59. Cambridge University Press, 1996.
2. Klaus Ambos-Spies, Hans Fleischhack and Hagen Huwig. Diagonalizations over polynomial time computable sets. *Theoretical Computer Science*, 51:177–204, 1987.
3. Klaus Ambos-Spies, Hans-Christian Neis, and Sebastiaan A. Terwijn. Genericity and measure for exponential time. *Theoretical Computer Science*, 168:3–19, 1996.
4. Klaus Ambos-Spies, Sebastiaan A. Terwijn, and Xizhong Zheng. Resource bounded randomness and weakly complete problems. *Theoretical Computer Science*, 172:195–207, 1997.
5. Balcázar and José L. Simplicity, relativizations and nondeterminism. *SIAM Journal on Computing*, 14(1):148–157, 1985.
6. José L. Balcázar, Josep Diaz, and Joaquim Gabarró. *Structural Complexity, vols. I and II*. Springer, Berlin, 1988.

7. Richard Beigel, Martin Kummer, and Frank Stephan. Approximable sets. *Information and Computation*, 120:304–314, 1995.
8. Leonard Berman. On the structure of complete sets: Almost everywhere complexity and infinitely often speedup. In *17th Annual Symposium on Foundations of Computer Science*, pages 76–80, Houston, Texas, 25–27 October 1976. IEEE.
9. Leonard Berman and Juris Hartmanis. On isomorphisms and density of NP and other complete sets. *SIAM Journal on Computing*, 6(2):305–322, June 1977.
10. Harry Buhrman. *Resource Bounded Reductions*. PhD thesis, University of Amsterdam, 1993.
11. Harry Buhrman. Complete sets are not simple. Unpublished manuscript, 1997.
12. Leo Harrington and Robert I. Soare. Post's program and incomplete recursively enumerable sets. *Proceedings of the National Academy of Science, USA*, 88:10242–10246, 1991.
13. Steven Homer and Wolfgang Maass. Oracle-dependent properties of the lattice of NP-sets. *Theoretical Computer Science*, 24:279–289. 1983.
14. Juris Hartmanis, Ming Li, and Yaacov Yesha. Containment, separation, complete sets, and immunity of complexity classes. In *Automata, Languages and Programming, 13th International Colloquium*, volume 226 of *Lecture Notes in Computer Science*, pages 136–145, Rennes, France, 15–19 July 1986. Springer-Verlag.
15. Jack H. Lutz. Category and measure in complexity classes. *SIAM Journal on Computing*, 19:1100–1131, 1990.
16. Jack H. Lutz. Almost everywhere high non-uniform complexity. *Journal of Computer and System Sciences*, 44:220–258, 1992.
17. Piergiorgio Odifreddi. *Classical recursion theory*. North-Holland, Amsterdam, 1989.
18. Emil Post. Recursively enumerable sets of positive integers and their decision problems. *Bulletin of the American Mathematical Society*, 50:284–316, 1944.
19. Marcus Schaefer. *Completeness and Incompleteness*. PhD thesis, University of Chicago, June 1999.
20. Marcus Schaefer and Stephen Fenner. Simplicity and strong reductions. Available from http://www.cse.sc.edu/~fenner/papers/simplicity.ps, 1998.
21. Nicholas Tran. On P-immunity of nondeterministic complete sets. In *Proceedings of the 10th Annual Conference on Structure in Complexity Theory 1995*, pages 262–263. IEEE Computer Society Press, June 1995.
22. Nikolai K. Vereshchagin. NP-sets are coNP-immune relative to a random oracle. *Third Israel Symposium on Theory of Computing and Systems, ISTCS 1995, Tel Aviv, Israel, January 4–6, 1995, Proceedings.* pages 40–45, IEEE Computer Society, 1995.
23. Christopher B. Wilson. Relativized circuit complexity. *Journal of Computer and System Sciences*, 31(2):169–181, October 1985.

The Complexity of Membership Problems for Circuits over Sets of Natural Numbers

Pierre McKenzie[1] and Klaus W. Wagner[2]

[1] Informatique et recherche opérationnelle, Université de Montréal,
C.P. 6128, Succ. Centre-Ville, Montréal (Québec), H3C 3J7 Canada
mckenzie@iro.umontreal.ca
[2] Theoretische Informatik, Bayerische Julius-Maximilians-Universität Würzburg,
Am Hubland, D-97074 Würzburg, Germany
wagner@informatik.uni-wuerzburg.de

Abstract. The problem of testing membership in the subset of the natural numbers produced at the output gate of a $\{\cup, \cap, ^-, +, \times\}$ combinational circuit is shown to capture a wide range of complexity classes. Although the general problem remains open, the case $\{\cup, \cap, +, \times\}$ is shown NEXPTIME-complete, the cases $\{\cup, \cap, ^-, \times\}$, $\{\cup, \cap, \times\}$, $\{\cup, \cap, +\}$ are shown PSPACE-complete, the case $\{\cup, +\}$ is shown NP-complete, the case $\{\cap, +\}$ is shown $C_=L$-complete, and several other cases are resolved. Interesting auxiliary problems are used, such as testing nonemptyness for union-intersection-concatenation circuits, and expressing each integer, drawn from a set given as input, as powers of relatively prime integers of one's choosing. Our results extend in nontrivial ways past work by Stockmeyer and Meyer (1973), Wagner (1984) and Yang (2000).

Classification: Computational complexity

1 Introduction

Combinational circuits permeate complexity theory. Countless lower bounds, complexity class characterizations, and completeness results involve circuits over the boolean semiring (see [Vo00] among many others). Circuits and formulas over more general structures have been studied as well (see for a few examples [BCGR92,BM95,CMTV98,BMPT97,AJMV98,AAD00]).

In this paper, we study circuits operating on the natural numbers. These include 0 and are simply called the set $\mathbb{N}$ of *numbers* from now on. Next to the boolean semiring, the semiring of numbers is certainly the most fundamental, and two results involving number arithmetic have appeared recently: iterated number multiplication was finally shown to belong to uniform TC^0 [He01,ABH01,CDL] and $(\cup, +, \times)$-circuit evaluation was shown PSPACE-hard [Ya00].

In the boolean setting, the AND and the OR operations combine to capture alternation (in, say, simulations of alternating Turing machines by circuits). In the setting of numbers, perhaps the closest analogs to the AND and the OR become the $\cap$ and the $\cup$. To make sense, this requires an adjustment: since gates

H. Alt and M. Habib (Eds.): STACS 2003, LNCS 2607, pp. 571–582, 2003.

will now compute sets of numbers, a +-gate and a ×-gate having input gates computing $S_1 \subseteq \mathbb{N}$ and $S_2 \subseteq \mathbb{N}$ then compute $\{a + b : a \in S_1, b \in S_2\}$ and $\{a \times b : a \in S_1, b \in S_2\}$ respectively.

Another reason to study such circuits over $\{\cup, \cap, +, \times\}$ is that they obey some form of monotonicity condition: if the set $S \subset \mathbb{N}$ carried by an input gate is replaced by a larger set $S' \supset S$, then the set computed at the output gate of the circuit can only become larger (never smaller). This is reminiscent of monotone boolean functions (a boolean function is monotone if flipping an input from 0 to 1 can never change the output from 1 to 0), for which significant complexity bounds are known. How are the circuit and formula evaluation problems over subsets of $\{\cup, \cap, +, \times\}$ related to monotone boolean function complexity?

Here we study the following problem, which we think of as combining number arithmetic with some form of alternation: given a number b and a $\{\cup, \cap, {}^-, +, \times\}$-circuit C with number inputs, does b belong to the set computed by the output gate of C? We call this problem $\mathrm{MC}(\cup, \cap, {}^-, +, \times)$, and we call $\mathrm{MF}(\cup, \cap, {}^-, +, \times)$ the same problem restricted to formulas. Note that a complement gate ${}^-$ applied to a finite set $S \subset \mathbb{N}$ computes the infinite set $\mathbb{N} \setminus S$. Hence, in the presence of complement gates, the brute force strategy which would exhaustively compute all the sets encountered in the circuit fails. The notation for restricted versions of these problems, for instance $\mathrm{MC}(\cup, +)$, is self-explanatory (see section 2).

Beyond the results stated above, it was known prior to this work that the problem $\mathrm{MF}(\cup, +)$ is NP-complete [SM73], that $\mathrm{MF}(\cup, \cap, {}^-, +)$ and $\mathrm{MF}(\cup, {}^-, +)$ are PSPACE-complete [SM73], and that $\mathrm{MC}(\cup, +)$ is in PSPACE [Wa84].

Adding results from the present paper, some of whose proofs will be deferred to the full version of the paper for lack of space, we obtain Table 1. We highlight here some of the interesting results or techniques:

- The problem $\mathrm{MC}(\cup, \cap, +, \times)$ is NEXPTIME-complete. As an intermediate step, we prove that determining whether a union-intersection-concatenation circuit over a finite alphabet produces a nonempty set is also NEXPTIME-complete.
- The problem $\mathrm{MC}(\cup, \cap, \times)$ is PSPACE-complete.
- The problem $\mathrm{MC}(\cup, \cap, \times)$ reduces in polynomial time to $\mathrm{MC}(\cup, \cap, +)$, which is thus also PSPACE-complete. This reduction is possible because the following problem is solvable in polynomial time: given any set S of numbers excluding 0, compute a set T of pairwise relatively prime numbers and express each $m \in S$ as a product of powers of the numbers in T.
- The problem $\mathrm{MC}(\cup, +)$ is NP-complete. This is a nontrivial improvement over the former PSPACE upper bound.
- The problem $\mathrm{MC}(\cap, +)$ is $\mathrm{C}_{=}\mathrm{L}$-complete, and so is the problem of testing whether two $\mathrm{MC}(+)$-circuits are equivalent.

The hardest problem we consider is $\mathrm{MC}(\cup, \cap, {}^-, +, \times)$. As will be seen, we do not have an upper bound for this problem and it may well be undecidable. On the other hand, Table 1 shows that its various restrictions hit upon a wealth of complexity classes.

2   Definitions and Known Results

A *circuit* $C = (G, E, g_C)$ is a finite directed acyclic graph (G, E) with a specified node g_C, the *output gate*. The gates with indegree 0 are called the *input gates*.

We consider different types of arithmetic circuits. Let $\mathcal{O} \subseteq \{\cup, \cap, ^-, +, \times\}$. An $\mathcal{O}$-*circuit* $C = (G, E, g_C, \alpha)$ is a circuit (G, E, g_C) whose gates have indegree 0, 1, or 2 and are labelled by the function $\alpha : G \mapsto \mathcal{O} \cup \mathbb{N}$ in the following way: Every input gate g has a label $\alpha(g) \in \mathbb{N}$, every gate g with indegree 1 has the label $\alpha(g) = ^-$, and every gate g with indegree 2 has a label $\alpha(g) \in \{\cup, \cap, +, \times\}$. For each of its gates g the arithmetic circuit C computes a set $I(g) \subseteq \mathbb{N}$ inductively defined as follows:

- If g is an input gate with label a then $I(g) =_{\text{def}} \{a\}$.
- If g is +-gate with pred g_1, g_2 then $I(g) =_{\text{def}} \{k+m : k \in I(g_1) \wedge m \in I(g_2)\}$.
- If g is ×-gate with pred g_1, g_2 then $I(g) =_{\text{def}} \{k \cdot m : k \in I(g_1) \wedge m \in I(g_2)\}$.
- If g is a $\cup$-gate with predecessors g_1, g_2 then $I(g) =_{\text{def}} I(g_1) \cup I(g_2)$.
- If g is a $\cap$-gate with predecessors g_1, g_2 then $I(g) =_{\text{def}} I(g_1) \cap I(g_2)$.
- If g is a $^-$-gate with predecessor g_1 then $I(g) =_{\text{def}} \mathbb{N} \setminus I(g_1)$.

The set computed by C is $I(C) =_{\text{def}} I(g_C)$. If $I(g) = \{a\}$ then we also write $I(g) = a$. An $\mathcal{O}$-*formula* is an $\mathcal{O}$-circuit with maximal outdegree 1. For $\mathcal{O} \subseteq \{\cup, \cap, ^-, +, \times\}$ the *membership problems* for $\mathcal{O}$-circuits and $\mathcal{O}$-formulae are

$$\text{MC}(\mathcal{O}) =_{\text{def}} \{(C, b) : C \text{ is an } \mathcal{O}\text{-circuit and } b \in \mathbb{N} \text{ such that } b \in I(C)\}$$

and

$$\text{MF}(\mathcal{O}) =_{\text{def}} \{(F, b) : F \text{ is an } \mathcal{O}\text{-formula and } b \in \mathbb{N} \text{ such that } b \in I(F)\}.$$

For simplicity we write $\text{MC}(o_1, \ldots o_r)$ instead of $\text{MC}(\{o_1, \ldots o_r\})$, and we write $\text{MF}(o_1, \ldots o_r)$ instead of $\text{MF}(\{o_1, \ldots o_r\})$.

Examples. A circuit PRIMES such that $I(\text{PRIMES})$ is the set of prime numbers is obtained by defining the subcircuit GE2 as $\overline{0 \cup 1}$ and defining PRIMES as $\text{GE2} \cap \overline{(\text{GE2} \times \text{GE2})}$. This circuit could easily be turned into a formula. Hence the problem of primality testing easily reduces to $\text{MF}(\cup, \cap, ^-, \times)$. As another example, consider the circuit GOLDBACH defined as $(\text{GE2} \times 2) \cap \overline{(\text{PRIMES} + \text{PRIMES})}$. Then $I(\text{GOLDBACH})$ is empty iff every even number greater than 2 is expressible as a sum of two primes. Hence Goldbach's conjecture holds iff "$0 \in \overline{0 \times \text{GOLDBACH}}$" is a positive instance of $\text{MC}(\cup, \cap, ^-, +, \times)$.

If not otherwise stated, the hardness results in this paper are in terms of many-one logspace reducibility. We assume any circuit and formula encoding in which the gates are sorted topologically and in which immediate predecessors are readily available (say in AC^0). Viewed as graphs, circuits and formulas are not necessarily connected. Numbers are encoded in binary notation.

The following results are known from the literature:

Theorem 1 *1. [SM73] The problem $\text{MF}(\cup, +)$ is NP-complete.*
2. [SM73] The problems $\text{MF}(\cup, \cap, ^-, +)$ and $\text{MF}(\cup, ^-, +)$ are PSPACE-complete.
3. [Wa84] The problem $\text{MC}(\cup, +)$ is in PSPACE.
4. [Ya00] The problem $\text{MC}(\cup, +, \times)$ is PSPACE-complete.

In [Wa84] it is shown that the problem $\text{MC}(\cup, +)$ restricted to $(\cup, +)$-circuits for which every +-gate has at least one input gate as predecessor is NP-complete.

By De Morgan's laws we have

Proposition 2 *For every* $\mathcal{O} \subseteq \{+, \times\}$,

1. $\mathrm{MC}(\{\cup, \cap, ^- \} \cup \mathcal{O}) \equiv_m^P \mathrm{MC}(\{\cup, ^- \} \cup \mathcal{O}) \equiv_m^P \mathrm{MC}(\{\cap, ^- \} \cup \mathcal{O})$.
2. $\mathrm{MF}(\{\cup, \cap, ^- \} \cup \mathcal{O}) \equiv_m^P \mathrm{MF}(\{\cup, ^- \} \cup \mathcal{O}) \equiv_m^P \mathrm{MF}(\{\cap, ^- \} \cup \mathcal{O})$.

Hence we can omit $\mathrm{MC}(\{\cup, ^- \} \cup \mathcal{O})$, $\mathrm{MC}(\{\cap, ^- \} \cup \mathcal{O})$, $\mathrm{MF}(\{\cup, ^- \} \cup \mathcal{O})$, and $\mathrm{MF}(\{\cap, ^- \} \cup \mathcal{O})$ from our exhaustive study.

3   Multiplication versus Addition

In this section we will establish a relationship between the complexity of the membership problems for $(\mathcal{O} \cup \{\times\})$-circuits and $(\mathcal{O} \cup \{+\})$-circuits, for $\mathcal{O} \subseteq \{\cup, \cap, ^- \}$. To this end we need the following problem. Let $\gcd(a, b)$ be the greatest common divisor of the numbers $a, b \geq 1$.

Gcd-Free Basis (GFB)

Given: Numbers $a_1, a_2, \ldots, a_n \geq 1$.
Compute: Numbers $m \geq 1$, $q_1, \ldots, q_m \geq 2$, and $e_{11}, \ldots, e_{nm} \geq 0$ such that
$$\gcd(q_i, q_j) = 1 \text{ for } i \neq j \text{ and } a_i = \prod_{j=1}^{m} q_j^{e_{ij}} \text{ for } i = 1, \ldots, n.$$
Despite the fact that factoring may not be possible in polynomial time, the following is known (see [BS96]):

Proposition 3 *Gcd-Free Basis can be computed in polynomial time.*

As an auxiliary tool we need the generalized membership problems $\mathrm{MC}^*(\mathcal{O})$ and $\mathrm{MF}^*(\mathcal{O})$ for arithmetic circuit and formulae, resp., with addition. These problems deal with elements of $\mathbb{N}^m \cup \{\infty\}$, for an $m \geq 1$ prescribed on input, where the addition on m-tuples is defined componentwise and $a + \infty = \infty + a = \infty + \infty = \infty$ for every $a \in \mathbb{N}^m$. Note that polynomial space many-one reducibility is understood to be performed by polynomial space computable polynomially bounded functions.

Lemma 4 1. *For* $\mathcal{O} \subseteq \{\cup, \cap\}$, *the problem* $\mathrm{MC}(\mathcal{O} \cup \{\times\})$ *is polynomial time many-one reducible to the problem* $\mathrm{MC}^*(\mathcal{O} \cup \{+\})$.
  2. *For* $\mathcal{O} \subseteq \{\cup, \cap\}$, *the problem* $\mathrm{MF}(\mathcal{O} \cup \{\times\})$ *is polynomial time many-one reducible to the problem* $\mathrm{MF}^*(\mathcal{O} \cup \{+\})$.
  3. *For* $\mathcal{O} \subseteq \{\cup, \cap, ^- \}$, *the problem* $\mathrm{MC}(\mathcal{O} \cup \{\times\})$ *is polynomial space many-one reducible to the problem* $\mathrm{MC}^*(\mathcal{O} \cup \{+\})$.
  4. *For* $\mathcal{O} \subseteq \{\cup, \cap, ^- \}$, *the problem* $\mathrm{MF}(\mathcal{O} \cup \{\times\})$ *is polynomial space many-one reducible to the problem* $\mathrm{MF}^*(\mathcal{O} \cup \{+\})$.

Proof. 1. Let $\mathcal{O} \subseteq \{\cup, \cap\}$, let C be a $(\mathcal{O} \cup \{\times\})$-circuit with the input gates $u_1, \ldots, u_s$, and let $b \in \mathbb{N}$. Observe that the absence of $+$ in C entails that any number in $I(C)$ is expressible as a monomial in the inputs. Compute in polynomial time (Lemma 3) numbers $q_1, \ldots, q_m \geq 2$ and $e_{11}, \ldots, e_{sm}, e_1, \ldots, e_m \geq 1$ such that $\gcd(q_i, q_j) = 1$ for $i \neq j$, $\alpha(u_i) = \prod_{j=1}^{m} q_j^{e_{ij}}$ for $i = 1, \ldots, s$ such that $\alpha(u_i) > 0$ and $b = \prod_{j=1}^{m} q_j^{e_j}$ if $b > 0$. Let $M =_{\mathrm{def}} \{ \prod_{j=1}^{m} q_j^{f_j} : f_1, \ldots, f_m \geq 0 \} \subseteq \mathbb{N}$ and let $\sigma : M \cup \{0\} \to \mathbb{N}^m \cup \{\infty\}$

be defined by $\sigma(\prod_{j=1}^{m} q_j^{f_j}) =_{\text{def}} (f_1, \ldots, f_m)$ and $\sigma(0) = \infty$. Obviously, σ is a monoid isomorphism between $(M \cup \{0\}, \times)$ and $(\mathbb{N}^m \cup \{\infty\}, +)$ where we define $\infty + \infty = \infty + a = a + \infty = \infty$ for every $a \in \mathbb{N}^m$. Because $M \cup \{0\}$ is closed under $\times$, the set of numbers computed by any gate in C is included in $M \cup \{0\}$. Furthermore, the following holds for any $\emptyset \subseteq S_1, S_2 \subseteq M \cup \{0\}$:

$$\sigma(S_1 \times S_2) = \sigma(S_1) + \sigma(S_2),$$
$$\sigma(S_1 \cup S_2) = \sigma(S_1) \cup \sigma(S_2),$$
$$\sigma(S_1 \cap S_2) = \sigma(S_1) \cap \sigma(S_2).$$

The reduction therefore consists of converting C into a $(\mathcal{O} \cup \{+\})$-circuit C' which has the same structure as C where a $\times$-gate in C becomes a $+$-gate in C' and an input gate u_i gets label $\sigma(\alpha(u_i))$. An induction using the three identities above shows that for all $a \in M \cup \{0\}$ the following holds: $a \in I(v)$ in $C \Leftrightarrow \sigma(a) \in I(v)$ in C'. This concludes the proof because $b \in M \cup \{0\}$.

2. Same as above because the construction preserves the circuit structure.

3. and 4. If we have complementation then it is no longer true that every number computed within C has a decomposition into $q_1, \ldots, q_m$. To salvage the above construction, the isomorphism σ must therefore be extended to convey information about $\mathbb{N} \setminus M$. A slick way to do this is to begin from the *full* prime decomposition of the numbers $u_1, \ldots, u_s, b$ and to trade the former isomorphism for a homomorphism. Indeed let $q_1, \ldots, q_m$ exhaust the distinct prime divisors of $u_1, \ldots, u_s, b$, and let $q_1, \ldots, q_m, q_{m+1}, q_{m+2}, \ldots$ be the sequence of all primes (in some order). For every number $\prod_{j=1}^{\infty} q_j^{d_j} \in \mathbb{N} \setminus \{0\}$ define $\sigma(\prod_{j=1}^{\infty} q_j^{d_j}) =_{\text{def}} (d_1, \ldots, d_m, \sum_{j>m} d_j)$, and define $\sigma(0) =_{\text{def}} \infty$. Let $M =_{\text{def}} \{ \prod_{j=1}^{m} q_j^{f_j} : f_1, \ldots, f_m \geq 0 \} \subseteq \mathbb{N}$. Because the full prime decomposition was used, σ is a well-defined monoid homomorphism from $(\mathbb{N}, \times)$ onto $(\mathbb{N}^{m+1} \cup \{\infty\}, +)$, where $\sigma(M \cup \{0\}) = (\mathbb{N}^m \otimes \{0\}) \cup \{\infty\}$ (σ is one-one on this part; $\otimes$ denotes direct product) and $\sigma(\mathbb{N} \setminus (M \cup \{0\})) = (\mathbb{N}^m \otimes (\mathbb{N} \setminus \{0\}))$.

The reduction then again consists of converting C into a $(\mathcal{O} \cup \{+\})$-circuit C' having the same structure as C where a $\times$-gate in C becomes a $+$-gate in C' and an input gate u_i gets label $\sigma(\alpha(u_i))$. An induction proves that for any $a \in \mathbb{N}$ and any gate v in C, $a \in I(v)$ in $C \Leftrightarrow \sigma(a) \in I(v)$ in C'. This implies that $b \in I(C) \Leftrightarrow \sigma(b) \in I(C')$, completing the proof. The polynomial space is needed to perform the prime decomposition (if needed, a possibly weaker reducibility, like a many-one polynomial time reduction with an NP oracle, would suffice). $\square$

In some cases the generalized membership problems used above are logspace equivalent to their standard versions:

Lemma 5 *Let* $\{\cup\} \subseteq \mathcal{O} \subseteq \{\cup, \cap\}$.

1. $\text{MC}^*(\mathcal{O} \cup \{+\}) \equiv_m^{\log} \text{MC}(\mathcal{O} \cup \{+\})$.
2. $\text{MF}^*(\mathcal{O} \cup \{+\}) \equiv_m^{\log} \text{MF}(\mathcal{O} \cup \{+\})$.

4   NP-Complete Membership Problems

Lemma 6 *The problem* $\mathrm{MF}(\cup, \cap, +, \times)$ *is in* NP.

Proof sketch. An NP-algorithm can guess a proof that $b \in I(F)$ and can check that the input gates used in the proof carry the required values.   □

The following is a nontrivial improvement over the known PSPACE upper bound for $\mathrm{MC}(\cup, +)$:

Lemma 7 *The problems* $\mathrm{MC}(\cup, +)$ *and* $\mathrm{MC}(\cup, \times)$ *are in* NP.

Proof. Let C be a $\{\cup, +\}$-circuit, and let T_C be the tree which is the result of unfolding C into a tree. A subtree T of T_C is called *computation tree* of C iff

- the output gate of T_C is in T,
- both predecessors of a +-gate of T are in T, and
- exactly one predecessor of a $\cup$-gate of T is in T.

Hence T describes one of the many ways to compute a number from $I(C)$.

A gate g in C corresponds to several copies of it in T_C (and hence also in a computation tree T of C). Let $\beta_{C,T}(g)$ be the number of copies of g in T. In the same way, an edge in C corresponds to several copies of it in T_C (and hence also in a computation tree T of C). Let $\beta_{C,T}(e)$ be the number of copies of e in T.

Defining $s(C,T) =_{\mathrm{def}} \sum\limits_{\text{input gate } g \text{ of } C} \alpha(g) \cdot \beta_{C,T}(g)$

we obtain immediately $I(C) = \{s(C,T) : T \text{ is a computation tree of } C\}$.

A function $\beta : G \cup E \mapsto \mathbb{N}$ is a *valuation function* of the $\{\cup, +\}$-circuit $C = (G, E, \alpha)$ if the following holds:

- $\beta(g_C) = 1$,
- if g is a +-gate with the incoming edges e_1 and e_2 then $\beta(g) = \beta(e_1) = \beta(e_2)$,
- if g is a $\cup$-gate with the incoming edges e_1 and e_2 then $\beta(g) = \beta(e_1) + \beta(e_2)$,
- if g is a gate with the outgoing edges $e_1, \ldots e_k$ then $\beta(g) = \beta(e_1) + \ldots + \beta(e_k)$.

For a valuation function β of C define $s(C, \beta) =_{\mathrm{def}} \sum\limits_{\text{input gate } g \text{ of } C} \alpha(g) \cdot \beta(g)$.

See the full paper for the proof of the following two claims:

Claim 1: If T is a computation tree of C then there exists a valuation function β of C such that $s(C, \beta) = s(C, T)$.

Claim 2: If β is a valuation function of C then there exists a computation tree T of C such that $s(C, T) = s(C, \beta)$.

Then we obtain $I(C) = \{s(C, \beta) : \beta \text{ is a valuation function of } C\}$, and hence $a \in I(C) \Leftrightarrow \exists \beta (\beta \text{ is a valuation function of } C \text{ and } s(C, \beta) = a)$. However, the latter property is in NP. This completes the proof of $\mathrm{MC}(\cup, +) \in$ NP. From this, Lemma 4, and Lemma 5 we obtain $\mathrm{MC}(\cup, \times) \in$ NP.   □

Lemma 8 $\mathrm{MF}(\cup, +)$ *and* $\mathrm{MF}(\cup, \times)$ *are* NP-*hard.*

Proof. The NP-hardness of $\mathrm{MF}(\cup, +)$ is known ([SM73], cf. Theorem 1). We prove in the full paper that 3-SAT $\leq^{\log}_{\mathrm{m}} \mathrm{MF}(\cup, \times)$.   □

As an immediate consequence of the preceding lemmas we obtain:

Theorem 9 *The problems* $\mathrm{MF}(\cup, \cap, +, \times)$, $\mathrm{MF}(\cup, \cap, +)$, $\mathrm{MF}(\cup, \cap, \times)$, $\mathrm{MF}(\cup, +, \times)$, $\mathrm{MF}(\cup, +)$, $\mathrm{MF}(\cup, \times)$, $\mathrm{MC}(\cup, +)$, *and* $\mathrm{MC}(\cup, \times)$ *are* NP-*complete.*

5 PSPACE-Complete Membership Problems

Theorem 10 *1.* [Ya00] *The problem* $\mathrm{MC}(\cup, +, \times)$ *is* PSPACE-*complete.*
2. [SM73] *The problem* $\mathrm{MF}(\cup, \cap, ^-, +)$ *is* PSPACE-*complete.*

Lemma 11 *The problems* $\mathrm{MF}(\cup, \cap, ^-, \times)$, $\mathrm{MC}(\cup, \cap, \times)$, *and* $\mathrm{MC}(\cup, \cap, +)$ *are* PSPACE-*hard, the latter w.r.t. polytime reducibility.*

Proof. A single proof in the full paper shows that the quantified boolean 3-CNF formula problem, which is known to be PSPACE-complete [SM73], is logspace many-one reducible to the problems $\mathrm{MF}(\cup, \cap, ^-, \times)$ and $\mathrm{MC}(\cup, \cap, \times)$. The hardness of $\mathrm{MC}(\cup, \cap, +)$ then follows by Lemma 4 and Lemma 5.   □

Lemma 12 *For every gate g of a generalized* $(\cup, \cap, ^-, +)$-*circuit, if* $I(g) \neq \emptyset$ *then* $I(g) \cap (\{0, 1, \dots, 2^{|C|+1}\}^m \cup \{\infty\}) \neq \emptyset$.

Lemma 13 *The problems* $\mathrm{MC}(\cup, \cap, ^-, +)$ *and* $\mathrm{MC}(\cup, \cap, ^-, \times)$ *are in* PSPACE.

Proof. Note that $\mathrm{MC}(\cup, \cap, ^-, +)$ reduces to $\mathrm{MC}^*(\cup, \cap, ^-, +)$ simply by replacing every negation gate $\bar{g}$ with $\bar{g} \cap \overline{\infty}$. To prove Lemma 13, it then suffices by Lemma 4 to show that $\mathrm{MC}^*(\cup, \cap, ^-, +) \in$ PSPACE.

For a generalized $\{\cup, \cap, ^-, +\}$-circuit C and $b \in \mathbb{N}^m \cup \{\infty\}$, the idea is to use alternating polynomial time to guess an (alternating) proof that $b \in I(C)$. A subtlety arises when a $+$-gate g is encountered and it is guessed that ∞ is fed into g by one of its two inputs. Then we seek a witness to the fact that the other input to g carries a nonempty set. Now how large can such a witness be? Lemma 12 ensures that if a witness exists at all, then a witness exists of polynomial length. The details are given in the full paper.

As a direct consequence of the preceding lemmas we obtain:

Theorem 14 *The five problems* $\mathrm{MC}(\cup, \cap, ^-, +)$, $\mathrm{MC}(\cup, \cap, +)$, $\mathrm{MC}(\cup, \cap, ^-, \times)$, $\mathrm{MC}(\cup, \cap, \times)$, *and* $\mathrm{MF}(\cup, \cap, ^-, \times)$ *are* PSPACE-*complete, the latter w.r.t. polytime reducibility. The problem* $\mathrm{MF}(\cup, \cap, ^-, +, \times)$ *is* PSPACE-*hard.*

6 Beyond PSPACE

As an auxiliary tool we introduce the following $(\cup, \cap, \cdot)$-circuits which compute finite sets of words. Such a circuit C only has gates of indegrees 0 and 2. Every input gate (i.e., gate of indegree 0) g is labelled with a word from a given alphabet Σ^*, and every gate of in degree 2 is labelled with $\cup$, $\cap$, or $\cdot$ (concatenation). For each of its gates g, the circuit computes a set $I(g) \subseteq \Sigma^*$ inductively defined as follows: If g is an input gate with label v then $I(g) =_{\mathrm{def}} \{v\}$. If g is an ω-gate ($\omega \in \{\cup, \cap, \cdot\}$) with predecessors g_l, g_r then $I(g) =_{\mathrm{def}} I(g_l) \, \omega \, I(g_r)$. Finally, $I(C) =_{\mathrm{def}} I(g_C)$ where g_C is the output gate of C. A $(\cup, \cap, \cdot)$-circuit C is called *special* if for every gate g of C there exists a $k \geq 0$ such that $I(g) \subseteq \Sigma^k$. Let $\mathrm{NE}(\cup, \cap, \cdot)$ be the nonemptyness problem for special $(\cup, \cap, \cdot)$-circuits, i.e., $\mathrm{NE}(\cup, \cap, \cdot) =_{\mathrm{def}} \{C : C \text{ is a special } (\cup, \cap, \cdot)\text{-circuit such that } I(C) \neq \emptyset\}$.

Lemma 15 *The problem* $\mathrm{NE}(\cup, \cap, \cdot)$ *is* NEXPTIME-*hard.*

Proof. A delicate generic reduction is given in the full paper. Care is needed to recursively construct sets of words capable of ensuring the match between equal length subwords representing successive machine configurations.   □

Theorem 16 *The problem* $\mathrm{MC}(\cup, \cap, +, \times)$ *is* NEXPTIME-*complete.*

Proof. 1. To prove $\mathrm{MC}(\cup, \cap, +, \times)$ is NEXPTIME-hard we show $\mathrm{NE}(\cup, \cap, \cdot)$ $\leq^{\log}_m \mathrm{MC}(\cup, \cap, +, \times)$. For $w \in \{0,1\}^*$ let $\mathrm{bin}^{-1}(w)$ be that natural number whose binary description is w (possibly with leading zeros), and for $L \subseteq \{0,1\}^*$ let $\mathrm{bin}^{-1}(L) =_{\mathrm{def}} \{\mathrm{bin}^{-1}(w) : w \in L\}$.

Given a special $(\cup, \cap, \cdot)$-circuit C such that $I(C) \subseteq \{0,1\}^k$ (using a block encoding this can be assumed without loss of generality) we construct in logarithmic space a $(\cup, \cap, +, \times)$-circuit C' such that $I(C') = \mathrm{bin}^{-1}(I(C))$. The circuit C' basically has the same structure as C: An input gate in C with label w becomes an input gate in C' with label $\mathrm{bin}^{-1}(w)$, a $\cup$-gate in C becomes a $\cup$-gate in C', and a $\cap$-gate in C becomes a $\cap$-gate in C'. A $\cdot$-gate g in C with predecessor g_1, g_2 such that $I(g_2) \subseteq \{0,1\}^k$ is replaced in C' by a subcircuit which computes $\mathrm{bin}^{-1}(I(g_1) \cdot I(g_2)) = (2^k \times \mathrm{bin}^{-1}(I(g_1))) + \mathrm{bin}^{-1}(I(g_2))$. (Here it is important that C is a *special* $(\cup, \cap, \cdot)$-circuit.)

Now, $I(C) \neq \emptyset \Leftrightarrow I(C') \neq \emptyset \Leftrightarrow 0 \in (\{0\} \times I(C'))$.

2. To see that $\mathrm{MC}(\cup, \cap, +, \times)$ is in NEXPTIME, simply unfold a given $(\cup, \cap, +, \times)$-circuit into a (possibly exponentially larger) $(\cup, \cap, +, \times)$-formula and apply the NP-algorithm from Lemma 6.   □

Corollary 17 *The problem* $\mathrm{NE}(\cup, \cap, \cdot)$ *is* NEXPTIME-*complete.*

As an immediate consequence of Theorem 16 we obtain also:

Theorem 18 *The problem* $\mathrm{MC}(\cup, \cap, ^-, +, \times)$ *is* NEXPTIME-*hard.*

Remark. Since $\mathrm{MC}(\cup, \cap, ^-, +, \times) \equiv^{\log}_m \overline{\mathrm{MC}(\cup, \cap, ^-, +, \times)}$ these problems cannot be in NEXPTIME unless NEXPTIME = co-NEXPTIME. In fact, there is evidence suggesting that $\mathrm{MF}(\cup, \cap, ^-, +, \times)$ might not be decidable. Indeed, Christian Glaßer in Würzburg was the first to observe that there is a simple $\{\cup, \cap, ^-, +, \times\}$-formula G (see the examples given in Section 2) having the property that $(G, 0) \in \mathrm{MF}(\cup, \cap, ^-, +, \times)$ if and only if Goldbach's Conjecture is true. Hence, a decision procedure for $\mathrm{MF}(\cup, \cap, ^-, +, \times)$ would provide a terminating algorithm to test Goldbach's Conjecture; this would be surprising.

7 P-Complete Membership Problems

Theorem 19 *The problem* $\mathrm{MC}(+, \times)$ *is* P-*complete.*

Theorem 20 *The problems* $\mathrm{MC}(\cup, \cap)$ *and* $\mathrm{MC}(\cup, \cap, ^-)$ *are* P-*complete.*

Proof. To prove $\mathrm{MC}(\cup, \cap, ^-) \in$ P, let C be a $(\cup, \cap, ^-)$-circuit and $b \in \mathbb{N}$. Define $S =_{\mathrm{def}} \bigcup_{v \text{ input gate}} I(v)$. We prove that (*) for every gate v there are sets

$P, N \subseteq S$ such that $I(v) = P \cup \overline{N}$. Then we can compute in polynomial time all $I(v)$ from the inputs down to the output by storing only the sets P and N.

To see (*) let $P_1, P_2, N_1, N_2 \subseteq S$ and observe $(P_1 \cup \overline{N_1}) \cup (P_2 \cup \overline{N_2}) = (P_1 \cup P_2) \cup \overline{(N_1 \cap N_2)}$, $(P_1 \cup \overline{N_1}) \cap (P_2 \cup \overline{N_2}) = ((P_1 \cap P_2) \cup (P_1 \setminus N_2) \cup (P_2 \setminus N_1)) \cup \overline{(N_1 \cup N_2)}$, and $\overline{P_1 \cup \overline{N_1}} = N_1 \setminus P_1$.

The hardness proof is by showing that the P-complete *monotone boolean circuit value problem* can be reduced to $MC(\cup, \cap)$. To do so we convert a monotone boolean circuit C into a $(\cup, \cap)$-circuit C' of almost the same structure where every input gate in C with boolean value 1 becomes an input gate in C' with integer value 1, every input gate in C with boolean value 0 becomes an $\cap$-gate in C' with two input gates with labels 0 and 1, resp., as predecessors, every $\vee$-gate in C becomes a $\cup$-gate in C', and every $\wedge$-gate in C becomes a $\cap$-gate in C'. It is easy to see that v evaluates to 0 in C if and only if $I(v) = \emptyset$ in C', and v evaluates to 1 in C if and only if $I(v) = \{1\}$ in C'. Hence, C evaluates to 1 if and only if $1 \in I(C')$. $\square$

8   Circuits with Intersection as the Only Set Operation

Circuits with intersection as the only set operation are special in the sense that every node computes a singleton or the empty set. Thus these circuits bear some relationship to circuits of the same type without intersection. For $\mathcal{O} \subseteq \{+, \times\}$ define $EQ(\mathcal{O}) =_{\text{def}} \{(C_1, C_2) : C_1, C_2 \text{ are } \mathcal{O}\text{-circuits such that } I(C_1) = I(C_2)\}$.

Lemma 21   *1.* $MC(\cap, +) \leq_m^{\log} EQ(+)$
2. $MC(\cap, +, \times) \equiv_m^{\log} EQ(+, \times)$

Proof. Note that an empty set computed at any accessible gate of a $\{\cap, +, \times\}$-circuit propagates to the output. The reductions from left to right consist of progressively bypassing $\cap$-gates, creating for each such gate g a pair (C_g, C_g') of $\cap$-free subcircuits corresponding to the inputs to g, and rigging in the end two $\cap$-free circuits that are equivalent iff C_g is equivalent to C_g' in all the pairs created from accessible gates. See the full paper for details. $\square$

For the following we need the complexity classes #L and $C_=L$. For a nondeterministic logarithmic space machine M, define $n_M(x)$ as the number of accepting paths of M on input x. The class #L precisely consists of these functions n_M. A set A is in $C_=L$ if and only if there exist $f \in \#L$ and a logarithmic space computable function g such that $x \in A \Leftrightarrow f(x) = g(x)$ for every x. For a survey on these and other counting classes see [All97].

Observe that this definition is equivalent to: A set A is in $C_=L$ if and only if there exist $f, g \in \#L$ such that $x \in A \Leftrightarrow f(x) = g(x)$ for every x.

Theorem 22 $MC(\cap, +)$, $MC(+)$, *and* $EQ(+)$ *are* $\leq_m^{\log}$-*complete for* $C_=L$.

Theorem 23   *1. The problem* $MC(\cap, +, \times)$ *is in co-NP.* [1]
2. The problem $MC(\cap, +, \times)$ *is P-hard.*

---

[1] Christian Glaßer, Würzburg, recently proved that $MC(\cap, +, \times)$ is in co-R.

Theorem 24   *1. The problem* $MC(\cap, \times)$ *is in* P.
  2. The problem $MC(\cap, \times)$ *is* $C_=L$-*hard.*

Theorem 25 *The problem* $MC(\times)$ *is* NL-*complete.*

Proof. Hardness is argued in the full paper. Here we describe a many-one reduction from $MC(\times)$ to the iterated multiplication decision problem IMD $=_{def}$ $\{(a_1, \ldots a_r, b) : a_1, \ldots a_r, b \in \mathbb{N} \wedge \prod_{i=1}^{r} a_i = b\}$ via a function which is logspace computable with an oracle from NL. Since IMD $\in$ L ([CDL], or better yet, uniform TC^0 [ABH01,He01]) we obtain $MC(\times) \in L^{NL}$. However, $L^{NL} = NL$.

Let C be a $(\times)$-circuit and $b \in \mathbb{N}$. Let $g_1, \ldots g_r$ be the input gates of C, let g_C be the output gate of C, and let $n(C, g, g')$ be the number of different paths in C from gate g to gate g'. Obviously, $n \in \#L$, and we obtain $I(C) = \prod_{i=1}^{r} I(g_i)^{n(C, g_i, g_C)}$. Defining $s(i) =_{def} \min\{n(C, g_i, g_C), |b|+1\}$ for $i = 1, \ldots, r$ we obtain

$$b \in I(C) \Leftrightarrow b = \prod_{i=1}^{r} I(g_i)^{n(C, g_i, g_C)} \Leftrightarrow b = \prod_{i=1}^{r} I(g_i)^{s(i)}$$

$$\Leftrightarrow (\underbrace{I(g_1), \ldots, I(g_1)}_{s(1)}, \ldots, \underbrace{I(g_r), \ldots, I(g_r)}_{s(r)}, b) \in \text{IMD}$$

The latter tuple of numbers is generated as follows: for $i = 1, \ldots r$ and $k = 1, \ldots, |b|+1$ ask $n(C, g_i, g_C) \geq k$, which are queries to a NL-set [ARZ99]. (This owes to the fact that only small k, i.e. k whose values are polynomially bounded in the length of the input, are considered.) If such a query $n(C, g_i, g_C) \geq k$ is answered in the affirmative, output $I(g_i)$, and finally output b. $\square$

9   Further Results

The following results are argued in the full paper:

Theorem 26 *The problems* $MC(\cup)$ *and* $MC(\cap)$ *are* NL-*complete.*

Theorem 27 *The problems* $MF(\cap, +, \times)$ *and* $MF(+, \times)$ *are in* DLOGCFL.

Theorem 28 *The problems* $MF(\cup, \cap, ^-)$, $MF(\cup, \cap)$, $MF(\cup)$, $MF(\cap)$, $MF(\cap, +)$, $MF(\cap, \times)$, $MF(+)$ *and* $MF(\times)$ *are* L-*complete under* AC^0-*reducibility.*

The L-hardness of the MF problems considered in this section owe to our choice of formula encoding. At such low complexity levels, a more appropriate choice is infix notation; then $MF(\cup, \cap, ^-)$ becomes NC^1-complete by equivalence with the Boolean formula value problem [Bu87], and some of the other restrictions considered in Theorem 28 drop down into yet smaller classes.

The Complexity of Membership Problems for Circuits 581

Table 1. State of the art. The results on MF($\cup$, +), MF($\cup$, $\cap$,$^-$, +) as well as on MC($\cup$, +, ×) were already known from the literature, please refer to the relevant sections for the appropriate credit. Lower bounds of course refer to hardness results.

$\mathcal{O}$	MC($\mathcal{O}$) lower bound	MC($\mathcal{O}$) upper bound	Th.	MF($\mathcal{O}$) low. bound	MF($\mathcal{O}$) upp. bound	Th.
$\cup,\cap,^-,+,\times$	NEXPTIME	?	18	PSPACE	?	14
$\cup,\cap,+,\times$	NEXPTIME	NEXPTIME	16	NP	NP	9
$\cup,+,\times$	PSPACE	PSPACE	10	NP	NP	9
$\cap,+,\times$	P	co-R	23	L	DLOGCFL	27
$+,\times$	P	P	19	L	DLOGCFL	27
$\cup,\cap,^-,+$	PSPACE	PSPACE	14	PSPACE	PSPACE	10
$\cup,\cap,+$	PSPACE	PSPACE	14	NP	NP	9
$\cup,+$	NP	NP	9	NP	NP	9
$\cap,+$	$C_{=}L$	$C_{=}L$	22	L	L	28
$+$	$C_{=}L$	$C_{=}L$	22	L	L	28
$\cup,\cap,^-,\times$	PSPACE	PSPACE	14	PSPACE	PSPACE	14
$\cup,\cap,\times$	PSPACE	PSPACE	14	NP	NP	9
$\cup,\times$	NP	NP	9	NP	NP	9
$\cap,\times$	$C_{=}L$	P	24	L	L	28
$\times$	NL	NL	25	L	L	28
$\cup,\cap,^-$	P	P	20	L	L	28
$\cup,\cap$	P	P	20	L	L	28
$\cup$	NL	NL	26	L	L	28
$\cap$	NL	NL	26	L	L	28

10 Conclusion

Table 1 summarizes the known complexity status of the membership problems for arithmetic circuits over subsets of N. Several open questions are apparent from the table, most notably that of finding an upper bound (if one exists) on the complexity of MC($\cup$, $\cap$,$^-$, +, ×).

We observe that the problems MC(×) and MC(+), complete for NL and for the $C_{=}L$ respectively, offer an interesting new perspective on these two classes. If one could reduce MC(+) to MC(×), then it would follow that NL = $C_{=}L$.

Acknowledgements. We are grateful to Eric Allender helping us with the upper bound in Theorem 25, to Heribert Vollmer and Christian Glaßer for very useful discussions, and to the anonymous referees for valuable suggestions, including the need to correct our choice of formula encoding and to clarify its ramifications.

References

[AAD00]   M. Agrawal, E. Allender, and S. Datta, On TC^0, AC^0, and arithmetic circuits, *J. Computer and System Sciences* 60 (2000), pp. 395–421.

[All97]   E. Allender, Making computation count: Arithmetic circuits in the Nineties, in the Complexity Theory Column, *SIGACT NEWS* 28 (4) (1997) pp. 2–15.

[ARZ99]   E. Allender, K. Reinhardt, S. Zhou, Isolation, matching, and counting: Uniform and nonuniform upper bounds, *J. Computer and System Sciences* 59(1999), pp. 164–181.

[ABH01]   E. Allender, D. Barrington, and W. Hesse, Uniform constant-depth threshold circuits for division and iterated multiplication, *Proceedings 16th Conference on Computational Complexity*, 2001, pp. 150–159.

[AJMV98]   E. Allender, J. Jiao, M. Mahajan and V. Vinay, Non-commutative arithmetic circuits: depth-reduction and depth lower bounds, *Theoretical Computer Science* Vol. 209 (1,2) (1998), pp. 47–86.

[BS96]   E. Bach, J. Shallit, Algorithmic Number Theory, Volume I: Efficient Algorithms, MIT Press 1996.

[BM95]   M. Beaudry and P. McKenzie, Circuits, matrices and nonassociative computation, *J. Computer and System Sciences* 50 (1995), pp. 441–455.

[BMPT97]   M. Beaudry, P. McKenzie, P. Péladeau, D. Thérien, Finite monoids: from word to circuit evaluation, *SIAM J. Computing* 26 (1997), pp. 138–152.

[BCGR92]   S. Buss, S. Cook, A. Gupta, V. Ramachandran, An optimal parallel algorithm for formula evaluation, *SIAM J. Computing* 21 (1992), pp. 755–780.

[Bu87]   S. R. Buss, The boolean formula value problemis in ALOGTIME, Proceedings 19th ACM Symp, on the Theory of Computing, 1987, pp. 123–131.

[CMTV98]   H. Caussinus, P. McKenzie, D. Thérien, H. Vollmer, Nondeterministic NC^1 computation, *J. Computer and System Sciences*, 57 (2), 1998, pp. 200–212.

[CSV84]   A. K. Chandra, L. Stockmeyer, U. Vishkin, Constant depth reducibility, *SIAM Journal on Computing*, 13, 1984, pp. 423–439.

[CDL]   A. Chiu, G. Davida, and B. Litow, NC^1 *division*, available at http://www.cs.jcu.edu.au/ bruce/papers/cr00_3.ps.gz

[Go77]   L. M. Goldschlager, The monotone and planar circuit value problems are logspace complete for P, *SIGACT News*, 9, 1977, pp. 25–29.

[He01]   W. Hesse, Division in uniform TC^0, Proceedings of the 28th International Colloquium on Automata, Languages, and Programming 2001, Lecture Notes in Computer Science 2076, pp. 104–114

[Ga84]   J. von zur Gathen, Parallel algorithms for algebraic problems, *SIAM J. on Computing* 13(4), (1984), pp. 802–824.

[GHR95]   R. Greenlaw, J. Hoover and L. Ruzzo, *Limits to parallel computation, P-completeness theory,* Oxford University Press, 1995, 311 pages.

[Og98]   M. Ogihara, The PL hierarchy collapses, *SIAM Journal of Computing*, 27, 1998, pp. 1430–1437.

[SM73]   L. J. Stockmeyer, A. R. Meyer, Word Problems Requiring Exponential Time, Proceedings 5th ACM Symposium on the Theory of Computing, 1973, pp. 1–9.

[Vo00]   H. Vollmer, Circuit complexity, Springer, 2000.

[Wa84]   K. W. Wagner, The complexity of problems concerning graphs with regularities, Proceedings 11th Mathematical Foundations of Computer Science 1984, Lecture Notes in Computer Science 176, pp. 544–552. Full version as TR N/84/52, Friedrich-Schiller-Universität Jena, 1984.

[Ya00]   K. Yang, Integer circuit evaluation is PSPACE-complete, Proceedings 15th Conference on Computational Complexity, 2000, pp. 204–211.

Performance Ratios for the Differencing Method Applied to the Balanced Number Partitioning Problem

Wil Michiels[1,2], Jan Korst[2], Emile Aarts[1,2], and Jan van Leeuwen[3]

[1] Philips Research Laboratories, Prof. Holstlaan 4, 5656 AA Eindhoven, The Netherlands
michiels@natlab.research.philips.com
[2] Eindhoven University of Technology, P.O. Box 513, 5600 MB Eindhoven, The Netherlands
[3] Utrecht University, P.O. Box 80.089, 3508 TB Utrecht, The Netherlands

Abstract. We consider the problem of partitioning a set of n numbers into m subsets of cardinality $k = \lceil n/m \rceil$ or $\lfloor n/m \rfloor$, such that the maximum subset sum is minimal. We prove that the performance ratios of the Differencing Method of Karmarkar and Karp for $k = 3, 4, 5$, and 6 are precisely 4/3, 19/12, 103/60, and 643/360, respectively, by means of a novel approach in which the ratios are explicitly calculated using mixed integer linear programming. Moreover, we show that for $k \geq 7$ the performance ratio lies between $2 - 2/k$ and $2 - 1/(k-1)$. For the case that m is given instead of k, we prove a performance ratio of precisely $2 - 1/m$. The results settle the problem of determining the worst-case performance of the Differencing Method.

1 Introduction

A classical problem in complexity theory is the number partitioning problem: given a set of n numbers, partition it into m subsets such that the maximum subset sum is minimal. In this paper, we study this problem with the additional constraint that the cardinality of the subsets is balanced, which means that each subset contains either $\lceil n/m \rceil$ or $\lfloor n/m \rfloor$ numbers. For the number partitioning problem and this so-called balanced number partitioning problem, which are both strongly NP-hard [7], the Differencing Method of Karmarkar and Karp [10] outperforms other existing polynomial-time approximation algorithms from an average-case perspective [5,15,19] and it has remained a challenging open problem to come with tight bounds on the worst-case performance ratio of this approach. By means of a novel technique in which the ratios are explicitly calculated using mixed integer linear programming (MILP), we will largely settle this question for the balanced number partitioning problem.

Definition (Balanced Number Partitioning). A problem instance I is given by an integer m and a set $A = \{1, 2, \ldots, n\}$ of n items, where each item $j \in A$ has a nonnegative size a_j with $a_1 \leq a_2 \leq \cdots \leq a_n$. Find a partition $\mathcal{A} = (A_1, A_2, \ldots, A_m)$ of A into m subsets of cardinality $k - 1$ or k with $k = \lceil \frac{n}{m} \rceil$, such that

$$f_I(\mathcal{A}) = \max_{1 \leq i \leq m} S(A_i)$$

is minimal, where $S(A_i) = \sum_{j \in A_i} a_j$. □

H. Alt and M. Habib (Eds.): STACS 2003, LNCS 2607, pp. 583–595, 2003.
© Springer-Verlag Berlin Heidelberg 2003

As applications of the balanced number partitioning problem, Tsai [21] mentions the allocation of component types to pick-and-place machines for printed circuit board assembly [2] and the assignment of tools to machines in flexible manufacturing systems [20].

The Differencing Method works as follows, where for simplicity we assume that $m = 2$. In Section 2, we discuss how the approach generalizes to $m \geq 2$. The approach starts with a sequence of all n numbers. First, it selects two numbers a_i and a_j and commits both numbers to different subsets without deciding yet to which subsets the two numbers are actually assigned. This decision is equivalent to deciding to which subset we assign the absolute difference $|a_i - a_j|$. Therefore, the approach replaces the numbers a_i and a_j in the sequence by $|a_i - a_j|$. This operation, called differencing, leaves a sequence with one number less. The described process is now repeated until a single number remains. By backtracing through the successive differencing operations, we can easily determine the two subsets A_1 and A_2 that partition the n numbers and for which the difference in sum equals the last remaining number in the sequence.

Various strategies have been proposed for selecting the pair of numbers to be differenced. For example, the Paired Differencing Method (PDM) works in successive phases as follows. In each phase, the numbers of the sequence are ordered non-increasingly and the largest two numbers, the third and fourth largest number, etc. are differenced. Note that such a phase halves the sequence length. PDM iterates these phases until the sequence consists of one number. Alternatively, the Largest Differencing Method (LDM) each time selects the largest two numbers for differencing.

Although LDM has a better average-case performance than PDM [23], the latter has the advantage that it is guaranteed to give a balanced partition. To exploit the strengths of both algorithms, Yakir [23] combined them in the Balanced Largest Differencing Method (BLDM). This algorithm performs the first phase of PDM and proceeds with LDM. Whereas the average-case performance of differencing methods is well studied in the literature, this is not the case for the worst-case performance. In this paper, we derive worst-case performance results for BLDM. We say that an algorithm has a *performance bound U*, if it always delivers a solution with a cost at most U times the optimal cost. If bound U is tight, then U is called a *performance ratio*.

Our results. In Section 2 we present BLDM, which we informally discussed above for $m = 2$. Next, in Sections 3 and 4 we analyze the worst-case performance of the algorithm for the case that k is given and m is arbitrary. To prevent case distinctions that distract from the essence of the derivation, we restrict ourselves to $k \geq 4$. Nevertheless, a similar analysis can be given for $k = 3$ resulting in a performance ratio of 4/3. For $k = 2$, the problem is easy as can be seen as follows. Assume that we are given a problem instance with $2m$ items. Otherwise, we constuct such a problem instance by adding one item with size zero. Then an optimal partition is obtained by assigning the $n/2$ smallest items increasingly and the $n/2$ largest items decreasingly to $A_1, A_2, \ldots, A_m$, i.e., $A_i = \{i, n - i + 1\}$ for $1 \leq i \leq m$.

In Section 3 we prove that the performance ratio of BLDM is bounded between $2 - 2/k$ and $2 - 1/(k - 1)$ for any $k \geq 4$ by transforming the algorithm and showing that the set of relevant problem instances can be reduced. This analysis is extended in Section 4, where we formulate the problem of determining a performance ratio for any $k \geq 4$ as an

MILP problem. By using branch-and-bound techniques, we obtain performance ratios for $k = 4, 5$, and 6 of precisely $19/12, 103/60$, and $643/360$, respectively. To our knowledge, this is the first time that an MILP formulation is used to explicitly calculate performance ratios of an algorithm.

Whereas for some applications the performance of BLDM for a given cardinality k of the subsets is particularly interesting, for others the performance as a function of m can be of more interest. In Section 5, we prove that for given $m \geq 2$, BLDM has a performance ratio of precisely $2 - 1/m$. In the full paper, we also derive performance ratios for the case that both k and m are given. We end with some concluding remarks in Section 6.

Related work. Several approaches to the balanced number partitioning problem have been studied in the past. For $k = 3$ and fixed $m \geq 2$, Kellerer and Woeginger [12] prove a worst-case performance ratio of $4/3 - 1/(3m)$ for the well-known Largest Processing Time (LPT) algorithm [9] adapted to the balanced number partitioning problem. An algorithm with a performance ratio of $7/6$ is presented for $k = 3$ by Kellerer and Kotov [11]. For $k \geq 3$, Babel et al. [1] analyze several approximation algorithms, of which a mixture of LPT and Multifit [4] achieves the best performance bound, namely $4/3$. However, all these algorithms are outperformed by differencing methods such as BLDM and PDM with respect to their average-case performance.

We now discuss some related work on the Differencing Method. Consider LDM when applied to number partitioning. Fischetti and Martello [6] prove that the algorithm has a performance ratio of $7/6$ for $m = 2$. Furthermore, we prove in a next paper [18] that for $m \geq 3$ the performance ratio of LDM is bounded between $\frac{4}{3} - \frac{1}{3(m-1)}$ and $\frac{4}{3} - \frac{1}{3m}$. There, we also analyze the performance ratio if in addition to the number of subsets m, the number of items n is fixed as well. The two most popular polynomial-time algorithms for number partitioning are LPT and Multifit. It follows that although LDM has a better average-case performance than LPT and Multifit, Multifit outperforms LDM from a worst-case perspective. However, the worst-case performance of LDM is at least as good as that of LPT [18].

Yakir [23] proves that if $m = 2$ and the item sizes are uniformly distributed on $[0, 1]$, then the expected difference between the sum of the two subsets in a partition generated by either LDM or BLDM is given by $\mathcal{O}(n^{-c \log n})$ for some constant c. This implies that also the expected deviation of the cost of such a partition from the cost of an optimal partition, either balanced or not necessarily balanced, is $\mathcal{O}(n^{-c \log n})$. If the partition is given by PDM, Lueker [14] proves that the expected difference is $\mathcal{O}(1/n)$. In [3,21], a probabilistic analysis is given for two alternative differencing methods for balanced number partitioning with $m = 2$. However, both algorithms have a worse average-case performance than BLDM.

For given $m \geq 2$, Karmarkar and Karp [10] present a rather elaborate differencing method that does, as LDM, not necessarily give a balanced partition. The algorithm uses some randomization in selecting the pair that is to be differenced so as to facilitate its probabilistic analysis. For the algorithm, they prove that the difference between the maximum and minimum sum of any subset is at most $\mathcal{O}(n^{-c \log n})$, almost surely, when the item sizes are in $[0, 1]$ and the density function is reasonably smooth. Tsai [19] proposes a modification of the algorithm that preserves this probabilistic result but enforces that balanced partitions are obtained.

Korf [13] presents a branch-and-bound algorithm, which starts with LDM and then tries to find a better solution until it ultimately finds and proves the optimal solution to the number partitioning problem. By running BLDM instead of LDM and by modifying the search for better solutions, Mertens [15] changes the algorithm into an optimal algorithm for balanced number partitioning. Although both algorithms are practically useful for $m = 2$, they are less interesting for $m > 2$.

2 Balanced Largest-First Differencing Method: The Algorithm

We discuss how BLDM can be generalized to any $m \geq 2$. It is the algorithm that we will analyze in the sequel. We assume that m and k divide n, i.e., $n = mk$. Otherwise, we add $l = m - (n \bmod m)$ items with size zero. Let G_1 be the set containing the m smallest items, G_2 the set containing the m smallest remaining items, and so on. Hence, $G_i = \{(i-1)m+r \mid 1 \leq r \leq m\}$ for $1 \leq i \leq k$. Initially, BLDM starts with a sequence L of k partial solutions $\mathcal{A}_1, \mathcal{A}_2, \ldots, \mathcal{A}_k$, where $\mathcal{A}_i = \{A_{i1}, A_{i2}, \ldots, A_{im}\}$ is obtained by assigning each item of G_i to a different subset. More precisely, $A_{ij} = \{(i-1)m+j\}$. This is a generalization of the first phase of PDM. Next, the algorithm selects the two partial solutions from L for which $d(\mathcal{A})$ is maximal, where $d(\mathcal{A})$ is defined as the difference between the maximum and minimum subset sum in $\mathcal{A}$. These two solutions, denoted by $\mathcal{A}'$ and $\mathcal{A}''$, are combined into a new partial solution $\mathcal{A}$ by joining the subset with the smallest sum in $\mathcal{A}'$ with the subset with the largest sum in $\mathcal{A}''$, the subset with the second smallest sum in $\mathcal{A}'$ with the subset with the second largest sum in $\mathcal{A}''$, and so on. Hence, $\mathcal{A}$ is formed by the m subsets $A'_j \cup A''_{m-j+1}$ with $1 \leq j \leq m$. This replaces $\mathcal{A}'$ and $\mathcal{A}''$ in L. We iterate this differencing operation until only one solution in L remains, which is the balanced solution called $\mathcal{A}^{\text{BLDM}}$. Note that if m and k do not divide n in the original problem instance, then $\mathcal{A}^{\text{BLDM}}$ has at least l subsets containing an item with size zero. Removing these items yields a balanced partition for I with the same cost as $\mathcal{A}^{\text{BLDM}}$.

In this paper, we will define sets of items by giving the sequence of the sizes instead of the items themselves. For example, instead of $A = \{1, 2, 3\}$ with $a_i = 5+i$ we write $A = 6, 7, 8$. Furthermore, $A_1 - A_2 - \ldots - A_m$ denotes the partition $\{A_1, A_2, \ldots, A_m\}$ and A_i^l is a short-hand notation for the partition $A_j - A_j - \ldots - A_j$ (l times). We illustrate the algorithm by means of an example. Let $m = 3$, $k = 4$, and the twelve item sizes be 1,2,4,5,8,10,11,11,16,17,17,21. Initially L consists of the partial solutions $\mathcal{A}_1 = 1 - 2 - 4$, $\mathcal{A}_2 = 5 - 8 - 10$, $\mathcal{A}_3 = 11 - 11 - 16$, and $\mathcal{A}_4 = 17 - 17 - 21$. As $d(\mathcal{A}_1) = 3$, $d(\mathcal{A}_2) = 5$, $d(\mathcal{A}_3) = 5$, and $d(\mathcal{A}_4) = 4$, the first iteration of BLDM replaces $\mathcal{A}_2$ and $\mathcal{A}_3$ in L by $\mathcal{A}_5 = 5, 16 - 8, 11 - 10, 11$ with $d(\mathcal{A}_5) = 2$. In the next two iterations $\mathcal{A}_1$ and $\mathcal{A}_4$ are replaced by $\mathcal{A}_6 = 1, 21 - 2, 17 - 4, 17$ and $\mathcal{A}_5$ and $\mathcal{A}_6$ by $\mathcal{A}^{\text{BLDM}} = 2, 5, 16, 17 - 1, 8, 11, 21 - 4, 10, 11, 17$, which is the final balanced solution. This solution is not optimal as in the solution the three subsets have a sum of 40, 41, and 42, respectively, whereas in solution $1, 8, 11, 21 - 2, 5, 17, 17 - 4, 10, 11, 16$ the sums of the three subsets are all 41.

3 Performance Ratios: Basic Analysis

Consider a problem instance I in which m and k do not divide n and let I' be obtained from I by adding $m - (n \bmod m)$ items with size zero. As indicated in Section 2, the cost of the partition given by BLDM for I equals the cost of the partition it constructs I'. Furthermore, the cost of an optimal partition for I' is at most the cost of an optimal partition for I. Hence, the performance ratio of BLDM for I' is at least the performance ratio of BLDM for I. Hence, as I' contains km items, we can assume without loss of generality that $n = mk$ in our analysis of the worst-case performance of BLDM as a function of k and/or m.

In this section, we prove that the performance ratio R of BLDM satisfies $2 - 2/k \leq R \leq 2 - 1/(k-1)$ for $k \geq 4$. Since the value of k will be clear from the context, k is not explicitly given as index of R. Our main challenge will be to derive a sharp upper bound U. A lower bound L can simply be proved by giving a problem instance with a performance ratio of L. Besides proving lower and upper bounds on R for any $k \geq 4$, this section also serves as a basis for Section 4.

Let $k \geq 4$. By definition, U is a performance bound of BLDM if and only if for each instance I, $f_I(\mathcal{A}^{\text{BLDM}}) \leq U \cdot f_I^*$, where f_I^* denotes the cost of an optimal partition. For proving $f_I(\mathcal{A}^{\text{BLDM}}) \leq U \cdot f_I^*$, we may substitute $f_I(\mathcal{A}^{\text{BLDM}})$ by a larger expression and f_I^* by a smaller expression. The first two steps of the analysis given in this section concentrate on the nontrivial task of finding such expressions, where we successively change the algorithm and the problem instance. These steps are summarized in Figure 1. We use this in the third step to prove lower and upper bounds on R.

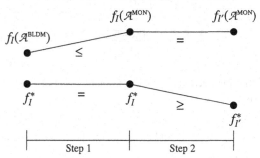

Fig. 1. Outline of the first two steps in Section 3.

In our analysis, we can disregard problem instances with a performance ratio at most $3/2$ as for each $k \geq 4$ an instance with a ratio of $19/12$ exists. For $k = 4$ this is the case as partition $12, 4, 3, 0 - (4, 4, 3, 0)^7$ with objective value 19 can be obtained by BLDM, whereas an optimal partition of these numbers is $12, 0, 0, 0 - (4, 4, 4, 0)^5 - (3, 3, 3, 3)^2$, which has objective value 12. Furthermore, when $k = 4 + i$ for $i \geq 1$ an instance with performance ratio of $19/12$ can be obtained from this instance by adding im items with size zero. Then in both the partition given by BLDM and an optimal partition, i items with size zero are added to each subset resulting in partitions with the same objective value.

Step 1: construction of a partition $\mathcal{A}^{\text{MON}}$ with $f_I(\mathcal{A}^{\text{BLDM}}) \leq f_I(\mathcal{A}^{\text{MON}})$. Partition $\mathcal{A}^{\text{BLDM}}$ has the property that each subset contains exactly one item from each set G_i. The same property holds for the so-called monotone partition $\mathcal{A}^{\text{MON}} = (A_1^{\text{MON}}, A_2^{\text{MON}}, \ldots, A_m^{\text{MON}})$, where subset A_j^{MON} contains the jth largest item from G_k and the jth smallest items from $G_1, G_2, \ldots, G_{k-1}$, i.e., the subset contains item $n - j + 1$ from G_k and item $(i-1)m + j$

from G_i with $1 \leq i < k$. The next lemma, which is proved in the full paper, implies that indeed $f_I(\mathcal{A}^{\text{BLDM}}) \leq f_I(\mathcal{A}^{\text{MON}})$.

Lemma 1. *If $f_I(\mathcal{A}^{\text{BLDM}})/f_I^* > 3/2$ for a given instance I, then $f_I(\mathcal{A}^{\text{BLDM}}) \leq f_I(\mathcal{A}^{\text{MON}})$.*

Step 2: construction of a new problem instance I' with $f_I(\mathcal{A}^{\text{MON}}) = f_{I'}(\mathcal{A}^{\text{MON}})$ and $f_I^* \geq f_{I'}^*$. We construct instance I' from I by changing the item sizes. Let A_j^{MON} be the subset of $\mathcal{A}^{\text{MON}}$ with largest sum. Furthermore, let x_i with $1 \leq i \leq k$ be defined as the size of the only item that is both in A_j^{MON} and G_i. Note that $f_I(\mathcal{A}^{\text{MON}}) = \sum_{i=1}^k x_i$ and $x_1 \leq x_2 \leq \cdots \leq x_k$.

Instance I' is now constructed out of I by decreasing the size of each item i that is not in A_j^{MON} until it is either equal to the size of the largest item i' in A_j^{MON} with $i' < i$ or equal to zero; see Figure 2. More precisely, subset A_l^{MON} with $l \neq j$ is set to

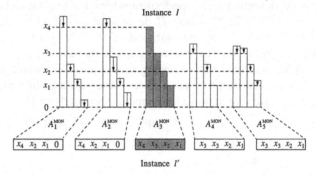

Fig. 2. Example of the construction of problem instance I' out of I, where A_j^{MON} is the subset in $\mathcal{A}^{\text{MON}}$ with largest sum.

$x_k, x_{k-2}, x_{k-3}, \ldots, x_1, 0$ if $l < j$ and to $x_{k-1}, x_{k-1}, x_{k-2}, \ldots, x_2, x_1$ otherwise. Hence, $\mathcal{A}^{\text{MON}}$ is given by

$$(x_k, x_{k-2}, x_{k-3}, \ldots, x_1, 0)^{j-1} - x_k, x_{k-1}, x_{k-2}, \ldots, x_2, x_1 - (x_{k-1}, x_{k-1}, x_{k-2}, \ldots, x_2, x_1)^{m-j}.$$

As we only decrease sizes and do not affect the sizes in A_j^{MON}, both $f_I(\mathcal{A}^{\text{MON}}) = f_{I'}(\mathcal{A}^{\text{MON}})$ and $f_I^* \geq f_{I'}^*$.

Step 3: bounding the performance ratio. If we combine the results from Steps 1 and 2, we get $f_I(\mathcal{A}^{\text{BLDM}}) \leq f_{I'}(\mathcal{A}^{\text{MON}})$ and $f_I^* \geq f_{I'}^*$; see Figure 1. Hence, if we prove for some U that $f_{I'}(\mathcal{A}^{\text{MON}}) \leq U \cdot f_{I'}^*$, then it follows that $f_I(\mathcal{A}^{\text{BLDM}}) \leq U \cdot f_I^*$, i.e., U is a performance bound for instance I. Consequently, to prove a performance bound U for BLDM, it suffices to prove that for each $x_k \geq x_{k-1} \geq \cdots \geq x_1 \geq 0$, $m_l \geq 0$, and $m_r \geq 0$, we have $f_I(\mathcal{A}^{\text{MON}}) \leq U \cdot f_I^*$, where I is defined such that $\mathcal{A}^{\text{MON}}$ is given by

$$(x_k, x_{k-2}, x_{k-3}, \ldots, x_1, 0)^{m_l} - x_k, x_{k-1}, x_{k-2}, \ldots, x_2, x_1 - (x_{k-1}, x_{k-1}, x_{k-2}, \ldots, x_2, x_1)^{m_r}.$$

$$(1)$$

Using this observation, we can prove the following theorem.

Theorem 1. *For given $k \geq 4$, the performance ratio R of* BLDM *satisfies*

$$2 - \frac{2}{k} \leq R \leq 2 - \frac{1}{k-1}.$$

Proof. First, we show that $2 - \frac{1}{k-1}$ is a performance bound for $k \geq 4$ by proving that

$$f_I(\mathcal{A}^{\text{MON}}) \leq \left(1 + \frac{k-2}{k-1}\right) f_I^*, \tag{2}$$

where I is defined such that $\mathcal{A}^{\text{MON}}$ is given by (1) for some $x_k \geq x_{k-1} \geq \cdots \geq x_1 \geq 0$, $m_l \geq 0$, and $m_r \geq 0$. Let ω_1, ω_2, and ω_3 be defined as the sum of the three types of subsets in $\mathcal{A}^{\text{MON}}$, i.e.,

$$\begin{aligned} \omega_1 &= \omega_2 - x_{k-1}, \\ \omega_2 &= \sum_{i=1}^{k} x_i, \text{ and} \\ \omega_3 &= \omega_2 - x_k + x_{k-1}. \end{aligned}$$

Since ω_1, ω_2, and ω_3 can not all three be strictly larger than f_I^* and since ω_2 is at least as large as both ω_1 and ω_3, we have $\omega_1 \leq f_I^*$ or $\omega_3 \leq f_I^*$. We consider these two cases separately.

Case 1: $\omega_1 \leq f_I^*$.
By definition, $f_I(\mathcal{A}^{\text{MON}}) = \omega_2 = \omega_1 + x_{k-1}$. Hence, (2) follows when $x_{k-1} \leq \frac{k-2}{k-1} f_I^*$. Since I contains $m_l + 2m_r + 2$ items with a size at least x_{k-1} and since these items have to be assigned to $m = m_l + m_r + 1$ subsets, an optimal partition has at least one subset containing two items with size at least x_{k-1}. Consequently, $x_{k-1} \leq \frac{1}{2} f_I^*$, which yields that $x_{k-1} \leq \frac{k-2}{k-1} f_I^*$ since $\frac{1}{2} \leq \frac{k-2}{k-1}$ for $k \geq 4$.

Case 2: $\omega_3 \leq f_I^*$.
If $\omega_3 = 0$, then $f_I(\mathcal{A}^{\text{MON}}) = x_k = f_I^*$, which clearly implies (2). Next, assume that $\omega_3 > 0$, which implies that $x_{k-1} > 0$. Since the number of items with size x_k is larger than the number of items with size 0, $x_k + x_1 \leq f_I^*$. Hence, for proving (2), it suffices to show that $x_{k-1} + x_{k-2} + \cdots + x_2 \leq \frac{k-2}{k-1} f_I^*$. By the definition of ω_3 and since $\omega_3 \leq f_I^*$, this is implied by

$$\frac{x_{k-1} + x_{k-2} + \cdots + x_2}{x_{k-1} + x_{k-1} + x_{k-2} + \cdots + x_1} \leq \frac{k-2}{k-1}.$$

It can be verified that the left-hand side is maximal whenever $x_1 = 0$ and $x_i = x_{k-1}$ for $2 \leq i < k$. Hence, the left-hand side is at most $\frac{(k-2)x_{k-1}}{(k-1)x_{k-1}} = \frac{k-2}{k-1}$. This proves the performance bound $2 - \frac{1}{k-1}$ for BLDM. The lower bound $2 - \frac{2}{k}$ follows as BLDM can generate partition $k, 1, 1, \ldots, 1, 0$ - $(1, 1, 1, \ldots, 1, 0)^{k-2}$ with cost $2k - 2$, whereas partition $k, 0, 0, \ldots, 0, 0$ - $(1, 1, 1, \ldots, 1, 1)^{k-2}$ has cost k.

4   An MILP Approach to Determining Performance Ratios

In Section 3, we proved the essential fact that

$$\sup_{(\mathcal{A}, I) \in \mathcal{I}^{(1)}} \frac{f_I(\mathcal{A}^{\text{MON}})}{f_I(\mathcal{A})} \tag{3}$$

is a performance bound of BLDM for given $k \geq 4$. Here $(\mathcal{A}, I) \in \mathcal{I}^{(1)}$ if and only if $0 \leq x_1 \leq x_2 \leq \cdots \leq x_k$, $m_l \geq 0$, and $m_r \geq 0$ exist such that for problem instance I partition $\mathcal{A}^{\text{MON}}$ is given by (1). In the full paper, we prove that we can restrict ourselves to the subset $\mathcal{I}^{(2)}$ of $\mathcal{I}^{(1)}$, i.e., that (3) equals

$$\sup_{(\mathcal{A},I) \in \mathcal{I}^{(2)}} \frac{f_I(\mathcal{A}^{\text{MON}})}{f_I(\mathcal{A})} \tag{4}$$

where $\mathcal{I}^{(2)}$ contains $(\mathcal{A}, I) \in \mathcal{I}^{(1)}$ if the following constraints are satisfied:

- $m_l = 0$, i.e., $\mathcal{A}^{\text{MON}} = x_k, x_{k-1}, \ldots, x_1 - (x_{k-1}, x_{k-1}, x_{k-2}, \ldots, x_1)^{m-1}$.
- the subset in $\mathcal{A}$ containing x_k does, besides x_k, only contain zeros.
- $m \geq k-1$. Hence, we can assume that $A_1 = x_k, 0, 0, \ldots, 0 = x_k, x_1, x_1, \ldots, x_1$.

For an $(\mathcal{A}, I) \in \mathcal{I}^{(2)}$, partition $\mathcal{A}^{\text{MON}} = x_k, x_{k-1}, \ldots, x_1 - (x_{k-1}, x_{k-1}, x_{k-2}, \ldots, x_1)^{m-1}$ is the same partition as the one given by BLDM. Furthermore, $f_I(\mathcal{A}) \geq f_I^*$. This implies that the performance bound given by (4) is tight. Hence, although it may seem that by deriving (4) in successive steps we lost the tightness of the bound, this is not the case. In addition, a worst-case example for BLDM is given by an $(\mathcal{A}, I) \in \mathcal{I}^{(2)}$ for which (4) is attained.

Theorem 2. *For any $k \geq 4$, the performance ratio of BLDM is given by (4).*

In this section, we formulate the problem of determining (4) for a given $k \geq 4$ as an MILP problem. By using standard branch-and-bound techniques it will enable us to obtain performance ratios for $k = 4, 5$, and 6 and a lower bound for $k = 7$ that is better than the one given by Theorem 1. However, we first introduce some definitions.

We define $\mathcal{W}$ as the set of all $(k-1)$-tuples $\bar{b}_j = (b_{1j}, b_{2j}, \ldots, b_{k-1,j}) \in \mathbb{N}^{k-1}$ with $\sum_{i=1}^{k-1} b_{ij} = k$. We let these tuples be numbered from 1 to t. Let $\bar{b}_j \in \mathcal{W}$ define the subset V_j containing b_{ij} items with size x_i for $1 \leq i < k$. For example, for $k = 5$ tuple $\bar{b}_j = (1, 2, 1, 1)$ defines subset $V_j = x_1, x_2, x_2, x_3, x_4$. Note that x_k does not occur in a subset corresponding to a tuple in $\mathcal{W}$. We say that partition $\mathcal{A}$ and instance I are *characterized* by $\bar{z} = (z_1, z_2, \ldots, z_t)$ if

$$\mathcal{A} = x_k, x_1, x_1, \ldots, x_1 - V_1^{z_1} - V_2^{z_2} - \ldots - V_t^{z_t},$$

where subset V_j is defined by $\bar{b}_j$. In addition, if for instance I, partition $\mathcal{A}^{\text{MON}}$ is given by (1) with $m_l = 0$ and $m \geq k-1$, i.e., $(\mathcal{A}, I) \in \mathcal{I}^{(2)}$ for any $0 = x_1 \leq x_2 \leq \cdots \leq x_k$, then $\bar{z}$ is called *feasible*.

Now, it can be verified that the following formulation, in which we indicate two related subproblems, describes the problem of determining the performance ratio given by (4).

$$\text{Maximize } \frac{x_k + x_{k-1} + \cdots + x_1}{f_I(\mathcal{A})},$$

$$\text{such that } \mathcal{A}, I \text{ are characterized by } \bar{z}$$

$$0 = x_1 \leq x_2 \leq \cdots \leq x_k$$

$$\bar{x} \geq \bar{0}$$

$$\bar{z} \text{ feasible}$$

$$\bar{z} \geq \bar{0}, \ \bar{z} \text{ integer}$$

Subproblem 2

Subproblem 1

The remainder of this section is organized as follows. First, we formulate Subproblem 1 as an Integer Linear Programming (ILP) problem and Subproblem 2 as a programming problem that is linear in $\bar{x}$ but conditional in $\bar{z}$. By replacing the conditional constraints by linear constraints at the cost of binary variables, we obtain an MILP formulation of the problem to determine the performance ratio for a given k. Finally, we discuss some results obtained by solving this MILP problem.

Subproblem 1. We will derive linear constraints for determining whether a given $\bar{z} \in \mathbb{N}^t$ is feasible, which yields an ILP formulation of Subproblem 1. Let problem instance I and partition $\mathcal{A}$ be characterized by $\bar{z}$. By definition, $\bar{z}$ is feasible if and only if the following two conditions are satisfied.

1. $m \geq k - 1$, i.e., $\sum_{j=1}^{t} z_j \geq k - 2$ as $m = 1 + \sum_{j=1}^{t} z_j$, and
2. $\mathcal{A}$ contains x_i once for $i = k$, $2m - 1$ times for $i = k - 1$, and m times for $1 \leq i \leq k - 2$. This is equivalent to stating that $\mathcal{A} = x_k, x_{k-1}, \ldots, x_1 \cdot (x_{k-1}, x_{k-1}, x_{k-2}, \ldots, x_1)^{m-1}$.

Consider Condition 2. By definition, we have that $\mathcal{A}$ contains x_k exactly once. Furthermore, to check the number of occurrences of the remaining $k - 1$ variables, it suffices to check the number of occurrences of only $k - 2$ of them. Hence, Condition 2 is equivalent to the condition that $\mathcal{A}$ contains $2m - 1$ times x_{k-1} and m times x_i for $2 \leq i \leq k - 2$. Since the total number of occurrences of x_i in $\mathcal{A}$ is given by $\sum_{j=1}^{t} z_j \cdot b_{ij}$, by definition, the condition is formalized by $\sum_{j=1}^{t} z_j \cdot b_{k-1,j} = 2m - 1$ and $\sum_{j=1}^{t} z_j \cdot b_{ij} = m$ for $1 \leq i \leq k - 2$. As a result, $\bar{z}$ is feasible, i.e., Conditions 1 and 2 hold, if and only if the linear constraints

$$\sum_{j=1}^{t} z_j \geq k - 2$$
$$D\bar{z} = \bar{1},$$

are satisfied, where matrix D is given by

$$\begin{pmatrix} b_{21} - 1 & b_{22} - 1 & \cdots & b_{2t} - 1 \\ b_{31} - 1 & b_{32} - 1 & \cdots & b_{3t} - 1 \\ & & \vdots & \\ b_{k-2,1} - 1 & b_{k-2,2} - 1 & \cdots & b_{k-2,t} - 1 \\ b_{k-1,1} - 2 & b_{k-1,2} - 2 & \cdots & b_{k-1,t} - 2 \end{pmatrix}.$$

Subproblem 2. Next, we write Subproblem 2 as a programming problem with constraints that are conditional in $\bar{z}$ and linear in all other variables. We start with eliminating $f_I(\mathcal{A})$ from the problem formulation.

If partition $\mathcal{A}$ and instance I are characterized by $\bar{z}$, then $\mathcal{A}$ contains subset V_j if $z_j > 0$ and it does not contain V_j if $z_j = 0$. Hence, $f_I(\mathcal{A})$ equals the maximum sum of subset $x_k, x_1, x_1, \ldots, x_1$ and of any V_j with $z_j > 0$. As $x_1 = 0$, the sum of V_j is given by $(\sum_{i=2}^{k-1} b_{ij} x_i)$ and the sum of subset $x_k, x_1, x_1, \ldots, x_1$ by x_k. Hence, $f_I(\mathcal{A})$ is given by the minimum C satisfying

$$\sum_{i=2}^{k-1} b_{ij} x_i \leq C, \text{ for all } j \text{ with } z_j > 0$$
$$x_k \leq C$$

As a result, Subproblem 2 is equivalent to

$$\text{Maximize } \frac{x_k + x_{k-1} + \cdots + x_2}{C},$$

$$\text{such that } \left(\sum_{i=2}^{k-1} b_{ij} x_i / C \right) \leq 1, \quad \text{for all } 1 \leq j \leq t \text{ with } z_j > 0$$

$$x_k / C \leq 1$$

$$0 \leq x_2 \leq x_3 \leq \cdots \leq x_k$$

$$x_i \geq 0, \qquad \text{for } 1 < i \leq k$$

$$C > 0$$

We introduce new decision variables y_j with $1 < j < k$, which we substitute for x_j / C, i.e., y_j represents $x_j / f_I(\mathcal{A})$. Hence, instead of having a formulation that depends on both the exact cost of partition $\mathcal{A}$ and the exact values of $x_2, x_3, \ldots, x_k$, we have a formulation that only depends on the values of $x_2, x_3, \ldots, x_k$ expressed as a fraction of the cost of partition $\mathcal{A}$. We now get the following formalization of Subproblem 2. We thereby use that $y_k = x_k / C = 1$ in an optimal solution, which can easily be verified.

$$\text{Maximize } 1 + \sum_{i=2}^{k-1} y_i,$$

$$\text{such that } \left(\sum_{i=2}^{k-1} b_{ij} y_i \right) \leq 1, \qquad \text{for all } j \text{ with } z_j > 0$$

$$0 \leq y_2 \leq y_3 \leq \cdots \leq y_{k-1} \leq 1$$

$$y_i \geq 0, \qquad \text{for } 1 < i < k$$

To illustrate the first type of constraint, assume that $V_j = x_3\, x_3\, x_2\, x_2\, x_2$ occurs in partition $\mathcal{A}$ characterized by $\bar{z}$, i.e., $z_j > 0$. Then the constraint $\sum_{i=2}^{k-1} b_{ij} y_i \leq 1$ corresponds to $2y_3 + 3y_2 \leq 1$. Hence, the constraint gives a necessary condition on y_3 and y_2 to represent values x_3 and x_2 as a fraction of the cost of partition $\mathcal{A}$.

Subproblems 1 and 2 combined. The following problem formulation combines the derived constraints for Subproblems 1 and 2. As we prove in the full paper, the integrality constraint on $\bar{z}$ can be removed.

$$\text{Maximize } 1 + \sum_{i=2}^{k-1} y_i,$$

$$\text{such that } \left(\sum_{i=2}^{k-1} b_{ij} y_i \right) \leq 1, \qquad \text{for all } 1 \leq j \leq t \text{ with } z_j > 0$$

$$\sum_{j=1}^{t} z_j \geq k - 2$$

$$D\bar{z} = \bar{1} \qquad\qquad\qquad\qquad\qquad\qquad (\text{P})$$

$$0 \leq y_2 \leq y_3 \leq \cdots \leq y_{k-1} \leq 1$$

$$y_i \geq 0, \qquad \text{for all } 1 < i < k$$

$$\bar{z} \geq \bar{0}$$

To transform this problem into an MILP formulation, we have to eliminate the conditional constraints. In [22], an approach is presented for replacing conditional constraints by linear constraints at the cost of binary variables.

As mentioned above, an MILP problem can be solved to optimality by using standard branch-and-bound techniques. When applying these techniques, we obtain for $k = 4, 5$, and 6 the performance ratios $19/12$, $103/60$, and $643/360$, respectively; see Table 1. Furthermore, we obtain a lower bound of $L = 268/147$ for $k = 7$, whereas, by Theorem 1,

Table 1. Performance ratios for $k = 3, 4, 5$, and 6 and lower and upper bounds for $k = 7$. Furthermore, problem instances are defined with performance ratio R for $3 \leq k \leq 6$ and with performance ratio L for $k = 7$ by giving the partition $\mathcal{A}^{\mathrm{BLDM}}$ given by BLDM and an optimal partition $\mathcal{A}^*$.

k	R	$\mathcal{A}^{\mathrm{BLDM}}$ $\mathcal{A}^*$
3	4/3	$3, 1, 0 - 1, 1, 0$ $3, 0, 0 - 1, 1, 1$
4	19/12	$12, 4, 3, 0 - (4, 4, 3, 0)^7$ $12, 0, 0, 0 - (4, 4, 4, 0)^5 - (3, 3, 3, 3)^2$
5	103/60	$60, 16, 15, 12, 0 - (16, 16, 15, 12, 0)^{43}$ $60, 0, 0, 0, 0 - (16, 16, 16, 12, 0)^{29} - (15, 15, 15, 15, 0)^{11} - (12, 12, 12, 12, 12)^3$
6	643/360	$360, 76, 75, 72, 60, 0 - (76, 76, 75, 72, 60, 0)^{283}$ $360, 0, 0, 0, 0, 0 - (76, 76, 76, 72, 60, 0)^{189} - (75, 75, 75, 75, 60, 0)^{71} - (72, 72, 72, 72, 72, 0)^{19}$ $- (60, 60, 60, 60, 60, 60)^4$
7	$[268/147, 11/6]$	$294, 51, 51, 50, 48, 41, 0 - (51, 51, 51, 50, 48, 41, 0)^{242}$ $294, 0, 0, 0, 0, 0, 0 - (51, 51, 51, 51, 48, 42, 0)^{127} - (51, 51, 50, 50, 50, 42, 0)^{81}$ $- (51, 51, 48, 48, 48, 48, 0)^{29} - (42, 42, 42, 42, 42, 42, 42)^5$

$U = 1 + (k - 2)/(k - 1) = 11/6$ is an upper bound for this case. It can be verified that $L = (1 - \varepsilon)U$, where $\varepsilon = 5.6 \cdot 10^{-3}$, i.e., the lower bound is only 0.56% smaller than the upper bound. By construction, we can derive an $(\mathcal{A}, I) \in \mathcal{I}^{(2)}$ for which (4) is attained from an optimal solution of MILP and, as mentioned above, such an $(\mathcal{A}, I)$ gives a worst-case instance for BLDM. These worst-case instances are also given in Table 1. For the sake of completeness, we also add the results for $k = 3$, which are proved in the full paper.

Theorem 3. *For given $k \geq 4$, the performance ratio R of BLDM is given by the optimal objective value of (P). Using this, we can derive $R = 19/12, 103/60$, and $643/360$ for $k = 4, 5$, and 6, respectively, and $268/147 \leq R \leq 11/6$ for $k = 7$.*

5 Performance Ratios Depending on m

We now focus on the performance ratio of BLDM as a function of m. We show that for any $m \geq 2$, the performance ratio is given by $2 - 1/m$. A well-known algorithm for the number partitioning problem is List Scheduling (LS), which assigns the next item in some prespecified list l to the subset with the smallest sum. Assume that l is defined such that the first m items are from G_1, the next m items from G_2, and so on. We call the assignment of all items from a set G_i by LS with $1 \leq i \leq k$ a phase and we say that a partition is group-based if each subset contains exactly one item from each G_i.

If the partial partition derived by LS at the end of phase $i - 1$ is group-based then the difference between the largest and smallest subset sum cannot be larger than the smallest item size in G_i. This implies that in phase i the items from G_i are assigned in such a way that at the end of phase i the partition is still group-based. Based on this observation, it can be verified that any group-based partition can be obtained by a proper choice of l. As BLDM always returns a group-based partition and as LS has a performance bound of $2 - 1/m$ for the number partitioning problem [8], $2 - 1/m$ is also a performance bound for BLDM when applied to the number partitioning problem. Clearly, the optimal cost of this problem is at most the optimal cost of the balanced partitioning problem. Hence, the performance bound is also valid for the latter problem. Moreover, this performance bound is tight as can be seen as follows. Let M be a multiple of m. Then partition

$$M, \overbrace{1,1,\ldots,1}^{\frac{(m-1)M}{m}}, \overbrace{0,0,\ldots,0}^{\frac{M}{m}} - (\overbrace{1,1,\ldots,1}^{\frac{(m-1)M}{m}+1} \overbrace{0,0,\ldots,0}^{\frac{M}{m}})^{m-1},$$

with objective value $(2 - 1/m)M$ can be given by BLDM, whereas the optimal partition $M, 0, 0, \ldots, 0, 0 - (1,1,1,\ldots,1,1)^{m-1}$ only has objective value $M + 1$. For $M \to \infty$, the performance ratio of this problem instance approaches $2 - 1/m$.

Theorem 4. *For given $m \geq 2$, the performance ratio of* BLDM *is given by* $2 - 1/m$.

6   Concluding Remarks

In this paper, we analyzed the performance ratio of BLDM both as a function of the cardinality k of the subsets and as a function of the number of subsets m. For this we employed an MILP formulation. Similar results are derived in the full paper and in [16] for the case that both k and m are assumed to be fixed. There, we also prove that using another strategy for determining which solutions are to be differenced in an iteration of BLDM cannot improve the worst-case performance of the algorithm.

Finally, we remark that a similar analysis can be used to derive worst-case performance results for the Grouped Best-Fit Algorithm [17] when applied to the Min-Max Subsequence problem.

References

1. L. Babel, H. Kellerer, and V. Kotov. The k-partitioning problem. *Mathematical Methods of Operations Research*, 47, pages 59–82, 1998.
2. M. Ball and M. Magazine. Sequencing of insertions in printed circuit board assembly. *Operations Research*, 36, pages 192–201, 1988.
3. E. Coffman, G. Frederickson, and G. Lueker. A note on expected makespans for largest-first sequences of independent tasks on two processors. *Mathematics of Operations Research*, 9, pages 260–266, 1984.
4. E.G. Coffman Jr, M.R. Garey, and D.S. Johnson. An application of bin-packing to multiprocessor scheduling. *SIAM Journal on Computing*, 7(1):1–17, 1978.
5. E.G. Coffman Jr. and W. Whitt. Recent asymptotic results in the probabilistic analysis of schedule makespans. In P. Chretienne, E.G. Coffman Jr., J.K. Lenstra, and Z. Liu, editors, *Scheduling Theory and its Applications*, pages 15–31. Wiley, 1995.
6. M. Fischetti and S. Martello. Worst-case analysis of the differencing method for the partition problem. *Mathematical Programming*, 37, pages 117–120, 1987.
7. M. Garey and D. Johnson. *Computers and Intractability: A Guide to the Theory of NP-Completeness*. W.H. Freeman and Company, 1979.
8. R. Graham. Bounds for certain multiprocessing anomalies. *Bell System Technical Journal*, 45, pages 1563–1581, 1966.

9. R. Graham. Bounds on multiprocessing timing anomalies. *SIAM Journal on Applied Mathematics*, 17, pages 416–429, 1969.
10. N. Karmarkar and R. Karp. The differencing method of set partitioning. Technical Report UCB/CSD 82/113, University of California, Berkeley, 1982.
11. H. Kellerer and V. Kotov. A 7/6-approximation algorithm for 3-partitioning and its application to multiprocessor scheduling. *INFOR*, 37(1), pages 48–56, 1999.
12. H. Kellerer and G. Woeginger. A tight bound for 3-partitioning. *Discrete Applied Mathematics*, 45, pages 249–259, 1993.
13. R. Korf. A complete anytime algorithm for number partitioning. *Artificial Intelligence*, 106, pages 181–203, 1998.
14. G. Lueker. A note on the average-case behavior of a simple differencing method for partitioning. *Operations Research Letters*, 6, pages 285–288, 1987.
15. S. Mertens. A complete anytime algorithm for balanced number partitioning, 1999. preprint xxx.lanl.gov/abs/cs.DS/9903011.
16. W. Michiels. *Performance Ratios for Differencing Methods*. PhD thesis, Technische Universiteit Eindhoven, 2003.
17. W. Michiels and J. Korst. Min-max subsequence problems in multi-zone disk recording. *Journal of Scheduling*, 4, pages 271–283, 2001.
18. W. Michiels, J. Korst, E. Aarts, and J. van Leeuwen. Performance ratios for the Karmarkar-Karp Differencing Method. Technical Report 02/17, Technische Universiteit Eindhoven, 2002. Submitted to Journal of Algorithms.
19. L. Tasi. The modified differencing method for the set partitioning problem with cardinality constraints. *Discrete Applied Mathematics*, 63, pages 175–180, 1995.
20. L. Tsai. *The loading and scheduling problems in flexible manufacturing systems*. PhD thesis, University of California, Berkeley, 1987.
21. L. Tsai. Asymptotic analysis of an algorithm for balanced parallel processor scheduling. *SIAM Journal on Computing*, 21, pages 59–64, 1992.
22. H. Williams. *Model Building in Mathematical Programming*. Wiley, 1978.
23. B. Yakir. The differencing algorithm LDM for partitioning: A proof of karp's conjecture. *Mathematics of Operations Research*, 21, pages 85–99, 1996.

Cake-Cutting Is Not a Piece of Cake

Malik Magdon-Ismail, Costas Busch, and Mukkai S. Krishnamoorthy

Department of Computer Science
Rensselaer Polytechnic Institute
110 8th Street, Troy, NY 12180
{magdon, buschc, moorthy}@cs.rpi.edu

Abstract. Fair cake-cutting is the division of a cake or resource among N users so that each user is content. Users may value a given piece of cake differently, and information about how a user values different parts of the cake can only be obtained by requesting users to "cut" pieces of the cake into specified ratios. One of the most interesting open questions is to determine the minimum number of cuts required to divide the cake fairly. It is known that $O(N \log N)$ cuts suffices, however, it is not known whether one can do better.

We show that sorting can be reduced to cake-cutting: *any* algorithm that performs fair cake-division can sort. For a general class of cake-cutting algorithms, which we call linearly-labeled, we obtain an $\Omega(N \log N)$ lower bound on their computational complexity. All the known cake-cutting algorithms fit into this general class, which leads us to conjecture that every cake-cutting algorithm is linearly-labeled. If in addition, the number of comparisons per cut is bounded (comparison-bounded algorithms), then we obtain an $\Omega(N \log N)$ lower bound on the number of cuts. All known algorithms are comparison-bounded.

We also study variations of envy-free cake-division, where each user feels that they have more cake than every other user. We construct utility functions for which any algorithm (including continuous algorithms) requires $\Omega(N^2)$ cuts to produce such divisions. These are the the first known general lower bounds for envy-free algorithms. Finally, we study another general class of algorithms called *phased* algorithms, for which we show that even if one is to simply guarantee each user a piece of cake with positive value, then $\Omega(N \log N)$ cuts are needed in the worst case. Many of the existing cake-cutting algorithms are phased.

1 Introduction

Property sharing problems, such as chore division, inheritance allocation, and room selection, have been extensively studied in economics and game-theory [1, 5,7,9,11]. Property sharing problems arise often in everyday computing when different users compete for the same resources. Typical examples of such problems are: job scheduling; sharing the CPU time of a multiprocessor machine; sharing the bandwidth of a network connection; etc. The resource to be shared can be viewed as a cake, and the problem of sharing such a resource is called

H. Alt and M. Habib (Eds.): STACS 2003, LNCS 2607, pp. 596–607, 2003.

cake-cutting or *cake-division*. In this work, we study fair cake-division; we are interested in quantifying how hard a computational problem this is. In the original formulation of the cake-division problem, introduced in the 1940's by Steinhaus [15], N users wish to share a cake in such a way that each user gets a portion of the cake that she is content with (a number of definitions of content can be considered, and we will discuss these more formally later). Users may value a given piece of the cake differently. For example, some users may prefer the part of the cake with the chocolate topping, and others may prefer the part without the topping. Suppose there are five users, and lets consider the situation from the first user's point of view. The result of a cake-division is that every user gets a portion of the cake, in particular user 1 gets a portion. User 1 will certainly not be content if the portion that she gets is worth (in her opinion) less than one fifth the value (in her opinion) of the entire cake. So in order to make user 1 content, we must give her a portion that she considers to be worth *at least* one fifth the value of the cake, and similarly for the other users. If we succeed in finding such a division for which all the users are content, we say that it is a *fair cake-division*.

More formally, we represent the cake as an interval $I = [0, 1]$. A piece of the cake corresponds to some sub-interval of this interval, and a portion of cake can be viewed as a collection of pieces. The user only knows how to value pieces as specified by her utility function, and has no knowledge about the utility functions of the other users. The cake-division process (an assignment of portions to users) is to be effected by a superuser S who initially has no knowledge about the utility functions of the users. We point out that this may appear to diverge from the accepted mentality in the field, where protocols are viewed as self-enforcing – players are advised as to how they should cut, and are guaranteed that if they cut according to the advice, then the result will be equitable to them. The way such algorithms usually proceed is that the players make cuts, and somehow based on these cuts, portions are assigned to the players. Some computing body needs to do this assignment, and perform the necessary calculations so that the resulting assignment is guaranteed to be equitable to all the players (provided that they followed the advice). It is exactly this computing body that we envision as the superuser, because we would ultimately like to quantify all the computation that takes place in the cake division process. In order to construct an appropriate division, the superuser may request users to cut pieces into ratios that the superuser may specify. Based on the information learned from a number of such cuts, the superuser must now make an appropriate assignment of portions, such that each user is content. A simple example will illustrate the process. Suppose that two users wish to share the cake. The superuser can ask one of the users to cut the entire cake into two equal parts. The superuser now asks the other user to evaluate the two resulting parts. The second user is then assigned the part that she had higher value for, and the first user gets the remaining part. This well known division scheme, sometimes termed "I cut, you choose", clearly leaves both users believing they have at least half the cake, and it is thus a successful fair division algorithm. From this example we see that one cut suffices to perform fair division for two users. An interesting question to ask is: what is the minimum number of cuts required to perform a fair division when there are N users.

The cake-division problem has been extensively studied in the literature [4,6, 8,12,13,14,16,17]. From a computational point of view, we want to minimize the number of cuts needed, since this leads to a smaller number of computational steps performed by the algorithm. Most of the algorithms proposed in the literature require $O(N^2)$ cuts for N users (see for example Algorithm A, Section 2.3), while the best known cake-cutting algorithm, which is based on a divide-and-conquer procedure, uses $O(N \log N)$ cuts (see for example Algorithm B, Section 2.3). More examples can be found in [4,14]. It is not known whether one can do better than $O(N \log N)$. In fact, it is conjectured that there is no algorithm that uses $o(N \log N)$ cuts in the worst case. The problem of determining what the minimum number of cuts required to guarantee a fair cake-division seems to be a very hard one. We quote from Robertson and Webb [14, Chapter 2.7]:

> "The problem of determining in general the fewest number of cuts required for fair division seems to be a very hard one. ... We have lost bets before, but if we were asked to gaze into the crystal ball, we would place our money against finding a substantial improvement on the $N \log N$ bound."

Of course, a well known continuous protocol that uses $N - 1$ cuts is the moving knife algorithm (see Section 2.3). Certainly $N-1$ cuts cannot be beaten, however, such continuous algorithms are excluded from our present discussion. Our results apply only to discrete protocols except when explicitly stated otherwise.

Our main result is that sorting can be reduced to cake-cutting: *any* fair cake-cutting algorithm can be converted to an equivalent one that can sort an arbitrary sequence of distinct positive integers. Further, this new algorithm uses no more cuts (for any set of N users) than the original one did. Therefore, cake-cutting should be at least as hard as sorting. The heart of this reduction lies in a mechanism for labeling the pieces of the cake. Continuing, we define the class of linearly-labeled cake-cutting algorithms as those for which the extra cost of labeling is linear in the number of cuts. Essentially, the converted algorithm is as efficient as the original one. For the class of linearly-labeled algorithms, we obtain an $\Omega(N \log N)$ lower bound on their computational complexity. To our knowledge, all the known fair cake-cutting algorithms fit into this general class, which leads us to conjecture that every fair cake-cutting algorithm is linearly-labeled, a conjecture that we have not yet settled. From the practical point of view, the computational power of the superuser can be a limitation, and so we introduce the class of algorithms that allow the super user a budget, in terms of computation, for every cut that is made. Thus, the computation that the superuser performs can only grow linearly with the number of cuts performed. Such algorithms we term comparison-bounded. All the known algorithms are comparison-bounded. If in addition to being linearly-labeled, the algorithm is also comparison-bounded, then we obtain an $\Omega(N \log N)$ lower bound on the number of cuts required in the worst case. Thus the conjecture of Robertson and Webb is true within the class of linearly-labeled & comparison-bounded algorithms. To our knowledge, this class includes all the known algorithms, which makes it a very interesting and natural class of cake-cutting algorithms. These are the first "hardness" results for a general class of cake-cutting algorithms that are applicable to a general number of users.

We also provide lower bounds for some types of *envy-free* cake-division. A cake-division is envy-free if each user believes she has at least as large a portion (in her opinion) as every other user, i.e., no user is envious of another user. The two person fair division scheme presented earlier is also an envy-free division scheme. Other envy-free algorithms for more users can be found in [3,4,8,11,14]. It is known that for any set of utility functions there are envy-free solutions [14]. However, there are only a few envy-free algorithms known in the literature for N users [14]. Remarkably, all these algorithms are *unbounded*, in the sense that there exist utility functions for which the number of cuts is finite, but arbitrarily large. Again, no lower bounds on the number cuts required for envy-free division exist. We give the first such lower bounds for two variations of the envy-free problem, and our bounds are applicable to both discrete and continuous algorithms. A division is *strong envy-free*, if each user believes she has *more* cake than the other users, i.e., each user believes the other users will be envious of her. We show that $\Omega(N^2)$ cuts are required in the worst case to guarantee a strong envy-free division when it exists. A division is *super envy-free*, if every user believes that every other user has at most a fair share of the cake (see for example [2, 14]. We show that $\Omega(N^2)$ cuts are required in the worst case to guarantee super envy-free division when it exists. These lower bounds give a first explanation of why the problem of envy-free cake-division is harder than fair cake-division for general N.

The last class of cake-cutting algorithms that we consider are called *phased* algorithms. In phased algorithms, the execution of the algorithm is partitioned into phases. At each phase all the "active" users make a cut. At the end of a phase users may be assigned portions, in which case they become "inactive" for the remainder of the algorithm. Many known cake-cutting algorithms are phased. We show that there are utility functions for which *any* phased cake-cutting algorithm requires $\Omega(N \log N)$ cuts to guarantee every user a portion they believe to be of positive value (a much weaker condition than fair). For such algorithms, assigning positive portions alone is hard, so requiring the portions to also be fair or envy-free can only make the problem harder. In particular, Algorithm B (see Section 2.3), a well known divide and conquer algorithm is phased, and obtains a fair division using $O(N \log N)$ cuts. Therefore this algorithm is optimal among the class of phased algorithms *even* if we compare to algorithms that merely assign positive value portions. The issue of determining the maximum value that can be guaranteed to every user with K cuts has been studied in the literature [14, Chapter 9]. We have that for phased algorithms, this maximum value is zero for $K = o(N \log N)$.

The outline of the remainder of the paper is as follows. In the next section, we present the formal definitions of the cake-division model that we use, and what constitutes a cake-cutting algorithm, followed by some example algorithms. Next, we present the lower bounds. In Section 3 we introduce phased algorithms and give lower bounds for the number of cuts needed. Section 4 discusses labeled algorithms and the connection to sorting. In Section 5 we give the lower bounds for envy-free division, and finally, we make some concluding remarks in Section 6. Due to space constraints, we refer to the accompanying technical report [10] for most of the proofs.

2 Preliminaries

2.1 Cake-Division

We denote the cake as the interval $I = [0, 1]$. A *piece* of the cake is any interval $P = [l, r]$, $0 \le l \le r \le 1$, where l is the left end point and r the right end point of P. The *width* of P is $r - l$, and we take the width of the empty set $\emptyset$ to be zero. $P_1 = [l_1, r_1]$ and $P_2 = [l_2, r_2]$ are *separated* if the width of $P_1 \cap P_2$ is 0, otherwise we say that P_1 and P_2 *overlap*. P_1 *contains* P_2 if $P_2 \subseteq P_1$. If P_1 and P_2 are separated, then we say that P_1 is left of P_2 if $l_1 < l_2$. The *concatenation* of the $M > 1$ pieces $\{[l, s_1], [s_1, s_2], [s_2, s_3], \dots, [s_{M-1}, r]\}$ is the piece $[l, r]$.

A *portion* is a non-empty set of separated pieces $W = \{P_1, P_2, \dots, P_k\}$, $k \ge 1$. Note that a portion may consist of pieces which are not adjacent (i.e. a portion might be a collection of "crumbs" from different parts of the cake). Two portions W_1 and W_2 are *separated* if every piece in W_1 is separated from every piece in W_2. An *N-partition* of the cake is a collection of separated portions $W_1, \dots, W_N$ whose union is the entire cake I.

Suppose that the N users $u_1, \dots, u_N$ wish to share the cake. Each user u_i has a *utility function* $F_i(x)$, which determines how user u_i values the piece $[0, x]$, where $0 \le x \le 1$. Each user u_i knows only its own utility function $F_i(x)$, and has no information regarding the utility functions of other users. The functions $F_i(x)$ are monotonically non-decreasing with $F_i(0) = 0$ and $F_i(1) = 1$, for every user u_i. We require that the value of a portion is the sum of the values of the individual pieces in that portion[1]. Thus, the value of piece $[l, r]$ to user u_i is $F_i([l, r]) = F_i(r) - F_i(l)$, and for any portion $W = \{P_1, P_2, \dots P_k\}$, $F_i(W) = \sum_{i=1}^{k} F_i(P_i)$.

The goal of cake-division is to partition the entire cake I into N separated portions, assigning each user to a portion. Formally, a *cake-division* is an N-partition $W_1, \dots, W_N$ of cake I, with an assignment of portion W_i to user u_i, for all $1 \le i \le N$. Two cake-divisions $W_1, \dots, W_N$ and $W_1', \dots, W_N'$ are *equivalent* if $\bigcup_{P_i \in W_j} P_i = \bigcup_{P_i \in W_j'} P_i$ for all j, i.e., every user gets the same part of cake in both divisions (but perhaps divided into different pieces).

The cake-division is *fair* or *proportional* if $F_i(W_i) \ge 1/N$, for all $1 \le i \le N$, i.e., each user u_i gets what she considers to be at least $1/N$ of the cake according to her own utility function F_i. We obtain the following interesting variations of

---

[1] This is a commonly made technical assumption. Practically, there could be situations where a pound of crumbs is not equivalent to a pound of cake.

fair cake-division, if, in addition to fair, we impose further restrictions or *fairness constraints* on the relationship between the assigned portions:

Envy-free: $F_i(\mathcal{W}_i) \geq F_i(\mathcal{W}_j)$ for all i, j; *strong envy-free* if $F_i(\mathcal{W}_i) > F_i(\mathcal{W}_j)$ for all $i \neq j$.

Super envy-free: $F_i(\mathcal{W}_j) \leq 1/n$ for all $i \neq j$; *strong super envy-free* if $F_i(\mathcal{W}_i) < 1/n$ for all $i \neq j$.

These definitions are standard and found in [14].

2.2   Cake-Cutting Algorithms

We now move on to defining a cake-cutting protocol/algorithm. Imagine the existence of some administrator or superuser $\mathcal{S}$ who is responsible for the cake-division. The superuser $\mathcal{S}$ has limited computing power, namely she can perform basic operations such as comparisons, additions and multiplications. We assume that each such basic operation requires one time step.

Superuser $\mathcal{S}$ can ask the users to cut pieces of the cake in order to get information regarding their utility functions. A cut is composed of the following steps: superuser $\mathcal{S}$ specifies to user u_i a piece $[l, r]$ and a ratio R with $0 \leq R \leq 1$; the user then returns the point C in $[l, r]$ such that $F_i([l, C])/F_i([l, r]) = R$. Thus, a cut can be represented by the four-tuple $\langle u_i; [l, r]; R; C \rangle$. We call C the *position* of the cut. It is possible that a cut could yield multiple cut positions, i.e. when some region of the cake evaluates to zero; in such a case we require that the cut position returned is the left-most. In cake-cutting algorithms, the endpoints of the piece to be cut must be either 0, 1, or cut positions that have been produced by earlier cuts. So for example, the first cut has to be of the form $\langle u_{i_1}; [0, 1]; R_1; C_1 \rangle$. The second cut could then be made on $[0, 1]$, $[0, C_1]$ or $[C_1, 1]$. From now every piece will be of this form. We assume that a user can construct a cut in constant time[2]. A cake-cutting algorithm (implemented by the superuser $\mathcal{S}$) is a sequence of cuts that $\mathcal{S}$ constructs in order to output the desired cake-division.

Definition 1 (Cake-Cutting Algorithm).

Input: *The N utility functions, $F_1(x), \ldots, F_N(x)$ for the users $u_1, \ldots, u_N$.*

Output: *A cake-division satisfying the necessary fairness constraint.*

Computation: *The algorithm is a sequence of steps, $t = 1 \ldots K$. At every step t, the superuser requests a user u_{i_t} to perform a cut on a piece $[l_t, r_t]$ with ratio R_t: $\langle u_{i_t}; [l_t, r_t]; R_t; C_t \rangle$. In determining what cut to make, the superuser may use her limited computing power and the information contained in all previous cuts. A single cut conveys to the super user an amount of information that the superuser would otherwise need to obtain using some comparisons.*

---

[2] From the computational point of view, this may be a strong assumption, for example dividing a piece by an irrational ratio is a non-trivial computational task, however it is a standard assumption made in the literature, and so we continue with the tradition.

These comparisons need to also be taken into account in the computational complexity of the algorithm.

We say that this algorithm uses K cuts. K can depend on N and the utility functions F_i. A correct cake-division must take into account the utility functions of all the users, however, the superuser does not know these utility functions. The superuser implicitly infers the necessary information about each user's utility function from the cuts made. The history of all the cuts represents the entire knowledge that $\mathcal{S}$ has regarding the utility functions of the users. By a suitable choice of cuts, $\mathcal{S}$ then outputs a correct cake-division. An algorithm is named according to the fairness constraint the cake-division must satisfy. For example, if the output is fair (envy-free) then the algorithm is called a *fair (envy-free) cake-cutting algorithm*.

A cut as we have defined it is equivalent to a constant number of comparisons. A number of additional requirements can be placed on the model for cake-cutting given above. For example, when a cut is made, a common assumption in the literature is that *every* user evaluates the resulting two pieces for the superuser. Computationally, this assumes that utility function evaluation is a negligible cost operation. For the most part, our lower bounds do not require such additional assumptions. In our discussion we will make clear what further assumptions we make when necessary.

2.3   Particular Algorithms

We briefly present some well known cake-cutting algorithms. More details can be found in [14]. Algorithms A and B are both fair cake-cutting algorithms.

In algorithm A, all the users cut at $1/N$ of the whole cake. The user who cut the smallest piece is given that piece, and the remaining users recursively divide the remainder of the cake fairly. The value of the remainder of the cake to each of the remaining users is at least $1 - 1/N$, and so the resulting division is fair. This algorithm requires $\frac{1}{2}N(N+1) - 1$ cuts.

In algorithm B, for simplicity assume that there are 2^M users (although the algorithm is general). All the users cut the cake at $1/2$. The users who made the smallest $N/2$ cuts recursively divide the left "half" of the cake up to and including the median cut, and the users who cut to the right of the median cut recursively divide the right "half" of the cake. Since all the left users value the left part of the cake at $\geq 1/2$ and all the right users value the right part of the cake at $\geq 1/2$, the algorithm produces a fair division. This algorithm requires $N\lceil \log_2 N \rceil - 2^{\lceil \log_2 N \rceil} + 1$ cuts.

A perfectly legitimate cake-cutting algorithm that does not fit within this framework is the *moving knife fair division algorithm*. The superuser moves a knife continuously from the left end of the cake to the right. The first user (without loss of generality u_1) who is happy with the piece to the left of the current position of the knife yells "cut" and is subsequently given that piece. User u_1 is happy with that piece, and the remaining users were happy to give up that piece. Thus the remaining users must be happy with a fair division of the remaining of the cake. The process is then repeated with the remaining cake and the remaining $N - 1$ users. This algorithm makes $N - 1$ cuts which cannot be improved upon, since at least $N - 1$ cuts need to be made to generate N pieces. However, this algorithm does not fit within the framework we have described, and is an example of a *continuous algorithm*: there is no way to simulate the moving knife with any sequence of discrete cuts. Further, each cut in this algorithm is not equivalent to a constant number of comparisons, for example the first cut conveys the information in $\Omega(N)$ comparisons. Hence, such an algorithm is not of much interest from the computational point of view. The types of algorithms that our framework admits are usually termed *finite* or *discrete* algorithms. More details, including algorithms for envy-free can be found in [14].

3   A Lower Bound for Phased Algorithms

We consider a general class of cake-cutting algorithms, that we call "phased". We find a lower bound on the number of cuts required by phased algorithms that guarantee every user a positive valued portion. *Phased* cake-cutting algorithms have the following properties.

– The steps of the algorithm are divided into *phases*.
– In each phase, every *active* user cuts a piece, the endpoints of which are defined using cuts made during *previous* phases only. In the first phase, each user cuts the whole cake.
– Once a user is assigned a portion, that user becomes inactive for the remainder of the algorithm. (Assigned portions are not considered for the remainder of the algorithm.)

Many cake-cutting algorithms fit into the class of phased algorithms. Typical examples are Algorithms A and B. There also exist algorithms that are not phased, for example Steinhaus' original algorithm.

We say that two algorithms are *equivalent* if they use the same number of cuts for any set of utility functions, and produce equivalent cake-divisions. A piece is *solid* if it does not contain any cut positions – a non-solid piece is the union of two or more separated solid pieces. Our first observation is that any cut by a user on a non-solid piece P giving cut position C can be replaced with a cut by the same user on a solid piece contained in P, yielding the *same* cut position.

Lemma 1. *Suppose that P is the concatenation of separated solid pieces $P_1, \ldots, P_k$, for $k \geq 2$, and that the cut $\langle u_i; P; R; C \rangle$ produces a cut position C. Then, for suitably chosen R' and some solid piece P_m, the cut $\langle u_i; P_m; R'; C' \rangle$ produces the same cut position ($C' = C$). Further, R' and m depend only on R and $F_i(P_1), \ldots, F_i(P_k)$.*

Lemma 1 allows us to restrict our attention to solid piece phased algorithms. The following lemma then gives that an initially solid piece to be cut by some users may be cut in the same spot by each of these users.

Lemma 2. *For any phased algorithm, there are utility functions for which all users who are to cut the same (initially solid) piece will cut at the same position.*

We now give our lower bound for phased algorithms, which applies to any algorithm that guarantees each user a portion of positive value.

Theorem 1 (Lower bound for phased algorithms). *Any phased algorithm that guarantees each of N users a portion of positive value for any set of utility functions, requires $\Omega(N \log N)$ cuts in the worst case.*

The lower bound of $\Omega(N \log N)$ cuts for phased algorithms, demonstrates that even the problem of assigning positive portions to users is non-trivial. This lower bound immediately applies to fair and envy-free algorithms, since these algorithms assign positive portions to users.

4   A Lower Bound for Labeled Algorithms

We present a lower bound on the number of cuts required for a general class of fair algorithms that we refer to as "linearly-labeled & comparison-bounded". The proofs are by reducing sorting to cake-cutting. First, we show that any cake-cutting algorithm can be converted to a *labeled* algorithm which labels every piece in the cake-division. Then, by appropriately choosing utility functions, we use the labels of the pieces to sort a given sequence of integers.

First, we define labeled algorithms and then show how any cake-cutting algorithm can be converted to a labeled one. A *full binary tree* is a binary tree in which every node is either a leaf or the parent of two nodes. A *labeling tree* is a full binary tree in which every left edge has label 0 and every right edge has label 1. Every leaf is labeled with the binary number obtained by concatenating the labels of every edge on the path from the root to that leaf. An example labeling tree is shown in Figure 1. Let v be the deepest common ancestor of two leaves v_1 and v_2. If v_1 belongs to the left subtree of v and v_2 belongs to the right subtree of v, then v_1 is *left* of v_2.

Consider an N-partition $\mathcal{W}_1, \ldots, \mathcal{W}_N$ of the cake. The partition is *labeled* if the following hold:

- For some labeling tree, every (separated) piece P_i in the partition has a distinct label b_i that is a leaf on this tree, and every leaf on this tree labels some piece.
- P_i is left of P_j in the cake if and only if leaf b_i is left of leaf b_j in the labeling tree.

A cake-cutting algorithm is *labeled* if it always produces an N-partition that is labeled. An example of a labeled partition is shown in Figure 1. In general, there are many ways to label a partition, and the algorithm need only output one of those ways. Next, we show that any cake-cutting algorithm can be converted to an equivalent labeled algorithm.

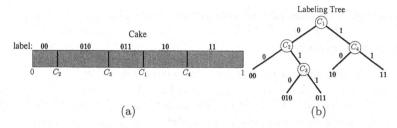

Fig. 1. (a) A labeled partition. (b) Corresponding labeling tree.

Theorem 2. *Every cake-cutting algorithm is equivalent to a labeled cake-cutting algorithm.*

We now show that a labeled cake-cutting algorithm can be used to sort N positive distinct integers $x_1, \ldots, x_N$. To relate sorting to cake-cutting, we first define a "less than" relation for pieces. If P_1 and P_2 are separated, then $P_1 < P_2$ if P_1 is on the left of P_2. Clearly, this "<" relation imposes a total order on any set of separated pieces. Our approach is to show that given N positive distinct integers, we can construct utility functions such that any fair division will allow us to sort the integers *quickly*. Define the utility functions $F_i(x) = \min(1, N^{x_i}x)$, for user u_i. In what follows, F_i will always refer to the utility functions defined above. Let $V_i = 1/N^{x_i}$. Only pieces that overlap $[0, V_i]$ have positive value for user u_i.

Consider any N-partition $\mathcal{W}_1, \ldots, \mathcal{W}_N$, such that each $\mathcal{W}_i$ has a non-zero value for the respective user u_i. Let $R_i \in \mathcal{W}_i$ be the rightmost piece of $\mathcal{W}_i$ that overlaps $[0, V_i]$. The ordering relation on pieces now induces an ordering on portions: $\mathcal{W}_i < \mathcal{W}_j$ if and only if $R_i < R_j$. Next, we show that the order of the portions $\mathcal{W}_i$ is related with the order of the integers x_i.

Lemma 3. *Let $\mathcal{W}_1, \ldots, \mathcal{W}_N$ be a fair cake-division for the utility functions $F_1, \ldots, F_N$. Then, $x_i < x_j$ if and only if $\mathcal{W}_j < \mathcal{W}_i$.*

The ordering relation on portions can be used to sort the N-partition $\mathcal{W}_1, \ldots, \mathcal{W}_N$, i.e., find the sequence of indices $i_1, \ldots, i_N$, such that $\mathcal{W}_{i_1} < \mathcal{W}_{i_2} < \cdots < \mathcal{W}_{i_N}$. An application of Lemma 3 then gives that $x_{i_1} > x_{i_2} > \cdots > x_{i_N}$, thus sorting the partition is equivalent to sorting the integers. We now show that if the partition is labeled, we can use the labels to sort the portions $\mathcal{W}_i$ quickly, which in turn will allow us to sort the integers quickly.

Lemma 4. *Any labeled N-partition $\mathcal{W}_1, \ldots, \mathcal{W}_N$, can be sorted in $O(K)$ time, where K is the total number of pieces in the partition.*

By Lemma 3, sorting the partition $\mathcal{W}_1, \ldots, \mathcal{W}_N$ is equivalent to (reverse) sorting the integers $x_1, \ldots, x_N$. From Lemma 4, we know that if the fair cake-division is labeled, then we can sort the partition in $O(K)$ time, where K is the number of pieces in the partition. Thus, we obtain the following theorem, which reduces sorting to cake-cutting:

Theorem 3 (Reduction of sorting to cake-cutting). *Given a K-piece, labeled, fair cake-division for utility functions $F_1, \ldots, F_N$, we can sort the numbers $x_1, \ldots, x_N$ in $O(K)$ time.*

Theorem 2 showed that every cake-cutting algorithm can be converted to an equivalent labeled cake-cutting algorithm. Of importance is the complexity of this conversion. We say that a cake-cutting algorithm H that outputs a cake-division with K pieces is *linearly-labeled* if it can be converted to a labeled algorithm H' that outputs an equivalent cake division with $O(K)$ pieces using at most $O(K)$ extra time, i.e., if it can be converted to an equally efficient algorithm that outputs essentially the same division. To our knowledge, all the known cake-cutting algorithms are linearly-labeled. In particular, Algorithms A and B can be easily converted to labeled algorithms using at most $O(K)$ additional operations to output an equivalent cake-division. Since sorting is reducible to labeled cake-cutting, labeled cake-cutting cannot be faster than sorting. We have the following result.

Theorem 4 (Lower bound for labeled algorithms). *For any linearly-labeled fair cake-cutting algorithm H, there are utility functions for which $\Omega(N \log N)$ comparisons will be required.*

From the practical point of view one might like to limit the amount of computation the superuser is allowed to use in order to determine what cuts are to be made. Each step in the algorithm involves a cut, and computations necessary for performing the cut. Among these computations might be comparisons, i.e., the superuser might compare cut positions. At step t, let K_t denote the number of comparisons performed. The algorithm is *comparison-bounded* if $\sum_{t=1}^{T} K_t \leq \alpha T$ for a constant α and all T. Essentially, the number of comparisons is linear in the number of cuts. The labeled algorithms A are B are easily shown to be comparison-bounded. We now give our lower bound on the number of cuts required for linearly-labeled comparison-bounded algorithms.

Theorem 5 (Lower bound for comparison-bounded algorithms). *For any linearly-labeled comparison-bounded fair algorithm H, utility functions exist for which $\Omega(N \log N)$ cuts will be made.*

5   Lower Bounds for Envy-Free Algorithms

We give lower bounds on the number of cuts required for strong and super envy-free division, when such divisions exist. We show that there exist utility functions that admit acceptable divisions for which $\Omega(N^2)$ cuts are needed.

Theorem 6 (Lower bound for strong envy-free division). *There exist utility functions for which a strong envy-free division requires $\Omega(0.086N^2)$ cuts.*

Theorem 7 (Lower bound for super envy-free division). *There exist utility functions for which a super envy-free division requires $\Omega(0.25N^2)$ cuts.*

6 Concluding Remarks

The most general results are that any cake-cutting algorithm can be converted to a solid piece algorithm, and then to a labeled algorithm. We then showed that any labeled fair cake-cutting algorithm can be used to sort, therefore any fair cake-cutting algorithm can be used to sort. This provided the connection between sorting and cake-cutting. We also provided an independent strong result for phased algorithms, namely that $\Omega(N \log N)$ cuts are needed to guarantee each user a positive valued portion, and we also obtained $\Omega(N^2)$ bounds for two types of envy-free division. Important open questions remain and we refer the reader to the technical report [10] and the literature for more details.

References

1. J. Barbanel. Game-theoretic algorithms for fair and strongly fair cake division with entitlements. *Colloquium Math.*, 69:59–53, 1995.
2. J. Barbanel. Super envy-free cake division and independence of measures. *J. Math. Anal. Appl.*, 197:54–60, 1996.
3. Steven J. Brams and Allan D Taylor. An envy-free cake division protocol. *Am. Math. Monthly*, 102:9–18, 1995.
4. Steven J. Brams and Allan D. Taylor. *Fair Division: From Cake-Cutting to Dispute Resolution.* Cambridge University Press, New York, NY, 1996.
5. Stephen Demko and Theodore P. Hill. Equitable distribution of indivisible objects. *Mathematical Social Sciences*, 16(2):145–58, October 1988.
6. L. E. Dubins and E. H. Spanier. How to cut a cake fairly. *Am. Math. Monthly*, 68:1–17, 1961.
7. Jacob Glazer and Ching-to Albert Ma. Efficient allocation of a 'prize' – King Solomon's dilemma. *Games and Economic Behavior*, 1(3):223–233, 1989.
8. C-J Haake, M. G. Raith, and F. E. Su. Bidding for envy-freeness: A procedural approach to n-player fair-division problems. *Social Choice and Welfare*, To appear.
9. Jerzy Legut and Wilczyński. Optimal partitioning of a measuarble space. *Proceedings of the American Mathematical Society*, 104(1):262–264, September 1988.
10. Malik Magdon-Ismail, Costas Busch, and Mukkai Krishnamoorthy. Cake-cutting is not a piece of cake. Technical Report 02-12, Rensselaer Polytechnic Institute, Troy, NY 12180, USA, 2002.
11. Elisa Peterson and F. E. Su. Four-person envy-free chore division. *Mathematics Magazine*, April 2002.
12. K. Rebman. How to get (at least) a fair share of the cake. *in Mathematical Plums (Edited by R. Honsberger), The Mathematical Association of America*, pages 22–37, 1979.
13. Jack Robertson and William Webb. Approximating fair division with a limited number of cuts. *J. Comp. Theory*, 72(2):340–344, 1995.
14. Jack Robertson and William Webb. *Cake-Cutting Algorithms: Be Fair If You Can.* A. K. Peters, Nattick, MA, 1998.
15. H. Steinhaus. The problem of fair division. *Econometrica*, 16:101–104, 1948.
16. F. E. Su. Rental harmony: Sperner's lemma in fair division. *American Mathematical Monthly*, 106:930–942, 1999.
17. Gerhard J. Woeginger. An approximation scheme for cake division with a linear number of cuts. In *European Symposium on Algorithms (ESA)*, pages 896–901, 2002.

The Price of Truth: Frugality in Truthful Mechanisms

Kunal Talwar*

University of California, Berkeley CA 94720, USA,
kunal@cs.berkeley.edu

Abstract. The celebrated Vickrey-Clarke-Grove(VCG) mechanism induces selfish agents to behave truthfully by paying them a premium. In the process, it may end up paying more than the actual cost to the agents. For the minimum spanning tree problem, if the market is "competitive", one can show that VCG never pays too much. On the other hand, for the shortest s-t path problem, Archer and Tardos [5] showed that VCG can overpay by a factor of $\Omega(n)$. A natural question that arises then is: For what problems does VCG overpay by a lot? We quantify this notion of overpayment, and show that the class of instances for which VCG never overpays is a natural generalization of matroids, that we call *frugoids*. We then give some sufficient conditions to upper bound and lower bound the overpayment in other cases, and apply these to several important combinatorial problems. We also relate the overpayment in an suitable model to the *locality ratio* of a natural local search procedure.

Classification: Current challenges, Mechanism design.

1 Introduction

Many problems require the co-operation of multiple participants, e.g. several autonomous systems participate to route packets in the Internet. Often these participants or *agents* have their own selfish motives, which may conflict with social welfare goals. In particular, it may be in an agent's interest to misrepresent her utilities/costs. The field of *mechanism design* deals with the design of protocols which ensure that the designer's goals are achieved by incentivizing selfish agents to be truthful. A *mechanism* is a protocol that takes the announced preferences of a set of agents and returns an *outcome*. A mechanism is *truthful* or *strategy proof* if for every agent, it is most beneficial to reveal her true preferences.

Consider, for example the problem of choosing one of several contractors for a particular task. Each contractor bids an amount representing her cost for performing the task. If we were to choose the lowest bidder, and pay her what her bid was, it might be in her interest to bid higher than her true cost, and make a large profit. We may, on the other hand, use the *VCG mechanism* [24,8, 11] which in this case, selects the lowest bidder, and pays her an amount equal

---

* Supported in part by NSF grants CCR-0105533 and CCR-9820897.

H. Alt and M. Habib (Eds.): STACS 2003, LNCS 2607, pp. 608–619, 2003.
© Springer-Verlag Berlin Heidelberg 2003

to the second lowest bid. It can be shown that in this case, it is in the best interest of every contractor to reveal her true cost (assuming that agents don't collude). Moreover, the task get completed at the lowest possible cost. The VCG mechanism is strategy proof, and minimizes the *social cost*. On the other hand, the cost to the mechanism itself may be large. In this case, for example, while the task gets completed by the most efficient contractor, the amount the mechanism has to pay to extract the truth is *more* than the true cost. In this paper, we address the question: How much more?

We look at a more general setting where some task can be accomplished by hiring a team of agents. Each agent performs a fixed service and incurs a fixed cost for performing that service. There are some given teams of agents, such that any one team can accomplish the complete task. The cost of each agent is known only to the agent herself, and agents are selfish. While the VCG mechanism selects a team that performs the task with minimum cost to itself, the amount that the mechanism pays to the agents may be high. We are interested in characterizing problems where this payment is *not too large*. Archer and Tardos [5] showed that in general, this payment can be $\Omega(n)$ times the real cost, even if there is sufficient competition.

It is not immediately clear what this payment should be compared to. Ideally, we would like to say that we never pay too much more than the real cost of the optimal. However, in a monopolistic situation, VCG can do pretty badly. So we impose the condition that the market is sufficiently competitive(a notion elucidated later), and we would be satisfied if our mechanism did well under these constraints. Qualitatively, we want to say that a mechanism is *frugal* for an instance if, in the presence of competition, the amount that the mechanism pays is not too much more than the true cost. Concretely let opt be the most cost-effective team for the task. We compare the amount that the mechanism pays to the agents in opt to the cost of the best rival solution opt′, i.e. the best solution to the instance that does not use any agent in opt. If it is the case that opt and opt′ have exactly the same cost, it can be shown that VCG does not over-pay, i.e. it pays exactly the cost of opt. However, if opt′ is a little costlier than opt, the performance may degrade rapidly. We define the *frugality ratio* of VCG on an instance to be the worst possible ratio of the payment to the cost of opt′.[1] This definition turns out to be equivalent to the *agents are substitutes* condition defined independently by [7] in a different context. We discuss the implications of this in section 5.

Under this definition of frugality, we characterize exactly the class of problems with frugality ratio 1. This class is a natural generalization of the class of matroids, which we call *frugoids*. We also give more general upper and lower bounds on the frugality in terms of some parameters of the problem. Using these, we classify several interesting problems by their frugality ratio. We also note a very interesting connection between the notion of frugality, and the *locality gap* of a natural local search procedure for the facility location problem; we discuss this in section 4.

---

[1] We show in section 3.1 how this definition of frugality relates well to the notion of overpayment in the presence of competition.

Related Work

The problem of mechanism design has classically been a part of game theory and economics (see, e.g. [21], [18]). In the past few years, there has been a lot of work the border of computer science, economics and game theory (see [22] for a survey). Nisan and Ronen [19] applied the mechanism design framework to some optimization problems in computer science. Computational issues in such mechanisms have also been considered in various scenarios (see, e.g.[20], [10], [15], [7],[6]). Work has also been done is studying efficienlty computable mechanisms other than VCG for specific problems (e.g. [16],[4],[17],[3]).

The issue of frugality is raised in [4] who look at a scheduling problem and compare the payment to the actual cost incurred by the machines under competitiveness assumptions. In [5,9], the authors show that for the shortest path problem, the overpayment is large even in the presence of competition. On the positive side, [6] show that when the underlying optimization problem is a matroid, the frugality ratio is 1. [12] relates frugality to the core of a cooperative game.

The rest of the paper is organized as follows. In section 2, we define the framework formally. We state and prove our results on frugality in section 3. In section 4, we apply the results to several combinatorial problems, and in section 5, we consider some interesting implications. We conclude in section 6 with some open problems.

2   Definitions and Notation

We first define formally the problem for which we analyze the performance of VCG. We are given a set E of elements and a family $\mathcal{F} \subseteq 2^E$ of *feasible* subsets of E. We shall call $(E, \mathcal{F})$ a *set system*. A set system $(E, \mathcal{F})$ is *upwards closed* if for every $S \in \mathcal{F}$ and every superset $T : S \subseteq T \subseteq E$, it is the case that $T \in \mathcal{F}$. We also have a non-negative cost function $\hat{c} : E \to \Re_0$. For a set $S \subseteq E$, let $\hat{c}(S) = \sum_{e \in S} \hat{c}(e)$. The goal is to find a feasible set of minimum total cost. Clearly for the minimization problem, there is no loss of generality in assuming that $(E, \mathcal{F})$ is upward closed. $(E, \mathcal{F})$ is therefore fully defined by the minimal sets in $\mathcal{F}$. We shall henceforth represent a set system by its minimal feasible sets (*bases*) $\mathcal{B} = \{S \in \mathcal{F} : $ no proper subset of S is in $\mathcal{F}\}$. Note that the family $\mathcal{B}$ may be exponential in size and given implicitly. Several problems such as minimum spanning tree, shortest path, etc. can be put in this framework.

Now suppose that each element $e \in E$ is owned by a selfish agent, and she alone knows the true cost $\hat{c}(e)$ of this element. A *mechanism* is denoted as $m = (\mathcal{A}, p)$ where

- $\mathcal{A}$ is an allocation algorithm that takes as input the instance $(E, \mathcal{B})$ and the revealed costs $c(e)$ for each agent e, and outputs a set $S \in \mathcal{F}$.
- p is a payment scheme that defines the payment $p(e)$ to be made to each agent.

For the VCG mechanism, $\mathcal{A}$ selects the minimum cost feasible set. The payment scheme p is as follows. Let $OPT(c)$ be the cost of an optimal solution under cost function c, i.e. $OPT(c) = \min_{S \in \mathcal{B}} c(S)$ and let $\text{opt}(c)$ denote the set achieving this minimum[2]. Let $c[e \mapsto x]$ denote the cost function c' which is equal to c everywhere except at e, where it takes the value x. Then

$$p(e, c) = \begin{cases} 0 & \text{if } e \notin \text{opt}(c) \\ OPT(c[e \mapsto \infty]) - OPT(c[e \mapsto 0]) & \text{if } e \in \text{opt}(c) \end{cases}$$

It can be shown that the mechanism is *truthful*(see for example [18]). Hence assuming that agents are rational, $c(e) = \hat{c}(e)$ for all agents. We denote by $OPT'(c)$ the cost of the cheapest feasible set disjoint from $\text{opt}(c)$, i.e. $OPT'(c) = \min_{S \in \mathcal{B}, S \cap \text{opt}(c) = \phi} \{c(S)\}$. Note that such a set may not always exist, in which case this minimum is infinite. If $OPT'(c)$ is finite, let $\text{opt}'(c)$ denote the set achieving the minimum. For any set $S \subseteq E$, let $p(S, c) = \sum_{e \in S} p(e, c)$. For an instance $(E, \mathcal{B}, c)$, such that $OPT'(c) > 0$, define the *frugality ratio* $\phi(E, \mathcal{B}, c) = p(\text{opt}(c), c)/OPT'(c)$. (If $OPT'(c) = 0$, in which case $p(\text{opt}(c), c) = 0$ as well, we shall define the frugality ratio to be 1.) For a set system $(E, \mathcal{B})$, define the frugality ratio of the the set system $\phi(E, \mathcal{B}) = \sup_c \phi(E, \mathcal{B}, c)$.

An alternative definition of frugality, what we call the *marginal frugality*, is the ratio of VCG overpayment to the difference in the costs of opt and opt'. Formally, $\phi'(E, \mathcal{B}) = \sup_c (p(\text{opt}(c), c) - OPT(c))/(OPT'(c) - OPT(c))$. In the next section, we show how these two definitions are related. The following theorem shows that the marginal frugality of a set system bounds the rate of change of the payment with respect to the value of OPT'. Thus the marginal frugality behaves like the derivative of the payment with respect to $OPT'(c)$. We omit the proofs of the following from this extended abstract.

Theorem 1. *Let $(E, \mathcal{B})$ be a set system and let c and c' be cost functions such that $\text{opt}(c) = \text{opt}(c') = A$, $\text{opt}'(c) = \text{opt}'(c') = B$, and $c(e) = c'(e)$ for all $e \notin B$. Then*

$$p(A, c') - p(A, c) \le \phi'(E, \mathcal{B}) \cdot (c'(B) - c(B))$$

Corollary 1.

$$\phi'(E, \mathcal{B}) = \lim_{\epsilon \to 0^+} \sup(p(A, c') - p(A, c))/\epsilon$$

where the sup is over all c, c' satisfying the conditions in theorem 1 such that $c'(B) - c(B) = \epsilon$.

Thus, the "slope" of the payment function with respect to OPT' is always less than ϕ', and is equal to ϕ' at $OPT' = OPT$.

3 Frugality

3.1 Canonical Cost Functions

Call a cost function c *canonical* if the following hold:

- $c(e) = 0$ for all $e \in \text{opt}(c)$.
- $c(e) = \infty$ for all $e \notin \text{opt}(c) \cup \text{opt}'(c)$.

---

[2] We assume that ties are broken in a particular way, say $\text{opt}(c)$ is the lexicographically smallest set $S \in \mathcal{B}$ such that $c(S) = OPT(c)$, so that $\text{opt}(c)$ is well defined.

The following lemma shows that for every set system $(E, \mathcal{B})$, and every cost function c, there exists a cost function c' which is canonical, and has a higher frugality ratio. Hence the definition of frugality ratio can be modified so as to consider only canonical cost functions.

Lemma 1. *Let $(E, \mathcal{B})$ be a set system, and c be a non negative cost function on E. Then there is a cost function c' such that the following hold:*

- *c' is canonical.*
- *$\phi(E, \mathcal{B}, c') \geq \phi(E, \mathcal{B}, c)$*

We omit the proof from this extended abstract.

It is easy to see that VCG payments satisfy the following properties.

Property 1. For every cost function c such that $OPT(c) = OPT'(c)$, $p(\text{opt}(c), c) = OPT(c)$.

Property 2. For every cost function c and every constant $\alpha > 0$, $p(\text{opt}(\alpha c), \alpha c) = \alpha p(\text{opt}(c), c)$.

The following lemma shows that under some constraints, VCG payments are super-additive.

Lemma 2. *Let c_1 be an arbitrary cost function and c_2 be a canonical cost function such that $\text{opt}(c_1) = \text{opt}(c_2)$ and $\text{opt}'(c_1) = \text{opt}'(c_2)$. Let $c = c_1 + c_2$. Then*

$$p(\text{opt}(c), c) \geq p(\text{opt}(c_1), c_1) + p(\text{opt}(c_2), c_2)$$

Proof. First note that $\text{opt}(c) = \text{opt}(c_1) = \text{opt}(c_2)$ and $\text{opt}'(c) = \text{opt}'(c_1) = \text{opt}'(c_2)$. Now let e be any element in $\text{opt}(c)$. Then

$$p(e, c) = OPT(c[e \mapsto \infty]) - OPT(c[e \mapsto 0])$$
$$\text{[by definition]}$$
$$\geq c_1(\text{opt}(c[e \mapsto \infty])) + c_2(\text{opt}(c[e \mapsto \infty])) - c_1(\text{opt}(c)) - c_2(\text{opt}(c))$$
$$\text{[Since } c[e \mapsto 0](\text{opt}(c)) \leq c(\text{opt}(c))]$$
$$\geq c_1(\text{opt}(c_1[e \mapsto \infty])) + c_2(\text{opt}(c_2[e \mapsto \infty])) - c_1(\text{opt}(c))$$
$$\text{[Since } c(\text{opt}(c)) \leq c(\text{opt}(c')), c_2 \text{ is canonical]}$$
$$= (c_1(\text{opt}(c_1[e \mapsto \infty])) - c_1(\text{opt}(c_1[e \mapsto 0]))) + c_2(\text{opt}(c_2[e \mapsto \infty]))$$
$$\text{[Since } \text{opt}(c_1[e \mapsto 0]) = \text{opt}(c)]$$
$$= p(e, c_1) + p(e, c_2)$$
$$\text{[} c_2 \text{ is canonical]}$$

Summing over all e in $\text{opt}(c)$, we get the desired result.

Lemmas 1 and 2 imply:

Corollary 2. *Let $(E, \mathcal{B})$ be a set system. Then for any $\epsilon > 0$, there is a cost function c such that $OPT(c) = 1$, $OPT'(c) = (1 + \epsilon)$ and $p(\text{opt}(c), c) \geq 1 + \phi(E, \mathcal{B})\epsilon$.*

Proof. (Sketch) Let c be a canonical cost function achieving [3] a frugality ratio of $\phi(E, \mathcal{B})$. Let $A = \text{opt}(c)$ and $B = \text{opt}'(c)$. Since c is canonical, by observation 1, we can assume that $c(B) = 1$. Let c' be a cost function such that $c'(a) = \frac{1}{|A|}$ for $a \in A$, $c'(b) = \frac{1}{|B|}$ for $b \in B$ and $c'(x) = 1$ otherwise. Finally, consider the cost function $c' + \epsilon c$. By lemma 2, the claim follows.

Corollary 3. *For any set system* $(E, \mathcal{B})$, $\phi'(E, \mathcal{B}) \geq \phi(E, \mathcal{B})$. *Moreover,* $\phi(E, \mathcal{B}) = 1$ *iff* $\phi'(E, \mathcal{B}) = 1$.

Lemma 1 shows that the worst frugality ratio is attained at a canonical cost function, which is very non competitive. Thus it might seem that even when the frugality ratio is large, the VCG mechanism could do well for cost functions we care about. The above corollary however shows that if the frugality ratio is high, there are competitive cost functions where the overpayment is large. On the other hand, a low frugality ratio clearly implies that the overpayment is never large. Thus our definition of frugality is a robust one.

We now note that for any canonical cost function c, the VCG payments have a nice structure. The proof is simple and omitted from this extended abstract.

Proposition 1. *Let c be any canonical cost function. Then*

$$p(e, c) = \min_{T \subseteq \text{opt}'(c) : \text{opt}(c) \setminus \{e\} \cup T \in \mathcal{F}} c(T)$$

3.2   Frugoids

For disjoint sets $A, B \in \mathcal{B}$, and any $Y \subseteq B$, we say that $x \in A$ is *dependent* on Y with respect to A, B if x cannot be replaced in A by some element of $B \setminus Y$, i.e. if for any $y \in B \setminus Y, A \setminus \{x\} \cup \{y\}$ is not feasible. Define the *set of dependents* of Y with respect to A, B as $D^{A,B}(Y) = \{x \in A : A \setminus \{x\} \cup \{y\}$ is not feasible for any $y \in B \setminus Y\}$.

Call a set system $(E, \mathcal{B})$ a *frugoid* if for every pair of disjoint sets $A, B \in \mathcal{B}$ and every $Y \subseteq B$, $|D^{A,B}(Y)| \leq |Y|$. The following proposition shows that matroids satisfy the above condition for every pair of bases A and B (not necessarily disjoint).

Proposition 2. *Let $(E, \mathcal{B})$ be a matroid. Then for any $A, B \in \mathcal{B}$ and any $Y \subseteq B$, $|D^{A,B}(Y)| \leq |Y|$.*

Proof. From the definition of matroids, it follows that any two base sets have the same cardinality. Also for any base sets S and T and any set $T' \subseteq T$ satisfying $|T'| < |T|$, there exists $S' \subseteq S \setminus T'$ with $|S'| = |S| - |T'|$ such that $T' \cup S'$ is a base set. Now consider any $A, B \in \mathcal{B}$ and let $Y \subseteq B$ be arbitrary. Since $|B \setminus Y| < |B|$ and $A, B \in \mathcal{B}$, there exists $X \subseteq A$ with $|X| = |Y|$ such that $(B \setminus Y) \cup X \in \mathcal{B}$. We shall show that $D^{A,B}(Y) \subseteq X$, from which the claim

---

[3] We assume for simplicity that there is a cost function attaining the frugality. Similar results would hold if this was not the case.

follows. Consider any $a \in A \setminus X$. Since $|A \setminus \{a\}| < |A|$ and $(B \setminus Y) \cup X \in \mathcal{B}$, there exists a $b \in ((B \setminus Y) \cup X) \setminus (A \setminus \{a\})$ such that $A \setminus \{a\} \cup \{b\}$ is a base set. However, then b must belong to $B \setminus Y$. Thus $a \notin D^{A,B}(Y)$. Since a was arbitrary, $D^{A,B}(Y)$ contains no elements from $A \setminus X$, and hence must be contained in X.

Thus every matroid is a frugoid. We give below an example of a set system which is a frugoid but not a matroid.

Example 1. Consider the set system $(E_1, \mathcal{B}_1)$ where $E_1 = \{a_1, a_2, b_1, b_2, c_1, c_2\}$ and $\mathcal{B}_1$ contains the following sets:

$$\{a_1, b_1, c_1\} \ \{a_2, b_1, c_1\} \ \{a_1, b_2, c_1\} \ \{a_1, b_1, c_2\}$$
$$\{a_2, b_2, c_2\} \ \{a_1, b_2, c_2\} \ \{a_2, b_1, c_2\} \ \{a_2, b_2, c_1\}$$
$$\{a_1, a_2\}$$

Since any pair of disjoint sets in $\mathcal{B}_1$ come from the matroid $(E_1, \mathcal{B}_1 \setminus \{\{a_1, a_2\}\})$, it is a frugoid.

The following theorem gives an exact characterization of set systems that have frugality ratio at most one.

Theorem 2. *A set system $(E, \mathcal{B})$ is a frugoid iff $\phi(E, \mathcal{B}) \leq 1$.*

Proof. (proof of $\Rightarrow$) From Lemma 1, it suffices to show that for all canonical cost functions c, $\phi(E, \mathcal{B}, c) \leq 1$. Let c be a canonical cost function, let $A = \mathrm{opt}(c)$ and $B = \mathrm{opt}'(c)$. We need to show that $p(A) \leq c(B)$. We shall find a one-one mapping π from A to B such that for all $a \in A, p(a)$ will be no more than $c(\pi(a))$. The claim would then follow. Consider a bipartite graph with vertex set $A \cup B$, and an edge between a and b if $(A \setminus \{a\}) \cup \{b\}$ is feasible (and hence $p(a) \leq c(b)$). The condition $|D^{A,B}(Y)| \leq |Y|$ implies that there is no Hall set in A (recall that a hall set is a set A such whose neighbourhood is of size strictly smaller than A itself). Hall's theorem [13](see [14] for a proof) then implies that we can find such a mapping. The claim follows.

(Proof of $\Leftarrow$) We first show that any two disjoint base sets must have the same cardinality. Assume the contrary. Let A and B be disjoint sets with different cardinalities. Without loss of generality, $|A| > |B|$. Consider the cost function c defined as follows:

$$c(e) = \begin{cases} 0 & \text{if } e \in A \\ 1 & \text{if } e \in B \\ \infty & \text{otherwise} \end{cases}$$

Now consider any $e \in A$. Since $A \setminus \{e\}$ is not feasible, $p(e, c) \geq 1$. Thus $p(A, c) \geq |A|$. On the other hand, $c(B) = |B| < |A|$. This contradicts the fact that frugality ratio of $(E, \mathcal{B})$ is at most 1.

Now suppose the condition is violated for some A, B, Y. Consider the canonical cost function c defined as follows:

$$c(e) = \begin{cases} 0 & \text{if } e \in A \\ 1 & \text{if } e \in B \setminus Y \\ 2 & \text{if } e \in Y \\ \infty & \text{otherwise} \end{cases}$$

Again, for any $e \in A$, $p(e, c) \geq 1$. Moreover, from the definition of dependency, for any $e \in A$ which is 1-dependent on Y, $p(e, c) \geq 2$. Thus $p(A, c) \geq |A| + |D^{A,B}(Y)|$. On the other hand, $c(B) = |B| + |Y|$. Since $|A| = |B|$ and $|D^{A,B}(Y)| > |Y|$, $\phi(E, \mathcal{B}, c) > 1$, which is a contradiction. Hence the claim follows.

3.3   Non Frugoids

For systems which are not frugoids, we give some upper bounds on the frugality ratios.

Analogous to the definition of dependence, we define *k-dependence* as follows. For disjoint sets $A, B \in \mathcal{B}$, and any $Y \subseteq B$, we say that $x \in A$ is *k-dependent* on Y with respect to A, B if x cannot be replaced in A by at most k elements of $B \setminus Y$, i.e. if for any $X \subseteq B \setminus Y : |X| \leq k$, $A \setminus \{x\} \cup X$ is not feasible. Define the *set of k-dependents* of Y with respect to A, B as $D_k^{A,B}(Y) = \{x \in A : A \setminus \{x\} \cup X$ is not feasible for any $X \subseteq B \setminus Y, |X| \leq k\}$.

Theorem 3. *The following hold:*
(i)   Let $(E, \mathcal{B})$ be a set system such that for every pair of disjoint sets A and B, and for some positive integer k, and for all $Y \subseteq B$, it is the case that $k \frac{|D_k^{A,B}(Y)|}{|Y|} \leq f$. Then the frugality ratio $\phi(E, \mathcal{B}) \leq f$.
(ii) Let $(E, \mathcal{B})$ be a set system such that the size of each set in $\mathcal{B}$ is at most l. Then $\phi(E, \mathcal{B}) \leq l$.
(iii) Let $(E, \mathcal{B})$ be a set system instance derived from a set cover problem where each set is of size at most k. Then $\phi(E, \mathcal{B}) \leq k$.

Proof. The proof of (i) is analogous to theorem 2. (ii) is immediate from the fact that for any $e \in \text{opt}(c)$, $p(e, c) \leq OPT'(c)$. We prove (iii) below. Let c be a canonical cost function. Consider any edge e in $\text{opt}(c)$. Since c is canonical, $p(e, c) \leq c(T_e)$ for any $T_e \subseteq \text{opt}'(c)$ such that $\text{opt}(c) \setminus \{e\} \cup T_e$ is feasible. We construct one such T_e as follows. For each $v' \in e$ that is covered only by e in $\text{opt}(c)$, add to T_e an arbitrary set e' from $\text{opt}'(c)$ that covers v. We say then that e *requires* e' to cover v'. We now show that $\sum_{e \in \text{opt}(c)} c(T_e) \leq kOPT'(c)$. Let $e'' = \{v_1', v_2', \dots, v_l'\}$ be any set in $\text{opt}'(c)$. If v_i' is covered more than once in $\text{opt}(c)$, no set in $\text{opt}(c)$ requires e'' to cover v_i'. Otherwise, for each v_i', there is at most one set in $\text{opt}(c)$ that requires e'' to cover v_i'. Hence in all, at most k T_e's for $e \in \text{opt}(c)$ contain e''. Now $p(\text{opt}(c), e) \leq \sum_{e \in \text{opt}(c)} c(T_e) = \sum_{e'' \in \text{opt}'(c)} \sum_{e \in \text{opt}(c): e'' \in T_e} c(e'') \leq kc(\text{opt}'(c))$. Hence the claim follows.

The same upper bounds can be shown on the marginal frugality ϕ' as well. We omit the proof from this extended abstract.

We now state some simple lower bounds on the frugality of set systems.

Theorem 4. *The following hold:*
(i)   Let $(E, \mathcal{B})$ be a set system such that $\mathcal{B}$ contains two disjoint sets A and B such that $|A| = \alpha|B|$. Then $\phi(E, \mathcal{B}) \geq \alpha$.

(ii) Let $(E, \mathcal{B})$ be a set system such that $\mathcal{B}$ contains two disjoint sets A and B such that for all $a \in A$, for all $Y \subseteq B$ of cardinality less than k, $A \setminus \{a\} \cup Y$ is not feasible. Then $\phi(E, \mathcal{B}) \geq k \frac{|A|}{|B|}$.

Proof. Consider the cost function:

$$c(e) = \begin{cases} 0 & \text{if } e \in A \\ 1 & \text{if } e \in B \\ \infty & \text{otherwise} \end{cases}$$

It is easy to check that this cost function gives the lower bounds in both cases.

Note that because of corollary 3, these lower bounds apply to the marginal frugality ϕ' as well.

In the next section, we use these results to estimate the frugality ratio of several important combinatorial problems.

4   Examples

For a class of set system instances, we define the frugality ratio of the class as the largest frugality ratio for an instance from the class. For example, from theorem 2, it follows that the minimum spanning tree problem on graphs (being a matroid) has frugality 1. Figure 1 tabulates some simple consequences of theorems 3 and 4.

Problem	Frugality ratio	Theorems used
Any matroid problem	1	Prop.2 and Thm. 2
Vertex Cover with maximum degree d	d	Thms. 2(ii) and 3(i)
Edge Cover	2	Thms. 2(ii) and 3(ii)
Bipartite graph matching	$\Theta(n)$	Thms. 2(ii) and 3(ii)
Minimum cut	$\Theta(n)$	Thms. 2(ii) and 3(i)
Minimum vertex cut	$\Theta(n)$	Thms. 2(ii) and 3(i)
Dominating set	$\Theta(n)$	Thms. 2(ii) and 3(i)
Set cover - each set of size k	k	Thms. 2(iii) and 3(ii)
Uncapacitated facility location	4	See discussion below

Fig. 1. Frugality ratios of some combinatorial problems

Frugality and Locality Ratio

We note an interesting connection between frugality and *locality ratio* of a natural local search procedure for the facility location problem. The notion of frugality here is slightly different. We assume that the facilities are owned by agents, and only they know the facility cost. The distances however are well known. Analogous to the locality ratio analysis (theorem 4.3) of Arya et.al. [2], we can show that the payment to the facilities in opt is no more than

$(cost_f(\text{opt}) + 2cost_f(\text{opt}') + 3cost_s(\text{opt}'))$ where $cost_f()$ and $cost_s()$ denote the facility and the service costs of a solution. Thus the total "expenditure" in this case is $(cost_f(\text{opt}) + 2cost_f(\text{opt}') + 3cost_s(\text{opt}')) + cost_s(\text{opt}) \leq 4cost(\text{opt}')$. We omit the details from this extended abstract.

5 Discussion

In this section, we look at some interesting implications of the positive and negative results of the previous sections.

We first look at the problem of computing the VCG payments. In general, this requires solving an optimization problem for finding the optimal solution, and then solving one optimization problem for each agent in the optimal solution. [7] show that whenever the "agents are substitutes" condition holds, all the VCG payments can be computed using the variables in the dual of a linear programming formulation of the underlying optimization problem. (Intuitively, the dual variables correspond to the effect of the corresponding primal constraint on the value of the optimal, which is precisely the "bonus" to the agent in VCG). Since the "agents are substitutes" condition is equivalent[4] to the set system having frugality ratio 1, the VCG payment for frugoids can be computed by solving a single linear program and its dual.

One criticism of VCG is that it requires all agents to divulge their true values to the auctioneer (and hence trust the auctioneer to not somehow use this information, e.g. in similar cases in the future). A suggested solution to this is to design *iterative mechanisms*, where agents respond to a sequence of offers made by the auctioneer. This has the advantages of simplicity and privacy (see e.g. [1], [23],etc. for further arguments in favour of iterative auctions). [7] also show that under the "agents are substitutes" condition, an iterative mechanism can be designed that gives the same outcome as the VCG mechanism. This, then holds for all frugoids. Moreover [6] conjecture that if the substitutes condition does not hold, there is no iterative mechanism yielding the Vickrey outcome. Assuming this conjecture then, frugoids is exactly the class of minimization problems which have iterative mechanisms implementing the social optimum.

We also note that in general, VCG is the only truthful mechanism that selects the optimal allocation and satisfies individual rationality. Moreover, since every dominant strategy mechanism has an equivalent truthful mechanism (the revelation principle), better frugality ratios cannot be achieved by any mechanism while maintaining optimality. However by imposing some additional restrictions, the class of truthful mechanisms can be made larger and more frugal mechanisms can be designed.

While we have addressed the questions of frugality of exact mechanisms, it turns out that approximate mechanisms need not even be approximately frugal, even for frugoids. In fact, there is a constant factor approximation algorithm for the minimum spanning tree problem that can be implemented truthfully, but the resulting mechanisms has frugality ratio $\Omega(n)$.

---

[4] We omit the simple proof from this extended abstract

6   Conclusion and Further Work

We have defined the frugality ratio which is a robust measure of the economic performance of a mechanism in the presence of competition. We show that frugoids, a natural generalization of matroids is the exact class of problems with frugality ratio 1. We also give lower and upper bounds on the frugality ratio of set systems. We use these to estimate the frugality of several interesting combinatorial problems. An exact characterization of set systems with frugality ratio exactly k for $k > 1$ is an interesting open problem. Further, while it turns out that some important problems have large frugality ratios in the worst case, it would be interesting to see if we can get better positive results by restricting the instances and/or restricting the cost functions in some reasonable way.

We also define the notion of *marginal frugality* and show that it is always lower bounded by the frugality ratio. We leave open the intriguing question of whether or not they are equal.

Acknowledgements. I would like to thank Christos Papadimitriou for several useful discussions and comments. I would also like to thank Ranjit Jhala, Amin Saberi and Nikhil Devanur for useful discussions. Many thanks to Amir Ronen, Aaron Archer and Vijay Vazirani for discussions that led to the frugality questions addressed here. I would also like to thank the anonymous refrees for their several helpful comments.

References

1. LAWRENCE M. AUSUBEL, "An Efficient Ascending-Bid Auction for Multiple Objects," Working Paper, University of Maryland, Department of Economics, 1997.
2. VIJAY ARYA, NAVEEN GARG, ROHIT KHANDEKAR, ADAM MEYERSON, KAMESH MUNAGALA, VINAYAKA PANDIT, "Local search heuristics for k-median and facility location problems", *Proceedings of the 33rd ACM symposium on the theory of computing, 2001.*
3. AARON ARCHER, CHRISTOS PAPADIMITRIOU, KUNAL TALWAR, ÉVA TARDOS, "An approximate truthful mechanism for combinatorial auctions with single parameter agents", *To appear in SODA 2003*
4. AARON ARCHER, ÉVA TARDOS "Truthful mechanisms for one-parameter agents",*Proceedings of the 42nd IEEE annual symposium on Foundations of Computer Science, 2001.*
5. AARON ARCHER, ÉVA TARDOS "Frugal Path Mechanisms", *SODA 2002.*
6. SUSHIL BIKHCHANDANI, SVEN DE VRIES, JAMES SCHUMMER, RAKESH VOHRA "Linear Programming and Vickrey Auctions," *IMA Volumes in Mathematics and its Applications, Mathematics of the Internet: E-Auction and Markets*, 127:75–116, 2001.
7. SUSHIL BIKHCHANDANI, JOSEPH M. OSTROY, "The package assignment model", working paper, 2001.
8. E. H. CLARKE "Multipart pricing of public goods", *Public Choice*, 8:17–33, 1971.
9. EDITH ELKIND, AMIT SAHAI, "Shortest Paths are costly", *manuscript*, October 2002.

10. JOAN FEIGENBAUM, CHRISTOS PAPADIMITRIOU, AND SCOTT SHENKER, "Sharing the cost of multicast transmissions", *Journal of Computer and System Sciences* 63:21–41, 2001.

11. THEODORE GROVES "Incentives in teams", *Econometrica*, 41(4):617–631, 1973.

12. RAHUL GARG, VIJAY KUMAR, ATRI RUDRA AND AKSHAT VERMA, "When can we devise frugal mechanisms for network design problems?", *manuscript*, 2002.

13. PHILIP HALL, "On representatives of subsets," *Journal of London Mathematics Society*, 10:26–30, 1935.

14. FRANK HARARY, "Graph Theory", Addison Wesley, 1971.

15. JOHN HERSHBERGER, SUBHASH SURI, ". Vickrey Pricing in network routing: Fast payment computation", *Proceedings of the 42nd IEEE Symposium on Foundations of Computer Science, 2001.*

16. DANIEL LEHMANN, LIADAN ITA O'CALLAGHAN, YOAV SHOHAM "Truth revelation in rapid approximately efficient combinatorial auctions", *1st ACM conference on electronic commerce, 1999*

17. AHUVA MU'ALEM, NOAM NISAN "Truthful approximation mechanism for restricted combinatorial auctions", to appear in *AAAI 2002*

18. ANDREU MAS-COLELL, MICHAEL D. WHINSTON, JERRY R. GREEN", Microeconomic Theory", Oxford University Press, 1995.

19. NOAM NISAN, AMIR RONEN "Algorithmic mechanism design", *Proceedings of the Thirty-First Annual ACM Symposium on Theory of Computing, 1999.*

20. NOAM NISAN, AMIR RONEN "Computationally feasible VCG mechanisms", *ACM conference on electronic commerce*, 242–252, 2000.

21. MARTIN J. OSBORNE, ARIEL RUBINSTEIN"A course in game theory ", MIT Press, Cambridge, 1994.

22. CHRISTOS PAPADIMITRIOU "Algorithms, games and the Internet", *Proceedings of the 33rd Annual ACM Symposium on Theory of Computation*, 749–753, 2001.

23. DAVID PARKES, LYLE UNGAR, "Iterative Combinatorial Auctions: Theory and Practice", *In Proc. 17th National Conference on Artificial Intelligence, (AAAI-00)* pp. 74–81, 2000.

24. WILLIAM VICKREY "Counterspeculation, auctions and competitive sealed tenders", *Journal of Finance*, 16:8–37, 1961.

Untameable Timed Automata!

(Extended Abstract)

Patricia Bouyer[1,2*]

[1] LSV – CNRS UMR 8643 & ENS de Cachan
61, Av. du Président Wilson
94235 Cachan Cedex – France
[2] BRICS[***] – Aalborg University
Fredrik Bajers Vej 7E
9220 Aalborg Ø – Denmark
bouyer@lsv.ens-cachan.fr

Abstract. Timed automata are a widely studied model for real-time systems. Since 8 years, several tools implement this model and are successfully used to verify real-life examples. In spite of this well-established framework, we prove that the forward analysis algorithm implemented in these tools is not correct! However, we also prove that it is correct for a restricted class of timed automata, which has been sufficient for modeling numerous real-life systems.

1 Introduction

Real-Time Systems – Since their introduction by Alur and Dill in [AD94], timed automata are one of the most studied models for real-time systems. Numerous works have been devoted to the "theoretical" comprehension of timed automata: determinization [AFH94], minimization [ACH+92], power of ε-transitions [BDGP98], power of clocks [ACH94,HKWT95], extensions of the model [DZ98,HRS98,CG00,BDFP00,BFH+01], logical characterizations [Wil94,HRS98]... have in particular been investigated. Practical aspects of the model have also been studied and several model-checkers are now available (HyTech[1] [HHWT97], Kronos[2] [DOTY96], Uppaal[3] [LPY97]). Timed automata afford to modelize many real-time systems and the existing model-checkers have allowed to verify a lot of industrial case studies (see the web pages of the tools or, for example, [HSLL97,TY98]).

Implementation of Timed Automata – The decidability of the timed automata model has been proved by Alur and Dill in [AD94]. It is based on the construction of the so-called region automaton: it abstracts finitely and in a correct way the behaviours of timed automata. However, such a construction does not support a natural implementation.

---

* Partly supported by the French RNRT Project Calife
*** Basic Research in Computer Science (www.brics.dk),
funded by the Danish National Research Foundation.
[1] http://www-cad.eecs.berkeley.edu:80/~tah/HyTech/
[2] http://www-verimag.imag.fr/TEMPORISE/kronos/
[3] http://www.uppaal.com/

Therefore, instead of this construction, algorithms glancing on-the-fly through the automaton are implemented. These last algorithms are based on the notion of zones and are efficiently implemented using data structures like DBMs [Dil89] and CDDs [BLP+99]. There are two classes of such algorithms, the class of forward analysis algorithms and the class of backward analysis algorithms. KRONOS implements the two kinds of algorithms, whereas UPPAAL implements only forward analysis algorithms, because it allows the feature of bounded integer variables, for which backward analysis is not well-appropriate.

Our contribution – In this paper, we are interested in the forward analysis algorithms. Our main result is that the forward analysis algorithm implemented in many tools is not correct for the whole class of timed automata. This might appear as very surprising because tools implementing this algorithm are successfully used since 8 years. The problem is due to the fact that we can compare the values of two clocks in the model. We then propose several subclasses of timed automata for which we can safely use a forward analysis algorithm. These subclasses contain in particular the large class of timed automata in which we can not compare two clocks, which might explain why this problem has not been detected before.

Outline of the paper – The structure of the paper is the following: after presenting basic definitions (Section 2), we recall some aspects of the implementation of timed automata and present in particular the algorithm we will study (Section 3). We then discuss the correctness of the algorithm and prove that it is surprisingly not correct (Section 4). However, it is correct for some subclasses of timed automata (Section 5). We conclude with a small discussion (Section 6). For lack of space, technical proofs are not presented in this paper but can be found in [Bou02b].

2   Preliminaries

If Z is any set, let Z^* be the set of *finite* sequences of elements in Z. We consider as time domain $\mathbb{T}$ the set Q^+ of non-negative rationals or the set R^+ of non-negative reals and Σ as a finite set of *actions*. A *time sequence* over $\mathbb{T}$ is a finite non decreasing sequence $\tau = (t_i)_{1 \leq i \leq p} \in \mathbb{T}^*$. A *timed word* $\omega = (a_i, t_i)_{1 \leq i \leq p}$ is an element of $(\Sigma \times \mathbb{T})^*$, also written as a pair $\omega = (\sigma, \tau)$, where $\sigma = (a_i)_{1 \leq i \leq p}$ is a word in Σ^* and $\tau = (t_i)_{1 \leq i \leq p}$ a time sequence in $\mathbb{T}^*$ of same length.

Clock Valuations – We consider a finite set X of variables, called *clocks*. A *clock valuation* over X is a mapping $v : X \to \mathbb{T}$ that assigns to each clock a time value. The set of all clock valuations over X is denoted $\mathbb{T}^X$. Let $t \in \mathbb{T}$, the valuation $v + t$ is defined by $(v + t)(x) = v(x) + t, \forall x \in X$. We also use the notation $(\alpha_i)_{1 \leq i \leq n}$ for the valuation v such that $v(x_i) = \alpha_i$. For a subset C of X, we denote by $[C \leftarrow 0]v$ the valuation such that for each $x \in C$, $([C \leftarrow 0]v)(x) = 0$ and for each $x \in X \setminus C$, $([C \leftarrow 0]v)(x) = v(x)$.

Clock Constraints – Given a set of clocks X, we introduce two sets of clock constraints over X. The most general one, denoted by $\mathcal{C}(X)$, is defined by the following grammar:

$$g ::= x \sim c \mid x - y \sim c \mid g \wedge g \mid true$$
$$\text{where } x, y \in X, c \in \mathbb{Z} \text{ and } \sim \in \{<, \leq, =, \geq, >\}.$$

We also use the proper subset of *diagonal-free* constraints where the comparison between two clocks is not allowed. This set is defined by the grammar:

$$g ::= x \sim c \mid g \wedge g \mid true,$$

where $x \in X$, $c \in \mathbb{Z}$ and $\sim \in \{<, \leq, =, \geq, >\}$.

A *k-bounded clock constraint* is a clock constraint that involves only constants between $-k$ and $+k$.

If v is a clock valuation we write $v \models g$ when v satisfies the clock constraint g and we say that v satisfies $x \sim c$ (resp. $x - y \sim c$) whenever $v(x) \sim c$ (resp. $v(x) - v(y) \sim c$).

Timed Automata – A *timed automaton* over $\mathbb{T}$ is a tuple $\mathcal{A} = (\Sigma, Q, T, I, F, X)$, where Σ is a finite alphabet of actions, Q is a finite set of states, X is a finite set of clocks, $T \subseteq Q \times [\mathcal{C}(X) \times \Sigma \times 2^X] \times Q$ is a finite set of transitions[4], $I \subseteq Q$ is the subset of initial states and $F \subseteq Q$ is the subset of final states.

A *path* in $\mathcal{A}$ is a finite sequence of consecutive transitions:

$$P = q_0 \xrightarrow{g_1, a_1, C_1} q_1 \dots q_{p-1} \xrightarrow{g_p, a_p, C_p} q_p$$

where $(q_{i-1}, g_i, a_i, C_i, q_i) \in T$ for each $1 \leq i \leq p$.

The path is said to be *accepting* if it starts in an initial state ($q_0 \in I$) and ends in a final state ($q_p \in F$). A *run* of the automaton through the path P is a sequence of the form:

$$(q_0, v_0) \xrightarrow[t_1]{g_1, a_1, C_1} (q_1, v_1) \dots \xrightarrow[t_p]{g_p, a_p, C_p} (q_p, v_p)$$

where $\tau = (t_i)_{1 \leq i \leq p}$ is a time sequence and $(v_i)_{1 \leq i \leq p}$ are clock valuations defined by:

$$\begin{cases} v_0(x) = 0, \ \forall x \in X, \\ v_{i-1} + (t_i - t_{i-1}) \models g_i, \\ v_i = [C_i \leftarrow 0] \left(v_{i-1} + (t_i - t_{i-1})\right). \end{cases}$$

The label of the run is the timed word $w = (a_1, t_1) \dots (a_p, t_p)$. If the path P is accepting then the timed word w is said to be accepted by the timed automaton. The set of all timed words accepted by $\mathcal{A}$ is denoted by $L(\mathcal{A})$.

3   Implementation of Timed Automata

For verification purposes, a fundamental question about timed automata is to decide whether the accepted language is empty. This problem is called the *emptiness problem*. A class of models is said *decidable* if the emptiness problem is decidable for these models. Note that this problem is equivalent to the *reachability problem* which tests whether a state can be reached in a model.

---

[4] For more readability, a transition will often be written as $q \xrightarrow{g, a, C} q'$ or even as $q \xrightarrow{g, a, C := 0} q'$ instead of simply the tuple (q, g, a, C, q').

3.1   From Decidability to Implementation

Alur and Dill proved in [AD94] that the emptiness problem is decidable for timed automata. The proof of this result is based on a "region automaton construction". This construction suffers from an enormous combinatorics explosion. The idea of the region automaton is to construct a finite simulation graph for the automaton based on an equivalence relation (of finite index) defined on the set of clock valuations and which abstracts finitely and correctly (with respect to reachability) the behaviours of timed automata. Some works have been done to reduce the size of this simulation graph by enlarging the equivalence relation, see for example [ACD+92,YL97,TY01]. However, in practice, such graphs are not constructed and on-the-fly zone algorithms glancing symbolically through the graph are implemented.

One of the advantages of these algorithms is that they can easily be implemented using the *Difference Bounded Matrices* data structure (DBM for short), initially proposed by [Dil89]. There are two families of algorithms, the one performing a forward analysis and the one performing a backward analysis. The tool KRONOS [BTY97,Daw97,Yov98] implements the two kinds of algorithms whereas the tool UPPAAL [BL96,LPY97] implements only forward analysis procedures, because it is more appropriate for dealing also with (bounded) integer variables, a feature proposed by UPPAAL. In this work, we will prove that the forward analysis algorithms implemented in the two previous tools are not correct!

We will now present the basic notions of zones and DBMs, a data structure adapted for representing zones. We will then be able to present the forward analysis algorithm.

3.2   Zones

A *zone* is a subset of $\mathbb{T}^n$ defined by a general clock constraint. Let k be a constant. A *k-bounded zone* is a zone defined by a k-bounded clock constraint. Let Z be a zone. The set of k-bounded zones containing Z is finite and not empty, the intersection of these k-bounded zones is a k-bounded zone containing Z, and is thus the smallest one having this property. It is called the *k-approximation* of Z and is denoted $Approx_k(Z)$. In the following, such operators will be called *extrapolation operators*.

Example 1. Consider the zone Z drawn with ▨ on the figure beside: Z is defined by the clock constraint

$$1 < x < 4 \wedge 2 < y < 4 \wedge x - y < 1.$$

Taking $k = 2$, the k-approximation of Z is drawn adding the part ▨ ; it is defined by the clock constraint

$$1 < x \wedge 2 < y \wedge x - y < 1.$$

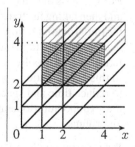

3.3   The DBM Data Structure

A *difference bounded matrice* (say *DBM* for short) for n clocks is an $(n + 1)$-square matrix of pairs

$$(m; \prec) \in \mathbb{V} = (\mathbb{Z} \times \{<, \leq\}) \cup \{(\infty; <)\}.$$

A DBM $M = (m_{i,j}, \prec_{i,j})_{i,j=1...n}$ defines the following subset of $\mathbb{T}^n$ (the clock x_0 is supposed to be always equal to zero, *i.e.* for each valuation v, $v(x_0) = 0$):

$$\{v : \{x_1, \ldots, x_n\} \longrightarrow \mathbb{T} \mid \forall\, 0 \le i, j \le n,\ v(x_i) - v(x_j) \prec_{i,j} m_{i,j}\}$$

where $\gamma < \infty$ means that γ is some real (there is no bound on it).

This subset of $\mathbb{T}^n$ is a zone and will be denoted, in what follows, by $[\![M]\!]$. Each DBM on n clocks represents a zone of $\mathbb{T}^n$. Note that several DBMs can define the same zone.

Example 2. The zone defined by the equations $x_1 > 3 \wedge x_2 \le 5 \wedge x_1 - x_2 < 4$ can be represented by the two DBMs

$$\begin{pmatrix} (0; \le) & (-3; <) & (\infty; <) \\ (\infty; <) & (0; \le) & (4; <) \\ (5; \le) & (\infty; <) & (0; \le) \end{pmatrix} \quad \text{and} \quad \begin{pmatrix} (\infty; <) & (-3; <) & (\infty; <) \\ (\infty; \le) & (\infty; <) & (4; <) \\ (5; \le) & (\infty; <) & (0; \le) \end{pmatrix}.$$

Thus the DBMs are not a canonical representation of zones. Moreover, it isn't possible to test syntactically whether $[\![M_1]\!] = [\![M_2]\!]$. A *normal form* has thus been defined for representing zones. Its computation uses the Floyd algorithm (see [Dil89,CGP99] for a description of this procedure). We will not detail the nice and numerous properties of the DBMs, but we can notice that this data structure allows to compute many operations on zones and also to test for inclusion of zones. More details about this can be found in [Ben02,Bou02b].

3.4 Forward Analysis Algorithm

Let $\mathcal{A}$ be a classical timed automaton. If $e = (q \xrightarrow{g,a,C:=0} q')$ is a transition of $\mathcal{A}$ and if Z is a zone, then $\mathsf{Post}(Z, e)$ denotes the set $[C \leftarrow 0](g \cap \overrightarrow{Z})$ where $\overrightarrow{Z}$ represents the *future* of Z and is defined by

$$\overrightarrow{Z} = \{v + t \mid v \in Z \text{ and } t \ge 0\}$$

$\mathsf{Post}(Z, e)$ is the set of valuations which can be reached by waiting in the current state, q, and then taking the transition e. Forward analysis consists in computing the successors of the initial configuration(s) by iterating the previous Post function. The exact computation does not always terminate, an extrapolation operator on zones is thus used to enforce the termination. In tools, some other abstractions are used [DT98], but they are "orthogonal" to this extrapolation operator in the sense that they are used together with this operator to reduce time and space consumption of the verification process. The algorithm using the extrapolation operator is thus the basis of all the implemented forward analysis algorithms.

We associate with $\mathcal{A}$ the largest constant, k, appearing in $\mathcal{A}$ (*i.e.* the largest constant c such that there is a constraint $x \sim c$ for some clock x or $x - y \sim c$ for some clocks x and y in $\mathcal{A}$). A maximal constant can be computed for each clock x (in a similar way), but for our purpose, it does not change anything, the presentation would just be a bit more complicated. The basis forward analysis algorithm which is implemented in tools is presented as Algorithm 1 (q_0 is the initial location of the automaton whereas Z_0

Algorithm 1 Forward Analysis Algorithm for Timed Automata

```
# Forward Analysis Algorithm (𝒜: TA ; k: integer) {
#     Visited := ∅;                              (* Visited stores the visited states *)
#     Waiting := {(q₀,Approxₖ(Z₀))};
#     Repeat
#        Get and Remove (q,Z) from Waiting;
#        If q is final
#           then {Return "Yes, a final state is reachable";}
#           else {If there is no (q,Z') ∈ Visited such that Z ⊆ Z'
#                   then {Visited := Visited ∪ {(q,Z)};
#                         Successor := {(q',Approxₖ(Post(Z, e))) | e transition from q to q'};
#                         Waiting := Waiting ∪ Successor;}}
#     Until (Waiting = ∅);
#     Return "No, no final state is reachable"; }
```

represents the initial zone, it is most of the time the valuation where all the clocks are set to zero) and is applied taking the maximal constant k as second parameter.

This algorithm terminates because there are finitely many k-bounded zones and thus finitely many k-approximations of zones that can be computed for each control state of the automaton. Moreover, the DBM data structure can be used to compute all the operations that appear in Algorithm 1, see [Ben02,Bou02b] for details. Let us just point out how it is easy to compute the k-approximation of zones, which may explain why this operator has immediately been adopted in implementations, even before its formalization in [DT98]. Let k be an integer and Z a zone represented by a DBM in normal form $M = (m_{i,j}; \prec_{i,j})_{i,j=0...n}$. We define the DBM $M' = (m'_{i,j}; \prec'_{i,j})_{i,j=0...n}$ by

$$(m'_{i,j}; \prec'_{i,j}) = \begin{cases} (\infty; <) & \text{if } m_{i,j} > k, \\ (-k; <) & \text{if } m_{i,j} < -k, \\ (m_{i,j}; \prec_{i,j}) & \text{otherwise.} \end{cases}$$

The DBM M' may not be in normal form, but $[\![M']\!] = Approx_k(Z)$.

Algorithm 1 computes step-by-step an overapproximation of the set of reachable states and tests whether this approximation intersects the set of final states, or not. Thus, if the answer of the algorithm is "No", it is sure that no final state can be reached. However, if the answer is "Yes", it can *a priori* be the case that the algorithm does a mistake. The algorithm will be said *correct* (with respect to reachability) whenever it never does such a mistake. We will now discuss in details the correctness of Algorithm 1.

4   A Correctness Problem

Algorithm 1 is implemented in tools like UPPAAL and KRONOS. The correctness of this algorithm is asserted in many papers in the literature and several attempts of proofs can be found [WT94,DT98,Daw98,Tri98,Pet99,Bou02a]. All these attempts of proofs are however incomplete or buggy (*cf* [Bou02b] for more details), and these proofs can not be corrected, because we will prove that Algorithm 1 is indeed not correct!

A Surprising Observation...

In trying to write a complete proof for the correctness of Algorithm 1, we have had some troubles, and studying precisely what were the problems we were confronted to, we have been forced to face the facts that **Algorithm 1 is not correct!** whatever is the choice of the parameter k. Consider the automaton C depicted on Fig. 1.

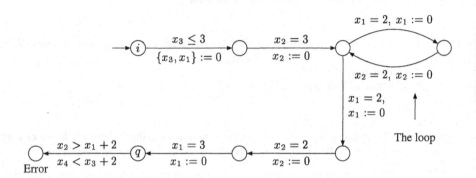

Fig. 1. A surprising timed automaton, C

Consider a path from i to q in the automaton C. If d is the date the first transition is taken and if α is the number of loops taken along the run, the valuation v of the clocks when arriving in q is defined by:

$$v(x_1) = 0 \qquad v(x_3) = d$$
$$v(x_2) = 2\alpha + 5 \qquad v(x_4) = d + 2\alpha + 5$$

Thus, applying an exact computation of the successors of the initial state i, with all the clocks set to 0, the set of valuations that can be reached in state q, when the loop is taken α times, is defined by the relations (1) in **Table 1**. This set of valuations is of course a zone and is denoted by Z_α. It can be depicted by the scheme on Fig. 2.

Table 1. Equations of the zones Z_α (on the left) and $Approx_k(Z_\alpha)$ (on the right)

$$
\begin{cases}
x_2 \geq 1 \\
x_3 \geq 2\alpha + 5 \\
x_4 \geq 2\alpha + 6 \\
1 \leq x_2 - x_1 \leq 3 \\
1 \leq x_4 - x_3 \leq 3 \\
x_3 - x_1 = 2\alpha + 5 \\
x_4 - x_2 = 2\alpha + 5
\end{cases}
\quad (1)
\qquad\qquad
\begin{cases}
x_2 \geq 1 \\
x_3 > k \\
x_4 > k \\
1 \leq x_2 - x_1 \leq 3 \\
1 \leq x_4 - x_3 \leq 3 \\
x_3 - x_1 > k \\
x_4 - x_2 > k
\end{cases}
\quad (2)
$$

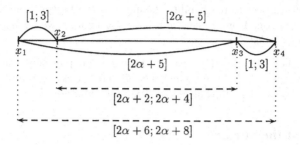

Fig. 2. The zone Z_α

Even if it is not explicit in the description of Z_α through the equations (1), we can easily deduce, and it appears clearly on the representation of the zone on Fig. 2, that if v is a valuation of Z_α, we have the very strong constraint that

$$v(x_4) - v(x_3) = v(x_2) - v(x_1) \tag{3}$$

In particular, we get that the state "Error" of $\mathcal{C}$ is not reachable.

However, the condition (3) is not explicit in the definition of Z_α, and this condition will be forgotten when we will apply an extrapolation operator to the zone. Let k be a constant and let α be such that $2\alpha + 2 > k$. Applying the extrapolation operator "$Approx_k$" to Z_α, the zone $Approx_k(Z_\alpha)$ we obtain is defined by the relations (2) in Table 1.

We thus get that Algorithm 1 will compute, whatever is the choice of the parameter k, that the state "Error" is reachable, which is, as said before, wrong. Thus, we conclude that for the automaton $\mathcal{C}$, no extrapolation operator is correct with respect to reachability: there is no way to choose correctly a constant k such that Algorithm 1 applied to $\mathcal{C}$ with the parameter k gives a correct answer.

Remark 1. Note that, for simplicity, we only applied the extrapolation operator to the zone computed for the state q. However, as it is an increasing operator, if we use it at each state along the computation, the zone that will be computed when arriving in state q will contain Z_α (if the loop is taken α times), and thus, the state "Error" will of course also be computed as reachable.

From this counter-example, we get that it is impossible to propose "good" constants which will allow to apply safely Algorithm 1. This is really a big surprise! To sum up, we get the following result:

Theorem 1. *Algorithm 1 is not correct with respect to reachability, whatever is the choice of the parameter k.*

Moreover, assume we want to modify slightly Algorithm 1 and we want to find a finite abstraction operator abs[5] such that, if Z is a zone, $abs(Z)$ is a zone containing Z. Then, a simple corollary of Theorem 1 is that Algorithm 1 modified in such a way that the extrapolation operator $Approx_k$ is replaced by the abstraction operator abs, is also not

---

[5] An abstraction operator is said *finite* whenever $\{abs(Z) \mid Z \text{ is a zone}\}$ is a finite set.

correct with respect to reachability (taking k as the biggest constant involved in the definition of a zone from the set $\{abs(Z) \mid Z \text{ is a zone}\}$, we get that for every zone Z, $Approx_k(Z) \subseteq abs(Z))$.

The aim of the remainder of this paper is to study more precisely where are the problems and to bring out subclasses for which we can find a correct extrapolation operator. This will partly explain why the bug has not been found before.

5   Out of the Surface

What precedes might appear as a very bad news for the verification of timed systems, because one of the bases of the model-checking of timed automata collapses. However, it is now 8 years that this algorithm is implemented in tools like KRONOS and UPPAAL and used for checking the correctness of many timed systems and this "bug" has not been found yet. If we refine the previous result and if we focus on the counter-example we constructed, we get that Algorithm 1 is not correct for timed automata that use diagonal constraints (the last transition of C is labeled by the constraint $x_2 > x_1 + 2 \wedge x_4 < x_3 + 2$) and that use more than four clocks. For lack of space, the following results can not be detailed in this paper, see [Bou02b] for more details.

Diagonal-Free Timed Automata – We restrict to diagonal-free timed automata, *i.e.* timed automata which use only clock constraints of the form $x \sim c$. The very nice following theorem then holds.

Theorem 2. *Algorithm 1 is correct for diagonal-free timed automata (where the constant used for the extrapolation operator is the maximum constant used in one of the clock constraints of the automaton).*

This theorem, even if it is not as general as we could expect, is already interesting because in many practical cases, the automata built for real systems are diagonal-free (see as examples [HSLL97] or [BBP02]). This might explain why the bug has not been found before. Moreover, from a theoretical point of view, every timed automaton can be transformed into a diagonal-free timed automaton (see [BDGP98] for a proof), but the transformation suffers from an exponential blow-up of the size of the automaton.

General Timed Automata – We have seen that it is hopeless to find a very wide class of general timed automata for which Algorithm 1 is correct. However, the following result completes the puzzle.

Proposition 3 *Algorithm 1 is correct for timed automata which use no more than three clocks (the constant used as a parameter for the extrapolation operator is for example the product $n.k$ where k is the maximal constant appearing in the automaton and n is the number of clocks which is used [6]).*

The reason why this theorem holds for three clocks, but not for four clocks, apart from the counter-example of Fig. 1, is that resetting clocks is similar to being in a two-dimensional time-space, and the two-dimensional time-space has very nice properties that one can **visualize** but that can not be extended to other dimensions.

---

[6] However, this constant can be tightened, see for example [Bou02b].

We have thus characterized in a very thin way the classes of automata for which we can safely use Algorithm 1: if the timed automaton is diagonal-free or if it has no more than three clocks, then we can use Algorithm 1, but if the timed automaton has both diagonal clock constraints and more than four clocks, then we can not use safely Algorithm 1, it may do a mistake.

6 Conclusion and Discussion

In this paper, we studied the forward analysis algorithm which is implemented in several tools. This algorithm belongs to the "basic knowledge" of all the people working on the verification of timed systems. However, in spite of the success stories of several implementations of this algorithm and their applications to real case studies, we did prove that it is not correct!

This algorithm is based on an abstraction operator, which is a very natural operator on DBMs, the data structure used to represent zones. However, when we apply this operator, we can lose some important relations in zones, like, for example, the equality of two differences of clocks. An alternative to this abstraction operator has to be proposed, in order to be able to verify safely timed automata that also use diagonal constraints on clocks. However, as said page 628, a corollary of our non-correctness result is that no finite abstraction on zones (which transforms a zone into a larger zone) is correct for a forward analysis. Some alternative propositions are done in [Bou02b], but it is not really satisfactory because the solutions suffer from a big combinatorics explosion. Zones are maybe not as appropriate as what one could think for analyzing timed automata...

Acknowledgments. I would like to thank Antoine Petit for his careful reading of some drafts of this paper and François Laroussinie for some interesting discussions on DBMs. I would also like to thank Kim G. Larsen, Emmanuel Fleury and Gerd Behrmann for our discussions on the implementation of timed automata during our so-called UPPAAL meetings.

References

[ACD$^+$92]   Rajeev Alur, Costas Courcoubetis, David Dill, Nicolas Halbwachs and Howard Wong-Toi. *An Implementation of Three Algorithms for Timing Verification Based on Automata Emptiness.* In *Proc. 13th IEEE Real-Time Systems Symposium (RTSS'92)*, pp. 157–166. IEEE Computer Society Press, 1992.

[ACH$^+$92]   Rajeev Alur, Costas Courcoubetis, Nicolas Halbwachs, David Dill and Howard Wong-Toi. *Minimization of Timed Transition Systems.* In *Proc. 3rd International Conference on Concurrency Theory (CONCUR'92)*, vol. 630 of *Lecture Notes in Computer Science*, pp. 340–354. Springer, 1992.

[ACH94]   Rajeev Alur, Costas Courcoubetis and Thomas A. Henzinger. *The Observational Power of Clocks.* In *Proc. 5th International Conference on Concurrency Theory (CONCUR'94)*, vol. 836 of *Lecture Notes in Computer Science*, pp. 162–177. Springer, 1994.

[AD94]   Rajeev Alur and David Dill. *A Theory of Timed Automata.* Theoretical Computer Science (TCS), vol. 126(2):pp. 183–235, 1994.

[AFH94]   Rajeev Alur, Limor Fix and Thomas A. Henzinger. *A Determinizable Class of Timed Automata*. In *Proc. 6th International Conference on Computer Aided Verification (CAV'94)*, vol. 818 of *Lecture Notes in Computer Science*, pp. 1–13. Springer, 1994.

[BBP02]   Béatrice Bérard, Patricia Bouyer and Antoine Petit. *Analysing the PGM Protocol with* Uppaal. In *Proc. 2nd Workshop on Real-Time Tools (RT-TOOLS'02)*. 2002. Proc. published as Technical Report 2002-025, Uppsala University, Sweden.

[BDFP00]   Patricia Bouyer, Catherine Dufourd, Emmanuel Fleury and Antoine Petit. *Are Timed Automata Updatable?*. In *Proc. 12th International Conference on Computer Aided Verification (CAV'2000)*, vol. 1855 of *Lecture Notes in Computer Science*, pp. 464–479. Springer, 2000.

[BDGP98]   Béatrice Bérard, Volker Diekert, Paul Gastin and Antoine Petit. *Characterization of the Expressive Power of Silent Transitions in Timed Automata*. Fundamenta Informaticae, vol. 36(2–3):pp. 145–182, 1998.

[Ben02]   Johan Bengtsson. *Clocks, DBMs ans States in Timed Systems*. Ph.D. thesis, Department of Information Technology, Uppsala University, Uppsala, Sweden, 2002.

[BFH$^+$01] Gerd Behrmann, Ansgar Fehnker, Thomas Hune, Kim G. Larsen, Paul Pettersson, Judi Romijn and Frits Vaandrager. *Minimum-Cost Reachability for Priced Timed Automata*. In *Proc. 4th International Workshop on Hybrid Systems: Computation and Control (HSCC'01)*, vol. 2034 of *Lecture Notes in Computer Science*, pp. 147–161. Springer, 2001.

[BL96]   Johan Bengtsson and Fredrik Larsson. Uppaal, *a Tool for Automatic Verification of Real-Time Systems*. Master's thesis, Department of Computer Science, Uppsala University, Sweden, 1996.

[BLP$^+$99] Gerd Behrmann, Kim G. Larsen, Justin Pearson, Carsten Weise and Wang Yi. *Efficient Timed Reachability Analysis Using Clock Difference Diagrams*. In *Proc. 11th International Conference on Computer Aided Verification (CAV'99)*, vol. 1633 of *Lecture Notes in Computer Science*, pp. 341–353. Springer, 1999.

[Bou02a]   Patricia Bouyer. *Modèles et algorithmes pour la vérification des systèmes temporisés*. Ph.D. thesis, École Normale Supérieure de Cachan, Cachan, France, 2002.

[Bou02b]   Patricia Bouyer. *Timed Automata May Cause Some Troubles*. Research Report LSV–02–9, Laboratoire Spécification et Vérification, ENS de Cachan, France, 2002. Also Available as *BRICS Research Report RS-02-35*, Aalborg University, Denmark, 2002.

[BTY97]   Ahmed Bouajjani, Stavros Tripakis and Sergio Yovine. *On-the-Fly Symbolic Model-Checking for Real-Time Systems*. In *Proc. 18th IEEE Real-Time Systems Symposium (RTSS'97)*, pp. 25–35. IEEE Computer Society Press, 1997.

[CG00]   Christian Choffrut and Massimiliano Goldwurm. *Timed Automata with Periodic Clock Constraints*. Journal of Automata, Languages and Combinatorics (JALC), vol. 5(4):pp. 371–404, 2000.

[CGP99]   Edmund Clarke, Orna Grumberg and Doron Peled. *Model-Checking*. The MIT Press, Cambridge, Massachusetts, 1999.

[Daw97]   Conrado Daws. *Analyse par simulation symbolique des systèmes temporisés avec* Kronos. *Research report*, Verimag, 1997.

[Daw98]   Conrado Daws. *Méthodes d'analyse de systèmes temporisés : de la théorie à la pratique*. Ph.D. thesis, Institut National Polytechnique de Grenoble, Grenoble, France, 1998.

[Dil89]   David Dill. *Timing Assumptions and Verification of Finite-State Concurrent Systems*. In *Proc. of the Workshop on Automatic Verification Methods for Finite State Systems*, vol. 407 of *Lecture Notes in Computer Science*, pp. 197–212. Springer, 1989.

[DOTY96]   Conrado Daws, Alfredo Olivero, Stavros Tripakis and Sergio Yovine. *The Tool* KRONOS. In *Proc. Hybrid Systems III: Verification and Control (1995)*, vol. 1066 of *Lecture Notes in Computer Science*, pp. 208–219. Springer, 1996.

[DT98]   Conrado Daws and Stavros Tripakis. *Model-Checking of Real-Time Reachability Properties using Abstractions*. In *Proc. 4th International Conference on Tools and Algorithms for the Construction and Analysis of Systems (TACAS'98)*, vol. 1384 of *Lecture Notes in Computer Science*, pp. 313–329. Springer, 1998.

[DZ98]   François Demichelis and Wieslaw Zielonka. *Controlled Timed Automata*. In *Proc. 9th International Conference on Concurrency Theory (CONCUR'98)*, vol. 1466 of *Lecture Notes in Computer Science*, pp. 455–469. Springer, 1998.

[HHWT97]   Thomas A. Henzinger, Pei-Hsin Ho and Howard Wong-Toi. HYTECH: *A Model-Checker for Hybrid Systems*. Journal on Software Tools for Technology Transfer (STTT), vol. 1(1–2):pp. 110–122, 1997.

[HKWT95]   Thomas A. Henzinger, Peter W. Kopke and Howard Wong-Toi. *The Expressive Power of Clocks*. In *Proc. 22nd International Colloquium on Automata, Languages and Programming (ICALP'95)*, vol. 944 of *Lecture Notes in Computer Science*, pp. 417–428. Springer, 1995.

[HRS98]   Thomas A. Henzinger, Jean-François Raskin and Pierre-Yves Schobbens. *The Regular Real-Time Languages*. In *Proc. 25th International Colloquium on Automata, Languages and Programming (ICALP'98)*, vol. 1443 of *Lecture Notes in Computer Science*, pp. 580–591. Springer, 1998.

[HSLL97]   Klaus Havelund, Arne Skou, Kim G. Larsen and Kristian Lund. *Formal Modeling and Analysis of an Audio/Video Protocol: An Industrial Case Study Using* UPPAAL. In *Proc. 18th IEEE Real-Time Systems Symposium (RTSS'97)*, pp. 2–13. IEEE Computer Society Press, 1997.

[LPY97]   Kim G. Larsen, Paul Pettersson and Wang Yi. UPPAAL *in a Nutshell*. Journal of Software Tools for Technology Transfer (STTT), vol. 1(1–2):pp. 134–152, 1997.

[Pet99]   Paul Pettersson. *Modelling and Verification of Real-Time Systems Using Timed Automata: Theory and Practice*. Ph.D. thesis, Department of Computer Systems, Uppsala University, Uppsala, Sweden, 1999. Available as DoCS Technical Report 99/101.

[Tri98]   Stavros Tripakis. *L'analyse formelle des systèmes temporisés en pratique*. Ph.D. thesis, Université Joseph Fourier, Grenoble, France, 1998.

[TY98]   Stavros Tripakis and Sergio Yovine. *Verification of the Fast Reservation Protocol with Delayed Transmission using the Tool* KRONOS. In *Proc. 4th IEEE Real-Time Technology and Applications Symposium (RTAS'98)*, pp. 165–170. IEEE Computer Society Press, 1998.

[TY01]   Stavros Tripakis and Sergio Yovine. *Analysis of Timed Systems using Time-Abstracting Bisimulations*. Formal Methods in System Design, vol. 18(1):pp. 25–68, 2001.

[Wil94]   Thomas Wilke. *Specifying Timed State Sequences in Powerful Decidable Logics and Timed Automata*. In *Proc. 3rd International Symposium on Formal Techniques in Real-Time and Fault-Tolerant Systems (FTRTFT'94)*, vol. 863 of *Lecture Notes in Computer Science*, pp. 694–715. Springer, 1994.

[WT94]   Howard Wong-Toi. *Symbolic Approximations for Verifying Real-Time Systems*. Ph.D. thesis, Stanford University, USA, 1994.

[YL97]   Mihalis Yannakakis and David Lee. *An Efficient Algorithm for Minimizing Real-Time Transition Systems*. Formal Methods in System Design, vol. 11(2):pp. 113–136, 1997.

[Yov98]   Sergio Yovine. *Model-Checking Timed Automata*. In *School on Embedded Systems*, vol. 1494 of *Lecture Notes in Computer Science*, pp. 114–152. Springer, 1998.

The Intrinsic Universality Problem of One-Dimensional Cellular Automata

Nicolas Ollinger

LIP, École Normale Supérieure de Lyon, 46, allée d'Italie
69 364 Lyon Cedex 07, France
Nicolas.Ollinger@ens-lyon.fr

Abstract. Undecidability results of cellular automata properties usually concern one time step or long time behavior of cellular automata. Intrinsic universality is a dynamical property of another kind. We prove the undecidability of this property for one-dimensional cellular automata. The construction used in this proof may be extended to other properties.

Cellular automata are simple discrete dynamical systems given by a triple of objects: a *regular lattice of cells*, a *neighborhood vector* on this space, and a finitely described *local transition function* defining how the state of a cell of the lattice evolves according to the states of its neighbors. A *configuration* of a cellular automaton is a mapping from the lattice of cells to the finite set of states which assigns a state to each cell. The *global transition function* of the cellular automaton, which defines its *dynamics*, transforms a configuration into another one by applying the local transition function uniformly, and in parallel, to each cell. A *space-time diagram* is an infinite sequence of configurations obtained by iteration of the global transition function starting from an initial configuration. A main concern of the study of cellular automata is to understand the links between local and global properties of cellular automata: how can very simple local transition functions provide very rich dynamics?

Recently, an algebraic framework was proposed by J. Mazoyer and I. Rapaport [9] to induce an order on cellular automata which is somehow relevant from the point of view of global behavior study. A cellular automaton is said to simulate another one if, up to some rescaling of both cellular automata, the set of space-time diagrams of the former includes the set of space-time diagrams of the later. This relation is a quasi-order on the set of cellular automata and the induced equivalent classes and order on these classes provide a natural way to compare cellular automata. In [9], this order was proven to admit a global minimum and simple known dynamical properties were used to characterize classes at the bottom of the order. In [10], we introduced a generalization of this order by allowing a more general notion of cellular automata rescaling. Thanks to these new geometrical transformations on cellular automata space-time diagrams, we were able to take new phenomena into account. In particular, contrary to the original one, our new order admits a global maximum. This maximum

H. Alt and M. Habib (Eds.): STACS 2003, LNCS 2607, pp. 632–641, 2003.

corresponds to the set of intrinsically universal cellular automata, that is cellular automaton which can simulate any other cellular automaton step by step, already present in the work of E. R. Banks [3] and formalized for the first time by J. Albert and K. Čulik [1].

In the present article, we prove that the intrinsic universality property is undecidable. Contrary to the computation universality for Turing machines, this result is not a direct consequence of some Rice's theorem because cellular automata behavior lacks such tool. This result explains the difficulty to exhibit intrinsically universal cellular automata with a few states as discussed in [11]. Moreover, when interpreted in the framework of cellular automata comparison, this implies that there exist no class of cellular automata that lie at the limit downside intrinsically universal ones.

1   Definitions

A *cellular automaton* $\mathcal{A}$ is a quadruple $\left(\mathbb{Z}^d, S, N, \delta\right)$ such that $\mathbb{Z}^d$ is the d-dimensional regular grid, S is a finite set of states, N is a finite set of ν vectors of $\mathbb{Z}^d$ called the neighborhood of $\mathcal{A}$ and δ is the local transition function of $\mathcal{A}$ which maps S^ν to S. Two classical neighborhoods for one-dimensional cellular automata are the one-way neighborhood $N_{\text{OCA}} = \{-1, 0\}$ and the first-neighbors neighborhood $N_{\text{vN}} = \{-1, 0, 1\}$.

A *configuration* $\mathcal{C}$ of a cellular automaton $\mathcal{A}$ maps $\mathbb{Z}^d$ to the set of states of $\mathcal{A}$. The state of the i-th cell of $\mathcal{C}$ is denoted as $\mathcal{C}_i$. A configuration $\mathcal{C}$ is *periodic* if there exists a basis $(v_1, \dots, v_d)$ of $\mathbb{Z}^d$ such that for any index k and any cell i of $\mathbb{Z}^d$, $\mathcal{C}_{i+v_k} = \mathcal{C}_i$. In the case of one-dimensional configurations, the smallest strictly positive value for v_1 is the *period* of the configuration.

The local transition function δ of $\mathcal{A}$ is naturally extended to a *global transition function* $G_\mathcal{A}$ which maps a configuration $\mathcal{C}$ of $\mathcal{A}$ to a configuration $\mathcal{C}'$ of $\mathcal{A}$ satisfying, for each cell i, the equation $\mathcal{C}'_i = \delta\left(\mathcal{C}_{i+v_1}, \dots, \mathcal{C}_{i+v_\nu}\right)$, where $\{v_1, \dots, v_\nu\}$ is the neighborhood of $\mathcal{A}$. A *space-time diagram* of a cellular automaton $\mathcal{A}$ is an infinite sequence of configurations $(\mathcal{C}_t)_{t\in\mathbb{N}}$ such that, for every time t, $\mathcal{C}_{t+1} = G_\mathcal{A}(\mathcal{C}_t)$. The usual way to represent space-time diagrams is to draw the sequence of configurations successively, from bottom to top.

The *limit set* $\Omega(\mathcal{A})$ of a d-dimensional cellular automaton $\mathcal{A}$ with set of states S is the non-empty set of configurations of $\mathcal{A}$ that can appear at any time step in a space-time diagram. Formally,

$$\Omega(\mathcal{A}) = \bigcap_{t\in\mathbb{N}} G_\mathcal{A}^t\left(S^{\mathbb{Z}^d}\right).$$

A *fixed point* $\mathcal{C}$ of a cellular automaton $\mathcal{A}$ is a configuration of $\mathcal{A}$ such that $G_\mathcal{A}(\mathcal{C}) = \mathcal{C}$. Thus, we say that a configuration $\mathcal{C}$ evolves to a fixed point if there exists a time t such that $G_\mathcal{A}^t(\mathcal{C})$ is a fixed point.

A cellular automaton $\mathcal{A}$ is *nilpotent* if any configuration of $\mathcal{A}$ evolves to a same fixed point, *i.e.* $\Omega(\mathcal{A})$ is a singleton. Symmetrically, a cellular automaton $\mathcal{A}$ is *nilpotent for periodic configurations* if any periodic configuration of $\mathcal{A}$ evolves to a same fixed point. Notice that in both cases, the fixed point configuration has to be monochromatic, *i.e.* consisting of only one state s. To emphasize the choice of s, we will speak of s-nilpotency.

A *sub-automaton*[1] of a cellular automaton corresponds to a stable restriction on the set of states. A cellular automaton is a sub-automaton of another cellular automaton if (up to a renaming of states) the space-time diagrams of the first one are space-time diagrams of the second one. To compare cellular automata, we introduce a notion of rescaling space-time diagrams. To formalize this idea, we introduce the following notations:

σ^k. Let S be a finite set of states and k be a vector of $\mathbb{Z}^d$. The shift σ^k is the bijective map from $S^{\mathbb{Z}^d}$ onto $S^{\mathbb{Z}^d}$ which maps a configuration $\mathcal{C}$ to the configuration $\mathcal{C}'$ such that, for each cell i, the equation $\mathcal{C}'_{i+k} = \mathcal{C}_i$ is satisfied.

o^m. Let S be a finite set of states and $m = (m_1, \ldots, m_d)$ be a finite sequence of strictly positive integers. The *packing map* o^m is the bijective map from $S^{\mathbb{Z}^d}$ onto $(S^{m_1 \cdots m_d})^{\mathbb{Z}^d}$ which maps a configuration $\mathcal{C}$ to the configuration $\mathcal{C}'$ such that, for each cell i, the equation $\mathcal{C}'_i = (\mathcal{C}_{mi}, \ldots, \mathcal{C}_{m(i+1)-1})$ is satisfied. The principle of $o^{(3,2)}$ is depicted on Fig. 1.

Fig. 1. The way $o^{(3,2)}$ cuts $\mathbb{Z}^2$ space

Definition 1. *Let $\mathcal{A}$ be a d-dimensional cellular automaton with set of states S. A $\langle m, n, k \rangle$-rescaling of $\mathcal{A}$ is a cellular automaton $\mathcal{A}^{\langle m,n,k \rangle}$ with set of states $S^{m_1 \cdots m_d}$ and global transition function $G_{\mathcal{A}}^{\langle m,n,k \rangle} = \sigma^k \circ o^m \circ G_{\mathcal{A}}^n \circ o^{-m}$.*

Definition 2. *Let $\mathcal{A}$ and $\mathcal{B}$ be two cellular automata. Then $\mathcal{B}$ simulates $\mathcal{A}$ if there exists a rescaling of $\mathcal{A}$ which is a sub-automaton of a rescaling of $\mathcal{B}$.*

The relation of simulation is a quasi-order on cellular automata. It is a generalization of the order introduced by Mazoyer and Rapaport [9]. In [10], we motivate the introduction of this relation and discuss its main properties. In particular, it induces a maximal equivalence class which exactly corresponds to the set of intrinsically universal cellular automata as described by Banks [3] and Albert and Čulik II [1].

---

[1] The prefix *sub* emphasizes the fact that $(S'^{\mathbb{Z}}, G)$ is an (algebraic) sub-structure of $(S^{\mathbb{Z}}, G)$. One could have also used the terminology *divisor* as the set of space-time diagrams of one automaton is included into the one of the other.

Definition 3. *A cellular automaton* $\mathcal{A}$ *is* intrinsically universal *if, for each cellular automaton* $\mathcal{B}$, *there exists a rescaling of* $\mathcal{A}$ *of which* $\mathcal{B}$ *is a sub-automaton.*

As any one-dimensional cellular automaton can be simulated by a one-way cellular automaton, that is a cellular automaton with neighborhood N_{OCA}, there exist intrinsically universal one-way cellular automata. Therefore, to prove that a particular one-dimensional cellular automaton is intrinsically universal, it is sufficient to prove that it can simulate any one-way cellular automaton. In particular, the intrinsically universal cellular automaton of section 3 is constructed by mean of one-way cellular automata simulation.

2   Deciding Properties of Cellular Automata

One time step behavior of cellular automata, like the injectivity or the surjectivity of the global transition function (for cellular automata, injectivity implies bijectivity) have been well studied. Both properties have been proved decidable for one-dimensional cellular automata by Amoroso and Patt [2]. However, in the case of higher dimensions, both properties have been proved undecidable by Kari [6]. There is a gap of complexity between dimensions one and two. The proofs of Kari rely on reductions to tiling problems of the plane and aperiodic tilings studied by Robinson [12].

Long time behavior of cellular automata deals with the limit set of cellular automata: the set of configurations that can appear after any possible number of time steps. This set is known to be either a singleton either an infinite set. In the first case, the unique configuration of the limit set is monochromatic and the cellular automaton is said nilpotent. Nilpotency has been proved undecidable for two-dimensional cellular automata by Čulik II, Pachl, and Yu [4]. This result has been extended to one-dimensional cellular automata by Kari [5] using a reduction to the tiling problem of the plane for NW-deterministic tile sets. Eventually, Kari [7] has generalized his result to an analog of Rice's theorem for limit set properties of cellular automata: every non-trivial long time behavior properties of cellular automata are undecidable.

Long time behavior of cellular automata on periodic configurations has also its literature because previous results do not automatically extend to periodic configurations. In the case of one-dimensional cellular automata, Sutner [13] proved that it is undecidable to know whether each periodic configuration of a one-dimensional cellular automaton evolves to a fixed point. This result was recently extended by Mazoyer and Rapaport [8] as follows: it is undecidable to know whether each periodic configuration of a one-dimensional cellular automaton evolves to a same fixed point. The proof still uses a reduction to a particular tiling problem. In the present article, we prove an undecidability result by reduction to a simple variant of Mazoyer and Rapaport's result, CA-1D-NIL-PER.

CA-1D-NIL-PER

Input	A one-dimensional cellular automaton $\mathcal{A}$ and a particular state s from $\mathcal{A}$
Question	Is $\mathcal{A}$ s-nilpotent for periodic configurations ?

Some dynamical properties of cellular automata are neither one time step nor long time properties. An example of such properties is the intrinsic universality. Briefly, a cellular automaton is intrinsically universal if it can simulate, in a particular sense, any other cellular automaton step by step. In the classical case of Turing machines, computational universality, even if not formally defined, is undecidable considering Rice's theorem. In the case of cellular automata, there is no such tool. We will now prove that the problem CA-1D-UNIV is undecidable using a new technique which should work for other dynamical properties.

CA-1D-UNIV

Input	A one-dimensional cellular automaton $\mathcal{A}$
Question	Is $\mathcal{A}$ intrinsically universal ?

3   An Intrinsically Universal Cellular Automaton

Our proof of the undecidability of the intrinsic universality problem of one-dimensional cellular automata proceeds by reduction to the nilpotency problem for periodic configurations and relies on the existence of a particular intrinsically universal cellular automaton $\mathcal{U}$. We briefly describe its structure and properties.

The cellular automaton $\mathcal{U}$ is defined by simulation of a multi-head Turing machine $\mathcal{M} = (Q_u, \Sigma_u, \pi)$ with set of states Q_u, alphabet Σ_u and whose transition function π maps $Q_u \times \Sigma_u$ to $Q_u \times \Sigma_u \times \{\leftarrow, \downarrow, \rightarrow\}$. Notice that the behavior of $\mathcal{M}$ on configurations where several heads share a same position is undefined. By simulation, we mean here that $\mathcal{U}$ is defined as

$$\mathcal{U} = (\mathbb{Z}, (\{\cdot\} \cup Q_u) \times \Sigma_u, N_{\text{vN}}, \delta_u) .$$

A state of $\mathcal{U}$ is a pair constituted of a head or a blank and a letter. A configuration of $\mathcal{U}$ looks like a configuration of $\mathcal{M}$:

$$a\,a\,a\,a\,b\,b\,a\,b\,a\,b\,a\,a\,a\,b\,b\,a\,a\,a\,b\,a\,b\,b$$
$$\cdot\,\cdot\,\cdot\,\cdot\,\cdot\,s\,\cdot\,\cdot\,\cdot\,\cdot\,\cdot\,\cdot\,\cdot\,\cdot\,\cdot\,s'\,\cdot\,\cdot\,\cdot\,\cdot\,\cdot$$

The local transition function δ_u is defined in order to emulate $\mathcal{M}$ according to π on locally valid configurations of $\mathcal{M}$. By locally valid configurations of $\mathcal{M}$ we mean configurations where no two heads are in the von Neumann neighborhood of each other. To fully define δ_u, we simply ask that no new head is created, for example by destroying heads that locally invalidate a configuration.

We also give some constraints on intrinsic simulation. For each one-way cellular automaton $\mathcal{A}$, there must exist a positive integer m and an injective map

φ from $S_{\mathcal{A}}$ into $S_{\mathcal{U}}^m$ such that $\mathcal{U}$ simulates $\mathcal{A}$ according to φ without shift: there exists some n such that $\overline{\varphi} \circ G_{\mathcal{A}} = G_{\mathcal{U}}^{\langle m,n,0 \rangle} \circ \overline{\varphi}$ where $\overline{\varphi}$ is the extension of φ to $S_{\mathcal{A}}^{\mathbb{Z}}$, i.e. the following diagram commutes:

$$
\begin{array}{ccc}
C & \xrightarrow{\ o^{-m} \circ \overline{\varphi}\ } & (o^{-m} \circ \overline{\varphi})(C) \\[2pt]
{\scriptstyle G_{\mathcal{A}}}\Big\downarrow & & \Big\downarrow{\scriptstyle G_{\mathcal{U}}^n} \\[2pt]
G_{\mathcal{A}}(C) & \xrightarrow{\ o^{-m} \circ \overline{\varphi}\ } & (o^{-m} \circ \overline{\varphi})(G_{\mathcal{A}}(C))
\end{array}
\quad .
$$

Moreover φ has the following three properties:

First, each macro-cell (a block of cells encoding one cell of the simulated cellular automaton) is driven by a Turing head. Formally, there exists a state s_0 of $\mathcal{M}$ such that, for each state s of $\mathcal{A}$, its image $\varphi(s)$ contains s_0 as the head component of its first cell and no head elsewhere.

Second, during the simulation, the Turing heads move like a comb. Formally, for each configuration C of $\mathcal{A}$ and for each time t, the head components of the configuration $C^{(t)} = G_{\mathcal{U}}^t (\overline{\varphi}(C))$ of $\mathcal{U}$ contains heads exactly at positions $(mi + l_t)_{i \in \mathbb{Z}}$ for some l_t.

Third, we can extend the simulation macro-cells to bigger m by padding them and still have the same properties. Formally, there exists a letter a of $\mathcal{M}$ such that, for each positive integer l, the sum $m + l$ and the injective map

$$
\begin{aligned}
\varphi_l : S_{\mathcal{A}} &\longrightarrow S_{\mathcal{U}}^{m+l} \\
s &\longmapsto (\varphi(s))(\cdot, a)^l
\end{aligned}
$$

are valid choices to replace m and φ and keep the same simulation properties.

Let us now sketch what the behavior of such a cellular automaton $\mathcal{U}$ could be. The following ideas are depicted on Fig. 2, where the head movements during the times that are not depicted are represented by straight segments.

The first idea of the construction is to cut the line of cells regularly into blocks of cells. Each block corresponds to a macro-cell which encodes one cell of the simulated one-way cellular automaton. The border line between two such blocks is materialized by two border letters # separated by a void of letters ..

Inside the block, several regions are distinguished and separated by a letter @. In a first region is stored the state of the macro-cell. A second region is used during the transition to store the state of the neighbor macro-cell state. A third region is used to temporarily store the next state of the macro-cell. Finally, a last region is required to store the transition table of the simulated cellular automaton.

One transition of the simulated macro-cell is operated thanks to the following steps. First, the head copies the state of the neighbor macro-cell in the appropriate storage region. Next, the transition table is read entirely. At each step, the values of the stored states are compared to the current position inside the

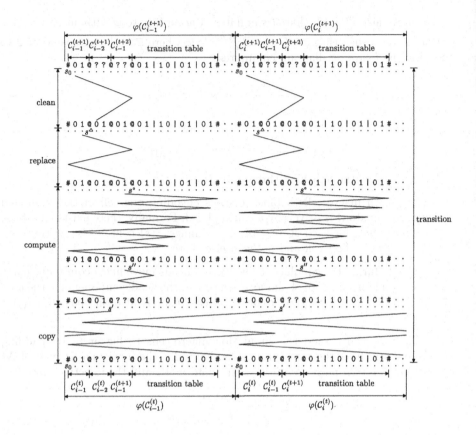

Fig. 2. Typical behavior of the intrinsically universal cellular automaton $\mathcal{U}$

transition table and the result of the transition is copied into the next state storage region, or this copy is emulated if the values do not match. Once the reading of the transition table is achieved, the head replaces the content of the state storage region by the content of the next state storage region and cleans up the storage areas. Afterwards, the head goes back to its initial position and enters state s_0.

4   Undecidability of the Intrinsic Universality Problem

Once convinced that such a $\mathcal{U}$ exists, consider the following transformation which is the key idea of our reduction. For each cellular automaton $\mathcal{A}$ and each state s of $\mathcal{A}$, we define a product cellular automaton $\mathcal{A} \circledast_s \mathcal{U}$ by

$$\mathcal{A} \circledast_s \mathcal{U} = (\mathbb{Z}, S_{\mathcal{A}} \times \{\cdot, *\} \times S_{\mathcal{U}}, N_{\mathcal{A}} \cup N_{\mathrm{OCA}} \cup N_{\mathcal{U}}, \delta),$$

a three layers automaton whose local transition rule δ is described layer by layer.

The bottom layer is the energy production layer. It consists of a configuration of $\mathcal{A}$ which evolves according to the local transition function $\delta_{\mathcal{A}}$ of $\mathcal{A}$.

The middle layer is the energy diffusion layer. It consists of a configuration on states $\cdot$ (no energy) and $*$ (an energy dot). The evolution of this layer is the one of a shift but: if the bottom layer of a cell does not contain state s the middle layer produces an energy dot $*$; if the upper layer of a cell contains a head state and its middle layer receives an energy dot $*$ from its neighbor, the middle layer dissipates the incoming energy dot and receives a no energy state.

The upper layer is the energy consumption layer. It consists of a configuration of $\mathcal{U}$ which evolves according to the local transition function of $\mathcal{U}$ under the control of the middle layer. If the upper layer of a cell contains a head state in a cell of its neighborhood then, if the middle layer of the cell containing the head receives an energy dot $*$, the upper layer evolves according to $\mathcal{U}$.

Lemma 1. *For any cellular automaton $\mathcal{A}$, the automaton $\mathcal{A} \circledast_s \mathcal{U}$ is intrinsically universal if and only if $\mathcal{A}$ is not s-nilpotent on periodic configurations.*

Proof. Let $\mathcal{A}$ be a cellular automaton and s a state of $\mathcal{A}$. The proof is by discrimination on the s-nilpotency on periodic configurations of $\mathcal{A}$.

If $\mathcal{A}$ is not s-nilpotent on periodic configurations, there exists a configuration $\mathcal{C}$ of $\mathcal{A}$ which is periodic both in space and time and is not s-monochromatic. Let p be a spatial period of $\mathcal{C}$. To prove that $\mathcal{A} \circledast_s \mathcal{U}$ is intrinsically universal, we prove that it is in the maximal equivalence class for the simulation relation by proving that it can simulate any one-way cellular automaton. Consider any one-way cellular automaton $\mathcal{B}$. Let m, n and φ be the parameters of one of our particular intrinsic simulations of $\mathcal{B}$ by $\mathcal{U}$ such that p divides m (use padding property if necessary). Then, $\mathcal{A} \circledast_s \mathcal{U}$ simulates $\mathcal{B}^T$ for some strictly positive T with parameters m, Tn and ψ where ψ is obtained from φ using the following ideas. The bottom layer of ψ consists of periods of $\mathcal{C}$. The upper layer of ψ is directly given by φ. As $\mathcal{C}$ is periodic and its period divide m, it periodically produces energy for the universal simulation, preserving the comb structure, thus allowing the simulation to take place with a given slowdown. By a simple pigeonhole reasoning, it is possible to choose a middle layer for ψ such that during any simulation this layer appears on a periodic duration basis t (remember that each macro-cell produces and consums energy the same way whatever state it encodes). As it is straightforward to see, by choosing for T a value such that t divides Tn, the following two properties hold: ψ is injective and $\mathcal{A} \circledast_s \mathcal{U}$ simulates $\mathcal{B}^T$ with parameters m, Tn and ψ.

If $\mathcal{A}$ is s-nilpotent on periodic configurations, we prove that $\mathcal{A} \circledast_s \mathcal{U}$ cannot simulate the cellular automaton $\mathcal{B} = (\mathbb{Z}, \{\circ, \bullet\}, N_{\mathrm{OCA}}, \oplus)$ where $(\{\circ, \bullet\}, \oplus)$ is the cyclic group $(\mathbb{Z}_2, +)$ where $\circ$ corresponds to 0 and $\bullet$ to 1. Assume that $\mathcal{A} \circledast_s \mathcal{U}$ is intrinsically universal. In particular, it simulates $\mathcal{B}$: let $w_\circ$ and $w_\bullet$ be the respective encoding of its states. The automaton $\mathcal{B}$ admits a space-time diagram Δ periodic in both space and time with the following filling pattern:

As $\mathcal{A}$ is s-nilpotent on periodic configurations, the bottom layer of the space-time diagram of the simulation of $\mathcal{B}$ on Δ stops creating energy on the middle layer after a finite time. Thus, either $w_\circ$ and $w_\bullet$ cannot contain both energy cells on their middle layer and heads on their upper layer (because the heads would consume the whole energy in finite time which would be a problem to preserve the injectivity of the encoding).

If there is a head in the upper layer of $w_\circ$ or $w_\bullet$ then the middle layer is empty of energy cells: the bottom layer behaves periodically and the other layers are constant, thus $\mathcal{A} \circledast_s \mathcal{U}$ behaves like a periodic automaton on the simulation configurations.

If there is no head in the upper layer of $w_\circ$ and $w_\bullet$ then the first layer is periodic, the second layer is a shift and the third layer is constant, thus $\mathcal{A} \circledast_s \mathcal{U}$ behaves like a cartesian product of a shift and a periodic automaton on the simulation configurations.

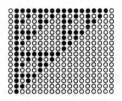

Fig. 3. Pascal triangle modulo two produced by $\mathcal{B}$

In both cases, these behaviors fail to capture the behavior of $\mathcal{B}$ as the behavior of $\mathcal{B}$ on the configuration ○-monochromatic but on cell 0 which has state ● cannot be simulated by such cellular automaton as it produces a Pascal triangle modulo two (as depicted on Fig. 3) which cannot be obtained by the product of a shift and a periodic cellular automaton. ∎

Theorem 1. *The intrinsic universality problem of one-dimensional cellular automata is undecidable.*

Proof. This proposition is a corollary of the previous lemma. As the computation of the cellular automaton $\mathcal{A} \circledast_s \mathcal{U}$ from the cellular automaton $\mathcal{A}$ is recursive, the decidability of the intrinsic universality problem of one-dimensional cellular automata would imply the decidability of s-nilpotency problem on periodic configurations. ∎

5   Cellular Automata Dynamics and Computation

The dynamical properties of cellular automata known to be undecidable are one time step or long time behavior properties. We were able to prove the undecidability of the intrinsic universality of one-dimensional cellular automata, which is a dynamical property of another kind, using a particular reduction to a long time behavior property.

The key idea of our proof is to combine an intrinsically universal cellular automaton with a second cellular automaton which acts as an energy provider. The first automaton consumes energy provided by the second one. The product cellular automaton is intrinsically universal if and only if the energy provider is not nilpotent for periodic configurations. It should be possible to replace intrinsic universality by other dynamical properties to prove other undecidability results.

Intrinsic universality plays the same role for cellular automata dynamics as computation universality for Turing machines. A formalization of this idea seems worth studying as it would certainly lead to a better understanding of the way computation occurs inside space-time diagrams and provide undecidability results of the same kind as Rice's theorem.

References

1. J. Albert and K. Čulik II, A simple universal cellular automaton and its one-way and totalistic version, *Complex Systems*, **1**(1987), no. 1, 1–16.
2. S. Amoroso and Y. N. Patt, Decision procedures for surjectivity and injectivity of parallel maps for tessellation structures, *J. Comput. System Sci.*, **6**(1972), 448–464.
3. E. R. Banks, Universality in cellular automata, in *Conference Record of 1970 Eleventh Annual Symposium on Switching and Automata Theory*, pages 194–215, IEEE, 1970.
4. K. Čulik II, J. K. Pachl, and S. Yu, On the limit sets of cellular automata, *SIAM J. Comput.*, **18**(1989), no. 4, 831–842.
5. J. Kari, The nilpotency problem of one-dimensional cellular automata, *SIAM J. Comput.*, **21**(1992), 571–586.
6. J. Kari, Reversibility and surjectivity problems of cellular automata, *J. Comput. System Sci.*, **48**(1994), 149–182.
7. J. Kari, Rice's theorem for the limit sets of cellular automata, *Theoretical Computer Science*, **127**(1994), no. 2, 229–254.
8. J. Mazoyer and I. Rapaport, Global fixed point attractors of circular cellular automata and periodic tilings of the plane: undecidability results, *Discrete Math.*, **199**(1999), 103–122.
9. J. Mazoyer and I. Rapaport, Inducing an order on cellular automata by a grouping operation, *Discrete Appl. Math.*, **218**(1999), 177–196.
10. N. Ollinger, Toward an algorithmic classification of cellular automata dynamics, 2001, LIP RR2001-10, http://www.ens-lyon.fr/LIP.
11. N. Ollinger, The quest for small universal cellular automata, in P. Widmayer, F. Triguero, R. Morales, M. Hennessy, S. Eidenbenz, and R. Conejo, editors, *Automata, languages and programming (Málaga, Spain, 2002)*, volume 2380 of *Lecture Notes in Computer Science*, pages 318–329, Springer, Berlin, 2002.
12. R. Robinson, Undecidability and nonperiodicity for tilings of the plane, *Invent. Math.*, **12**(1971), 177–209.
13. K. Sutner, Classifying circular cellular automata, *Physica D*, **45**(1990), 386–395.

On Sand Automata

Julien Cervelle[1] and Enrico Formenti[2]

[1] Laboratoire d'Informatique de l'Institut Gaspard-Monge
77454 Marne-la-Vallée Cedex 2
cervelle@univ-mlv.fr
[2] LIF-CMI, 39 rue F. Joliot Curie, 13453 Marseille cedex 13
eforment@cmi.univ-mrs.fr

Abstract. In this paper we introduce sand automata in order to give a common and useful framework for the study of most of the models of sandpiles. Moreover we give the possibility to have sources and sinks of "sand grains". We prove a result which shows that the class of sand automata is rich enough to simulate any reasonable model of sandpiles based on local interaction rules. We also give an algorithm to find the fixed points of the evolutions of the sandpiles. Finally we prove that reversibility is equivalent to bijectivity.

1 Introduction

Self-organized criticality (SOC) have received great attention since it was proposed as a paradigm for the description of a wide variety of dynamical processes in physics, biology and computer science [13,2,7,1,3].

Roughly speaking a SOC system evolves to a "critical state" after some transient. Small perturbations to this critical state cause the system to step to another critical state by some internal reorganization sub-process. Sandpiles are a simple example of this phenomenon.

A sandpile can be built by dropping grains of sand on a table, one by one. This sandpile will steepen until the slope of its edges exceeds some critical value. At this point, further addition of grains will cause cascades of sand (technically called "avalanches") to topple down the sandpile.

No external parameter needs to be tuned since the new critical state is reached by a self-organization of grains and their redistribution by avalanches.

In [8], an interesting mathematical model for sandpiles, SPM, have been introduced. It is based on the iteration of a simple local rule. Notwithstanding, it has a complex behavior and all the characteristics of a SOC system. The authors showed that, under some reasonable conditions, SPM has fixed point dynamics and gave a formula for the length of the transient behavior. In [4], Brylawski presents a more general model by adding a non-local rule. He also proves that the space of configurations is a lattice. This lattice structure helps in proving most of the above mentioned results. Recently, several variants to these seminal models have been introduced in order to study how local rules

H. Alt and M. Habib (Eds.): STACS 2003, LNCS 2607, pp. 642–653, 2003.

interact with the underlying lattice structure and how small perturbations in the local rules still present a model with a lattice structure [14,6,9,11,12,10].

In this paper we try to give to all the above cited models a common and useful "playground" for their study in the discrete dynamical system approach. For this reason, we introduce a topology on distributions of sand grains and prove that it has some useful properties such as local compactness, perfectness and that it is totally disconnected.

We define the class of *sand automata* and we prove that it generalizes all sandpiles models based on local interaction rules (see Theorem 1).

As we have already seen above, in SOC systems, fixed points play a central role. In this paper, we introduce special graphs (Π-graphs) in order to find easy polynomial algorithms (in the size of the automaton) for finding fixed points.

In our approach we allow *sources i.e.* sites which contain an infinite number of sand grains, and *sinks i.e.* sites where sand grains disappear forever.

Furthermore, we address the problem of reversibility *i.e.* to study when a SA admit an inverse map which is still a SA. Proposition 5 proves that reversibility is equivalent to bijectivity.

Finally, we point out as all the results in the paper can be generalized to lattice of dimension bigger than 1. The only exception is the algorithm based on the notion of Π-graph for finding fixed points. This kind of "graph based" approach has the same drawbacks as De Brujin graphs in cellular automata theory (see [5] for more on this subject).

The study of lattices of dimension bigger than 1 can be very interesting also from a physical point of view. In fact, they can use one dimension for the grain content and the others to take into account different physical quantities such as potential energy, velocity and so on.

2 A Topology for Sand Grains

A *configuration* is a distribution of sand grains on a regular lattice; here for the sake of simplicity we will assume that the lattice is one-dimensional *i.e.* a regular lattice indexed by $\mathbb{Z}$. Each site is labeled by the number of sand grains it contains.

Assume there is an operator who has a set of tools for measuring the heights of sand piles in a configuration. Each tool has a measuring limit and can make measurements only on a finite range of sites. Moreover, heights bigger than the tool limit are declared to be infinite. If the operator estimates that more precision is needed, he may decide to change the tool with a more powerful one.

The problem is how to measure the distance between two configurations. This is a hard problem because we want to choose our metric in order to obtain a compact or at least locally compact space and, at the same time, not to lose intuitiveness and adequacy to the sandpiles context.

We propose the following behavior for the operator when measuring the distance between two configurations x and y. First of all, he stands on the reference point: the site of index 0. If the number of grains at index 0 are different, then the operator declares those configurations completely different, *i.e.* at distance 1. Otherwise, the configurations will be observed putting a measuring device on top of the pile of the reference point. The height of this pile will be referred to as *the reference height*. Measuring devices are square windows of side $2l$. From the reference point, the operator will note the differences of height between the $2l$ neighboring sites (l to the left, and l to the right). This difference is declared infinite if it is greater than l, and therefore out of sight of the measuring device. If the current device is precise enough to point out a difference between x and y, then the distance between x and y is 2^{-l}. Otherwise the operator starts the process again using a more powerful measuring device *i.e.* a box of size $l + 1$. The process continues until he can distinguish between x and y.

Before giving the formal definition of the distance between configurations we need some notation.

Let $\widetilde{\mathbb{Z}} = \mathbb{Z} \cup \{+\infty, -\infty\}$, and $\forall a, b \in \mathbb{Z}$ $(a \leqslant b)$ let $[\![a, b]\!] = \{a, a+1, \dots, b\}$ and $\widetilde{[\![a, b]\!]} = [\![a, b]\!] \cup \{+\infty, -\infty\}$. "Measuring devices" of size $l \in \mathbb{N}$ and reference height $m \in \widetilde{\mathbb{Z}}$ are nothing but functions from $\widetilde{\mathbb{Z}}$ to $\widetilde{[\![-l, l]\!]}$ defined as follows

$$\beta_l^m(n) = \begin{cases} +\infty & \text{if } m + l < n \text{ ,} \\ -\infty & \text{if } m - l > n \text{ ,} \\ n - m & \text{otherwise.} \end{cases}$$

For the sake of simplicity, in the above definition we assume $(+\infty) - (+\infty) = (-\infty) - (-\infty) = 0$.

For any configuration $z \in \widetilde{\mathbb{Z}}^{\mathbb{Z}}$, $l \in \mathbb{N} \setminus \{0\}$ and $i \in \mathbb{Z}$ let $d_l^i(z)$ be the finite sequence:

$$d_l^i(z) = (\beta_l^{z_i}(z_{i-l}), \ \dots, \beta_l^{z_i}(z_{i-1}), \beta_l^{z_i}(z_{i+1}), \ \dots, \beta_l^{z_i}(z_{i+l})) \in \widetilde{[\![-l, l]\!]}^{2l} \ .$$

These are the measures observed by the operator using the device β_l and site i as reference point. In the sequel, the word $w = d_l^i(z)$ will be called the *range* of radius l and center i. In order to simplify the notation in the proofs, subscripts of w take values in $[\![-l, -1]\!] \cup [\![1, l]\!]$ *i.e.* $w = w_{-l} \dots w_{-1} w_1 \dots w_l$.

Definition 1. *The distance between two configurations x and y is defined as $d(x, y) = 2^{-l}$, where l is the least integer such that $d_l^0(x) \neq d_l^0(y)$.*

For $w \in \widetilde{\mathbb{Z}}$, let $\underline{w}$ be the configuration such that $\forall i \in \mathbb{Z}$, $\underline{w}_i = w$.

Let X be the space of configurations endowed with the topology induced by d. Open balls in X are called *boxes* and can be defined as follows.

Definition 2 (Boxes). *For any positive integer l, let $u \in \widetilde{\mathbb{Z}}$ and let w a range of radius l. The box $\langle\!\langle u, w \rangle\!\rangle_l$, centered on u, of range w and radius l is defined as follows*

$$\langle\!\langle u, w \rangle\!\rangle_l = \left\{ x \in \widetilde{\mathbb{Z}}^{\mathbb{Z}} \mid x_0 = u, \ \forall i \in [\![-l, -1]\!] \cup [\![1, l]\!], \ \beta_l^u(x_i) = w_i \right\} \ .$$

Proposition 1. *The space X is perfect (i.e. it has no isolated points).*

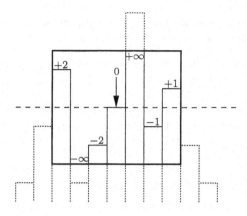

Fig. 1. Illustration of a "measuring device" of size 3. Here $d_3^0(x) = (2, -\infty, -2, +\infty, -1, 1)$ is the "range" of radius 3 and center 0.

Proof. Choose an arbitrary configuration $x \in X$. For any $l \in \mathbb{N}$, build a configuration $z \in X$ defined as follows

$$\forall j \in \mathbb{Z}, z_j = \begin{cases} x_j & \text{if } j \neq l+1 \\ x_0 & \text{if } j = l+1 \text{ and } x_{l+1} \neq x_0 \\ 0 & \text{if } j = l+1 \text{ and } |x_0| = \infty \\ x_0 + 1 & \text{otherwise.} \end{cases}$$

By definition, $d(x, z) = 2^{-l-1}$. $\qquad\qquad\qquad\qquad\qquad\qquad\qquad\qquad$ $\square$

Many classical results in discrete dynamical systems dynamics relay on the compactness of the space. Unfortunately X is not compact. In fact, it is easy to see that the sequence $(n)_{n \in \mathbb{N}}$ has no converging subsequence. Corollary 1 proves that X is at least locally compact. A central role in the proof of this result is played by the sets $E_u = \left\{ x \in \widetilde{\mathbb{Z}}^{\mathbb{Z}}, x_0 = u \right\}$, for $u \in \widetilde{\mathbb{Z}}$, as one can see in the following proposition.

Proposition 2. *For all $u \in \widetilde{\mathbb{Z}}$, the set E_u is compact.*

Proof. Consider a sequence $(x^n)_{n \in \mathbb{N}}$ of configurations in E_u. We are going to build a configuration z such that for all $\varepsilon > 0$, there exists an i such that $d(x^i, z) \leqslant \varepsilon$. Let $z_0 = u$. We are going to assign the values for z at positions 1, -1, 2, -2, 3 etc. Let $U_0 = \mathbb{N}$. For each value $i \neq 0$, let U_ℓ the set defined at the previous step ($-i$ if $i > 0$ and $-i + 1$ if $i < 0$), or U_0 if $i = 1$. Consider the sequence $(x_i^j)_{j \in U_\ell}$; there are three possible cases:

i)   there is a value k_i which occurs infinitely many times, set $z_i = k_i$, and let U_i the set of indices j such that $x_i^j = k_i$;

ii) there exists a strictly increasing subsequence, set $z_i = +\infty$, and let $U_i \subset U_\ell$ be an infinite set of indices j such that $(x_i^j)_{j \in U_i}$ is strictly increasing;

iii) there exists a strictly decreasing subsequence; then set $z_i = -\infty$, and $U_i \subset U_\ell$ be an infinite set of indices j such that $(x_i^j)_{j \in U_i}$ is strictly decreasing.

Let us prove that z has the required property. Let l be a positive integer. We want to find an integer n such that $d(x^n, z) \leqslant 2^{-l}$. Consider the set U_{-l}. For all configurations x^i, the integer $i \in U_{-l}$, and for all k between $-l$ and l, either $z_k = x_k^i$ or $z_k = +\infty$ [resp. $-\infty$], and the sequence $(x_k^j)_{j \in U_{-l}}$ is strictly increasing [resp. decreasing]. Choose $n \in U_{-l}$ such that, for all k between $-l$ and l such that $z_k = +\infty$ implies $x_k^n > l$ and $z_k = -\infty$ implies $x_k^n < l$. We have $d_l^0(x) = d_l^0(z)$ and hence $d(x^n, z) \leqslant 2^{-l}$.   □

Corollary 1. *The space X is locally compact.*

Proof. Each point x belongs to E_{x_0}. By Proposition 2, the set E_{x_0} is compact.   □

Corollary 2. *Boxes are clopen (i.e. closed and open) sets.*

Proof. Any box $\langle\!\langle u, w \rangle\!\rangle_l$ is, by definition, open. Let $W = \overbrace{[\![-l, l]\!]}^{2l} \setminus \{w\}$. We have that $\langle\!\langle u, w \rangle\!\rangle_l = E_u \setminus \bigcup_{\bar{w} \in W} \langle\!\langle u, \bar{w} \rangle\!\rangle_l$. By Proposition 2, E_u is compact and hence closed. We conclude that $\langle\!\langle u, w \rangle\!\rangle_l$ is a closed set minus some open sets, therefore it is closed.   □

Corollary 3. *The topological space X is totally disconnected.*

Proof. Consider two configurations x and y. Let B_x be the open balls of center x and radius $d(x, y)$. Since B_x is clopen, its complementary is open. Hence, X is the union of B_x which contains x and its complementary (which contains y). Both of these sets are open. We conclude that X is totally disconnected.   □

3   Sand Automata

A sand automaton is a deterministic finite automaton working on configurations. Each site is updated according to a local rule which computes the new sand content for the site taking into account its current sand content and the one of a fixed number of neighboring sites. All sites are updated synchronously and in parallel. The maximal number of grains that an automaton can add or delete from a site is called the *radius* of the automaton. More formally, one can give the following definition.

Definition 3 (Sand automaton). *A SA   is a couple $\langle r, \lambda \rangle$, where r is the radius and $\lambda : \overbrace{[\![-r, r]\!]}^{2r} \to [\![-r, r]\!]$ the local rule of the automaton. By means of the local rule, one can define the global rule $f : \widetilde{\mathbb{Z}}^{\mathbb{Z}} \to \widetilde{\mathbb{Z}}^{\mathbb{Z}}$ as follows*

$$\forall x \in \widetilde{\mathbb{Z}}^{\mathbb{Z}} \, \forall i \in \mathbb{Z}, \; f(x)_i = \begin{cases} x_i & \text{if } x_i = \infty \text{ or } x_i = -\infty \\ x_i + \lambda(d_r^i(x)) & \text{otherwise.} \end{cases}$$

When no misunderstanding is possible, we will make no distinction between the global rule and the SA itself.

Example 1 (SA version of SPM). Consider a SA $\mathcal{A} = \langle 1, \lambda \rangle$, where λ is defined as follows

$$\forall a, b \in \widetilde{[\![-1,1]\!]}, \ \lambda(a,b) = \begin{cases} 1 \text{ if } a = +\infty \text{ and } b \neq -\infty \\ -1 \text{ if } a \neq \infty \text{ and } b = -\infty \\ 0 \text{ otherwise.} \end{cases}$$

This rule simulates sand piles acting under the rules of SPM: a grain falls down to the right if the pile to the right has two or more grain less.

In a way similar to Example 1, one can simulate all variants of sandpiles models that are based on a local rule (for more on variants of sandpiles models see [14]). This will be formally proved in Theorem 1 but before we need some more definitions and a technical lemma.

In the sequel we often refer to the *shift map* σ that is defined as follows

$$\forall x \in \widetilde{\mathbb{Z}}^{\mathbb{Z}} \forall i \in \mathbb{Z}, \ \sigma(x)_i = x_{i+1} \ ,$$

and the *raising map* ρ which is defined as follows

$$\forall x \in \widetilde{\mathbb{Z}}^{\mathbb{Z}} \forall i \in \mathbb{Z}, \ \rho(x)_i = x_i + 1 \ .$$

One can immediately verify that for any $x \in \widetilde{\mathbb{Z}}^{\mathbb{Z}}$, $d_r^{i+1}(x) = d_r^i(\sigma(x))$ and $d_r^i(x) = d_r^i(\rho(x))$.

Definition 4. *A function $f : X \to X$ is* shift-invariant *if it commutes with the shift i.e. $f \circ \sigma = \sigma \circ f$. The function f is* vertical invariant *if it commutes with the raising map i.e. $f \circ \rho = \rho \circ f$. We say that f is* infiniteness conserving *if $\forall x \in \widetilde{\mathbb{Z}}^{\mathbb{Z}}$, $f(x)_i = x_i$ if $x_i = +\infty$ or $-\infty$ and $|f(x)_i| < \infty$ if $|x_i| < \infty$.*

Notice that, by definition, all SA are infiniteness conserving.

Lemma 1. *Let $f : X \to X$ be a continuous, vertical invariant and infiniteness conserving function. Then, $\forall u \in \widetilde{\mathbb{Z}}$, $f^{-1}(E_u)$ is compact.*

Proof. Let f be a continuous, vertical invariant and infiniteness conserving function.

By infiniteness conservation, if $u \in \widetilde{\mathbb{Z}}^{\mathbb{Z}}$ is infinite, we have $f^{-1}(E_u) \subseteq E_u$. By Proposition 2, E_u is compact and, by continuity of f, $f^{-1}(E_u)$ is closed. We conclude that $f^{-1}(E_u)$ is compact since it is a closed subset of compact set.

Now, assume that u is finite. Let $U = f^{-1}(E_u)$. By the continuity of f, U is clopen. Consider a partition $(U_i)_{i \in I}$ of U such that $\emptyset \subsetneq U_i = U \cap E_i$. Since all E_i are clopens then each U_i is clopen too.

By contradiction, assume $|I| = \infty$. First of all, recall that for all configuration x and all integer l, $d_l^0(x) = d_l^0(\rho(x))$. Let $(u_n)_{n \in \mathbb{N}}$ be a sequence of distinct integers contained in I. For all $n \in \mathbb{N}$, let $B_n = \langle\!\langle u_n, w_n \rangle\!\rangle_{-d_n}$ be an open ball included in U_{u_n}. Let n_0 be the integer such that $d_{n_0} = \min_{i \in \mathbb{N}} d_{n_i}$. Let w_n' be the range of radius d_{n_0} such that $\langle\!\langle u_n, w_n \rangle\!\rangle_{d_n} \subset \langle\!\langle u_n, w_n' \rangle\!\rangle_{d_{n_0}}$. Remark that w_n' is the central portion of radius d_{n_0} of w_n, whose values greater than d_{n_0} are replaced by ∞ and those lower than $-d_{n_0}$ are replaced by $-\infty$. By the hypothesis, for all $n \in \mathbb{N}$, $f(B_n) \subset E_u$. Hence, since f is vertical invariant, for all $i \in \mathbb{Z}$, $f(\rho^i(B_n)) \subset E_{u+i}$, which can be rewritten as $f(\langle\!\langle u_n + i, w_n \rangle\!\rangle_{d_{n_0}}) \subset E_{u+i}$. For $i = u_{n_0} - u_n$, $f(\langle\!\langle u_{n_0}, w_n \rangle\!\rangle_{d_{n_0}}) \subset E_{u+u_{n_0}-u_n}$. Note that for all n and m such that $n \neq m$, the sets $E_{u+u_{n_0}-u_n}$ and $E_{u+u_{n_0}-u_m}$ are disjoint. This fact implies that $w'_n \neq w'_m$, otherwise we have a common member in $\langle\!\langle u_{n_0}, w_n \rangle\!\rangle_{d_{n_0}}$ and $\langle\!\langle u_{n_0}, w_m \rangle\!\rangle_{d_{n_0}}$ which have disjoint image sets. The fact that ranges $(w'_k)_{k \in \mathbb{Z}}$ have finitely many possible values (they are taken from $[\![-d_{n_0}, d_{n_0}]\!]^{\overbrace{2d_{n_0}}}$) leads to a contradiction.

Therefore I has finite cardinality. Since $U \subset \bigcup_{i \in I} E_i$ and U is closed, we conclude that U is compact.   □

The following theorem is a strong representation result that characterizes a wide class of functions that have finite description on X. The advantage of such functions is that they are really suitable for computer simulations. The finite description allows a faultless computation of the values of the function diminishing the sensibility to approximations errors which can completely bias simulations.

Theorem 1. *A function $f : X \to X$ is a SA if and only if following conditions hold:*

i)   *f is continuous;*
ii) *f is shift invariant;*
iii) *f is vertical invariant;*
iv) *f is infiniteness conserving.*

Proof.
Consider a SA $\mathcal{A} = \langle r, \lambda \rangle$ with global rule f. It follows immediately by the definition of SA that f is shift-invariant, vertical invariant and infiniteness conserving.

It remains to prove that f is continuous. For any configuration x and any positive integer l, we have to find an integer m such that $d(x, y) < 2^{-m}$ implies $d(f(x), f(y)) < 2^{-l}$ for all configurations y. Choose $m = 4l + r$. Let y be a configuration such that $d(x, y) < 2^{-m}$. We have $x_0 = y_0$ and $d_m^0(x) = d_m^0(y)$, which implies, on the one hand, $f(x)_0 = f(y)_0$, and on the other hand, $\beta_m^{x_0}(x_i) = \beta_m^{y_0}(y_i)$, for all integer $i \in [\![-m, m]\!]$.

We claim that for all integer $j \in [\![-l, l]\!]$, $j \neq 0$ we have two possible cases:

i) $|x_j - x_0| > 3l$ and hence $|y_j - y_0| > 3l$;
ii) $x_j = y_j$ and for all integer $k \neq j$ with $k \in [\![j - r, j + r]\!]$, $\beta_l^{x_j}(x_k) = \beta_l^{y_j}(y_k)$.

In fact, suppose that, on the one hand, $|x_j - x_0| > 3l$. Then, it holds that $|y_j - y_0| > 2l$, since $\beta_m^{x_0}(x_j) = \beta_m^{y_0}(y_j)$.

On the other hand, assume that $|x_j - x_0| \leqslant 3l$. We have that $x_j = y_j$, since $\beta_m^{x_0}(x_j) = \beta_m^{y_0}(y_j)$. Then, for all integer $k \neq j$ with $k \in [\![j - r, j + r]\!]$, we have three possible subcases (since $\beta_m^{x_0}(x_k) = \beta_m^{y_0}(y_k)$):

a) $x_k = y_k$ then $\underline{\beta_l^{x_j}(x_k) = \beta_l^{y_j}(y_k)}$;

b) $x_k - x_0 > m > 4l$ and $y_k - y_0 > m > 4l$. As $|x_j - x_0| \leqslant 3l$ and $|y_j - y_0| \leqslant 3l$ (recall that $x_j = y_j$ and $x_0 = y_0$), then $x_k - x_j > l$ and $y_k - y_j > l$ which implies that $\underline{\beta_l^{x_j}(x_k)} = \underline{\beta_l^{y_j}(y_k)}$;

c) else $x_k - x_0 < -m < -4l$ and $y_k - y_0 < -m < -4l$. Using the same chain of inequalities, it holds that $\underline{\beta_l^{x_j}(x_k)} = \underline{\beta_l^{y_j}(y_k)}$

(in the formulas above and the sequel of the proof, we have underlined some parts to stress that they are equals).

We conclude that, for all integers $j \in [\![-l, l]\!]$ with $j \neq 0$, if case i) occurs, since the local rule can only increase or decrease a value by at most l, it holds that

$$|f(x)_j - f(x)_0| \geqslant |x_j - x_0| - |f(x)_0 - x_0| - |f(x)_l - x_l| > 3l - l - l = l$$

and, by using the same argument, one finds $|f(y)_j - f(y)_0| > l$. Therefore it holds that $\beta_l^{f(x)_0}(f(x)_j) = \beta_l^{f(y)_0}(f(y)_j)$.

If case ii) occurs, then $x_j = y_j$ and $d_r^j(x) = d_r^j(y)$. That means that $f(x)_j = f(y)_j$, and so $\underline{\beta_l^{f(x)_0}(f(x)_j)} = \underline{\beta_l^{f(y)_0}(f(y)_j)}$.

Hence, it holds that $d_l^0(f(x)) = d_l^0(f(y))$, and then $d(f(x), f(y)) < 2^{-l}$.

For the second part of the proof, let $f : X \to X$ be a continuous, shift-invariant, vertical invariant and infiniteness conserving function. We are going to prove that it is the global rule of a suitable SA.

Consider the clopen set E_0. Let $U = f^{-1}(E_0)$. By Lemma 1, the set U is compact, and, hence, it is an union of finitely many open balls: $U = \bigcup_{i \in I} \langle\!\langle u_i, w^i \rangle\!\rangle_{r_i}$ with $|I| < \infty$. Since each box can be decomposed into finitely many boxes of larger radius, without loss of generality, one can suppose that each range w^i has the same radius r. Remaind that boxes of larger radius are indeed spheres of smaller radius. We claim that every range must appear exactly once in the sequence $(w^i)_{i \in I}$. In fact, if a range appears twice, it would contradict vertical invariance (recall that $d_r^0(x) = d_r^0(\rho(x))$) since two distinct centers with the same range cannot both lead to E_0. If a range w does not appear, let x be a configuration such that $x_0 = 0$ and $d_r^0(x) = w$. The configuration $f(x)$ belongs to an E_j, for some finite j since f is infiniteness conserving. Hence, by vertical invariance, $f(\rho^{-j}(x)) \in E_0$. Since $d_r^0(\rho^{-j}(x)) = d_r^0(x) = w$, it means that w appears in $(w^i)_{i \in I}$. Hence, it is natural to define λ as follows: $\lambda(w^i) = -u_i$. Let f' be the global rule of the SA $\langle r, \lambda \rangle$. Let us prove that $f = f'$. For all configurations x, for all integers n, let $i \in I$ be such that $w^i = d_r^j(x)$. We have that $f'(x)_n = x_n - u_i$. Let us compute $f(x)_n$. Since f is vertical invariant and shift invariant, we have that

$$f(x)_n = f(\sigma^n(x))_0 = x_n - u_i + f(\rho^{u_i - x_n}(\sigma^n(x)))_0$$

Let $y = \rho^{u_i - x_n}(\sigma^n(x))$. We have $y_0 = u_i - x_n + x_n = u_i$ and $d_r^0(y) = w_i$. Hence, by definition of u_i and w_i, $f(y)_0 = 0$. Hence, $f(x)_n = x_n - u_i + f(y)_0 = x_n - u_i = f'(x)$. We conclude that $f = f'$. $\qquad\square$

Cellular automata are often used as a paradigmatic example for modeling phenomena ruled by local interaction rules. Proposition 3 says that SA can be used as well. Cellular automata can be formally defined as follows.

A *cellular automaton* is a map $F : S^{\mathbb{Z}} \to S^{\mathbb{Z}}$ defined for any $x \in S^{\mathbb{Z}}$ and $i \in \mathbb{Z}$ by $F(x)_i = \mu(x_{i-h} \ldots x_i \ldots x_{i+h})$, where $\mu : S^{2h+1} \to S$ is the *local rule* and h is the *radius*. The set S is finite and it is usually called the set of *states* of the cellular automaton. The function F is called the *global rule* of the cellular automaton (for more on cellular automata, for example, see [15]).

Proposition 3. *Any cellular automaton can be simulated by a suitable SA.*

4   Reversibility

Given a SA $\mathcal{A} = \langle r, \lambda \rangle$ of global rule f, we say that $\mathcal{A}$ is injective (resp. bijective) if f is injective (resp. bijective).

Proposition 4. *Consider a SA $\mathcal{A}$ of global rule f. If $\mathcal{A}$ is injective then f is open.*

Proof. Consider an injective SA $\mathcal{A} = \langle r, \lambda \rangle$ of global rule f. Let A be the box $\langle\!\langle u, w \rangle\!\rangle_l$, with $l > r$. By definition of SA, $f(A) \subset E_{u+i}$ where i is the result of the application of the λ to w. Let $C = f^{-1}(E_{u+1})$ and $B = C \setminus A$. Since f is injective, we have that $f(C) = E_{u+1}$ is the disjoint union of $f(A)$ and $f(B)$. Since f is continuous, C is a clopen. As A is also a clopen, B is a clopen. Using Lemma 1, C is included in a finite union of sets E_i hence, by Proposition 2, it is compact. We deduce that, since B is closed, it is compact, and, by the continuity of f, $f(B)$ is compact too. We conclude that $f(B)$ is closed, and hence, that $f(A) = f(C) - f(B)$ is open since $f(C) = E_{u+1}$ is open.   □

Proposition 5. *Consider a SA $\mathcal{A}$ of global rule f. If $\mathcal{A}$ is a bijective, then f^{-1} is a SA.*

Proof. Consider an injective SA $\mathcal{A}$ of global rule f. By Proposition 4, f is open and, hence, f^{-1} is continuous. Since f is vertical invariant so is f^{-1}: for all configuration x, and $y = f^{-1}(x)$, $f(\rho(x)) = \rho(f(x)) \Rightarrow f(\rho(f^{-1}(y))) = \rho(y) \Rightarrow \rho(f^{-1}(y)) = f^{-1}(\rho(y))$. Replacing ρ by σ gives that f^{-1} is shift invariant too. It is clear that f^{-1} is infiniteness conserving. Hence by Theorem 1, f^{-1} is the global rule of a SA.   □

5 Π-Graphs and Fixed Points

The study of sandpiles have revealed that, although at first they seem to be governed by some intrinsic instability, they show many examples of local stability. For this reason, in literature, many papers give characterizations of the set of fixed points and show formulas for computing the length of the transient trajectory to a fixed point. Similar results can be found for SA. In this section, we give an algorithm to find the set of fixed points of any SA.

A necessary and sufficient condition for a SA $\mathcal{A} = \langle r, \lambda \rangle$ to admit a configuration $c \in X$ as a fixed point is that $\forall i \in \mathbb{Z}$, $\lambda(d_r^i(x)) = 0$. This condition gives an easy algorithm to find out all fixed points of a SA. In order to better illustrate this algorithm we introduce a new tool: Π-graphs.

For sake of simplicity, we will describe the case where the radius is 1. The Π-graph of a SA $\mathcal{A} = \langle 1, \lambda \rangle$ is a labeled directed graph $\langle V, E \rangle$, where $V = \widehat{[\![-1,1]\!]}$ and $E = V \times V$. Any edge (a, b) is labeled $\lambda(-a, b)$.

Figure 2 shows the Π-graph for the SA described in Example 1. Π-graphs for $r > 1$ are a little bit difficult to define and represent since one should take into account much more information on the structure of the neighborhood of the SA. They will be further developped in the long version of the paper.

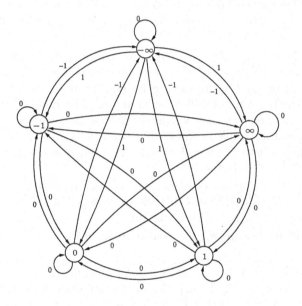

Fig. 2. The Π-graph for the SA in Example 1

Given a configuration x and a Π-graph, one can compute $f(x)$ as follows. Compute the bi-infinite sequence $v(x)$ such that $\forall i \in \mathbb{Z}, v(x)_{i+1} = \beta_1^{x_i}(x_{i+1})$.

Then start, for instance, at position 0 and vertex $v(x)_0$. Follow the path $v(x)_1, v(x)_2$ etc. For each edge, add its value to the corresponding site of the configuration: going from $v(x)_i$ to $v(x)_{i+1}$, add the label of this edge to x_i; this computes the value of $f(x)_i$ since

$$x_i + \text{label}(v(x)_i, v(x)_{i+1}) = x_i + \lambda(-\beta_1^{x_{i-1}}(x_i), \beta_1^{x_i}(x_{i+1}))$$
$$= x_i + \lambda(\beta_1^{x_i}(x_{i-1}), \beta_1^{x_i}(x_{i+1})) = f(x)_i$$

At this point it is easy to see that fixed points are given by bi-infinite path with all edges labeled by 0.

However, one should take care of two difficulties. First, from a path $v(x)$, there exists infinitely many y such that $v(y) = v(x)$. In order to reconstruct y from $v(x)$, one can fix a value at any position i and compute the values at position $i + 1$, $i + 2$, etc., by adding the values of $v(x)$, paying attention that if it is ∞ (resp. $-\infty$), any value strictly greater than 1 can be added (resp. subtracted).

Second, we should pay attention to values ∞ and $-\infty$ in configurations. Since the rule is not applied (one simply uses infiniteness conservation) when a site has these values, for this reason one can jump between vertices ∞ and $-\infty$ even if the edges between them are not labeled by 0, if you put ∞ or $-\infty$ at the corresponding sites in x. For instance, if the 0 labeled edged are $(-\infty, -1), (-1, -\infty), (\infty, 0), (0, \infty), (0, 0)$, then a path $-1, -\infty, \infty, 0, 0, \ldots$ is correct though the label of $(-\infty, \infty)$ is not 0 since it can correspond to $0, -1, -\infty, \infty, 40, 40, \ldots$. Anyway this sequence is a portion of a fixed point.

By the definition of SA, it is trivial to deduce that configuration which are exclusively made by sinks and/or sources are fixed points. They are a real element of stability of the system. In fact, in Proposition 6, we prove that for all $x \in \{\infty, -\infty\}$, the trajectory of all point in any neighborhood U of x never leaves U. The notion of equicontinuity point formalized this behavior.

A configuration $x \in X$ is an *equicontinuity* (or *Lyapunov stable*) point if for all $\varepsilon > 0$ there exists $\delta > 0$ such that for all configurations y, $d(x, y) \leqslant \delta \implies \forall n \in \mathbb{N}, d(f^n(x), f^n(y)) \leqslant \varepsilon$.

It is easy to see that any configuration of $\{-\infty, \infty\}^{\mathbb{Z}}$ are not only fixed points, but the next proposition proves that they are Lyapunov stable, that it to say all SA have a certain degree of local stability.

Proposition 6. *For any SA the configurations of $\{-\infty, \infty\}^{\mathbb{Z}}$ are equicontinuity points.*

Proof. Consider a SA $\mathcal{A} = \langle r, \lambda \rangle$. Let x be a configuration of $\{-\infty, \infty\}^{\mathbb{Z}}$. Fix $\varepsilon > 0$ and choose $\delta = 2^{-h}$ such that $\delta \leqslant \varepsilon$ for some $h \in \mathbb{N}$. For any y such that $d(x, y) \leqslant \delta$, it holds that for all integer $i \neq 0$ between $-h - 1$ and $h + 1$, $x_i = \infty \Rightarrow y_i = \infty$ and $x_i = -\infty \Rightarrow y_i < \infty$. As f is a SA a is ruled by a local rule, if $x_i = y_i = \infty$, $\forall n \in \mathbb{N}$, $f^n(x)_i = f^n(y)_i = \infty$ and if $x_i = -\infty$ and $y_i < \infty$, then $\forall n \in \mathbb{N}$, $f^n(x)_i = \infty$ and $f^n(y_i) < \infty$. Hence, $d(f^n(x), f^n(y)) < 2^{-h} \leqslant \varepsilon$. $\square$

6   Greater Dimensions

There are two ways to increase the dimension of the underlying lattice which SA work on. The first one consists in increasing the dimension of the values taken by each site. The new working space would be $\left(\widetilde{\mathbb{Z}^n} \right)^{\mathbb{Z}}$. The first coordinate could be the height of the sand pile, as usual, and the others could be used to describe physical quantities about the pile, such as energy, velocity, or, using cellular automata emulation, any finite state variable. The measuring tool used to compute the distance would compare two elements of $\widetilde{\mathbb{Z}}^n$, coordinate by coordinate, up to its limit. All our results can be extended to this n-dimensional case. The second one consists in increasing the dimension of the number of sites. The new working space would be $X^m = \widetilde{\mathbb{Z}}^{\mathbb{Z}^m}$. For instance, with $m = 2$, a 3-dimensional sand hill could be simulated. The measuring tool used to compute the distance would compare all piles around the site under computation in a box of dimension m. For this m-dimensional case, all our result extend, except for the construction of Π-graphs, due to the appearence of problems related to the fact that one has to deal with multiple paths leading from a point to another.

References

1. R. Anderson, L. Lovász, P. Shor, J. Spencer, E. Tardos, and S. Winograd. Disks, balls and walls: analysis of a combinatorial game. *American mathematical monthly*, 96:481–493, 1989.
2. P. Bak. *How nature works – The science of SOC*. Oxford University Press, 1997.
3. A. Bjorner and L. Lovász. Chip firing games on directed graphs. *Journal of algebraic combinatorics*, 1:305–328, 1992.
4. T. Brylawski. The lattice of integer partitions. *Discrete mathematics*, 6:201–219, 1973.
5. B. Durand. Global properties of cellular automata. In E. Goles and S. Martinez, editors, *Cellular Automata and Complex Systems*. Kluwer, 1998.
6. J. Durand-Lose. *Automates cellulaires, automates à partition et tas de sable*. PhD thesis, Université de Bordeaux – LABRI, 1996.
7. E. Goles. Sand pile automata. *Annales Insitut Henri Poincaré, Physique Théorique*, 56(1):75–90, 1992.
8. E. Goles and M. A. Kiwi. Game on line graphs and sandpile automata. *Theoretical computer science*, 115:321–349, 1993.
9. E. Goles and M. A. Kiwi. Sandpiles dynamics in a one-dimensional bounded lattice. *Theoretical computer science*, 136(2):527–532, 1994.
10. E. Goles, M. Morvan, and H. D. Phan. Lattice structure and convergence of a game of cards. *Annals of Combinatorics*, 2002. To appear.
11. E. Goles, M. Morvan, and H. D. Phan. Sandpiles and order structure of integer partitions. *Discrete Applied Mathematics*, 117(1–3):51–64, 2002.
12. E. Goles, M. Morvan, and H. D. Phan. The structure of linear chip firing game and related models. *Theoretical Computer Science*, 270:827–841, 2002.
13. H. J. Jensen. *Self-organized criticality*. Cambridge University Press, 1998.
14. H. D. Phan. *Structures ordonnées et dynamique de pile de sable*. PhD thesis, Université Denis Diderot Paris VII – LIAFA, 1999.
15. S. Wolfram. *Theory and application of cellular automata*. Wold Scientific Publishing Co., 1986.

Adaptive Sorting and the Information Theoretic Lower Bound

Amr Elmasry[1] and Michael L. Fredman[2*]

[1] Dept. of Computer Science
Alexandria University
Alexandria, Egypt
[2] Dept. of Computer Science
Rutgers University
New Brunswick, NJ 08903

Abstract. We derive a variation of insertion sort that is near optimal with respect to the number of inversions present in the input. The number of comparisons performed by our algorithms, on an input sequence of length n that has I inversions, is at most $n \log_2 \left(\frac{I}{n} + 1 \right) + O(n)$. Moreover, we give an implementation of the algorithm that runs in time $O(n \log_2 \left(\frac{I}{n} + 1 \right) + n)$. All previously known algorithms require at least $cn \log_2 (\frac{I}{n} + 1)$ comparisons for some $c > 1$.

Category: **Algorithms and Data Structures**

1 Introduction

In many applications of sorting the lists to be sorted do not consist of randomly ordered elements, but are already partially sorted. An adaptive sorting algorithm benefits from the existing presortedness in the input sequence. One of the commonly recognized measures of presortedness is the number of inversions in the input sequence [11]. The number of inversions is the number of pairs of input items in the wrong order. More precisely, for an input sequence X of length n, the number of inversions, $\text{Inv}(X)$, is defined

$$\text{Inv}(X) = |\{(i,j) \mid 1 \le i < j \le n \text{ and } x_i > x_j\}|.$$

Let $P(I)$ denote the number of permutations having I or fewer inversions. It is readily demonstrated that $\log P(I) = n \log(\frac{I}{n} + 1) + O(n)$ [10]. (Throughout this paper all logarithms are taken to base 2.) Appealing to the information-theoretic lower bound applicable to comparison based algorithms, it is therefore a worthy goal to construct a sorting algorithm that sorts an input X in time $O(n \log(\frac{\text{Inv}(X)}{n} + 1) + n)$, and indeed there are a number of such algorithms that accomplish this as will be summarized shortly. Our focus, however, concerns a theoretical exploration of the complexity limits, with the goal of obtaining more precise results than can otherwise be inferred from the existing algorithms.

* Supported in part by NSF grant CCR-9732689

H. Alt and M. Habib (Eds.): STACS 2003, LNCS 2607, pp. 654–662, 2003.
© Springer-Verlag Berlin Heidelberg 2003

Appealing to a general information-theoretic theorem of Fredman [9], *if we know in advance* that $\text{Inv}(X) \leq I$, then X can be sorted with only $\log P(I) + O(n)$ comparisons. Combining this with the above estimate for $\log P(I)$, we obtain a near optimal bound of $n \log(\frac{I}{n} + 1) + O(n)$ comparisons for sorting under these circumstances. This observation, however, is deficient in two respects: First, the information-theoretic result is not accompanied by an implementable algorithm. Second, there is a possibility that no *single* algorithm can serve to sort an arbitrary input X *uniformly* utilizing only $n \log(\frac{\text{Inv}(X)}{n} + 1) + O(n)$ comparisons; in other words, achieving the information-theoretic bound may require an apriori estimate for $\text{Inv}(X)$.

A simple example illustrates this second issue in an analogous, if more limited setting. Consider the task of inserting a value y into a sorted sequence, $x_1 < x_2 < \cdots < x_{n-1}$. The method of binary insertion attains the optimal number of required comparisons, $\lceil \log n \rceil$. Moreover, with advance knowledge that y will arrive in one of the first i positions, $\lceil \log i \rceil$ comparisons suffice. The question can thus be raised: Can the insertion of y be accomplished so that, *uniformly*, only $\log i$ comparisons are required should y arrive in the ith position? A result of Bentley and Yao [3] shows that this is not possible. In fact, even $\log i + \log \log i$ comparisons will not suffice [3]. (This work of Bentley and Yao actually concerns the closely related problem of unbounded searching.)

Besides providing an analogous setting, this insertion problem is directly relevant to our sorting problem. Setting aside the issue of implementation, *if we could* accomplish this insertion uniformly with $\log i + O(1)$ comparisons, then using the strategy of successively inserting the items $x_1, \cdots, x_n$ in reverse order (into an initially empty list), the total numbers of comparisons required to sort would be bounded by $n \log(\frac{\text{Inv}(X)}{n} + 1) + O(n)$, which is asymptotically optimal. (Let i_j denote the position in which x_j arrives when it is inserted in the sorted list consisting of $x_{j+1}, \cdots, x_n$. Then $\sum i_j = \text{Inv}(X) + n$ and $\sum \log i_j \leq n \log \frac{\sum i_j}{n}$.) The work of Bentley-Yao [3] also includes a positive result for the insertion problem, providing a construction that accomplishes the insertion with at most $\log i + 2 \log \log i$ comparisons. (Thus, the inherent complexity of this insertion problem is in some sense $\log i + \Theta(\log \log i)$. In fact, the Bentley-Yao results include much tighter estimates [3].) It follows that a sorting by insertion strategy can achieve a bound of $n \log(\frac{\text{Inv}(X)}{n} + 1) + O(n \log \log(\frac{\text{Inv}(X)}{n} + 1))$ comparisons, which is near optimal but still leaves a gap relative to our bound of $n \log(\frac{\text{Inv}(X)}{n} + 1) + O(n)$ comparisons, attainable with prior knowledge of $\text{Inv}(X)$.

Almost paradoxically, our main result shows that sorting by insertion can be utilized to sort with $n \log(\frac{\text{Inv}(X)}{n} + 1) + O(n)$ comparisons! Moreover, we provide an implementation for which the overall run time is $O(n \log(\frac{\text{Inv}(X)}{n} + 1) + n)$, staying within the $n \log(\frac{\text{Inv}(X)}{n} + 1) + O(n)$ bound for the number of comparisons. The space utilized is $O(n)$.

Related Work

There exists a fair amount of prior work on the subject of adaptive sorting. The first adaptive sorting algorithm with an $O(n \log(\frac{\text{Inv}(X)}{n} + 1) + n)$ run time is due to Guibas, McCreight, Plass and Roberts [10]. This algorithm provides an implementation of sorting by insertion, utilizing a finger-based balanced search tree. Mehlhorn [18] describes a similar approach. Other implementations of finger trees appear in [2,17]. As a consequence of the dynamic finger theorem for splay trees (see Cole [5]), the splay trees of Sleator and Tarjan [21] provide a simplified substitute for finger trees to achieve the same run time. The work of Moffat, Petersson, and Wormald [19] utilizes a merge sort strategy to attain these same bounds. Some other adaptive sorting algorithms include Blocksort [15] which runs in-place, Splitsort [12], and Adaptive Heapsort [13]. Of these, Splitsort [12] and Adaptive Heapsort [13] are the most promising from the practical point of view. The more recently introduced Trinomialsort of Elmasry [6] requires $1.89n \log(\frac{\text{Inv}(X)}{n} + 1)) + O(n)$ comparisons in the worst case, and provides a practical, efficient algorithm for adaptive sorting.

Several authors have proposed other measures of presortedness and proposed optimal algorithms with respect to these measures [8,4,13,15]. Mannila [16] formalized the concept of presortedness. He studied several measures of presortedness and introduced the concept of optimality with respect to these measures. Petersson and Moffat [20] related all of the various known measures in a partial order and established new definitions with respect to the optimality of adaptive algorithms.

2   Adaptive Insertion Sort

Our initial focus exclusively concerns comparison complexity, ignoring the question of implementation. Consider the following method for inserting y into a sorted sequence, $x_1 < x_2 < \cdots < x_{n-1}$. For a specified value r we first perform a linear search among the items, $x_r, x_{2r}, \cdots, x_{r\lfloor n/r \rfloor}$, to determine the interval of length r among $x_1 < x_2 < \cdots < x_{n-1}$ into which y falls. Next, we perform a binary search within the resulting interval of length r to determine the precise location for y. If y ends up in position i, then $\frac{i}{r} + \log r + O(1)$ comparisons suffice for this insertion. Using the notation from the Introduction and choosing $r = \frac{\sum i_j}{n} = \frac{\text{Inv}(X)}{n} + 1$, we conclude that by using this method of insertion as part of an insertion sort algorithm, the total number of comparisons performed in sorting a sequence X is bounded by $n \log r + \frac{\sum i_j}{r} + O(n) = n \log(\frac{\text{Inv}(X)}{n} + 1) + O(n)$. This would solve our problem were it not for the fact that we are unable to determine the appropriate value for r in advance. To circumvent this deficiency, we choose r to be a dynamic quantity that is maintained as insertions take place; r is initially chosen to be $r_1 = 1$, and during the kth insertion, $k > 1$, r is given by $r_k = \frac{1}{k-1} \sum_{1 \le j < k} i_j$.

Lemma 1. *The preceding insertion sort algorithm performs at most* $n \log(\frac{Inv(X)}{n} + 1) + O(n)$ *comparisons to sort an input* X *of length* n.

Proof: Define $E(k)$, the excess number of comparisons performed during the first k insertions, to be the actual number performed minus $k \log(\frac{1}{k} \sum_{1 \le j \le k} i_j)$. We demonstrate that $E(k) = O(n)$ when $k = n$. We proceed to estimate $E(k + 1) - E(k)$.

Let r' denote the average, $\frac{1}{k+1} \sum_{1 \le j \le k+1} i_j$, and let r denote the corresponding quantity, $\frac{1}{k} \sum_{1 \le j \le k} i_j$. Then

$$E(k+1) - E(k) = \log r + i_{k+1}/r - (k+1) \log r' + k \log r + O(1)$$
$$= i_{k+1}/r + (k+1)(\log r - \log r') + O(1). \tag{1}$$

Now write $r' = (k \cdot r + i_{k+1})/(k+1) = (k/(k+1)) \cdot r \cdot g$, where $g = 1 + i_{k+1}/(k \cdot r)$. Substituting into (1) this expression for r', we obtain

$$E(k+1) - E(k) = i_{k+1}/r + (k+1)(\log(k+1)/k) - (k+1) \log g + O(1).$$

The term $(k+1)(\log(k+1)/k)$ is $O(1)$, leaving us to estimate $i_{k+1}/r - (k+1) \log g$. We have two cases: (i) $i_{k+1} \le k \cdot r$ and (ii) $i_{k+1} > k \cdot r$

For case (i), using the fact that $\log(1 + x) \ge x$ for $0 \le x \le 1$, we find that $i_{k+1}/r - (k+1) \log g \le 0$ (since $\log g \ge i_{k+1}/(k \cdot r)$ for this case). For case (ii), we bound $i_{k+1}/r - (k+1) \log g$ from above using i_{k+1}. But the condition for case (ii), namely $i_{k+1} > k \cdot r = \sum_{1 \le j \le k} i_j$, implies that the sum of these i_{k+1} estimates (over those k for which case (ii) applies) is at most twice the last such term, which is bounded by n. Since $E(1) = 0$, we conclude that $E(n) = O(n)$.

$\square$

Efficient Implemention

To convert the above construction to an implementable algorithm with total running time $O(n \log(\frac{Inv(X)}{n} + 1) + n)$, without sacrificing the bound on the number of comparisons performed, we utilize the considerable freedom available in the choice of the r_k values in the above construction, while preserving the result of the preceding Lemma. Let $\alpha \ge 1$ be an arbitrary constant. If we replace our choice for r_k in the above algorithm by any quantity s_k satisfying $r_k \le s_k \le \alpha \cdot r_k$, then the above lemma still holds; the cost $i_k/s_k + \log s_k + O(1)$ of a single insertion cannot grow by more than $O(1)$ as s_k deviates from its initial value r_k while remaining in the indicated range.

An overview of the implementation: At each insertion point the previously inserted items are organized into a list of consecutive intervals, but with varying numbers of items in a given interval. The items belonging to a given interval are organized as a search tree to facilitate efficient insertion once the interval that contains the next inserted item has been identified. The relevant interval is first identified by employing a linear search through the interval list. After each insertion, the interval list may require reorganization, though on a relatively infrequent basis. The details of this implementation is facilitated by defining a system of *interval mechanics*.

A system of interval mechanics refers to a collection of procedures that support the following six operations on a list of intervals, with the indicated performance characteristics. (The specified time bounds are both amortized and worst case, with the exception of insertion, for which the bound is amortized only.) The space utilized for an interval of size s is $O(s)$.

1. Split: An interval can be split into two consecutive intervals, each one-half the size of the original interval, in time $O(s)$, where s is the size of the original interval. (No comparisons take place.)
2. Combine: Two (or three) consecutive intervals can be combined into a single interval in time $O(s)$, where s is the size of the resulting interval. (No comparisons take place.)
3. Find largest: The value of the largest member of a given interval can be accessed in constant time. (No comparisons take place.)
4. Insertion: An insertion of a new value into an interval of size s can be performed with $\log s + O(1)$ comparisons, and in time $O(\log s)$.
5. Initialize: An interval of size 4 can be constructed in constant time.
6. Extraction: A sorted sequence of the members of an interval can be formed in $O(s)$ time, where s is the size of the interval. (No comparisons take place.)

The work of Andersson and Lai [1] concerning near optimal binary search trees provides an implementation of interval mechanics. Consider now the cost of the single operation: inserting y into a sorted sequence $S = x_1 < x_2 < \cdots < x_{n-1}$. If S is organized via interval mechanics into a list of intervals, and y belongs to the kth interval, which is of size m, then the insertion requires no more than $k + \log m + O(1)$ comparisons.

Define the nominal size of an interval to be the largest power of two that does not exceed its actual size. The intervals of a list are to be maintained so that their nominal sizes satisfy: (i) they form a non-decreasing sequence, with each term at least 4; (ii) the smallest size (leftmost) occurs at least twice (unless there is only one interval); and (iii) if r and r' are the nominal sizes of two consecutive intervals, then $\log r' - \log r \leq 1$. The three conditions are referred to as the *monotonicity conditions*. To maintain these conditions, whenever the actual size of an interval grows to the next power of two, as a result of an insertion, we immediately apply a splitting operation to the interval; the nominal sizes of the daughter intervals are the same as that of the original prior to the insertion.

As in the preceding lemma, we define $r_k = \frac{1}{k-1} \sum_{1 \leq j < k} i_j$, for $k > 1$. Besides enforcing the monotonicity conditions, interval operations will be utilized to maintain the following additional condition: Just prior to the kth insertion the nominal size r of the leftmost interval satisfies

$$r_k < r < 4r_k \quad \text{when } r \geq 8, \text{ and} \tag{2}$$

$$1 \leq r_k < r \quad \text{when } r = 4. \tag{3}$$

(The quantity r_k is always at least 1.) Now assume $k < n$. If (as a result of an insertion) r_k drops to $r/4$ and $r \geq 8$, then we split the leftmost interval into two intervals, each having nominal size $r/2$. This is referred to as a reduction operation and should not be confused with a splitting operation which preserves nominal size. If r_k grows to r $(r \geq 4)$, then we execute a coalescing operation defined as follows. We proceed from left to right through all of the intervals of nominal size r, combining them in pairs to form intervals of nominal size $2r$. (In case there are an odd number of intervals of nominal size r, we combine the last three into one (or two) interval(s) of nominal size $2r$, using a splitting operation in case two intervals are required. For the purpose of our subsequent analysis, we view this as an atomic operation whose total cost is proportional to the sizes of the resulting interval(s).) We observe that the three operations, splitting, reduction, and the coalescing operation, serve to preserve the monotonicity conditions as well as the inequalities (2) and (3).

Observe that a single insertion does not change r_k by more than 1. This implies that immediately after a coalescing or reduction operation, $r_k \leq r/2 + 1$, where r is the new nominal size of the leftmost interval. (It also implies that after a reduction operation the required condition, $r_k < r < 4r_k$, holds for $r \geq 8$.) We view the process to be initialized so that there is one interval of size 4; at this point r_k $(k = 5)$ is at most $2.5 \leq r/2 + 1$.

Analysis

Lemma 2. *Our algorithm performs at most* $n \log(\frac{Inv(X)}{n} + 1) + O(n)$ *comparisons to sort an input X of length n.*

Proof: In view of Lemma 1, it suffices to show that a given insertion, arriving in position L, reqires at most

$$\log r_k + L/r_k + O(1) \tag{4}$$

comparisons. Let $r = 2^h$ be the nominal size of the leftmost interval, let 2^j be the nominal size of the interval into which the newly inserted item falls, and let i be the position of this interval, so that the total insertion cost is $i + j + O(1)$. We argue that

$$i + j < \frac{L}{r} + h + 2m - 2^m + 2 \text{ (where } m = j - h \geq 0) \tag{5}$$

$$\leq \frac{L}{r} + h + O(1) \tag{6}$$

The inequalities (2), (3) and (6) establish (4). To establish (5) we observe that for a given choice of i and j, to minimize L subject to the monotonicity conditions, the sequence consisting of the nominal sizes of the first i intervals has the form:

$$\underbrace{r, r, \cdots, r,}_{i - m \text{ terms}} 2r, 4r, \cdots, 2^j$$

(Note the crucial role of the third monotonicity condition.) A calculation shows that $L > (i - (m + 2))r + 2^j$, from which (5) follows.   □

In considering the processing costs apart from comparisons, we note that the total cost for insertions into intervals is reflected by our corresponding estimate for the number of comparisons involved. What remains, therefore, is the task of estimating the total costs of the interval processing tasks, splitting and combining.

The total cost of reduction operations matches (within a constant factor) the total cost of the coalescing operations. This follows from the observation that each instance of a reduction operation initially involving an interval of nominal size r can be uniquely matched against the most recent coalescing operation resulting in intervals of nominal size r.

The total cost of splitting operations that preserve nominal size is bounded by $O(n) + O$(the total cost of coalescing operations), argued as follows. Define the *excess size* of an interval to be its actual size minus its nominal size. A splitting operation that preserves nominal size can be charged against the decrease in total excess size of the intervals (multiplied by factor of 2). Each insertion operation increases excess size by 1 unit. When two intervals are combined in a coalescing operation, resulting in an interval of increased nominal size, there is no net change in the total excess sizes of the intervals. Similarly when three intervals are combined to form two intervals during a coalescing operation, there is no net increase in total excess size. However, if three intervals are combined into one interval, there is a net increase in total excess size, but this increase is dominated by the cost of the coalescing step. Finally, reduction operations do not affect the net total excess sizes of the intervals. Combining these cases, we conclude that the total cost of these splitting operations is bounded by $O(n) + O$(the total cost of coalescing operations), as claimed.

What remains to be shown is that the cost of coalescing intervals is bounded by our estimate for the number of comparisons executed. At the onset of a coalescing operation taking place just after t items have been inserted, consider the most recent preceding insertion point t' that triggers either a reduction or a coalescing operation; and let r be the nominal size of the leftmost interval at this point, upon completion of the triggered operation. (If there have been no preceding coalescing or reduction operations, then we choose $t' = 4$.) This will be the nominal size of the intervals participating in the upcoming coalescing operation triggered by the tth insertion. Let m' be the combined sizes of the intervals of nominal size r at the point immediately preceding the $(t' + 1)$st insertion and let m be the combined sizes of the intervals of nominal size r at the point following the tth insertion just before the coalescing operation gets underway. We trivially have $m \geq m'$, and it is also clear that at least $m - m'$ insertions have taken place, each of which involves at least one comparison.

If $m' < m/2$, then the number of comparisons executed subsequent to the t'th insertion is $\Omega(m) = \Omega$(time to execute the coalescing operation), and we are justified in charging this cost to those comparisons.

Next suppose that $m' \geq m/2$. We proceed to show that the number of comparisons that take place subsequent to the t'th insertion in this case is likewise $\Omega(m)$, the cost of the coalescing operation. Immediately preceding the $(t'+1)$st insertion we have $r_{t'+1} \leq r/2+1$, since this was the point immediately following the previous coalesing or reduction operation. The sum, $\sum_{t'<j\leq t} i_j$ is at least $r \cdot t/4$, since $r \geq 4$ and r_j will have increased from at most $r/2 + 1$ to at least r (the condition that triggers to next coalescing operation). Now consider the number of items belonging to intervals of nominal size r that are in front of the item constituting the jth insertion (just after the insertion takes place) and let h_j be this same quantity plus 1. For j in the range $t' < j \leq t$, $h_j = \min(i_j, m'+1)$. Since $i_j \leq t$, we have $h_j/i_j \geq (m'+1)/t$, and it follows that

$$\sum_{t'<j\leq t} h_j \geq \frac{m'+1}{t} \sum_{t'<j\leq t} i_j \geq \frac{m'+1}{t} \cdot r \cdot t/4 \geq r \cdot m'/4. \tag{7}$$

Now if c_j comparisons take place during the jth insertion, then since each comparison accounts for at most one interval of nominal size r, and each such interval has fewer than $2r$ items, we conclude that

$$h_j \leq 2r \cdot c_j + 1 \leq 3r \cdot c_j \tag{8}$$

(since $c_j \geq 1$). Combining (7) and (8) we conclude that

$$3r \sum_{t'<j\leq t} c_j \geq r \cdot m'/4 = \Omega(r \cdot m)$$

since $m' \geq m/2$.

Thus for each coalescing operation, the number of comparisons taking place prior to the operation, but subsequent to the preceding coalescing operation, effectively bounds the amount of work performed for the operation. We have thus established the following theorem.

Theorem 1. *The preceding insertion sort algorithm sorts an input X of length n in time $O(n \log(\frac{Inv(X)}{n}+1) + n)$, and performs at most $n \log(\frac{Inv(X)}{n}+1) + O(n)$ comparisons. The space requirement for the algorithm is $O(n)$.*

3   Discussion

We have also shown that the same bounds can be derived using a merge sort approach, although the construction is perhaps less interesting. An insertion sort also has the advantage of facilitating on-line sorting. It remains an open problem whether there are practical adaptive sorting algorithms that are competitive in the worst case with the best non-adaptive algorithms.

References

1. A. Andersson and T. W. Lai. *Fast updating of well-balanced trees*. In Proc. Scandinavian Workshop on Algorithm Theory, Springer Verlag (1990), 111–121.

2. M. Brown and R. Tarjan. *Design and analysis of data structures for representing sorted lists.* SIAM J. Comput. 9(1980), 594–614.
3. J. Bentley and A. Yao. *An almost optimal algorithm for unbounded searching.* Information Processing Letters 5(3) (1976), 82–87.
4. S. Carlsson, C. Levcopoulos and O. Petersson. *Sublinear merging and natural Mergesort.* Algorithmica 9 (1993), 629–648.
5. R. Cole. *On the dynamic finger conjecture for splay trees. Part II: The proof.* SIAM J. Comput. 30 (2000), 44–85.
6. A. Elmasry. *Priority queues, pairing and adaptive sorting.* 29th ICALP. In LNCS 2380 (2002), 183–194.
7. V. Estivill-Castro. *A survey of adaptive sorting algorithms.* ACM Comput. Surv. vol 24(4) (1992), 441–476.
8. V. Estivill-Castro and D. Wood. *A new measure of presortedness.* Infor. and Comput. 83 (1989), 111–119.
9. M. Fredman. *How good is the information theory bound in sorting?.* Theoretical Computer Science 1 (1976), 355–361.
10. L. Guibas, E. McCreight, M. Plass and J. Roberts. *A new representation of linear lists.* ACM Symp. on Theory of Computing 9 (1977), 49–60.
11. D. Knuth. *The art of Computer programming. Vol III: Sorting and Searching.* Addison-wesley, second edition (1998).
12. C. Levcopoulos and O. Petersson. *Splitsort - An adaptive sorting algorithm.* Information Processing Letters 39 (1991), 205–211.
13. C. Levcopoulos and O. Petersson. *Adaptive Heapsort.* J. of Algorithms 14 (1993), 395–413.
14. C. Levcopoulos and O. Petersson. *Sorting shuffled monotone sequences.* Inform. and Comput. 112 (1994), 37–50.
15. C. Levcopoulos and O. Petersson. *Exploiting few inversions when sorting: Sequential and parallel algorithms.* Theoretical Computer Science 163 (1996), 211–238.
16. H. Mannila. *Measures of presortedness and optimal sorting algorithms.* IEEE Trans. Comput. C-34 (1985), 318–325.
17. K. Mehlhorn *Data structures and algorithms. Vol.1. Sorting and Searching.* Springer-Verlag, Berlin/Heidelberg. (1984)
18. K. Mehlhorn. *Sorting presorted files.* Proc. Of the 4th GI Conference on Theory of Computer Science. LNCS 67 (1979), 199–212.
19. A. Moffat, O. Petersson and N. Wormald *A tree-based Mergesort.* Acta Informatica, Springer-Verlag (1998), 775–793.
20. O. Petersson and A. Moffat. *A framework for adaptive sorting.* Discrete App. Math. 59 (1995), 153–179.
21. D. Sleator and R. Tarjan. *Self-adjusting binary search trees.* J. ACM 32(3) (1985), 652–686.

A Discrete Subexponential Algorithm for Parity Games[*]

Henrik Björklund, Sven Sandberg, and Sergei Vorobyov

Computing Science Department, Uppsala University, Sweden

Abstract. We suggest a new randomized algorithm for solving parity games with worst case time complexity roughly

$$\min\left(O\left(n^3 \cdot \left(\frac{n}{k} + 1\right)^k \right),\ 2^{O(\sqrt{n \log n})} \right),$$

where n is the number of vertices and k the number of colors of the game. This is comparable with the previously known algorithms when the number of colors is small. However, the subexponential bound is an advantage when the number of colors is large, $k = \Omega(n^{1/2+\varepsilon})$.

1 Introduction

Parity games are infinite games played on finite directed bipartite leafless graphs, with vertices colored by integers. Two players alternate moving a pebble along edges. The goal of Player 0 is to ensure that the biggest color visited by the pebble infinitely often is even, whereas Player 1 tries to make it odd. The complexity of determining a winner in parity games, equivalent to the Rabin chain tree automata non-emptiness, as well as to the μ-calculus[1] model checking [5,3], is a fundamental open problem in complexity theory [11]. The problem belongs to NP∩coNP, but its PTIME-membership status remains widely open. All known algorithms for the problem are exponential, with an exception of [12] when the number of colors is large and games are binary.

In this paper we present a new discrete, randomized, subexponential algorithm for parity games. It combines ideas from iterative strategy improvement based on randomized techniques of Kalai [9] for Linear Programming and of Ludwig [10] for simple stochastic games, with discrete strategy evaluation similar to that of Vöge and Jurdziński [15]. Generally, algorithms for parity games are *exponential* in the number of colors k, which may be as big as the number n of vertices. For most, exponentially hard input instances are known [4,3,2,14, 8]. Our algorithm is subexponential in n. Earlier we suggested a subexponential algorithm [12], similar to [10], but based on graph optimization rather than linear programming subroutines. Both algorithms [10,12] become exponential

---

[*] Supported by Swedish Research Council Grants "Infinite Games: Algorithms and Complexity", "Interior-Point Methods for Infinite Games".

[1] One of the most expressive temporal logics of programs [3].

H. Alt and M. Habib (Eds.): STACS 2003, LNCS 2607, pp. 663–674, 2003.

for graphs with unbounded vertex outdegree. The present paper eliminates this drawback. There is a well-known reduction from parity to mean payoff games, but the best known algorithms for the latter [7,16,13] are known to be exponential (pseudopolynomial). Reducing parity to simple stochastic games [16] leads to manipulating high-precision arithmetic and to algorithms invariably subexponential in the number of vertices, which is worse than an exponential dependence on colors when colors are few.

A recent iterative strategy improvement algorithm [15] uses a discrete strategy evaluation involving game graph characteristics like colors, sets of vertices, and path lengths. Despite a reportedly good practical behavior, the only known worst-case bound for this algorithm is exponential in the number of vertices, independently of the number of colors.

Our new algorithm avoids any reductions and directly applies to parity games of arbitrary outdegree. We use a discrete strategy evaluation measure similar to, but more economical than the one used in [15]. Combined with Kalai's and Ludwig's randomization schemes this provides for a worst case bound that is simultaneously subexponential in the number of vertices and exponential in the number of colors. This is an advantage when the colors are few.

Outline. After preliminaries on parity games, we start by presenting a simpler, Ludwig-style randomized algorithm in combination with an abstract discrete measure on strategies. This simplifies motivation, exposition, and definitions for the specific tight discrete measure we build upon. We then proceed to a more involved Kalai-style randomized algorithm allowing for arbitrary vertex outdegrees. All proofs can be found in [1].

2   Parity Games

Definition 1 (Parity Games). *A parity game is an infinite game played on a finite directed bipartite leafless graph $G[n,k] = (V_0, V_1, E, c)$, where $n = |V_0 \cup V_1|$, $E \subseteq (V_0 \times V_1) \cup (V_1 \times V_0)$, $k \in \mathbb{N}$, and $c : V_0 \cup V_1 \to \{1, \ldots, k\}$ is a coloring function. The sizes of V_0 and V_1 are denoted by n_0 and n_1, respectively. Starting from a vertex, Player 0 and 1 alternate moves constructing an infinite sequence of vertices; Player i moves from a vertex in V_i by selecting one of its successors. Player 0 wins if the highest color encountered infinitely often in this sequence is even, while Player 1 wins otherwise.*[2]   □

Parity games are known to be *determined*: from each vertex exactly one player has a winning positional strategy, selecting a unique successor to every vertex [5]. All our results straightforwardly generalize to the non-bipartite case.

A *binary* parity game is a game where the vertex outdegree is at most two.

---

[2] We systematically use n for the number of vertices and k for the number of colors; consequently we usually skip $[n,k]$ in $G[n,k]$.

3   Ludwig-Style Algorithm with a Well-Behaved Measure

Every positional strategy of Player 0 in a binary parity game can be associated with a corner of the n_0-dimensional boolean hypercube. If there is an appropriate way of assigning values to strategies, then we can apply an algorithm similar to [10] to find the best strategy as follows.

1. Start with some strategy σ_0 of Player 0.
2. Randomly choose a facet F of the hypercube, containing σ_0.
3. Recursively find the best strategy σ' on F.
4. Let σ'' be the neighbor of σ' on the opposite facet $\overline{F}$. If σ' is better than σ'', then return σ'. Else recursively find the optimum on $\overline{F}$, starting from σ''.

To guarantee correctness and subexponentiality, the assignment cannot be completely unstructured. Also, evaluating strategies is costly, so a full evaluation should only be performed for strategies that are really better than the current one. In subsequent sections, we present a function EVALUATE that given a strategy σ returns an assignment ν_σ of values to vertices of the game that meets the following criteria (where $\prec$ is a comparison operator on the values).

Stability. Let $\sigma(v)$ be the successor of vertex v selected by strategy σ and let $\overline{\sigma}(v)$ be the other successor of v. If $\nu_\sigma(\sigma(v)) \succeq \nu_\sigma(\overline{\sigma}(v))$ for all vertices v of Player 0, then σ is optimal (maximizes the winning set of Player 0).

Uniqueness of optimal values. All optimal strategies have the same valuation. (This is essential for a subexponential bound.)

Profitability. Suppose that $\nu_\sigma(\sigma(u)) \succeq \nu_\sigma(\overline{\sigma}(u))$ for every vertex $u \in V_0 \setminus v$ and $\nu_\sigma(\sigma(v)) \prec \nu_\sigma(\overline{\sigma}(v))$ (attractiveness). Let σ' be the strategy obtained by changing σ only at v (single switch), and let $\nu_{\sigma'}$ be its valuation. Then $\nu_\sigma(v) \prec \nu_{\sigma'}(v)$ and $\nu_\sigma(u) \preceq \nu_{\sigma'}(u)$ for all other vertices u (profitability).

The Ludwig-style algorithm with EVALUATE applies to solving binary parity games. The evaluation function has the benefit that in step 4 of the algorithm, σ'' does not have to be evaluated, unless the recursive call is needed.

   Ludwig [10] shows that his algorithm for simple stochastic games has a $2^{O(\sqrt{n_0})}$ upper bound on the expected number of improvement steps. With only minor modifications, the same proof shows that the Ludwig-style algorithm together with our EVALUATE function has the same bound for parity games.

   The value space of EVALUATE allows at most $O(n^3 \cdot (n/k+1)^k)$ improvement steps. Since the algorithm makes only improving switches, the upper bound on the number of switches of the combined approach is

$$\min \left(O\left(n^3 \cdot \left(\frac{n}{k} + 1 \right)^k \right), 2^{O(\sqrt{n_0})} \right).$$

Any parity game reduces to a binary one. This allows for a subexponential algorithm for games with subquadratic total outdegree. For arbitrary games the reduction gives a quadratic explosion in the number of vertices and the Ludwig-style algorithm becomes exponential. In Section 10 we achieve a subexponential bound by employing a more involved randomization scheme from [9].

4   Strategies and Values

For technical reasons, each vertex is assigned a unique value, called a *tint*.

Definition 2 (Tints). *A bijection* $\mathbf{t} : V \to \{1, \ldots, n\}$ *such that* $c(u) \leq c(v) \Rightarrow$ $\mathbf{t}(u) \leq \mathbf{t}(v)$ *assigns tints to vertices. The color of a tint* $s \in \{1, \ldots, n\}$ *equals* $c(\mathbf{t}^{-1}(s))$. □

Note that tints of vertices of the same color form a consecutive segment of natural numbers. Subsequently we identify vertices with their tints, and slightly abuse notation by writing $c(t)$ for the color of the vertex with tint t.

Definition 3 (Winning and Losing Colors and Tints). *Color i is* winning for Player 0 *(Player 1 resp.) if it is even (odd resp.). Tint t is* winning for Player 0 *(Player 1 resp.) if its color $c(t)$ is. A color or tint is* losing *for a player if it is winning for his adversary.* □

Note that tints of different colors are ordered as these colors. Within the same winning (resp. losing) color the bigger (resp. smaller) tint is better for Player 0.

In this section we define the 'value' of a strategy – the target to be iteratively improved. An elementary improvement step is as follows: given a strategy σ of Player 0, its value is a vector of values of all vertices of the game, assuming that the adversary Player 1 applies an 'optimal' response counterstrategy τ against σ. The value of each vertex is computed with respect to the pair of strategies (σ, τ), where the optimality of τ is essential for guiding Player 0 in improving σ. We delay the issue of constructing optimal counterstrategies until Section 9, assuming for now that Player 1 always responds with an optimal counterstrategy.

Definition 4. *A positional strategy for Player 0 is a function* $\sigma : V_0 \to V_1$, *such that if* $\sigma(v) = v'$, *then* $(v, v') \in E$. *Saying that Player 0 fixes his positional strategy means that he deterministically chooses the successor $\sigma(v)$ each time the play comes to v, independently of the history of the play. Positional strategies for Player 1 are defined symmetrically.* □

Assumption. From now on we restrict our attention to positional strategies only. The iterative improvement proceeds by improving positional strategies for Player 0, and this is justified by Profitability, Stability, and Uniqueness Theorems 20, 22, and 23 below. The fact that Player 1 may also restrict himself to positional strategies is demonstrated in Section 9.

Definition 5 (Single Switch). *A* single switch *in a positional strategy σ of Player 0 is a change of successor assignment of σ in exactly one vertex.* □

When the players fix their positional strategies, the trace of any play is a simple path leading to a simple loop. Roughly speaking, the value of a vertex with respect to a pair of positional strategies consists of a loop value (major tint) and a path value (a record of the numbers of more significant colors on the path to the major, plus the length of this path), as defined below.

Notation 6 *Denote by V^i the set of vertices of color i and by $V^{>t}$ the set of vertices with tints numerically bigger than t.* □

Definition 7 (Traces, Values). *Suppose the players fix positional strategies σ and τ, respectively. Then from every vertex u_0 the* trace *of the play takes a simple δ-shape form: an initial simple path (of length $q \geq 0$, possibly empty) ending in a loop:*

$$u_0, u_1, \ldots u_q, \ldots, u_r, \ldots u_s = u_q, \tag{1}$$

where all vertices u_i are distinct, except $u_q = u_s$. The vertex u_r with the maximal tint t on the loop $u_q, \ldots, u_r, \ldots, u_s = u_q$ in (1) is called principal *or* major.

VALUES FOR NON-PRINCIPAL VERTICES. *If the vertex u_0 is non-principal, then its value $\nu_{\sigma,\tau}(u_0)$ with respect to the pair of strategies (σ, τ) has the form (L, P, p) and consists of:*

LOOP VALUE (TINT) *L equal to the principal tint t;*
PATH COLOR HIT RECORD RELATIVE TO t *defined as a vector*

$$P = (m_k, m_{k-1}, \ldots, m_l, \underbrace{0, \ldots, 0}_{l-1\ times}),$$

where $l = c(t)$ is the color of the principal tint t, and

$$m_i = \left| \{u_0, u_1, \ldots u_{r-1}\} \cap V^i \cap V^{>t} \right|$$

is the number of vertices of color $i \geq l$ on the path to from u_0 to the major u_r (except that for the color l of the major we account only for the vertices with tint bigger than t.)
PATH LENGTH *$p = r$.*

VALUES FOR PRINCIPAL VERTICES. *If the vertex u_0 is principal (case $q = r = 0$ in (1)) then its value $\nu_{\sigma,\tau}(u_0)$ with respect to the pair of strategies (σ, τ) is defined as $(t, \bar{0}, s)$, where $\bar{0}$ is a k-dimensional vector of zeros.*

PATH VALUE *is a pair (P, p), where (t, P, p) is a vertex value.*   □

The reason of the complexity of this definition is to meet the criteria enumerated in Section 3 and simultaneously obtain the 'tightest possible' bound on the number of iterative improvements. It is clear that such a bound imposed by the value measure from Definition 7 is $O(n^3 \cdot (n/k + 1)^k)$.

5   Value Comparison and Attractive Switches

Definition 8 (Preference Orders). *The* preference order *on colors (as seen by Player 0) is as follows: $c \prec c'$ iff $(-1)^c \cdot c < (-1)^{c'} \cdot c'$.*
The preference order *on tints (as seen by Player 0) is as follows:*

$$t \prec t' \ iff \ (-1)^{c(t)} \cdot t < (-1)^{c(t')} \cdot t'.$$ □

We thus have $\ldots \prec 5 \prec 3 \prec 1 \prec 0 \prec 2 \prec 4 \prec \ldots$ on colors.

Definition 9 ('Lexicographic' Ordering). *Given two vectors (indexed in descending order from the maximal color k to some $l \geq 1$)*

$$P = (m_k, m_{k-1}, \ldots, m_{l+1}, m_l),$$
$$P' = (m'_k, m'_{k-1}, \ldots, m'_{l+1}, m'_l),$$

define $P \prec P'$ if the vector

$$\left((-1)^k \cdot m_k, (-1)^{k-1} \cdot m_{k-1}, \ldots, (-1)^{l+1} \cdot m_{l+1}, (-1)^l \cdot m_l\right)$$

is lexicographically smaller (assuming the usual ordering of integers) than the vector $\left((-1)^k \cdot m'_k, (-1)^{k-1} \cdot m'_{k-1}, \ldots, (-1)^{l+1} \cdot m'_{l+1}, (-1)^l \cdot m'_l\right)$. □

Definition 10 (Path Attractiveness). *For two vertex values (t, P_1, p_1) and (t, P_2, p_2), where t is a tint, $l = c(t)$ is its color, and*

$$P_1 = (m_k, m_{k-1}, \ldots, m_{l+1}, m_l, \ldots, m_1),$$
$$P_2 = (m'_k, m'_{k-1}, \ldots, m'_{l+1}, m'_l, \ldots, m'_1),$$

say that the path value (P_2, p_2) is more attractive[3] modulo t than the path value (P_1, p_1), symbolically $(P_1, p_1) \prec_t (P_2, p_2)$, if:

1. *either $(m_k, m_{k-1}, \ldots, m_{l+1}, m_l) \prec (m'_k, m'_{k-1}, \ldots, m'_{l+1}, m'_l)$,*
2. *or $(m_k, m_{k-1}, \ldots, m_{l+1}, m_l) = (m'_k, m'_{k-1}, \ldots, m'_{l+1}, m'_l)$ and*

$$(-1)^l \cdot p_1 > (-1)^l \cdot p_2. \tag{2}$$

Remark 11. Note that (2) means that shorter (resp. longer) paths are better for Player 0 when the loop tint t is winning (resp. losing) for him. □

Definition 12 (Value Comparison). *For two vertex values define $(t_1, P_1, p_1) \prec (t_2, P_2, p_2)$ if*

1. *either $t_1 \prec t_2$,*
2. *or $t_1 = t_2 = t$, and $(P_1, p_1) \prec_t (P_2, p_2)$.* □

Definition 13 (Vertex Values). *The value $\nu_\sigma(v)$ of a vertex v with respect to a strategy σ of Player 0 is the minimum of the values $\nu_{\sigma,\tau}(v)$, taken over all strategies τ of Player 1.* □

In Section 9 we show that the 'minimum' in this definition can be achieved in all vertices simultaneously by a positional strategy τ of Player 1.

Definition 14. *The value of a strategy σ of Player 0 is a vector of values of all vertices with respect to the pair of strategies (σ, τ), where τ is an optimal response counterstrategy of Player 1 against σ; see Section 9.* □

---

[3] In the sequel, when saying "attractive", "better", "worse", etc., we consistently take the viewpoint of Player 0.

Definition 15. *A strategy σ' improves σ, symbolically $\sigma \prec \sigma'$, if $\nu_\sigma(v) \preceq \nu_{\sigma'}(v)$ for all vertices v and there is at least one vertex u with $\nu_\sigma(u) \prec \nu_{\sigma'}(u)$.* □

Proposition 16. *The relations $\prec$ on colors, tints, values, and strategies, and $\prec_t$ on path values (for each t) are transitive.* □

Our algorithms proceed by single attractive switches only.

Definition 17 (Attractive Switch). *Let (t_1, P_1, p_1) and (t_2, P_2, p_2) be the values with respect to σ of vertices v_1 and v_2, respectively. Consider a single switch in strategy σ of Player 0, consisting in changing the successor of v with respect to σ from v_1 to v_2. The switch is* attractive *if $(t_1, P_1, p_1) \prec (t_2, P_2, p_2)$.* □

Remark 18. Note that deciding whether a switch is attractive (when comparing values of its successors) we do not directly account for the color/tint of the current vertex. However, this color/tint may be eventually included in the values of successors possibly dependent on the current vertex.

6   Profitability of Attractive Switches

Our algorithms proceed by making single attractive switches. Attractiveness is established locally, by comparing values of a vertex successors with respect to a current strategy; see Definition 17.

Definition 19. *Say that a single switch from σ to σ' is* profitable *if $\sigma \prec \sigma'$.* □

Profitability of attractive switches is crucial for the efficiency, correctness, and termination of our algorithms, as explained in Sections 3 and 10. Profitability is a consequence of the the preceding complicated definitions of values, value comparison, and strategy evaluation.

Theorem 20 (Profitability). *Every attractive switch is profitable:*

1. *it increases the value of the vertex where it is made, and*
2. *all other vertices either preserve or increase their values,*

i.e., the switch operator is monotone. □

7   Stability Implies Optimality

The Main Theorem 22 of this section guarantees that iterative improvement can terminate once a strategy with no attractive switches is found. In more general terms it states that every local optimum is global. This is one of the main motivations for the complex strategy evaluation definitions.

Definition 21. *Say that a strategy σ is* stable *if it does not have attractive switches with respect to $\tau(\sigma)$, an optimal counterstrategy of Player 1.* □

In Section 9 we show that all optimal counterstrategies provide for the *same* values. Thus stability of σ in the previous definition may be checked after computing any optimal counterstrategy $\tau(\sigma)$.

Theorem 22 (Stability). *Any stable strategy of Player 0 is optimal: vertices with loop values of even colors form the winning set of Player 0.* □

8   Uniqueness of Optimal Values

Theorem 22 does not guarantee that different stable (hence optimal) strategies provide for the same (or even comparable) vectors of values for the game vertices. The uniqueness of optimal values is however crucial for the subexponential complexity analysis of Sections 3 and 10, and is provided by the following

Theorem 23 (Uniqueness). *Any two stable strategies of Player 0 give the same values for all vertices of the game.*   □

9   Computing Optimal Counterstrategies

Let G_σ be the game graph induced by a positional strategy σ of Player 0 (delete all edges of Player 0 not used by σ). Partition vertices of G_σ into classes L_t containing the vertices from which Player 1 can ensure the loop tint t, but cannot guarantee any worse loop tint. This can be done by using finite reachability in G_σ as follows. For each tint t in $\prec$-ascending order, check whether t can be reached from itself without passing any tint $t' > t$. If so, Player 1 can form a loop with t as major. Since the tints are considered in $\prec$-ascending order, t will be the best loop value Player 1 can achieve for all vertices from which t is reachable. Remove them from the graph, place them in class L_t, and proceed with the next tint.

For each class L_t, use dynamic programming to calculate the values of 1-optimal paths of different lengths from each vertex to t. For each vertex, the algorithm first computes the optimal color hit record (abbreviated *chr* in the algorithm) over all paths of length 0 to the loop major (∞ for each vertex except t). Then it calculates the color hit record of optimal paths of length one, length two, and so forth. It uses the values from previous steps in each step except the initial one.[4]

Algorithm 1: Computing path values within a class L_t.

PATHVALUES(L_t)

(1) $t.chr[0] \leftarrow (0,\dots,0)$
(2)   **foreach** vertex $v \in L_t$ except t
(3) $v.chr[0] \leftarrow \infty$
(4)   **for** $i \leftarrow 1$ **to** $|L_t| - 1$
(5)   **foreach** vertex $v \in L_t$ except t
(6) $v.chr[i] \leftarrow \min_{\prec_t}\{\text{ADDCOLOR}(t, v'.chr[i-1], \mathbf{t}(v)) : v' \in L_t \text{ is a successor of } v)\}$
(7)   **foreach** vertex $v \in L_t$ except t
(8) $v.pathvalue \leftarrow \min_{\prec_t}\{(v.chr[i], i) : 0 \leq i < |L_t|\}$
(9) $t.pathvalue \leftarrow \min_{\prec_t}\{v.pathvalue : v \in L_t \text{ is a successor of } t\}$
(10) $t.pathvalue.pathlength \leftarrow t.pathvalue.pathlength + 1$

---

[4] The algorithm assumes the game is bipartite; in particular, t in line (9) cannot be a successor of itself. It can be straightforwardly generalized for the non-bipartite case.

The function ADDCOLOR takes a tint, a color hit record, and a second tint. If the second tint is bigger than the first one, then ADDCOLOR increases the position in the vector representing the color of the second tint. The function always returns ∞ when the second argument has value ∞.

The algorithm also handles non-binary games.

Lemma 24 (Algorithm Correctness). *The algorithm correctly computes values of optimal paths. Moreover:*

1. *optimal paths are simple;*
2. *the values computed are consistent with an actual positional strategy that guarantees loop value t.* □

Lemma 25 (Algorithm Complexity). *The algorithm for computing an optimal counterstrategy runs in time* $O(|V| \cdot |E| \cdot k)$, *where* $|V|$ *is the number of vertices of the graph,* $|E|$ *is the number of edges, and* k *is the number of colors.*

10   Kalai-Style Randomization for Games with Unbounded Outdegree

As discussed in Section 3, any non-binary parity game reduces to a binary one, and the Ludwig-style algorithm applies. However, the resulting complexity gets worse and may become exponential (rather than subexponential) due to a possibly quadratic blow-up in the number of vertices. In this section we describe a different approach relying on the randomization scheme of Kalai [9,6] used for Linear Programming. This results in a subexponential randomized algorithm directly applicable to parity games of arbitrary outdegree, without any preliminary translations. When compared with reducing to the binary case combined with the Ludwig-style algorithm of Section 3, the algorithm of this section provides for a better complexity when the total number of edges is roughly $\Omega(n \log n)$.

Games, Subgames, and Facets. Let $\mathcal{G}(d, m)$ be the class of parity games with vertices of Player 0 partitioned into two sets U_1 of outdegree one and U_2 of an arbitrary outdegree $\delta(v) \geq 1$, with $|U_2| = d$ and $m \geq \sum_{v \in U_2} \delta(v)$. Informally, d is the dimension (number of variables to determine), and m is a bound on the number of edges (constraints) to choose from. The numbers of vertices and edges of Player 1 are unrestricted.

Given a game $G \in \mathcal{G}(d, m)$, a vertex $v \in U_2$ of Player 0, and an edge e leaving v, consider the (sub)game F obtained by fixing e and deleting all other edges leaving v. Obviously, $F \in \mathcal{G}(d-1, m-\delta(v))$ and also, by definition, $F \in \mathcal{G}(d, m)$, which is convenient when we need to consider a strategy in the subgame F as a strategy in the full game G in the sequel. Call the game F a *facet* of G.

If σ is a positional strategy and e is an edge leaving a vertex v of Player 0, then we define $\sigma[e]$ as the strategy coinciding with σ in all vertices, except possibly v, where the choice is e. If σ is a strategy in $G \in \mathcal{G}(d, m)$, then a facet F is σ-*improving* if some *witness* strategy σ' in the game F (considered as a member of $\mathcal{G}(d, m)$) satisfies $\sigma \prec \sigma'$.

The Algorithm takes a game $G \in \mathcal{G}(d, m)$ and an initial strategy σ_0 as inputs, and proceeds in three steps.

1. Collect a set M containing r pairs (F, σ) of σ_0-improving facets F of G and corresponding witness strategies $\sigma \succ \sigma_0$.
   (The parameter r specified later depends on d and m. Different choices of r give different algorithms. The subroutine to find σ_0-improving facets is described below. This subroutine may find an optimal strategy in G, in which case the algorithm returns it immediately.)
2. Select one pair $(F, \sigma_1) \in M$ uniformly at random. Find an optimal strategy σ in F by applying the algorithm recursively, taking σ_1 as the initial strategy.[5]
3. If σ is an optimal strategy also in G, then return σ. Otherwise, let σ' be a strategy differing from σ by an attractive switch. Restart from step 1 using the new strategy σ' and the same game $G \in \mathcal{G}(d, m)$.

The algorithm terminates because each solved subproblem starts from a strictly better strategy. It is correct because it can only terminate by returning an optimal strategy.

How to Find Many Improving Facets. In step 1 the algorithm above needs to find either r different σ_0-improving facets or an optimal strategy in G. To this end we construct a sequence $(G^0, G^1, \ldots, G^{r-d})$ of games, with $G^i \in \mathcal{G}(d, d+i)$. All the $d + i$ facets of G^i are σ_0-improving; we simultaneously determine the corresponding witness strategies σ^j optimal in G^j. The subroutine returns r facets of G, each one obtained by fixing one of the r edges in $G^{r-d} \in \mathcal{G}(d, r)$. All these are σ_0-improving by construction.

Let e be the target edge of an attractive switch from σ_0. (If no attractive switch exists, then σ_0 is optimal in G and we are done.) Set G^0 to the game where all choices are fixed as in $\sigma_0[e]$, and all other edges of Player 0 in G are deleted. Let σ^0 be the unique, hence optimal, strategy $\sigma_0[e]$ in G^0. Fixing any of the d edges of σ^0 in G defines a σ_0-improving facet of G with σ^0 as a witness.

To construct G^{i+1} from G^i, let e be the target edge of an attractive switch from σ^i in G. (Note that σ^i is optimal in G^i but not necessarily in the full game G. If it is, we terminate.) Let G^{i+1} be the game G^i with e added, and compute σ^{i+1} as an optimal strategy in G^{i+1}, by a recursive application of the algorithm above. Note that fixing any of the $i+1$ added target edges defines a σ_0-improving facet of G. Therefore, upon termination we have r such facets.

Complexity Analysis. The following recurrence bounds the expected number of calls to the algorithm solving a game in $\mathcal{G}(d, m)$ in the worst case:

$$T(d, m) \leq \sum_{i=d}^{r} T(d, i) + T(d-1, m-2) + \frac{1}{r} \sum_{i=1}^{r} T(d, m-i) + 1$$

---

[5] Rather than computing all of M, we may select a random number $x \in \{1, \ldots, r\}$ before step 1 and compute only x improving facets. This is crucial in order for the computed sequence of strategies to be strictly improving, and saves some work.

The first term represents the work of finding r different σ_0-improving facets in step 1. The second term comes from the recursive call in step 2. [6] The last term accounts for step 3 and can be understood as an average over the r equiprobable choices made in step 2, as follows. All facets of G are partially ordered by the values of their optimal strategies (although this order is unknown to the algorithm). Optimal values in facets are unique by Theorem 23, and the algorithm visits only improving strategies. It follows that all facets possessing optimal strategies that are worse, equal, or incomparable to the strategy σ of step 2 will never be visited in the rest of the algorithm. In the *worst* case, the algorithm selects the r worst possible facets in step 1. Thus, in the worst case, in step 3 it solves a game in $\mathcal{G}(d, m - i)$ for $i = 1, \ldots, r$, with probability $1/r$. This justifies the last term.

Kalai uses $r = \max(d, m/2)$ in step 1 to get the best solution of the recurrence. The result is subexponential, $m^{O(\sqrt{d/\log d})}$. By symmetry, we can choose to optimize a strategy of the player possessing fewer vertices.

Let n_i denote the number of vertices of player i. Since m is bounded above by the maximal number of edges, $(n_0 + n_1)^2$, and $d \leq \min(n_0, n_1)$, we get

$$\min \left\{ 2^{O\left((\log n_1) \cdot \sqrt{n_0 / \log n_0}\right)}, 2^{O\left((\log n_0) \cdot \sqrt{n_1 / \log n_1}\right)} \right\}$$

as the bound on the number of calls to the algorithm. Combining it with the bound on the maximal number of improving steps allowed by our measure yields

$$\min \left\{ 2^{O\left((\log n_1) \cdot \sqrt{n_0 / \log n_0}\right)}, 2^{O\left((\log n_0) \cdot \sqrt{n_1 / \log n_1}\right)}, O\left(n^3 \cdot \left(\frac{n}{k} + 1\right)^k\right) \right\}.$$

If $n_0 = O(poly(n_1))$ and $n_1 = O(poly(n_0))$ then this reduces to

$$\min \left\{ 2^{O(\sqrt{n_0 \log n_0})}, 2^{O(\sqrt{n_1 \log n_1})}, O\left(n^3 \cdot \left(\frac{n}{k} + 1\right)^k\right) \right\}.$$

These are the bounds on the number of recursive calls to the algorithm. Within each recursive call, the auxiliary work is dominated by time to compute a strategy value, multiplying the running time by $O(n \cdot |E| \cdot k)$.

Acknowledgements. We thank anonymous referees for valuable remarks and suggestions.

References

1. H. Björklund, S. Sandberg, and S. Vorobyov. A discrete subexponential algorithm for parity games. Technical Report 2002-026, Department of Information Technology, Uppsala University, September 2002.
   http://www.it.uu.se/research/reports/.

---

[6] Actually, if δ is the outdegree in the vertex where we fix an edge, then the second term is $T(d - 1, m - \delta)$. We consider the worst case of $\delta = 2$.

2. A. Browne, E. M. Clarke, S. Jha, D. E Long, and W Marrero. An improved algorithm for the evaluation of fixpoint expressions. *Theor. Comput. Sci.*, 178:237–255, 1997. Preliminary version in CAV'94, LNCS'818.
3. E. A. Emerson. Model checking and the Mu-calculus. In N. Immerman and Ph. G. Kolaitis, editors, *DIMACS Series in Discrete Mathematics*, volume 31, pages 185–214, 1997.
4. E. A. Emerson, C. Jutla, and A. P. Sistla. On model-checking for fragments of μ-calculus. In *Computer Aided Verification, Proc. 5th Int. Conference*, volume 697, pages 385–396. Lect. Notes Comput. Sci., 1993.
5. E. A. Emerson and C. S. Jutla. Tree automata, μ-calculus and determinacy. In *Annual IEEE Symp. on Foundations of Computer Science*, pages 368–377, 1991.
6. Goldwasser. A survey of linear programming in randomized subexponential time. *SIGACTN: SIGACT News (ACM Special Interest Group on Automata and Computability Theory)*, 26:96–104, 1995.
7. V. A. Gurvich, A. V. Karzanov, and L. G. Khachiyan. Cyclic games and an algorithm to find minimax cycle means in directed graphs. *U.S.S.R. Computational Mathematics and Mathematical Physics*, 28(5):85–91, 1988.
8. M. Jurdzinski. Small progress measures for solving parity games. In *17th STACS*, volume 1770 of *Lect. Notes Comput. Sci.*, pages 290–301. Springer-Verlag, 2000.
9. G. Kalai. A subexponential randomized simplex algorithm. In *24th ACM STOC*, pages 475–482, 1992.
10. W. Ludwig. A subexponential randomized algorithm for the simple stochastic game problem. *Information and Computation*, 117:151–155, 1995.
11. C. Papadimitriou. Algorithms, games, and the internet. In *ACM Annual Symposium on Theory of Computing*, pages 749–753. ACM, July 2001.
12. V. Petersson and S. Vorobyov. A randomized subexponential algorithm for parity games. *Nordic Journal of Computing*, 8:324–345, 2001.
13. N. Pisaruk. Mean cost cyclical games. *Mathematics of Operations Research*, 24(4):817–828, 1999.
14. H Seidl. Fast and simple nested fixpoints. *Information Processing Letters*, 59(3):303–308, 1996.
15. J. Vöge and M. Jurdzinski. A discrete strategy improvement algorithm for solving parity games. In *CAV'00: Computer-Aided Verification*, volume 1855 of *Lect. Notes Comput. Sci.*, pages 202–215. Springer-Verlag, 2000.
16. U. Zwick and M. Paterson. The complexity of mean payoff games on graphs. *Theor. Comput. Sci.*, 158:343–359, 1996.

Cryptographically Sound and Machine-Assisted Verification of Security Protocols

Michael Backes[1] and Christian Jacobi[2]

[1] IBM Zurich Research Laboratory, Rüschlikon, Switzerland[†] mbc@zurich.ibm.com
[2] IBM Deutschland Entwicklung GmbH, Processor Development 2, Böblingen, Germany[†]
cjacobi@de.ibm.com

Abstract. We consider machine-aided verification of suitably constructed abstractions of security protocols, such that the verified properties are valid for the concrete implementation of the protocol with respect to cryptographic definitions. In order to link formal methods and cryptography, we show that integrity properties are preserved under step-wise refinement in asynchronous networks with respect to cryptographic definitions, so formal verifications of our abstractions carry over to the concrete counterparts. As an example, we use the theorem prover PVS to formally verify a system for ordered secure message transmission, which yields the first example ever of a formally verified but nevertheless cryptographically sound proof of a security protocol. We believe that a general methodology for verifying cryptographic protocols cryptographically sound can be derived by following the ideas of this example.

Keywords: cryptography, specification, verification, PVS, semantics, simulatability

1 Introduction

Nowadays, formal analysis and verification of security protocols is getting more and more attention in both theory and practice. One main goal of protocol verification is to consider the cryptographic aspects of protocols in order to obtain complete and mathematically rigorous proofs with respect to cryptographic definitions. We speak of (cryptographically) sound proofs in this case. Ideally, these proofs should be performed machine-aided in order to eliminate (or at least minimize) human inaccuracies. As formally verifying cryptographic protocols presupposes abstractions of them, which are suitable for formal methods, it has to be ensured that properties proved for these abstract specifications carry over to the concrete implementations.

Both formal verification and cryptographically sound proofs have been investigated very well from their respective communities during the last years. Especially the formal verification has been subject of lots of papers in the literature, e.g., [20,13,21,1,16]. The underlying abstraction is almost always based on the Dolev-Yao model [7]. Here cryptographic operations, e.g., E for encryption and D for decryption, are considered as operators in a free algebra where only certain predefined cancellation rules hold,

---

[†] Work was done while both authors were affiliated with Saarland University.

H. Alt and M. Habib (Eds.): STACS 2003, LNCS 2607, pp. 675–686, 2003.

i.e., twofold encryption of a message m does not yield another message from the basic message space but the term $E(E(m))$. A typical cancellation rule is $D(E(m)) = m$.

This abstraction simplifies proofs of larger protocols considerably. Unfortunately, these formal proofs lack a link to the rigorous security definitions in cryptography. The main problem is that the abstraction requires that *no* equations hold except those that can be derived within the algebra. Cryptographic definitions do not aim at such statements. For example, encryption is only required to keep cleartexts secret, but there is no restriction on structure in the ciphertexts. Hence a protocol that is secure in the Dolev-Yao framework is not necessarily secure in the real world even if implemented with provably secure cryptographic primitives, cf. [17] for a concrete counterexample. Thus, the appropriateness of this approach is at least debatable.

On the other hand, we have the computational view whose definitions are based on complexity theory, e.g., [8,9,5]. Here, protocols can be rigorously proven with respect to cryptographic definitions, but these proofs are not feasible for formal verification because of the occurrence of probabilities and complexity-theoretic restrictions, so they are both prone to errors and simply too complex for larger systems.

Unfortunately, achieving both points at the same time seems to be a very difficult task. To the best of our knowledge, no cryptographic protocol has been formally verified such that this verification is valid for the concrete implementation with respect to the strongest possible cryptographic definitions, cf. the Related Literature for more details.

Our goal is to link both approaches to get the best overall result: proofs of cryptographic protocols that allow abstraction and the use of formal methods, but retain a sound cryptographic semantics. Moreover, these proofs should be valid for completely asynchronous networks and the strongest cryptographic definitions possible, e.g., security against adaptive chosen ciphertext attack [5] in case of asymmetric encryption.

In this paper, we address the verification of integrity properties such that these properties automatically carry over from the abstract specification to the concrete implementation. Our work is motivated by the work of Pfitzmann and Waidner which have already shown in [18] that integrity properties are preserved under refinement for a synchronous timing model. However, a synchronous definition of time is difficult to justify in the real world since no notion of rounds is naturally given there and it seems to be very difficult to establish them for the Internet, for example. In contrast to that, asynchronous scenarios are attractive, because no assumptions are made about network delays and the relative execution speed of the parties. Moreover, [18] solely comprises the theoretical background, since they neither investigated the use of nor actually used formal proof tools for the verification of a concrete example.

Technically, the first part of our work can be seen as an extension of the results of [18] to asynchronous scenarios as presented in [19]. This extension is not trivial since synchronous time is much easier to handle; moreover, both models do not only differ in the definition of time but also in subtle, but important details. The second part of this paper is dedicated to the actual verification of a concrete cryptographic protocol: secure message transmission with ordered channels [4]. This yields the first example of a machine-aided proof of a cryptographic protocol in asynchronous networks such that the proven security is equivalent to the strongest cryptographic definition possible.

Related Literature. An extended version of this work is available as an IBM Research Report [3]. The goal of retaining a sound cryptographic semantics and nevertheless provide abstract interfaces for formal methods is pursued by several researchers: our approach is based on the model for reactive systems in asynchronous networks recently introduced in [19], which we believe to be really close to this goal. As we already mentioned above, Pfitzmann and Waidner have shown in [18] that integrity properties are preserved for reactive systems, but only under a synchronous timing model and they have neither investigated the use of formal methods nor the verification of a concrete example. Other possible ways to achieve this goal have been presented in [20,21,13,16, 12,1], e.g., but these either do not provide abstractions for using formal methods, or they are based on unfaithful abstractions – following the approach of Dolev and Yao [7] – in the sense that no secure cryptographic implementation of them is known.

In [2], Abadi and Rogaway have shown that a slight variation of the standard Dolev-Yao abstraction is cryptographically faithful specifically for symmetric encryption. However, their results hold only for passive adversaries and for a synchronous timing model, but the authors already state that active adversaries and an asynchronous definition of time are important goals to strive for. Another interesting approach has been presented by Guttman et. al. [11], which starts adapting the strand space theory to concrete cryptographic definitions. However, their results are specific for the Wegman-Carter system so far. Moreover, as this system is information-theoretically secure, its security proof is much easier to handle than asymmetric primitives since no reduction proofs against underlying number-theoretic assumptions have to be made.

2   Reactive Systems in Asynchronous Networks

In this section we briefly sketch the model for reactive systems in asynchronous networks from [19]. All details not necessary for understanding are omitted. Machines are represented by probabilistic state-transition machines, similar to probabilistic I/O automata [14]. For complexity we consider every automaton to be implemented as a probabilistic Turing machine; complexity is measured in the length of its initial state, i.e., the initial worktape content (often a security parameter k in unary representation).

2.1   General System Model and Simulatability

A *system* consists of several possible *structures*. A structure is a pair $(\hat{M}, S)$ of a set $\hat{M}$ of connected correct machines and a subset S of the free ports[1], called *specified ports*. Roughly, specified ports provide certain services to the honest users. In a *standard cryptographic system*, the structures are derived from one intended structure and a trust model. The trust model consists of an access structure $\mathcal{ACC}$ and a channel model χ. Here $\mathcal{ACC}$ contains the possible sets $\mathcal{H}$ of indices of uncorrupted machines (among the intended ones), and χ designates whether each channel is secure, reliable (authentic but not private) or insecure. Each structure can be completed to a *configuration* by adding an arbitrary *user* machine H and *adversary* machine A. H connects only to ports in S and A to the rest, and they may interact. The general scheduling model in [19] gives each connection c a buffer, and the machine with the corresponding clock port $c^{\lhd}!$ can schedule a message when it makes a transition. In real asynchronous cryptographic systems, all

---

[1] A port is called *free* if its corresponding port does not belong to a machine in $\hat{M}$. These ports are connected to the users and the adversary.

connections are typically scheduled by A. Thus a configuration is a runnable system, i.e., one gets a probability space of *runs* and *views* of individual machines in these runs. For a configuration $conf$, we denote the random variables over this probability space by $run_{conf,k}$ and $view_{conf,k}$, respectively. For a polynomial l, we further obtain random variables for l-step prefixes of the runs, denoted by $run_{conf,k,l(k)}$. Moreover, a run r can be restricted to a set S of ports which is denoted by $r\lceil_S$.

Simulatability essentially means that whatever can happen to certain users in the real system can also happen to the same users in the ideal (abstract) system: For every structure $struc_1 = (\hat{M}_1, S_1)$ of the real system, every user H, and every adversary A_1 there exists an adversary A_2 on a corresponding ideal structure $struc_2 = (\hat{M}_2, S_2)$, such that the view of H in the two configurations is indistinguishable, cf. Figure 1. Indistinguishability is a well-defined cryptographic notion from [22]. We write this $Sys_{real} \geq_{sec} Sys_{id}$ and

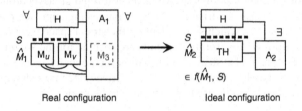

Real configuration   Ideal configuration

Fig. 1. Overview of the simulatability definition. The view of H must be indistinguishable in the two configurations. In this example, $\mathcal{H} = \{1, 2\}$.

say that Sys_{real} is *at least as secure as* Sys_{id}. In general, a mapping f may denote the correspondence between ideal and real structures and one writes $\geq_{sec}^{f}$, but with the further restriction that f maps identical sets of specified ports. An important feature of the system model is transitivity of $\geq_{sec}$, i.e., the preconditions $Sys_1 \geq_{sec} Sys_2$ and $Sys_2 \geq_{sec} Sys_3$ together imply $Sys_1 \geq_{sec} Sys_3$ [19].

In a typical ideal system, each structure contains only one machine TH called *trusted host*, whereas structures of real systems typically consist of several machines M_i, one for each honest user.

3   Integrity Properties

In this section, we show how the relation "at least as secure as" relates to integrity properties a system should fulfill, e.g., safety properties expressed in temporal logic. As a rather general version of integrity properties, independent of the concrete formal language, we consider those that have a linear-time semantics, i.e., that correspond to a set of allowed traces of in- and outputs. We allow different properties for different sets of specified ports, since different requirements of various parties in cryptography are often made for different trust assumptions.

Definition 1 (Integrity Properties). An integrity property Req for a system Sys is a function that assigns a set of valid traces at the ports in S to each set S with $(\hat{M}, S) \in Sys$. More precisely such a trace is a sequence $(v_i)_{i \in \mathbb{N}}$ of values over port names and Σ^*, so that v_i is of the form $v_i := \bigcup_{p \in S}\{p : v_{p,i}\}$ and $v_{p,i} \in \Sigma^*$. We say that Sys fulfills Req

a) *perfectly* (written $Sys \models^{perf} Req$) if for any configuration $conf = (\hat{M}, S, \mathsf{H}, \mathsf{A}) \in$ Conf(Sys), the restrictions $r \lceil_S$ of all runs of this configuration to the specified ports S lie in $Req(S)$. In formulas, $[(run_{conf,k} \lceil_S)] \subseteq Req(S)$ for all k, where $[\cdot]$ denotes the carrier set of a probability distribution.

b) *statistically* for a class $SMALL$ ($Sys \models^{SMALL} Req$) if for any configuration $conf = (\hat{M}, S, \mathsf{H}, \mathsf{A}) \in$ Conf(Sys), the probability that $Req(S)$ is not fulfilled is small, i.e., for all polynomials l (and as a function of k),

$$P(run_{conf,k,l(k)} \lceil_S \notin Req(S)) \in SMALL.$$

The class $SMALL$ must be closed under addition and contain any function g' less than or equal to any function $g \in SMALL$.

c) *computationally* ($Sys \models^{poly} Req$) if for any polynomial configuration $conf = (\hat{M}, S, \mathsf{H}, \mathsf{A}) \in$ Conf$_{poly}$(Sys), the probability that $Req(S)$ is not fulfilled is negligible, i.e.,

$$P(run_{conf,k} \lceil_S \notin Req(S)) \in NEGL.$$

For the computational and statistical case, the trace has to be finite. Note that a) is normal fulfillment. We write "$\models$" if we want to treat all three cases together. $\Diamond$

Obviously, perfect fulfillment implies statistical fulfillment for every non-empty class $SMALL$ and statistical fulfillment for a class $SMALL$ implies fulfillment in the computational case if $SMALL \subseteq NEGL$.

We now prove that integrity properties of abstract specifications carry over to their concrete counterparts in the sense of simulatability, i.e., if the properties are valid for a specification, the concrete implementation also fulfills concrete versions of these goals. As specifications are usually built by only one idealized, deterministic machine TH, they are quite easy to verify using formal proof systems, e.g., PVS. Now, our result implies that these verified properties automatically carry over to the (usually probabilistic) implementation without any further work.

The actual proof will be done by contradiction, i.e., we will show that if the real system does not fulfill its goals, the two systems can be distinguished. However, in order to exploit simulatability, we have to consider an honest user that connects to *all* specified ports. Otherwise, the contradiction might stem from those specified ports which are connected to the adversary, but those ports are not considered by simulatability. The following lemma will help us to circumvent this problem:

Lemma 1. *Let a system Sys be given. For every configuration $conf = (\hat{M}, S, \mathsf{H}, \mathsf{A}) \in$ Conf(Sys), there is a configuration $conf_s = (\hat{M}, S, \mathsf{H_s}, \mathsf{A_s}) \in$ Conf(Sys) with $S \subseteq$ ports($\mathsf{H_s}$), such that $run_{conf} \lceil_S = run_{conf_s} \lceil_S$, i.e., the probability of the runs restricted to the set S of specified ports is identical in both configurations. If $conf$ is polynomial-time, then $conf_s$ is also polynomial-time.* $\square$

We omit the proof due to space constraints and refer to [3].

Lemma 2. *The statistical distance $\Delta(\phi(var_k), \phi(var'_k))$ between a function ϕ of two random variables is at most $\Delta(var_k, var'_k)$.* $\square$

This is a well-known fact, hence we omit the easy proof.

Theorem 1 (Conservation of Integrity Properties). Let a system Sys_2 be given that fulfills an integrity property Req, i.e., $Sys_2 \models Req$, and let $Sys_1 \geq^f_{sec} Sys_2$ for an arbitrary mapping f. Then also $Sys_1 \models Req$. This holds in the perfect and statistical sense, and in the computational sense if membership in the set $Req(S)$ is decidable in polynomial time for all S.   □

Proof. Req is well-defined on Sys_1, since simulatability implies that for each $(\hat{M}_1, S_1) \in Sys_1$ there exists $(\hat{M}_2, S_2) \in f(\hat{M}_1, S_1)$ with $S_1 = S_2$. We will now prove that if Sys_1 does not fulfill the property, the two systems can be distinguished yielding a contradiction.

Assume that a configuration $conf_1 = (\hat{M}_1, S_1, \mathsf{H}, \mathsf{A}_1)$ of Sys_1 contradicts the theorem. As already described above, we need an honest user that connects to all specified ports. This is precisely what Lemma 1 does, i.e., there is a configuration $conf_{s,1}$ in which the user connects to all specified ports, with $run_{conf_{s,1}} \lceil S_1 = run_{conf_1} \lceil S_1$, so $conf_{s,1}$ also contradicts the theorem. Note that all specified ports are now connected to the honest user; thus, we can exploit simulatability.[2] Because of our precondition $Sys_1 \geq^f_{sec} Sys_2$, there exists an indistinguishable configuration $conf_{s,2} = (\hat{M}, S, \mathsf{H}_s, \mathsf{A}_2)$ of Sys_2, i.e., $view_{conf_{s,1}}(\mathsf{H}_s) \approx view_{conf_{s,2}}(\mathsf{H}_s)$. By assumption, the property is fulfilled for this configuration (perfectly, statistically, or computationally). Furthermore, the view of H_s in both configurations contains the trace at $S := S_1 = S_2$, i.e., the trace is a function $\lceil S$ of the view.

In the perfect case, the distribution of the views is identical. This contradicts the assumption that $[(run_{conf_{s,1},k} \lceil S)] \not\subseteq Req(S)$ while $[(run_{conf_{s,2},k} \lceil S)] \subseteq Req(S)$.

In the statistical case, let any polynomial l be given. The statistical distance $\Delta(view_{conf_{s,1},k,l(k)}(\mathsf{H}_s), view_{conf_{s,2},k,l(k)}(\mathsf{H}_s))$ is a function $g(k) \in SMALL$. We apply Lemma 2 to the characteristic function $1_{v\lceil S \notin Req(S)}$ on such views v. This gives $|P(run_{conf_{s,1},k,l(k)} \lceil S \notin Req(S)) - P(run_{conf_{s,2},k,l(k)} \lceil S \notin Req(S))| \leq g(k)$. As $SMALL$ is closed under addition and under making functions smaller, this gives the desired contradiction.

In the computational case, we define a distinguisher Dis: Given the view of machine H_s, it extracts the run restricted to S and verifies whether the result lies in $Req(S)$. If yes, it outputs 0, otherwise 1. This distinguisher is polynomial-time (in the security parameter k) because the view of H_s is of polynomial length, and membership in $Req(S)$ was required to be polynomial-time decidable. Its advantage in distinguishing is $|P(\mathsf{Dis}(1^k, view_{conf_{s,1},k}) = 1) - P(\mathsf{Dis}(1^k, view_{conf_{s,2},k}) = 1)| = |P(run_{conf_{s,1},k} \lceil S \notin Req(S)) - P(run_{conf_{s,2},k} \lceil S \notin Req(S))|$. Since the second term is negligible by assumption, and $NEGL$ is closed under addition, the first term also has to be negligible, yielding the desired contradiction.   ∎

In order to apply this theorem to integrity properties formulated in a logic, e.g., temporal logic, we have to show that abstract derivations in the logic are valid with respect to the cryptographic sense. This can be proven similar to the version with synchronous time, we only include it for reasons of completeness (without proof).

---

[2] In the proof for the synchronous timing model, this problem was avoided by combining the honest user and the adversary to the new honest user. However, this combination would yield an invalid configuration in the asynchronous model.

Theorem 2.

a) If $Sys \models Req_1$ and $Req_1 \subseteq Req_2$, then also $Sys \models Req_2$.
b) If $Sys \models Req_1$ and $Sys \models Req_2$, then also $Sys \models Req_1 \cap Req_2$.

Here "$\subseteq$" and "$\cap$" are interpreted pointwise, i.e., for each S. This holds in the perfect and statistical sense, and in the computational sense if for a) membership in $Req_2(S)$ is decidable in polynomial time for all S.   □

If we now want to apply this theorem to concrete logics, we have to show that the common deduction rules hold. For modus ponens, e.g., if one has derived that a and $a \rightarrow b$ are valid in a given model, then b is also valid in this model. If Req_a etc. denote the semantics of the formulas, i.e., the trace sets they represent, we have to show that

$$(Sys \models Req_a \text{ and } Sys \models Req_{a \rightarrow b}) \text{ implies } Sys \models Req_b.$$

From Theorem 2b we conclude $Sys \models Req_a \cap Req_{a \rightarrow b}$. Obviously, $Req_a \cap Req_{a \rightarrow b} = Req_{a \wedge b} \subseteq Req_b$ holds, so the claim follows from Theorem 2a.

4   Verification of the Ordered Channel Specification

In this section we review the specification for secure message transmission with ordered channels [4], and we formally verify that message reordering is in fact prevented.

4.1   Secure Message Transmission with Ordered Channels

Let n and $\mathcal{M} := \{1, \dots, n\}$ denote the number of participants and the set of indices respectively. The specification is of the typical form $Sys^{\mathsf{spec}} = \{(\{\mathsf{TH}_\mathcal{H}\}, S_\mathcal{H}) | \mathcal{H} \subseteq \mathcal{M}\}$, i.e., there is one structure for every subset of the machines, denoting the honest users. The remaining machines are corrupted, i.e., they are absorbed into the adversary.

The ideal machine $\mathsf{TH}_\mathcal{H}$ models initialization, sending and receiving of messages. A user u can initialize communications with other users by inputting a command of the form (snd_init) to the port $\mathsf{in}_u?$ of $\mathsf{TH}_\mathcal{H}$. In real systems, initialization corresponds to key generation and authenticated key exchange. Sending of messages to a user v is triggered by a command (send, m, v). If v is honest, the message is stored in an internal array $deliver_{u,v}^{\mathsf{spec}}$ of $\mathsf{TH}_\mathcal{H}$ together with a counter indicating the number of the message. After that, a command (send_blindly, i, l, v) is output to the adversary, l and i denote the length of the message m and its position in the array, respectively. This models that the adversary will notice in the real world that a message has been sent and he might also be able to know the length of that message. Because of the underlying asynchronous timing model, $\mathsf{TH}_\mathcal{H}$ has to wait for a special term (receive_blindly, u, i) or (rec_init, u) sent by the adversary, signaling, that the message stored at the ith position of $deliver_{u,v}^{\mathsf{spec}}$ should be delivered to v, or that a connection between u and v should be initialized. In the first case, $\mathsf{TH}_\mathcal{H}$ reads $(m, j) := deliver_{u,v}^{\mathsf{spec}}[i]$ and checks whether $msg\_out_{u,v}^{\mathsf{spec}} \leq j$ holds for a message counter $msg\_out_{u,v}^{\mathsf{spec}}$. If the test is successful the message is delivered at $\mathsf{out}_v!$ and the counter is set to $j + 1$, otherwise $\mathsf{TH}_\mathcal{H}$ outputs nothing. The condition $msg\_out_{u,v}^{\mathsf{spec}} \leq j$ ensures that messages can only be delivered in the order they have been received by $\mathsf{TH}_\mathcal{H}$, i.e., neither replay attacks nor reordering messages is possible for the adversary; cf. [4] for details. The user will receive inputs of the form (receive, u, m) and (rec_init, u), respectively. If v is dishonest, $\mathsf{TH}_{\mathcal{H}}$ will simply output (send, m, v) to the adversary. Finally, the adversary can send a message m to a user u by sending a command (receive, v, m) to the port from_adv$_u$? of $\mathsf{TH}_{\mathcal{H}}$ for a corrupted user v, and he can also stop the machine of any user by sending a command (stop) to a corresponding port of $\mathsf{TH}_{\mathcal{H}}$, which corresponds to exceeding the machine's runtime bound in the real world.

In contrast to the concrete implementation, which we will review later on, the machine $\mathsf{TH}_{\mathcal{H}}$ is completely deterministic, hence it can be expressed very well within a formal proof system, which supports the required data structures.

4.2 The Integrity Property

The considered specification has been designed to fulfill the property that the order of messages is maintained during every trace of the configuration. Thus, for arbitrary traces, arbitrary users $u, v \in \mathcal{H}$, $u \neq v$, and any point in time, the messages from v received so far by u via $\mathsf{TH}_{\mathcal{H}}$ should be a sublist of the messages sent from v to $\mathsf{TH}_{\mathcal{H}}$ aimed for forwarding to u. The former list is called *receive-list*, the latter *send-list*.

In order to obtain trustworthy proofs, we formally verify the integrity property in the theorem proving system PVS [15]. This will be described in the following. For reasons of readability and brevity, we use standard mathematical notation instead of PVS syntax. The PVS sources are available online.[3]

The formalization of the machine $\mathsf{TH}_{\mathcal{H}}$ in PVS is described in [4]. We assume that the machine operates on an input set $\mathcal{I}_{\mathsf{TH}_{\mathcal{H}}}$ (short $\mathcal{I}$), a state set $States_{\mathsf{TH}_{\mathcal{H}}}$ (short S), and an output set $\mathcal{O}_{\mathsf{TH}_{\mathcal{H}}}$ (short $\mathcal{O}$). For convenience, the transition function $\delta_{\mathsf{TH}_{\mathcal{H}}}: \mathcal{I} \times S \to S \times \mathcal{O}$ is split into $\delta: \mathcal{I} \times S \to S$ and $\omega: \mathcal{I} \times S \to \mathcal{O}$, which denote the next-state and output part of $\delta_{\mathsf{TH}_{\mathcal{H}}}$, respectively. The function $\delta_{\mathsf{TH}_{\mathcal{H}}}$ is defined in PVS's specification language, which contains a complete functional programming language. PVS provides natural, rational, and real numbers, arithmetic, lists, arrays, etc. Furthermore, custom datatypes (including algebraic abstract datatypes) are supported.

In order to formulate the property, we need a PVS-suited, formal notation of (infinite) runs of a machine, of lists, of what it means that a list l_1 is a sublist of a list l_2, and we need formalizations of the *receive-list* and *send-list*.

Definition 2 (Input sequence, state trace, output sequence). *Let* M *be a machine with input set* $\mathcal{I}_{\mathsf{M}}$, *state set* $States_{\mathsf{M}}$, *output set* $\mathcal{O}_{\mathsf{M}}$, *state transition function* δ, *and output transition function* ω. *Call* $s_{init} \in States_{\mathsf{M}}$ *the initial state. An* input sequence $i: \mathbb{N} \to \mathcal{I}_{\mathsf{M}}$ *for machine* M *is a function mapping the time (modeled as the set* $\mathbb{N}$*) to inputs* $i(t) \in \mathcal{I}_{\mathsf{M}}$. *A given input sequence* i *defines a sequence of states* $s^i: \mathbb{N} \to States_{\mathsf{M}}$ *of the machine* M *by the following recursive construction:*

$$s^i(0) := s_{init},$$
$$s^i(t+1) := \delta(i(t), s^i(t)).$$

The sequence s^i *is called* state-trace *of* M *under* i. *The* output sequence $o^i : \mathbb{N} \to O$ *of the run is defined as*

$$o^i(t) := \omega(i(t), s^i(t)).$$

[3] http://www.zurich.ibm.com/~mbc/OrdSecMess.tgz

We omit the index i if the input sequence is clear from the context. For components x of the state type, we write $x(t)$ for the content of x in $s(t)$, e.g., we write $deliver_{u,v}^{\text{spec}}(t)$ to denote the content at time t of the list $deliver_{u,v}^{\text{spec}}$, which is part of the state of $\mathsf{TH}_{\mathcal{H}}$. ◇

These definitions precisely match the model-intern definition of views, cf. [19]. In the context of $\mathsf{TH}_{\mathcal{H}}$, the input sequence i consists of the messages that the honest users and the adversary send to $\mathsf{TH}_{\mathcal{H}}$. In the following, a list l_1 being a sublist of a list l_2 is expressed by $l_1 \subseteq l_2$, l_1 being a sublist of the k-prefix of l_2 by $l_1 \subseteq^k l_2$.

Lemma 3. *Let l_1, l_2 be lists over some type T, let $k \in \mathbb{N}_0$. It holds:*

$$k < length(l_2) \wedge l_1 \subseteq^k l_2 \implies append(nth(l_2, k), l_1) \subseteq^{k+1} l_2,$$

that is, one may append the kth element (counted from 0) of l_2 to l_1 while preserving the prefix-sublist property. □

Definition 3 (Receive- and send-list). *Let i be an input sequence for machine $\mathsf{TH}_{\mathcal{H}}$, and let s and o be the corresponding state-trace of $\mathsf{TH}_{\mathcal{H}}$ and the output sequence, respectively. Let $u, v \in \mathcal{H}$. The receive-list is obtained by appending a new element m whenever v receives a message $(receive, m, u)$ from $\mathsf{TH}_{\mathcal{H}}$. The send-list is obtained by appending m whenever u sends a message $(send, m, v)$ to $\mathsf{TH}_{\mathcal{H}}$. Formally, this is captured in the following recursive definitions:*

$$recvlist_{u,v}^i(t) := \begin{cases} null & \text{if } t = -1, \\ append(m, recvlist_{u,v}^i(t-1)) & \text{if } t \geq 0 \wedge o^i(t) = (receive, m, u) \\ & \text{at } out_v!. \\ recvlist_{u,v}^i(t-1) & \text{otherwise} \end{cases}$$

$$sendlist_{u,v}^i(t) := \begin{cases} null & \text{if } t = -1, \\ append(m, sendlist_{u,v}^i(t-1)) & \text{if } t \geq 0 \wedge i(t) = (send, m, v) \\ & \text{at } in_u?. \\ sendlist_{u,v}^i(t-1) & \text{otherwise} \end{cases}$$

◇

We now are ready to give a precise, PVS-suited formulation of the integrity property we are aiming to prove:

Theorem 3. *For any $\mathsf{TH}_{\mathcal{H}}$ input sequence i, for any $u, v \in \mathcal{H}$, $u \neq v$, and any point in time $t \in \mathbb{N}$, it holds*

$$recvlist_{u,v}^i(t) \subseteq sendlist_{u,v}^i(t). \tag{1}$$

In the following, we omit the index i. □

Proof (sketch). The proof is split into two parts: we prove $recvlist_{u,v}(t-1) \subseteq deliver_{u,v}^{\text{spec}}(t)$ and $deliver_{u,v}^{\text{spec}}(t) \subseteq sendlist_{u,v}(t-1)$. The claim of the theorem then follows from transitivity of sublists.

The claim $deliver_{u,v}^{\text{spec}}(t) \subseteq sendlist_{u,v}(t-1)$ is proved by induction on t. Both induction base and step are proved in PVS by the built-in strategy ⟨grind⟩, which performs automatic definition expanding and rewriting with sublist-related lemmas.

The claim $recvlist_{u,v}(t-1) \subseteq deliver_{u,v}^{spec}(t)$ is more complicated. The claim is also proved by induction on t. However, it is easy to see that the claim is not inductive: in case of a (receive_blindly, u, i) at from_adv$_v$?, $\mathsf{TH}_\mathcal{H}$ outputs (receive, m, u) to out$_v$!, where $(m, j) := deliver_{u,v}^{spec}[i]$, i.e., m is the ith message of the $deliver_{u,v}^{spec}$ list (cf. Section 4.1, or [4] for more details). By the definition of the receive-list, the message m is appended to $recvlist_{u,v}$. In order to prove that $recvlist_{u,v} \subseteq deliver_{u,v}^{spec}$ is preserved during this transition, it is necessary to know that the receive list was a sublist of the prefix of the $deliver_{u,v}^{spec}$ list that does not reach to m. It would suffice to know that

$$recvlist_{u,v}(t-1) \subseteq^i deliver_{u,v}^{spec}(t).$$

Then the claim follows from Lemma 3.

We therefore strengthen the invariant to comprise the prefix-sublist property. However, the value i in the above prefix-sublist relation stems from the input (receive_blindly, u, i), and hence is not suited to state the invariant. To circumvent this problem, we recursively construct a sequence $last\_rcv\_blindly_{u,v}(t)$ which holds the parameter i of the last valid (receive_blindly, u, i) received by $\mathsf{TH}_\mathcal{H}$ on from_adv$_v$?; then

$$recvlist_{u,v}(t-1) \subseteq^l deliver_{u,v}^{spec}(t) \text{ with } l = last\_rcv\_blindly_{u,v}(t)$$

is an invariant of the system. We further strengthen this invariant by asserting that $last\_rcv\_blindly_{u,v}(t)$ and the j's stored in the $deliver_{u,v}^{spec}$ list grow monotonically. Together this yields the inductive invariant. We omit the details and again refer the to the PVS files available online.[3]   ∎

Applying Definition 1 of integrity properties, we can now define that the property Req holds for an arbitrary trace tr if and only if Equation 1 holds for all $u, v \in \mathcal{H}$, $u \neq v$, and the input sequence i of the given trace tr. Thus, Theorem 3 can be rewritten in the notation of Definition 1 as $[(run_{conf,k}\lceil s\rceil) \subseteq Req(S)]$ for all k, i.e., we have shown that the specification Sys^{spec} perfectly fulfills the integrity property Req.

4.3   The Concrete Implementation

For understanding it is sufficient to give a brief review of the concrete implementation Sys^{impl}, a detailed description can be found in [4]. Sys^{impl} is of the typical form $Sys^{impl} = \{(\hat{M}_\mathcal{H}, S_\mathcal{H}) \mid \mathcal{H} \subseteq \mathcal{M}\}$, where $\hat{M}_\mathcal{H} = \{M_u \mid u \in \mathcal{H}\}$, i.e., there is one machine for each honest participant. It uses asymmetric encryption and digital signatures as cryptographic primitives, which satisfy the strongest cryptographic definition possible, i.e., security against adaptive chosen-ciphertext attack in case of encryption (e.g., [6]) and security against existential forgery under adaptive chosen-message attacks in case of digital signatures (e.g., [10]).

A user u can let his machine create signature and encryption keys that are sent to other users over authenticated channels. Messages sent from user u to user v are signed and encrypted by M_u and sent to M_v over an insecure channel, representing a real network. Similar to $\mathsf{TH}_\mathcal{H}$ each machine maintains internal counters used for discarding messages that are out of order. The adversary schedules the communication between correct machines and it can send arbitrary messages m to arbitrary users.

Now the validity of the integrity property of the concrete implementation immediately follows from the Preservation Theorem and the verification of the abstract specification. More precisely, we have shown that the specification fulfills its integrity property of Theorem 3 perfectly, which especially implies computational fulfillment. As $Sys^{impl} \geq^{poly}_{sec} Sys^{spec}$ has already been shown in [4], our proof of integrity carries over to the concrete implementation for the computational case according to Theorem 1.

5 Conclusion

In this paper, we have addressed the problem how cryptographic protocols in asynchronous networks can be verified both machine-aided and sound with respect to the definitions of cryptography. We have shown that the verification of integrity properties of our abstract specifications automatically carries over to the cryptographic implementations, and that logic derivations among integrity properties are valid for the concrete systems in the cryptographic sense, which makes them accessible to theorem provers. As an example, we have formally verified the scheme for ordered secure message transmission [4] using the theorem proving system PVS [15]. This yields the first formal verification of an integrity property of a cryptographic protocol whose security is equivalent to the underlying cryptography with respect to the strongest cryptographic definitions possible.

References

1. M. Abadi and A. D. Gordon. A calculus for cryptographic protocols: The spi calculus. *Information and Computation*, 148(1):1–70, 1999.
2. M. Abadi and P. Rogaway. Reconciling two views of cryptography: The computational soundness of formal encryption. In *Proc. 1st IFIP International Conference on Theoretical Computer Science*, volume 1872 of *Lecture Notes in Computer Science*, pages 3–22. Springer, 2000.
3. M. Backes and C. Jacobi. Cryptographically sound and machine-assisted verification of security protocols. Research Report RZ 3468, IBM Research, 2002.
4. M. Backes, C. Jacobi, and B. Pfitzmann. Deriving cryptographically sound implementations using composition and formally verified bisimulation. In *Proc. 11th Symposium on Formal Methods Europe (FME 2002)*, volume 2391 of *Lecture Notes in Computer Science*, pages 310–329. Springer, 2002.
5. M. Bellare, A. Desai, D. Pointcheval, and P. Rogaway. Relations among notions of security for public-key encryption schemes. In *Advances in Cryptology: CRYPTO '98*, volume 1462 of *Lecture Notes in Computer Science*, pages 26–45. Springer, 1998.
6. R. Cramer and V. Shoup. Practical public key cryptosystem provably secure against adaptive chosen ciphertext attack. In *Advances in Cryptology: CRYPTO '98*, volume 1462 of *Lecture Notes in Computer Science*, pages 13–25. Springer, 1998.
7. D. Dolev and A. C. Yao. On the security of public key protocols. *IEEE Transactions on Information Theory*, 29(2):198–208, 1983.
8. S. Goldwasser and S. Micali. Probabilistic encryption. *Journal of Computer and System Sciences*, 28:270–299, 1984.
9. S. Goldwasser, S. Micali, and C. Rackoff. The knowledge complexity of interactive proof systems. *SIAM Journal on Computing*, 18(1):186–207, 1989.

686   M. Backes and C. Jacobi

10. S. Goldwasser, S. Micali, and R. L. Rivest. A digital signature scheme secure against adaptive chosen-message attacks. *SIAM Journal on Computing*, 17(2):281–308, 1988.
11. J. D. Guttman, F. J. Thayer Fabrega, and L. Zuck. The faithfulness of abstract protocol analysis: Message authentication. In *Proc. 8th ACM Conference on Computer and Communications Security*, pages 186–195, 2001.
12. P. Lincoln, J. Mitchell, M. Mitchell, and A. Scedrov. A probabilistic poly-time framework for protocol analysis. In *Proc. 5th ACM Conference on Computer and Communications Security*, pages 112–121, 1998.
13. G. Lowe. Breaking and fixing the Needham-Schroeder public-key protocol using FDR. In *Proc. 2nd International Conference on Tools and Algorithms for the Construction and Analysis of Systems (TACAS)*, volume 1055 of *Lecture Notes in Computer Science*, pages 147–166. Springer, 1996.
14. N. Lynch. *Distributed Algorithms*. Morgan Kaufmann Publishers, San Francisco, 1996.
15. S. Owre, N. Shankar, and J. M. Rushby. PVS: A prototype verification system. In *Proc. 11th International Conference on Automated Deduction (CADE)*, volume 607 of *Lecture Notes in Computer Science*, pages 748–752. springer, 1992.
16. L. Paulson. The inductive approach to verifying cryptographic protocols. *Journal of Cryptology*, 6(1):85–128, 1998.
17. B. Pfitzmann, M. Schunter, and M. Waidner. Cryptographic security of reactive systems. Presented at the DERA/RHUL Workshop on Secure Architectures and Information Flow, Electronic Notes in Theoretical Computer Science (ENTCS), March 2000. http://www.elsevier.nl/cas/tree/store/tcs/free/noncas/pc/menu.htm.
18. B. Pfitzmann and M. Waidner. Composition and integrity preservation of secure reactive systems. In *Proc. 7th ACM Conference on Computer and Communications Security*, pages 245–254, 2000.
19. B. Pfitzmann and M. Waidner. A model for asynchronous reactive systems and its application to secure message transmission. In *Proc. 22nd IEEE Symposium on Security & Privacy*, pages 184–200, 2001.
20. A. W. Roscoe. Modelling and verifying key-exchange protocols using CSP and FDR. In *Proc. 8th IEEE Computer Security Foundations Workshop (CSFW)*, pages 98–107, 1995.
21. S. Schneider. Security properties and CSP. In *Proc. 17th IEEE Symposium on Security & Privacy*, pages 174–187, 1996.
22. A. C. Yao. Theory and applications of trapdoor functions. In *Proc. 23rd IEEE Symposium on Foundations of Computer Science (FOCS)*, pages 80–91, 1982.

Durations, Parametric Model-Checking in Timed Automata with Presburger Arithmetic*

Véronique Bruyère[1], Emmanuel Dall'Olio[2], and Jean-François Raskin[2]

[1] Institut d'Informatique,
Université de Mons-Hainaut, Le Pentagone,
Avenue du Champ de Mars 6, B-7000 Mons, Belgium
Veronique.Bruyere@umh.ac.be
[2] Département d'Informatique, Université Libre de Bruxelles,
Blvd du Triomphe, CP 212. B-1050-Bruxelles, Belgium
{Emmanuel.Dallolio,Jean-Francois.Raskin}@ulb.ac.be

Abstract. We consider the problem of model-checking a parametric extension of the logic TCTL over timed automata and establish its decidability. Given a timed automaton, we show that the set of durations of runs starting from a region and ending in another region is definable in the arithmetic of Presburger (when the time domain is discrete) or in the theory of the reals (when the time domain is dense). With this logical definition, we show that the parametric model-checking problem for the logic TCTL can easily be solved. More generally, we are able to effectively characterize the values of the parameters that satisfy the parametric TCTL formula.

1 Introduction

For more than twenty years, temporal logics and automata theory play a central role in the foundations for the formal verification of reactive systems. Reactive systems often have to meet real-time constraints. As a consequence, in the early nineties, temporal logics and automata theory have been extended to explicitly refer to real-time. Timed automata [2], an extension of usual automata with clocks, have been proposed as a natural model for systems that must respect real-time constraints. Temporal logics have also been extended to express quantitative real-time properties [1,10].

The model-checking problem, given a timed automaton $\mathcal{A}$ and a real-time logic formula ϕ, consists in answering by YES or NO if the executions of $\mathcal{A}$ verify the property expressed by ϕ. So the model-checking problem tries to answer the question $\mathcal{A} \models^? \phi$. If the answer is NO then the model-checking algorithm is able to exhibit an example of execution where the property is falsified.

Response property is a typical example of a real-time property: "if every request reaches the server within two time units, then every request must be

---

* Supported by the FRFC project "Centre Fédéré en Vérification" funded by the Belgian National Science Fundation (FNRS) under grant nr 2.4530.02

H. Alt and M. Habib (Eds.): STACS 2003, LNCS 2607, pp. 687–698, 2003.
© Springer-Verlag Berlin Heidelberg 2003

accepted or rejected within four time units". This property is easily expressed in TCTL as follows:

$$\forall\Box(\text{send} \rightarrow \forall\Diamond_{\leq 2}\text{request}) \rightarrow \forall\Box(\text{send} \rightarrow \forall\Diamond_{\leq 4}\text{granted} \vee \text{rejected})$$

The main drawback with this formulation of the response property is that it refers to fixed constants. In the context of real-time systems, it is often more natural to use parameters instead of constants to expressed the delays involved in the model. In particular, we would like to formalize the above property in a more abstract way like "if every request reaches the server within α time units, then there should exist a bound of time γ within which every request must be accepted or rejected, that bound should be less than 2α". In this way, the property is less dependent on the particular model on which it is checked. This is particularly important when the model is not completely known a priori. The parametric property above can be formalized using an extension of TCTL with parameters and quantification over parameters as follows:

$$\forall\alpha\exists\gamma \cdot \gamma \leq 2\alpha \wedge [\forall\Box(\text{send} \rightarrow \forall\Diamond_{\leq\alpha}\text{request})$$
$$\rightarrow \forall\Box(\text{send} \rightarrow \forall\Diamond_{\leq\gamma}\text{granted} \vee \text{rejected})]$$

The use of parameters can also be very useful to *learn* about a model. For example, parameters in conjunction with temporal logics can be used to express questions that are not expressible in usual temporal logics. As an example, consider the following question: "Is there a bound on the delay to reach a σ state from an initial state ?". This property is easily expressed using the TCTL syntax extended with parameters: $\exists\alpha \cdot \forall\Diamond_{\leq\alpha}\sigma$. The answer to the question is YES if the formula is true. We can even be more ambitious and ask to characterize the set of parameter valuations for which a formula is true on a given timed model. As an example, consider the following TCTL formula with parameter γ: $\forall\Diamond_{\leq\gamma}\sigma$. If there is no bound on the time needed to reach a σ state from an initial state, then the answer to the parameter valuation problem would be "empty", while it would be all the valuations v such that $v(\alpha) \geq c$ if all paths starting from an initial state reach a σ state within c time units.

We show in this paper that the problems outlined above are decidable for discrete- and dense-timed automata (if parameters range over natural numbers). When those results were partially known, see [11,12,8], the main contributions of this paper are as follows. First, we show that the *natural* and *intuitive* notion of *duration* (of runs of a timed automaton starting from a region and ending to another region) is central and sufficient to answer the parametric TCTL verification problem. Second, we show that Presburger arithmetic (or the additive theory of reals) and classical automata are *elegant* tools to formalize this notion of run durations. Finally, using those simple and clean concepts from logic and automata theory, we are able to prove the decidability of the model-checking problem for a parametric version of TCTL that strictly subsumes, to the best of our knowledge, all the parametric extensions of TCTL proposed so far. Furthermore, we are able to effectively characterize the values of the parameters that satisfy the parametric TCTL formula. Finally, it should be noted that the technique to establish these results is very similar if the time domain is discrete or dense.

Due to space limitations, this paper does not contain important proofs. The interested reader is refered to [6] for full details.

Related Works. The work by Wang et al [11,12] is closely related to our work. The logic they consider is a strict subset of the logic considered in this paper: in their works the parameters are all implicitly quantified existentially (when we also consider universal quantification). The technique they use to establish the decidability result while ingenious is more complex than the technique that we propose in this paper. So the main contribution with regard to their work is, we feel, a simpler proof of decidability of a generalization of their logic. For the existential fragment of our logic, we obtain a similar bound on the complexity of the model-checking problem. Emerson et al have also studied an extension of TCTL with parameters in [8] but they make strong assumptions on the timed models used and strong restrictions on the use of parameters in the way they constrain the scope of the temporal operators. Alur et al [4] have also studied the extension of real-time logics with parameters but in the context of linear time. Alur et al [3] have studied the introduction of parameters in timed automata. Other researchers have also proposed the use of Presburger arithmetic (or Real arithmetic) in the context of timed automata. In particular, Comon et al [7] have studied the use of the Real arithmetic to express the reachability relation of timed automata. Their work is more ambitious, since they do not only consider durations between regions but also durations between individual clock valuations. Nevertheless they do not consider properties expressible in temporal logics.

2   Timed Automata

In this section, we recall the classical notion of timed automaton [2]. Throughout the paper, we denote by $X = \{x_1, \ldots, x_n\}$ a set of n clocks. We use the same notation $x = (x_1, \ldots, x_n)$ for the *variables* and for an *assignment* of values to these variables. Depending on whether the timed automata are *dense*-timed or *discrete*-timed, the values of the variables are taken in domain $\mathbb{T}$ equal to the set $\mathbb{R}^+$ of nonnegative reals or to the set $\mathbb{N}$ of natural numbers. Given a clock assignment x and $\tau \in \mathbb{T}$, $x + \tau$ is the clock assignment $(x_1 + \tau, \ldots, x_n + \tau)$. A *simple guard* is an expression of the form $x_i \sim c$ where x_i is a clock, $c \in \mathbb{N}$ is an integer constant, and $\sim$ is one of the symbols $\{<, \leq, =, >, \geq\}$. A *guard* is any finite conjunction of simple guards. We denote by $\mathcal{G}$ the set of guards. Let g be a guard and x be a clock assignment, notation $x \models g$ means that x satisfies g. A *reset* function r indicates which clocks are reset to 0. Thus either $r(x)_i = 0$ or $r(x)_i = x_i$. The set of reset functions is denoted by $\mathcal{R}$. We use notation Σ for the set of *atomic propositions*.

Definition 1. *A timed automaton is a quadruple $\mathcal{A} = (L, X, E, I, \mathcal{L})$ with the following components: (i) L is a finite set of locations, (ii) X is a set of clocks,* *(iii)* $E \subseteq L \times \mathcal{G} \times \mathcal{R} \times L$ *is a finite set of* edges, *(iv)* $I : L \to \mathcal{G}$ *assigns an* invariant *to each location, (v)* $\mathcal{L} : L \to 2^{\Sigma}$ *is the* labeling *function.*

Definition 2. *A* timed automaton $\mathcal{A} = (L, X, E, I, \mathcal{L})$ *generates a* transition system $T_{\mathcal{A}} = (Q, \to)$ *with a set of* states Q *equal to* $\{(l, x) \mid l \in L, x \in \mathbb{T}^n, x \models I(l)\}$ *and a transition relation* $\to = \bigcup_{\tau \in \mathbb{T}} \overset{\tau}{\to}$ *defined as follows:* $(l, x) \overset{\tau}{\to} (l', x')$ *(i) if* $\tau > 0$*, then* $l = l'$ *and* $x' = x + \tau$ *(elapse of time at location* l*) (ii) if* $\tau = 0$*, then* $(l, g, r, l') \in E$*,* $x \models g$ *and* $r(x) = x'$ *(instantaneous switch)*

The states (l, x) of $T_{\mathcal{A}}$ are shortly denoted by q. Given $q = (l, x) \in Q$ and $\tau \in \mathbb{T}$, we denote by $q + \tau$ the state $(l, x + \tau)$. Note that if $\mathcal{A}$ is a discrete-timed automaton, the elapse of time is discrete with $\tau = 1, 2, 3, \cdots$. If $\mathcal{A}$ is a dense-timed automaton, then τ is any positive real number. Note also that given a transition $q \to q'$ of $T_{\mathcal{A}}$, it is easy to compute the unique τ such that $q \overset{\tau}{\to} q'$.

Definition 3. *Given a transition system* $T_{\mathcal{A}}$*, a* run $\rho = (q_i)_{i \geq 0}$ *is an infinite path in* $T_{\mathcal{A}}$ $\rho = q_0 \overset{\tau_0}{\to} q_1 \overset{\tau_1}{\to} q_2 \cdots q_i \overset{\tau_i}{\to} q_{i+1} \cdots$ *such that* $\Sigma_{i \geq 0} \tau_i = \infty$ *(Non-Zenoness property). A* finite run $\rho = (q_i)_{0 \leq i \leq j}$ *is any finite path in* $T_{\mathcal{A}}$*. A* position *in* ρ *is a state* $q_i + \tau$*, where* $i \geq 0$*,* $\tau \in \mathbb{T}$ *and either* $\tau = 0$*, or* $0 < \tau < \tau_i$*. The* duration $t = D_{\rho}(q)$ *at position* $q = q_i + \tau$ *is equal to* $t = \tau + \Sigma_{0 \leq i' < i} \tau_{i'}$*.*

So, if $\mathcal{A}$ is a discrete-timed automaton, then D_{ρ} is a function which assigns a natural number to any position of the run ρ. If $\mathcal{A}$ is a dense-timed automaton, the assignment is in $\mathbb{R}^+$. As we allow *several* consecutive instantaneous switches in the definition of a run, different positions q can have the same duration $D_{\rho}(q)$. The set of positions in a run ρ can be totally ordered as follows. Let $q = q_i + \tau$ and $q' = q_{i'} + \tau'$ be two positions of ρ. Then $q < q'$ iff either $i < i'$ or $i = i'$ and $\tau < \tau'$.

3   Parametric Timed CTL Logic

Let $P = \{\theta_1, \ldots, \theta_m\}$ be a fixed set of *parameters*. A *parameter valuation* for P is a function $v : P \to \mathbb{N}$ which assigns a natural number to each parameter $\theta \in P$. In the sequel, α means any element of $P \cup \mathbb{N}$ and β means any linear term $\Sigma_{j \in J} c_j \theta_j + c$, with $c_j, c \in \mathbb{N}$ and $J \subseteq \{1, \ldots m\}$. A parameter valuation v is naturally extended to linear terms by defining $v(c) = c$ for any $c \in \mathbb{N}$.

   The *syntax* of *Parametric Timed CTL logic*, PTCTL logic for short, is defined in two steps: we first define formulae of first type, and then formulae of second type. We propose two logics, the discrete PTCTL logic and the dense PTCTL logic. The first one is dedicated to discrete-timed automata whereas the second one is dedicated to dense-timed automata. Notation σ means any atomic proposition $\sigma \in \Sigma$.

Definition 4. *Discrete* PTCTL *formulae* φ *of first type are given by the following grammar* $\varphi ::= \sigma \mid \neg \varphi \mid \varphi \vee \varphi \mid \exists \bigcirc \varphi \mid \varphi \exists U_{\sim \alpha} \varphi \mid \varphi \forall U_{\sim \alpha} \varphi$ *and formulae* f *of second type by* $f ::= \varphi \mid \theta \sim \beta \mid \neg f \mid f \vee f \mid \exists \theta f$*. Dense* PTCTL *formulae* *are defined in the same way, except that operator $\exists\bigcirc$ is forbidden. The fragment* $\exists$PTCTL *is the fragment where formulae are of the form* $\exists\theta_1,\ldots,\theta_n g$ *where* g *is a* PTCTL *formula without quantifiers.*

In this definition, in second type formula $\exists\theta f$, it is assumed that θ is a free parameter of formula f. The set of free parameters of f is denoted by P_f. Note that usual operators $\exists U$ and $\forall U$ are obtained as $\exists U_{\geq 0}$ and $\forall U_{\geq 0}$. We now give the *semantics* of PTCTL logic.

Definition 5. *Let* $T_{\mathcal{A}}$ *be the transition graph of a discrete-timed automaton* $\mathcal{A}$ *and* $q = (l, x)$ *be a state of* $T_{\mathcal{A}}$. *Let* f *be a discrete* PTCTL *formula and* v *be a parameter valuation on* P_f. *Then the* satisfaction *relation* $q \models_v f$ *is defined inductively as indicated below. If* $\mathcal{A}$ *is a dense-timed automaton and* f *is a dense* PTCTL *formula, then the satisfaction relation is defined in the same way, except that* $q \models_v \exists\bigcirc\varphi$ *does not exist.*

(1) *First type formulae: (i)* $q \models_v \sigma$ *iff* $\sigma \in \mathcal{L}(l)$ *(ii)* $q \models_v \neg\varphi$ *iff* $q \not\models_v \varphi$ *(iii)* $q \models_v \varphi \vee \psi$ *iff* $q \models_v \varphi$ *or* $q \models_v \psi$ *(iv)* $q \models_v \exists\bigcirc \varphi$ *iff there exists a run* $\rho = (q_i)_{i\geq 0}$ *in* $T_{\mathcal{A}}$ *with* $q = q_0$ *and* $q_0 \xrightarrow{\tau} q_1$ *satisfying* $\tau = 0$ *or* $\tau = 1$, *such that* $q_1 \models_v \varphi$ *(v)* $q \models_v \varphi\exists U_{\sim\alpha}\psi$ *iff there exists a run* $\rho = (q_i)_{i\geq 0}$ *in* $T_{\mathcal{A}}$ *with* $q = q_0$, *there exists a position* p *in* ρ *such that* $D_\rho(p) \sim v(\alpha)$, $p \models_v \psi$ *and* $p' \models_v \varphi$ *for all* $p' < p$ *(vi)* $q \models_v \varphi\forall U_{\sim\alpha}\psi$ *iff for any run* $\rho = (q_i)_{i\geq 0}$ *in* $T_{\mathcal{A}}$ *with* $q = q_0$, *there exists a position* p *in* ρ *such that* $D_\rho(p) \sim v(\alpha)$, $p \models_v \psi$ *and* $p' \models_v \varphi$ *for all* $p' < p$

(2) *Second type formulae: (i)* $q \models_v \theta \sim \beta$ *iff* $v(\theta) \sim v(\beta)$ *(ii)* $q \models_v \neg f$ *iff* $q \not\models_v f$ *(iii)* $q \models_v f \vee g$ *iff* $q \models_v f$ *or* $q \models_v g$ *(iv)* $q \models_v \exists\theta f$ *iff there exists* $c \in \mathbb{N}$ *such that* $q \models_{v'} f$ *where* v' *is defined on* P_f *by* $v' = v$ *on* $P_{\exists\theta f}$ *and* $v'(\theta) = c$

For convenience, we use the following abbreviations: *(i)* $\exists\Diamond_{\sim c}\varphi \equiv \top\exists U_{\sim c}\varphi$, *(ii)* $\forall\Diamond_{\sim c}\varphi \equiv \top\forall U_{\sim c}\varphi$, *(iii)* $\exists\Box_{\sim c}\varphi \equiv \neg\forall\Diamond_{\sim c}\neg\varphi$, and *(iv)* $\forall\Box_{\sim c}\varphi \equiv \neg\exists\Diamond_{\sim c}\neg\varphi$. When necessary, we will use the alternative grammar given in the next proposition.

Proposition 1. *An alternative grammar for the first type* PTCTL *formulae is given by:* $\varphi ::= \sigma \mid \neg\varphi \mid \varphi \vee \varphi \mid \exists\bigcirc\varphi \mid \varphi\exists U_{\sim\alpha}\varphi \mid \exists\Box_{\sim\alpha}\varphi$.

Problem 1. The *model-checking* problem is the following. Given a timed automaton $\mathcal{A}$ and a state q of $T_{\mathcal{A}}$, given a PTCTL formula f, is there a parameter valuation v on P_f such that $q \models_v f$? The model-checking problem is called *discrete* if $\mathcal{A}$ is a discrete-timed automaton and f a discrete PTCTL formula. It is called *dense* if $\mathcal{A}$ is a dense-timed automaton and f a dense PTCTL formula.

Problem 2. The *parameter synthesis* problem is the following. Given a timed automaton $\mathcal{A}$ and a state q of $T_{\mathcal{A}}$, given a PTCTL formula f, compute a symbolic representation of the set of parameter valuations v on P_f such that $q \models_v f$. This symbolic representation should support boolean operations, projections and checking emptiness.

4   Region Graphs

In this section we recall the definition of region graph [2]. It is usually given for dense-timed automata. It can also be applied to discrete-timed automata. Thus in the sequel $\mathcal{A} = (L, X, E, I, \mathcal{L})$ is a discrete- or a dense-timed automaton.

We first recall the usual equivalence on clock assignments and its extension to states of the transition system T_A generated by $\mathcal{A}$. For clock x_i, let c_i be the largest constant that x_i is compared with in any guard g of E. For $\tau \in \mathbb{T}$, $\text{fract}(\tau)$ denotes its fractional part and $\lfloor \tau \rfloor$ denotes its integral part.

Definition 6. *Two clock assignements x and x' are equivalent, $x \approx x'$, iff the following conditions hold (i) For any i, $1 \leq i \leq n$, either $\lfloor x_i \rfloor = \lfloor x'_i \rfloor$ or $x_i, x'_i > c_i$. (ii) For any $i \neq j$, $1 \leq i,j \leq n$ such that $x_i \leq c_i$ and $x_j \leq c_j$, (a) $\text{fract}(x_i) \leq \text{fract}(x_j)$ iff $\text{fract}(x'_i) \leq \text{fract}(x'_j)$, (b) $\text{fract}(x_i) = 0$ iff $\text{fract}(x'_i) = 0$. The equivalence relation $\approx$ is extended to the states of T_A as follows: $q = (l, x) \approx q' = (l', x')$ iff $l = l'$ and $x \approx x'$.*

We use $[x]$ (resp. $[q]$) to denote the equivalence class to which x (resp. q) belongs. A *region* is an equivalence class $[q]$. The set of all regions is denoted by R. A region $[q]$ is called *boundary* if $q + \tau \not\approx q$ for any $\tau > 0$. It is called *unbounded* if it satisfies $q = (l, x)$ with $x_i > c_i$ for all i. We note $B(r)$ the fact that the region r is boundary. It is well-known [2] that $\approx$ is *back-stable* on Q: if $q \approx q'$ and $p \to q$, then $\exists p' \approx p$ such that $p' \to q'$. It is proved in [1] that states belonging to the same region satisfy the same set of TCTL formulae. The proof is easily adapted to PTCTL formulae.

Proposition 2. *Let q, q' be two states of T_A such that $q \approx q'$. Let f be a PTCTL formula and v be a parameter valuation. Then $q \models_v f$ iff $q' \models_v f$.*

As a consequence, Problems 1 and 2 can be restated as follows. The interest is obviously the finite number of regions.

Problem 3. Given a timed automaton $\mathcal{A}$ and a region r of R, given a PTCTL formula f, is there a parameter valuation v on P_f such that $r \models_v f$? Is it possible to compute a symbolic representation of the set of parameter valuations v on P_f such that $q \models_v f$?

We proceed with the definition of region graph. Given two regions $r = [q]$, $r' = [q']$ such that $r \neq r'$, we say that r' is a *successor* of r, $r' = \text{succ}(r)$, if $\exists \tau \in \mathbb{T}$, $q + \tau \in r'$, and $\forall \tau' < \tau$, $q + \tau' \in r \cup r'$.

Definition 7. *Let $\mathcal{A}$ be a timed automaton. The region graph $R_A = (R, F)$ is a finite graph with R as vertex set and its edge set F defined as follows. Given two regions $r, r' \in R$, the edge (r, r') belongs to F if: (i) $q \xrightarrow{\tau} q'$ in T_A, with $\tau = 0$, $r = [q]$ and $r' = [q']$, or (ii) $r' = \text{succ}(r)$, or (iii) $r = r'$ is an unbounded region. We recall that the size $|R_A|$ of the region automaton R_A is in $\mathcal{O}(2^{|A|})$ [2].*

There is a simple correspondence between runs of T_A and paths in R_A. More precisely, let $\rho = (q_i)_{i \geq 0}$ be a run. Consider $q_i \xrightarrow{\tau_i} q_{i+1}$. If $\tau_i = 0$ or if $[q_i] = [q_{i+1}]$ is an unbounded region, then $([q_i], [q_{i+1}])$ is an edge of R_A. Otherwise, there is a path $(r_{i,k})_{1 \leq k \leq n_i}$ in R_A such that $r_{i,1} = [q_i]$, $r_{i,n_i} = [q_{i+1}]$ and $r_{i,k+1} = \mathrm{succ}(r_{i,k})$ for all k, $1 \leq k < n_i$. This path can be empty (when $[q_i] = [q_{i+1}]$). In this way, an infinite path denoted $\pi(\rho)$ of R_A corresponds to the run ρ of T_A. We say that $\pi(\rho)$ is the path *associated* to ρ.

Definition 8. *Any path $\pi(\rho)$ associated to a run ρ is called* progressive *since time progresses without bound along ρ (see Definition 3).*

On the other hand, for any progressive path of R_A, we can find a corresponding run of T_A [2]. This run is not unique.

In Definition 6, only condition (i) is useful when A is discrete-timed. Thus given a clock x_i, the possible values for an equivalence class are 1, 2, ..., c_i and $c_i^+ = \{c \in \mathbb{N} \mid c > c_i\}$. If r, r' are two regions such that $r' = \mathrm{succ}(r)$, then any clock assignment has been increased by 1.

5   Durations

To solve the model-checking problem (see Problem 1), we want an algorithm that, given a timed automaton A and a region r of R_A, given a PTCTL formula f, tests whether there exists a parameter valuation v on P_f such that $r \models_v f$.

Our approach is the following. Let $P_f = \{\theta_1, \ldots, \theta_m\}$. In the case of the discrete model-checking problem, we are going to construct a formula $\Delta(f, r)$ of Presburger arithmetic with free variables $\theta_1, \ldots, \theta_m$ such that $r \models_v f$ for some valuation v iff the sentence $\exists \theta_1 \cdots \exists \theta_m \Delta(f, r)$ is true. Our algorithm follows because Presburger arithmetic has a decidable theory.

The *main tool* of our approach is a description, given two regions s, s' of R_A, of all the possible values of duration $D_\rho(q_j)$ for any finite run ρ from q_0 to q_j in T_A such that $[q_0] = s$, $[q_j] = s'$ (see Definition 9 below). The description is given by a Presburger arithmetic formula. The construction of $\Delta(f, r)$ is then easily performed by induction on the formula f. The approach is the same for the dense model-checking problem but with a real arithmetic instead of Presburger arithmetic.

Definition 9. *Let A be a timed automaton and $R_A = (R, F)$ be its region graph. Let $s, s' \in R$ and $S \subseteq R$. Then $\lambda^S_{s,s'}$ is the set of $t \in \mathbb{T}$ such that (i) there exists a finite run $\rho = (q_i)_{0 \leq i \leq j}$ in T_A with duration $t = D_\rho(q_j)$, (ii) let $\pi(\rho) = (r_l)_{0 \leq l \leq k}$ be the path in R_A associated with ρ, then $s = r_0$, $s' = r_k$ and $r_l \in S$ for any l, $0 \leq l < k$. In this definition, s belongs to S and s' may not belong to S.*

Discrete Time. We recall that Presburger arithmetic, PA for short, is the set of first-order formulae of $\langle \mathbb{N}, +, <, 0, 1 \rangle$. Terms are built from variables, the constants 0, 1, and the symbol $+$. Atomic formulae are either equations $t_1 = t_2$ or inequations $t_1 < t_2$ between terms t_1, t_2. Formulae are built from atomic formulae using first-order quantifiers and the usual connectives. Formulae are interpreted over the natural numbers, with the usual interpretation of $+$, $<$, 0 and 1. Presburger *theory* is the set of all the sentences of Presburger arithmetic i.e., formulae without free variables. It is well-known that Presburger theory is decidable with a complexity in 3EXPTIME in the size of the sentence and in NP if only existential quantification is used [9]. A set $X \subseteq \mathbb{N}$ is *definable* by a PA formula $\varphi(x)$ if X is exactly the set of assignments of variable x making formula φ true.

Proposition 3. *Let* $\mathcal{A} = (L, X, E, I, \mathcal{L})$ *be a discrete-timed automaton and* $R_\mathcal{A} = (R, F)$ *be its region graph. Let* $s, s' \in R$ *and* $S \subseteq R$. *Then set* $\lambda_{s,s'}^S$ *is definable by a PA formula. The construction of the formula is effective.*

Proof. As $\mathcal{A}$ is a discrete-timed automaton, its region graph satisfies the following property. Consider the edge $(r, r') \in F$. Either it corresponds to an instantaneous switch, which takes no time. Either $r' = \text{succ}(r)$ and any clock has been increased by 1 from r to r'. Or $r' = r$ is an unbounded region for which we can suppose that any clock is increased by 1 along (r, r').

Let a be a fixed symbol meaning an increment by 1 of the clocks. We define a classical automaton $\mathcal{C}$, as a particular subgraph of $R_\mathcal{A}$. It has $S \cup \{s'\}$ as set of states and $F \cap S \times (S \cup \{s'\})$ as set of transitions. Any of its transitions (r, r') is labeled by ϵ (the empty word) if it corresponds to an instantaneous switch, and by a otherwise. It has a unique initial state equal to s and a unique final state equal to s'. The standard subset construction is then applied to $\mathcal{C}$ to get a deterministic automaton without ϵ-transitions. The resulting automaton $\mathcal{C}'$ has the special structure of "frying pan" automaton (see figure 1) since the only symbol labeling the transitions is a. Denote its states by $\{\zeta_0, \cdots, \zeta_k, \cdots, \zeta_l\}$, with ζ_0 the initial state. Note that $\mathcal{C}'$ could be without cycle.

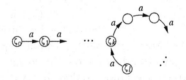

Fig. 1. Frying pan automaton

It is not difficult to check that t belongs to $\lambda_{s,s'}^S$ iff t is the length of a path in $\mathcal{C}'$ starting at ζ_0 and ending at some final state ζ_m. Hence if $m < k$, then $t = m$ and if $k \leq m \leq l$, then $\exists z \in \mathbb{N}, t = m + cz$ where $c = l - k + 1$ is the length of the cycle. Therefore $\lambda_{s,s'}^S$ is definable by a PA formula given by a disjunction of terms like $t = m$ or $\exists z \ t = m + cz$ where m, c are constants.

Dense Time. To deal with the dense time, we work with the Real arithmetic, RA, instead of Presburger arithmetic. It is the set of first-order formulae of $\langle \mathbb{R}, +, <,$

$\mathbb{N}, 0, 1\rangle$ where $\mathbb{N}$ is a unary predicate. Formulae are interpreted over the reals numbers. The interpretation of $\mathbb{N}$ is defined such that $\mathbb{N}(x)$ holds iff x is a natural number. As for Presburger arithmetic, the Real arithmetic has a decidable theory with a complexity in 3ExpTime in the size of the sentence [13,5]. Note that any PA formula is a RA formula thanks to predicate $\mathbb{N}(x)$.

Proposition 4. *Suppose that $\mathcal{A}$ is a dense-timed automaton. Then set $\lambda^S_{s,s'}$ is definable by a* RA *formula. The construction of the formula is effective.*

The proof of Proposition 4 is in the same vein as the proof of Proposition 3, however with some fitting due to dense time.

Additional Sets. To solve the model-checking problem, we need two auxiliary sets that we present now. We begin with the definition of set $\chi^S_{s,s'}$ which is very close to the definition of $\lambda^S_{s,s'}$.

Definition 10. *Let $\mathcal{A}$ be a timed automaton and $R_{\mathcal{A}} = (R, F)$ be its region graph. Let $s, s' \in R$ and $S \subseteq R$. Then $\chi^S_{s,s'}$ is the set of $t \in \mathbb{N}$ such that (i) there exists a finite run $\rho = (q_i)_{0 \leq i \leq j}$ in $T_{\mathcal{A}}$ with duration $t = D_\rho(q_j)$, (ii) let $\pi(\rho) = (r_l)_{0 \leq l \leq k}$ be the path in $R_{\mathcal{A}}$ associated with ρ, then $s = r_0$, $s' = r_k$ and $r_l \in S$ for any l, $0 \leq l < k$, (iii) any position $p < q_j$ satisfies $D_\rho(p) < t$.*

The differences with $\lambda^S_{s,s'}$ are the following. The duration t is necessarily a natural number. A third condition has been added which means that q_j is the first position in ρ with duration equal to t (remember that several positions can have the same duration in a run).

Proposition 5. *Set $\chi^S_{s,s'}$ is definable by a* PA *formula[1]. The construction of the formula is effective.*

We end this paragraph with the definition of set Pr_S.

Definition 11. *Let $\mathcal{A}$ be a timed automaton and $R_{\mathcal{A}} = (R, F)$ be its region graph. Let $S \subseteq R$ and $s \in S$. Then $s \in \mathrm{Pr}_S$ iff there exists a progressive path in $R_{\mathcal{A}}$ with all its vertices in S and its first vertex equal to s.*

Proposition 6. *It is decidable whether $s \in \mathrm{Pr}_S$.*

6   Model-Checking

Discrete Model-Checking. To solve the discrete model-checking problem, we rely on the duration formulae defined in the previous section. The inductive definition of the PA formula $\Delta(f, r)$ is given in Table 1. This table only refers to formulae with temporal operators following the grammar of Proposition 1. The definition of $\Delta(f, r)$ is immediate for the other formulae. Note that due to discrete time, symbol $\sim$ is restricted to $\{<, \geq, =\}$. This contruction allows us to obtain the following theorem.

---
[1] This is true both for the discrete and dense case (RA is not needed here)

Theorem 1. *Let $\mathcal{A}$ be a discrete-timed automaton and r be a region of $R_{\mathcal{A}} = (R, F)$. Let f be a PTCTL formula with $P_f = \{\theta_1, \ldots, \theta_m\}$. Then there exists a PA formula $\Delta(f, r)$ such that $r \models_v f$ for some valuation v iff the sentence $\exists \theta_1 \cdots \exists \theta_m \Delta(f, r)$ is true. The construction of $\Delta(f, r)$ is effective.*

Corollary 1. *The discrete model-checking problem is decidable.*

The following two statements characterise the complexity of our method.

Proposition 7. *The PA formulae $\lambda_{r,r'}^S(t)$ and $\chi_{r,r'}^S(t)$ have a size, and can be constructed in time, bounded by $\mathcal{O}(2^{2 \cdot |R_{\mathcal{A}}|})$. The PA formula $\Delta(f, r)$ has a size, and can be constructed in time, bounded by $\mathcal{O}(2^{6 \cdot |R_{\mathcal{A}}| \cdot |f|})$.*

Corollary 2. *The discrete model-checking problem is in 4ExpTime in the product of the sizes of $R_{\mathcal{A}}$ and f, and in 5ExpTime in the product of the sizes of $\mathcal{A}$ and f. If f is a formula of $\exists$PTCTL then the model-checking problem is in 3ExpTime in the product of the sizes of $\mathcal{A}$ and f.*

Dense Model-Checking. To solve the dense model-checking, we proceed in the same way and obtain the following theorem.

Theorem 2. *Let $\mathcal{A}$ be a dense-timed automaton and r be a region of $R_{\mathcal{A}} = (R, F)$. Let f be a PTCTL formula with $P_f = \{\theta_1, \ldots, \theta_m\}$. Then there exists a RA formula $\Delta(f, r)$ such that $r \models_v f$ for some valuation v iff the sentence $\exists \theta_1 \cdots \exists \theta_m \Delta(f, r)$ is true. The construction of $\Delta(f, r)$ is effective.*

The inductive definition for formula $\Delta(f, r)$ is given in Table 2.

Corollary 3. *The dense model-checking problem is decidable.*

The complexity of the method is given by the following theorem:

Theorem 3. *The dense model-checking problem for a PTCTL is in 4ExpTime in the sizes of $R_{\mathcal{A}}$ and f, and in 5ExpTime in the product of the sizes of $\mathcal{A}$ and f. If f is a formula of $\exists$PTCTL then the model-checking problem is in 3ExpTime in the product of the sizes of $\mathcal{A}$ and f.*

Parameter Synthesis. The proofs given for Theorems 1 and 2 also solve the parameter synthesis problem (see Problem 2). Indeed the formula $\Delta(f, r)$ is a PA or RA formula with free variables in P_f that exactly defines the values of the parameters of P_f that satisfy formula f at region r. As its construction is effective, we get a symbolic representation of these values, that supports boolean operations, projections and checking emptiness.

Theorem 4. *Let $\mathcal{A}$ be a timed automaton and r be a region of $R_{\mathcal{A}} = (R, F)$. Let f be a PTCTL formula with $P_f = \{\theta_1, \ldots, \theta_m\}$. Then the formula $\Delta(f, r)(\theta_1, \ldots, \theta_m)$ with free variables $\theta_1, \ldots, \theta_m$ is an effective characterization is the set of valuations v for P_f such that $r \models_v f$.*

Table 1. Formulae $\Delta(\varphi, r)$ for discrete time.

φ	$\Delta(\varphi, r)$
$\exists \bigcirc \psi$	$\bigvee_{(r,r') \in F} (\Delta(\psi, r') \wedge \Pr_R(r'))$
$\psi \exists U_{\sim \alpha} \phi$	$(\Delta(\phi, r) \wedge \Pr_R(r) \wedge 0 \sim \alpha)$
	$\qquad \vee \bigvee_{r' \in R} \bigvee_{S \subseteq R} \left(\exists t \sim \alpha \; \lambda_{r,r'}^S(t) \wedge \Delta(\phi, r') \wedge \bigwedge_{s \in S} \Delta(\psi, s) \wedge \Pr_R(r') \right)$
$\exists \Box_{< \alpha} \psi$	$\bigvee_{r' \in R} \bigvee_{S \subseteq R} \left(\chi_{r,r'}^S(\alpha) \wedge \bigwedge_{s \in S} \Delta(\psi, s) \wedge \Pr_R(r') \right)$
$\exists \Box_{\geq \alpha} \psi$	$\bigvee_{r' \in R} \bigvee_{S \subseteq R} \left(\chi_{r,r'}^R(\alpha) \wedge \bigwedge_{s \in S} \Delta(\psi, s) \wedge \Pr_S(r') \right)$
$\exists \Box_{= \alpha} \psi$	$\bigvee_{r',r'' \in R} \bigvee_{S \subseteq R} \left(\chi_{r,r'}^R(\alpha) \wedge \chi_{r',r''}^S(1) \wedge \bigwedge_{s \in S} \Delta(\psi, s) \wedge \Pr_R(r'') \right)$

Table 2. Formulae $\Delta(\varphi, r)$ for dense time.

φ	$\Delta(\varphi, r)$
$\psi \exists U_{\sim \alpha} \phi$	$(\Delta(\phi, r) \wedge \Pr_R(r) \wedge 0 \sim \alpha) \qquad \vee$
	$\bigvee_{r' \in R} \bigvee_{S \subseteq R} (\exists t \sim \alpha \; \lambda_{r,r'}^S(t) \wedge \Delta(\phi, r') \wedge \bigwedge_{s \in S} \Delta(\psi, s)$
	$\qquad \qquad \wedge \Pr_R(r') \wedge (\neg B(r') \rightarrow \Delta(\psi, r')))$
$\exists \Box_{\geq \alpha} \psi$	$\bigvee_{r' \in R} \bigvee_{S \subseteq R} \left(\chi_{r,r'}^R(\alpha) \wedge \bigwedge_{s \in S} \Delta(\psi, s) \wedge \Pr_S(r') \right)$
$\exists \Box_{< \alpha} \psi$	$\bigvee_{r' \in R} \bigvee_{S \subseteq R} \left(\chi_{r,r'}^S(\alpha) \wedge \bigwedge_{s \in S} \Delta(\psi, s) \wedge \Pr_R(r') \wedge (\neg B(r') \rightarrow \Delta(\psi, r')) \right)$
$\exists \Box_{\leq \alpha} \psi$	$\bigvee_{S \subseteq R} \bigvee_{r' \in S} \bigvee_{r'',(r',r'') \in F^>} \left(\lambda_{r,r'}^S(\alpha) \wedge \bigwedge_{s \in S} \Delta(\psi, s) \wedge \Pr_R(r'') \right)$
$\exists \Box_{> \alpha} \psi$	$\bigvee_{S \subseteq R} \bigvee_{r' \in R} \bigvee_{r'',(r',r'') \in F^>}$
	$\qquad \left(\lambda_{r,r'}^R(\alpha) \wedge \bigwedge_{s \in S} \Delta(\psi, s) \wedge \Pr_S(r'') \wedge (\neg B(r') \rightarrow \Delta(\psi, r')) \right)$
$\exists \Box_{= \alpha} \psi$	$\bigvee_{S \subseteq R} \bigvee_{r',r'' \in S} \bigvee_{r''',(r'',r''') \in F^>}$
	$\qquad \left(\chi_{r,r'}^R(\alpha) \wedge \lambda_{r',r''}^S(0) \wedge \bigwedge_{s \in S} \Delta(\psi, s) \wedge \Pr_R(r''') \right)$

Let us denote by $V(\mathcal{A}, f, r)$ the set of valuations v for P_f such that $r \models_v f$. Since PA and RA have a decidable theory, any question formulated in these logics about the set $V(\mathcal{A}, f, r)$ is therefore decidable. For instance, the question "Is the set $V(\mathcal{A}, f, r)$ not empty ?" is decidable when formulated in PA or RA by $\exists \theta_1, \ldots, \exists \theta_m \Delta(f, r)(\theta_1, \ldots, \theta_m)$. It is nothing else than the model-checking problem. Other questions can be the following ones. "Does the set $V(\mathcal{A}, f, r)$ contain every valuation ?", that is, $\forall \theta_1, \ldots, \forall \theta_m \Delta(f, r)(\theta_1, \ldots, \theta_m)$. "Is the set $V(\mathcal{A}, f, r)$ finite ?", that is, $\exists z \forall \theta_1, \ldots, \forall \theta_m \Delta(f, r)(\theta_1, \ldots, \theta_m) \Rightarrow \bigwedge_{1 \leq i \leq m} \theta_i < z$.

We could also be interested in optimisation problems. Assume for example that there is only one parameter θ like in a formula f such $\exists \Box_{\leq \theta} \sigma$. It is interesting to know what is the maximum value z of θ for which this formula is satisfied. Such a maximum value can be defined by $\Delta(f, r)(z) \wedge \forall \theta (\Delta(f, r)(\theta) \Rightarrow \theta \leq z)$. Now, PA and RA have an effective quantifier elimination, by adding the congruences $\equiv_n$ for each natural number n in PA and by adding the congruences $\equiv_n$ and the integer part operation $\lfloor \ \rfloor$ in RA (see [13]). It follows that we can compute the value of z. More generally, if an optimisation problem can be formulated in these logics, then its solutions can be effectively symbolically represented by a quantifier-free formula.

Remark on the complexity. Concerning lower bounds, there is a gap which needs more research effort. Model-checking for ∃PTCTL logic is at least PSPACE-HARD since the model-checking problem defined in [1] is a particular case of our problem. Model-checking problem for PTCTL logic is at least 3ExpTime-Hard since full Presburger arithmetic is already present.

References

1. R. Alur, C. Courcoubetis, and D.L. Dill. Model checking for real-time systems. In *Proceedings of the Fifth Annual Symposium on Logic in Computer Science*, pages 414–425. IEEE Computer Society Press, 1990.
2. R. Alur and D.L. Dill. A theory of timed automata. *Theoretical Computer Science*, 126:183–235, 1994.
3. R. Alur, T.A. Henzinger, and M.Y. Vardi. Parametric real-time reasoning. In *Proceedings of the 25th Annual Symposium on Theory of Computing*, pages 592–601. ACM Press, 1993.
4. Rajeev Alur, Kousha Etessami, Salvatore La Torre, and Doron Peled. Parametric temporal logic for "model measuring". *Lecture Notes in Computer Science*, 1644:159–173, 1999.
5. Bernard Boigelot, Sébastien Jodogne, and Pierre Wolper. On the use of weak automata for deciding linear arithmetic with integer and real variables. In *Proc. International Joint Conference on Automated Reasoning*, volume 2083, pages 611–625. Springer-Verlag, 2001.
6. V. Bruyère, E. Dall'olio, and J.-F. Raskin. Durations, parametric model-checking in timed automata with Presburger arithmetic. Technical Report CFV-2002-1, Centre Fédéré en Vérification (Belgique), 2002.
   http://www.ulb.ac.be/di/ssd/cfv/TechReps/TechRep_CFV_2002_1.ps.
7. H. Comon and Y. Jurski. Timed automata and the theory of real numbers. In *Proc. 10th Int. Conf. Concurrency Theory (CONCUR'99), Eindhoven, The Netherlands, Aug. 1999*, volume 1664 of *Lecture Notes in Computer Science*, pages 242–257. Springer, 1999.
8. E. Allen Emerson and Richard J. Trefler. Parametric quantitative temporal reasoning. In *Logic in Computer Science*, pages 336–343, 1999.
9. Jeanne Ferrante and Charles W. Rackoff. *The computational complexity of logical theories*. Number 718 in Lecture Notes in Mathematics. Springer-Verlag, 1979.
10. T.A. Henzinger. It's about time: Real-time logics reviewed. In D. Sangiorgi and R. de Simone, editors, *CONCUR 98: Concurrency Theory*, Lecture Notes in Computer Science 1466, pages 439–454. Springer-Verlag, 1998.
11. Farn Wang. Timing behavior analysis for real-time systems. In *In Proceedings of the Tenth IEEE Symposium on Logic in Computer Science*, pages 112–122, 1995.
12. Farn Wang and Pao-Ann Hsiung. Parametric analysis of computer systems. In *Algebraic Methodology and Software Technology*, pages 539–553, 1997.
13. Volker Weispfenning. Mixed real-integer linear quantifier elimination. In *ISSAC: Proceedings of the ACM SIGSAM International Symposium on Symbolic and Algebraic Computation*, 1999.

Table 1. Formulae $\Delta(\varphi, r)$ for discrete time.

φ	$\Delta(\varphi, r)$
$\exists\bigcirc\psi$	$\bigvee_{(r,r')\in F}(\Delta(\psi, r') \wedge \text{Pr}_R(r'))$
$\psi\exists U_{\sim\alpha}\phi$	$(\Delta(\phi, r) \wedge \text{Pr}_R(r) \wedge 0 \sim \alpha)$
	$\vee \bigvee_{r'\in R}\bigvee_{S\subseteq R}(\exists t \sim \alpha\, \lambda^S_{r,r'}(t) \wedge \Delta(\phi, r') \wedge \bigwedge_{s\in S}\Delta(\psi, s) \wedge \text{Pr}_R(r'))$
$\exists\square_{<\alpha}\psi$	$\bigvee_{r'\in R}\bigvee_{S\subseteq R}(\chi^S_{r,r'}(\alpha) \wedge \bigwedge_{s\in S}\Delta(\psi, s) \wedge \text{Pr}_R(r'))$
$\exists\square_{\geq\alpha}\psi$	$\bigvee_{r'\in R}\bigvee_{S\subseteq R}(\chi^R_{r,r'}(\alpha) \wedge \bigwedge_{s\in S}\Delta(\psi, s) \wedge \text{Pr}_S(r'))$
$\exists\square_{=\alpha}\psi$	$\bigvee_{r',r''\in R}\bigvee_{S\subseteq R}(\chi^R_{r,r'}(\alpha) \wedge \chi^S_{r',r''}(1) \wedge \bigwedge_{s\in S}\Delta(\psi, s) \wedge \text{Pr}_R(r''))$

Table 2. Formulae $\Delta(\varphi, r)$ for dense time.

φ	$\Delta(\varphi, r)$
$\psi\exists U_{\sim\alpha}\phi$	$(\Delta(\phi, r) \wedge \text{Pr}_R(r) \wedge 0 \sim \alpha) \qquad \vee$
	$\bigvee_{r'\in R}\bigvee_{S\subseteq R}(\exists t \sim \alpha\, \lambda^S_{r,r'}(t) \wedge \Delta(\phi, r') \wedge \bigwedge_{s\in S}\Delta(\psi, s)$
	$\wedge\, \text{Pr}_R(r') \wedge (\neg\text{B}(r') \to \Delta(\psi, r')))$
$\exists\square_{\geq\alpha}\psi$	$\bigvee_{r'\in R}\bigvee_{S\subseteq R}(\chi^R_{r,r'}(\alpha) \wedge \bigwedge_{s\in S}\Delta(\psi, s) \wedge \text{Pr}_S(r'))$
$\exists\square_{<\alpha}\psi$	$\bigvee_{r'\in R}\bigvee_{S\subseteq R}(\chi^S_{r,r'}(\alpha) \wedge \bigwedge_{s\in S}\Delta(\psi, s) \wedge \text{Pr}_R(r') \wedge (\neg\text{B}(r') \to \Delta(\psi, r')))$
$\exists\square_{\leq\alpha}\psi$	$\bigvee_{S\subseteq R}\bigvee_{r'\in S}\bigvee_{r'',(r',r'')\in F>}(\lambda^S_{r,r'}(\alpha) \wedge \bigwedge_{s\in S}\Delta(\psi, s) \wedge \text{Pr}_R(r''))$
$\exists\square_{>\alpha}\psi$	$\bigvee_{S\subseteq R}\bigvee_{r'\in R}\bigvee_{r'',(r',r'')\in F>}$
	$(\lambda^R_{r,r'}(\alpha) \wedge \bigwedge_{s\in S}\Delta(\psi, s) \wedge \text{Pr}_S(r'') \wedge (\neg\text{B}(r') \to \Delta(\psi, r')))$
$\exists\square_{=\alpha}\psi$	$\bigvee_{S\subseteq R}\bigvee_{r',r''\in S}\bigvee_{r''',(r'',r''')\in F>}$
	$(\chi^R_{r,r'}(\alpha) \wedge \lambda^S_{r',r''}(0) \wedge \bigwedge_{s\in S}\Delta(\psi, s) \wedge \text{Pr}_R(r'''))$

Let us denote by $V(\mathcal{A}, f, r)$ the set of valuations v for P_f such that $r \models_v f$. Since PA and RA have a decidable theory, any question formulated in these logics about the set $V(\mathcal{A}, f, r)$ is therefore decidable. For instance, the question "Is the set $V(\mathcal{A}, f, r)$ not empty ?" is decidable when formulated in PA or RA by $\exists\theta_1,\ldots,\exists\theta_m\Delta(f, r)(\theta_1,\ldots,\theta_m)$. It is nothing else than the model-checking problem. Other questions can be the following ones. "Does the set $V(\mathcal{A}, f, r)$ contain every valuation ?", that is, $\forall\theta_1,\ldots,\forall\theta_m\Delta(f, r)(\theta_1,\ldots,\theta_m)$. "Is the set $V(\mathcal{A}, f, r)$ finite ?", that is, $\exists z\forall\theta_1,\ldots,\forall\theta_m\Delta(f, r)(\theta_1,\ldots,\theta_m) \Rightarrow \bigwedge_{1\leq i\leq m}\theta_i < z$.

We could also be interested in optimisation problems. Assume for example that there is only one parameter θ like in a formula f such $\exists\square_{\leq\theta}\sigma$. It is interesting to know what is the maximum value z of θ for which this formula is satisfied. Such a maximum value can be defined by $\Delta(f, r)(z) \wedge \forall\theta(\Delta(f, r)(\theta) \Rightarrow \theta \leq z)$. Now, PA and RA have an effective quantifier elimination, by adding the congruences $\equiv_n$ for each natural number n in PA and by adding the congruences $\equiv_n$ and the integer part operation $\lfloor\ \rfloor$ in RA (see [13]). It follows that we can compute the value of z. More generally, if an optimisation problem can be formulated in these logics, then its solutions can be effectively symbolically represented by a quantifier-free formula.

Remark on the complexity. Concerning lower bounds, there is a gap which needs more research effort. Model-checking for ∃PTCTL logic is at least PSPACE-HARD since the model-checking problem defined in [1] is a particular case of our problem. Model-checking problem for PTCTL logic is at least 3EXPTIME-HARD since full Presburger arithmetic is already present.

References

1. R. Alur, C. Courcoubetis, and D.L. Dill. Model checking for real-time systems. In *Proceedings of the Fifth Annual Symposium on Logic in Computer Science*, pages 414–425. IEEE Computer Society Press, 1990.
2. R. Alur and D.L. Dill. A theory of timed automata. *Theoretical Computer Science*, 126:183–235, 1994.
3. R. Alur, T.A. Henzinger, and M.Y. Vardi. Parametric real-time reasoning. In *Proceedings of the 25th Annual Symposium on Theory of Computing*, pages 592–601. ACM Press, 1993.
4. Rajeev Alur, Kousha Etessami, Salvatore La Torre, and Doron Peled. Parametric temporal logic for "model measuring". *Lecture Notes in Computer Science*, 1644:159–173, 1999.
5. Bernard Boigelot, Sébastien Jodogne, and Pierre Wolper. On the use of weak automata for deciding linear arithmetic with integer and real variables. In *Proc. International Joint Conference on Automated Reasoning*, volume 2083, pages 611–625. Springer-Verlag, 2001.
6. V. Bruyère, E. Dall'olio, and J.-F. Raskin. Durations, parametric model-checking in timed automata with Presburger arithmetic. Technical Report CFV-2002-1, Centre Fédéré en Vérification (Belgique), 2002.
   http://www.ulb.ac.be/di/ssd/cfv/TechReps/TechRep_CFV_2002_1.ps.
7. H. Comon and Y. Jurski. Timed automata and the theory of real numbers. In *Proc. 10th Int. Conf. Concurrency Theory (CONCUR'99), Eindhoven, The Netherlands, Aug. 1999*, volume 1664 of *Lecture Notes in Computer Science*, pages 242–257. Springer, 1999.
8. E. Allen Emerson and Richard J. Trefler. Parametric quantitative temporal reasoning. In *Logic in Computer Science*, pages 336–343, 1999.
9. Jeanne Ferrante and Charles W. Rackoff. *The computational complexity of logical theories*. Number 718 in Lecture Notes in Mathematics. Springer-Verlag, 1979.
10. T.A. Henzinger. It's about time: Real-time logics reviewed. In D. Sangiorgi and R. de Simone, editors, *CONCUR 98: Concurrency Theory*, Lecture Notes in Computer Science 1466, pages 439–454. Springer-Verlag, 1998.
11. Farn Wang. Timing behavior analysis for real-time systems. In *In Proceedings of the Tenth IEEE Symposium on Logic in Computer Science*, pages 112–122, 1995.
12. Farn Wang and Pao-Ann Hsiung. Parametric analysis of computer systems. In *Algebraic Methodology and Software Technology*, pages 539–553, 1997.
13. Volker Weispfenning. Mixed real-integer linear quantifier elimination. In *ISSAC: Proceedings of the ACM SIGSAM International Symposium on Symbolic and Algebraic Computation*, 1999.

Author Index

Lecture Notes in Computer Science

For information about Vols. 1–2501

please contact your bookseller or Springer-Verlag

Vol. 2541: T. Barkowsky, Mental Representation and Processing of Geographic Knowledge. X, 174 pages. 2002. (Subseries LNAI).

Vol. 2542: I. Dimov, I. Lirkov, S. Margenov, Z. Zlatev (Eds.), Numerical Methods and Applications. Proceedings, 2002. X, 174 pages. 2003.

Vol. 2544: S. Bhalla (Ed.), Databases in Networked Information Systems. Proceedings 2002. X, 285 pages. 2002.

Vol. 2545: P. Forbrig, Q, Limbourg, B. Urban, J. Vanderdonckt (Eds.), Interactive Systems. Proceedings 2002. XII, 574 pages. 2002.

Vol. 2546: J. Sterbenz, O. Takada, C. Tschudin, B. Plattner (Eds.), Active Networks. Proceedings, 2002. XIV, 267 pages. 2002.

Vol. 2547: R. Fleischer, B. Moret, E. Meineche Schmidt (Eds.), Experimental Algorithmics. XVII, 279 pages. 2002.

Vol. 2548: J. Hernández, Ana Moreira (Eds.), Object-Oriented Technology. Proceedings, 2002. VIII, 223 pages. 2002.

Vol. 2549: J. Cortadella, A. Yakovlev, G. Rozenberg (Eds.), Concurrency and Hardware Design. XI, 345 pages. 2002.

Vol. 2550: A. Jean-Marie (Ed.), Advances in Computing Science – ASIAN 2002. Proceedings, 2002. X, 233 pages. 2002.

Vol. 2551: A. Menezes, P. Sarkar (Eds.), Progress in Cryptology – INDOCRYPT 2002. Proceedings, 2002. XI, 437 pages. 2002.

Vol. 2552: S. Sahni, V.K. Prasanna, U. Shukla (Eds.), High Performance Computing – HiPC 2002. Proceedings, 2002. XXI, 735 pages. 2002.

Vol. 2553: B. Andersson, M. Bergholtz, P. Johannesson (Eds.), Natural Language Processing and Information Systems. Proceedings, 2002. X, 241 pages. 2002.

Vol. 2554: M. Beetz, Plan-Based Control of Robotic Agents. XI, 191 pages. 2002. (Subseries LNAI).

Vol. 2555: E.-P. Lim, S. Foo, C. Khoo, H. Chen, E. Fox, S. Urs, T. Costantino (Eds.), Digital Libraries: People, Knowledge, and Technology. Proceedings, 2002. XVII, 535 pages. 2002.

Vol. 2556: M. Agrawal, A. Seth (Eds.), FST TCS 2002: Foundations of Software Technology and Theoretical Computer Science. Proceedings, 2002. XI, 361 pages. 2002.

Vol. 2557: B. McKay, J. Slaney (Eds.), AI 2002: Advances in Artificial Intelligehce. Proceedings, 2002. XV, 730 pages. 2002. (Subseries LNAI).

Vol. 2558: P. Perner, Data Mining on Multimedia Data. X, 131 pages. 2002.

Vol. 2559: M. Oivo, S. Komi-Sirviö (Eds.), Product Focused Software Process Improvement. Proceedings, 2002. XV, 646 pages. 2002.

Vol. 2560: S. Goronzy, Robust Adaptation to Non-Native Accents in Automatic Speech Recognition. Proceedings, 2002. XI, 144 pages. 2002. (Subseries LNAI).

Vol. 2561: H.C.M. de Swart (Ed.), Relational Methods in Computer Science. Proceedings, 2001. X, 315 pages. 2002.

Vol. 2562: V. Dahl, P. Wadler (Eds.), Practical Aspects of Declarative Languages. Proceedings, 2003. X, 315 pages. 2002.

Vol. 2566: T.Æ. Mogensen, D.A. Schmidt, I.H. Sudborough (Eds.), The Essence of Computation. XIV, 473 pages. 2002.

Vol. 2567: Y.G. Desmedt (Ed.), Public Key Cryptography – PKC 2003. Proceedings, 2003. XI, 365 pages. 2002.

Vol. 2568: M. Hagiya, A. Ohuchi (Eds.), DNA Computing. Proceedings, 2002. XI, 338 pages. 2003.

Vol. 2569: D. Gollmann, G. Karjoth, M. Waidner (Eds.), Computer Security – ESORICS 2002. Proceedings, 2002. XIII, 648 pages. 2002. (Subseries LNAI).

Vol. 2570: M. Jünger, G. Reinelt, G. Rinaldi (Eds.), Combinatorial Optimization – Eureka, You Shrink!. Proceedings, 2001. X, 209 pages. 2003.

Vol. 2571: S.K. Das, S. Bhattacharya (Eds.), Distributed Computing. Proceedings, 2002. XIV, 354 pages. 2002.

Vol. 2572: D. Calvanese, M. Lenzerini, R. Motwani (Eds.), Database Theory – ICDT 2003. Proceedings, 2003. XI, 455 pages. 2002.

Vol. 2574: M.-S. Chen, P.K. Chrysanthis, M. Sloman, A. Zaslavsky (Eds.), Mobile Data Management. Proceedings, 2003. XII, 414 pages. 2003.

Vol. 2575: L.D. Zuck, P.C. Attie, A. Cortesi, S. Mukhopadhyay (Eds.), Verification, Model Checking, and Abstract Interpretation. Proceedings, 2003. XI, 325 pages. 2003.

Vol. 2576: S. Cimato, C. Galdi, G. Persiano (Eds.), Security in Communication Networks. Proceedings, 2002. IX, 365 pages. 2003.

Vol. 2578: F.A.P. Petitcolas (Ed.), Information Hiding. Proceedings, 2002. IX, 427 pages. 2003.

Vol. 2580: H. Erdogmus, T. Weng (Eds.), COTS-Based Software Systems. Proceedings, 2003. XVIII, 261 pages. 2003.

Vol. 2581: J.S. Sichman, F. Bousquet, P. Davidsson (Eds.), Multi-Agent-Based Simulation II. Proceedings, 2002. X, 195 pages. 2003. (Subseries LNAI).

Vol. 2583: S. Matwin, C. Sammut (Eds.), Inductive Logic Programming. Proceedings, 2002. X, 351 pages. 2003. (Subseries LNAI).

Vol. 2588: A. Gelbukh (Ed.), Computational Linguistics and Intelligent Text Processing. Proceedings, 2003. XV, 648 pages. 2003.

Vol. 2589: E. Börger, A. Gargantini, E. Riccobene (Eds.), Abstract State Machines 2003. Proceedings, 2003. XI, 427 pages. 2003.

Vol. 2594: A. Asperti, B. Buchberger, J.H. Davenport (Eds.), Mathematical Knowledge Management. Proceedings, 2003. X, 225 pages. 2003.

Vol. 2598: R. Klein, H.-W. Six, L. Wegner (Eds.), Computer Science in Perspective. X, 357 pages. 2003.

Vol. 2600: S. Mendelson, A.J. Smola, Advanced Lectures on Machine Learning. Proceedings, 2002. IX, 259 pages. 2003. (Subseries LNAI).

Vol. 2601: M. Ajmone Marsan, G. Corazza, M. Listanti, A. Roveri (Eds.) Quality of Service in Multiservice IP Networks. Proceedings, 2003. XV, 759 pages. 2003.

Vol. 2602: C. Priami (Ed.), Computational Methods in Systems Biology. Proceedings, 2003. IX, 214 pages. 2003.

Vol. 2607: H. Alt, M. Habib (Eds.), STACS 2003. Proceedings, 2003. XVII, 700 pages. 2003.